STUDENT'S SOLUTIONS MANUAL

ELKA M. BLOCK
FRANK PURCELL

FINITE MATHEMATICS AND CALCULUS WITH APPLICATIONS

NINTH EDITION

Margaret L. Lial
American River College

Raymond N. Greenwell
Hofstra University

Nathan P. Ritchey
Youngstown State University

PEARSON

Boston Columbus Indianapolis New York San Francisco Upper Saddle River
Amsterdam Cape Town Dubai London Madrid Milan Munich Paris Montreal Toronto
Delhi Mexico City Sao Paulo Sydney Hong Kong Seoul Singapore Taipei Tokyo

Reproduced by Pearson from electronic files supplied by the author.

Publishing as Pearson, 75 Arlington Street, Boston, MA 02116.

ISBN-13: 978-0-321-74623-8
ISBN-10: 0-321-74623-6

1 2 3 4 5 6 BRR 15 14 13 12 11

www.pearsonhighered.com
PEARSON

CONTENTS

CHAPTER 7 SETS AND PROBABILITY

CHAPTER 8 COUNTING PRINCIPLES

CHAPTER 9 STATISTICS

CHAPTER 10 NONLINEAR FUNCTIONS

CHAPTER 11 THE DERIVATIVE

CHAPTER 12 CALCULATING THE DERIVATIVE

CHAPTER 13 GRAPHS AND THE DERIVATIVE

CHAPTER 14 APPLICATIONS OF THE DERIVATIVE

CHAPTER 15 INTEGRATION

CHAPTER 16 FURTHER TECHNIQUES AND APPLICATIONS OF INTEGRATION

CHAPTER 17 MULTIVARIABLE CALCULUS

CHAPTER 18 PROBABILITY AND CALCULUS

Chapter R

ALGEBRA REFERENCE

R.1 Polynomials

Your Turn 1

$3(x^2 - 4x - 5) - 4(3x^2 - 5x - 7)$

$= 3x^2 - 12x - 15 - 12x^2 + 20x + 28$

$= -9x^2 + 8x + 13$

Your Turn 2

$(3y + 2)(4y^2 - 2y - 5)$

$= (3y)(4y^2 - 2y - 5) + (2)(4y^2 - 2y - 5)$

$= 12y^3 - 6y^2 - 15y + 8y^2 - 4y - 10$

$= 12y^3 + 2y^2 - 19y - 10$

R.1 Exercises

1. $(2x^2 - 6x + 11) + (-3x^2 + 7x - 2)$

$= 2x^2 - 6x + 11 - 3x^2 + 7x - 2$

$= (2 - 3)x^2 + (7 - 6)x + (11 - 2)$

$= -x^2 + x + 9$

3. $-6(2q^2 + 4q - 3) + 4(-q^2 + 7q - 3)$

$= (-12q^2 - 24q + 18) + (-4q^2 + 28q - 12)$

$= (-12q^2 - 4q^2) + (-24q + 28q) + (18 - 12)$

$= -16q^2 + 4q + 6$

5. $(0.613x^2 - 4.215x + 0.892) - 0.47(2x^2 - 3x + 5)$

$= 0.613x^2 - 4.215x + 0.892 - 0.94x^2 + 1.41x - 2.35$

$= -0.327x^2 - 2.805x - 1.458$

7. $-9m(2m^2 + 3m - 1)$

$= -9m(2m^2) - 9m(3m) - 9m(-1)$

$= -18m^3 - 27m^2 + 9m$

9. $(3t - 2y)(3t + 5y)$

$= (3t)(3t) + (3t)(5y) + (-2y)(3t) + (-2y)(5y)$

$= 9t^2 + 15ty - 6ty - 10y^2$

$= 9t^2 + 9ty - 10y^2$

11. $(2 - 3x)(2 + 3x)$

$= (2)(2) + (2)(3x) + (-3x)(2) + (-3x)(3x)$

$= 4 + 6x - 6x - 9x^2$

$= 4 - 9x^2$

13. $\left(\frac{2}{5}y + \frac{1}{8}z\right)\left(\frac{3}{5}y + \frac{1}{2}z\right)$

$= \left(\frac{2}{5}y\right)\left(\frac{3}{5}y\right) + \left(\frac{2}{5}y\right)\left(\frac{1}{2}z\right) + \left(\frac{1}{8}z\right)\left(\frac{3}{5}y\right) + \left(\frac{1}{8}z\right)\left(\frac{1}{2}z\right)$

$= \frac{6}{25}y^2 + \frac{1}{5}yz + \frac{3}{40}yz + \frac{1}{16}z^2$

$= \frac{6}{25}y^2 + \left(\frac{8}{40} + \frac{3}{40}\right)yz + \frac{1}{16}z^2$

$= \frac{6}{25}y^2 + \frac{11}{40}yz + \frac{1}{16}z^2$

15. $(3p - 1)(9p^2 + 3p + 1)$

$= (3p - 1)(9p^2) + (3p - 1)(3p) + (3p - 1)(1)$

$= 3p(9p^2) - 1(9p^2) + 3p(3p) - 1(3p) + 3p(1) - 1(1)$

$= 27p^3 - 9p^2 + 9p^2 - 3p + 3p - 1$

$= 27p^3 - 1$

17. $(2m + 1)(4m^2 - 2m + 1)$

$= 2m(4m^2 - 2m + 1) + 1(4m^2 - 2m + 1)$

$= 8m^3 - 4m^2 + 2m + 4m^2 - 2m + 1$

$= 8m^3 + 1$

19. $(x + y + z)(3x - 2y - z)$

$= x(3x) + x(-2y) + x(-z) + y(3x) + y(-2y) + y(-z) + z(3x) + z(-2y) + z(-z)$

$= 3x^2 - 2xy - xz + 3xy - 2y^2 - yz + 3xz - 2yz - z^2$

$= 3x^2 + xy + 2xz - 2y^2 - 3yz - z^2$

21. $(x + 1)(x + 2)(x + 3)$

$= [x(x + 2) + 1(x + 2)](x + 3)$

$= [x^2 + 2x + x + 2](x + 3)$

$= [x^2 + 3x + 2](x + 3)$

$= x^2(x + 3) + 3x(x + 3) + 2(x + 3)$

$= x^3 + 3x^2 + 3x^2 + 9x + 2x + 6$

$= x^3 + 6x^2 + 11x + 6$

23. $(x + 2)^2 = (x + 2)(x + 2)$

$= x(x + 2) + 2(x + 2)$

$= x^2 + 2x + 2x + 4$

$= x^2 + 4x + 4$

25. $(x - 2y)^3$

$= [(x - 2y)(x - 2y)](x - 2y)$

$= (x^2 - 2xy - 2xy + 4y^2)(x - 2y)$

$= (x^2 - 4xy + 4y^2)(x - 2y)$

$= (x^2 - 4xy + 4y^2)x + (x^2 - 4xy + 4y^2)(-2y)$

$= x^3 - 4x^2y + 4xy^2 - 2x^2y + 8xy^2 - 8y^3$

$= x^3 - 6x^2y + 12xy^2 - 8y^3$

R.2 Factoring

Your Turn 1

Factor $4z^4 + 4z^3 + 18z^2$.

$4z^4 + 4z^3 + 18z^2$

$= (2z^2) \cdot 2z^2 + (2z^2) \cdot 2z + (2z^2) \cdot 9$

$= (2z^2)(2z^2 + 2z + 9)$

Your Turn 2

Factor $6a^2 + 5ab - 4b^2$.

$6a^2 + 5ab - 4b^2 = (2a - b)(3a + 4b)$

R.2 Exercises

1. $7a^3 + 14a^2 = 7a^2 \cdot a + 7a^2 \cdot 2$

$= 7a^2(a + 2)$

3. $13p^4q^2 - 39p^3q + 26p^2q^2$

$= 13p^2q \cdot p^2q - 13p^2q \cdot 3p + 13p^2q \cdot 2q$

$= 13p^2q(p^2q - 3p + 2q)$

5. $m^2 - 5m - 14 = (m - 7)(m + 2)$

since $(-7)(2) = -14$ and $-7 + 2 = -5$.

7. $z^2 + 9z + 20 = (z + 4)(z + 5)$

since $4 \cdot 5 = 20$ and $4 + 5 = 9$.

9. $a^2 - 6ab + 5b^2 = (a - b)(a - 5b)$

since $(-b)(-5b) = 5b^2$ and $-b + (-5b) = -6b$.

11. $y^2 - 4yz - 21z^2 = (y + 3z)(y - 7z)$

since $(3z)(-7z) = -21z^2$ and $3z + (-7z) = -4z$.

13. $3a^2 + 10a + 7$

The possible factors of $3a^2$ are $3a$ and a and the possible factors of 7 are 7 and 1. Try various combinations until one works.

$3a^2 + 10a + 7 = (a + 1)(3a + 7)$

15. $21m^2 + 13mn + 2n^2 = (7m + 2n)(3m + n)$

17. $3m^3 + 12m^2 + 9m = 3m(m^2 + 4m + 3)$

$= 3m(m + 1)(m + 3)$

19. $24a^4 + 10a^3b - 4a^2b^2$

$= 2a^2(12a^2 + 5ab - 2b^2)$

$= 2a^2(4a - b)(3a + 2b)$

21. $x^2 - 64 = x^2 - 8^2$

$= (x + 8)(x - 8)$

23. $10x^2 - 160 = 10(x^2 - 16)$

$= 10(x^2 - 4^2)$

$= 10(x + 4)(x - 4)$

25. $z^2 + 14zy + 49y^2 = z^2 + 2 \cdot 7zy + 7^2y^2$

$= (z + 7y)^2$

27. $9p^2 - 24p + 16 = (3p)^2 - 2 \cdot 3p \cdot 4 + 4^2$

$= (3p - 4)^2$

29. $27r^3 - 64s^3 = (3r)^3 - (4s)^3$

$= (3r - 4s)(9r^2 + 12rs + 16s^2)$

31. $x^4 - y^4 = (x^2)^2 - (y^2)^2$

$= (x^2 + y^2)(x^2 - y^2)$

$= (x^2 + y^2)(x + y)(x - y)$

R.3 Rational Expressions

Your Turn 1

Write in lowest terms $\frac{z^2 + 5z + 6}{2z^2 + 7z + 3}$.

$$\frac{z^2 + 5z + 6}{2z^2 + 7z + 3} = \frac{(z + 3)(z + 2)}{(z + 3)(2z + 1)}$$

$$= \frac{z + 2}{2z + 1}$$

Your Turn 2

Perform each of the following operations.

(a) $\frac{z^2 + 5z + 6}{2z^2 - 5z - 3} \cdot \frac{2z^2 - z - 1}{z^2 + 2z - 3}$

$$= \frac{(z + 2)(z + 3)}{(2z + 1)(z - 3)} \cdot \frac{(2z + 1)(z - 1)}{(z + 3)(z - 1)}$$

$$= \frac{(z + 2)\cancel{(z + 3)}\cancel{(2z + 1)}\cancel{(z - 1)}}{\cancel{(2z + 1)}(z - 3)\cancel{(z + 3)}\cancel{(z - 1)}}$$

$$= \frac{z + 2}{z - 3}$$

(b) $\frac{a - 3}{a^2 + 3a + 2} + \frac{5a}{a^2 - 4}$

$$= \frac{a - 3}{(a + 2)(a + 1)} + \frac{5a}{(a - 2)(a + 2)}$$

$$= \frac{a - 3}{(a + 2)(a + 1)} \cdot \frac{(a - 2)}{(a - 2)} + \frac{5a}{(a - 2)(a + 2)} \cdot \frac{(a + 1)}{(a + 1)}$$

$$= \frac{(a^2 - 5a + 6) + (5a^2 + 5a)}{(a - 2)(a + 2)(a + 1)}$$

$$= \frac{6a^2 + 6}{(a - 2)(a + 2)(a + 1)}$$

$$= \frac{6(a^2 + 1)}{(a - 2)(a + 2)(a + 1)}$$

R.3 Exercises

1. $\frac{5v^2}{35v} = \frac{5 \cdot v \cdot v}{5 \cdot 7 \cdot v} = \frac{v}{7}$

3. $\frac{8k + 16}{9k + 18} = \frac{8(k + 2)}{9(k + 2)} = \frac{8}{9}$

5. $\frac{4x^3 - 8x^2}{4x^2} = \frac{4x^2(x - 2)}{4x^2} = x - 2$

7. $\frac{m^2 - 4m + 4}{m^2 + m - 6} = \frac{(m - 2)(m - 2)}{(m - 2)(m + 3)}$

$$= \frac{m - 2}{m + 3}$$

9. $\frac{3x^2 + 3x - 6}{x^2 - 4} = \frac{3(x + 2)(x - 1)}{(x + 2)(x - 2)} = \frac{3(x - 1)}{x - 2}$

11. $\frac{m^4 - 16}{4m^2 - 16} = \frac{(m^2 + 4)(m + 2)(m - 2)}{4(m + 2)(m - 2)}$

$$= \frac{m^2 + 4}{4}$$

13. $\frac{9k^2}{25} \cdot \frac{5}{3k} = \frac{3 \cdot 3 \cdot 5k^2}{5 \cdot 5 \cdot 3k} = \frac{3k^2}{5k} = \frac{3k}{5}$

15. $\frac{3a + 3b}{4c} \cdot \frac{12}{5(a + b)} = \frac{3(a + b)}{4c} \cdot \frac{3 \cdot 4}{5(a + b)}$

$$= \frac{3 \cdot 3}{c \cdot 5}$$

$$= \frac{9}{5c}$$

17. $\frac{2k - 16}{6} \div \frac{4k - 32}{3} = \frac{2k - 16}{6} \cdot \frac{3}{4k - 32}$

$$= \frac{2(k - 8)}{6} \cdot \frac{3}{4(k - 8)}$$

$$= \frac{1}{4}$$

19. $\frac{4a + 12}{2a - 10} \div \frac{a^2 - 9}{a^2 - a - 20}$

$$= \frac{4(a + 3)}{2(a - 5)} \cdot \frac{(a - 5)(a + 4)}{(a - 3)(a + 3)}$$

$$= \frac{2(a + 4)}{a - 3}$$

21. $\frac{k^2 + 4k - 12}{k^2 + 10k + 24} \cdot \frac{k^2 + k - 12}{k^2 - 9}$

$$= \frac{(k + 6)(k - 2)}{(k + 6)(k + 4)} \cdot \frac{(k + 4)(k - 3)}{(k + 3)(k - 3)}$$

$$= \frac{k - 2}{k + 3}$$

23. $\dfrac{2m^2-5m-12}{m^2-10m+24} \div \dfrac{4m^2-9}{m^2-9m+18}$

$= \dfrac{2m^2-5m-12}{m^2-10m+24} \cdot \dfrac{m^2-9m+18}{4m^2-9}$

$= \dfrac{(2m+3)(m-4)(m-6)(m-3)}{(m-6)(m-4)(2m-3)(2m+3)}$

$= \dfrac{m-3}{2m-3}$

25. $\dfrac{a+1}{2} - \dfrac{a-1}{2} = \dfrac{(a+1)-(a-1)}{2}$

$= \dfrac{a+1-a+1}{2}$

$= \dfrac{2}{2} = 1$

27. $\dfrac{6}{5y} - \dfrac{3}{2} = \dfrac{6\cdot 2}{5y\cdot 2} - \dfrac{3\cdot 5y}{2\cdot 5y} = \dfrac{12-15y}{10y}$

29. $\dfrac{1}{m-1} + \dfrac{2}{m} = \dfrac{m}{m}\left(\dfrac{1}{m-1}\right) + \dfrac{m-1}{m-1}\left(\dfrac{2}{m}\right)$

$= \dfrac{m+2m-2}{m(m-1)}$

$= \dfrac{3m-2}{m(m-1)}$

31. $\dfrac{8}{3(a-1)} + \dfrac{2}{a-1} = \dfrac{8}{3(a-1)} + \dfrac{3}{3}\left(\dfrac{2}{a-1}\right)$

$= \dfrac{8+6}{3(a-1)}$

$= \dfrac{14}{3(a-1)}$

33. $\dfrac{4}{x^2+4x+3} + \dfrac{3}{x^2-x-2}$

$= \dfrac{4}{(x+3)(x+1)} + \dfrac{3}{(x-2)(x+1)}$

$= \dfrac{4(x-2)}{(x-2)(x+3)(x+1)}$

$+ \dfrac{3(x+3)}{(x-2)(x+3)(x+1)}$

$= \dfrac{4(x-2)+3(x+3)}{(x-2)(x+3)(x+1)}$

$= \dfrac{4x-8+3x+9}{(x-2)(x+3)(x+1)}$

$= \dfrac{7x+1}{(x-2)(x+3)(x+1)}$

35. $\dfrac{3k}{2k^2+3k-2} - \dfrac{2k}{2k^2-7k+3}$

$= \dfrac{3k}{(2k-1)(k+2)} - \dfrac{2k}{(2k-1)(k-3)}$

$= \left(\dfrac{k-3}{k-3}\right)\dfrac{3k}{(2k-1)(k+2)}$

$- \left(\dfrac{k+2}{k+2}\right)\dfrac{2k}{(2k-1)(k-3)}$

$= \dfrac{(3k^2-9k)-(2k^2+4k)}{(2k-1)(k+2)(k-3)}$

$= \dfrac{k^2-13k}{(2k-1)(k+2)(k-3)}$

$= \dfrac{k(k-13)}{(2k-1)(k+2)(k-3)}$

37. $\dfrac{2}{a+2} + \dfrac{1}{a} + \dfrac{a-1}{a^2+2a}$

$= \dfrac{2}{a+2} + \dfrac{1}{a} + \dfrac{a-1}{a(a+2)}$

$= \left(\dfrac{a}{a}\right)\dfrac{2}{a+2} + \left(\dfrac{a+2}{a+2}\right)\dfrac{1}{a} + \dfrac{a-1}{a(a+2)}$

$= \dfrac{2a+a+2+a-1}{a(a+2)}$

$= \dfrac{4a+1}{a(a+2)}$

R.4 Equations

Your Turn 1

Solve $3x-7 = 4(5x+2)-7x$.

$3x-7 = 20x+8-7x$

$3x-7 = 13x+8$

$-10x = 15$

$x = -\dfrac{15}{10}$

$x = -\dfrac{3}{2}$

Your Turn 2

Solve $2m^2+7m = 15$.

$2m^2+7m-15 = 0$

$(2m-3)(m+5) = 0$

$2m-3 = 0$ or $m+5 = 0$

$m = \dfrac{3}{2}$ or $m = -5$

Your Turn 3

Solve $z^2 + 6 = 8z$.

$z^2 - 8z + 6 = 0$

Use the quadratic formula with $a = 1$, $b = -8$, and $c = 6$.

$$z = \frac{-(-8) \pm \sqrt{(-8)^2 - 4(1)(6)}}{2(1)}$$
$$= \frac{8 \pm \sqrt{64 - 24}}{2}$$
$$= \frac{8 \pm \sqrt{40}}{2}$$
$$= \frac{8 \pm \sqrt{4 \cdot 10}}{2}$$
$$= \frac{8 \pm 2\sqrt{10}}{2}$$
$$= 4 \pm \sqrt{10}$$

Your Turn 4

Solve $\frac{1}{x^2 - 4} + \frac{2}{x - 2} = \frac{1}{x}$.

$$\frac{1}{(x-2)(x+2)} + \frac{2}{x-2} = \frac{1}{x}$$
$$(x-2)(x+2)(x) \cdot \frac{1}{(x-2)(x+2)}$$
$$+ (x-2)(x+2)(x) \cdot \frac{2}{x-2}$$
$$= (x-2)(x+2)(x) \cdot \frac{1}{x}$$
$$x + 2x^2 + 4x = x^2 - 4$$
$$x^2 + 5x + 4 = 0$$
$$(x+1)(x+4) = 0$$
$$x = -1 \text{ or } x = -4$$

Neither of these values makes a denominator equal to zero, so both are solutions.

R.4 Exercises

1. $2x + 8 = x - 4$
$$x + 8 = -4$$
$$x = -12$$

The solution is -12.

3. $0.2m - 0.5 = 0.1m + 0.7$
$$10(0.2m - 0.5) = 10(0.1m + 0.7)$$
$$2m - 5 = m + 7$$
$$m - 5 = 7$$
$$m = 12$$

The solution is 12.

5. $3r + 2 - 5(r + 1) = 6r + 4$
$$3r + 2 - 5r - 5 = 6r + 4$$
$$-3 - 2r = 6r + 4$$
$$-3 = 8r + 4$$
$$-7 = 8r$$
$$-\frac{7}{8} = r$$

The solution is $-\frac{7}{8}$.

7. $2[3m - 2(3 - m) - 4] = 6m - 4$
$$2[3m - 6 + 2m - 4] = 6m - 4$$
$$2[5m - 10] = 6m - 4$$
$$10m - 20 = 6m - 4$$
$$4m - 20 = -4$$
$$4m = 16$$
$$m = 4$$

The solution is 4.

9. $x^2 + 5x + 6 = 0$
$$(x + 3)(x + 2) = 0$$
$$x + 3 = 0 \quad \text{or} \quad x + 2 = 0$$
$$x = -3 \quad \text{or} \quad x = -2$$

The solutions are -3 and -2.

11. $m^2 = 14m - 49$
$$m^2 - 14m + 49 = 0$$
$$(m)^2 - 2(7m) + (7)^2 = 0$$
$$(m - 7)^2 = 0$$
$$m - 7 = 0$$
$$m = 7$$

The solution is 7.

13. $12x^2 - 5x = 2$
$$12x^2 - 5x - 2 = 0$$
$$(4x + 1)(3x - 2) = 0$$
$$4x + 1 = 0 \quad \text{or} \quad 3x - 2 = 0$$
$$4x = -1 \quad \text{or} \quad 3x = 2$$
$$x = -\frac{1}{4} \quad \text{or} \quad x = \frac{2}{3}$$

The solutions are $-\frac{1}{4}$ and $\frac{2}{3}$.

15. $4x^2 - 36 = 0$

Divide both sides of the equation by 4.

$$x^2 - 9 = 0$$
$$(x + 3)(x - 3) = 0$$
$$x + 3 = 0 \quad \text{or} \quad x - 3 = 0$$
$$x = -3 \quad \text{or} \quad x = 3$$

The solutions are -3 and 3.

17. $12y^2 - 48y = 0$

$$12y(y) - 12y(4) = 0$$
$$12y(y - 4) = 0$$
$$12y = 0 \quad \text{or} \quad y - 4 = 0$$
$$y = 0 \quad \text{or} \quad y = 4$$

The solutions are 0 and 4.

19. $2m^2 - 4m = 3$

$$2m^2 - 4m - 3 = 0$$
$$m = \frac{-(-4) \pm \sqrt{(-4)^2 - 4(2)(-3)}}{2(2)}$$
$$= \frac{4 \pm \sqrt{40}}{4} = \frac{4 \pm \sqrt{4 \cdot 10}}{4}$$
$$= \frac{4 \pm \sqrt{4}\sqrt{10}}{4}$$
$$= \frac{4 \pm 2\sqrt{10}}{4} = \frac{2 \pm \sqrt{10}}{2}$$

The solutions are $\frac{2+\sqrt{10}}{2} \approx 2.5811$ and $\frac{2-\sqrt{10}}{2} \approx -0.5811$.

21. $k^2 - 10k = -20$

$$k^2 - 10k + 20 = 0$$
$$k = \frac{-(-10) \pm \sqrt{(-10)^2 - 4(1)(20)}}{2(1)}$$
$$k = \frac{10 \pm \sqrt{100 - 80}}{2}$$
$$k = \frac{10 \pm \sqrt{20}}{2}$$
$$k = \frac{10 \pm 2\sqrt{5}}{2}$$
$$k = \frac{2(5 \pm \sqrt{5})}{2}$$
$$k = 5 \pm \sqrt{5}$$

The solutions are $5 + \sqrt{5} \approx 7.2361$ and $5 - \sqrt{5} \approx 2.7639$.

23. $2r^2 - 7r + 5 = 0$

$$(2r - 5)(r - 1) = 0$$
$$r - 1 = 0 \quad \text{or} \quad r = 1$$
$$2r - 5 = 0 \quad \text{or} \quad r = \frac{5}{2}$$

The solutions are $\frac{5}{2}$ and 1.

25. $3k^2 + k = 6$

$$3k^2 + k - 6 = 0$$
$$k = \frac{-1 \pm \sqrt{1 - 4(3)(-6)}}{2(3)}$$
$$= \frac{-1 \pm \sqrt{73}}{6}$$

The solutions are $\frac{-1+\sqrt{73}}{6} \approx 1.2573$ and $\frac{-1-\sqrt{73}}{6} \approx -1.5907$.

27. $\frac{3x - 2}{7} = \frac{x + 2}{5}$

$$35\left(\frac{3x - 2}{7}\right) = 35\left(\frac{x + 2}{2}\right)$$
$$5(3x - 2) = 7(x + 2)$$
$$15x - 10 = 7x + 14$$
$$8x = 24$$
$$x = 3$$

The solution is $x = 3$.

29. $\frac{4}{x - 3} - \frac{8}{2x + 5} + \frac{3}{x - 3} = 0$

$$\frac{4}{x - 3} + \frac{3}{x - 3} - \frac{8}{2x + 5} = 0$$
$$\frac{7}{x - 3} - \frac{8}{2x + 5} = 0$$

Multiply both sides by $(x - 3)(2x + 5)$. Note that $x \neq 3$ and $x \neq \frac{5}{2}$.

$$(x - 3)(2x + 5)\left(\frac{7}{x - 3} - \frac{8}{2x + 5}\right)$$
$$= (x - 3)(2x + 5)(0)$$
$$7(2x + 5) - 8(x - 3) = 0$$
$$14x + 35 - 8x + 24 = 0$$
$$6x + 59 = 0$$
$$6x = -59 \quad \text{or} \quad x = -\frac{59}{6}$$

Note: It is especially important to check solutions of equations that involve rational expressions. Here, a check shows that $-\frac{59}{6}$ is a solution.

31. $\dfrac{2m}{m-2} - \dfrac{6}{m} = \dfrac{12}{m^2 - 2m}$

$\dfrac{2m}{m-2} - \dfrac{6}{m} = \dfrac{12}{m(m-2)}$

Multiply both sides by $m(m-2)$. Note that $m \neq 0$ and $m \neq 2$.

$$m(m-2)\left(\frac{2m}{m-2} - \frac{6}{m}\right) = m(m-2)\left(\frac{12}{m(m-2)}\right)$$

$$\begin{aligned} m(2m) - 6(m-2) &= 12 \\ 2m^2 - 6m + 12 &= 12 \\ 2m^2 - 6m &= 0 \\ 2m(m-3) &= 0 \end{aligned}$$

$$2m = 0 \quad \text{or} \quad m - 3 = 0$$
$$m = 0 \quad \text{or} \quad m = 3$$

Since $m \neq 0$, 0 is not a solution. The solution is 3.

33. $\dfrac{1}{x-2} - \dfrac{3x}{x-1} = \dfrac{2x+1}{x^2 - 3x + 2}$

$\dfrac{1}{x-2} - \dfrac{3x}{x-1} = \dfrac{2x+1}{(x-2)(x-1)}$

Multiply both sides by $(x-2)(x-1)$. Note that $x \neq 2$ and $x \neq 1$.

$$(x-2)(x-1)\left(\frac{1}{x-2} - \frac{3x}{x-1}\right) = (x-2)(x-1)\cdot\left[\frac{2x+1}{(x-2)(x-1)}\right]$$

$$(x-2)(x-1)\left(\frac{1}{x-2}\right) - (x-2)(x-1)\cdot\left(\frac{3x}{x-1}\right) = \frac{(x-2)(x-1)(2x+1)}{(x-2)(x-1)}$$

$$\begin{aligned} (x-1) - (x-2)(3x) &= 2x + 1 \\ x - 1 - 3x^2 + 6x &= 2x + 1 \\ -3x^2 + 7x - 1 &= 2x + 1 \\ -3x^2 + 5x - 2 &= 0 \\ 3x^2 - 5x + 2 &= 0 \\ (3x-2)(x-1) &= 0 \end{aligned}$$

$$3x - 2 = 0 \quad \text{or} \quad x - 1 = 0$$
$$x = \frac{2}{3} \quad \text{or} \quad x = 1$$

1 is not a solution since $x \neq 1$. The solution is $\frac{2}{3}$.

35. $\dfrac{5}{b+5} - \dfrac{4}{b^2 + 2b} = \dfrac{6}{b^2 + 7b + 10}$

$\dfrac{5}{b+5} - \dfrac{4}{b(b+2)} = \dfrac{6}{(b+5)(b+2)}$

Multiply both sides by $b(b+5)(b+2)$. Note that $b \neq 0, b \neq -5$, and $b \neq -2$.

$$b(b+5)(b+2)\left(\frac{5}{b+5} - \frac{4}{b(b+2)}\right) = b(b+5)(b+2)\left(\frac{6}{(b+5)(b+2)}\right)$$

$$\begin{aligned} 5b(b+2) - 4(b+5) &= 6b \\ 5b^2 + 10b - 4b - 20 &= 6b \\ 5b^2 - 20 &= 0 \\ b^2 - 4 &= 0 \\ (b+2)(b-2) &= 0 \end{aligned}$$

$$b + 2 = 0 \quad \text{or} \quad b - 2 = 0$$
$$b = -2 \quad \text{or} \quad b = 2$$

Since $b \neq -2, -2$ is not a solution. The solution is 2.

37. $\dfrac{4}{2x^2 + 3x - 9} + \dfrac{2}{2x^2 - x - 3} = \dfrac{3}{x^2 + 4x + 3}$

$\dfrac{4}{(2x-3)(x+3)} + \dfrac{2}{(2x-3)(x+1)} = \dfrac{3}{(x+3)(x+1)}$

Multiply both sides by $(2x-3)(x+3)(x+1)$. Note that $x \neq \frac{3}{2}$, $x \neq -3$, and $x \neq -1$.

$$(2x-3)(x+3)(x+1)\cdot\left(\frac{4}{(2x-3)(x+3)} + \frac{2}{(2x-3)(x+1)}\right) = (2x-3)(x+3)(x+1)\left(\frac{3}{(x+3)(x+1)}\right)$$

$$\begin{aligned} 4(x+1) + 2(x+3) &= 3(2x-3) \\ 4x + 4 + 2x + 6 &= 6x - 9 \\ 6x + 10 &= 6x - 9 \\ 10 &= -9 \end{aligned}$$

This is a false statement. Therefore, there is no solution.

R.5 Inequalities

Your Turn 1

Solve $3z - 2 > 5z + 7.$

$$\begin{aligned} 3z - 2 &> 5z + 7 \\ 3z - 2 + 2 &> 5z + 7 + 2 \\ 3z &> 5z + 9 \\ 3z - 5z &> 5z - 5z + 9 \\ -2z &> 9 \\ \frac{-2z}{-2} &< \frac{9}{-2} \\ z &< -\frac{9}{2} \end{aligned}$$

Your Turn 2

Solve $3y^2 \leq 16y + 12.$

$$3y^2 \leq 16y + 12$$

$$3y^2 - 16y - 12 \leq 0$$

First solve the equation $3y^2 - 16y - 12 = 0.$

$$3y^2 - 16y - 12 = 0$$

$$(3y + 2)(y - 6) = 0$$

$$3y + 2 = 0 \quad \text{or} \quad y - 6 = 0$$

$$y = -\frac{2}{3} \quad \text{or} \quad y = 6$$

Determine three intervals on the number line and choose a test point in each interval.

A B C

$-\frac{2}{3}$ 6

Choose -1 from interval A: $3(-1)^2 - 16(-1) - 12 > 0$

Choose 0 from interval B: $3(0)^2 - 16(0) - 12 < 0$

Choose 7 from interval C: $3(7)^2 - 16(7) - 12 > 0$

The numbers in interval B satisfy the inequality, and since the sign was less than or equal to, the boundary points of interval B are also part of the solution. The solution is $[-2/3, 6]$.

Your Turn 3

Solve $\frac{k^2 - 35}{k} \geq 2.$

First solve the corresponding equation $\frac{k^2 - 35}{k} = 2.$

$$\frac{k^2 - 35}{k} = 2$$

$$k^2 - 35 = 2k$$

$$k^2 - 2k - 35 = 0$$

$$(k - 7)(k + 5) = 0$$

$$k = 7 \text{ or } k = -5$$

The denominator is 0 when $k = 0,$ so there are four intervals to consider:

$$(-\infty, -5),\ (-5, 0),\ (0, 7), \text{ and } (7, \infty).$$

Choose a test point in each interval.

$$k = -8;\ \frac{(-8)^2 - 35}{-8} < 2$$

$$k = -1;\ \frac{(-1)^2 - 35}{-1} > 2$$

$$k = 5;\ \frac{(-5)^2 - 35}{5} < 2$$

$$k = 10;\ \frac{(10)^2 - 35}{10} > 2$$

The second and fourth intervals are part of the solution. Since the inequality is greater than or equal to, we can include the endpoints -5 and 7 but not the endpoint 0, which makes the denominator 0. The solution is $[-5, 0) \cup [7, \infty)$.

R.5 Exercises

1. $x < 4$

Because the inequality symbol means "less than," the endpoint at 4 is not included. This inequality is written in interval notation as $(-\infty, 4)$. To graph this interval on a number line, place an open circle at 4 and draw a heavy arrow pointing to the left.

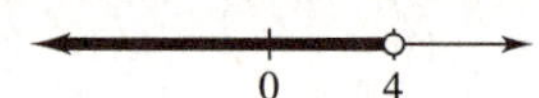

3. $1 \leq x < 2$

The endpoint at 1 is included, but the endpoint at 2 is not. This inequality is written in interval notation as $[1, 2)$. To graph this interval, place a closed circle at 1 and an open circle at 2; then draw a heavy line segment between them.

5. $-9 > x$

This inequality may be rewritten as $x < -9,$ and is written in interval notation as $(-\infty, -9)$. Note that the endpoint at -9 is not included. To graph this interval, place an open circle at -9 and draw a heavy arrow pointing to the left.

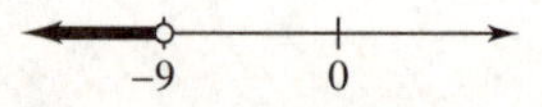

7. $[-7, -3]$

This represents all the numbers between -7 and -3, including both endpoints. This interval can be written as the inequality $-7 \le x \le -3$.

9. $(-\infty, -1]$

This represents all the numbers to the left of -1 on the number line and includes the endpoint. This interval can be written as the inequality $x \le -1$.

11. Notice that the endpoint -2 is included, but 6 is not. The interval shown in the graph can be written as the inequality $-2 \le x < 6$.

13. Notice that both endpoints are included. The interval shown in the graph can be written as $x \le -4$ or $x \ge 4$.

15.
$$6p + 7 \le 19$$
$$6p \le 12$$
$$\left(\frac{1}{6}\right)(6p) \le \left(\frac{1}{6}\right)(12)$$
$$p \le 2$$

The solution in interval notation is $(-\infty, 2]$.

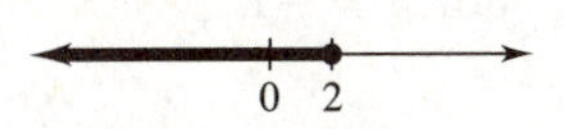

17.
$$m - (3m - 2) + 6 < 7m - 19$$
$$m - 3m + 2 + 6 < 7m - 19$$
$$-2m + 8 < 7m - 19$$
$$-9m + 8 < -19$$
$$-9m < -27$$
$$-\frac{1}{9}(-9m) > -\frac{1}{9}(-27)$$
$$m > 3$$

The solution is $(3, \infty)$.

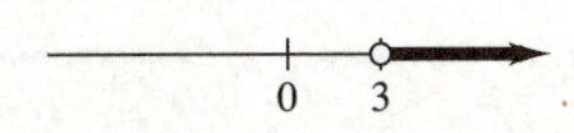

19.
$$3p - 1 < 6p + 2(p - 1)$$
$$3p - 1 < 6p + 2p - 2$$
$$3p - 1 < 8p - 2$$
$$-5p - 1 < -2$$
$$-5p < -1$$
$$-\frac{1}{5}(-5p) > -\frac{1}{5}(-1)$$
$$p > \frac{1}{5}$$

The solution is $\left(\frac{1}{5}, \infty\right)$.

21.
$$-11 < y - 7 < -1$$
$$-11 + 7 < y - 7 + 7 < -1 + 7$$
$$-4 < y < 6$$

The solution is $(-4, 6)$.

23.
$$-2 < \frac{1 - 3k}{4} \le 4$$
$$4(-2) < 4\left(\frac{1 - 3k}{4}\right) \le 4(4)$$
$$-8 < 1 - 3k \le 16$$
$$-9 < -3k \le 15$$
$$-\frac{1}{3}(-9) > -\frac{1}{3}(-3k) \ge -\frac{1}{3}(15)$$

Rewrite the inequalities in the proper order.

$-5 \le k < 3$

The solution is $[-5, 3)$.

25.
$$\frac{3}{5}(2p + 3) \ge \frac{1}{10}(5p + 1)$$
$$10\left(\frac{3}{5}\right)(2p + 3) \ge 10\left(\frac{1}{10}\right)(5p + 1)$$
$$6(2p + 3) \ge 5p + 1$$
$$12p + 18 \ge 5p + 1$$
$$7p \ge -17$$
$$p \ge -\frac{17}{7}$$

The solution is $\left[-\frac{17}{7}, \infty\right)$.

27. $(m - 3)(m + 5) < 0$

Solve $(m - 3)(m + 5) = 0$.
$$(m - 3)(m + 5) = 0$$
$$m = 3 \quad \text{or} \quad m = -5$$

Intervals: $(-\infty, -5), (-5, 3), (3, \infty)$

For $(-\infty, -5)$, choose -6 to test for m.
$$(-6 - 3)(-6 + 5) = -9(-1) = 9 \not< 0$$

For $(-5, 3)$, choose 0.
$$(0 - 3)(0 + 5) = -3(5) = -15 < 0$$

For $(3, \infty)$, choose 4.
$$(4 - 3)(4 + 5) = 1(9) = 9 \not< 0$$

The solution is $(-5, 3)$.

29. $y^2 - 3y + 2 < 0$

$(y - 2)(y - 1) < 0$

Solve $(y - 2)(y - 1) = 0$.

$y = 2$ or $y = 1$

Intervals: $(-\infty, 1), (1, 2), (2, \infty)$

For $(-\infty, 1)$, choose $y = 0$.

$$0^2 - 3(0) + 2 = 2 \not< 0$$

For (1, 2), choose $y = \frac{3}{2}$.

$$\left(\frac{3}{2}\right)^2 - 3\left(\frac{3}{2}\right) + 2 = \frac{9}{4} - \frac{9}{2} + 2$$

$$= \frac{9 - 18 + 8}{4}$$

$$= -\frac{1}{4} < 0$$

For $(2, \infty)$, choose 3.

$$3^2 - 3(3) + 2 = 2 \not< 0$$

The solution is (1, 2).

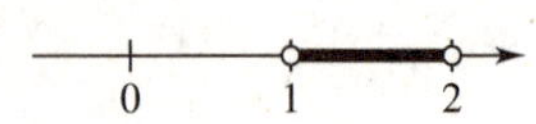

31. $x^2 - 16 > 0$

Solve $x^2 - 16 = 0$.

$$x^2 - 16 = 0$$

$$(x + 4)(x - 4) = 0$$

$x = -4$ or $x = 4$

Intervals: $(-\infty, -4), (-4, 4), (4, \infty)$

For $(-\infty, -4)$, choose -5.

$$(-5)^2 - 16 = 9 > 0$$

For $(-4, 4)$, choose 0.

$$0^2 - 16 = -16 \not> 0$$

For $(4, \infty)$, choose 5.

$$5^2 - 16 = 9 > 0$$

The solution is $(-\infty, -4) \cup (4, \infty)$.

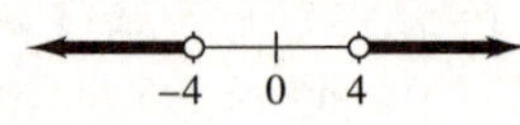

33. $x^2 - 4x \geq 5$

Solve $x^2 - 4x = 5$.

$$x^2 - 4x = 5$$

$$x^2 - 4x - 5 = 0$$

$$(x + 1)(x - 5) = 0$$

$x + 1 = 0$ or $x - 5 = 0$

$x = -1$ or $x = 5$

Intervals: $(-\infty, -1), (-1, 5), (5, \infty)$

For $(-\infty, -1)$, choose -2.

$$(-2)^2 - 4(-2) = 12 \geq 5$$

For $(-1, 5)$, choose 0.

$$0^2 - 4(0) = 0 \not\geq 5$$

For $(5, \infty)$, choose 6.

$$(6)^2 - 4(6) = 12 \geq 5$$

The solution is $(-\infty, -1] \cup [5, \infty)$.

35. $3x^2 + 2x > 1$

Solve $3x^2 + 2x = 1$.

$$3x^2 + 2x = 1$$

$$3x^2 + 2x - 1 = 0$$

$$(3x - 1)(x + 1) = 0$$

$x = \frac{1}{3}$ or $x = -1$

Intervals: $(-\infty, -1), \left(-1, \frac{1}{3}\right), \left(\frac{1}{3}, \infty\right)$

For $(-\infty, -1)$, choose -2.

$$3(-2)^2 + 2(-2) = 8 > 1$$

For $\left(-1, \frac{1}{3}\right)$, choose 0.

$$3(0)^2 + 2(0) = 0 \not> 1$$

For $\left(\frac{1}{3}, \infty\right)$, choose 1.

$$3(1)^2 + 2(1) = 5 > 1$$

The solution is $(-\infty, -1) \cup \left(\frac{1}{3}, \infty\right)$.

37. $9 - x^2 \leq 0$

Solve $9 - x^2 = 0$.

$$9 - x^2 = 0$$

$$(3 + x)(3 - x) = 0$$

$x = -3$ or $x = 3$

Intervals: $(-\infty, -3), (-3, 3), (3, \infty)$

For $(-\infty, -3)$, choose -4.

$$9 - (-4)^2 = -7 \leq 0$$

For $(-3, 3)$, choose 0.

$$9 - (0)^2 = 9 \not\leq 0$$

For $(3, \infty)$, choose 4.

$$9 - (4)^2 = -7 \leq 0$$

The solution is $(-\infty, -3] \cup [3, \infty)$.

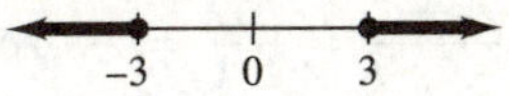

39. $x^3 - 4x \geq 0$

Solve $x^3 - 4x = 0$.

$$x^3 - 4x = 0$$
$$x(x^2 - 4) = 0$$
$$x(x + 2)(x - 2) = 0$$
$$x = 0, \quad \text{or} \quad x = -2, \quad \text{or} \quad x = 2$$

Intervals: $(-\infty, -2), (-2, 0), (0, 2), (2, \infty)$

For $(-\infty, -2)$, choose -3.

$$(-3)^3 - 4(-3) = -15 \not\geq 0$$

For $(-2, 0)$, choose -1.

$$(-1)^3 - 4(-1) = 3 \geq 0$$

For $(0, 2)$, choose 1.

$$(1)^3 - 4(1) = -3 \not\geq 0$$

For $(2, \infty)$, choose 3.

$$(3)^3 - 4(3) = 15 \geq 0$$

The solution is $[-2, 0] \cup [2, \infty)$.

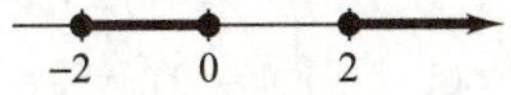

41. $2x^3 - 14x^2 + 12x < 0$

Solve $2x^3 - 14x^2 + 12x = 0$.

$$2x^3 - 14x^2 + 12x = 0$$
$$2x(x^2 - 7x + 6) = 0$$
$$2x(x - 1)(x - 6) = 0$$
$$x = 0, \quad \text{or} \quad x = 1, \quad \text{or} \quad x = 6$$

Intervals: $(-\infty, 0), (0, 1), (1, 6), (6, \infty)$

For $(-\infty, 0)$, choose -1.

$$2(-1)^3 - 14(-1)^2 + 12(-1) = -28 < 0$$

For $(0, 1)$, choose $\frac{1}{2}$.

$$2\left(\frac{1}{2}\right)^3 - 14\left(\frac{1}{2}\right)^2 + 12\left(\frac{1}{2}\right) = \frac{11}{4} \not< 0$$

For $(1, 6)$, choose 2.

$$2(2)^3 - 14(2)^2 + 12(2) = -16 < 0$$

For $(6, \infty)$, choose 7.

$$2(7)^3 - 14(7)^2 + 12(7) = 84 \not< 0$$

The solution is $(-\infty, 0) \cup (1, 6)$.

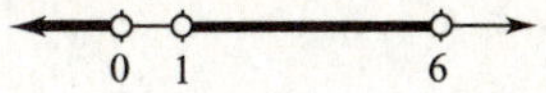

43. $\dfrac{m - 3}{m + 5} \leq 0$

Solve $\dfrac{m - 3}{m + 5} = 0$.

$$(m + 5)\frac{m - 3}{m + 5} = (m + 5)(0)$$
$$m - 3 = 0$$
$$m = 3$$

Set the denominator equal to 0 and solve.

$$m + 5 = 0$$
$$m = -5$$

Intervals: $(-\infty, -5), (-5, 3), (3, \infty)$

For $(-\infty, -5)$, choose -6.

$$\frac{-6 - 3}{-6 + 5} = 9 \not\leq 0$$

For $(-5, 3)$, choose 0.

$$\frac{0 - 3}{0 + 5} = -\frac{3}{5} \leq 0$$

For $(3, \infty)$, choose 4.

$$\frac{4 - 3}{4 + 5} = \frac{1}{9} \not\leq 0$$

Although the $\leq$ symbol is used, including -5 in the solution would cause the denominator to be zero.

The solution is $(-5, 3]$.

45. $\dfrac{k - 1}{k + 2} > 1$

Solve $\dfrac{k - 1}{k + 2} = 1$.

$$k - 1 = k + 2$$
$$-1 \neq 2$$

The equation has no solution. Solve $k + 2 = 0$.

$$k = -2$$

Intervals: $(-\infty, -2), (-2, \infty)$

For $(-\infty, -2)$, choose -3.

$$\frac{-3 - 1}{-3 + 2} = 4 > 1$$

For $(-2, \infty)$, choose 0.

$$\frac{0 - 1}{0 + 2} = -\frac{1}{2} \not> 1$$

The solution is $(-\infty, -2)$.

47. $\frac{2y+3}{y-5} \le 1$

Solve $\frac{2y+3}{y-5} = 1.$

$$2y + 3 = y - 5$$
$$y = -8$$

Solve $y - 5 = 0.$

$$y = 5$$

Intervals: $(-\infty, -8), (-8, 5), (5, \infty)$

For $(-\infty, -8)$, choose $y = -10$.

$$\frac{2(-10)+3}{-10-5} = \frac{17}{15} \not\le 1$$

For $(-8, 5)$, choose $y = 0$.

$$\frac{2(0)+3}{0-5} = -\frac{3}{5} \le 1$$

For $(5, \infty)$, choose $y = 6$.

$$\frac{2(6)+3}{6-5} = \frac{15}{1} \not\le 1$$

The solution is $[-8, 5)$.

49. $\frac{2k}{k-3} \le \frac{4}{k-3}$

Solve $\frac{2k}{k-3} = \frac{4}{k-3}$.

$$\frac{2k}{k-3} = \frac{4}{k-3}$$
$$\frac{2k}{k-3} - \frac{4}{k-3} = 0$$
$$\frac{2k-4}{k-3} = 0$$
$$2k - 4 = 0$$
$$k = 2$$

Set the denominator equal to 0 and solve for k.

$$k - 3 = 0$$
$$k = 3$$

Intervals: $(-\infty, 2), (2, 3), (3, \infty)$

For $(-\infty, 2)$, choose 0.

$$\frac{2(0)}{0-3} = 0 \text{ and } \frac{4}{0-3} = -\frac{4}{3}, \text{ so}$$
$$\frac{2(0)}{0-3} \not\le \frac{4}{0-3}.$$

For $(2, 3)$, choose $\frac{5}{2}$.

$$\frac{2\left(\frac{5}{2}\right)}{\frac{5}{2}-3} = \frac{5}{-\frac{1}{2}} = -10$$
$$\text{and } \frac{4}{\frac{5}{2}-3} = \frac{4}{-\frac{1}{2}} = -8, \text{ so}$$
$$\frac{2\left(\frac{5}{2}\right)}{\frac{5}{2}-3} \le \frac{4}{\frac{5}{2}-3}.$$

For $(3, \infty)$, choose 4.

$$\frac{2(4)}{4-3} = 8 \text{ and } \frac{4}{4-3} = 4, \text{ so}$$
$$\frac{2(4)}{4-3} \not\le \frac{4}{4-3}.$$

The solution is $[2, 3)$.

51. $\frac{2x}{x^2 - x - 6} \ge 0$

Solve $\frac{2x}{x^2 - x - 6} = 0.$

$$\frac{2x}{x^2 - x - 6} = 0$$
$$2x = 0$$
$$x = 0$$

Set the denominator equal to 0 and solve for x.

$$x^2 - x - 6 = 0$$
$$(x+2)(x-3) = 0$$
$$x + 2 = 0 \quad \text{or} \quad x - 3 = 0$$
$$x = -2 \quad \text{or} \quad x = 3$$

Intervals: $(-\infty, -2), (-2, 0), (0, 3), (3, \infty)$

For $(-\infty, -2)$, choose -3.

$$\frac{2(-3)}{(-3)^2 - (-3) - 6} = -1 \not\ge 0$$

For $(-2, 0)$, choose -1.

$$\frac{2(-1)}{(-1)^2 - (-1) - 6} = \frac{1}{2} \ge 0$$

For $(0, 3)$, choose 2.

$$\frac{2(2)}{2^2 - 2 - 6} = -1 \not\ge 0$$

For $(3, \infty)$, choose 4.

$$\frac{2(4)}{4^2 - 4 - 6} = \frac{4}{3} \ge 0$$

The solution is $(-2, 0] \cup (3, \infty)$.

53. $\frac{z^2 + z}{z^2 - 1} \geq 3$

Solve

$$\frac{z^2 + z}{z^2 - 1} = 3.$$
$$z^2 + z = 3z^2 - 3$$
$$-2z^2 + z + 3 = 0$$
$$-1(2z^2 - z - 3) = 0$$
$$-1(z + 1)(2z - 3) = 0$$
$$z = -1 \quad \text{or} \quad z = \frac{3}{2}$$

Set $z^2 - 1 = 0.$

$$z^2 = 1$$
$$z = -1 \quad \text{or} \quad z = 1$$

Intervals: $(-\infty, -1), (-1, 1), \left(1, \frac{3}{2}\right), \left(\frac{3}{2}, \infty\right)$

For $(-\infty, -1)$, choose $x = -2$.

$$\frac{(-2)^2 + 3}{(-2)^2 - 1} = \frac{7}{3} \not\geq 3$$

For $(-1, 1)$, choose $x = 0$.

$$\frac{0^2 + 3}{0^2 - 1} = -3 \not\geq 3$$

For $\left(1, \frac{3}{2}\right)$, choose $x = \frac{3}{2}$.

$$\frac{\left(\frac{3}{2}\right)^2 + 3}{\left(\frac{3}{2}\right)^2 - 1} = \frac{21}{5} \geq 3$$

For $\left(\frac{3}{2}, \infty\right)$, choose $x = 2$.

$$\frac{2^2 + 3}{2^2 - 1} = \frac{7}{3} \not\geq 3$$

The solution is $\left(1, \frac{3}{2}\right]$.

R.6 Exponents

Your Turn 1

Simplify $\left(\frac{y^2 z^{-4}}{y^{-3} z^4}\right)^{-2}$.

$$\left(\frac{y^2 z^{-4}}{y^{-3} z^4}\right)^{-2} = \frac{\left(y^2 z^{-4}\right)^{-2}}{\left(y^{-3} z^4\right)^{-2}} = \frac{y^{(2)(-2)} z^{(-4)(-2)}}{y^{(-3)(-2)} z^{(4)(-2)}}$$
$$= \frac{y^{-4} z^8}{y^6 z^{-8}} = \frac{z^{8-(-8)}}{y^{6-(-4)}} = \frac{z^{16}}{y^{10}}$$

Your Turn 2

Factor $5z^{1/3} + 4z^{-2/3}$.

$$5z^{1/3} + 4z^{-2/3} = z^{-2/3}\left(5z^{(1/3)+(2/3)} + 4z^{(-2/3)+(2/3)}\right)$$
$$= z^{-2/3}(5z + 4)$$

R.6 Exercises

1. $8^{-2} = \frac{1}{8^2} = \frac{1}{64}$

3. $5^0 = 1$, by definition.

5. $-(-3)^{-2} = -\frac{1}{(-3)^2} = -\frac{1}{9}$

7. $\left(\frac{1}{6}\right)^{-2} = \frac{1}{\left(\frac{1}{6}\right)^2} = \frac{1}{\frac{1}{36}} = 36$

9. $\frac{4^{-2}}{4} = 4^{-2-1} = 4^{-3} = \frac{1}{4^3} = \frac{1}{64}$

11. $\frac{10^8 \cdot 10^{-10}}{10^4 \cdot 10^2}$

$$= \frac{10^{8+(-10)}}{10^{4+2}} = \frac{10^{-2}}{10^6}$$
$$= 10^{-2-6} = 10^{-8}$$
$$= \frac{1}{10^8}$$

13. $\frac{x^4 \cdot x^3}{x^5} = \frac{x^{4+3}}{x^5} = \frac{x^7}{x^5} = x^{7-5} = x^2$

15. $\frac{(4k^{-1})^2}{2k^{-5}} = \frac{4^2 k^{-2}}{2k^{-5}} = \frac{16k^{-2-(-5)}}{2}$

$$= 8k^{-2+5} = 8k^3$$
$$= 2^3 k^3$$

17. $\frac{3^{-1} \cdot x \cdot y^2}{x^{-4} \cdot y^5} = 3^{-1} \cdot x^{1-(-4)} \cdot y^{2-5}$

$$= 3^{-1} \cdot x^{1+4} \cdot y^{-3}$$
$$= \frac{1}{3} \cdot x^5 \cdot \frac{1}{y^3}$$
$$= \frac{x^5}{3y^3}$$

19. $\left(\frac{a^{-1}}{b^2}\right)^{-3} = \frac{(a^{-1})^{-3}}{(b^2)^{-3}} = \frac{a^{(-1)(-3)}}{b^{2(-3)}}$

$= \frac{a^3}{b^{-6}} = a^3b^6$

21. $a^{-1} + b^{-1} = \frac{1}{a} + \frac{1}{b}$

$= \left(\frac{b}{b}\right)\left(\frac{1}{a}\right) + \left(\frac{a}{a}\right)\left(\frac{1}{b}\right)$

$= \frac{b}{ab} + \frac{a}{ab}$

$= \frac{b+a}{ab}$

$= \frac{a+b}{ab}$

23. $\frac{2n^{-1} - 2m^{-1}}{m + n^2} = \frac{\frac{2}{n} - \frac{2}{m}}{m + n^2}$

$= \frac{\frac{2}{n} \cdot \frac{m}{m} - \frac{2}{m} \cdot \frac{n}{n}}{(m + n^2)}$

$= \frac{2m - 2n}{mn(m + n^2)}$

or $\frac{2(m - n)}{mn(m + n^2)}$

25. $(x^{-1} - y^{-1})^{-1} = \frac{1}{\frac{1}{x} - \frac{1}{y}}$

$= \frac{1}{\frac{1}{x} \cdot \frac{y}{y} - \frac{1}{y} \cdot \frac{x}{x}}$

$= \frac{1}{\frac{y}{xy} - \frac{x}{xy}}$

$= \frac{1}{\frac{y-x}{xy}}$

$= \frac{xy}{y - x}$

27. $121^{1/2} = (11^2)^{1/2} = 11^{2(1/2)} = 11^1 = 11$

29. $32^{2/5} = (32^{1/5})^2 = 2^2 = 4$

31. $\left(\frac{36}{144}\right)^{1/2} = \frac{36^{1/2}}{144^{1/2}} = \frac{6}{12} = \frac{1}{2}$

This can also be solved by reducing the fraction first.

$\left(\frac{36}{144}\right)^{1/2} = \left(\frac{1}{4}\right)^{1/2} = \frac{1^{1/2}}{4^{1/2}} = \frac{1}{2}$

33. $8^{-4/3} = (8^{1/3})^{-4} = 2^{-4} = \frac{1}{2^4} = \frac{1}{16}$

35. $\left(\frac{27}{64}\right)^{-1/3} = \frac{27^{-1/3}}{64^{-1/3}} = \frac{64^{1/3}}{27^{1/3}} = \frac{4}{3}$

37. $3^{2/3} \cdot 3^{4/3} = 3^{(2/3)+(4/3)} = 3^{6/3} = 3^2 = 9$

39. $\frac{4^{9/4} \cdot 4^{-7/4}}{4^{-10/4}} = 4^{9/4-7/4-(-10/4)}$

$= 4^{12/4} = 4^3 = 64$

41. $\left(\frac{x^6y^{-3}}{x^{-2}y^5}\right)^{1/2} = (x^{6-(-2)}y^{-3-5})^{1/2}$

$= (x^8y^{-8})^{1/2}$

$= (x^8)^{1/2}(y^{-8})^{1/2}$

$= x^4y^{-4}$

$= \frac{x^4}{y^4}$

43. $\frac{7^{-1/3} \cdot 7r^{-3}}{7^{2/3} \cdot (r^{-2})^2} = \frac{7^{-1/3+1}r^{-3}}{7^{2/3} \cdot r^{-4}}$

$= 7^{-1/3+3/3-2/3}r^{-3-(-4)}$

$= 7^0r^{-3+4} = 1 \cdot r^1 = r$

45. $\frac{3k^2 \cdot (4k^{-3})^{-1}}{4^{1/2} \cdot k^{7/2}}$

$= \frac{3k^2 \cdot 4^{-1}k^3}{2 \cdot k^{7/2}}$

$= 3 \cdot 2^{-1} \cdot 4^{-1}k^{2+3-(7/2)}$

$= \frac{3}{8} \cdot k^{3/2}$

$= \frac{3k^{3/2}}{8}$

47. $\frac{a^{4/3}}{a^{2/3}} \cdot \frac{b^{1/2}}{b^{-3/2}} = a^{4/3-2/3}b^{1/2-(-3/2)}$

$= a^{2/3}b^2$

49. $\frac{k^{-3/5} \cdot h^{-1/3} \cdot t^{2/5}}{k^{-1/5} \cdot h^{-2/3} \cdot t^{1/5}}$

$= k^{-3/5-(-1/5)}h^{-1/3-(-2/3)}t^{2/5-1/5}$

$= k^{-3/5+1/5}h^{-1/3+2/3}t^{2/5-1/5}$

$= k^{-2/5}h^{1/3}t^{1/5} = \frac{h^{1/3}t^{1/5}}{k^{2/5}}$

51. $3x^3(x^2+3x)^2 - 15x(x^2+3x)^2$
$= 3x \cdot x^2(x^2+3x)^2 - 3x \cdot 5(x^2+3x)^2$
$= 3x(x^2+3x)^2(x^2-5)$

53. $10x^3(x^2-1)^{-1/2} - 5x(x^2-1)^{1/2}$
$= 5x \cdot 2x^2(x^2-1)^{-1/2} - 5x(x^2-1)^{-1/2}(x^2-1)^1$
$= 5x(x^2-1)^{-1/2}[2x^2-(x^2-1)]$
$= 5x(x^2-1)^{-1/2}(x^2+1)$

55. $x(2x+5)^2(x^2-4)^{-1/2} + 2(x^2-4)^{1/2}(2x+5)$
$= (2x+5)^2(x^2-4)^{-1/2}(x)$
$+ (x^2-4)^1(x^2-4)^{-1/2}(2)(2x+5)$
$= (2x+5)(x^2-4)^{-1/2}$
$\cdot [(2x+5)(x) + (x^2-4)(2)]$
$= (2x+5)(x^2-4)^{-1/2} \cdot (2x^2+5x+2x^2-8)$
$= (2x+5)(x^2-4)^{-1/2}(4x^2+5x-8)$

R.7 Radicals

Your Turn 1

Simplify $\sqrt{28x^9y^5}$.

$\sqrt{28x^9y^5} = \sqrt{4 \cdot x^8 \cdot y^4 \cdot 7xy}$
$= 2x^4y^2\sqrt{7xy}$

Your Turn 2

Rationalize the denominator in $\dfrac{5}{\sqrt{x}-\sqrt{y}}$.

$\dfrac{5}{\sqrt{x}-\sqrt{y}} = \dfrac{5}{\sqrt{x}-\sqrt{y}} \cdot \dfrac{\sqrt{x}+\sqrt{y}}{\sqrt{x}+\sqrt{y}}$
$= \dfrac{5\left(\sqrt{x}+\sqrt{y}\right)}{x-y}$

R.7 Exercises

1. $\sqrt[3]{125} = 5$ because $5^3 = 125$.

3. $\sqrt[5]{-3125} = -5$ because $(-5)^5 = -3125$.

5. $\sqrt{2000} = \sqrt{4 \cdot 100 \cdot 5}$
$= 2 \cdot 10\sqrt{5}$
$= 20\sqrt{5}$

7. $\sqrt{27} \cdot \sqrt{3} = \sqrt{27 \cdot 3} = \sqrt{81} = 9$

9. $7\sqrt{2} - 8\sqrt{18} + 4\sqrt{72}$
$= 7\sqrt{2} - 8\sqrt{9 \cdot 2} + 4\sqrt{36 \cdot 2}$
$= 7\sqrt{2} - 8(3)\sqrt{2} + 4(6)\sqrt{2}$
$= 7\sqrt{2} - 24\sqrt{2} + 24\sqrt{2}$
$= 7\sqrt{2}$

11. $4\sqrt{7} - \sqrt{28} + \sqrt{343}$
$= 4\sqrt{7} - \sqrt{4}\sqrt{7} + \sqrt{49}\sqrt{7}$
$= 4\sqrt{7} - 2\sqrt{7} + 7\sqrt{7}$
$= (4-2+7)\sqrt{7}$
$= 9\sqrt{7}$

13. $\sqrt[3]{2} - \sqrt[3]{16} + 2\sqrt[3]{54}$
$= \sqrt[3]{2} - (\sqrt[3]{8 \cdot 2}) + 2(\sqrt[3]{27 \cdot 2})$
$= \sqrt[3]{2} - \sqrt[3]{8}\sqrt[3]{2} + 2(\sqrt[3]{27}\sqrt[3]{2})$
$= \sqrt[3]{2} - 2\sqrt[3]{2} + 2(3\sqrt[3]{2})$
$= \sqrt[3]{2} - 2\sqrt[3]{2} + 6\sqrt[3]{2}$
$= 5\sqrt[3]{2}$

15. $\sqrt{2x^3y^2z^4} = \sqrt{x^2y^2z^4 \cdot 2x} = xyz^2\sqrt{2x}$

17. $\sqrt[3]{128x^3y^8z^9} = \sqrt[3]{64x^3y^6z^9 \cdot 2y^2}$
$= \sqrt[3]{64x^3y^6z^9}\sqrt[3]{2y^2}$
$= 4xy^2z^3\sqrt[3]{2y^2}$

19. $\sqrt{a^3b^5} - 2\sqrt{a^7b^3} + \sqrt{a^3b^9}$
$= \sqrt{a^2b^4ab} - 2\sqrt{a^6b^2ab} + \sqrt{a^2b^8ab}$
$= ab^2\sqrt{ab} - 2a^3b\sqrt{ab} + ab^4\sqrt{ab}$
$= (ab^2 - 2a^3b + ab^4)\sqrt{ab}$
$= ab\sqrt{ab}(b - 2a^2 + b^3)$

21. $\sqrt{a} \cdot \sqrt[3]{a} = a^{1/2} \cdot a^{1/3}$
$= a^{1/2+(1/3)}$
$= a^{5/6} = \sqrt[6]{a^5}$

23. $\sqrt{16 - 8x + x^2}$
$= \sqrt{(4-x)^2}$
$= |4-x|$

25. $\sqrt{4-25z^2} = \sqrt{(2+5z)(2-5z)}$

This factorization does not produce a perfect square, so the expression $\sqrt{4-25z^2}$ cannot be simplified.

27. $\frac{5}{\sqrt{7}} = \frac{5}{\sqrt{7}} \cdot \frac{\sqrt{7}}{\sqrt{7}} = \frac{5\sqrt{7}}{7}$

29. $$\frac{-3}{\sqrt{12}} = \frac{-3}{\sqrt{4 \cdot 3}}$$
$$= \frac{-3}{2\sqrt{3}} \cdot \frac{\sqrt{3}}{\sqrt{3}} = \frac{-3\sqrt{3}}{6} = -\frac{\sqrt{3}}{2}$$

31. $$\frac{3}{1-\sqrt{2}} = \frac{3}{1-\sqrt{2}} \cdot \frac{1+\sqrt{2}}{1+\sqrt{2}}$$
$$= \frac{3(1+\sqrt{2})}{1-2}$$
$$= -3(1+\sqrt{2})$$

33. $$\frac{6}{2+\sqrt{2}} = \frac{6}{2+\sqrt{2}} \cdot \frac{2-\sqrt{2}}{2-\sqrt{2}}$$
$$= \frac{6(2-\sqrt{2})}{4-2\sqrt{2}+2\sqrt{2}-\sqrt{4}}$$
$$= \frac{6(2-\sqrt{2})}{4-2} = \frac{6(2-\sqrt{2})}{2}$$
$$= 3(2-\sqrt{2})$$

35. $$\frac{1}{\sqrt{r}-\sqrt{3}} = \frac{1}{\sqrt{r}-\sqrt{3}} \cdot \frac{\sqrt{r}+\sqrt{3}}{\sqrt{r}+\sqrt{3}}$$
$$= \frac{\sqrt{r}+\sqrt{3}}{r-3}$$

37. $$\frac{y-5}{\sqrt{y}-\sqrt{5}} = \frac{y-5}{\sqrt{y}-\sqrt{5}} \cdot \frac{\sqrt{y}+\sqrt{5}}{\sqrt{y}+\sqrt{5}}$$
$$= \frac{(y-5)(\sqrt{y}+\sqrt{5})}{y-5}$$
$$= \sqrt{y}+\sqrt{5}$$

39. $$\frac{\sqrt{x}+\sqrt{x+1}}{\sqrt{x}-\sqrt{x+1}} = \frac{\sqrt{x}+\sqrt{x+1}}{\sqrt{x}-\sqrt{x+1}} \cdot \frac{\sqrt{x}+\sqrt{x+1}}{\sqrt{x}+\sqrt{x+1}}$$
$$= \frac{x+2\sqrt{x(x+1)}+(x+1)}{x-(x+1)}$$
$$= \frac{2x+2\sqrt{x(x+1)}+1}{-1}$$
$$= -2x-2\sqrt{x(x+1)}-1$$

41. $$\frac{1+\sqrt{2}}{2} = \frac{\left(1+\sqrt{2}\right)\left(1-\sqrt{2}\right)}{2\left(1-\sqrt{2}\right)}$$
$$= \frac{1-2}{2\left(1-\sqrt{2}\right)}$$
$$= -\frac{1}{2\left(1-\sqrt{2}\right)}$$

43. $$\frac{\sqrt{x}+\sqrt{x+1}}{\sqrt{x}-\sqrt{x+1}}$$
$$= \frac{\sqrt{x}+\sqrt{x+1}}{\sqrt{x}-\sqrt{x+1}} \cdot \frac{\sqrt{x}-\sqrt{x+1}}{\sqrt{x}-\sqrt{x+1}}$$
$$= \frac{x-(x+1)}{x-2\sqrt{x}\cdot\sqrt{x+1}+(x+1)}$$
$$= \frac{-1}{2x-2\sqrt{x(x+1)}+1}$$

Chapter 1

LINEAR FUNCTIONS

1.1 Slopes and Equations of Lines

Your Turn 1

Find the slope of the line through $(1,5)$ and $(4,6)$.

Let $(x_1, y_1) = (1,5)$ and $(x_2, y_2) = (4,6)$.

$$m = \frac{6-5}{4-1} = \frac{1}{3}$$

Your Turn 2

Find the equation of the line with x-intercept -4 and y-intercept 6.

We know that $b = 6$ and that the line crosses the axes at $(-4,0)$ and $(0,6)$. Use these two intercepts to find the slope m.

$$m = \frac{6-0}{0-(-4)} = \frac{6}{4} = \frac{3}{2}$$

Thus the equation for the line in slope-intercept form is $y = \frac{3}{2}x + 6.$

Your Turn 3

Find the slope of the line whose equation is $8x + 3y = 5$.

Solve the equation for y.

$$\begin{aligned} 8x + 3y &= 5 \\ 3y &= -8x + 5 \\ y &= -\frac{8}{3}x + \frac{5}{3} \end{aligned}$$

The slope is $-8/3$.

Your Turn 4

Find the equation (in slope-intercept form) of the line through $(2,9)$ and $(5,3)$.

First find the slope.

$$m = \frac{3-9}{5-2} = \frac{-6}{3} = -2$$

Now use the point-slope form, with $(x_1, y_1) = (5,3)$.

$$\begin{aligned} y - y_1 &= m(x - x_1) \\ y - 3 &= -2(x-5) \\ y - 3 &= -2x + 10 \\ y &= -2x + 13 \end{aligned}$$

Your Turn 5

Find (in slope-intercept form) the equation of the line that passes through the point $(4,5)$ and is parallel to the line $3x - 6y = 7$.

First find the slope of the line $3x - 6y = 7$ by solving this equation for y.

$$\begin{aligned} 3x - 6y &= 7 \\ 6y &= 3x - 7 \\ y &= \frac{3}{6}x - \frac{7}{6} \\ y &= \frac{1}{2}x - \frac{7}{6} \end{aligned}$$

Since the line we are to find is parallel to this line, it will also have slope $1/2$. Use the point-slope form with $(x_1, y_1) = (4,5)$.

$$\begin{aligned} y - y_1 &= m(x - x_1) \\ y - 5 &= \frac{1}{2}(x-4) \\ y - 5 &= \frac{1}{2}x - 2 \\ y &= \frac{1}{2}x + 3 \end{aligned}$$

Your Turn 6

Find (in slope-intercept form) the equation of the line that passes through the point $(3,2)$ and is perpendicular to the line $2x + 3y = 4$.

First find the slope of the line $2x + 3y = 4$ by solving this equation for y.

$$\begin{aligned} 2x + 3y &= 4 \\ 3y &= -2x + 4 \\ y &= -\frac{2}{3}x + \frac{4}{3} \end{aligned}$$

Since the line we are to find is perpendicular to a line with slope $-2/3$, , it will have slope $3/2$. (Note that $(-2/3)(3/2) = -1$.)

Use the point-slope form with $(x_1, y_1) = (3,2)$.

$$y - y_1 = m(x - x_1)$$
$$y - 2 = \frac{3}{2}(x - 3)$$
$$y - 2 = \frac{3}{2}x - \frac{9}{2}$$
$$y = \frac{3}{2}x - \frac{5}{2}$$

1.1 Exercises

1. Find the slope of the line through (4, 5) and (−1, 2).

$$m = \frac{5-2}{4-(-1)}$$
$$= \frac{3}{5}$$

3. Find the slope of the line through (8, 4) and (8, −7).

$$m = \frac{4-(-7)}{8-8} = \frac{11}{0}$$

The slope is undefined; the line is vertical.

5. $y = x$

Using the slope-intercept form, $y = mx + b$, we see that the slope is 1.

7. $5x - 9y = 11$

Rewrite the equation in slope-intercept form.

$$9y = 5x - 11$$
$$y = \frac{5}{9}x - \frac{11}{9}$$

The slope is $\frac{5}{9}$.

9. $x = 5$

This is a vertical line. The slope is undefined.

11. $y = 8$

This is a horizontal line, which has a slope of 0.

13. Find the slope of a line parallel to $6x - 3y = 12$.

Rewrite the equation in slope-intercept form.

$$-3y = -6x + 12$$
$$y = 2x - 4$$

The slope is 2, so a parallel line will also have slope 2.

15. The line goes through (1, 3), with slope $m = -2$. Use point-slope form.

$$y - 3 = -2(x - 1)$$
$$y = -2x + 2 + 3$$
$$y = -2x + 5$$

17. The line goes through (−5, −7) with slope $m = 0$. Use point-slope form.

$$y - (-7) = 0[x - (-5)]$$
$$y + 7 = 0$$
$$y = -7$$

19. The line goes through (4, 2) and (1, 3). Find the slope, then use point-slope form with either of the two given points.

$$m = \frac{3-2}{1-4} = -\frac{1}{3}$$
$$y - 3 = -\frac{1}{3}(x - 1)$$
$$y = -\frac{1}{3}x + \frac{1}{3} + 3$$
$$y = -\frac{1}{3}x + \frac{10}{3}$$

21. The line goes through $\left(\frac{2}{3}, \frac{1}{2}\right)$ and $\left(\frac{1}{4}, -2\right)$.

$$m = \frac{-2 - \frac{1}{2}}{\frac{1}{4} - \frac{2}{3}} = \frac{-\frac{4}{2} - \frac{1}{2}}{\frac{3}{12} - \frac{8}{12}}$$
$$m = \frac{\frac{-5}{2}}{-\frac{5}{12}} = \frac{60}{10} = 6$$
$$y - (-2) = 6\left(x - \frac{1}{4}\right)$$
$$y + 2 = 6x - \frac{3}{2}$$
$$y = 6x - \frac{3}{2} - 2$$
$$y = 6x - \frac{3}{2} - \frac{4}{2}$$
$$y = 6x - \frac{7}{2}$$

23. The line goes through (−8, 4) and (−8, 6).

$$m = \frac{4-6}{-8-(-8)} = \frac{-2}{0};$$

which is undefined.

This is a vertical line; the value of x is always −8.

The equation of this line is $x = -8$.

25. The line has x-intercept −6 and y-intercept −3. Two points on the line are (−6, 0) and (0, −3). Find the slope; then use slope-intercept form.

$$m = \frac{-3-0}{0-(-6)} = \frac{-3}{6} = -\frac{1}{2}$$
$$b = -3$$

$$y = -\frac{1}{2}x - 3$$
$$2y = -x - 6$$
$$x + 2y = -6$$

27. The vertical line through $(-6, 5)$ goes through the point $(-6, 0)$, so the equation is $x = -6$.

29. Write an equation of the line through $(-4, 6)$, parallel to $3x + 2y = 13$.

Rewrite the equation of the given line in slope-intercept form.

$$3x + 2y = 13$$
$$2y = -3x + 13$$
$$y = -\frac{3}{2}x + \frac{13}{2}$$

The slope is $-\frac{3}{2}$.

Use $m = -\frac{3}{2}$ and the point $(-4, 6)$ in the point-slope form.

$$y - 6 = -\frac{3}{2}[x - (-4)]$$
$$y = -\frac{3}{2}(x + 4) + 6$$
$$y = -\frac{3}{2}x - 6 + 6$$
$$y = -\frac{3}{2}x$$
$$2y = -3x$$
$$3x + 2y = 0$$

31. Write an equation of the line through $(3, -4)$, perpendicular to $x + y = 4$.

Rewrite the equation of the given line as

$$y = -x + 4.$$

The slope of this line is -1. To find the slope of a perpendicular line, solve

$$-1m = -1.$$
$$m = 1$$

Use $m = 1$ and $(3, -4)$ in the point-slope form.

$$y - (-4) = 1(x - 3)$$
$$y = x - 3 - 4$$
$$y = x - 7$$
$$x - y = 7$$

33. Write an equation of the line with y-intercept 4, perpendicular to $x + 5y = 7$.

Find the slope of the given line.

$$x + 5y = 7$$
$$5y = -x + 7$$
$$y = -\frac{1}{5}x + \frac{7}{5}$$

The slope is $-\frac{1}{5}$, so the slope of the perpendicular line will be 5. If the y-intercept is 4, then using the slope-intercept form we have

$$y = mx + b$$
$$y = 5x + 4, \quad \text{or} \quad 5x - y = -4$$

35. Do the points $(4, 3)$, $(2, 0)$, and $(-18, -12)$ lie on the same line?

Find the slope between $(4, 3)$ and $(2, 0)$.

$$m = \frac{0 - 3}{2 - 4} = \frac{-3}{-2} = \frac{3}{2}$$

Find the slope between $(4, 3)$ and $(-18, -12)$.

$$m = \frac{-12 - 3}{-18 - 4} = \frac{-15}{-22} = \frac{15}{22}$$

Since these slopes are not the same, the points do not lie on the same line.

37. A parallelogram has 4 sides, with opposite sides parallel. The slope of the line through $(1, 3)$ and $(2, 1)$ is

$$m = \frac{3 - 1}{1 - 2}$$
$$= \frac{2}{-1}$$
$$= -2.$$

The slope of the line through $\left(-\frac{5}{2}, 2\right)$ and $\left(-\frac{7}{2}, 4\right)$ is

$$m = \frac{2 - 4}{-\frac{5}{2} - \left(-\frac{7}{2}\right)} = \frac{-2}{1} = -2.$$

Since these slopes are equal, these two sides are parallel.

The slope of the line through $\left(-\frac{7}{2}, 4\right)$ and $(1, 3)$ is

$$m = \frac{4 - 3}{-\frac{7}{2} - 1} = \frac{1}{-\frac{9}{2}} = -\frac{2}{9}.$$

Slope of the line through $\left(-\frac{5}{2}, 2\right)$ and $(2, 1)$ is

$$m = \frac{2 - 1}{-\frac{5}{2} - 2} = \frac{1}{-\frac{9}{2}} = -\frac{2}{9}.$$

Since these slopes are equal, these two sides are parallel.

Since both pairs of opposite sides are parallel, the quadrilateral is a parallelogram.

39. The line goes through (0, 2) and (−2, 0)

$$m = \frac{2-0}{0-(-2)} = \frac{2}{2} = 1$$

The correct choice is (a).

41. The line appears to go through (0, 0) and (−1, 4).

$$m = \frac{4-0}{-1-0} = \frac{4}{-1} = -4$$

43. (a) See the figure in the textbook.

Segment MN is drawn perpendicular to segment PQ. Recall that MQ is the length of segment MQ.

$$m_1 = \frac{\Delta y}{\Delta x} = \frac{MQ}{PQ}$$

From the diagram, we know that $PQ = 1$.

Thus, $m_1 = \frac{MQ}{1}$, so MQ has length m_1.

(b) $m_2 = \frac{\Delta y}{\Delta x} = \frac{-QN}{PQ} = \frac{-QN}{1}$

$QN = -m_2$

(c) Triangles MPQ, PNQ, and MNP are right triangles by construction. In triangles MPQ and MNP, angle M = angle M, and in the right triangles PNQ and MNP,

angle N = angle N.

Since all right angles are equal, and since triangles with two equal angles are similar, triangle MPQ is similar to triangle MNP and triangle PNQ is similar to triangle MNP.

Therefore, triangles MNQ and PNQ are similar to each other.

(d) Since corresponding sides in similar triangles are proportional,

$$MQ = k \cdot PQ \quad \text{and} \quad PQ = k \cdot QN.$$

$$\frac{MQ}{PQ} = \frac{k \cdot PQ}{k \cdot QN}$$

$$\frac{MQ}{PQ} = \frac{PQ}{QN}$$

From the diagram, we know that $PQ = 1$.

$$MQ = \frac{1}{QN}$$

From (a) and (b), $m_1 = MQ$ and $-m_2 = QN$.

Substituting, we get $m_1 = \frac{1}{-m_2}$.

Multiplying both sides by m_2, we have $m_1 m_2 = -1$.

45. $y = x - 1$

Three ordered pairs that satisfy this equation are (0, −1), (1, 0), and (4, 3). Plot these points and draw a line through them.

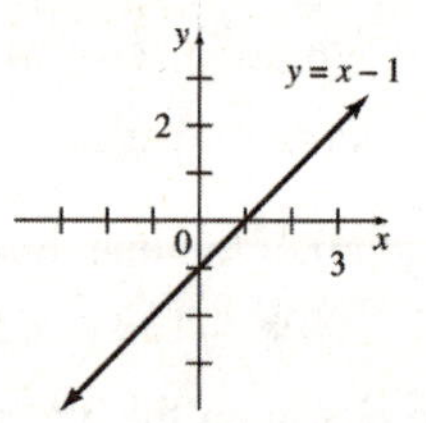

47. $y = -4x + 9$

Three ordered pairs that satisfy this equation are (0, 9), (1, 5), and (2, 1). Plot these points and draw a line through them.

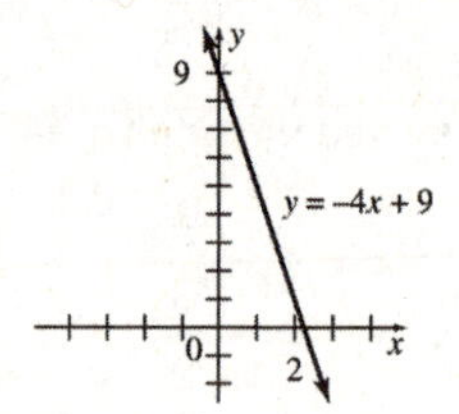

49. $2x - 3y = 12$

Find the intercepts.

If $y = 0$, then

$$\begin{aligned} 2x - 3(0) &= 12 \\ 2x &= 12 \\ x &= 6 \end{aligned}$$

so the x-intercept is 6.

If $x = 0$, then

$$\begin{aligned} 2(0) - 3y &= 12 \\ -3y &= 12 \\ y &= -4 \end{aligned}$$

so the y-intercept is −4.

Plot the ordered pairs (6, 0) and (0, −4) and draw a line through these points. (A third point may be used as a check.)

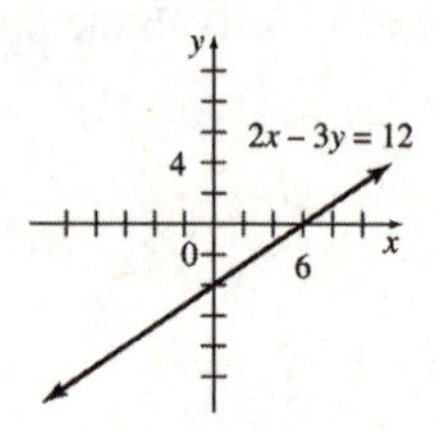

51. $3y - 7x = -21$

Find the intercepts.

If $y = 0$, then

$$3(0) + 7x = -21$$
$$-7x = -21$$
$$x = 3$$

so the x-intercept is 3.

If $x = 0$, then

$$3y - 7(0) = -21$$
$$3y = -21$$
$$y = -7$$

So the y-intercepts is -7.

Plot the ordered pairs $(3, 0)$ and $(0, -7)$ and draw a line through these points. (A third point may be used as a check.)

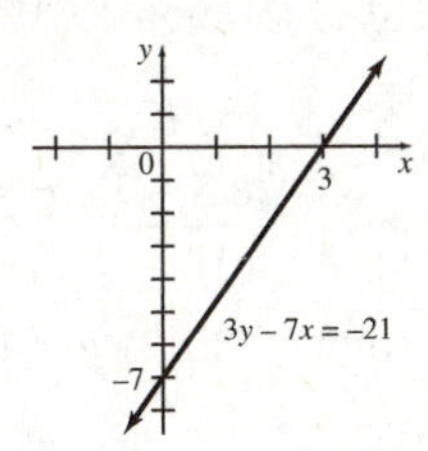

53. $y = -2$

The equation $y = -2$, or, equivalently, $y = 0x - 2$, always gives the same y-value, -2, for any value of x. The graph of this equation is the horizontal line with y-intercept -2.

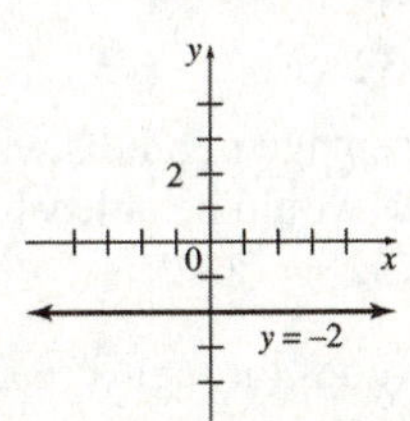

55. $x + 5 = 0$

This equation may be rewritten as $x = -5$. For any value of y, the x-value is -5. Because all ordered pairs that satisfy this equation have the same first number, this equation does not represent a function. The graph is the vertical line with x-intercept -5.

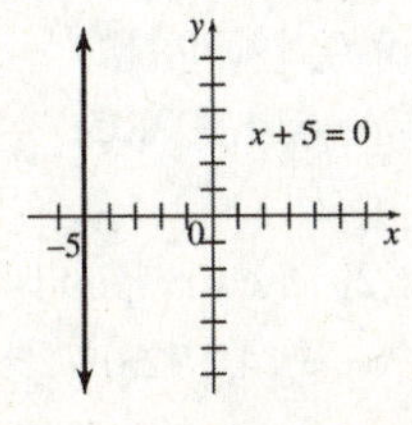

57. $y = 2x$

Three ordered pairs that satisfy this equation are $(0, 0)$, $(-2, -4)$, and $(2, 4)$. Use these points to draw the graph.

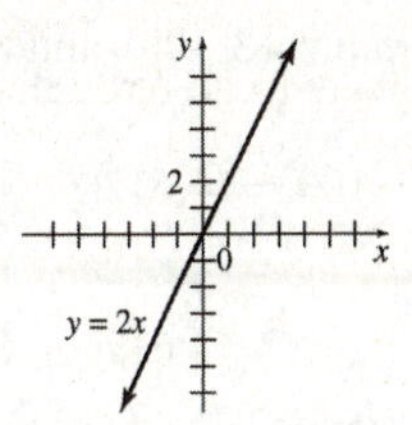

59. $x + 4y = 0$

If $y = 0$, then $x = 0$, so the x-intercept is 0. If $x = 0$, then $y = 0$, so the y-intercept is 0. Both intercepts give the same ordered pair, $(0, 0)$. To get a second point, choose some other value of x (or y). For example if $x = 4$, then

$$x + 4y = 0$$
$$4 + 4y = 0$$
$$4y = -4$$
$$y = -1,$$

giving the ordered pair $(4, -1)$. Graph the line through $(0, 0)$ and $(4, -1)$.

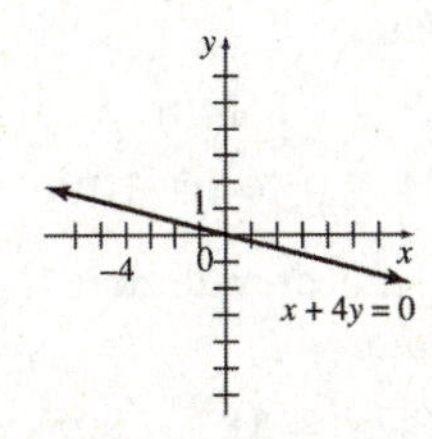

61. **(a)** The line goes through $(2, 27{,}000)$ and $(5, 63{,}000)$.

$$m = \frac{63{,}000 - 27{,}000}{5 - 2}$$
$$= 12{,}000$$
$$y - 27{,}000 = 12{,}000(x - 2)$$
$$y - 27{,}000 = 12{,}000x - 24{,}000$$
$$y = 12{,}000x + 3000$$

(b) Let $y = 100{,}000$; find x.

$$100{,}000 = 12{,}000x + 3000$$
$$97{,}000 = 12{,}000x$$
$$8.08 = x$$

Sales would surpass \$100,000 after 8 years, 1 month.

63. **(a)** The line goes through $(3, 100)$ and $(28, 215.3)$.

$$m = \frac{215.3 - 100}{28 - 3} \approx 4.612$$

Use the point $(3, 100)$ and the point-slope form.

$$y - 100 = 4.612(t - 3)$$
$$y = 4.612t - 13.836 + 100$$
$$y = 4.612t + 86.164$$

(b) The year 2000 corresponds to $t = 2000 - 1980 = 20$.

$$y = 4.612(20) + 86.164$$
$$y \approx 178.4$$

The predicted value is slightly more than the actual CPI of 172.2.

(c) The annual CPI is increasing at a rate of

65. **(a)** Let $x =$ age.

$$u = 0.85(220 - x) = 187 - 0.85x$$
$$l = 0.7(200 - x) = 154 - 0.7x$$

(b) $u = 187 - 0.85(20) = 170$

$l = 154 - 0.7(20) = 140$

The target heart rate zone is 140 to 170 beats per minute.

(c) $u = 187 - 0.85(40) = 153$

$l = 154 - 0.7(40) = 126$

The target heart rate zone is 126 to 153 beats per minute.

(d)
$$154 - 0.7x = 187 - 0.85(x + 36)$$
$$154 - 0.7x = 187 - 0.85x - 30.6$$
$$154 - 0.7x = 156.4 - 0.85x$$
$$0.15x = 2.4$$
$$x = 16$$

The younger woman is 16; the older woman is $16 + 36 = 52$. $l = 0.7(220 - 16) \approx 143$ beats per minute.

67. Let $x = 0$ correspond to 1900. Then the "life expectancy from birth" line contains the points (0, 46) and (104, 77.8).

$$m = \frac{77.8 - 46}{104 - 0} = \frac{31.3}{102} = 0.306$$

Since (0, 46) is one of the points, the line is given by the equation.

$$y = 0.306x + 46.$$

The "life expectancy from age 65" line contains the points (0, 76) and (104, 83.7).

$$m = \frac{83.7 - 76}{104 - 0} = \frac{7.7}{104} \approx 0.074$$

Since (0, 76) is one of the points, the line is given by the equation

$$y = 0.07x + 76.$$

Set the two expressions for y equal to determine where the lines intersect. At this point, life expectancy should increase no further.

$$0.306x + 46 = 0.074x + 76$$
$$0.232x = 30$$
$$x \approx 129$$

Determine the y-value when $x = 129$. Use the first equation.

$$y = 0.306(129) + 46$$
$$= 39.474 + 46$$
$$= 85.474$$

Thus, the maximum life expectancy for humans is about 86 years.

69. **(a)** The line goes through $(9, 17.2)$ and $(18, 20.3)$.

$$m = \frac{20.3 - 17.2}{18 - 9} \approx 0.344$$

Use the point $(9, 17.2)$ and the point-slope form.

$$y - 17.2 = 0.344(t - 9)$$
$$y = 0.344t - 3.096 + 17.2$$
$$y = 0.344t + 14.1$$

(b) Let $y = 25$.

$$25 = 0.344t + 14.1$$
$$10.9 = 0.344t$$
$$32 \approx t$$

The percentage of adults without health insurance would be at least 25% in the year $1990 + 32 = 2022$.

71. **(a)** The line goes through $(50, 249{,}187)$ and $(108, 1{,}107{,}126)$.

$$m = \frac{1{,}107{,}126 - 249{,}187}{108 - 50}$$
$$\approx 14{,}792.05$$

Use the point $(50, 249{,}187)$ and the point-slope form.

$$y - 249{,}187 = 14{,}792.05(t - 50)$$
$$y = 14{,}792.05t - 739{,}602.5 + 249{,}187$$
$$y = 14{,}792.05t - 490{,}416$$

(b) The year 2015 corresponds to $t = 115$.

$$y = 14{,}792.05(115) - 490{,}416$$
$$y \approx 1{,}210{,}670$$

The number of immigrants admitted to the United States in 2015 will be about 1,210,670.

(c) The equation $y = 14{,}792.05t - 490{,}416$ has $-490{,}416$ for the y-intercept, indicating that the number of immigrants admitted in the year 1900 was $-490{,}416$. Realistically, the number of immigrants cannot be a negative value, so the equation cannot be used for valid predicted values.

73. (a) Plot the points $(15, 1600)$, $(200, 15{,}000)$, $(290, 24{,}000)$, and $(520, 40{,}000)$.

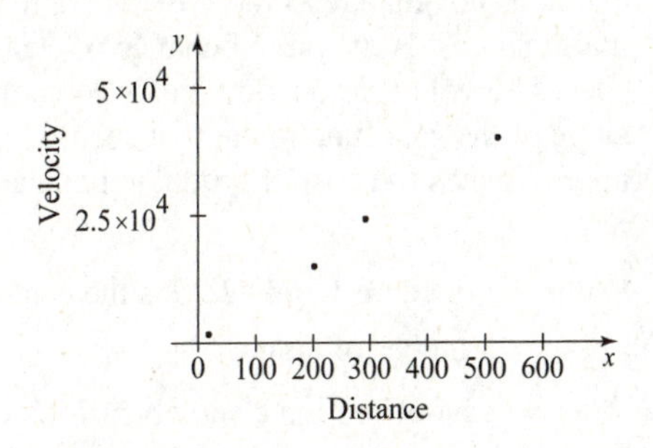

The points lie approximately on a line, so there appears to be a linear relationship between distance and time.

(b) The graph of any equation of the form $y = mx$ goes through the origin, so the line goes through $(520, 40{,}000)$ and $(0, 0)$.

$$m = \frac{40{,}000 - 0}{520 - 0} \approx 76.9$$
$$b = 0$$
$$y = 76.9x + 0$$
$$y = 76.9x$$

(c) Let $y = 60{,}000$; solve for x.

$$60{,}000 = 76.9x$$
$$780.23 \approx x$$

Hydra is about 780 megaparsecs from earth.

(d) $A = \dfrac{9.5 \times 10^{11}}{m}, m = 76.9$

$$A = \frac{9.5 \times 10^{11}}{76.9}$$
$$= 12.4 \text{ billion years}$$

75. (a)

y
3×10^4
2.5×10^4
2×10^4
1.5×10^4
1×10^4
5000
Tuition and Fees
0 2 4 6 8 10 t
Year

Yes, the data appear to lie roughly along a straight line.

(b) The line goes through $(0, 16{,}072)$ and $(9, 26{,}273)$.

$$m = \frac{26{,}273 - 16{,}072}{9 - 0} \approx 1133.4$$
$$b = 16{,}072$$
$$y = 1133.4t + 16{,}072$$

The slope 1133.4 indicates that tuition and fees have increased approximately \$1133 per year.

(c) The year 2025 is too far in the future to rely on this equation to predict costs; too many other factors may influence these costs by then.

1.2 Linear Functions and Applications

Your Turn 1

For $g(x) = -4x + 5$, calculate $g(-5)$.

$$g(x) = -4x + 5$$
$$g(-5) = -4(-5) + 5$$
$$= 20 + 5$$
$$= 25$$

Your Turn 2

For the demand and supply functions given in Example 2, find the quantity of watermelon demanded and supplied at a price of \$3.30 per watermelon.

$$p = D(q) = 9 - 0.75q$$
$$3.30 = 9 - 0.75q$$
$$0.75q = 5.7$$
$$q = \frac{5.7}{0.75} = 7.6$$

Since the quantity is in thousands, 7600 watermelon are demanded at a price of \$3.30.

$$p = S(q) = 0.75q$$
$$3.30 = 0.75q$$
$$q = \frac{3.3}{0.75} = 4.4$$

Since the quantity is in thousands, 4400 watermelon are supplied at a price of \$3.30.

Your Turn 3

Set the two price expressions equal and solve for the equilibrium quantity q.

$$10 - 0.85q = 0.4q$$
$$10 = 1.25q$$
$$q = \frac{10}{1.25} = 8$$

The equilibrium quantity is 8000 watermelon. Use either price expression to find the equilibrium price p.

$$p = 0.4q$$
$$p = 0.4(8) = 3.2$$

The equilibrium price is \$3.20 per watermelon.

Your Turn 4

The marginal cost is the slope of the cost function $C(x)$, so this function has the form $C(x) = 15x + b$. To find b, use the fact that producing 80 batches costs \$1930.

$$C(x) = 15x + b$$
$$C(80) = 15(80) + b$$
$$1930 = 1200 + b$$
$$b = 730$$

Thus the cost function is $C(x) = 15x + 730$.

Your Turn 5

The cost function is $C(x) = 35x + 250$ and the revenue function is $R(x) = 58x$. Thus the profit function is

$$P(x) = R(x) - C(x)$$
$$= 58x - (35x + 250)$$
$$= 23x - 250$$

The profit is to be \$8030.

$$P(x) = 23x - 250$$
$$8030 = 23x - 250$$
$$23x = 8280$$
$$x = \frac{8280}{23} = 360$$

Sale of 360 units will produce \$8030 profit.

1.2 Exercises

1. $f(2) = 7 - 5(2) = 7 - 10 = -3$

3. $f(-3) = 7 - 5(-3) = 7 + 15 = 22$

5. $g(1.5) = 2(1.5) - 3 = 3 - 3 = 0$

7. $g\left(-\frac{1}{2}\right) = 2\left(-\frac{1}{2}\right) - 3 = -1 - 3 = -4$

9. $f(t) = 7 - 5(t) = 7 - 5t$

11. This statement is true.

When we solve $y = f(x) = 0$, we are finding the value of x when $y = 0$, which is the x-intercept. When we evaluate $f(0)$, we are finding the value of y when $x = 0$, which is the y-intercept.

13. This statement is true.

Only a vertical line has an undefined slope, but a vertical line is not the graph of a function. Therefore, the slope of a linear function cannot be undefined.

15. The fixed cost is constant for a particular product and does not change as more items are made. The marginal cost is the rate of change of cost at a specific level of production and is equal to the slope of the cost function at that specific value; it approximates the cost of producing one additional item.

19. \$10 is the fixed cost and \$2.25 is the cost per hour.

Let x = number of hours;

$R(x)$ = cost of renting a snowboard for x hours.

Thus,

$$R(x) = \text{fixed cost} + (\text{cost per hour}) \cdot (\text{number of hours})$$
$$R(x) = 10 + (2.25)(x)$$
$$= 2.25x + 10$$

21. \$2 is the fixed cost and \$0.75 is the cost per half-hour.

Let x = the number of half-hours;

$C(x)$ = the cost of parking a car for x half-hours.

Thus,

$$C(x) = 2 + 0.75x$$
$$= 0.75x + 2$$

23. Fixed cost, \$100; 50 items cost \$1600 to produce.

Let $C(x)$ = cost of producing x items.

$C(x) = mx + b$, where b is the fixed cost.

$$C(x) = mx + 100$$

Now,

$C(x) = 1600$ when $x = 50$, so

$$1600 = m(50) + 100$$
$$1500 = 50m$$
$$30 = m.$$

Thus, $C(x) = 30x + 100$.

25. Marginal cost: \$75; 50 items cost \$4300.

$$C(x) = 75x + b$$

Now, $C(x) = 4300$ when $x = 50$.

$$4300 = 75(50) + b$$
$$550 = b$$

Thus, $C(x) = 75x + 550$.

27. $D(q) = 16 - 1.25q$

(a) $D(0) = 16 - 1.25(0) = 16 - 0 = 16$

When 0 watches are demanded, the price is $16.

(b) $D(4) = 16 - 1.25(4) = 16 - 5 = 11$

When 400 watches are demanded, the price is $11.

(c) $D(8) = 16 - 1.25(8) = 16 - 10 = 6$

When 800 watches are demanded, the price is $6.

(d) Let $D(q) = 8$. Find q.

$$8 = 16 - 1.25q$$
$$\frac{5}{4}q = 8$$
$$q = 6.4$$

When the price is $8, the number of watches demanded is 640.

(e) Let $D(q) = 10$. Find q.

$$10 = 16 - 1.25q$$
$$\frac{5}{4}q = 6$$
$$q = 4.8$$

When the price is $10, the number of watches demanded is 480.

(f) Let $D(q) = 12$. Find q.

$$12 = 16 - 1.25q$$
$$\frac{5}{4}q = 4$$
$$q = 3.2$$

When the price is $12, the number of watches demanded is 320.

(g)

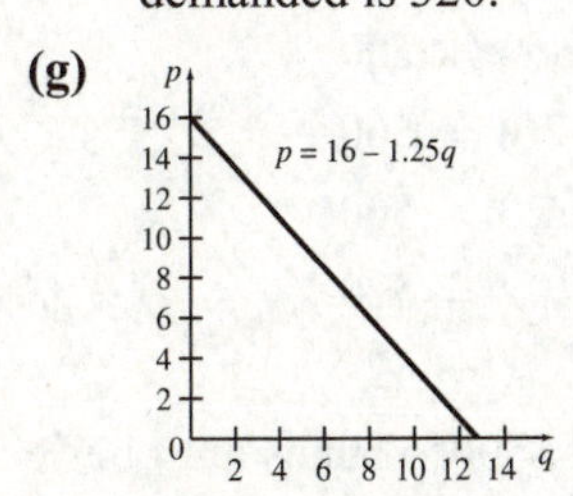

(h) $S(q) = 0.75q$

Let $S(q) = 0$. Find q.

$$0 = 0.75q$$
$$0 = q$$

When the price is $0, the number of watches supplied is 0.

(i) Let $S(q) = 10$. Find q.

$$10 = 0.75q$$
$$\frac{40}{3} = q$$
$$q = 13.\overline{3}$$

When the price is $10, The number of watches supplied is about 1333.

(j) Let $S(q) = 20$. Find q.

$$20 = 0.75q$$
$$\frac{80}{3} = q$$
$$q = 26.\overline{6}$$

When the price is $20, the number of watches demanded is about 2667.

(k)

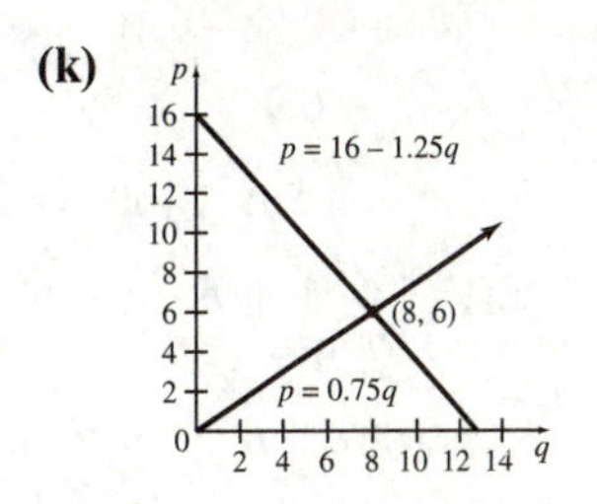

(l)

$$D(q) = S(q)$$
$$16 - 1.25q = 0.75q$$
$$16 = 2q$$
$$8 = q$$

$$S(8) = 0.75(8) = 6$$

The equilibrium quantity is 800 watches, and the equilibrium price is $6.

29. $p = S(q) = \frac{2}{5}q;\ p = D(q) = 100 - \frac{2}{5}q$

(a)

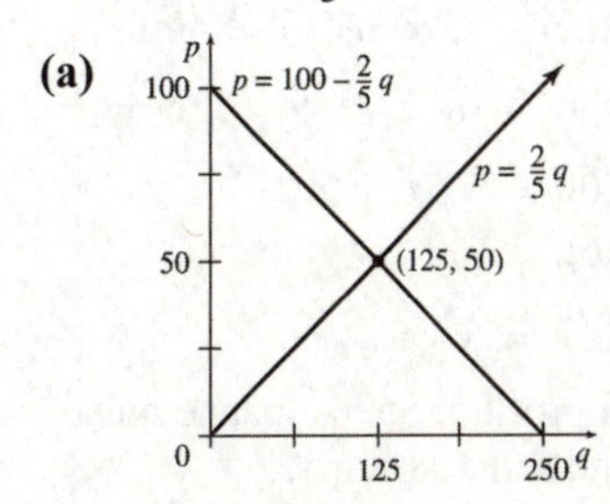

(b) $S(q) = D(q)$

$$\frac{2}{5}q = 100 - \frac{2}{5}q$$
$$\frac{4}{5}q = 100$$
$$q = 125$$

$$S(125) = \frac{2}{5}(125) = 50$$

The equilibrium quantity is 125, the equilibrium price is $50.

31. Use the supply function to find the equilibrium quantity that corresponds to the given equilibrium price of $4.50.

$$\begin{aligned} S(q) = p &= 0.3q + 2.7 \\ 4.50 &= 0.3q + 2.7 \\ 1.8 &= 0.3q \\ 6 &= q \end{aligned}$$

The line that represents the demand function goes through the given point $(2, 6.10)$ and the equilibrium point $(6, 4.50)$.

$$m = \frac{4.50 - 6.10}{6 - 2} = -0.4$$

Use point-slope form and the point $(2, 6.10)$.

$$\begin{aligned} D(q) - 6.10 &= -0.4(q - 2) \\ D(q) &= -0.4q + 0.8 + 6.10 \\ D(q) &= -0.4q + 6.9 \end{aligned}$$

33. **(a)** $C(x) = mx + b;\ m = 3.50;\ C(60) = 300$

$$C(x) = 3.50x + b$$

Find b.

$$\begin{aligned} 300 &= 3.50(60) + b \\ 300 &= 210 + b \\ 90 &= b \end{aligned}$$

$$C(x) = 3.50x + 90$$

(b) $R(x) = 9x$

$C(x) = R(x)$

$$\begin{aligned} 3.50x + 90 &= 9x \\ 90 &= 5.5x \\ 16.36 &= x \end{aligned}$$

Joanne must produce and sell 17 shirts.

(c) $P(x) = R(x) - C(x); P(x) = 500$

$$\begin{aligned} 500 &= 9x - (3.50x + 90) \\ 500 &= 5.5x - 90 \\ 590 &= 5.5x \\ 107.27 &= x \end{aligned}$$

To make a profit of $500, Joanne must produce and sell 108 shirts.

35. **(a)** Using the points (100, 11.02) and (400, 40.12),

$$\begin{aligned} m &= \frac{40.12 - 11.02}{400 - 100} \\ &= \frac{29.1}{300} = 0.097. \end{aligned}$$

$$\begin{aligned} y - 11.02 &= 0.097(x - 100) \\ y - 11.02 &= 0.097x - 9.7 \\ y &= 0.097x + 1.32 \\ C(x) &= 0.097x + 1.32 \end{aligned}$$

(b) The fixed cost is given by the constant in $C(x)$. It is $1.32.

(c)
$$\begin{aligned} C(1000) &= 0.097(1000) + 1.32 \\ &= 97 + 1.32 \\ &= 98.32 \end{aligned}$$

The total cost of producing 1000 cups is $98.32.

(d)
$$\begin{aligned} C(1001) &= 0.097(1001) + 1.32 \\ &= 97.097 + 1.32 \\ &= 98.417 \end{aligned}$$

The total cost of producing 10001 cups is $98.42.

(e) Marginal cost $= 98.417 - 98.32$

$= \$0.097$ or $9.7¢$

(f) The marginal cost for *any* cup is the slope, $0.097 or 9.7¢. This means the cost of producing one additional cup of coffee would be 9.7¢.

37. $C(x) = 5x + 20;\ R(x) = 15x$

(a)
$$\begin{aligned} C(x) &= R(x) \\ 5x + 20 &= 15x \\ 20 &= 10x \\ 2 &= x \end{aligned}$$

The break-even quantity is 2 units.

(b)
$$\begin{aligned} P(x) &= R(x) - C(x) \\ P(x) &= 15x - (5x + 20) \\ P(100) &= 15(100) - (5 \cdot 100 + 20) \\ &= 1500 - 520 \\ &= 980 \end{aligned}$$

The profit from 100 units is $980.

(c)
$$\begin{aligned} P(x) &= 500 \\ 15x - (5x + 20) &= 500 \\ 10x - 20 &= 500 \\ 10x &= 520 \\ x &= 52 \end{aligned}$$

For a profit of $500, 52 units must be produced.

39. $C(x) = 85x + 900$

$R(x) = 105x$

Set $C(x) = R(x)$ to find the break-even quantity.

$$85x + 900 = 105x$$
$$900 = 20x$$
$$45 = x$$

The break-even quantity is 45 units. You should decide not to produce since no more than 38 units can be sold.

$$P(x) = R(x) - C(x)$$
$$= 105x - (85x + 900)$$
$$= 20x - 900$$

The profit function is $P(x) = 20x - 900$.

41. $C(x) = 70x + 500$

$R(x) = 60x$

$$70x + 500 = 60x$$
$$10x = -500$$
$$x = -50$$

This represents a break-even quantity of -50 units. It is impossible to make a profit when the break-even quantity is negative. Cost will always be greater than revenue.

$$P(x) = R(x) - C(x)$$
$$= 60x - (70x + 500)$$
$$= -10x - 500$$

The profit function is $P(x) = -10x - 500$.

43. Since the fixed cost is \$400, the cost function is $C(x) = mx + 100$, where m is the cost per unit. The revenue function is $R(x) = px$, where p is the price per unit.

The profit $P(x) = R(x) - C(x)$ is 0 at the given break-even quantity of 80.

$$P(x) = px - (mx + 400)$$
$$P(x) = px - mx - 400$$
$$P(x) = Mx - 400 \qquad (\text{Let } M = p - m.)$$
$$P(80) = M \cdot 80 - 400$$
$$0 = 80M - 400$$
$$400 = 80M$$
$$5 = M$$

So, the linear profit function is $P(x) = 5x - 400$, and the marginal profit is 5.

45. Use the formula derived in Example 7 in this section of the textbook.

$$F = \frac{9}{5}C + 32 \quad \text{or} \quad C = \frac{5}{9}(F - 32)$$

(a) $F = 58$; find C.

$$C = \frac{5}{9}(58 - 32)$$
$$C = \frac{5}{9}(26) = 14.4$$

The temperature is 14.4°C.

(b) $F = -20$; find C.

$$C = \frac{5}{9}(F - 32)$$
$$C = \frac{5}{9}(-20 - 32)$$
$$C = \frac{5}{9}(-52) = -28.9$$

The temperature is -28.9°C.

(c) $C = 50$; find F.

$$F = \frac{9}{5}C + 32$$
$$F = \frac{9}{5}(50) + 32$$
$$F = 90 + 32 = 122$$

The temperature is 122°F.

47. If the temperatures are numerically equal, then $F = C$.

$$F = \frac{9}{5}C + 32$$
$$C = \frac{9}{5}C + 32$$
$$-\frac{4}{5}C = 32$$
$$C = -40$$

The Celsius and Fahrenheit temperatures are numerically equal at $-40°$.

1.3 The Least Squares Line

Your Turn 1

Rather than recompute all the numbers in the solution table for Example 1, we record the changes to the totals. Note that we have $(90)(40.2) = 3618$, which replaces the next to last value in the xy column. Also note that we have $40.2^2 = 1616.04$, which replaces the next to last value in the y^2 column. The new totals are as follows:

$\Sigma x = 550 - 100 = 450$

$\Sigma y = 595.5 - 34.0 - 36.9 + 40.2 = 564.8$

$\Sigma xy = 28{,}135 - 3400 - 3321 + 3618 = 25{,}032$

$\Sigma x^2 = 38{,}500 - 10{,}000 = 28{,}500$

$$\Sigma y^2 = 38{,}249.41 - 1156.00 - 1361.61 + 1616.04 = 37{,}347.84$$

The number of data points n is now 9 rather than 10. Put the new column totals into the formulas for the slope and intercept.

$$m = \frac{n(\Sigma xy) - (\Sigma x)(\Sigma y)}{n(\Sigma x^2) - (\Sigma x)^2}$$

$$m = \frac{9(25{,}032) - (450)(564.8)}{9(28{,}500) - (450)^2} \approx -0.534667$$

$m \approx -0.535$

$$b = \frac{\Sigma y - m(\Sigma x)}{n} = \frac{564.8 - (-0.534667)(450)}{9} \approx 89.5$$

The least squares line is $Y = -0.535x + 89.5$.

Your Turn 2

Use the new column totals computed in Your Turn 1.

$$r = \frac{n(\Sigma xy) - (\Sigma x)(\Sigma y)}{\sqrt{n(\Sigma x^2) - (\Sigma x)^2} \cdot \sqrt{n(\Sigma y^2) - (\Sigma y)^2}}$$

$$= \frac{9(25{,}032) - (450)(564.8)}{\sqrt{9(28{,}500) - (450)^2} \cdot \sqrt{9(37{,}347.84) - (564.8)^2}}$$

≈ -0.949

1.3 Exercises

3. (a)

(b)

x	y	xy	x^2	y^2
1	0	0	1	0
2	0.5	1	4	0.25
3	1	3	9	1
4	2	8	16	4
5	2.5	12.5	25	6.25
6	3	18	36	9
7	3	21	49	9
8	4	32	64	16
9	4.5	40.5	81	20.25
10	5	50	100	25
55	25.5	186	385	90.75

$$r = \frac{n(\Sigma xy) - (\Sigma x)(\Sigma y)}{\sqrt{n(\Sigma x^2) - (\Sigma x)^2} \cdot \sqrt{n(\Sigma y^2) - (\Sigma y)^2}}$$

$$= \frac{10(186) - (55)(25.5)}{\sqrt{10(385) - (55)^2} \cdot \sqrt{10(90.75) - (25.5)^2}}$$

≈ 0.993

(c) The least squares line is of the form $Y = mx + b$. First solve for m.

$$m = \frac{n(\Sigma xy) - (\Sigma x)(\Sigma y)}{n(\Sigma x^2) - (\Sigma x)^2} = \frac{10(186) - (55)(25.5)}{10(385) - (55)^2} = 0.5545454545 \approx 0.555$$

Now find b.

$$b = \frac{\Sigma y - m(\Sigma x)}{n} = \frac{25.5 - 0.5545454545(55)}{10} = -0.5$$

Thus, $Y = 0.555x - 0.5$.

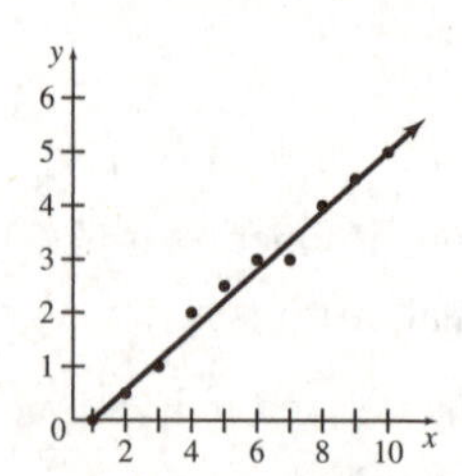

(d) Let $x = 11$. Find Y.

$Y = 0.55(11) - 0.5 = 5.6$

5.

x	y	xy	x^2	y^2
1	1	1	1	1
1	2	2	1	4
2	1	2	4	1
2	2	4	4	4
9	9	81	81	81
15	15	90	91	91

(a) $n = 5$

$$m = \frac{n(\sum xy) - (\sum x)(\sum y)}{n(\sum x^2) - (\sum x)^2} = \frac{5(90) - (15)(15)}{5(91) - (15)^2} = 0.9782608 \approx 0.9783$$

$$b = \frac{\sum y - m(\sum x)}{n} = \frac{15 - (0.9782608)(15)}{5} \approx 0.0652$$

Thus, $Y = 0.9783x + 0.0652$.

$$r = \frac{n(\sum xy) - (\sum x)(\sum y)}{\sqrt{n(\sum x^2) - (\sum x)^2} \cdot \sqrt{n(\sum y^2) - (\sum y)^2}} = \frac{5(90) - (15)(15)}{\sqrt{5(91) - (15)^2} \cdot \sqrt{5(91) - (15)^2}} \approx 0.9783$$

(b)

x	y	xy	x^2	y^2
1	1	1	1	1
1	2	2	1	4
2	1	2	4	1
2	2	4	4	4
6	6	9	10	10

$n = 4$

$$m = \frac{n(\sum xy) - (\sum x)(\sum y)}{n(\sum x^2) - (\sum x)^2} = \frac{4(9) - (6)(6)}{4(10) - (6)^2} = 0$$

$$b = \frac{\sum y - m(\sum x)}{n} = \frac{6 - (0)(6)}{4} = 1.5$$

Thus, $Y = 0x + 1.5$, or $Y = 1.5$.

$$r = \frac{n(\sum xy) - (\sum x)(\sum y)}{\sqrt{n(\sum x^2) - (\sum x)^2} \cdot \sqrt{n(\sum y^2) - (\sum y)^2}} = \frac{4(9) - (6)(6)}{\sqrt{4(10) - (6)^2} \cdot \sqrt{4(10) - (6)^2}} = 0$$

(c)

The point $(9, 9)$ is an outlier that has a strong effect on the least squares line and the correlation coefficient.

7.

x	y	xy	x^2	y^2
1	1	1	1	1
2	1	2	4	1
3	1	3	9	1
4	1.1	4.4	16	1.21
10	4.1	10.4	30	4.21

(a) $n = 4$

$$r = \frac{n(\sum xy) - (\sum x)(\sum y)}{\sqrt{n(\sum x^2) - (\sum x)^2} \cdot \sqrt{n(\sum y^2) - (\sum y)^2}} = \frac{4(10.4) - (10)(4.1)}{\sqrt{4(30) - (10)^2} \cdot \sqrt{4(4.21) - (4.1)^2}} = 0.7745966 \approx 0.7746$$

(b)

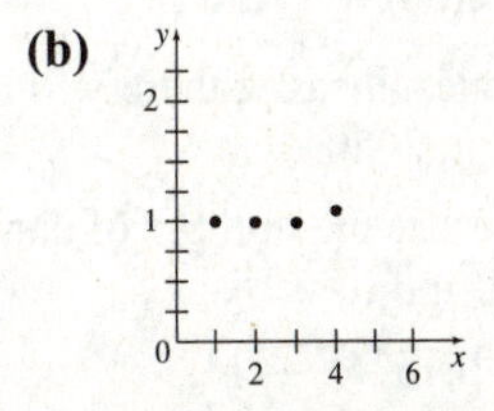

(c) Yes; because the data points are either on or very close to the horizontal line $y = 1$, it seems that the data should have a strong linear relationship. The correlation coefficient does not describe well a linear relationship if the data points fit a horizontal line.

9. $nb + (\sum x)m = \sum y$

$$(\sum x)b + (\sum x^2)m = \sum xy$$

$$nb + (\sum x)m = \sum y$$

$$nb = (\sum y) - (\sum x)m$$

$$b = \frac{\sum y - m(\sum x)}{n}$$

$$(\sum x)\left(\frac{\sum y - m(\sum x)}{n}\right) + (\sum x^2)m = \sum xy$$

$$(\sum x)[(\sum y) - m(\sum x)] + nm(\sum x^2) = n(\sum xy)$$

$$(\sum x)(\sum y) - m(\sum x)^2 + nm(\sum x^2) = n(\sum xy)$$

$$nm(\sum x^2) - m(\sum x)^2 = n(\sum xy) - (\sum x)(\sum y)$$

$$m[n(\sum x^2) - (\sum x)^2] = n(\sum xy) - (\sum x)(\sum y)$$

$$m = \frac{n(\sum xy) - (\sum x)(\sum y)}{n(\sum x^2) - (\sum x)^2}$$

11. (a) $m = \frac{n(\sum xy) - (\sum x)(\sum y)}{n(\sum x^2) - (\sum x)^2}$

$$= \frac{10(1810.095) - (235)(77.564)}{10(5605) - (235)^2}$$

$$\approx -0.1534$$

$$b = \frac{\sum y - m(\sum x)}{n}$$

$$= \frac{77.564 - (-0.1534)(235)}{10}$$

$$\approx 11.36$$

Thus, $Y = -0.1534x + 11.36$.

(b) The year 2020 corresponds to $x = 2020 - 1990 = 30$.

$$Y = -0.1534(30) + 11.36 = 6.758$$

If the trend continues linearly, there will be about 6760 banks in 2020.

(c) Let $Y = 6$ (since Y is the number of banks in thousands) and find x.

$$6 = -0.1534x + 11.36$$

$$-5.36 = -0.1534x$$

$$34.94 = x$$

$$35 \approx x$$

The number of U.S. banks will drop below 6000 in the year $1990 + 35 = 2025$.

(d) $r = \frac{n(\sum xy) - (\sum x)(\sum y)}{\sqrt{n(\sum x^2) - (\sum x)^2} \cdot \sqrt{n(\sum y^2) - (\sum y)^2}}$

$$= \frac{10(1810.095) - (235)(77.564)}{\sqrt{10(5605) - (235)^2} \cdot \sqrt{10(603.60324) - (77.564)^2}}$$

$$\approx -0.9890$$

This means that the least squares line fits the data points very well. The negative sign indicates that the number of banks is decreasing as the years increase.

(e) $r = \frac{n(\sum xy) - (\sum x)(\sum y)}{\sqrt{n(\sum x^2) - (\sum x)^2} \cdot \sqrt{n(\sum y^2) - (\sum y)^2}}$

$$= \frac{5(659) - (20)(126.2)}{\sqrt{5(120) - (20)^2} \cdot \sqrt{5(3784.82) - (126.2)^2}}$$

$$\approx 0.9957$$

This means that the least squares line fits the data points extremely well.

13.

x	y	xy	x^2	y^2
4	2219.5	8878.0	16	4,926,180.25
5	2319.8	11,599.0	25	5,381,472.04
6	2415.0	14,490.0	36	5,832,225.00
7	2551.9	17,863.3	49	6,512,193.61
8	2592.1	20,786.8	64	6,718,982.41
30	12,098.3	73,567.1	190	29,371,053.31

(a) $m = \frac{n(\sum xy) - (\sum x)(\sum y)}{n(\sum x^2) - (\sum x)^2}$

$$= \frac{5(73,567.1) - (30)(12,098.3)}{5(190) - (30)^2} \approx 97.73$$

$$b = \frac{\sum y - m(\sum x)}{n}$$

$$= \frac{12,098.3 - (97.73)(30)}{5} \approx 1833.3$$

Thus, $Y = 97.73x + 1833.3$.

(b) The total amount of consumer credit is increasing at a rate of about $97.73 billion per year.

(c) The year 2015 corresponds to $x = 15$.

$$Y = 97.73(15) + 1833.3 = 3299.25$$

If the trend continues linearly, the total amount of consumer credit will be about $3299 billion in 2015.

(d) Let $Y = 4000$ and find x.

$$4000 = 97.73x + 1833.3$$
$$2166.7 = 97.73x$$
$$22.17 \approx x$$

The total debt will exceed \$4000 billion in the year $2000 + 23 = 2023$.

(e) $$r = \frac{n(\sum xy) - (\sum x)(\sum y)}{\sqrt{n(\sum x^2) - (\sum x)^2} \cdot \sqrt{n(\sum y^2) - (\sum y)^2}}$$

$$= \frac{5(73,567.1) - (30)(12,098.3)}{\sqrt{5(190) - (30)^2} \cdot \sqrt{5(29,371,053.31) - (12,098.3)^2}}$$

$$\approx 0.9909$$

This means that the least squares line fits the data points extremely well.

15. (a)

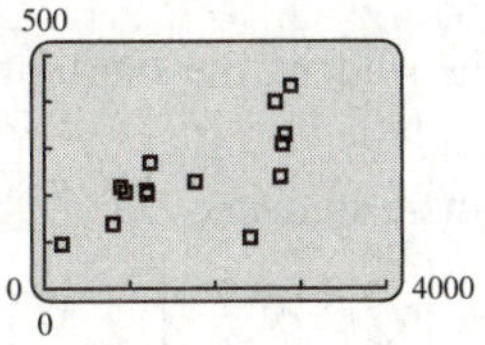

Yes, the data points lie in a linear pattern.

(b)

x	y	xy	x^2	y^2
206	95	19,570	42,436	9025
802	138	110,676	643,204	19,044
1771	228	403,788	3,136,441	51,984
1198	209	250,382	1,435,204	43,681
1238	269	333,022	1,532,644	72,361
2786	309	860,874	1,761,796	95,481
1207	202	243,814	1,456,849	40,804
892	217	193,564	795,664	47,089
2411	109	262,799	5,812,921	11,881
2885	434	1,252,090	8,323,225	188,356
2705	399	1,079,295	7,317,025	159,201
948	206	195,288	898,704	42,436
2762	239	660,118	7,628,644	57,121
2815	329	926,135	7,.924,255	108,241
24,626	3383	6,791,415	54,708,982	946,705

$$r = \frac{14(6,791,415) - (24,626)(3383)}{\sqrt{14(54,708,982) - 24,626^2} \cdot \sqrt{14(946,705) - 3383^2}}$$

$$\approx 0.693$$

There is a positive correlation between the price and the distance.

(c) $$m = \frac{n(\sum xy) - (\sum x)(\sum y)}{n(\sum x^2) - (\sum x)^2}$$

$$m = \frac{14(6,791,415) - (24,626)(3383)}{14(54,708,982) - 24,626^2}$$

$$m = 0.0737999664 \approx 0.0738$$

$$b = \frac{\sum y - m(\sum x)}{n}$$

$$b = \frac{3383 - (0.0737999664)(24,626)}{14}$$

$$\approx 111.83$$

$$Y = 0.0738x + 111.83$$

The marginal cost is about 7.38 cents per mile.

(d) In 2000, the marginal cost was 2.43 cents per mile. It increased to 7.38 cents per mile by 2006.

(e) Phoenix is the outlier.

17. (a)

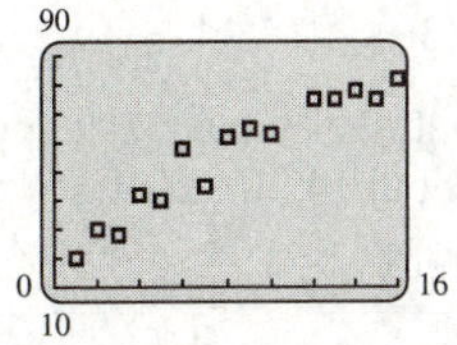

Yes, the points lie in a linear pattern.

(b) Using a calculator's STAT feature, the correlation coefficient is found to be $r \approx 0.959$. This indicates that the percentage of successful hunts does trend to increase with the size of the hunting party.

(c) $Y = 3.98x + 22.7$

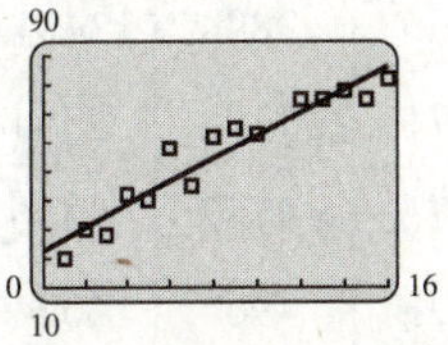

19.

x	y	xy	x^2	y^2
0	17.4	0	0	302.76
4	17.7	70.8	16	313.29
8	16.9	135.2	64	285.61
12	16.2	194.4	144	262.44
16	15.8	252.8	256	249.64
40	84.0	653.2	480	1413.74

(a) $$m = \frac{n(\sum xy) - (\sum x)(\sum y)}{n(\sum x^2) - (\sum x)^2} = \frac{5(653.2) - (40)(84)}{5(480) - (40)^2} \approx -0.1175$$

$$b = \frac{\sum y - m(\sum x)}{n} = \frac{84 - (-0.1175)(40)}{5} \approx 17.74$$

Thus, $Y = -0.1175x + 17.74$.

(b) The year 2020 corresponds to $x = 2020 - 1990 = 30$.

$Y = -0.1175(30) + 17.74 = 14.215 \approx 14.2$

If the trend continues linearly, the pupil-teacher ratio will be about 14.2 in 2020.

(c) $$r = \frac{n(\sum xy) - (\sum x)(\sum y)}{\sqrt{n(\sum x^2) - (\sum x)^2} \cdot \sqrt{n(\sum y^2) - (\sum y)^2}} = \frac{5(653.2) - (40)(84)}{\sqrt{5(480) - (40)^2} \cdot \sqrt{5(1413.74) - (84)^2}} \approx -0.9326$$

The value indicates a strong linear correlation.

21.

x	y	xy	x^2	y^2
59	66	3894	3481	4356
62	71	4402	3844	5041
66	72	4752	4356	5184
68	73	4964	4624	5329
71	75	5325	5041	5625
67	63	4221	4489	3969
70	63	4410	4900	3969
71	67	4757	5041	4489
73	66	4818	5329	4356
75	66	4950	5625	4356
682	682	46,493	46,730	46,674

(a) $$m = \frac{n(\sum xy) - (\sum x)(\sum y)}{n(\sum x^2) - (\sum x)^2} = \frac{10(46{,}493) - (682)(682)}{10(46{,}730) - (682)^2} \approx -0.08915$$

$$b = \frac{\sum y - m(\sum x)}{n} = \frac{682 - (-0.08915)(682)}{10} \approx 74.28$$

Thus, $Y = -0.08915x + 74.28$.

$$r = \frac{n(\sum xy) - (\sum x)(\sum y)}{\sqrt{n(\sum x^2) - (\sum x)^2} \cdot \sqrt{n(\sum y^2) - (\sum y)^2}} = \frac{10(46{,}493) - (682)(682)}{\sqrt{10(46{,}730) - (682)^2} \cdot \sqrt{10(46{,}674) - (682)^2}} \approx -0.1035$$

The taller the student, the shorter the ideal partner's height is.

(b) Data for female students:

x	y	xy	x^2	y^2
59	66	3894	3481	4356
62	71	4402	3844	5041
66	72	4752	4356	5184
68	73	4964	4624	5329
71	75	5325	5041	5625
326	357	23,337	21,346	25,535

$$m = \frac{n(\sum xy) - (\sum x)(\sum y)}{n(\sum x^2) - (\sum x)^2} = \frac{5(23{,}337) - (326)(357)}{5(21{,}346) - (326)^2} \approx 0.6674$$

$$b = \frac{\sum y - m(\sum x)}{n} = \frac{357 - (0.6674)(326)}{5} \approx 27.89$$

Thus, $Y = 0.6674x + 27.89$.

$$r = \frac{n(\sum xy) - (\sum x)(\sum y)}{\sqrt{n(\sum x^2) - (\sum x)^2} \cdot \sqrt{n(\sum y^2) - (\sum y)^2}} = \frac{5(23{,}337) - (326)(357)}{\sqrt{5(21{,}346) - (326)^2} \cdot \sqrt{5(25{,}535) - (357)^2}} \approx 0.9459$$

Data for male students:

x	y	xy	x^2	y^2
67	63	4221	4489	3969
70	63	4419	4900	3969
71	67	4757	5041	4489
73	66	4818	5329	4356
75	66	4950	5625	4356
356	325	23,156	25,384	21,139

$$m = \frac{n(\sum xy) - (\sum x)(\sum y)}{n(\sum x^2) - (\sum x)^2}$$

$$= \frac{5(23{,}156) - (356)(325)}{5(25{,}384) - (356)^2} \approx 0.4348$$

$$b = \frac{\sum y - m(\sum x)}{n}$$

$$= \frac{325 - (0.4348)(356)}{5} \approx 34.04$$

Thus, $Y = 0.4348x + 34.04$.

$$r = \frac{n(\sum xy) - (\sum x)(\sum y)}{\sqrt{n(\sum x^2) - (\sum x)^2} \cdot \sqrt{n(\sum y^2) - (\sum y)^2}}$$

$$= \frac{5(23{,}156) - (356)(325)}{\sqrt{5(25{,}384) - (356)^2} \cdot \sqrt{5(21{,}139) - (325)^2}}$$

$$\approx 0.7049$$

(c)

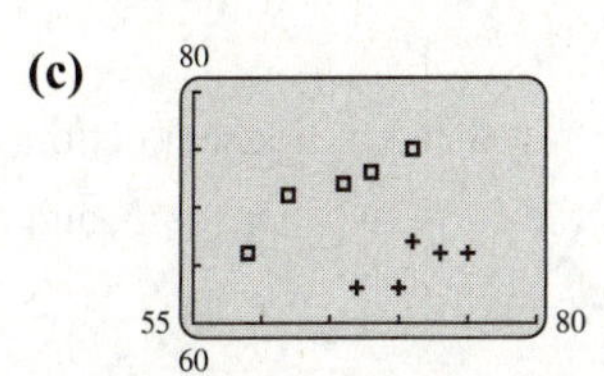

There is no linear relationship among all 10 data pairs. However, there is a linear relationship among the first five data pairs (female students) and a separate linear relationship among the second five data pairs (male students).

23. (a)

(b)

L	T	LT	L^2	T^2
1.0	1.11	1.11	1	1.2321
1.5	1.36	2.04	2.25	1.8496
2.0	1.57	3.14	4	2.4649
2.5	1.76	4.4	6.25	3.0976
3.0	1.92	5.76	9	3.6864
3.5	2.08	7.28	12.25	4.3264
4.0	2.22	8.88	16	4.9844
17.5	12.02	32.61	50.75	21.5854

$$m = \frac{n(\sum xy) - (\sum x)(\sum y)}{n(\sum x^2) - (\sum x)^2}$$

$$m = \frac{7(32.61) - (17.5)(12.02)}{7(50.75) - 17.5^2}$$

$$m = 0.3657142857$$

$$\approx 0.366$$

$$b = \frac{\sum T - m(\sum L)}{n}$$

$$b = \frac{12.02 - 0.3657142857(17.5)}{7}$$

$$\approx 0.803$$

$$Y = 0.366x + 0.803$$

The line seems to fit the data.

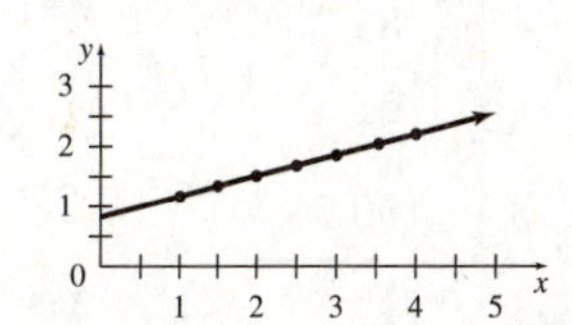

(c)

$$r = \frac{7(32.61) - (17.5)(12.02)}{\sqrt{7(50.75) - 17.5^2} \cdot \sqrt{7(21.5854) - 12.02^2}}$$

$$= 0.995,$$

which is a good fit and confirms the conclusion in part (b).

25. (a)

$$r = \frac{10(399.16) - (500)(20.668)}{\sqrt{10(33{,}250) - 500^2} \cdot \sqrt{10(91.927042) - (20.668)^2}}$$

$$= -0.995$$

Yes, there does appear to be a linear correlation.

(b) $$m = \frac{n(\sum xy) - (\sum x)(\sum y)}{n(\sum x^2) - (\sum x)^2}$$

$$m = \frac{10(399.16) - (500)(20.668)}{10(33{,}250) - 500^2}$$

$$m = -0.0768775758 \approx -0.0769$$

$$b = \frac{\sum y - m(\sum x)}{n}$$

$$b = \frac{20.668 - (-0.0768775758)(500)}{10}$$

$$\approx 5.91$$

$$Y = -0.0769x + 5.91$$

(c) Let $x = 50$

$$Y = -0.0769(50) + 5.91 \approx 2.07$$

The predicted number of points expected when a team is at the 50 yard line is 2.07 points.

27.

x	y
0.00	0.0
2.3167	11.5
3.7167	18.9
5.6000	27.8
7.0833	32.8
7.5000	36.0
8.5000	43.9
10.6000	51.5
11.9333	58.4
15.2333	71.8
17.8167	80.9
18.9667	85.2
20.8333	91.3
23.3833	100.5
153.4833	710.5

(a) Skaggs' average speed was $100.5/23.3833 \approx 4.298$ miles per hour.

(b)

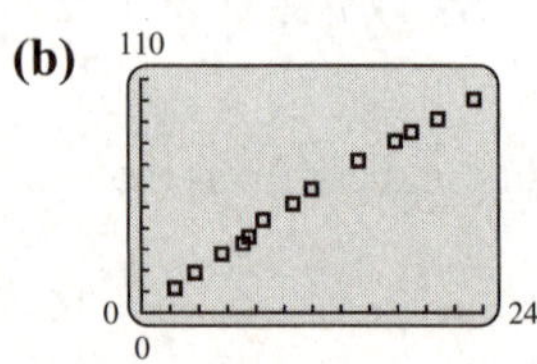

The data appear to lie approximately on a straight line.

(c) Using a graphing calculator,

$$Y = 4.317x + 3.419.$$

(d) Using a graphing calculator,

$$r \approx 0.9971$$

Yes, the least squares line is a very good fit to the data.

(e) A good value for Skaggs' average speed would be the slope of the least squares line, or

$$m = 4.317 \text{ miles per hour.}$$

This value is faster than the average speed found in part (a). The value 4.317 miles per hour is most likely the better value because it takes into account all 14 data pairs.

Chapter 1 Review Exercises

1. False; a line can have only one slant, so its slope is unique.

2. False; the equation $y = 3x + 4$ has slope 3.

3. True; the point $(3, -1)$ is on the line because $-1 = -2(3) + 5$ is a true statement.

4. False; the points $(2, 3)$ and $(2, 5)$ do not have the same y-coordinate.

5. True; the points $(4, 6)$ and $(5, 6)$ do have the same y-coordinate.

6. False; the x-intercept of the line $y = 8x + 9$ is $-\frac{9}{8}$.

7. True; $f(x) = \pi x + 4$ is a linear function because it is in the form $y = mx + b$, where m and b are real numbers.

8. False; $f(x) = 2x^2 + 3$ is not linear function because it isn't in the form $y = mx + b$, and it is a second-degree equation.

9. False; the line $y = 3x + 17$ has slope 3, and the line $y = -3x + 8$ has slope -3. Since $3 \cdot -3 \neq -1$, the lines cannot be perpendicular.

10. False; the line $4x + 3y = 8$ has slope $-\frac{4}{3}$, and the line $4x + y = 5$ has slope -4. Since the slopes are not equal, the lines cannot be parallel.

11. False; a correlation coefficient of zero indicates that there is no linear relationship among the data.

12. True; a correlation coefficient always will be a value between -1 and 1.

13. Marginal cost is the rate of change of the cost function; the fixed cost is the initial expenses before production begins.

15. Through $(-3, 7)$ and $(2, 12)$

$$m = \frac{12 - 7}{2 - (-3)} = \frac{5}{5} = 1$$

17. Through the origin and $(11,-2)$

$$m=\frac{-2-0}{11-0}=-\frac{2}{11}$$

19. $4x+3y=6$

$$3y=-4x+6$$
$$y=-\frac{4}{3}x+2$$

Therefore, the slope is $m=-\frac{4}{3}$.

21. $y+4=9$

$$y=5$$
$$y=0x+5$$
$$m=0$$

23. $y=5x+4$

$$m=5$$

25. Through $(5,-1)$; slope $\frac{2}{3}$

Use point-slope form.

$$y-(-1)=\frac{2}{3}(x-5)$$
$$y+1=\frac{2}{3}(x-5)$$
$$3(y+1)=2(x-5)$$
$$3y+3=2x-10$$
$$3y=2x-13$$
$$y=\frac{2}{3}x-\frac{13}{3}$$

27. Through $(-6,3)$ and $(2,-5)$

$$m=\frac{-5-3}{2-(-6)}=\frac{-8}{8}=-1$$

Use point-slope form.

$$y-3=-1[x-(-6)]$$
$$y-3=-x-6$$
$$y=-x-3$$

29. Through $(2,-10)$, perpendicular to a line with undefined slope

A line with undefined slope is a vertical line. A line perpendicular to a vertical line is a horizontal line with equation of the form $y=k$. The desired line passed through $(2,-10)$, so $k=-10$. Thus, an equation of the desired line is $y=-10$.

31. Through $(3,-4)$ parallel to $4x-2y=9$

Solve $4x-2y=9$ for y.

$$-2y=-4x+9$$
$$y=2x-\frac{9}{2} \quad \text{so } m=2$$

The desired line has the same slope. Use the point-slope form.

$$y-(-4)=2(x-3)$$
$$y+4=2x-6$$
$$y=2x-10$$

Rearrange.

$$2x-y=10$$

33. Through $(-1,4)$; undefined slope

Undefined slope means the line is vertical.

The equation of the vertical line through $(-1,4)$ is $x=-1$.

35. Through $(3,-5)$, parallel to $y=4$

Find the slope of the given line.

$y=0x+4$, so $m=0$, and the required line will also have slope 0.

Use the point-slope from.

$$y-(-5)=0(x-3)$$
$$y+5=0$$
$$y=-5$$

37. $y=4x+3$

Let $x=0$: $\quad y=4(0)+3$

$$y=3$$

Let $y=0$: $\quad 0=4x+3$

$$-3=4x$$
$$-\frac{3}{4}=x$$

Draw the line through $(0,3)$ and $\left(-\frac{3}{4},0\right)$.

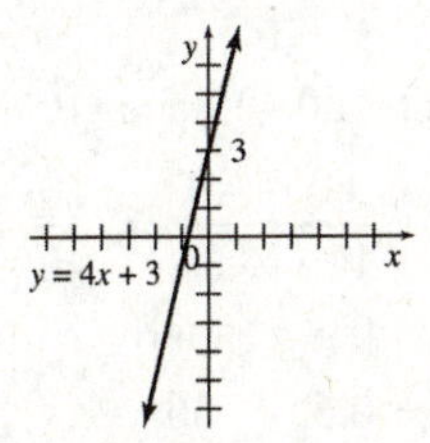

39. $3x-5y=15$

$$-5y=-3x+15$$
$$y=\frac{3}{5}x-3$$

When $x=0, y=-3$; when $y=0, x=5$.

Draw the line through $(0,-3)$ and $(5,0)$.

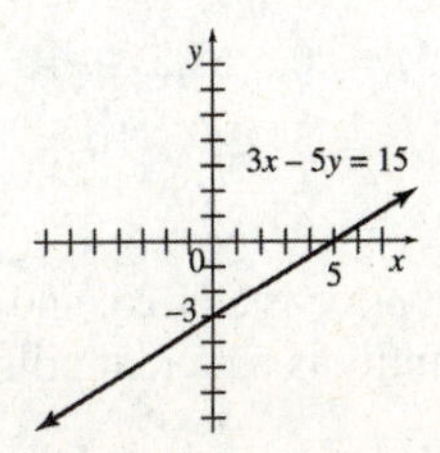

41. $x - 3 = 0$

$x = 3$

This is the vertical line through (3, 0).

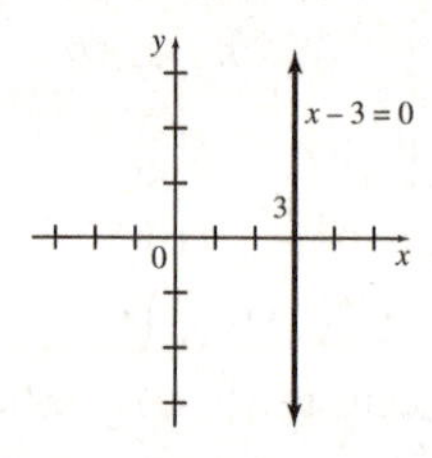

43. $y = 2x$

When $x = 0, y = 0$.

When $x = 1, y = 2$.

Draw the line through (0, 0) and (1, 2).

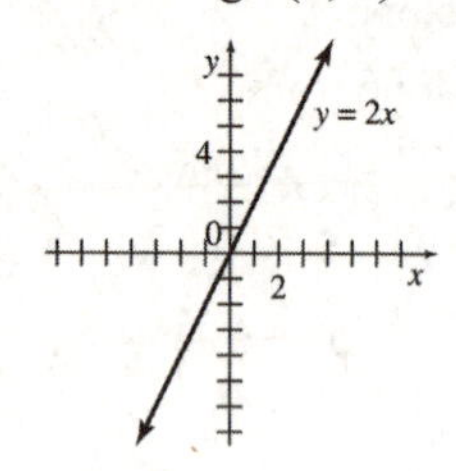

45. **(a)** $E = 352 + 42x$ (where x is in thousands)

(b) $R = 130x$ (where x is in thousands)

(c) $R > E$

$130x > 352 + 42x$

$88x > 352$

$x > 4$

For a profit to be made, more than 4000 chips must be sold.

47. Using the points (60, 40) and (100, 60),

$$m = \frac{60 - 40}{100 - 60} = \frac{20}{40} = 0.5.$$

$$p - 40 = 0.5(q - 60)$$
$$p - 40 = 0.5q - 30$$
$$p = 0.5q + 10$$
$$S(q) = 0.5q + 10$$

49. $S(q) = D(q)$

$$0.5q + 10 = -0.5q + 72.50$$
$$q = 62.5$$

$$S(62.5) = 0.5(62.5) + 10$$
$$= 31.25 + 10$$
$$= 41.25$$

The equilibrium price is \$41.25, and the equilibrium quantity is 62.5 diet pills.

51. Fixed cost is \$2000; 36 units cost \$8480.

Two points on the line are (0, 2000) and (36, 8480), so

$$m = \frac{8480 - 2000}{36 - 0} = \frac{6480}{36} = 180.$$

Use point-slope form.

$$y = 180x + 2000$$
$$C(x) = 180x + 2000$$

53. Thirty units cost \$1500; 120 units cost \$5640. Two points on the line are (30, 1500), (120, 5640), so

$$m = \frac{5640 - 1500}{120 - 30} = \frac{4140}{90} = 46.$$

Use point-slope form.

$$y - 1500 = 46(x - 30)$$
$$y = 46x - 1380 + 1500$$
$$y = 46x + 120$$
$$C(x) = 46x + 120$$

55. **(a)** $C(x) = 3x + 160; R(x) = 7x$

$$C(x) = R(x)$$
$$3x + 160 = 7x$$
$$160 = 4x$$
$$40x = x$$

The break-even quantity is 40 pounds.

(b) $R(40) = 7 \cdot 40 = \$280$

The revenue for 40 pounds is \$280.

57. Let y represent imports to China in billions of dollars. Using the points (1, 19.1) and (8, 69.7)

$$m = \frac{69.7 - 19.1}{8 - 1} = \frac{50.6}{7} \approx 7.23$$

$$y - 19.1 = 7.23(t - 1)$$
$$y - 19.1 = 7.23t - 7.23$$
$$y = 7.23t + 11.9$$

59.

x	y
0	7500
5	12,000
10	16,000
15	20,450
20	24,900
25	28,400

(a) Use the points (0, 7500) and (25, 28,400) to find the slope.

$$m = \frac{28,400 - 7500}{25 - 0} = 836$$

$$b = 7500$$

The linear equation for the average new car cost since 1980 is $y = 836x + 7500$.

(b) Use the points (15, 20,450) and (25, 28,400) to find the slope.

$$m = \frac{28,400 - 20,450}{25 - 15} = 795$$

$$y - 20,450 = 795(x - 15)$$

$$y - 20,450 = 795x - 11,925$$

$$y = 8525$$

The linear equation for the average new car cost since 1980 is $y = 795x + 8525$.

(c) Using a graphing calculator, the least square line is $Y = 843.7x + 7662$.

(d)

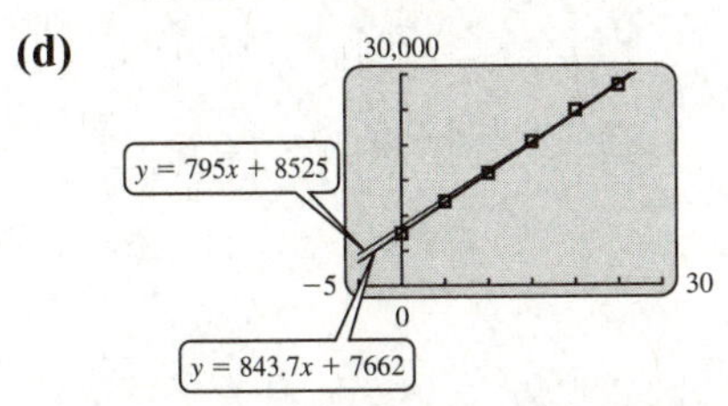

(e) The least squares lines best describes the data. Since the data seems to fit a straight line, a linear model describes the data very well.

(f) Using a graphing calculator, $r \approx 0.9995$.

61. (a)

x	y	xy	x^2	y^2
130	170	22,100	16,900	28,900
138	160	22,080	19,044	25,600
142	173	24,566	20,164	29,929
159	181	28,779	25,281	32,761
165	201	33,165	27,225	40,401
200	192	38,400	40,000	36,864
210	240	50,400	44,100	57,600
250	290	72,500	62,500	84,100
1394	1607	291,990	255,214	336,155

$$m = \frac{n(\sum xy) - (\sum x)(\sum y)}{n(\sum x^2) - (\sum x)^2}$$

$$m = \frac{8(291,990) - (1394)(1607)}{8(225,214) - 1394^2}$$

$$m = 0.9724399854 \approx 0.9724$$

$$b = \frac{\sum y - m(\sum x)}{n}$$

$$b = \frac{1607 - 0.9724(1394)}{8} \approx 31.43$$

$$Y = 0.9724x + 31.43$$

(b) Let $x = 190$; find Y.

$$Y = 0.9724(190) + 31.43$$

$$Y = 216.19 \approx 216$$

The cholesterol level for a person whose blood sugar level is 190 would be about 216.

(c)
$$r = \frac{8(291,990) - (1394)(1607)}{\sqrt{8(255,214) - 1394^2} \cdot \sqrt{8(336,155) - 1607^2}}$$

$$= 0.933814 \approx 0.93$$

63. Using the points (5, 55) and (19, 72.1),

$$m = \frac{72.1 - 55}{19 - 5} = \frac{17.1}{14} \approx 1.22$$

$$y - 55 = 1.22(t - 5)$$

$$y - 55 = 1.22t - 6.1$$

$$y = 1.22t + 48.9$$

65. (a) Using a graphing calculator, $r = 0.6998$. The data seem to fit a line but the fit is not very good.

(b)

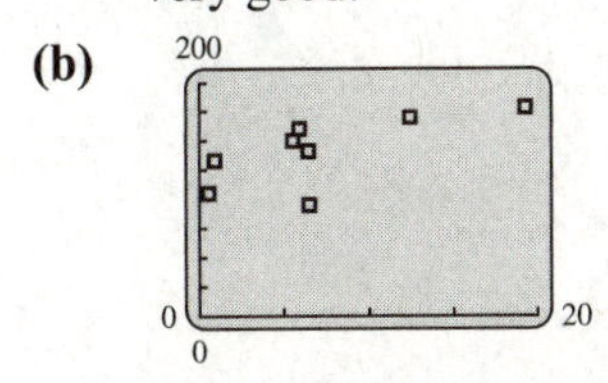

(c) Using a graphing calculator,

$$Y = 3.396x + 117.2$$

(d) The slope is 3.396 thousand (or 3396). On average, the governor's salary increases \$3396 for each additional million in population.

Chapter 2

SYSTEMS OF LINEAR EQUATIONS AND MATRICES

2.1 Solution of Linear Systems by the Echelon Method

Your Turn 1

$$2x + 3y = 12 \quad (1)$$
$$3x - 4y = 1 \quad (2)$$

Use row transformations to eliminate x in equation (2).

$$2x + 3y = 12$$
$$3R_1 + (-2)R_2 \to R_2 \qquad 17y = 34$$

Now make the coefficient of the first term in each equation equal to 1.

$$\tfrac{1}{2}R_1 \to R_1 \qquad x + \frac{3}{2}y = 6$$
$$\tfrac{1}{17}R_2 \to R_2 \qquad y = 2$$

Back-substitute to solve for x.

$$x + \frac{3}{2}(2) = 6$$
$$x + 3 = 6$$
$$x = 3$$

The solution of the system is (3, 2).

Your Turn 2

Let $x =$ flight time eastward

$y =$ difference in time zones

$$x + y = 16 \quad (1)$$
$$x - y = 2 \quad (2) \text{ (no wind)}$$

$$x + y = 16$$
$$R_1 + (-1)R_2 \to R_2 \qquad 2y = 14$$
$$y = 7$$
$$x + 7 = 16$$
$$x = 9$$

The flight time eastward is 9 hours and the difference in time zones is 7 hours.

Your Turn 3

Since the permissible values of z are 5, 6, 7, . . . , 16, the largest value of z is $z = 16$.

$$x = 2z - 9 = 2(16) - 9 = 23$$
$$y = 49 - 3z = 49 - 3(16) = 1$$

The solution with the largest number of spoons is 23 knives, 1 fork, and 16 spoons.

2.1 Exercises

In Exercises 1–16 and 19–28, check each solution by substituting it in the original equations of the system.

1.
$$x + y = 5 \quad (1)$$
$$2x - 2y = 2 \quad (2)$$

To eliminate x in equation (2), multiply equation (1) by -2 and add the result to equation (2). The new system is

$$x + y = 5 \quad (1)$$
$$-2R_1 + R_2 \to R_2 \qquad -4y = -8. \quad (3)$$

Now make the coefficient of the first term in each row equal 1. To accomplish this, multiply equation (3) by $-\frac{1}{4}$.

$$x + y = 5 \quad (1)$$
$$-\tfrac{1}{4}R_2 \to R_2 \qquad y = 2 \quad (4)$$

Substitute 2 for y in equation (1).

$$x + 2 = 5$$
$$y = 3$$

The solution is (3, 2).

3.
$$3x - 2y = -3 \quad (1)$$
$$5x - y = 2 \quad (2)$$

To eliminate x in equation (2), multiply equation (1) by -5 and equation (2) by 3. Add the results. The new system is

$$3x - 2y = -3 \quad (1)$$
$$-5R_1 + 3R_2 \to R_2 \qquad 7y = 21. \quad (3)$$

Now make the coefficient of the first term in each row equal 1. To accomplish this, multiply equation (1) by $\frac{1}{3}$ and equation (3) by $\frac{1}{7}$.

$$\tfrac{1}{3}R_1 \to R_1 \quad x - \frac{2}{3}y = -1 \quad (4)$$
$$\tfrac{1}{7}R_2 \to R_2 \qquad y = 3 \quad (5)$$

Back-substitution of 3 for y in equation (4) gives

$$x - \frac{2}{3}(3) = -1$$
$$x - 2 = -1$$
$$x = 1.$$

The solution is (1, 3).

5. $3x + 2y = -6 \quad (1)$

$5x - 2y = -10 \quad (2)$

Eliminate x in equation (2) to get the system

$$3x + 2y = -6 \quad (1)$$
$$5R_1 + (-3)R_2 \to R_2 \qquad 16y = 0. \quad (3)$$

Make the coefficient of the first term in each equation equal 1.

$$\tfrac{1}{3}R_1 \to R_1 \quad x + \frac{2}{3}y = -2 \quad (4)$$
$$\tfrac{1}{16}R_2 \to R_2 \qquad y = 0 \quad (5)$$

Substitute 0 for y in equation (4) to get $x = -2$. The solution is $(-2, 0)$.

7. $6x - 2y = -4 \quad (1)$

$3x + 4y = 8 \quad (2)$

Eliminate x in equation (2).

$$6x - 2y = -4 \quad (1)$$
$$-1R_1 + 2R_2 \to R_2 \qquad 10y = 20 \quad (3)$$

Make the coefficient of the first term in each row equal 1.

$$\tfrac{1}{6}R_1 \to R_1 \quad x - \frac{1}{3}y = -\frac{2}{3} \quad (4)$$
$$\tfrac{1}{10}R_2 \to R_2 \qquad y = 2 \quad (5)$$

Substitute 2 for y in equation (4) to get $x = 0$. The solution is (0, 2).

9. $5p + 11q = -7 \quad (1)$

$3p - 8q = 25 \quad (2)$

Eliminate p in equation (2).

$$5p + 11q = -7 \quad (1)$$
$$-3R_1 + 5R_2 \to R_2 \qquad -73q = 146 \quad (3)$$

Make the coefficient of the first term in each row equal 1.

$$\tfrac{1}{5}R_1 \to R_1 \quad p + \frac{11}{5}q = -\frac{7}{5} \quad (4)$$
$$-\tfrac{1}{73}R_2 \to R_2 \qquad q = -1 \quad (5)$$

Substitute -2 for q in equation (4) to get $p = 3$. The solution is $(3, -2)$.

11. $6x + 7y = -2 \quad (1)$

$7x - 6y = 26 \quad (2)$

Eliminate x in equation (2).

$$6x + 7y = -2 \quad (1)$$
$$7R_1 + (-6)R_2 \to R_2 \qquad 85y = -170 \quad (3)$$

Make the coefficient of the first term in each equation equal 1.

$$\tfrac{1}{6}R_1 \to R_1 \quad x + \frac{7}{6}y = -\frac{1}{3} \quad (4)$$
$$\tfrac{1}{85}R_2 \to R_2 \qquad y = -2 \quad (5)$$

Substitute -2 for y in equation (4) to get $x = 2$. The solution is $(2, -2)$.

13. $3x + 2y = 5 \quad (1)$

$6x + 4y = 8 \quad (2)$

Eliminate x in equation (2).

$$3x + 2y = 5 \quad (1)$$
$$-2R_1 + R_2 \to R_2 \qquad 0 = -2 \quad (3)$$

Equation (3) is a false statement.

The system is inconsistent and has no solution.

15. $3x - 2y = -4 \quad (1)$

$-6x + 4y = 8 \quad (2)$

Eliminate x in equation (2).

$$3x - 2y = -4 \quad (1)$$
$$2R_1 + R_2 \to R_2 \qquad 0 = 0 \quad (3)$$

The true statement in equation (3) indicates that there are an infinite number of solutions for the system. Solve equation (1) for x.

$$3x - 2y = -4 \quad (1)$$
$$3x = 2y - 4$$
$$x = \frac{2y - 4}{3} \quad (4)$$

For each value of y, equation (4) indicates that $x = \frac{2y-4}{3}$, and all ordered pairs of the form $\left(\frac{2y-4}{3}, y\right)$ are solutions.

17. $x - \dfrac{3y}{2} = \dfrac{5}{2} \quad (1)$

$\dfrac{4x}{3} + \dfrac{2y}{3} = 6 \quad (2)$

Rewrite the equations without fractions.

$$2R_1 \to R_1 \quad 2x - 3y = 5 \quad (3)$$
$$3R_2 \to R_2 \quad 4x + 2y = 18 \quad (4)$$

Eliminate x in equation (4).

$$2x - 3y = 5 \quad (3)$$
$$-2R_1 + R_2 \to R_2 \qquad 8y = 8 \quad (5)$$

Make the coefficient of the first term in each equation equal 1.

$$\tfrac{1}{2}R_1 \to R_1 \quad x - \frac{3}{2}y = \frac{5}{2} \quad (6)$$
$$\tfrac{1}{8}R_2 \to R_2 \qquad y = 1 \quad (7)$$

Substitute 1 for y in equation (6) to get $x = 4$. The solution is (4, 1).

19. $\frac{x}{2} + y = \frac{3}{2}$ (*1*)

$\frac{x}{3} + y = \frac{1}{3}$ (*2*)

Rewrite the equations without fractions.

$$2R_1 \to R_1 \quad x + 2y = 3 \quad (3)$$
$$3R_2 \to R_2 \quad x + 3y = 1 \quad (4)$$

Eliminate x in equation (4).

$$x + 2y = 3 \quad (3)$$
$$-1R_1 + R_2 \to R_2 \quad y = -2 \quad (5)$$

Substitute -2 for y in equation (3) to get $x = 7$. The solution is $(7, -2)$.

21. An inconsistent system has *no* solutions.

23. $x + 2y + 3z = 90$ (*1*)

$3y + 4z = 36$ (*2*)

Let z be the parameter and solve equation (2) for y in terms of z.

$$3y + 4z = 36$$
$$3y = 36 - 4z$$
$$y = 12 - \frac{4}{3}z$$

Substitute this expression for y in equation (1) to solve for x in terms of z.

$$x + 2\left(12 - \frac{4}{3}z\right) + 3z = 90$$
$$x + 24 - \frac{8}{3}z + 3z = 90$$
$$x + \frac{1}{3}z = 66$$
$$x = 66 - \frac{1}{3}z$$

Thus the solutions are $\left(66 - \frac{1}{3}z, 12 - \frac{4}{3}z, z\right)$, where z is any real number. Since the solutions have to be nonnegative integers, set $66 - \frac{1}{3}z \geq 0$. Solving for z gives $z \leq 198$.

Since y must be nonnegative, we have $12 - \frac{4}{3}z \geq 0$. Solving for z gives $z \leq 9$.

Since z must be a multiple of 3 for x and y to be integers, the permissible values of z are 0, 3, 6, and 9, which gives 4 solutions.

25. $3x + 2y + 4z = 80$ (*1*)

$y - 3z = 10$ (*2*)

Let z be the parameter and solve equation (2) for y in terms of z.

$$y - 3z = 10$$
$$y = 3z + 10$$

Substitute this expression for y in equation (1) to solve for x in terms of z.

$$3x + 2(3z + 10) + 4z = 80$$
$$3x + 6z + 20 + 4z = 80$$
$$3x + 10z = 60$$
$$3x = 60 - 10z$$
$$x = 20 - \frac{10}{3}z$$

Thus the solutions are $\left(20 - \frac{10}{3}z, 3z + 10, z\right)$, where z is any real number. Since the solutions have to be nonnegative integers, set $20 - \frac{10}{3}z \geq 0$. Solving for z gives $z \leq 6$.

Since y must be nonnegative, we have $3z + 10 \geq 0$. Solving for z gives $z \geq -10\frac{1}{3}$.

Since z must be a multiple of 3 for x to be an integer, the permissible values of z are 0, 3, and, 6, which gives 3 solutions.

29. $2x + 3y - z = 1$ (*1*)

$3x + 5y + z = 3$ (*2*)

Eliminate x in equation (2).

$$2x + 3y - z = 1 \quad (1)$$
$$-3R_1 + 2R_2 \to R_2 \quad y + 5z = 3 \quad (3)$$

Since there are only two equations, it is not possible to continue with the echelon method as in the previous exercises involving systems with three equations and three variables. To complete the solution, make the coefficient of the first term in the each equation equal 1.

$$\tfrac{1}{2}R_1 \to R_1 \quad x + \frac{3}{2}y + \frac{1}{2}z = \frac{1}{2} \quad (4)$$
$$y + 5z = 3 \quad (3)$$

Solve equation (3) for y in terms of the parameter z.

$$y + 5z = 3$$
$$y = 3 - 5z$$

Substitute this expression for y in equation (4) to solve for x in terms of the parameter z.

$$x + \frac{3}{2}(3 - 5z) - \frac{1}{2}z = \frac{1}{2}$$
$$x + \frac{9}{2} - \frac{15}{2}z - \frac{1}{2}z = \frac{1}{2}$$
$$x - 8z = -4$$
$$x = 8z - 4$$

The solution is $(8z - 4, 3 - 5z, z)$.

31. $x + 2y + 3z = 11$ (1)
$2x - y + z = 2$ (2)

$$x + 2y + 3z = 11 \quad (1)$$
$$-2R_1 + R_2 \rightarrow R_2 \quad -5y - 5z = -20 \quad (3)$$

$$x + 2y + 3z = 11 \quad (1)$$
$$-\tfrac{1}{5}R_2 \rightarrow R_2 \quad y + z = 4 \quad (4)$$

Since there are only two equations, it is not possible to continue with the echelon method. To complete the solution, solve equation (4) for y in terms of the parameter z.

$$y = 4 - z$$

Now substitute $4 - z$ for y in equation (1) and solve for x in terms of z.

$$x + 2(4 - z) + 3z = 11$$
$$x + 8 - 2z + 3z = 11$$
$$x = 3 - z$$

The solution is $(3 - z, 4 - z, z)$.

33. $nb + (\sum x)m = \sum y$ (1)
$(\sum x)b + (\sum x^2)m = \sum xy$ (2)

Multiply equation (1) by $\frac{1}{n}$.

$$b + \frac{\sum x}{n}m = \frac{\sum y}{n} \quad (3)$$
$$(\sum x)b + (\sum x^2)m = \sum xy \quad (2)$$

Eliminate b from equation (2).

$$b + \frac{\sum x}{n}m = \frac{\sum y}{n} \quad (3)$$

$(-\sum x)R_1 + R_2 \rightarrow R_2$

$$\left[-\frac{(\sum x)^2}{n} + \sum x^2\right]m = \frac{-(\sum x)(\sum y)}{n} + \sum xy \quad (4)$$

Multiply equation (4) by $\dfrac{1}{-\frac{(\sum x)^2}{n} + \sum x^2}$.

$$b + \frac{\sum x}{n}m = \frac{\sum y}{n} \quad (3)$$

$$m = \left[\frac{-(\sum x)(\sum y)}{n} + \sum xy\right]\left[\frac{1}{-\frac{(\sum x)^2}{n} + \sum x^2}\right] \quad (5)$$

Simplify the right side of equation (5).

$$m = \left[\frac{-(\sum x)(\sum y) + n\sum xy}{n}\right]\left[\frac{n}{-(\sum x)^2 + n(\sum x^2)}\right]$$

$$m = \frac{n\sum xy - (\sum x)(\sum y)}{n(\sum x^2) - (\sum x)^2}$$

From equation (3) we have

$$b = \frac{\sum y}{n} - \frac{\sum x}{n}m$$
$$b = \frac{\sum y - m(\sum x)}{n}.$$

35. Let x = the cost per pound of rice, and
y = the cost per pound of potatoes.

The system to be solved is

$$20x + 10y = 16.20 \quad (1)$$
$$30x + 12y = 23.04. \quad (2)$$

Multiply equation (1) by $\frac{1}{20}$.

$$\tfrac{1}{20}R_1 \rightarrow R_1 \quad x + 0.5y = 0.81 \quad (3)$$
$$30x + 12y = 23.04 \quad (2)$$

Eliminate x in equation (2).

$$x + 0.5y = 0.81 \quad (3)$$
$$-30R_1 + R_2 \rightarrow R_2 \quad -3y = -1.26 \quad (4)$$

Multiply equation (4) by $-\frac{1}{3}$.

$$x + 0.5y = 0.81 \quad (3)$$
$$-\tfrac{1}{3}R_2 \rightarrow R_2 \quad y = 0.42 \quad (5)$$

Substitute 0.42 for y in equation (3).

$$x + 0.5(0.42) = 0.81$$
$$x + 0.21 = 0.81$$
$$x = 0.60$$

The cost of 10 pounds of rice and 50 pounds of potatoes is

$$10(0.60) + 50(0.42) = 27,$$

that is, \$27.

37. Let x = the number of seats on the main floor, and
y = the number of seats in the balcony.

The system to be solved is

$$8x + 5y = 4200 \quad (1)$$
$$0.25(8x) + 0.40(5y) = 1200. \quad (2)$$

Make the coefficient of the first term in equation (1) equal 1.

$$\tfrac{1}{8}R_1 \rightarrow R_1 \quad x + \frac{5}{8}y = 525 \quad (3)$$
$$2x + 2y = 1200 \quad (2)$$

Eliminate x in equation (2).

$$x + \frac{5}{8}y = 525 \quad (3)$$

$$-2R_1 + R_2 \to R_2 \qquad \frac{6}{8}y = 150 \quad (4)$$

Make the coefficient of the first term in equation (4) equal 1.

$$x + \frac{5}{8}y = 525 \quad (3)$$

$$\frac{8}{6}R_2 \to R_2 \qquad y = 200 \quad (5)$$

Substitute 200 for y in equation (3).

$$x + \frac{5}{8}(200) = 525$$

$$x + 125 = 525$$

$$x = 400$$

There are 400 main floor seats and 200 balcony seats.

39. Let $x =$ the number of model 201 to make each day, and
$y =$ the number of model 301 to make each day.

The system to be solved is

$$2x + 3y = 34 \quad (1)$$
$$18x + 27y = 335. \quad (2)$$

Make the coefficient of the first term in equation (1) equal 1.

$$\frac{1}{2}R_1 \to R_1 \qquad x + \frac{3}{2}y = 17 \quad (3)$$
$$18x + 27y = 335 \quad (2)$$

Eliminate x in equation (2).

$$x + \frac{3}{2}y = 17 \quad (3)$$
$$-18R_1 + R_2 \to R_2 \qquad 0 = 29 \quad (4)$$

Since equation (4) is false, the system is inconsistent. Therefore, this situation is impossible.

41. Let $x =$ the number of buffets produced each week,
$y =$ the number of chairs produced each week,
$z =$ the number of tables produced each week.

Make a table.

	Buffet	Chair	Table	Totals
Construction	30	10	10	350
Finishing	10	10	30	150

The system to be solved is

$$30x + 10y + 10z = 350 \quad (1)$$
$$10x + 10y + 30z = 150. \quad (2)$$

Make the coefficient of the first term in equation (1) equal 1.

$$\frac{1}{30}R_1 \to R_1 \quad x + \frac{1}{3}y + \frac{1}{3}z = \frac{35}{3} \quad (3)$$
$$10x + 10y + 30z = 150 \quad (2)$$

Eliminate x from equation (2).

$$x + \frac{1}{3}y + \frac{1}{3}z = \frac{35}{3} \quad (3)$$
$$-10R_1 + R_2 \to R_2 \qquad \frac{20}{3}y + \frac{80}{3}z = \frac{100}{3} \quad (4)$$

Solve equation (4) for y. Multiply by 3.

$$20y + 80z = 100$$
$$y + 4z = 5$$
$$y = 5 - 4z$$

Substitute $5 - 4z$ for y in equation (1) and solve for x.

$$30x + 10(5 - 4z) + 10z = 350$$
$$30x + 50 - 40z + 10z = 350$$
$$30x = 300 + 30z$$
$$x = 10 + z$$

The solution is $(10 + z, 5 - 4z, z)$. All variables must be nonnegative integers. Therefore,

$$5 - 4z \geq 0$$
$$5 \geq 4z$$
$$z \leq \frac{5}{4},$$

so $z = 0$ or $z = 1$. (Any larger value of z would cause y to be negative, which would make no sense in the problem.) If $z = 0$, then the solution is $(10, 5, 0)$. If $z = 1$, then the solution is $(11, 1, 1)$. Therefore, the company should make either 10 buffets, 5 chairs, and no tables or 11 buffets, 1 chair, and 1 table each week.

43. Let $x =$ the number of long-sleeve blouses,
$y =$ the number of short-sleeve blouses, and
$z =$ the number of sleeveless blouses.

Make a table.

	Long Sleeve	Short Sleeve	Sleeveless	Totals
Cutting	1.5	1	0.5	380
Sewing	1.2	0.9	0.6	330

The system to be solved is

$$1.5x + y + 0.5z = 380 \quad (1)$$
$$1.2x + 0.9y + 0.6z = 330. \quad (2)$$

Simplify the equations. Multiply equation (1) by 2 and equation (2) by $\frac{10}{3}$.

$$3x + 2y + z = 760 \quad (3)$$
$$4x + 3y + 2z = 1100 \quad (4)$$

Make the leading coefficient of equation (3) equal 1.

$$\tfrac{1}{3}R_1 \to R_1 \quad x + \frac{2}{3}y + \frac{1}{3}z = \frac{760}{3} \quad (5)$$
$$4x + 3y + 2z = 1100 \quad (4)$$

Eliminate x from equation (4).

$$x + \frac{2}{3}y + \frac{1}{3}z = \frac{760}{3} \quad (5)$$
$$-4R_1 + R_2 \to R_2 \quad \frac{1}{3}y + \frac{2}{3}z = \frac{260}{3} \quad (6)$$

Make the leading coefficient of equation (6) equal 1.

$$x + \frac{2}{3}y + \frac{1}{3}z = \frac{760}{3} \quad (5)$$
$$3R_2 \to R_2 \quad y + 2z = 260 \quad (7)$$

From equation (7), $y = 260 - 2z$. Substitute this into equation (5).

$$x + \frac{2}{3}(260 - 2z) + \frac{1}{3}z = \frac{760}{3}$$
$$x + \frac{520}{3} - \frac{4}{3}z + \frac{1}{3}z = \frac{760}{3}$$
$$x - z = \frac{240}{3}$$
$$x = z + 80$$

The solution is $(z + 80, 260 - 2z, z)$. In this problem x, y, and z must be nonnegative, so

$$260 - 2z \geq 0$$
$$-2z \geq -260$$
$$z \leq 130.$$

Therefore, the plant should make $z = 80$ long-sleeve blouses, $260 - 2z$ short-sleeve blouses, and z sleeveless blouses with $0 \leq z \leq 130$.

45. **(a)** For the first equation, the first sighting in 2000 was on day $y = 759 - 0.338(2000) = 83$, or during the eighty-third day of the year. Since 2000 was a leap year, the eighty-third day fell on March 23.

For the second equation, the first sighting in 2000 was on day $y = 1637 - 0.779(2000) = 79$, or during the seventy-ninth day of the year. Since 2000 was a leap year, the seventy-ninth day fell on March 19.

(b) $y = 759 - 0.338x \quad (1)$

$y = 1637 - 0.779x \quad (2)$

Rewrite equations so that variables are on the left side and constant term is on the right side.

$$0.338x + y = 759 \quad (3)$$
$$0.779x + y = 1637 \quad (4)$$

Eliminate y from equation (4).

$$0.338x + y = 759 \quad (3)$$
$$-1R_1 + R_2 \to R_2 \quad 0.441x = 878 \quad (5)$$

Make leading coefficient for equation (5) equal 1.

The two estimates agree in the year closest to $x = \frac{878}{0.441} \approx 1990.93$, so they agree in 1991.

The estimated number of days into the year when a robin can be expected is

$$0.338\left(\frac{878}{0.441}\right) + y = 759$$
$$y \approx 86.$$

47. Let x = number of field goals, and
y = number of foul shots.

Then
$$x + y = 64 \quad (1)$$
$$2x + y = 100 \quad (2).$$

Eliminate x in equation (2).

$$x + y = 64 \quad (1)$$
$$-2R_1 + R_2 \to R_2 \quad -y = -28 \quad (3)$$

Make the coefficients of the first term of each equation equal 1.

$$x + y = 64 \quad (1)$$
$$-1R_2 \to R_2 \quad y = 28 \quad (4)$$

Substitute 28 for y in equation (1) to get $x = 36$. Wilt Chamberlain made 36 field goals and 28 foul shots.

2.2 Solution of Linear Systems by the Gauss-Jordan Method

Your Turn 1

$$4x + 5y = 10$$
$$7x + 8y = 19$$

Write the augmented matrix and transform the matrix.

$$\left[\begin{array}{cc|c} 4 & 5 & 10 \\ 7 & 8 & 19 \end{array}\right]$$

$$\tfrac{1}{4}R_1 \to R_1 \left[\begin{array}{cc|c} 1 & \frac{5}{4} & \frac{5}{2} \\ 7 & 8 & 19 \end{array}\right]$$

$$-7R_1 + R_2 \to R_2 \left[\begin{array}{cc|c} 1 & \frac{5}{4} & \frac{5}{2} \\ 0 & -\frac{3}{4} & \frac{3}{2} \end{array}\right]$$

$$-\tfrac{4}{3}R_2 \to R_2 \quad \left[\begin{array}{cc|c} 1 & \frac{5}{4} & \frac{5}{2} \\ 0 & 1 & -2 \end{array}\right]$$

$$-\tfrac{5}{4}R_2 + R_1 \to R_1 \quad \left[\begin{array}{cc|c} 1 & 0 & 5 \\ 0 & 1 & -2 \end{array}\right]$$

The solution is $x = 5$ and $y = -2$, or $(5, -2)$.

Your Turn 2

$$\begin{aligned} x + 2y + 3z &= 2 \\ 2x + 2y - 3z &= 27 \\ 3x + 2y + 5z &= 10 \end{aligned}$$

Write the augmented matrix and transform the matrix.

$$\left[\begin{array}{ccc|c} 1 & 2 & 3 & 2 \\ 2 & 2 & -3 & 27 \\ 3 & 2 & 5 & 10 \end{array}\right]$$

$$\begin{array}{r} \\ -2R_1 + R_2 \to R_2 \\ -3R_1 + R_3 \to R_3 \end{array} \left[\begin{array}{ccc|c} 1 & 2 & 3 & 2 \\ 0 & -2 & -9 & 23 \\ 0 & -4 & -4 & 4 \end{array}\right]$$

$$\begin{array}{r} \\ -\frac{1}{2}R_2 \to R_2 \\ \\ \end{array} \left[\begin{array}{ccc|c} 1 & 2 & 3 & 2 \\ 0 & 1 & \frac{9}{2} & -\frac{23}{2} \\ 0 & -4 & -4 & 4 \end{array}\right]$$

$$\begin{array}{r} -2R_2 + R_1 \to R_1 \\ \\ 4R_2 + R_3 \to R_3 \end{array} \left[\begin{array}{ccc|c} 1 & 0 & -6 & 25 \\ 0 & 1 & \frac{9}{2} & -\frac{23}{2} \\ 0 & 0 & 14 & -42 \end{array}\right]$$

$$\begin{array}{r} \\ \\ \frac{1}{14}R_3 \to R_3 \end{array} \left[\begin{array}{ccc|c} 1 & 0 & -6 & 25 \\ 0 & 1 & \frac{9}{2} & -\frac{23}{2} \\ 0 & 0 & 1 & -3 \end{array}\right]$$

$$\begin{array}{r} 6R_3 + R_1 \to R_1 \\ -\frac{9}{2}R_3 + R_2 \to R_2 \\ \\ \end{array} \left[\begin{array}{ccc|c} 1 & 0 & 0 & 7 \\ 0 & 1 & 0 & 2 \\ 0 & 0 & 1 & -3 \end{array}\right]$$

The solution is $x = 7$, $y = 2$, and $z = -3$, or $(7, 2, -3)$.

Your Turn 3

$$\begin{aligned} 2x - 2y + 3z - 4w &= 6 \\ 3x + 2y + 5z - 3w &= 7 \\ 4x + y + 2z - 2w &= 8 \end{aligned}$$

Write the augmented matrix and transform the matrix.

$$\left[\begin{array}{cccc|c} 2 & -2 & 3 & -4 & 6 \\ 3 & 2 & 5 & -3 & 7 \\ 4 & 1 & 2 & -2 & 8 \end{array}\right]$$

$$\begin{array}{r} \frac{1}{2}R_1 \to R_1 \\ \\ \\ \end{array} \left[\begin{array}{cccc|c} 1 & -1 & \frac{3}{2} & -2 & 3 \\ 3 & 2 & 5 & -3 & 7 \\ 4 & 1 & 2 & -2 & 8 \end{array}\right]$$

$$\begin{array}{r} \\ -3R_1 + R_2 \to R_2 \\ -4R_1 + R_3 \to R_3 \end{array} \left[\begin{array}{cccc|c} 1 & -1 & \frac{3}{2} & -2 & 3 \\ 0 & 5 & \frac{1}{2} & 3 & -2 \\ 0 & 5 & -4 & 6 & -4 \end{array}\right]$$

$$\begin{array}{r} \\ \frac{1}{5}R_2 \to R_2 \\ \\ \end{array} \left[\begin{array}{cccc|c} 1 & -1 & \frac{3}{2} & -2 & 3 \\ 0 & 1 & \frac{1}{10} & \frac{3}{5} & -\frac{2}{5} \\ 0 & 5 & -4 & 6 & -4 \end{array}\right]$$

$$\begin{array}{r} R_2 + R_1 \to R_1 \\ \\ -5R_2 + R_3 \to R_3 \end{array} \left[\begin{array}{cccc|c} 1 & 0 & \frac{8}{5} & -\frac{7}{5} & \frac{13}{5} \\ 0 & 1 & \frac{1}{10} & \frac{3}{5} & -\frac{2}{5} \\ 0 & 0 & -\frac{9}{2} & 3 & -2 \end{array}\right]$$

$$\begin{array}{r} \\ \\ -\frac{2}{9}R_3 \to R_3 \end{array} \left[\begin{array}{cccc|c} 1 & 0 & \frac{8}{5} & -\frac{7}{5} & \frac{13}{5} \\ 0 & 1 & \frac{1}{10} & \frac{3}{5} & -\frac{2}{5} \\ 0 & 0 & 1 & -\frac{2}{3} & \frac{4}{9} \end{array}\right]$$

$$\begin{array}{r} -\frac{8}{5}R_3 + R_1 \to R_1 \\ -\frac{1}{10}R_3 + R_2 \to R_2 \\ \\ \end{array} \left[\begin{array}{cccc|c} 1 & 0 & 0 & -\frac{1}{3} & \frac{17}{9} \\ 0 & 1 & 0 & \frac{2}{3} & -\frac{4}{9} \\ 0 & 0 & 1 & -\frac{2}{3} & \frac{4}{9} \end{array}\right]$$

We cannot change the values in column 4 without changing the form of the other three columns. So, let w be the parameter. The last matrix gives these equations.

$$x - \frac{1}{3}w = \frac{17}{9}, \quad \text{or} \quad x = \frac{17}{9} + \frac{1}{3}w$$

$$y + \frac{2}{3}w = -\frac{4}{9}, \quad \text{or} \quad y = -\frac{4}{9} - \frac{2}{3}w$$

$$z - \frac{2}{3}w = \frac{4}{9}, \quad \text{or} \quad z = \frac{4}{9} + \frac{2}{3}w$$

The solution is $\left(\frac{17}{9} + \frac{1}{3}w, -\frac{4}{9} - \frac{2}{3}w, \frac{4}{9} + \frac{2}{3}w, w\right)$, where w is a real number.

2.2 Exercises

1. $3x + y = 6$
$2x + 5y = 15$

The equations are already in proper form. The augmented matrix obtained from the coefficients and the constants is

$$\left[\begin{array}{cc|c} 3 & 1 & 6 \\ 2 & 5 & 15 \end{array}\right].$$

3. $2x + y + z = 3$
$3x - 4y + 2z = -7$
$x + y + z = 2$

leads to the augmented matrix

$$\left[\begin{array}{ccc|c} 2 & 1 & 1 & 3 \\ 3 & -4 & 2 & -7 \\ 1 & 1 & 1 & 2 \end{array}\right].$$

5. We are given the augmented matrix

$$\left[\begin{array}{cc|c} 1 & 0 & 2 \\ 0 & 1 & 3 \end{array}\right].$$

This is equivalent to the system of equations

$$x = 2$$
$$y = 3,$$

or $x = 2, y = 3$.

7. $$\left[\begin{array}{ccc|c} 1 & 0 & 0 & 4 \\ 0 & 1 & 0 & -5 \\ 0 & 0 & 1 & 1 \end{array}\right]$$

The system associated with this matrix is

$$x = 4$$
$$y = -5$$
$$z = 1,$$

or $x = 4, \ y = -5, \ z = 1$.

9. *Row operations* on a matrix correspond to transformations of a system of equations.

11. $$\left[\begin{array}{ccc|c} 3 & 7 & 4 & 10 \\ 1 & 2 & 3 & 6 \\ 0 & 4 & 5 & 11 \end{array}\right]$$

Find $R_1 + (-3)R_2$.

In row 2, column 1,

$$3 + (-3)1 = 0.$$

In row 2, column 2,

$$7 + (-3)2 = 1.$$

In row 2, column 3,

$$4 + (-3)3 = -5.$$

In row 2, column 4,

$$10 + (-3)6 = -8.$$

Replace R_2 with these values. The new matrix is

$$\left[\begin{array}{ccc|c} 3 & 7 & 4 & 10 \\ 0 & 1 & -5 & -8 \\ 0 & 4 & 5 & 11 \end{array}\right].$$

13. $$\left[\begin{array}{ccc|c} 1 & 6 & 4 & 7 \\ 0 & 3 & 2 & 5 \\ 0 & 5 & 3 & 7 \end{array}\right]$$

Find $(-2)R_2 + R_1 \to R_1$

$$\left[\begin{array}{ccc|c} (-2)0+1 & (-2)3+6 & (-2)2+4 & (-2)5+7 \\ 0 & 3 & 2 & 5 \\ 0 & 5 & 3 & 7 \end{array}\right]$$

$$= \left[\begin{array}{ccc|c} 1 & 0 & 0 & -3 \\ 0 & 3 & 2 & 5 \\ 0 & 5 & 3 & 7 \end{array}\right]$$

15. $$\left[\begin{array}{ccc|c} 3 & 0 & 0 & 18 \\ 0 & 5 & 0 & 9 \\ 0 & 0 & 4 & 8 \end{array}\right]$$

$$\frac{1}{3}R_1 \to R_1 \left[\begin{array}{ccc|c} \frac{1}{3}(3) & \frac{1}{3}(0) & \frac{1}{3}(0) & \frac{1}{3}(18) \\ 0 & 5 & 0 & 9 \\ 0 & 0 & 4 & 8 \end{array}\right] = \left[\begin{array}{ccc|c} 1 & 0 & 0 & 6 \\ 0 & 5 & 0 & 9 \\ 0 & 0 & 4 & 8 \end{array}\right]$$

17. $x + y = 5$
$3x + 2y = 12$

Write the augmented matrix and use row operations.

$$\left[\begin{array}{cc|c} 1 & 1 & 5 \\ 3 & 2 & 12 \end{array}\right]$$

$$-3R_1 + R_2 \to R_2 \left[\begin{array}{cc|c} 1 & 1 & 5 \\ 0 & -1 & -3 \end{array}\right]$$

$$-1R_2 \to R_2 \left[\begin{array}{cc|c} 1 & 1 & 5 \\ 0 & 1 & 3 \end{array}\right]$$

$$-1R_2 + R_1 \to R_1 \left[\begin{array}{cc|c} 1 & 0 & 2 \\ 0 & 1 & 3 \end{array}\right]$$

The solution is (2, 3).

19. $x + y = 7$
$4x + 3y = 22$

Write the augmented matrix and use row operations.

$$\left[\begin{array}{rr|r}1&1&7\\4&3&22\end{array}\right]$$

$$-4R_1+R_2\to R_2\left[\begin{array}{rr|r}1&1&7\\0&-1&-6\end{array}\right]$$

$$-1R_2\to R_2\left[\begin{array}{rr|r}1&1&7\\0&1&6\end{array}\right]$$

$$-1R_2+R_1\to R_1\left[\begin{array}{rr|r}1&0&1\\0&1&6\end{array}\right]$$

The solution is (1, 6).

21. $2x-3y=2$
$4x-6y=1$

Write the augmented matrix and use row operations.

$$\left[\begin{array}{rr|r}2&-3&2\\4&-6&1\end{array}\right]$$

$$-2R_1+R_2\to R_2\left[\begin{array}{rr|r}2&-3&2\\0&0&-3\end{array}\right]$$

The system associated with the last matrix is

$$2x-3y=2$$
$$0x+0y=-3.$$

Since the second equation, $0=-3$, is false, the system is inconsistent and therefore has no solution.

23. $6x-3y=1$
$-12x+6y=-2$

Write the augmented matrix of the system and use row operations.

$$\left[\begin{array}{rr|r}6&-3&1\\-12&6&-2\end{array}\right]$$

$$2R_1+R_2\to R_2\left[\begin{array}{rr|r}6&-3&1\\0&0&0\end{array}\right]$$

$$\tfrac{1}{6}R_1\to R_1\left[\begin{array}{rr|r}1&-\frac{1}{2}&\frac{1}{6}\\0&0&0\end{array}\right]$$

This is as far as we can go with the Gauss-Jordan method. To complete the solution, write the equation that corresponds to the first row of the matrix.

$$x-\frac{1}{2}y=\frac{1}{6}$$

Solve this equation for x in terms of y.

$$x=\frac{1}{2}y+\frac{1}{6}=\frac{3y+1}{6}$$

The solution is $\left(\frac{3y+1}{6},y\right)$, y any real number.

25. $y=x-3$
$y=1+z$
$z=4-x$

First write the system in proper form.

$$\begin{aligned}-x+y\quad&=-3\\ y-z&=1\\ x\quad+z&=4\end{aligned}$$

Write the augmented matrix and use row operations.

$$\left[\begin{array}{rrr|r}-1&1&0&-3\\0&1&-1&1\\1&0&1&4\end{array}\right]$$

$$-1R_1\to R_1\left[\begin{array}{rrr|r}1&-1&0&3\\0&1&-1&1\\1&0&1&4\end{array}\right]$$

$$-1R_1+R_3\to R_3\left[\begin{array}{rrr|r}1&-1&0&3\\0&1&-1&1\\0&1&1&1\end{array}\right]$$

$$\begin{array}{r}R_2+R_1\to R_1\\ \\ -1R_2+R_3\to R_3\end{array}\left[\begin{array}{rrr|r}1&0&-1&4\\0&1&-1&1\\0&0&2&0\end{array}\right]$$

$$\begin{array}{r}R_3+2R_1\to R_1\\ R_3+2R_2\to R_2\\ \\ \end{array}\left[\begin{array}{rrr|r}2&0&0&8\\0&2&0&2\\0&0&2&0\end{array}\right]$$

$$\begin{array}{r}\frac{1}{2}R_1\to R_1\\ \frac{1}{2}R_2\to R_2\\ \frac{1}{2}R_3\to R_3\end{array}\left[\begin{array}{rrr|r}1&0&0&4\\0&1&0&1\\0&0&1&0\end{array}\right]$$

The solution is (4, 1, 0).

27. $2x-2y=-5$
$2y+z=0$
$2x+z=-7$

Write the augmented matrix and use row operations.

$$\left[\begin{array}{rrr|r}2&-2&0&-5\\0&2&1&0\\2&0&1&-7\end{array}\right]$$

$$-1R_1+R_3\to R_3\left[\begin{array}{rrr|r}2&-2&0&-5\\0&2&1&0\\0&2&1&-2\end{array}\right]$$

$$\begin{array}{r}R_2+R_1\to R_1\\ \\ -1R_2+R_3\to R_3\end{array}\left[\begin{array}{rrr|r}2&0&1&-5\\0&2&1&0\\0&0&0&-2\end{array}\right]$$

This matrix corresponds to the system of equations

$$\begin{aligned} 2x + z &= -5 \\ 2y + z &= 0 \\ 0 &= -2. \end{aligned}$$

This false statement $0 = -2$ indicates that the system is inconsistent and therefore has no solution.

29. $4x + 4y - 4z = 24$
$2x - y + z = -9$
$x - 2y + 3z = 1$

Write the augmented matrix and use row operations.

$$\left[\begin{array}{rrr|r} 4 & 4 & -4 & 24 \\ 2 & -1 & 1 & -9 \\ 1 & -2 & 3 & 1 \end{array}\right]$$

$$\begin{array}{r} \\ R_1 + (-2)R_2 \to R_2 \\ R_1 + (-4)R_3 \to R_3 \end{array} \left[\begin{array}{rrr|r} 4 & 4 & -4 & 24 \\ 0 & 6 & -6 & 42 \\ 0 & 12 & -16 & 20 \end{array}\right]$$

$$\begin{array}{r} 2R_2 + (-3)R_1 \to R_1 \\ \\ -2R_2 + R_3 \to R_3 \end{array} \left[\begin{array}{rrr|r} -12 & 0 & 0 & 12 \\ 0 & 6 & -6 & 42 \\ 0 & 0 & -4 & -64 \end{array}\right]$$

$$\begin{array}{r} \\ -3R_3 + 2R_2 \to R_2 \\ \\ \end{array} \left[\begin{array}{rrr|r} -12 & 0 & 0 & 12 \\ 0 & 12 & 0 & 276 \\ 0 & 0 & -4 & -64 \end{array}\right]$$

$$\begin{array}{r} -\frac{1}{12}R_1 \to R_1 \\ \frac{1}{12}R_2 \to R_2 \\ -\frac{1}{4}R_3 \to R_3 \end{array} \left[\begin{array}{rrr|r} 1 & 0 & 0 & -1 \\ 0 & 1 & 0 & 23 \\ 0 & 0 & 1 & 16 \end{array}\right]$$

The solution is $(-1, 23, 16)$.

31. $3x + 5y - z = 0$
$4x - y + 2z = 1$
$7x + 4y + z = 1$

Write the augmented matrix and use row operations.

$$\left[\begin{array}{rrr|r} 3 & 5 & -1 & 0 \\ 4 & -1 & 2 & 1 \\ 7 & 4 & 1 & 1 \end{array}\right]$$

$$\begin{array}{r} \\ 4R_1 + (-3)R_2 \to R_2 \\ 7R_1 + (-3)R_3 \to R_3 \end{array} \left[\begin{array}{rrr|r} 3 & 5 & -1 & 0 \\ 0 & 23 & -10 & -3 \\ 0 & 23 & -10 & -3 \end{array}\right]$$

$$\begin{array}{r} 23R_1 + (-5)R_2 \to R_1 \\ \\ R_2 + (-1)R_3 \to R_3 \end{array} \left[\begin{array}{rrr|r} 69 & 0 & 27 & 15 \\ 0 & 23 & -10 & -3 \\ 0 & 0 & 0 & 0 \end{array}\right]$$

$$\begin{array}{r} \frac{1}{69}R_1 \to R_1 \\ \frac{1}{23}R_2 \to R_2 \\ \\ \end{array} \left[\begin{array}{rrr|r} 1 & 0 & \frac{9}{23} & \frac{5}{23} \\ 0 & 1 & -\frac{10}{23} & -\frac{3}{23} \\ 0 & 0 & 0 & 0 \end{array}\right]$$

The row of zeros indicates dependent equations. Solve the first two equations respectively for x and y in terms of z to obtain

$$x = -\frac{9}{23}z + \frac{5}{23} = \frac{-9z + 5}{23}$$

and

$$y = \frac{10}{23}z - \frac{3}{23} = \frac{10z - 3}{23}.$$

The solution is $\left(\frac{-9z+5}{23}, \frac{10z-3}{23}, z\right)$.

33. $5x - 4y + 2z = 6$
$5x + 3y - z = 11$
$15x - 5y + 3z = 23$

Write the augmented matrix and use row operations.

$$\left[\begin{array}{rrr|r} 5 & -4 & 2 & 6 \\ 5 & 3 & -1 & 11 \\ 15 & -5 & 3 & 23 \end{array}\right]$$

$$\begin{array}{r} \\ -1R_1 + R_2 \to R_2 \\ -3R_1 + R_3 \to R_3 \end{array} \left[\begin{array}{rrr|r} 5 & -4 & 2 & 6 \\ 0 & 7 & -3 & 5 \\ 0 & 7 & -3 & 5 \end{array}\right]$$

$$\begin{array}{r} 4R_2 + 7R_1 \to R_1 \\ \\ -1R_2 + R_3 \to R_3 \end{array} \left[\begin{array}{rrr|r} 35 & 0 & 2 & 62 \\ 0 & 7 & -3 & 5 \\ 0 & 0 & 0 & 0 \end{array}\right]$$

$$\begin{array}{r} \frac{1}{35}R_1 \to R_1 \\ \frac{1}{7}R_2 \to R_2 \\ \\ \end{array} \left[\begin{array}{rrr|r} 1 & 0 & \frac{2}{35} & \frac{62}{35} \\ 0 & 1 & -\frac{3}{7} & \frac{5}{7} \\ 0 & 0 & 0 & 0 \end{array}\right]$$

The row of zeros indicates dependent equations. Solve the first two equations respectively for x and y in terms of z to obtain

$$x = -\frac{2}{35}z + \frac{62}{35} = \frac{-2z + 62}{35}$$

and

$$y = \frac{3}{7}z + \frac{5}{7} = \frac{3z + 5}{7}.$$

The solution is $\left(\frac{-2z+62}{35}, \frac{3z+5}{7}, z\right)$.

35. $2x + 3y + z = 9$
$4x + 6y + 2z = 18$
$-\frac{1}{2}x - \frac{3}{4}y - \frac{1}{4}z = -\frac{9}{4}$

Write the augmented matrix and use row operations.

$$\left[\begin{array}{rrr|r} 2 & 3 & 1 & 9 \\ 4 & 6 & 2 & 18 \\ -\frac{1}{2} & -\frac{3}{4} & -\frac{1}{4} & -\frac{9}{4} \end{array}\right]$$

$$\begin{array}{r} \\ -2R_1 + R_2 \to R_2 \\ \frac{1}{4}R_1 + R_3 \to R_3 \end{array} \left[\begin{array}{rrr|r} 2 & 3 & 1 & 9 \\ 0 & 0 & 0 & 0 \\ 0 & 0 & 0 & 0 \end{array}\right]$$

The rows of zeros indicate dependent equations. Since the equation involves x, y, and z, let y and z be parameters. Solve the equation for x to obtain $x = \frac{9 - 3y - z}{2}$.

The solution is $\left(\frac{9z - 3y - z}{2}, y, z\right)$, where y and z are any real numbers.

37.
$$\begin{aligned} x + 2y \quad\;\; - w &= 3 \\ 2x + \quad 4z + 2w &= -6 \\ x + 2y - z \quad\;\; &= 6 \\ 2x - y + z + w &= -3 \end{aligned}$$

Write the augmented matrix and use row operations.

$$\left[\begin{array}{rrrr|r} 1 & 2 & 0 & -1 & 3 \\ 2 & 0 & 4 & 2 & -6 \\ 1 & 2 & -1 & 0 & 6 \\ 2 & -1 & 1 & 1 & -3 \end{array}\right]$$

$$\begin{array}{r} \\ -2R_1 + R_2 \to R_2 \\ -1R_1 + R_3 \to R_3 \\ -2R_1 + R_4 \to R_4 \end{array} \left[\begin{array}{rrrr|r} 1 & 2 & 0 & -1 & 3 \\ 0 & -4 & 4 & 4 & -12 \\ 0 & 0 & -1 & 1 & 3 \\ 0 & -5 & 1 & 3 & -9 \end{array}\right]$$

$$\begin{array}{r} R_2 + 2R_1 \to R_1 \\ \\ \\ -5R_2 + 4R_4 \to R_4 \end{array} \left[\begin{array}{rrrr|r} 2 & 0 & 4 & 2 & -6 \\ 0 & -4 & 4 & 4 & -12 \\ 0 & 0 & -1 & 1 & 3 \\ 0 & 0 & -16 & -8 & 24 \end{array}\right]$$

$$\begin{array}{r} 4R_3 + R_1 \to R_1 \\ 4R_3 + R_2 \to R_2 \\ \\ 16R_3 + (-1)R_4 \to R_4 \end{array} \left[\begin{array}{rrrr|r} 2 & 0 & 0 & 6 & 6 \\ 0 & -4 & 0 & 8 & 0 \\ 0 & 0 & -1 & 1 & 3 \\ 0 & 0 & 0 & 24 & 24 \end{array}\right]$$

$$\begin{array}{r} R_4 + (-4)R_1 \to R_1 \\ R_4 + (-3)R_2 \to R_2 \\ R_4 + (-24)R_3 \to R_3 \\ \\ \end{array} \left[\begin{array}{rrrr|r} -8 & 0 & 0 & 0 & 0 \\ 0 & 12 & 0 & 0 & 24 \\ 0 & 0 & 24 & 0 & -48 \\ 0 & 0 & 0 & 24 & 24 \end{array}\right]$$

$$\begin{array}{r} -\frac{1}{8}R_1 \to R_1 \\ \frac{1}{12}R_2 \to R_2 \\ \frac{1}{24}R_3 \to R_3 \\ \frac{1}{24}R_4 \to R_4 \end{array} \left[\begin{array}{rrrr|r} 1 & 0 & 0 & 0 & 0 \\ 0 & 1 & 0 & 0 & 2 \\ 0 & 0 & 1 & 0 & -2 \\ 0 & 0 & 0 & 1 & 1 \end{array}\right]$$

The solution is $x = 0$, $y = 2$, $z = -2$, $w = 1$, or $(0, 2, -2, 1)$.

39.
$$\begin{aligned} x + y - z + 2w &= -20 \\ 2x - y + z + w &= 11 \\ 3x - 2y + z - 2w &= 27 \end{aligned}$$

$$\left[\begin{array}{rrrr|r} 1 & 1 & -1 & 2 & -20 \\ 2 & -1 & 1 & 1 & 11 \\ 3 & -2 & 1 & -2 & 27 \end{array}\right]$$

$$\begin{array}{r} \\ -2R_1 + R_2 \to R_2 \\ -3R_1 + R_3 \to R_3 \end{array} \left[\begin{array}{rrrr|r} 1 & 1 & -1 & 2 & -20 \\ 0 & -3 & 3 & -3 & 51 \\ 0 & -5 & 4 & -8 & 87 \end{array}\right]$$

$$\begin{array}{r} \\ -\frac{1}{3}R_2 \to R_2 \\ \\ \end{array} \left[\begin{array}{rrrr|r} 1 & 1 & -1 & 2 & -20 \\ 0 & 1 & -1 & 1 & -17 \\ 0 & -5 & 4 & -8 & 87 \end{array}\right]$$

$$\begin{array}{r} -1R_2 + R_1 \to R_1 \\ \\ 5R_2 + R_3 \to R_3 \end{array} \left[\begin{array}{rrrr|r} 1 & 0 & 0 & 1 & -3 \\ 0 & 1 & -1 & 1 & -17 \\ 0 & 0 & -1 & -3 & 2 \end{array}\right]$$

$$\begin{array}{r} \\ \\ -1R_3 \to R_3 \end{array} \left[\begin{array}{rrrr|r} 1 & 0 & 0 & 1 & -3 \\ 0 & 1 & -1 & 1 & -17 \\ 0 & 0 & 1 & 3 & -2 \end{array}\right]$$

$$\begin{array}{r} \\ R_3 + R_2 \to R_2 \\ \\ \end{array} \left[\begin{array}{rrrr|r} 1 & 0 & 0 & 1 & -3 \\ 0 & 1 & 0 & 4 & -19 \\ 0 & 0 & 1 & 3 & -2 \end{array}\right]$$

This is as far as we can go using row operations. To complete the solution, write the equations that correspond to the matrix.

$$\begin{aligned} x + w &= -3 \\ y + 4w &= -19 \\ z + 3w &= -2 \end{aligned}$$

Let w be the parameter and express x, y, and z in terms of w. From the equations above, $x = -w - 3$, $y = -4w - 19$, and $z = -3w - 2$.

The solution is $(-w - 3, -4w - 19, -3w - 2, w)$, where w is any real number.

41. $10.47x + 3.52y + 2.58z - 6.42w = 218.65$

$8.62x - 4.93y - 1.75z + 2.83w = 157.03$

$4.92x + 6.83y - 2.97z + 2.65w = 462.3$

$2.86x + 19.10y - 6.24z - 8.73w = 398.4$

Write the augmented matrix of the system.

$$\left[\begin{array}{cccc|c} 10.47 & 3.52 & 2.58 & -6.42 & 218.65 \\ 8.62 & -4.93 & -1.75 & 2.83 & 157.03 \\ 4.92 & 6.83 & -2.97 & 2.65 & 462.3 \\ 2.86 & 19.10 & -6.24 & -8.73 & 398.4 \end{array}\right]$$

This exercise should be solved by graphing calculator or computer methods. The solution, which may vary slightly, is $x \approx 28.9436$, $y \approx 36.6326$, $z \approx 9.6390$, and $w \approx 37.1036$, or

$(28.9436, 36.6326, 9.6390, 37.1036)$.

43. Insert the given values, introduce variables, and the table is as follows.

$\frac{3}{8}$	a	b
c	d	$\frac{1}{4}$
e	f	g

From this, we obtain the following system of equations.

$$\begin{array}{l} a + b + \frac{3}{8} = 1 \\ c + d + \frac{1}{4} = 1 \\ e + f + g = 1 \\ c + e + \frac{3}{8} = 1 \\ a + d + f = 1 \\ b + g + \frac{1}{4} = 1 \\ d + g + \frac{3}{8} = 1 \\ b + d + e = 1 \end{array}$$

The augmented matrix and the final form after row operations are as follows.

$$\left[\begin{array}{ccccccc|c} 1 & 1 & 0 & 0 & 0 & 0 & 0 & \frac{5}{8} \\ 0 & 0 & 1 & 1 & 0 & 0 & 0 & \frac{3}{4} \\ 0 & 0 & 0 & 0 & 1 & 1 & 1 & 1 \\ 0 & 0 & 1 & 0 & 1 & 0 & 0 & \frac{5}{8} \\ 1 & 0 & 0 & 1 & 0 & 1 & 0 & 1 \\ 0 & 1 & 0 & 0 & 0 & 0 & 1 & \frac{3}{4} \\ 0 & 0 & 0 & 1 & 0 & 0 & 1 & \frac{5}{8} \\ 0 & 1 & 0 & 1 & 1 & 0 & 0 & 1 \end{array}\right] \rightarrow \left[\begin{array}{ccccccc|c} 1 & 0 & 0 & 0 & 0 & 0 & 0 & \frac{1}{6} \\ 0 & 1 & 0 & 0 & 0 & 0 & 0 & \frac{11}{24} \\ 0 & 0 & 1 & 0 & 0 & 0 & 0 & \frac{5}{12} \\ 0 & 0 & 0 & 1 & 0 & 0 & 0 & \frac{1}{3} \\ 0 & 0 & 0 & 0 & 1 & 0 & 0 & \frac{5}{24} \\ 0 & 0 & 0 & 0 & 0 & 1 & 0 & \frac{1}{2} \\ 0 & 0 & 0 & 0 & 0 & 0 & 1 & \frac{7}{12} \\ 0 & 0 & 0 & 0 & 0 & 0 & 0 & 0 \end{array}\right]$$

The solution to the system is read from the last column.

$$a = \frac{1}{6}, b = \frac{11}{24}, c = \frac{5}{12}, d = \frac{1}{3},$$
$$e = \frac{5}{24}, f = \frac{1}{2}, \text{ and } g = \frac{7}{24}$$

So the magic square is:

$\frac{3}{8}$	$\frac{1}{6}$	$\frac{11}{24}$
$\frac{5}{12}$	$\frac{1}{3}$	$\frac{1}{4}$
$\frac{5}{24}$	$\frac{1}{2}$	$\frac{7}{24}$

45. Let x = amount invested in U.S. savings bonds,
y = amount invested in mutual funds, and
z = amount invested in a money market account.

Since the total amount invested was \$10,000, $x + y + z = 10{,}000$.

Katherine invested twice as much in mutual funds as in savings bonds, so $y = 2x$.

The total return on her investments was \$470, so $0.025x + 0.06y + 0.045z = 470$.

The system to be solved is

$$\begin{array}{lr} x + y + z = 10{,}000 & (1) \\ 2x - y = 0 & (2) \\ 0.025x + 0.06y + 0.045z = 470 & (3). \end{array}$$

Multiply equation (3) by 1000.

$$\begin{array}{llr} & x + y + z = 10{,}000 & (1) \\ & 2x - y = 0 & (2) \\ 1000R_3 \rightarrow R_3 & 25x + 60y + 45z = 470{,}000 & (4) \end{array}$$

Eliminate x in equations (2) and (4).

$$\begin{array}{llr} & x + y + z = 10{,}000 & (1) \\ -2R_1 + R_2 \rightarrow R_2 & -3y - 2z = -20{,}000 & (5) \\ -25R_1 + R_3 \rightarrow R_3 & 35y + 20z = 220{,}000 & (6) \end{array}$$

Eliminate y in equation (6).

$$\begin{array}{llr} & x + y + z = 10{,}000 & (1) \\ & -3y - 2z = -20{,}000 & (5) \\ 35R_2 + 3R_3 \rightarrow R_3 & -10z = -40{,}000 & (7) \end{array}$$

Make each leading coefficient equal 1.

$$\begin{array}{llr} & x + y + z = 10{,}000 & (1) \\ -\frac{1}{3}R_2 \rightarrow R_2 & y + \frac{2}{3}z = \frac{20{,}000}{3} & (8) \\ -\frac{1}{10}R_3 \rightarrow R_3 & z = 4000 & (9) \end{array}$$

Substitute 4000 for z in equation (8) to get $y = 4000$. Finally, substitute 4000 for 2 and 4000 for y in equation (1) to get $x = 2000$. Ms. Chong invested \$2000 in U.S. savings bonds, \$4000 in mutual funds, and \$4000 in a money market account.

47. Let x = the number of chairs produced each week,
y = the number of cabinets produced each week, and
z = the number of buffets produced each week.

Make a table to organize the information.

	Chair	Cabinet	Buffet	Totals
Cutting	0.2	0.5	0.3	1950
Assembly	0.3	0.4	0.1	1490
Finishing	0.1	0.6	0.4	2160

The system to be solved is

$$\begin{aligned} 0.2x + 0.5y + 0.3z &= 1950 \\ 0.3x + 0.4y + 0.1z &= 1490 \\ 0.1x + 0.6y + 0.4z &= 2160. \end{aligned}$$

Write the augmented matrix of the system

$$\left[\begin{array}{ccc|c} 0.2 & 0.5 & 0.3 & 1950 \\ 0.3 & 0.4 & 0.1 & 1490 \\ 0.1 & 0.6 & 0.4 & 2160 \end{array}\right]$$

$$\begin{array}{r} 10R_1 \to R_1 \\ 10R_2 \to R_2 \\ 10R_3 \to R_3 \end{array} \left[\begin{array}{ccc|c} 2 & 5 & 3 & 19{,}500 \\ 3 & 4 & 1 & 14{,}900 \\ 1 & 6 & 4 & 21{,}600 \end{array}\right]$$

Interchange rows 1 and 3.

$$\left[\begin{array}{ccc|c} 1 & 6 & 4 & 21{,}600 \\ 3 & 4 & 1 & 14{,}900 \\ 2 & 5 & 3 & 19{,}500 \end{array}\right]$$

$$\begin{array}{r} \\ -3R_1 + R_2 \to R_2 \\ -2R_1 + R_3 \to R_3 \end{array} \left[\begin{array}{ccc|c} 1 & 6 & 4 & 21{,}600 \\ 0 & -14 & -11 & -49{,}900 \\ 0 & -7 & -5 & -23{,}700 \end{array}\right]$$

$$\begin{array}{r} \\ -\frac{1}{14}R_2 \to R_2 \\ \\ \end{array} \left[\begin{array}{ccc|c} 1 & 6 & 4 & 21{,}600 \\ 0 & 1 & \frac{11}{14} & \frac{24{,}950}{7} \\ 0 & -7 & -5 & -23{,}700 \end{array}\right]$$

$$\begin{array}{r} -6R_2 + R_1 \to R_1 \\ \\ 7R_2 + R_3 \to R_3 \end{array} \left[\begin{array}{ccc|c} 1 & 0 & -\frac{5}{7} & \frac{1500}{7} \\ 0 & 1 & \frac{11}{14} & \frac{24{,}950}{7} \\ 0 & 0 & \frac{1}{2} & 1250 \end{array}\right]$$

$$\begin{array}{r} \\ \\ 2R_3 \to R_3 \end{array} \left[\begin{array}{ccc|c} 1 & 0 & -\frac{5}{7} & \frac{1500}{7} \\ 0 & 1 & \frac{11}{14} & \frac{24{,}950}{7} \\ 0 & 0 & 1 & 2500 \end{array}\right]$$

$$\begin{array}{r} \frac{5}{7}R_3 + R_1 \to R_1 \\ -\frac{11}{14}R_3 + R_2 \to R_2 \\ \\ \end{array} \left[\begin{array}{ccc|c} 1 & 0 & 0 & 2000 \\ 0 & 1 & 0 & 1600 \\ 0 & 0 & 1 & 2500 \end{array}\right]$$

The solution is (2000, 1600, 2500). Therefore, 2000 chairs, 1600 cabinets, and 2500 buffets should be produced.

49. **(a)** Let x be the number of trucks used, y be the number of vans, and z be the number of SUVs. We first obtain the equations given here.

$$\begin{aligned} 2x + 3y + 3z &= 25 \\ 2x + 4y + 5z &= 33 \\ 3x + 2y + z &= 22 \end{aligned}$$

Write the augmented matrix and use row operations.

$$\left[\begin{array}{ccc|c} 2 & 3 & 3 & 25 \\ 2 & 4 & 5 & 33 \\ 3 & 2 & 1 & 22 \end{array}\right]$$

$$\begin{array}{r} \\ -1R_1 + R_2 \to R_2 \\ -3R_1 + 2R_3 \to R_3 \end{array} \left[\begin{array}{ccc|c} 2 & 3 & 3 & 25 \\ 0 & 1 & 2 & 8 \\ 0 & -5 & -7 & -31 \end{array}\right]$$

$$\begin{array}{r} -3R_2 + R_1 \to R_1 \\ \\ 5R_2 + R_3 \to R_3 \end{array} \left[\begin{array}{ccc|c} 2 & 0 & -3 & 1 \\ 0 & 1 & 2 & 8 \\ 0 & 0 & 3 & 9 \end{array}\right]$$

$$\begin{array}{r} R_3 + R_1 \to R_1 \\ -2R_3 + 3R_2 \to R_2 \\ \\ \end{array} \left[\begin{array}{ccc|c} 2 & 0 & 0 & 10 \\ 0 & 3 & 0 & 6 \\ 0 & 0 & 3 & 9 \end{array}\right]$$

$$\begin{array}{r} \frac{1}{2}R_1 \to R_1 \\ \frac{1}{3}R_2 \to R_2 \\ \frac{1}{3}R_3 \to R_3 \end{array} \left[\begin{array}{ccc|c} 1 & 0 & 0 & 5 \\ 0 & 1 & 0 & 2 \\ 0 & 0 & 1 & 3 \end{array}\right]$$

Read the solution from the last column of the matrix. The solution is 5 trucks, 2 vans, and 3 SUVs.

(b) The system of equations is now

$$\begin{aligned} 2x + 3y + 3z &= 25 \\ 2x + 4y + 5z &= 33. \end{aligned}$$

Write the augmented matrix and use row operations.

$$\left[\begin{array}{ccc|c} 2 & 3 & 3 & 25 \\ 2 & 4 & 5 & 33 \end{array}\right]$$

$$\begin{array}{r} \\ -R_1 + R_2 \to R_2 \end{array} \left[\begin{array}{ccc|c} 2 & 3 & 3 & 25 \\ 0 & 1 & 2 & 8 \end{array}\right]$$

$$-3R_2 + R_1 \to R_1 \left[\begin{array}{ccc|c} 2 & 0 & -3 & 1 \\ 0 & 1 & 2 & 8 \end{array}\right]$$

Obtain a one in row 1, column 1.

$$\tfrac{1}{2}R_1 \to R_1 \left[\begin{array}{ccc|c} 1 & 0 & -\frac{3}{2} & \frac{1}{2} \\ 0 & 1 & 2 & 8 \end{array}\right]$$

The last row indicates multiple solutions are possible. The remaining equations are

$$x - \frac{3}{2}z = \frac{1}{2} \text{ and } y + 2z = 8.$$

Solving these for x and y, we have

$$x = \frac{3}{2}z + \frac{1}{2} \text{ and } y = -2z + 8.$$

The form of the solution is

$$\left(\tfrac{3}{2}z + \tfrac{1}{2}, -2z + 8, z\right).$$

Since the solutions must be whole numbers,

$$\frac{3}{2}z + \frac{1}{2} \geq 0 \quad \text{and} \quad -2z + 8 \geq 0$$
$$\frac{3}{2}z \geq -\frac{1}{2} \qquad -2z \geq -8$$
$$z \geq -\frac{1}{3} \qquad z \leq 4$$

Thus, there are 4 possible solutions but each must be checked to determine if they produce whole numbers for x and y.

When $z = 0, \left(\frac{1}{2}, 8, 0\right)$ which is not realistic.

When $z = 1, (2, 6, 1)$.

When $z = 2, \left(\frac{7}{2}, 4, 2\right)$ which is not realistic.

When $z = 3, (5, 2, 3)$.

When $z = 4, \left(\frac{13}{2}, 0, 4\right)$ which is not realistic.

The company has 2 options. Either use 2 trucks, 6 vans, and 1 SUV or use 5 trucks, 2 vans, and 3 SUVs.

51. Let x = the amount borrowed at 8%,
y = the amount borrowed at 9%, and
z = the amount borrowed at 10%.

(a) The system to be solved is

$$x + y + z = 25{,}000$$
$$0.08x + 0.09y + 0.10z = 2190$$
$$y = z + 1000$$

Multiply the second equation by 100 and rewrite the equations in standard form.

$$x + y + z = 25{,}000$$
$$8x + 9y + 10z = 219{,}000$$
$$y - z = 1000.$$

Write the augmented matrix and use row operations to solve

$$\left[\begin{array}{ccc|c} 1 & 1 & 1 & 25{,}000 \\ 8 & 9 & 10 & 219{,}000 \\ 0 & 1 & -1 & 1000 \end{array}\right]$$

$$-8R_1 + R_2 \to R_2 \left[\begin{array}{ccc|c} 1 & 1 & 1 & 25{,}000 \\ 0 & 1 & 2 & 19{,}000 \\ 0 & 1 & -1 & 1000 \end{array}\right]$$

$$\begin{array}{r} -1R_2 + R_1 \to R_1 \\ \\ -1R_2 + R_3 \to R_3 \end{array} \left[\begin{array}{ccc|c} 1 & 0 & -1 & 6000 \\ 0 & 1 & 2 & 19{,}000 \\ 0 & 0 & -3 & -18{,}000 \end{array}\right]$$

$$\begin{array}{r} -1R_3 + 3R_1 \to R_1 \\ 2R_3 + 3R_2 \to R_2 \\ \\ \end{array} \left[\begin{array}{ccc|c} 3 & 0 & 0 & 36{,}000 \\ 0 & 3 & 0 & 21{,}000 \\ 0 & 0 & -3 & -18{,}000 \end{array}\right]$$

$$\begin{array}{r} \frac{1}{3}R_1 \to R_1 \\ \frac{1}{3}R_2 \to R_2 \\ \frac{1}{3}R_3 \to R_3 \end{array} \left[\begin{array}{ccc|c} 1 & 0 & 0 & 12{,}000 \\ 0 & 1 & 0 & 7000 \\ 0 & 0 & 1 & 6000 \end{array}\right]$$

The solution is (12,000, 7000, 6000). The company borrowed \$12,000 at 8%, \$7000 at 9%, and \$6000 at 10%.

(b) If the condition is dropped, the initial augmented matrix and solution is found as before.

$$\left[\begin{array}{ccc|c} 1 & 1 & 1 & 25{,}000 \\ 8 & 9 & 10 & 219{,}000 \end{array}\right]$$

$$-8R_1 + R_2 \to R_2 \left[\begin{array}{ccc|c} 1 & 1 & 1 & 25{,}000 \\ 0 & 1 & 2 & 19{,}000 \end{array}\right]$$

$$-1R_2 + R_1 \to R_1 \left[\begin{array}{ccc|c} 1 & 0 & -1 & 6000 \\ 0 & 1 & 2 & 19{,}000 \end{array}\right]$$

This gives the system of equations

$$x = z + 6000$$
$$y = -2x + 19{,}000$$

Since all values must be nonnegative,

$$z + 6000 \geq 0 \quad \text{and} \quad -2z + 19{,}000 \geq 0$$
$$z \geq -6000 \qquad z \leq 9500.$$

The second inequality produces the condition that the amount borrowed at 10% must be less than or equal to \$9500. If \$5000 is borrowed at 10%, $z = 5000$, and

$$x = 500 + 6000 = 11{,}000$$
$$y = -2(5000) + 19{,}000 = 9000.$$

This means \$11,000 is borrowed at 8% and \$9000 is borrowed at 9%.

(c) The original conditions resulted in \$12,000 borrowed at 8%. So, if the bank sets a maximum of \$10,000 at the 8% rate, no solution is possible.

(d) The total interest would be

$$\begin{aligned} &0.08(10{,}000) + 0.09(8000) + 0.10(7000) \\ &= 800 + 720 + 700 \\ &= 2220 \end{aligned}$$

or \$2220, which is not the \$2190 interest as specified as one of the conditions of the problem.

53. Let x_1 = the number of units from first supplier for Roseville,

x_2 = the number of units from first supplier for Akron,

x_3 = the number of units from second supplier for Roseville, and

x_4 = the number of units from second supplier for Akron.

Roseville needs 40 units so

$$x_1 + x_3 = 40.$$

Akron needs 75 units so

$$x_2 + x_4 = 75.$$

The manufacturer orders 75 units from the first supplier so

$$x_1 + x_2 = 75.$$

The total cost is \$10,750 so

$$70x_1 + 90x_2 + 80x_3 + 120x_4 = 10{,}750.$$

The system to be solved is

$$\begin{aligned} x_1 \qquad + x_3 \qquad &= 40 \\ x_2 \qquad + x_4 &= 75 \\ x_1 + x_2 \qquad\qquad &= 75 \\ 70x_1 + 90x_2 + 80x_3 + 120x_4 &= 10{,}750. \end{aligned}$$

Write augmented matrix and use row operations.

$$\left[\begin{array}{cccc|c} 1 & 0 & 1 & 0 & 40 \\ 0 & 1 & 0 & 1 & 75 \\ 1 & 1 & 0 & 0 & 75 \\ 70 & 90 & 80 & 120 & 10{,}750 \end{array}\right]$$

$$\begin{array}{r} \\ \\ -1R_1 + R_3 \rightarrow R_3 \\ -70R_1 + R_4 \rightarrow R_4 \end{array} \left[\begin{array}{cccc|c} 1 & 0 & 1 & 0 & 40 \\ 0 & 1 & 0 & 1 & 75 \\ 0 & 1 & -1 & 0 & 35 \\ 0 & 90 & 10 & 120 & 7950 \end{array}\right]$$

$$\begin{array}{r} \\ \\ -1R_2 + R_3 \rightarrow R_3 \\ -90R_2 + R_4 \rightarrow R_4 \end{array} \left[\begin{array}{cccc|c} 1 & 0 & 1 & 0 & 40 \\ 0 & 1 & 0 & 1 & 75 \\ 0 & 0 & -1 & -1 & -40 \\ 0 & 0 & 10 & 30 & 1200 \end{array}\right]$$

$$\begin{array}{r} \\ \\ -1R_3 \rightarrow R_3 \\ 10R_3 + R_4 \rightarrow R_4 \end{array} \left[\begin{array}{cccc|c} 1 & 0 & 1 & 0 & 40 \\ 0 & 1 & 0 & 1 & 75 \\ 0 & 0 & 1 & 1 & 40 \\ 0 & 0 & 0 & 20 & 800 \end{array}\right]$$

$$\begin{array}{r} \\ \\ \\ \frac{1}{20}R_4 \rightarrow R_4 \end{array} \left[\begin{array}{cccc|c} 1 & 0 & 1 & 0 & 40 \\ 0 & 1 & 0 & 1 & 75 \\ 0 & 0 & 1 & 1 & 40 \\ 0 & 0 & 0 & 1 & 40 \end{array}\right]$$

$$\begin{array}{r} \\ -1R_4 + R_2 \rightarrow R_2 \\ -1R_4 + R_3 \rightarrow R_3 \\ \\ \end{array} \left[\begin{array}{cccc|c} 1 & 0 & 1 & 0 & 40 \\ 0 & 1 & 0 & 0 & 35 \\ 0 & 0 & 1 & 0 & 0 \\ 0 & 0 & 0 & 1 & 40 \end{array}\right]$$

$$\begin{array}{r} -1R_3 + R_1 \rightarrow R_1 \\ \\ \\ \\ \end{array} \left[\begin{array}{cccc|c} 1 & 0 & 0 & 0 & 40 \\ 0 & 1 & 0 & 0 & 35 \\ 0 & 0 & 1 & 0 & 0 \\ 0 & 0 & 0 & 1 & 40 \end{array}\right]$$

The solution of the system is $x_1 = 40$, $x_2 = 35$, $x_3 = 0$, $x_4 = 40$, or (40, 35, 0, 40). The first supplier should send 40 units to Roseville and 35 units to Akron. The second supplier should send 0 units to Roseville and 40 units to Akron.

55. Let x = the number of two-person tents,

y = the number of four-person tents, and

z = the number of six-person tents that were ordered.

(a) The problem is to solve the following system of equations.

$$\begin{aligned} 2x + 4y + 6z &= 200 \\ 40x + 64y + 88z &= 3200 \\ 129x + 179y + 229z &= 8950 \end{aligned}$$

Write the augmented matrix and use row operations to solve.

$$\left[\begin{array}{ccc|c} 2 & 4 & 6 & 200 \\ 40 & 64 & 88 & 3200 \\ 129 & 179 & 229 & 8950 \end{array}\right]$$

$$\begin{array}{r} \\ 20R_1 + (-1)R_2 \rightarrow R_2 \\ 129R_1 + (-2)R_3 \rightarrow R_3 \end{array} \left[\begin{array}{ccc|c} 2 & 4 & 6 & 200 \\ 0 & 16 & 32 & 800 \\ 0 & 158 & 316 & 7900 \end{array}\right]$$

$$\begin{array}{l} R_2 + (-4)R_1 \to R_1 \\ \\ 79R_2 + (-8)R_3 \to R_3 \end{array} \left[\begin{array}{ccc|c} -8 & 0 & 8 & 0 \\ 0 & 16 & 32 & 800 \\ 0 & 0 & 0 & 0 \end{array}\right]$$

$$\begin{array}{l} -\frac{1}{8}R_1 \to R_1 \\ \\ \frac{1}{16}R_2 \to R_2 \end{array} \left[\begin{array}{ccc|c} 1 & 0 & -1 & 0 \\ 0 & 1 & 2 & 50 \\ 0 & 0 & 0 & 0 \end{array}\right]$$

Since the last row is all zeros, there is more than one solution. Let z be the parameter. The matrix gives

$$x - z = 0$$
$$y + 2z = 50.$$

Solving these equations for x and y, the solution is $(z, -2z + 50, z)$. The numbers in the solution must be nonnegative integers. Therefore,

$$y \geq 0$$
$$-2z + 50 \geq 0$$
$$z \leq 25.$$

Thus, $z \in \{0, 1, 2, 3, \ldots, 25\}$ In other words, depending on the number of six-person tents, there are 26 solutions to this problem.

(b) The number of four-person tents is given by the value of the variable y. Since $y = -2z + 50$, the most four-person tents will result when z is as small as possible, or 0. When this occurs, $y = -2(0) + 50 = 50$. And since $x = z$, the solution with the most four-person tents is 0 two-person tents, 50 four-person tents, and 0 six-person tents.

(c) The number of two-person tents is given by the value of the variable x. Since $x = z$, the most two-person tents will result when y is as small as possible, or 0. When this occurs,

$$2x + 4y + 6z = 200$$
$$2(z) + 4(0) + 6(z) = 200$$
$$8z = 200$$
$$z = 25.$$

The solution with the most two-person tents is 25 two-person tents, 0 four-person tents, and 25 six-person tents.

57. Let $x =$ the number of grams of group A, $y =$ the number of grams of group B, and $z =$ the number of grams of group C.

(a) The system to be solved is

$$x + y + z = 400 \quad (1)$$
$$x = \frac{1}{3}y \quad (2)$$
$$x + z = 2y. \quad (3)$$

Rewrite equations (2) and (3) in proper form and multiply both sides of equation (2) by 3.

$$x + y + z = 400$$
$$3x - y = 0$$
$$x - 2y + z = 0$$

Write the augmented matrix.

$$\left[\begin{array}{ccc|c} 1 & 1 & 1 & 400 \\ 3 & -1 & 0 & 0 \\ 1 & -2 & 1 & 0 \end{array}\right]$$

$$\begin{array}{l} \\ -3R_1 + R_2 \to R_2 \\ -1R_1 + R_3 \to R_3 \end{array} \left[\begin{array}{ccc|c} 1 & 1 & 1 & 400 \\ 0 & -4 & -3 & -1200 \\ 0 & -3 & 0 & -400 \end{array}\right]$$

$$\begin{array}{l} \\ \\ -\frac{1}{3}R_3 \to R_3 \end{array} \left[\begin{array}{ccc|c} 1 & 1 & 1 & 400 \\ 0 & -4 & -3 & -1200 \\ 0 & 1 & 0 & \frac{400}{3} \end{array}\right]$$

Interchange rows 2 and 3.

$$\left[\begin{array}{ccc|c} 1 & 1 & 1 & 400 \\ 0 & 1 & 0 & \frac{400}{3} \\ 0 & -4 & -3 & -1200 \end{array}\right]$$

$$\begin{array}{l} -1R_2 + R_1 \to R_1 \\ \\ 4R_2 + R_3 \to R_3 \end{array} \left[\begin{array}{ccc|c} 1 & 0 & 1 & \frac{800}{3} \\ 0 & 1 & 0 & \frac{400}{3} \\ 0 & 0 & -3 & -\frac{2000}{3} \end{array}\right]$$

$$\begin{array}{l} \\ \\ -\frac{1}{3}R_3 \to R_3 \end{array} \left[\begin{array}{ccc|c} 1 & 0 & 1 & \frac{800}{3} \\ 0 & 1 & 0 & \frac{400}{3} \\ 0 & 0 & 1 & \frac{2000}{9} \end{array}\right]$$

$$-1R_3 + R_1 \to R_1 \left[\begin{array}{ccc|c} 1 & 0 & 0 & \frac{400}{9} \\ 0 & 1 & 0 & \frac{400}{3} \\ 0 & 0 & 1 & \frac{2000}{9} \end{array}\right]$$

The solution is $\left(\frac{400}{9}, \frac{400}{3}, \frac{2000}{9}\right)$. Include $\frac{400}{9}$ g of group A, $\frac{400}{3}$ g of group B, and $\frac{2000}{9}$ g of group C.

(b) If the requirement that the diet include one-third as much of A as of B is dropped, refer to the first two rows of the fifth augmented matrix in part (a).

$$\left[\begin{array}{ccc|c} 1 & 0 & 1 & \frac{800}{3} \\ 0 & 1 & 0 & \frac{400}{3} \end{array}\right]$$

This gives

$$x = \frac{800}{3} - z$$

$$y = \frac{400}{3}.$$

Therefore, for any positive number z of grams of group C, there should be z grams less than $\frac{800}{3}$ g of group A and $\frac{400}{3}$ g of group B.

(c) Since there was a unique solution for the original problem, by adding an additional condition. the only possible solution would be the one from part (a). However, by substituting those values of A, B, and C for x, y, and z in the equation for the additional condition, $0.02x + 0.02y + 0.03z = 8.00$, the values do not work. Thus, no solution is possible.

59. Let $x =$ the number of species A,
$y =$ the number of species B, and
$z =$ the number of species C.

Use a chart to organize the information.

		Species			
		A	B	C	Totals
Food	I	1.32	2.1	0.86	490
	II	2.9	0.95	1.52	897
	III	1.75	0.6	2.01	653

The system to be solved is

$$1.32x + 2.1y + 0.86z = 490$$
$$2.9x + 0.95y + 1.52z = 897$$
$$1.75x + 0.6y + 2.01z = 653.$$

Use graphing calculator or computer methods to solve this system. The solution, which may vary slightly, is to stock about 244 fish of species A, 39 fish of species B, and 101 fish of species C.

61. (a) Bulls:

The number of white ones was one half plus one third the number of black greater than the brown.

$$X = \left(\frac{1}{2} + \frac{1}{3}\right)Y + T$$

$$X = \frac{5}{6}Y + T$$

$$6X - 5Y = 6T$$

The number of the black, one quarter plus one fifth the number of the spotted greater than the brown.

$$Y = \left(\frac{1}{4} + \frac{1}{5}\right)Z + T$$

$$Y = \frac{9}{20}Z + T$$

$$20Y = 9Z + 20T$$

$$20Y - 9Z = 20T$$

The number of the spotted, one sixth and one seventh the number of the white greater than the brown.

$$Z = \left(\frac{1}{6} + \frac{1}{7}\right)X + T$$

$$Z = \frac{13}{42}X + T$$

$$42Z = 13X + 42T$$

$$42Z - 13X = 42T$$

So the system of equations for the bulls is

$$6X - 5Y = 6T$$
$$20Y - 9Z = 20T$$
$$42Z - 13X = 42T.$$

Cows:

The number of white ones was one third plus one quarter of the total black cattle.

$$x = \left(\frac{1}{3} + \frac{1}{4}\right)(Y + y)$$

$$x = \frac{7}{12}(Y + y)$$

$$12x = 7Y + 7y$$

$$12x - 7y = 7Y$$

The number of the black, one quarter plus one fifth the total of the spotted cattle.

$$y = \left(\frac{1}{4} + \frac{1}{5}\right)(Z + z)$$

$$y = \frac{9}{20}(Z + z)$$

$$20y = 9Z + 9z$$

$$20y - 9z = 9Z$$

The number of the spotted, one fifth plus one sixth the total of the brown cattle.

$$z = \left(\frac{1}{5} + \frac{1}{6}\right)(T + t)$$

$$z = \frac{11}{30}(T + t)$$

$$30z = 11T + 11t$$

$$30z - 11t = 11T$$

The number of the brown, one sixth plus one seventh the total of the white cattle.

$$t = \left(\frac{1}{6} + \frac{1}{7}\right)(X + x)$$

$$t = \frac{13}{42}(X + x)$$

$$42t = 13X + 13x$$

$$42t - 13x = 13X$$

So the system of equations for the cows is

$$12x - 7y = 7Y$$
$$20y - 9z = 9Z$$
$$30z - 11t = 11T$$
$$-13x + 42t = 13X$$

(b) For $T = 4{,}149{,}387$, the 3×3 system to be solved is

$$\begin{aligned} 6X - 5Y \quad &= 24{,}896{,}322 \\ 20Y - 9Z &= 82{,}987{,}740 \\ -13X \quad + 42Z &= 174{,}274{,}254 \end{aligned}$$

Write the augmented matrix of the system.

$$\left[\begin{array}{ccc|c} 6 & -5 & 0 & 24{,}896{,}322 \\ 0 & 20 & -9 & 82{,}987{,}740 \\ -13 & 0 & 42 & 174{,}274{,}254 \end{array}\right]$$

This exercise should be solved by graphing calculator or computer methods. The solution is $X = 10{,}366{,}482$ white bulls, $Y = 7{,}460{,}514$ black bulls, and $Z = 7{,}358{,}060$ spotted bulls.

For $X = 10{,}366{,}482$, $Y = 7{,}460{,}514$, and $Z = 7{,}358{,}060$, the 4×4 system to be solved is

$$12x - 7y = 52{,}223{,}598$$
$$20y - 9z = 66{,}222{,}540$$
$$30z - 11t = 45{,}643{,}257$$
$$-13x + 42t = 134{,}764{,}266$$

Write the augmented matrix of the system.

$$\left[\begin{array}{cccc|c} 12 & -7 & 0 & 0 & 52{,}223{,}598 \\ 0 & 20 & -9 & 0 & 66{,}222{,}540 \\ 0 & 0 & 30 & -11 & 45{,}643{,}257 \\ -13 & 0 & 0 & 42 & 134{,}764{,}266 \end{array}\right]$$

This exercise should be solved by graphing calculator or computer methods. The solution is $x = 7{,}206{,}360$ white cows, $y = 4{,}893{,}246$ black cows, $z = 3{,}515{,}820$ spotted cows, and $t = 5{,}439{,}213$ brown cows.

63. (a) The system to be solved is

$$0 = 200{,}000 - 0.5r - 0.3b$$
$$0 = 350{,}000 - 0.5r - 0.7b.$$

First, write the system in proper form.

$$0.5r + 0.3b = 200{,}000$$
$$0.5r + 0.7b = 350{,}000$$

Write the augmented matrix and use row operations.

$$\left[\begin{array}{cc|c} 0.5 & 0.3 & 200{,}000 \\ 0.5 & 0.7 & 350{,}000 \end{array}\right]$$

$$\begin{array}{r} 10R_1 \rightarrow R_1 \\ 10R_2 \rightarrow R_2 \end{array} \left[\begin{array}{cc|c} 5 & 3 & 2{,}000{,}000 \\ 5 & 7 & 3{,}500{,}000 \end{array}\right]$$

$$-1R_1 + R_2 \rightarrow R_2 \left[\begin{array}{cc|c} 5 & 3 & 2{,}000{,}000 \\ 0 & 4 & 1{,}500{,}000 \end{array}\right]$$

$$-\tfrac{3}{4}R_2 + R_1 \rightarrow R_1 \left[\begin{array}{cc|c} 5 & 0 & 875{,}000 \\ 0 & 4 & 1{,}500{,}000 \end{array}\right]$$

$$\begin{array}{r} \tfrac{1}{5}R_1 \rightarrow R_1 \\ \tfrac{1}{4}R_2 \rightarrow R_2 \end{array} \left[\begin{array}{cc|c} 1 & 0 & 175{,}000 \\ 0 & 1 & 375{,}000 \end{array}\right]$$

The solution is (175,000, 375,000). When the rate of increase for each is zero, there are 175,000 soldiers in the Red Army and 375,000 soldiers in the Blue Army.

65. Let x = the number of calories in each gram of fat
y = the number of calories in each gram of carbohydrates
z = the number of calories in each gram of protein

We want to solve the following system.

$$10x + 36y + 2z = 240$$
$$14x + 37y + 3z = 280$$
$$20x + 23y + 11z = 295$$

Write the augmented matrix and transform the matrix.

$$\left[\begin{array}{ccc|c} 10 & 36 & 2 & 240 \\ 14 & 37 & 3 & 280 \\ 20 & 23 & 11 & 295 \end{array}\right]$$

$$\tfrac{1}{10}R_1 \rightarrow R_1 \left[\begin{array}{ccc|c} 1 & 3.6 & 0.2 & 24 \\ 14 & 37 & 3 & 280 \\ 20 & 23 & 11 & 295 \end{array}\right]$$

$$\begin{array}{r} \\ -14R_1 + R_2 \rightarrow R_2 \\ -20R_1 + R_3 \rightarrow R_3 \end{array} \left[\begin{array}{ccc|c} 1 & 3.6 & 0.2 & 24 \\ 0 & -13.4 & 0.2 & -56 \\ 0 & -49 & 7 & -185 \end{array}\right]$$

$$-\tfrac{1}{13.4}R_2 \rightarrow R_2 \left[\begin{array}{ccc|c} 1 & 3.6 & 0.2 & 24 \\ 0 & 1 & -0.014925 & 4.179104 \\ 0 & -49 & 7 & -185 \end{array}\right]$$

$$\begin{array}{r} -3.6R_2 + R_1 \rightarrow R_1 \\ \\ 49R_2 + R_3 \rightarrow R_3 \end{array} \left[\begin{array}{ccc|c} 1 & 0 & 0.25373 & 8.95522 \\ 0 & 1 & -0.014925 & 4.179104 \\ 0 & 0 & 6.268675 & 19.7761 \end{array}\right]$$

$$\frac{1}{6.268675}R_3 \to R_3 \left[\begin{array}{ccc|c} 1 & 0 & 0.25373 & 8.95522 \\ 0 & 1 & -0.014925 & 4.179104 \\ 0 & 0 & 1 & 3.15475 \end{array}\right]$$

$$\begin{array}{r} -0.25373R_3 + R_1 \to R_1 \\ 0.014925R_3 + R_2 \to R_2 \\ \end{array} \left[\begin{array}{ccc|c} 1 & 0 & 0 & 8.15476 \\ 0 & 1 & 0 & 4.22619 \\ 0 & 0 & 1 & 3.15475 \end{array}\right]$$

The solution is $(8.15, 4.23, 3.15)$. There are 8.15 calories in a gram of fat, 4.23 calories in a gram of carbohydrates, and 3.15 calories in a gram of protein.

67. Let x = the number of balls,
y = the number of dolls, and
z = the number of cars.

(a) The system to be solved is

$$\begin{aligned} x + y \quad z &= 100 \\ 2x + 3y + 4z &= 295 \\ 12x + 16y + 18z &= 1542. \end{aligned}$$

Write the augmented matrix of the system.

$$\left[\begin{array}{ccc|c} 1 & 1 & 1 & 100 \\ 2 & 3 & 4 & 295 \\ 12 & 16 & 18 & 1542 \end{array}\right]$$

$$\begin{array}{r} \\ -2R_1 + R_2 \to R_2 \\ -12R_1 + R_3 \to R_3 \end{array} \left[\begin{array}{ccc|c} 1 & 1 & 1 & 100 \\ 0 & 1 & 2 & 95 \\ 0 & 4 & 6 & 342 \end{array}\right]$$

$$\begin{array}{r} -1R_2 + R_1 \to R_1 \\ \\ -4R_2 + R_3 \to R_3 \end{array} \left[\begin{array}{ccc|c} 1 & 0 & -1 & 5 \\ 0 & 1 & 2 & 95 \\ 0 & 0 & -2 & -38 \end{array}\right]$$

$$\begin{array}{r} \\ \\ -\frac{1}{2}R_3 \to R_3 \end{array} \left[\begin{array}{ccc|c} 1 & 0 & -1 & 5 \\ 0 & 1 & 2 & 95 \\ 0 & 0 & 1 & 19 \end{array}\right]$$

$$\begin{array}{r} R_3 + R_1 \to R_1 \\ -2R_3 + R_2 \to R_2 \\ \\ \end{array} \left[\begin{array}{ccc|c} 1 & 0 & 0 & 24 \\ 0 & 1 & 0 & 57 \\ 0 & 0 & 1 & 19 \end{array}\right]$$

The solution is (24, 57, 19). There were 24 balls, 57 dolls, and 19 cars.

(b) The augmented matrix becomes

$$\left[\begin{array}{ccc|c} 1 & 1 & 1 & 100 \\ 2 & 3 & 4 & 295 \\ 11 & 15 & 19 & 1542 \end{array}\right].$$

$$\begin{array}{r} \\ -2R_1 + R_2 \to R_2 \\ -11R_1 + R_3 \to R_3 \end{array} \left[\begin{array}{ccc|c} 1 & 1 & 1 & 100 \\ 0 & 1 & 2 & 95 \\ 0 & 4 & 8 & 442 \end{array}\right]$$

$$\begin{array}{r} -1R_2 + R_1 \to R_1 \\ \\ -4R_2 + R_3 \to R_3 \end{array} \left[\begin{array}{ccc|c} 1 & 0 & -1 & 5 \\ 0 & 1 & 2 & 95 \\ 0 & 0 & 0 & 62 \end{array}\right]$$

Since row 3 yields a false statement, $0 = 62$, there is no solution.

(c) The augmented matrix becomes

$$\left[\begin{array}{ccc|c} 1 & 1 & 1 & 100 \\ 2 & 3 & 4 & 295 \\ 11 & 15 & 19 & 1480 \end{array}\right].$$

$$\begin{array}{r} \\ -2R_1 + R_2 \to R_2 \\ -11R_1 + R_3 \to R_3 \end{array} \left[\begin{array}{ccc|c} 1 & 1 & 1 & 100 \\ 0 & 1 & 2 & 95 \\ 0 & 4 & 8 & 380 \end{array}\right]$$

$$\begin{array}{r} -1R_2 + R_1 \to R_1 \\ \\ -4R_2 + R_3 \to R_3 \end{array} \left[\begin{array}{ccc|c} 1 & 0 & -1 & 5 \\ 0 & 1 & 2 & 95 \\ 0 & 0 & 0 & 0 \end{array}\right]$$

Since the last row is all zeros, there are infinitely many solutions Let z be the parameter. The matrix gives

$$\begin{aligned} x - z &= 5 \\ y + 2z &= 95. \end{aligned}$$

Solving these equations for x and y, the solution is $(5 + z, 95 - 2z, z)$. The numbers in the solution must be nonnegative integers. Therefore,

$$\begin{aligned} 95 - 2z &\geq 0 \\ -2z &\geq -95 \\ z &\leq 47.5. \end{aligned}$$

Thus, $z \in \{0, 1, 2, 3, ..., 47\}$. There are 48 possible solutions.

(d) For the smallest number of cars, $z = 0$, the solution is (5, 95, 0). This means 5 balls, 95 dolls, and no cars.

(e) For the largest number of cars, $z = 47$, the solution is (52, 1, 47). This means 52 balls, 1 doll, and 47 cars.

69. (a)
$$\begin{aligned} x_{11} + x_{12} + x_{21} &= 1 \\ x_{11} + x_{12} + x_{22} &= 1 \\ x_{11} + x_{21} + x_{22} &= 1 \\ x_{12} + x_{21} + x_{22} &= 1 \end{aligned}$$

Write the augmented matrix of the system.

$$\left[\begin{array}{cccc|c} 1 & 1 & 1 & 0 & 1 \\ 1 & 1 & 0 & 1 & 1 \\ 1 & 0 & 1 & 1 & 1 \\ 0 & 1 & 1 & 1 & 1 \end{array}\right]$$

$$\begin{array}{l} \\ \\ -1R_1 + R_2 \rightarrow R_2 \\ -1R_1 + R_3 \rightarrow R_3 \end{array} \left[\begin{array}{cccc|c} 1 & 1 & 1 & 0 & 1 \\ 0 & 0 & -1 & 1 & 0 \\ 0 & -1 & 0 & 1 & 0 \\ 0 & 1 & 1 & 1 & 1 \end{array}\right]$$

Since $-1 = 1$ modulo 2, replace -1 with 1.

$$\left[\begin{array}{cccc|c} 1 & 1 & 1 & 0 & 1 \\ 0 & 0 & 1 & 1 & 0 \\ 0 & 1 & 0 & 1 & 0 \\ 0 & 1 & 1 & 1 & 1 \end{array}\right]$$

Interchange rows 2 and 3.

$$\left[\begin{array}{cccc|c} 1 & 1 & 1 & 0 & 1 \\ 0 & 1 & 0 & 1 & 0 \\ 0 & 0 & 1 & 1 & 0 \\ 0 & 1 & 1 & 1 & 1 \end{array}\right]$$

$$\begin{array}{l} -1R_2 + R_1 \rightarrow R_1 \\ \\ \\ -R_2 + R_4 \rightarrow R_4 \end{array} \left[\begin{array}{cccc|c} 1 & 0 & 1 & -1 & 1 \\ 0 & 1 & 0 & 1 & 0 \\ 0 & 0 & 1 & 1 & 0 \\ 0 & 0 & 1 & 0 & 1 \end{array}\right]$$

Again, replace -1 with 1.

$$\left[\begin{array}{cccc|c} 1 & 0 & 1 & 1 & 1 \\ 0 & 1 & 0 & 1 & 0 \\ 0 & 0 & 1 & 1 & 0 \\ 0 & 0 & 1 & 0 & 1 \end{array}\right]$$

$$\begin{array}{l} -1R_3 + R_1 \rightarrow R_1 \\ \\ \\ -1R_3 + R_4 \rightarrow R_4 \end{array} \left[\begin{array}{cccc|c} 1 & 0 & 0 & 0 & 1 \\ 0 & 1 & 0 & 1 & 0 \\ 0 & 0 & 1 & 1 & 0 \\ 0 & 0 & 0 & -1 & 1 \end{array}\right]$$

Replace -1 with 1.

$$\left[\begin{array}{cccc|c} 1 & 0 & 0 & 0 & 1 \\ 0 & 1 & 0 & 1 & 0 \\ 0 & 0 & 1 & 1 & 0 \\ 0 & 0 & 0 & 1 & 1 \end{array}\right]$$

$$\begin{array}{l} \\ -1R_4 + R_2 \rightarrow R_2 \\ -1R_4 + R_3 \rightarrow R_3 \\ \\ \end{array} \left[\begin{array}{cccc|c} 1 & 0 & 0 & 0 & 1 \\ 0 & 1 & 0 & 0 & -1 \\ 0 & 0 & 1 & 0 & -1 \\ 0 & 0 & 0 & 1 & 1 \end{array}\right]$$

Finally, replace -1 with 1.

$$\left[\begin{array}{cccc|c} 1 & 0 & 0 & 0 & 1 \\ 0 & 1 & 0 & 0 & 1 \\ 0 & 0 & 1 & 0 & 1 \\ 0 & 0 & 0 & 1 & 1 \end{array}\right]$$

The solution (1, 1, 1, 1) corresponds to $x_{11} = 1,\ x_{12} = 1,\ x_{21} = 1$, and $x_{22} = 1$. Since 1 indicates that a button is pushed, the strategy required to turn all the lights out is to push every button one time.

(b)
$$\begin{aligned} x_{11} + x_{12} + x_{21} &= 0 \\ x_{11} + x_{12} + x_{22} &= 1 \\ x_{11} + x_{21} + x_{22} &= 1 \\ x_{12} + x_{21} + x_{22} &= 0 \end{aligned}$$

Write the augmented matrix of the system.

$$\left[\begin{array}{cccc|c} 1 & 1 & 1 & 0 & 0 \\ 1 & 1 & 0 & 1 & 1 \\ 1 & 0 & 1 & 1 & 1 \\ 0 & 1 & 1 & 1 & 0 \end{array}\right]$$

$$\begin{array}{l} \\ \\ -1R_1 + R_2 \rightarrow R_2 \\ -1R_1 + R_3 \rightarrow R_3 \end{array} \left[\begin{array}{cccc|c} 1 & 1 & 1 & 0 & 0 \\ 0 & 0 & -1 & 1 & 1 \\ 0 & -1 & 0 & 1 & 1 \\ 0 & 1 & 1 & 1 & 0 \end{array}\right]$$

Replace -1 with 1.

$$\left[\begin{array}{cccc|c} 1 & 1 & 1 & 0 & 0 \\ 0 & 0 & 1 & 1 & 1 \\ 0 & 1 & 0 & 1 & 1 \\ 0 & 1 & 1 & 1 & 0 \end{array}\right]$$

Interchange rows 2 and 3.

$$\left[\begin{array}{cccc|c} 1 & 1 & 1 & 0 & 0 \\ 0 & 1 & 0 & 1 & 1 \\ 0 & 0 & 1 & 1 & 1 \\ 0 & 1 & 1 & 1 & 0 \end{array}\right]$$

$$-1R_2 + R_1 \to R_1 \quad -1R_2 + R_4 \to R_4 \quad \left[\begin{array}{cccc|c} 1 & 0 & 1 & -1 & -1 \\ 0 & 1 & 0 & 1 & 1 \\ 0 & 0 & 1 & 1 & 1 \\ 0 & 0 & 1 & 0 & -1 \end{array}\right]$$

Replace -1 with 1.

$$\left[\begin{array}{cccc|c} 1 & 0 & 1 & 1 & 1 \\ 0 & 1 & 0 & 1 & 1 \\ 0 & 0 & 1 & 1 & 1 \\ 0 & 0 & 1 & 0 & 1 \end{array}\right]$$

$$-1R_3 + R_1 \to R_1 \quad -1R_3 + R_4 \to R_4 \quad \left[\begin{array}{cccc|c} 1 & 0 & 0 & 0 & 0 \\ 0 & 1 & 0 & 1 & 1 \\ 0 & 0 & 1 & 1 & 1 \\ 0 & 0 & 0 & -1 & 0 \end{array}\right]$$

Replace -1 with 1.

$$\left[\begin{array}{cccc|c} 1 & 0 & 0 & 0 & 0 \\ 0 & 1 & 0 & 1 & 1 \\ 0 & 0 & 1 & 1 & 1 \\ 0 & 0 & 0 & 1 & 0 \end{array}\right]$$

$$-1R_4 + R_2 \to R_2 \quad -1R_4 + R_3 \to R_3 \quad \left[\begin{array}{cccc|c} 1 & 0 & 0 & 0 & 0 \\ 0 & 1 & 0 & 0 & 1 \\ 0 & 0 & 1 & 0 & 1 \\ 0 & 0 & 0 & 1 & 0 \end{array}\right]$$

The solution (0, 1, 1, 0) corresponds to $x_{11} = 0$, $x_{12} = 1$, $x_{21} = 1$, and $x_{22} = 0$. Since 1 indicates that a button is pushed and 0 indicates that it is not, the strategy required to turn all the lights out is to push the button in the first row, second column, and push the button in the second row first column.

2.3 Addition and Subtraction of Matrices

Your Turn 1

(a) It is not possible to add a 2×4 matrix and a 2×3 matrix.

(b)

$$\begin{bmatrix} 3 & 4 & 5 \\ 1 & 2 & 3 \end{bmatrix} + \begin{bmatrix} 1 & -2 & 4 \\ -2 & -4 & 8 \end{bmatrix} = \begin{bmatrix} 4 & 2 & 9 \\ -1 & -2 & 11 \end{bmatrix}$$

Your Turn 2

$$\begin{bmatrix} 3 & 4 & 5 \\ 1 & 2 & 3 \end{bmatrix} - \begin{bmatrix} 1 & -2 & 4 \\ -2 & -4 & 8 \end{bmatrix} = \begin{bmatrix} 2 & 6 & 1 \\ 3 & 6 & -5 \end{bmatrix}$$

2.3 Exercises

1. $\begin{bmatrix} 1 & 3 \\ 5 & 7 \end{bmatrix} = \begin{bmatrix} 1 & 5 \\ 3 & 7 \end{bmatrix}$

This statement is false, since not all corresponding elements are equal.

3. $\begin{bmatrix} x \\ y \end{bmatrix} = \begin{bmatrix} -2 \\ 8 \end{bmatrix}$ if $x = -2$ and $y = 8$.

This statement is true. The matrices are the same size and corresponding elements are equal.

5. $\begin{bmatrix} 1 & 9 & -4 \\ 3 & 7 & 2 \\ -1 & 1 & 0 \end{bmatrix}$ is a square matrix.

This statement is true. The matrix has 3 rows and 3 columns.

7. $\begin{bmatrix} -4 & 8 \\ 2 & 3 \end{bmatrix}$ is a 2×2 square matrix.

9. $\begin{bmatrix} -6 & 8 & 0 & 0 \\ 4 & 1 & 9 & 2 \\ 3 & -5 & 7 & 1 \end{bmatrix}$ is a 3×4 matrix.

11. $\begin{bmatrix} -7 \\ 5 \end{bmatrix}$ is a 2×1 column matrix.

13. Undefined

15. $\begin{bmatrix} 3 & 4 \\ -8 & 1 \end{bmatrix} = \begin{bmatrix} 3 & x \\ y & z \end{bmatrix}$

Corresponding elements must be equal for the matrices to be equal. Therefore, $x = 4$, $y = -8$, and $z = 1$.

17. $\begin{bmatrix} s-4 & t+2 \\ -5 & 7 \end{bmatrix} = \begin{bmatrix} 6 & 2 \\ -5 & r \end{bmatrix}$

Corresponding elements must be equal

$$s - 4 = 6 \qquad t + 2 = 2 \qquad r = 7.$$
$$s = 10 \qquad t = 0$$

Thus, $s = 10$, $t = 0$, and $r = 7$.

19. $\begin{bmatrix} a+2 & 3b & 4c \\ d & 7f & 8 \end{bmatrix} + \begin{bmatrix} -7 & 2b & 6 \\ -3d & -6 & -2 \end{bmatrix} = \begin{bmatrix} 15 & 25 & 6 \\ -8 & 1 & 6 \end{bmatrix}$

Add the two matrices on the left side to obtain

$$\begin{bmatrix} a+2 & 3b & 4c \\ d & 7f & 8 \end{bmatrix} + \begin{bmatrix} -7 & 2b & 6 \\ -3d & -6 & -2 \end{bmatrix}$$

$$= \begin{bmatrix} (a+2)+(-7) & 3b+2b & 4c+6 \\ d+(-3d) & 7f+(-6) & 8+(-2) \end{bmatrix}$$

$$= \begin{bmatrix} a-5 & 5b & 4c+6 \\ -2d & 7f-6 & 6 \end{bmatrix}$$

Corresponding elements of this matrix and the matrix on the right side of the original equation must be equal.

$$a-5=15 \quad 5b=25 \quad 4c+6=6$$
$$a=20 \quad b=5 \quad c=0$$

$$-2d=-8 \quad 7f-6=1$$
$$d=4 \quad f=1$$

Thus, $a=20$, $b=5$, $c=0$, $d=4$, and $f=1$.

21. $\begin{bmatrix} 2 & 4 & 5 & -7 \\ 6 & -3 & 12 & 0 \end{bmatrix} + \begin{bmatrix} 8 & 0 & -10 & 1 \\ -2 & 8 & -9 & 11 \end{bmatrix}$

$$= \begin{bmatrix} 2+8 & 4+0 & 5+(-10) & -7+1 \\ 6+(-2) & -3+8 & 12+(-9) & 0+11 \end{bmatrix}$$

$$= \begin{bmatrix} 10 & 4 & -5 & -6 \\ 4 & 5 & 3 & 11 \end{bmatrix}$$

23. $\begin{bmatrix} 1 & 3 & -2 \\ 4 & 7 & 1 \end{bmatrix} + \begin{bmatrix} 3 & 0 \\ 6 & 4 \\ -5 & 2 \end{bmatrix}$

These matrices cannot be added since the first matrix has size 2×3, while the second has size 3×2. Only matrices that are the same size can be added.

25. The matrices have the same size, so the subtraction can be done. Let A and B represent the given matrices.

$A - B =$

$$= \begin{bmatrix} 2-1 & 8-3 & 12-6 & 0-9 \\ 7-2 & 4-(-3) & -1-(-3) & 5-4 \\ 1-8 & 2-0 & 0-(-2) & 10-17 \end{bmatrix}$$

$$= \begin{bmatrix} 1 & 5 & 6 & -9 \\ 5 & 7 & 2 & 1 \\ -7 & 2 & 2 & -7 \end{bmatrix}$$

27. $\begin{bmatrix} 2 & 3 \\ -2 & 4 \end{bmatrix} + \begin{bmatrix} 4 & 3 \\ 7 & 8 \end{bmatrix} - \begin{bmatrix} 3 & 2 \\ 1 & 4 \end{bmatrix}$

$$= \begin{bmatrix} 2+4-3 & 3+3-2 \\ -2+7-1 & 4+8-4 \end{bmatrix} = \begin{bmatrix} 3 & 4 \\ 4 & 8 \end{bmatrix}$$

29. $\begin{bmatrix} 2 & -1 \\ 0 & 13 \end{bmatrix} - \begin{bmatrix} 4 & 8 \\ -5 & 7 \end{bmatrix} + \begin{bmatrix} 12 & 7 \\ 5 & 3 \end{bmatrix}$

$$= \begin{bmatrix} 2-4+12 & -1-8+7 \\ 0-(-5)+5 & 13-7+3 \end{bmatrix} = \begin{bmatrix} 10 & -2 \\ 10 & 9 \end{bmatrix}$$

31. $\begin{bmatrix} -4x+2y & -3x+y \\ 6x-3y & 2x-5y \end{bmatrix} + \begin{bmatrix} -8x+6y & 2x \\ 3y-5x & 6x+4y \end{bmatrix}$

$$= \begin{bmatrix} (-4x+2y)+(-8x+6y) & (-3x+y)+2x \\ (6x-3y)+(3y-5x) & (2x-5y)+(6x+4y) \end{bmatrix}$$

$$= \begin{bmatrix} -12x+8y & -x+y \\ x & 8x-y \end{bmatrix}$$

33. $O - X = \begin{bmatrix} 0 & 0 \\ 0 & 0 \end{bmatrix} - \begin{bmatrix} x & y \\ z & w \end{bmatrix}$

$$= \begin{bmatrix} 0-x & 0-y \\ 0-z & 0-w \end{bmatrix} = \begin{bmatrix} -x & -y \\ -z & -w \end{bmatrix}$$

35. Show that $X + (T + P) = (X + T) + P$.

On the left side, the sum $T + P$ is obtained first, and then

$$X + (T + P).$$

This gives the matrix

$$\begin{bmatrix} x+(r+m) & y+(s+n) \\ z+(t+p) & w+(u+q) \end{bmatrix}.$$

For the right side, first the sum $X + T$ is obtained, and then

$$(X + T) + P.$$

This gives the matrix

$$\begin{bmatrix} (x+r)+m & (y+s)+n \\ (z+t)+p & (w+u)+q \end{bmatrix}.$$

Comparing corresponding elements, we see that they are equal by the associative property of addition of real numbers. Thus,

$$X + (T + P) = (X + T) + P.$$

37. Show that $P + O = P$.

$$P + O = \begin{bmatrix} m & n \\ p & q \end{bmatrix} + \begin{bmatrix} 0 & 0 \\ 0 & 0 \end{bmatrix} = \begin{bmatrix} m+0 & n+0 \\ p+0 & q+0 \end{bmatrix}$$

$$= \begin{bmatrix} m & n \\ p & q \end{bmatrix} = P$$

Thus, $P + O = P$.

39. **(a)** The production cost matrix for Chicago is

$$\begin{matrix} & \text{Phones} & \text{Calculators} \\ \text{Material} & 4.05 & 7.01 \\ \text{Labor} & 3.27 & 3.51 \end{matrix}$$

The production cost matrix for Seattle is

$$\begin{matrix} & \text{Phones} & \text{Calculators} \\ \text{Material} & 4.40 & 6.90 \\ \text{Labor} & 3.54 & 3.76 \end{matrix}$$

(b) The new production cost matrix for Chicago is

$$\begin{matrix} & \text{Phones} & \text{Calculators} \\ \text{Material} & 4.05 + 0.37 & 7.01 + 0.42 \\ \text{Labor} & 3.27 + 0.11 & 3.51 + 0.11 \end{matrix}$$

or $\begin{bmatrix} 4.42 & 7.43 \\ 3.38 & 3.62 \end{bmatrix}$.

41. **(a)** There are four food groups and three meals. To represent the data by a 3×4 matrix, we must use the rows to correspond to the meals, breakfast, lunch, and dinner, and the columns to correspond to the four food groups. Thus, we obtain the matrix

$$\begin{bmatrix} 2 & 1 & 2 & 1 \\ 3 & 2 & 2 & 1 \\ 4 & 3 & 2 & 1 \end{bmatrix}.$$

(b) There are four food groups. These will correspond to the four rows. There are three components in each food group: fat, carbohydrates, and protein. These will correspond to the three columns. The matrix is

$$\begin{bmatrix} 5 & 0 & 7 \\ 0 & 10 & 1 \\ 0 & 15 & 2 \\ 10 & 12 & 8 \end{bmatrix}.$$

(c) The matrix is

$$\begin{bmatrix} 8 \\ 4 \\ 5 \end{bmatrix}.$$

43.

$$\begin{matrix} & \text{Obtained Pain Relief} & \\ & \text{Yes} & \text{No} \\ \text{Painfree} & 22 & 3 \\ \text{Placebo} & 8 & 17 \end{matrix}$$

(a) Of the 25 patients who took the placebo, 8 got relief.

(b) Of the 25 patients who took Painfree, 3 got no relief.

(c)

$$\begin{bmatrix} 22 & 3 \\ 8 & 17 \end{bmatrix} + \begin{bmatrix} 21 & 4 \\ 6 & 19 \end{bmatrix} + \begin{bmatrix} 19 & 6 \\ 10 & 15 \end{bmatrix} + \begin{bmatrix} 23 & 2 \\ 3 & 22 \end{bmatrix} = \begin{bmatrix} 85 & 15 \\ 27 & 73 \end{bmatrix}$$

(d) Yes, it appears that Painfree is effective. Of the 100 patients who took the medication, 85% got relief.

45. **(a)** The matrix for the life expectancy of African Americans is

$$\begin{matrix} & M & F \\ 1970 & 60.0 & 68.3 \\ 1980 & 63.8 & 72.5 \\ 1990 & 64.5 & 73.6 \\ 2000 & 68.3 & 75.2 \end{matrix}$$

(b) The matrix for the life expectancy of White Americans is

$$\begin{matrix} & M & F \\ 1970 & 68.0 & 75.6 \\ 1980 & 70.7 & 78.1 \\ 1990 & 72.7 & 79.4 \\ 2000 & 74.9 & 80.1 \end{matrix}$$

(c) The matrix showing the difference between the life expectancy between the two groups is

$$\begin{bmatrix} 60.0 & 68.3 \\ 63.8 & 72.5 \\ 64.5 & 73.6 \\ 68.3 & 75.2 \end{bmatrix} - \begin{bmatrix} 68.0 & 75.6 \\ 70.7 & 78.1 \\ 72.7 & 79.4 \\ 74.9 & 80.1 \end{bmatrix} = \begin{bmatrix} -8.0 & -7.3 \\ -6.9 & -5.6 \\ -8.2 & -5.8 \\ -6.6 & -4.9 \end{bmatrix}$$

47. **(a)** The matrix for the educational attainment of African Americans is

	4 Years of High School or More	4 Years of College or More
1980	51.2	7.9
1985	59.8	11.1
1990	66.2	11.3
1995	73.8	13.2
2000	78.5	16.5
2008	83.0	19.6

(b) The matrix for the educational attainment of Hispanic Americans is

	4 Years of High School or More	4 Years of College or More
1980	45.3	7.9
1985	47.9	8.5
1990	50.8	9.2
1995	53.4	9.3
2000	57.0	10.6
2008	62.3	13.3

(c) The matrix showing the difference in the educational attainment between African and Hispanic Americans is

$$\begin{bmatrix} 51.2 & 7.9 \\ 59.8 & 11.1 \\ 66.2 & 11.3 \\ 73.8 & 13.2 \\ 78.5 & 16.5 \\ 83.0 & 19.6 \end{bmatrix} - \begin{bmatrix} 45.3 & 7.9 \\ 47.9 & 8.5 \\ 50.8 & 9.2 \\ 53.4 & 9.3 \\ 57.0 & 10.6 \\ 62.3 & 13.3 \end{bmatrix} = \begin{bmatrix} 5.9 & 0 \\ 11.9 & 2.6 \\ 15.4 & 2.1 \\ 20.4 & 3.9 \\ 21.5 & 5.9 \\ 20.7 & 6.3 \end{bmatrix}$$

2.4 Multiplication of Matrices

Your Turn 1

$$AB = \begin{bmatrix} 3 & 4 \\ 1 & 2 \end{bmatrix} \begin{bmatrix} 1 & -2 \\ -2 & -4 \end{bmatrix}$$

$$= \begin{bmatrix} 3(1) + 4(-2) & 3(-2) + 4(-4) \\ 1(1) + 2(-2) & 1(-2) + 2(-4) \end{bmatrix}$$

$$= \begin{bmatrix} -5 & -22 \\ -3 & -10 \end{bmatrix}$$

Your Turn 2

$$AB = \begin{bmatrix} 3 & 5 & -1 \\ 2 & 4 & -2 \end{bmatrix} \begin{bmatrix} 3 & -4 \\ -5 & -3 \end{bmatrix}$$

AB does not exist because a 2×3 matrix cannot be multiplied by a 2×2 matrix.

$$BA = \begin{bmatrix} 3 & -4 \\ -5 & -3 \end{bmatrix} \begin{bmatrix} 3 & 5 & -1 \\ 2 & 4 & -2 \end{bmatrix}$$

$$= \begin{bmatrix} 3(3) - 4(2) & 3(5) - 4(4) & 3(-1) - 4(-2) \\ -5(3) - 3(2) & -5(5) - 3(4) & -5(-1) - 3(-2) \end{bmatrix}$$

$$= \begin{bmatrix} 1 & -1 & 5 \\ -21 & -37 & 11 \end{bmatrix}$$

2.4 Exercises

In Exercises 1-6, let

$$A = \begin{bmatrix} -2 & 4 \\ 0 & 3 \end{bmatrix} \text{ and } B = \begin{bmatrix} -6 & 2 \\ 4 & 0 \end{bmatrix}.$$

1. $2A = 2\begin{bmatrix} -2 & 4 \\ 0 & 3 \end{bmatrix} = \begin{bmatrix} -4 & 8 \\ 0 & 6 \end{bmatrix}$

3. $-6A = -6\begin{bmatrix} -2 & 4 \\ 0 & 3 \end{bmatrix} = \begin{bmatrix} 12 & -24 \\ 0 & -18 \end{bmatrix}$

5. $-4A + 5B = -4\begin{bmatrix} -2 & 4 \\ 0 & 3 \end{bmatrix} + 5\begin{bmatrix} -6 & 2 \\ 4 & 0 \end{bmatrix}$

$$= \begin{bmatrix} 8 & -16 \\ 0 & -12 \end{bmatrix} + \begin{bmatrix} -30 & 10 \\ 20 & 0 \end{bmatrix}$$

$$= \begin{bmatrix} -22 & -6 \\ 20 & -12 \end{bmatrix}$$

7. Matrix A size Matrix B size

$2 \times \underline{2}$ $\qquad \underline{2} \times 2$

The number of columns of A is the same as the number of rows of B, so the product AB exists. The size of the matrix AB is 2×2.

Matrix B size Matrix A size

$2 \times \underline{2}$ $\qquad \underline{2} \times 2$

Since the number of columns of B is the same as the number of rows of A, the product BA also exists and has size 2×2.

9. Matrix A size Matrix B size

$3 \times \underline{4}$ $\qquad \underline{4} \times 4$

Since matrix A has 4 columns and matrix B has 4 rows, the product AB exists and has size 3×4.

Matrix B size Matrix A size

$4 \times \underline{4}$ $\qquad \underline{3} \times 4$

Since B has 4 columns and A has 3 rows, the product BA does not exist.

11. Matrix A size Matrix B size

$4 \times \underline{2}$ $\qquad \underline{3} \times 4$

The number of columns of A is not the same as the number of rows of B, so the product AB does not exist.

Matrix B size Matrix A size

$3 \times \underline{4}$ $\qquad \underline{4} \times 2$

The number of columns of B is the same as the number of rows of A, so the product BA exists and has size 3×2.

13. To find the product matrix AB, the number of *columns* of A must be the same as the number of *rows* of B.

15. Call the first matrix A and the second matrix B. The product matrix AB will have size 2×1.

Step 1: Multiply the elements of the first row of A by the corresponding elements of the column of B and add.

$$\begin{bmatrix} \mathbf{2} & \mathbf{-1} \\ 5 & 8 \end{bmatrix}\begin{bmatrix} \mathbf{3} \\ \mathbf{-2} \end{bmatrix} \quad 2(3) + (-1)(-2) = 8$$

Therefore, 8 is the first row entry of the product matrix AB.

Step 2: Multiply the elements of the second row of A by the corresponding elements of the column of B and add.

$$\begin{bmatrix} 2 & -1 \\ \mathbf{5} & \mathbf{8} \end{bmatrix}\begin{bmatrix} \mathbf{3} \\ \mathbf{-2} \end{bmatrix} \quad 5(3) + 8(-2) = -1$$

The second row entry of the product is -1.

Step 3: Write the product using the two entries found above.

$$AB = \begin{bmatrix} 2 & -1 \\ 5 & 8 \end{bmatrix}\begin{bmatrix} 3 \\ -2 \end{bmatrix} = \begin{bmatrix} 8 \\ -1 \end{bmatrix}$$

17. $$\begin{bmatrix} 2 & -1 & 7 \\ -3 & 0 & -4 \end{bmatrix}\begin{bmatrix} 5 \\ 10 \\ 2 \end{bmatrix}$$

$$= \begin{bmatrix} 2 \cdot 5 + (-1) \cdot 10 + 7 \cdot 2 \\ (-3) \cdot 5 + 0 \cdot 10 + (-4) \cdot 2 \end{bmatrix}$$

$$= \begin{bmatrix} 14 \\ -23 \end{bmatrix}$$

19. $$\begin{bmatrix} 2 & -1 \\ 3 & 6 \end{bmatrix}\begin{bmatrix} -1 & 0 & 4 \\ 5 & -2 & 0 \end{bmatrix}$$

$$= \begin{bmatrix} 2 \cdot (-1) + (-1) \cdot 5 & 2 \cdot 0 + (-1) \cdot (-2) & 2 \cdot 4 + (-1) \cdot 0 \\ 3 \cdot (-1) + 6 \cdot 5 & 3 \cdot 0 + 6 \cdot (-2) & 3 \cdot 4 + 6 \cdot 0 \end{bmatrix}$$

$$= \begin{bmatrix} -7 & 2 & 8 \\ 27 & -12 & 12 \end{bmatrix}$$

21. $$\begin{bmatrix} 2 & 2 & -1 \\ 3 & 0 & 1 \end{bmatrix}\begin{bmatrix} 0 & 2 \\ -1 & 4 \\ 0 & 2 \end{bmatrix}$$

$$= \begin{bmatrix} 2 \cdot 0 + 2(-1) + (-1)0 & 2 \cdot 2 + 2 \cdot 4 + (-1)2 \\ 3 \cdot 0 + 0(-1) + 1(0) & 3 \cdot 2 + 0 \cdot 4 + 1 \cdot 2 \end{bmatrix}$$

$$= \begin{bmatrix} -2 & 10 \\ 0 & 8 \end{bmatrix}$$

23. $$\begin{bmatrix} 1 & 2 \\ 3 & 4 \end{bmatrix}\begin{bmatrix} -1 & 5 \\ 7 & 0 \end{bmatrix}$$

$$= \begin{bmatrix} 1(-1) + 2 \cdot 7 & 1 \cdot 5 + 2 \cdot 0 \\ 3(-1) + 4 \cdot 7 & 3 \cdot 5 + 4 \cdot 0 \end{bmatrix}$$

$$= \begin{bmatrix} 13 & 5 \\ 25 & 15 \end{bmatrix}$$

25. $$\begin{bmatrix} -2 & -3 & 7 \\ 1 & 5 & 6 \end{bmatrix}\begin{bmatrix} 1 \\ 2 \\ 3 \end{bmatrix} = \begin{bmatrix} -2(1) + (-3)2 + 7 \cdot 3 \\ 1 \cdot 1 + 5 \cdot 2 + 6 \cdot 3 \end{bmatrix}$$

$$= \begin{bmatrix} 13 \\ 29 \end{bmatrix}$$

27. $$\left(\begin{bmatrix} 2 & 1 \\ -3 & -6 \\ 4 & 0 \end{bmatrix}\begin{bmatrix} 1 & -2 \\ 2 & -1 \end{bmatrix}\right)\begin{bmatrix} 3 \\ 1 \end{bmatrix} = \begin{bmatrix} 4 & -5 \\ -15 & 12 \\ 4 & -8 \end{bmatrix}\begin{bmatrix} 3 \\ 1 \end{bmatrix}$$

$$= \begin{bmatrix} 7 \\ -33 \\ 4 \end{bmatrix}$$

29. $$\begin{bmatrix} 2 & -2 \\ 1 & -1 \end{bmatrix}\left(\begin{bmatrix} 4 & 3 \\ 1 & 2 \end{bmatrix} + \begin{bmatrix} 7 & 0 \\ -1 & 5 \end{bmatrix}\right)$$

$$= \begin{bmatrix} 2 & -2 \\ 1 & -1 \end{bmatrix}\begin{bmatrix} 11 & 3 \\ 0 & 7 \end{bmatrix}$$

$$= \begin{bmatrix} 22 & -8 \\ 11 & -4 \end{bmatrix}$$

31. **(a)** $AB = \begin{bmatrix} -2 & 4 \\ 1 & 3 \end{bmatrix}\begin{bmatrix} -2 & 1 \\ 3 & 6 \end{bmatrix} = \begin{bmatrix} 16 & 22 \\ 7 & 19 \end{bmatrix}$

(b) $BA = \begin{bmatrix} -2 & 1 \\ 3 & 6 \end{bmatrix}\begin{bmatrix} -2 & 4 \\ 1 & 3 \end{bmatrix} = \begin{bmatrix} 5 & -5 \\ 0 & 30 \end{bmatrix}$

(c) No, AB and BA are not equal here.

(d) No, AB does not always equal BA.

33. Verify that $P(X+T) = PX + PT$.

Find $P(X+T)$ and $PX + PT$ separately and compare their values to see if they are the same.

$$P(X+T) = \begin{bmatrix} m & n \\ p & q \end{bmatrix}\left(\begin{bmatrix} x & y \\ z & w \end{bmatrix} + \begin{bmatrix} r & s \\ t & u \end{bmatrix}\right) = \begin{bmatrix} m & n \\ p & q \end{bmatrix}\left(\begin{bmatrix} x+r & y+s \\ z+t & w+u \end{bmatrix}\right)$$

$$= \begin{bmatrix} m(x+r)+n(z+t) & m(y+s)+n(w+u) \\ p(x+r)+q(z+t) & p(y+s)+q(w+u) \end{bmatrix} = \begin{bmatrix} mx+mr+nz+nt & my+ms+nw+nu \\ px+pr+qz+qt & py+ps+qw+qu \end{bmatrix}$$

$$PX + PT = \begin{bmatrix} m & n \\ p & q \end{bmatrix}\begin{bmatrix} x & y \\ z & w \end{bmatrix} + \begin{bmatrix} m & n \\ p & q \end{bmatrix}\begin{bmatrix} r & s \\ t & u \end{bmatrix} = \begin{bmatrix} mx+nz & my+nw \\ px+qz & py+qw \end{bmatrix} + \begin{bmatrix} mr+nt & ms+nu \\ pr+qt & ps+qu \end{bmatrix}$$

$$= \begin{bmatrix} (mx+nz)+(mr+nt) & (my+nw)+(ms+nu) \\ (px+qz)+(pr+qt) & (py+qw)+(ps+qu) \end{bmatrix} = \begin{bmatrix} mx+nz+mr+nt & my+nw+ms+nu \\ px+qz+pr+qt & py+qw+ps+qu \end{bmatrix}$$

$$= \begin{bmatrix} mx+mr+nz+nt & my+ms+nw+nu \\ px+pr+qz+qt & py+ps+qw+qu \end{bmatrix}$$

Observe that the two results are identical. Thus, $P(X+T) = PX + PT$.

35. Verify that $(k+h)P = kP + hP$ for any real numbers k and h.

$$(k+h)P = (k+h)\begin{bmatrix} m & n \\ p & q \end{bmatrix}$$

$$= \begin{bmatrix} (k+h)m & (k+h)n \\ (k+h)p & (k+h)q \end{bmatrix}$$

$$= \begin{bmatrix} km+hm & kn+hn \\ kp+hp & kq+hq \end{bmatrix}$$

$$= \begin{bmatrix} km & kn \\ kp & kq \end{bmatrix} + \begin{bmatrix} hm & hn \\ hp & hq \end{bmatrix}$$

$$= k\begin{bmatrix} m & n \\ p & q \end{bmatrix} + h\begin{bmatrix} m & n \\ p & q \end{bmatrix}$$

$$= kP + hP$$

Thus, $(k+h)P = kP + hP$ for any real numbers k and h.

37. $$\begin{bmatrix} 2 & 3 & 1 \\ 1 & -4 & 5 \end{bmatrix}\begin{bmatrix} x_1 \\ x_2 \\ x_3 \end{bmatrix} = \begin{bmatrix} 2x_1+3x_2+x_3 \\ x_1-4x_2+5x_3 \end{bmatrix},$$

and $\begin{bmatrix} 2x_1+3x_2+x_3 \\ x_1-4x_2+5x_3 \end{bmatrix} = \begin{bmatrix} 5 \\ 8 \end{bmatrix}.$

This is equivalent to

$$2x_1 + 3x_2 + x_3 = 5$$
$$x_1 - 4x_2 + 5x_3 = 8$$

since corresponding elements of equal matrices must be equal. Reversing this, observe that the given system of linear equations can be written as the matrix equation

$$\begin{bmatrix} 2 & 3 & 1 \\ 1 & -4 & 5 \end{bmatrix}\begin{bmatrix} x_1 \\ x_2 \\ x_3 \end{bmatrix} = \begin{bmatrix} 5 \\ 8 \end{bmatrix}.$$

39. **(a)** Use a graphing calculator or a computer to find the product matrix. The answer is

$$AC = \begin{bmatrix} 6 & 106 & 158 & 222 & 28 \\ 120 & 139 & 64 & 75 & 115 \\ -146 & -2 & 184 & 144 & -129 \\ 106 & 94 & 24 & 116 & 110 \end{bmatrix}.$$

(b) CA does not exist.

(c) AC and CA are clearly not equal, since CA does not even exist.

41. Use a graphing calculator or computer to find the matrix products and sums. The answers are as follows.

(a) $$C + D = \begin{bmatrix} -1 & 5 & 9 & 13 & -1 \\ 7 & 17 & 2 & -10 & 6 \\ 18 & 9 & -12 & 12 & 22 \\ 9 & 4 & 18 & 10 & -3 \\ 1 & 6 & 10 & 28 & 5 \end{bmatrix}$$

(b) $(C + D)B = \begin{bmatrix} -2 & -9 & 90 & 77 \\ -42 & -63 & 127 & 62 \\ 413 & 76 & 180 & -56 \\ -29 & -44 & 198 & 85 \\ 137 & 20 & 162 & 103 \end{bmatrix}$

(c) $CB = \begin{bmatrix} -56 & -1 & 1 & 45 \\ -156 & -119 & 76 & 122 \\ 315 & 86 & 118 & -91 \\ -17 & -17 & 116 & 51 \\ 118 & 19 & 125 & 77 \end{bmatrix}$

(d) $DB = \begin{bmatrix} 54 & -8 & 89 & 32 \\ 114 & 56 & 51 & -60 \\ 98 & -10 & 62 & 35 \\ -12 & -27 & 82 & 34 \\ 19 & 1 & 37 & 26 \end{bmatrix}$

(e) $CB + DB = \begin{bmatrix} -2 & -9 & 90 & 77 \\ -42 & -63 & 127 & 62 \\ 413 & 76 & 180 & -56 \\ -29 & -44 & 198 & 85 \\ 137 & 20 & 162 & 103 \end{bmatrix}$

(f) Yes, $(C + D)B$ and $CB + DB$ are equal, as can be seen by observing that the answers to parts (b) and (e) are identical.

43. (a) $\begin{bmatrix} 10 & 4 & 3 & 5 & 6 \\ 7 & 2 & 2 & 3 & 8 \\ 4 & 5 & 1 & 0 & 10 \\ 0 & 3 & 4 & 5 & 5 \end{bmatrix} \begin{bmatrix} 2 & 3 \\ 1 & 1 \\ 4 & 3 \\ 3 & 3 \\ 1 & 2 \end{bmatrix}$

$$= \begin{array}{c} \\ \text{Dept. 1} \\ \text{Dept. 2} \\ \text{Dept. 3} \\ \text{Dept. 4} \end{array} \begin{array}{c} \begin{matrix} \text{A} & \text{B} \end{matrix} \\ \begin{bmatrix} 57 & 70 \\ 41 & 54 \\ 27 & 40 \\ 39 & 40 \end{bmatrix} \end{array}$$

(b) The total cost to buy from supplier A is $57 + 41 + 27 + 39 = \$164$, and the total cost to buy from supplier B is $70 + 54 + 40 + 40 = \$204$. The company should make the purchase from supplier A, since \$164 is a lower total cost than \$204.

45. (a) To find the average, add the matrices. Then multiply the resulting matrix by $\frac{1}{3}$. (Multiplying by $\frac{1}{3}$ is the same as dividing by 3.)

$$\frac{1}{3}\left(\begin{bmatrix} 4.27 & 6.94 \\ 3.45 & 3.65 \end{bmatrix} + \begin{bmatrix} 4.05 & 7.01 \\ 3.27 & 3.51 \end{bmatrix} + \begin{bmatrix} 4.40 & 6.90 \\ 3.54 & 3.76 \end{bmatrix}\right)$$

$$= \frac{1}{3}\begin{bmatrix} 12.72 & 20.85 \\ 10.26 & 10.92 \end{bmatrix} = \begin{bmatrix} 4.24 & 6.95 \\ 3.42 & 3.64 \end{bmatrix}$$

(b) To find the new average, add the new matrix for the Chicago plant and the matrix for the Seattle plant. Since there are only two matrices now, multiply the resulting matrix by $\frac{1}{2}$ to get the average. (Multiplying by $\frac{1}{2}$ is the same as dividing by 2.)

$$\frac{1}{2}\left(\begin{bmatrix} 4.42 & 7.43 \\ 3.38 & 3.62 \end{bmatrix} + \begin{bmatrix} 4.40 & 6.90 \\ 3.54 & 3.76 \end{bmatrix}\right)$$

$$= \frac{1}{2}\begin{bmatrix} 8.82 & 14.33 \\ 6.92 & 7.38 \end{bmatrix} = \begin{bmatrix} 4.41 & 7.17 \\ 3.46 & 3.69 \end{bmatrix}$$

47. (a)

$$P = \begin{array}{c} \\ \text{Sal's} \\ \text{Fred's} \end{array} \begin{array}{c} \begin{matrix} \text{Sh} & \text{Sa} & \text{B} \end{matrix} \\ \begin{bmatrix} 80 & 40 & 120 \\ 60 & 30 & 150 \end{bmatrix} \end{array}$$

(b)

$$F = \begin{array}{c} \\ \text{Sh} \\ \text{Sa} \\ \text{B} \end{array} \begin{array}{c} \begin{matrix} \text{CA} & \text{AR} \end{matrix} \\ \begin{bmatrix} \frac{1}{2} & \frac{1}{5} \\ \frac{1}{4} & \frac{1}{5} \\ \frac{1}{4} & \frac{3}{5} \end{bmatrix} \end{array}$$

(c) $PF = \begin{bmatrix} 80 & 40 & 120 \\ 60 & 30 & 150 \end{bmatrix} \begin{bmatrix} \frac{1}{2} & \frac{1}{5} \\ \frac{1}{4} & \frac{1}{5} \\ \frac{1}{4} & \frac{3}{5} \end{bmatrix}$

$$= \begin{bmatrix} 80\left(\frac{1}{2}\right) + 40\left(\frac{1}{4}\right) + 120\left(\frac{1}{4}\right) & 80\left(\frac{1}{5}\right) + 40\left(\frac{1}{5}\right) + 120\left(\frac{3}{5}\right) \\ 60\left(\frac{1}{2}\right) + 30\left(\frac{1}{4}\right) + 150\left(\frac{1}{4}\right) & 60\left(\frac{1}{5}\right) + 30\left(\frac{1}{5}\right) + 150\left(\frac{3}{5}\right) \end{bmatrix}$$

$$= \begin{bmatrix} 80 & 96 \\ 75 & 108 \end{bmatrix}$$

The rows give the average price per pair of footwear sold by each store, and the columns give the state.

(d) The average price of footwear at a Fred's outlet in Arizona is \$108.

49. (a)

$$XY = \begin{bmatrix} 2 & 1 & 2 & 1 \\ 3 & 2 & 2 & 1 \\ 4 & 3 & 2 & 1 \end{bmatrix} \begin{bmatrix} 5 & 0 & 7 \\ 0 & 10 & 1 \\ 0 & 15 & 2 \\ 10 & 12 & 8 \end{bmatrix} = \begin{bmatrix} 20 & 52 & 27 \\ 25 & 62 & 35 \\ 30 & 72 & 43 \end{bmatrix}$$

The rows give the amounts of fat, carbohydrates, and protein, respectively, in each of the daily meals.

(b) $$YZ = \begin{bmatrix} 5 & 0 & 7 \\ 0 & 10 & 1 \\ 0 & 15 & 2 \\ 10 & 12 & 8 \end{bmatrix} \begin{bmatrix} 8 \\ 4 \\ 5 \end{bmatrix} = \begin{bmatrix} 75 \\ 45 \\ 70 \\ 168 \end{bmatrix}$$

The rows give the number of calories in one exchange of each of the food groups.

(c) Use the matrices found for XY and YZ from parts (a) and (b).

$$(XY)Z = \begin{bmatrix} 20 & 52 & 27 \\ 25 & 62 & 35 \\ 30 & 72 & 43 \end{bmatrix} \begin{bmatrix} 8 \\ 4 \\ 5 \end{bmatrix} = \begin{bmatrix} 503 \\ 623 \\ 743 \end{bmatrix}$$

$$X(YZ) = \begin{bmatrix} 2 & 1 & 2 & 1 \\ 3 & 2 & 2 & 1 \\ 4 & 3 & 2 & 1 \end{bmatrix} \begin{bmatrix} 75 \\ 45 \\ 70 \\ 168 \end{bmatrix} = \begin{bmatrix} 503 \\ 623 \\ 743 \end{bmatrix}$$

The rows give the number of calories in each meal.

51. $$\frac{1}{6}\begin{bmatrix} 60.0 & 68.3 \\ 63.8 & 72.5 \\ 64.5 & 73.6 \\ 68.3 & 75.2 \end{bmatrix} + \frac{5}{6}\begin{bmatrix} 68.0 & 75.6 \\ 70.7 & 78.1 \\ 72.7 & 79.4 \\ 74.9 & 80.1 \end{bmatrix}$$

$$= \frac{1}{6}\left(\begin{bmatrix} 60.0 & 68.3 \\ 63.8 & 72.5 \\ 64.5 & 73.6 \\ 68.3 & 75.2 \end{bmatrix} + 5\begin{bmatrix} 68.0 & 75.6 \\ 70.7 & 78.1 \\ 72.7 & 79.4 \\ 74.9 & 80.1 \end{bmatrix}\right)$$

$$= \frac{1}{6}\left(\begin{bmatrix} 60.0 & 68.3 \\ 63.8 & 72.5 \\ 64.5 & 73.6 \\ 68.3 & 75.2 \end{bmatrix} + \begin{bmatrix} 340.0 & 378.0 \\ 353.5 & 390.5 \\ 363.5 & 397.0 \\ 374.5 & 400.5 \end{bmatrix}\right)$$

$$= \frac{1}{6}\begin{bmatrix} 400.0 & 446.3 \\ 417.3 & 463.0 \\ 428.0 & 470.6 \\ 442.8 & 475.7 \end{bmatrix} = \begin{bmatrix} 66.7 & 74.4 \\ 69.6 & 77.2 \\ 71.3 & 78.4 \\ 73.8 & 79.3 \end{bmatrix}$$

53. (a) The matrices are

$$A = \begin{bmatrix} 0.0346 & 0.0118 \\ 0.0174 & 0.0073 \\ 0.0189 & 0.0059 \\ 0.0135 & 0.0083 \\ 0.0099 & 0.0103 \end{bmatrix}$$

$$B = \begin{bmatrix} 361 & 2038 & 286 & 227 & 460 \\ 473 & 2494 & 362 & 252 & 484 \\ 627 & 2978 & 443 & 278 & 499 \\ 803 & 3435 & 524 & 314 & 511 \\ 1013 & 3824 & 591 & 344 & 522 \end{bmatrix}$$

(b) The total number of births and deaths each year is found by multiplying matrix B by matrix A.

$$BA = \begin{bmatrix} 361 & 2038 & 286 & 227 & 460 \\ 473 & 2494 & 362 & 252 & 484 \\ 627 & 2978 & 443 & 278 & 499 \\ 803 & 3435 & 524 & 314 & 511 \\ 1013 & 3824 & 591 & 344 & 522 \end{bmatrix} \begin{bmatrix} 0.0346 & 0.0118 \\ 0.0174 & 0.0073 \\ 0.0189 & 0.0059 \\ 0.0135 & 0.0083 \\ 0.0099 & 0.0103 \end{bmatrix}$$

$$= \begin{array}{c} \\ 1970 \\ 1980 \\ 1990 \\ 2000 \\ 2010 \end{array}\begin{array}{c} \text{Births} \quad \text{Deaths} \\ \begin{bmatrix} 60.98 & 27.45 \\ 74.80 & 33.00 \\ 90.58 & 39.20 \\ 106.75 & 45.51 \\ 122.57 & 51.59 \end{bmatrix} \end{array}$$

2.5 Matrix Inverses

Your Turn 1

Use row operations to transform the augmented matrix so that the identity matrix is the first three columns.

$$A = \left[\begin{array}{ccc|ccc} 2 & 3 & 1 & 1 & 0 & 0 \\ 1 & -2 & -1 & 0 & 1 & 0 \\ 3 & 3 & 2 & 0 & 0 & 1 \end{array}\right]$$

$$\text{R}_1 \leftrightarrow \text{R}_2 \left[\begin{array}{ccc|ccc} 1 & -2 & -1 & 0 & 1 & 0 \\ 2 & 3 & 1 & 1 & 0 & 0 \\ 3 & 3 & 2 & 0 & 0 & 1 \end{array}\right]$$

$$\begin{array}{r} \\ -2\text{R}_1 + \text{R}_2 \to \text{R}_2 \\ -3\text{R}_1 + \text{R}_3 \to \text{R}_3 \end{array}\left[\begin{array}{ccc|ccc} 1 & -2 & -1 & 0 & 1 & 0 \\ 0 & 7 & 3 & 1 & -2 & 0 \\ 0 & 9 & 5 & 0 & -3 & 1 \end{array}\right]$$

$$\begin{array}{l} \\ \frac{1}{7}R_2 \to R_2 \\ \\ \end{array}\left[\begin{array}{ccc|ccc} 1 & -2 & -1 & 0 & 1 & 0 \\ 0 & 1 & \frac{3}{7} & \frac{1}{7} & -\frac{2}{7} & 0 \\ 0 & 9 & 5 & 0 & -3 & 1 \end{array}\right]$$

$$\begin{array}{r} 2R_2+R_1 \to R_1 \\ \\ -9R_2+R_3 \to R_3 \end{array}\left[\begin{array}{ccc|ccc} 1 & 0 & -\frac{1}{7} & \frac{2}{7} & \frac{3}{7} & 0 \\ 0 & 1 & \frac{3}{7} & \frac{1}{7} & -\frac{2}{7} & 0 \\ 0 & 0 & \frac{8}{7} & -\frac{9}{7} & -\frac{3}{7} & 1 \end{array}\right]$$

$$\begin{array}{r} \\ \\ \frac{7}{8}R_3 \to R_3 \end{array}\left[\begin{array}{ccc|ccc} 1 & 0 & -\frac{1}{7} & \frac{2}{7} & \frac{3}{7} & 0 \\ 0 & 1 & \frac{3}{7} & \frac{1}{7} & -\frac{2}{7} & 0 \\ 0 & 0 & 1 & -\frac{9}{8} & -\frac{3}{8} & \frac{7}{8} \end{array}\right]$$

$$\begin{array}{r} \frac{1}{7}R_3+R_1 \to R_1 \\ -\frac{3}{7}R_3+R_2 \to R_2 \\ \\ \end{array}\left[\begin{array}{ccc|ccc} 1 & 0 & 0 & \frac{1}{8} & \frac{3}{8} & \frac{1}{8} \\ 0 & 1 & 0 & \frac{5}{8} & -\frac{1}{8} & -\frac{3}{8} \\ 0 & 0 & 1 & -\frac{9}{8} & -\frac{3}{8} & \frac{7}{8} \end{array}\right]$$

Thus, $A^{-1} = \begin{bmatrix} \frac{1}{8} & \frac{3}{8} & \frac{1}{8} \\ \frac{5}{8} & -\frac{1}{8} & -\frac{3}{8} \\ -\frac{9}{8} & -\frac{3}{8} & \frac{7}{8} \end{bmatrix}$.

Your Turn 2

Solve the linear system

$$\begin{aligned} 5x + 4y &= 23 \\ 4x - 3y &= 6 \end{aligned}.$$

Let $A = \begin{bmatrix} 5 & 4 \\ 4 & -3 \end{bmatrix}$ be the coefficient matrix,

$B = \begin{bmatrix} 23 \\ 6 \end{bmatrix}$, and $X = \begin{bmatrix} x \\ y \end{bmatrix}$.

First find A^{-1}.

$$\left[\begin{array}{cc|cc} 5 & 4 & 1 & 0 \\ 4 & -3 & 0 & 1 \end{array}\right]$$

$$\begin{array}{r} \frac{1}{5}R_1 \to R_1 \\ \\ \end{array}\left[\begin{array}{cc|cc} 1 & \frac{4}{5} & \frac{1}{5} & 0 \\ 4 & -3 & 0 & 1 \end{array}\right]$$

$$\begin{array}{r} \\ -4R_1+R_2 \to R_2 \end{array}\left[\begin{array}{cc|cc} 1 & \frac{4}{5} & \frac{1}{5} & 0 \\ 0 & -\frac{31}{5} & -\frac{4}{5} & 1 \end{array}\right]$$

$$\begin{array}{r} \\ -\frac{5}{31}R_2 \to R_2 \end{array}\left[\begin{array}{cc|cc} 1 & \frac{4}{5} & \frac{1}{5} & 0 \\ 0 & 1 & \frac{4}{31} & -\frac{5}{31} \end{array}\right]$$

$$\begin{array}{r} -\frac{4}{5}R_2+R_1 \to R_1 \\ \\ \end{array}\left[\begin{array}{cc|cc} 1 & 0 & \frac{3}{31} & \frac{4}{31} \\ 0 & 1 & \frac{4}{31} & -\frac{5}{31} \end{array}\right]$$

$$A^{-1} = \begin{bmatrix} \frac{3}{31} & \frac{4}{31} \\ \frac{4}{31} & -\frac{5}{31} \end{bmatrix}$$

$$X = A^{-1}B = \begin{bmatrix} \frac{3}{31} & \frac{4}{31} \\ \frac{4}{31} & -\frac{5}{31} \end{bmatrix}\begin{bmatrix} 23 \\ 6 \end{bmatrix} = \begin{bmatrix} 3 \\ 2 \end{bmatrix}$$

The solution to the system is $(3, 2)$.

Your Turn 3

(a) The word "behold" gives two 3×1 matrices:

$$\begin{bmatrix} 2 \\ 5 \\ 8 \end{bmatrix} \text{ and } \begin{bmatrix} 15 \\ 12 \\ 4 \end{bmatrix}$$

Use the coding matrix $A = \begin{bmatrix} 1 & 3 & 4 \\ 2 & 1 & 3 \\ 4 & 2 & 1 \end{bmatrix}$ to find the product of A with each column matrix.

$$\begin{bmatrix} 1 & 3 & 4 \\ 2 & 1 & 3 \\ 4 & 2 & 1 \end{bmatrix}\begin{bmatrix} 2 \\ 5 \\ 8 \end{bmatrix} = \begin{bmatrix} 49 \\ 33 \\ 26 \end{bmatrix} \text{ and } \begin{bmatrix} 1 & 3 & 4 \\ 2 & 1 & 3 \\ 4 & 2 & 1 \end{bmatrix}\begin{bmatrix} 15 \\ 12 \\ 4 \end{bmatrix} = \begin{bmatrix} 67 \\ 54 \\ 88 \end{bmatrix}$$

The coded message is 49, 33, 26, 67, 54, 88.

(b) Use the inverse of the coding matrix to decode the message 96, 87, 74, 141, 117, 114.

$$A^{-1} = \begin{bmatrix} -0.2 & 0.2 & 0.2 \\ 0.4 & -0.6 & 0.2 \\ 0 & 0.4 & -0.2 \end{bmatrix}, \begin{bmatrix} 96 \\ 87 \\ 74 \end{bmatrix}, \text{ and } \begin{bmatrix} 141 \\ 117 \\ 114 \end{bmatrix}$$

$$\begin{bmatrix} -0.2 & 0.2 & 0.2 \\ 0.4 & -0.6 & 0.2 \\ 0 & 0.4 & -0.2 \end{bmatrix}\begin{bmatrix} 96 \\ 87 \\ 74 \end{bmatrix} = \begin{bmatrix} 13 \\ 1 \\ 20 \end{bmatrix}$$

$$\begin{bmatrix} -0.2 & 0.2 & 0.2 \\ 0.4 & -0.6 & 0.2 \\ 0 & 0.4 & -0.2 \end{bmatrix}\begin{bmatrix} 141 \\ 117 \\ 114 \end{bmatrix} = \begin{bmatrix} 18 \\ 9 \\ 24 \end{bmatrix}$$

The message is the word "matrix."

2.5 Exercises

1. $$\begin{bmatrix} 2 & 1 \\ 5 & 3 \end{bmatrix}\begin{bmatrix} 3 & -1 \\ -5 & 2 \end{bmatrix} = \begin{bmatrix} 6-5 & -2+2 \\ 15-15 & -5+6 \end{bmatrix} = \begin{bmatrix} 1 & 0 \\ 0 & 1 \end{bmatrix} = I$$

$$\begin{bmatrix} 3 & -1 \\ -5 & 2 \end{bmatrix}\begin{bmatrix} 2 & 1 \\ 5 & 3 \end{bmatrix} = \begin{bmatrix} 6-5 & 3-3 \\ -10+10 & -5+6 \end{bmatrix} = \begin{bmatrix} 1 & 0 \\ 0 & 1 \end{bmatrix} = I$$

Since the products obtained by multiplying the matrices in either order are both the 2×2 identity matrix, the given matrices are inverses of each other.

3. $$\begin{bmatrix} 2 & 6 \\ 2 & 4 \end{bmatrix}\begin{bmatrix} -1 & 2 \\ 2 & -4 \end{bmatrix} = \begin{bmatrix} 10 & -20 \\ 6 & -12 \end{bmatrix} \neq I$$

No, the matrices are not inverses of each other since their product matrix is not I.

5. $$\begin{bmatrix} 2 & 0 & 1 \\ 1 & 1 & 2 \\ 0 & 1 & 0 \end{bmatrix}\begin{bmatrix} 1 & 1 & -1 \\ 0 & 1 & 0 \\ -1 & -2 & 2 \end{bmatrix}$$

$$= \begin{bmatrix} 2+0-1 & 2+0-2 & -2+0+2 \\ 1+0-2 & 1+1-4 & -1+0+4 \\ 0+0+0 & 0+1+0 & 0+0+0 \end{bmatrix}$$

$$= \begin{bmatrix} 1 & 0 & 0 \\ -1 & -2 & 3 \\ 0 & 1 & 0 \end{bmatrix} \neq I$$

No, the matrices are not inverses of each other since their product matrix is not I.

7. $$\begin{bmatrix} 1 & 3 & 3 \\ 1 & 4 & 3 \\ 1 & 3 & 4 \end{bmatrix}\begin{bmatrix} 7 & -3 & -3 \\ -1 & 1 & 0 \\ -1 & 0 & 1 \end{bmatrix} = \begin{bmatrix} 1 & 0 & 0 \\ 0 & 1 & 0 \\ 0 & 0 & 1 \end{bmatrix} = I$$

$$\begin{bmatrix} 7 & -3 & -3 \\ -1 & 1 & 0 \\ -1 & 0 & 1 \end{bmatrix}\begin{bmatrix} 1 & 3 & 3 \\ 1 & 4 & 3 \\ 1 & 3 & 4 \end{bmatrix} = \begin{bmatrix} 1 & 0 & 0 \\ 0 & 1 & 0 \\ 0 & 0 & 1 \end{bmatrix} = I$$

Yes, these matrices are inverses of each other.

9. No, a matrix with a row of all zeros does not have an inverse; the row of all zeros makes it impossible to get all the 1's in the main diagonal of the identity matrix.

11. Let $A = \begin{bmatrix} 1 & -1 \\ 2 & 0 \end{bmatrix}$.

Form the augmented matrix $[A|I]$.

$$[A|I] = \left[\begin{array}{cc|cc} 1 & -1 & 1 & 0 \\ 2 & 0 & 0 & 1 \end{array}\right]$$

Perform row operations on $[A|I]$ to get a matrix of the form $[I|B]$.

$$\left[\begin{array}{cc|cc} 1 & -1 & 1 & 0 \\ 2 & 0 & 0 & 1 \end{array}\right]$$

$$-2R_1 + R_2 \rightarrow R_2 \quad \left[\begin{array}{cc|cc} 1 & -1 & 1 & 0 \\ 0 & 2 & -2 & 1 \end{array}\right]$$

$$2R_1 + R_2 \rightarrow R_1 \quad \left[\begin{array}{cc|cc} 2 & 0 & 0 & 1 \\ 0 & 2 & -2 & 1 \end{array}\right]$$

$$\begin{array}{l} \frac{1}{2}R_1 \rightarrow R_1 \\ \frac{1}{2}R_2 \rightarrow R_2 \end{array} \quad \left[\begin{array}{cc|cc} 1 & 0 & 0 & \frac{1}{2} \\ 0 & 1 & -1 & \frac{1}{2} \end{array}\right] = [I|B]$$

The matrix B in the last transformation is the desired multiplicative inverse.

$$A^{-1} = \begin{bmatrix} 0 & \frac{1}{2} \\ -1 & \frac{1}{2} \end{bmatrix}$$

This answer may be checked by showing that $AA^{-1} = I$ and $A^{-1}A = I$.

13. Let $$A = \begin{bmatrix} 3 & -1 \\ -5 & 2 \end{bmatrix}.$$

$$[A|I] = \left[\begin{array}{cc|cc} 3 & -1 & 1 & 0 \\ -5 & 2 & 0 & 1 \end{array}\right]$$

$$5R_1 + 3R_2 \rightarrow R_2 \quad \left[\begin{array}{cc|cc} 3 & -1 & 1 & 0 \\ 0 & 1 & 5 & 3 \end{array}\right]$$

$$R_1 + R_2 \rightarrow R_1 \quad \left[\begin{array}{cc|cc} 3 & 0 & 6 & 3 \\ 0 & 1 & 5 & 3 \end{array}\right]$$

$$\frac{1}{3}R_1 \rightarrow R_1 \quad \left[\begin{array}{cc|cc} 1 & 0 & 2 & 1 \\ 0 & 1 & 5 & 3 \end{array}\right] = [I|B]$$

The desired inverse is

$$A^{-1} = \begin{bmatrix} 2 & 1 \\ 5 & 3 \end{bmatrix}.$$

15. Let $A = \begin{bmatrix} 1 & -3 \\ -2 & 6 \end{bmatrix}$.

$$[A|I] = \left[\begin{array}{rr|rr} 1 & -3 & 1 & 0 \\ -2 & 6 & 0 & 1 \end{array}\right]$$

$$2R_1 + R_2 \to R_2 \quad \left[\begin{array}{rr|rr} 1 & -3 & 1 & 0 \\ 0 & 0 & 2 & 1 \end{array}\right]$$

Because the last row has all zeros to the left of the vertical bar, there is no way to complete the desired transformation. A has no inverse.

17. Let $A = \begin{bmatrix} 1 & 0 & 0 \\ 0 & -1 & 0 \\ 1 & 0 & 1 \end{bmatrix}$.

$$[A|I] = \left[\begin{array}{rrr|rrr} 1 & 0 & 0 & 1 & 0 & 0 \\ 0 & -1 & 0 & 0 & 1 & 0 \\ 1 & 0 & 1 & 0 & 0 & 1 \end{array}\right]$$

$$-1R_1 + R_3 \to R_3 \quad \left[\begin{array}{rrr|rrr} 1 & 0 & 0 & 1 & 0 & 0 \\ 0 & -1 & 0 & 0 & 1 & 0 \\ 0 & 0 & 1 & -1 & 0 & 1 \end{array}\right]$$

$$-1R_2 \to R_2 \quad \left[\begin{array}{rrr|rrr} 1 & 0 & 0 & 1 & 0 & 0 \\ 0 & 1 & 0 & 0 & -1 & 0 \\ 0 & 0 & 1 & -1 & 0 & 1 \end{array}\right]$$

$$A^{-1} = \begin{bmatrix} 1 & 0 & 0 \\ 0 & -1 & 0 \\ -1 & 0 & 1 \end{bmatrix}$$

19. Let $A = \begin{bmatrix} -1 & -1 & -1 \\ 4 & 5 & 0 \\ 0 & 1 & -3 \end{bmatrix}$.

$$[A|I] = \left[\begin{array}{rrr|rrr} -1 & -1 & -1 & 1 & 0 & 0 \\ 4 & 5 & 0 & 0 & 1 & 0 \\ 0 & 1 & -3 & 0 & 0 & 1 \end{array}\right]$$

$$4R_1 + R_2 \to R_2 \quad \left[\begin{array}{rrr|rrr} -1 & -1 & -1 & 1 & 0 & 0 \\ 0 & 1 & -4 & 4 & 1 & 0 \\ 0 & 1 & -3 & 0 & 0 & 1 \end{array}\right]$$

$$\begin{array}{l} R_2 + R_1 \to R_1 \\ \\ -1R_2 + R_3 \to R_3 \end{array} \quad \left[\begin{array}{rrr|rrr} -1 & 0 & -5 & 5 & 1 & 0 \\ 0 & 1 & -4 & 4 & 1 & 0 \\ 0 & 0 & 1 & -4 & -1 & 1 \end{array}\right]$$

$$\begin{array}{l} 5R_3 + R_1 \to R_1 \\ 4R_3 + R_2 \to R_2 \end{array} \quad \left[\begin{array}{rrr|rrr} -1 & 0 & 0 & -15 & -4 & 5 \\ 0 & 1 & 0 & -12 & -3 & 4 \\ 0 & 0 & 1 & -4 & -1 & 1 \end{array}\right]$$

$$-1R_1 \to R_1 \quad \left[\begin{array}{rrr|rrr} 1 & 0 & 0 & 15 & 4 & -5 \\ 0 & 1 & 0 & -12 & -3 & 4 \\ 0 & 0 & 1 & -4 & -1 & 1 \end{array}\right]$$

$$A^{-1} = \begin{bmatrix} 15 & 4 & -5 \\ -12 & -3 & 4 \\ -4 & -1 & 1 \end{bmatrix}$$

21. Let $A = \begin{bmatrix} 1 & 2 & 3 \\ -3 & -2 & -1 \\ -1 & 0 & 1 \end{bmatrix}$.

$$[A|I] = \left[\begin{array}{rrr|rrr} 1 & 2 & 3 & 1 & 0 & 0 \\ -3 & -2 & -1 & 0 & 1 & 0 \\ -1 & 0 & 1 & 0 & 0 & 1 \end{array}\right]$$

$$\begin{array}{l} \\ 3R_1 + R_2 \to R_2 \\ R_1 + R_3 \to R_3 \end{array} \quad \left[\begin{array}{rrr|rrr} 1 & 2 & 3 & 1 & 0 & 0 \\ 0 & 4 & 8 & 3 & 1 & 0 \\ 0 & 2 & 4 & 1 & 0 & 1 \end{array}\right]$$

$$\begin{array}{l} R_2 + (-2R_1) \to R_1 \\ \\ R_2 + (-2R_3) \to R_3 \end{array} \quad \left[\begin{array}{rrr|rrr} -2 & 0 & 2 & 1 & 1 & 0 \\ 0 & 4 & 8 & 3 & 1 & 0 \\ 0 & 0 & 0 & 1 & 1 & -2 \end{array}\right]$$

Because the last row has all zeros to the left of the vertical bar, there is no way to complete the desired transformation. A has no inverse.

23. Find the inverse of $A = \begin{bmatrix} 1 & 3 & -2 \\ 2 & 7 & -3 \\ 3 & 8 & -5 \end{bmatrix}$, if it exists.

$$[A|I] = \left[\begin{array}{rrr|rrr} 1 & 3 & -2 & 1 & 0 & 0 \\ 2 & 7 & -3 & 0 & 1 & 0 \\ 3 & 8 & -5 & 0 & 0 & 1 \end{array}\right]$$

$$\begin{array}{l} \\ -2R_1 + R_2 \to R_2 \\ -3R_1 + R_3 \to R_3 \end{array} \quad \left[\begin{array}{rrr|rrr} 1 & 3 & -2 & 1 & 0 & 0 \\ 0 & 1 & 1 & -2 & 1 & 0 \\ 0 & -1 & 1 & -3 & 0 & 1 \end{array}\right]$$

$$\begin{array}{l} -3R_2 + R_1 \to R_1 \\ \\ R_2 + R_3 \to R_3 \end{array} \quad \left[\begin{array}{rrr|rrr} 1 & 0 & -5 & 7 & -3 & 0 \\ 0 & 1 & 1 & -2 & 1 & 0 \\ 0 & 0 & 2 & -5 & 1 & 1 \end{array}\right]$$

$$\begin{array}{l} 5R_3 + 2R_1 \to R_1 \\ -1R_3 + 2R_2 \to R_2 \\ \\ \end{array} \quad \left[\begin{array}{rrr|rrr} 2 & 0 & 0 & -11 & -1 & 5 \\ 0 & 2 & 0 & 1 & 1 & -1 \\ 0 & 0 & 2 & -5 & 1 & 1 \end{array}\right]$$

$$\begin{array}{l} \frac{1}{2}R_1 \to R_1 \\ \frac{1}{2}R_2 \to R_2 \\ \frac{1}{2}R_3 \to R_3 \end{array} \left[\begin{array}{ccc|ccc} 1 & 0 & 0 & -\frac{11}{2} & -\frac{1}{2} & \frac{5}{2} \\ 0 & 1 & 0 & \frac{1}{2} & \frac{1}{2} & -\frac{1}{2} \\ 0 & 0 & 1 & -\frac{5}{2} & \frac{1}{2} & \frac{1}{2} \end{array}\right]$$

$$A^{-1} = \begin{bmatrix} -\frac{11}{2} & -\frac{1}{2} & \frac{5}{2} \\ \frac{1}{2} & \frac{1}{2} & -\frac{1}{2} \\ -\frac{5}{2} & \frac{1}{2} & \frac{1}{2} \end{bmatrix}$$

25. Let $A = \begin{bmatrix} 1 & -2 & 3 & 0 \\ 0 & 1 & -1 & 1 \\ -2 & 2 & -2 & 4 \\ 0 & 2 & -3 & 1 \end{bmatrix}$.

$$[A|I] = \left[\begin{array}{cccc|cccc} 1 & -2 & 3 & 0 & 1 & 0 & 0 & 0 \\ 0 & 1 & -1 & 1 & 0 & 1 & 0 & 0 \\ -2 & 2 & -2 & 4 & 0 & 0 & 1 & 0 \\ 0 & 2 & -3 & 1 & 0 & 0 & 0 & 1 \end{array}\right]$$

$$2R_1 + R_3 \to R_3 \left[\begin{array}{cccc|cccc} 1 & -2 & 3 & 0 & 1 & 0 & 0 & 0 \\ 0 & 1 & -1 & 1 & 0 & 1 & 0 & 0 \\ 0 & -2 & 4 & 4 & 2 & 0 & 1 & 0 \\ 0 & 2 & -3 & 1 & 0 & 0 & 0 & 1 \end{array}\right]$$

$$\begin{array}{l} 2R_2 + R_1 \to R_1 \\ \\ 2R_2 + R_3 \to R_3 \\ -2R_2 + R_4 \to R_4 \end{array} \left[\begin{array}{cccc|cccc} 1 & 0 & 1 & 2 & 1 & 2 & 0 & 0 \\ 0 & 1 & -1 & 1 & 0 & 1 & 0 & 0 \\ 0 & 0 & 2 & 6 & 2 & 2 & 1 & 0 \\ 0 & 0 & -1 & -1 & 0 & -2 & 0 & 1 \end{array}\right]$$

$$\begin{array}{l} R_3 + (-2)R_1 \to R_1 \\ R_3 + 2R_2 \to R_2 \\ \\ R_3 + 2R_4 \to R_4 \end{array} \left[\begin{array}{cccc|cccc} -2 & 0 & 0 & 2 & 0 & -2 & 1 & 0 \\ 0 & 2 & 0 & 8 & 2 & 4 & 1 & 0 \\ 0 & 0 & 2 & 6 & 2 & 2 & 1 & 0 \\ 0 & 0 & 0 & 4 & 2 & -2 & 1 & 2 \end{array}\right]$$

$$\begin{array}{l} R_4 + (-2)R_1 \to R_1 \\ -2R_4 + R_2 \to R_2 \\ -3R_4 + 2R_3 \to R_3 \\ \\ \end{array} \left[\begin{array}{cccc|cccc} 4 & 0 & 0 & 0 & 2 & 2 & -1 & 2 \\ 0 & 2 & 0 & 0 & -2 & 8 & -1 & -4 \\ 0 & 0 & 4 & 0 & -2 & 10 & -1 & -6 \\ 0 & 0 & 0 & 4 & 2 & -2 & 1 & 2 \end{array}\right]$$

$$\begin{array}{l} \frac{1}{4}R_1 \to R_1 \\ \frac{1}{2}R_2 \to R_2 \\ \frac{1}{4}R_3 \to R_3 \\ \frac{1}{4}R_4 \to R_4 \end{array} \left[\begin{array}{cccc|cccc} 1 & 0 & 0 & 0 & \frac{1}{2} & \frac{1}{2} & -\frac{1}{4} & \frac{1}{2} \\ 0 & 1 & 0 & 0 & -1 & 4 & -\frac{1}{2} & -2 \\ 0 & 0 & 1 & 0 & -\frac{1}{2} & \frac{5}{2} & -\frac{1}{4} & -\frac{3}{2} \\ 0 & 0 & 0 & 1 & \frac{1}{2} & -\frac{1}{2} & \frac{1}{4} & \frac{1}{2} \end{array}\right]$$

$$A^{-1} = \begin{bmatrix} \frac{1}{2} & \frac{1}{2} & -\frac{1}{4} & \frac{1}{2} \\ -1 & 4 & -\frac{1}{2} & -2 \\ -\frac{1}{2} & \frac{5}{2} & -\frac{1}{4} & -\frac{3}{2} \\ \frac{1}{2} & -\frac{1}{2} & \frac{1}{4} & \frac{1}{2} \end{bmatrix}$$

27.
$$\begin{aligned} 2x + 5y &= 15 \\ x + 4y &= 9 \end{aligned}$$

First, write the system in matrix form.

$$\begin{bmatrix} 2 & 5 \\ 1 & 4 \end{bmatrix} \begin{bmatrix} x \\ y \end{bmatrix} = \begin{bmatrix} 15 \\ 9 \end{bmatrix}$$

Let $A = \begin{bmatrix} 2 & 5 \\ 1 & 4 \end{bmatrix}$, $X = \begin{bmatrix} x \\ y \end{bmatrix}$, and $B = \begin{bmatrix} 15 \\ 9 \end{bmatrix}$.

The system in matrix form is $AX = B$. We wish to find $X = A^{-1}AX = A^{-1}B$. Use row operations to find A^{-1}.

$$[A|I] = \left[\begin{array}{cc|cc} 2 & 5 & 1 & 0 \\ 1 & 4 & 0 & 1 \end{array}\right]$$

$$-1R_1 + 2R_2 \to R_2 \left[\begin{array}{cc|cc} 2 & 5 & 1 & 0 \\ 0 & 3 & -1 & 2 \end{array}\right]$$

$$-5R_2 + 3R_1 \to R_1 \left[\begin{array}{cc|cc} 6 & 0 & 8 & -10 \\ 0 & 3 & -1 & 2 \end{array}\right]$$

$$\begin{array}{l} \frac{1}{6}R_1 \to R_1 \\ \frac{1}{3}R_2 \to R_2 \end{array} \left[\begin{array}{cc|cc} 1 & 0 & \frac{4}{3} & -\frac{5}{3} \\ 0 & 1 & -\frac{1}{3} & \frac{2}{3} \end{array}\right]$$

$$A^{-1} = \begin{bmatrix} \frac{4}{3} & -\frac{5}{3} \\ -\frac{1}{3} & \frac{2}{3} \end{bmatrix} = \frac{1}{3}\begin{bmatrix} 4 & -5 \\ -1 & 2 \end{bmatrix}$$

Next find the product $A^{-1}B$.

$$\begin{aligned} X = A^{-1}B &= \begin{bmatrix} \frac{4}{3} & -\frac{5}{3} \\ -\frac{1}{3} & \frac{2}{3} \end{bmatrix} \begin{bmatrix} 15 \\ 9 \end{bmatrix} \\ &= \frac{1}{3}\begin{bmatrix} 4 & -5 \\ -1 & 2 \end{bmatrix} \begin{bmatrix} 15 \\ 9 \end{bmatrix} \\ &= \frac{1}{3}\begin{bmatrix} 15 \\ 3 \end{bmatrix} = \begin{bmatrix} 5 \\ 1 \end{bmatrix} \end{aligned}$$

Thus, the solution is (5, 1).

29. $2x + y = 5$
$5x + 3y = 13$

Let $A = \begin{bmatrix} 2 & 1 \\ 5 & 3 \end{bmatrix}, \quad X = \begin{bmatrix} x \\ y \end{bmatrix}, \quad B = \begin{bmatrix} 5 \\ 13 \end{bmatrix}.$

Use row operations to obtain

$$A^{-1} = \begin{bmatrix} 3 & -1 \\ -5 & 2 \end{bmatrix}.$$

$$X = A^{-1}B = \begin{bmatrix} 3 & -1 \\ -5 & 2 \end{bmatrix}\begin{bmatrix} 5 \\ 13 \end{bmatrix} = \begin{bmatrix} 2 \\ 1 \end{bmatrix}$$

The solution is (2, 1).

31. $3x - 2y = 3$
$7x - 5y = 0$

First, write the system in matrix form.

$$\begin{bmatrix} 3 & -2 \\ 7 & -5 \end{bmatrix}\begin{bmatrix} x \\ y \end{bmatrix} = \begin{bmatrix} 3 \\ 0 \end{bmatrix}$$

Let $A = \begin{bmatrix} 3 & -2 \\ 7 & -5 \end{bmatrix}, \quad X = \begin{bmatrix} x \\ y \end{bmatrix},$ and $B = \begin{bmatrix} 3 \\ 0 \end{bmatrix}.$

The system is in matrix form $AX = B$. We wish to find $X = A^{-1}\ AX = A^{-1}B$. Use row operations to find A^{-1}.

$$[A|I] = \left[\begin{array}{cc|cc} 3 & -2 & 1 & 0 \\ 7 & -5 & 0 & 1 \end{array}\right]$$

$$-7R_1 + 3R_2 \rightarrow R_2 \quad \left[\begin{array}{cc|cc} 3 & -2 & 1 & 0 \\ 0 & -1 & -7 & 3 \end{array}\right]$$

$$-2R_2 + R_1 \rightarrow R_1 \quad \left[\begin{array}{cc|cc} 3 & 0 & 15 & -6 \\ 0 & -1 & -7 & 3 \end{array}\right]$$

$$\begin{array}{r} \frac{1}{3}R_1 \rightarrow R_1 \\ -1R_2 \rightarrow R_2 \end{array} \quad \left[\begin{array}{cc|cc} 1 & 0 & 5 & -2 \\ 0 & 1 & 7 & -3 \end{array}\right]$$

$$A^{-1} = \begin{bmatrix} 5 & -2 \\ 7 & -3 \end{bmatrix}$$

Next find the product $A^{-1}B$.

$$X = A^{-1}B = \begin{bmatrix} 5 & -2 \\ 7 & -3 \end{bmatrix}\begin{bmatrix} 3 \\ 0 \end{bmatrix} = \begin{bmatrix} 15 \\ 21 \end{bmatrix}$$

Thus, the solution is (15, 21).

33. $-x - 8y = 12$
$3x + 24y = -36$

Let $A = \begin{bmatrix} -1 & -8 \\ 3 & 24 \end{bmatrix}, \quad X = \begin{bmatrix} x \\ y \end{bmatrix}, \quad B = \begin{bmatrix} 12 \\ -36 \end{bmatrix}.$

Using row operations on $[A|I]$ leads to the matrix

$$\left[\begin{array}{cc|cc} 1 & 8 & -1 & 0 \\ 0 & 0 & 3 & 1 \end{array}\right],$$

but the zeros in the second row indicate that matrix A does not have an inverse. We cannot complete the solution by this method.

Since the second equation is a multiple of the first, the equations are dependent. Solve the first equation of the system for x.

$$-x - 8y = 12$$
$$-x = 8y + 12$$
$$x = -8y - 12$$

The solution is $(-8y - 12, y)$, where y is any real number.

35. $-x - y - z = 1$
$4x + 5y = -2$
$y - 3z = 3$

has coefficient matrix

$$A = \begin{bmatrix} -1 & -1 & -1 \\ 4 & 5 & 0 \\ 0 & 1 & -3 \end{bmatrix}.$$

In Exercise 19, it was found that

$$A^{-1} = \begin{bmatrix} -1 & -1 & -1 \\ 4 & 5 & 0 \\ 0 & 1 & 3 \end{bmatrix}^{-1}$$

$$= \begin{bmatrix} 15 & 4 & -5 \\ -12 & -3 & 4 \\ -4 & -1 & 1 \end{bmatrix}.$$

Since $X = A^{-1}B$,

$$\begin{bmatrix} x \\ y \\ z \end{bmatrix} = \begin{bmatrix} 15 & 4 & -5 \\ -12 & -3 & 4 \\ -4 & -1 & 1 \end{bmatrix}\begin{bmatrix} 1 \\ -2 \\ 3 \end{bmatrix} = \begin{bmatrix} -8 \\ 6 \\ 1 \end{bmatrix}.$$

The solution is (−8, 6, 1).

37.
$$\begin{aligned} x + 3y - 2z &= 4 \\ 2x + 7y - 3z &= 8 \\ 3x + 8y - 5z &= -4 \end{aligned}$$

has coefficient matrix

$$A = \begin{bmatrix} 1 & 3 & -2 \\ 2 & 7 & -3 \\ 3 & 8 & -5 \end{bmatrix}.$$

In Exercise 23, it was calculated that

$$A^{-1} = \begin{bmatrix} 1 & 3 & -2 \\ 2 & 7 & -3 \\ 3 & 8 & -5 \end{bmatrix}^{-1} = \begin{bmatrix} -\frac{11}{2} & -\frac{1}{2} & \frac{5}{2} \\ \frac{1}{2} & \frac{1}{2} & -\frac{1}{2} \\ -\frac{5}{2} & \frac{1}{2} & \frac{1}{2} \end{bmatrix}$$

$$= \frac{1}{2}\begin{bmatrix} -11 & -1 & 5 \\ 1 & 1 & -1 \\ -5 & 1 & 1 \end{bmatrix}.$$

Since $X = A^{-1}B$.

$$\begin{bmatrix} x \\ y \\ z \end{bmatrix} = \frac{1}{2}\begin{bmatrix} -11 & -1 & 5 \\ 1 & 1 & -1 \\ -5 & 1 & 1 \end{bmatrix}\begin{bmatrix} 4 \\ 8 \\ -4 \end{bmatrix}$$

$$= \frac{1}{2}\begin{bmatrix} -72 \\ 16 \\ -16 \end{bmatrix} = \begin{bmatrix} -36 \\ 8 \\ -8 \end{bmatrix}$$

39.
$$\begin{aligned} 2x - 2y &= 5 \\ 4y + 8z &= 7 \\ x + 2z &= 1 \end{aligned}$$

has coefficient matrix

$$A = \begin{bmatrix} 2 & -2 & 0 \\ 0 & 4 & 8 \\ 1 & 0 & 2 \end{bmatrix}.$$

However, using row operations on $[A|I]$ shows that A does not have an inverse, so another method must be used.

Try the Gauss-Jordan method. The augmented matrix is

$$\left[\begin{array}{ccc|c} 2 & -2 & 0 & 5 \\ 0 & 4 & 8 & 7 \\ 1 & 0 & 2 & 1 \end{array}\right].$$

After several row operations, we obtain the matrix

$$\left[\begin{array}{ccc|c} 1 & 0 & 2 & \frac{17}{4} \\ 0 & 1 & 2 & \frac{7}{4} \\ 0 & 0 & 0 & 13 \end{array}\right].$$

The bottom row of this matrix shows that the system has no solution, since $0 = 13$ is a false statement.

41.
$$\begin{aligned} x - 2y + 3z &= 4 \\ y - z + w &= -8 \\ -2x + 2y - 2z + 4w &= 12 \\ 2y - 3z + w &= -4 \end{aligned}$$

has coefficient matrix

$$A = \begin{bmatrix} 1 & -2 & 3 & 0 \\ 0 & 1 & -1 & 1 \\ -2 & 2 & -2 & 4 \\ 0 & 2 & -3 & 1 \end{bmatrix}.$$

In Exercise 25, it was found that

$$A^{-1} = \begin{bmatrix} \frac{1}{2} & \frac{1}{2} & -\frac{1}{4} & \frac{1}{2} \\ -1 & 4 & -\frac{1}{2} & -2 \\ -\frac{1}{2} & \frac{5}{2} & -\frac{1}{4} & -\frac{3}{2} \\ \frac{1}{2} & -\frac{1}{2} & \frac{1}{4} & \frac{1}{2} \end{bmatrix}.$$

Since $X = A^{-1}B$,

$$\begin{bmatrix} x \\ y \\ z \\ w \end{bmatrix} = \begin{bmatrix} \frac{1}{2} & \frac{1}{2} & -\frac{1}{4} & \frac{1}{2} \\ -1 & 4 & -\frac{1}{2} & -2 \\ -\frac{1}{2} & \frac{5}{2} & -\frac{1}{4} & -\frac{3}{2} \\ \frac{1}{2} & -\frac{1}{2} & \frac{1}{4} & \frac{1}{2} \end{bmatrix}\begin{bmatrix} 4 \\ -8 \\ 12 \\ -4 \end{bmatrix} = \begin{bmatrix} -7 \\ -34 \\ -19 \\ 7 \end{bmatrix}.$$

The solution is $(-7, -34, -19, 7)$.

In Exercises 43–48, let $A = \begin{bmatrix} a & b \\ c & d \end{bmatrix}$.

43. $IA = \begin{bmatrix} 1 & 0 \\ 0 & 1 \end{bmatrix}\begin{bmatrix} a & b \\ c & d \end{bmatrix} = \begin{bmatrix} a & b \\ c & d \end{bmatrix} = A$

Thus, $IA = A$.

45. $A \cdot O = \begin{bmatrix} a & b \\ c & d \end{bmatrix}\begin{bmatrix} 0 & 0 \\ 0 & 0 \end{bmatrix} = \begin{bmatrix} 0 & 0 \\ 0 & 0 \end{bmatrix} = O$

Thus, $A \cdot O = O$.

47. In Exercise 46, it was found that

$$A^{-1} = \frac{1}{ad - bc}\begin{bmatrix} d & -b \\ -c & a \end{bmatrix}.$$

$$A^{-1}A = \left(\frac{1}{ad - bc}\begin{bmatrix} d & -b \\ -c & a \end{bmatrix}\right)\begin{bmatrix} a & b \\ c & d \end{bmatrix}$$

$$= \frac{1}{ad - bc}\left(\begin{bmatrix} d & -b \\ -c & a \end{bmatrix}\begin{bmatrix} a & b \\ c & d \end{bmatrix}\right)$$

$$= \frac{1}{ad - bc}\begin{bmatrix} ad - bc & 0 \\ 0 & ad - bc \end{bmatrix}$$

$$= \begin{bmatrix} 1 & 0 \\ 0 & 1 \end{bmatrix} = I$$

Thus, $A^{-1}A = I$.

49.

$$AB = O$$
$$A^{-1}(AB) = A^{-1} \cdot O$$
$$(A^{-1}A)B = O$$
$$I \cdot B = O$$
$$B = O$$

Thus, if $AB = O$ and A^{-1} exists, then $B = O$.

51. This exercise should be solved by graphing calculator or computer methods. The solution, which may vary slightly, is

$$C^{-1} = \begin{bmatrix} -0.0477 & -0.0230 & 0.0292 & 0.0895 & -0.0402 \\ 0.0921 & 0.0150 & 0.0321 & 0.0209 & -0.0276 \\ -0.0678 & 0.0315 & -0.0404 & 0.0326 & 0.0373 \\ 0.0171 & -0.0248 & 0.0069 & -0.0003 & 0.0246 \\ -0.0208 & 0.0740 & 0.0096 & -0.1018 & 0.0646 \end{bmatrix}.$$

(Entries are rounded to 4 places.)

53. This exercise should be solved by graphing calculator or computer methods. The solution, which may vary slightly, is

$$D^{-1} = \begin{bmatrix} 0.0394 & 0.0880 & 0.0033 & 0.0530 & -0.1499 \\ -0.1492 & 0.0289 & 0.0187 & 0.1033 & 0.1668 \\ -0.1330 & -0.0543 & 0.0356 & 0.1768 & 0.1055 \\ 0.1407 & 0.0175 & -0.0453 & -0.1344 & 0.0655 \\ 0.0102 & -0.0653 & 0.0993 & 0.0085 & -0.0388 \end{bmatrix}.$$

(Entries are rounded to 4 places.)

55. This exercise should be solved by graphing calculator or computer methods. The solution may vary slightly. The answer is, yes, $D^{-1}C^{-1} = (CD)^{-1}$.

57. This exercise should be solved by graphing calculator or computer methods. The solution, which may vary slightly, is

$$\begin{bmatrix} 1.51482 \\ 0.053479 \\ -0.637242 \\ 0.462629 \end{bmatrix}.$$

59. **(a)** The matrix is $B = \begin{bmatrix} 72 \\ 48 \\ 60 \end{bmatrix}$.

(b) The matrix equation is

$$\begin{bmatrix} 2 & 4 & 2 \\ 2 & 1 & 2 \\ 2 & 1 & 3 \end{bmatrix}\begin{bmatrix} x_1 \\ x_2 \\ x_3 \end{bmatrix} = \begin{bmatrix} 72 \\ 48 \\ 60 \end{bmatrix}.$$

(c) To solve the system, begin by using row operations to find A^{-1}.

$$[A|I] = \left[\begin{array}{ccc|ccc} 2 & 4 & 2 & 1 & 0 & 0 \\ 2 & 1 & 2 & 0 & 1 & 0 \\ 2 & 1 & 3 & 0 & 0 & 1 \end{array}\right]$$

$$\begin{array}{r} \\ R_1 + (-1)R_2 \rightarrow R_2 \\ R_1 + (-1)R_3 \rightarrow R_3 \end{array} \left[\begin{array}{ccc|ccc} 2 & 4 & 2 & 1 & 0 & 0 \\ 0 & 3 & 0 & 1 & -1 & 0 \\ 0 & 3 & -1 & 1 & 0 & -1 \end{array}\right]$$

$$\begin{array}{r} -4R_2 + 3R_1 \rightarrow R_1 \\ \\ R_2 + (-1)R_3 \rightarrow R_3 \end{array} \left[\begin{array}{ccc|ccc} 6 & 0 & 6 & -1 & 4 & 0 \\ 0 & 3 & 0 & 1 & -1 & 0 \\ 0 & 0 & 1 & 0 & -1 & 1 \end{array}\right]$$

$$\begin{array}{r} -6R_3 + R_1 \rightarrow R_1 \\ \\ \\ \end{array} \left[\begin{array}{ccc|ccc} 6 & 0 & 0 & -1 & 10 & -6 \\ 0 & 3 & 0 & 1 & -1 & 0 \\ 0 & 0 & 1 & 0 & -1 & 1 \end{array}\right]$$

$$\begin{array}{r} \frac{1}{6}R_1 \rightarrow R_1 \\ \frac{1}{3}R_2 \rightarrow R_2 \\ \\ \end{array} \left[\begin{array}{ccc|ccc} 1 & 0 & 0 & -\frac{1}{6} & \frac{5}{3} & -1 \\ 0 & 1 & 0 & \frac{1}{3} & -\frac{1}{3} & 0 \\ 0 & 0 & 1 & 0 & -1 & 1 \end{array}\right]$$

The inverse matrix is

$$A^{-1} = \begin{bmatrix} -\frac{1}{6} & \frac{5}{3} & -1 \\ \frac{1}{3} & -\frac{1}{3} & 0 \\ 0 & -1 & 1 \end{bmatrix}.$$

Since $X = A^{-1}B$,

$$\begin{bmatrix} x_1 \\ x_2 \\ x_3 \end{bmatrix} = \begin{bmatrix} -\frac{1}{6} & \frac{5}{3} & -1 \\ \frac{1}{3} & -\frac{1}{3} & 0 \\ 0 & -1 & 1 \end{bmatrix} \begin{bmatrix} 72 \\ 48 \\ 60 \end{bmatrix} = \begin{bmatrix} 8 \\ 8 \\ 12 \end{bmatrix}.$$

There are 8 daily orders for type I, 8 for type II, and 12 for type III.

61. Let x = the amount invested in AAA bonds, y = the amount invested in A bonds, and z = amount invested in B bonds.

(a) The total investment is $x + y + z = 25{,}000$. The annual return is $0.06x + 0.065y + 0.08z = 1650$. Since twice as much is invested in AAA bonds as in B bonds, $x = 2z$.

The system to be solved is

$$\begin{aligned} x + y + z &= 25{,}000 \\ 0.06x + 0.065y + 0.08z &= 1650 \\ x - 2z &= 0 \end{aligned}$$

Let $A = \begin{bmatrix} 1 & 1 & 1 \\ 0.06 & 0.065 & 0.08 \\ 1 & 0 & -2 \end{bmatrix}$, $B = \begin{bmatrix} 25{,}000 \\ 1650 \\ 0 \end{bmatrix}$,

and $X = \begin{bmatrix} x \\ y \\ z \end{bmatrix}$.

Use a graphing calculator to obtain

$$A^{-1} = \begin{bmatrix} -26 & 400 & 3 \\ 40 & -600 & -4 \\ -13 & 200 & 1 \end{bmatrix}.$$

Use a graphing calculator again to solve the matrix equation $X = A^{-1}B$.

$$\begin{bmatrix} x \\ y \\ z \end{bmatrix} = \begin{bmatrix} -26 & 400 & 3 \\ 40 & -600 & -4 \\ -13 & 200 & 1 \end{bmatrix} \begin{bmatrix} 25{,}000 \\ 1650 \\ 0 \end{bmatrix} = \begin{bmatrix} 10{,}000 \\ 10{,}000 \\ 5000 \end{bmatrix}$$

\$10,000 should be invested at 6% in AAA bonds, \$10,000 at 6.5% in A bonds, and \$5000 at 8% in B bonds.

(b) The matrix of constants is changed to

$$B = \begin{bmatrix} 30{,}000 \\ 1985 \\ 0 \end{bmatrix}.$$

$$\begin{bmatrix} x \\ y \\ z \end{bmatrix} = \begin{bmatrix} -26 & 400 & 3 \\ 40 & -600 & -4 \\ -13 & 200 & 1 \end{bmatrix} \begin{bmatrix} 30{,}000 \\ 1985 \\ 0 \end{bmatrix} = \begin{bmatrix} 14{,}000 \\ 9{,}000 \\ 7000 \end{bmatrix}$$

\$14,000 should be invested at 6% in AAA bonds, \$9000 at 6.5% in A bonds, and \$7000 at 8% in B bonds.

(c) The matrix of constants is changed to

$$B = \begin{bmatrix} 40{,}000 \\ 2660 \\ 0 \end{bmatrix}.$$

$$\begin{bmatrix} x \\ y \\ z \end{bmatrix} = \begin{bmatrix} -26 & 400 & 3 \\ 40 & -600 & -4 \\ -13 & 200 & 1 \end{bmatrix} \begin{bmatrix} 40{,}000 \\ 2660 \\ 0 \end{bmatrix} = \begin{bmatrix} 24{,}000 \\ 4000 \\ 12{,}000 \end{bmatrix}$$

\$24,000 should be invested at 6% in AAA bonds, \$4000 at 6.5% in A bonds, and \$12,000 at 8% in B bonds.

63. Let x = the number of Super Vim tablets, y = the number of Multitab tablets, and z = the number of Mighty Mix tablets.

The total number of vitamins is

$$x + y + z.$$

The total amount of niacin is

$$15x + 20y + 25z.$$

The total amount of Vitamin E is

$$12x + 15y + 35z.$$

(a) The system to be solved is

$$\begin{aligned} x + y + z &= 225 \\ 15x + 20y + 25z &= 4750 \\ 12x + 15y + 35z &= 5225. \end{aligned}$$

Let $A = \begin{bmatrix} 1 & 1 & 1 \\ 15 & 20 & 25 \\ 20 & 15 & 35 \end{bmatrix}$, $X = \begin{bmatrix} x \\ y \\ z \end{bmatrix}$, $B = \begin{bmatrix} 225 \\ 4750 \\ 5225 \end{bmatrix}$.

Thus, $AX = B$ and

$$\begin{bmatrix} 1 & 1 & 1 \\ 15 & 20 & 25 \\ 12 & 15 & 35 \end{bmatrix} \begin{bmatrix} x \\ y \\ z \end{bmatrix} = \begin{bmatrix} 225 \\ 4750 \\ 5225 \end{bmatrix}.$$

Use row operations to obtain the inverse of the coefficient matrix.

$$A^{-1} = \begin{bmatrix} \frac{65}{17} & -\frac{4}{17} & \frac{1}{17} \\ -\frac{45}{17} & \frac{23}{85} & -\frac{2}{17} \\ -\frac{3}{17} & -\frac{3}{85} & \frac{1}{17} \end{bmatrix}$$

Since $X = A^{-1}B$,

$$\begin{bmatrix} x \\ y \\ z \end{bmatrix} = \begin{bmatrix} \frac{65}{17} & -\frac{4}{17} & \frac{1}{17} \\ -\frac{45}{17} & \frac{23}{85} & -\frac{2}{17} \\ -\frac{3}{17} & -\frac{3}{85} & \frac{1}{17} \end{bmatrix} \begin{bmatrix} 225 \\ 4750 \\ 5225 \end{bmatrix} = \begin{bmatrix} 50 \\ 75 \\ 100 \end{bmatrix}.$$

There are 50 Super Vim tablets, 75 Multitab tablets, and 100 Mighty Mix tablets.

(b) The matrix of constants is changed to

$$B = \begin{bmatrix} 185 \\ 3625 \\ 3750 \end{bmatrix}.$$

$$\begin{bmatrix} x \\ y \\ z \end{bmatrix} = \begin{bmatrix} \frac{65}{17} & -\frac{4}{17} & \frac{1}{17} \\ -\frac{45}{17} & \frac{23}{85} & -\frac{2}{17} \\ -\frac{3}{17} & -\frac{3}{85} & \frac{1}{17} \end{bmatrix} \begin{bmatrix} 185 \\ 3625 \\ 3750 \end{bmatrix} = \begin{bmatrix} 75 \\ 50 \\ 60 \end{bmatrix}$$

There are 75 Super Vim tablets, 50 Multitab tablets, and 60 Mighty Mix tablets.

(c) The matrix of constants is changed to

$$B = \begin{bmatrix} 230 \\ 4450 \\ 4210 \end{bmatrix}.$$

$$\begin{bmatrix} x \\ y \\ z \end{bmatrix} = \begin{bmatrix} \frac{65}{17} & -\frac{4}{17} & \frac{1}{17} \\ -\frac{45}{17} & \frac{23}{85} & -\frac{2}{17} \\ -\frac{3}{17} & -\frac{3}{85} & \frac{1}{17} \end{bmatrix} \begin{bmatrix} 230 \\ 4450 \\ 4210 \end{bmatrix} = \begin{bmatrix} 80 \\ 100 \\ 50 \end{bmatrix}$$

There are 80 Super Vim tablets, 100 Multitab tablets, and 50 Mighty Mix tablets.

65. (a) First, divide the letters and spaces of the message into groups of three, writing each group as a column vector.

$$\begin{bmatrix} T \\ o \\ \text{(space)} \end{bmatrix}, \begin{bmatrix} b \\ e \\ \text{(space)} \end{bmatrix}, \begin{bmatrix} o \\ r \\ \text{(space)} \end{bmatrix}, \begin{bmatrix} n \\ o \\ t \end{bmatrix}, \begin{bmatrix} \text{(space)} \\ t \\ o \end{bmatrix}, \begin{bmatrix} \text{(space)} \\ b \\ e \end{bmatrix}$$

Next, convert each letter into a number, assigning 1 to A, 2 to B, and so on, with the number 27 used to represent each space between words.

$$\begin{bmatrix} 20 \\ 15 \\ 27 \end{bmatrix}, \begin{bmatrix} 2 \\ 5 \\ 27 \end{bmatrix}, \begin{bmatrix} 15 \\ 18 \\ 27 \end{bmatrix}, \begin{bmatrix} 14 \\ 15 \\ 20 \end{bmatrix}, \begin{bmatrix} 27 \\ 20 \\ 15 \end{bmatrix}, \begin{bmatrix} 27 \\ 2 \\ 5 \end{bmatrix}$$

Now, find the product of the coding matrix B and each column vector. This produces a new set of vectors, which represents the coded message.

$$\begin{bmatrix} 262 \\ -161 \\ -12 \end{bmatrix}, \begin{bmatrix} 186 \\ -103 \\ -22 \end{bmatrix}, \begin{bmatrix} 264 \\ -168 \\ -9 \end{bmatrix}, \begin{bmatrix} 208 \\ -134 \\ -5 \end{bmatrix}, \begin{bmatrix} 224 \\ -152 \\ 5 \end{bmatrix}, \begin{bmatrix} 92 \\ -50 \\ -3 \end{bmatrix}$$

This message will be transmitted as 262, −161, −12, 186, −103, −22, 264, −168, −9, 208, −134, −5, 224, −152, 5, 92, −50, −3.

(b) Use row operations or a graphing calculator to find the inverse of the coding matrix B.

$$B^{-1} = \begin{bmatrix} 1.75 & 2.5 & 3 \\ -0.25 & -0.5 & 0 \\ -0.25 & -0.5 & -1 \end{bmatrix}$$

(c) First, divide the coded message into groups of three numbers and form each group into a column vector.

$$\begin{bmatrix} 116 \\ -60 \\ -15 \end{bmatrix}, \begin{bmatrix} 294 \\ -197 \\ -2 \end{bmatrix}, \begin{bmatrix} 148 \\ -92 \\ -9 \end{bmatrix}, \begin{bmatrix} 96 \\ -64 \\ -4 \end{bmatrix}, \begin{bmatrix} 264 \\ -182 \\ -2 \end{bmatrix}$$

Next, find the product of the decoding matric B^{-1} and each of the column vectors. This produces a new set of vectors, which represents the decoded message.

$$\begin{bmatrix} 8 \\ 1 \\ 16 \end{bmatrix}, \begin{bmatrix} 16 \\ 25 \\ 27 \end{bmatrix}, \begin{bmatrix} 2 \\ 9 \\ 18 \end{bmatrix}, \begin{bmatrix} 20 \\ 8 \\ 4 \end{bmatrix}, \begin{bmatrix} 1 \\ 25 \\ 27 \end{bmatrix}$$

Last, convert each number into a letter, assigning A to 1, B to 2, and so on, with the number 27 used to represent a space between words. The decoded message is HAPPY BIRTHDAY.

2.6 Input-Output Models

Your Turn 1

$$X = (I - A)^{-1}D$$

$$X = \begin{bmatrix} 1.395 & 0.496 & 0.589 \\ 0.837 & 1.364 & 0.620 \\ 0.558 & 0.465 & 1.302 \end{bmatrix}\begin{bmatrix} 322 \\ 447 \\ 133 \end{bmatrix} = \begin{bmatrix} 749.239 \\ 961.682 \\ 560.697 \end{bmatrix}$$

The productions of 749 units of agriculture, 962 units of manufacturing, and 561 units of transportation are required to satisfy the demands of 322, 447, and 133 units, respectively.

Your Turn 2

$$A = \begin{bmatrix} \frac{1}{2} & \frac{1}{4} & \frac{1}{6} \\ 0 & \frac{1}{4} & \frac{1}{6} \\ \frac{1}{2} & \frac{1}{2} & \frac{2}{3} \end{bmatrix} \quad I - A = \begin{bmatrix} \frac{1}{2} & -\frac{1}{4} & -\frac{1}{6} \\ 0 & \frac{3}{4} & -\frac{1}{6} \\ -\frac{1}{2} & -\frac{1}{2} & \frac{1}{3} \end{bmatrix}$$

$$(I - A)X = \begin{bmatrix} \frac{1}{2}x_1 & -\frac{1}{4}x_2 & -\frac{1}{6}x_3 \\ 0x_1 & \frac{3}{4}x_2 & -\frac{1}{6}x_3 \\ -\frac{1}{2}x_1 & -\frac{1}{2}x_2 & \frac{1}{3}x_3 \end{bmatrix} = \begin{bmatrix} 0 \\ 0 \\ 0 \end{bmatrix}$$

We get the following system.

$$\frac{1}{2}x_1 - \frac{1}{4}x_2 - \frac{1}{6}x_3 = 0$$
$$\frac{3}{4}x_2 - \frac{1}{6}x_3 = 0$$
$$-\frac{1}{2}x_1 - \frac{1}{2}x_2 + \frac{1}{3}x_3 = 0$$

Clearing fractions gives the following system.

$$6x_1 - 3x_2 - 2x_3 = 0$$
$$9x_2 - 2x_3 = 0$$
$$-3x_1 - 3x_2 + 2x_3 = 0$$

$$9x_2 - 2x_3 = 0$$
$$9x_2 = 2x_3$$
$$x_2 = \frac{2}{9}x_3$$

$$6x_1 - 3\left(\frac{2}{9}x_3\right) - 2x_3 = 0$$
$$6x_1 - \frac{2}{3}x_3 - 2x_3 = 0$$
$$6x_1 - \frac{8}{3}x_3 = 0$$
$$6x_1 = \frac{8}{3}x_3$$
$$x_1 = \frac{4}{9}x_3$$

The solution is $\left(\frac{4}{9}x_3, \frac{2}{9}x_3, x_3\right)$, where x_3 is any real number. For $x_3 = 9$, the solution is (4, 2, 9). So, the production of the three commodities should be in the ratio 4:2:9.

2.6 Exercises

1. $A = \begin{bmatrix} 0.8 & 0.2 \\ 0.2 & 0.7 \end{bmatrix}, D = \begin{bmatrix} 2 \\ 3 \end{bmatrix}$

To find the production matrix, first calculate $I - A$.

$$I - A = \begin{bmatrix} 1 & 0 \\ 0 & 1 \end{bmatrix} - \begin{bmatrix} 0.8 & 0.2 \\ 0.2 & 0.7 \end{bmatrix} = \begin{bmatrix} 0.2 & -0.2 \\ -0.2 & 0.3 \end{bmatrix}$$

Using row operations, find the inverse of $I - A$.

$$[I - A|I] = \left[\begin{array}{cc|cc} 0.2 & -0.2 & 1 & 0 \\ -0.2 & 0.3 & 0 & 1 \end{array}\right]$$

$$\begin{array}{r} 10R_1 \to R_1 \\ 10R_2 \to R_2 \end{array} \left[\begin{array}{cc|cc} 2 & -2 & 10 & 0 \\ -2 & 3 & 0 & 10 \end{array}\right]$$

$$R_1 + R_2 \to R_2 \left[\begin{array}{cc|cc} 2 & -2 & 10 & 0 \\ 0 & 1 & 10 & 10 \end{array}\right]$$

$$2R_2 + R_1 \to R_1 \left[\begin{array}{cc|cc} 2 & 0 & 30 & 20 \\ 0 & 1 & 10 & 10 \end{array}\right]$$

$$\frac{1}{2}R_1 \to R_1 \left[\begin{array}{cc|cc} 1 & 0 & 15 & 10 \\ 0 & 1 & 10 & 10 \end{array}\right]$$

$$(I - A)^{-1} = \begin{bmatrix} 15 & 10 \\ 10 & 10 \end{bmatrix}$$

Since $X = (I - A)^{-1}D$, the product matrix is

$$X = \begin{bmatrix} 15 & 10 \\ 10 & 10 \end{bmatrix}\begin{bmatrix} 2 \\ 3 \end{bmatrix} = \begin{bmatrix} 60 \\ 50 \end{bmatrix}.$$

3. $A = \begin{bmatrix} 0.1 & 0.03 \\ 0.07 & 0.6 \end{bmatrix}, D = \begin{bmatrix} 5 \\ 10 \end{bmatrix}$

First, calculate $I - A$.

$$I - A = \begin{bmatrix} 0.9 & -0.03 \\ -0.07 & 0.4 \end{bmatrix}$$

Use row operations to find the inverse of $I - A$, which is

$$(I - A)^{-1} \approx \begin{bmatrix} 1.118 & 0.084 \\ 0.196 & 2.515 \end{bmatrix}.$$

Since $X = (I - A)^{-1}D$, the production matrix is

$$X = \begin{bmatrix} 1.118 & 0.084 \\ 0.196 & 2.515 \end{bmatrix} \begin{bmatrix} 5 \\ 10 \end{bmatrix} = \begin{bmatrix} 6.43 \\ 26.12 \end{bmatrix}.$$

5. $A = \begin{bmatrix} 0.8 & 0 & 0.1 \\ 0.1 & 0.5 & 0.2 \\ 0 & 0 & 0.7 \end{bmatrix}, D = \begin{bmatrix} 1 \\ 6 \\ 3 \end{bmatrix}$

To find the production matrix, first calculate $I - A$.

$$I - A = \begin{bmatrix} 1 & 0 & 0 \\ 0 & 1 & 0 \\ 0 & 0 & 1 \end{bmatrix} - \begin{bmatrix} 0.8 & 0 & 0.1 \\ 0.1 & 0.5 & 0.2 \\ 0 & 0 & 0.7 \end{bmatrix}$$

$$= \begin{bmatrix} 0.2 & 0 & -0.1 \\ -0.1 & 0.5 & -0.2 \\ 0 & 0 & 0.3 \end{bmatrix}$$

Using row operations, find the inverse of $I - A$.

$$[I - A \mid I] = \left[\begin{array}{ccc|ccc} 0.2 & 0 & -0.1 & 1 & 0 & 0 \\ -0.1 & 0.5 & -0.2 & 0 & 1 & 0 \\ 0 & 0 & 0.3 & 0 & 0 & 1 \end{array}\right]$$

$$\begin{array}{r} 10R_1 \to R_1 \\ 10R_2 \to R_2 \\ 10R_3 \to R_3 \end{array} \left[\begin{array}{ccc|ccc} 2 & 0 & -1 & 10 & 0 & 0 \\ -1 & 5 & -2 & 0 & 10 & 0 \\ 0 & 0 & 3 & 0 & 0 & 10 \end{array}\right]$$

$$\begin{array}{r} \\ R_1 + 2R_2 \to R_2 \\ \end{array} \left[\begin{array}{ccc|ccc} 2 & 0 & -1 & 10 & 0 & 0 \\ 0 & 10 & -5 & 10 & 20 & 0 \\ 0 & 0 & 3 & 0 & 0 & 10 \end{array}\right]$$

$$\begin{array}{r} R_3 + 3R_1 \to R_1 \\ 5R_3 + 3R_2 \to R_2 \\ \end{array} \left[\begin{array}{ccc|ccc} 6 & 0 & 0 & 30 & 0 & 10 \\ 0 & 30 & 0 & 30 & 60 & 50 \\ 0 & 0 & 3 & 0 & 0 & 10 \end{array}\right]$$

$$\begin{array}{r} \frac{1}{6}R_1 \to R_1 \\ \frac{1}{30}R_2 \to R_2 \\ \frac{1}{3}R_3 \to R_3 \end{array} \left[\begin{array}{ccc|ccc} 1 & 0 & 0 & 5 & 0 & \frac{5}{3} \\ 0 & 1 & 0 & 1 & 2 & \frac{5}{3} \\ 0 & 0 & 1 & 0 & 0 & \frac{10}{3} \end{array}\right]$$

$$(I - A)^{-1} = \begin{bmatrix} 5 & 0 & \frac{5}{3} \\ 1 & 2 & \frac{5}{3} \\ 0 & 0 & \frac{10}{3} \end{bmatrix}$$

Since $X = (I - A)^{-1}D$, the product matrix is

$$X = \begin{bmatrix} 5 & 0 & \frac{5}{3} \\ 1 & 2 & \frac{5}{3} \\ 0 & 0 & \frac{10}{3} \end{bmatrix} \begin{bmatrix} 1 \\ 6 \\ 3 \end{bmatrix} = \begin{bmatrix} 10 \\ 18 \\ 10 \end{bmatrix}.$$

7.

$$\begin{array}{c} \\ A \\ B \\ C \end{array} \begin{array}{c} \begin{array}{ccc} A & B & C \end{array} \\ \begin{bmatrix} 0.3 & 0.1 & 0.8 \\ 0.5 & 0.6 & 0.1 \\ 0.2 & 0.3 & 0.1 \end{bmatrix} \end{array} = A$$

$$I - A = \begin{bmatrix} 0.7 & -0.1 & -0.8 \\ -0.5 & 0.4 & -0.1 \\ -0.2 & -0.3 & 0.9 \end{bmatrix}$$

Set $(I - A)X = O$ to obtain the following.

$$\begin{bmatrix} 0.7 & -0.1 & -0.8 \\ -0.5 & 0.4 & -0.1 \\ -0.2 & -0.3 & 0.9 \end{bmatrix} \begin{bmatrix} x_1 \\ x_2 \\ x_3 \end{bmatrix} = \begin{bmatrix} 0 \\ 0 \\ 0 \end{bmatrix}$$

$$\begin{bmatrix} 0.7x_1 - 0.1x_2 - 0.8x_3 \\ -0.5x_1 + 0.4x_2 - 0.1x_3 \\ -0.2x_1 - 0.3x_2 + 0.9x_3 \end{bmatrix} = \begin{bmatrix} 0 \\ 0 \\ 0 \end{bmatrix}$$

Rewrite this matrix equation as a system of equations.

$$\begin{aligned} 0.7x_1 - 0.1x_2 - 0.8x_3 &= 0 \\ -0.5x_1 + 0.4x_2 - 0.1x_3 &= 0 \\ -0.2x_1 - 0.3x_2 + 0.9x_3 &= 0 \end{aligned}$$

Rewrite the equations without decimals.

$$\begin{aligned} 7x_1 - x_2 - 8x_3 &= 0 \quad (1) \\ -5x_1 + 4x_2 - x_3 &= 0 \quad (2) \\ -2x_1 - 3x_2 + 9x_3 &= 0 \quad (3) \end{aligned}$$

Use row operations to solve this system of equations. Begin by eliminating x_1 in equations (2) and (3)

$$\begin{array}{rr} & 7x_1 - x_2 - 8x_3 = 0 \quad (1) \\ 5R_1 + 7R_2 \to R_1 & 23x_2 - 47x_3 = 0 \quad (4) \\ 2R_1 + 7R_3 \to R_3 & -23x_2 + 47x_3 = 0 \quad (5) \end{array}$$

Eliminate x_2 in equations (1) and (5).

$$\begin{array}{rr} 23R_1 + R_2 \to R_1 & 161x_1 \qquad - 231x_3 = 0 \quad (6) \\ & 23x_2 - 47x_3 = 0 \quad (4) \\ R_2 + R_3 \to R_3 & 0 = 0 \quad (7) \end{array}$$

The true statement in equation (7) indicates that the equations are dependent. Solve equation (6) for x_1 and equation (4) for x_2, each in terms of x_3.

$$x_1 = \frac{231}{161}x_3 = \frac{33}{23}x_3$$

$$x_2 = \frac{47}{23}x_3$$

The solution of the system is

$$\left(\frac{33}{23}x_3, \frac{47}{23}x_3, x_3\right).$$

If $x_3 = 23$, then $x_1 = 33$ and $x_2 = 47$, so the production of the three commodities should be in the ratio 33:47:23.

9. Use a graphing calculator or a computer to find the production matrix $X = (I - A)^{-1}D$. The answer is

$$X = \begin{bmatrix} 7697 \\ 4205 \\ 6345 \\ 4106 \end{bmatrix}.$$

Values have been rounded.

11. In Example 4, it was found that

$$(I - A)^{-1} \approx \begin{bmatrix} 1.3882 & 0.1248 \\ 0.5147 & 1.1699 \end{bmatrix}.$$

Since $X = (I - A)^{-1}D$, the production matrix is

$$X = \begin{bmatrix} 1.3882 & 0.1248 \\ 0.5147 & 1.1699 \end{bmatrix}\begin{bmatrix} 925 \\ 1250 \end{bmatrix} = \begin{bmatrix} 1440.085 \\ 1938.473 \end{bmatrix}.$$

Thus, about 1440. metric tons of wheat and 1938 metric tons of oil should be produced.

13. In Example 3, it was found that

$$(I - A)^{-1} \approx \begin{bmatrix} 1.3953 & 0.4961 & 0.5891 \\ 0.8372 & 1.3643 & 0.6202 \\ 0.5581 & 0.4651 & 1.3023 \end{bmatrix}.$$

Since $X = (I - A)^{-1}D$, the production matrix is

$$X = \begin{bmatrix} 1.3953 & 0.4961 & 0.5891 \\ 0.8372 & 1.3643 & 0.6202 \\ 0.5581 & 0.4651 & 1.3023 \end{bmatrix}\begin{bmatrix} 607 \\ 607 \\ 607 \end{bmatrix} = \begin{bmatrix} 1505.66 \\ 1712.77 \\ 1411.58 \end{bmatrix}.$$

Thus, about 1506 units of agriculture, 1713 units of manufacturing, and 1412 units of transportation should be produced.

15. From the given data, we get the input-output matrix

$$A = \begin{bmatrix} 0 & \frac{1}{2} & \frac{1}{4} \\ \frac{1}{4} & 0 & \frac{1}{4} \\ \frac{1}{2} & \frac{1}{4} & 0 \end{bmatrix}.$$

$$I - A = \begin{bmatrix} 1 & -\frac{1}{2} & -\frac{1}{4} \\ -\frac{1}{4} & 1 & -\frac{1}{4} \\ -\frac{1}{2} & -\frac{1}{4} & 1 \end{bmatrix}$$

Use row operations to find the inverse of $I - A$, which is

$$(I - A)^{-1} \approx \begin{bmatrix} 1.538 & 0.923 & 0.615 \\ 0.615 & 1.436 & 0.513 \\ 0.923 & 0.821 & 1.436 \end{bmatrix}.$$

Since $X = (I - A)^{-1}D$, the production matrix is

$$X = \begin{bmatrix} 1.538 & 0.923 & 0.615 \\ 0.615 & 1.436 & 0.513 \\ 0.923 & 0.821 & 1.436 \end{bmatrix}\begin{bmatrix} 1000 \\ 1000 \\ 1000 \end{bmatrix} \approx \begin{bmatrix} 3077 \\ 2564 \\ 3179 \end{bmatrix}.$$

Thus, the production should be about 3077 units of agriculture, 2564 units of manufacturing, and 3179 units of transportation.

17. From the given data, we get the input-output matrix

$$A = \begin{bmatrix} \frac{1}{4} & \frac{1}{6} \\ \frac{1}{2} & 0 \end{bmatrix}.$$

$$I - A = \begin{bmatrix} \frac{3}{4} & -\frac{1}{6} \\ -\frac{1}{2} & 1 \end{bmatrix}$$

Use row operations to find the inverse of $I - A$, which is

$$(I - A)^{-1} = \begin{bmatrix} \frac{3}{2} & \frac{1}{4} \\ \frac{3}{4} & \frac{9}{8} \end{bmatrix}.$$

(a) The production matrix is

$$X = (I - A)^{-1}D = \begin{bmatrix} \frac{3}{2} & \frac{1}{4} \\ \frac{3}{4} & \frac{9}{8} \end{bmatrix}\begin{bmatrix} 1 \\ 1 \end{bmatrix} = \begin{bmatrix} \frac{7}{4} \\ \frac{15}{8} \end{bmatrix}.$$

Thus, $\frac{7}{4}$ bushels of yams and $\frac{15}{8} \approx 2$ pigs should be produced.

(b) The production matrix is

$$X = (I - A)^{-1}D = \begin{bmatrix} \frac{3}{2} & \frac{1}{4} \\ \frac{3}{4} & \frac{9}{8} \end{bmatrix} \begin{bmatrix} 100 \\ 70 \end{bmatrix} = \begin{bmatrix} 167.5 \\ 153.75 \end{bmatrix}.$$

Thus, 167.5 bushels of yams and $153.75 \approx 154$ pigs should be produced.

19. Use a graphing calculator or a computer to find the production matrix $X = (I - A)^{-1}D$. The answer is

$$\begin{bmatrix} 848 \\ 516 \\ 2970 \end{bmatrix}.$$

Values have been rounded.

Produce 848 units of agriculture, 516 units of manufacturing, and 2970 units of households.

21. Use a graphing calculator or a computer to find the production matrix $X = (I - A)^{-1}D$. The answer is

$$\begin{bmatrix} 195{,}492 \\ 25{,}933 \\ 13{,}580 \end{bmatrix}.$$

Values have been rounded. Change from thousands of pounds to millions of pounds.

Produce about 195 million Israeli pounds of agriculture, 26 million Israeli pounds of manufacturing, and 13.6 million Israeli pounds of energy.

23. Use a graphing calculator or a computer to find the production matrix $X = (I - A)^{-1}D$. The answer is

$$\begin{bmatrix} 532 \\ 481 \\ 805 \\ 1185 \end{bmatrix}.$$

Values have been rounded.

Produce about 532 units of natural resources, 481 manufacturing units, 805 trade and service units, and 1185 personal consumption units. Units are millions of dollars.

25. (a) Use a graphing calculator or a computer to find the matrix $(I - A)^{-1}$. The answer is

$$\begin{bmatrix} 1.67 & 0.56 & 0.56 \\ 0.19 & 1.17 & 0.06 \\ 3.15 & 3.27 & 4.38 \end{bmatrix}.$$

Values have been rounded.

(b) These multipliers imply that if the demand for one community's output increases by \$1 then the output in the other community will increase by the amount in the row and column of this matrix. For example, if the demand for Hermitage's output increases by \$1, then output from Sharon will increase \$0.56, from Farrell by \$0.06, and from Hermitage by \$4.38.

27. Calculate $I - A$, and then set $(I - A)X = O$ to find X.

$$(I - A)X = \left(\begin{bmatrix} 1 & 0 \\ 0 & 1 \end{bmatrix} - \begin{bmatrix} \frac{3}{4} & \frac{1}{3} \\ \frac{1}{4} & \frac{2}{3} \end{bmatrix} \right) \begin{bmatrix} x_1 \\ x_2 \end{bmatrix}$$

$$= \begin{bmatrix} \frac{1}{4} & -\frac{1}{3} \\ -\frac{1}{4} & \frac{1}{3} \end{bmatrix} \begin{bmatrix} x_1 \\ x_2 \end{bmatrix}$$

$$= \begin{bmatrix} \frac{1}{4}x_1 - \frac{1}{3}x_2 \\ -\frac{1}{4}x_1 + \frac{1}{3}x_2 \end{bmatrix} \begin{bmatrix} 0 \\ 0 \end{bmatrix}$$

Thus,

$$\frac{1}{4}x_1 - \frac{1}{3}x_2 = 0$$

$$\frac{1}{4}x_1 = \frac{1}{3}x_2$$

$$x_1 = \frac{4}{3}x_2$$

If $x_2 = 3, x_1 = 4$. Therefore, produce 4 units of steel for every 3 units of coal.

29. For this economy,

$$A = \begin{bmatrix} \frac{1}{5} & \frac{3}{5} & 0 \\ \frac{2}{5} & \frac{1}{5} & \frac{4}{5} \\ \frac{2}{5} & \frac{1}{5} & \frac{1}{5} \end{bmatrix}.$$

Find the value of $I - A$, then set $(I - A)X = O$.

$$(I - A)X = \left(\begin{bmatrix} 1 & 0 & 0 \\ 0 & 1 & 0 \\ 0 & 0 & 1 \end{bmatrix} - \begin{bmatrix} \frac{1}{5} & \frac{3}{5} & 0 \\ \frac{2}{5} & \frac{1}{5} & \frac{4}{5} \\ \frac{2}{5} & \frac{1}{5} & \frac{1}{5} \end{bmatrix} \right) \begin{bmatrix} x_1 \\ x_2 \\ x_3 \end{bmatrix}$$

$$= \begin{bmatrix} \frac{4}{5} & -\frac{3}{5} & 0 \\ -\frac{2}{5} & \frac{4}{5} & -\frac{4}{5} \\ -\frac{2}{5} & -\frac{1}{5} & \frac{4}{5} \end{bmatrix} \begin{bmatrix} x_1 \\ x_2 \\ x_3 \end{bmatrix}$$

$$= \begin{bmatrix} \frac{4}{5}x_1 - \frac{3}{5}x_2 \\ -\frac{2}{5}x_1 + \frac{4}{5}x_2 - \frac{4}{5}x_3 \\ -\frac{2}{5}x_1 - \frac{1}{5}x_2 + \frac{4}{5}x_3 \end{bmatrix} = \begin{bmatrix} 0 \\ 0 \\ 0 \end{bmatrix}$$

The system to be solved is

$$\begin{aligned} \frac{4}{5}x_1 - \frac{3}{5}x_2 \qquad &= 0 \\ -\frac{2}{5}x_1 + \frac{4}{5}x_2 - \frac{4}{5}x_3 &= 0 \\ -\frac{2}{5}x_1 - \frac{1}{5}x_2 + \frac{4}{5}x_3 &= 0. \end{aligned}$$

Write the augmented matrix of the system.

$$\left[\begin{array}{ccc|c} \frac{4}{5} & -\frac{3}{5} & 0 & 0 \\ -\frac{2}{5} & \frac{4}{5} & -\frac{4}{5} & 0 \\ -\frac{2}{5} & -\frac{1}{5} & \frac{4}{5} & 0 \end{array}\right]$$

$$\begin{array}{l} \frac{5}{4}R_1 \to R_1 \\ 5R_2 \to R_2 \\ 5R_3 \to R_3 \end{array} \left[\begin{array}{ccc|c} 1 & -\frac{3}{4} & 0 & 0 \\ -2 & 4 & -4 & 0 \\ -2 & -1 & 4 & 0 \end{array}\right]$$

$$\begin{array}{l} \\ 2R_1 + R_2 \to R_2 \\ 2R_1 + R_3 \to R_3 \end{array} \left[\begin{array}{ccc|c} 1 & -\frac{3}{4} & 0 & 0 \\ 0 & \frac{5}{2} & -4 & 0 \\ 0 & -\frac{5}{2} & 4 & 0 \end{array}\right]$$

$$\begin{array}{l} \\ \frac{2}{5}R_2 \to R_2 \\ \\ \end{array} \left[\begin{array}{ccc|c} 1 & -\frac{3}{4} & 0 & 0 \\ 0 & 1 & -\frac{8}{5} & 0 \\ 0 & -\frac{5}{2} & 4 & 0 \end{array}\right]$$

$$\begin{array}{l} \frac{3}{4}R_2 + R_1 \to R_1 \\ \\ \frac{5}{2}R_2 + R_3 \to R_3 \end{array} \left[\begin{array}{ccc|c} 1 & 0 & -\frac{6}{5} & 0 \\ 0 & 1 & -\frac{8}{5} & 0 \\ 0 & 0 & 0 & 0 \end{array}\right]$$

Use x_3 as the parameter. Therefore, $x_1 = \frac{6}{5}x_3$ and $x_2 = \frac{8}{5}x_3$, and the solution is $\left(\frac{6}{5}x_3, \frac{8}{5}x_3, x_3\right)$. If $x_3 = 5$, then $x_1 = 6$ and $x_2 = 8$.

Produce 6 units of mining for every 8 units of manufacturing and 5 units of communication.

Chapter 2 Review Exercises

1. False; a system with three equations and four unknowns has an infinite number of solutions.
2. False; matrix A is a 2×2 matrix and matrix B is a 3×2 matrix. Only matrices having the same dimension can be added.
3. True
4. True
5. False; only matrices having the same dimension can be added.
6. True
7. False; in general, matrix multiplication is not commutative.
8. False; only square matrices can have inverses.
9. True; for example, $0 \cdot A = A \cdot 0 = 0$
10. False; only row operations can be used.
11. False; any $n \times n$ zero matrix does not have an inverse, and the matrix $\begin{bmatrix} 2 & -4 \\ 1 & -2 \end{bmatrix}$ is an example of another square matrix that doesn't have an inverse.
12. False; if $AB = C$ and A has an inverse, then $B = A^{-1}C$.
13. True
14. True
15. False; $AB = CB$ implies $A = C$ only if B is the identity matrix or B has an inverse.
16. True
17. For a system of m linear equations in n unknowns and $m = n$, there could be one, none, or an infinite number of solutions. If $m < n$, there are an infinite number of solutions. If $m > n$, there could be one, none, or an infinite number of solutions.

19.
$$\begin{aligned} 2x - 3y &= 14 \quad (1) \\ 3x + 2y &= -5 \quad (2) \end{aligned}$$

Eliminate x in equation (2).

$$\begin{array}{lrl} & 2x - 3y &= 14 \quad (1) \\ -3R_1 + 2R_2 \to R_2 & 13y &= -52 \quad (3) \end{array}$$

Make each leading coefficient equal 1.

$$\begin{array}{lrl} \frac{1}{2}R_1 \to R_1 & x - \frac{3}{2}y = & 7 \quad (4) \\ \frac{1}{13}R_2 \to R_2 & y = & -4 \quad (5) \end{array}$$

Substitute –4 for y in equation (4) to get $x = 1$.

The solution is $(1, -4)$.

21. $2x - 3y + z = -5 \quad (1)$
$5x + 5y + 3z = 14 \quad (2)$

Eliminate x in equation (2).

$$\begin{array}{rl} & 2x - 3y + z = -5 \\ 5R_1 + (-2)R_2 \to R_2 & -25y - z = -53 \end{array}$$

Let z be the parameter. Solve for y and for x in terms of z.

$$\begin{aligned} -25y - z &= -53 \\ -25y &= -53 + z \\ y &= \frac{53 - z}{25} \end{aligned}$$

$$\begin{aligned} 2x - 3\left(\frac{53 - z}{25}\right) + z &= -5 \\ 2x - \frac{159}{25} + \frac{3z}{25} + z &= -5 \\ 2x + \frac{28}{25}z &= \frac{34}{25} \\ 2x &= \frac{34}{25} - \frac{28}{25}z \\ x &= \frac{34 - 28z}{50} \end{aligned}$$

The solutions are $\left(\frac{34 - 28z}{50}, \frac{53 - z}{25}, z\right)$, where z is any real number.

23. $2x + 4y = -6$
$-3x - 5y = 12$

Write the augmented matrix and use row operations.

$$\left[\begin{array}{rr|r} 2 & 4 & -6 \\ -3 & -5 & 12 \end{array}\right]$$

$$3R_1 + 2R_2 \to R_2 \quad \left[\begin{array}{rr|r} 2 & 4 & -6 \\ 0 & 2 & 6 \end{array}\right]$$

$$-2R_2 + R_1 \to R_1 \quad \left[\begin{array}{rr|r} 2 & 0 & -18 \\ 0 & 2 & 6 \end{array}\right]$$

$$\begin{array}{r} \frac{1}{2}R_1 \to R_1 \\ \frac{1}{2}R_2 \to R_2 \end{array} \quad \left[\begin{array}{rr|r} 1 & 0 & -9 \\ 0 & 1 & 3 \end{array}\right]$$

The solution is $(-9, 3)$.

25. $x - y + 3z = 13$
$4x + y + 2z = 17$
$3x + 2y + 2z = 1$

Write the augmented matrix and use row operations.

$$\left[\begin{array}{rrr|r} 1 & -1 & 3 & 13 \\ 4 & 1 & 2 & 17 \\ 3 & 2 & 2 & 1 \end{array}\right]$$

$$\begin{array}{r} \\ -4R_1 + R_2 \to R_2 \\ -3R_1 + R_3 \to R_3 \end{array} \left[\begin{array}{rrr|r} 1 & -1 & 3 & 13 \\ 0 & 5 & -10 & -35 \\ 0 & 5 & -7 & -38 \end{array}\right]$$

$$\begin{array}{r} R_2 + 5R_1 \to R_1 \\ \\ -1R_2 + R_3 \to R_3 \end{array} \left[\begin{array}{rrr|r} 5 & 0 & 5 & 30 \\ 0 & 5 & -10 & -35 \\ 0 & 0 & 3 & -3 \end{array}\right]$$

$$\begin{array}{r} 5R_3 + (-3R_1) \to R_1 \\ 10R_3 + 3R_2 \to R_2 \\ \\ \end{array} \left[\begin{array}{rrr|r} -15 & 0 & 0 & -105 \\ 0 & 15 & 0 & -135 \\ 0 & 0 & 3 & -3 \end{array}\right]$$

$$\begin{array}{r} -\frac{1}{15}R_1 \to R_1 \\ \frac{1}{15}R_2 \to R_2 \\ \frac{1}{3}R_3 \to R_3 \end{array} \left[\begin{array}{rrr|r} 1 & 0 & 0 & 7 \\ 0 & 1 & 0 & -9 \\ 0 & 0 & 1 & -1 \end{array}\right]$$

The solution is $(7, -9, -1)$.

27. $3x - 6y + 9z = 12$
$-x + 2y - 3z = -4$
$x + y + 2z = 7$

Write the augmented matrix and use row operations.

$$\left[\begin{array}{rrr|r} 3 & -6 & 9 & 12 \\ -1 & 2 & -3 & -4 \\ 1 & 1 & 2 & 7 \end{array}\right]$$

$$\begin{array}{r} \\ R_1 + 3R_2 \to R_2 \\ -1R_1 + 3R_3 \to R_3 \end{array} \left[\begin{array}{rrr|r} 3 & -6 & 9 & 12 \\ 0 & 0 & 0 & 0 \\ 0 & 9 & -3 & 9 \end{array}\right]$$

The zero in row 2, column 2 is an obstacle. To proceed, interchange the second and third rows.

$$\left[\begin{array}{rrr|r} 3 & -6 & 9 & 12 \\ 0 & 9 & -3 & 9 \\ 0 & 0 & 0 & 0 \end{array}\right]$$

$$3R_1 + 2R_2 \to R_1 \left[\begin{array}{rrr|r} 9 & 0 & 21 & 54 \\ 0 & 9 & -3 & 9 \\ 0 & 0 & 0 & 0 \end{array}\right]$$

$$\begin{array}{l} \frac{1}{9}R_1 \to R_1 \\ \frac{1}{9}R_2 \to R_2 \\ \end{array} \left[\begin{array}{ccc|c} 1 & 0 & \frac{7}{3} & 6 \\ 0 & 1 & -\frac{1}{3} & 1 \\ 0 & 0 & 0 & 0 \end{array}\right]$$

The row of zeros indicates dependent equations. Solve the first two equations respectively for x and y in terms of z to obtain

$$x = 6 - \frac{7}{3}z \text{ and } y = 1 + \frac{1}{3}z$$

The solution of the system is

$$\left(6 - \frac{7}{3}z, 1 + \frac{1}{3}z, z\right),$$

where z is any real number.

In Exercises 29–32, corresponding elements must be equal.

29. $\begin{bmatrix} 2 & 3 \\ 5 & q \end{bmatrix} = \begin{bmatrix} a & b \\ c & 9 \end{bmatrix}$

Size: 2×2; $a = 2, b = 3, c = 5, q = 9$; square matrix

31. $\begin{bmatrix} 2m & 4 & 3z & -12 \end{bmatrix}$

$= \begin{bmatrix} 12 & k+1 & -9 & r-3 \end{bmatrix}$

Size: 1×4; $m = 6, k = 3, z = -3, r = -9$; row matrix

33. $A + C = \begin{bmatrix} 4 & 10 \\ -2 & -3 \\ 6 & 9 \end{bmatrix} + \begin{bmatrix} 5 & 0 \\ -1 & 3 \\ 4 & 7 \end{bmatrix}$

$= \begin{bmatrix} 9 & 10 \\ -3 & 0 \\ 10 & 16 \end{bmatrix}$

35. $3C + 2A = 3\begin{bmatrix} 5 & 0 \\ -1 & 3 \\ 4 & 7 \end{bmatrix} + 2\begin{bmatrix} 4 & 10 \\ -2 & -3 \\ 6 & 9 \end{bmatrix}$

$= \begin{bmatrix} 15 & 0 \\ -3 & 9 \\ 12 & 21 \end{bmatrix} + \begin{bmatrix} 8 & 20 \\ -4 & -6 \\ 12 & 18 \end{bmatrix}$

$= \begin{bmatrix} 23 & 20 \\ -7 & 3 \\ 24 & 39 \end{bmatrix}$

37. $2A - 5C = 2\begin{bmatrix} 4 & 10 \\ -2 & -3 \\ 6 & 9 \end{bmatrix} - 5\begin{bmatrix} 5 & 0 \\ -1 & 3 \\ 4 & 7 \end{bmatrix}$

$= \begin{bmatrix} 8 & 20 \\ -4 & -6 \\ 12 & 18 \end{bmatrix} - \begin{bmatrix} 25 & 0 \\ -5 & 15 \\ 20 & 35 \end{bmatrix}$

$= \begin{bmatrix} -17 & 20 \\ 1 & -21 \\ -8 & -17 \end{bmatrix}$

39. A is 3×2 and C is 3×2, so finding the product AC is not possible.

$$\begin{array}{cc} A & C \\ 3 \times \mathbf{2} & \mathbf{3} \times 2 \end{array}$$

(The inner two numbers must match.)

41. $ED = [1 \quad 3 \quad -4]\begin{bmatrix} 6 \\ 1 \\ 0 \end{bmatrix}$

$= [1.6 + 3.1 + (-4)0] = [9]$

43. $EC = \begin{bmatrix} 1 & 3 & -4 \end{bmatrix}\begin{bmatrix} 5 & 0 \\ -1 & 3 \\ 4 & 7 \end{bmatrix}$

$= [1 \cdot 5 + 3(-1) + (-4) \cdot 4 \quad 1 \cdot 0$
$+ 3 \cdot 3 + (-4) \cdot 7]$
$= \begin{bmatrix} -14 & -19 \end{bmatrix}$

45. Find the inverse of $B = \begin{bmatrix} 2 & 3 & -2 \\ 2 & 4 & 0 \\ 0 & 1 & 2 \end{bmatrix}$, if it exists.

Write the augmented matrix to obtain

$$\begin{bmatrix} B|I \end{bmatrix} = \left[\begin{array}{ccc|ccc} 2 & 3 & -2 & 1 & 0 & 0 \\ 2 & 4 & 0 & 0 & 1 & 0 \\ 0 & 1 & 2 & 0 & 0 & 1 \end{array}\right].$$

$$-1R_1 + R_2 \to R_2 \left[\begin{array}{ccc|ccc} 2 & 3 & -2 & 1 & 0 & 0 \\ 0 & 1 & 2 & -1 & 1 & 0 \\ 0 & 1 & 2 & 0 & 0 & 1 \end{array}\right]$$

$$\begin{array}{l} -3R_2 + R_1 \to R_1 \\ \\ -1R_2 + R_3 \to R_3 \end{array} \left[\begin{array}{ccc|ccc} 2 & 0 & -8 & 4 & -3 & 0 \\ 0 & 1 & 2 & -1 & 1 & 0 \\ 0 & 0 & 0 & 1 & -1 & 1 \end{array}\right]$$

No inverse exists, since the third row is all zeros to the left of the vertical bar.

47. Find the inverse of $A = \begin{bmatrix} 1 & 3 \\ 2 & 7 \end{bmatrix}$, if it exists.

Write the augmented matrix $[A|I]$.

$$[A \mid I] = \left[\begin{array}{cc|cc} 1 & 3 & 1 & 0 \\ 2 & 7 & 0 & 1 \end{array}\right]$$

Perform row operations on $[A|I]$ to get a matrix of the form $[I|B]$.

$$-2R_1 + R_2 \to R_2 \quad \left[\begin{array}{cc|cc} 1 & 3 & 1 & 0 \\ 0 & 1 & -2 & 1 \end{array}\right]$$

$$-3R_2 + R_1 \to R_1 \quad \left[\begin{array}{cc|cc} 1 & 0 & 7 & -3 \\ 0 & 1 & -2 & 1 \end{array}\right]$$

The last augmented matrix is of the form $[I|B]$, so the desired inverse is

$$A^{-1} = \begin{bmatrix} 7 & -3 \\ -2 & 1 \end{bmatrix}.$$

49. Find the inverse of $A = \begin{bmatrix} 3 & -6 \\ -4 & 8 \end{bmatrix}$, if it exists.

Write the augmented matrix $[A|I]$.

$$[A|I] = \left[\begin{array}{cc|cc} 3 & -6 & 1 & 0 \\ -4 & 8 & 0 & 1 \end{array}\right]$$

Perform row operations on $[A|I]$ to get a matrix of the form $[I|B]$.

$$4R_1 + 3R_2 \to R_2 \quad \left[\begin{array}{cc|cc} 3 & -6 & 1 & 0 \\ 0 & 0 & 4 & 3 \end{array}\right]$$

Since the entries left of the vertical bar in the second row are zeros, no inverse exists.

51. Find the inverse of $A = \begin{bmatrix} 2 & -1 & 0 \\ 1 & 0 & 1 \\ 1 & -2 & 0 \end{bmatrix}$, if it exists.

The augmented matrix is

$$[A|I] = \left[\begin{array}{ccc|ccc} 2 & -1 & 0 & 1 & 0 & 0 \\ 1 & 0 & 1 & 0 & 1 & 0 \\ 1 & -2 & 0 & 0 & 0 & 1 \end{array}\right].$$

$$\begin{array}{r} \\ R_1 + (-2)R_2 \to R_2 \\ R_1 + (-2)R_3 \to R_3 \end{array} \left[\begin{array}{ccc|ccc} 2 & -1 & 0 & 1 & 0 & 0 \\ 0 & -1 & -2 & 1 & -2 & 0 \\ 0 & 3 & 0 & 1 & 0 & -2 \end{array}\right]$$

$$\begin{array}{r} -1R_2 + R_1 \to R_1 \\ \\ 3R_2 + R_3 \to R_3 \end{array} \left[\begin{array}{ccc|ccc} 2 & 0 & 2 & 0 & 2 & 0 \\ 0 & -1 & -2 & 1 & -2 & 0 \\ 0 & 0 & -6 & 4 & -6 & -2 \end{array}\right]$$

$$\begin{array}{r} R_3 + 3R_1 \to R_1 \\ R_3 + (-3)R_2 \to R_2 \\ \\ \end{array} \left[\begin{array}{ccc|ccc} 6 & 0 & 0 & 4 & 0 & -2 \\ 0 & 3 & 0 & 1 & 0 & -2 \\ 0 & 0 & -6 & 4 & -6 & -2 \end{array}\right]$$

$$\begin{array}{r} \frac{1}{6}R_1 \to R_1 \\ \frac{1}{3}R_2 \to R_2 \\ -\frac{1}{6}R_3 \to R_3 \end{array} \left[\begin{array}{ccc|ccc} 1 & 0 & 0 & \frac{2}{3} & 0 & -\frac{1}{3} \\ 0 & 1 & 0 & \frac{1}{3} & 0 & -\frac{2}{3} \\ 0 & 0 & 1 & -\frac{2}{3} & 1 & \frac{1}{3} \end{array}\right]$$

$$A^{-1} = \begin{bmatrix} \frac{2}{3} & 0 & -\frac{1}{3} \\ \frac{1}{3} & 0 & -\frac{2}{3} \\ -\frac{2}{3} & 1 & \frac{1}{3} \end{bmatrix}$$

53. Find the inverse of $A = \begin{bmatrix} 1 & 3 & 6 \\ 4 & 0 & 9 \\ 5 & 15 & 30 \end{bmatrix}$, if it exists

$$[A|I] = \left[\begin{array}{ccc|ccc} 1 & 3 & 6 & 1 & 0 & 0 \\ 4 & 0 & 9 & 0 & 1 & 0 \\ 5 & 15 & 30 & 0 & 0 & 1 \end{array}\right]$$

$$\begin{array}{r} \\ -4R_1 + R_2 \to R_2 \\ -5R_1 + R_3 \to R_3 \end{array} \left[\begin{array}{ccc|ccc} 1 & 3 & 6 & 1 & 0 & 0 \\ 0 & -12 & -15 & -4 & 1 & 0 \\ 0 & 0 & 0 & -5 & 0 & 1 \end{array}\right]$$

The last row is all zeros to the left of the bar, so no inverse exists.

55. $A = \begin{bmatrix} 5 & 1 \\ -1 & -2 \end{bmatrix}, B = \begin{bmatrix} -8 \\ 24 \end{bmatrix}$

The matrix equation to be solved is $AX = B$, or

$$\begin{bmatrix} 5 & 1 \\ -1 & -2 \end{bmatrix}\begin{bmatrix} x \\ y \end{bmatrix} = \begin{bmatrix} -8 \\ 24 \end{bmatrix}.$$

Calculate the inverse of the coefficient matrix A to obtain

$$\begin{bmatrix} 5 & 1 \\ -1 & -2 \end{bmatrix}^{-1} = \begin{bmatrix} \frac{1}{4} & \frac{1}{8} \\ -\frac{1}{4} & -\frac{5}{8} \end{bmatrix}.$$

Now $X = A^{-1}B$, so

$$\begin{bmatrix} x \\ y \end{bmatrix} = \begin{bmatrix} \frac{1}{4} & \frac{1}{8} \\ -\frac{1}{4} & -\frac{5}{8} \end{bmatrix}\begin{bmatrix} -8 \\ 24 \end{bmatrix} = \begin{bmatrix} 1 \\ -13 \end{bmatrix}.$$

57. $A = \begin{bmatrix} 1 & 0 & 2 \\ -1 & 1 & 0 \\ 3 & 0 & 4 \end{bmatrix}, B = \begin{bmatrix} 8 \\ 4 \\ -6 \end{bmatrix}$

By the usual method, we find that the inverse of the coefficient matrix is

$$A^{-1} = \begin{bmatrix} -2 & 0 & 1 \\ -2 & 1 & 1 \\ \frac{3}{2} & 0 & -\frac{1}{2} \end{bmatrix}.$$

Since $X = A^{-1}B$,

$$X = \begin{bmatrix} -2 & 0 & 1 \\ -2 & 1 & 1 \\ \frac{3}{2} & 0 & -\frac{1}{2} \end{bmatrix} \begin{bmatrix} 8 \\ 4 \\ -6 \end{bmatrix} = \begin{bmatrix} -22 \\ -18 \\ 15 \end{bmatrix}.$$

59. $x + 2y = 4$
$2x - 3y = 1$

The coefficient matrix is

$$A = \begin{bmatrix} 1 & 2 \\ 2 & -3 \end{bmatrix}.$$

Calculate the inverse of A.

$$A^{-1} = \begin{bmatrix} \frac{3}{7} & \frac{2}{7} \\ \frac{2}{7} & -\frac{1}{7} \end{bmatrix}$$

Use $X = A^{-1}B$ to solve.

$$\begin{bmatrix} x \\ y \end{bmatrix} = \begin{bmatrix} \frac{3}{7} & \frac{2}{7} \\ \frac{2}{7} & -\frac{1}{7} \end{bmatrix} \begin{bmatrix} 4 \\ 1 \end{bmatrix} = \begin{bmatrix} 2 \\ 1 \end{bmatrix}$$

The solution is (2, 1).

61. $x + y + z = 1$
$2x + y = -2$
$3y + z = 2$

The coefficient matrix is

$$A = \begin{bmatrix} 1 & 1 & 1 \\ 2 & 1 & 0 \\ 0 & 3 & 1 \end{bmatrix}.$$

Find that the inverse of A is

$$A^{-1} = \begin{bmatrix} \frac{1}{5} & \frac{2}{5} & -\frac{1}{5} \\ -\frac{2}{5} & \frac{1}{5} & \frac{2}{5} \\ \frac{6}{5} & -\frac{3}{5} & -\frac{1}{5} \end{bmatrix}.$$

Now $X = A^{-1}B$, so

$$\begin{bmatrix} x \\ y \\ z \end{bmatrix} = \begin{bmatrix} \frac{1}{5} & \frac{2}{5} & -\frac{1}{5} \\ -\frac{2}{5} & \frac{1}{5} & \frac{2}{5} \\ \frac{6}{5} & -\frac{3}{5} & -\frac{1}{5} \end{bmatrix} \begin{bmatrix} 1 \\ -2 \\ 2 \end{bmatrix} = \begin{bmatrix} -1 \\ 0 \\ 2 \end{bmatrix}.$$

The solution is $(-1, 0, 2)$.

63. $A = \begin{bmatrix} 0.01 & 0.05 \\ 0.04 & 0.03 \end{bmatrix}, D = \begin{bmatrix} 200 \\ 300 \end{bmatrix}$

$$X = (I - A)^{-1}D$$

$$I - A = \begin{bmatrix} 1 & 0 \\ 0 & 1 \end{bmatrix} - \begin{bmatrix} 0.01 & 0.05 \\ 0.04 & 0.03 \end{bmatrix} = \begin{bmatrix} 0.99 & -0.05 \\ -0.04 & 0.97 \end{bmatrix}$$

Use row operations to find the inverse of $I - A$, which is

$$(I - A)^{-1} = \begin{bmatrix} 1.0122 & 0.0522 \\ 0.0417 & 1.0331 \end{bmatrix}.$$

Since $X = (I - A)^{-1}D$, the production matrix is

$$X = \begin{bmatrix} 1.0122 & 0.0522 \\ 0.0417 & 1.0331 \end{bmatrix} \begin{bmatrix} 200 \\ 300 \end{bmatrix} = \begin{bmatrix} 218.1 \\ 318.3 \end{bmatrix}.$$

65. $x + 2y + z = 7$ (1)
$2x - y - z = 2$ (2)
$3x - 3y + 2z = -5$ (3)

(a) To solve the system by the echelon method, begin by eliminating x in equations (2) and (3).

$x + 2y + z = 7$ (1)
$-2R_1 + R_2 \rightarrow R_2$ $\quad -5y - 3z = -12$ (4)
$-3R_1 + R_3 \rightarrow R_3$ $\quad -9y - z = -26$ (5)

Eliminate y in equation (5).

$x + 2y + z = 7$ (1)
$-5y - 3z = -12$ (4)
$-9R_2 + 5R_3 \rightarrow R_3$ $\quad 22z = -22$ (6)

Make each leading coefficient equal 1.

$$\begin{aligned} & x + 2y + z = 7 && (1) \\ -\tfrac{1}{5}R_2 \to R_2 \quad & y + \tfrac{3}{5}z = \tfrac{12}{5} && (7) \\ \tfrac{1}{22}R_3 \to R_3 \quad & z = -1 && (8) \end{aligned}$$

Substitute -1 for z in equation (7) to get $y = 3$. Substitute -1 for z and 3 for y in equation (1) to get $x = 2$.

The solution is $(2, 3, -1)$.

(b) The same system is to be solved using the Gauss-Jordan method. Write the augmented matrix and use row operations.

$$\left[\begin{array}{ccc|c} 1 & 2 & 1 & 7 \\ 2 & -1 & -1 & 2 \\ 3 & -3 & 2 & -5 \end{array}\right]$$

$$\begin{array}{r} \\ -2R_1 + R_2 \to R_2 \\ -3R_1 + R_3 \to R_3 \end{array} \left[\begin{array}{ccc|c} 1 & 2 & 1 & 7 \\ 0 & -5 & -3 & -12 \\ 0 & -9 & -1 & -26 \end{array}\right]$$

$$\begin{array}{r} 2R_2 + 5R_1 \to R_1 \\ \\ -9R_2 + 5R_3 \to R_3 \end{array} \left[\begin{array}{ccc|c} 5 & 0 & -1 & 11 \\ 0 & -5 & -3 & -12 \\ 0 & 0 & 22 & -22 \end{array}\right]$$

$$\begin{array}{r} R_3 + 22R_1 \to R_1 \\ 3R_3 + 22R_2 \to R_2 \\ \\ \end{array} \left[\begin{array}{ccc|c} 110 & 0 & 0 & 220 \\ 0 & -110 & 0 & -330 \\ 0 & 0 & 22 & -22 \end{array}\right]$$

$$\begin{array}{r} \frac{1}{110}R_1 + R_1 \\ -\frac{1}{110}R_2 \to R_2 \\ \frac{1}{22}R_3 \to R_3 \end{array} \left[\begin{array}{ccc|c} 1 & 0 & 0 & 2 \\ 0 & 1 & 0 & 3 \\ 0 & 0 & 1 & -1 \end{array}\right]$$

The corresponding system is $\begin{aligned} x &= 2 \\ y &= 3 \\ z &= -1. \end{aligned}$

The solution is $(2, 3, -1)$

(c) The system can be written as a matrix equation $AX = B$ by writing

$$\begin{bmatrix} 1 & 2 & 1 \\ 2 & -1 & -1 \\ 3 & -3 & 2 \end{bmatrix} \begin{bmatrix} x \\ y \\ z \end{bmatrix} = \begin{bmatrix} 7 \\ 2 \\ -5 \end{bmatrix}.$$

(d) The inverse of the coefficient matrix A can be found by using row operations.

$$[A|I] = \left[\begin{array}{ccc|ccc} 1 & 2 & 1 & 1 & 0 & 0 \\ 2 & -1 & -1 & 0 & 1 & 0 \\ 3 & -3 & 2 & 0 & 0 & 1 \end{array}\right]$$

$$\begin{array}{r} \\ -2R_1 + R_2 \to R_2 \\ -3R_1 + R_3 \to R_3 \end{array} \left[\begin{array}{ccc|ccc} 1 & 2 & 1 & 1 & 0 & 0 \\ 0 & -5 & -3 & -2 & 1 & 0 \\ 0 & -9 & -1 & -3 & 0 & 1 \end{array}\right]$$

$$\begin{array}{r} 2R_2 + 5R_1 \to R_1 \\ \\ -9R_2 + 5R_3 \to R_3 \end{array} \left[\begin{array}{ccc|ccc} 5 & 0 & -1 & 1 & 2 & 0 \\ 0 & -5 & -3 & -2 & 1 & 0 \\ 0 & 0 & 22 & 3 & -9 & 5 \end{array}\right]$$

$$\begin{array}{r} R_3 + 22R_1 \to R_1 \\ 3R_3 + 22R_2 \to R_2 \\ \\ \end{array} \left[\begin{array}{ccc|ccc} 110 & 0 & 0 & 25 & 35 & 5 \\ 0 & -110 & 0 & -35 & -5 & 15 \\ 0 & 0 & 22 & 3 & -9 & 5 \end{array}\right]$$

$$\begin{array}{r} \frac{1}{110}R_1 + R_1 \\ -\frac{1}{110}R_2 \to R_2 \\ \frac{1}{22}R_3 \to R_3 \end{array} \left[\begin{array}{ccc|ccc} 1 & 0 & 0 & \frac{5}{22} & \frac{7}{22} & \frac{1}{22} \\ 0 & 1 & 0 & \frac{7}{22} & \frac{1}{22} & -\frac{3}{22} \\ 0 & 0 & 1 & \frac{3}{22} & -\frac{9}{22} & \frac{5}{22} \end{array}\right]$$

The inverse of matrix A is

$$A^{-1} = \begin{bmatrix} \frac{5}{22} & \frac{7}{22} & \frac{1}{22} \\ \frac{7}{22} & \frac{1}{22} & -\frac{3}{22} \\ \frac{3}{22} & -\frac{9}{22} & \frac{5}{22} \end{bmatrix} \approx \begin{bmatrix} 0.23 & 0.32 & 0.05 \\ 0.32 & 0.05 & -0.14 \\ 0.14 & -0.41 & 0.23 \end{bmatrix}.$$

(e) Since $X = A^{-1}B$,

$$\begin{bmatrix} x \\ y \\ z \end{bmatrix} = \begin{bmatrix} \frac{5}{22} & \frac{7}{22} & \frac{1}{22} \\ \frac{7}{22} & \frac{1}{22} & -\frac{3}{22} \\ \frac{3}{22} & -\frac{9}{22} & \frac{5}{22} \end{bmatrix} \begin{bmatrix} 7 \\ 2 \\ -5 \end{bmatrix} = \begin{bmatrix} 2 \\ 3 \\ -1 \end{bmatrix}.$$

Once again, the solution is $(2, 3, -1)$.

67. Let $x_1 =$ the number of blankets,

$x_2 =$ the number of rugs, and

$x_3 =$ the number of skirts.

The given information leads to the system

$$\begin{aligned} 24x_1 + 30x_2 + 12x_3 &= 306 && (1) \\ 4x_1 + 5x_2 + 3x_3 &= 59 && (2) \\ 15x_1 + 18x_2 + 9x_3 &= 201. && (3) \end{aligned}$$

Simplify equations (1) and (3).

$$\frac{1}{6}R_1 \to R_1 \quad 4x_1 + 5x_2 + 2x_3 = 51 \quad (4)$$
$$4x_1 + 5x_2 + 3x_3 = 59 \quad (2)$$
$$\frac{1}{3}R_3 \to R_3 \quad 5x_1 + 6x_2 + 3x_3 = 67 \quad (5)$$

Solve this system by the Gauss-Jordan method. Write the augmented matrix and use row operations.

$$\left[\begin{array}{ccc|c} 4 & 5 & 2 & 51 \\ 4 & 5 & 3 & 59 \\ 5 & 6 & 3 & 67 \end{array}\right]$$

$$\begin{array}{r} \\ -1R_1 + R_2 \to R_2 \\ -4R_3 + 5R_1 \to R_3 \end{array} \left[\begin{array}{ccc|c} 4 & 5 & 2 & 51 \\ 0 & 0 & 1 & 8 \\ 0 & 1 & -2 & -13 \end{array}\right]$$

Interchange the second and third rows.

$$\left[\begin{array}{ccc|c} 4 & 5 & 2 & 51 \\ 0 & 1 & -2 & -13 \\ 0 & 0 & 1 & 8 \end{array}\right]$$

$$-5R_2 + R_1 \to R_1 \left[\begin{array}{ccc|c} 4 & 0 & 12 & 116 \\ 0 & 1 & -2 & -13 \\ 0 & 0 & 1 & 8 \end{array}\right]$$

$$\begin{array}{r} -12R_3 + R_1 \to R_1 \\ 2R_3 + R_2 \to R_2 \\ \\ \end{array} \left[\begin{array}{ccc|c} 4 & 0 & 0 & 20 \\ 0 & 1 & 0 & 3 \\ 0 & 0 & 1 & 8 \end{array}\right]$$

$$\frac{1}{4}R_1 \to R_1 \left[\begin{array}{ccc|c} 1 & 0 & 0 & 5 \\ 0 & 1 & 0 & 3 \\ 0 & 0 & 1 & 8 \end{array}\right]$$

The solution of the system is $x = 5$, $y = 3$, $z = 8$. So, 5 blankets, 3 rugs, and 8 skirts can be made.

69. The 4×5 matrix of stock reports is

	div	*ratio*	*sales*	*price*	*change*
AT & T	1.33	17.6	152,000	26.75	1.88
GE	1.00	20	238,200	32.36	−1.50
SaraLee	0.79	25.4	39,110	16.51	−0.89
Disney	0.27	21.2	122,500	28.60	0.75

71. (a) The input-output matrix is

$$A = \begin{bmatrix} 0 & \frac{1}{2} \\ \frac{2}{3} & 0 \end{bmatrix}.$$

(b) $I - A = \begin{bmatrix} 1 & -\frac{1}{2} \\ -\frac{2}{3} & 1 \end{bmatrix}$, $D = \begin{bmatrix} 400 \\ 800 \end{bmatrix}$

Use row operations to find the inverse of $I - A$, which is

$$(I - A)^{-1} = \begin{bmatrix} \frac{3}{2} & \frac{3}{4} \\ 1 & \frac{3}{2} \end{bmatrix}.$$

Since $X = (I - A)^{-1}D$,

$$X = \begin{bmatrix} \frac{3}{2} & \frac{3}{4} \\ 1 & \frac{3}{2} \end{bmatrix} \begin{bmatrix} 400 \\ 800 \end{bmatrix} = \begin{bmatrix} 1200 \\ 1600 \end{bmatrix}.$$

The production required is 1200 units of cheese and 1600 units of goats.

73. The given information can be written as the following 4×3 matrix.

$$\begin{bmatrix} 8 & 8 & 8 \\ 10 & 5 & 9 \\ 7 & 10 & 7 \\ 8 & 9 & 7 \end{bmatrix}$$

75. (a) $a + b = 0.60 \quad (1)$

$c + d = 0.75 \quad (2)$

$a + c = 0.65 \quad (3)$

$b + d = 0.70 \quad (4)$

The augmented matrix of the system is

$$\left[\begin{array}{cccc|c} 1 & 1 & 0 & 0 & 0.60 \\ 0 & 0 & 1 & 1 & 0.75 \\ 1 & 0 & 1 & 0 & 0.65 \\ 0 & 1 & 0 & 1 & 0.70 \end{array}\right].$$

$$\begin{array}{r} \\ \\ -1R_1 + R_3 \to R_3 \\ \\ \end{array} \left[\begin{array}{cccc|c} 1 & 1 & 0 & 0 & 0.60 \\ 0 & 0 & 1 & 1 & 0.75 \\ 0 & -1 & 1 & 0 & 0.05 \\ 0 & 1 & 0 & 1 & 0.70 \end{array}\right]$$

Interchange rows 2 and 4.

$$\left[\begin{array}{cccc|c} 1 & 1 & 0 & 0 & 0.60 \\ 0 & 1 & 0 & 1 & 0.70 \\ 0 & -1 & 1 & 0 & 0.05 \\ 0 & 0 & 1 & 1 & 0.75 \end{array}\right]$$

$$\begin{array}{r} -1R_2 + R_1 \to R_1 \\ \\ R_2 + R_3 \to R_3 \\ \\ \end{array} \left[\begin{array}{cccc|c} 1 & 0 & 0 & -1 & -0.10 \\ 0 & 1 & 0 & 1 & 0.70 \\ 0 & 0 & 1 & 1 & 0.75 \\ 0 & 0 & 1 & 1 & 0.75 \end{array}\right]$$

Since R_3 and R_4 are identical, there will be infinitely many solutions. We do not have enough information to determine the values of a, b, c, and d.

(b) i. If $d = 0.33$, the system of equations in part (a) becomes

$$\begin{aligned} a + b &= 0.60 \quad (1) \\ c + 0.33 &= 0.75 \quad (2) \\ a + c &= 0.65 \quad (3) \\ b + 0.33 &= 0.70. \quad (4) \end{aligned}$$

Equation (2) gives $c = 0.42$, and equation (4) gives $b = 0.37$. Substituting $c = 0.42$ into equation (3) gives $a = 0.23$. Therefore, $a = 0.23$, $b = 0.37$, $c = 0.42$, and $d = 0.33$.

Thus, A is healthy, B and D are tumorous, and C is bone.

ii. If $d = 0.43$, the system of equations in part (a) becomes

$$\begin{aligned} a + b &= 0.60 \quad (1) \\ c + 0.43 &= 0.75 \quad (2) \\ a + c &= 0.65 \quad (3) \\ b + 0.43 &= 0.70. \quad (4) \end{aligned}$$

Equation (2) gives $c = 0.32$, and equation (4) gives $b = 0.27$. Substituting $c = 0.32$ into equation (3) gives $a = 0.33$. Therefore, $a = 0.33$, $b = 0.27$, $c = 0.32$, and $d = 0.43$.

Thus, A and C are tumorous, B could be healthy or tumorous, and D is bone.

(c) The original system now has two additional equations.

$$\begin{aligned} a + b &= 0.60 \quad (1) \\ c + d &= 0.75 \quad (2) \\ a + c &= 0.65 \quad (3) \\ b + d &= 0.70 \quad (4) \\ b + c &= 0.85 \quad (5) \\ a + d &= 0.50 \quad (6) \end{aligned}$$

The augmented matrix of this system is

$$\left[\begin{array}{cccc|c} 1 & 1 & 0 & 0 & 0.60 \\ 0 & 0 & 1 & 1 & 0.75 \\ 1 & 0 & 1 & 0 & 0.65 \\ 0 & 1 & 0 & 1 & 0.70 \\ 0 & 1 & 1 & 0 & 0.85 \\ 1 & 0 & 0 & 1 & 0.50 \end{array}\right].$$

Using the Gauss-Jordan method we obtain

$$\left[\begin{array}{cccc|c} 1 & 0 & 0 & 0 & 0.20 \\ 0 & 1 & 0 & 0 & 0.40 \\ 0 & 0 & 1 & 0 & 0.45 \\ 0 & 0 & 0 & 1 & 0.30 \\ 0 & 0 & 0 & 0 & 0 \\ 0 & 0 & 0 & 0 & 0 \end{array}\right].$$

Therefore, $a = 0.20$, $b = 0.40$, $c = 0.45$, and $d = 0.30$. Thus, A is healthy, B and C are bone, and D is tumorous.

(d) As we saw in part (c), the six equations reduced to four independent equations. We need only four beams, correctly chosen, to obtain a solution. The four beams must pass through all four cells and must lead to independent equations. One such choice would be beams 1, 2, 3, and 6. Another choice would be beams 1, 2, 4, and 5.

77. $$\begin{aligned} \frac{\sqrt{3}}{2}(W_1 + W_2) &= 100 \quad (1) \\ W_1 - W_2 &= 0 \quad (2) \end{aligned}$$

Equation (2) gives $W_1 = W_2$. Substitute W_1 for W_2 in equation (1).

$$\begin{aligned} \frac{\sqrt{3}}{2}(W_1 + W_1) &= 100 \\ \frac{\sqrt{3}}{2}(2W_1) &= 100 \\ \sqrt{3}W_1 &= 100 \\ W_1 &= \frac{100}{\sqrt{3}} = \frac{100\sqrt{3}}{3} \approx 58 \end{aligned}$$

Therefore, $W_1 = W_2 \approx 58$ lb.

79. $C = at^2 + by + c$

Use the values for C from the table.

(a) For 1960, $t = 0$ and $C = 317$.

$$\begin{aligned} 317 &= a(0)^2 + b(0) + c \\ 317 &= c \end{aligned}$$

For 1980, $t = 20$ and $C = 339$.

$$\begin{aligned} 339 &= a(20)^2 + b(20) + 317 \\ 22 &= 400a + 20b \end{aligned}$$

For 2004, $t = 44$ and $C = 377$.

$$\begin{aligned} 377 &= a(44)^2 + b(44) + 317 \\ 60 &= 1936a + 44b \end{aligned}$$

Thus, we need to solve the system

$$\begin{bmatrix} 400 & 20 \\ 1936 & 44 \end{bmatrix} = \begin{bmatrix} 22 \\ 60 \end{bmatrix}$$

$$\begin{bmatrix} a \\ b \end{bmatrix} = \begin{bmatrix} 400 & 20 \\ 1936 & 44 \end{bmatrix} \begin{bmatrix} 22 \\ 60 \end{bmatrix}$$

$$= \begin{bmatrix} -\frac{1}{480} & \frac{1}{1056} \\ \frac{11}{120} & -\frac{5}{264} \end{bmatrix} \begin{bmatrix} 22 \\ 60 \end{bmatrix}$$

$$= \begin{bmatrix} \frac{29}{2640} \\ \frac{581}{660} \end{bmatrix} \approx \begin{bmatrix} 0.010985 \\ 0.8830 \end{bmatrix}$$

Therefore,

$C = 0.010985t^2 + 0.8803t + 317.$

(b) In 1960, $C = 317$. So, double that level would be $C = 634$.

$$634 = 0.010985t^2 + 0.8803t + 317$$
$$0 = 0.010985t^2 + 0.8803t - 317$$

Multiply this equation by 2640 to clear the decimal values.

$$0 = 29t^2 + 2324t - 836{,}880$$

Use the quadratic formula with $a = 29$, $b = 2324$, and $c = -836{,}880$.

$$t = \frac{-2324 \pm \sqrt{2324^2 - 4(29)(-836{,}880)}}{2(29)}$$
$$= \frac{-2324 \pm \sqrt{5{,}400{,}976 - (-97{,}078{,}080)}}{2(29)}$$
$$\approx \frac{-2324 \pm 10{,}123}{58}$$
$$\approx 134.5 \text{ or} - 214.6$$

Ignore the negative value. If $t = 134.5$, then $1960 + 134.5 = 2094.5$. The 1960 CO_2 level will double in the year 2095.

81. Let $x =$ the number of boys and $y =$ the number of girls.

$$0.2x + 0.3y = 500 \quad (1)$$
$$0.6x + 0.9y = 1500 \quad (2)$$

The augmented matrix is

$$\left[\begin{array}{cc|c} 0.2 & 0.3 & 500 \\ 0.6 & 0.9 & 1500 \end{array}\right].$$

$$5R_1 \rightarrow R_1 \left[\begin{array}{cc|c} 1 & 1.5 & 2500 \\ 0.6 & 0.9 & 1500 \end{array}\right]$$

$$-0.6R_1 + R_2 \rightarrow R_2 \left[\begin{array}{cc|c} 1 & 1.5 & 2500 \\ 0 & 0 & 0 \end{array}\right]$$

Thus,

$$x + 1.5y = 2500$$
$$x = 2500 - 1.5y.$$

There are y girls and $2500 - 1.5y$ boys, where y is any even integer between 0 and 1666 since $y \geq 0$ and

$$2500 - 1.5y \geq 0$$
$$-1.5y \geq -2500$$
$$y \leq 1666.\overline{6}.$$

83. Let $x =$ the weight of a single chocolate wafer and
$y =$ the weight of a single layer of vanilla creme.

A serving of regular Oreo cookies is three cookies so that $3(2x + y) = 34$.

A serving of Double Stuf is two cookies so that $2(2x + 2y) = 29$.

Write the equations in proper form, obtain the augmented matrix, and use row operations to solve.

$$\left[\begin{array}{cc|c} 6 & 3 & 34 \\ 4 & 4 & 29 \end{array}\right]$$

$$-2R_1 + 3R_2 \rightarrow R_2 \left[\begin{array}{cc|c} 6 & 3 & 34 \\ 0 & 6 & 19 \end{array}\right]$$

$$-1R_2 + 2R_1 \rightarrow R_1 \left[\begin{array}{cc|c} 12 & 0 & 49 \\ 0 & 6 & 19 \end{array}\right]$$

$$\begin{array}{l} \frac{1}{12}R_1 \rightarrow R_1 \\ \frac{1}{6}R_2 \rightarrow R_2 \end{array} \left[\begin{array}{cc|c} 1 & 0 & \frac{49}{12} \\ 0 & 1 & \frac{19}{6} \end{array}\right]$$

The solution is $\left(\frac{49}{12}, \frac{19}{6}\right)$, or about (4.08,3.17).

A choclate wafer weighs 4.08 g and a single layer of vanilla creme weighs 3.17g.

Chapter 3

LINEAR PROGRAMMING: THE GRAPHICAL METHOD

3.1 Graphing Linear Inequalities

Your Turn 1

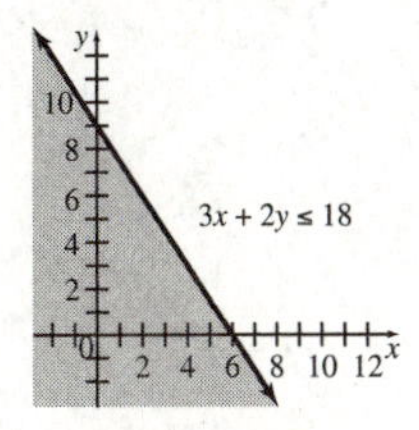

Your Turn 2

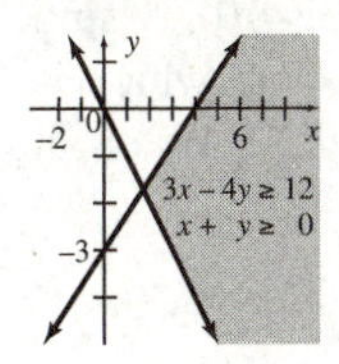

3.1 Exercises

1. $x + y \le 2$

First graph the boundary line $x + y = 2$ using the points $(2, 0)$ and $(0, 2)$. Since the points on this line satisfy $x + y \le 2$, draw a solid line. To find the correct region to shade, choose any point not on the line. If $(0, 0)$ is used as the test point, we have

$$x + y \le 2$$
$$0 + 0 \le 2$$
$$0 \le 2,$$

which is a true statement. Shade the half-plane containing $(0, 0)$, or all points below the line.

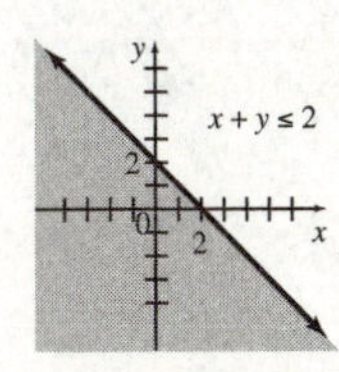

3. $x \ge 2 - y$

First graph the boundary line $x = 2 - y$ using the points $(0, 2)$ and $(2, 0)$. This will be a solid line. Choose $(0, 0)$ as a test point.

$$x \ge 2 - y$$
$$0 \ge 2 - 0$$
$$0 \ge 2,$$

which is a false statement. Shade the half-plane that does not contain $(0, 0)$, or all points below the line.

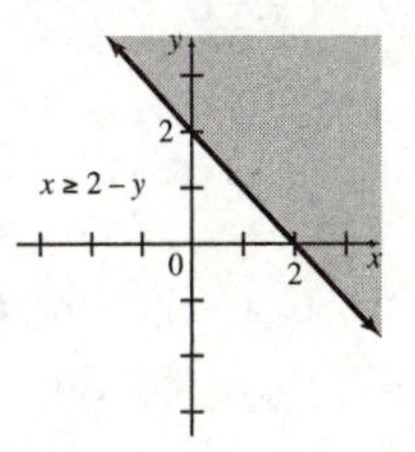

5. $4x - y < 6$

Graph $4x - y = 6$ as a dashed line, since the points on the line are not part of the solution; the line passes through the points $(0, -6)$ and $\left(\frac{3}{2}, 0\right)$. Using the test point $(0, 0)$, we have $0 - 0 < 6$ or $0 < 6$, a true statement. Shade the half-plane containing $(0, 0)$, or all points above the line.

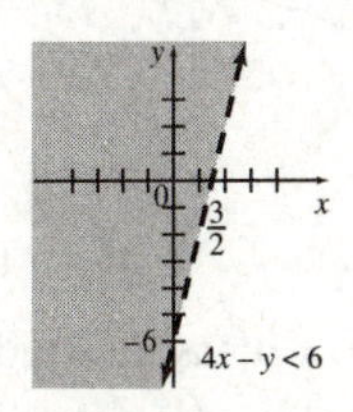

7. $4x + y < 8$

Graph $4x + y = 8$ as a dashed line through $(2, 0)$ and $(0, 8)$. Using the test point $(0, 0)$, we get $4 \cdot (0) + 0 < 8$ or $0 < 8$, 0 a true statement. Shade the half-plane containing $(0, 0)$, or all points below the line.

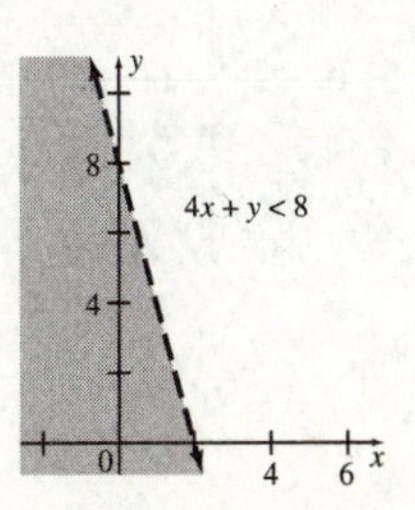

9. $x + 3y \geq -2$

The graph includes the line $x + 3y = -2$, whose intercepts are the points $\left(0, -\frac{2}{3}\right)$ and $(-2, 0)$. Graph $x + 3y = -2$ as a solid line and use the origin as a test point. Since $0 + 3(0) \geq -2$ is true, shade the half-plane containing $(0, 0)$, or all points above the line.

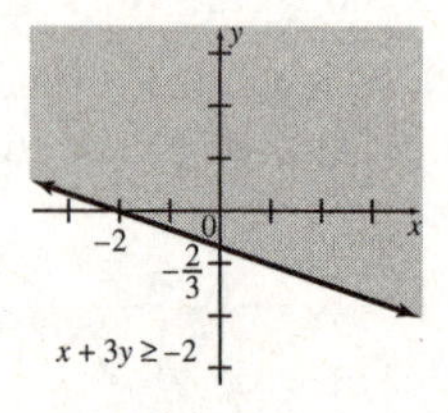

11. $x \leq 3y$

Graph $x = 3y$ as a solid line through the points $(0, 0)$ and $(3, 1)$. Since this line contains the origin, some point other than $(0, 0)$ must be used as a test point. If we use the point $(1, 2)$, we obtain $1 \leq 3(2)$ or $1 \leq 6$, a true statement. Shade the half-plane containing $(1, 2)$, or all points above the line.

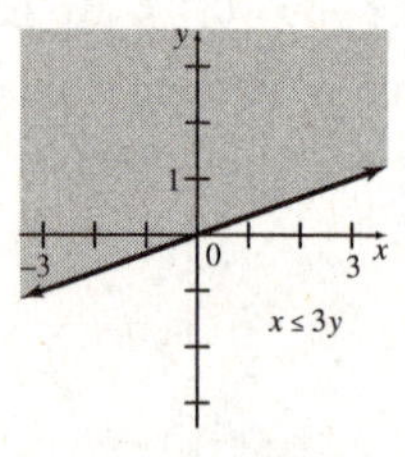

13. $x + y \leq 0$

Graph $x + y = 0$ as a solid line through the points $(0, 0)$ and $(1, -1)$. This line contains $(0, 0)$. If we use $(-1, 0)$ as a test point, we obtain $-1 + 0 \leq 0$ or $-1 \leq 0$, a true statement. Shade the half-plane containing $(-1, 0)$, or all points below the line.

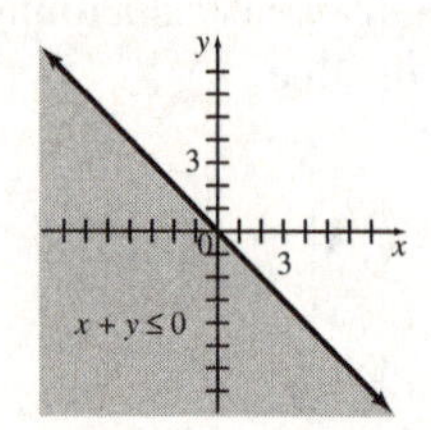

15. $y < x$

Graph $y = x$ as a dashed line through the points $(0, 0)$ and $(1, 1)$. Since this line contains the origin, choose a point other than $(0, 0)$ as a test point. If we use $(2, 3)$, we obtain $3 < 2$, which is false. Shade the half-plane that does not contain $(2, 3)$, or all points below the line.

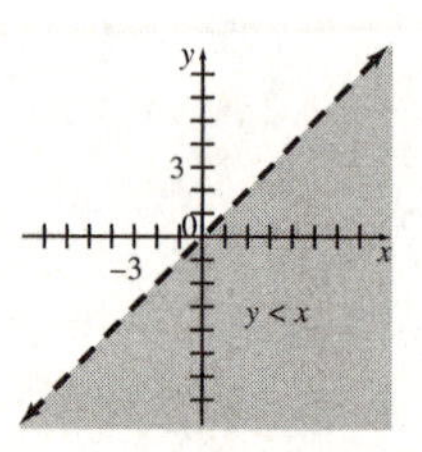

17. $x < 4$

Graph $x = 4$ as a dashed line. This is the vertical line crossing the x-axis at the point $(4, 0)$. Using $(0, 0)$ as a test point, we obtain $0 < 4$, which is true. Shade the half-plane containing $(0, 0)$, or all points to the left of the line.

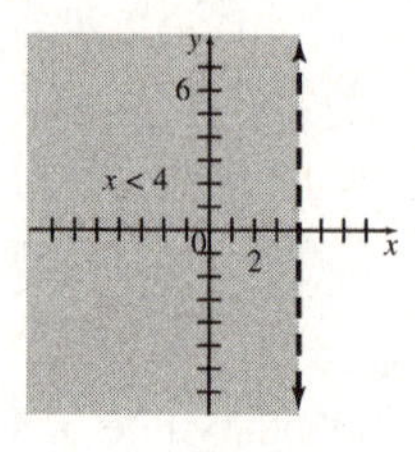

19. $y \leq -2$

Graph $y = -2$ as a solid horizontal line through the point $(0, -2)$. Using the origin as a test point, we obtain $0 \leq -2$, which is false. Shade the half-plane that does not contain $(0, 0)$, or all points below the line.

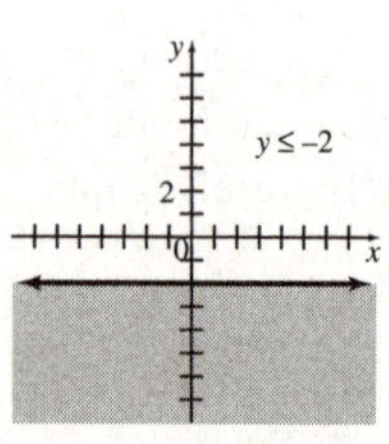

21. $x + y \leq 1$
$x - y \geq 2$

Graph the solid lines

$$x + y = 1 \quad \text{and}$$
$$x - y = 2.$$

$0 + 0 \leq 1$ is true, and $0 - 0 \geq 2$ is false. In each case, the graph is the region below the line. Shade the overlapping part of these two half-planes, which is the region below both lines. The shaded region is the feasible region for this system.

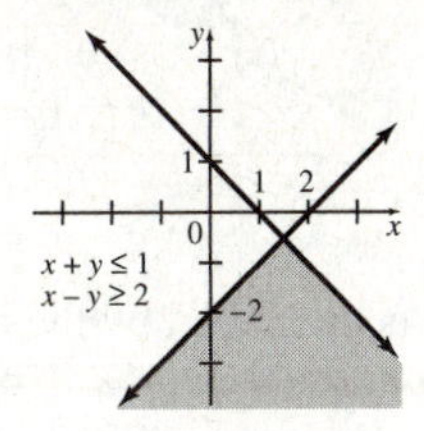

Unbounded

23. $x + 3y \leq 6$
$2x + 4y \geq 7$

Graph the solid lines $x + 3y = 6$ and $2x + 4y = 7$. Use $(0, 0)$ as a test point. $0 + 0 \leq 6$ is true, and $0 + 0 \geq 7$ is false. Shade all points below $x + 3y = 6$ and above $2x + 4y = 7$. The feasible region is the overlap of the two half-planes.

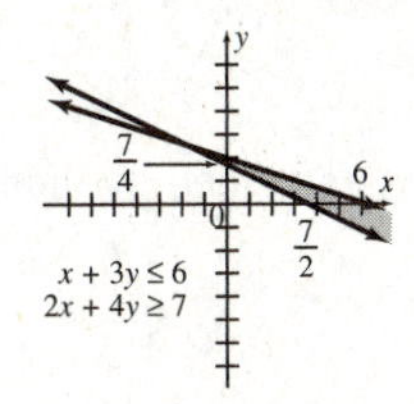

Unbounded

25. $x + y \leq 7$
$x - y \leq -4$
$4x + y \geq 0$

The graph of $x + y \leq 7$ consists of the solid line $x + y = 7$ and all the points below it. The graph of $x - y \leq -4$ consists of the solid line $x - y = -4$ and all the points above it. The graph of $4x + y \geq 0$ consists of the solid line $4x + y = 0$ and all the points above it. The feasible region is the overlapping part of these three half-planes.

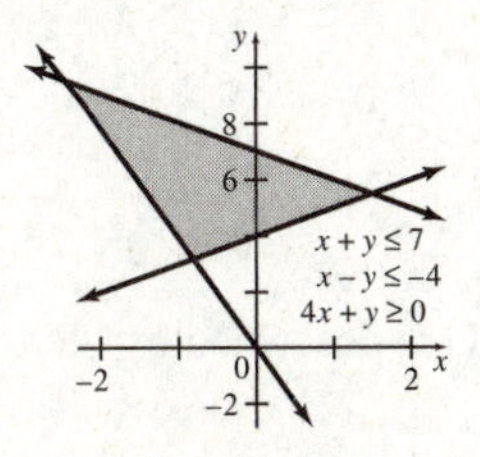

Bounded

27. $-2 < x < 3$
$-1 \leq y \leq 5$
$2x + y < 6$

The graph of $-2 < x < 3$ is the region between the vertical lines $x = -2$ and $x = 3$, but not including the lines themselves (so the two vertical boundaries are drawn as dashed lines). The graph of $-1 \leq y \leq 5$ is the region between the horizontal lines $y = -1$ and $y = 5$, including the lines (so the two horizontal boundaries are drawn as solid lines). The graph of $2x + y < 6$ is the region below the line $2x + y = 6$ (so the boundary is drawn as a dashed line). Shade the region common to all three graphs to show the feasible region.

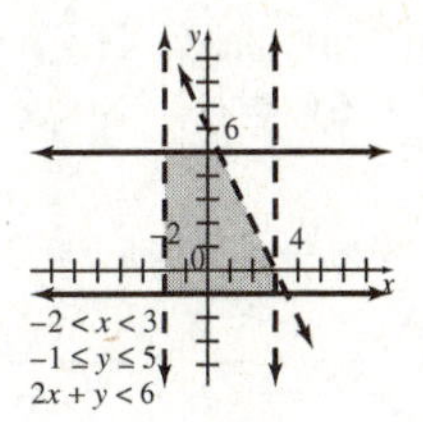

Bounded

29. $y - 2x \leq 4$
$y \geq 2 - x$
$x \geq 0$
$y \geq 0$

The graph of $y - 2x \leq 4$ consists of the boundary line $y - 2x = 4$ and the region below it. The graph of $y \geq 2 - x$ consists of the boundary line $y = 2 - x$ and the region above it. The inequalities $x \geq 0$ and $y \geq 0$ restrict the feasible region to the first quadrant. Shade the region in the first quadrant where the first two graphs overlap to show the feasible region.

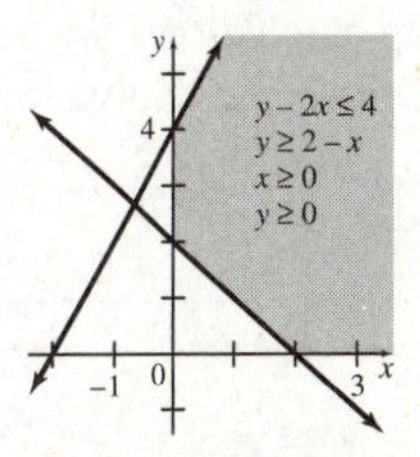

Unbounded

31. $3x + 4y > 12$
$2x - 3y < 6$
$0 \leq y \leq 2$
$x \geq 0$

$3x + 4y > 12$ is the set of points above the dashed line $3x + 4y = 12$; $2x - 3y < 6$ is the set of points above the dashed line $2x - 3y = 6$; $0 \leq y \leq 2$ is the set of points lying on or between the horizontal lines $y = 0$ and $y = 2$; and $x \geq 0$ consists of all the points on or to the right of the y-axis. Shade the feasible region, which is the triangular region satisfying all of the inequalities.

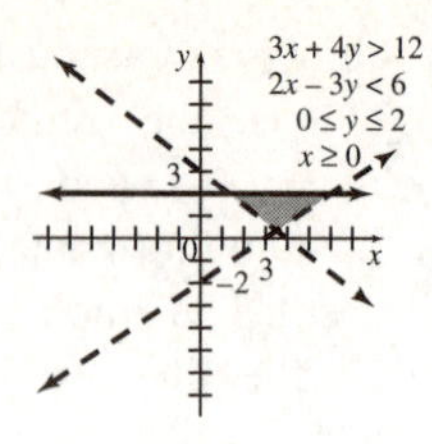

Bounded

33. $2x - 6y > 12$

Use a graphing calculator. The boundary line is the graph of $2x - 6y = 12$. Solve this equation for y.

$$-6y = -2x + 12$$

$$y = \frac{-2}{-6}x + \frac{12}{-6}$$

$$y = \frac{1}{3}x - 2$$

Enter $y_1 = \frac{1}{3}x - 2$ and graph it. Using the origin as a test point, we obtain $0 > 12$, which is false. Shade the region that does not contain the origin.

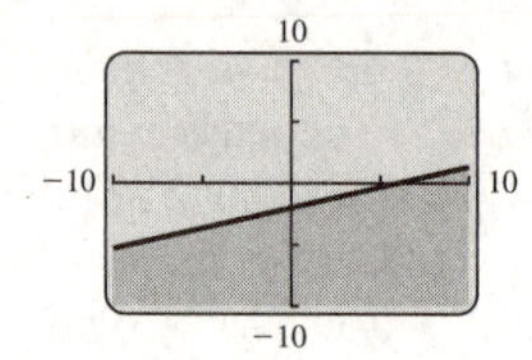

35. $3x - 4y < 6$
$2x + 5y > 15$

Use a graphing calculator. One boundary line is the graph of $3x - 4y = 6$. Solve this equation for y.

$$-4y = -3x + 6$$

$$y = \frac{-3}{-4}x + \frac{6}{-4}$$

$$y = \frac{3}{4}x - \frac{3}{2}$$

Enter $y_1 = \frac{3}{4}x - \frac{3}{2}$ and graph it. Using the origin as a test point, we obtain $0 < 6$, which is true. Shade the region that contains the origin.

The other boundary line is the graph of $2x + 5y = 15$. Solve this equation for y.

$$5y = -2x + 15$$

$$y = -\frac{2}{5}x + 3$$

Enter $y_2 = -\frac{2}{5}x + 3$ and graph it. Using the origin as a test point, we obtain $0 > 15$, which is false. Shade the region that does not contain the origin. The overlap of the two graphs is the feasible region.

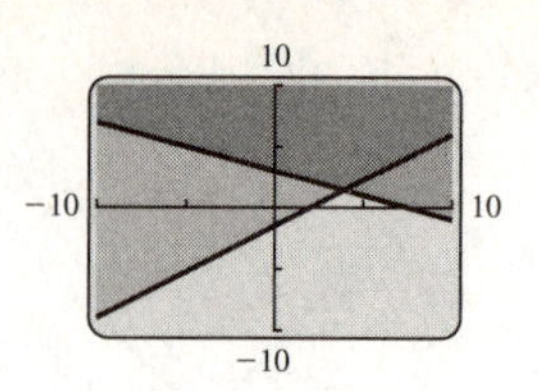

37. The region B is described by the inequalities

$$x + 3y \le 6$$
$$x + y \le 3$$
$$x - 2y \le 2$$
$$x \ge 0$$
$$y \ge 0.$$

The region C is described by the inequalities

$$x + 3y \ge 6$$
$$x + y \ge 3$$
$$x - 2y \le 2$$
$$x \ge 0$$
$$y \ge 0.$$

The region D is described by the inequalities

$$x + 3y \le 6$$
$$x + y \ge 3$$
$$x - 2y \le 2$$
$$x \ge 0$$
$$y \ge 0.$$

The region E is described by the inequalities

$$x + 3y \le 6$$
$$x + y \le 3$$
$$x - 2y \ge 2$$
$$x \ge 0$$
$$y \ge 0.$$

The region F is described by the inequalities

$$x + 3y \le 6$$
$$x + y \ge 3$$
$$x - 2y \ge 2$$
$$x \ge 0$$
$$y \ge 0.$$

The region G is described by the inequalities

$$x + 3y \ge 6$$
$$x + y \ge 3$$
$$x - 2y \ge 2$$
$$x \ge 0$$
$$y \ge 0.$$

39. (a)

	Shawls	Afghans	Total
Number Made	x	y	
Spinning Time	1	2	≤ 8
Dyeing Time	1	1	≤ 6
Weaving Time	1	4	≤ 14

(b) $x + 2y \leq 8$ *Spinning inequality*

$x + y \leq 6$ *Dyeing inequality*

$x + 4y \leq 14$ *Weaving inequality*

$x \geq 0$ *Ensures a nonnegative*

$y \geq 0$ *number of each*

Graph the solid lines $x + 2y = 8$, $x + y = 6$, $x + 4y = 14$, $x = 0$ and $y = 0$, and shade the appropriate half-planes to get the feasible region.

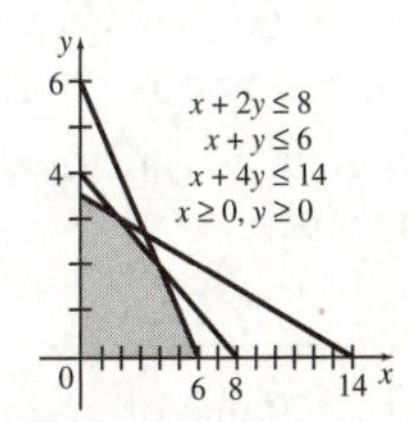

(c) Yes, 3 shawls and 2 afghans can be made because this corresponds to the point $(3, 2)$, which is in the feasible region.

No, 4 shawls and 3 afghans cannot be made because this corresponds to the point $(4, 3)$, which is not in the feasible region.

41. (a) The first sentence of the problem tells us that a total of \$30 million or $x + y \leq 30$ has been set aside for loans. The second sentence of the problem gives $x \geq 4y$. The third and fourth sentences give

$$0.06x + 0.08y \geq 1.6$$

Also, $x \geq 0$ and $y \geq 0$ ensure nonnegative numbers. Thus,

$$x + y \leq 30$$
$$x \geq 4y$$
$$0.06x + 0.08y \geq 1.6$$
$$x \geq 0$$
$$y \geq 0$$

(b) Using the above system, graph solid lines and shade appropriate half-planes to get the feasible region.

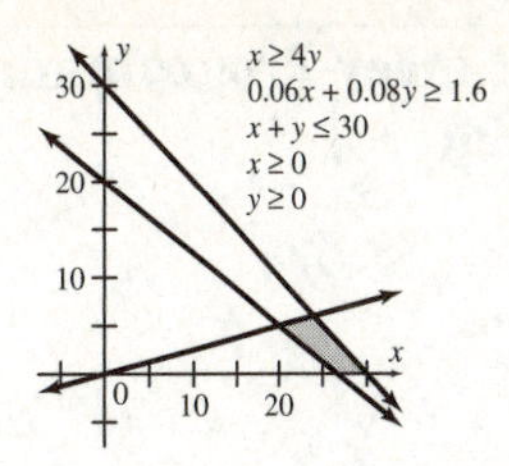

43. (a) The second sentence of the problem tells us that the number of M3 Power™ razors is never more than half the number of Fusion Power™ razors or $x \leq \frac{1}{2}y$. The third sentence tells us that a total of at most 800 razors can be produced per week or $x + y \leq 800$. The inequalities $x \geq 0$ and $y \geq 0$ ensure non-negative numbers. Thus,

$$x \leq \frac{1}{2}y$$
$$x + y \leq 800$$
$$x \geq 0$$
$$y \geq 0.$$

(b)

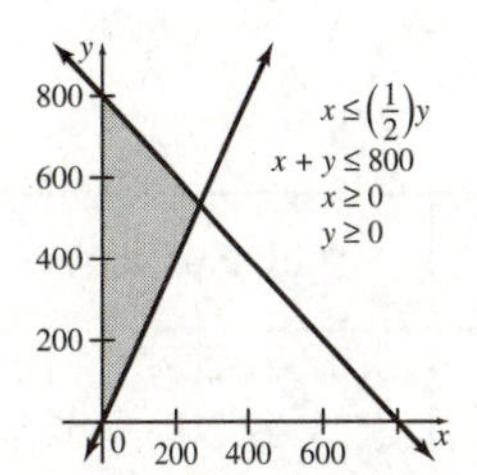

45. (a) The problem tells us that each ounce of fruit supplies 1 unit of protein and each ounce of nuts supplies 1 unit of protein. Thus, $1x + 1y$ is the number of units of protein per package. Since each package must provide at least 7 units of protein, we get the inequality $x + y \geq 7$. Similarly, we get the inequalities $2x + y \geq 10$ for carbohydrates and $x + y \leq 9$ for fat. The inequalities $x \geq 0$ and $y \geq 0$ ensure nonnegative numbers. Thus,

$$x + y \geq 7$$
$$2x + y \geq 10$$
$$x + y \leq 9$$
$$x \geq 0$$
$$y \geq 0.$$

(b)

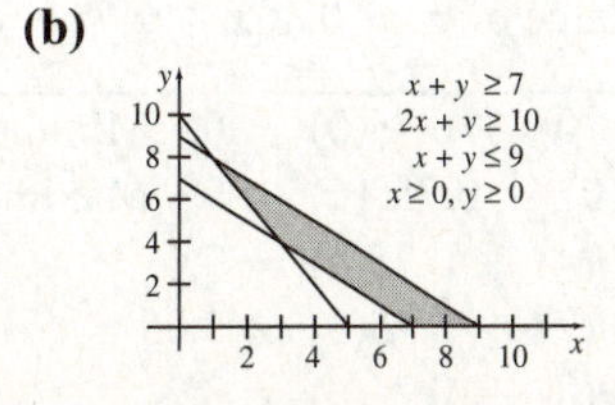

3.2 Solving Linear Programming Problems Graphically

Your Turn 1

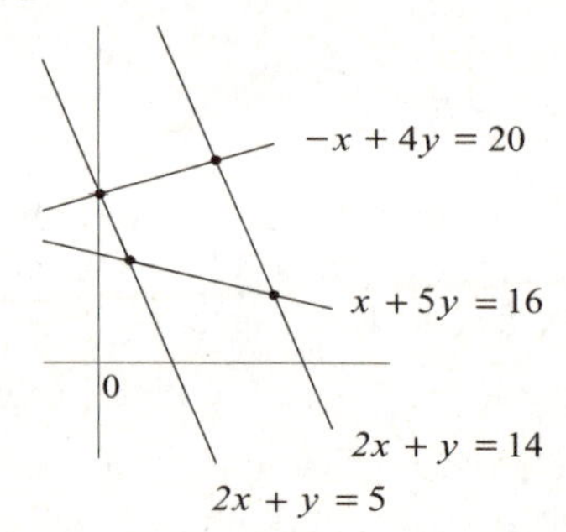

Corner Point	Value of $z = 3x + 4y$
(0, 5)	$3(0) + 4(5) = 20$
(1, 3)	$3(1) + 4(3) = 15$ Minimum
(4, 6)	$3(4) + 4(6) = 36$ Maximum
(6, 2)	$3(6) + 4(2) = 26$

The maximum value of 36 occurs at (4, 6).
The minimum value of 15 occurs at (1, 3).

3.2 Exercises

1. (a)

Corner Point	Value of $z = 3x + 2y$
(0, 5)	$3(0) + 2(5) = 10$ Minimum
(3, 8)	$3(3) + 2(8) = 25$
(7, 4)	$3(7) + 2(4) = 29$ Maximum
(4, 1)	$3(4) + 2(1) = 14$

The maximum value of 29 occurs at (7, 4). The minimum value of 10 occurs at (0, 5).

(b)

Corner Point	Value of $z = x + 4y$
(0, 5)	$0 + 4(5) = 20$
(3, 8)	$3 + 4(8) = 35$ Maximum
(7, 4)	$7 + 4(4) = 23$
(4, 1)	$4 + 4(1) = 8$ Minimum

The maximum value of 35 occurs at (3, 8). The minimum value of 8 occurs at (4, 1).

3. (a)

Corner Point	Value of $z = 0.40x + 0.75y$
(0, 0)	$0.40(0) + 0.75(0) = 0$ Minimum
(0, 12)	$0.40(0) + 0.75(12) = 9$ Maximum
(4, 8)	$0.40(4) + 0.75(8) = 7.6$
(7, 3)	$0.40(7) + 0.75(3) = 5.05$
(8, 0)	$0.40(8) + 0.75(0) = 3.2$

The maximum value of 9 occurs at (0, 12). The minimum value of 0 occurs at (0, 0).

(b)

Corner Point	Value of $z = 1.50x + 0.25y$
(0, 0)	$1.50.(0) + 0.25(0) = 0$ Minimum
(0, 12)	$1.50(0) + 0.25(12) = 3$
(4, 8)	$1.50(4) + 0.25(8) = 8$
(7, 3)	$1.50(7) + 0.25(3) = 11.25$
(8, 0)	$1.50(8) + 0.25(0) = 12$ Maximum

The maximum value of 12 occurs at (8, 0). The minimum value of 0 occurs at (0, 0).

5. (a)

Corner Point	Value of $z = 4x + 2y$
(0, 8)	$4(0) + 2(8) = 16$ Minimum
(3, 4)	$4(3) + 2(4) = 20$
$\left(\frac{13}{2}, 2\right)$	$4\left(\frac{13}{2}\right) + 2(2) = 30$
(12, 0)	$4(12) + 2(0) = 48$

The minimum value is 16 at (0, 8). Since the feasible region is unbounded, there is no maximum value.

(b)

Corner Point	Value of $z = 2x + 3y$
(0, 8)	$2(0) + 3(8) = 24$
(3, 4)	$2(3) + 3(4) = 18$ Minimum
$\left(\frac{13}{2}, 2\right)$	$2\left(\frac{13}{2}\right) + 3(2) = 19$
(12, 0)	$2(12) + 3(0) = 24$

The minimum value is 18 at (3, 4); there is no maximum value since the feasible region is unbounded.

(c)

Corner Point	Value of $z = 2x + 4y$
(0, 8)	$2(0) + 4(8) = 32$
(3, 4)	$2(3) + 4(4) = 22$
$\left(\frac{13}{2}, 2\right)$	$2\left(\frac{13}{2}\right) + 4(2) = 21$ Minimum
(12, 0)	$2(12) + 4(0) = 24$

The minimum value is 21 at $\left(\frac{13}{2}, 2\right)$; there is no maximum value since the feasible region is unbounded.

(d)

Corner Point	Value of $z = x + 4y$
(0, 8)	$0 + 4(8) = 32$
(3, 4)	$3 + 4(4) = 19$
$\left(\frac{13}{2}, 2\right)$	$\frac{13}{2} + 4(2) = \frac{29}{2}$
(12, 0)	$12 + 4(0) = 12$ Minimum

The minimum value is 12 at (12, 0); there is no maximum value since the feasible region is unbounded.

7. Minimize $z = 4x + 7y$

subject to:

$$\begin{aligned} x - y &\geq 1 \\ 3x + 2y &\geq 18 \\ x &\geq 0 \\ y &\geq 0. \end{aligned}$$

Sketch the feasible region.

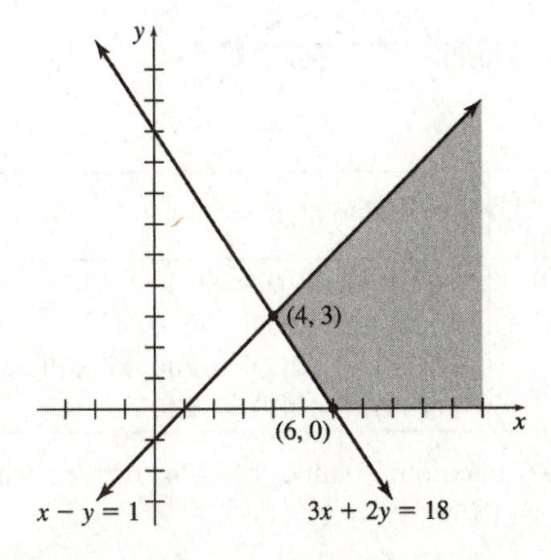

The sketch shows that the feasible region is unbounded. The corner points are (4, 3) and (6, 0). The corner point (4, 3) can be found by solving the system

$$\begin{aligned} x - y &= 1 \\ 3x + 2y &= 18. \end{aligned}$$

Use the corner points to find the minimum value of the objective function.

Corner Point	Value of $z = 4x + 7y$
(4, 3)	$4(4) + 7(3) = 37$
(6, 0)	$4(6) + 7(0) = 24$ Minimum

The minimum value is 24 when $x = 6$ and $y = 0$.

9. Maximize $z = 5x + 2y$

subject to:

$$\begin{aligned} 4x - y &\leq 16 \\ 2x + y &\geq 11 \\ x &\geq 3 \\ y &\leq 8. \end{aligned}$$

Sketch the feasible region.

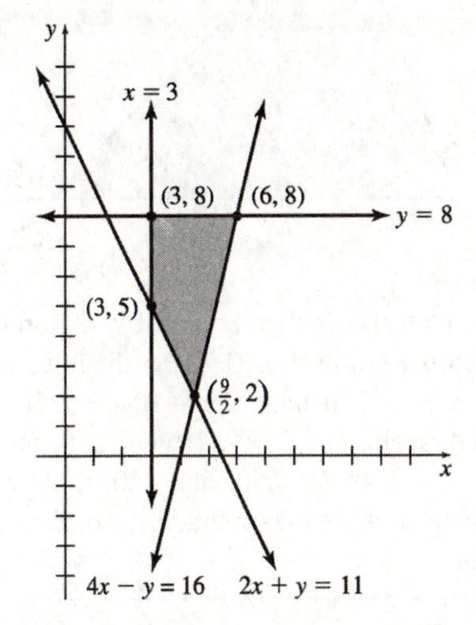

The sketch shows that the feasible region is bounded. The corner points are: (3, 5), the intersection of $2x + y = 11$ and $x = 3$; (3, 8), the intersection of $x = 3$ and $y = 8$; (6, 8), the intersection of $y = 8$ and $4x - y = 16$; and $(9/2, 2)$, the intersection of $4x - y = 16$ and $2x + y = 11$. Use the corner points to find the maximum value of the objective function.

Corner Point	Value of $z = 5x + 2y$
(3, 5)	$5(3) + 2(5) = 25$
(3, 8)	$5(3) + 2(8) = 31$
(6, 8)	$5(6) + 2(8) = 46$ Maximum
$\left(\frac{9}{2}, 2\right)$	$5\left(\frac{9}{2}\right) + 2(2) = 26.5$

The maximum value is 46 when $x = 6$ and $y = 8$.

11. Maximize $z = 10x + 10y$

subject to:

$$\begin{aligned} 5x + 8y &\geq 200 \\ 25x - 10y &\geq 250 \\ x + y &\leq 150 \\ x &\geq 0 \\ y &\geq 0. \end{aligned}$$

Sketch the feasible region.

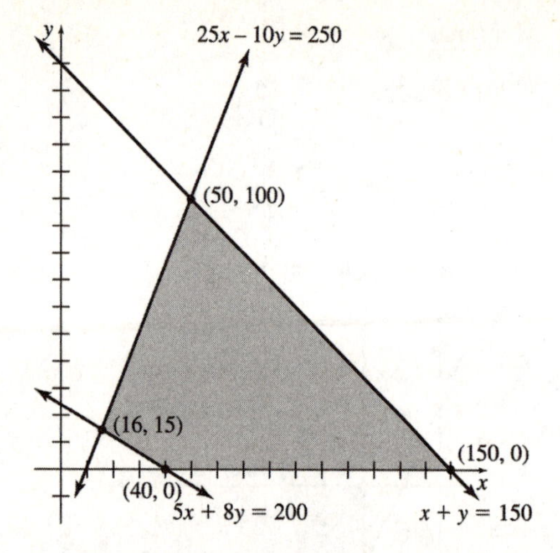

The sketch shows that the feasible region is bounded. The corner points are: $(16, 15)$, the intersection of $5x + 8y = 200$ and $25x - 10y = 250$; $(50, 100)$, the intersection of $25x - 10y = 250$ and $x + y = 150$; $(150, 0)$; and $(40, 0)$. Use the corner points to find the maximum value of the objective function.

Corner Point	Value of $z = 10x + 10y$
(16, 15)	$10(16) + 10(15) = 310$
(50, 100)	$10(50) + 10(100) = 1500$ Maximum
(150, 0)	$10(150) + 10(0) = 1500$ Maximum
(40, 0)	$10(40) + 10(0) = 400$

The maximum value is 1500 when $x = 50$ and $y = 100$, as well as when $x = 150$ and $y = 0$ and all points on the line between.

13. Maximize $z = 3x + 6y$

subject to: $2x - 3y \leq 12$

$x + y \geq 5$

$3x + 4y \geq 24$

$x \geq 0$

$y \geq 0.$

Sketch the feasible region.

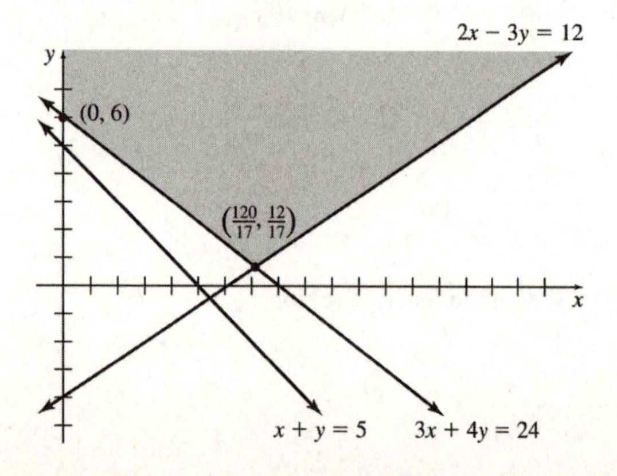

The graph shows that the feasible region is unbounded. Therefore, there is no maximum value of the objective function on the feasible region, hence, no solution.

15. Maximize $z = 10x + 12y$ subject to the following sets of constraints, with $x \geq 0$ and $y \geq 0$.

(a) $x + y \leq 20$

$x + 3y \leq 24$

Sketch the feasible region in the first quadrant, and identify the corner points at $(0, 0)$, $(0, 8)$, $(18, 2)$, which is the intersection of $x + y = 20$ and $x + 3y = 24$, and $(20, 0)$.

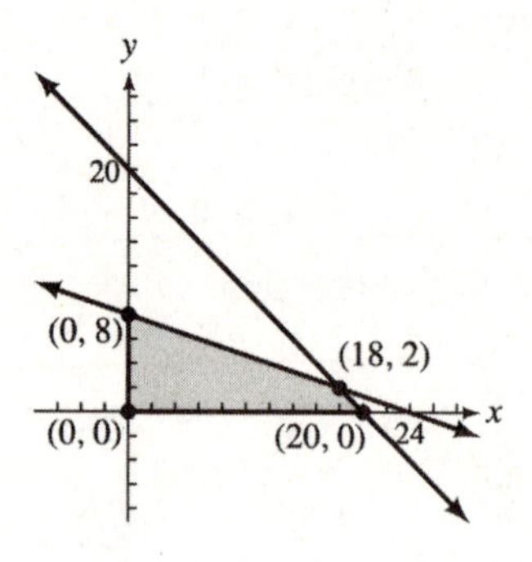

Corner Point	Value of $z = 10x + 12y$
(0, 0)	$10(0) + 12(0) = 0$
(0, 8)	$10(0) + 12(8) = 96$
(18, 2)	$10(18) + 12(2) = 204$ Maximum
(20, 0)	$10(20) + 12(0) = 200$

The maximum value of 204 occurs when $x = 18$ and $y = 2$.

(b) $3x + y \leq 15$

$x + 2y \leq 18$

Sketch the feasible region in the first quadrant, and identify the corner points. The corner point $\left(\frac{12}{5}, \frac{39}{5}\right)$ can be found by solving the system

$3x + y = 15$

$x + 2y = 18.$

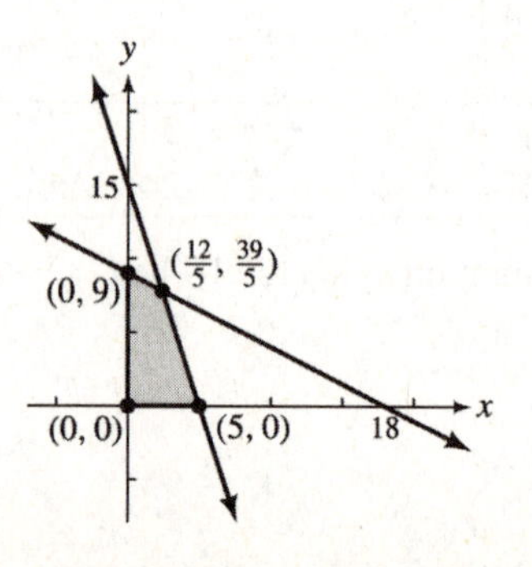

Corner Point	Value of $z = 10x + 12y$
$(0, 0)$	$10(0) + 12(0) = 0$
$(0, 9)$	$10(0) + 12(9) = 108$
$\left(\frac{12}{5}, \frac{39}{5}\right)$	$10\left(\frac{12}{5}\right) + 12\left(\frac{39}{5}\right) = \frac{588}{5} = 117\frac{3}{5}$ Maximum
$(5, 0)$	$10(5) + 12(0) = 50$

The maximum value of $\frac{588}{5}$ occurs when $x = \frac{12}{5}$ and $y = \frac{39}{5}$.

(c) $2x + 5y \geq 22$
$4x + 3y \leq 28$
$2x + 2y \leq 17$

Sketch the feasible region in the first quadrant, and identify the corner points. The corner point $\left(\frac{5}{2}, 6\right)$ can be found by solving the system

$$4x + 3y = 28$$
$$2x + 2y = 17,$$

and the corner point $\left(\frac{37}{7}, \frac{16}{7}\right)$ can be found by solving the system

$$2x + 5y = 22$$
$$4x + 3y = 28.$$

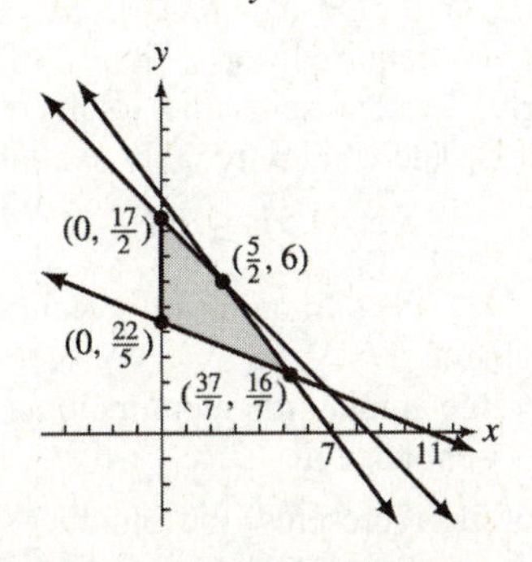

Corner Point	Value of $z = 10x + 12y$
$\left(0, \frac{22}{5}\right)$	$10(0) + 12\left(\frac{22}{5}\right) = \frac{264}{5} = 52.8$
$\left(0, \frac{17}{2}\right)$	$10(0) + 12\left(\frac{17}{2}\right) = 102$ Maximum
$\left(\frac{5}{2}, 6\right)$	$10\left(\frac{5}{2}\right) + 12(6) = 97$
$\left(\frac{37}{7}, \frac{16}{7}\right)$	$10\left(\frac{37}{7}\right) + 12\left(\frac{16}{7}\right) = \frac{562}{7} \approx 80.3$

The maximum value of 102 occurs when $x = 0$ and $y = \frac{17}{2}$.

17. Maximize $z = c_1x_1 + c_2x_2$

subject to: $2x_1 + x_2 \leq 11$
$-x_1 + 2x_2 \leq 2$
$x_1 \geq 0, \quad x_2 \geq 0.$

Sketch the feasible region.

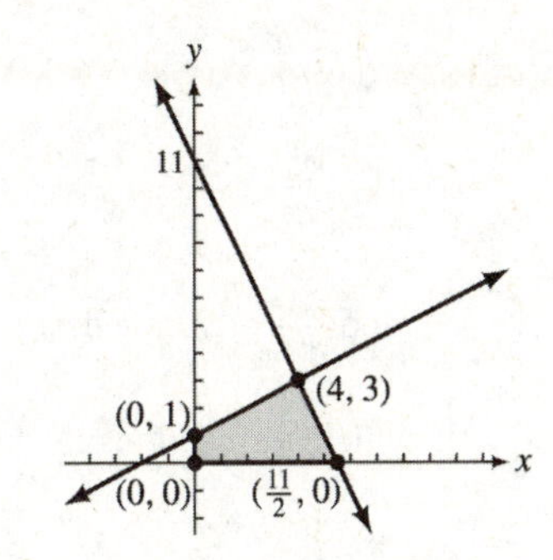

The region is bounded, with corner points $(0, 0)$, $(0, 1)$,, $(4, 3)$, and $\left(\frac{11}{2}, 0\right)$.

Corner Point	Value of $z = c_1x_1 + c_2x_2$
$(0, 0)$	$c_1(0) + c_2(0) = 0$
$(0, 1)$	$c_1(0) + c_2(1) = c_2$
$(4, 3)$	$c_1(4) + c_2(3) = 4c_1 + 3c_2$
$\left(\frac{11}{2}, 0\right)$	$c_1\left(\frac{11}{2}\right) + c_2(0) = \frac{11}{2}c_1$

If we are to have $(x_1, x_2) = (4, 3)$ as an optimal solution, then it must be true that both $4c_1 + 3c_2 \geq c_2$ and $4c_1 + 3c_2 \geq \frac{11}{2}c_1$, because the value of z at $(4, 3)$ cannot be smaller than the other values of z in the table. Manipulate the symbols in these two inequalities in order to isolate $\frac{c_1}{c_2}$ in each; keep in mind the given information that $c_2 > 0$ when performing division by c_2. First,

$$4c_1 + 3c_2 \geq c_2$$
$$4c_1 \geq -2c_2$$
$$\frac{4c_1}{4c_2} \geq \frac{-2c_2}{4c_2}$$
$$\frac{c_1}{c_2} \geq -\frac{1}{2}.$$

Then,

$$4c_1 + 3c_2 \geq \frac{11}{2}c_1$$
$$-\frac{3}{2}c_1 + 3c_2 \geq 0$$
$$3c_1 - 6c_2 \leq 0$$
$$3c_1 \leq 6c_2$$
$$\frac{3c_1}{3c_2} \leq \frac{6c_2}{3c_2}$$
$$\frac{c_1}{c_2} \leq 2.$$

Since $\frac{c_1}{c_2} \geq -\frac{1}{2}$ and $\frac{c_1}{c_2} \leq 2$, the desired range for $\frac{c_1}{c_2}$ is $\left[-\frac{1}{2}, 2\right]$, which corresponds to choice (b).

Your Turn 3

The new resource constraint is $25x + 75y \leq 1050$, or $x + 3y \leq 42$.

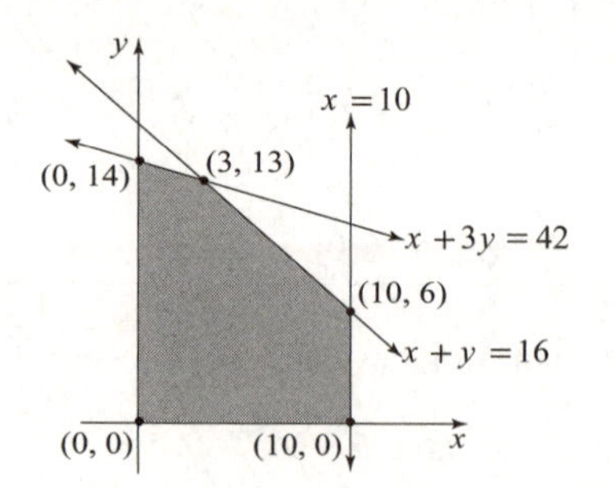

Corner Point	Value of $z = 12x + 40y$
(0, 0)	$12(0) + 40(0) = 0$
(0, 14)	$12(0) + 40(14) = 560$
(3, 13)	$12(3) + 40(13) = 556$
(10, 6)	$12(10) + 40(6) = 360$
(10, 0)	$12(10) + 40(0) = 120$

14 pigs and 0 goats produces a maximum profit of $560.

3.3 Applications of Linear Programming

Your Turn 1

Corner Point	Value of $z = 25x + 35y$
(0, 0)	$25(0) + 35(0) = 0$
(65, 0)	$25(65) + 35(0) = 1625$
(0, 60)	$25(0) + 35(60) = 2100$
(25, 40)	$25(0) + 35(40) = 2025$

He should buy 60 kayaks and no canoes for a maximum revenue of $2100 a day.

Your Turn 2

Corner Point	Value of $z = 20x + 30y$
(0, 0)	$20(0) + 30(0) = 0$
(0, 6)	$20(0) + 30(6) = 180$
(6, 0)	$20(6) + 30(0) = 120$
(4, 4)	$20(4) + 30(4) = 200$

The company should make 4 batches of each for a maximum profit of $200.

3.3 Exercises

1. Let x represent the number of product A made and y represent the number of product B. Each item of A uses 3 hr on the machine, so $3x$ represents the total hours required for x items of product A. Similarly, $5y$ represents the total hours used for product B. There are only 60 hr available, so

$$3x + 5y \leq 60.$$

3. Let x = the amount of calcium carbonate supplement
and y = the amount of calcium citrate supplement.

Then $600x$ represents the number of units of calcium provided by the calcium carbonate supplement and $250y$ represents the number of units provided by the calcium citrate supplement. Since at least 1500 units are needed per day,

$$600x + 250y \geq 1500.$$

5. Let x represent the number of pounds of $8 coffee and y represent the number of pounds of $10 coffee. Since the mixture must weigh at least 40 lb,

$$x + y \geq 40.$$

(Notice that the price per pound is not used in setting up this inequality.)

7. Let x = the number of engines to ship to plant I and y = the number of engines to ship to plant II.

Minimize $z = 30x + 40y$

subject to:
$$x \geq 45$$
$$y \geq 32$$
$$x + y \leq 120$$
$$20x + 15y \geq 1500$$
$$x \geq 0.$$
$$y \geq 0.$$

Sketch the feasible region in quadrant I, and identify the corner points.

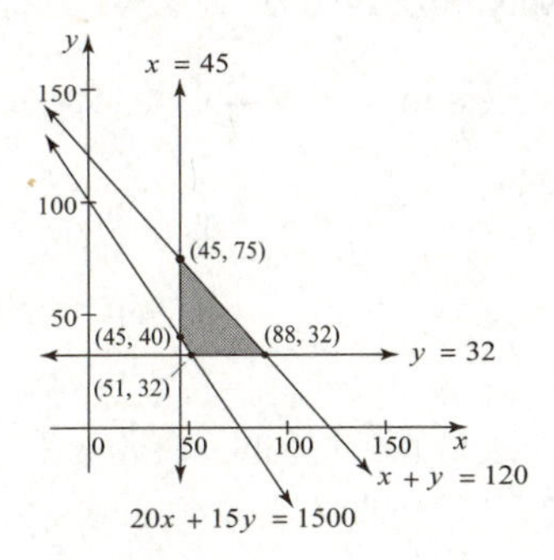

The corner points are:

$(45, 75)$, the intersection of $x = 45$ and $x + y = 120$,

$(45, 40)$, the intersection of $x = 45$ and $20x + 15y = 1500$,

$(88, 32)$, the intersection of $y = 32$ and $x + y = 120$, and

$(51, 32)$, the intersection of $y = 32$ and $20x + 15y = 1500$.

Use the corner points to find the minimum value of the objective function.

Corner Point	Value of $z = 30x + 40y$
(45, 75)	$30(45) + 40(75) = 4350$
(45, 40)	$30(45) + 40(40) = 2950$
(88, 32)	$30(88) + 40(32) = 3920$
(51, 32)	$30(51) + 40(32) = 2810$

The minimum value is \$2810, which occurs when 51 engines are shipped to plant I and 32 engines are shipped to plant II.

9. **(a)** Let x = the number of units of policy A and y = the number of units of policy B.

Minimize $z = 50x + 40y$

subject to:
$$10{,}000x + 15{,}000y \geq 300{,}000$$
$$180{,}000x + 120{,}000y \geq 3{,}000{,}000$$
$$x \geq 0$$
$$y \geq 0.$$

Sketch the feasible region in quadrant I, and identify the corner points. The corner point $(6, 16)$ can be found by solving the system

$$10{,}000x + 15{,}000y = 300{,}000$$
$$180{,}000x + 120{,}000y = 3{,}000{,}000,$$

which can be simplified as

$$2x + 3y = 60$$
$$3x + 2y = 50.$$

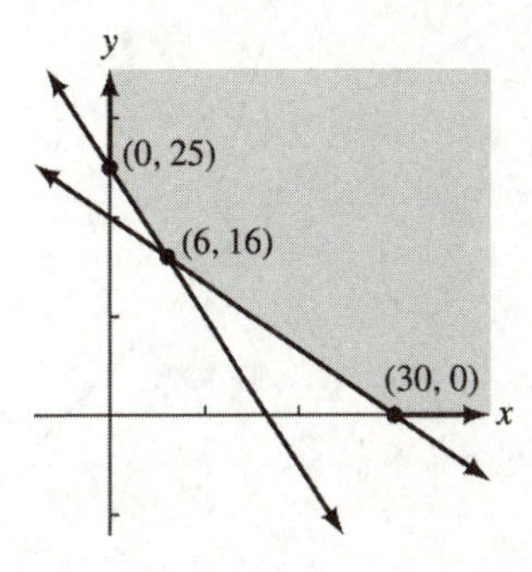

Corner Point	Value of $z = 50x + 40y$	
(0, 25)	$50(0) + 40(25) = 1000$	
(6, 16)	$50(6) + 40(16) = 940$	Minimum
(30, 0)	$50(30) + 40(0) = 1500$	

The minimum cost is \$940, which occurs when 6 units of policy A and 16 units of policy B are purchased.

(b) The objective function changes to $z = 25x + 40y$, but the constraints remain the same. Use the same corner points as in part (a).

Corner Point	Value of $z = 25x + 40y$	
(0, 25)	$25(0) + 40(25) = 1000$	
(6, 16)	$25(6) + 40(16) = 790$	
(30, 0)	$25(30) + 40(0) = 750$	Minimum

The minimum cost is \$750. which occurs when 30 units of policy A and no units of policy B are purchased.

11. (a) Let x = the number of type I bolts and y = the number of type II bolts.

Maximize $z = 0.15x + 0.20y$

subject to:
$$0.2x + 0.2y \le 300$$
$$0.6x + 0.2y \le 720$$
$$0.04x + 0.08y \le 100$$
$$x \ge 0$$
$$y \ge 0.$$

Graph the feasible region and identify the corner points.

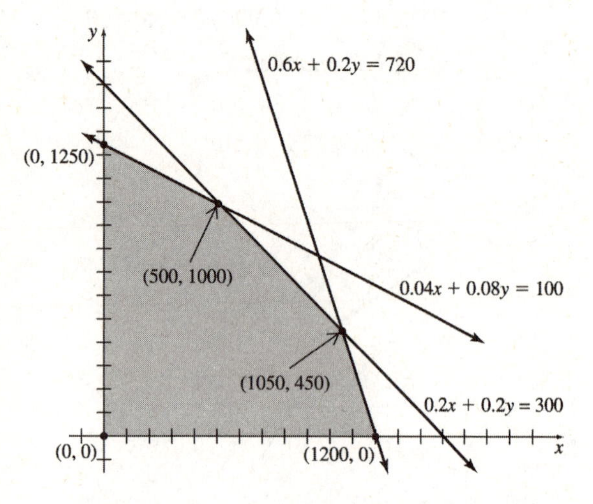

The corner points are $(0, 0)$; $(0, 1250)$; $(500, 1000)$, the intersection of $0.04x + 0.08y = 100$ and $0.02x + 0.2y = 300$; $(1050, 450)$, the intersection of $0.02x + 0.2y = 300$ and $0.6x + 0.2y = 720$; and $(1200, 0)$. Use the corner points to find the maximum value of the objective function.

Corner Point	Value of $z = 0.15x + 0.90y$
(0, 0)	$0.15(0) + 0.20(0) = 0$
(0, 1250)	$0.15(0) + 0.20(1250) = 250$
(500, 1000)	$0.15(500) + 0.20(1000) = 275$ Maximum
(1050, 450)	$0.15(1050) + 0.20(450) = 247.5$
(1200, 0)	$0.15(1200) + 0.20(0) = 180$

The shop should manufacture 500 type I bolts and 1000 type II bolts to maximize revenue.

(b) The maximum revenue is \$275.

(c) When the slope of the line $z = px + 0.20y$ for constant z matches the slope of the line joining the corner points $(500, 1000)$ and $(1050, 450)$, all points in the feasible region on this line produce the same maximum value of the objective function. This happens when

$$\frac{-p}{0.20} = \frac{1000 - 450}{500 - 1050} = -1,$$

or $p = 0.20$. As we increase the price of type I bolts beyond 20¢, the slope of $z = px + 0.20y$ becomes more negative and the corner point $(1050, 450)$ takes over and produces the maximum profit. So the answer is a price of 20¢ per bolt for type I bolts.

13. (a) Let x = the number of kg of the half-and-half mixture and y = the number of kg of the second mixture.

Maximize $z = 7x + 9.5y$

subject to:
$$\frac{1}{2}x + \frac{3}{4}y \le 150$$
$$\frac{1}{2}x + \frac{1}{4}y \le 90$$
$$x \ge 0$$
$$y \ge 0.$$

Sketch the feasible region and identify the corner points.

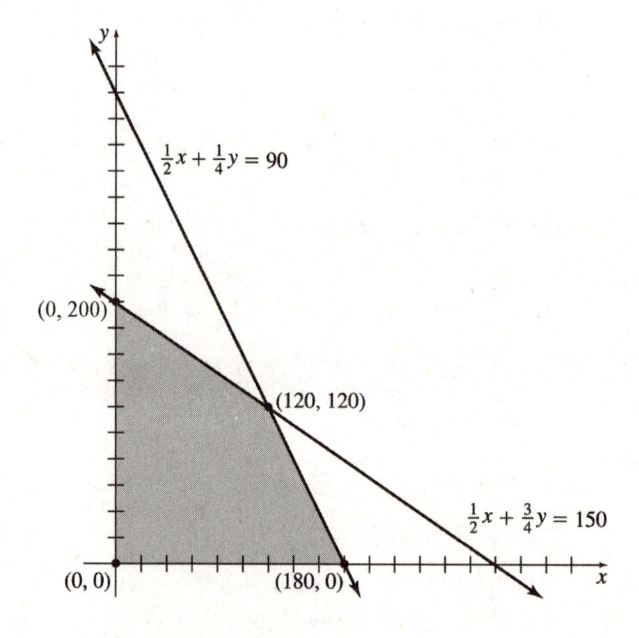

The corner points are $(0, 0)$; $(0, 200)$; $(120, 120)$, the intersection of $\frac{1}{2}x + \frac{3}{4}y = 150$ and $\frac{1}{2}x + \frac{1}{4}y = 90$; and $(180, 0)$. Use the corner points to find the maximum value of the objective function.

Corner Point	Value of $z = 7x + 9.5y$
(0, 0)	$7(0) + 9.5(0) = 0$
(0, 200)	$7(0) + 9.5(200) = 1900$
(120, 120)	$7(120) + 9.5(120) = 1980$ Maximum
(180, 0)	$7(180) + 9.5(0) = 1260$

The candy company should prepare 120 kg of the half-and-half mixture and 120 kg of the second mixture for a maximum revenue of \$1980.

(b) The objective function to be maximized is now $z = 7x + 11y$. The corner points remain the same.

Corner Point	Value of $z = 7x + 11y$
(0, 0)	$7(0) + 11(0) = 0$
(0, 200)	$7(0) + 11(200) = 2200$
(120, 120)	$7(120) + 11(120) = 2160$ Maximum
(180, 0)	$7(180) + 11(0) = 1260$

In order to maximize the revenue under the altered conditions, the candy company should prepare 0 kg of the half-and-half mixture and 200 kg of the second mixture for a maximum revenue of \$2200.

15. (a) Let x = the number of gallons from dairy I and y = the number of gallons from dairy II.

Maximize $z = 0.037x + 0.032y$

subject to:
$$0.60x + 0.20y \le 36$$
$$x \le 50$$
$$y \le 80$$
$$x + y \le 100$$
$$x \ge 0$$
$$y \ge 0.$$

Sketch the feasible region, and identify the corner points.

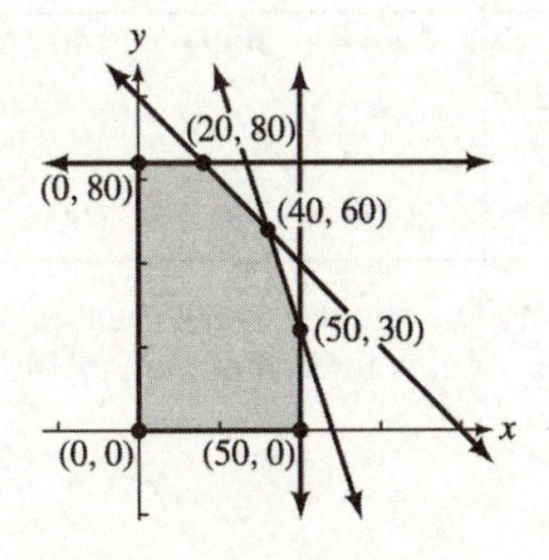

Corner Point	Value of $z = 0.037x + 0.032y$
(0, 0)	0
(0, 80)	2.56
(20, 80)	3.30
(40, 60)	3.40 Maximum
(50, 30)	2.81
(50, 0)	1.85

The maximum amount of butterfat is 3.4 gal, which occurs when 40 gal are purchased from dairy I and 60 gal are purchased from dairy II.

(b) In the solution to part (a), Mostpure uses 40 gallons from dairy I with a capacity for 50 gallons. Therefore, there is an excess capacity of 10 gallons from dairy I. Similarly, there is an excess capacity of 80 − 60 or 20 gallons from dairy II. No, the excess capacity cannot be used.

17. Let x = the amount (in millions) invested in U.S. Treasury bonds and y = the amount (in millions) invested in mutual funds.

Maximize $z = 0.04x + 0.08y$

subject to:
$$x + y \le 30$$
$$x \ge 5$$
$$y \ge 10$$
$$100x + 200y \ge 5000$$
$$x \ge 0$$
$$y \ge 0.$$

Sketch the feasible region, and identify the corner points.

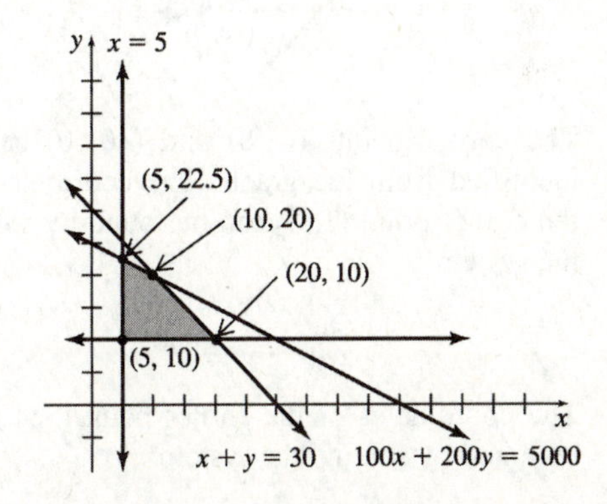

The corner points are (5, 10); (5, 22.5), the intersection of $x = 5$ and $100x + 200y = 5000$; (10, 20), the intersection of $100x + 200y = 5000$ and $x + y = 30$; and (20, 10). Use the corner points to find the maximum value of the objective function.

Corner Point	Value of $z = 0.04x + 0.08y$	
(5, 10)	$0.04(5) + 0.08(10) = 1$	
(5, 22.5)	$0.04(5) + 0.08(22.5) = 2$	Maximum
(10, 20)	$0.04(10) + 0.08(20) = 2$	Maximum
(20, 10)	$0.04(20) + 0.08(10) = 1.6$	

The maximum annual interest of $2 million can be achieved by investing $5 million in U.S. Treasury bonds and $22.5 million in mutual funds, or $10 million in bonds and $20 million in mutual funds (or in any solution on the line between those two points).

19. Beta is limited to 400 units per day, so Beta $\leq$ 400. The correct answer is choice (a).

21. **(a)** Let $x =$. the number of pill 1 and $y =$ the number of pill 2.

Minimize $z = 0.15x + 0.30y$

subject to:
$$240x + 60y \geq 480$$
$$x + y \geq 5$$
$$2x + 7y \geq 20$$
$$x \geq 0$$
$$y \geq 0.$$

Sketch the feasible region in quadrant I.

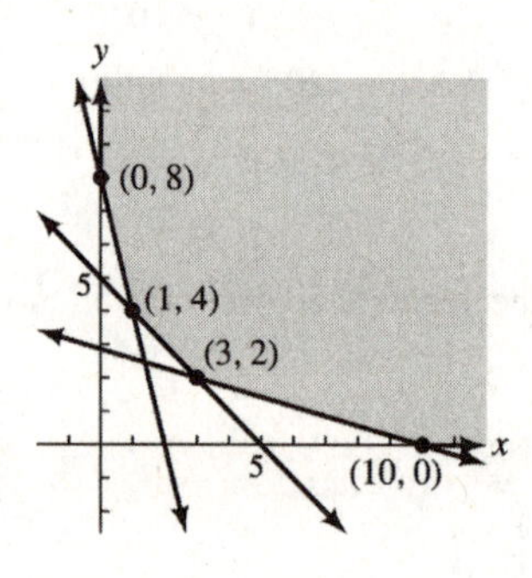

The corner points (0, 8) and (10, 0) can be identified from the graph. The coordinates of the corner point (1, 4) can be found by solving the system

$$240x + 60y = 480$$
$$x + y = 5.$$

The coordinates of the corner point (3, 2) can be found by solving the system

$$2x + 7y = 20$$
$$x + y = 5.$$

Corner Point	Value of $z = 0.15x + 0.30y$
(1, 4)	1.35
(3, 2)	1.05 Minimum
(0, 8)	2.40
(10, 0)	1.50

A minimum daily cost of $1.05 is incurred by taking three of pill 1 and two of pill 2.

(b) In the solution to part (a), Mark receives $240(3) + 60(2)$ or 840 units of vitamin A.

This is a surplus of $840 - 480$ or 360 units of vitamin A. No, there is no way for him to avoid receiving a surplus.

23. Let $x =$ the number of ounces of fruit and $y =$ the number of ounces of nuts.

Minimize $z = 20x + 30y$

subject to:
$$3y \geq 6$$
$$2x + y \geq 10$$
$$x + 2y \leq 9$$
$$x \geq 0$$
$$y \geq 0.$$

Sketch the feasible region, and identify the corner points.

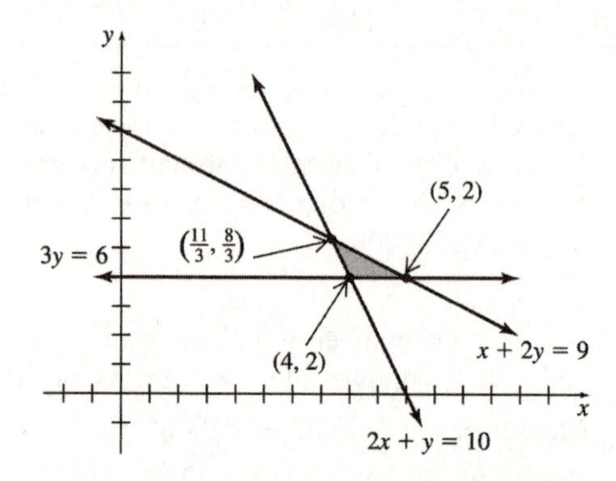

The corner points are: (4, 2), the intersection of $y = 2$ and $2x + y = 10$; $\left(\frac{11}{3}, \frac{8}{3}\right)$, the intersection of $2x + y = 10$ and $x + 2y = 9$; and (5, 2), the intersection of $x + 2y = 9$ and $y = 2$. Use the corner points to find the minimum value of the objective function.

Corner Point	Value of $z = 20x + 30y$
(4, 2)	$20(4) + 30(2) = 140$ Minimum
$\left(\frac{11}{3}, \frac{8}{3}\right)$	$20\left(\frac{11}{3}\right) + 30\left(\frac{8}{3}\right) = \frac{460}{3} \approx 153$
(5, 2)	$20(5) + 30(2) = 160$

The dietician should use 4 ounces of fruit and 2 ounces of nuts for a minimum of 140 calories.

25. Let x = the number of units of plants and y = the number of animals.

Minimize $z = 30x + 15y$

subject to: $30x + 20y \geq 360$

$$10x + 25y \geq 300$$
$$y \geq 8$$
$$0 \leq x \leq 25$$
$$0 \leq y \leq 25.$$

Sketch the feasible region in quadrant I.

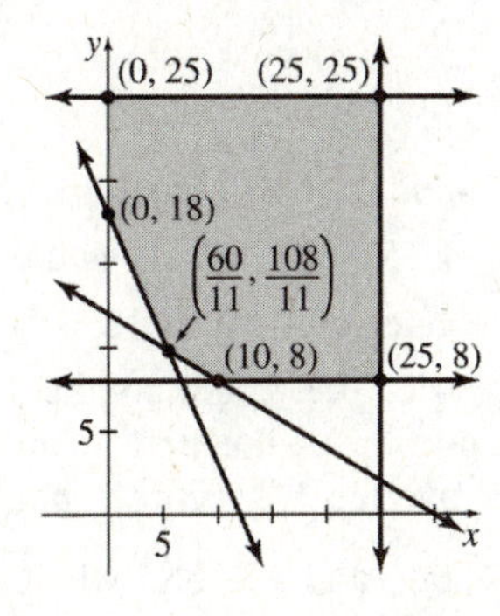

The corner points (0, 18), (0, 25), (25, 25), and (25, 8) can be determined from the graph. The corner point $\left(\frac{60}{11}, \frac{108}{11}\right)$ can be found by solving the system

$$30x + 20y = 360$$
$$10x + 25y = 300.$$

The corner point (10, 8) can be found by solving the system

$$10x + 25y = 300$$
$$y = 8.$$

Corner Point	Value of $z = 30x + 15y$	
(0, 18)	270	Minimum
(0, 25)	375	
(25, 25)	1125	
(25, 8)	870	
$\left(\frac{60}{11}, \frac{108}{11}\right)$	$\frac{3420}{11} \approx 310.91$	
(10, 8)	420	

The minimum labor is 270 hours and is achieved when 0 units of plants and 18 animals are collected.

Chapter 3 Review Exercises

1. False; the graphical method is impossible with four or more variables.

2. True

3. False; $x \leq 2y$ means that the number of acres of wheat will be at most twice the number of acres of corn planted.

4. False; the total amount of time cannot exceed 60 hours, so the constraint is represented by $2x + y \leq 60$.

5. False; the objective function to maximize profit is $10x + 14y$.

6. False; a corner point of the feasible region is $\left(\frac{30}{13}, \frac{75}{26}\right)$, or (2.308, 2.885), not (2, 3).

7. True

8. False; the optimal solution occurs at a corner point, and (2, 3) is not a corner point.

9. False; the half-planes of both sides of a linear inequality cannot contain the same points.

10. True

11. True

12. True

13. True

15. $y \geq 2x + 3$

Graph $y = 2x + 3$ as a solid line, using the intercepts (0, 3) and $\left(-\frac{3}{2}, 0\right)$. Using the origin as a test point, we get $0 \geq 2(0) + 3$ or $0 \geq 3$, which is false. Shade the region that does not contain the origin, that is, the half-plane above the line.

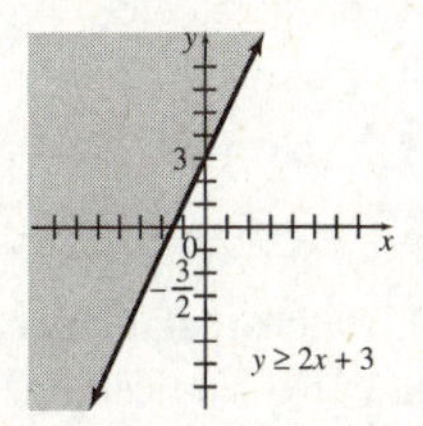

17. $2x + 6y \leq 8$

Graph $2x + 6y = 8$ as a solid line, using the intercepts $\left(0, \frac{4}{3}\right)$ and (4, 0). Using the origin as a test point, we get $0 \leq 8$, which is true. Shade the region that contains the origin, that is, the half-plane below the line.

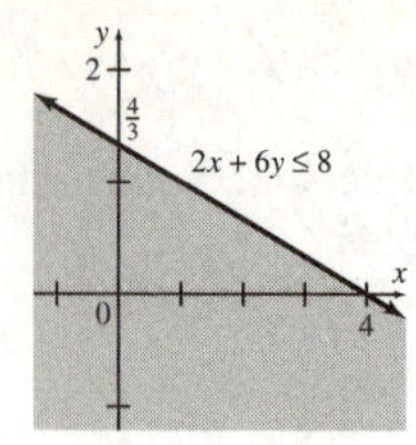

19. $y \geq x$

Graph $y = x$ as a solid line. Since this line contains the origin, choose a point other than (0, 0) as a test point. If we use (1, 4), we get $4 \geq 1$, which is true. Shade the region that contains the test point, that is, the half-plane above the line.

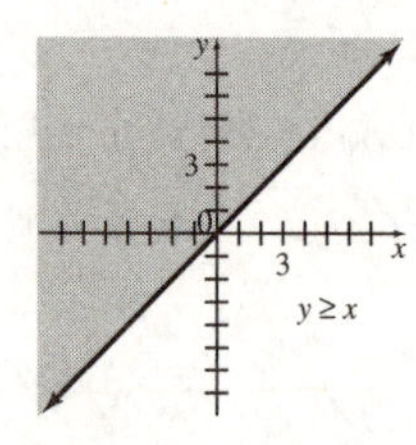

21. $x + y \leq 6$
$2x - y \geq 3$

$x + y \leq 6$ is the half-plane on or below the line $x + y = 6$; $2x - y \geq 3$ is the half-plane on or below the line $2x - y = 3$. Shade the overlapping part of these two half-planes, which is the region below both lines. The only corner point is the intersection of the two boundary lines, the point (3, 3).

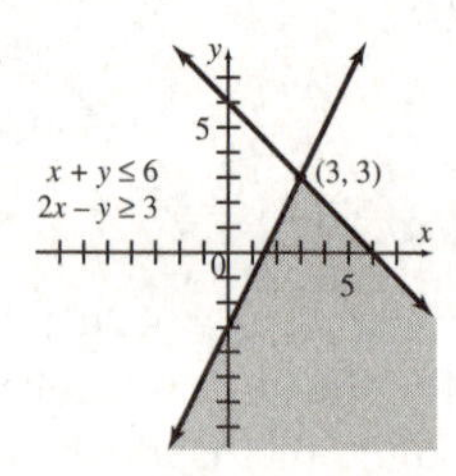

Unbounded

23. $-4 \leq x \leq 2$
$-1 \leq y \leq 3$
$x + y \leq 4$

$-4 \leq x \leq 2$ is the rectangular region lying on or between the two vertical lines, $x = -4$ and $x = 2$; $-1 \leq y \leq 3$ is the rectangular region lying on or between the two horizontal lines, $y = -1$ and $y = 3$; $x + y \leq 4$ is the half-plane lying on or below the line $x + y = 4$. Shade the overlapping part of these three regions. The corner points are $(-4, -1)$, $(-4, 3)$, $(1, 3)$, $(2, 2)$, and $(2, -1)$.

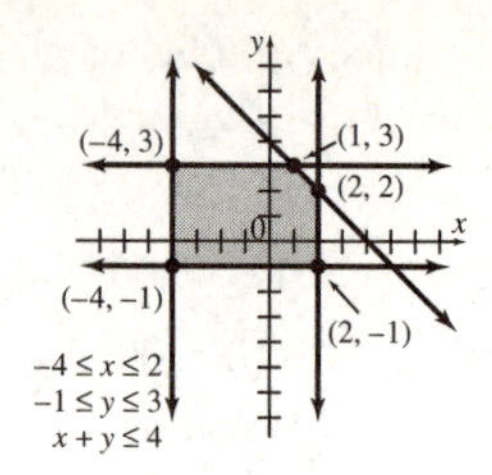

Bounded

25. $x + 2y \leq 4$
$5x - 6y \leq 12$
$x \geq 0$
$y \geq 0$

$x + 2y \leq 4$ is the half-plane on or below the line $x + 2y = 4$; $5x - 6y \leq 12$ is the half-plane on or above the line $5x - 6y = 12$; $x \geq 0$ and $y \geq 0$ together restrict the graph to the first quadrant. Shade the portion of the first quadrant where the half-planes overlap. The corner points are (0, 0), (0, 2), $\left(\frac{12}{5}, 0\right)$, and $\left(3, \frac{1}{2}\right)$, which can be found by solving the system

$$x + 2y = 4$$
$$5x - 6y = 12.$$

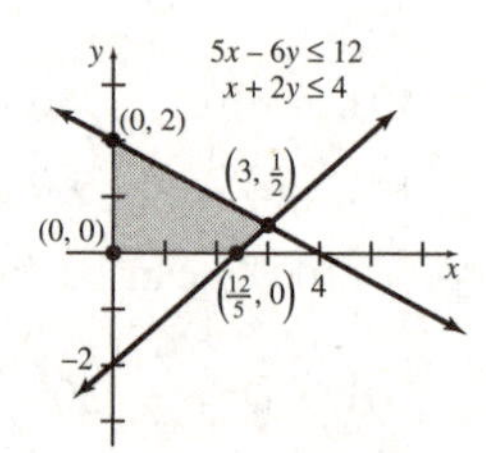

Bounded

27. Evaluate the objective function $z = 2x + 4y$ at each corner point.

Corner Point	Value of $z = 2x + 4y$	
(0, 0)	$2(0) + 4(0) = 0$	Minimum
(0, 4)	$2(0) + 4(4) = 16$	
(3, 4)	$2(3) + 4(4) = 22$	Maximum
(6, 2)	$2(6) + 4(2) = 20$	
(4, 0)	$2(4) + 4(0) = 8$	

The maximum value 22 occurs at (3, 4), and the minimum value of 0 occurs at (0, 0).

29. Maximize $z = 2x + 4y$

subject to: $3x + 2y \le 12$

$$5x + y \ge 5$$
$$x \ge 0$$
$$y \ge 0.$$

Sketch the feasible region in quadrant I.

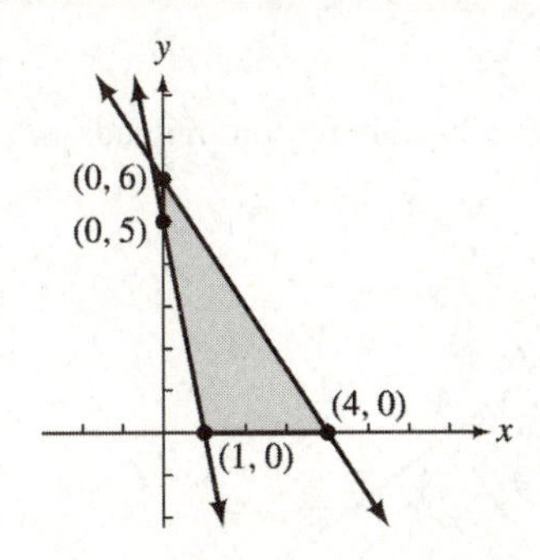

The corner points are (0, 5), (0, 6), (4, 0), and (1, 0).

Corner Point	Value of $z = 2x + 4y$
(0, 5)	20
(0, 6)	24 Maximum
(4, 0)	8
(1, 0)	2

The maximum value is 24 at (0, 6).

31. Minimize $z = 4x + 2y$

subject to: $x + y \le 50$

$$2x + y \ge 20$$
$$x + 2y \ge 30$$
$$x \ge 0$$
$$y \ge 0.$$

Sketch the feasible region.

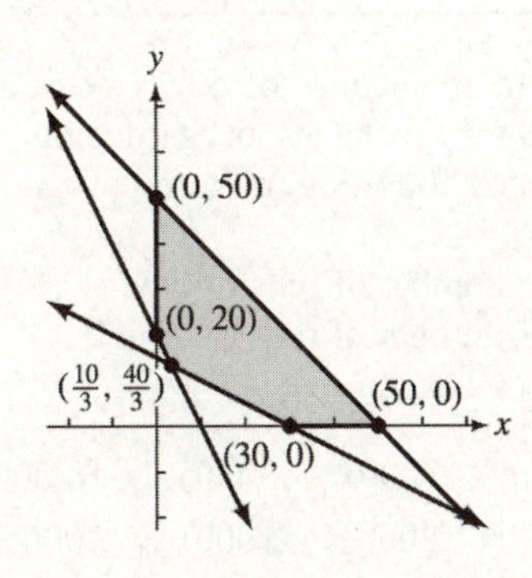

The corner points are (0, 20), $\left(\frac{10}{3}, \frac{40}{3}\right)$, (30, 0), (50, 0), and (0, 50). The corner point $\left(\frac{10}{3}, \frac{40}{3}\right)$ can be found by solving the system

$$2x + y = 20$$
$$x + 2y = 30.$$

Corner Point	Value of $z = 4x + 2y$
(0, 20)	40 Minimum
$\left(\frac{10}{3}, \frac{40}{3}\right)$	40 Minimum
(30, 0)	120
(50, 0)	200
(0, 50)	100

Thus, the minimum value is 40 and occurs at every point on the line segment joining (0, 20) and $\left(\frac{10}{3}, \frac{40}{3}\right)$.

35. Maximize $z = 3x + 4y$

subject to: $2x + y \le 4$

$$-x + 2y \le 4$$
$$x \ge 0$$
$$y \ge 0.$$

(a) Sketch the feasible region. All corner points except one can be read from the graph. Solving the system

$$2x + y = 4$$
$$-x + 2y = 4$$

gives the final corner point, $\left(\frac{4}{5}, \frac{12}{5}\right)$.

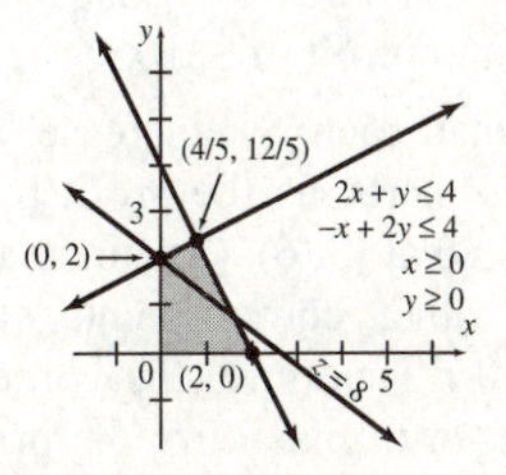

(b)

37. Let $x =$ the number of batches of cakes and $y =$ the number of batches of cookies.
Then we have the following inequalities.

$$2x + \frac{3}{2}y \leq 15 \quad \text{(oven time)}$$

$$3x + \frac{2}{3}y \leq 13 \quad \text{(decorating)}$$

$$x \geq 0$$

$$y \geq 0$$

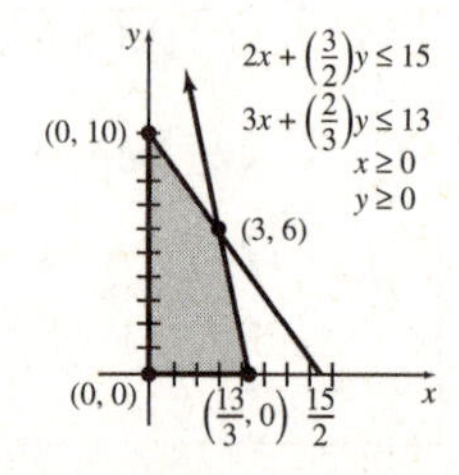

39. **(a)** From the graph for Exercise 37, the corner points are (0, 10), (3, 6), $\left(\frac{13}{3}, 0\right)$, and (0, 0).

Since x was the number of batches of cakes and y the number of batches of cookies, the revenue function is

$$z = 30x + 20y.$$

Evaluate this objective function at each corner point.

Corner Point	Value of $z = 30x + 20y$
(0, 10)	200
(3, 6)	210 Maximum
$\left(\frac{13}{3}, 0\right)$	130
(0, 0)	0

Therefore, 3 batches of cakes and 6 batches of cookies should be made to produce a maximum profit of $210.

(b) Note that each $1 increase in the price of cookies increases the profit by $10 at (0, 10) and by $6 at (3, 6). (The increase has no effect at the other corner points since, for those, $y = 0$.) Therefore, the corner point (0, 10) will begin to maximize the profit when x, the number of one-dollar increases, is larger than the solution to the following equation.

$$30(0) + (20 + 1x)(10) = 30(3) + (20 + 1x)(6)$$

$$(20 + x)(4) = 90$$

$$4x = 10$$

$$x = 2.5$$

If the profit per batch of cookies increases by more than $2.50 (to $22.50), then it will be more profitable to make 10 batches of cookies and no batches of cake.

41. Let $x =$ number of packages of gardening mixture and $y =$ number of packages of potting mixture.

Maximize $z = 3x + 5y$

subject to:
$$2x + y \leq 16$$
$$x + 2y \leq 11$$
$$x + 3y \leq 15$$
$$x \geq 0$$
$$y \geq 0.$$

Sketch the feasible region in quadrant I.

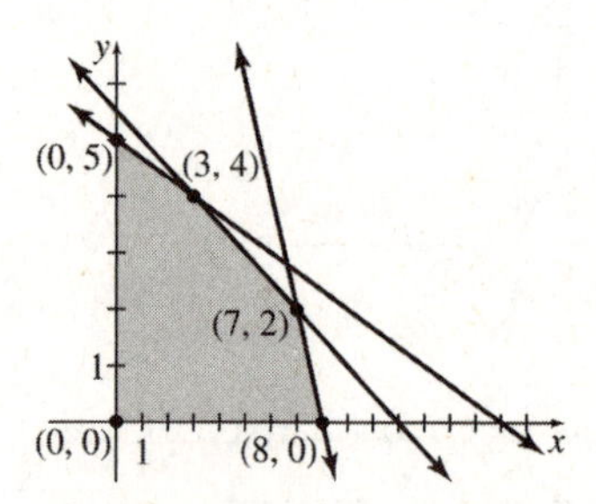

The corner points (0, 0), (0, 5), and (8, 0) can be identified from the graph. The corner point (3, 4) can be found by solving the system

$$x + 2y = 11$$
$$x + 3y = 15.$$

The corner point (7, 2) can be found by solving the system

$$2x + y = 16$$
$$x + 2y = 11.$$

Corner Point	Value of $z = 3x + 5y$
(0, 0)	0
(0, 5)	25
(3, 4)	29
(7, 2)	31 Maximum
(8, 0)	24

A maximum income of $31 can be achieved by preparing 7 packages of gardening mixture and 2 packages of potting mixture.

43. Let $x =$ number of runs of type I and $y =$ number of runs of type II.

Minimize $z = 15{,}000x + 6000y$

subject to:
$$3000x + 3000y \geq 18{,}000$$
$$2000x + 1000y \geq 7000$$
$$2000x + 3000y \geq 14{,}000$$
$$x \geq 0$$
$$y \geq 0.$$

Sketch the feasible region in quadrant I.

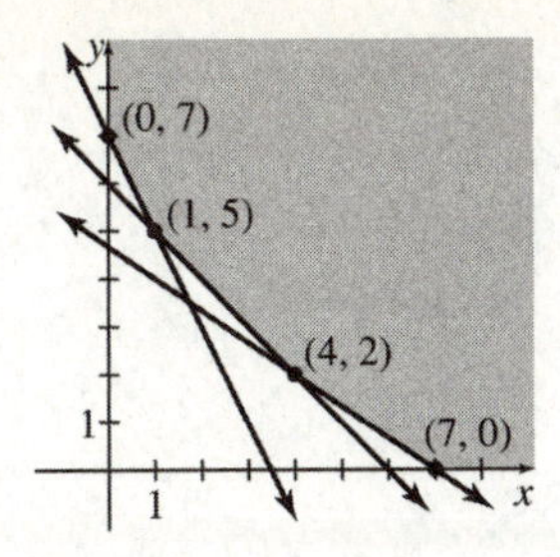

The corner points (0, 7) and (7, 0) can be identified from the graph. The corner point (1, 5) can be found by solving the system

$$3000x + 3000y = 18{,}000$$
$$2000x + 1000y = 7000.$$

The corner point (4, 2) can be found by solving the system

$$3000x + 3000y = 18{,}000$$
$$2000x + 3000y = 14{,}000.$$

Corner Point	Value of $z = 15{,}000x + 6000y$	
(0, 7)	42,000	Minimum
(1, 5)	45,000	
(4, 2)	72,000	
(7, 0)	105,000	

The company should produce 0 runs of type I and 7 runs of type II for a minimum cost of $42,000.

45. Let x = the number of acres devoted to millet and y = the number of acres devoted to wheat.

Maximize $z = 400x + 800y$

subject to: $36x + 8y \leq 48$

$$x + y \leq 2$$
$$x \geq 0$$
$$y \geq 0.$$

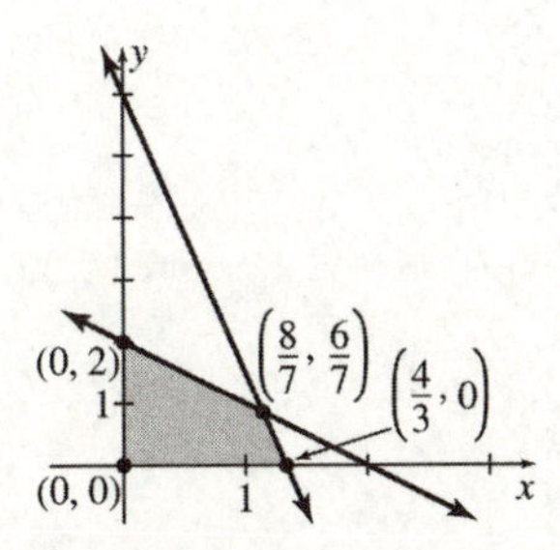

The corner points (0, 0) and (0, 2) can be identified from the graph. The corner point $\left(\frac{4}{3}, 0\right)$ can be found by solving the system

$$36x + 8y = 48$$
$$y = 0.$$

The corner point $\left(\frac{8}{7}, \frac{6}{7}\right)$ can be found by solving the system

$$36x + 8y = 48$$
$$x + y = 2.$$

Corner Point	Value of $z = 400x + 800y$	
(0, 0)	0	
(0, 2)	1600	Maximum
$\left(\frac{8}{7}, \frac{6}{7}\right)$	$\frac{8000}{7}$	
$\left(\frac{4}{3}, 0\right)$	$\frac{1600}{3}$	

The maximum amount of grain is 1600 pounds and can be obtained by planting 2 acres of wheat and no millet.

Chapter 4

LINEAR PROGRAMMING: THE SIMPLEX METHOD

4.1 Slack Variables and the Pivot

Your Turn 1

The two new equations are:

$$\begin{aligned} 300x_1 + 60x_2 + 180x_3 &\le 20{,}000 \\ 5x_1 + 10x_2 + 15x_3 &\le 900 \end{aligned}$$

The new answer tableau is:

$$\begin{array}{c} \begin{matrix} x_1 & x_2 & x_3 & s_1 & s_2 & s_3 & z \end{matrix} \\ \left[\begin{array}{rrrrrrr|r} 1 & 1 & 1 & 1 & 0 & 0 & 0 & 100 \\ 15 & 3 & 9 & 0 & 1 & 0 & 0 & 1000 \\ 1 & 2 & 3 & 0 & 0 & 1 & 0 & 180 \\ \hline -120 & -40 & -60 & 0 & 0 & 0 & 1 & 0 \end{array}\right] \end{array}$$

Your Turn 2

Pivot around the indicated 6.

$$\begin{array}{c} \begin{matrix} x_1 & x_2 & x_3 & s_1 & s_2 & s_3 & z \end{matrix} \\ \left[\begin{array}{rrrrrrr|r} 3 & \boxed{6} & 2 & 1 & 0 & 0 & 0 & 60 \\ 8 & 5 & 4 & 0 & 1 & 0 & 0 & 80 \\ 3 & 6 & 7 & 0 & 0 & 1 & 0 & 120 \\ \hline -30 & -50 & -15 & 0 & 0 & 0 & 1 & 0 \end{array}\right] \end{array}$$

The result is:

$$\begin{array}{r} \\ \\ -5R_1 + 6R_2 \to R_2 \\ -R_1 + R_3 \to R_3 \\ 25R_1 + 3R_4 \to R_4 \end{array} \begin{array}{c} \begin{matrix} x_1 & x_2 & x_3 & s_1 & s_2 & s_3 & z \end{matrix} \\ \left[\begin{array}{rrrrrrr|r} 3 & \boxed{6} & 2 & 1 & 0 & 0 & 0 & 60 \\ 33 & 0 & 14 & -5 & 6 & 0 & 0 & 180 \\ 0 & 0 & 5 & -1 & 0 & 1 & 0 & 60 \\ \hline -15 & 0 & 5 & 25 & 0 & 0 & 3 & 1500 \end{array}\right] \end{array}$$

$$\begin{array}{r} \\ \frac{1}{6}R_1 \to R_1 \\ \frac{1}{6}R_2 \to R_2 \\ \\ \frac{1}{3}R_4 \to R_4 \end{array} \begin{array}{c} \begin{matrix} x_1 & x_2 & x_3 & s_1 & s_2 & s_3 & z \end{matrix} \\ \left[\begin{array}{rrrrrrr|r} 1/2 & 1 & 1/3 & 1/6 & 0 & 0 & 0 & 10 \\ 11 & 0 & 7/3 & -5/6 & 1 & 0 & 0 & 30 \\ 0 & 0 & 5 & -1 & 0 & 1 & 0 & 60 \\ \hline -5 & 0 & 5/3 & 25/3 & 0 & 0 & 1 & 500 \end{array}\right] \end{array}$$

The solution given by this tableau is:

$$\begin{aligned} &x_1 = 0, x_2 = 10, x_3 = 0, s_1 = 0, \\ &s_2 = 30, s_3 = 60, z = 500 \end{aligned}$$

4.1 Exercises

1. $x_1 + 2x_2 \le 6$

Add s_1 to the given inequality to obtain

$$x_1 + 2x_2 + s_1 = 6.$$

3. $2.3x_1 + 5.7x_2 + 1.8x_3 \le 17$

Add s_1 to the given inequality to obtain

$$2.3x_1 + 5.7x_2 + 1.8x_3 + s_1 = 17.$$

5. (a) Since there are three constraints to be converted into equations we need three slack variables.

(b) We use s_1, s_2, and s_3 for the slack variables.

(c) The equations are

$$\begin{aligned} 2x_1 + 3x_2 + s_1 \phantom{{}+ s_2 + s_3} &= 15 \\ 4x_1 + 5x_2 \phantom{{}+ s_1} + s_2 \phantom{{}+ s_3} &= 35 \\ x_1 + 6x_2 \phantom{{}+ s_1 + s_2} + s_3 &= 20. \end{aligned}$$

7. (a) There are two constraints to be converted into equations, so we must introduce two slack variables.

(b) Call the slack variables s_1 and s_2.

(c) The equations are

$$\begin{aligned} 7x_1 + 6x_2 + 8x_3 + s_1 \phantom{{}+ s_2} &= 118 \\ 4x_1 + 5x_2 + 10x_3 \phantom{{}+ s_1} + s_2 &= 220. \end{aligned}$$

9. Find $x_1 \ge 0$ and $x_2 \ge 0$ such that

$$\begin{aligned} 4x_1 + 2x_2 &\le 5 \\ x_1 + 2x_2 &\le 4 \end{aligned}$$

and $z = 7x_1 + x_2$ is maximized.

We need two slack variables, s_1 and s_2. Then the problem can be restated as:

Find $x_1 \ge 0, x_2 \ge 0, s_1 \ge 0$, and $s_2 \ge 0$ such that

$$\begin{aligned} 4x_1 + 2x_2 + s_1 \phantom{{}+ s_2} &= 5 \\ x_1 + 2x_2 \phantom{{}+ s_1} + s_2 &= 4. \end{aligned}$$

and $z = 7x_1 + x_2$ is maximized.

Rewrite the objective function as

$$-7x_1 - x_2 + z = 0.$$

The initial simplex tableau is

$$\begin{array}{c} \begin{matrix} x_1 & x_2 & s_1 & s_2 & z \end{matrix} \\ \left[\begin{array}{rrrrr|r} 4 & 2 & 1 & 0 & 0 & 5 \\ 1 & 2 & 0 & 1 & 0 & 4 \\ \hline -7 & -1 & 0 & 0 & 1 & 0 \end{array}\right]. \end{array}$$

11. Find $x_1 \geq 0$ and $x_2 \geq 0$ such that

$$\begin{aligned} x_1 + x_2 &\leq 10 \\ 5x_1 + 2x_2 &\leq 20 \\ x_1 + 2x_2 &\leq 36 \end{aligned}$$

and $z = x_1 + 3x_2$ is maximized.

Using slack variables $s_1, s_2,$ and s_3, the problem can be restated as:

Find $x_1 \geq 0, x_2 \geq 0, s_1 \geq 0,$ $s_2 \geq 0,$ and $s_3 \geq 0$ such that

$$\begin{aligned} x_1 + x_2 + s_1 \qquad\qquad &= 10 \\ 5x_1 + 2x_2 \qquad + s_2 \qquad &= 20 \\ x_1 + 2x_2 \qquad\qquad + s_3 &= 36 \end{aligned}$$

and $z = x_1 + 3x_2$ is maximized.

Rewrite the objective function as

$$-x_1 - 3x_2 + z = 0.$$

The initial simplex tableau is

$$\begin{array}{c} \begin{matrix} x_1 & x_2 & s_1 & s_2 & s_3 & z \end{matrix} \\ \left[\begin{array}{rrrrrr|r} 1 & 1 & 1 & 0 & 0 & 0 & 10 \\ 5 & 2 & 0 & 1 & 0 & 0 & 20 \\ 1 & 2 & 0 & 0 & 1 & 0 & 36 \\ \hline -1 & -3 & 0 & 0 & 0 & 1 & 0 \end{array}\right]. \end{array}$$

13. Find $x_1 \geq 0$ and $x_2 \geq 0$ such that

$$\begin{aligned} 3x_1 + x_2 &\leq 12 \\ x_1 + x_2 &\leq 15 \end{aligned}$$

and $z = 2x_1 + x_2$ is maximized.

Using slack variables s_1 and s_2, the problem can be restated as:

Find $x_1 \geq 0, x_2 \geq 0, s_1 \geq 0,$ and $s_2 \geq 0$ such that

$$\begin{aligned} 3x_1 + x_2 + s_1 \qquad &= 12 \\ x_1 + x_2 \qquad + s_2 &= 15 \end{aligned}$$

and $z = 2x_1 + x_2$ is maximized.

Rewrite the objective function as

$$-2x_1 - x_2 + z = 0.$$

The initial simplex tableau is

$$\begin{array}{c} \begin{matrix} x_1 & x_2 & s_1 & s_2 & z \end{matrix} \\ \left[\begin{array}{rrrrr|r} 3 & 1 & 1 & 0 & 0 & 12 \\ 1 & 1 & 0 & 1 & 0 & 15 \\ \hline -2 & -1 & 0 & 0 & 1 & 0 \end{array}\right]. \end{array}$$

15.

$$\begin{array}{c} \begin{matrix} x_1 & x_2 & x_3 & s_1 & s_2 & z \end{matrix} \\ \left[\begin{array}{rrrrrr|r} 1 & 0 & 4 & 5 & 1 & 0 & 8 \\ 3 & 1 & 1 & 2 & 0 & 0 & 4 \\ \hline -2 & 0 & 2 & 3 & 0 & 1 & 28 \end{array}\right] \end{array}$$

The variables x_2 and s_2 are basic variables, because the columns for these variables have all zeros except for one nonzero entry. If the remaining variables x_1, x_3, and s_1 are zero, then $x_2 = 4$ and $s_2 = 8$. From the bottom row, $z = 28$. The basic feasible solution is $x_1 = 0,$ $x_2 = 4,$ $x_3 = 0,$ $s_1 = 0,$ $s_2 = 8,$ and $z = 28$.

17.

$$\begin{array}{c} \begin{matrix} x_1 & x_2 & x_3 & s_1 & s_2 & s_3 & z \end{matrix} \\ \left[\begin{array}{rrrrrrr|r} 6 & 2 & 2 & 3 & 0 & 0 & 0 & 16 \\ 2 & 2 & 0 & 1 & 0 & 5 & 0 & 35 \\ 2 & 1 & 0 & 3 & 1 & 0 & 0 & 6 \\ \hline -3 & -2 & 0 & 2 & 0 & 0 & 3 & 36 \end{array}\right] \end{array}$$

The basic variables are x_3, s_2, and s_3. If x_1, x_2, and s_1 are zero, then $2x_3 = 16$, so $x_3 = 8$. Similarly, $s_2 = 6$ and $5s_3 = 35$, so $s_3 = 7$. From the bottom row, $3z = 36$, so $z = 12$. The basic feasible solution is $x_1 = 0, x_2 = 0, x_3 = 8, s_1 = 0,$ $s_2 = 6, s_3 = 7,$ and $z = 12$.

19.

$$\begin{array}{c} \begin{matrix} x_1 & x_2 & x_3 & s_1 & s_2 & z \end{matrix} \\ \left[\begin{array}{rrrrrr|r} 1 & 2 & 4 & 1 & 0 & 0 & 56 \\ 2 & \boxed{2} & 1 & 0 & 1 & 0 & 40 \\ \hline -1 & -3 & -2 & 0 & 0 & 1 & 0 \end{array}\right] \end{array}$$

Clear the x_2 column.

$$-R_2 + R_1 \rightarrow R_1 \begin{array}{c} \begin{matrix} x_1 & x_2 & x_3 & s_1 & s_2 & z \end{matrix} \\ \left[\begin{array}{rrrrrr|r} -1 & 0 & 3 & 1 & -1 & 0 & 16 \\ 2 & \boxed{2} & 1 & 0 & 1 & 0 & 40 \\ \hline -1 & -3 & -2 & 0 & 0 & 1 & 0 \end{array}\right] \end{array}$$

$$\begin{array}{r} \\ \\ 3R_2 + 2R_3 \rightarrow R_3 \end{array} \left[\begin{array}{cccccc|c} x_1 & x_2 & x_3 & s_1 & s_2 & z & \\ -1 & 0 & 3 & 1 & -1 & 0 & 16 \\ 2 & 2 & 1 & 0 & 1 & 0 & 40 \\ \hline 4 & 0 & -1 & 0 & 3 & 2 & 120 \end{array}\right]$$

x_2 and s_1 are now basic. The solution is $x_1 = 0$, $x_2 = 20$, $x_3 = 0$, $s_1 = 16$, $s_2 = 0$, and $z = 60$.

21.

$$\left[\begin{array}{ccccccc|c} x_1 & x_2 & x_3 & s_1 & s_2 & s_3 & z & \\ 2 & 2 & \boxed{1} & 1 & 0 & 0 & 0 & 12 \\ 1 & 2 & 3 & 0 & 1 & 0 & 0 & 45 \\ 3 & 1 & 1 & 0 & 0 & 1 & 0 & 20 \\ \hline -2 & -1 & -3 & 0 & 0 & 0 & 1 & 0 \end{array}\right]$$

Clear the x_3 column.

$$\begin{array}{r} \\ \\ -3R_1 + R_2 \rightarrow R_2 \\ -R_1 + R_3 \rightarrow R_3 \\ 3R_1 + R_4 \rightarrow R_4 \end{array} \left[\begin{array}{ccccccc|c} x_1 & x_2 & x_3 & s_1 & s_2 & s_3 & z & \\ 2 & 2 & 1 & 1 & 0 & 0 & 0 & 12 \\ -5 & -4 & 0 & -3 & 1 & 0 & 0 & 9 \\ 1 & -1 & 0 & -1 & 0 & 1 & 0 & 8 \\ \hline 4 & 5 & 0 & 3 & 0 & 0 & 1 & 36 \end{array}\right]$$

$x_3, s_2,$ and s_3 are now basic. The solution is $x_1 = 0$, $x_2 = 0$, $x_3 = 12$, $s_1 = 0$, $s_2 = 9$, $s_3 = 8$, and $z = 36$.

23.

$$\left[\begin{array}{ccccccc|c} x_1 & x_2 & x_3 & s_1 & s_2 & s_3 & z & \\ 2 & \boxed{2} & 3 & 1 & 0 & 0 & 0 & 500 \\ 4 & 1 & 1 & 0 & 1 & 0 & 0 & 300 \\ 7 & 2 & 4 & 0 & 0 & 1 & 0 & 700 \\ \hline -3 & -4 & -2 & 0 & 0 & 0 & 1 & 0 \end{array}\right]$$

Clear the x_2 column.

$$\begin{array}{r} \\ \\ -R_1 + 2R_2 \rightarrow R_2 \\ -R_1 + R_3 \rightarrow R_3 \\ 2R_1 + R_4 \rightarrow R_4 \end{array} \left[\begin{array}{ccccccc|c} x_1 & x_2 & x_3 & s_1 & s_2 & s_3 & z & \\ 2 & \boxed{2} & 3 & 1 & 0 & 0 & 0 & 500 \\ 6 & 0 & -1 & -1 & 2 & 0 & 0 & 100 \\ 5 & 0 & 1 & -1 & 0 & 1 & 0 & 200 \\ \hline 1 & 0 & 4 & 2 & 0 & 0 & 1 & 1000 \end{array}\right]$$

$x_2, s_2,$ and s_3 are now basic. Thus, the solution is $x_1 = 0$, $x_2 = 250$, $x_3 = 0$, $s_1 = 0$, $s_2 = 50$, $s_3 = 200$, and $z = 1000$.

25. A slack variable (a nonnegative quantity), converts a linear inequality into a linear equation by adding the amount needed in an expression to be equal to a specific value.

27. Let x_1 represent the number of simple figures, x_2 the number of figures with additions, and x_3 the number of computer-drawn sketches. Organize the information in a table.

	Simple Figures	Figures with Additions	Computer -Drawn Sketches	Maximum Allowed
Cost	20	35	60	2200
Royalties	95	200	325	

The cost constraint is

$$20x_1 + 35x_2 + 60x_3 \le 2200.$$

The limit of 400 figures leads to the constraint

$$x_1 + x_2 + x_3 \le 400.$$

The other stated constraints are

$$x_3 \le x_1 + x_2 \text{ and } x_1 \ge 2x_2,$$

and these can be rewritten in standard form as

$$-x_1 - x_2 + x_3 \le 0 \text{ and } -x_1 + 2x_2 \le 0$$

respectively. The problem may be stated as:

Find $x_1 \ge 0$, $x_2 \ge 0$, and $x_3 \ge 0$ such that

$$\begin{aligned} 20x_1 + 35x_2 + 60x_3 &\le 2200 \\ x_1 + x_2 + x_3 &\le 400 \\ -x_1 - x_2 + x_3 &\le 0 \\ -x_1 + 2x_2 &\le 0 \end{aligned}$$

and $z = 95x_1 + 200x_2 + 325x_3$ is maximized.

Introduce slack variables s_1, s_2, s_3, and s_4, and the problem can be restated as:

Find $x_1 \ge 0$, $x_2 \ge 0$, $x_3 \ge 0$, $s_1 \ge 0$, $s_2 \ge 0$, $s_3 \ge 0$, and $s_4 \ge 0$ such that

$$\begin{aligned} 20x_1 + 35x_2 + 60x_3 + s_1 &= 2200 \\ x_1 + x_2 + x_3 + s_2 &= 400 \\ -x_1 - x_2 + x_3 + s_3 &= 0 \\ -x_1 + 2x_2 + s_4 &= 0 \end{aligned}$$

and $z = 95x_1 + 200x_2 + 325x_3$ is maximized.

Rewrite the objective function as

$$-95x_1 - 200x_2 - 325x_3 + z = 0.$$

The initial simlpex tableau is

$$\begin{array}{cccccccc|c} x_1 & x_2 & x_3 & s_1 & s_2 & s_3 & s_4 & z & \\ 20 & 35 & 60 & 1 & 0 & 0 & 0 & 0 & 2200 \\ 1 & 1 & 1 & 0 & 1 & 0 & 0 & 0 & 400 \\ -1 & -1 & 1 & 0 & 0 & 1 & 0 & 0 & 0 \\ -1 & 2 & 0 & 0 & 0 & 0 & 1 & 0 & 0 \\ \hline -95 & -200 & -325 & 0 & 0 & 0 & 0 & 1 & 0 \end{array}.$$

29. Let x_1 represent the number of redwood tables, x_2 the number of stained Douglas fir tables, and x_3 the number of stained white spruce tables. Organize the information in a table.

	Redwood	Douglas Fir	White Spruce	Maximum Available
Assembly Time	8	7	8	90 8-hr days = 720 hr
Staining Time	0	2	2	60 8-hr days = 480 hr
Cost	\$159	\$138.85	\$129.35	\$15,000

The limit of 720 hr for carpenters leads to the constraint

$$8x_1 + 7x_2 + 8x_3 \le 720.$$

The limit of 480 hr for staining leads to the constraint

$$2x_2 + 2x_3 \le 480.$$

The cost constraint is

$$159x_1 + 138.85x_2 + 129.35x_3 \le 15{,}000.$$

The problem may be stated as:

Find $x_1 \ge 0$, $x_2 \ge 0$, and $x_3 \ge 0$ such that

$$\begin{aligned} 8x_1 + 7x_2 + 8x_3 &\le 720 \\ 2x_2 + 2x_3 &\le 480 \\ 159x_1 + 138.85x_2 + 129.35x_3 &\le 15{,}000 \end{aligned}$$

and $z = x_1 + x_2 + x_3$ is maximized.

Introduce slack variables s_1, s_2, and s_3, and the problem can be restated as:

Find $x_1 \ge 0$, $x_2 \ge 0$, $x_3 \ge 0$, $s_1 \ge 0$, $s_2 \ge 0$, and $s_3 \ge 0$ such that

$$\begin{aligned} 8x_1 + 7x_2 + 8x_3 + s_1 &= 720 \\ 2x_2 + 2x_3 + s_2 &= 480 \\ 159x_1 + 138.85x_2 + 129.35x_3 + s_3 &= 15{,}000 \end{aligned}$$

and $z = x_1 + x_2 + x_3$ is maximized.

Rewrite the objective function as

$$-x_1 - x_2 - x_3 + z = 0.$$

The initial simplex tableau is

$$\begin{array}{ccccccc|c} x_1 & x_2 & x_3 & s_1 & s_2 & s_3 & z & \\ 8 & 7 & 8 & 1 & 0 & 0 & 0 & 720 \\ 0 & 2 & 2 & 0 & 1 & 0 & 0 & 480 \\ 159 & 138.85 & 129.35 & 0 & 0 & 1 & 0 & 15{,}000 \\ \hline -1 & -1 & -1 & 0 & 0 & 0 & 1 & 0 \end{array}.$$

31. Let x_1 = the number of newspaper ads,
x_2 = the number of internet banners,
and x_3 = the number of TV ads.

Organize the information in a table.

	Newspaper Ads	Internet Banners	TV Ads
Cost per Ad	400	20	2000
Maximum Number	30	60	10
Women Seeing Ad	4000	3000	10,000

The cost constraint is

$$400x_1 + 20x_2 + 2000x_3 \le 8000$$

The constraints on the numbers of ads is

$$x_1 \le 30$$
$$x_2 \le 60$$
$$x_3 \le 10.$$

The problem may be stated as:

Find $x_1 \ge 0$, $x_2 \ge 0$, and $x_3 \ge 0$ such that

$$\begin{aligned} 400x_1 + 20x_2 + 2000x_3 &\le 8000 \\ x_1 &\le 30 \\ x_2 &\le 60 \\ x_3 &\le 10 \end{aligned}$$

and $z = 4000x_1 + 3000x_2 + 10{,}000x_3$ is maximized.

Introduce slack variables s_1, s_2, s_3 and s_4, and the problem can be restated as:

Find $x_1 \ge 0$, $x_2 \ge 0$, $x_3 \ge 0$, $s_1 \ge 0$, $s_2 \ge 0$, $s_3 \ge 0$, and $s_4 \ge 0$ such that

$$\begin{aligned} 400x_1 + 20x_2 + 2000x_3 + s_1 &= 8000 \\ x_1 + s_2 &= 30 \\ x_2 + s_3 &= 60 \\ x_3 + s_4 &= 10 \end{aligned}$$

and $z = 4000x_1 + 3000x_2 + 10{,}000x_3$ is maximized.

Rewrite the objective function as

$$-4000x_1 - 3000x_2 - 10{,}000x_3 + z = 0.$$

The initial simplex tableau is

x_1	x_2	x_3	s_1	s_2	s_3	s_4	z	
400	20	2000	1	0	0	0	0	8000
1	0	0	0	1	0	0	0	30
0	1	0	0	0	1	0	0	60
0	0	1	0	0	0	1	0	10
−4000	−3000	−10,000	0	0	0	0	1	0

4.2 Maximization Problems

Your Turn 1

Example 2 of Section 3.3 yields the following linear programming problem, where we have renamed x and y as x_1 and x_2.

Maximize $\quad z = 12x_1 + 40x_2$

subject to:

$$\begin{aligned} x_1 + x_2 &\le 16 \\ x_1 + 3x_2 &\le 36 \\ x_1 &\le 10 \\ x_1 &\ge 0 \\ x_2 &\ge 0 \end{aligned}$$

Add a slack variable to each of the first three constraints:

$$\begin{aligned} x_1 + x_2 + s_1 &\le 16 \\ x_1 + 3x_2 + s_2 &\le 36 \\ x_1 + s_3 &\le 10 \end{aligned}$$

with $x_1 \ge 0,\ x_2 \ge 0,\ s_1 \ge 0,\ s_2 \ge 0,\ s_3 \ge 0$

The corresponding initial tableau is

x_1	x_2	s_1	s_2	s_3	z	
1	1	1	0	0	0	16
1	[3]	0	1	0	0	36
1	0	0	0	1	0	10
−12	−40	0	0	0	1	0

Since the most negative indicator is −40 and the quotient 36/3 is smaller than 16/1, we pivot on the 3 in column 2:

	x_1	x_2	s_1	s_2	s_3	z	
$-R_2 + 3R_1 \to R_1$	2	0	3	−1	0	0	12
	1	3	0	1	0	0	36
	1	0	0	0	1	0	10
$40R_2 + 3R_4 \to R_4$	4	0	0	40	0	3	1440

There are now no negative indicators, so we can read the solution:

$$x_1 = 0,\ x_2 = \frac{36}{3} = 12,\ z = \frac{1440}{3} = 480$$

Your Turn 2

Pivot on the 4 in column 2 of the following tableau.

x_1	x_2	s_1	s_2	s_3	z	
1	−2	1	0	0	0	100
3	[4]	0	1	0	0	200
5	0	0	0	1	0	150
−10	−25	0	0	0	1	0

	x_1	x_2	s_1	s_2	s_3	z	
$R_2 + 2R_1 \to R_1$	5	0	2	1	0	0	400
	3	4	0	1	0	0	200
	5	0	0	0	1	0	150
$25R_2 + 4R_4 \to R_4$	35	0	0	25	0	4	5000

There are no negative indicators, so the optimal solution is:

$$x_1 = 0,\ x_2 = \frac{200}{4} = 50,\ s_1 = 200,$$

$$s_2 = 0,\ s_3 = 150,\ z = \frac{5000}{4} = 1250$$

4.2 Exercises

1.

x_1	x_2	x_3	s_1	s_2	z	
1	4	4	1	0	0	16
2	1	5	0	1	0	20
−3	−1	−2	0	0	1	0

The most negative indicator is −3, in the first column. Find the quotients $\frac{16}{1} = 16$ and $\frac{20}{2} = 10$; since 10 is the smaller quotient, 2 in row 2, column 1 is the pivot.

	x_1	x_2	x_3	s_1	s_2	z	
$\frac{16}{1} = 16$	1	4	4	1	0	0	16
$\frac{20}{2} = 10$	[2]	1	5	0	1	0	20
	−3	−1	−2	0	0	1	0

Performing row transformations, we get the following tableau.

$$\begin{array}{r} \\ -R_2+2R_1 \rightarrow R_1 \\ \\ 3R_2+2R_3 \rightarrow R_3 \end{array} \begin{array}{cccccc|c} x_1 & x_2 & x_3 & s_1 & s_2 & z & \\ \hline 0 & 7 & 3 & 2 & -1 & 0 & 12 \\ 2 & 1 & 5 & 0 & 1 & 0 & 20 \\ \hline 0 & 1 & 11 & 0 & 3 & 2 & 60 \end{array}$$

All of the numbers in the last row are nonnegative, so we are finished pivoting. Create a 1 in the columns corresponding to x_1, s_1 and z.

$$\begin{array}{r} \\ \frac{1}{2}R_1 \rightarrow R_1 \\ \frac{1}{2}R_2 \rightarrow R_2 \\ \frac{1}{2}R_3 \rightarrow R_3 \end{array} \begin{array}{cccccc|c} x_1 & x_2 & x_3 & s_1 & s_2 & z & \\ \hline 0 & \frac{7}{2} & \frac{3}{2} & 1 & -\frac{1}{2} & 0 & 6 \\ 1 & \frac{1}{2} & \frac{5}{2} & 0 & \frac{1}{2} & 0 & 10 \\ \hline 0 & \frac{1}{2} & \frac{11}{2} & 0 & \frac{3}{2} & 1 & 30 \end{array}$$

The maximum value is 30 and occurs when $x_1 = 10,\ x_2 = 0,\ x_3 = 0,\ s_1 = 6$, and $s_2 = 0$.

3.

$$\begin{array}{cccccc|c} x_1 & x_2 & s_1 & s_2 & s_3 & z & \\ \hline 1 & 3 & 1 & 0 & 0 & 0 & 12 \\ 2 & 1 & 0 & 1 & 0 & 0 & 10 \\ 1 & 1 & 0 & 0 & 1 & 0 & 4 \\ \hline -2 & -1 & 0 & 0 & 0 & 1 & 0 \end{array}$$

The most negative indicator is -2, in the first column. Find the quotients $\frac{12}{1} = 12$, $\frac{10}{2} = 5$, and $\frac{4}{1} = 4$; since 4 is the smallest quotient, 1 in row 3, column 1 is the pivot.

$$\begin{array}{cccccc|c} x_1 & x_2 & s_1 & s_2 & s_3 & z & \\ \hline 1 & 3 & 1 & 0 & 0 & 0 & 12 \\ 2 & 1 & 0 & 1 & 0 & 0 & 10 \\ \boxed{1} & 1 & 0 & 0 & 1 & 0 & 4 \\ \hline -2 & -1 & 0 & 0 & 0 & 1 & 0 \end{array}$$

$$\begin{array}{r} \\ -R_3+R_1 \rightarrow R_1 \\ -2R_3+R_2 \rightarrow R_2 \\ \\ 2R_3+R_4 \rightarrow R_4 \end{array} \begin{array}{cccccc|c} x_1 & x_2 & s_1 & s_2 & s_3 & z & \\ \hline 0 & 2 & 1 & 0 & -1 & 0 & 8 \\ 0 & -1 & 0 & 1 & -2 & 0 & 2 \\ 1 & 1 & 0 & 0 & 1 & 0 & 4 \\ \hline 0 & 1 & 0 & 0 & 2 & 1 & 8 \end{array}$$

This is a final tableau since all of the numbers in the last row are nonnegative. The maximum value is 8 when $x_1 = 4,\ x_2 = 0,\ s_1 = 8,\ s_2 = 2$, and $s_3 = 0$.

5.

$$\begin{array}{ccccccc|c} x_1 & x_2 & x_3 & s_1 & s_2 & s_3 & z & \\ \hline 2 & 2 & 8 & 1 & 0 & 0 & 0 & 40 \\ 4 & -5 & 6 & 0 & 1 & 0 & 0 & 60 \\ 2 & -2 & 6 & 0 & 0 & 1 & 0 & 24 \\ \hline -14 & -10 & -12 & 0 & 0 & 0 & 1 & 0 \end{array}$$

The most negative indicator is -14, in the first column. Find the quotients $\frac{40}{2} = 20$, $\frac{60}{4} = 15$, and $\frac{24}{2} = 12$; since 12 is the smallest quotient, 2 in row 3, column 1 is the pivot.

$$\begin{array}{ccccccc|c} x_1 & x_2 & x_3 & s_1 & s_2 & s_3 & z & \\ \hline 2 & 2 & 8 & 1 & 0 & 0 & 0 & 40 \\ 4 & -5 & 6 & 0 & 1 & 0 & 0 & 60 \\ \boxed{2} & -2 & 6 & 0 & 0 & 1 & 0 & 24 \\ \hline -14 & -10 & -12 & 0 & 0 & 0 & 1 & 0 \end{array}$$

Performing row transformations, we get the following tableau.

$$\begin{array}{r} \\ -R_3+R_1 \rightarrow R_1 \\ -2R_3+R_2 \rightarrow R_2 \\ \\ 7R_3+R_4 \rightarrow R_4 \end{array} \begin{array}{ccccccc|c} x_1 & x_2 & x_3 & s_1 & s_2 & s_3 & z & \\ \hline 0 & \boxed{4} & 2 & 1 & 0 & -1 & 0 & 16 \\ 0 & -1 & -6 & 0 & 1 & -2 & 0 & 12 \\ 2 & -2 & 6 & 0 & 0 & 1 & 0 & 24 \\ \hline 0 & -24 & 30 & 0 & 0 & 7 & 1 & 168 \end{array}$$

Since there is still a negative indicator, we must repeat the process. The second pivot is the 4 in column 2, since $\frac{16}{4}$ is the only nonnegative quotient in the only column with a negative indicator. Performing row transformations again, we get the following tableau.

$$\begin{array}{r} \\ \\ R_1+4R_2 \rightarrow R_2 \\ R_1+2R_3 \rightarrow R_3 \\ 6R_1+R_4 \rightarrow R_4 \end{array} \begin{array}{ccccccc|c} x_1 & x_2 & x_3 & s_1 & s_2 & s_3 & z & \\ \hline 0 & 4 & 2 & 1 & 0 & -1 & 0 & 16 \\ 0 & 0 & -22 & 1 & 4 & -9 & 0 & 64 \\ 4 & 0 & 14 & 1 & 0 & 1 & 0 & 64 \\ \hline 0 & 0 & 42 & 6 & 0 & 1 & 1 & 264 \end{array}$$

All of the numbers in the last row are nonnegative, so we are finished pivoting. Create a 1 in the columns corresponding to x_1, x_2, and s_2.

$$\begin{array}{r} \\ \frac{1}{4}R_1 \rightarrow R_1 \\ \frac{1}{4}R_2 \rightarrow R_2 \\ \frac{1}{4}R_3 \rightarrow R_3 \\ \\ \end{array} \begin{array}{ccccccc|c} x_1 & x_2 & x_3 & s_1 & s_2 & s_3 & z & \\ \hline 0 & 1 & \frac{1}{2} & \frac{1}{4} & 0 & -\frac{1}{4} & 0 & 4 \\ 0 & 0 & -\frac{11}{2} & \frac{1}{4} & 1 & -\frac{9}{4} & 0 & 16 \\ 1 & 0 & \frac{7}{2} & \frac{1}{4} & 0 & \frac{1}{4} & 0 & 16 \\ \hline 0 & 0 & 42 & 6 & 0 & 1 & 1 & 264 \end{array}$$

The maximum value is 264 and occurs when $x_1 = 16,\ x_2 = 4,\ x_3 = 0,\ s_1 = 0,\ s_2 = 16$, and $s_3 = 0$.

7. Maximize $z = 3x_1 + 5x_2$

subject to: $4x_1 + x_2 \leq 25$

$2x_1 + 3x_2 \leq 15$

with $x_1 \geq 0, x_2 \geq 0.$

Two slack variables, s_1 and s_2, need to be introduced. The problem can be restated as:

Maximize $z = 3x_1 + 5x_2$

subject to: $4x_1 + x_2 + s_1 = 25$

$2x_1 + 3x_2 + s_2 = 15$

with $x_1 \geq 0, x_2 \geq 0, s_1 \geq 0, s_2 \geq 0.$

Rewrite the objective function as

$$-3x_1 - 5x_2 + z = 0.$$

The initial simplex tableau follows.

$$\begin{array}{c} \begin{array}{ccccc} x_1 & x_2 & s_1 & s_2 & z \end{array} \\ \left[\begin{array}{ccccc|c} 4 & 1 & 1 & 0 & 0 & 25 \\ 2 & 3 & 0 & 1 & 0 & 15 \\ \hline -3 & -5 & 0 & 0 & 1 & 0 \end{array}\right] \end{array}$$

The most negative indicator is -5, in the second column. To select the pivot from column 2, find the quotients $\frac{25}{1} = 25$ and $\frac{15}{3} = 5$. The smaller quotient is 5, so 3 is the pivot.

$$\begin{array}{c} \begin{array}{ccccc} x_1 & x_2 & s_1 & s_2 & z \end{array} \\ \left[\begin{array}{ccccc|c} 4 & 1 & 1 & 0 & 0 & 25 \\ 2 & \boxed{3} & 0 & 1 & 0 & 15 \\ \hline -3 & -5 & 0 & 0 & 1 & 0 \end{array}\right] \end{array}$$

$$\begin{array}{r} \\ -R_2 + 3R_1 \to R_1 \\ \\ 5R_2 + 3R_3 \to R_3 \end{array} \begin{array}{c} \begin{array}{ccccc} x_1 & x_2 & s_1 & s_2 & z \end{array} \\ \left[\begin{array}{ccccc|c} 10 & 0 & 3 & -1 & 0 & 60 \\ 2 & 3 & 0 & 1 & 0 & 15 \\ \hline 1 & 0 & 0 & 5 & 3 & 75 \end{array}\right] \end{array}$$

All of the indicators are nonnegative. Create a 1in the columns corresponding to x_2, s_1, and z.

$$\begin{array}{r} \\ \frac{1}{3}R_1 \to R_1 \\ \frac{1}{3}R_2 \to R_2 \\ \frac{1}{3}R_3 \to R_3 \end{array} \begin{array}{c} \begin{array}{ccccc} x_1 & x_2 & s_1 & s_2 & z \end{array} \\ \left[\begin{array}{ccccc|c} \frac{10}{3} & 0 & 1 & -\frac{1}{3} & 0 & 20 \\ \frac{2}{3} & 1 & 0 & \frac{1}{3} & 0 & 5 \\ \hline \frac{1}{3} & 0 & 0 & \frac{5}{3} & 1 & 25 \end{array}\right] \end{array}$$

The maximum value is 25 when $x_1 = 0$, $x_2 = 5$, $s_1 = 20$, and $s_2 = 0$.

9. Maximize $z = 10x_1 + 12x_2$

subject to: $4x_1 + 2x_2 \leq 20$

$5x_1 + x_2 \leq 50$

$2x_1 + 2x_2 \leq 24$

with $x_1 \geq 0, x_2 \geq 0.$

Three slack variables, s_1, s_2, and s_3, need to be introduced. The initial tableau is as follows.

$$\begin{array}{c} \begin{array}{cccccc} x_1 & x_2 & s_1 & s_2 & s_3 & z \end{array} \\ \left[\begin{array}{cccccc|c} 4 & 2 & 1 & 0 & 0 & 0 & 20 \\ 5 & 1 & 0 & 1 & 0 & 0 & 50 \\ 2 & 2 & 0 & 0 & 1 & 0 & 24 \\ \hline -10 & -12 & 0 & 0 & 0 & 1 & 0 \end{array}\right] \end{array}$$

The most negative indicator is -12, in column 2.

The quotients are $\frac{20}{2} = 10$, $\frac{50}{1} = 50$, and $\frac{24}{2} = 12$; the smallest is 10, so 2 in row 1, column 2 is the pivot.

$$\begin{array}{c} \begin{array}{cccccc} x_1 & x_2 & s_1 & s_2 & s_3 & z \end{array} \\ \left[\begin{array}{cccccc|c} 4 & \boxed{2} & 1 & 0 & 0 & 0 & 20 \\ 5 & 1 & 0 & 1 & 0 & 0 & 50 \\ 2 & 2 & 0 & 0 & 1 & 0 & 24 \\ \hline -10 & -12 & 0 & 0 & 0 & 1 & 0 \end{array}\right] \end{array}$$

$$\begin{array}{r} \\ \\ -R_1 + 2R_2 \to R_2 \\ -R_1 + R_3 \to R_3 \\ 6R_1 + R_4 \to R_4 \end{array} \begin{array}{c} \begin{array}{cccccc} x_1 & x_2 & s_1 & s_2 & s_3 & z \end{array} \\ \left[\begin{array}{cccccc|c} 4 & 2 & 1 & 0 & 0 & 0 & 20 \\ 6 & 0 & -1 & 2 & 0 & 0 & 80 \\ -2 & 0 & -1 & 0 & 1 & 0 & 4 \\ \hline 14 & 0 & 6 & 0 & 0 & 1 & 120 \end{array}\right] \end{array}$$

All of the indicators are nonnegative, so we are finished pivoting. Create a 1 in the columns corresponding to x_2 and s_2.

$$\begin{array}{r} \\ \\ \frac{1}{2}R_1 \to R_1 \\ \frac{1}{2}R_2 \to R_2 \\ \\ \end{array} \begin{array}{c} \begin{array}{cccccc} x_1 & x_2 & s_1 & s_2 & s_3 & z \end{array} \\ \left[\begin{array}{cccccc|c} 2 & 1 & \frac{1}{2} & 0 & 0 & 0 & 10 \\ 3 & 0 & -\frac{1}{2} & 1 & 0 & 0 & 40 \\ -2 & 0 & -1 & 0 & 1 & 0 & 4 \\ \hline 14 & 0 & 6 & 0 & 0 & 1 & 120 \end{array}\right] \end{array}$$

The maximum value is 120 when $x_1 = 0$, $x_2 = 10$, $s_1 = 0$, $s_2 = 40$, and $s_3 = 4$.

11. Maximize $z = 8x_1 + 3x_2 + x_3$

subject to: $x_1 + 6x_2 + 8x_3 \le 118$
$x_1 + 5x_2 + 10x_3 \le 220$

with $x_1 \ge 0, x_2 \ge 0, x_3 \ge 0.$

Two slack variables, s_1 and s_2, need to be introduced. The initial simplex tableau is as follows.

x_1	x_2	x_3	s_1	s_2	z	
[1]	6	8	1	0	0	118
1	5	10	0	1	0	220
−8	−3	−1	0	0	1	0

The most negative indicator is -8, in the first column. The quotients are $\frac{118}{1} = 118$ and $\frac{220}{1} = 220$; since 118 is the smaller, 1 in row 1, column 1 is the pivot. Performing row transformations, we get the following tableau.

	x_1	x_2	x_3	s_1	s_2	z	
	1	6	8	1	0	0	118
$-R_1 + R_2 \to R_2$	0	−1	2	−1	1	0	102
$8R_1 + R_3 \to R_3$	0	45	63	8	0	1	944

All of the indicators are nonnegative, so we are finished pivoting. The maximum value is 944 when $x_1 = 118,\ x_2 = 0,\ x_3 = 0,\ s_1 = 0,$ and $s_2 = 102.$

13. Maximize $z = 10x_1 + 15x_2 + 10x_3 + 5x_4$

subject to: $x_1 + x_2 + x_3 + x_4 \le 300$
$x_1 + 2x_2 + 3x_3 + x_4 \le 360$

with $x_1 \ge 0, x_2 \ge 0, x_3 \ge 0, x_4 \ge 0.$

The initial tableau is as follows.

x_1	x_2	x_3	x_4	s_1	s_2	z	
1	1	1	1	1	0	0	300
1	[2]	3	1	0	1	0	360
−10	−15	−10	−5	0	0	1	0

In the column with the most negative indicator, -15, the quotients are $\frac{300}{1} = 300$ and $\frac{360}{2} = 180$. The smaller quotient is 180, so the 2 in row 2, column 2, is the pivot.

	x_1	x_2	x_3	x_4	s_1	s_2	z	
$-R_2 + 2R_1 \to R_1$	[1]	0	−1	1	2	−1	0	240
	1	2	3	1	0	1	0	360
$15R_2 + 2R_3 \to R_3$	−5	0	25	5	0	15	2	5400

Pivot on the 1 in row 1, column 1.

	x_1	x_2	x_3	x_4	s_1	s_2	z	
	1	0	−1	1	2	−1	0	240
$-R_1 + R_2 \to R_2$	0	2	4	0	−2	2	0	120
$5R_1 + R_3 \to R_3$	0	0	20	10	10	10	2	6600

Create a 1 in the columns corresponding to x_2 and z.

	x_1	x_2	x_3	x_4	s_1	s_2	z	
	1	0	−1	1	2	−1	0	240
$\frac{1}{2}R_2 \to R_2$	0	1	2	0	−1	1	0	60
$\frac{1}{2}R_3 \to R_3$	0	0	10	5	5	5	1	3300

The maximum value is 3300 when $x_1 = 240,$ $x_2 = 60,\ x_3 = 0,\ x_4 = 0,\ s_1 = 0,$ and $s_2 = 0.$

15. Maximize $z = 4x_1 + 6x_2$

subject to: $x_1 - 5x_2 \le 25$
$4x_1 - 3x_2 \le 12$

with $x_1 \ge 0, x_2 \ge 0.$

x_1	x_2	s_1	s_2	z	
1	−5	1	0	0	25
4	−3	0	1	0	12
−4	−6	0	0	1	0

The most negative indicator is -6. The negative quotients $25/(-5)$ and $12/(-3)$ indicate an unbounded feasible region, so there is no unique optimum solution.

17. Maximize

$z = 37x_1 + 34x_2 + 36x_3 + 30x_4 + 35x_5$

subject to:

$$16x_1 + 19x_2 + 23x_3 + 15x_4 + 21x_5 \le 42{,}000$$
$$15x_1 + 10x_2 + 19x_3 + 23x_4 + 10x_5 \le 25{,}000$$
$$9x_1 + 16x_2 + 14x_3 + 12x_4 + 11x_5 \le 23{,}000$$
$$18x_1 + 20x_2 + 15x_3 + 17x_4 + 19x_5 \le 36{,}000$$

with $x_1 \ge 0, x_2 \ge 0, x_3 \ge 0, x_4 \ge 0, x_5 \ge 0.$

Four slack variables, s_1, s_2, s_3, and s_4, need to be introduced. The initial simplex tableau follows.

x_1	x_2	x_3	x_4	x_5	s_1	s_2	s_3	s_4	z	
16	19	23	15	21	1	0	0	0	0	42,000
15	10	19	23	10	0	1	0	0	0	25,000
9	16	14	12	11	0	0	1	0	0	23,000
18	20	15	17	19	0	0	0	1	0	36,000
−37	−34	−36	−30	−35	0	0	0	0	1	0

Using a graphing calculator or computer program, the maximum value is found to be 70,818.18 when $x_1 = 181.82$, $x_2 = 0$, $x_3 = 454.55$, $x_4 = 0$, $x_5 = 1363.64$, $s_1 = 0$, $s_2 = 0$, $s_3 = 0$, and $s_4 = 0$.

23. Organize the information in a table.

	Church Group	Labor Union	Maximum Time Available
Letter Writing	2	2	16
Follow-up	1	3	12
Money Raised	\$100	\$200	

Let x_1 and x_2 be the number of church groups and labor unions contacted respectively. We need two slack variables, s_1 and s_2.

Maximize $z = 100x_1 + 200x_2$

subject to: $2x_1 + 2x_2 + s_1 = 16$

$x_1 + 3x_2 + s_2 = 12$

with $x_1 \ge 0, x_2 \ge 0, s_1 \ge 0, s_2 \ge 0.$

The initial simplex tableau is as follows.

x_1	x_2	s_1	s_2	z	
2	2	1	0	0	16
1	[3]	0	1	0	12
−100	−200	0	0	1	0

Pivot on the 3 in row 2, column 2.

	x_1	x_2	s_1	s_2	z	
$-2R_2 + 3R_1 \to R_1$	[4]	0	3	−2	0	24
	1	3	0	1	0	12
$200R_2 + 3R_3 \to R_3$	−100	0	0	200	3	2400

Pivot on the 4 in row 1, column 1.

	x_1	x_2	s_1	s_2	z	
	4	0	3	−2	0	24
$-R_1 + 4R_2 \to R_2$	0	12	−3	6	0	24
$25R_1 + R_3 \to R_3$	0	0	75	150	3	3000

This is a final tableau, since all of the indicators are nonnegative. Create a 1 in the columns corresponding to x_1, x_2, and z.

	x_1	x_2	s_1	s_2	z	
$\frac{1}{4}R_1 \to R_1$	1	0	$\frac{3}{4}$	$-\frac{1}{2}$	0	6
$\frac{1}{12}R_2 \to R_2$	0	1	$-\frac{1}{4}$	$\frac{1}{2}$	0	2
$\frac{1}{3}R_3 \to R_3$	0	0	25	50	1	1000

The maximum amount of money raised is \$1000/mo when $x_1 = 6$ and $x_2 = 2$, that is, when 6 churches and 2 labor unions are contacted.

25. (a) Let x_1 be the number of Royal Flush poker sets, x_2 be the number of Deluxe Diamond sets, and x_3 be the number of Full House sets. The problem can be stated as follows.

Maximize $z = 38x_1 + 22x_2 + 12x_3$

subject to:

$$1000x_1 + 600x_2 + 300x_3 \le 2{,}800{,}000$$
$$4x_1 + 2x_2 + 2x_3 \le 10{,}000$$
$$10x_1 + 5x_2 + 5x_3 \le 25{,}000$$
$$2x_1 + x_2 + x_3 \le 6000$$

with $x_1 \ge 0, x_2 \ge 0, x_3 \ge 0.$

Since there are four constraints, introduce slack variables, s_1, s_2, s_3, and s_4 and set up the initial simplex tableau.

$$\begin{array}{c} \\ \end{array}\begin{array}{rrrrrrrr|r} x_1 & x_2 & x_3 & s_1 & s_2 & s_3 & s_4 & z & \\ \hline 1000 & 600 & 300 & 1 & 0 & 0 & 0 & 0 & 2{,}800{,}000 \\ 4 & 2 & 2 & 0 & 1 & 0 & 0 & 0 & 10{,}000 \\ 10 & 5 & 5 & 0 & 0 & 1 & 0 & 0 & 25{,}000 \\ 2 & 1 & 1 & 0 & 0 & 0 & 1 & 0 & 6000 \\ \hline -38 & -22 & -12 & 0 & 0 & 0 & 0 & 1 & 0 \end{array}$$

Using a graphing calculator or computer program, the maximum profit is \$104,000 and is obtained when 1000 Royal Flush poker sets, 3000 Deluxe Diamond poker sets, and no Full House poker sets are assembled.

(b) According to the poker chip constraint:

$$1000(1000) + 600(3000) + 300(0) + s_2$$
$$= 2{,}800{,}000 \quad s_1 = 0.$$

So all of the poker chips are used. Checking the card constraint:

$$4(1000) + 2(3000) + 2(0) + s_2 = 10{,}000$$
$$s_2 = 0.$$

So all of the cards are used. Checking the dice constraint:

$$10(1000) + 5(3000) + 5(0) + s_3 = 25{,}000$$
$$s_3 = 0.$$

So all of the dice are used. Finally, checking the dealer button constraint:

$$2(1000) + 3000 + 0 + s_4 = 6000$$
$$s_4 = 1000.$$

This means there are 1000 unused dealer buttons.

27. (a) Let x_1 represent the number of racing bicycles, x_2 the number of touring bicycles, and x_3 the number of mountain bicycles.

From Exercise 28 in Section 4.1, the initial simplex tableau is as follows.

$$\begin{array}{rrrrrr|r} x_1 & x_2 & x_3 & s_1 & s_2 & z & \\ \hline 17 & 27 & \boxed{34} & 1 & 0 & 0 & 91{,}800 \\ 12 & 21 & 15 & 0 & 1 & 0 & 42{,}000 \\ \hline -8 & -12 & -22 & 0 & 0 & 1 & 0 \end{array}$$

Pivot on the 34 in row 1, column 3.

$$\begin{array}{r|rrrrrr|r} & x_1 & x_2 & x_3 & s_1 & s_2 & z & \\ \hline & 17 & 27 & 34 & 1 & 0 & 0 & 91{,}800 \\ -15R_1 + 34R_2 \to R_2 & 153 & 309 & 0 & -15 & 34 & 0 & 51{,}000 \\ \hline 11R_1 + 17R_3 \to R_3 & 51 & 93 & 0 & 11 & 0 & 17 & 1{,}009{,}800 \end{array}$$

This is a final tableau, since all of the indicators are nonnegative. Create a 1 in the columns corresponding to x_3, s_2, and z.

$$\begin{array}{r|rrrrrr|r} & x_1 & x_2 & x_3 & s_1 & s_2 & z & \\ \hline \frac{1}{34}R_1 \to R_1 & \frac{1}{2} & \frac{27}{34} & 1 & \frac{1}{34} & 0 & 0 & 2700 \\ \frac{1}{34}R_2 \to R_2 & \frac{9}{2} & \frac{309}{34} & 0 & -\frac{15}{34} & 1 & 0 & 1500 \\ \hline \frac{1}{17}R_3 \to R_3 & 3 & \frac{93}{17} & 0 & \frac{11}{17} & 0 & 1 & 59{,}400 \end{array}$$

From the tableau, $x_1 = 0$, $x_2 = 0$, and $x_3 = 2700$. The company should make no racing or touring bicycles and 2700 mountain bicycles.

(b) From the third row of the final tableau, the maximum profit is \$59,400.

(c) When $x_1 = 0$, $x_2 = 0$, and $x_3 = 2700$, the number of units of steel used is

$$17(0) + 27(0) + 34(2700) = 91{,}800$$

which is all the steel available. The number of units of aluminum used is

$$12(0) + 21(0) + 15(2700) = 40{,}500$$

which leaves $42{,}000 - 40{,}500 = 1500$ units of aluminum unused.

Checking the second constraint:

$$12x_1 + 21x_2 + 15x_3 + s_2 = 42{,}000$$
$$12(0) + 21(0) + 15(2700) + s_2 = 42{,}000$$
$$s_2 = 1500.$$

29. (a) Let x_1 be the number of newspaper ads, x_2 be the number of Internet banner ads, and x_3 be the number of TV ads. Here is the initial tableau:

$$\begin{array}{c} \\ \end{array}\begin{array}{cccccccc|c} x_1 & x_2 & x_3 & s_1 & s_2 & s_3 & s_4 & z & \\ \hline 400 & 20 & \boxed{2000} & 1 & 0 & 0 & 0 & 0 & 8000 \\ 1 & 0 & 0 & 0 & 1 & 0 & 0 & 0 & 30 \\ 0 & 1 & 0 & 0 & 0 & 1 & 0 & 0 & 60 \\ 0 & 0 & 1 & 0 & 0 & 0 & 1 & 0 & 10 \\ \hline -4000 & -3000 & -10{,}000 & 0 & 0 & 0 & 0 & 1 & 0 \end{array}$$

Pivot on the 2000 in row 1, column 3.

$$\begin{array}{r} \\ \\ \\ \\ -R_1 + 2000R_4 \to R_4 \\ 5R_1 + R_5 \to R_5 \end{array}\begin{array}{cccccccc|c} x_1 & x_2 & x_3 & s_1 & s_2 & s_3 & s_4 & z & \\ \hline 400 & 20 & 2000 & 1 & 0 & 0 & 0 & 0 & 8000 \\ 1 & 0 & 0 & 0 & 1 & 0 & 0 & 0 & 30 \\ 0 & \boxed{1} & 0 & 0 & 0 & 1 & 0 & 0 & 60 \\ -400 & -20 & 0 & -1 & 0 & 0 & 2000 & 0 & 12{,}000 \\ \hline -2000 & -2900 & 0 & 5 & 0 & 0 & 0 & 1 & 40{,}000 \end{array}$$

Pivot on the 1 in row 3, column 2.

$$\begin{array}{r} \\ -20R_3 + R_1 \to R_1 \\ \\ \\ 20R_3 + R_4 \to R_4 \\ 2900R_3 + R_5 \to R_5 \end{array}\begin{array}{cccccccc|c} x_1 & x_2 & x_3 & s_1 & s_2 & s_3 & s_4 & z & \\ \hline \boxed{400} & 20 & 2000 & 1 & 0 & -20 & 0 & 0 & 6800 \\ 1 & 0 & 0 & 0 & 1 & 0 & 0 & 0 & 30 \\ 0 & 1 & 0 & 0 & 0 & 1 & 0 & 0 & 60 \\ -400 & 0 & 0 & -1 & 0 & 20 & 2000 & 0 & 13{,}200 \\ \hline -2000 & 0 & 0 & 5 & 0 & 2900 & 0 & 1 & 214{,}000 \end{array}$$

Pivot on the 400 in row 1, column 1.

$$\begin{array}{r} \\ \\ -R_1 + 400R_2 \to R_2 \\ \\ R_1 + R_4 \to R_4 \\ 5R_1 + R_5 \to R_5 \end{array}\begin{array}{cccccccc|c} x_1 & x_2 & x_3 & s_1 & s_2 & s_3 & s_4 & z & \\ \hline 400 & 0 & 2000 & 1 & 0 & -20 & 0 & 0 & 6800 \\ 0 & 0 & -2000 & -1 & 400 & 20 & 0 & 0 & 5200 \\ 0 & 1 & 0 & 0 & 0 & 1 & 0 & 0 & 60 \\ 0 & 0 & 2000 & 0 & 0 & 0 & 2000 & 0 & 20{,}000 \\ \hline 0 & 0 & 10{,}000 & 10 & 0 & 2800 & 0 & 1 & 248{,}000 \end{array}$$

Create a 1 in the columns corresponding to $x_1, s_2,$ and s_4.

$$\begin{array}{r} \\ \frac{1}{400}R_1 \to R_1 \\ \frac{1}{400}R_2 \to R_2 \\ \\ \frac{1}{2000}R_4 \to R_4 \\ \\ \end{array}\begin{array}{cccccccc|c} x_1 & x_2 & x_3 & s_1 & s_2 & s_3 & s_4 & z & \\ \hline 1 & 0 & 5 & \frac{1}{400} & 0 & -\frac{1}{20} & 0 & 0 & 17 \\ 0 & 0 & -5 & -\frac{1}{400} & 1 & \frac{1}{20} & 0 & 0 & 13 \\ 0 & 1 & 0 & 0 & 0 & 1 & 0 & 0 & 60 \\ 0 & 0 & 1 & 0 & 0 & 0 & 1 & 0 & 10 \\ \hline 0 & 0 & 10{,}000 & 10 & 0 & 2800 & 0 & 1 & 248{,}000 \end{array}$$

This is the final tableau. The maximum exposure is 248,000 women when 17 newspaper ads, 60 Internet banner ads, and no TV ads are used.

31. **(a)** The coefficients of the objective function are the profit coefficients from the table: 5, 4, and 3; choice (3) is correct.

(b) The constraints are the available man-hours for the 2 departments, 400 and 600; choice (4) is correct.

(c) $2X_1 + 3X_2 + 1X_3 \leq 400$ is the constraint on department 1; choice (3) is correct.

33. Maximize $z = 100x + 200y$

subject to: $2x + 2y \leq 16$

$x + 3y \leq 12$

with $x \geq 0, y \geq 0.$

Using Excel, we enter the variables x and y in cells Al and Bl, respectively. Enter the x- and y-coordinates of the initial corner point of the feasible region, (0, 0), in cells A2 and B2, respectively, and NAME these cells x and y, respectively. In cells C2, C4, C5, C6, and C7, enter the formula for the function to maximize and each of the constraints: $100x + 200y$, $2x + 2y$, $x + 3y, x$, and y. Since x and y have been set to 0, all the cells containing formulas should also show the value 0, as below.

	A	B	C
1	x	y	
2	0	0	0
3			
4			0
5			0
6			0
7			0

Using the SOLVER, ask Excel to maximize the value in cell C2 subject to the constraints C4 $\leq$ 16, C5 $\leq$ 12, C6 $\geq$ 0, C7 $\geq$ 0. Make sure you have checked off the box *Assume Linear Model* in SOLVER OPTIONS.

Excel returns the following values and allows you to choose a report.

	A	B	C
1	x	y	
2	6	2	1000
3			
4			16
5			12
6			6
7			2

Select the sensitivity report. The report will appear on a new sheet of the spread sheet.

Adjustable Cells

Cell	Name	Final Value	Reduced Cost	Objective Coefficient	Allowable Increase	Allowable Decrease
A2	x	6	0	100	100	33.33333333
B2	y	2	0	200	100	100

Constraints

Cell	Name	Final Value	Shadow Price	Constraint R.H. Side	Allowable Increase	Allowable Decrease
C4		16	25	16	8	8
C5		12	50	12	12	4
C6		6	0	0	6	1E + 30
C7		2	0	0	2	1E + 30

The church group's allowable increase is \$100 and the allowable decrease is \$33.33. So their contribution can be as high as $\$100 + \$100 = \$200$ or as low as $\$100 - \$33.33 = \$66.67$ and the original solution is still optimal. The unions' allowable increase is \$100 and the allowable decrease is \$100. So their contribution can be as high as $\$200 + \$100 = \$300$ or as low as $\$200 - \$100 = \$100$ and the original solution is still optimal.

35. Let $x_1 =$ number of hours running, x_2 be the number of hours biking, and x_3 be the number hours walking. The problem can be stated as follows.

Maximize $z = 531x_1 + 472x_2 + 354x_3$

subject to:
$$x_1 + x_2 + x_3 \le 15$$
$$x_1 \le 3$$
$$2x_2 - x_3 \le 0$$

with $x_1 \ge 0, x_2 \ge 0, x_3 \ge 0.$

We need three slack variables, s_1, s_2, and s_3. The initial simplex tableau as follows.

$$\begin{array}{c} \\ \\ \\ \\ \\ \end{array}\begin{array}{ccccccc|c} x_1 & x_2 & x_3 & s_1 & s_2 & s_3 & z & \\ 1 & 1 & 1 & 1 & 0 & 0 & 0 & 15 \\ \boxed{1} & 0 & 0 & 0 & 1 & 0 & 0 & 3 \\ 0 & 2 & -1 & 0 & 0 & 1 & 0 & 0 \\ \hline -531 & -472 & -354 & 0 & 0 & 0 & 1 & 0 \end{array}$$

Pivot on the 1 in row 2, column 1.

$$\begin{array}{r} \\ -R_2 + R_1 \to R_1 \\ \\ \\ 531R_2 + R_4 \to R_4 \end{array}\begin{array}{ccccccc|c} x_1 & x_2 & x_3 & s_1 & s_2 & s_3 & z & \\ 0 & 1 & 1 & 1 & -1 & 0 & 0 & 12 \\ 1 & 0 & 0 & 0 & 1 & 0 & 0 & 3 \\ 0 & \boxed{2} & -1 & 0 & 0 & 1 & 0 & 0 \\ \hline 0 & -472 & -354 & 0 & 351 & 0 & 1 & 1593 \end{array}$$

Pivot on the 2 in row 3, column 2.

$$\begin{array}{r} \\ -R_3 + 2R_1 \to R_1 \\ \\ \\ 236R_3 + R_4 \to R_4 \end{array}\begin{array}{ccccccc|c} x_1 & x_2 & x_3 & s_1 & s_2 & s_3 & z & \\ 0 & 0 & \boxed{3} & 2 & -2 & -1 & 0 & 24 \\ 1 & 0 & 0 & 0 & 1 & 0 & 0 & 3 \\ 0 & 2 & -1 & 0 & 0 & 1 & 0 & 0 \\ \hline 0 & 0 & -590 & 0 & 531 & 236 & 1 & 1593 \end{array}$$

Finally pivot on the 3 in row 1, column 3.

$$\begin{array}{r} \\ \\ \\ R_1 + 3R_3 \to R_3 \\ 590R_1 + 3R_4 \to R_4 \end{array}\begin{array}{ccccccc|c} x_1 & x_2 & x_3 & s_1 & s_2 & s_3 & z & \\ 0 & 1 & 3 & 2 & -2 & -1 & 0 & 24 \\ 1 & 0 & 0 & 0 & 1 & 0 & 0 & 3 \\ 0 & 6 & 0 & 2 & -2 & 2 & 0 & 24 \\ \hline 0 & 0 & 0 & 1180 & 413 & 118 & 3 & 18{,}939 \end{array}$$

Create a 1 in the columns corresponding to x_2, x_3, and z.

$$
\begin{array}{r}
\\
\frac{1}{3}R_1 \rightarrow R_1 \\
\\
\frac{1}{6}R_3 \rightarrow R_3 \\
\frac{1}{3}R_4 \rightarrow R_4
\end{array}
\begin{array}{ccccccc|c}
x_1 & x_2 & x_3 & s_1 & s_2 & s_3 & z & \\
\hline
0 & 0 & 1 & \frac{2}{3} & -\frac{2}{3} & -\frac{1}{3} & 0 & 8 \\
1 & 0 & 0 & 0 & 1 & 0 & 0 & 3 \\
0 & 1 & 0 & \frac{1}{3} & -\frac{1}{3} & \frac{1}{3} & 0 & 4 \\
\hline
0 & 0 & 0 & \frac{1180}{3} & \frac{413}{3} & \frac{118}{3} & 1 & 6313
\end{array}
$$

Rachel should run 3 hours, bike 4 hours, and walk 8 hours for a maximum calorie expenditure of 6313 calories.

37. (a) Let x_1 represent the number of species A, x_2 represent the number of species B, and x_3 represent the number of species C.

Maximize $\quad z = 1.62x_1 + 2.14x_2 + 3.01x_3$

subject to:
$$
\begin{aligned}
1.32x_1 + 2.1x_2 + 0.86x_3 &\le 490 \\
2.9x_1 + 0.95x_2 + 1.52x_3 &\le 897 \\
1.75x_1 + 0.6x_2 + 2.01x_3 &\le 653
\end{aligned}
$$

with $\quad x_1 \ge 0, x_2 \ge 0, x_3 \ge 0.$

Use a graphing calculator or computer to solve this problem and find that the answer is to stock none of species A, 114 of species B, and 291 of species C for a maximum combined weight of 1119.72 kg.

(b) When $x_1 = 0$, $x_2 = 114$, and $x_3 = 291$, the number of units used are as follows.

$$\text{Food I: } 1.32(0) + 2.1 + (114) + 0.86(291) = 489.66$$

or 490 units, which is the total amount available of Food I.

$$\text{Food II: } 2.9(0) + 0.95(114) + 1.52(291) = 550.62$$

or 551 units, which leaves 897 – 551, or 346 units of Food II available.

$$\text{Food III: } 1.75(0) + 0.6(114) + 2.01(291) = 653.31$$

or 653 units, which is the total amount available of Food III.

(c) Many answers are possible. The idea is to choose average weights for species B and C that are considerably smaller than the average weight chosen for species A, so that species A dominates the objective function.

(d) Many answers are possible. The idea is to choose average weights for species A and B that are considerably smaller than the average weight chosen for species C.

39. Let x_1 represent the number of minutes for the senator, x_2 the number of minutes for the congresswoman, and x_3 the number of minutes for the governor.

Of the half-hour show's time, at most only $30 - 3 = 27$ min are available to be allotted to the politicians. The given information leads to the inequality

$$x_1 + x_2 + x_3 \le 27$$

and the inequalities

$$x_1 \ge 2x_3 \quad \text{and} \quad x_1 + x_3 \ge 2x_2,$$

and we are to maximize the objective function

$$z = 35x_1 + 40x_2 + 45x_3.$$

Rewrite the equation as

$$x_3 \le 27 - x_1 - x_2$$

and the inequalities as

$$-x_1 + 2x_3 \le 0 \quad \text{and} \quad -x_1 + 2x_2 - x_3 \le 0.$$

Substitute $27 - x_1 - x_2$ for x_3 in the objective function and the inequalities, and the problem is as follows.

Maximize $\quad z = 35x_1 + 40x_2 + 45x_3$

$$\text{subject to:} \quad \begin{aligned} -x_1 \phantom{{}+2x_2} + 2x_3 &\le 0 \\ -x_1 + 2x_2 - x_3 &\le 0 \\ x_1 + x_2 + x_3 &\le 27 \end{aligned}$$

with $x_1 \ge 0,\ x_2 \ge 0,\ x_3 \ge 0.$

We need three slack variables. The initial simplex tableau is as follows.

$$\begin{array}{c} \begin{array}{ccccccc|c} x_1 & x_2 & x_3 & s_1 & s_2 & s_3 & z & \end{array} \\ \left[\begin{array}{ccccccc|c} -1 & 0 & 2 & 1 & 0 & 0 & 0 & 0 \\ -1 & \boxed{2} & -1 & 0 & 1 & 0 & 0 & 0 \\ 1 & 1 & 1 & 0 & 0 & 1 & 0 & 27 \\ \hline -35 & -40 & -45 & 0 & 0 & 0 & 1 & 0 \end{array}\right] \end{array}$$

Pivot on the 2 in row 2, column 2, and then pivot on the 9 in row 3, column 1.

$$\begin{array}{r} \\ \\ -R_2 + 2R_3 \to R_3 \\ 20R_2 + R_4 \to R_4 \end{array} \left[\begin{array}{ccccccc|c} x_1 & x_2 & x_3 & s_1 & s_2 & s_3 & z & \\ -1 & 0 & \boxed{2} & 1 & 0 & 0 & 0 & 0 \\ -1 & 2 & -1 & 0 & 1 & 0 & 0 & 0 \\ 3 & 0 & 3 & 0 & -1 & 2 & 0 & 54 \\ \hline -55 & 0 & -65 & 0 & 20 & 0 & 1 & 0 \end{array}\right]$$

$$\begin{array}{r} \\ \\ R_1 + 2R_2 \to R_2 \\ -3R_1 + 2R_3 \to R_3 \\ 65R_1 + 2R_4 \to R_4 \end{array} \left[\begin{array}{ccccccc|c} x_1 & x_2 & x_3 & s_1 & s_2 & s_3 & z & \\ -1 & 0 & 2 & 1 & 0 & 0 & 0 & 0 \\ -3 & 4 & 0 & 1 & 2 & 0 & 0 & 0 \\ \boxed{9} & 0 & 0 & -3 & -2 & 4 & 0 & 108 \\ \hline -175 & 0 & 0 & 65 & 40 & 0 & 2 & 0 \end{array}\right]$$

Pivot on the 9 in row 3, column 1.

$$\begin{array}{r} \\ R_3 + 9R_1 \to R_1 \\ R_3 + 3R_2 \to R_2 \\ \\ 175R_3 + 9R_4 \to R_4 \end{array} \left[\begin{array}{ccccccc|c} x_1 & x_2 & x_3 & s_1 & s_2 & s_3 & z & \\ 0 & 0 & 18 & 6 & -2 & 4 & 0 & 108 \\ 0 & 12 & 0 & 0 & 4 & 4 & 0 & 108 \\ 9 & 0 & 0 & -3 & -2 & 4 & 0 & 108 \\ \hline 0 & 0 & 0 & 60 & 10 & 700 & 18 & 18{,}900 \end{array}\right]$$

Create a 1 in the columns corresponding to x_1, x_2, x_3, and z.

$$\begin{array}{r} \\ \frac{1}{18}R_1 \to R_1 \\ \frac{1}{12}R_2 \to R_2 \\ \frac{1}{9}R_3 \to R_3 \\ \frac{1}{18}R_4 \to R_4 \end{array} \left[\begin{array}{ccccccc|c} x_1 & x_2 & x_3 & s_1 & s_2 & s_3 & z & \\ 0 & 0 & 1 & \frac{1}{3} & -\frac{1}{9} & \frac{2}{9} & 0 & 6 \\ 0 & 1 & 0 & 0 & \frac{1}{3} & \frac{1}{3} & 0 & 9 \\ 1 & 0 & 0 & -\frac{1}{3} & -\frac{2}{9} & \frac{4}{9} & 0 & 12 \\ \hline 0 & 0 & 0 & \frac{10}{3} & \frac{5}{9} & \frac{350}{9} & 1 & 1050 \end{array}\right]$$

The maximum value of z is 1050 when $x_1 = 12$, $x_2 = 9$, and $x_3 = 6$. That is, for a maximum of 1,050,000 viewers, the time allotments should be 12 minutes for the senator, 9 minutes for the congresswoman, and 6 minutes for the governor.

4.3 Minimization Problems; Duality

Your Turn 1

Write the augmented matrix.

$$\left[\begin{array}{ccc|c} 3 & 3 & 4 & 24 \\ 5 & 1 & 3 & 27 \\ \hline 25 & 12 & 27 & 0 \end{array}\right]$$

Transpose to get the matrix for the dual problem.

$$\left[\begin{array}{cc|c} 3 & 5 & 25 \\ 3 & 1 & 12 \\ 4 & 3 & 27 \\ \hline 24 & 27 & 0 \end{array}\right]$$

Write the dual problem.

Maximize $z = 24x_1 + 27x_2$

subject to: $3x_1 + 5x_2 \leq 25$

$3x_1 + x_2 \leq 12$

$4x_1 + 3x_2 \leq 27$

with $x_1 \geq 0,\ x_2 \geq 0$

Your Turn 2

Write the augmented matrix.

$$\left[\begin{array}{cc|c} 3 & 5 & 20 \\ 3 & 1 & 18 \\ \hline 15 & 12 & 0 \end{array}\right]$$

Transpose to get the matrix for the dual problem.

$$\left[\begin{array}{cc|c} 3 & 3 & 15 \\ 5 & 1 & 12 \\ \hline 20 & 18 & 0 \end{array}\right]$$

Write the dual problem.

Maximize $z = 20x_1 + 18x_2$

subject to: $3x_1 + 3x_2 \leq 15$

$5x_1 + x_2 \leq 12$

with $x_1 \geq 0,\ x_2 \geq 0.$

The initial tableau for this problem is

$$\begin{array}{c} \\ \\ \\ \end{array}\begin{array}{c} x_1 \quad x_2 \quad s_1 \quad s_2 \quad z \\ \left[\begin{array}{ccccc|c} 3 & 3 & 1 & 0 & 0 & 15 \\ \boxed{5} & 1 & 0 & 1 & 0 & 12 \\ \hline -20 & -18 & 0 & 0 & 1 & 0 \end{array}\right] \end{array}$$

Pivot around the indicated 5.

$$\begin{array}{r} \\ -3R_2 + 5R_1 \to R_1 \\ \\ 4R_2 + R_3 \to R_3 \end{array} \begin{array}{c} x_1 \quad x_2 \quad s_1 \quad s_2 \quad z \\ \left[\begin{array}{ccccc|c} 0 & \boxed{12} & 5 & -3 & 0 & 39 \\ 5 & 1 & 0 & 1 & 0 & 12 \\ \hline 0 & -14 & 0 & 4 & 1 & 48 \end{array}\right] \end{array}$$

Now pivot around the indicated 12:

$$\begin{array}{r} \\ \\ -R_1 + 12R_2 \to R_2 \\ 7R_1 + 6R_3 \to R_3 \end{array} \begin{array}{c} x_1 \quad x_2 \quad s_1 \quad s_2 \quad z \\ \left[\begin{array}{ccccc|c} 0 & 12 & 5 & -3 & 0 & 39 \\ 60 & 0 & -5 & 15 & 0 & 105 \\ \hline 0 & 0 & 35 & 3 & 6 & 561 \end{array}\right] \end{array}$$

Finally divide the last row by 6 to produce a 1 in the z column:

$$\begin{array}{r} \\ \\ \\ R_3/6 \to R_3 \end{array} \begin{array}{c} x_1 \quad x_2 \quad s_1 \quad s_2 \quad z \\ \left[\begin{array}{ccccc|c} 0 & 12 & 5 & -3 & 0 & 39 \\ 60 & 0 & -5 & 15 & 0 & 105 \\ \hline 0 & 0 & \frac{35}{6} & \frac{1}{2} & 1 & \frac{187}{2} \end{array}\right] \end{array}$$

In the original problem, w has a minimum of $\frac{187}{2}$ when $y_1 = \frac{35}{6}$ and $y_2 = \frac{1}{2}$.

4.3 Exercises

1. To form the transpose of a matrix, the rows of the original matrix are written as the columns of the transpose. The transpose of

$$\begin{bmatrix} 1 & 2 & 3 \\ 3 & 2 & 1 \\ 1 & 10 & 0 \end{bmatrix}$$

is

$$\begin{bmatrix} 1 & 3 & 1 \\ 2 & 2 & 10 \\ 3 & 1 & 0 \end{bmatrix}.$$

3. The transpose of

$$\begin{bmatrix} 4 & 5 & -3 & 15 \\ 7 & 14 & 20 & -8 \\ 5 & 0 & -2 & 23 \end{bmatrix}$$

is

$$\begin{bmatrix} 4 & 7 & 5 \\ 5 & 14 & 0 \\ -3 & 20 & -2 \\ 15 & -8 & 23 \end{bmatrix}.$$

5. Maximize $z = 4x_1 + 3x_2 + 2x_3$

subject to:
$$\begin{aligned} x_1 + x_2 + x_3 &\le 5 \\ x_1 + x_2 \quad\;\; &\le 4 \\ 2x_1 + x_2 + 3x_3 &\le 15 \end{aligned}$$

with $x_1 \ge 0, x_2 \ge 0, x_3 \ge 0.$

To form the dual, first write the augmented matrix for the given problem.

$$\left[\begin{array}{ccc|c} 1 & 1 & 1 & 5 \\ 1 & 1 & 0 & 4 \\ 2 & 1 & 3 & 15 \\ \hline 4 & 3 & 2 & 0 \end{array}\right]$$

Then form the transpose of this matrix.

$$\left[\begin{array}{ccc|c} 1 & 1 & 2 & 4 \\ 1 & 1 & 1 & 3 \\ 1 & 0 & 3 & 2 \\ \hline 5 & 4 & 15 & 0 \end{array}\right]$$

The dual problem is stated from this second matrix (using y instead of x).

Minimize $w = 5y_1 + 4y_2 + 15y_3$

subject to:
$$\begin{aligned} y_1 + y_2 + 2y_3 &\ge 4 \\ y_1 + y_2 + y_3 &\ge 3 \\ y_1 \qquad + 3y_3 &\ge 2 \end{aligned}$$

with $y_1 \ge 0, y_2 \ge 0, y_3 \ge 0.$

7. Minimize $w = 3y_1 + 6y_2 + 4y_3 + y_4$

subject to:
$$\begin{aligned} y_1 + y_2 + y_3 + y_4 &\ge 150 \\ 2y_1 + 2y_2 + 3y_3 + 4y_4 &\ge 275 \end{aligned}$$

with $y_1 \ge 0, y_2 \ge 0, y_3 \ge 0, y_4 \ge 0.$

To find the dual problem, first write the augmented matrix for the problem.

$$\left[\begin{array}{cccc|c} 1 & 1 & 1 & 1 & 150 \\ 2 & 2 & 3 & 4 & 275 \\ \hline 3 & 6 & 4 & 1 & 0 \end{array}\right]$$

Then form the transpose of this matrix.

$$\left[\begin{array}{cc|c} 1 & 2 & 3 \\ 1 & 2 & 6 \\ 1 & 3 & 4 \\ 1 & 4 & 1 \\ \hline 150 & 275 & 0 \end{array}\right]$$

The dual problem is

Maximize $z = 150x_1 + 275x_2$

subject to:
$$\begin{aligned} x_1 + 2x_2 &\le 3 \\ x_1 + 2x_2 &\le 6 \\ x_1 + 3x_2 &\le 4 \\ x_1 + 4x_2 &\le 1 \end{aligned}$$

with $x_1 \ge 0,\ x_2 \ge 0.$

9. Find $y_1 \ge 0$ and $y_2 \ge 0$ such that

$$\begin{aligned} 2y_1 + 3y_2 &\ge 6 \\ 2y_1 + y_2 &\ge 7 \end{aligned}$$

and $w = 5y_1 + 2y_2$ is minimized.

Write the augmented matrix for this problem.

$$\left[\begin{array}{cc|c} 2 & 3 & 6 \\ 2 & 1 & 7 \\ \hline 5 & 2 & 0 \end{array}\right]$$

Form the transpose of this matrix.

$$\left[\begin{array}{cc|c} 2 & 2 & 5 \\ 3 & 1 & 2 \\ \hline 6 & 7 & 0 \end{array}\right]$$

Use this matrix to write the dual problem.

Find $x_1 \ge 0$ and $x_2 \ge 0$ such that

$$\begin{aligned} 2x_1 + 2x_2 &\le 5 \\ 3x_1 + x_2 &\le 2 \end{aligned}$$

and $z = 6x_1 + 7x_2$ is maximized.

Introduce slack variables s_1 and s_2. The initial tableau is as follows.

$$\begin{array}{c} \begin{array}{ccccc} x_1 & x_2 & s_1 & s_2 & z \end{array} \\ \left[\begin{array}{ccccc|c} 2 & 2 & 1 & 0 & 0 & 5 \\ 3 & \boxed{1} & 0 & 1 & 0 & 2 \\ \hline -6 & -7 & 0 & 0 & 1 & 0 \end{array}\right] \end{array}$$

Pivot on the 1 in row 2, column 2, since that column has the most negative indicator and that row has the smallest nonnegative quotient.

$$\begin{array}{r} \\ -2R_2 + R_1 \to R_1 \\ \\ 7R_2 + R_3 \to R_3 \end{array} \begin{array}{c} \begin{array}{ccccc} x_1 & x_2 & s_1 & s_2 & z \end{array} \\ \left[\begin{array}{ccccc|c} -4 & 0 & 1 & -2 & 0 & 1 \\ 3 & 1 & 0 & 1 & 0 & 2 \\ \hline 15 & 0 & 0 & 7 & 1 & 14 \end{array}\right] \end{array}$$

The minimum value of w is the same as the maximum value of z. The minimum value of w is 14 when $y_1 = 0$ and $y_2 = 7$. (Note that the values of y_1 and y_2 are given by the entries in the bottom row of the columns corresponding to the slack variables in the final tableau.)

11. Find $y_1 \geq 0$ and $y_2 \geq 0$ such that

$$10y_1 + 5y_2 \geq 100$$
$$20y_1 + 10y_2 \geq 150$$

and $w = 4y_1 + 5y_2$ is minimized.

Write the augmented matrix for this problem.

$$\left[\begin{array}{cc|c} 10 & 5 & 100 \\ 20 & 10 & 150 \\ \hline 4 & 5 & 0 \end{array}\right]$$

Form the transpose of this matrix.

$$\left[\begin{array}{cc|c} 10 & 20 & 4 \\ 5 & 10 & 5 \\ \hline 100 & 150 & 0 \end{array}\right]$$

Write the dual problem from this matrix.

Find $x_1 \geq 0$ and $x_2 \geq 0$ such that

$$10x_1 + 20x_2 \leq 4$$
$$5x_1 + 10x_2 \leq 5$$

and $z = 100x_1 + 150x_2$ is maximized.

The initial simplex tableau is as follows.

$$\begin{array}{c} \begin{array}{ccccc} x_1 & x_2 & s_1 & s_2 & z \end{array} \\ \left[\begin{array}{ccccc|c} 10 & \boxed{20} & 1 & 0 & 0 & 4 \\ 5 & 10 & 0 & 1 & 0 & 5 \\ \hline -100 & -150 & 0 & 0 & 1 & 0 \end{array}\right] \end{array}$$

Pivot on the 20 in row 1, column 2.

$$\begin{array}{r} \\ -R_1 + 2R_1 \rightarrow R_2 \\ 15R_1 + 2R_3 \rightarrow R_3 \end{array} \left[\begin{array}{ccccc|c} \boxed{10} & 20 & 1 & 0 & 0 & 4 \\ 0 & 0 & -1 & 2 & 0 & 6 \\ \hline -50 & 0 & 15 & 0 & 2 & 60 \end{array}\right]$$

(columns: x_1, x_2, s_1, s_2, z)

Pivot on the 10 in row 1, column 1.

$$\begin{array}{r} \\ \\ 5R_1 + R_3 \rightarrow R_3 \end{array} \left[\begin{array}{ccccc|c} 10 & 20 & 1 & 0 & 0 & 4 \\ 0 & 0 & -1 & 2 & 0 & 6 \\ \hline 0 & 100 & 20 & 0 & 2 & 80 \end{array}\right]$$

(columns: x_1, x_2, s_1, s_2, z)

Create a 1 in the columns corresponding to x_1, s_2, and z.

$$\begin{array}{r} \frac{1}{10}R_1 \rightarrow R_1 \\ \frac{1}{2}R_2 \rightarrow R_2 \\ \frac{1}{2}R_3 \rightarrow R_3 \end{array} \left[\begin{array}{ccccc|c} 1 & 2 & \frac{1}{10} & 0 & 0 & \frac{2}{5} \\ 0 & 0 & -\frac{1}{2} & 1 & 0 & 3 \\ \hline 0 & 50 & 10 & 0 & 1 & 40 \end{array}\right]$$

(columns: x_1, x_2, s_1, s_2, z)

The minimum value of w is 40 when $y_1 = 10$ and $y_2 = 0$. (These values of y_1 and y_2 are read from the last row of the columns corresponding to s_1 and s_2 in the final tableau.)

13. Minimize $\quad w = 6y_1 + 10y_2$

subject to: $\quad 3y_1 + 5y_2 \geq 15$

$\quad\quad 4y_1 + 7y_2 \geq 20$

with $\quad y_1 \geq 0,\ y_2 \geq 0.$

Write the augmented matrix.

$$\left[\begin{array}{cc|c} 3 & 5 & 15 \\ 4 & 7 & 20 \\ \hline 6 & 10 & 0 \end{array}\right]$$

Transpose to get the matrix for the dual problem.

$$\left[\begin{array}{cc|c} 3 & 4 & 6 \\ 5 & 7 & 10 \\ \hline 15 & 20 & 0 \end{array}\right]$$

Write the dual problem.

Maximize $\quad z = 15x_1 + 20x_2$

subject to: $\quad 3x_1 + 4x_2 \leq 6$

$\quad\quad 5x_1 + 7x_2 \leq 10$

with $\quad x_1 \geq 0,\ x_2 \geq 0.$

Write the initial tableau for this problem.

$$\begin{array}{c} \\ \\ \\ \end{array}\begin{array}{cccccc} x_1 & x_2 & s_1 & s_2 & z & \\ \left[\begin{array}{c}\boxed{3}\\5\\ \hline -15\end{array}\right. & \begin{array}{c}4\\7\\ \hline -20\end{array} & \begin{array}{c}1\\0\\ \hline 0\end{array} & \begin{array}{c}0\\1\\ \hline 0\end{array} & \begin{array}{c}0\\0\\ \hline 1\end{array} & \left|\begin{array}{c}6\\10\\ \hline 0\end{array}\right] \end{array}$$

Pivot around the indicated 3 to obtain this final tableau:

$$\begin{array}{r} \\ -5R_1 + 3R_2 \rightarrow R_2 \\ 5R_1 + R_3 \rightarrow R_3 \end{array} \left[\begin{array}{ccccc|c} 3 & 4 & 1 & 0 & 0 & 6 \\ 0 & 1 & -5 & 3 & 0 & 0 \\ \hline 0 & 0 & 5 & 0 & 1 & 30 \end{array}\right] \quad (x_1\ x_2\ s_1\ s_2\ z)$$

Instead we could pivot around the 5 in the first row, second column. This produces the following final tableau:

$$\begin{array}{r} -3R_2 + 5R_1 \rightarrow R_1 \\ \\ 3R_2 + R_3 \rightarrow R_3 \end{array} \left[\begin{array}{ccccc|c} 0 & -1 & 5 & -3 & 0 & 6 \\ 5 & 7 & 0 & 1 & 0 & 10 \\ \hline 0 & 1 & 0 & 3 & 1 & 30 \end{array}\right] \quad (x_1\ x_2\ s_1\ s_2\ z)$$

In the original problem, w has a minimum of 30 when $y_1 = 5$ and $y_2 = 0$ (reading from the first final tableau) or when $y_1 = 0$ and $y_2 = 3$ (reading from the second final tableau). Any point on the line segment between $(5, 0)$ and $(0, 3)$ also gives the minimum of 30.

15. Minimize $w = 2y_1 + y_2 + 3y_3$

subject to: $y_1 + y_2 + y_3 \geq 100$
$2y_1 + y_2 \geq 50$

with $y_1 \geq 0,\ y_2 \geq 0,\ y_3 \geq 0.$

Write the augmented matrix.

$$\left[\begin{array}{ccc|c} 1 & 1 & 1 & 100 \\ 2 & 1 & 0 & 50 \\ \hline 2 & 1 & 3 & 0 \end{array}\right]$$

Form the transpose of this matrix.

$$\left[\begin{array}{cc|c} 1 & 2 & 2 \\ 1 & 1 & 1 \\ 1 & 0 & 3 \\ \hline 100 & 50 & 0 \end{array}\right]$$

The dual problem is as follows.

Maximize $z = 100x_1 + 50x_2$

subject to: $x_1 + 2x_2 \leq 2$
$x_1 + x_2 \leq 1$
$x_1 \leq 3$

with $x_1 \geq 0, x_2 \geq 0.$

The initial simplex tableau is as follows.

$$\left[\begin{array}{cccccc|c} 1 & 2 & 1 & 0 & 0 & 0 & 2 \\ \boxed{1} & 1 & 0 & 1 & 0 & 0 & 1 \\ 1 & 0 & 0 & 0 & 1 & 0 & 3 \\ \hline -100 & -50 & 0 & 0 & 0 & 1 & 0 \end{array}\right] \quad (x_1\ x_2\ s_1\ s_2\ s_3\ z)$$

Pivot on the 1 in row 2, column 1.

$$\begin{array}{r} -R_2 + R_1 \rightarrow R_1 \\ \\ -R_2 + R_3 \rightarrow R_3 \\ 100R_2 + R_4 \rightarrow R_4 \end{array} \left[\begin{array}{cccccc|c} 0 & 1 & 1 & -1 & 0 & 0 & 1 \\ 1 & 1 & 0 & 1 & 0 & 0 & 1 \\ 0 & -1 & 0 & -1 & 1 & 0 & 2 \\ 0 & 50 & 0 & 100 & 0 & 1 & 100 \end{array}\right] \quad (x_1\ x_2\ s_1\ s_2\ s_3\ z)$$

The minimum value of w is 100 when $y_1 = 0$, $y_2 = 100$, and $y_3 = 0$.

17. Minimize $z = x_1 + 2x_2$

subject to: $-2x_1 + x_2 \geq 1$
$x_1 - 2x_2 \geq 1$

with $x_1 \geq 0,\ x_2 \geq 0.$

A quick sketch of the constraints $-2x_1 + x_2 \geq 1$ and $x_1 - 2x_2 \geq 1$ will verify that the two corresponding half planes do not overlap in the first quadrant of the x_1x_2-plane. Therefore, this problem (P) has no feasible solution. The dual of the given problem is as follows:

Maximize $w = y_1 + y_2$

subject to: $-2y_1 + y_2 \leq 1$
$y_1 - 2y_2 \leq 2$

with $y_1 \geq 0, y_2 \geq 0.$

A quick sketch here will verify that there is a feasible region in the y_1y_2-plane, and it is unbounded. Therefore, there is no maximum value of w in this problem (D).

(P) has no feasible solution and the objective function of (D) is unbounded; this is choice (a).

19. **(a)** Let y_1 = the number of units of regular beer

and y_2 = the number of units of light beer.

Minimize $w = 32{,}000y_1 + 50{,}000y_2$

subject to:

$$\begin{aligned} y_1 \qquad\qquad\qquad\quad &\geq 10 \\ y_2 &\geq 15 \\ y_1 + \qquad\qquad\quad y_2 &\geq 45 \\ 120{,}000y_1 + 300{,}000y_2 &\geq 9{,}000{,}000 \\ y_1 + \qquad\quad y_2 &\geq 20 \end{aligned}$$

with $y_1 \geq 0,\ y_2 \geq 0.$

Write the augmented matrix for this problem, and form the transpose to give the matrix for the dual problem.

$$\left[\begin{array}{cc|c} 1 & 0 & 10 \\ 0 & 1 & 15 \\ 1 & 1 & 45 \\ 120{,}000 & 300{,}000 & 9{,}000{,}000 \\ 1 & 1 & 20 \\ \hline 32{,}000 & 50{,}000 & 0 \end{array}\right]$$

$$\left[\begin{array}{ccccc|c} 1 & 0 & 1 & 120{,}000 & 1 & 32{,}000 \\ 0 & 1 & 1 & 300{,}000 & 1 & 50{,}000 \\ \hline 10 & 15 & 45 & 9{,}000{,}000 & 20 & 0 \end{array}\right]$$

The dual problem is

Maximize $z = 10x_1 + 15x_2 + 45x_3 + 9{,}000{,}000x_4 + 20x_5$

subject to:
$$\begin{aligned} x_1 + x_3 + 120{,}000x_4 + x_5 &\leq 32{,}000 \\ x_2 + x_3 + 300{,}000x_4 + x_5 &\leq 50{,}000 \end{aligned}$$

with $x_1 \geq 0,\ x_2 \geq 0,\ x_3 \geq 0,\ x_4 \geq 0,\ x_5 \geq 0.$

Write the initial simplex tableau.

$$\begin{array}{c} \begin{array}{cccccccc} x_1 & x_2 & x_3 & x_4 & x_5 & s_1 & s_2 & z \end{array} \\ \left[\begin{array}{cccccccc|c} 1 & 0 & 1 & 120{,}000 & 1 & 1 & 0 & 0 & 32{,}000 \\ 0 & 1 & 1 & \boxed{300{,}000} & 1 & 0 & 1 & 0 & 50{,}000 \\ \hline -10 & -15 & -45 & -9{,}000{,}000 & -20 & 0 & 0 & 1 & 0 \end{array}\right] \end{array}$$

Pivot on the 300,000 in row 2, column 3.

$$\begin{array}{r} \\ -2\text{R}_2 + 5\text{R}_1 \to \text{R}_1 \\ \\ 30\text{R}_2 + \text{R}_3 \to \text{R}_3 \end{array} \begin{array}{c} \begin{array}{cccccccc} x_1 & x_2 & x_3 & x_4 & x_5 & s_1 & s_2 & z \end{array} \\ \left[\begin{array}{cccccccc|c} 5 & -2 & \boxed{3} & 0 & 3 & 5 & -2 & 0 & 60{,}000 \\ 0 & 1 & 1 & 300{,}000 & 1 & 0 & 1 & 0 & 50{,}000 \\ \hline -10 & 15 & -15 & 0 & 10 & 0 & 30 & 1 & 1{,}500{,}000 \end{array}\right] \end{array}$$

Pivot on the 3 in row 1, column 3.

$$\begin{array}{r} \\ \\ -R_1 + 3R_2 \to R_2 \\ 5R_1 + R_3 \to R_3 \end{array} \begin{array}{cccccccc|c} x_1 & x_2 & x_3 & x_4 & x_5 & s_1 & s_2 & z & \\ 5 & -2 & 3 & 0 & 3 & 5 & -2 & 0 & 60{,}000 \\ -5 & 5 & 0 & 900{,}000 & 0 & -5 & 5 & 0 & 90{,}000 \\ \hline 15 & 5 & 0 & 0 & 25 & 25 & 20 & 1 & 1{,}800{,}000 \end{array}$$

Create a 1 in the columns corresponding to x_3 and x_4.

$$\begin{array}{r} \\ \frac{1}{3}R_1 \to R_1 \\ \frac{1}{900{,}000}R_3 \to R_3 \\ \\ \end{array} \begin{array}{cccccccc|c} x_1 & x_2 & x_3 & x_4 & x_5 & s_1 & s_2 & z & \\ \frac{5}{3} & -\frac{2}{3} & 1 & 0 & 1 & \frac{5}{3} & -\frac{2}{3} & 0 & 20{,}000 \\ -\frac{1}{180{,}000} & \frac{1}{180{,}000} & 0 & 1 & 0 & -\frac{1}{180{,}000} & \frac{1}{180{,}000} & 0 & \frac{1}{10} \\ \hline 15 & 5 & 0 & 0 & 25 & 25 & 20 & 1 & 1{,}800{,}000 \end{array}$$

The minimum value of w is 1,800,000 when $y_1 = 25$ and $y_2 = 20$.

Therefore, 25 units of regular beer and 20 units of light beer should be made for a minimum cost of $1,800,000.

(b) The shadow cost for revenue is $\frac{1}{10}$ dollar or $0.10. An increase in $500,000 in revenue will increase costs to

$$\$1{,}800{,}000 + \$0.10(500{,}000) = \$1{,}850{,}000.$$

21. (a) The initial matrix for the original problem is

$$\left[\begin{array}{ccc|c} 1 & 1 & 1 & 100 \\ 400 & 160 & 280 & 20{,}000 \\ \hline 120 & 40 & 60 & 0 \end{array}\right].$$

The transposed matrix, for the dual problem, is

$$\left[\begin{array}{cc|c} 1 & 400 & 120 \\ 1 & 160 & 40 \\ 1 & 280 & 60 \\ \hline 100 & 20{,}000 & 0 \end{array}\right].$$

Minimize $w = 100y_1 + 20{,}000y_2$

subject to:
$$\begin{aligned} y_1 + 400y_2 &\geq 120 \\ y_1 + 160y_2 &\geq 40 \\ y_1 + 280y_2 &\geq 60 \end{aligned}$$

with $y_1 \geq 0, y_2 \geq 0.$

(b) We apply the simplex algorithm to the original maximization problem. The initial tableau is

$$\left[\begin{array}{cccccc|c} x_1 & x_2 & x_3 & s_1 & s_2 & z & \\ 1 & 1 & 1 & 1 & 0 & 0 & 100 \\ \boxed{400} & 160 & 280 & 0 & 1 & 0 & 20{,}000 \\ \hline -120 & -40 & -60 & 0 & 0 & 1 & 0 \end{array}\right]$$

Pivot on the 400 in row 2, column 1.

$$\begin{array}{r} \\ -R_2 + 400R_1 \rightarrow R_1 \\ \\ \frac{3}{10}R_2 + R_3 \rightarrow R_3 \end{array} \begin{array}{c} \begin{matrix} x_1 & x_2 & x_3 & s_1 & s_2 & z \end{matrix} \\ \left[\begin{array}{cccccc|c} 0 & 240 & 120 & 400 & -1 & 0 & 20{,}000 \\ 400 & 160 & 280 & 0 & 1 & 0 & 20{,}000 \\ \hline 0 & 8 & 24 & 0 & 0.3 & 1 & 6000 \end{array}\right] \end{array}$$

Create a 1 in the columns corresponding to x_1 and s_1.

$$\begin{array}{r} \\ \frac{1}{400}R_1 \rightarrow R_1 \\ \frac{1}{400}R_2 \rightarrow R_2 \\ \\ \end{array} \begin{array}{c} \begin{matrix} x_1 & x_2 & x_3 & s_1 & s_2 & z \end{matrix} \\ \left[\begin{array}{cccccc|c} 0 & 0.6 & 0.3 & 1 & -\frac{1}{400} & 0 & 50 \\ 1 & 0.4 & 0.7 & 0 & \frac{1}{400} & 0 & 50 \\ \hline 0 & 8 & 24 & 0 & 0.3 & 1 & 6000 \end{array}\right] \end{array}$$

This solution is optimal. A maximum profit of \$6000 is achieved by planting 50 acres of potatoes, 0 acres of corn, and 0 acres of cabbage.

From the dual solution, the shadow cost of acreage is 0 and of capital is $\frac{3}{10}$.

$$\begin{aligned} \text{New profit} &= 6000 + 0(-10) + \left(\frac{3}{10}\right)1000 \\ &= \$6300 \end{aligned}$$

Now calculate the number of acres of each:

$$\begin{aligned} \text{Profit} &= 120P + 40C + 60B \\ 6300 &= 120P + 40(0) + 60(0) \\ P &= 52.5. \end{aligned}$$

The farmer will make a profit of \$6300 by planting 52.5 acres of potatoes and no corn or cabbage.

(c)
$$\begin{aligned} \text{New profit} &= 6000 + 0(10) + \left(\frac{3}{10}\right)(-1000) \\ &= \$5700 \end{aligned}$$

Calculate the number of acres of each:

$$\begin{aligned} \text{Profit} &= 120P + 40C + 60B \\ 5700 &= 120P + 40(0) + 60(0) \\ P &= 47.5. \end{aligned}$$

The farmer will make a profit of \$5700 by planting 47.5 acres of potatoes and no corn or cabbage.

23. Let y_1 = the number of political interviews conducted

and y_2 = the number of market interviews conducted.

The problem is:

Minimize $w = 45y_1 + 55y_2$

subject to:

$$\begin{aligned} y_1 + y_2 &\geq 8 \\ 8y_1 + 10y_2 &\geq 60 \\ 6y_1 + 5y_2 &\geq 40 \end{aligned}$$

with $y_1 \geq 0, y_2 \geq 0$.

Write the augmented matrix.

$$\left[\begin{array}{cc|c} 1 & 1 & 8 \\ 8 & 10 & 60 \\ 6 & 5 & 40 \\ \hline 45 & 55 & 0 \end{array}\right]$$

Transpose to get the matrix for the dual problem.

$$\left[\begin{array}{ccc|c} 1 & 8 & 6 & 45 \\ 1 & 10 & 5 & 55 \\ \hline 8 & 60 & 40 & 0 \end{array}\right]$$

Write the dual problem:

Maximize $z = 8x_1 + 60x_2 + 40x_3$

subject to:

$$\begin{aligned} x_1 + 8x_2 + 6x_3 &\leq 45 \\ x_1 + 10x_2 + 5x_3 &\leq 55 \end{aligned}$$

with $x_1 \geq 0, x_2 \geq 0, x_3 \geq 0$.

Write the initial tableau.

$$\begin{array}{c} \\ \\ \\ \end{array}\begin{array}{cccccc|c} x_1 & x_2 & x_3 & s_1 & s_2 & z & \\ \hline 1 & 8 & 6 & 1 & 0 & 0 & 45 \\ 1 & \boxed{10} & 5 & 0 & 1 & 0 & 55 \\ \hline -8 & -60 & -40 & 0 & 0 & 1 & 0 \end{array}$$

Pivot on the 10 in row 2, column 2.

$$\begin{array}{r} \\ -4R_2 + 5R_1 \to R_1 \\ \\ 6R_2 + R_3 \to R_3 \end{array}\begin{array}{cccccc|c} x_1 & x_2 & x_3 & s_1 & s_2 & z & \\ \hline 1 & 0 & \boxed{10} & 5 & -4 & 0 & 5 \\ 1 & 10 & 5 & 0 & 1 & 0 & 55 \\ \hline -2 & 0 & -10 & 0 & 6 & 1 & 330 \end{array}$$

Pivot on the 10 in row 1, column 3.

$$\begin{array}{r} \\ \\ -R_1 + 2R_2 \to R_2 \\ R_1 + R_3 \to R_3 \end{array}\begin{array}{cccccc|c} x_1 & x_2 & x_3 & s_1 & s_2 & z & \\ \hline \boxed{1} & 0 & 10 & 5 & -4 & 0 & 5 \\ 1 & 20 & 0 & -5 & 6 & 0 & 105 \\ \hline -1 & 0 & 0 & 5 & 2 & 1 & 335 \end{array}$$

Pivot on the 1 in row 1, column 1.

$$\begin{array}{r} \\ \\ -R_1 + R_2 \to R_2 \\ R_1 + R_3 \to R_3 \end{array}\begin{array}{cccccc|c} x_1 & x_2 & x_3 & s_1 & s_2 & z & \\ \hline 1 & 0 & 10 & 5 & -4 & 0 & 5 \\ 0 & 20 & -10 & -10 & \boxed{10} & 0 & 100 \\ \hline 0 & 0 & 10 & 10 & -2 & 1 & 340 \end{array}$$

Pivot on the 10 in row 2, column 5.

$$\begin{array}{r} \\ 2R_2 + 5R_1 \to R_1 \\ \\ R_2 + 5R_3 \to R_3 \end{array}\begin{array}{cccccc|c} x_1 & x_2 & x_3 & s_1 & s_2 & z & \\ \hline 5 & 40 & 30 & 5 & 0 & 0 & 225 \\ 0 & 20 & -10 & -10 & 10 & 0 & 100 \\ \hline 0 & 20 & 40 & 40 & 0 & 5 & 1800 \end{array}$$

Create a 1 in the columns corresponding to x_1, s_2, and z.

$$\begin{array}{r} \\ \frac{1}{5}R_1 \to R_1 \\ \frac{1}{10}R_2 \to R_2 \\ \frac{1}{5}R_3 \to R_3 \end{array}\begin{array}{cccccc|c} x_1 & x_2 & x_3 & s_1 & s_2 & z & \\ \hline 1 & 8 & 6 & 1 & 0 & 0 & 45 \\ 0 & 2 & -1 & -1 & 1 & 0 & 10 \\ \hline 0 & 4 & 8 & 8 & 0 & 1 & 360 \end{array}$$

The minimum time spent is 360 min when $y_1 = 8$ and $y_2 = 0$, that is, when 8 political interviews and no market interviews are done.

25. Organize the information in a table.

	Units of Nutrient A (per bag)	Units of Nutrient B (per bag)	Cost (per bag)
Feed 1	1	2	$3
Feed 2	3	1	$2
Minimum	7	4	

Let y_1 = the number of bags of feed 1

and y_2 = the number of bags of feed 2.

(a) We want the cost to equal $7 for 7 units of A and 4 units of B exactly. Therefore, use a system of equations rather than a system of inequalities.

$$3y_1 + 2y_2 = 7$$
$$y_1 + 3y_2 = 7$$
$$2y_1 + y_2 = 4$$

Use Gauss-Jordan elimination to solve this system of equations.

$$\left[\begin{array}{cc|c} 3 & 2 & 7 \\ 1 & 3 & 7 \\ 2 & 1 & 4 \end{array}\right]$$

$$\begin{array}{r} \\ -R_1 + 3R_2 \rightarrow R_2 \\ -2R_1 + 3R_3 \rightarrow R_3 \end{array} \left[\begin{array}{cc|c} 3 & 2 & 7 \\ 0 & 7 & 14 \\ 0 & -1 & -2 \end{array}\right]$$

$$\begin{array}{r} -2R_2 + 7R_1 \rightarrow R_1 \\ \\ R_2 + 7R_3 \rightarrow R_3 \end{array} \left[\begin{array}{cc|c} 21 & 0 & 21 \\ 0 & 7 & 14 \\ 0 & 0 & 0 \end{array}\right]$$

$$\begin{array}{r} \frac{1}{21}R_1 \rightarrow R_1 \\ \frac{1}{7}R_2 \rightarrow R_2 \\ \\ \end{array} \left[\begin{array}{cc|c} 1 & 0 & 1 \\ 0 & 1 & 2 \\ 0 & 0 & 0 \end{array}\right]$$

Thus, $y_1 = 1$ and $y_2 = 2$, so use 1 bag of feed 1 and 2 bags of feed 2. The cost will be $3(1) + 2\ (2) = \$7$ as desired. The number of units of A is $1(1) + 3(2) = 7$, and the number of units of B is $2(1) + 1(2) = 4$.

(b)

	Units of Nutrient A (per bag)	Units of Nutrient B (per bag)	Cost (per bag)
Feed 1	1	2	\$3
Feed 2	3	1	\$2
Minimum	5	4	

The problem is:

Minimize $w = 3y_1 + 2y_2$

subject to:
$$y_1 + 3y_2 \geq 5$$
$$2y_1 + y_2 \geq 4$$

with $y_1 \geq 0, y_2 \geq 0$.

The dual problem is as follows.

Maximize $z = 5x_1 + 4x_2$

subject to:
$$x_1 + 2x_2 \leq 3$$
$$3x_1 + x_2 \leq 2$$

with $x_1 \geq 0,\ x_2 \geq 0$.

The initial tableau is as follows.

$$\begin{array}{c} \begin{array}{cccccc} x_1 & x_2 & s_1 & s_2 & z & \end{array} \\ \left[\begin{array}{ccccc|c} 1 & 2 & 1 & 0 & 0 & 3 \\ \boxed{3} & 1 & 0 & 1 & 0 & 2 \\ \hline -5 & -4 & 0 & 0 & 1 & 0 \end{array}\right] \end{array}$$

Pivot as indicated.

$$\begin{array}{r} -R_2 + 3R_1 \rightarrow R_1 \\ \\ 5R_2 + 3R_3 \rightarrow R_3 \end{array} \begin{array}{c} \begin{array}{cccccc} x_1 & x_2 & s_1 & s_2 & z & \end{array} \\ \left[\begin{array}{ccccc|c} 0 & \boxed{5} & 3 & -1 & 0 & 7 \\ 3 & 1 & 0 & 1 & 0 & 2 \\ \hline 0 & -7 & 0 & 5 & 3 & 10 \end{array}\right] \end{array}$$

$$\begin{array}{r} \\ -R_1 + 5R_2 \rightarrow R_2 \\ 7R_1 + 5R_3 \rightarrow R_3 \end{array} \begin{array}{c} \begin{array}{cccccc} x_1 & x_2 & s_1 & s_2 & z & \end{array} \\ \left[\begin{array}{ccccc|c} 0 & 5 & 3 & -1 & 0 & 7 \\ 15 & 0 & -3 & 6 & 0 & 3 \\ \hline 0 & 0 & 21 & 18 & 15 & 99 \end{array}\right] \end{array}$$

Create a 1 in the columns corresponding to x_1, x_2, and z.

$$\begin{array}{r} \frac{1}{5}R_1 \rightarrow R_1 \\ \frac{1}{15}R_2 \rightarrow R_2 \\ \frac{1}{15}R_3 \rightarrow R_3 \end{array} \begin{array}{c} \begin{array}{cccccc} x_1 & x_2 & s_1 & s_2 & z & \end{array} \\ \left[\begin{array}{ccccc|c} 0 & 1 & \frac{3}{5} & -\frac{1}{5} & 0 & \frac{7}{5} \\ 1 & 0 & -\frac{1}{5} & \frac{2}{5} & 0 & \frac{1}{5} \\ \hline 0 & 0 & \frac{7}{5} & \frac{6}{5} & 1 & \frac{33}{5} \end{array}\right] \end{array}$$

Reading from the final column of the final tableau, $x_2 = \$1.40$ is the cost of nutrient B and $x_1 = \$0.20$ is the cost of nutrient A. With 5 units of A and 4 units of B, this gives a minimum cost of

$$5(\$0.20) + 4(\$1.40) = \$6.60$$

as given in the lower right corner. 1.4 $\left(\text{or } \frac{7}{5}\right)$ bags of feed 1 and 1.2 $\left(\text{or } \frac{6}{5}\right)$ bags of feed 2 should be used.

27. Let $y_1 =$ the number of minutes spent walking,

$y_2 =$ the number of minutes spent cycling,

and $y_3 =$ the number of minutes spent swimming.

Minimize $w = y_1 + y_2 + y_3$

subject to:
$$3.5y_1 + 4y_2 + 8y_3 \geq 1500$$
$$y_1 + y_2 \geq 3y_3$$
$$y_1 \geq 30$$

with $y_1 \geq 0, y_2 \geq 0, y_3 \geq 0.$

The second constraint can be written as

$$y_1 + y_2 - 3y_3 \geq 0.$$

Write the augmented matrix for this problem.

$$\left[\begin{array}{ccc|c} 3.5 & 4 & 8 & 1500 \\ 1 & 1 & -3 & 0 \\ 1 & 0 & 0 & 30 \\ \hline 1 & 1 & 1 & 0 \end{array}\right]$$

Transpose to get the matrix for the dual problem.

$$\left[\begin{array}{ccc|c} 3.5 & 1 & 1 & 1 \\ 4 & 1 & 0 & 1 \\ 8 & -3 & 0 & 1 \\ \hline 1500 & 0 & 30 & 0 \end{array}\right]$$

Write the dual problem.

Maximize $z = 1500x_1 + 30x_3$

subject to:
$$\begin{aligned} 3.5x_1 + x_2 + x_3 &\leq 1 \\ 4x_1 + x_2 &\leq 1 \\ 8x_1 - 3x_2 &\leq 1 \end{aligned}$$

with $x_1 \geq 0, s_2 \geq 0, x_3 \geq 0.$

Write the initial simplex tableau.

$$\begin{array}{c} \begin{array}{ccccccc} x_1 & x_2 & x_3 & x_4 & s_1 & s_2 & z \end{array} \\ \left[\begin{array}{ccccccc|c} 3.5 & 1 & 1 & 1 & 0 & 0 & 0 & 1 \\ 4 & 1 & 0 & 0 & 1 & 0 & 0 & 1 \\ 8 & -3 & 0 & 0 & 0 & 1 & 0 & 1 \\ \hline -1500 & 0 & -30 & 0 & 0 & 0 & 1 & 0 \end{array}\right] \end{array}$$

Using a graphing calculator or computer program, such as Solver in Microsoft Excel, we obtain the optimal answer: 30 minutes walking, 197.25 minutes cycling, and 75.75 minutes swimming for a total minimum time of 303 minutes per week.

29. Let y_1 = the number of units of ingredient I;

y_2 = the number of units of ingredient II;

and y_3 = the number of units of ingredient III.

The problem is:

Minimize $w = 4y_1 + 7y_2 + 5y_3$

subject to:
$$\begin{aligned} 4y_1 + y_2 + 10y_3 &\geq 10 \\ 3y_1 + 2y_2 + y_3 &\geq 12 \\ 4y_2 + 5y_3 &\geq 20 \end{aligned}$$

with $y_1 \geq 0, y_2 \geq 0, y_3 \geq 0.$

The dual problem is as follows.

Maximize $z = 10x_1 + 12x_2 + 20x_3$

subject to:
$$\begin{aligned} 4x_1 + 3x_2 &\leq 4 \\ x_1 + 2x_2 + 4x_3 &\leq 7 \\ 10x_1 + x_2 + 5x_3 &\leq 5 \end{aligned}$$

with $x_1 \geq 0, x_2 \geq 0, x_3 \geq 0.$

The initial tableau is as follows.

$$\begin{array}{c} \begin{array}{ccccccc} x_1 & x_2 & x_3 & s_1 & s_2 & s_3 & z \end{array} \\ \left[\begin{array}{ccccccc|c} 4 & 3 & 0 & 1 & 0 & 0 & 0 & 4 \\ 1 & 2 & 4 & 0 & 1 & 0 & 0 & 7 \\ 10 & 1 & \boxed{5} & 0 & 0 & 1 & 0 & 5 \\ \hline -10 & -12 & -20 & 0 & 0 & 0 & 1 & 0 \end{array}\right] \end{array}$$

Pivot as indicated.

$$\begin{array}{r} \\ -4R_3 + 5R_2 \to R_2 \\ \\ 4R_3 + R_4 \to R_4 \end{array} \left[\begin{array}{ccccccc|c} 4 & \boxed{3} & 0 & 1 & 0 & 0 & 0 & 4 \\ -35 & 6 & 0 & 0 & 5 & -4 & 0 & 15 \\ 10 & 1 & 5 & 0 & 0 & 1 & 0 & 5 \\ \hline 30 & -8 & 0 & 0 & 0 & 4 & 1 & 20 \end{array}\right]$$

(columns: x_1, x_2, x_3, s_1, s_2, s_3, z)

$$\begin{array}{r} \\ -2R_1 + R_2 \to R_2 \\ -R_1 + 3R_3 \to R_3 \\ 8R_1 + 3R_4 \to R_4 \end{array} \left[\begin{array}{ccccccc|c} 4 & 3 & 0 & 1 & 0 & 0 & 0 & 4 \\ -43 & 0 & 0 & -2 & 5 & -4 & 0 & 7 \\ 26 & 0 & 15 & -1 & 0 & 3 & 0 & 11 \\ \hline 122 & 0 & 0 & 8 & 0 & 12 & 3 & 92 \end{array}\right]$$

(columns: x_1, x_2, x_3, s_1, s_2, s_3, z)

Create a 1 in the columns corresponding to x_2, x_3, and z.

$$\begin{array}{r} \frac{1}{3}R_1 \to R_1 \\ \\ \frac{1}{15}R_3 \to R_3 \\ \frac{1}{3}R_4 \to R_4 \end{array} \left[\begin{array}{ccccccc|c} \frac{4}{3} & 1 & 0 & \frac{1}{3} & 0 & 0 & 0 & \frac{4}{3} \\ -43 & 0 & 0 & -2 & 5 & -4 & 0 & 7 \\ \frac{26}{15} & 0 & 1 & -\frac{1}{15} & 0 & \frac{1}{5} & 0 & \frac{11}{5} \\ \hline \frac{122}{3} & 0 & 0 & \frac{8}{3} & 0 & 4 & 1 & \frac{92}{3} \end{array}\right]$$

(columns: x_1, x_2, x_3, s_1, s_2, s_3, z)

From the last row, the minimum value is $\frac{92}{3}$ when $y_1 = \frac{8}{3}$, $y_2 = 0$, and $y_3 = 4$. The biologist can meet his needs at a minimum cost of \$30.67 by using $\frac{8}{3}$ units of ingredient I and 4 units of ingredient III. (Ingredient II should not be used at all.)

4.4 Nonstandard Problems

Your Turn 1

Minimize $w = 6y_1 + 4y_2$

subject to: $3y_1 + 4y_2 \ge 10$
$9y_1 + 7y_2 \le 18$

with $y_1 \ge 0,\ y_2 \ge 0.$

Instead we maximize $z = -w = -6y_1 - 4y_2$ subject to the same constraints. Inserting slack and surplus variables produces the following initial tableau.

$$\begin{array}{c} \\ \\ \\ \\ \end{array}\begin{array}{ccccc|c} y_1 & y_2 & s_1 & s_2 & z & \\ 3 & 4 & -1 & 0 & 0 & 10 \\ \boxed{9} & 7 & 0 & 1 & 0 & 18 \\ \hline 6 & 4 & 0 & 0 & 1 & 0 \end{array}$$

Because s_1 is negative, we choose the positive entry farthest to the left in row 1, which is the 3 in column 1. The entry 9 in this column gives the smallest quotient so we choose 9 as the pivot.

$$\begin{array}{r} \\ -R_2 + 3R_1 \to R_1 \\ \\ -2R_2 + 3R_3 \to R_3 \end{array}\begin{array}{ccccc|c} y_1 & y_2 & s_1 & s_2 & z & \\ 0 & \boxed{5} & -3 & -1 & 0 & 12 \\ 9 & 7 & 0 & 1 & 0 & 18 \\ \hline 0 & -2 & 0 & -2 & 3 & -36 \end{array}$$

s_2 is still negative, so we pivot on the 5 in column 2.

$$\begin{array}{r} \\ \\ -7R_1 + 5R_2 \to R_2 \\ 2R_1 + 5R_3 \to R_3 \end{array}\begin{array}{ccccc|c} y_1 & y_2 & s_1 & s_2 & z & \\ 0 & 5 & -3 & -1 & 0 & 12 \\ 45 & 0 & 21 & \boxed{12} & 0 & 6 \\ \hline 0 & 0 & -6 & -12 & 15 & -156 \end{array}$$

Now we work on the largest negative indicator and pivot on the 12 in column 4.

$$\begin{array}{r} \\ R_2 + 12R_1 \to R_1 \\ \\ R_2 + R_3 \to R_3 \end{array}\begin{array}{ccccc|c} y_1 & y_2 & s_1 & s_2 & z & \\ 45 & 60 & -15 & 0 & 0 & 150 \\ 45 & 0 & 21 & 12 & 0 & 6 \\ \hline 45 & 0 & 15 & 0 & 15 & -150 \end{array}$$

From this we can read the solution: The minimum is $-\left(\frac{-150}{15}\right) = 10$ when $y_1 = 0$ and $y_2 = \frac{150}{60} = \frac{5}{2}$.

Your Turn 2

We start with this tableau.

$$\begin{array}{ccccccccc|c} y_1 & y_2 & y_3 & y_4 & s_1 & s_2 & s_3 & s_4 & z & \\ 0 & 1 & 0 & 1 & 0 & 0 & 0 & -1 & 0 & 16 \\ 0 & 0 & 0 & 0 & 1 & 1 & 1 & 1 & 0 & 0 \\ 1 & 0 & 0 & -1 & 1 & 0 & 0 & 1 & 0 & 12 \\ 0 & 0 & 1 & \boxed{1} & -1 & 0 & -1 & -1 & 0 & 8 \\ \hline 0 & 0 & 0 & -300 & 180 & 0 & 400 & 480 & 1 & -10{,}640 \end{array}$$

We pivot on the 1 in row 4 of column 4.

$$\begin{array}{r} \\ -R_4 + R_1 \to R_1 \\ \\ R_4 + R_3 \to R_3 \\ \\ 300R_4 + R_5 \to R_5 \end{array}\begin{array}{ccccccccc|c} y_1 & y_2 & y_3 & y_4 & s_1 & s_2 & s_3 & s_4 & z & \\ 0 & 1 & -1 & 0 & 1 & 0 & 1 & 0 & 0 & 8 \\ 0 & 0 & 0 & 0 & \boxed{1} & 1 & 1 & 1 & 0 & 0 \\ 1 & 0 & 1 & 0 & 0 & 0 & -1 & 0 & 0 & 20 \\ 0 & 0 & 1 & 1 & -1 & 0 & -1 & -1 & 0 & 8 \\ \hline 0 & 0 & 300 & 0 & -120 & 0 & 100 & 180 & 1 & -8240 \end{array}$$

Finally we pivot on the 1 in row 2 of column 5.

$$\begin{array}{r} \\ -R_2 + R_1 \to R_1 \\ \\ \\ R_2 + R_4 \to R_4 \\ 120R_2 + R_5 \to R_5 \end{array}\begin{array}{ccccccccc|c} y_1 & y_2 & y_3 & y_4 & s_1 & s_2 & s_3 & s_4 & z & \\ 0 & 1 & -1 & 0 & 0 & -1 & 0 & -1 & 0 & 8 \\ 0 & 0 & 0 & 0 & 1 & 1 & 1 & 1 & 0 & 0 \\ 1 & 0 & 1 & 0 & 0 & 0 & -1 & 0 & 0 & 20 \\ 0 & 0 & 1 & 1 & 0 & 1 & 0 & 0 & 0 & 8 \\ \hline 0 & 0 & 300 & 0 & 0 & 120 & 220 & 300 & 1 & -8240 \end{array}$$

Since there are now no negative indicators this tableau gives the solution:
$y_1 = 20,\ y_2 = 8,\ y_3 = 0,\ y_4 = 8$, with a minimum cost of \$8240.

4.4 Exercises

1. $2x_1 + 3x_2 \le 8$
$x_1 + 4x_2 \ge 7$

Introduce the slack variable s_1 and the surplus variable s_2 to obtain the following equations:

$$\begin{aligned} 2x_1 + 3x_2 + s_1 \quad &= 8 \\ x_1 + 4x_2 \quad - s_2 &= 7. \end{aligned}$$

3. $2x_1 + x_2 + 2x_3 \le 50$
$x_1 + 3x_2 + x_3 \ge 35$
$x_1 + 2x_2 \ge 15$

Introduce the slack variable s_1 and the surplus variables s_2 and s_3 to obtain the following equations:

$$\begin{aligned} 2x_1 + x_2 + 2x_3 + s_1 \qquad\qquad &= 50 \\ x_1 + 3x_2 + x_3 \qquad - s_2 \qquad &= 35 \\ x_1 + 2x_2 \qquad\qquad\qquad - s_3 &= 15. \end{aligned}$$

5. Minimize $w = 3y_1 + 4y_2 + 5y_3$

subject to:
$$y_1 + 2y_2 + 3y_3 \geq 9$$
$$y_2 + 2y_3 \geq 8$$
$$2y_1 + y_2 + 2y_3 \geq 6$$

with $y_1 \geq 0, y_2 \geq 0, y_3 \geq 0.$

Change this to a maximization problem by letting $z = -w$. The problem can now be stated equivalently as follows:

Maximize $z = -3y_1 - 4y_2 - 5y_3$

subject to:
$$y_1 + 2y_2 + 3y_3 \geq 9$$
$$y_2 + 2y_3 \geq 8$$
$$2y_1 + y_2 + 2y_3 \geq 6$$

with $y_1 \geq 0, y_2 \geq 0, y_3 \geq 0.$

7. Minimize $w = y_1 + 2y_2 + y_3 + 5y_4$

subject to:
$$y_1 + y_2 + y_3 + y_4 \geq 50$$
$$3y_1 + y_2 + 2y_3 + y_4 \geq 100$$

with $y_1 \geq 0, y_2 \geq 0, y_3 \geq 0, y_4 \geq 0.$

Change this to a maximization problem by letting $z = -w$. The problem can now be stated equivalently as follows:

Maximize $z = -y_1 - 2y_2 - y_3 - 5y_4$

subject to:
$$y_1 + y_2 + y_3 + y_4 \geq 50$$
$$3y_1 + y_2 + 2y_3 + y_4 \geq 100$$

with $y_1 \geq 0, y_2 \geq 0, y_3 \geq 0, y_4 \geq 0.$

9. Find $x_1 \geq 0$ and $x_2 \geq 0$ such that

$$x_1 + 2x_2 \geq 24$$
$$x_1 + x_2 \leq 40$$

and $z = 12x_1 + 10x_2$ is maximized.

Subtracting the surplus variable s_1 and adding the slack variable s_2 leads to the equations

$$x_1 + 2x_2 - s_1 = 24$$
$$x_1 + x_2 + s_2 = 40.$$

The initial simplex tableau is as follows.

x_1	x_2	s_1	s_2	z	
[1]	2	−1	0	0	24
1	1	0	1	0	40
−12	−10	0	0	1	0

The initial basic solution is not feasible since $s_1 = -24$ is negative, so row transformations must be used. Pivot on the 1 in row 1, column 1, since it is the positive entry that is farthest to the left in the first row (the row containing the −1) and since, in the first column, $\frac{24}{1} = 24$ is a smaller quotient than $\frac{40}{1} = 40$. After row transformations, we obtain the following tableau.

	x_1	x_2	s_1	s_2	z	
	1	2	−1	0	0	24
$-R_1 + R_2 \rightarrow R_2$	0	−1	[1]	1	0	16
$12R_1 + R_3 \rightarrow R_3$	0	14	−12	0	1	288

The basic solution is now feasible, but the problem is not yet finished since there is a negative indicator. Continue in the usual way. The 1 in column 3 is the next pivot. After row transformations, we get the following tableau.

	x_1	x_2	s_1	s_2	z	
$R_1 + R_2 \rightarrow R_1$	1	1	0	1	0	40
	0	−1	1	1	0	16
$12R_2 + R_3 \rightarrow R_3$	0	2	0	12	1	480

This is a final tableau since the entries in the last row are all nonnegative. The maximum value is 480 when $x_1 = 40$ and $x_2 = 0$.

11. Find $x_1 \geq 0$, $x_2 \geq 0$, and $x_3 \geq 0$ such that

$$x_1 + x_2 + x_3 \leq 150$$
$$x_1 + x_2 + x_3 \geq 100$$

and $z = 2x_1 + 5x_2 + 3x_3$ is maximized.

The initial tableau is as follows.

x_1	x_2	x_3	s_1	s_2	z	
1	1	1	1	0	0	150
[1]	1	1	0	−1	0	100
−2	−5	−3	0	0	1	0

Note that s_1 is a slack variable, while s_2 is a surplus variable. The initial basic solution is not feasible, since $s_2 = -100$ is negative. Pivot on the 1 in row 2, column 1.

	x_1	x_2	x_3	s_1	s_2	z	
$-R_2 + R_1 \rightarrow R_1$	0	0	0	1	1	0	50
	1	[1]	1	0	−1	0	100
$2R_2 + R_3 \rightarrow R_3$	0	−3	−1	0	−2	1	200

Pivot on the 1 in row 2, column 2.

$$
\begin{array}{r} \\ \\ \\ 3R_2 + R_3 \to R_3 \end{array}
\begin{array}{c}
\begin{array}{cccccc|c} x_1 & x_2 & x_3 & s_1 & s_2 & z & \end{array} \\
\left[\begin{array}{cccccc|c}
0 & 0 & 0 & 1 & \boxed{1} & 0 & 50 \\
1 & 1 & 1 & 0 & -1 & 0 & 100 \\ \hline
3 & 0 & 2 & 0 & -5 & 1 & 500
\end{array}\right]
\end{array}
$$

Pivot on the 1 in row 1, column 5.

$$
\begin{array}{r} \\ \\ R_1 + R_2 \to R_2 \\ 5R_1 + R_3 \to R_3 \end{array}
\begin{array}{c}
\begin{array}{cccccc|c} x_1 & x_2 & x_3 & s_1 & s_2 & z & \end{array} \\
\left[\begin{array}{cccccc|c}
0 & 0 & 0 & 1 & 1 & 0 & 50 \\
1 & 1 & 1 & 1 & 0 & 0 & 150 \\ \hline
3 & 0 & 2 & 5 & 0 & 1 & 750
\end{array}\right]
\end{array}
$$

This is a final tableau. The maximum value is 750 when $x_1 = 0$, $x_2 = 150$, and $x_3 = 0$.

13. Find $x_1 \geq 0$ and $x_2 \geq 0$ such that

$$
\begin{aligned}
x_1 + x_2 &\leq 100 \\
2x_1 + 3x_2 &\leq 75 \\
x_1 + 4x_2 &\geq 50
\end{aligned}
$$

and $z = 5x_1 - 3x_2$ is maximized.

The initial simplex tableau is

$$
\begin{array}{c}
\begin{array}{cccccc|c} x_1 & x_2 & s_1 & s_2 & s_3 & z & \end{array} \\
\left[\begin{array}{cccccc|c}
1 & 1 & 1 & 0 & 0 & 0 & 100 \\
\boxed{2} & 3 & 0 & 1 & 0 & 0 & 75 \\
1 & 4 & 0 & 0 & -1 & 0 & 50 \\ \hline
-5 & 3 & 0 & 0 & 0 & 1 & 0
\end{array}\right]
\end{array}.
$$

The initial basic solution is not feasible since $s_3 = -50$. Pivot on the 2 in row 2, column 1.

$$
\begin{array}{r} \\ -R_2 + 2R_1 \to R_1 \\ \\ -R_2 + 2R_3 \to R_3 \\ 5R_2 + 2R_4 \to R_4 \end{array}
\begin{array}{c}
\begin{array}{cccccc|c} x_1 & x_2 & x_3 & s_1 & s_2 & z & \end{array} \\
\left[\begin{array}{cccccc|c}
0 & -1 & 2 & -1 & 0 & 0 & 125 \\
2 & 3 & 0 & 1 & 0 & 0 & 75 \\
0 & \boxed{5} & 0 & -1 & -2 & 0 & 25 \\ \hline
0 & 21 & 0 & 5 & 0 & 2 & 375
\end{array}\right]
\end{array}.
$$

This solution is still not feasible since $s_3 = -\frac{25}{2}$. Pivot on the 5 in row 3, column 2.

$$
\begin{array}{r} \\ R_3 + 5R_1 \to R_1 \\ -3R_3 + 5R_2 \to R_2 \\ \\ -21R_3 + 5R_4 \to R_4 \end{array}
\begin{array}{c}
\begin{array}{cccccc|c} x_1 & x_2 & x_3 & s_1 & s_2 & z & \end{array} \\
\left[\begin{array}{cccccc|c}
0 & 0 & 10 & -6 & -2 & 0 & 650 \\
10 & 0 & 0 & 8 & 6 & 0 & 300 \\
0 & \boxed{5} & 0 & -1 & -2 & 0 & 25 \\ \hline
0 & 0 & 0 & 46 & 42 & 10 & 1350
\end{array}\right]
\end{array}
$$

Create a 1 in the columns corresponding to x_1, x_2, s_1, and z.

$$
\begin{array}{r} \\ \frac{1}{10}R_1 \to R_1 \\ \frac{1}{10}R_2 \to R_2 \\ \frac{1}{5}R_3 \to R_3 \\ \frac{1}{10}R_4 \to R_4 \end{array}
\begin{array}{c}
\begin{array}{cccccc|c} x_1 & x_2 & x_3 & s_1 & s_2 & z & \end{array} \\
\left[\begin{array}{cccccc|c}
0 & 0 & 1 & -\frac{3}{5} & -\frac{1}{5} & 0 & 65 \\
1 & 0 & 0 & \frac{4}{5} & \frac{3}{5} & 0 & 30 \\
0 & 1 & 0 & -\frac{1}{5} & -\frac{2}{5} & 0 & 5 \\ \hline
0 & 0 & 0 & \frac{23}{5} & \frac{21}{5} & 1 & 135
\end{array}\right]
\end{array}.
$$

This is a final tableau. The maximum is 135 when $x_1 = 30$, $x_2 = 5$.

15. Find $y_1 \geq 0$, $y_2 \geq 0$, and $y_3 \geq 0$ such that

$$
\begin{aligned}
5y_1 + 3y_2 + 2y_3 &\leq 150 \\
5y_1 + 10y_2 + 3y_3 &\geq 90
\end{aligned}
$$

and $w = 10y_1 + 12y_2 + 10y_3$ is minimized.

Let $z = -w = -10y - 12y_2 - 10y_3$. Maximize z.

The initial simplex tableau is

$$
\begin{array}{c}
\begin{array}{cccccc|c} y_1 & y_2 & y_3 & s_1 & s_2 & z & \end{array} \\
\left[\begin{array}{cccccc|c}
5 & 3 & 2 & 1 & 0 & 0 & 150 \\
\boxed{5} & 10 & 3 & 0 & -1 & 0 & 90 \\ \hline
10 & 12 & 10 & 0 & 0 & 1 & 0
\end{array}\right]
\end{array}
$$

The initial basic solution is not feasible since $s_2 = -90$. Pivot on the 5 in row 2, column 1.

$$
\begin{array}{r} \\ -R_2 + R_1 \to R_1 \\ \\ -2R_2 + R_3 \to R_3 \end{array}
\begin{array}{c}
\begin{array}{cccccc|c} y_1 & y_2 & y_3 & s_1 & s_2 & z & \end{array} \\
\left[\begin{array}{cccccc|c}
0 & -7 & -1 & 1 & 1 & 0 & 60 \\
5 & \boxed{10} & 3 & 0 & -1 & 0 & 90 \\ \hline
0 & -8 & 4 & 0 & 2 & 1 & -180
\end{array}\right]
\end{array}
$$

Pivot on the 10 in row 2, column 2.

$$
\begin{array}{r} \\ 7R_2 + 10R_1 \to R_1 \\ \\ 8R_2 + 10R_3 \to R_3 \end{array}
\begin{array}{c}
\begin{array}{cccccc|c} y_1 & y_2 & y_3 & s_1 & s_2 & z & \end{array} \\
\left[\begin{array}{cccccc|c}
35 & 0 & 11 & 10 & 3 & 0 & 1230 \\
5 & 10 & 3 & 0 & -1 & 0 & 90 \\ \hline
40 & 0 & 64 & 0 & 12 & 10 & -1080
\end{array}\right]
\end{array}
$$

Create a 1 in the columns corresponding to y_2, s_1, and z.

$$\begin{array}{r} \\ \frac{1}{10}R_1 \to R_1 \\ \frac{1}{10}R_2 \to R_2 \\ \frac{1}{10}R_3 \to R_3 \end{array} \begin{array}{cccccc|c} y_1 & y_2 & y_3 & s_1 & s_2 & z & \\ \frac{7}{2} & 0 & \frac{11}{10} & 1 & \frac{3}{10} & 0 & 123 \\ \frac{1}{2} & 1 & \frac{3}{10} & 0 & -\frac{1}{10} & 0 & 9 \\ \hline 4 & 0 & \frac{32}{5} & 0 & \frac{6}{5} & 1 & -108 \end{array}$$

This is a final tableau. The minimum is 108 when $y_1 = 0, y_2 = 9,$ and $y_3 = 0$.

17. Maximize $z = 3x_1 + 2x_1$

subject to:
$$x_1 + x_2 = 50$$
$$4x_1 + 2x_2 \geq 120$$
$$5x_1 + 2x_2 \leq 200$$

with $x_1 \geq 0, x_2 \geq 0.$

The artificial variable a_1 is used to rewrite $x_1 + x_2 = 50$ as $x_1 + x_2 + a_1 = 50$; note that a_1 must equal 0 for this equation to be a true statement. Also the surplus variable s_1 and the slack variable s_2 are needed. The initial tableau is as follows.

$$\begin{array}{cccccc|c} x_1 & x_2 & a_1 & s_1 & s_2 & z & \\ 1 & 1 & 1 & 0 & 0 & 0 & 50 \\ \boxed{4} & 2 & 0 & -1 & 0 & 0 & 120 \\ 5 & 2 & 0 & 0 & 1 & 0 & 200 \\ \hline -3 & -2 & 0 & 0 & 0 & 1 & 0 \end{array}$$

The initial basic solution is not feasible. Pivot on the 4 in row 2, column 1.

$$\begin{array}{r} \\ -R_2 + 4R_1 \to R_1 \\ \\ -5R_2 + 4R_3 \to R_3 \\ 3R_2 + 4R_4 \to R_4 \end{array} \begin{array}{cccccc|c} x_1 & x_2 & a_1 & s_1 & s_2 & z & \\ 0 & 2 & 4 & 1 & 0 & 0 & 80 \\ 4 & 2 & 0 & -1 & 0 & 0 & 120 \\ 0 & -2 & 0 & \boxed{5} & 4 & 0 & 200 \\ \hline 0 & -2 & 0 & -3 & 0 & 4 & 360 \end{array}$$

The basic solution is now feasible, but there are negative indicators. Pivot on the 5 in row 3, column 4 (which is the column with the most negative indicator and the row with the smallest nonnegative quotient).

$$\begin{array}{r} \\ -R_3 + 5R_1 \to R_1 \\ R_3 + 5R_2 \to R_2 \\ \\ 3R_3 + 5R_4 \to R_4 \end{array} \begin{array}{cccccc|c} x_1 & x_2 & a_1 & s_1 & s_2 & z & \\ 0 & \boxed{12} & 20 & 0 & -4 & 0 & 200 \\ 20 & 8 & 0 & 0 & 4 & 0 & 800 \\ 0 & -2 & 0 & 5 & 4 & 0 & 200 \\ \hline 0 & -16 & 0 & 0 & 12 & 20 & 2400 \end{array}$$

Pivot on the 12 in row 1, column 2.

$$\begin{array}{r} \\ \\ -2R_1 + 3R_2 \to R_2 \\ R_1 + 6R_3 \to R_3 \\ 4R_1 + 3R_4 \to R_4 \end{array} \begin{array}{cccccc|c} x_1 & x_2 & a_1 & s_1 & s_2 & z & \\ 0 & 12 & 20 & 0 & -4 & 0 & 200 \\ 60 & 0 & -40 & 0 & 20 & 0 & 2000 \\ 0 & 0 & 20 & 30 & 20 & 0 & 1400 \\ \hline 0 & 0 & 80 & 0 & 20 & 60 & 8000 \end{array}$$

We now have $a_1 = 0$, so drop the a_1 column.

$$\begin{array}{ccccc|c} x_1 & x_2 & s_1 & s_2 & z & \\ 0 & 12 & 0 & -4 & 0 & 200 \\ 60 & 0 & 0 & 20 & 0 & 2000 \\ 0 & 0 & 30 & 20 & 0 & 1400 \\ \hline 0 & 0 & 0 & 20 & 60 & 8000 \end{array}$$

We are finished pivoting. Create a 1 in the columns corresponding to $x_1, x_2, s_1,$ and z.

$$\begin{array}{r} \\ \frac{1}{12}R_1 \to R_1 \\ \frac{1}{60}R_2 \to R_2 \\ \frac{1}{30}R_3 \to R_3 \\ \frac{1}{60}R_4 \to R_4 \end{array} \begin{array}{ccccc|c} x_1 & x_2 & s_1 & s_2 & z & \\ 0 & 1 & 0 & -\frac{1}{3} & 0 & \frac{50}{3} \\ 1 & 0 & 0 & \frac{1}{3} & 0 & \frac{100}{3} \\ 0 & 0 & 1 & \frac{2}{3} & 0 & \frac{140}{3} \\ \hline 0 & 0 & 0 & \frac{1}{3} & 1 & \frac{400}{3} \end{array}$$

The maximum value is $\frac{400}{3}$ when $x_1 = \frac{100}{3}$ and $x_2 = \frac{50}{3}$.

19. Minimize $w = 32y_1 + 40y_2 + 48y_3$

subject to:
$$20y_1 + 10y_2 + 5y_3 = 200$$
$$25y_1 + 40y_2 + 50y_3 \leq 500$$
$$18y_1 + 24y_2 + 12y_3 \geq 300$$

with $y_1 \geq 0, y_2 \geq 0, y_3 \geq 0$

With artificial, slack, and surplus variables, this problem becomes

Maximize $z = -32y_1 - 40y_2 - 48y_3$

subject to:

$$20y_1 + 10y_2 + 5y_3 + a_1 = 200$$
$$25y_1 + 40y_2 + 50y_3 + s_1 = 500$$
$$18y_1 + 24y_2 + 12y_3 - s_2 = 300.$$

The initial tableau is as follows.

$$\begin{array}{ccccccc|c} y_1 & y_2 & y_3 & a_1 & s_1 & s_2 & z & \\ \boxed{20} & 10 & 5 & 1 & 0 & 0 & 0 & 200 \\ 25 & 40 & 50 & 0 & 1 & 0 & 0 & 500 \\ 18 & 24 & 12 & 0 & 0 & -1 & 0 & 300 \\ \hline 32 & 40 & 48 & 0 & 0 & 0 & 1 & 0 \end{array}$$

The initial basic tableau is not feasible. Pivot on the 20 in row 1, column 1.

$$\begin{array}{r} \\ \\ -5R_1 + 4R_2 \to R_2 \\ -9R_1 + 10R_3 \to R_3 \\ -8R_1 + 5R_4 \to R_4 \end{array} \begin{array}{ccccccc|c} y_1 & y_2 & y_3 & a_1 & s_1 & s_2 & z & \\ 20 & 10 & 5 & 1 & 0 & 0 & 0 & 200 \\ 0 & 110 & 175 & -5 & 4 & 0 & 0 & 1000 \\ 0 & 150 & 75 & -9 & 0 & -10 & 0 & 1200 \\ \hline 0 & 120 & 200 & -8 & 0 & 0 & 5 & -1600 \end{array}$$

Eliminate the a_1 column.

$$\begin{array}{cccccc|c} y_1 & y_2 & y_3 & s_1 & s_2 & z & \\ 20 & 10 & 5 & 0 & 0 & 0 & 200 \\ 0 & 110 & 175 & 4 & 0 & 0 & 1000 \\ 0 & \boxed{150} & 75 & 0 & -10 & 0 & 1200 \\ \hline 0 & 120 & 200 & 0 & 0 & 5 & -1600 \end{array}$$

Pivot on the 150 in row 3, column 2.

$$\begin{array}{r} \\ -R_3 + 15R_1 \to R_1 \\ -11R_3 + 15R_2 \to R_2 \\ \\ -4R_3 + 5R_4 \to R_4 \end{array} \begin{array}{cccccc|c} y_1 & y_2 & y_3 & s_1 & s_2 & z & \\ 300 & 0 & 0 & 0 & 10 & 0 & 1800 \\ 0 & 0 & 1800 & 60 & 110 & 0 & 1800 \\ 0 & 150 & 75 & 0 & -10 & 0 & 1200 \\ \hline 0 & 0 & 700 & 0 & 40 & 25 & -12{,}800 \end{array}$$

Create ones in the columns corresponding to y_1, y_2, s_1, and z.

$$\begin{array}{r} \\ \frac{1}{300}R_1 \to R_1 \\ \frac{1}{60}R_2 \to R_2 \\ \frac{1}{150}R_3 \to R_3 \\ \frac{1}{25}R_4 \to R_4 \end{array} \begin{array}{cccccc|c} y_1 & y_2 & y_3 & s_1 & s_2 & z & \\ 1 & 0 & 0 & 0 & \frac{1}{30} & 0 & 6 \\ 0 & 0 & 30 & 1 & \frac{11}{6} & 0 & 30 \\ 0 & 1 & \frac{1}{2} & 0 & -\frac{1}{15} & 0 & 8 \\ \hline 0 & 0 & 28 & 0 & \frac{8}{5} & 1 & -512 \end{array}$$

This is a final tableau. The minimum value is 512 when $y_1 = 6$, $y_2 = 8$, and $y_3 = 0$.

23. (a) Let y_1 = amount shipped from S_1 to D_1,

y_2 = amount shipped from S_1 to D_2,

y_3 = amount shipped from S_2 to D_1,

and y_4 = amount shipped from S_2 to D_2.

Minimize $w = 30y_1 + 20y_2 + 25y_3 + 22y_4$

subject to:

$$\begin{aligned} y_1 + y_3 &\geq 3000 \\ y_2 + y_4 &\geq 5000 \\ y_1 + y_2 &\leq 5000 \\ y_3 + y_4 &\leq 5000 \\ 2y_1 + 6y_2 + 5y_3 + 4y_4 &\leq 40{,}000 \end{aligned}$$

with $y_1 \geq 0,\ y_2 \geq 0,\ y_3 \geq 0,\ y_4 \geq 0.$

Maximize $z = -w = -30y_1 - 20y_2 - 25y_3 - 22y_4.$

$$\begin{array}{c} \\ \\ \\ \\ \\ \\ \end{array}
\begin{array}{cccccccccc|c} y_1 & y_2 & y_3 & y_4 & s_1 & s_2 & s_3 & s_4 & s_5 & z & \\ \boxed{1} & 0 & 1 & 0 & -1 & 0 & 0 & 0 & 0 & 0 & 3000 \\ 0 & 1 & 0 & 1 & 0 & -1 & 0 & 0 & 0 & 0 & 5000 \\ 1 & 1 & 0 & 0 & 0 & 0 & 1 & 0 & 0 & 0 & 5000 \\ 0 & 0 & 1 & 1 & 0 & 0 & 0 & 1 & 0 & 0 & 5000 \\ 2 & 6 & 5 & 4 & 0 & 0 & 0 & 0 & 1 & 0 & 40{,}000 \\ \hline 30 & 20 & 25 & 22 & 0 & 0 & 0 & 0 & 0 & 1 & 0 \end{array}$$

Pivot on the 1 in row 1, column 1 since the feasible solution has a negative value, $s_1 = -3000$.

$$\begin{array}{r} \\ \\ \\ -R_1 + R_3 \to R_3 \\ \\ -2R_1 + R_5 \to R_5 \\ -30R_1 + R_6 \to R_6 \end{array}
\begin{array}{cccccccccc|c} y_1 & y_2 & y_3 & y_4 & s_1 & s_2 & s_3 & s_4 & s_5 & z & \\ 1 & 0 & 1 & 0 & -1 & 0 & 0 & 0 & 0 & 0 & 3000 \\ 0 & 1 & 0 & 1 & 0 & -1 & 0 & 0 & 0 & 0 & 5000 \\ 0 & \boxed{1} & -1 & 0 & 1 & 0 & 1 & 0 & 0 & 0 & 2000 \\ 0 & 0 & 1 & 1 & 0 & 0 & 0 & 1 & 0 & 0 & 5000 \\ 0 & 6 & 3 & 4 & 2 & 0 & 0 & 0 & 1 & 0 & 34{,}000 \\ \hline 0 & 20 & -5 & 22 & 30 & 0 & 0 & 0 & 0 & 1 & -90{,}000 \end{array}$$

Since the feasible solution has a negative value ($s_2 = -5000$), pivot on the 1 in row 3, column 2.

$$\begin{array}{r} \\ \\ -R_3 + R_2 \to R_2 \\ \\ \\ -6R_3 + R_5 \to R_5 \\ -20R_3 + R_6 \to R_6 \end{array}
\begin{array}{cccccccccc|c} y_1 & y_2 & y_3 & y_4 & s_1 & s_2 & s_3 & s_4 & s_5 & z & \\ 1 & 0 & 1 & 0 & -1 & 0 & 0 & 0 & 0 & 0 & 3000 \\ 0 & 0 & 1 & 1 & -1 & -1 & -1 & 0 & 0 & 0 & 3000 \\ 0 & 1 & -1 & 0 & 1 & 0 & 1 & 0 & 0 & 0 & 2000 \\ 0 & 0 & 1 & 1 & 0 & 0 & 0 & 1 & 0 & 0 & 5000 \\ 0 & 0 & \boxed{9} & 4 & -4 & 0 & -6 & 0 & 1 & 0 & 22{,}000 \\ \hline 0 & 0 & 15 & 22 & 10 & 0 & -20 & 0 & 0 & 1 & -130{,}000 \end{array}$$

Since the feasible solution has a negative value ($s_2 = -3000$), pivot on the 9 in row 5, column 3.

	y_1	y_2	y_3	y_4	s_1	s_2	s_3	s_4	s_5	z	
$-R_5 + 9R_1 \rightarrow R_1$	9	0	0	−4	−5	0	6	0	−1	0	5000
$-R_5 + 9R_2 \rightarrow R_2$	0	0	0	[5]	−5	−9	−3	0	−1	0	5000
$R_5 + 9R_3 \rightarrow R_3$	0	9	0	4	5	0	3	0	1	0	40,000
$-R_5 + 9R_4 \rightarrow R_4$	0	0	0	5	4	0	6	9	−1	0	23,000
	0	0	9	4	−4	0	−6	0	1	0	22,000
$-5R_5 + 3R_6 \rightarrow R_6$	0	0	0	46	50	0	−30	0	−5	3	−500,000

Pivot on the 5 in row 2, column 4.

	y_1	y_2	y_3	y_4	s_1	s_2	s_3	s_4	s_5	z	
$4R_2 + 5R_1 \rightarrow R_1$	45	0	0	0	−45	−36	18	0	−9	0	45,000
	0	0	0	5	−5	−9	−3	0	−1	0	5000
$-4R_2 + 5R_3 \rightarrow R_3$	0	45	0	0	45	36	27	0	9	0	180,000
$-R_2 + R_4 \rightarrow R_4$	0	0	0	0	9	9	[9]	9	0	0	18,000
$-4R_2 + 5R_5 \rightarrow R_5$	0	0	45	0	0	36	−18	0	9	0	90,000
$-46R_2 + 5R_6 \rightarrow R_6$	0	0	0	0	480	414	−12	0	21	15	−2,730,000

Pivot on the 9 in row 4, column 7.

	y_1	y_2	y_3	y_4	s_1	s_2	s_3	s_4	s_5	z	
$-2R_4 + R_1 \rightarrow R_1$	45	0	0	0	−63	−54	0	−18	−9	0	9000
$R_4 + 3R_2 \rightarrow R_2$	0	0	0	15	−6	−18	0	9	−3	0	33,000
$-3R_4 + R_3 \rightarrow R_3$	0	45	0	0	18	9	0	−27	9	0	126,000
	0	0	0	0	9	9	9	9	0	0	18,000
$2R_4 + R_5 \rightarrow R_5$	0	0	45	0	18	54	0	18	9	0	126,000
$4R_4 + 3R_6 \rightarrow R_6$	0	0	0	0	1476	450	0	36	63	45	−8,118,000

Create a 1 in the columns corresponding to y_1, y_2, y_3, y_4, and z.

	y_1	y_2	y_3	y_4	s_1	s_2	s_3	s_4	s_5	z	
$\frac{1}{45}R_1 \rightarrow R_1$	1	0	0	0	$-\frac{7}{5}$	$-\frac{6}{5}$	0	$-\frac{2}{5}$	$-\frac{1}{5}$	0	200
$\frac{1}{15}R_2 \rightarrow R_2$	0	0	0	1	$-\frac{2}{5}$	$-\frac{6}{5}$	0	$\frac{3}{5}$	$-\frac{1}{5}$	0	2200
$\frac{1}{45}R_3 \rightarrow R_3$	0	1	0	0	$\frac{2}{5}$	$\frac{1}{5}$	0	$-\frac{3}{5}$	$\frac{1}{5}$	0	2800
	0	0	0	0	9	9	9	9	0	0	18,000
$\frac{1}{45}R_5 \rightarrow R_5$	0	0	1	0	$\frac{2}{5}$	$\frac{6}{5}$	0	$\frac{2}{5}$	$\frac{1}{5}$	0	2800
$\frac{1}{45}R_6 \rightarrow R_6$	0	0	0	0	$\frac{164}{5}$	10	0	$\frac{4}{5}$	$\frac{7}{5}$	1	−180,400

Here, $y_1 = 200$, $y_2 = 2800$, $y_3 = 2800$, $y_4 = 2200$, and $-z = w = 180,400$. So, ship 200 barrels of oil from supplier S_1 to distributor D_1. Ship 2800 barrels of oil from supplier S_1 to distributor D_2. Ship 2800 barrels of oil from supplier S_2 to distributor D_1. Ship 2200 barrels of oil from supplier S_2 to distributor D_2. The minimum cost is \$180,400.

(b) From the final tableau, $9s_3 = 18,000$, so $s_3 = 2000$. Therefore, S_1 could furnish 2000 more barrels of oil.

25. Let $x_1 =$ the number of million dollars for home loans

and $x_2 =$ the number of million dollars for commercial loans.

Maximize $z = 0.12x_1 + 0.10x_2$

subject to:

$$\begin{aligned} x_1 \geq 4x_2 \text{ or } x_1 - 4x_2 &\geq 0 \\ x_1 + x_2 &\geq 10 \\ 3x_1 + 2x_2 &\leq 72 \\ x_1 + x_2 &\leq 25 \end{aligned}$$

with $x_1 \geq 0, x_2 \geq 0.$

$$\begin{array}{c} \begin{matrix} x_1 & x_2 & s_1 & s_2 & s_3 & s_4 & z \end{matrix} \\ \left[\begin{array}{ccccccc|c} 1 & -4 & -1 & 0 & 0 & 0 & 0 & 0 \\ 1 & 1 & 0 & -1 & 0 & 0 & 0 & 10 \\ 3 & 2 & 0 & 0 & 1 & 0 & 0 & 72 \\ 1 & 1 & 0 & 0 & 0 & 1 & 0 & 25 \\ \hline -0.12 & -0.10 & 0 & 0 & 0 & 0 & 1 & 0 \end{array}\right] \end{array}$$

Eliminate the decimals in the last row by multiplying by 100

$$\begin{array}{c} \begin{matrix} x_1 & x_2 & s_1 & s_2 & s_3 & s_4 & z \end{matrix} \\ \left[\begin{array}{ccccccc|c} 1 & -4 & -1 & 0 & 0 & 0 & 0 & 0 \\ \boxed{1} & 1 & 0 & -1 & 0 & 0 & 0 & 10 \\ 3 & 2 & 0 & 0 & 1 & 0 & 0 & 72 \\ 1 & 1 & 0 & 0 & 0 & 1 & 0 & 25 \\ \hline -12 & -10 & 0 & 0 & 0 & 0 & 100 & 0 \end{array}\right] \end{array}$$

Pivot on the 1 in row 2, column 1.

$$\begin{array}{r} \\ -R_2 + R_1 \to R_1 \\ \\ -3R_2 + R_3 \to R_3 \\ -R_2 + R_4 \to R_4 \\ 12R_2 + R_5 \to R_5 \end{array} \begin{array}{c} \begin{matrix} x_1 & x_2 & s_1 & s_2 & s_3 & s_4 & z \end{matrix} \\ \left[\begin{array}{ccccccc|c} 0 & -5 & -1 & 1 & 0 & 0 & 0 & -10 \\ 1 & 1 & 0 & -1 & 0 & 0 & 0 & 10 \\ 0 & -1 & 0 & \boxed{3} & 1 & 0 & 0 & 42 \\ 0 & 0 & 0 & 1 & 0 & 1 & 0 & 15 \\ \hline 0 & 2 & 0 & -12 & 0 & 0 & 100 & 120 \end{array}\right] \end{array}$$

Pivot on the 3 in row 3, column 4.

$$\begin{array}{r} \\ -R_3 + 3R_1 \to R_1 \\ R_3 + 3R_2 \to R_2 \\ \\ -R_3 + 3R_4 \to R_4 \\ 4R_3 + R_5 \to R_5 \end{array} \begin{array}{c} \begin{matrix} x_1 & x_2 & s_1 & s_2 & s_3 & s_4 & z \end{matrix} \\ \left[\begin{array}{ccccccc|c} 0 & -14 & -3 & 0 & -1 & 0 & 0 & -72 \\ 3 & 2 & 0 & 0 & -2 & 0 & 0 & 72 \\ 0 & -1 & 0 & 3 & 1 & 0 & 0 & 42 \\ 0 & \boxed{1} & 0 & 0 & -1 & 3 & 0 & 3 \\ \hline 0 & -2 & 0 & 0 & 4 & 0 & 100 & 288 \end{array}\right] \end{array}$$

Pivot on the 1 in row 4, column 2.

$$\begin{array}{r} \\ 14R_4 + R_1 \to R_1 \\ -2R_4 + R_2 \to R_2 \\ R_4 + R_3 \to R_3 \\ \\ 2R_4 + R_5 \to R_5 \end{array} \begin{array}{c} \begin{matrix} x_1 & x_2 & s_1 & s_2 & s_3 & s_4 & z \end{matrix} \\ \left[\begin{array}{ccccccc|c} 0 & 0 & -3 & 0 & -15 & 42 & 0 & -30 \\ 3 & 0 & 0 & 0 & 0 & -6 & 0 & 66 \\ 0 & 0 & 0 & 3 & 0 & 3 & 0 & 45 \\ 0 & 1 & 0 & 0 & -1 & 3 & 0 & 3 \\ \hline 0 & 0 & 0 & 0 & 2 & 6 & 100 & 294 \end{array}\right] \end{array}$$

Create a 1 in the columns corresponding to x_1 and z.

$$\begin{array}{r} \\ \\ \frac{1}{3}R_2 \to R_2 \\ \\ \\ \frac{1}{100}R_5 \to R_5 \end{array} \begin{array}{c} \begin{matrix} x_1 & x_2 & s_1 & s_2 & s_3 & s_4 & z \end{matrix} \\ \left[\begin{array}{ccccccc|c} 0 & 0 & -3 & 0 & -15 & 42 & 0 & -30 \\ 1 & 0 & 0 & 0 & 0 & -2 & 0 & 22 \\ 0 & 0 & 0 & 3 & 0 & 3 & 0 & 45 \\ 0 & 1 & 0 & 0 & -1 & 3 & 0 & 3 \\ \hline 0 & 0 & 0 & 0 & 0.02 & 0.06 & 1 & 2.94 \end{array}\right] \end{array}$$

Here, $x_1 = 22$, $x_2 = 3$, and $z = 2.94$. Make \$22 million (\$22,000,000) in home loans and \$3 million (\$3,000,000) in commercial loans for a maximum return of \$2.94 million, or \$2,940,000.

27. Let $x_1 =$ the number of pounds of bluegrass seed,

$x_2 =$ the number of pounds of rye seed,

and $x_3 =$ the number of pounds of Bermuda seed.

If each batch must contain at least 25% bluegrass seed, then

$$y_1 \geq 0.25(y_1 + y_2 + y_3)$$

$$0.75y_1 - 0.25y_2 - 0.25y_3 \geq 0.$$

And if the amount of Bermuda must be no more than $\frac{2}{3}$ the amount of rye, then

$$y_3 \leq \tfrac{2}{3}y_2$$

$$-2y_2 + 3y_3 = 0.$$

Using these forms for our constraints, we can now state the problem as follows.

Minimize $w = 16y_1 + 14y_2 + 12y_3$

subject to:

$$\begin{aligned} 0.75y_1 - 0.25y_2 - 0.25y_3 &\geq 0 \\ -\quad 2y_2 + \quad 3y_3 &\leq 0 \\ y_1 + \quad y_2 + \quad y_3 &\geq 6000 \end{aligned}$$

with $y_1 \geq 0, y_2 \geq 0, y_3 \geq 0.$

The initial simplex tableau is

y_1	y_2	y_3	s_1	s_2	a	z	
0.75	−0.25	−0.25	−1	0	0	0	0
0	−2	3	0	1	0	0	0
$\boxed{1}$	1	1	0	0	1	0	6000
16	14	12	0	0	0	1	0

First eliminate the artificial variable a. Pivot on the 1 in row 3, column 1.

	y_1	y_2	y_3	s_1	s_2	a	z	
$0.75R_3 - R_1 \to R_1$	0	1	1	1	0	0.75	0	4500
	0	−2	3	0	1	0	0	0
	1	1	1	0	0	1	0	6000
$-16R_3 + R_4 \to R_4$	0	−2	−4	0	0	−16	1	−96,000

Since $a = 0$, we can drop the a column.

y_1	y_2	y_3	s_1	s_2	z	
0	1	1	1	0	0	4500
0	−2	$\boxed{3}$	0	1	0	0
1	1	1	0	0	0	6000
0	−2	−4	0	0	1	−96,000

Pivot on the 3 in row 2, column 3.

	y_1	y_2	y_3	s_1	s_2	z	
$-R_2 + 3R_1 \to R_1$	0	$\boxed{5}$	0	3	−1	0	13,500
	0	−2	3	0	1	0	0
$-R_2 + 3R_3 \to R_3$	3	5	0	0	−1	0	18,000
$4R_2 + 3R_4 \to R_4$	0	−14	0	0	4	3	−288,000

Pivot on the 5 in row 1, column 2.

	y_1	y_2	y_3	s_1	s_2	z	
	0	5	0	3	−1	0	13,500
$2R_1 + 5R_2 \to R_2$	0	0	15	6	3	0	27,000
$-R_1 + R_3 \to R_3$	3	0	0	−3	0	0	4500
$14R_1 + 5R_4 \to R_4$	0	0	0	42	6	15	−1,251,000

Create a 1 in the columns corresponding to y_1, y_2, y_3, and z.

	y_1	y_2	y_3	s_1	s_2	z	
$\frac{1}{5}R_1 \to R_1$	0	1	0	0.6	−0.2	0	2700
$\frac{1}{15}R_2 \to R_2$	0	0	1	0.4	0.2	0	1800
$\frac{1}{3}R_3 \to R_3$	1	0	0	−1	0	0	1500
$\frac{1}{15}R_4 \to R_4$	0	0	0	2.8	0.4	1	−83,400

Here, $y_1 = 1500$, $y_2 = 2700$, $y_3 = 1800$, and $z = -w = 83{,}400$. Therefore, use 1500 lb of bluegrass, 2700 lb of rye, and 1800 lb of Bermuda for a minimum cost of \$834.

29. (a) Let $x_1 =$ the number of computers shipped from W_1 to D_1,
$x_2 =$ the number of computers shipped from W_1 to D_2,
$x_3 =$ the number of computers shipped from W_2 to D_1,
and $x_4 =$ the number of computers shipped from W_2 to D_2.

Minimize $w = 14x_1 + 12x_2 + 12x_3 + 10x_4$

subject to:
$$x_1 + x_3 \geq 32$$
$$x_2 + x_4 \geq 20$$
$$x_1 + x_2 \leq 25$$
$$x_3 + x_4 \leq 30$$

with $x_1 \geq 0,\ x_2 \geq 0,\ x_3 \geq 0,\ x_4 \geq 0.$

Maximize
$$z = -w = -14x_1 - 12x_2 - 12x_3 - 10x_4.$$

The initial tableau looks like the following.

x_1	x_2	x_3	x_4	s_1	s_2	s_3	s_4	z	
1	0	1	0	−1	0	0	0	0	32
0	1	0	1	0	−1	0	0	0	20
$\boxed{1}$	1	0	0	0	0	1	0	0	25
0	0	1	1	0	0	0	1	0	30
14	12	12	10	0	0	0	0	1	0

The variable s_1 is negative; we pivot on the 1 in row 3 of column 1.

	x_1	x_2	x_3	x_4	s_1	s_2	s_3	s_4	z	
$-R_3 + R_1 \to R_1$	0	−1	$\boxed{1}$	0	−1	0	−1	0	0	7
	0	1	0	1	0	−1	0	0	0	20
	1	1	0	0	0	0	1	0	0	25
	0	0	1	1	0	0	0	1	0	30
$-14R_3 + R_5 \to R_5$	0	−2	12	10	0	0	−14	0	1	−350

The variable s_1 is still negative; we pivot on the 1 in row 1 of column 3.

	x_1	x_2	x_3	x_4	s_1	s_2	s_3	s_4	z	
	0	−1	1	0	−1	0	−1	0	0	7
	0	$\boxed{1}$	0	1	0	−1	0	0	0	20
	1	1	0	0	0	0	1	0	0	25
$-R_1 + R_4 \to R_4$	0	1	0	1	1	0	1	1	0	23
$-12R_1 + R_5 \to R_5$	0	10	0	10	12	0	−2	0	1	−434

The variable s_2 is still negative; we pivot on the 1 in row 2 of column 2.

$$\begin{array}{r} \\ R_2 + R_1 \to R_1 \\ \\ -R_2 + R_3 \to R_3 \\ -R_2 + R_4 \to R_4 \\ -10R_2 + R_5 \to R_5 \end{array} \begin{array}{c} \begin{array}{ccccccccc|c} x_1 & x_2 & x_3 & x_4 & s_1 & s_2 & s_3 & s_4 & z & \end{array} \\ \left[\begin{array}{ccccccccc|c} 0 & 0 & 1 & 1 & -1 & -1 & -1 & 0 & 0 & 27 \\ 0 & 1 & 0 & 1 & 0 & -1 & 0 & 0 & 0 & 20 \\ 1 & 0 & 0 & -1 & 0 & 1 & 1 & 0 & 0 & 5 \\ 0 & 0 & 0 & 0 & 1 & 1 & \boxed{1} & 1 & 0 & 3 \\ \hline 0 & 0 & 0 & 0 & 12 & 10 & -2 & 0 & 1 & -634 \end{array}\right] \end{array}$$

Now we eliminate the only negative indicator by pivoting on the 1 in row 4 of column 7.

$$\begin{array}{r} \\ R_4 + R_1 \to R_1 \\ \\ -R_4 + R_3 \to R_3 \\ \\ 2R_4 + R_5 \to R_5 \end{array} \begin{array}{c} \begin{array}{ccccccccc|c} x_1 & x_2 & x_3 & x_4 & s_1 & s_2 & s_3 & s_4 & z & \end{array} \\ \left[\begin{array}{ccccccccc|c} 0 & 0 & 1 & 1 & 0 & 0 & 0 & 1 & 0 & 30 \\ 0 & 1 & 0 & 1 & 0 & 0 & 0 & 0 & 0 & 20 \\ 1 & 0 & 0 & -1 & -1 & 0 & 0 & -1 & 0 & 2 \\ 0 & 0 & 0 & 0 & 1 & 1 & 1 & 1 & 0 & 3 \\ \hline 0 & 0 & 0 & 0 & 14 & 12 & 0 & 2 & 1 & -628 \end{array}\right] \end{array}$$

From this we can read the solution: Ship 2 computers from W_1 to D_1, ship 20 computers from W_1 to D_2, ship 30 computers from W_2 to D_1, and 0 computers from W_2 to D_2. The resulting minimum cost is $628.

(b) From the final tableau, $s_3 = 3$. Therefore, warehouse W_1 has three more computers that it could ship.

31. Let x_1 = the amount of chemical I,

x_2 = the amount of chemical II,

and x_3 = the amount of chemical III.

Minimize $w = 1.09x_1 + 0.87x_2 + 0.65x_3$

subject to:

$$\begin{aligned} x_1 + x_2 + x_3 &\geq 750 \\ 0.09x_1 + 0.04x_2 + 0.03x_3 &\geq 30 \\ 3x_2 &= 4x_3 \end{aligned}$$

with $x_1 \geq 0,\ x_2 \geq 0,\ x_3 \geq 0.$

We follow the suggestion in the note in the text to reduce the number of variables by using the fact that $x_3 = 0.75x_2$ to express our constraints as follows:

Minimize $w = 1.09x_1 + 1.3575x_2$

subject to

$$\begin{aligned} x_1 + 1.75x_2 &\geq 750 \\ 0.09x_1 + 0.0625x_2 &\geq 30 \end{aligned}$$

We maximize $z = -w$ and after multiplying the second constraint through by 100, our initial tableau is the following:

$$\begin{array}{c} \begin{array}{ccccc|c} x_1 & x_2 & s_1 & s_2 & z & \end{array} \\ \left[\begin{array}{ccccc|c} 1 & 1.75 & -1 & 0 & 0 & 750 \\ \boxed{9} & 6.25 & 0 & -100 & 0 & 3000 \\ \hline 1.09 & 1.3575 & 0 & 0 & 1 & 0 \end{array}\right] \end{array}$$

Since s_1 is negative, we look for a pivot in the first column, and choose 9 because it has the smallest ratio with the corresponding entry in the last column.

$$\begin{array}{r} \\ -R_2 + 9R_1 \to R_1 \\ \\ -1.09R_2 + 9R_3 \to R_3 \end{array} \begin{array}{c} \begin{array}{ccccc|c} x_1 & x_2 & s_1 & s_2 & z & \end{array} \\ \left[\begin{array}{ccccc|c} 0 & \boxed{9.5} & -9 & 100 & 0 & 3750 \\ 9 & 6.25 & 0 & -100 & 0 & 3000 \\ \hline 0 & 5.405 & 0 & 109 & 9 & -3270 \end{array}\right] \end{array}$$

s_1 is still negative so we pivot on the 9.5 in column 2.

$$\begin{array}{r} \\ \\ -6.25R_1 + 9.5R_2 \to R_2 \\ -5.405R_1 + 9.5R_3 \to R_3 \end{array} \begin{array}{c} \begin{array}{ccccc|c} x_1 & x_2 & s_1 & s_2 & z & \end{array} \\ \left[\begin{array}{ccccc|c} 0 & 9.5 & -9 & 100 & 0 & 3750 \\ 85.5 & 0 & 56.25 & -1575 & 0 & 5062.5 \\ \hline 0 & 0 & 48.645 & 495 & 85.5 & -51{,}333.75 \end{array}\right] \end{array}$$

This tableau yields the following solution.

$$x_1 = \frac{5062.5}{85.5} = 59.21,\ x_2 = \frac{3750}{9.5} = 394.74,$$

$$x_3 = \frac{3750}{9.5} \cdot \frac{3}{4} = 296.05$$

$$\text{Minimum} = -\left(\frac{-51{,}333.75}{85.5}\right) = 600.39$$

So use 59.21 kg of chemical I, 394.74 kg of chemical II, and 296.05 kg of chemical III, for a minimum cost of $600.39.

33. Let y_1 = the number of ounces of ingredient I,

y_2 = the number of ounces of ingredient II,

and y_3 = the number of ounces of ingredient III.

Expressing the problem in cents, the problem is:

Minimize $w = 30y_1 + 9y_2 + 27y_3$

subject to

$$\begin{aligned} y_1 + y_2 + y_3 &\geq 10 \\ y_1 + y_2 + y_3 &\leq 15 \\ y_1 &\geq \frac{1}{4}y_2 \\ y_3 &\geq y_1 \end{aligned}$$

with $y_1 \geq 0,\ y_2 \geq 0,\ y_3 \geq 0.$

Rewrite the last two inequalities so that the problem becomes:

Minimize $w = 30y_1 + 9y_2 + 27y_3$

subject to:

$$\begin{aligned} y_1 + y_2 + y_3 &\geq 10 \\ y_1 + y_2 + y_3 &\leq 15 \\ -4y_1 + y_2 &\leq 0 \\ y_1 - y_3 &\leq 0 \end{aligned}$$

with $y_1 \geq 0, y_2 \geq 0, y_3 \geq 0.$

We maximize $z = -w$ and have the following initial tableau.

$$\begin{array}{c} \begin{array}{cccccccc} y_1 & y_2 & y_3 & s_1 & s_2 & s_3 & s_4 & z \end{array} \\ \left[\begin{array}{cccccccc|c} 1 & 1 & 1 & -1 & 0 & 0 & 0 & 0 & 10 \\ 1 & 1 & 1 & 0 & 1 & 0 & 0 & 0 & 15 \\ -4 & 1 & 0 & 0 & 0 & 1 & 0 & 0 & 0 \\ \boxed{1} & 0 & -1 & 0 & 0 & 0 & 1 & 0 & 0 \\ \hline 30 & 9 & 27 & 0 & 0 & 0 & 0 & 1 & 0 \end{array}\right] \end{array}$$

Because the solution is not feasible $(s_1 = -10)$, pivot on the 1 in row 4, column 1.

$$\begin{array}{r} \\ -R_4 + R_1 \to R_1 \\ -R_4 + R_2 \to R_2 \\ 4R_4 + R_3 \to R_3 \\ \\ -30R_4 + R_5 \to R_5 \end{array} \begin{array}{c} \begin{array}{cccccccc} y_1 & y_2 & y_3 & s_1 & s_2 & s_3 & s_4 & z \end{array} \\ \left[\begin{array}{cccccccc|c} 0 & 1 & 2 & -1 & 0 & 0 & -1 & 0 & 10 \\ 0 & 1 & 2 & 0 & 1 & 0 & -1 & 0 & 15 \\ 0 & \boxed{1} & -4 & 0 & 0 & 1 & 4 & 0 & 0 \\ 1 & 0 & -1 & 0 & 0 & 0 & 1 & 0 & 0 \\ \hline 0 & 9 & 57 & 0 & 0 & 0 & -30 & 1 & 0 \end{array}\right] \end{array}$$

Because the solution is still not feasible $(s_1 = -10)$, pivot on the 1 in row 3, column 2.

$$\begin{array}{r} \\ -R_3 + R_1 \to R_1 \\ -R_3 + R_2 \to R_2 \\ \\ \\ -9R_3 + R_5 \to R_5 \end{array} \begin{array}{c} \begin{array}{cccccccc} y_1 & y_2 & y_3 & s_1 & s_2 & s_3 & s_4 & z \end{array} \\ \left[\begin{array}{cccccccc|c} 0 & 1 & \boxed{6} & -1 & 0 & -1 & -5 & 0 & 10 \\ 0 & 1 & 6 & 0 & 1 & -1 & -5 & 0 & 15 \\ 0 & 1 & -4 & 0 & 0 & 1 & 4 & 0 & 0 \\ 1 & 0 & -1 & 0 & 0 & 0 & 1 & 0 & 0 \\ \hline 0 & 0 & 93 & 0 & 0 & -9 & -66 & 1 & 0 \end{array}\right] \end{array}$$

Because the solution is still not feasible $(s_1 = -10)$, pivot on the 6 in row 1, column 3.

$$\begin{array}{r} \\ \\ -R_1 + R_2 \to R_2 \\ 2R_1 + 3R_3 \to R_3 \\ R_1 + 6R_4 \to R_4 \\ -31R_3 + 2R_5 \to R_5 \end{array} \begin{array}{c} \begin{array}{cccccccc} y_1 & y_2 & y_3 & s_1 & s_2 & s_3 & s_4 & z \end{array} \\ \left[\begin{array}{cccccccc|c} 0 & 0 & 6 & -1 & 0 & -1 & -5 & 0 & 10 \\ 0 & 0 & 0 & 1 & 1 & 0 & 0 & 0 & 5 \\ 0 & 3 & 0 & -2 & 0 & 1 & 2 & 0 & 20 \\ 6 & 0 & 0 & -1 & 0 & -1 & 1 & 0 & 10 \\ \hline 0 & 0 & 0 & 31 & 0 & 13 & 23 & 2 & -310 \end{array}\right] \end{array}$$

Create a 1 in the columns corresponding to y_1, y_2, y_2, and z.

$$\begin{array}{r} \\ \\ \frac{1}{6}R_2 \to R_2 \\ \frac{1}{3}R_3 \to R_3 \\ \frac{1}{6}R_4 \to R_4 \\ \frac{1}{2}R_5 \to R_5 \end{array} \begin{array}{c} \begin{array}{cccccccc} y_1 & y_2 & y_3 & s_1 & s_2 & s_3 & s_4 & z \end{array} \\ \left[\begin{array}{cccccccc|c} 0 & 0 & 1 & -\frac{1}{6} & 0 & -\frac{1}{6} & -\frac{5}{6} & 0 & \frac{5}{3} \\ 0 & 0 & 0 & 1 & 1 & 0 & 0 & 0 & 5 \\ 0 & 1 & 0 & -\frac{2}{3} & 0 & \frac{1}{3} & \frac{2}{3} & 0 & \frac{20}{3} \\ 1 & 0 & 0 & -\frac{1}{6} & 0 & -\frac{1}{6} & \frac{1}{6} & 0 & \frac{5}{3} \\ \hline 0 & 0 & 0 & \frac{31}{2} & 0 & \frac{13}{2} & \frac{23}{2} & 1 & -155 \end{array}\right] \end{array}$$

Here $y_1 = \frac{5}{3}, y_2 = \frac{20}{3}, y_3 = \frac{5}{3},$ and $w = -z = 155.$

Therefore, the additive should consist of $\frac{5}{3}$ oz of ingredient I, $\frac{20}{3}$ oz of ingredient II, and $\frac{5}{3}$ oz of ingredient III, for a minimum cost of 155¢/gal, or \$1.55/gal. The amount of additive that should be used per gallon of gasoline is $\frac{5}{3} + \frac{20}{3} + \frac{5}{3} = 10$ oz.

Chapter 4 Review Exercises

1. True
2. False
3. True
4. False
5. False
6. True
7. True
8. False
9. False
10. True
11. False
12. True
13. False
14. True
15. The simplex method should be used for problems with more than two variables or problems with two variables and many constants.

17. (a) Maximize $z = 2x_1 + 7x_2$

subject to:
$$\begin{aligned} 4x_1 + 6x_2 &\le 60 \\ 3x_1 + x_2 &\le 18 \\ 2x_1 + 5x_2 &\le 20 \\ x_1 + x_2 &\le 15 \end{aligned}$$

with $x_1 \ge 0, x_2 \ge 0.$

Adding slack variables s_1, s_2, s_3, and s_4, we obtain the following equations.

$$\begin{aligned} 4x_1 + 6x_2 + s_1 \qquad\qquad &= 60 \\ 3x_1 + x_2 \quad + s_2 \qquad &= 18 \\ 2x_1 + 5x_2 \qquad + s_3 \quad &= 20 \\ x_1 + x_2 \qquad\qquad + s_4 &= 15. \end{aligned}$$

(b) The initial simplex tableau is as follows.

$$\begin{array}{c} \begin{array}{ccccccc} x_1 & x_2 & s_1 & s_2 & s_3 & s_4 & z \end{array} \\ \left[\begin{array}{rrrrrrr|r} 4 & 6 & 1 & 0 & 0 & 0 & 0 & 60 \\ 3 & 1 & 0 & 1 & 0 & 0 & 0 & 18 \\ 2 & 5 & 0 & 0 & 1 & 0 & 0 & 20 \\ 1 & 1 & 0 & 0 & 0 & 1 & 0 & 15 \\ \hline -2 & -7 & 0 & 0 & 0 & 0 & 1 & 0 \end{array}\right]. \end{array}$$

19. Maximize $z = 5x_1 + 8x_2 + 6x_3$

subject to:
$$\begin{aligned} x_1 + x_2 + x_3 &\le 90 \\ 2x_1 + 5x_2 + x_3 &\le 120 \\ x_1 + 3x_2 \qquad &\ge 80 \end{aligned}$$

with $x_1 \ge 0, x_2 \ge 0, x_3 \ge 0.$

(a) Adding the slack variables s_1 and s_2 and subtracting the surplus variable s_3, we obtain the following equations:

$$\begin{aligned} x_1 + x_2 + x_3 + s_1 \qquad\qquad &= 90 \\ 2x_1 + 5x_2 + x_3 \qquad + s_2 \qquad &= 120 \\ x_1 + 3x_2 \qquad\qquad\qquad - s_3 &= 80. \end{aligned}$$

(b) The initial tableau is

$$\begin{array}{c} \begin{array}{ccccccc} x_1 & x_2 & x_3 & s_1 & s_2 & s_3 & z \end{array} \\ \left[\begin{array}{rrrrrrr|r} 1 & 1 & 1 & 1 & 0 & 0 & 0 & 90 \\ 2 & 5 & 1 & 0 & 1 & 0 & 0 & 120 \\ 1 & 3 & 0 & 0 & 0 & -1 & 0 & 80 \\ \hline -5 & -8 & -6 & 0 & 0 & 0 & 1 & 0 \end{array}\right]. \end{array}$$

21.

$$\begin{array}{c} \begin{array}{cccccc} x_1 & x_2 & x_3 & s_1 & s_2 & z \end{array} \\ \left[\begin{array}{rrrrrr|r} 4 & 5 & 2 & 1 & 0 & 0 & 18 \\ 2 & 8 & \boxed{6} & 0 & 1 & 0 & 24 \\ \hline -5 & -3 & -6 & 0 & 0 & 1 & 0 \end{array}\right]. \end{array}$$

The most negative entry in the last row is -6, and the smaller of the two quotients is $\frac{24}{6} = 4$. Hence, the 6 in row 2, column 3, is the first pivot. Performing row transformations leads to the following tableau.

$$\begin{array}{r} \\ -R_2 + 3R_1 \to R_1 \\ \\ R_2 + R_3 \to R_3 \end{array} \begin{array}{c} \begin{array}{cccccc} x_1 & x_2 & x_3 & s_1 & s_2 & z \end{array} \\ \left[\begin{array}{rrrrrr|r} \boxed{10} & 7 & 0 & 3 & -1 & 0 & 30 \\ 2 & 8 & 6 & 0 & 1 & 0 & 24 \\ \hline -3 & 5 & 0 & 0 & 1 & 1 & 24 \end{array}\right]. \end{array}$$

Pivot on the 10 in row 1, column 1.

$$\begin{array}{r} \\ \\ -R_1 + 5R_2 \to R_2 \\ 3R_1 + 10R_3 \to R_3 \end{array} \begin{array}{c} \begin{array}{cccccc} x_1 & x_2 & x_3 & s_1 & s_2 & z \end{array} \\ \left[\begin{array}{rrrrrr|r} 10 & 7 & 0 & 3 & -1 & 0 & 30 \\ 0 & 33 & 30 & -3 & 6 & 0 & 90 \\ \hline 0 & 71 & 0 & 9 & 7 & 10 & 330 \end{array}\right]. \end{array}$$

Create a 1 in the columns corresponding to x_1, x_3, and z.

$$\begin{array}{r} \\ \frac{1}{10}R_1 \to R_1 \\ \frac{1}{30}R_2 \to R_2 \\ \frac{1}{10}R_3 \to R_3 \end{array} \begin{array}{c} \begin{array}{cccccc} x_1 & x_2 & x_3 & s_1 & s_2 & z \end{array} \\ \left[\begin{array}{cccccc|c} 1 & \frac{7}{10} & 0 & \frac{3}{10} & -\frac{1}{10} & 0 & 3 \\ 0 & \frac{11}{10} & 1 & -\frac{1}{10} & \frac{1}{5} & 0 & 3 \\ \hline 0 & \frac{71}{10} & 0 & \frac{9}{10} & \frac{7}{10} & 10 & 33 \end{array}\right]. \end{array}$$

The maximum value is 33 when $x_1 = 3, x_2 = 0, x_3 = 3, s_1 = 0,$ and $s_2 = 0$.

23.

$$\begin{array}{c} \begin{array}{ccccccc} x_1 & x_2 & x_3 & s_1 & s_2 & s_3 & z \end{array} \\ \left[\begin{array}{rrrrrrr|r} 1 & 2 & 2 & 1 & 0 & 0 & 0 & 50 \\ \boxed{3} & 1 & 0 & 0 & 1 & 0 & 0 & 20 \\ 1 & 0 & 2 & 0 & 0 & -1 & 0 & 15 \\ \hline -5 & -3 & -2 & 0 & 0 & 0 & 1 & 0 \end{array}\right] \end{array}$$

The initial basic solution is not feasible since $s_3 = -15$. In the third row where the negative coefficient appears, the nonnegative entry that appears farthest to the left is the 1 in the first column. In the first column, the smallest nonnegative quotient is $\frac{20}{3}$. Pivot on the 3 in row 2, column 1.

	x_1	x_2	x_3	s_1	s_2	s_3	z	
$-R_2 + 3R_1 \to R_1$	0	5	6	3	−1	0	0	130
	3	1	0	0	1	0	0	20
$-R_2 + 3R_3 \to R_3$	0	−1	[6]	0	−1	−3	0	25
$5R_2 + 3R_4 \to R_4$	0	−4	−6	0	5	0	3	100

Continue by pivoting on each boxed entry.

	x_1	x_2	x_3	s_1	s_2	s_3	z	
$-R_3 + R_2 \to R_1$	0	[6]	0	3	0	3	0	105
	3	1	0	0	1	0	0	20
	0	−1	6	0	−1	−3	0	25
$R_3 + R_4 \to R_4$	0	−5	0	0	4	−3	3	125

The basic solution is now feasible, but there are negative indicators.

Continue pivoting.

	x_1	x_2	x_3	s_1	s_2	s_3	z	
	0	6	0	3	0	[3]	0	105
$-R_1 + 6R_2 \to R_2$	18	0	0	−3	6	−3	0	15
$R_1 + 6R_3 \to R_3$	0	0	36	3	0	−15	0	255
$5R_1 + 6R_4 \to R_4$	0	0	0	15	24	−3	18	1275

	x_1	x_2	x_3	s_1	s_2	s_3	z	
	0	6	0	3	0	3	0	105
$R_1 + R_2 \to R_2$	18	6	0	0	6	0	0	120
$5R_1 + R_3 \to R_3$	0	30	36	18	0	0	0	780
$R_1 + R_4 \to R_4$	0	6	0	18	24	0	18	1380

Create a 1 in the columns corresponding to x_1, x_3, s_3, and z.

	x_1	x_2	x_3	s_1	s_2	s_3	z	
$\frac{1}{3}R_1 \to R_1$	0	2	0	1	0	1	0	35
$\frac{1}{18}R_2 \to R_2$	1	.33	0	0	.33	0	0	6.67
$\frac{1}{36}R_3 \to R_3$	0	.83	1	.5	0	0	0	21.67
$\frac{1}{18}R_4 \to R_4$	0	.33	0	1	1.33	0	1	76.67

The maximum value is about 76.67 when $x_1 \approx 6.67$, $x_2 = 0$, $x_3 \approx 21.67$, $s_1 = 0$, $s_2 = 0$, and $s_3 = 35$.

25. Minimize $w = 10y_1 + 15y_2$

subject to:
$$y_1 + y_2 \geq 17$$
$$5y_1 + 8y_2 \geq 42$$

with $y_1 \geq 0, y_2 \geq 0.$

Using the dual method:

To form the dual, write the augmented matrix for the given problem.

$$\left[\begin{array}{cc|c} 1 & 1 & 17 \\ 5 & 8 & 42 \\ \hline 10 & 15 & 0 \end{array}\right]$$

Form the transpose of this matrix.

$$\left[\begin{array}{cc|c} 1 & 5 & 10 \\ 1 & 8 & 15 \\ \hline 17 & 42 & 0 \end{array}\right]$$

Write the dual problem.

Maximize $z = 17x_1 + 42x_2$

subject to:
$$x_1 + 5x_2 \leq 10$$
$$x_1 + 8x_2 \leq 15$$

with $x_1 \geq 0, x_2 \geq 0.$

The initial simplex tableau is as follows.

x_1	x_2	s_1	s_2	z	
1	5	1	0	0	10
1	[8]	0	1	0	15
−17	−42	0	0	1	0

Pivot on the 8 in row 2 column 2.

	x_1	x_2	s_1	s_2	z	
$-5R_2 + 8R_1 \to R_1$	[3]	0	8	−5	0	5
	1	8	0	1	0	15
$21R_2 + 4R_3 \to R_3$	−47	0	0	21	4	315

Pivot on the 3 in row 1, column 1.

	x_1	x_2	s_1	s_2	z	
	3	0	8	−5	0	5
$-R_1 + 3R_2 \to R_2$	0	24	−8	[8]	0	40
$47R_1 + 3R_3 \to R_3$	0	0	376	−172	12	1180

Pivot on the 8 in row 2, column 4.

	x_1	x_2	s_1	s_2	z	
$5R_2 + 8R_1 \to R_1$	24	120	24	0	0	240
	0	24	−8	8	0	40
$43R_2 + 2R_3 \to R_3$	0	1032	408	0	24	4080

Create a 1 in the columns corresponding to x_1, x_2, and z.

$$\begin{array}{r} \\ \frac{1}{24}R_1 \to R_1 \\ \frac{1}{8}R_2 \to R_2 \\ \frac{1}{24}R_3 \to R_3 \end{array} \begin{array}{c} \begin{array}{ccccc|c} x_1 & x_2 & s_1 & s_2 & z & \end{array} \\ \left[\begin{array}{ccccc|c} 1 & 5 & 1 & 0 & 0 & 10 \\ 0 & 3 & -1 & 1 & 0 & 5 \\ \hline 0 & 43 & 17 & 0 & 1 & 170 \end{array}\right] \end{array}$$

The minimum value is 170 when $y_1 = 17$ and $y_2 = 0$.

Using the method of 4.4:

Change the objective function to

$$\text{Maximize } z = -w = -10y_1 - 15y_2.$$

The constraints are not changed.

The initial simplex tableau is as follows.

$$\begin{array}{c} \begin{array}{ccccc|c} y_1 & y_2 & s_1 & s_2 & z & \end{array} \\ \left[\begin{array}{ccccc|c} 1 & 1 & -1 & 0 & 0 & 17 \\ \boxed{5} & 8 & 0 & -1 & 0 & 42 \\ \hline 10 & 15 & 0 & 0 & 1 & 0 \end{array}\right] \end{array}$$

The solution is not feasible since $s_1 = -17$ and $s_2 = -42$. Pivot on the 5 in row 2, column 1.

$$\begin{array}{r} \\ -R_2 + 5R_1 \to R_1 \\ \\ -2R_2 + R_3 \to R_3 \end{array} \begin{array}{c} \begin{array}{ccccc|c} y_1 & y_2 & s_1 & s_2 & z & \end{array} \\ \left[\begin{array}{ccccc|c} 0 & -3 & -5 & 1 & 0 & 43 \\ 5 & 8 & 0 & -1 & 0 & 42 \\ \hline 0 & -1 & 0 & 2 & 1 & -84 \end{array}\right] \end{array}$$

The solution is still not feasible since $s_1 = -\frac{43}{5}$. But there are no positive entries to the left of the -5 in column 3 so it is not possible to choose a pivot element. The method of 4.4 fails to provide a solution in this case.

27. Minimize $w = 7y_1 + 2y_2 + 3y_3$

subject to:
$$\begin{aligned} y_1 + y_2 + 2y_3 &\geq 48 \\ y_1 + y_2 &\geq 12 \\ y_3 &\geq 10 \\ 3y_1 + y_3 &\geq 30 \end{aligned}$$

with $y_1 \geq 0,\ y_2 \geq 0,\ y_3 \geq 0$.

Using the dual method:

To form the dual, write the augmented matrix for the given problem.

$$\left[\begin{array}{ccc|c} 1 & 1 & 2 & 48 \\ 1 & 1 & 0 & 12 \\ 0 & 0 & 1 & 10 \\ 3 & 0 & 1 & 30 \\ \hline 7 & 2 & 3 & 0 \end{array}\right]$$

Form the transpose of this matrix.

$$\left[\begin{array}{cccc|c} 1 & 1 & 0 & 3 & 7 \\ 1 & 1 & 0 & 0 & 2 \\ 2 & 0 & 1 & 1 & 3 \\ \hline 48 & 12 & 10 & 30 & 0 \end{array}\right]$$

Write the dual problem.

Maximize $z = 48x_1 + 12x_2 + 10x_3 + 30x_4$

subject to:
$$\begin{aligned} x_1 + x_2 + 3x_4 &\leq 7 \\ x_1 + x_2 &\leq 2 \\ 2x_1 + x_3 + x_4 &\leq 3 \end{aligned}$$

with $x_1 \geq 0,\ x_2 \geq 0,\ x_3 \geq 0,\ x_4 \geq 0$.

The initial simplex tableau is as follows.

$$\begin{array}{c} \begin{array}{cccccccc|c} x_1 & x_2 & x_3 & x_4 & s_1 & s_2 & s_3 & z & \end{array} \\ \left[\begin{array}{cccccccc|c} 1 & 1 & 0 & 3 & 1 & 0 & 0 & 0 & 7 \\ 1 & 1 & 0 & 0 & 0 & 1 & 0 & 0 & 2 \\ \boxed{2} & 0 & 1 & 1 & 0 & 0 & 1 & 0 & 3 \\ \hline -48 & -12 & -10 & -30 & 0 & 0 & 0 & 1 & 0 \end{array}\right] \end{array}$$

Pivot on the 2 in row 3, column 1.

$$\begin{array}{r} \\ -R_3 + 2R_1 \to R_1 \\ -R_3 + 2R_2 \to R_2 \\ \\ 24R_3 + R_4 \to R_4 \end{array} \begin{array}{c} \begin{array}{cccccccc|c} x_1 & x_2 & x_3 & x_4 & s_1 & s_2 & s_3 & z & \end{array} \\ \left[\begin{array}{cccccccc|c} 0 & 2 & -1 & 5 & 2 & 0 & -1 & 0 & 11 \\ 0 & \boxed{2} & -1 & -1 & 0 & 2 & -1 & 0 & 1 \\ 2 & 0 & 1 & 1 & 0 & 0 & 1 & 0 & 3 \\ \hline 0 & -12 & 14 & -6 & 0 & 0 & 24 & 1 & 72 \end{array}\right] \end{array}$$

Pivot on the 2 in row 2, column 2.

$$\begin{array}{r} \\ -R_2 + R_1 \to R_1 \\ \\ \\ 6R_2 + R_4 \to R_4 \end{array} \begin{array}{c} \begin{array}{cccccccc|c} x_1 & x_2 & x_3 & x_4 & s_1 & s_2 & s_3 & z & \end{array} \\ \left[\begin{array}{cccccccc|c} 0 & 0 & 0 & \boxed{6} & 2 & -2 & 0 & 0 & 10 \\ 0 & 2 & -1 & -1 & 0 & 2 & -1 & 0 & 1 \\ 2 & 0 & 1 & 1 & 0 & 0 & 1 & 0 & 3 \\ \hline 0 & 0 & 8 & -12 & 0 & 12 & 18 & 1 & 78 \end{array}\right] \end{array}$$

Pivot on the 6 in row 1, column 4.

$$\begin{array}{r|cccccccc|c} & x_1 & x_2 & x_3 & x_4 & s_1 & s_2 & s_3 & z & \\ \hline & 0 & 0 & 0 & 6 & 2 & -2 & 0 & 0 & 10 \\ R_1+6R_2\to R_2 & 0 & 12 & -6 & 0 & 2 & 10 & -6 & 0 & 16 \\ -R_1+6R_3\to R_3 & 12 & 0 & 6 & 0 & -2 & 2 & 6 & 0 & 8 \\ \hline 2R_1+R_4\to R_4 & 0 & 0 & 8 & 0 & 4 & 8 & 18 & 1 & 98 \end{array}$$

Create a 1 in the columns corresponding to x_1, x_2, and x_4.

$$\begin{array}{r|cccccccc|c} & x_1 & x_2 & x_3 & x_4 & s_1 & s_2 & s_3 & z & \\ \hline \frac{1}{6}R_1\to R_1 & 0 & 0 & 0 & 1 & \frac{1}{3} & -\frac{1}{3} & 0 & 0 & \frac{5}{3} \\ \frac{1}{12}R_2\to R_2 & 0 & 1 & -\frac{1}{2} & 0 & \frac{1}{6} & \frac{5}{6} & -\frac{1}{2} & 0 & \frac{4}{3} \\ \frac{1}{12}R_3\to R_3 & 1 & 0 & \frac{1}{2} & 0 & -\frac{1}{6} & \frac{1}{6} & \frac{1}{2} & 0 & \frac{2}{3} \\ \hline & 0 & 0 & 8 & 0 & 4 & 8 & 18 & 1 & 98 \end{array}$$

The minimum value is 98 when $y_1 = 4$, $y_2 = 8$, and $y_3 = 18$.

Using the method of 4.4:

Change the objective function to

$$\text{Maximize } z = -w = -7y_1 - 2y_2 - 3y_3.$$

The constraints are not changed.

The initial simplex tableau is as follows.

$$\begin{array}{cccccccc|c} y_1 & y_2 & y_3 & s_1 & s_2 & s_3 & s_4 & z & \\ \hline 1 & 1 & 2 & -1 & 0 & 0 & 0 & 0 & 48 \\ 1 & 1 & 0 & 0 & -1 & 0 & 0 & 0 & 12 \\ 0 & 0 & 1 & 0 & 0 & -1 & 0 & 0 & 10 \\ \boxed{3} & 0 & 1 & 0 & 0 & 0 & -1 & 0 & 30 \\ \hline 7 & 2 & 3 & 0 & 0 & 0 & 0 & 1 & 0 \end{array}$$

The solution is not feasible since $s_1 = -48$, $s_2 = -12$, $s_3 = -10$, and $s_4 = -30$. Pivot on the 3 in row 4, column 1.

$$\begin{array}{r|cccccccc|c} & y_1 & y_2 & y_3 & s_1 & s_2 & s_3 & s_4 & z & \\ \hline -R_4+3R_1\to R_1 & 0 & \boxed{3} & 5 & -3 & 0 & 0 & 1 & 0 & 114 \\ -R_4+3R_2\to R_2 & 0 & 3 & -1 & 0 & -3 & 0 & 1 & 0 & 6 \\ & 0 & 0 & 1 & 0 & 0 & -1 & 0 & 0 & 10 \\ & 3 & 0 & 1 & 0 & 0 & 0 & -1 & 0 & 30 \\ \hline -7R_4+3R_5\to R_5 & 0 & 6 & 2 & 0 & 0 & 0 & 7 & 3 & -210 \end{array}$$

The solution is still not feasible since $s_1 = -38$, $s_2 = -2$, and $s_3 = -10$. Pivot on the 3 in row 1, column 2.

$$\begin{array}{r|cccccccc|c} & y_1 & y_2 & y_3 & s_1 & s_2 & s_3 & s_4 & z & \\ \hline & 0 & 3 & 5 & -3 & 0 & 0 & 1 & 0 & 114 \\ R_1-R_2\to R_2 & 0 & 0 & 6 & -3 & 3 & 0 & 0 & 0 & 108 \\ & 0 & 0 & \boxed{1} & 0 & 0 & -1 & 0 & 0 & 10 \\ & 3 & 0 & 1 & 0 & 0 & 0 & -1 & 0 & 30 \\ \hline -2R_1+R_5\to R_5 & 0 & 0 & -8 & 6 & 0 & 0 & 5 & 3 & -438 \end{array}$$

Again, the solution is not feasible since $s_3 = -10$ and $s_4 = -30$. Pivot on the 1 in row 3, column 3.

$$\begin{array}{r|cccccccc|c} & y_1 & y_2 & y_3 & s_1 & s_2 & s_3 & s_4 & z & \\ \hline -5R_3+R_1\to R_1 & 0 & 3 & 0 & -3 & 0 & 5 & 1 & 0 & 64 \\ -6R_3+R_2\to R_2 & 0 & 0 & 0 & -3 & 3 & \boxed{6} & 0 & 0 & 48 \\ & 0 & 0 & 1 & 0 & 0 & -1 & 0 & 0 & 10 \\ -R_3+R_4\to R_4 & 3 & 0 & 0 & 0 & 0 & 1 & -1 & 0 & 20 \\ \hline 8R_3+R_5\to R_5 & 0 & 0 & 0 & 6 & 0 & -8 & 5 & 3 & -358 \end{array}$$

The solution is feasible because all variables are nonnegative. But it is still not optimal. Pivot on the 6 in row 2, column 6.

$$\begin{array}{r|cccccccc|c} & y_1 & y_2 & y_3 & s_1 & s_2 & s_3 & s_4 & z & \\ \hline -5R_2+6R_1\to R_1 & 0 & 18 & 0 & -3 & -15 & 0 & 6 & 0 & 144 \\ & 0 & 0 & 0 & -3 & 3 & 6 & 0 & 0 & 48 \\ R_2+6R_3\to R_3 & 0 & 0 & 6 & -3 & 3 & 0 & 0 & 0 & 108 \\ -R_2+6R_4\to R_4 & 18 & 0 & 0 & 3 & -3 & 0 & -6 & 0 & 72 \\ \hline 4R_2+3R_5\to R_5 & 0 & 0 & 0 & 6 & 12 & 0 & 15 & 9 & -882 \end{array}$$

Create a 1 in the columns corresponding to y_1, y_2, y_3, s_3 and z.

$$\begin{array}{r|cccccccc|c} & y_1 & y_2 & y_3 & s_1 & s_2 & s_3 & s_4 & z & \\ \hline \frac{1}{18}R_1\to R_1 & 0 & 1 & 0 & -\frac{1}{6} & -\frac{5}{6} & 0 & \frac{1}{3} & 0 & 8 \\ \frac{1}{6}R_2\to R_2 & 0 & 0 & 0 & -\frac{1}{2} & \frac{1}{2} & 1 & 0 & 0 & 8 \\ \frac{1}{6}R_3\to R_3 & 0 & 0 & 1 & -\frac{1}{2} & \frac{1}{2} & 0 & 0 & 0 & 18 \\ \frac{1}{18}R_4\to R_4 & 1 & 0 & 0 & \frac{1}{6} & -\frac{1}{6} & 0 & -\frac{1}{3} & 0 & 4 \\ \hline \frac{1}{9}R_5\to R_5 & 0 & 0 & 0 & \frac{2}{3} & \frac{4}{3} & 0 & \frac{5}{3} & 1 & -98 \end{array}$$

Since $z = -w = -98$, the minimum value is 98 when $y_1 = 4$, $y_2 = 8$, and $y_3 = 18$.

29.

$$\begin{array}{ccccc|c} x_1 & x_2 & s_1 & s_2 & z & \\ \hline 5 & 10 & 1 & 0 & 0 & 120 \\ \boxed{10} & 15 & 0 & -1 & 0 & 200 \\ \hline -20 & -30 & 0 & 0 & 1 & 0 \end{array}$$

The initial tableau is not feasible. Pivot on the 10 in row 2, column 1.

$$\begin{array}{r|ccccc|c|} & x_1 & x_2 & s_1 & s_2 & z & \\ \hline -R_2+2R_1\to R_1 & 0 & 5 & 2 & \boxed{1} & 0 & 40 \\ & 10 & 15 & 0 & -1 & 0 & 200 \\ \hline 2R_2+R_3\to R_3 & 0 & 0 & 0 & -2 & 1 & 400 \end{array}$$

The basic solution is feasible, but there are negative indicators. Pivot on the 1 in row 1, column 4.

$$\begin{array}{r|ccccc|c|} & x_1 & x_2 & s_1 & s_2 & z & \\ \hline & 0 & 5 & 2 & 1 & 0 & 40 \\ R_2+2R_1\to R_2 & 10 & 20 & 2 & 0 & 0 & 240 \\ \hline 2R_1+R_3\to R_3 & 0 & 10 & 4 & 0 & 1 & 480 \end{array}$$

Create a one in the column corresponding to x_1.

$$\begin{array}{r|ccccc|c|} & x_1 & x_2 & s_1 & s_2 & z & \\ \hline & 0 & 5 & 2 & 1 & 0 & 40 \\ \frac{1}{10}R_2\to R_2 & 1 & 2 & \frac{1}{5} & 0 & 0 & 24 \\ \hline & 0 & 10 & 4 & 0 & 1 & 480 \end{array}$$

The maximum value is $z = 480$ when $x_1 = 24$ and $x_2 = 0$.

31. Maximize $z = 10x_1 + 12x_2$

subject to: $2x_1 + 2x_2 = 17$

$2x_1 + 5x_2 \geq 22$

$4x_1 + 3x_2 \leq 28$

with $x_1 \geq 0, x_2 \geq 0.$

Introduce artificial variable a, surplus variable s_1, and slack variable s_2. The initial simplex tableau as follows.

$$\begin{array}{cccccc|c} x_1 & x_2 & a & s_1 & s_2 & z & \\ \hline \boxed{2} & 2 & 1 & 0 & 0 & 0 & 17 \\ 2 & 5 & 0 & -1 & 0 & 0 & 22 \\ 4 & 3 & 0 & 0 & 1 & 0 & 28 \\ \hline -10 & -12 & 0 & 0 & 0 & 1 & 0 \end{array}$$

First, eliminate the artificial variable a. Pivot on the 2 in row 1, column 1.

$$\begin{array}{r|cccccc|c|} & x_1 & x_2 & a & s_1 & s_2 & z & \\ \hline & 2 & 2 & 1 & 0 & 0 & 0 & 17 \\ -R_1+R_2\to R_2 & 0 & 3 & -1 & -1 & 0 & 0 & 5 \\ 2R_1-R_3\to R_3 & 0 & 1 & 2 & 0 & -1 & 0 & 6 \\ \hline 5R_1+R_4\to R_4 & 0 & -2 & 5 & 0 & 0 & 1 & 85 \end{array}$$

Now $a = 0$, so we can drop the a column.

$$\begin{array}{ccccc|c} x_1 & x_2 & s_1 & s_2 & z & \\ \hline 2 & 2 & 0 & 0 & 0 & 17 \\ 0 & \boxed{3} & -1 & 0 & 0 & 5 \\ 0 & 1 & 0 & -1 & 0 & 6 \\ \hline 0 & -2 & 0 & 0 & 1 & 85 \end{array}$$

Because $s_1 = -5$, we choose the 3 in row 2, column 2, as the next pivot.

$$\begin{array}{r|ccccc|c|} & x_1 & x_2 & s_1 & s_2 & z & \\ \hline -2R_2+3R_1\to R_1 & 6 & 0 & 2 & 0 & 0 & 41 \\ & 0 & 3 & -1 & 0 & 0 & 5 \\ -R_2+3R_3\to R_3 & 0 & 0 & \boxed{1} & -3 & 0 & 13 \\ \hline 2R_2+3R_4\to R_4 & 0 & 0 & -2 & 0 & 3 & 265 \end{array}$$

The solution is still not feasible since $s_2 = -\frac{13}{3}$. Pivot on the 1 in row 3, column 3.

$$\begin{array}{r|ccccc|c|} & x_1 & x_2 & s_1 & s_2 & z & \\ \hline -2R_3+R_1\to R_1 & 6 & 0 & 0 & \boxed{6} & 0 & 15 \\ R_3+R_2\to R_2 & 0 & 3 & 0 & -3 & 0 & 18 \\ & 0 & 0 & 1 & -3 & 0 & 13 \\ \hline 2R_3+R_4\to R_4 & 0 & 0 & 0 & -6 & 3 & 291 \end{array}$$

The solution is now feasible but is not yet optimal. Pivot on the 6 in row 1, column 4.

$$\begin{array}{r|ccccc|c|} & x_1 & x_2 & s_1 & s_2 & z & \\ \hline & 6 & 0 & 0 & 6 & 0 & 15 \\ R_1+2R_2\to R_2 & 6 & 6 & 0 & 0 & 0 & 51 \\ R_1+2R_3\to R_3 & 6 & 0 & 2 & 0 & 0 & 41 \\ \hline R_1+R_4\to R_4 & 6 & 0 & 0 & 0 & 3 & 306 \end{array}$$

Create a 1 in the columns corresponding to x_2, s_1, s_2, and z.

$$\begin{array}{r|ccccc|c|} & x_1 & x_2 & s_1 & s_2 & z & \\ \hline \frac{1}{6}R_1\to R_1 & 1 & 0 & 0 & 1 & 0 & \frac{5}{2} \\ \frac{1}{6}R_2\to R_2 & 1 & 1 & 0 & 0 & 0 & \frac{17}{2} \\ \frac{1}{2}R_3\to R_3 & 3 & 0 & 1 & 0 & 0 & \frac{41}{2} \\ \hline \frac{1}{3}R_4\to R_4 & 2 & 0 & 0 & 0 & 1 & 102 \end{array}$$

The maximum is 102 when $x_1 = 0$ and $x_2 = \frac{17}{2}$.

33. Any maximizing or minimizing problems can be solved using slack, surplus, and artificial variables. Slack variables are used in problems involving "$\leq$" constraints. Surplus variables are used in problems involving "$\geq$" constraints. Artificial variables are used in problems involving "$=$" constraints.

35. $\left[\begin{array}{cccccc|c} 4 & 2 & 3 & 1 & 0 & 0 & 9 \\ 5 & 4 & 1 & 0 & 1 & 0 & 10 \\ \hline -6 & -7 & -5 & 0 & 0 & 1 & 0 \end{array}\right]$

(a) The 1 in column 4 and the 1 in column 5 indicate that the constraints involve $\leq$. The problem being solved with this tableau is:

Maximize $z = 6x_1 + 7x_2 + 5x_3$

subject to:

$$4x_1 + 2x_2 + 3x_3 \leq 9$$
$$5x_1 + 4x_2 + x_3 \leq 10$$

with $x_1 \geq 0, x_2 \geq 0, x_3 \geq 0.$

(b) If the 1 in row 1, column 4 was -1 rather than 1, then the first constraint would have a surplus variable rather than a slack variable, which means the first constraint would be $4x_1 + 2x_2 + 3x_3 \geq 9$ instead of $4x_1 + 2x_2 + 3x_3 \leq 9$.

(c)

$$\begin{array}{c} \begin{array}{cccccc} x_1 & x_2 & x_3 & s_1 & s_2 & z \end{array} \\ \left[\begin{array}{cccccc|c} 3 & 0 & 5 & 2 & -1 & 0 & 8 \\ 11 & 10 & 0 & -1 & 3 & 0 & 21 \\ \hline 47 & 0 & 0 & 13 & 11 & 10 & 227 \end{array}\right] \end{array}$$

From this tableau, the solution is $x_1 = 0$, $x_2 = \frac{21}{10} = 2.1$, $x_3 = \frac{8}{5} = 1.6$, and $z = \frac{227}{10} = 22.7$.

(d) The dual of the original problem is as follows:

Minimize $w = 9y_1 + 10y_2$

subject to:

$$4y_1 + 5y_2 \geq 6$$
$$2y_1 + 4y_2 \geq 7$$
$$3y_1 + y_2 \geq 5$$

with $y_1 \geq 0, \; y_2 \geq 0.$

(e) From the tableau in part (c), the solution of the dual in part (d) is $y_1 = \frac{13}{10} = 1.3$, $y_2 = \frac{11}{10} = 1.1$, and $w = \frac{227}{10} = 22.7$.

37. **(a)** Let x_1 = the number of cake plates,

x_2 = the number of bread plates,

and x_3 = the number of dinner plates.

(b) The objective function to maximize is $z = 15x_1 + 12x_2 + 5x_3$.

(c) The constraints are

$$15x_1 + 10x_2 + 8x_3 \leq 1500$$
$$5x_1 + 4x_2 + 4x_3 \leq 2700$$
$$6x_1 + 5x_2 + 5x_3 \leq 1200.$$

39. **(a)** Let x_1 = number of gallons of Fruity wine and x_2 = number of gallons of Crystal wine.

(b) The profit function is

$$z = 12x_1 + 15x_2.$$

(c) The ingredients available are the limitations; the constraints are

$$2x_1 + x_2 \leq 110$$
$$2x_1 + 3x_2 \leq 125$$
$$2x_1 + x_2 \leq 90.$$

41. Maximize $z = 15x_1 + 12x_2 + 5x_3$

subject to:

$$15x_1 + 10x_2 + 8x_3 \leq 1500$$
$$5x_1 + 4x_2 + 4x_3 \leq 2700$$
$$6x_1 + 5x_2 + 5x_3 \leq 1200$$

with $x_1 \geq 0, \; x_2 \geq 0, \; x_3 \geq 0.$

The initial tableau is as follows.

$$\begin{array}{c} \begin{array}{ccccccc} x_1 & x_2 & x_3 & s_1 & s_2 & s_3 & z \end{array} \\ \left[\begin{array}{ccccccc|c} \boxed{15} & 10 & 8 & 1 & 0 & 0 & 0 & 1500 \\ 5 & 4 & 4 & 0 & 1 & 0 & 0 & 2700 \\ 6 & 5 & 5 & 0 & 0 & 1 & 0 & 1200 \\ \hline -15 & -12 & -5 & 0 & 0 & 0 & 1 & 0 \end{array}\right] \end{array}.$$

Pivot on the 15 in row 1, column 1.

$$\begin{array}{r} \\ \\ -R_1 + 3R_2 \to R_2 \\ -2R_1 + 5R_3 \to R_3 \\ R_1 + R_4 \to R_4 \end{array} \begin{array}{c} \begin{array}{ccccccc} x_1 & x_2 & x_3 & s_1 & s_2 & s_3 & z \end{array} \\ \left[\begin{array}{ccccccc|c} 15 & \boxed{10} & 8 & 1 & 0 & 0 & 0 & 1500 \\ 0 & 2 & 4 & -1 & 3 & 0 & 0 & 6600 \\ 0 & 5 & 9 & -2 & 0 & 5 & 0 & 3000 \\ \hline 0 & -2 & 3 & 1 & 0 & 0 & 1 & 1500 \end{array}\right] \end{array}$$

Pivot on the 10 in row 1, column 2.

$$\begin{array}{r} \\ \\ -R_1 + 5R_2 \to R_2 \\ -R_1 + 2R_3 \to R_3 \\ R_1 + 5R_4 \to R_4 \end{array} \begin{array}{c} \begin{array}{ccccccc} x_1 & x_2 & x_3 & s_1 & s_2 & s_3 & z \end{array} \\ \left[\begin{array}{ccccccc|c} 15 & 10 & 8 & 1 & 0 & 0 & 0 & 1500 \\ -15 & 0 & 12 & -6 & 15 & 0 & 0 & 31{,}500 \\ -15 & 0 & 10 & -5 & 0 & 10 & 0 & 4500 \\ \hline 15 & 0 & 23 & 6 & 0 & 0 & 5 & 9000 \end{array}\right] \end{array}$$

Create a 1 in the columns corresponding to x_2, s_2, s_3, and z.

$$\begin{array}{r} \frac{1}{10}R_1 \to R_1 \\ \frac{1}{15}R_2 \to R_2 \\ \frac{1}{10}R_3 \to R_3 \\ \frac{1}{5}R_4 \to R_4 \end{array} \begin{array}{c} \begin{array}{ccccccc|c} x_1 & x_2 & x_3 & s_1 & s_2 & s_3 & z & \end{array} \\ \left[\begin{array}{ccccccc|c} \frac{3}{2} & 1 & \frac{4}{5} & \frac{1}{10} & 0 & 0 & 0 & 150 \\ -1 & 0 & \frac{4}{5} & -\frac{2}{5} & 1 & 0 & 0 & 2100 \\ -\frac{3}{2} & 0 & 1 & -\frac{1}{2} & 0 & 1 & 0 & 450 \\ \hline 3 & 0 & \frac{23}{5} & \frac{6}{5} & 0 & 0 & 1 & 1800 \end{array}\right] \end{array}$$

The maximum profit of \$1800 occurs when no cake plates, 150 bread plates, and no dinner plates are produced.

43. Based on Exercise 39, the initial tableau is

$$\begin{array}{c} \begin{array}{cccccc|c} x_1 & x_2 & s_1 & s_2 & s_3 & z & \end{array} \\ \left[\begin{array}{cccccc|c} 2 & 1 & 1 & 0 & 0 & 0 & 110 \\ 2 & \boxed{3} & 0 & 1 & 0 & 0 & 125 \\ 2 & 1 & 0 & 0 & 1 & 0 & 90 \\ \hline -12 & -15 & 0 & 0 & 0 & 1 & 0 \end{array}\right] \end{array}.$$

Locating the first pivot in the usual way, it is found to be the 3 in row 2, column 2. After row transformations, we get the next tableau.

$$\begin{array}{r} -R_2 + 3R_1 \to R_1 \\ \\ -R_2 + 3R_3 \to R_3 \\ 5R_2 + R_4 \to R_4 \end{array} \begin{array}{c} \begin{array}{cccccc|c} x_1 & x_2 & s_1 & s_2 & s_3 & z & \end{array} \\ \left[\begin{array}{cccccc|c} 4 & 0 & 3 & -1 & 0 & 0 & 205 \\ 2 & 3 & 0 & 1 & 0 & 0 & 125 \\ \boxed{4} & 0 & 0 & -1 & 3 & 0 & 145 \\ \hline -2 & 0 & 0 & 5 & 0 & 1 & 625 \end{array}\right] \end{array}.$$

Pivot on the 4 in row 3, column 1.

$$\begin{array}{r} -R_3 + R_1 \to R_1 \\ -R_3 + 2R_2 \to R_2 \\ \\ R_3 + 2R_4 \to R_4 \end{array} \begin{array}{c} \begin{array}{cccccc|c} x_1 & x_2 & s_1 & s_2 & s_3 & z & \end{array} \\ \left[\begin{array}{cccccc|c} 0 & 0 & 3 & 0 & -3 & 0 & 60 \\ 0 & 6 & 0 & 3 & -3 & 0 & 105 \\ 4 & 0 & 0 & -1 & 3 & 0 & 145 \\ \hline 0 & 0 & 0 & 9 & 3 & 2 & 1395 \end{array}\right] \end{array}$$

$$\begin{array}{r} \frac{1}{3}R_1 \to R_1 \\ \frac{1}{6}R_2 \to R_2 \\ \frac{1}{4}R_3 \to R_3 \\ \frac{1}{2}R_4 \to R_4 \end{array} \begin{array}{c} \begin{array}{cccccc|c} x_1 & x_2 & s_1 & s_2 & s_3 & z & \end{array} \\ \left[\begin{array}{cccccc|c} 0 & 0 & 1 & 0 & -1 & 0 & 20 \\ 0 & 1 & 0 & \frac{1}{2} & -\frac{1}{2} & 0 & \frac{35}{2} \\ 1 & 0 & 0 & -\frac{1}{4} & \frac{3}{4} & 0 & \frac{145}{4} \\ \hline 0 & 0 & 0 & \frac{9}{2} & \frac{3}{2} & 1 & \frac{1395}{2} \end{array}\right] \end{array}$$

The final tableau gives the solution $x_1 = \frac{145}{4}$, $x_2 = \frac{35}{2}$, and $z = \frac{1395}{2} = 697.5$. 36.25 gal of Fruity wine and 17.5 gal of Crystal wine should be produced for a maximum profit of \$697.50.

45. (a) Let $y_1 =$ the number of cases of corn,

$y_2 =$ the number of cases of beans

and $y_3 =$ the number of cases of carrots.

Minimize $w = 10y_1 + 15y_2 + 25y_3$

subject to:

$$y_1 + y_2 + y_3 \geq 1000$$
$$y_1 \geq 2y_2$$
$$y_3 \geq 340$$

with $y_1 \geq 0, y_2 \geq 0$.

The second constraint can be rewritten as $y_1 - 2y_2 \geq 0$. Change this to a maximization problem by letting $z = -w = -10y_1 - 15y_2 - 25y_3$. Now maximize $z = -10y_1 - 15y_2 - 25y_3$ subject to the constraints above. Begin by inserting surplus variables to set up the first tableau.

$$\begin{array}{c} \begin{array}{ccccccc|c} y_1 & y_2 & y_3 & s_1 & s_2 & s_3 & z & \end{array} \\ \left[\begin{array}{ccccccc|c} \boxed{1} & 1 & 1 & -1 & 0 & 0 & 0 & 1000 \\ 1 & -2 & 0 & 0 & -1 & 0 & 0 & 0 \\ 0 & 0 & 1 & 0 & 0 & -1 & 0 & 340 \\ \hline 10 & 15 & 25 & 0 & 0 & 0 & 1 & 0 \end{array}\right] \end{array}$$

Multiply row 2 by -1 so that s_2 is positive.

$$\begin{array}{r} \\ -R_2 \to R_2 \\ \\ \\ \end{array} \begin{array}{c} \begin{array}{ccccccc|c} y_1 & y_2 & y_3 & s_1 & s_2 & s_3 & z & \end{array} \\ \left[\begin{array}{ccccccc|c} \boxed{1} & 1 & 1 & -1 & 0 & 0 & 0 & 1000 \\ -1 & 2 & 0 & 0 & 1 & 0 & 0 & 0 \\ 0 & 0 & 1 & 0 & 0 & -1 & 0 & 340 \\ \hline 10 & 15 & 25 & 0 & 0 & 0 & 1 & 0 \end{array}\right] \end{array}$$

Pivot on the 1 in row 1, column 1.

$$\begin{array}{r} \\ R_1 + R_2 \to R_2 \\ \\ -10R_1 + R_4 \to R_4 \end{array} \begin{array}{c} \begin{array}{ccccccc|c} y_1 & y_2 & y_3 & s_1 & s_2 & s_3 & z & \end{array} \\ \left[\begin{array}{ccccccc|c} 1 & 1 & 1 & -1 & 0 & 0 & 0 & 1000 \\ 0 & 3 & 1 & -1 & 1 & 0 & 0 & 1000 \\ 0 & 0 & \boxed{1} & 0 & 0 & -1 & 0 & 340 \\ \hline 0 & 5 & 15 & 10 & 0 & 0 & 1 & -10{,}000 \end{array}\right] \end{array}$$

Pivot on the 1 in row 3, column 3.

$$\begin{array}{r} -R_3 + R_1 \to R_1 \\ -R_3 + R_2 \to R_2 \\ \\ -15R_3 + R_4 \to R_4 \end{array} \begin{array}{c} \begin{array}{ccccccc|c} y_1 & y_2 & y_3 & s_1 & s_2 & s_3 & z & \end{array} \\ \left[\begin{array}{ccccccc|c} 1 & 1 & 0 & -1 & 0 & 1 & 0 & 660 \\ 0 & 3 & 0 & -1 & 1 & 1 & 0 & 660 \\ 0 & 0 & 1 & 0 & 0 & -1 & 0 & 340 \\ \hline 0 & 5 & 0 & 10 & 0 & 15 & 1 & -15{,}100 \end{array}\right] \end{array}$$

The maximum value of z is $-15{,}100$ when $y_1 = 660$, $y_2 = 0$, and $y_3 = 340$. Hence the minimum value of w is 15,100 when $y_1 = 660$, $y_2 = 0$, and $y_3 = 340$. Produce 660 cases of corn and 340 cases of carrots for a minimum cost of \$15.000.

(b) The dual problem is as follows.

Maximize $\quad z = 1000x_1 + 340x_3$

subject to:
$$x_1 + x_2 \le 10$$
$$x_1 - 2x_2 \le 15$$
$$x_1 + x_3 \le 25$$

with $\quad x_1 \ge 0, x_2 \ge 0, x_3 \ge 0.$

The initial simplex tableau is as follows.

$$\begin{array}{c} \\ \\ \\ \\ \\ \end{array}\begin{array}{ccccccc|c} x_1 & x_2 & x_3 & s_1 & s_2 & s_3 & z & \\ \boxed{1} & 1 & 0 & 1 & 0 & 0 & 0 & 10 \\ 1 & -2 & 0 & 0 & 1 & 0 & 0 & 15 \\ 1 & 0 & 1 & 0 & 0 & 1 & 0 & 25 \\ \hline -1000 & 0 & -340 & 0 & 0 & 0 & 1 & 0 \end{array}$$

Pivot on the 1 in row 1, column 1.

$$\begin{array}{r} \\ \\ -R_1 + R_2 \to R_2 \\ -R_1 + R_3 \to R_3 \\ 1000R_1 + R_4 \to R_4 \end{array}\begin{array}{ccccccc|c} x_1 & x_2 & x_3 & s_1 & s_2 & s_3 & z & \\ 1 & 1 & 0 & 1 & 0 & 0 & 0 & 10 \\ 0 & -3 & 0 & -1 & 1 & 0 & 0 & 5 \\ 0 & -1 & \boxed{1} & -1 & 0 & 1 & 0 & 15 \\ \hline 0 & 1000 & -340 & 1000 & 0 & 0 & 1 & 10{,}000 \end{array}$$

Pivot on the 1 in row 3, column 3.

$$\begin{array}{r} \\ \\ \\ \\ 340R_3 + R_4 \to R_4 \end{array}\begin{array}{ccccccc|c} x_1 & x_2 & x_3 & s_1 & s_2 & s_3 & z & \\ 1 & 1 & 0 & 1 & 0 & 0 & 0 & 10 \\ 0 & -3 & 0 & -1 & 1 & 0 & 0 & 5 \\ 0 & -1 & 1 & -1 & 0 & 1 & 0 & 15 \\ \hline 0 & 660 & 0 & 660 & 0 & 340 & 1 & 15{,}100 \end{array}$$

The minimum value of w is 15,100 when $y_1 = 660$, $y_2 = 0$, and $y_3 = 340$, that is, 660 cases of corn, 0 cases of beans, and 340 cases of carrots should be produced to minimize costs, and the minimum cost is \$15,100.

(c) The final tableau for the dual solution shows that the shadow cost of acreage (x_1) is \$10 acre, so increasing the number of acres planted by 100 will increase the minimum cost by (\$10)(100) or \$1000, so the new minimum will be \$15,100 + \$1000 = \$16,100.

47. (a) Let $x_1 =$ the number of hours doing tai chi,

$x_2 =$ the number of hours riding a unicycle,

and $x_3 =$ the number of hours fencing.

If Ginger wants the total time doing tai chi to be at least twice as long as she rides a unicycle, then

$$x_1 \ge 2x_2$$
$$\text{or} \quad -x_1 + 2x_2 \le 0.$$

The problem can be stated as follows.

Maximize $\quad z = 236x_1 + 295x_2 + 354x_3$

subject to:
$$x_1 + x_2 + x_3 \le 10$$
$$x_3 \le 2$$
$$-x_1 + 2x_2 \le 0$$

with $\quad x_1 \ge 0, x_2 \ge 0, x_3 \ge 0.$

The initial simplex tableau is as follows.

$$\begin{array}{ccccccc|c} x_1 & x_2 & x_3 & s_1 & s_2 & s_3 & z & \\ 1 & 1 & 1 & 1 & 0 & 0 & 0 & 10 \\ 0 & 0 & \boxed{1} & 0 & 1 & 0 & 0 & 2 \\ -1 & 2 & 0 & 0 & 0 & 1 & 0 & 0 \\ \hline -236 & -295 & -354 & 0 & 0 & 0 & 1 & 0 \end{array}$$

Pivot on the 1 in row 2, column 3.

$$\begin{array}{r} \\ -R_2 + R_1 \to R_1 \\ \\ \\ 354R_2 + R_4 \to R_4 \end{array}\begin{array}{ccccccc|c} x_1 & x_2 & x_3 & s_1 & s_2 & s_3 & z & \\ 1 & 1 & 0 & 1 & -1 & 0 & 0 & 8 \\ 0 & 0 & 1 & 0 & 1 & 0 & 0 & 2 \\ -1 & \boxed{2} & 0 & 0 & 0 & 1 & 0 & 0 \\ \hline -236 & -295 & 0 & 0 & 354 & 0 & 1 & 708 \end{array}$$

Pivot on the 2 in row 3, column 2.

$$\begin{array}{r} \\ -R_3 + 2R_1 \to R_1 \\ \\ \\ 295R_3 + 2R_4 \to R_4 \end{array}\begin{array}{ccccccc|c} x_1 & x_2 & x_3 & s_1 & s_2 & s_3 & z & \\ \boxed{3} & 0 & 0 & 2 & -2 & -1 & 0 & 16 \\ 0 & 0 & 1 & 0 & 1 & 0 & 0 & 2 \\ -1 & 2 & 0 & 0 & 0 & 1 & 0 & 0 \\ \hline -767 & 0 & 0 & 0 & 708 & 295 & 2 & 1416 \end{array}$$

Pivot on the 3 in row 1, column 1.

$$\begin{array}{r} \\ \\ \\ R_1 + 3R_3 \to R_3 \\ 767R_1 + 3R_4 \to R_4 \end{array}\begin{array}{ccccccc|c} x_1 & x_2 & x_3 & s_1 & s_2 & s_3 & z & \\ 3 & 0 & 0 & 2 & -2 & -1 & 0 & 16 \\ 0 & 0 & 1 & 0 & 1 & 0 & 0 & 2 \\ 0 & 6 & 0 & 2 & -2 & 2 & 0 & 16 \\ \hline 0 & 0 & 0 & 1534 & 590 & 118 & 6 & 16{,}520 \end{array}$$

Create a 1 in the columns corresponding to $x_1, x_2,$ and z.

$$\begin{array}{r} \\ \frac{1}{3}R_1 \to R_1 \\ \\ \frac{1}{6}R_3 \to R_3 \\ \frac{1}{6}R_4 \to R_4 \end{array} \begin{array}{c} \begin{matrix} x_1 & x_2 & x_3 & s_1 & s_2 & s_3 & z & \end{matrix} \\ \left[\begin{array}{ccccccc|c} 1 & 0 & 0 & \frac{2}{3} & -\frac{2}{3} & -\frac{1}{3} & 0 & \frac{16}{3} \\ 0 & 0 & 1 & 0 & 1 & 0 & 0 & 2 \\ 0 & 1 & 0 & \frac{1}{3} & -\frac{1}{3} & \frac{1}{3} & 0 & \frac{8}{3} \\ \hline 0 & 0 & 0 & \frac{767}{3} & \frac{295}{3} & \frac{59}{3} & 1 & \frac{8260}{3} \end{array}\right] \end{array}$$

Ginger will burn a maximum of $2753\frac{1}{3}$ calories if she does $\frac{16}{3}$ hours of tai chi, $\frac{8}{3}$ hours riding a unicycle, and 2 hours fencing.

(b) Since fencing burns the most calories, she should do as much fencing as possible, which is 2 hours. This leaves 8 hours to divide between tai chi and the unicycle. The unicycle burns more calories, so she wants as much of unicycle as possible subject to the tai chi getting at least twice as much time as the unicycle. This requires devoting $\frac{1}{3}$ of the remaining 8 hours to the unicycle and $\frac{2}{3}$ of the 8 hours to tai chi. So the times are: $\frac{16}{3}$ hours of tai chi, $\frac{8}{3}$ hours of unicycle, and 2 hours of fencing.

Chapter 5

MATHEMATICS OF FINANCE

5.1 Simple and Compound Interest

Your Turn 1

Use the formula for maturity value, with $P = 3000$, $r = 0.058$, and $t = \frac{100}{360}$. We assume a year of 360 days.

$$A = P(1 + rt)$$
$$A = 3000\left[1 + 0.058\left(\frac{100}{360}\right)\right]$$
$$A = 3048.333$$

The maturity value is \$3048.33.

Your Turn 2

Use the formula $A = P(1 + rt)$ with $t = 0.75$ or three quarters of a year.

$$5243.75 = 5000\,[1 + r(0.75)]$$

Solve for r:

$$5243.75 = 5000 + 3750r$$
$$243.75 = 3750r$$
$$r = \frac{243.75}{3750} = 0.065$$

The interest rate is 6.5%.

Your Turn 3

For 7 years compounded quarterly there are $(7)(12) = 84$ periods. The interest rate per month is $\frac{0.042}{12} = 0.0035$. The interest earned on a principal of \$1600 is

$$1600(1 + 0.0035)^{84} - 1600 = 545.75 \text{ or } \$545.75.$$

Your Turn 4

The number of compounding periods is $(8)(12) = 96$. We need to solve the equation

$$6500\left(1 + \frac{r}{12}\right)^{96} = 8665.69$$

$$\left(1 + \frac{r}{12}\right)^{96} = \frac{8665.69}{6500} = 1.33318 \quad \text{Divide both sides by 6500.}$$

$$1 + \frac{r}{12} = 1.33318^{1/96} = 1.003 \quad \text{Raise both sides to the 1/96 power.}$$

$$\frac{r}{12} = 0.003 \quad \text{Subtract 1 from both sides.}$$

$$r = 0.036 \quad \text{Multiply both sides by 12.}$$

The annual interest rate is 3.6%.

Your Turn 5

Use the formula $r_E = \left(1 + \frac{r}{m}\right)^m - 1$ with $r = 0.027$ and $m = 12$.

$$r_E = \left(1 + \frac{0.027}{12}\right)^{12} - 1$$
$$r_E = 1.0273 - 1 = 0.0273$$

The effective rate is 2.73%.

Your Turn 6

The interest rate per quarter is $\frac{0.0425}{4} = 0.010625$. Use the formula for present value with $i = 0.010625$ and $n = (7)(4) = 28$.

$$P = \frac{A}{(1 + i)^n}$$
$$P = \frac{10{,}000}{(1 + 0.010625)^{28}}$$
$$P = 7438.39$$

The present value of the investment is \$7438.39.

Your Turn 7

The semiannual interest rate is $\frac{0.035}{2} = 0.0175$. Let n be the number of compounding periods. Then we want $7000 = 3800(1 + 0.0175)^n$.

Solving, we find

$$(1 + 0.0175)^n = \frac{7000}{3800} = 1.842$$

Using logarithms,

$$n\log(1.0175) = \log(1.842)$$
$$n = \frac{\log(1.842)}{\log(1.0175)} = 35.21$$

Since each period is half a year, this corresponds to $\frac{35.21}{2} = 17.605$ years. Rounding up to the next whole period we get an answer of 18 years.

Your Turn 8

Use the formula for continuous compounding with $P = 5000$, $r = 0.038$ and $t = 9$.

$$A = Pe^{rt}$$
$$A = 5000e^{(0.038)(9)}$$
$$A = 7038.80$$

Subtracting the initial investment we get $7038.80 - 5000 = 2038.80$, or \$2038.80 interest earned.

5.1 Exercises

5. \$25,000 at 3% for 9 mo

Use the formula for simple interest.

$$I = Prt$$
$$= 25{,}000(0.03)\left(\frac{9}{12}\right)$$
$$= 562.50$$

The simple interest is \$562.50.

7. \$1974 at 6.3% for 25 wk

Use the formula for simple interest.

$$I = Prt$$
$$= 1974(0.063)\left(\frac{25}{52}\right) \approx 59.79$$

The simple interest is \$59.79.

9. \$8192.17 at 3.1% for 72 days

Use the formula for simple interest.

$$I = Prt$$
$$= 8192.17(0.031)\left(\frac{72}{360}\right)$$
$$\approx 50.79$$

The simple interest is \$50.79.

11. Use the formula for future value for simple interest.

$$A = P(1 + rt)$$
$$= 3125\left[1 + 0.0285\left(\frac{7}{12}\right)\right]$$
$$\approx 3176.95$$

The maturity value is \$3176.95. The interest earned is $3176.95 - 3125 = \$51.95$.

13. Use the formula for simple interest.

$$I = Prt$$
$$56.25 = 1500r\left(\frac{6}{12}\right)$$
$$r = 0.075$$

The interest rate was 7.5%.

19. Use the formula for compound amount with $P = 1000$, $i = 0.06$, and $n = 8$.

$$A = P(1 + i)^n$$
$$= 1000(1 + 0.06)^8$$
$$\approx 1593.85$$

The compound amount is \$1593.85. The interest earned is $1593.85 - 1000 = \$593.85$.

21. Use the formula for compound amount with $P = 470$, $i = \frac{0.054}{2} = 0.027$, and $n = 12(2) = 24$.

$$A = P(1 + i)^n$$
$$= 470(1 + 0.027)^{24}$$
$$\approx 890.82$$

The compound amount is \$890.82. The interest earned is $890.82 - 470 = \$420.82$.

23. Use the formula for compound amount with $P = 8500$, $i = \frac{0.08}{4} = 0.02$, and $n = 5(4) = 20$.

$$A = P(1 + i)^n$$
$$= 8500(1 + 0.02)^{20}$$
$$\approx 12{,}630.55$$

The compound amount is \$12,630.55. The interest earned is $12{,}630.55 - 8500 = \$4130.55$.

25. The number of compounding periods is $(4)(8) = 32$.

$$8000\left(1 + \frac{r}{4}\right)^{32} = 11672.12$$
$$\left(1 + \frac{r}{4}\right)^{32} = \frac{11672.12}{8000} = 1.45902$$

$$1+\frac{r}{4} = 1.45902^{1/32} = 1.011875$$
$$\frac{r}{4} = 0.011875$$
$$r = 0.0475$$

The answer is 4.75%.

27. The number of compounding periods is $(12)(5) = 60$.

$$4500\left(1+\frac{r}{12}\right)^{60} = 5994.79$$
$$\left(1+\frac{r}{12}\right)^{60} = \frac{5994.79}{4500} = 1.332176$$
$$1+\frac{r}{12} = 1.332176^{1/60} = 1.004792$$
$$\frac{r}{12} = 0.004792$$
$$r = 0.0575$$

The answer is 5.75%

29. 4% compounded quarterly.

Use the formula for effective rate with $r = 0.04$ and $m = 4$.

$$r_E = \left(1+\frac{r}{m}\right)^m - 1$$
$$= \left(1+\frac{0.04}{4}\right)^4 - 1$$
$$\approx 0.04060$$

The effective rate is about 4.06%.

31. 7.25% compounded semiannually.

Use the formula for effective rate with $r = 0.0725$ and $m = 2$.

$$r_E = \left(1+\frac{r}{m}\right)^m - 1$$
$$= \left(1+\frac{0.0725}{2}\right)^2 - 1$$
$$\approx 0.07381$$

The effective rate is about 7.381%, or rounding to two decimal places, 7.38%.

33. Use the formula for present value for compound interest with $A = 12{,}820.77$, $i = 0.048$, and $n = 6$.

$$P = \frac{A}{(1+r)^n}$$
$$= \frac{12{,}820.77}{(1+0.048)^6}$$
$$\approx 9677.13$$

The present value is \$9677.13.

35. Use the formula for present value for compound interest with $A = 2000$, $i = \frac{0.06}{2} = 0.03$, and $n = 8(2) = 16$.

$$P = \frac{A}{(1+r)^n}$$
$$= \frac{2000}{(1+0.03)^{16}}$$
$$\approx 1246.33$$

The present value is \$1246.33.

37. Use the formula for present value for compound interest with $A = 8800$, $i = \frac{0.05}{4} = 0.0125$, and $n = 5(4) = 20$.

$$P = \frac{A}{(1+r)^n}$$
$$= \frac{8800}{(1+0.0125)^{20}}$$
$$\approx 6864.08$$

The present value is \$6864.08.

41. The quarterly interest rate is $\frac{0.04}{4} = 0.01$. Let n be the number of compounding periods. Then we want $9000 = 5000(1+0.01)^n$

or

$$(1+0.01)^n = \frac{9000}{5000} = 1.8$$

Using logarithms, we have

$$n\log(1.01) = \log(1.8)$$
$$n = \frac{\log(1.8)}{\log(1.01)}$$
$$n = 59.07$$

Since each period is one quarter of a year, the number of years is $\frac{59.07}{4} = 14.768$. Rounding up to the next whole quarter gives us an answer of 15 years.

43. The monthly interest rate is $\frac{0.036}{12} = 0.003$. Let n be the number of compounding periods. Then we want $11000 = 4500(1 + 0.003)^n$

or

$$(1 + 0.003)^n = \frac{11000}{4500} = 2.444$$

Using logarithms, we have

$$\begin{aligned} n\log(1.003) &= \log(2.444) \\ n &= \frac{\log(2.444)}{\log(1.003)} \\ n &= 298.325 \end{aligned}$$

Since each period is one twelfth of a year, the number of years is $\frac{298.325}{12} = 24.86$.

$$\frac{10}{12} = 0.833 \text{ and } \frac{11}{12} = 0.917$$

so rounding up to the next whole month gives us an answer of 24 years and 11 months.

45. **(a)** The doubling time for an inflation rate of 3.3% is the solution of $2 = (1.033)^n$. Taking logarithms on both sides we have

$$\begin{aligned} \log(2) &= n\log(1.033) \\ n &= \frac{\log(2)}{\log(1.033)} \\ n &= 21.349 \end{aligned}$$

The doubling time is about 21.35 years.

(b) Since $0.001 < 0.033 < 0.05$, this is a small growth rate and we may use the rule of 70 which estimates the doubling time as

$$\frac{70}{(100)(0.033)} = 21.212$$

or about 21.21 years.

47. **(a)** The future value is $5500e^{(0.031)(9)} = 7269.94$, or \$7269.94.

(b) The effective rate is $e^{0.031} - 1 = 0.0315$ or 3.15%.

(c) To find the time to reach \$10,000, we solve

$$\begin{aligned} 10{,}000 &= 5500e^{0.031t} \\ e^{0.031t} &= \frac{10{,}000}{5500} \end{aligned}$$

Using logarithms with base e we have

$$\begin{aligned} 0.031t &= \ln\left(\frac{10{,}000}{5500}\right) \\ t &= \frac{\ln\left(\frac{10{,}000}{5500}\right)}{0.031} \\ t &= 19.285 \end{aligned}$$

The time to reach \$10,000 is 19.29 years.

49. Start by finding the total amount repaid. Use the formula for future value for simple interest, with $P = 2700$, $r = 0.062$, and $t = \frac{9}{12}$.

$$\begin{aligned} A &= P(1 + rt) \\ &= 7200\left[1 + 0.062\left(\frac{9}{12}\right)\right] \\ &= 7534.80 \end{aligned}$$

Tanya repaid her father \$7534.80. To find the amount of this which was interest, subtract the original loan amount from the repayment amount.

$$7534.80 - 7200 = 334.80$$

Of the amount repaid, \$334.80 was interest.

51. The interest earned was

$$\$1521.25 - \$1500 = \$21.25$$

Use the formula for simple interest, with $I = 21.25$, $P = 1500$, and $t = \frac{75}{360}$.

$$\begin{aligned} I &= Prt \\ 21.25 &= 1500r\left(\frac{75}{360}\right) \\ 0.068 &= r \end{aligned}$$

The interest rate was 6.8%.

53. Start by finding the total interest earned.

$$I = (\$24 - \$22) + \$0.50 = \$2.50$$

Now use the formula for simple interest, with $I = 2.50$, $P = 22$, and $t = 1$.

$$\begin{aligned} I &= Prt \\ 2.50 &= 22r(1) \\ 0.11364 &\approx r \end{aligned}$$

The interest rate was about 11.36% or, rounding to one decimal place, 11.4%.

55. Use the formula for compound amount with $P = 40{,}000$, $i = \frac{0.0654}{12}$, and $n = 6$.

$$\begin{aligned} A &= P(1+i)^n \\ &= 40{,}000\left(1+\frac{0.0654}{12}\right)^6 \\ &\approx 41{,}325.95 \end{aligned}$$

When Kelly begins paying off his loan, he will owe \$41,325.95.

57. (a) Use the formula for compound amount to find the value of \$1000 in 5 yr.

$$\begin{aligned} A &= P(1+i)^n \\ &= 1000(1.06)^5 \\ &\approx 1338.23 \end{aligned}$$

In 5 yr, \$1000 will be worth \$1338.23. Since this is larger than the \$1210 one would receive in 5 yr, it would be more profitable to take the \$1000 now.

59. Let $P = 150{,}000$, $i = -2.4\% = -.024$, and $n = 4$.

$$\begin{aligned} A &= P(1+i)^n \\ &= 150{,}000[1+(-.024)]^4 \\ &= 150{,}000(.976)^4 \\ &\approx 136{,}110.16 \end{aligned}$$

After 4 yr, the amount on deposit will be \$136,110.16.

61. Use the formula

$$A = P(1+i)^n$$

with $P = \frac{2}{8}$ cent $= \$0.0025$ and $r = 0.04$ compounded quarterly for 2000 yr.

$$\begin{aligned} A &= 0.0025\left(1+\frac{0.04}{4}\right)^{4(2000)} \\ &= 0.0025(1.01)^{8000} \\ &\approx 9.31 \times 10^{31} \end{aligned}$$

2000 years later, the money would be worth $\$9.31 \times 10^{31}$.

63. Use the formula

$$A = P(1+i)^n$$

with $P = 10{,}000$ and $r = 0.05$ for 10 years777.

(a) If interest is compounding annually,

$$\begin{aligned} A &= 10{,}000(1+0.05)^{10} \\ &\approx 16{,}288.95. \end{aligned}$$

The future value is \$16,288.95.

(b) If interest is compounding quarterly,

$$\begin{aligned} A &= 10{,}000\left(1+\frac{0.05}{4}\right)^{40} \\ &\approx 16{,}436.19. \end{aligned}$$

The future value is \$16,436.19.

(c) If interest is compounding monthly,

$$\begin{aligned} A &= 10{,}000\left(1+\frac{0.05}{12}\right)^{120} \\ &\approx 16{,}470.09. \end{aligned}$$

The future value is \$16,470.09.

(d) If interest is compounding daily,

$$\begin{aligned} A &= 10{,}000\left(1+\frac{0.05}{365}\right)^{3650} \\ &\approx 16{,}486.65. \end{aligned}$$

The future value is \$16,486.65.

(e) If the interest is compounded continuously for 10 years at 5%, the future value is $10{,}000e^{(0.05)(10)} = 16{,}487.213$ or \$16,487.21.

65. First consider the case of earning interest at a rate of k per annnm compounded quarterly for all 8 yr and earning \$2203.76 on the \$1000 investment.

$$\begin{aligned} 2203.76 &= 1000\left(1+\frac{k}{4}\right)^{8(4)} \\ 2.20376 &= \left(1+\frac{k}{4}\right)^{32} \end{aligned}$$

Use a calculator to raise both sides to the power $\frac{1}{32}$.

$$\begin{aligned} 1.025 &= 1+\frac{k}{4} \\ 0.025 &= \frac{k}{4} \\ 0.1 &= k \end{aligned}$$

Next consider the actual investments. The \$1000 was invested for the first 5 yr at a rate of j per annum compounded semiannually.

$$A = 1000\left(1 + \frac{j}{2}\right)^{5(2)}$$

$$A = 1000\left(1 + \frac{j}{2}\right)^{10}$$

This amount was then invested for the remaining 3 yr at $k = .1$ per annum compounded quarterly for a final compound amount of \$1990.76.

$$1990.76 = A\left(1 + \frac{0.1}{4}\right)^{3(4)}$$

$$1990.76 = A(1.025)^{12}$$

$$1480.24 \approx A$$

Recall that $A = 1000\left(1 + \frac{j}{2}\right)^{10}$ and substitute this value into the above equation.

$$1480.24 = 1000\left(1 + \frac{j}{2}\right)^{10}$$

$$1.48024 = \left(1 + \frac{j}{2}\right)^{10}$$

Use a calculator to raise both sides to the power $\frac{1}{10}$.

$$1.04 \approx 1 + \frac{j}{2}$$

$$0.04 = \frac{j}{2}$$

$$0.08 = j$$

The ratio of k to j is

$$\frac{k}{j} = \frac{0.1}{0.08} = \frac{10}{8} = \frac{5}{4}.$$

67. For each quoted effective rate, find the corresponding nominal rate by using the formula for effective rate. Regardless of the CD's term, m always equals 4, since compounding is always quarterly.

For the 6-month CD, use $r_E = 0.025$.

$$r_E = \left(1 + \frac{r}{m}\right)^m - 1$$

$$0.025 = \left(1 + \frac{r}{4}\right)^4 - 1$$

$$(1 + 0.025)^{1/4} = 1 + \frac{r}{4}.$$

$$0.02477 \approx r$$

For the 6-month CD, the nominal rate is about 2.48%.

For the 9-month CD, use $r_E = 0.051$.

$$r_E = \left(1 + \frac{r}{m}\right)^m -1$$

$$0.051 = \left(1 + \frac{r}{4}\right)^4 -1$$

$$(1 + 0.051)^{1/4} = 1 + \frac{r}{4}.$$

$$0.050053 \approx r$$

For the 9-month CD, the nominal rate is about 5.01%.

For the 1-year CD, use $r_E = 0.0425$ and $m = 4$.

$$r_E = \left(1 + \frac{r}{m}\right)^m - 1$$

$$0.0425 = \left(1 + \frac{r}{4}\right)^4 - 1$$

$$(1 + 0.0425)^{1/4} = 1 + \frac{r}{4}.$$

$$0.04184 \approx r$$

For the 1-year CD, the nominal rate is about 4.18%.

For the 2-year CD, use $r_E = 0.045$.

$$r_E = \left(1 + \frac{r}{m}\right)^m - 1$$

$$0.045 = \left(1 + \frac{r}{4}\right)^4 - 1$$

$$(1 + 0.045)^{1/4} = 1 + \frac{r}{4}.$$

$$0.04426 \approx r$$

For the 2-year CD, the nominal rate is about 4.43%.

For the 3-year CD, use $r_E = 0.0525$.

$$r_E = \left(1 + \frac{r}{m}\right)^m - 1$$

$$0.0525 = \left(1 + \frac{r}{4}\right)^4 - 1$$

$$(1 + 0.0525)^{1/4} = 1 + \frac{r}{4}.$$

$$0.05150 \approx r$$

For the 3-year CD, the nominal rate is about 5.15%.

69. Start by finding the effective rate for the CD offered by Centennial Bank of Fountain Valley. Use the formula for effective rate, with $r = 0.055$ and $m = 12$.

$$r_E = \left(1 + \frac{r}{m}\right)^m - 1$$
$$= \left(1 + \frac{0.055}{12}\right)^{12} - 1$$
$$\approx 0.05641$$

The effective rate is about 5.64%.

Since the CD offered by First Source Bank of South Bend is compounded annually, the quoted rate of 5.63% is also the effective rate.

Centennial Bank of Fountain Valley pays a slightly higher effective rate.

71. Use the formula for present value for compound interest with $A = 30,000$, $i = \frac{0.055}{4} = 0.01375$, and $n = 5(4) = 20$.

$$P = \frac{A}{(1+r)^n}$$
$$= \frac{30,000}{(1 + 0.01375)^{20}}$$
$$\approx 22,829.89$$

The present value is $22,829.89, or rounding up to the nearest cent (to make sure that the investment really grows to $30,000), $22,829.90. That is how much of the inherited $25,000 Phyllis should invest in order to have $30,000 for a down payment in 5 years.

73. To find the number of years it will take prices to double at 4% annual inflation, find n in the equation

$$2 = (1 + 0.04)^n,$$

which simplifies to

$$2 = (1.04)^n.$$

By trying various values of n, find that $n = 18$ is approximately correct, because

$$1.04^{18} \approx 2.0258 \approx 2.$$

Prices will double in about 18 yr.

75. To find the number of years it will be until the generating capacity will need to be doubled, find n in the equation

$$2 = (1 + 0.06)^n,$$

which simplifies to

$$2 = (1.06)^n.$$

By trying various values of n, find that $n = 12$ is approximately correct, because

$$1.06^{12} \approx 2.0122 \approx 2.$$

The generating capacity will need to be doubled in about 12 yr.

77. **(a)** To find this rate of return we must solve $14 = 1(1 + r)^{14}$.

$$14 = 1(1 + r)^{14}$$
$$1 + r = 14^{1/14} = 1.207$$
$$r = 0.207$$

The required rate of return is 20.7%.

(b) With an annual rate of return of 113%, in 14 years an initial investment of $1 million would be worth $1(1 + 1.13)^{14} = 39,565.299$ million dollars or about $39.6 billion.

5.2 Future Value of an Annuity

Your Turn 1

For the geometric series 4, 12, 36,... the common ratio r is $\frac{12}{4} = 3$. The first term is $a = 4$, and to find the sum of the first nine terms we set $n = 9$ and use the formula for the sum of the first n terms of a geometric series:

$$S_n = \frac{a(r^n - 1)}{r - 1}.$$
$$S_9 = \frac{4(3^9 - 1)}{3 - 1}$$
$$S_9 = 39,364$$

Your Turn 2

Use the formula for the future value of an ordinary annuity,

$$S = R\left[\frac{(1+i)^n - 1}{i}\right],$$

with $R = 250$, $i = 0.033/12 = 0.00275$, and $n = (11)(12) = 132$.

$$S = 250\left[\frac{(1 + 0.00275)^{132} - 1}{0.00275}\right] = 39719.98$$

The accumulated amount after 11 years is $39,719.98.

Your Turn 3

Use the formula for a sinking fund payment,

$$R = \frac{Si}{(1+r)^n - 1},$$

with $S = 13{,}500$, $i = 0.0375/4 = 0.009375$, and $n = (4)(14) = 56$.

$$R = \frac{(13{,}500)(0.009375)}{(1+0.009375)^{56} - 1}$$

$$R = 184.41$$

The quarterly payment will be $184.41.

Your Turn 4

Use the formula for the future value of an annuity due,

$$S = R\left[\frac{(1+i)^{n+1} - 1}{i}\right] - R,$$

with $R = 325$, $i = 0.033/12 = 0.0025$, and $n = (12)(5) = 60$.

$$S = 325\left[\frac{(1+0.00275)^{61} - 1}{0.00275}\right] - 325 = 21{,}227.66$$

The future value of this annuity due is $21,227.66.

5.2 Exercises

1. $a = 3; r = 2$

The first five terms are

$$3, 3(2), 3(2)^2, 3(2)^3, 3(2)^4$$

or

$$3, 6, 12, 24, 48.$$

The fifth term is 48.

Or, use the formula $a_n = ar^{n-1}$ with $n = 5$.

$$a_5 = ar^{5-1} = 3(2)^4 = 3(16) = 48$$

3. $a = -8; r = 3; n = 5$

$$a_5 = ar^{5-1} = -8(3)^4 = -8(81) = -648$$

The fifth term is -648.

5. $a = 1; r = -3; n = 5$

$$a_5 = ar^{5-1} = 1(-3)^4 = 81$$

The fifth term is 81.

7. $a = 256; r = \frac{1}{4}; n = 5$

$$a_5 = ar^{5-1} = 256\left(\frac{1}{4}\right)^4 = 256\left(\frac{1}{256}\right) = 1$$

The fifth term is 1.

9. $a = 1; r = 2; n = 4$

To find the sum of the first 4 terms, S_4, use the formula for the sum of the first n terms of a geometric sequence.

$$S_n = \frac{a(r^n - 1)}{r - 1}$$

$$S_4 = \frac{1(2^4 - 1)}{2 - 1} = \frac{16 - 1}{1} = 15$$

11. $a = 5; r = \frac{1}{5}; n = 4$

$$S_n = \frac{a(r^n - 1)}{r - 1}$$

$$S_4 = \frac{5\left[\left(\frac{1}{5}\right)^4 - 1\right]}{\frac{1}{5} - 1} = \frac{5\left(-\frac{624}{625}\right)}{-\frac{4}{5}}$$

$$= \frac{-\frac{624}{125}}{-\frac{4}{5}} = \left(-\frac{624}{125}\right)\left(-\frac{5}{4}\right) = \frac{156}{25}$$

13. $a = 128; r = -\frac{3}{2}; n = 4$

$$S_n = \frac{a(r^n - 1)}{r - 1}$$

$$S_4 = \frac{128\left[\left(-\frac{3}{2}\right)^4 - 1\right]}{-\frac{3}{2} - 1} = \frac{128\left(\frac{65}{16}\right)}{-\frac{5}{2}}$$

$$= -208$$

17. $R = 100; i = 0.06; n = 4$

Use the formula for the future value of an ordinary annuity.

$$S = R\left[\frac{(1+i)^n - 1}{i}\right]$$

$$= 100\left[\frac{(1.06)^4 - 1}{0.06}\right]$$

$$= 100\left(\frac{1.262477 - 1}{0.06}\right)$$

$$\approx 437.46$$

The future value is $437.46.

19. $R = 25{,}000;\ i = 0.045;\ n = 36$

$$S = R\left[\frac{(1+i)^n - 1}{i}\right]$$
$$= 25{,}000\left[\frac{(1+0.045)^{36} - 1}{0.045}\right]$$
$$\approx 2{,}154{,}099.15$$

The future value is $\$2{,}154{,}099.15$.

21. $R = 9200$; 10% interest compounded semiannually for 7 yr

Interest of $\frac{10\%}{2} = 5\%$ is earned semiannually, so $i = 0.05$. In 7 yr, there are $7(2) = 14$ semiannual periods, so $n = 14$.

$$S = R\left[\frac{(1+i)^n - 1}{i}\right]$$
$$= 9200\left[\frac{(1.05)^{14} - 1}{0.05}\right]$$
$$\approx 180{,}307.41$$

The future value is $\$180{,}307.41$.

$\$9200$ is contributed in each of 14 periods. The total contribution is

$$\$9200(14) = \$128{,}800.$$

The amount from interest is

$$\$180{,}307.41 - 128{,}800 = \$51{,}507.41$$

23. $R = 800$; 6.51% interest compounded semiannually for 12 yr

Interest of $\frac{6.51\%}{2}$ is earned semiannually, so $i = \frac{0.0651}{2} = 0.03255$. In 12 yr, there are $12(2) = 24$ semiannual periods, so $n = 24$.

$$S = R\left[\frac{(1+i)^n - 1}{i}\right]$$
$$= 800\left[\frac{(1+0.03255)^{24} - 1}{0.03255}\right]$$
$$\approx 28{,}438.21$$

The future value is $\$28{,}438.21$.

$800 is contributed in each of 24 periods. The total contribution is

$$\$800(24) = \$19{,}200.$$

The amount from interest is

$$\$28{,}438.21 - 19{,}200 = \$9238.21.$$

25. $R = 12{,}000;\ i = \frac{0.048}{2} = 0.012;\ n = 16(4) = 64$

$$S = R\left[\frac{(1+i)^n - 1}{i}\right]$$
$$= 12{,}000\left[\frac{(1+0.012)^{64} - 1}{0.012}\right]$$
$$\approx 1{,}145{,}619.96$$

The future value is $\$1{,}145{,}619.96$.

$\$12{,}000$ is contributed in each of 64 periods. The total contribution is

$$\$12{,}000(64) = \$768{,}000.$$

The amount from interest is

$$\$1{,}145{,}619.96 - 768{,}000 = \$377{,}619.96.$$

29. Using the TMV Solver under the FINANCE menu on the TI-84 Plus calculator, set up the following input:

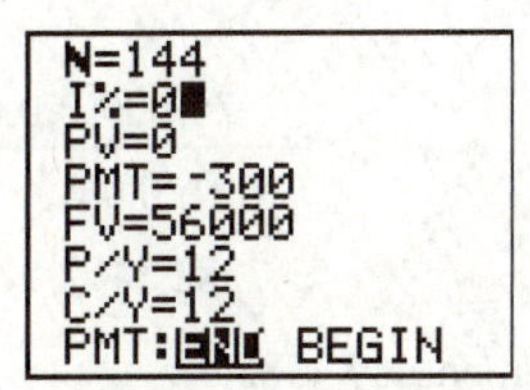

Put the cursor next to I% and press SOLVE to show the solution:

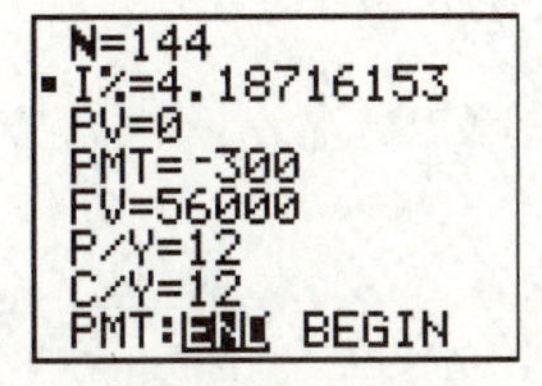

The required interest rate is 4.19%.

31. $S = \$10{,}000$; interest is 5% compounded annually; payments are made at the end of each year for 12 yr.

This is a sinking fund. Use the formula for an ordinary annuity with $S = 10{,}000$, $i = 0.05$, and $n = 12$ to find the value of R, the amount of each payment.

$$10{,}000 = Rs_{\overline{12}|0.05}$$
$$R = \frac{10{,}000}{s_{\overline{12}|0.05}} = \frac{10{,}000}{\frac{(1+0.05)^{12}-1}{0.05}} \approx 628.25$$

The required periodic payment is \$628.25.

33. $S = 8500;\ i = 0.08;\ n = 7$

$$R = \frac{S}{s_{\overline{n}|i}} = \frac{8500}{s_{\overline{7}|0.08}} = \frac{8500(0.08)}{(1+0.08)^7 - 1} \approx 952.62$$

The payment is \$952.62.

35. $S = 75{,}000;\ i = \frac{0.06}{2} = 0.03;\ n = 4\frac{1}{2}(2) = 9$

$$R = \frac{S}{s_{\overline{n}|i}} = \frac{75{,}000}{s_{\overline{9}|0.03}} = \frac{75{,}000(0.03)}{(1+0.03)^9 - 1} \approx 7382.54$$

The payment is \$7382.54.

37. \$65,000; money earns 7.5% compounded quarterly for $2\frac{1}{2}$ years

Thus, $i = \frac{0.075}{4} = 0.01875$ and $n = \left(2\frac{1}{2}\right)4 = 10.$

$$R = \frac{65{,}000}{s_{\overline{10}|0.01875}} = \frac{65{,}000}{\frac{(1+0.01875)^{10}-1}{0.01875}} \approx 5970.23$$

The amount of each payment is \$5970.23.

39. $R = 600;\ i = 0.06;\ n = 8$

To find the future value of an annuity due, use the formula for the future value of an ordinary annuity, but include one additional time period and subtract the amount of one payment.

$$S = R\left[\frac{(1+i)^{n+1} - 1}{i}\right] - R = 600\left[\frac{(1+0.06)^9 - 1}{0.06}\right] - 600 \approx 6294.79$$

The future value is \$6294.79.

41. $R = 16{,}000;\ i = 0.05;\ n = 7$

$$S = R\left[\frac{(1+i)^{n+1} - 1}{i}\right] - R = 16{,}000\left[\frac{(1+0.05)^8 - 1}{0.05}\right] - 16{,}000 \approx 136{,}785.74$$

The future value is \$136,785.74.

43. $R = 1000;\ i = \frac{0.0815}{2} = 0.04075;\ n = 9(2) = 18$

$$S = R\left[\frac{(1+i)^{n+1} - 1}{i}\right] - R = 1000\left[\frac{(1+0.04075)^{19} - 1}{0.04075}\right] - 1000 \approx 26{,}874.97$$

The future value is \$26,874.97.

\$1000 is contributed in each of 18 periods. The total contribution is

$$\$1000(18) = \$18{,}000.$$

The amount from interest is

$$\$26{,}874.97 - 18{,}000 = \$8874.97.$$

45. $R = 250;\ i = \frac{0.042}{2} = 0.0105;\ n = 12(4) = 48$

$$S = R\left[\frac{(1+i)^{n+1} - 1}{i}\right] - R = 250\left[\frac{(1+0.0105)^{49} - 1}{0.0105}\right] - 250 \approx 15{,}662.40$$

The future value is \$15,662.40.

\$250 is contributed in each of 48 periods. The total contribution is

$$\$250(48) = \$12{,}000.$$

The amount from interest is

$$15{,}662.40 - 12{,}000 = \$3662.40.$$

47. **(a)** $R = 12{,}000;\ i = 0.08;\ n = 9$

$$S = R\left[\frac{(1+i)^n - 1}{i}\right]$$
$$= 12{,}000\left[\frac{(1+0.08)^9 - 1}{0.08}\right]$$
$$\approx 149{,}850.69$$

The final amount is \$149,850.69.

(b) $R = 12{,}000;\ i = 0.06;\ n = 9$

$$S = R\left[\frac{(1+i)^n - 1}{i}\right]$$
$$= 12{,}000\left[\frac{(1+0.06)^9 - 1}{0.06}\right]$$
$$\approx 137{,}895.79$$

She will have \$137,895.79.

(c) The amount that would be lost is the difference between the two amounts in parts (a) and (b), which is

$$\$149{,}850.69 - 137{,}895.79 = \$11{,}954.90.$$

49. This is a future value problem with $R = 136.50$, $i = 0.048/12 = 0.004$, and $n = (12)(40) = 480$.

$$S = R\left[\frac{(1+i)^n - 1}{i}\right],$$
$$S = 136.50\left[\frac{(1.004)^{480} - 1}{0.004}\right]$$
$$S = 197{,}750.47$$

The account would be worth \$197,750.47.

51. From ages 50 to 60, we have an ordinary annuity with $R = 3000$, $i = \frac{0.05}{4} = 0.0125$, and $n = 10(4) = 40$. Use the formula for the future value of an ordinary annuity.

$$S = R\left[\frac{(1+i)^n - 1}{i}\right]$$
$$= 3000\left[\frac{(1.0125)^{40} - 1}{0.0125}\right]$$
$$\approx 154{,}468.67$$

At age 60, the value of the retirement account is \$154,468.67. This amount now earns 6.9% interest compounded monthly for 5 yr. Use the formula for compound amount with $P = 154{,}468.67$, $i = \frac{0.069}{12} = 0.00575$, and $n = 5(12) = 60$ to find the value of this amount after 5 yr.

$$A = P(1+i)^n$$
$$= 154{,}468.67(1.00575)^{60}$$
$$\approx 217{,}892.80$$

The value of the amount she withdraws from the retirement account will be \$217,892.80 when she reaches 65.

The deposits of \$300 at the end of each month into the mutual fund form another ordinary annuity. Use the formula for the future value of an ordinary annuity with $R = 300$, $i = \frac{0.069}{12} = 0.00575$, and $n = 12(5) = 60$.

$$S = R\left[\frac{(1+i)^n - 1}{i}\right]$$
$$= 300\left[\frac{(1.00575)^{60} - 1}{0.00575}\right]$$
$$\approx 21{,}422.37$$

The value of this annuity after 5 yr is \$21,422.37.

The total amount in the mutual fund account when the woman reaches age 65 will be

$$\$217{,}892.80 + 21{,}422.37 = \$239{,}315.17.$$

53. $R = 1000$, $i = \frac{0.08}{4} = 0.02$, and $n = 25(4) = 100$.

$$S = R\left[\frac{(1+i)^n - 1}{i}\right]$$
$$= 1000\left[\frac{(1+0.02)^{100} - 1}{0.02}\right]$$
$$\approx 312{,}232.31$$

There will be about \$312,232.31 in the IRA.

The total amount deposited was \$1000(100) = \$100,000. Thus, the amount of interest earned was

$$\$312{,}232.31 - 100{,}000 = \$212{,}232.31.$$

55. $R = 1000$, $i = \frac{0.10}{4} = 0.025$, and $n = 100$.

$$S = R\left[\frac{(1+i)^n - 1}{i}\right]$$
$$= 1000\left[\frac{(1+0.025)^{100} - 1}{0.025}\right]$$
$$\approx 432{,}548.65$$

There will be about \$432,548.65 in the IRA. The total amount deposited was \$100,000. Thus, the amount of interest earned was

$$\$432{,}548.65 - 100{,}000 = \$332{,}548.65.$$

57. This is a sinking fund with $S = 12{,}000$, $i = \frac{0.06}{2} = 0.03$, and $n = 4(2) = 8$.

$$\begin{aligned} R &= \frac{S}{s_{\overline{n}|i}} \\ &= \frac{12{,}000}{s_{\overline{8}|0.03}} \\ &= \frac{12{,}000(0.03)}{(1+0.03)^8 - 1} \\ &\approx 1349.48 \end{aligned}$$

Each payment should be \$1349.48.

59. $R = 80$; $i = \frac{0.025}{12}$; $n = 3(12) + 9 = 45$

Because the deposits are made at the beginning of each month, this is an annuity due.

$$\begin{aligned} S &= R\left[\frac{(1+i)^{n+1} - 1}{i}\right] - R \\ &= 80\left[\frac{\left(1 + \frac{0.025}{12}\right)^{46} - 1}{\frac{0.025}{12}}\right] - 80 \\ &\approx 3777.89 \end{aligned}$$

The account will have \$3777.89 in it.

61. For the first 8 yr, we have an annuity due with $R = 2435$, $i = \frac{0.06}{2} = 0.03$, and $n = 8(2) = 16$.

The amount on deposit after 8 yr is

$$\begin{aligned} S &= R\left[\frac{(1+i)^{n+1} - 1}{i}\right] - R \\ &= 2435\left[\frac{(1+0.03)^{17} - 1}{0.03}\right] - 2435 \\ &\approx 50{,}554.47. \end{aligned}$$

For the remaining 5 yr, this amount, \$50,554.47, earns compound interest at 6% compounded semi-annually. To find the final amount on deposit, use the formula for the compound amount with $P = 50{,}554.47$, $i = \frac{0.06}{2} = 0.03$, and $n = 5(2) = 10$.

$$\begin{aligned} A &= P(1+i)^n \\ &= 50{,}554.47(1.03)^{10} \\ &\approx 67{,}940.98 \end{aligned}$$

The final amount on deposit will be about \$67,940.98.

63. Let $x =$ the annual interest rate.

$$n = 20(12) = 240$$

Graph $y_1 = 147{,}126$ and

$$y_2 = 300\left[\frac{\left(1 + \frac{x}{12}\right)^{240} - 1}{\frac{x}{12}}\right].$$

The x-coordinate of the point of intersection is 0.06499984. Thus, the annual interest rate was about 6.5%.

65. (a) Compare the future amounts for an ordinary annuity with $R = 1{,}350{,}000$ and $i = 0.08$ to compound amounts with $P = 7{,}000{,}000$ and $i = .08$ for different values of n, starting with $n = 1$.

n	$S = R\left[\frac{(1+i)^n - 1}{i}\right]$	$A = P(1+i)^n$
1	$1{,}350{,}000\left[\frac{1.08 - 1}{0.08}\right] = \$1{,}350{,}000.00$	\$7,560,000.00
2	$1{,}350{,}000\left[\frac{(1.08)^2 - 1}{0.08}\right] = \$2{,}808{,}000.00$	\$8,164,800.00
3	$1{,}350{,}000\left[\frac{(1.08)^3 - 1}{0.08}\right] = \$4{,}382{,}640.00$	\$8,817,984.00
4	$1{,}350{,}000\left[\frac{(1.08)^4 - 1}{0.08}\right] = \$6{,}083{,}251.20$	\$9,523,422.72
5	$1{,}350{,}000\left[\frac{(1.08)^5 - 1}{0.08}\right] = \$7{,}919{,}911.30$	\$10,285,296.54
6	$1{,}350{,}000\left[\frac{(1.08)^6 - 1}{.08}\right] = \$9{,}903{,}504.20$	\$11,108,120.26
7	$1{,}350{,}000\left[\frac{(1.08)^7 - 1}{0.08}\right] = \$12{,}045{,}784.54$	\$11,996,769.88

After 7 yr, the investors would do better by winning the lottery.

(b) Repeat the calculations from part (a), but change the interest rate to $i = 0.12$.

n	$S = R\left[\dfrac{(1+i)^n - 1}{i}\right]$	$A = P(1+i)^n$
1	\$ 1,350,000.00	\$ 7,840,000.00
2	\$ 2,862,000.00	\$ 8,780,800.00
3	\$ 4,555,440.00	\$ 9,834,496.00
4	\$ 6,452,092.80	\$11,014,635.52
5	\$ 8,576,343.94	\$12,336,391.78
6	\$10,955,505.21	\$13,816,758.80
7	\$13,620,165.83	\$15,474,769.85
8	\$16,604,585.73	\$17,331,742.23
9	\$19,947,136.02	\$19,411,551.30

After 9 yr, the investors would do better by winning the lottery.

67. This exercise should be solved by graphing calculator or computer methods. The answers, which may vary slightly, are as follows.

(a) The amount of each interest payment is \$120.

(b) The amount of each payment is \$681.83, except the last payment, which is \$681.80. A table showing the amount in the sinking fund after each deposit is as follows.

Payment Number	Amount of Deposit	Interest Earned	Total
1	\$681.83	\$ 0	\$ 681.83
2	\$681.83	\$ 54.55	\$1418.21
3	\$681.83	\$113.46	\$2213.49
4	\$681.83	\$177.08	\$3072.40
5	\$681.81	\$245.79	\$4000.00

5.3 Present Value of an Annuity; Amortization

Your Turn 1

Use the formula for the present value of an annuity,

$$P = R\left[\frac{1-(1+i)^{-n}}{i}\right],$$

with $R = 500$, $i = 0.048/12 = 0.004$, and $n = (12)(5) = 60$.

$$P = 120\left[\frac{1-(1+0.004)^{-60}}{0.004}\right]$$

$$P = 6389.86$$

The present value is \$6389.86.

Your Turn 2

Compute the monthly payment using the formula

$$R = \frac{Pi}{1-(1+i)^{-n}}$$

with $P = 17{,}000$, $i = 0.054/12 = 0.0045$, and $n = 48$.

$$R = \frac{(17{,}000)(0.0045)}{1-(1+0.0045)^{-48}}$$

$$R = 394.59$$

The monthly car payment will be \$394.59.

Your Turn 3

The monthly payment will be given by

$$R = \frac{Pi}{1-(1+i)^{-n}}$$

with $R = 220{,}000$, $i = 0.07/12$, and $n = (12)(15) = 180$.

$$R = \frac{(220{,}000)\left(\dfrac{0.07}{12}\right)}{1-\left(1+\dfrac{0.07}{12}\right)^{-180}} = 1977.42$$

The monthly payment will be \$1977.42. The total of the 180 payments is

$$(1977.42)(180) = 355{,}935.60$$

To find the interest paid, subtract the principal from this payment total:

$$355{,}935.60 - 220{,}000 = 135{,}935.60$$

The total interest paid is \$135,935.60.

Your Turn 4

The only change in the calculation from Example 4 is that number of months remaining is 8 instead of 9, which gives a present value for the remaining balance of

$$88.8488\left[\frac{1-(1.01)^{-8}}{0.01}\right] = 679.84,$$

or \$679.84.

5.3 Exercises

3. Payments of $890 each year for 16 years at 6% compounded annually

Use the formula for present value of an annuity with $R = 890$, $i = 0.06$, and $n = 16$.

$$P = R\left[\frac{1-(1+i)^{-n}}{i}\right]$$

$$= 890\left[\frac{1-(1+0.06)^{-16}}{0.06}\right]$$

$$\approx 8994.25$$

The present value is $8994.25

5. Payments of $10,000 semiannually for 15 years at 5% compounded semiannually

Use the formula for present value of an annuity with $R = 10{,}000$, $i = \frac{0.05}{2} = 0.025$, and $n = 15(2) = 30$.

$$P = R\left[\frac{1-(1+i)^{-n}}{i}\right]$$

$$= 10{,}000\left[\frac{1-(1+0.025)^{-30}}{0.025}\right]$$

$$\approx 209{,}302.93$$

The present value is $209,302.93.

7. Payments of $15,806 quarterly for 3 years at 6.8% compounded quarterly

Use the formula for present value of an annuity with $R = 15{,}806$, $i = \frac{0.068}{4} = 0.017$, and $n = 3(4) = 12$.

$$P = R\left[\frac{1-(1+i)^{-n}}{i}\right]$$

$$= 15{,}806\left[\frac{1-(1+0.017)^{-12}}{0.017}\right]$$

$$\approx 170{,}275.47$$

The present value is $170,275.47.

9. 4% compounded annually

We want the present value, P, of an annuity with $R = 10{,}000$, $i = 0.04$, and $n = 15$.

$$P = R\left[\frac{1-(1+i)^{-n}}{i}\right]$$

$$= 10{,}000\left[\frac{1-(1.04)^{-15}}{0.04}\right]$$

$$\approx 111{,}183.87$$

The required lump sum is $111,183.87.

11. $P = 2500$, $i = \frac{0.06}{4} = 0.015$; $n = 6$

(a) To find the payment amount, use the formula for amortization payments.

$$R = \frac{Pi}{1-(1+i)^{-n}}$$

$$R = \frac{2500(0.015)}{1-(1+0.015)^{-6}}$$

$$\approx 438.81$$

Each payment is $438.81.

(b) To find the total payments, multiply the amount of one payment by $n = 6$.

$$438.81(6) = 2632.86$$

The total payments come out to $2632.86.

To find the total amount of interest paid, subtract the original loan amount from the total payments.

$$2632.86 - 2500 = 132.86$$

The total amount of interest paid is $132.86.

(c) Set $P = 2500$, $i = 0.015$, $n = 6$ and $R = 438.81$ and generate the following amortization table using software.

Payment Number	Amount of Payment	Interest for Period	Portion to Principal	Principal at End of Period
0	0.00	0.00	0.00	2500.00
1	438.81	37.50	401.31	2098.69
2	438.81	31.48	407.33	1691.36
3	438.81	25.37	413.44	1277.92
4	438.81	19.17	419.64	858.28
5	438.81	12.87	425.94	432.34
6	438.83	6.49	432.34	0.00

The sum of the Amount of Payment column gives the total payments, $2632.88.

The sum of the Interest column gives the total interest paid, $132.88.

13. $P = 90,000;\ i = 0.06;\ n = 12$

(a) To find the payment amount, use the formula for amortization payments.

$$R = \frac{Pi}{1-(1+i)^{-n}}$$

$$R = \frac{90,000(0.06)}{1-(1+0.06)^{-12}}$$

$$\approx 10,734.93$$

Each payment is $10,734.93.

(b) To find the total payments, multiply the amount of one payment by $n = 12$.

$$10734.93(12) = 128,819.16$$

The total payments come out to $128,819.16.

To find the total amount of interest paid, subtract the original loan amount from the total payments.

$$128,819.16 - 90,000 = 38,819.16$$

The total amount of interest paid is $38,819.16.

(c) Set $P = 90,000$, $i = 0.06$, $n = 12$ and $R = 10,734.93$ and generate the following amortization table using

Payment Number	Amount of Payment	Interest for Period	Portion to Principal	Principal at End of Period
0	0.00	0.00	0.00	90000.00
1	10734.93	5400.00	5334.93	84665.07
2	10734.93	5079.90	5655.03	79010.04
3	10734.93	4740.60	5994.33	73015.72
4	10734.93	4380.94	6353.99	66661.73
5	10734.93	3999.70	6735.23	59926.50
6	10734.93	3595.59	7139.34	52787.16
7	10734.93	3167.23	7567.70	45219.46
8	10734.93	2713.17	8021.76	37197.70
9	10734.93	2231.86	8503.07	28694.63
10	10734.93	1721.68	9013.25	19681.38
11	10734.93	1180.88	9554.05	10127.33
12	10734.97	607.64	10127.33	0.00

The sum of the Amount of Payment column gives the total payments, $128,819.20.

The sum of the Interest column gives the total interest paid, $38,819.20.

15. $P = 7400;\ i = \frac{0.062}{2} = 0.031;\ n = 18$

(a) To find the payment amount, use the formula for amortization payments.

$$R = \frac{Pi}{1-(1+i)^{-n}}$$

$$R = \frac{7400(0.031)}{1-(1+0.031)^{-18}}$$

$$\approx 542.60$$

Each payment is $542.60.

(b) To find the total payments, multiply the amount of one payment by $n = 18$.

$$542.60(18) = 9766.80$$

The total payments come out to $9766.80.

To find the total amount of interest paid, subtract the original loan amount from the total payments

$$9766.80 - 7400 = 2366.80$$

The total amount of interest paid is $2366.80.

(c) Set $P = 7400$, $i = 0.031$, $n = 18$ and $R = 542.60$ and generate the following amortization table using software.

Payment Number	Amount of Payment	Interest for Period	Portion to Principal	Principal at End of Period
0	0.00	0.00	0.00	7400.00
1	542.60	229.40	313.20	7086.80
2	542.60	219.69	322.91	6763.89
3	542.60	209.68	332.92	6430.97
4	542.60	199.36	343.24	6087.73
5	542.60	188.72	353.88	5733.85
6	542.60	177.75	364.85	5369.00
7	542.60	166.44	376.16	4992.84
8	542.60	154.78	387.82	4605.02
9	542.60	142.76	399.84	4205.17
10	542.60	130.36	412.24	3792.93
11	542.60	117.58	425.02	3367.91
12	542.60	104.41	438.19	2929.72
13	542.60	90.82	451.78	2477.94
14	542.60	76.82	465.78	2012.16
15	542.60	62.38	480.22	1531.93
16	542.60	47.49	495.11	1036.82
17	542.60	32.14	510.46	526.37
18	542.68	16.32	526.36	0.00

The sum of the Amount of Payment column gives the total payments, $9766.88.

The sum of the Interest column gives the total interest paid, $2366.88.

17. Using the first method in Example 4 and carrying more places in the payment amount we have

$$R = \frac{(90,000)(0.06)}{1-(1+0.06)^{-12}}$$
$$R = 10,734.9326$$

There are $12-3=9$ payments left, so the amount to pay off the loan is

$$10,734.9326\left[\frac{1-(1.06)^{-9}}{0.06}\right] = 73,015.71$$

or \$73,015.71.

19. Using the first method in Example 4 and carrying more places in the payment amount we have

$$R = \frac{(7400)(0.031)}{1-(1+0.031)^{-18}}$$
$$R = 542.6035$$

There are $18-6=12$ payments left, so the amount to pay off the loan is

$$542.6035\left[\frac{1-(1.031)^{-12}}{0.031}\right] = 5368.98$$

or \$5368.98.

21. Look at the entry for payment number 4 under the heading "Interest for Period." The amount of interest included in the fourth payment is \$7.61.

23. To find the amount of interest paid in the first 4 mo of the loan, add the entries for payment 1, 2, 3, and 4 under the heading "Interest for Period."

$$\$10.00 + 9.21 + 8.42 + 7.61 = \$35.24$$

In the first 4 mo of the loan, \$35.24 of interest is paid.

25. First, find the value of the annuity at the end of 8 yr. Use the formula for future value of an ordinary annuity.

$$S = R\left[\frac{(1+i)^n - 1}{i}\right]$$
$$= 1000\left[\frac{(1+0.06)^8 - 1}{0.06}\right]$$
$$\approx 9897.47$$

The future value of the annuity is \$9897.47.

Now find the present value of \$9897.47 at 5% compounded annually for 8 yr. Use the formula for present value for compound interest.

$$P = \frac{A}{(1+i)^n} = \frac{9897.47}{(1.05)^8} \approx 6699.00$$

The required amount is \$6699.

27. $P = 199,000;\ i = \frac{0.0701}{12};\ n = 25(12) = 300$

To find the payment amount, use the formula for amortization payments.

$$R = \frac{Pi}{1-(1+i)^{-n}}$$
$$R = \frac{199,000\left(\frac{0.0701}{12}\right)}{1-\left(1+\frac{0.0701}{12}\right)^{-300}}$$
$$\approx 1407.76$$

Each payment is \$1407.76.

To find the total payment, multiply the amount of one payment by $n=300$.

$$1407.76(300) = 422,328$$

The total payments come out to \$422,328.

To find the total amount of interest paid, subtract the original loan amount from the total payments.

$$422,328 - 199,000 = 223,328$$

The total amount of interest paid is \$223,328.

29. $P = 253,000,\ i = \frac{0.0645}{12},\ n = 30(12) = 360$

To find the payment amount, use the formula for amortization payments.

$$R = \frac{Pi}{1-(1+i)^{-n}}$$
$$R = \frac{253,000\left(\frac{0.0645}{12}\right)}{1-\left(1+\frac{0.0645}{12}\right)^{-360}}$$
$$\approx 1590.82$$

Each payment is \$1590.82.

To find the total payments, multiply the amount of one payment by $n = 360$.

$$1590.82(360) = 572,695.20$$

The total payments come out to \$572,695.20.

To find the total amount of interest paid, subtract the original loan amount from the total payments.

$$572{,}695.20 - 253{,}000 = 319{,}695.20$$

The total amount of interest paid is \$319,695.20.

31. **(a)** Solve as in Example 6:

$$90{,}000 = 16{,}000\left(\frac{1-1.06^{-n}}{0.06}\right)$$

$$\left(\frac{90{,}000}{16{,}000}\right)(0.06) = 0.3375$$

$$0.3375 = 1 - 1.06^{-n}$$

$$1.06^{-n} = 1 - 0.3375 = 0.6625$$

$$n = -\frac{\log(0.6625)}{\log(1.06)} = 7.066$$

Rounding to the next whole year, the loan will take 8 years to pay off.

(b) Use software to build an amortization table with $P = 90{,}000$, $i = 0.06$, $n = 8$, and $R = 16{,}000$.

Payment Number	Amount of Payment	Interest for Period	Portion to Principal	Principal at End of Period
0	0.00	0.00	0.00	90000.00
1	16000.00	5400.00	10600.00	79400.00
2	16000.00	4764.00	11236.00	68164.00
3	16000.00	4089.84	11910.16	56253.84
4	16000.00	3375.23	12624.77	43629.07
5	16000.00	2617.74	13382.26	30246.81
6	16000.00	1814.81	14185.19	16061.62
7	16000.00	963.70	15036.30	1025.32
8	1086.84	61.52	1025.32	0.00

The amortization table shows that the total of payments is \$113,068.84.

(c) Subtracting this value from the answer to Exercise 13(c), we find that the savings in interest is $128{,}819.20 - 113{,}086.84 = 15{,}732.36$ or \$15,732.36.

33. **(a)** Solve as in Example 6:

$$7400 = 850\left(\frac{1-1.031^{-n}}{0.031}\right)$$

$$\left(\frac{7400}{850}\right)(0.031) = 0.270$$

$$0.270 = 1 - 1.031^{-n}$$

$$1.031^{-n} = 1 - 0.270 = 0.730$$

$$n = -\frac{\log(0.73)}{\log(1.031)} = 10.309$$

Rounding to the next half year, the loan will take 11 semiannual periods to pay off.

(b) Use software to build an amortization table with $P = 7400$, $i = 0.031$, $n = 11$, and $R = 850$.

Payment Number	Amount of Payment	Interest for Period	Portion to Principal	Principal at End of Period
0	0.00	0.00	0.00	7400.00
1	850.00	229.40	620.60	6779.40
2	850.00	210.16	639.84	6139.56
3	850.00	190.33	659.67	5479.89
4	850.00	169.88	680.12	4799.76
5	850.00	148.79	701.21	4098.56
6	850.00	127.06	722.94	3375.61
7	850.00	104.64	745.36	2630.26
8	850.00	81.54	768.46	1861.79
9	850.00	57.72	792.28	1069.51
10	850.00	33.15	816.85	252.66
11	260.50	7.83	252.67	0.00

The amortization table shows that the total of payments is \$8760.50.

(c) Subtracting this value from the answer to Exercise 15(c), we find that the savings in interest is $9766.88 - 8760.50 = 1006.38$ or \$1006.38.

35. From Example 3, $P = 220{,}000$ and $i = \frac{0.06}{12} = 0.005$. For a 15-year loan, use $n = 15(12) = 180$.

$$R = \frac{Pi}{1-(1+i)^{-n}}$$

$$= \frac{220{,}000(0.005)}{1-(1+0.005)^{-180}}$$

$$\approx 1856.49$$

The monthly payments would be \$1856.49. The family makes 180 payments of \$1856.49 each, for a total of \$334,168.20. Since the amount of the loan was \$220,000, the total interest paid is

$$334{,}168.20 - 220{,}000 = 114{,}168.20.$$

The total amount of interest paid is \$114,168.20.

The payments for the 15-year loan are

$$1856.49 - \$1319.01 = \$537.48$$

more than those for the 30-year loan in Example 3. However, the total interest paid is

$$254{,}843.60 - \$114{,}168.20 = \$140{,}675.40$$

less than for the 30-year loan in Example 3.

37. **(a)** $P = 14{,}000,\ i = \frac{0.07}{12},\ n = 4(12) = 48$

$$\begin{aligned} R &= \frac{Pi}{1-(1+i)^{-n}} \\ &= \frac{14{,}000\left(\frac{0.07}{12}\right)}{1-\left(1+\frac{0.07}{12}\right)^{-48}} \\ &\approx 335.25 \end{aligned}$$

The amount of each payment is \$335.25.

(b) 48 payments of \$335.25 are made, and $48(\$335.25) = \$16{,}092$. The total amount of interest Le will pay is $\$16{,}092 - \$14{,}000 = \$2092$.

39. **(a)** Compute the monthly payment using the formula

$$R = \frac{Pi}{1-(1+i)^{-n}}$$

with $P = 30{,}000$, $i = 0.009/12$, and $n = 36$.

$$R = \frac{(30{,}000)\left(\frac{0.009}{12}\right)}{1-\left(1+\frac{0.009}{12}\right)^{-36}}$$

$$R = 844.95$$

The monthly payment is \$844.95 and the total paid will be $(844.95)(36) = 30{,}418.20$ or $\$30{,}418.20$.

(b) Compute the monthly payment with $P = 27{,}750$, $i = 0.0633/12$, and $n = 48$.

$$R = \frac{(27{,}750)\left(\frac{0.0633}{12}\right)}{1-\left(1+\frac{0.0633}{12}\right)^{-48}}$$

$$R = 655.92$$

The monthly payment is \$655.92 and the total paid will be $(655.92)(48) = 31{,}484.16$ or $\$31{,}484.16$.

41. For parts (a) and (b), if \$1 million is divided into 20 equal payments, each payment is \$50,000.

(a) $i = 0.05, n = 20$

$$\begin{aligned} P &= R\left[\frac{1-(1+i)^{-n}}{i}\right] \\ &= 50{,}000\left[\frac{1-(1+0.05)^{-20}}{0.05}\right] \\ &\approx 623{,}110.52 \end{aligned}$$

The present value is \$623,110.52.

(b) $i = 0.09, n = 20$

$$\begin{aligned} P &= R\left[\frac{1-(1+i)^{-n}}{i}\right] \\ &= 50{,}000\left[\frac{1-(1+0.09)^{-20}}{0.09}\right] \\ &\approx 456{,}427.28 \end{aligned}$$

The present value is \$456,427.28.

For parts (c) and (d), if \$1 million is divided into 25 equal payments, each payment is \$40,000.

(c) $i = 0.05, n = 25$

$$\begin{aligned} P &= R\left[\frac{1-(1+i)^{-n}}{i}\right] \\ &= 40{,}000\left[\frac{1-(1+0.05)^{-25}}{0.05}\right] \\ &\approx 563{,}757.78 \end{aligned}$$

The present value is \$563,757.78.

(d) $i = 0.09, n = 25$

$$\begin{aligned} P &= R\left[\frac{1-(1+i)^{-n}}{i}\right] \\ &= 40{,}000\left[\frac{1-(1+0.09)^{-25}}{0.09}\right] \\ &\approx 392{,}903.18 \end{aligned}$$

The present value is \$392,903.18.

43. Compute the monthly payment using the formula

$$R = \frac{Pi}{1-(1+i)^{-n}}$$

with $P = 55{,}000$, $i = 0.068/12$, and $n = 300$.

$$R = \frac{(55{,}000)\left(\frac{0.068}{12}\right)}{1-\left(1+\frac{0.068}{12}\right)^{-300}}$$

$$R = 381.74$$

The monthly payment will be \$381.74 and the total paid will be $(381.74)(300) = 114{,}522.00$ or $\$114{,}522$. The interest paid will be $\$114{,}522 - \$55{,}000 = \$59{,}522$.

45. $P = 110{,}000$, $i = \frac{0.08}{2} = 0.04$, $n = 9$

$$R = \frac{110{,}000}{a_{\overline{9}|0.04}} \approx \$14{,}794.23$$

is the amount of each payment.

Of the first payment, the company owes interest of

$$I = Prt = 110{,}000(0.08)\left(\tfrac{1}{2}\right) = \$4400.$$

Therefore, from the first payment, \$4400 goes to interest, and the balance.

$$\$14{,}794.23 - 4400 = \$10{,}394.23,$$

goes to principal. The principal at the end of this period is

$$\$110{,}000 - 10{,}394.23 = \$99{,}605.77.$$

The interest for the second payment is

$$I = Prt = 99{,}605.77(0.08)\left(\tfrac{1}{2}\right) \approx \$3984.23$$

Of the second payment, \$3984.23 goes to interest and

$$\$14{,}794.23 - 3984.23 = \$10{,}810.00$$

goes to principal. Continue in this fashion to complete the amortization schedule for the first four payments.

Payment Number	Amount of Payment	Interest for Period	Portion to Principal	Principal at End of Period
0	—	—	—	\$110,000.00
1	\$14,794.23	\$4400.00	\$10,394.23	\$ 99,605.77
2	\$14,794.23	\$3984.23	\$10,810.00	\$ 88,795.77
3	\$14,794.23	\$3551.83	\$11,242.40	\$ 77,553.37
4	\$14,794.23	\$3102.13	\$11,692.10	\$ 65,861.27

47. \$150,000 is the future value of an annuity over 79 yr compounded quarterly. So, there are 79(4) = 316 payment periods.

(a) The interest per quarter is $\frac{5.25\%}{4} = 1.3125\%$. Thus, $S = 150{,}000$, $n = 316$, $i = 0.013125$, and we must find the quarterly payment R in the formula

$$S = R\left[\frac{(1+i)^n - 1}{i}\right]$$

$$150{,}000 = R\left[\frac{(1.013125)^{316} - 1}{0.013125}\right]$$

$$R \approx 32.4923796$$

She would have to put \$32.49 into her savings at the end of every three months.

(b) For a 2% interest rate, the interest per quarter is $\frac{2\%}{4} = 0.5\%$. Thus, $S = 150{,}000$, $n = 316$, $i = 0.005$, and we must find the quarterly payment R in the formula

$$S = R\left[\frac{(1+i)^n - 1}{i}\right]$$

$$150{,}000 = R\left[\frac{(1.005)^{316} - 1}{0.005}\right]$$

$$R \approx 195.5222794$$

She would have to put \$195.52 into her savings at the end of every three months.

For a 7% interest rate, the interest per quarter is $\frac{7\%}{4} = 1.75\%$. Thus, $S = 150{,}000$, $n = 316$, $i = 0.0175$, and we must find the quarterly payment R in the formula

$$S = R\left[\frac{(1+i)^n - 1}{i}\right]$$

$$150{,}000 = R\left[\frac{(1.0175)^{316} - 1}{0.0175}\right]$$

$$R \approx 10.9663932$$

She would have to put \$10.97 into her savings at the end of every three months.

49. Throughout this exercise, $i = \frac{0.065}{12}$ and $P =$ the total amount financed, which is

$$\$285{,}000 - 60{,}000 = \$225{,}000.$$

(a) $n = 15(12) = 180$

$$R = \frac{Pi}{1-(1+i)^{-n}} = \frac{225{,}000\left(\frac{0.065}{12}\right)}{1-\left(1+\frac{0.065}{12}\right)^{-180}} \approx 1959.99$$

The monthly payment is \$1959.99.

$$\text{Total payments} = 180(\$1959.99) = \$352{,}798.20$$

$$\text{Total interest} = \$352{,}798.20 - 225{,}000 = \$127{,}798.20$$

(b) $n = 20(12) = 240$

$$R = \frac{Pi}{1-(1+i)^{-n}} = \frac{225{,}000\left(\frac{0.065}{12}\right)}{1-\left(1+\frac{0.065}{12}\right)^{-240}} \approx 1677.54$$

The monthly payment is \$1677.54.

$$\text{Total payments} = 240(\$1677.54) = \$402{,}609.60$$
$$\text{Total interest} = \$402{,}609.60 - 225{,}000 = \$177{,}609.60$$

(c) $n = 25(12) = 300$

$$R = \frac{Pi}{1-(1+i)^{-n}} = \frac{225{,}000\left(\frac{0.065}{12}\right)}{1-\left(1+\frac{0.065}{12}\right)^{-300}} \approx 1519.22$$

The monthly payment is \$1519.22.

$$\text{Total payments} = 300(\$1519.22) = \$455{,}766$$
$$\text{Total interest} = \$455{,}766 - 225{,}000 = \$230{,}766$$

(d) Graph

$$y_1 = 1677.54\left[\frac{1-\left(1+\frac{0.065}{12}\right)^{-(240-x)}}{\frac{0.065}{12}}\right] \text{ and}$$

$$y_2 = \frac{285{,}000 - 60{,}000}{2}.$$

The x-coordinate of the point of intersection is 156.44167, which rounds up to 157. Half the loan will be paid after 157 payments.

51. $P = 150{,}000$, $i = \frac{0.082}{12}$, and $n = 30(12) = 360$.

$$R = \frac{Pi}{1-(1+i)^{-n}} = \frac{150{,}000\left(\frac{0.082}{12}\right)}{1-\left(1+\frac{0.082}{12}\right)^{-360}} \approx 1121.63$$

The monthly payment is \$1121.63.

$$\text{Total payments} = 360(\$1121.63) = \$403{,}786.80$$
$$\text{Total interest} = \$403{,}786.80 - 150{,}000 = \$253{,}786.80$$

(b) 15 years of payments means $15(12) = 180$ payments.

$$y_{15} = 1121.63\left[\frac{1-\left(1+\frac{0.082}{12}\right)^{-(360-180)}}{\frac{0.082}{12}}\right] \approx 115{,}962.66$$

The unpaid balance after 15 years is approximately \$115,962.66.

The total of the remaining 180 payments is

$$180(\$1121.63) = \$201{,}893.40.$$

(c) The unpaid balance from part (b) is the new loan amount. Now $P = 115{,}962.66$, $i = \frac{0.065}{12}$, and again $n = 30(12) = 360$.

$$R = \frac{Pi}{1-(1+i)^{-n}} = \frac{115{,}962.66\left(\frac{0.065}{12}\right)}{1-\left(1+\frac{0.065}{12}\right)^{-360}} \approx 732.96$$

The new monthly payment would be \$732.96.

$$\text{Total payments} = 360(\$732.96) + \$3400 = \$267{,}265.60$$

(d) Again the unpaid balance from part (b) is the new loan amount. Again $P = 115{,}962.66$ and $i = \frac{0.065}{12}$, and this time $n = 15(12) = 180$.

$$R = \frac{Pi}{1-(1+i)^{-n}} = \frac{115{,}962.66\left(\frac{0.065}{12}\right)}{1-\left(1+\frac{0.065}{12}\right)^{-180}} \approx 1010.16$$

The new monthly payment would be \$1010.16.

$$\text{Total payments} = 180(\$1010.16) + \$4500 = \$186{,}328.80$$

53. This is just like a sinking fund in reverse.

(a) $P = 150{,}000$, $i = \frac{0.06}{2} = 0.03$, $n = 2(5) = 10$

$$R = \frac{Pi}{1-(1+i)^{-n}} = \frac{150{,}000(.03)}{1-(1+0.03)^{-10}} \approx 17{,}584.58$$

The amount of each withdrawal is \$17,584.58.

(b) $P = 150,000,\ i = \frac{0.06}{2} = 0.03,\ n = 2(6) = 12$

$$R = \frac{Pi}{1 - (1 + i)^{-n}}$$
$$= \frac{150,000(0.03)}{1 - (1 + 0.03)^{-12}}$$
$$\approx 15,069.31$$

If the money must last 6 yr, the amount of each withdrawal is \$15,069.31.

55. This exercise should be solved by graphing calculator or computer methods. The amortization schedule, which may vary slightly, is as follows.

Payment Number	Amount of Payment	Interest for Period	Portion to Principal	Principal at End of Period
0	—	—	—	\$4836.00
1	\$585.16	\$175.31	\$409.85	\$4426.15
2	\$585.16	\$160.45	\$424.71	\$4001.43
3	\$585.16	\$145.05	\$440.11	\$3561.32
4	\$585.16	\$129.10	\$456.06	\$3105.26
5	\$585.16	\$112.57	\$472.59	\$2632.67
6	\$585.16	\$ 95.43	\$489.73	\$2142.94
7	\$585.16	\$ 77.68	\$507.48	\$1635.46
8	\$585.16	\$ 59.29	\$525.87	\$1109.59
9	\$585.16	\$ 40.22	\$544.94	\$ 564.65
10	\$585.12	\$ 20.47	\$564.65	\$ 0.00

57. (a) Here $R = 1000$ and $i = 0.04$ and we have

$$P = \frac{R}{i} = \frac{100}{0.04} = 25,000$$

Therefore, the present value of the perpetuity is \$25,000.

(b) Here $R = 600$ and $i = \frac{0.06}{4} = 0.015$ and we have

$$P = \frac{R}{i} = \frac{600}{0.015} = 40,000$$

Therefore, the present value of the perpetuity is \$40,000.

Chapter 5 Review Exercises

1. True

2. False: The ratios of successive pairs of terms are not constant: For example, $\frac{4}{2} = 2$ but $\frac{6}{4} = 1.5$.

3. True

4. False: Both payments and interest on the accumulated value are added to a sinking fund at the end of each time period, so the value increases over time.

5. True

6. True

7. True

8. False: The effective rate formula gives an interest rate, not a present value.

9. False: The correct expression is

$$25,000\left[\frac{0.05/12}{1 - (1 + 0.05/12)^{-72}}\right].$$

10. True

11. $I = Prt$

$$= 15,903(0.06)\left(\frac{8}{12}\right)$$
$$= 636.12$$

The simple interest is \$636.12.

13. $I = Prt$

$$= 42,368(0.0522)\left(\frac{7}{12}\right)$$
$$\approx 1290.11$$

The simple interest is \$1290.11.

15. For a given amount of money at a given interest rate for a given time period greater than 1, compound interest produces more interest than simple interest.

17. \$19,456.11 at 8% compounded semiannually for 7 yr

Use the formula for compound amount with $P = 19,456.11$, $i = \frac{0.08}{2} = 0.04$, and $n = 7(2) = 14$.

$$A = P(1 + i)^n$$
$$= 19,456.11(1.04)^{14}$$
$$\approx 33,691.69$$

The compound amount is \$33,691.69.

19. \$57,809.34 at 6% compounded quarterly for 5 yr

Use the formula for compound amount with $P = 57{,}809.34$, $i = \frac{0.06}{4} = 0.015$, and $n = 5(4) = 20$.

$$\begin{aligned} A &= P(1+i)^n \\ &= 57{,}809.34(1.015)^{20} \\ &\approx 77{,}860.80 \end{aligned}$$

The compound amount is \$77,860.80.

21. \$12,699.36 at 5% compounded semiannually for 7 yr

Here $P = 12{,}699.36$, $i = \frac{0.05}{2} = 0.025$, and $n = 7(2) = 14$. First find the compound amount.

$$\begin{aligned} A &= P(1+i)^n \\ &= 12{,}699.36(1.025)^{14} \\ &\approx 17{,}943.86 \end{aligned}$$

The compound amount is \$17,943.86.

To find the amount of interest earned, subtract the initial deposit from the compound amount. The interest earned is

$$\$17{,}943.86 - 12{,}699.36 = \$5244.50.$$

23. \$34,677.23 at 4.8% compounded monthly for 32 mo

Here $P = 34{,}677.23$, $i = \frac{0.048}{12} = 0.004$, and $n = 32$.

$$\begin{aligned} A &= P(1+i)^n \\ &= 34{,}677.23(1.004)^{32} \\ &\approx 39{,}402.45 \end{aligned}$$

The compound amount is \$39,402.45

The interest earned is

$$\$39{,}402.45 - 34{,}677.23 = \$4725.22.$$

25. \$42,000 in 7 yr, 6% compounded monthly

Use the formula for present value for compound interest with $A = 42{,}000$, $i = \frac{0.06}{12} = 0.005$, and $n = 7(12) = 84$.

$$P = \frac{A}{(1+i)^n} = \frac{42{,}000}{(1.005)^{84}} \approx 27{,}624.86$$

The present value is \$27,624.86.

27. \$1347.89 in 3.5 yr, 6.77% compounded semiannually

Use the formula for present value for compound interest with $A = 1347.89$, $i = \frac{0.0677}{2} = 0.03385$, and $n = 3.5(2) = 7$.

$$P = \frac{A}{(1+i)^n} = \frac{1347.89}{(1.03385)^7} \approx 1067.71$$

The present value is \$1067.71.

29. $a = 2$; $r = 3$

The first five terms are

$$2,\ 2(3),\ 2(3)^2,\ 2(3)^3, \text{ and } 2(3)^4,$$

or

$$2,\ 6,\ 18,\ 54, \text{ and } 162.$$

31. $a = -3$; $r = 2$

To find the sixth term, use the formula $a_n = ar^{n-1}$ with $a = -3$, $r = 2$, and $n = 6$.

$$a_6 = ar^{6-1} = -3(2)^5 = -3(32) = -96$$

33. $a = -3$; $r = 3$

To find the sum of the first 4 terms of this geometric sequence, use the formula $S_n = \frac{a(r^n-1)}{r-1}$ with $n = 4$.

$$S_4 = \frac{-3(3^4-1)}{3-1} = \frac{-3(80)}{2} = \frac{-240}{2} = -120$$

35. $s_{\overline{n}|i} = \frac{(1+i)^n - 1}{i}$

$$s_{\overline{30}|0.02} = \frac{(1.02)^{30} - 1}{0.02} \approx 40.56808$$

39. $R = 1288$, $i = 0.04$, $n = 14$

This is an ordinary annuity.

$$\begin{aligned} S &= Rs_{\overline{n}|i} \\ S &= 1288s_{\overline{14}|0.04} \\ &= 1288\left[\frac{(1+0.04)^{14} - 1}{0.04}\right] \\ &\approx 23{,}559.98 \end{aligned}$$

The future value is \$23,559.98.

The total amount deposited is $\$1288(14) = \$18{,}032$.

Thus, the amount of interest is

$$\$23{,}559.98 - 18{,}032 = \$5527.98.$$

41. $R = 233,\ i = \frac{0.048}{12} = 0.004,\ n = 4(12) = 48$

This is an ordinary annuity.

$$S = R\left[\frac{(1+i)^n - 1}{i}\right]$$

$$S = 233\left[\frac{(1.004)^{48} - 1}{0.004}\right]$$

$$\approx 12{,}302.78$$

The future value is \$12,302.78.

The total amount deposited is \$233(48) = \$11,184.

Thus, the amount of interest is

$$\$12{,}302.78 - 11{,}184 = \$1118.78.$$

43. $R = 11{,}900,\ i = \frac{0.06}{12} = 0.005,\ n = 13$

This is an annuity due, so we use the formula for future value of an ordinary annuity, but include one additional time period and subtract the amount of one payment.

$$S = R\left[\frac{(1+i)^{n+1} - 1}{i}\right] - R$$

$$= 11{,}900\left[\frac{(1.005)^{14} - 1}{0.005}\right] - 11{,}900$$

$$\approx 160{,}224.29$$

The future value is \$160,224.29.

The total amount deposited is \$11,900(13) = \$154,700.

Thus, the amount of interest is

$$\$160{,}224.29 - 154{,}700 = \$5524.29.$$

45. \$6500; money earns 5% compounded annually; 6 annual payments

$$S = 6500,\ i = 0.05,\ n = 6$$

Let R be the amount of each payment.

$$S = Rs_{\overline{n}|i}$$

$$R = \frac{6500}{s_{\overline{6}|0.05}}$$

$$= \frac{6500(0.05)}{(1.05)^6 - 1}$$

$$\approx 955.61$$

The amount of each payment is \$955.61.

47. \$233,188; money earns 5.2% compounded quarterly for $7\frac{3}{4}$ years.

$$S = 233{,}188,\ i = \frac{0.052}{4} = 0.013,\ n = \left(7\tfrac{3}{4}\right)(4) = 31$$

Let R be the amount of each payment.

$$S = Rs_{\overline{n}|i}$$

$$R = \frac{233{,}188}{s_{\overline{31}|0.013}}$$

$$= \frac{233{,}188(0.013)}{(1.013)^{31} - 1}$$

$$\approx 6156.14$$

The amount of each payment is \$6156.14.

49. Deposits of \$850 annually for 4 years at 6% compounded annually

Use the formula for the present value of an annuity with $R = 850$, $i = 0.06$, and $n = 4$.

$$P = R\left[\frac{1 - (1+i)^{-n}}{i}\right]$$

$$= 850\left[\frac{1 - (1+0.06)^{-4}}{0.06}\right]$$

$$\approx 2945.34$$

The present value is \$2945.34.

51. Deposits of \$4210 semiannually for 8 years at 4.2% compounded annually

Use the formula for the present value of an annuity with $R = 4210$, $i = \frac{0.042}{2} = 0.021$, $n = 8(2) = 16$.

$$P = R\left[\frac{1 - (1+i)^{-n}}{i}\right]$$

$$= 4210\left[\frac{1 - (1.021)^{-16}}{0.021}\right]$$

$$\approx 56{,}711.93$$

The present value is \$56,711.93.

53. Two types of loans that are commonly amortized are home loans and auto loans.

55. $P = 3200,\ i = \frac{0.08}{4} = 0.02,\ n = 12$

$$\begin{aligned} R &= \frac{Pi}{1-(1+i)^{-n}} \\ &= \frac{3200(0.02)}{1-(1.02)^{-12}} \\ &\approx 302.59 \end{aligned}$$

The amount of each payment is \$302.59.

The total amount paid is \$302.59(12) = \$3631.08. Thus, the total interest paid is

$$\$3631.08 - 3200 = \$431.08.$$

57. $P = 51{,}607,\ i = \frac{0.08}{12} = 0.00\overline{6},\ n = 32$

$$\begin{aligned} R &= \frac{Pi}{1-(1+i)^{-n}} \\ &= \frac{51{,}607(0.00\overline{6})}{1-(1.00\overline{6})^{-32}} \\ &\approx 1796.20 \end{aligned}$$

The amount of each payment is \$1796.20.

The total amount paid is \$1796.20(32) = \$57,478.40. Thus, the total interest paid is

$$\$57{,}478.40 - 51{,}607 = \$5871.40.$$

59. $P = 177{,}110,\ i = \frac{0.0668}{12} = 0.0055\overline{6},$
$n = 30(12) = 360$

$$\begin{aligned} R &= \frac{Pi}{1-(1+i)^{-n}} \\ &= \frac{177{,}110(0.0055\overline{6})}{1-(1.0055\overline{6})^{-360}} \\ &\approx 1140.50 \end{aligned}$$

The amount of each payment is \$1140.50.
The total amount paid is \$1140.50(360) = \$410,580. Thus, the total interest paid is

$$\$410{,}580 - 177{,}110 = \$233{,}470.$$

61. The answer can be found in the table under payment number 12 in the column labeled "Portion to Principal." The amount of principal repayment included in the fifth payment is \$132.99.

63. The last entry in the column "Principal at End of Period," \$125,464.39, shows the debt remaining at the end of the first year (after 12 payments). Since the original debt (loan principal) was \$127,000, the amount by which the debt has been reduced at the end of the first year is

$$\$127{,}000 - 125{,}464.39 = \$1535.61.$$

65. Here $P = 9820$, $r = 6.7\% = 0.067$, and $t = \frac{7}{12}$.

$$\begin{aligned} I &= Prt \\ &= 9820(0.067)\left(\frac{7}{12}\right) \\ &\approx 383.80 \end{aligned}$$

The interest he will pay is \$383.80. The total amount he will owe in 7 mo is

$$\$9820 + 383.80 = \$10{,}203.80.$$

67. $P = 84{,}720,\ t = \frac{7}{12},\ I = 4055.46$

Substitute these values into the formula for simple interest to find the value of r.

$$\begin{aligned} I &= Prt \\ 4055.46 &= 84{,}720r\left(\frac{7}{12}\right) \\ 4055.46 &= 49{,}420r \\ 0.0821 &\approx r \end{aligned}$$

The interest rate is 8.21%.

69. In both cases use the formula for compound amount with $P = 500$ and $i = \frac{0.05}{4} = 0.0125$. For the investment at age 23 use $n = 42(4) = 168$.

$$\begin{aligned} A &= P(1+i)^n \\ &= 500(1+0.0125)^{168} \\ &\approx 4030.28 \end{aligned}$$

For the investment at age 40 use $n = 25(4) = 100$.

$$\begin{aligned} A &= P(1+i)^n \\ &= 500(1+0.0125)^{100} \\ &\approx 1731.70 \end{aligned}$$

The increased amount of money Tom will have if he invests now is

$$\$4030.28 - 1731.70 = \$2298.58.$$

71. $R = 5000,\ i = \frac{0.10}{2} = 0.05,\ n = 7\frac{1}{2}(2) = 15$

This is an ordinary annuity.

$$S = R\left[\frac{(1+i)^n - 1}{i}\right]$$

$$S = 5000\left[\frac{(1+0.05)^{15} - 1}{0.05}\right]$$

$$\approx 107{,}892.82$$

The future value is \$107,892.82. The amount of interest earned is

$$\$107{,}892.82 - 15(5000) = \$32{,}892.82.$$

73. Use the formula for amortization payments with $P = 48{,}000$, $i = 0.065$, and $n = 7$.

$$\begin{aligned} R &= \frac{Pi}{1-(1+i)^{-n}} \\ &= \frac{48{,}000(0.065)}{1-(1.065)^{-7}} \\ &\approx 8751.91 \end{aligned}$$

The owner should deposit \$8751.91 at the end of each year.

The total amount deposited is \$8751.91(7) = \$61,263.37. Thus, the total interest paid is

$$\$61{,}263.37 - 48{,}000 = \$13{,}263.37.$$

75. The effective rate paid by Ascencia would be

$$\left(1+\frac{0.0149}{12}\right)^{12} - 1 = 0.015$$

or 1.50%.

The effective rate paid by giantbank.com would be

$$\left(1+\frac{0.0145}{360}\right)^{360} - 1 = 0.0146$$

or 1.46%. Ascencia has the higher effective rate.

77. **(a)** For the 0% financing, the payments are simply 1/60 of the financed amount.

$$\frac{16{,}000}{60} = 266.67$$

Thus the rounded payments are \$266.67, and the total payments are equal to the financed amount of \$16,000. (In fact because of rounding the total payments are 20 cents more than this amount, but the final payment would be reduced by 20 cents to compensate.)

(b) For the 3.9% financing the monthly payment will be

$$\frac{(16{,}000)\left(\frac{0.039}{12}\right)}{1-\left(1+\frac{0.039}{12}\right)^{-72}} = 249.59$$

or \$249.59. The total payments will be 72 times this amount, or \$17,970.48.

(c) For the cash back option with a 6.33% interest rate for 489 months, the monthly payment will be

$$\frac{(12{,}000)\left(\frac{0.0633}{12}\right)}{1-\left(1+\frac{0.0633}{12}\right)^{-48}} = 283.64$$

or \$283.64. The total payments will be 48 times this amount, or \$13,614.72.

79.
$$\begin{aligned} \text{Amount of loan} &= \$191{,}000 - 40{,}000 \\ &= \$151{,}000 \end{aligned}$$

(a) Use the formula for amortization payments with $P = 151{,}000$, $i = \frac{0.065}{12} = 0.00541\overline{6}$, and $n = 30(12) = 360$.

$$\begin{aligned} R &= \frac{Pi}{1-(1+i)^{-n}} \\ &= \frac{151{,}000(0.00541\overline{6})}{1-(1.00541\overline{6})^{-360}} \\ &\approx 954.42 \end{aligned}$$

The monthly payment for this mortgage is \$954.42.

(b) To find the amount of the first payment that goes to interest, use $I = Prt$ with $P = 151{,}000$, $i = 0.00541\overline{6}$, and $t = 1$.

$$I = 151{,}000(0.065)\left(\frac{1}{12}\right) = 817.92$$

Of the first payment, \$817.92 is interest.

(c) Using method 1, since 180 of 360 payments were made, there are 180 remaining payments. The present value is

$$954.42\left[\frac{1-(1.00541\overline{6})^{-180}}{0.00541\overline{6}}\right] \approx 109{,}563.99,$$

so the remaining balance is \$109,563.99.

Using method 2, since 180 payments were already made, we have

$$954.42\left[\frac{1-(1.00541\overline{6})^{-180}}{0.00541\overline{6}}\right] \approx 109{,}563.99.$$

She still owes

$$\$151{,}000 - 109{,}563.99 = \$41{,}436.01.$$

Furthermore, she owes the interest on this amount for 180 mo, for a total remaining balance of

$$41{,}436.01(1.00541\overline{6})^{180} = 109{,}565.13.$$

(d) Closing costs $= 3700 + 0.025(238{,}000)$

$$= 3700 + 5950$$

$$= 9650$$

Closing costs are \$9650.

(e) Amount of money received

= Selling price − Closing costs − Current mortgage balance

Using method 1, the amount received is

\$238,000 − 9650 − 109,563.99 = \$118,786.01.

Using method 2, the amount received is

\$238,000 − 9650 − 109,565.13 = \$118,784.87.

81. (a) Use the formula for effective rate with $r_E = 0.10$ and $m = 12$.

$$r_E = \left(1 + \frac{r}{m}\right)^m - 1$$

$$0.10 = \left(1 + \frac{r}{12}\right)^{12} - 1$$

$$1.10 = \left(1 + \frac{r}{12}\right)^{12}$$

$$(1.10)^{1/12} = 1 + \frac{r}{12}$$

$$1.007974 \approx 1 + \frac{r}{12}$$

$$0.007974 \approx \frac{r}{12}$$

$$0.095688 \approx r$$

The annual interest rate is 9.569%.

(b) Use the formula for amortization payments with $P = 140{,}000$, $i = \frac{0.06625}{12}$, and $n = 30(12) = 360$.

$$R = \frac{Pi}{1 - (1 + i)^{-n}}$$

$$= \frac{140{,}000\left(\frac{0.06625}{12}\right)}{1 - \left(1 + \frac{0.06625}{12}\right)^{-360}}$$

$$\approx 896.44$$

Her monthly payment is \$896.44.

(c) This investment is an annuity with $R = 1200 - 896.44 = 303.56$, $i = \frac{0.09569}{12}$, and $n = 30(12) = 360$. The future value is

$$S = R\left[\frac{(1 + i)^n - 1}{i}\right]$$

$$= 303.56\left[\frac{\left(1 + \frac{0.09569}{12}\right)^{360} - 1}{\frac{0.09569}{12}}\right]$$

$$\approx 626{,}200.88$$

In 30 yr she will have \$626,200.88 in the fund.

(d) Use the formula for amortization payments with $P = 14{,}000$, $i = \frac{0.0625}{12}$, and $n = 15(12) = 180$.

$$R = \frac{Pi}{1 - (1 + i)^{-n}}$$

$$= \frac{140{,}000\left(\frac{0.0625}{12}\right)}{1 - \left(1 + \frac{0.0625}{12}\right)^{-180}}$$

$$\approx 1200.39$$

His monthly payment is \$1200.39.

(e) This investment is an annuity with $R = 1200$, $i = \frac{0.09569}{12}$, and $n = 15(12) = 180$. The future value is

$$S = R\left[\frac{(1 + i)^n - 1}{i}\right]$$

$$= 1200\left[\frac{\left(1 + \frac{0.09569}{12}\right)^{180} - 1}{\frac{0.09569}{12}}\right]$$

$$\approx 478{,}134.14$$

In 30 yr he will have \$478,134.14.

(f) Sue is ahead by

\$626,200.88 − 478,134.14 = \$148,066.74.

Chapter 6

LOGIC

6.1 Statements and Quantifiers

Your Turn 1

No; the statement "I bought Ben and Jerry's ice cream." is not compound because the "and" is not used as a logical connective that connects two simple statements.

Your Turn 2

The negative of the given statement is "Wal-Mart is the largest corporation in the USA."

Your Turn 3

The negation of $4x + 2y < 5$ is $4x + 2y \geq 5$.

Your Turn 4

$h \wedge \sim r$ represents "My backpack is heavy, and it is not going to rain."

Your Turn 5

p represents "$7 < 2$," which is false.

q represents "$4 > 3$," which is true.

$$\sim p \wedge q$$
$$\mathrm{T} \wedge \mathrm{T}$$
$$\mathrm{T}$$

The statement $\sim p \wedge q$ is true.

Your Turn 6

p is false, q is true, and r is false.

$$(\sim p \wedge q) \vee r$$
$$(\sim \mathrm{F} \wedge \mathrm{T}) \vee \mathrm{F}$$
$$(\mathrm{T} \wedge \mathrm{T}) \vee \mathrm{F}$$
$$\mathrm{T} \vee \mathrm{F}$$
$$\mathrm{T}$$

The statement $(\sim p \wedge q) \vee r$ is true.

Your Turn 7

p represents "$7 < 2$," q represents "$4 > 3$," and r represents "$2 > 8$."

$$p \vee (\sim q \wedge r)$$
$$\mathrm{F} \vee (\sim \mathrm{T} \wedge \mathrm{F})$$
$$\mathrm{F} \vee (\mathrm{F} \wedge \mathrm{F})$$
$$\mathrm{F} \vee \mathrm{F}$$
$$\mathrm{F}$$

The statement $p \vee (\sim q \wedge r)$ is false.

6.1 Exercises

1. Because the declarative sentence "Montevideo is the capital of Uruguay" has the property of being true or false, it is considered a statement. It is not compound.

3. "Don't feed the animals" is not a declarative sentence and does not have the property of being true or false. Hence, it is not considered a statement.

5. "$2 + 2 = 5$ and $3 + 3 = 7$" is a declarative sentence that is true or false and, therefore, is considered a statement. It is compound.

7. "Got milk?" is a question, not a declarative sentence, and, therefore, is not considered a statement.

9. "I am not a crook" is a compound statement because it contains the logical connective "not."

11. "She enjoyed the comedy team of Penn and Teller" is not compound because only one assertion is being made.

13. "If I get an A, I will celebrate" is a compound statement because it consists of two simple statements combined by the connective "if... then."

15. The negation of "My favorite flavor is chocolate" is "My favorite flavor is not chocolate."

17. A negation for "$y > 12$" (without using a slash sign) would be "$y \le 12$."

19. A negation for "$q \ge 5$" would be "$q < 5$."

23. A translation of "$\sim b$" is "I'm not getting better."

25. A translation of "$\sim b \vee d$" is "I'm not getting better or my parrot is dead."

27. A translation of "$\sim(b \wedge \sim d)$" is "It is not the case that both I'm getting better and my parrot is not dead."

29. If q is false, then $(p \wedge \sim q) \wedge q$ is false, since both parts of the conjunction must be true for the compound statement to be true.

31. If the conjunction $p \wedge q$ is true, then both p and q must be true. Thus, q must be true.

33. If $\sim(p \vee q)$ is true, then $p \vee q$ must be false, since a statement and its negation have opposite truth values. In order for the disjunction $p \vee q$ to be false, both component statements must be false. Thus, p and q are both false.

35. Since p is false, $\sim p$ is true, since a statement and its negation have opposite truth values.

37. Since p is false and q is true, we may consider the statement $p \vee q$ as

$$F \vee T,$$

which is true by the *or* truth table. That is, $p \vee q$ is true.

39. Since p is false and q is true, we may consider $p \vee \sim q$ as

$$F \vee \sim T$$
$$F \vee F$$
$$F.$$

That is, $p \vee \sim q$ is false.

41. With the given truth values for p and q, we may consider $\sim p \vee \sim q$ as

$$\sim F \vee \sim T$$
$$T \vee F$$
$$T.$$

Thus, $\sim p \vee \sim q$ is true.

43. Replacing p and q with the given truth values, we have

$$\sim(F \wedge \sim T)$$
$$\sim(F \wedge F)$$
$$\sim F$$
$$T.$$

Thus, the compound statement $\sim(p \wedge \sim q)$ is true.

45. Replacing p and q with the given truth values, we have

$$\sim[\sim F \wedge (\sim T \vee F)]$$
$$\sim[T \wedge (F \vee F)]$$
$$\sim[T \wedge F]$$
$$\sim F$$
$$T.$$

Thus, the compound statement $\sim[\sim p \wedge (\sim q \vee p)]$ is true.

47. The statement $3 \ge 1$ is a disjunction since it means "$3 > 1$" or "$3 = 1$."

49. Replacing p, q, and r with the given truth values, we have

$$(T \wedge F) \vee \sim F$$
$$F \vee T$$
$$T.$$

Thus, the compound statement $(p \wedge r) \vee \sim q$ is true.

51. Replacing p, q, and r with the given truth values, we have

$$T \wedge (F \vee F)$$
$$T \wedge F$$
$$F.$$

Thus, the compound statement $p \wedge (q \vee r)$ is false.

53. Replacing p, q, and r with the given truth values, we have

$$\sim(T \wedge F) \wedge (F \vee \sim F)$$
$$\sim F \wedge (F \vee T)$$
$$T \wedge T$$
$$T.$$

Thus, the compound statement $\sim(p \wedge q) \wedge (r \vee \sim q)$ is true.

55. Replacing p, q, and r with the given truth values, we have

$$\sim[(\sim T \wedge F) \vee F]$$
$$\sim[(F \wedge F) \vee F]$$
$$\sim[F \vee F]$$
$$\sim F$$
$$T.$$

Thus, the compound statement $\sim[(p \wedge q) \vee r]$ is true.

57. Since p is false and r is true, we have

$$F \wedge T$$
$$F.$$

The compound statement $p \wedge r$ is false.

59. Since q is false and r is true, we have

$$\sim F \vee \sim T$$
$$T \vee F$$
$$T.$$

The compound statement $\sim q \vee \sim r$ is true.

61. Since p and q are false and r is true, we have

$$(F \wedge F) \vee T$$
$$F \vee T$$
$$T.$$

The compound statement $(p \wedge q) \vee r$ is true.

63. Since p and q are false and r is true, we have

$$(\sim T \wedge F) \vee \sim F$$
$$(F \wedge F) \vee T$$
$$F \vee T$$
$$T.$$

The compound statement $(\sim r \wedge q) \vee \sim p$ is true.

65. **(b)**, **(c)**, and **(d)** are declarative sentences that are true or false and are therefore statements.

(a) is a command, not a declarative sentence, and is therefore not a statement.

67. An individual has to be your biological child to be a "qualifying" child.

69. $a \wedge j$

71. $\sim a \vee j$

73. Statements a and j are both true.

$a \wedge j$	$\sim a \wedge \sim j$	$\sim a \vee j$	$a \vee j$
$T \wedge T$	$F \wedge F$	$F \vee T$	$T \vee T$
T	F	T	T

Exercises 69, 71, and 72 are true.

75. **(a)** This is a question, not a declarative sentence, and therefore not a statement.

(b), (c), (d), (e) There are declarative sentences that are true or false and therefore statements.

77. The negation of "You may find that exercise helps you cope with stress" is "You may not find that exercise helps you cope with stress."

79. **(c)**, and **(d)** are compound statements that are formed by the disjunction *or*.

83. "New Orleans won the Super Bowl but Peyton Manning is not the best quarterback" may be symbolized as $n \wedge \sim m$.

85. "New Orleans did not win the Super Bowl or Peyton Manning is the best quarterback" may be symbolized as $\sim n \vee m$.

87. "Neither did New Orleans win the Super Bowl nor is Peyton Manning the best quarterback" may be symbolized as $\sim n \wedge \sim m$ or $\sim(n \vee m)$.

89. Assume that n is true and m is true. Under these conditions, the statements in Exercises 83–88 have the following truth values:

83. $n \wedge \sim m$: False, because n is true and $\sim m$ is false.

84. $\sim n \vee \sim m$: False, because $\sim n$ is false and $\sim m$ is false.

85. $\sim n \vee m$: True, because $\sim n$ is false and m is true.

86. $\sim n \wedge m$: False, because $\sim n$ is false and m is true.

87. $\sim n \wedge \sim m$: False, because $\sim n$ is false and $\sim m$ is false.

88. $n \vee m \wedge [\sim(n \wedge m)]$: False, since $(n \vee m)$ is true but $\sim(n \wedge m)$ is false.

Therefore, only Exercise 85 is a true statement.

6.2 Truth Tables and Equivalent Statements

Your Turn 1

p	q	$\sim p$	$\sim p \vee q$	$p \wedge (\sim p \vee q)$
T	T	F	T	T
T	F	F	F	F
F	T	T	T	F
F	F	T	T	F

Your Turn 2

Let p represent "I order pizza" and d represent "You make dinner." Then the statement "I do not order pizza, or you do not make dinner and I order pizza" can be represented by $\sim p \vee (\sim d \wedge p)$.

p	d	$\sim p$	$\sim d$	$\sim d \wedge p$	$\sim p \vee (\sim d \wedge p)$
T	T	F	F	F	F
T	F	F	T	T	T
F	T	T	F	F	T
F	F	T	T	F	T

Your Turn 3

Let d represent "You do make dinner" and p represent "I order pizza." "You do not make dinner or I order pizza" is symbolically $\sim d \vee p$. Applying DeMorgan's first law, the negation is $\sim(\sim d \vee p) \equiv d \wedge \sim p$. In words this reads "You make dinner and I do not order pizza."

6.2 Exercises

1. Since there are two simple statements (p and r), we have $2^2 = 4$ rows in the truth table.

3. Since there are four simple statements ($p, q, r,$ and s), we have $2^4 = 16$ rows in the truth table.

5. Since there are seven simple statements ($p, q, r, s, t, u,$ and v), we have $2^7 = 128$ rows in the truth table.

7. If the truth table for a certain compound statement has 64 rows, then there must be six distinct component statements since $2^6 = 64$.

9. $\sim p \wedge q$

p	q	$\sim p$	$\sim p \wedge q$
T	T	F	F
T	F	F	F
F	T	T	T
F	F	T	F

11. $\sim(p \wedge q)$

p	q	$p \wedge q$	$\sim(p \wedge q)$
T	T	T	F
T	F	F	T
F	T	F	T
F	F	F	T

13. $(q \vee \sim p) \vee \sim q$

p	q	$\sim p$	$\sim q$	$q \vee \sim p$	$(q \vee \sim p) \vee \sim q$
T	T	F	F	T	T
T	F	F	T	F	T
F	T	T	F	T	T
F	F	T	T	T	T

In Exercises 15–23 to save space we are using the alternative method, filling in columns in the order indicated by the numbers. Observe that columns with the same number are combined (by the logical definition of the connective) to get the next numbered column. Note that <u>*this is different*</u> *from the way the numbered columns are used in the textbook. Remember that the last column (highest numbered column) completed yields the truth values for the complete compound statement. Be sure to align truth values under the appropriate logical connective or simple statement.*

15. $\sim q \wedge (\sim p \vee q)$

p	q	$\sim q$	$\wedge$	$(\sim p$	$\vee$	$q)$
T	T	F	F	F	T	T
T	F	T	F	F	F	F
F	T	F	F	T	T	T
F	F	T	T	T	T	F
		1	4	2	3	2

17. $(p \vee \sim q) \wedge (p \wedge q)$

p	q	$(p$	$\vee$	$\sim q)$	$\wedge$	$(p$	$\wedge$	$q)$
T	T	T	T	F	T	T	T	T
T	F	T	T	T	F	T	F	F
F	T	F	F	F	F	F	F	T
F	F	F	T	T	F	F	F	F
		1	2	1	5	3	4	3

19. $(\sim p \wedge q) \wedge r$

p	q	r	$(\sim p$	$\wedge$	$q)$	$\wedge$	r
T	T	T	F	F	T	F	T
T	T	F	F	F	T	F	F
T	F	T	F	F	F	F	T
T	F	F	F	F	F	F	F
F	T	T	T	T	T	T	T
F	T	F	T	T	T	F	F
F	F	T	T	F	F	F	T
F	F	F	T	F	F	F	F
			1	2	1	4	3

21. $(\sim p \wedge \sim q) \vee (\sim r \vee \sim p)$

p	q	r	$(\sim p$	$\wedge$	$\sim q)$	$\vee$	$(\sim r$	$\vee$	$\sim p)$
T	T	T	F	F	F	F	F	F	F
T	T	F	F	F	F	T	T	T	F
T	F	T	F	F	T	F	F	F	F
T	F	F	F	F	T	T	T	T	F
F	T	T	T	F	F	T	F	T	T
F	T	F	T	F	F	T	T	T	T
F	F	T	T	T	T	T	F	T	T
F	F	F	T	T	T	T	T	T	T
			1	2	1	5	3	4	3

23. $\sim(\sim p \wedge \sim q) \vee (\sim r \vee \sim s)$

p	q	r	s	$\sim$	$(\sim p$	$\wedge$	$\sim q)$	$\vee$	$(\sim r$	$\vee$	$\sim s)$
T	T	T	T	T	F	F	F	T	F	F	F
T	T	T	F	T	F	F	F	T	F	T	T
T	T	F	T	T	F	F	F	T	T	T	F
T	T	F	F	T	F	F	F	T	T	T	T
T	F	T	T	T	F	F	T	T	F	F	F
T	F	T	F	T	F	F	T	T	F	T	T
T	F	F	T	T	F	F	T	T	T	T	F
T	F	F	F	T	F	F	T	T	T	T	T
F	T	T	T	T	T	F	F	T	F	F	F
F	T	T	F	T	T	F	F	T	F	T	T
F	T	F	T	T	T	F	F	T	T	T	F
F	T	F	F	T	T	F	F	T	T	T	T
F	F	T	T	F	T	T	T	F	F	F	F
F	F	T	F	F	T	T	T	T	F	T	T
F	F	F	T	F	T	T	T	T	T	T	F
F	F	F	F	F	T	T	T	T	T	T	T
				3	1	2	1	6	4	5	4

25. "It's vacation and I am having fun" has the symbolic form $p \wedge q$. The negation, $\sim(p \wedge q)$, is equivalent, by one of DeMorgan's laws, to $\sim p \vee \sim q$. The corresponding word statement is "It's not vacation or I am not having fun."

27. "Either the door was unlocked or the thief broke a window" has the symbolic form $p \vee q$. The negation, $\sim(p \vee q)$, is equivalent, by one of DeMorgan's laws, to $\sim p \wedge \sim q$. The corresponding word statement is "The door was locked and the thief didn't break a window."

29. "I'm ready to go, but Naomi Bahary isn't" has the symbolic form $p \wedge \sim q$. (The connective "but" is logically equivalent to "and.") The negation, $\sim(p \wedge \sim q)$, is equivalent, by one of DeMorgan's laws, to $\sim p \vee q$. The corresponding word statement is "I'm not ready to go, or Naomi Bahary is."

31. "$12 > 4$ or $8 = 9$" has the symbolic form $p \vee q$. The negation, $\sim(p \vee q)$, is equivalent, by one of DeMorgan's laws, to $\sim p \wedge \sim q$. The corresponding statement is "$12 \leq 4$ and $8 \neq 9$." (Note that the inequality "$\leq$" is logically equivalent to "$\not>$.")

33. "Larry or Moe is out sick today" has the symbolic form $p \vee q$. The negation, $\sim(p \vee q)$, is equivalent, by one of DeMorgan's laws, to $\sim p \wedge \sim q$. The corresponding word statement is "Neither Larry nor Moe is out sick today."

35. $p \veebar q$

p	q	$p \veebar q$
T	T	F
T	F	T
F	T	T
F	F	F

Observe that it is only the first line in the truth table that changes for "exclusive disjunction" since the component statements can not both be true at the same time.

37. "$(3 + 1 = 4) \veebar (2 + 5 = 9)$" is <u>true</u> since the first component statement is true and the second is false.

39. Store the truth values of the statements p, q, and s as P, Q, and S, respectively.

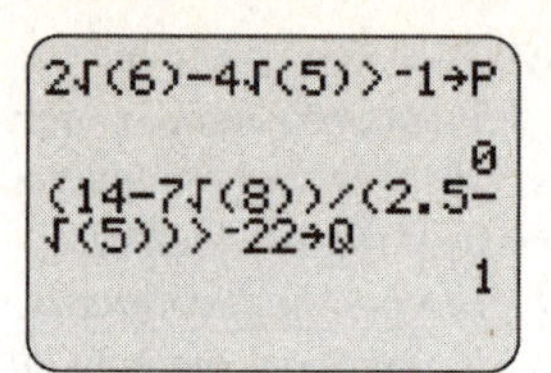

Use the stored values of P, Q, and S to find the truth values of each of the compound statements.

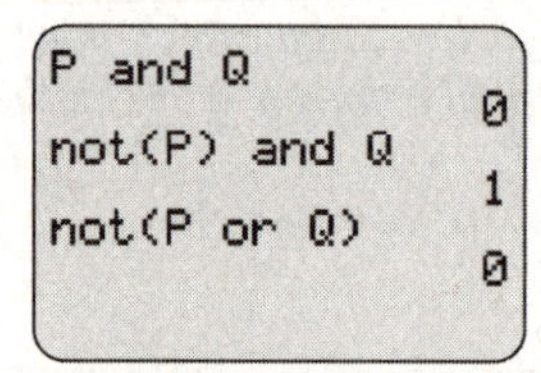

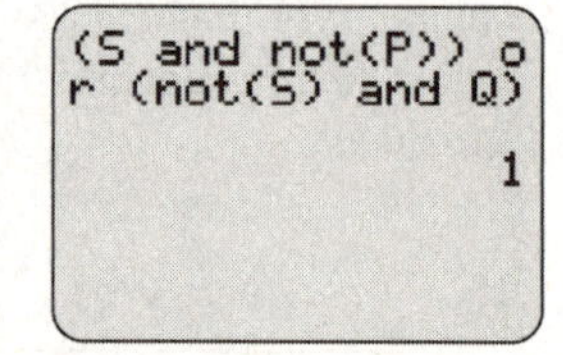

(a) P and Q returns 0, meaning $p \wedge q$ is false.

(b) not (P) and Q returns 1, meaning $\sim p \wedge q$ is true.

(c) not (P or Q) returns 0, meaning $\sim(p \vee q)$ is false.

(d) (S and not (P)) or (not (S) and Q) returns 1, meaning $(s \wedge \sim p) \vee (\sim s \wedge q)$ is true.

41. Service will not be performed at the location, and the store may not send the Covered Equipment to an Apple repair service location to be repaired.

43. Letting s represent "You will be completely satisfied," r represent "We will refund your money," and q represent "We will ask you questions," the guarantee translates into symbols as $s \vee (r \wedge \sim q)$. The truth table follows.

s	r	q	s	$\vee$	$(r$	$\wedge$	$\sim q)$
T	T	T	T	T	T	F	F
T	T	F	T	T	T	T	T
T	F	T	T	T	F	F	F
T	F	F	T	T	F	F	T
F	T	T	F	F	T	F	F
F	T	F	F	T	T	T	T
F	F	T	F	F	F	F	F
F	F	F	F	F	F	F	T
		1	1	4	2	3	2

The guarantee would be false in the three cases indicated as "F" in the column labeled "4." These would be if you are not completely satisfied, and they either don't refund your money or they ask you questions.

45. Inclusive, since the "and" case is allowed.

47. Letting p represent "Liberty without learning is always in peril" and q represent "Learning without liberty is always in vain," the quote translates symbolically as $p \wedge q$. Its negation is equivalent by DeMorgan's laws to $\sim p \vee \sim q$. A translation for the negation is "Liberty without learning is not always in peril, or learning without liberty is not always in vain."

49. p: You could reroll the die again for your Large Straight.

q: You could set aside the 2 Twos and roll for your Twos or for 3 of a Kind.

The statement is $p \vee q$.

Its negation is $\sim p \wedge \sim q$.

Negation: You cannot reroll the die again for your Large Straight and you cannot set aside the 2 Twos and roll for your Twos or for 3 of a Kind.

6.3 The Conditional and Circuits

Your Turn 1

If Little Rock is the capital of Arkansas, then New York City is the capital of New York.

The first component of this conditional is true, while the second is false. The given statement is false.

Your Turn 2

$$(4 > 5) \rightarrow \text{F}$$

The antecedent $4 > 5$ is false, while the consequent is false. The given statement is true.

Your Turn 3

p, q, and r are all false.

$$\sim q \rightarrow (p \rightarrow r)$$
$$\text{T} \rightarrow (\text{F} \rightarrow \text{F})$$
$$\text{T} \rightarrow \text{T}$$
$$\text{T}$$

Your Turn 4

p: You do the homework.

q: You will pass the quiz.

$$p \rightarrow q \equiv \sim p \vee q$$

"If you do the homework, then you will pass the quiz" is equivalent to "You do not do the homework, or you will pass the quiz."

Your Turn 5

Since the negation of $p \rightarrow q$ is $p \wedge \sim q$, the negation of "If you are on time, then we will be on time" is "You are on time, and we will not be on time."

6.3 Exercises

1. The statement "If the antecedent of a conditional statement is false, the conditional statement is true" is true, since a false antecedent will always yield a true conditional statement.

3. The statement "If q is true, then $(p \wedge q) \rightarrow q$ is true" is true, since with a true consequent the conditional statement is always true (even though the antecedent may be false).

5. "Given that $\sim p$ is true and q is false, the conditional $p \rightarrow q$ is true" is a true statement since the antecedent, p, must be false.

9. "$F \rightarrow (4 \neq 7)$" is a true, statement, since a false antecedent always yields a conditional statement which is true.

11. "$(4 = 11 - 7) \rightarrow (8 > 0)$" is true since the antecedent and the consequent are both true.

13. $d \rightarrow (e \wedge s)$ expressed in words becomes "If she dances tonight, then I'm leaving early and he sings loudly."

15. $\sim s \rightarrow (d \vee \sim e)$ expressed in words becomes "If he doesn't sing loudly, then she dances tonight or I'm not leaving early."

17. The statement "My dog ate my homework, or if I receive a failing grade then I'll run for governor" can be symbolized as $d \vee (f \rightarrow g)$.

19. The statement "I'll run for governor if I don't receive a failing grade" can be symbolized as $\sim f \rightarrow g$.

21. Replacing r and p with the given truth values, we have

$$\sim \text{F} \rightarrow \text{F}$$
$$\text{T} \rightarrow \text{F}$$
$$\text{F.}$$

Thus, the statement $\sim r \rightarrow p$ is false.

23. Replacing p, q, and r with the given truth values, we have

$$\sim F \rightarrow (T \wedge F)$$
$$T \rightarrow F$$
$$F.$$

Thus, the statement $\sim p \rightarrow (q \wedge r)$ is false.

25. Replacing p, q, and r with the given truth values, we have

$$\sim T \rightarrow (F \wedge F)$$
$$F \rightarrow F$$
$$T.$$

Thus, the statement $\sim q \rightarrow (p \wedge r)$ is true.

27. Replacing p, q, and r with the given truth values, we have

$$(\sim F \rightarrow \sim T) \rightarrow (\sim F \wedge \sim F)$$
$$(T \rightarrow F) \rightarrow (T \wedge T)$$
$$F \rightarrow T$$
$$T.$$

Thus, the statement $(p \rightarrow \sim q) \rightarrow (\sim p \wedge \sim r)$ is true.

31. $\sim p \rightarrow \sim q \equiv \sim(\sim p) \vee \sim q$ By Equivalent Statement 8

$\sim(\sim p) \vee \sim q \equiv p \vee \sim q$ By Equivalent Statement 7

So, $\sim p \rightarrow \sim q \equiv p \vee \sim q$ and $\sim p \rightarrow \sim q$ and $p \vee \sim q$ have the same truth values.

The negation of $p \vee \sim q$ is $\sim(p \vee \sim q)$.
Since $p \vee \sim q$ and $\sim(p \vee \sim q)$ cannot have the same truth value, $\sim p \rightarrow \sim q$ and $\sim(p \vee \sim q)$ cannot both be true. Thus, $(\sim p \rightarrow \sim q) \wedge \sim(p \vee \sim q)$ is a contradiction.

33. $\sim q \rightarrow p$

p	q	$\sim q$	$\rightarrow$	p
T	T	F	T	T
T	F	T	T	T
F	T	F	T	F
F	F	T	F	F

35. $(p \vee \sim p) \rightarrow (p \wedge \sim p)$

p	$(p$	$\vee$	$\sim p)$	$\rightarrow$	$(p$	$\wedge$	$\sim p)$
T	T	T	F	F	T	F	F
F	F	T	T	F	F	F	T
	1	2	1	5	3	4	3

Since the statement is always false (the truth values in column 5 are all false), it is a contradiction.

37. $(p \vee q) \rightarrow (q \vee p)$

p	q	$(p \vee q)$	$\rightarrow$	$(q \vee p)$
T	T	T T T	T	T T T
T	F	T T F	T	F T T
F	T	F T T	T	T T F
F	F	F F F	T	F F F
		1 2 1	5	3 4 3

Since this statement is always true (column 5), it is a tautology.

39. $r \rightarrow (p \wedge \sim q)$

p	q	r	r	$\rightarrow$	$(p \wedge \sim q)$
T	T	T	T	F	T F F
T	T	F	F	T	T F F
T	F	T	T	T	T T T
T	F	F	F	T	T T T
F	T	T	T	F	F F F
F	T	F	F	T	F F F
F	F	T	T	F	F F T
F	F	F	F	T	F F T
			1	4	2 3 2

41. $(\sim r \rightarrow s) \vee (p \rightarrow \sim q)$

p	q	r	s	$(\sim r \rightarrow s)$	$\vee$	$(p \rightarrow \sim q)$
T	T	T	T	F T T	T	T F F
T	T	T	F	F T F	T	T F F
T	T	F	T	T T T	T	T F F
T	T	F	F	T F F	F	T F F
T	F	T	T	F T T	T	T T T
T	F	T	F	F T F	T	T T T
T	F	F	T	T T T	T	T T T
T	F	F	F	T F F	T	T T T
F	T	T	T	F T T	T	F T F
F	T	T	F	F T F	T	F T F
F	T	F	T	T T T	T	F T F
F	T	F	F	T F F	T	F T F
F	F	T	T	F T T	T	F T T
F	F	T	F	F T F	T	F T T
F	F	F	T	T T T	T	F T T
F	F	F	F	T F F	T	F T T
				1 2 1	5	3 4 3

43. Let p represent "your eyes are bad" and q represent "your whole body will be full of darkness." The statement has the form $p \rightarrow q$. In words, the equivalent form $\sim p \vee q$ becomes "Your eyes are not bad or your whole body will be full of darkness."

45. Let p represent "I have the money" and q represent "I'd buy that car." The statement has the form $p \rightarrow q$. In words, the equivalent form $\sim p \vee q$ becomes "I don't have the money or I'd buy that car."

47. "If you ask me, I will do it" has the form $p \rightarrow q$. The negation has the form $p \wedge \sim q$, which translates as "You ask me, and I will not do it."

49. "If you don't love me, I won't be happy" has the form $\sim p \rightarrow \sim q$. The negation has the form $\sim p \wedge q$, which translates as "You don't love me and I will be happy."

51. The statements $p \rightarrow q$ and $\sim p \vee q$ are equivalent if they have the same truth tables.

p	q	p	$\rightarrow$	q	$\sim p$	$\vee$	q
T	T	T	T	T	F	T	T
T	F	T	F	F	F	F	F
F	T	F	T	T	T	T	T
F	F	F	T	F	T	T	F
		1	2	1	1	2	1

Since the truth values in the final columns for each statement are the same, the statements are equivalent.

53.

p	q	p	$\rightarrow$	q	q	$\rightarrow$	p
T	T	T	T	T	T	T	T
T	F	T	F	F	F	T	T
F	T	F	T	T	T	F	F
F	F	F	T	F	F	T	F
		1	2	1	1	2	1

Since the truth values in the final columns for each statement are not the same, the statements are not equivalent.

55.

p	q	p	$\rightarrow$	$\sim q$	$\sim p$	$\vee$	$\sim q$
T	T	T	F	F	F	F	F
T	F	T	T	T	F	T	T
F	T	F	T	F	T	T	F
F	F	F	T	T	T	T	T
		1	2	1	1	2	1

Since the truth values in the final columns for each statement are the same, the statements are equivalent.

57.

p	q	p	$\wedge$	$\sim q$	$\sim q$	$\rightarrow$	$\sim p$
T	T	T	F	F	F	T	F
T	F	T	T	T	T	F	F
F	T	F	F	F	F	T	T
F	F	F	F	T	T	T	T
		1	2	1	1	2	1

Since the truth values in the final columns for each statement are not the same, the statements are not equivalent. Observe that since they have opposite truth values, each statement is the negation of the other.

59.

p	q	p	$\wedge$	q	$\sim$	$(p$	$\rightarrow$	$\sim q)$
T	T	T	T	T	T	T	F	F
T	F	T	F	F	F	T	T	T
F	T	F	F	T	F	F	T	F
F	F	F	F	F	F	F	T	T
		1	2	1	5	3	4	3

The columns labeled 2 and 5 are identical.

61.

p	q	p	$\vee$	q	q	$\vee$	p
T	T	T	T	T	T	T	T
T	F	T	T	F	F	T	T
F	T	F	T	T	T	T	F
F	F	F	F	F	F	F	F
		1	2	1	3	4	3

The columns labeled 2 and 4 are identical.

63.

p	q	r	$(p$	$\vee$	$q)$	$\vee$	r	p	$\vee$	$(q$	$\vee$	$r)$
T	T	T	T	T	T	T	T	T	T	T	T	T
T	T	F	T	T	T	T	F	T	T	T	T	F
T	F	T	T	T	F	T	T	T	T	F	T	T
T	F	F	T	T	F	T	F	T	T	F	F	F
F	T	T	F	T	T	T	T	F	T	T	T	T
F	T	F	F	T	T	T	F	F	T	T	T	F
F	F	T	F	F	F	T	T	F	T	F	T	T
F	F	F	F	F	F	F	F	F	F	F	F	F
			1	2	1	4	3	5	8	6	7	6

The columns labeled 4 and 8 are identical.

65.

p	q	r	p	$\vee$	$(q$	$\wedge$	$r)$	$(p$	$\vee$	$q)$	$\wedge$	$(p$	$\vee$	$r)$
T	T	T	T	T	T	T	T	T	T	T	T	T	T	T
T	T	F	T	T	T	F	F	T	T	T	T	T	T	F
T	F	T	T	T	F	F	T	T	T	F	T	T	T	T
T	F	F	T	T	F	F	F	T	T	F	T	T	T	F
F	T	T	F	T	T	T	T	F	T	T	T	F	T	T
F	T	F	F	F	T	F	F	F	T	T	F	F	F	F
F	F	T	F	F	F	F	T	F	F	F	F	F	T	T
F	F	F	F	F	F	F	F	F	F	F	F	F	F	F
			1	4	2	3	2	5	6	5	9	7	8	7

The columns labeled 4 and 9 are identical.

67.

p	q	$(p$	$\wedge$	$q)$	$\vee$	p
T	T	T	T	T	T	T
T	F	T	F	F	T	T
F	T	F	F	T	F	F
F	F	F	F	F	F	F
		1	2	1	4	3

The p column and the column labeled 4 are identical.

69. In the diagram, two series circuits are shown, which correspond to $p \wedge q$ and $p \wedge \sim q$. These circuits, in turn, form a parallel circuit. Thus, the logical statement is

$$(p \wedge q) \vee (p \wedge \sim q).$$

One pair of equivalent statements listed in the text includes

$$(p \wedge q) \vee (p \wedge \sim q) \equiv p \wedge (q \vee \sim q)$$

Since $(q \vee \sim q)$ is always true, $p \wedge (q \vee \sim q)$ simplifies to

$$p \wedge \text{T} \equiv p.$$

71. In the diagram, a series circuit is shown, which corresponds to $\sim q \wedge r$. This circuit, in turn, forms a parallel circuit with p. Thus, the logical statement is

$$p \vee (\sim q \wedge r).$$

73. In the diagram, a parallel circuit corresponds to $p \vee q$. This circuit is parallel to $\sim p$. Thus, the total circuit corresponds to the logical statement

$$(p \vee q) \vee \sim p.$$

This statement, in turn, is equivalent to

$$(\sim p \vee p) \vee q.$$

Since $\sim p \vee p$ is always true, we have

$$\text{T} \vee q \equiv \text{T}.$$

75. In the diagram, series circuits corresponding to $p \wedge q$ and $p \wedge p$ form a parallel circuit. This parallel circuit is parallel to the series circuit corresponding to $r \wedge \sim r$. Thus, the logical statement is

$$[(p \wedge q) \vee (p \wedge p)] \vee (r \wedge \sim r).$$

This statement simplifies to p as follows:

$$\begin{aligned} &[(p \wedge q) \vee (p \wedge p)] \vee (r \wedge \sim r) \\ &\quad \equiv [(p \wedge q) \vee p] \vee (r \wedge \sim r) \\ &\quad \equiv p \vee (r \wedge \sim r) \\ &\quad \equiv p \vee \text{F} \\ &\quad \equiv p. \end{aligned}$$

77. The logical statement $p \wedge (q \vee \sim p)$ can be represented by the following circuit.

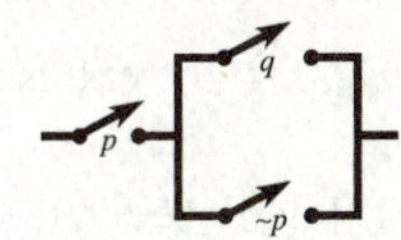

The statement $p \wedge (q \vee \sim p)$ simplifies to $p \wedge q$ as follows:

$$\begin{aligned} p \wedge (q \vee \sim p) &\equiv (p \wedge q) \vee (p \wedge \sim p) \\ &\equiv (p \wedge q) \vee \mathrm{F} \\ &\equiv p \wedge q. \end{aligned}$$

79. The logical statement $(p \vee q) \wedge (\sim p \wedge \sim q)$ can be represented by the following circuit.

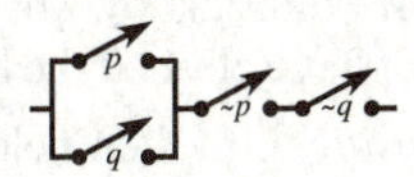

The statement $(p \vee q) \wedge (\sim p \wedge \sim q)$ simplifies to F as follows.

$$\begin{aligned} &(p \vee q) \wedge (\sim p \wedge \sim q) \\ &\equiv [p \wedge (\sim p \wedge \sim q)] \vee [q \wedge (\sim p \wedge \sim q)] \\ &\equiv [(p \wedge \sim p) \wedge \sim q)] \vee [q \wedge (\sim q \wedge \sim p)] \\ &\equiv [\mathrm{F} \wedge \sim q)] \vee [(q \wedge \sim q) \wedge \sim p] \\ &\equiv \mathrm{F} \vee (\mathrm{F} \wedge \sim p) \\ &\equiv \mathrm{F} \vee \mathrm{F} \\ &\equiv \mathrm{F} \end{aligned}$$

81. The logical statement $[(p \vee q) \wedge r] \wedge \sim p$ can be represented by the following circuit.

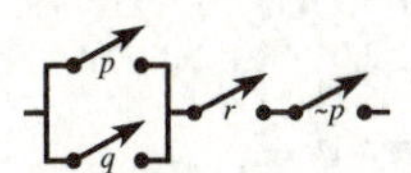

The statement $[(p \vee q) \wedge r] \wedge \sim p$ simplifies to $(r \wedge \sim p) \wedge q$ as follows:

$$\begin{aligned} &[(p \vee q) \wedge r] \wedge \sim p \\ &\equiv [(p \wedge r) \vee (q \wedge r)] \wedge \sim p \\ &\equiv [(p \wedge r) \wedge \sim p] \vee [(q \wedge r) \wedge \sim p] \\ &\equiv [p \wedge r \wedge \sim p] \vee [q \wedge r \wedge \sim p] \\ &\equiv [(p \wedge \sim p) \wedge r] \vee [(r \wedge \sim p) \wedge q] \\ &\equiv (\mathrm{F} \wedge \mathrm{r}) \vee [(r \wedge \sim p) \wedge q] \\ &\equiv \mathrm{F} \vee [(r \wedge \sim p) \wedge \mathrm{q}] \\ &\equiv (r \wedge \sim p) \wedge q \\ &\equiv r \wedge (\sim p \wedge q). \end{aligned}$$

83. The logical statement $\sim q \rightarrow (\sim p \rightarrow q)$ can be represented by the following circuit.

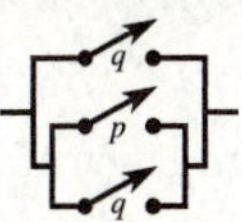

The statement $\sim q \rightarrow (\sim p \rightarrow q)$ simplifies to $p \vee q$ as follows:

$$\begin{aligned} \sim q \rightarrow (\sim p \rightarrow q) &\equiv \sim q \rightarrow (p \vee q) \\ &\equiv q \vee (p \vee q) \\ &\equiv q \vee p \vee q \\ &\equiv p \vee q \vee q \\ &\equiv p \vee (q \vee q) \\ &\equiv p \vee q. \end{aligned}$$

85. The logical statement $[(p \wedge q) \vee p] \wedge [(p \vee q) \wedge q]$ can be represented by the following circuit.

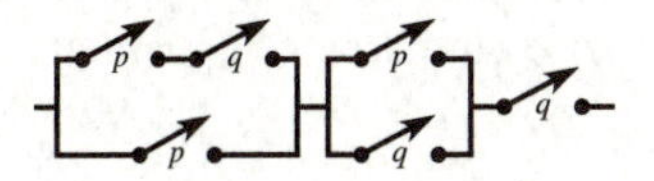

The statement simplifies to $p \wedge q$ as follows:

$$\begin{aligned} &[(p \wedge q) \vee p] \wedge [(p \vee q) \wedge q] \\ &\equiv p \wedge [(p \vee q) \wedge q] \\ &\equiv p \wedge q. \end{aligned}$$

89. Each statement has the form $p \rightarrow q$. The equivalent form using *or* is $\sim p \vee q$.

(a) You are not married at the end of the year, or you may file a joint return with your spouse.

(b) A bequest received by an executor from an estate is compensation for services, or it is tax free.

(c) A course does not improve your current job skills or does not lead to qualification for a new profession, or the course is not deductable.

91. **(a)** $(v \vee p) \rightarrow (s \wedge g)$

(b) The portfolio being worth $80,0000 means v is F, selling the Ford Motor stock at $56 per share means s is T and p is F, and keeping the proceeds means g is F.

$$\begin{gathered} (v \vee p) \rightarrow (s \wedge g) \\ (\mathrm{F} \vee \mathrm{F}) \rightarrow (\mathrm{T} \wedge \mathrm{F}) \\ \mathrm{F} \rightarrow \mathrm{F} \\ \mathrm{T} \end{gathered}$$

The statement is true.

(d) $\sim[(v \vee p) \rightarrow (s \wedge g)]$
$\equiv (v \vee p) \wedge \sim(s \wedge g)$
$\equiv (v \vee p) \wedge (\sim s \vee \sim g)$

In words, the negation of the given statement is:

The value of my portfolio exceeds $100,000 or the price of my stock in Ford Motor Company falls below $50 per share, and I will not sell all my shares of Ford stock or I will not give the proceeds to United Way.

93. $p \vee q \equiv \sim p \rightarrow q$

(a) If you cannot file a civil lawsuit yourself, then your attorney can do it for you.

(b) If your driver's license does not come with restrictions, then restrictions may sometimes be added on later.

(c) If you can marry when you're not at least 18 years old, then you have the permission of your parents or guardian.

6.4 More on the Conditional

Your Turn 1

(a) Being happy is sufficient for you to clap your hands is equivalent to If you are happy, then you clap your hands.

(b) All who seek shall find is equivalent to If you seek, then you shall find.

Your Turn 2

Given Statement: If I get another ticket, then I lose my license.
Converse: If I lose my license, then I get another ticket.
Inverse: If I didn't get another ticket, then I didn't lose my license.
Contrapositive: If I didn't lose my license, then I didn't get another ticket. The contrapositive is equivalent.

Your Turn 3

New Your City is the capital of the United States if and only if Paris is the capital of France. The statement is false because the two simple statements have different truth values.

6.4 Exercises

Wording may vary in the answers to Exercises 1–27.

1. *The direct statement:* If the exit is ahead, then I don't see it.

(a) *Converse:* If I don't see it, then the exit is ahead.

(b) *Inverse:* If the exit is not ahead, then I see it.

(c) *Contrapositive:* If I see it, then the exit is not ahead.

3. *The direct statement:* If I knew you were coming, I'd have cleaned the house.

(a) *Converse:* If I cleaned the house, then I knew you were coming.

(b) *Inverse:* If I didn't know you were coming, I wouldn't have cleaned the house.

(c) *Contrapositive:* If I didn't clean the house, then I didn't know you were coming.

5. *It is helpful to reword the given statement.*

The direct statement: If you are a mathematician, then you wear a pocket protector.

(a) *Converse:* If you wear a pocket protector, then you are a mathematician.

(b) *Inverse:* If you are not a mathematician, then you do not wear a pocket protector.

(c) *Contrapositive:* If you don't wear a pocket protector, then you are not a mathematician.

7. The direct statement: $p \rightarrow \sim q$.

(a) *Converse:* $\sim q \rightarrow p$.

(b) *Inverse:* $\sim p \rightarrow q$.

(c) *Contrapositive:* $q \rightarrow \sim p$.

9. The direct statement: $p \rightarrow (q \vee r)$.

(a) *Converse:* $(q \vee r) \rightarrow p$.

(b) *Inverse:* $\sim p \rightarrow \sim(q \vee r)$
or $\sim p \rightarrow (\sim q \wedge \sim r)$.

(c) *Contrapositive:* $(\sim q \wedge \sim r) \rightarrow \sim p$.

13. The statement "Your signature implies that you accept the conditions" becomes "If you sign, then you accept the conditions."

15. The statement "You can take this course pass/fail only if you have prior permission" becomes "If you can take this course pass/fail, then you have prior permission."

17. The statement "You can skate on the pond when the temperature is below 10°" becomes "If the temperature is below 10°, then you can skate on the pond."

19. The statement "Eating ten hot dogs is sufficient to make someone sick" becomes "If someone eats ten hot dogs, then he or she will get sick."

21. The statement "A valid passport is necessary for travel to France" becomes "If you travel to France, then you have a valid passport."

23. The statement "For a number to have a real square root, it is necessary that it be nonnegative" becomes "If a number has a real square root, then it is nonnegative."

25. The statement "All brides are beautiful" becomes "If someone is a bride, then she is beautiful."

27. The statement "A number is divisible by 3 if the sum of its digits is divisible by 3" becomes "If the sum of a number's digits is divisible by 3, then it is divisible by 3."

29. Option d is the answer since "r is necessary for s" represents the converse, $s \rightarrow r$, of all of the other statements.

33. The statement "$5 = 9 - 4$ if and only if $8 + 2 = 10$" is true, since this is a biconditional composed of two true statements.

35. The statement "$8 + 7 \neq 15$ if and only if $3 \times 5 \neq 9$" is false, since this is a biconditional consisting of one false statement and one true statement.

37. The statement "China is in Asia if and only if Mexico is in Europe" is false, since it is a biconditional consisting of a true statement and a false statement.

39.

p	q	$(\sim p$	$\wedge$	$q)$	$\leftrightarrow$	$(p$	$\rightarrow$	$q)$
T	T	F	F	T	F	T	T	T
T	F	F	F	F	T	T	F	F
F	T	T	T	T	T	F	T	T
F	F	T	F	F	F	F	T	F
		1	2	1	5	3	4	3

41. (a) If it is an employee contribution, then it must be reported on Form 8889.

(b) If certain tax benefits may be claimed by married persons, then they file jointly.

(c) If a child provides over half of his or her own support, then the child is not a qualifying child.

43. *Given statement:*

If your account is in default, then we may close your account without notice.

Converse:

If we close your account without notice, then your account is in default.

Inverse:

If your account is not in default, then we may not close your account without notice.

Contrapositive:

If we do not close your account without notice, then your account is not in default.

The original statement and the contrapositive are equivalent, and the converse and inverse are equivalent.

45. (a) Let p represent "there are triplets," let q represent "the most persistent stands to gain an extra meal," and let r represent "it may eat at the expense of another." Then the statement can be written as $p \rightarrow (q \wedge r)$.

(b) The contrapositive is $\sim(q \wedge r) \rightarrow p$, which is equivalent to $(\sim q \vee \sim r) \rightarrow p$: If the most persistent does not stand to gain an extra meal or it does not eat at the expense of another, then there are not triplets.

47. (a) *Converse:* If you can't get married again, then you are married.

Inverse: If you aren't married, then you can get married again.

Contrapositive: If you can get married again, then you are not married.

(b) *Converse:* If you are protected by the Fair Credit Billing Act, then you pay for your purchase with a credit card.

Inverse: If you do not pay for your purchase with a credit card, then you are not protected by the Fair Credit Billing Act.

Contrapositive: If you are not protected by the Fair Credit Billing Act, then you do not pay for your purchase with a credit card.

(c) *Converse:* If you're expected to make a reasonable effort to locate the owner, then you hit a parked car.

Inverse: If you did not hit a parked car, then you are not expected to make a reasonable effort to locate the owner.

Contrapositive: If you are not expected to make a reasonable effort to locate the owner, then you did not hit a parked car.

The original statement and the contrapositive are equivalent, and the converse and inverse are equivalent.

49. **(a)** Let d represent "political development in Western Europe will increase" and let a represent "social assimilation is increasing." Then the statement can be written as $d \leftrightarrow a$. The truth table for the statement is as follows.

d	a	$d \leftrightarrow a$
T	T	T
T	F	F
F	T	F
F	F	T

(b) If a is true and d is false then $d \leftrightarrow a$ is false.

51. If a country has democracy, then it has a high level of education.

Converse: If a country has a high level of education, then it has democracy.

Inverse: If a country does not have democracy, then it does not have a high level of education.
Contrapositive: If a country does not have a high level of education, then it does not have democracy.

The contrapositive is equivalent to the original.

53. The rule "If a card has a D on one side, then it must have a 3 on the other side" is violated when a card has a D on one side and the number on the other side is not 3. Thus, we only need to turn over cards that have a D or a number other than 3.

D card: This card must be turned over to see whether the rule has been violated. If the number on the other side is 3, the rule has not been violated; however, if the number is not 3, the rule has been violated.

F card: Since the premise of the rule is false for this card, the rule automatically holds. Thus, this card does not need to be turned over.

3 card: Since the conclusion of the rule is true, the rule automatically holds. Thus, this card does not need to be turned over.

7 card: This card must be turned over to see whether the rule has been violated. If the letter is D, then the rule has been violated; however, if the letter on the other side is not D, the rule has not been violated.

55. **(a)** If … then *form:* If nothing is ventured, then nothing is gained.
Contrapositive: If something is gained, then something is ventured.
Statement using or: Something is ventured or nothing is gained.

(b) If … then *form:* If something is one of the best things in life, then it is free.
Contrapositive: If something is not free, then it is not one of the best things in life.
Statement using or: Something is not one of the best things in life or it is free.

(c) If … then *form:* If something is a cloud, then it has a silver lining.
Contrapositive: If something doesn't have a silver lining, then it isn't a cloud.
Statement using or: Something is not a cloud or it doesn't have a silver lining.

57. **(a)** If you can score in this box, then the dice show any sequence of four numbers.
You cannot score in this box, or the dice show any sequence of four numbers.

(b) If two or more words are formed in the same play, then each is scored.
Two or more words are not formed in the same play, or each is scored.

(c) If words are labeled as a part of speech, then they are permitted.
Words are not labeled as parts of speech, or they are permitted.

6.5 Analyzing Arguments and Proofs

Your Turn 1

Let t represent "You watch television tonight," p represent "You write your paper tonight," and g represent "You get a good grade."

The given argument written symbolically is

1.	$t \vee p$	T
2.	$p \rightarrow g$	T
3.	g	T
	$\sim t$	F

Invalid argument; If t is true, p is true, and g is true, the premises are true, but the conclusion is false.

Your Turn 2

Let m represent "You put money in the parking meter", c represent "You buy a cup of coffee", and t represent "You get a ticket".

The given argument written symbolically as

1.	$\sim m \vee \sim c$	Premise
2.	$\sim m \rightarrow t$	Premise
3.	$\sim t$	Premise
4.	m	Statements 2, 3, Modus Tollens
5.	$\sim c$	Statements 1, 4, Disjunctive Syllogism

Conclusion: You did not buy a cup of coffee. Valid argument.

6.5 Exercises

1. Let p represent "she weighs the same as a duck," q represent "she's made of wood," and r represent "she's a witch." The argument is then represented symbolically by:

$$\begin{array}{l} p \rightarrow q \\ \underline{q \rightarrow r} \\ p \rightarrow r. \end{array}$$

This is the valid argument form Reasoning by Transitivity.

3. Let p represent "I had the money" and q represent "I'd go on vacation." The argument is then represented symbolically by:

$$\begin{array}{l} p \rightarrow q \\ \underline{p \qquad} \\ q. \end{array}$$

This is the valid argument form Modus Ponens.

5. Let p represent "you want to make trouble" and q represent "the door is that way." The argument is then represented symbolically by:

$$\begin{array}{l} p \rightarrow q \\ \underline{q \qquad} \\ p. \end{array}$$

Since this is the form Fallacy of the Converse, it is invalid and considered a fallacy.

7. Let p represent "Andrew Crowley plays" and q represent "the opponent gets shut out." The argument is then represented symbolically by:

$$\begin{array}{l} p \rightarrow q \\ \underline{\sim q \qquad} \\ \sim p. \end{array}$$

This is the valid argument form Modus Tollens.

9. Let p represent "we evolved a race of Isaac Newtons" and q represent "that would not be progress." The argument is then represented symbolically by:

$$\begin{array}{l} p \rightarrow q \\ \underline{\sim p \qquad} \\ \sim q. \end{array}$$

Note that since we let q represent "that would not be progress," $\sim q$ represents "that is progress." Since this is the form Fallacy of the Inverse, it is invalid and considered a fallacy.

11. Let p represent "Something is rotten in the state of Denmark" and q represent "my name isn't Hamlet." The argument is then represented symbolically by:

$$\begin{array}{l} p \vee q \text{ (or } q \vee p) \\ \underline{\sim q \qquad\qquad} \\ \sim p. \end{array}$$

Since this is the form Disjunctive Syllogism, it is a valid argument.

To show validity for the arguments in the following exercises we must show that the conjunction of the premises implies the conclusion. That is, the conditional statement $[P_1 \wedge P_2 \wedge \ldots \wedge P_n] \rightarrow C$ must be a tautology.

13. 1. $p \vee q$ T

2. $\underline{p}$ T

$\sim q$ F

The argument is <u>invalid</u>. When $p = \text{T}$ and $q = \text{T}$, the premises are true but the conclusion is false.

15. 1. $p \rightarrow q$ T

2. $\underline{q \rightarrow p}$ T

$p \wedge q$ F

The argument is <u>invalid</u>. When $p = \text{F}$ and $q = \text{F}$, the premises are true but the conclusion is false.

17. 1. $\sim p \rightarrow \sim q$ Premise

2. q Premise

3. p 1, 2, Modus Tollens

The argument is <u>valid</u>.

19. 1. $p \rightarrow q$ Premise

2. $\sim q$ Premise

3. $\sim p \rightarrow r$ Premise

4. $\sim p$ 1, 2, Modus Tollens

5. r 3, 4, Modus Ponens

The argument is <u>valid</u>.

21. 1. $p \rightarrow q$ Premise

2. $q \rightarrow r$ Premise

3. $\sim r$ Premise

4. $p \rightarrow r$ 1, 2, Transitivity

5. $\sim p$ 3, 4, Modus Tollens

The argument is <u>valid</u>.

23. 1. $p \rightarrow q$ Premise

2. $q \rightarrow \sim r$ Premise

3. p Premise

4. $r \vee s$ Premise

5. q 1, 3, Modus Ponens

6. $\sim r$ 2, 5, Modus Ponens

7. s 4, 6, Disjunctive Syllogism

The argument is <u>valid</u>.

25. Make a truth table for the statement $(p \wedge q) \rightarrow p$.

p	q	$(p$	$\wedge$	$q)$	$\rightarrow$	p
T	T	T	T	T	T	T
T	F	T	F	F	T	T
F	T	F	F	T	T	F
F	F	F	F	F	T	F
		1	2	1	3	2

Since the final column, 3, indicates that the conditional statement that represents the argument is true for all possible truth values of p and q, the statement is a tautology.

27. Make a truth table for the statement $(p \wedge q) \rightarrow (p \wedge q)$.

p	q	$(p$	$\wedge$	$q)$	$\rightarrow$	$(p$	$\wedge$	$q)$
T	T	T	T	T	T	T	T	T
T	F	T	F	F	T	T	F	F
F	T	F	F	T	T	F	F	T
F	F	F	F	F	T	F	F	F
		1	2	1	5	3	4	3

Since the final column, 3, indicates that the conditional statement that represents the argument is true for all possible truth values of p and q, the statement is a tautology.

29. Let a represent "Alex invests in AT&T," s represent "Sophia invests in Sprint Nextel," and v represent "Victor invests in Verizon".

The given argument written symbolically is

$$\begin{array}{c} a \rightarrow s \\ v \vee a \\ \underline{\sim v} \\ s \end{array}$$

1. $a \rightarrow s$ Premise
2. $v \vee a$ Premise
3. $\sim v$ Premise
4. a 2, 3, Disjunctive Syllogism
5. s 1, 4, Modus Ponens

The argument is <u>valid</u>.

31. Let b represent "it is a bearish market," p represent "prices are rising," and i represent "the investor sells stocks."

The given argument written symbolically is

$$\begin{array}{ll} b & \text{T} \\ p \to \sim b & \text{T} \\ \sim p \to i & \text{T} \\ \hline \sim i & \text{F} \end{array}$$

The argument is invalid. When $b = \text{T}$, $p = \text{F}$, and $i = \text{T}$, the premises are true but the conclusion is false.

33. Let s represent "animal is a spider," i represent "animal is an insect," l represent "animal has eight legs," and b represent "animal has two main body parts."

The given argument written symbolically is

$$\begin{array}{l} s \vee i \\ s \to (l \wedge b) \\ \sim l \vee \sim b \\ \hline i \end{array}$$

1. $s \vee i$ Premise
2. $s \to (l \wedge b)$ Premise
3. $\sim l \vee \sim b$ Premise
4. $\sim(l \wedge b)$ 3, DeMorgan's Law
5. $\sim s$ 2, 4, Modus Tollens
6. i 1, 5, Disjunctive Syllogism

The argument is valid.

35. Let m represent "I am married to you," o represent "we are one," and r represent "you are really a part of me."

The given argument written symbolically is

$$\begin{array}{ll} m \to o & \text{T} \\ o \vee \sim r & \text{T} \\ m \vee r & \text{T} \\ \hline m \to r & \text{F} \end{array}$$

The argument is invalid. When $m = \text{T}$, $o = \text{T}$, and $r = \text{F}$, the premises are true but the conclusion is false.

37. Let y represent "the Yankees will be in the World Series," p represent "the Phillies will be in the World Series," and n represent "the National League wins." The argument is then represented symbolically by:

$$\begin{array}{l} y \vee \sim p \\ \sim p \to \sim n \\ n \\ \hline y. \end{array}$$

The argument is valid.

1. $y \vee \sim p$ Premise
2. $\sim p \to \sim n$ Premise
3. n Premise
4. p 2, 3, Modus Tollens
5. y 1, 4, Disjunctive Syllogism

39. **(a)** $d \to \sim w$ **(b)** $o \to w$ or $w \to \sim o$
(c) $p \to d$ **(d)** $p \to \sim o$,

Conclusion: If it is my poultry, then it is not an officer. In Lewis Carroll's words, "My poultry are not officers."

41. **(a)** $b \to \sim t$ or $t \to \sim b$
(b) $w \to c$
(c) $\sim b \to h$
(d) $\sim w \to \sim p$ or $p \to w$
(e) $c \to t$
(f) $p \to h$,

Conclusion: If one is a pawnbroker, then one is honest. In Lewis Carroll's words, "No pawnbroker is dishonest."

43. **(a)** $d \to p$
(b) $\sim t \to \sim i$
(c) $r \to \sim f$ or $f \to \sim r$
(d) $o \to d$ or $\sim d \to \sim o$
(e) $\sim c \to i$
(f) $b \to s$
(g) $p \to f$
(h) $\sim o \to \sim c$ or $c \to o$
(i) $s \to \sim t$ or $t \to \sim s$
(j) $b \to \sim r$,

Conclusion: If it is written by Brown, then I can't read it. In Lewis Carroll's words, "I cannot read any of Brown's letters."

6.6 Analyzing Arguments with Quantifiers

Your Turn 1

(a) All college students study.

Let $c(x)$ represent "x is a college student" and $s(x)$ represent "x studies."

The statement can be written as

$$\forall x\,[c(x) \rightarrow s(x)]$$

Its negation is

$$\exists x\{\sim[c(x) \rightarrow s(x)]\},$$

which is equivalent to $\exists x\,[c(x) \wedge \sim s(x)]$.

In words, "Some college students do not study."

(b) Some professors are not organized.

Let $p(x)$ represent "x is a professor" and $o(x)$ represent "x is organized."

The statement can be written as

$$\exists x\,[p(x) \wedge \sim o(x)]$$

Its negation is

$$\forall x\,\{\sim[p(x) \wedge \sim o(x)]\},$$

which is equivalent to $\forall x\,[p(x) \rightarrow o(x)]$.

In words, "All professors are organized."

Your Turn 2

All insects are arthropods.
<u>A bee is an insect.</u>
A bee is an arthropod.

Let $i(x)$ represent "x is an insect," $a(x)$ represent "x is an arthropod," and b represent a "bee."

The argument becomes

$$\begin{array}{l} \forall x[i(x) \rightarrow a(x)] \\ \underline{i(b) \qquad\qquad} \\ a(b) \end{array}$$

The argument resembles Modus Ponens, so the argument is <u>valid</u>.

Your Turn 3

All birds have wings.
<u>Rover does not have wings.</u>
Rover is not a bird.

Let $b(x)$ represent "x is a bird," $w(x)$ represent "x has wings," and r represent "Rover."

The argument becomes

$$\begin{array}{l} \forall x\,[b(x) \rightarrow w(x)] \\ \underline{\sim w(r) \qquad\qquad} \\ \sim b(r) \end{array}$$

The argument resembles Modus Tollens, so the argument is <u>valid</u>.

Your Turn 4

Every man has his price.
<u>Sam has a price.</u>
Sam is a man.

Let $m(x)$ represent "x is a man," $p(x)$ represent "x has a price," and s represent "Sam."

The argument becomes

$$\begin{array}{l} \forall x[m(x) \rightarrow p(x)] \\ \underline{p(s) \qquad\qquad} \\ m(s) \end{array}$$

Draw an Euler diagram.

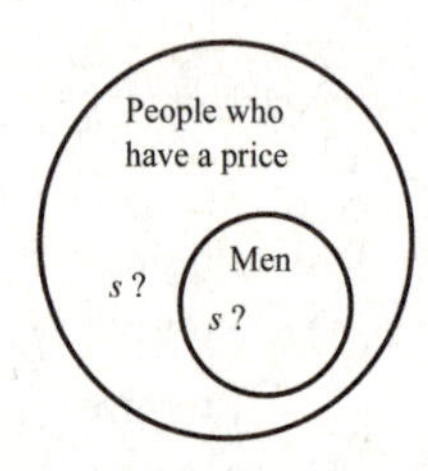

It is possible that Sam is not a man.

The argument resembles the Fallacy of the Converse, so the argument is <u>invalid</u>.

Your Turn 5

Some vegetarians eat eggs.
<u>Sarah is a vegetarian.</u>
Sarah eats eggs.

Let $v(x)$ represent "x is a vegetarian," $e(x)$ represent "x eats eggs," and s represent "Sarah."

The argument becomes

$$\begin{array}{l} \exists x[v(x) \wedge e(x)] \\ \underline{v(s) \qquad\qquad} \\ e(s) \end{array}$$

Draw an Euler diagram.

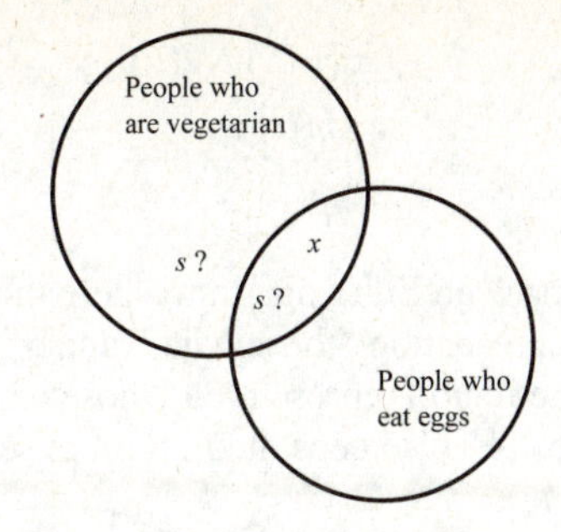

It's possible that Sarah is a vegetarian but does not eat eggs.

The argument is invalid.

6.6 Exercises

1. Let $b(x)$ represent "x is a book" and $s(x)$ represent" "x is a bestseller."
   - **(a)** $\exists x[b(x) \wedge s(x)]$
   - **(b)** $\forall x[b(x) \rightarrow \sim s(x)]$
   - **(c)** No books are bestsellers.

3. Let $c(x)$ represent "x is a CEO" and $s(x)$ represent "x sleeps well at night."
   - **(a)** $\forall x[c(x) \rightarrow \sim s(x)]$
   - **(b)** $\exists x[c(x) \wedge s(x)]$
   - **(c)** There is a CEO who sleeps well at night.

5. Let $l(x)$ represent "x is a leaf" and $b(x)$ represent "x is brown."
   - **(a)** $\forall x[l(x) \rightarrow b(x)]$
   - **(b)** $\exists x[l(x) \wedge \sim b(x)]$
   - **(c)** There is a leaf that's not brown.

7. **(a)** Let $g(x)$ represent "x is a graduate" and $f(x)$ represent "x wants to find a good job." Let t represent Theresa Cortesini. We can represent the argument symbolically as follows.

$$\begin{array}{l} \forall x[g(x) \rightarrow f(x)] \\ \underline{g(t)} \\ f(t) \end{array}$$

   (b) Draw an Euler diagram where the region representing "graduates" must be inside the region representing "people who want to find good jobs" so that the first premise is true.

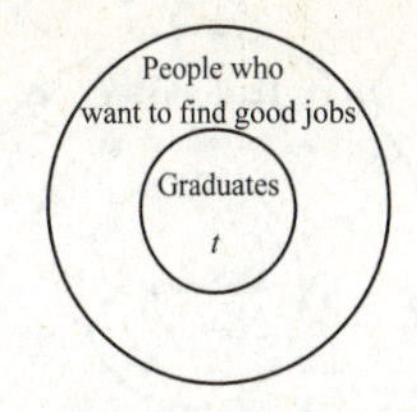

By the second premise, t must lie in the "graduates" region. Since this forces the conclusion to be true, the argument is valid.

9. **(a)** Let $p(x)$ represent "x is a professor" and $c(x)$ represent "x is covered with chalk dust." Let o represent Otis Taylor. We can represent the argument symbolically as follows.

$$\begin{array}{l} \forall x[p(x) \rightarrow c(x)] \\ \underline{c(o)} \\ p(o) \end{array}$$

   (b) Draw an Euler diagram where the region representing "professors" must be inside the region representing "those who are covered with chalk dust" so that the first premise is true.

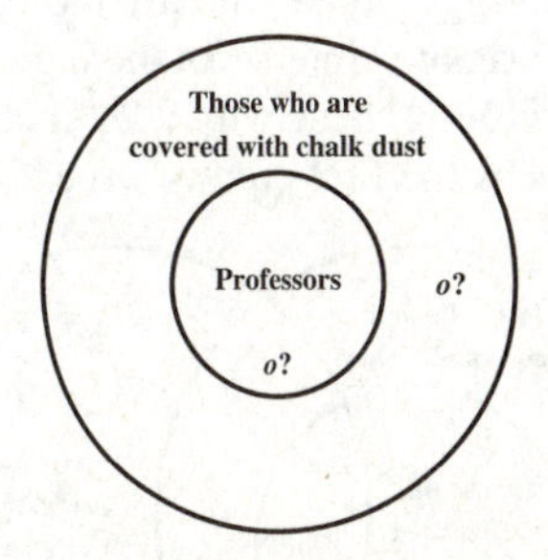

By the second premise, o must lie in the "those who are covered with chalk dust" region. Thus, o could be inside or outside the inner region "professors." Since this allows for a false conclusion (Otis doesn't have to be a professor to be covered with chalk dust), the argument is invalid.

11. **(a)** Let $c(x)$ represent "x is an accountant" and $p(x)$ represent "x uses a spreadsheet." Let n represent Nancy Hart. We can represent the argument symbolically as follows.

$$\begin{array}{l} \forall x[c(x) \rightarrow p(x)] \\ \underline{\sim p(n)} \\ \sim c(n) \end{array}$$

   (b) Draw an Euler diagram where the region representing "accountants" must be inside the region representing "those who use spreadsheets" so that the first premise is true.

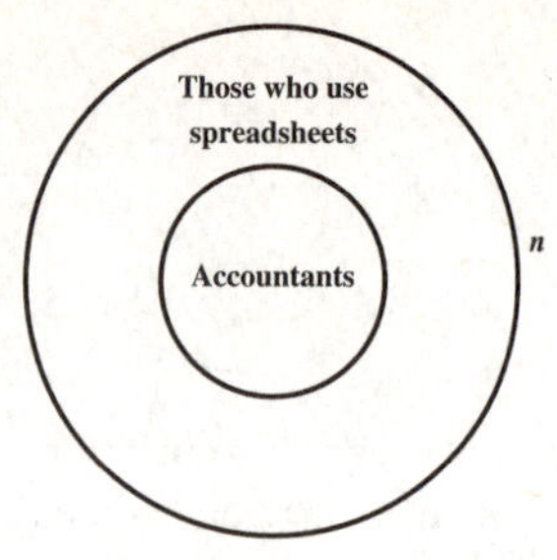

By the second premise, n must lie outside the region representing "those who use spreadsheets." Since this forces the conclusion to be true, the argument is valid.

13. (a) Let $t(x)$ represent "x is turned down for a mortgage," $s(x)$ represent "x has a second income," and $b(x)$ represent "x needs a mortgage broker." We can represent the argument symbolically as follows.

$$\exists x[t(x) \wedge s(x)]$$
$$\forall x[t(x) \rightarrow b(x)]$$
$$\overline{\exists x[s(x) \wedge b(x)]}$$

(b) Draw an Euler diagram where the region representing "Those who are turned down for a mortgage" intersects the region representing "Those with a 2nd income." This keeps the first premise true.

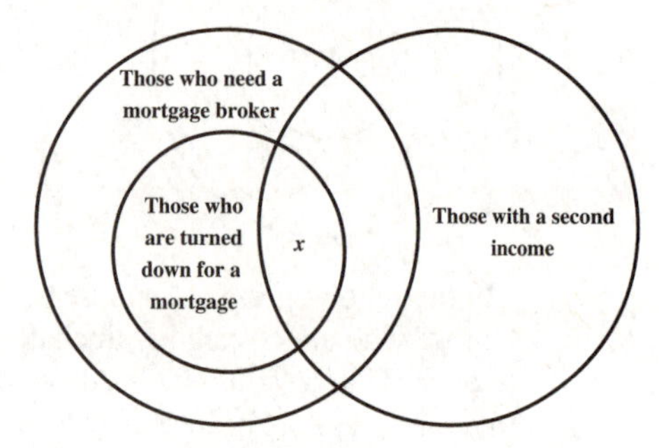

By the second premise the region representing "Those who are turned down for a mortgage" must be inside the region "Those who need a mortgage broker." Since x lies in both regions "These with a 2nd income" and "Those who need a mortgage broker," the conclusion is true and so the argument is valid.

15. (a) Let $w(x)$ represent "x wanders" and $l(x)$ represent "x is lost." Let m represent Marty McDonald. We can represent the argument symbolically as follows.

$$\exists x[w(x) \wedge l(x)]$$
$$w(m)$$
$$\overline{l(m)}$$

(b) Draw an Euler diagram where the region representing "those who wander" intersects the region representing "those who are lost." This keeps the first premise true.

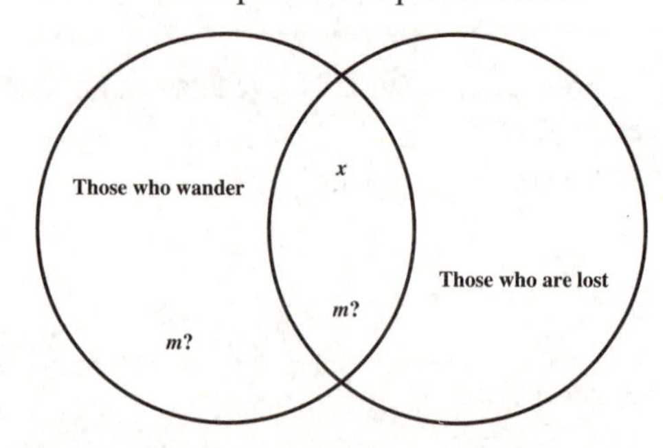

By the second premise, m must lie in the region representing "those who wander." But, m could be inside or outside the region representing "those who are lost." Since this allows for a false conclusion, the argument is invalid.

17. (a) Let $p(x)$ represent "x is a psychologist," $u(x)$ represent "x is a university professor," and $r(x)$ represent "x has a private practice." We can represent the argument symbolically as follows.

$$\exists x[p(x) \wedge u(x)]$$
$$\exists x[p(x) \wedge r(x)]$$
$$\overline{\exists x[u(x) \wedge r(x)]}$$

(b) Draw an Euler diagram where the region representing "psychologists" and "university professors" intersect each other to keep the first premise true. Then add a region representing "those with a private practice" intersecting the region "university professors" to keep the second premise true. In the most general case, this region should also intersect the region representing "psychologists."

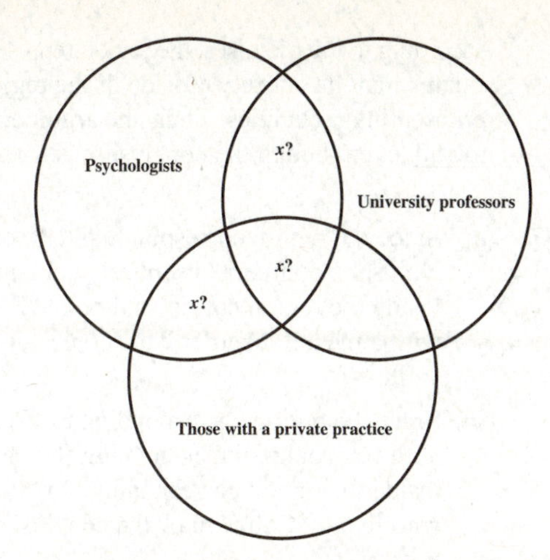

By the first premise, x must lie in the region shared by "psychologists" and "university professors"; by the second premise, x must lie in the region shared by "psychologists" and "those with a private practice." But x may not lie in the region shared by all three regions. Since this diagram shows true premises but a false conclusion, the argument is invalid.

19. **(a)** Let $a(x)$ represent "x is a saint" and $i(x)$ represent "x is a sinner." We can represent the argument symbolically as follows.

$$\forall x[a(x) \vee i(x)]$$
$$\underline{\exists x[\sim a(x)]}$$
$$\exists x[i(x)]$$

(b) Draw an Euler diagram where the region representing "saints" intersects the region representing "sinners" to keep the first premise true.

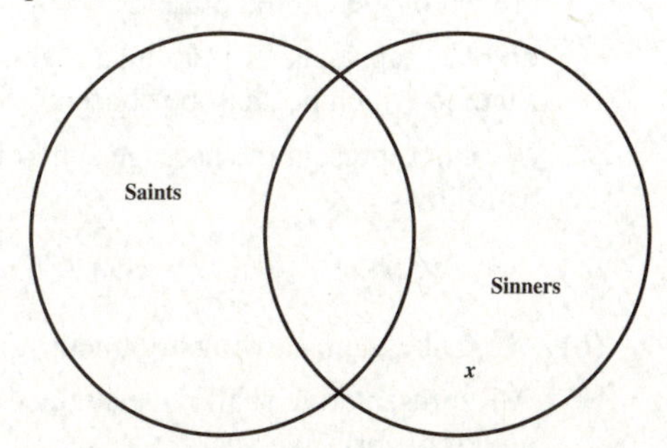

By the second premise, x must lie outside the region representing "saints." But that means x must lie in the part of the "sinners" region not shared by the "saints" region. Hence, the conclusion is true and so the argument is valid.

21. Interchanging the second premise and the conclusion of Example 4 yields the following argument.

All well-run businesses generate profits.
Monsters, Inc. is a well-run business.

Monsters, Inc. generates profits.

Draw an Euler diagram where the region representing "well-run businesses" must be inside the region representing "things that generate profits" so that the first premise is true.

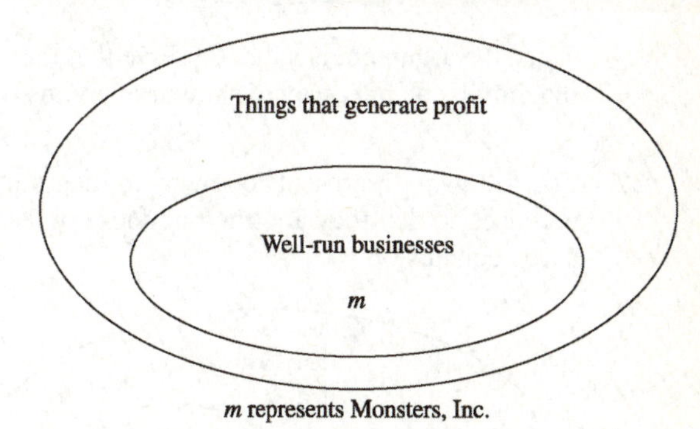

m represents Monsters, Inc.

Let m represent Monsters, Inc. By the second premise, m must lie inside the region representing "well-run businesses." Since this forces the conclusion to be true, the argument is valid, which makes the answer to the question "yes."

23. Since the region representing "major league baseball players" lies entirely inside the region representing "people who earn at least $300,000 a year," a possible first premise is

All major league baseball players earn at least $300,000 a year.

And if r represents Ryan Howard and r is inside the region representing "major league baseball players," then a possible second premise is

Ryan Howard is a major league baseball player.

A valid conclusion drawn from these two premises is that

Ryan Howard earns at least $300,000 a year.

Therefore, a valid argument based on the Euler diagram is as follows.

All major league baseball players earn at least $300,000 a year.
Ryan Howard is a major league baseball player.

Ryan Howard earns at least $300,000 a year.

25. The following diagram yields true premises. It also forces the conclusion to be true.

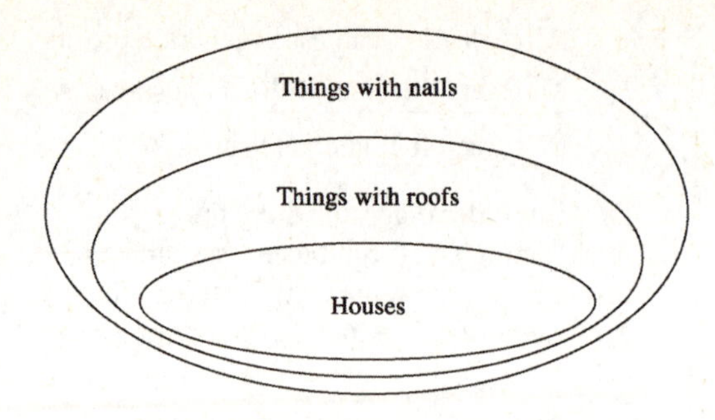

Thus, the argument is valid. Observe that the diagram is the only way to show true premises.

27. The following represents one way to diagram the premises so that they are true but does not lead to a true conclusion.

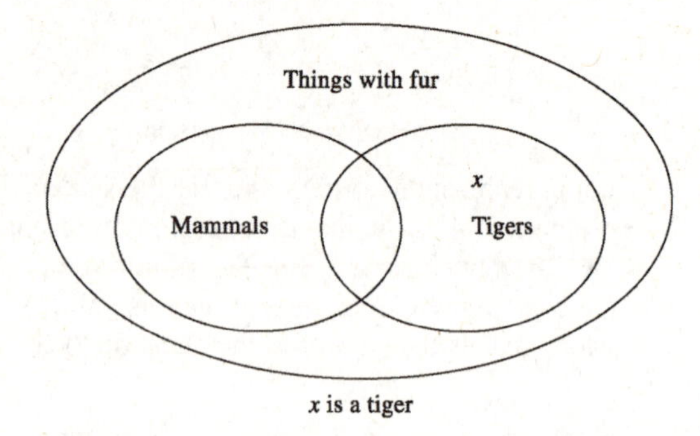

x is a tiger

If we let x be a tiger, according to the premises, x could also be a tiger but not a mammal. Thus, the argument is invalid.

29. The following Euler diagram illustrates that the conclusion is not forced to be true.

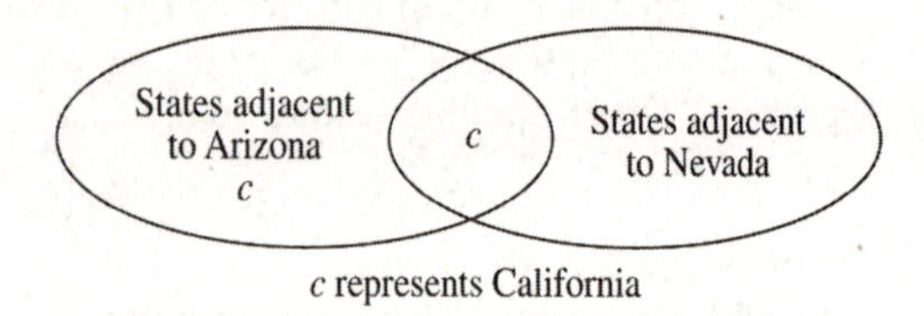

c represents California

The argument is invalid even though the conclusion is true.

31. The following Euler diagram represents true premises.

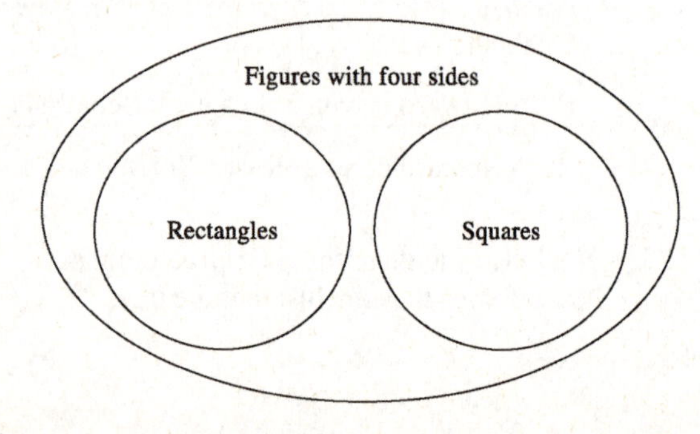

According to the premises, the region representing squares may lie entirely outside of the region representing rectangles. Thus, the argument is invalid, even though the conclusion is true.

37. **(a)** Since the region corresponding to "People with schizophrenia" is entirely contained inside the region corresponding to "People with a mental disorder," the conclusion is valid.

(b) Since the region corresponding to "People with schizophrenia" is not entirely contained inside the region corresponding to "People who live in California," the conclusion is invalid.

(c) Since some of the region corresponding to "People with schizophrenia" also lies in the region corresponding to "People who live in California," the conclusion is valid.

(d) Since some of the region corresponding to "People who live in California" lies outside the region corresponding to "People with schizophrenia" the conclusion is valid.

(e) Since some of the region corresponding to "People with a mental disorder" lies outside the region corresponding to "People with schizophrenia," the conclusion is invalid.

Thus, the answer is a, c, and d.

39. **(a)** Let $r(x)$ represent "x is a Representative,"

$a(x)$ represent "x has attained to the age of twenty-five years,"

$c(x)$ represent "x has been seven years a citizen of the United States,"

and $i(x)$ represent "x is an inhabitant of that State in which he shall be chosen."

We can represent the passage symbolically as follows.

$$\forall x\,\{r(x) \rightarrow [a(x) \wedge c(x) \wedge i(x)]\}$$

(b) We make the following argument.

A Representative shall have attained to the age of twenty-five years.

A Representative shall have been seven years a citizen of the United states.

A Representative shall, when elected, be an inhabitant of that State in which he shall be chosen.

John Boehner is a Representative.

John Boehner has attained to the age of twenty-five years, and been seven years a citizen of the United States, and was, when elected, an inhabitant of that State in which he was chosen.

The conclusion is true. Thus, the argument is valid.

(c) Draw an Euler diagram where the region representing "representative" must be inside the three other regions and the three other regions intersect each other. Let b represent John Boehner. By the fourth premise, b must lie in the region representing "Representative" and, thus, be in all of the regions.

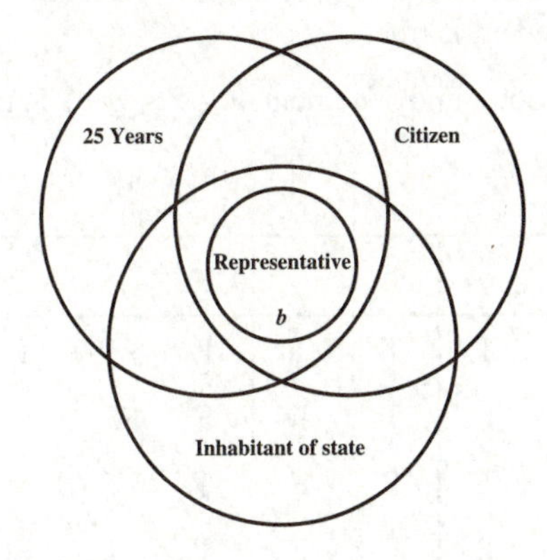

41. **(a)** Let $s(x)$ represent "x is a State,"

$t(x)$ represent "x enters into a treaty,"

$a(x)$ represent "x enters into an alliance,"

and $c(x)$ represent "x enters into a confederation."

We can represent the passage symbolically as follows.

$$\forall x \{s(x) \rightarrow \sim[t(x) \vee a(x) \vee c(x)]\}$$

(b) We make the following argument.

No State shall enter into any treaty, alliance, or confederatrion.

Texas is a state.

Texas shall not enter into any treaty, alliance, or confederation.

The conclusion is true. Thus, the argument is valid.

(c) Draw an Euler diagram where the region representing "states" must be outside the other three regions and the three other regions intersect. Let t represent Texas. By the second premise, t must lie in the region representing "states" and, thus, must be outside the other three regions.

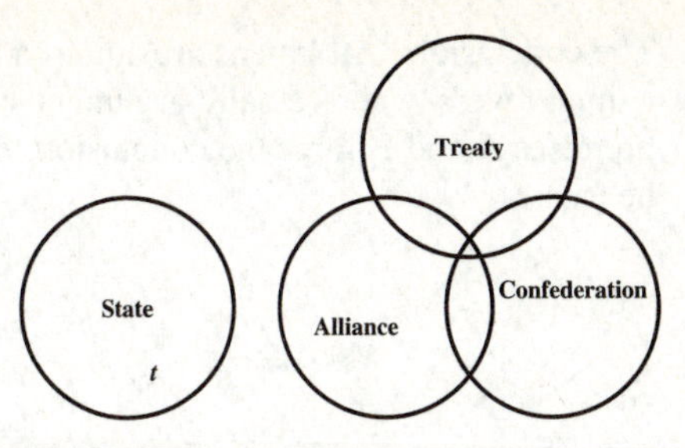

In Exercises 43–47, the premises marked A, B, and C are followed by several possible conclusions. Take each conclusion in turn, and check the resulting argument as valid *or* invalid.

A. *All kittens are cute animals.*

B. *All cute animals are admired by animal lovers.*

C. *Some dangerous animals are admired by animal lovers.*

Diagram the first two premises to be true. Then, notice that premise C is correctly represented by Case I, Case II, or Case III in the diagram.

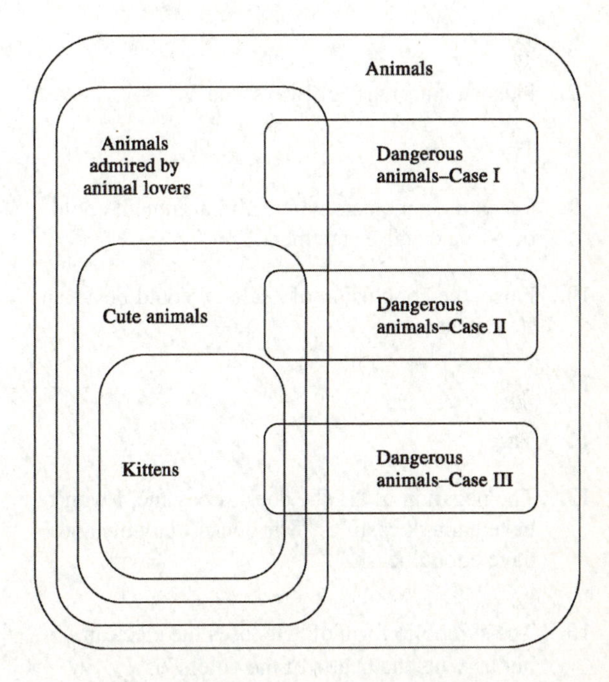

43. We are not forced into the conclusion "Some kittens are dangerous animals" since Case I and Case II represent true premises where this conclusion is false. Thus, the argument is invalid.

45. We are not forced into the conclusion "Some dangerous animals are cute" since Case I represents true premises where this conclusion is false. Thus, the argument is invalid.

47. The conclusion "All kittens are admired by animal lovers" yields a <u>valid</u> argument since premises A and B force the conclusion to be true.

Chapter 6 Review Exercises

1. True

2. False; a truth table with 5 variables has $2^5 = 32$ rows.

3. False; the number of rows any truth table has is always a power of 2, which is an even number.

4. False; the negation of a disjunction results in a conjunction.

5. False; the negation of a conditional statement is a conjunction.

6. True

7. False; a tautology is always true.

8. True

9. False; the conclusion of a valid argument could be a true or false statement.

10. False; the conclusion of a fallacy could be a true statement.

11. True

12. True

13. The negation of "If she doesn't pay me, I won't have enough cash" is "She doesn't pay me and I have enough cash."

15. The symbolic form of "He loses the election, but he wins the hearts of the voters" is $l \wedge w$.

17. The symbolic form of "He loses the election only if he doesn't win the hearts of the voters" is $l \rightarrow \sim w$.

19. Writing the symbolic form $\sim l \wedge w$ in words, we get "He doesn't lose the election and he wins the hearts of the voters."

21. Replacing q and r with the given truth values, we have

$$\sim F \wedge \sim F$$
$$T \wedge T$$
$$T$$

The compound statement $\sim q \wedge \sim r$ is true.

23. Replacing r with the given truth value (s not known), we have

$$F \rightarrow (s \wedge F)$$
$$T$$

since a conditional statement with a false antecedent is true.

The compound statement $r \rightarrow (s \vee r)$ is true.

27.

p	q	p	$\wedge$	$(\sim p$	$\vee$	$q)$
T	T	T	T	F	T	T
T	F	T	F	F	F	F
F	T	F	F	T	T	T
F	F	F	F	T	T	F
		1	4	2	3	2

The statement is not a tautology.

29. "All mathematicians are loveable" can be restated as "If someone is a mathematician, then that person is loveable."

31. "Having at least as many equations as unknowns is necessary for a system to have a unique solution" can be restated as "If a system has a unique solution, then it has at least as many equations as unknowns."

33. *The direct statement*: If the proposed regulations have been approved, then we need to change the way we do business.
   (a) *Converse:* If we need to change the way we do business, then the proposed regulations have been approved.
   (b) *Inverse:* If the proposed regulations have not been approved, then we do not need to change the way we do business.
   (c) *Contrapositive:* If we do not need to change the way we do business, then the proposed regulations have not been approved.

35. In the diagram, a series circuit corresponding to $p \wedge p$ is followed in series by a parallel circuit represented by $\sim p \vee q$. The logical statement is $(p \wedge p) \wedge (\sim p \vee q)$. This statement is equivalent to $p \wedge q$.

p	q	(p	∧	p)	∧	(~p	∨	q)	p	∧	q
T	T	T	T	T	T	F	T	T	T	T	T
T	F	T	T	T	F	F	F	F	T	F	F
F	T	F	F	F	F	T	T	T	F	F	T
F	F	F	F	F	F	T	T	F	F	F	F
		1	2	1	6	3	5	4	7	9	8

Columns 6 and 9 are identical.

37. The logical statement $(p \wedge q) \vee (p \wedge p)$ can be represented by the following circuit.

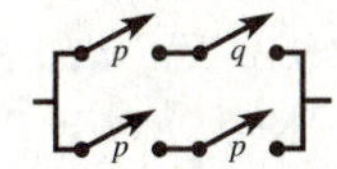

The statement simplifies to p as follows:

$$(p \wedge q) \vee (p \wedge p) = p \wedge (q \wedge p)$$
$$= p$$

39.

p	q	(p	⊻	q)	(p	∨	q)	∧	~	(p	∧	q)
T	T	T	F	T	T	T	T	F	F	T	T	T
T	F	T	T	F	T	T	F	T	T	T	F	F
F	T	F	T	T	F	T	T	T	T	F	F	T
F	F	F	F	F	F	F	F	F	T	F	F	F
		1	2	1	3	4	3	8	7	5	6	5

The columns labeled 2 and 8 are identical.

41. **(a)** Yes, the statement is true because "this year is 2010" is false and "$1 + 1 = 3$" is false and $\text{F} \rightarrow \text{F}$ is true.

(b) No, the statement was not true in 2010 because then the statement had the value $\text{T} \rightarrow \text{F}$, which is false.

43. Let l represent "you're late one more time" and d represent "you'll be docked." The argument is then presented symbolically as follows.

$$l \rightarrow d$$
$$\underline{l}$$
$$d$$

The argument is valid by Modus Ponens.

45. Let l represent "the instructor is late" and w represent "my watch is wrong." The argument is then presented symbolically as follows.

$$l \vee w$$
$$\underline{\sim w}$$
$$l$$

The argument is valid by Disjunctive Syllogism.

47. Let p represent "you play that song one more time" and n represent "I'm going nuts." The argument is then presented symbolically as follows.

$$p \rightarrow n$$
$$\underline{n}$$
$$p$$

The argument is invalid by Fallacy of the Converse.

49. Let h represent "we hire a new person," t represent "we'll spend more on training," and r represent "we rewrite the manual." The argument is then presented symbolically as follows.

$$h \rightarrow t$$
$$\underline{r \rightarrow \sim t}$$
$$\sim h$$

1. $h \rightarrow t$ Premise
2. $r \rightarrow \sim t$ Premise
3. r Premise
4. $\sim t$ 2, 3, Modus Ponens
5. $\sim h$ 1, 4, Modus Tollens

The argument is valid.

51.
1. $\sim p \rightarrow \sim q$ T
2. $\underline{q \rightarrow p}$ T

$p \vee q$ F.

The argument is <u>invalid</u>. When $p = \text{F}$ and $q = \text{F}$, the premises are true but the conculsion is false.

53. Let $d(x)$ represent "x is a dog" and $\ell(x)$ represent "x has a license".

(a) $\forall x\, [d(x) \rightarrow \ell(x)]$

(b) $\exists x\, [d(x) \wedge \sim\ell(x)]$

(c) There is a dog that doesn't have a license.

55. **(a)** Let $f(x)$ represent "x is a member of that fraternity," $w(x)$ represent "x does well academically," and j represent Jordan Enzor We can represent the argument symbolically as follows.

$$\forall x\, [f(x) \rightarrow w(x)]$$
$$\underline{f(j)}$$
$$w(j)$$

(b) Because of the first premise, the region representing "members of that fraternity" must be inside the region representing "those who do well academically." And j must be within the region representing "members of that fraternity" because of the second premise. Complete the Euler diagram as follows.

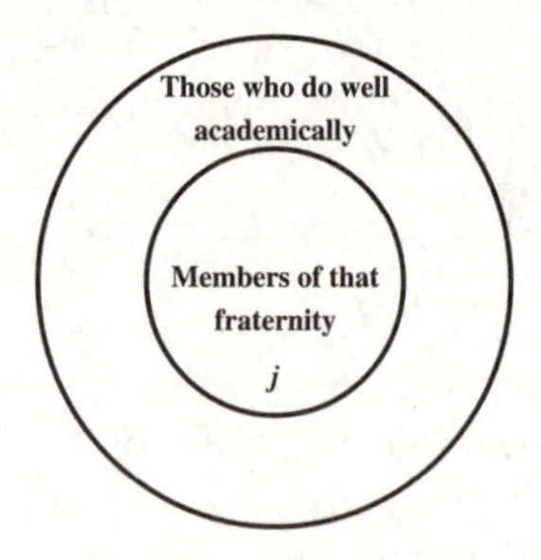

Since, when the premises are diagrammed as being true, we are forced into a true conclusion, the argument is valid.

57.

p	q	r	p	$\rightarrow$	$(q$	$\rightarrow$	$r)$	$(p$	$\rightarrow$	$q)$	$\rightarrow$	r
T	T	T	T	T	T	T	T	T	T	T	T	T
T	T	F	T	F	T	F	F	T	T	T	F	F
T	F	T	T	T	F	T	T	T	F	F	T	T
T	F	F	T	T	F	T	F	T	F	F	T	F
F	T	T	F	T	T	T	T	F	T	T	T	T
F	T	F	F	T	T	F	F	F	T	T	F	F
F	F	T	F	T	F	T	T	F	T	F	T	T
F	F	F	F	T	F	T	F	F	T	F	F	F
			1	4	2	3	2	5	6	5	8	7

To determine if the statements are equivalent, compare columns 4 and 8. Since they are not identical, the statements are not equivalent.

59. **(a)**

p	q	$(p$	$\wedge$	$\sim p)$	$\rightarrow$	q
T	T	T	F	F	T	T
T	F	T	F	F	T	F
F	T	F	F	T	T	T
F	F	F	F	T	T	F
		1	2	1	4	3

61. **(b)** and **(c)** are compound statements.

63. "If you use the Tax Table, then you do not have to compute your tax mathematically" is equivalent to "You do not use the Tax Table or you do not have to compute your tax mathematically."

65. **(a)** If you exercise regularly, then your heart becomes stronger and more efficient.

(b) If you are a teenager, then you need to be aware of the risks of drinking and driving.

(c) If you are visiting a country that has a high incidence of infectious diseases, then you may need extra immunizations.

(d) If you have good health, then you have food.

67. $(w \rightarrow d) \rightarrow v$

69. **(b), (c), (d)** These are declarative sentences and therefore are statements.

71. **(a)** $\sim s \rightarrow g$

(b) $l \rightarrow \sim g$

(c) $w \rightarrow l \equiv \sim l \rightarrow \sim w$

(d) $\sim s \rightarrow \sim w$

If the puppy does not lie still, it does not care to do worsted work. In Lewis Carroll's words, "Puppies that will not lie still never care to do worsted work."

73. **(a)** $f \rightarrow t \equiv \sim t \rightarrow \sim f$

(b) $\sim a \rightarrow \sim g \equiv g \rightarrow a$

(c) $w \rightarrow f$

(d) $t \rightarrow \sim g \equiv g \rightarrow \sim t$

(e) $a \rightarrow w \equiv \sim w \rightarrow \sim a$

(f) $g \rightarrow \sim e$

If the kitten will play with a gorilla, it does not have green eyes. In Lewis Carroll's words, "No kitten with green eyes will play with a gorilla."

Chapter 7

SETS AND PROBABILITY

7.1 Sets

Your Turn 1

There are three states whose names begin with O, Ohio Oklahoma, and Oregon Thus
$\{x \mid x$ is a state that begins with the letter O$\}$
$= \{$Ohio, Oklahoma, Oregon$\}$.

Your Turn 2

$\{2,4,6\} \subseteq \{6,2,4\}$ is true because each element of the first set is an element of the second set. (In this example the sets are in fact equal.)

Your Turn 3

A set of k distinct elements has 2^k subsets, so since there are four seasons, $\{x \mid x$ is a season of the year$\}$ has $2^4 = 16$ subsets.

Your Turn 4

$$U = \{0,1,2,...,10\}$$
$$A = \{3,6,9\}$$
$$B = \{2,4,6,8\}$$

Then $A' = \{0,1,2,4,5,7,8,10\}$. The elements common to A' and B are 2,4, and 8. Thus $A' \cap B = \{2,4,8\}$.

Your Turn 5

$$U = \{0,1,2,...,12\}$$
$$A = \{1,3,5,7,9,11\}$$
$$B = \{3,6,9,12\}$$
$$C = \{1,2,3,4,5\}$$

To find $A \cup (B \cap C')$ begin with the expression in parentheses and find C', which includes all the elements of the universal set that are not in C.

$$C' = \{6,7,8,9,10,11,12\}$$

The elements in C' that are also in B are 6, 9, and 12, so $B \cap C' = \{6,9,12\}$.

Now we list the elements of A and include any elements of $B \cap C'$ that are not already listed:

$$A \cup (B \cap C') = \{1,3,5,6,7,9,11,12\}$$

7.1 Exercises

1. $3 \in \{2,5,7,9,10\}$

The number 3 is not an element of the set, so the statement is false.

3. $9 \notin \{2,1,5,8\}$

Since 9 is not an element of the set, the statement is true.

5. $\{2,5,8,9\} = \{2,5,9,8\}$

The sets contain exactly the same elements, so they are equal. The statement is true.

7. {All whole numbers greater than 7 and less than 10} $= \{8,9\}$

Since 8 and 9 are the only such numbers, the statement is true.

9. $0 \in \emptyset$

The empty set has no elements. The statement is false.

In Exercises 11–22,

$$A = \{2,4,6,8,10,12\},$$
$$B = \{2,4,8,10\},$$
$$C = \{4,8,12\},$$
$$D = \{2,10\},$$
$$E = \{6\},$$
and $$U = \{2,4,6,8,10,12,14\}.$$

11. Since every element of A is also an element of U, A is a subset of U, written $A \subseteq U$.

13. A contains elements that do not belong to E, namely 2, 4, 8, 10, and 12, so A is not a subset of E, written $A \not\subseteq E$.

15. The empty set is a subset of every set, so $\emptyset \subseteq A$.

17. Every element of D is also an element of B, so D is a subset of B, $D \subseteq B$.

19. Since every element of A is also an element of U, and $A \neq U$, $A \boxed{\subset} U$.

Since every element of E is also an element of A, and $E \neq A$, $E \boxed{\subset} A$.

Since every element of A is not also an element of E, $A \boxed{\not\subseteq} E$.

Since every element of B is not also an element of C, $B \boxed{\not\subseteq} C$.

Since $\emptyset$ is a subset of every set, and $\emptyset \neq A$, $\emptyset \boxed{\subset} A$.

Since every element of $\{0,2\}$ is not also an element of D, $\{0,2\} \boxed{\not\subseteq} D$.

Since every element of D is not also an element of B, and $D \neq B$, $D \boxed{\subset} B$.

Since every element of A is not also an element of C, $A \boxed{\not\subseteq} C$.

21. A set with n distinct elements has 2^n subsets. A has $n = 6$ elements, so there are exactly $2^6 = 64$ subsets of A.

23. A set with n distinct elements has 2^n subsets, and C has $n = 3$ elements. Therefore, there are exactly $2^3 = 8$ subsets of C.

25. Since $\{7,9\}$ is the set of elements belonging to both sets, which is the intersection of the two sets, we write

$$\{5,7,9,19\} \cap \{7,9,11,15\} = \{7,9\}.$$

27. Since $\{1,2,5,7,9\}$ is the set of elements belonging to one or the other (or both) of the listed sets, it is their union.

$$\{2,1,7\} \cup \{1,5,9\} = \{1,2,5,7,9\}$$

29. Since $\emptyset$ contains no elements, there are no elements belonging to both sets. Thus, the intersection is the empty set, and we write

$$\{3,5,9,10\} \cap \emptyset = \emptyset.$$

31. $\{1,2,4\}$ is the set of elements belonging to both sets, and $\{1,2,4\}$ is also the set of elements in the first set or in the second set or possibly both. Thus,

$$\{1,2,4\} \cap \{1,2,4\} = \{1,2,4\}$$

and

$$\{1,2,4\} \cup \{1,2,4\} = \{1,2,4\}$$

are both true statements.

In Exercises 35–44,

$$U = \{1,2,3,4,5,6,7,8,9\},$$
$$X = \{2,4,6,8\},$$
$$Y = \{2,3,4,5,6\},$$

and

$$Z = \{1,2,3,8,9\}.$$

35. $X \cap Y$, the intersection of X and Y, is the set of elements belonging to both X and Y. Thus,

$$\begin{aligned} X \cap Y &= \{2,4,6,8\} \cap \{2,3,4,5,6\} \\ &= \{2,4,6\}. \end{aligned}$$

37. X', the complement of X, consists of those elements of U that are not in X. Thus,

$$X' = \{1,3,5,7,9\}.$$

39. From Exercise 37, $X' = \{1,3,5,7,9\}$; from Exercise 38, $Y' = \{1,7,8,9\}$. There are no elements common to both X' and Y' so

$$X' \cap Y' = \{1,7,9\}.$$

41. First find $X \cup Z$.

$$\begin{aligned} X \cup Z &= \{2,4,6,8\} \cup \{1,2,3,8,9\} \\ &= \{1,2,3,4,6,8,9\} \end{aligned}$$

Now find $Y \cap (X \cup Z)$.

$$\begin{aligned} Y \cap (X \cup Z) &= \{2,3,4,5,6\} \cap \{1,2,3,4,6,8,9\} \\ &= \{2,3,4,6\} \end{aligned}$$

43. $U = \{1,2,3,4,5,6,7,8,9\}$ and $Z = \{1,2,3,8,9\}$, so $Z' = \{4,5,6,7\}$.

From Exercise 38, $Y' = \{1,7,8,9\}$.

$$\begin{aligned} (X \cap Y') \cup (Z' \cap Y') &= (\{2,4,6,8\} \cap \{1,7,8,9\}) \\ &\quad \cup (\{4,5,6,7\} \cap \{1,7,8,9\}) \\ &= \{8\} \cup \{7\} \\ &= \{7,8\} \end{aligned}$$

45. $$\begin{aligned} (A \cap B) \cup (A \cap B') &= (\{3,6,9\} \cap \{2,4,6,8\}) \\ &\quad \cup (\{3,6,9\} \cap \{0,1,3,5,7,9,10\}) \\ &= \{6\} \cup \{3,9\} \\ &= \{3,6,9\} \\ &= A \end{aligned}$$

47. M' consists of all students in U who are not in M, so M' consists of all students in this school not taking this course.

49. $N \cap P$ is the set of all students in this school taking both accounting and zoology.

51. $A = \{2,4,6,8,10,12\}$,
$B = \{2,4,8,10\}$,
$C = \{4,8,12\}$,
$D = \{2,10\}$,
$E = \{6\}$,
$U = \{2,4,6,8,10,12,14\}$

A pair of sets is disjoint if the two sets have no elements in common. The pairs of these sets that are disjoint are B and E, C and E, D and E, and C and D.

53. B' is the set of all stocks on the list with a closing price below \$60 or above \$70.

$B' = \{$AT&T, Coca-Cola, FedEx, Disney$\}$

55. $(A \cap B)'$ is the set of all stocks on the list that do not have both a high price greater than \$50 and a closing price between \$60 and \$70.

$(A \cap B)' = \{$AT&T, Coca-Cola, FedEx, Disney$\}$

57. $A = \{1, 2, 3, \{3\}, \{1, 4, 7\}\}$

(a) $1 \in A$ is true.

(b) $\{3\} \in A$ is true.

(c) $\{2\} \in A$ is false. $(\{2\} \subseteq A)$

(d) $4 \in A$ is false. $(4 \in \{1, 4, 7\})$

(e) $\{\{3\}\} \subset A$ is true.

(f) $\{1,4,7\} \in A$ is true.

(g) $\{1, 4, 7\} \subseteq A$ is false. $(\{1, 4, 7\} \in A)$

For Exercises 59 through 61 refer to this abbreviated version of the table:

Vanguard 500 (V)	Fidelity New Millennium Fund (F)	Janus Perkins Large Cap Value (J)	Templeton Large Cap Value Fund (T)
Exxon	Pfizer	Exxon	IBM
Apple	Cisco	GE	GE
GE	Wal-Mart	Wal-Mart	HP
IBM	Apple	AT&T	Home Depot
JPMorgan	JPMorgan	JPMorgan	Aflac

59. $V \cap J$
$= \{$Exxon, Apple, GE, IBM, JPMorgan$\}$
$\cap \{$Exxon, GE, Wal-Mart, AT&T, JPMorgan$\}$
$= \{$Exxon, GE, JPMorgan$\}$

61.
$J \cup F$
$= \{$Exxon, GE, Wal-Mart, AT&T, JPMorgan$\}$
$\cup \{$Pfizer, Cisco,Wal-Mart, Apple, JPMorgan$\}$
$= \{$Exxon, GE, Wal-Mart, AT&T, JPMorgan, Pfizer, Cisco, Apple$\}$

$(J \cup F)' = \{$IBM, HP, Home Depot, Aflac$\}$

63. The number of subsets of a set with k elements is 2^k, so the number of possible sets of customers (including the empty set) is $2^9 = 512$.

65. $U = \{s, d, c, g, i, m, h\}$ and $N = \{s, d, c, g\}$, so

$$N' = \{i, m, h\}.$$

67. $N \cup O = \{s, d, c, g\} \cup \{i, m, h, g\}$
$= \{s, d, c, g, i, m, h\} = U$

69. The number of subsets of a set with 51 elements (50 states plus the District of Columbia) is

$$2^{51} \approx 2.252 \times 10^{15}.$$

For Exercises 71 through 75 refer to this table:

Network	Subscribers (millions)	Launch	Content
The Discovery Channel	98.0	1985	Nonfiction, nature, science
TNT	98.0	1988	Movies, sports, original programming
USA Network	97.5	1980	Sports, family entertainment
TLC	97.3	1980	Original programming, family entertainment
TBS	97.3	1976	Movies, sports, original programming

71. $F = \{$USA, TLC, TBS$\}$

73. $H = \{$Discovery, TNT$\}$

75. $G \cup H = \{$TNT, USA, TBS$\} \cup \{$Discovery, TNT$\}$
$= \{$TNT, USA, TBS, Discovery$\}$;
the set of networks that feature sports or that have more than 97.6 million viewers.

77. Joe should always first choose the complement of what Dorothy chose. This will leave only two sets to choose from, and Joe will get the last choice.

79. **(a)** $(A \cup B') \cap C$

$A \cup B$ is the set of states whose name contains the letter e or which have a population over 4,000,000. Therefore, $(A \cup B)'$ is the set of states which are not among those whose name contains the letter e or which have a population over 4,000,000. As a result, $(A \cup B)' \cap C$ is the set of states which are not among those whose name contains the letter e or which have a population over 4,000,000 and which also have an area over 40,000 square miles.

(b)

$$\begin{aligned}(A \cup B)' &= \{(\text{Kentucky, Maine, Nebraska, New Jersey}\} \\ &\quad \cup \{\text{Alabama, Colorado, Florida, Indiana, Kentucky, New Jersey}\})' \\ &= \{\text{Alabama, Colorado, Florida, Indiana, Kentucky, Maine, Nebraska, New Jersey}\}' \\ &= \{\text{Alaska, Hawaii}\}\end{aligned}$$

$$\begin{aligned}(A \cup B)' \cap C &= \{\text{Alaska, Hawai}\} \cap \{\text{Alabama, Alaska, Colorado, Florida, Kentucky, Nebraska}\} \\ &= \{\text{Alaska}\}\end{aligned}$$

7.2 Applications of Venn Diagrams

Your Turn 1

$A \cup B'$ is the set of elements in A or not in B or both in A and not in B.

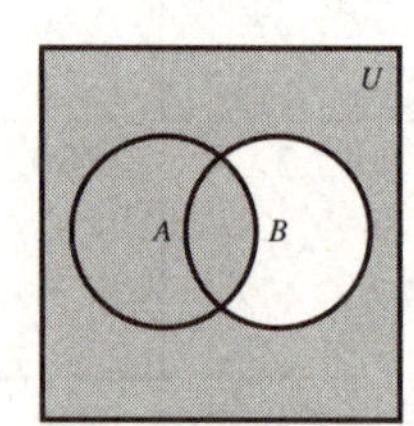

Your Turn 2

$A' \cap (B \cup C)$

First find A'.

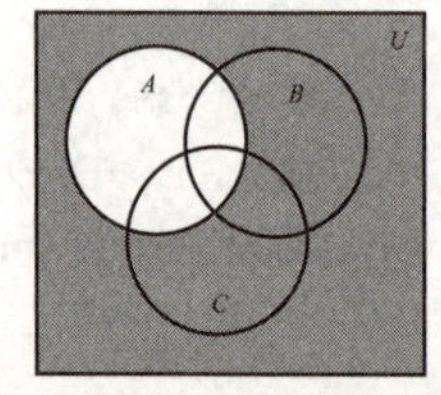

Then find $B \cup C$.

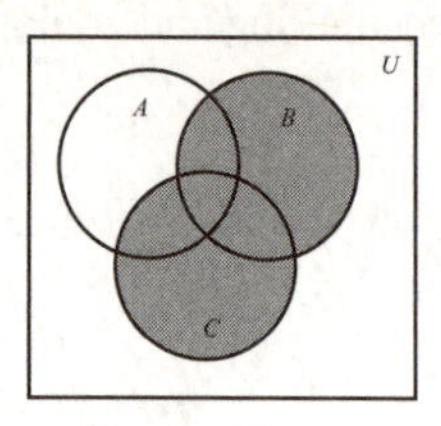

Then intersect these regions.

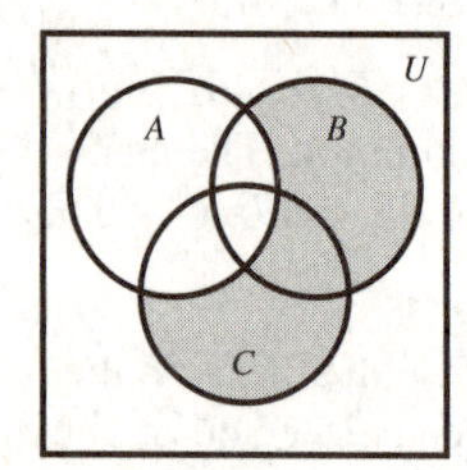

Your Turn 3

Start with $n(M \cap T \cap W) = 13$ and label the corresponding region with 13.

Since $n(M \cap T) = 17$, there are an additional 4 elements in $M \cap T$ but not in $M \cap T \cap W$. Label the corresponding region with 4.

Since $n(T \cap W) = 19$, there are an additional 6 elements in $T \cap W$ but not in $M \cap T \cap W$. Label the corresponding region with 6. The Venn diagram now looks like this:

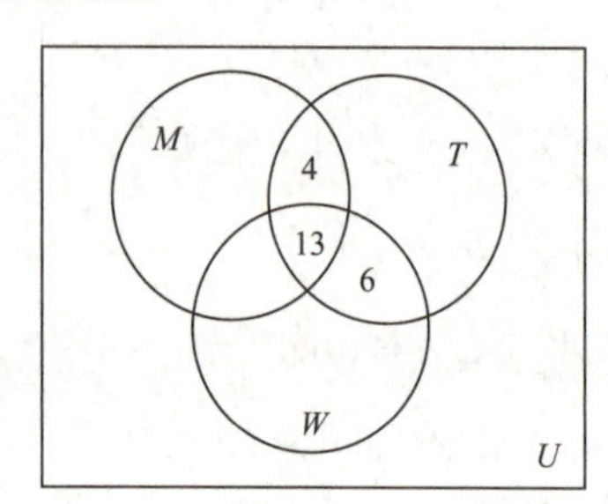

Since $n(T) = 46$, the number who ate only at Taco Bell is $46 - 4 - 13 - 6 = 23$.

Your Turn 4

Let T represent the set of those texting and M represent the set of those listening to music. The number in the lounge is $n(T \cup M) = n(T) + n(M) - n(T \cap M)$. We know that $n(T) = 15$, $n(M) = 11$, and $n(T \cap M) = 8$. Then $n(T \cup M) = 15 + 11 - 8 = 18$.

7.2 Exercises

1. $B \cap A'$ is the set of all elements in B *and* not in A.

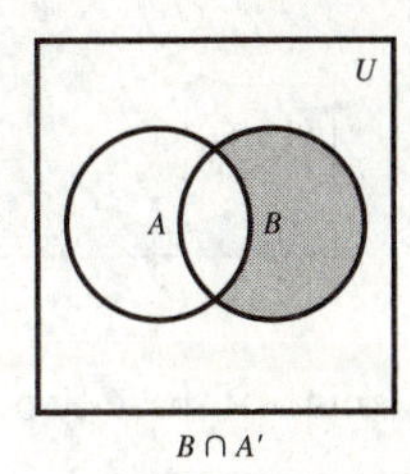

$B \cap A'$

3. $A' \cup B$ is the set of all elements that do not belong to A *or* that do belong to B, or both.

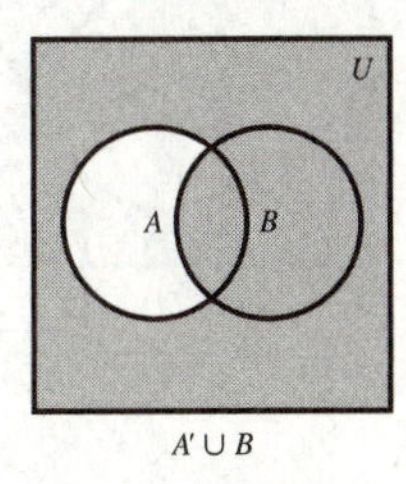

$A' \cup B$

5. $B' \cup (A' \cap B')$

First find $A' \cap B'$, the set of elements not in A *and* not in B.

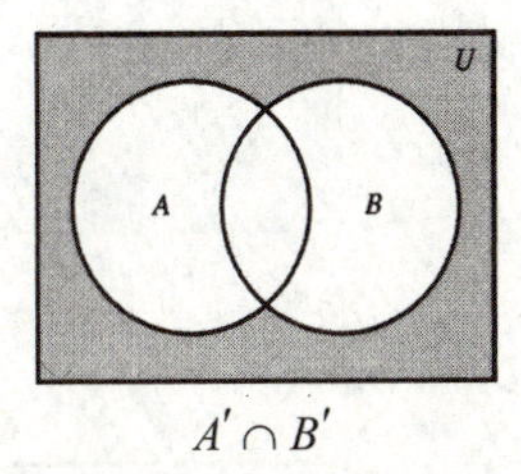

$A' \cap B'$

For the union, we want those elements in B' *or* $(A' \cap B')$, or both.

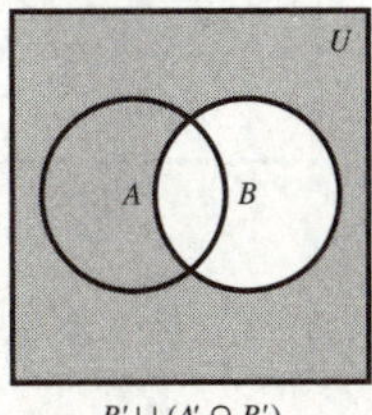

$B' \cup (A' \cap B')$

7. U' is the empty set $\emptyset$.

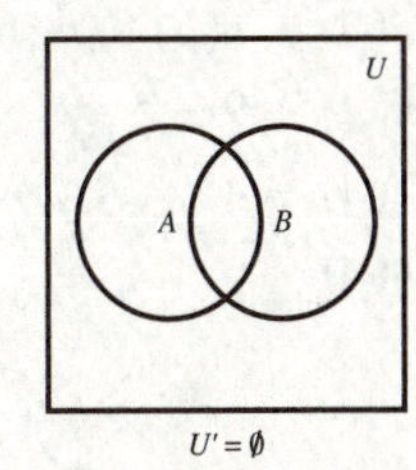

$U' = \emptyset$

9. Three sets divide the universal set into at most 8 regions. (Examples of this situation will be seen in Exercises 11–17.)

11. $(A \cap B) \cap C$

First form the intersection of A with B.

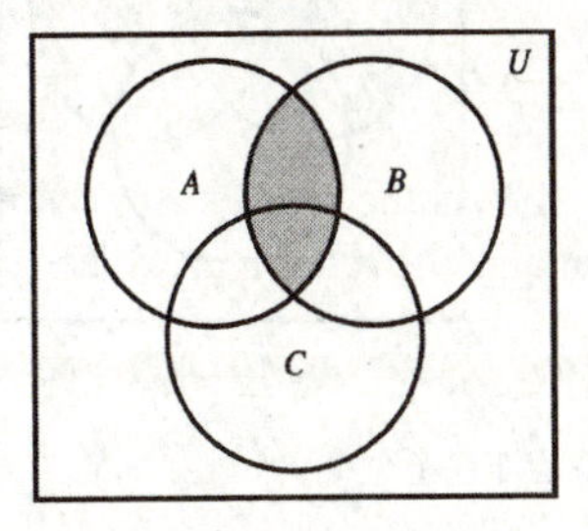

$A \cap B$

Now form the intersection of $A \cap B$ with C. The result will be the set of all elements that belong to all three sets.

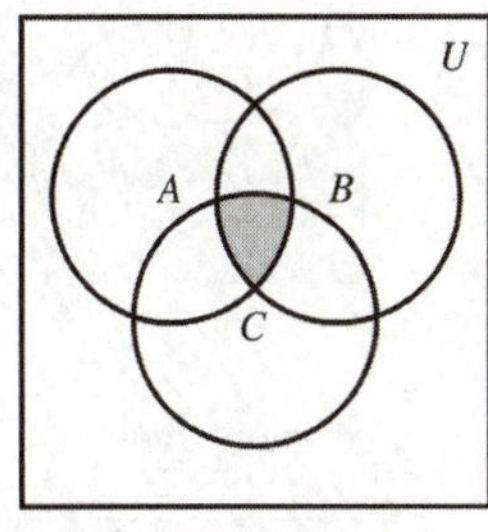

$(A \cap B) \cap C$

13. $A \cap (B \cup C')$

C' is the set of all elements in U that are not elements of C.

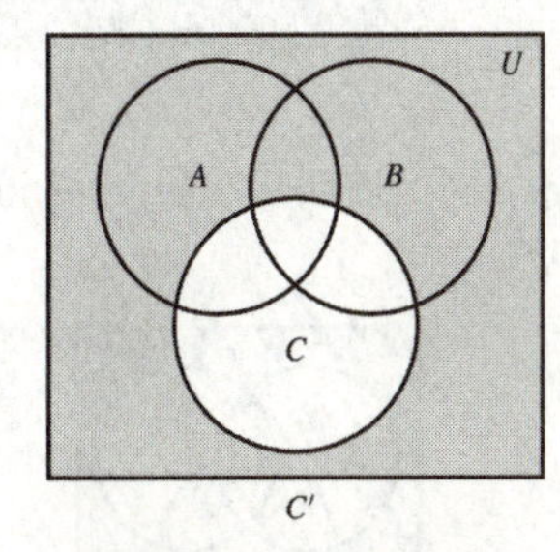

C'

Now form the union of C' with B.

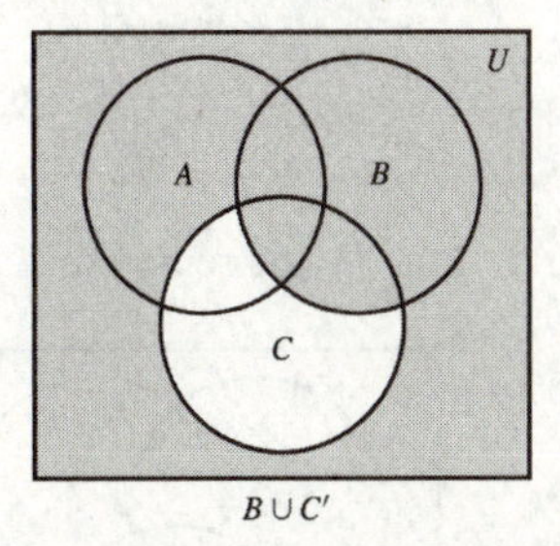

$B \cup C'$

Finally, find the intersection of this region with A.

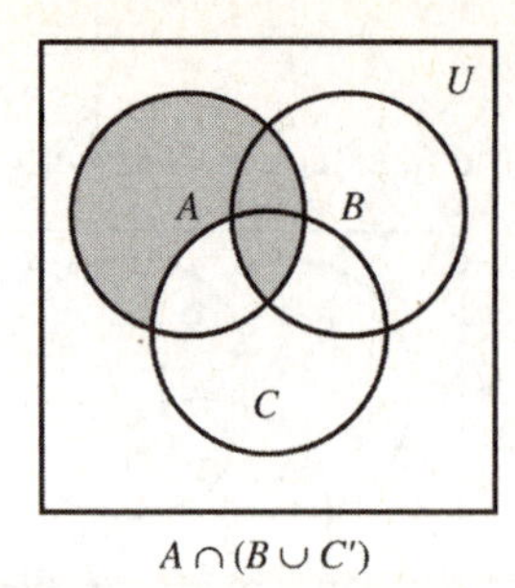

$A \cap (B \cup C')$

15. $(A' \cap B') \cap C'$

$A' \cap B'$ is the part of the universal set not in A *and* not in B:

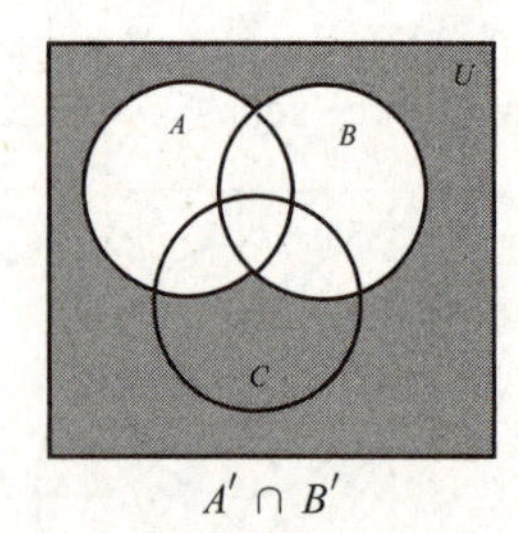

$A' \cap B'$

C' is the part of the universal set not in C:

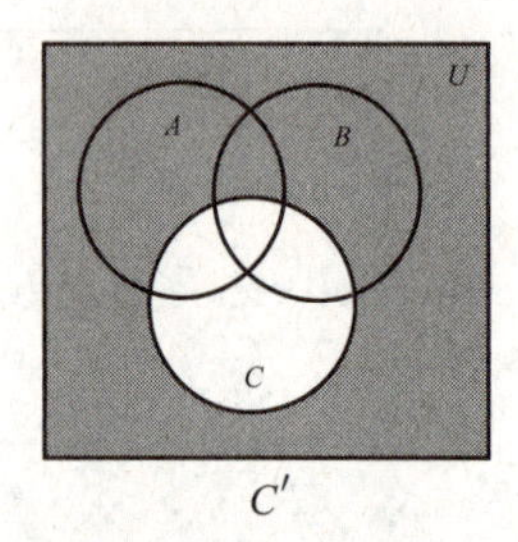

C'

Now intersect the shaded regions in these two diagrams:

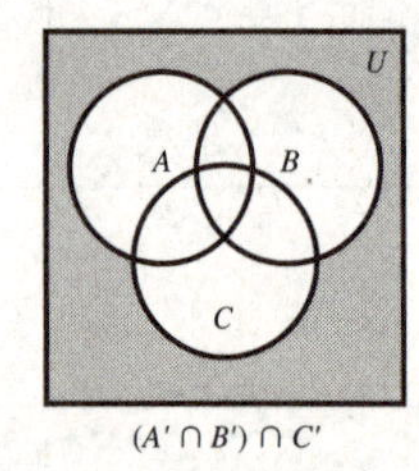

$(A' \cap B') \cap C'$

17. $(A \cap B') \cup C'$

First find $A \cap B'$, the region in A and not in B:

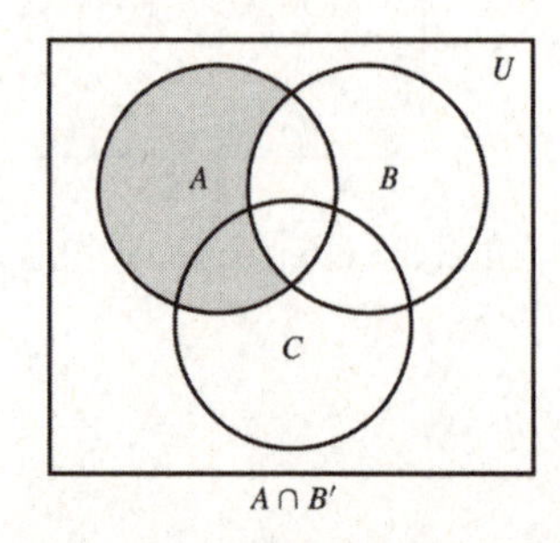

$A \cap B'$

C' is the region of the universal set not in C:

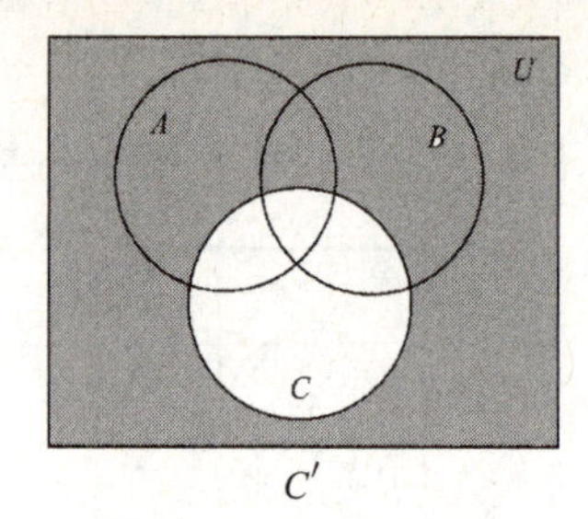

C'

Now form the union of these two regions.

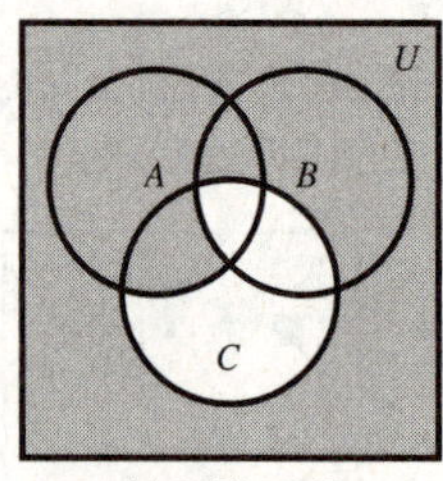

$(A \cap B') \cup C'$

19. $(A \cup B') \cap C$

First find $A \cup B'$, the region in A or B' or both.

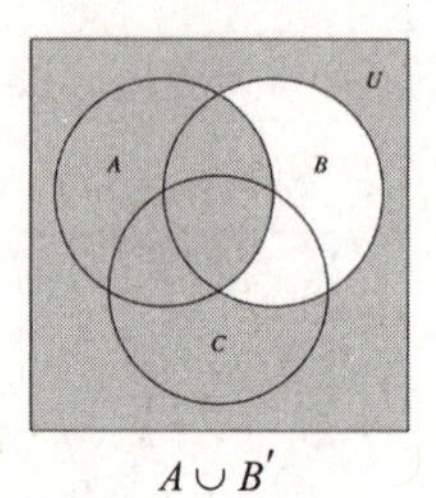

$A \cup B'$

Intersect this with C.

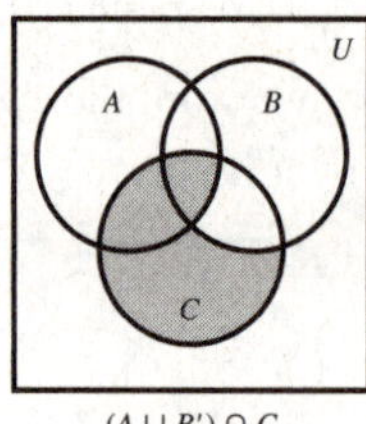

$(A \cup B') \cap C$

21.
$$\begin{aligned} n(A \cup B) &= n(A) + n(B) - n(A \cap B) \\ &= 5 + 12 - 4 \\ &= 13 \end{aligned}$$

23.
$$\begin{aligned} n(A \cup B) &= n(A) + n(B) - n(A \cap B) \\ 22 &= n(A) + 9 - 5 \\ 22 &= n(A) + 4 \\ 18 &= n(A) \end{aligned}$$

25. $n(U) = 41$

$n(A) = 16$

$n(A \cap B) = 12$

$n(B') = 20$

First put 12 in $A \cap B$. Since $n(A) = 16$, and 12 are in $A \cap B$, there must be 4 elements in A that are not in $A \cap B$. $n(B') = 20$, so there are 20 not in B. We already have 4 not in B (but in A), so there must be another 16 outside B *and* outside A. So far we have accounted for 32, and $n(U) = 41$, so 9 must be in B but not in any region yet identified. Thus $n(A' \cap B) = 9$.

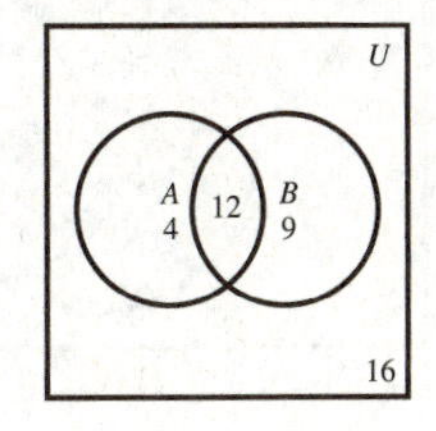

27. $n(A \cup B) = 24$

$n(A \cap B) = 6$

$n(A) = 11$

$n(A' \cup B') = 25$

Start with $n(A \cap B) = 6$. Since $n(A) = 11$, there must be 5 more in A not in B. $n(A \cup B) = 24$; we already have 11, so 13 more must be in B not yet counted. $A' \cup B'$ consists of all the region not in $A \cap B$, where we have 6. So far $5 + 13 = 18$ are in this region, so another $25 - 18 = 7$ must be outside both A and B.

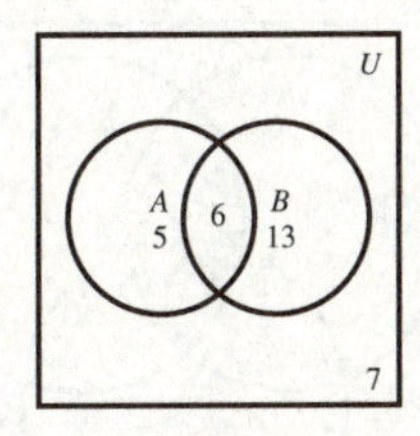

29. $n(A) = 28$ $\quad n(B) = 34$ $\quad n(C) = 25$

$n(A \cap B) = 14$ $\quad n(B \cap C) = 15$ $\quad n(A \cap C) = 11$

$n(A \cap B \cap C) = 9$

$n(U) = 59$

We start with $n(A \cap B \cap C) = 9$. If $n(A \cap B) = 14$, an additional 5 are in $A \cap B$ but not in $A \cap B \cap C$. Similarly, $n(B \cap C) = 15$, so $15 - 9 = 6$ are in $B \cap C$ but not in $A \cap B \cap C$. Also, $n(A \cap C) = 11$, so $11 - 9 = 2$ are in $A \cap C$ but not in $A \cap B \cap C$.

Now we turn our attention to $n(A) = 28$. So far we have $2 + 9 + 5 = 16$ in A; there must be another $28 - 16 = 12$ in A not yet counted. Similarly, $n(B) = 34$; we have $5 + 9 + 6 = 20$ so far, and $34 - 20 = 14$ more must be put in B.

For C, $n(C) = 25$; we have $2 + 9 + 6 = 17$ counted so far. Then there must be 8 more in C not yet counted. The count now stands at 56, and $n(U) = 59$, so 3 must be outside the three sets.

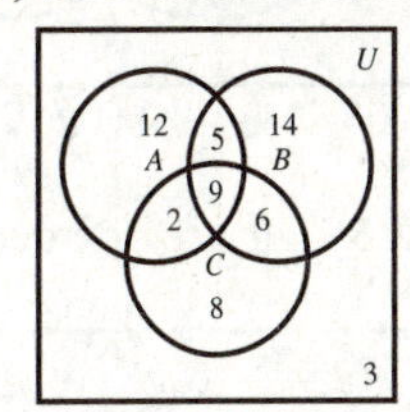

31. $n(A \cap B) = 6$ $\quad n(A \cap B \cap C) = 4$

$n(A \cap C) = 7$ $\quad n(B \cap C) = 4$

$n(A \cap C') = 11$ $\quad n(B \cap C') = 8$

$n(C) = 15$ $\quad n(A' \cap B' \cap C') = 5$

Start with $n(A \cap B) = 6$ and $n(A \cap B \cap C) = 4$ to get $6 - 4 = 2$ in that portion of $A \cap B$ outside of C. From $n(B \cap C) = 4$, there are $4 - 4 = 0$ elements in that portion of $B \cap C$ outside of A. Use $n(A \cap C) = 7$ to get $7 - 4 = 3$ elements in that portion of $A \cap C$ outside of B.

Since $n(A \cap C') = 11$, there are $11 - 2 = 9$ elements in that part of A outside of B and C. Use $n(B \cap C') = 8$ to get $8 - 2 = 6$ elements in that part of B outside of A and C. Since $n(C) = 15$, there are $15 - 3 - 4 - 0 = 8$ elements in C outside of A and B. Finally, 5 must be outside all three sets, since $n(A' \cap B' \cap C') = 5$.

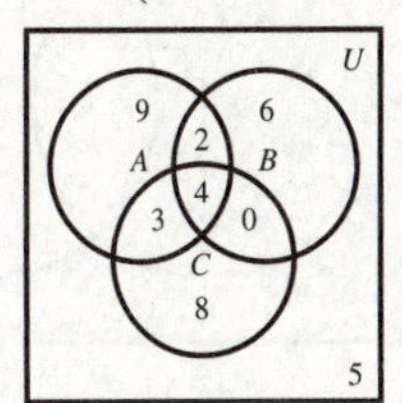

33. $(A \cup B)' = A' \cap B'$

For $(A \cup B)'$, first find $A \cup B$.

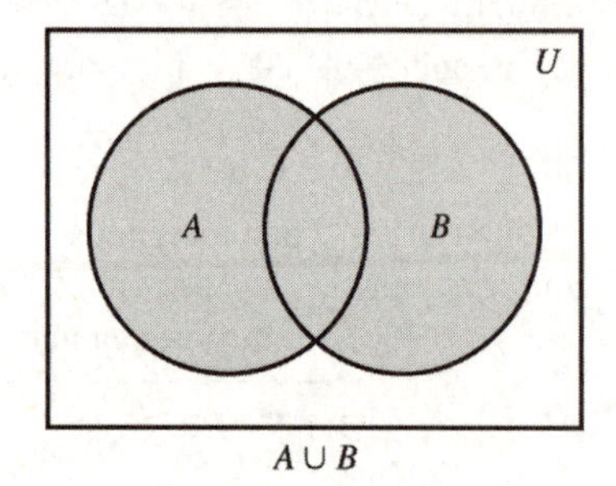

$A \cup B$

Now find $(A \cup B)'$, the region outside $A \cup B$.

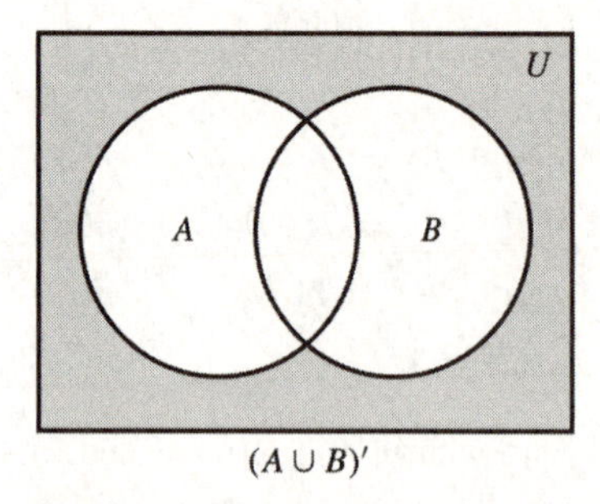

$(A \cup B)'$

For $A' \cap B'$, first find A' and B' individually.

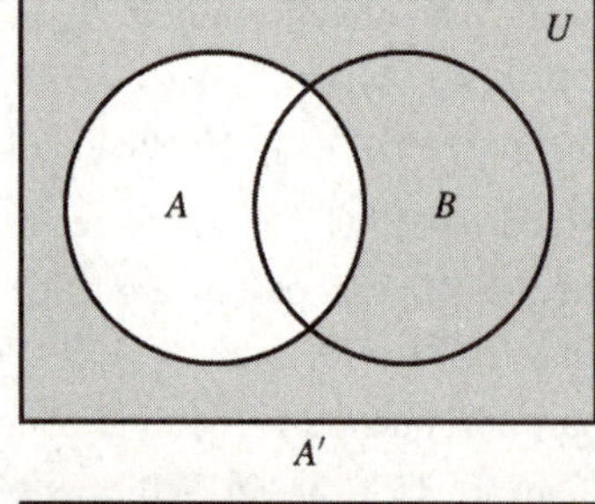

A'

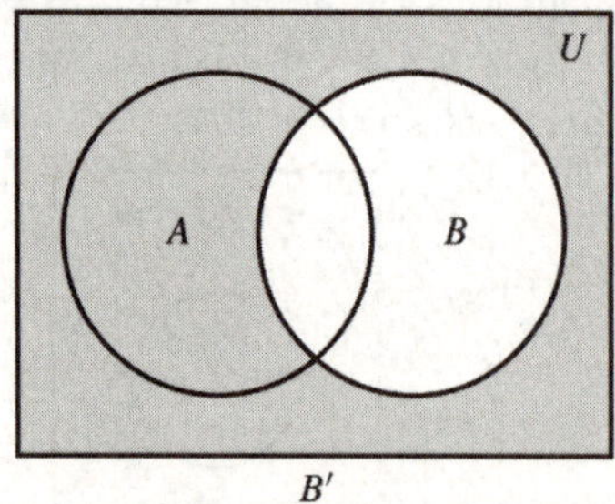

B'

Then $A' \cap B'$ is the region where A' and B' overlap, which is the entire region outside $A \cup B$ (the same result as in the second diagram). Therefore,

$$(A \cup B)' = A' \cap B'.$$

35. $A \cap (B \cup C) = (A \cap B) \cup (A \cap C)$

First find A and $B \cup C$ individually.

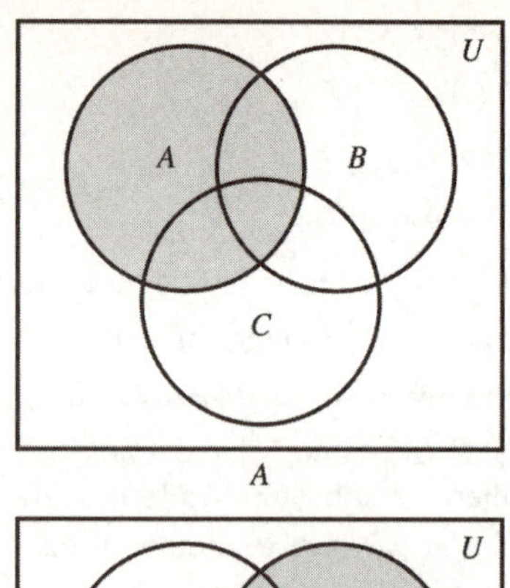

A

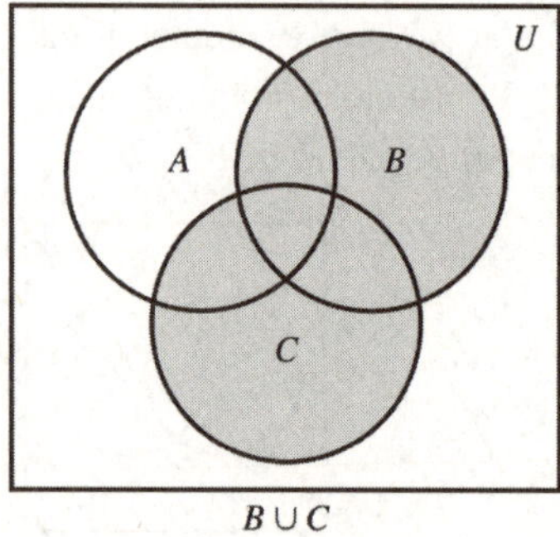

$B \cup C$

Then $A \cap (B \cup C)$ is the region where the above two diagram overlap.

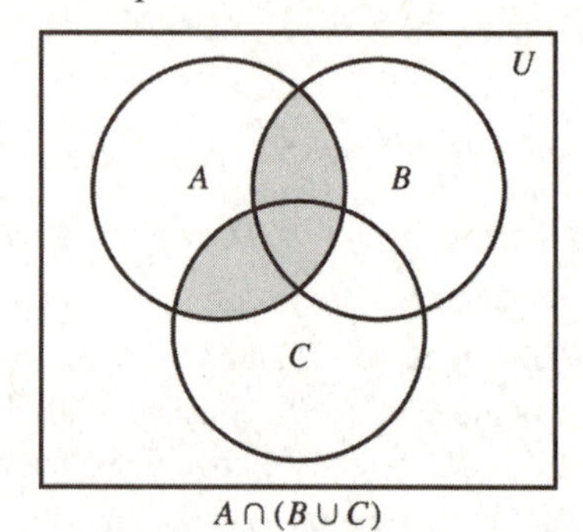

$A \cap (B \cup C)$

Next find $A \cap B$ and $A \cap C$ individually.

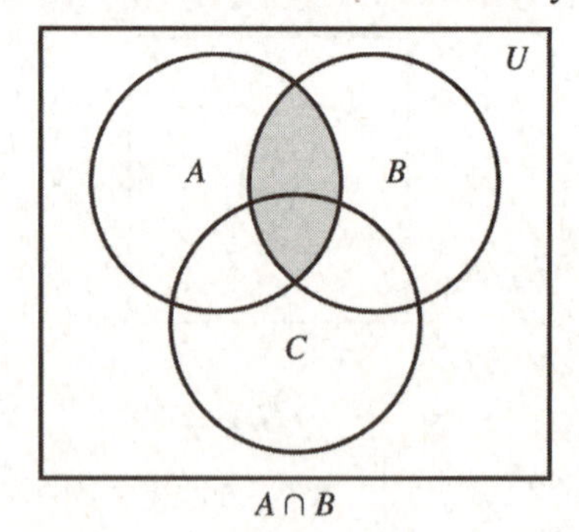

$A \cap B$

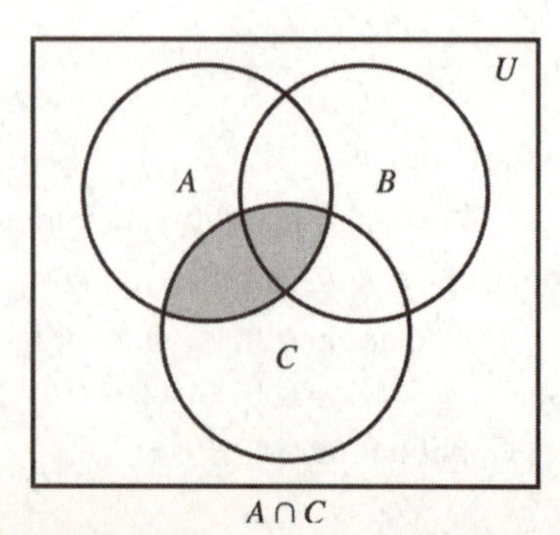

$A \cap C$

Then $(A \cap B) \cup (A \cap C)$ is the union of the above two diagrams.

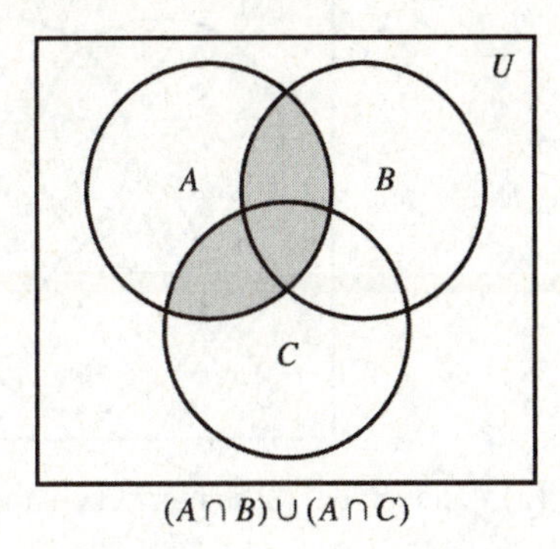

$(A \cap B) \cup (A \cap C)$

The Venn diagram for $A \cap (B \cup C)$ is identical to the Venn diagram for $(A \cap B) \cup (A \cap C)$, so conclude that

$$A \cap (B \cup C) = (A \cap B) \cup (A \cap C).$$

37. Prove

$$\begin{aligned} n(A \cup B \cup C) &= n(A) + n(B) + n(C) - n(A \cap B) - n(A \cap C) \\ &\quad - n(B \cap C) + n(A \cap B \cap C) \end{aligned}$$

$$\begin{aligned} n(A \cup B \cup C) &= n[A \cup (B \cup C)] \\ &= n(A) + n(B \cup C) - n[A \cap (B \cup C)] \\ &= n(A) + n(B) + n(C) - n(B \cap C) \\ &\quad - n[(A \cap B) \cup (A \cap C)] \\ &= n(A) + n(B) + n(C) - n(B \cap C) \\ &\quad - \{n(A \cap B) + n(A \cap C) \\ &\quad - n[(A \cap B) \cap (A \cap C)]\} \\ &= n(A) + n(B) + n(C) - n(B \cap C) - n(A \cap B) \\ &\quad - n(A \cap C) + n(A \cap B \cap C) \end{aligned}$$

39. Let A be the set of trucks that carried early peaches, B be the set of trucks that carried late peaches, and C be the set of trucks that carried extra late peaches. We are given the following information.

$n(A) = 34$ $\quad n(B) = 61$ $\quad n(C) = 50$

$n(A \cap B) = 25$

$n(B \cap C) = 30$

$n(A \cap C) = 8$

$n(A \cap B \cap C) = 6$

$n(A' \cap B' \cap C') = 9$

Start with $A \cap B \cap C$.

We know that $n(A \cap B \cap C) = 6$.

Since $n(A \cap B) = 25$, the number in $A \cap B$ but not in C is $25 - 6 = 19$.

Since $n(B \cap C) = 30$, the number in $B \cap C$ but not in A is $30 - 6 = 24$.

Since $n(A \cap C) = 8$, the number in $A \cap C$ but not in B is $8 - 6 = 2$.

Since $n(A) = 34$, the number in A but not in B or C is $34 - (19 + 6 + 2) = 7$.

Since $n(B) = 61$, the number in B but not in A or C is $61 - (19 + 6 + 24) = 12$.

Since $n(C) = 50$, the number in C but not in A or B is $50 - (24 + 6 + 2) = 18$.

Since $n(A' \cap B' \cap C') = 9$, the number outside $A \cup B \cup C$ is 9.

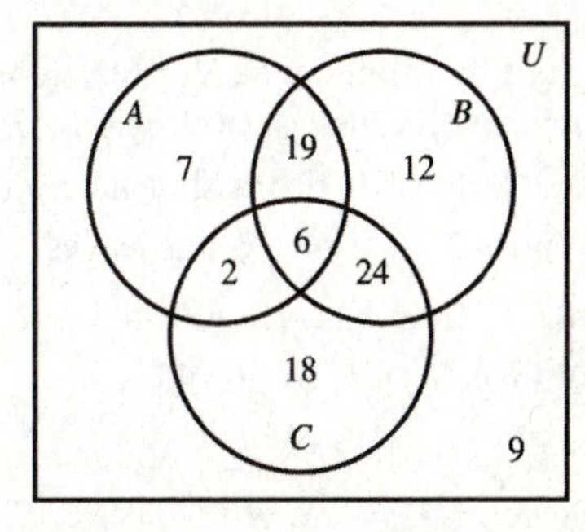

(a) From the Venn diagram, 12 trucks carried only late peaches.

(b) From the Venn diagram, 18 trucks carried only extra late peaches.

(c) From the Venn diagram, $7 + 12 + 18 = 37$ trucks carried only one type of peach.

(d) From the Venn diagram, $6 + 2 + 19 + 24 + 7 + 12 + 18 + 9 = 97$ trucks went out during the week.

41. **(a)** $n(Y \cap B) = 2$ since 2 is the number in the table where the Y row and the B column meet.

(b) $n(M \cup A) = n(M) + n(A) - n(M \cap A) = 33 + 41 - 14 = 60$

(c) $n[Y \cap (S \cup B)] = 6 + 2 = 8$

(d)

$$\begin{aligned} n[O' \cup (S \cup A)] &= n(O') + n(S \cup A) - n[O' \cap (S \cup A)] \\ &= (23 + 33) + (52 + 41) - (6 + 14 + 15 + 14) \\ &= 100 \end{aligned}$$

(e) Since is $M' \cup O'$ is the entire set, $(M' \cup O') \cap B = B$. Therefore,

$$n[(M' \cup O') \cap B] = n(B) = 27.$$

(f) $Y \cap (S \cup B)$ is the set of all bank customers who are of age 18–29 and who invest in stocks or bonds.

43. Let T be the set of all tall pea plants, G be the set of plants with green peas, and S be the set of plants with smooth peas. We are given the following information.

$n(U) = 50 \quad n(T) = 22 \quad n(G) = 25 \quad n(S) = 39$
$n(T \cap G) = 9$
$n(G \cap S) = 20$
$n(T \cap G \cap S) = 6$
$n(T' \cap G' \cap S') = 4$

Start by filling in the Venn Diagram with the numbers for the last two regions, $T \cap G \cap S$ and $T' \cap G' \cap S'$, as shown below.
With $n(T \cap G) = 9$, this leaves
$n(T \cap G \cap S') = 9 - 6 = 3.$
With $n(G \cap S) = 20$, this leaves
$n(T' \cap G \cap S) = 20 - 6 = 14.$
Since $n(G) = 25$, $n\,(T' \cap G \cap S')$
$= 25 - 3 - 6 - 14 = 2.$

With no other regions that we can calculate, denote by x the number in $T \cap G' \cap S'$. Then $n(T \cap G' \cap S') = 22 - 3 - 6 - x = 13 - x$, and $n(T' \cap G' \cap S) = 39 - 6 - 14 - x = 19 - x$, as shown. Summing the values for all eight regions,

$$\begin{aligned}(13 - x) + 3 + 2 + x + 6 + 14 + (19 - x) + 4 &= 50\\ 61 - x &= 50\\ x &= 11\end{aligned}$$

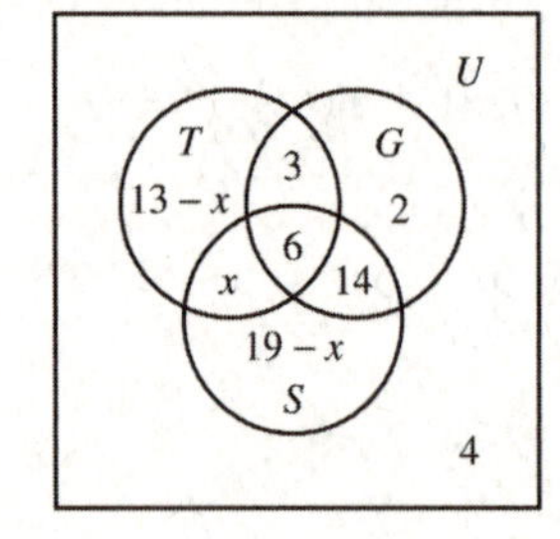

(a) $n(T \cap S) = 11 + 6 = 17$

(b) $n(T \cap G' \cap S') = 13 - x = 13 - 11 = 2$

(c) $n(T' \cap G \cap S) = 14$

45. First fill in the Venn diagram, starting with the region common to all three sets.

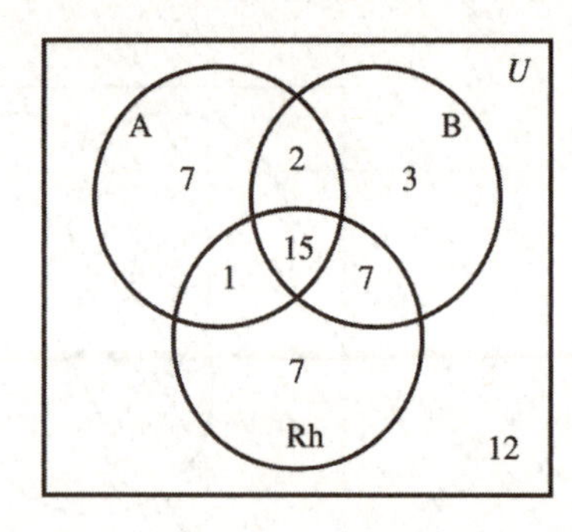

(a) The total of these numbers in the diagram is 54.

(b) $7 + 3 + 7 = 17$ had only one antigen.

(c) $1 + 2 + 7 = 10$ had exactly two antigens.

(d) A person with O-positive blood has only the Rh antigen, so this number is 7.

(e) A person with AB-positive blood has all three antigens, so this number is 15.

(f) A person with B-negative blood has only the B antigen, so this number is 3.

(g) A person with O-negative blood has none of the antigens. There are 12 such people.

(h) A person with A-positive blood has the A and Rh antigens, but not the B-antigen. The number is 1.

47. Extend the table to include totals for each row and each column.

	H	F	Total
A	95	34	129
B	41	38	79
C	9	7	16
D	202	150	352
Total	347	229	576

(a) $n(A \cap F)$ is the entry in the table that is in both row A and column F. Thus, there are 34 players in the set $A \cap F$.

(b) Since all players in the set C are either in set H or set $F, C \cap (H \cup F) = C$. Thus, $n(C \cap (H \cup F)) = n(C) = 16$, the total for row C. There are 16 players in the set $C \cap (H \cup F)$.

(c)
$$\begin{aligned} n(D \cup F) &= n(D) + n(F) - n(D \cap F)\\ &= 352 + 229 - 150\\ &= 431\end{aligned}$$

(d) $B' \cap C'$ is the set of players who are both *not* in B and *not* in C. Thus, $B' \cap C' = A \cup D$, and since A and D are disjoint, $n(A \cup D) = n(A) + n(D) = 129 + 352 = 481$. There are 481 players in the set $B' \cap C'$.

49. Reading directly from the table, $n(A \cap B) = 110.6$. Thus, there are 110.6 million people in the set $A \cap B$.

51.

$$\begin{aligned} n(G \cup (C \cap H)) &= n(G) + n(C \cap H) \\ &= 80.4 + 5.0 \\ &= 85.4 \end{aligned}$$

There are 85.4 million people in the set $G \cup (C \cap H)$.

53.

$$\begin{aligned} n(H \cup D) &= n(H) + n(D) - n(H \cap D) \\ &= 53.6 + 19.6 - 2.2 \\ &= 71.0 \end{aligned}$$

There are 71.0 million people in the set $H \cup D$.

For Exercises 55 through 58, use the following table, where the numbers are in thousands and the table has been extended to include row and column totals.

	W	B	H	A	Totals
N	54,205	13,547	12,021	3518	83,291
M	106,517	9577	16,111	6741	138,946
I	11,968	1740	1068	507	15,283
D	23,046	4590	3477	665	31,778
Totals	195,736	29,454	32,677	11,431	269,298

55. $N \cap (B \cup H)$ is the set of Blacks or Hispanics who never married. These people are located in the first row of the table, in the B and H columns, so $n(N \cap (B \cup H)) = 13,547 + 12,021 = 25,568$; since the table values are in thousands, this set contains 25,568,000 people.

57. $(D \cup W) \cap A'$ is the set of Whites or Divorced/separated people who are not Asian/Pacific Islanders. The number of these people is found by adding the total of the W column to the total of the D row and then subtracting the number of Divorced/separated Whites (so we don't count them twice) and subtracting the number of Divorced/separated Asians/Pacific Islanders (whom we don't want to count).

$$\begin{aligned} n((D \cup W) \cap A') &= 195,736 + 31,778 - 23,046 - 665 \\ &= 203,803; \end{aligned}$$

since the table values are in thousands, this set contains 203,803,000 people.

59. Let W be the set of women, C be the set of those who speak Cantonese, and F be the set of those who set off firecrackers. We are given the following information.

$n(W) = 120$ $\quad n(C) = 150$ $\quad n(F) = 170$

$n(W' \cap C) = 108$ $\quad n(W' \cap F') = 100$

$n(W \cap C' \cap F) = 18$

$n(W' \cap C' \cap F') = 78$

$n(W \cap C \cap F) = 30$

Note that

$$\begin{aligned} n(W' \cap C \cap F') &= n(W' \cap F') - n(W' \cap C' \cap F') \\ &= 100 - 78 \\ &= 22. \end{aligned}$$

Furthermore,

$$\begin{aligned} n(W' \cap C \cap F) &= n(W' \cap C) - n(W' \cap C \cap F') \\ &= 108 - 22 \\ &= 86. \end{aligned}$$

We now have

$$\begin{aligned} &n(W \cap C \cap F') \\ &\quad = n(C) - n(W' \cap C \cap F) - n(W \cap C \cap F) \\ &\qquad - n(W' \cap C \cap F') \\ &\quad = 150 - 86 - 30 - 22 = 12. \end{aligned}$$

With all of the overlaps of W, C, and F determined, we can now compute $n(W \cap C' \cap F') = 60$ and $n(W' \cap C' \cap F) = 36$.

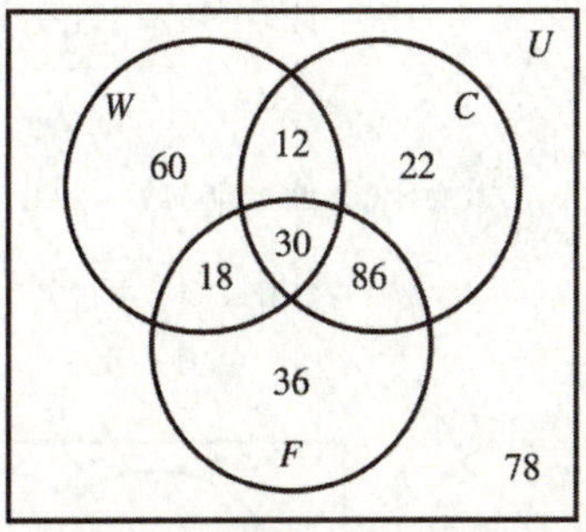

(a) Adding up the disjoint components, we find the total attendance to be

$$60 + 12 + 18 + 30 + 22 + 86 + 36 + 78 = 342.$$

(b) $n(C') = 342 - n(C) = 342 - 150 = 192$

(c) $n(W \cap F') = 60 + 12 = 72$

(d) $n(W' \cap C \cap F) = 86$

61. Let F be the set of fat chickens (so F' is the set of thin chickens), R be the set of red chickens (so R' is the set of brown chickens), and M be the set of male chickens, or roosters (so M' is the set of female chickens, or hens). We are given the following information.

$$n(F \cap R \cap M) = 9$$
$$n(F' \cap R' \cap M') = 13$$
$$n(R \cap M) = 15$$
$$n(F' \cap R) = 11$$
$$n(R \cap M') = 17$$
$$n(F) = 56$$
$$n(M) = 41$$
$$n(M') = 48$$

First, note that $n(M) + n(M') = n(U) = 89$, the total number of chickens.

Since $n(R \cap M) = 15$, $n(F' \cap R \cap M)$ $= 15 - 9 = 6$.

Since $n(F' \cap R) = 11$, $n(F' \cap R \cap M')$ $= 11 - 6 = 5$.

Since $n(R \cap M') = 17$, $n(F \cap R \cap M')$ $= 17 - 5 = 12$.

Since $n(M') = 48$, $n(F \cap R' \cap M')$ $= 48 - (12 + 5 + 13) = 18$.

Since $n(F) = 56$, $n(F \cap R' \cap M)$ $= 56 - (18 + 12 + 9) = 17$.

And, finally, since $n(M) = 41$,

$$n(F' \cap R' \cap M) = 41 - (17 + 9 + 6) = 9.$$

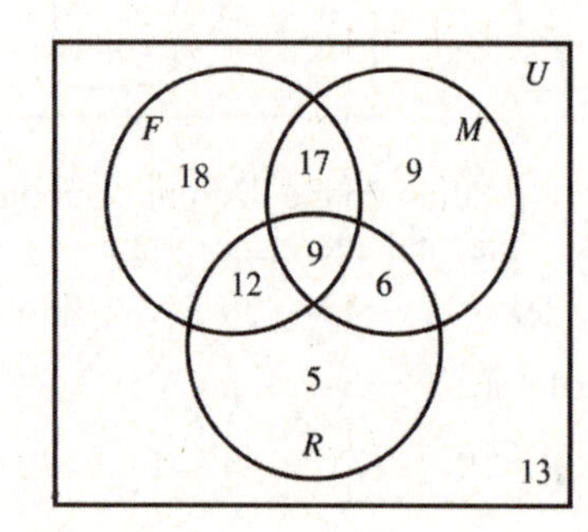

(a) $n(U) = 18 + 9 + 5 + 17 + 6 + 12 + 9 + 13 = 89$

(b) $n(R) = n(R \cap M) + n(R \cap M') = 15 + 17 = 32$

(c) $n(F \cap M) = n(F \cap R \cap M) + n(F \cap R' \cap M) = 17 + 9 = 26$

(d) $n(F \cap M') = n(F) - n(F \cap M) = 56 - 26 = 30$

(e) $n(F' \cap R') = n(F' \cap R' \cap M) + n(F' \cap R' \cap M') = 9 + 13 = 22$

(f) $n(F \cap R) = n(F \cap R \cap M) + n(F \cap R \cap M') = 9 + 12 = 21$

7.3 Introduction to Probability

Your Turn 1

The sample space of equally likely outcomes is $\{hh, ht, th, tt\}$.

Your Turn 2

The sample space for tossing two coins is $\{hh, ht, th, tt\}$. The event E: "the coins show exactly one head" is $E = \{ht, th\}$.

Your Turn 3

E: worker is under 20, so E': worker is 20 or over.

F: worker is white, so F': worker is not white.

Thus, $E' \cap F'$ is the event that the worker is 20 or over and is not white.

Your Turn 4

The sample space S is $S = \{1, 2, 3, 4, 5, 6\}$. The event H that the die shows a number less than 5 is $H = \{1, 2, 3, 4\}$. The probability of H is

$$P(H) = \frac{n(H)}{n(S)} = \frac{4}{6} = \frac{2}{3}.$$

Your Turn 5

In a standard 52-card deck there are 4 jacks and 4 kings, so $P(\text{jack or king}) = \frac{8}{52} = \frac{2}{13}$.

7.3 Exercises

3. The sample space is the set of the twelve months, {January, February, March, . . ., December}.

5. The possible number of points earned could be any whole number from 0 to 80. The sample space is the set

$$\{0,1,2,3,\ldots,79,80\}.$$

7. The possible decisions are to go ahead with a new oil shale plant or to cancel it. The sample space is the set {go ahead, cancel}.

9. Let h = heads and t = tails for the coin; the die can display 6 different numbers. There are 12 possible outcomes in the sample space, which is the set

$$\{(h,1),(h,2),(h,3),(h,4),(h,5),(h,6),\\ (t,1),(t,2),(t,3),(t,4),(t,5),(t,6)\}.$$

13. Use the first letter of each name. The sample space is the set

$$S = \{\text{AB},\text{AC},\text{AD},\text{AE},\text{BC},\text{BD},\text{BE},\text{CD},\text{CE},\text{DE}\}.$$

$n(S) = 10$. Assuming the committee is selected at random, the outcomes are equally likely.

(a) One of the committee members must be Chinn. This event is {AC, BC, CD, CE}.

(b) Alam, Bartolini, and Chinn may be on any committee; Dickson and Ellsberg may not be on the same committee. This event is

{AB, AC, AD, AE, BC, BD, BE, CD, CE}.

(c) Both Alam and Chinn are on the committee. This event is {AC}.

15. Each outcome consists of two of the numbers 1, 2, 3, 4, and 5, without regard for order. For example, let (2, 5) represent the outcome that the slips of paper marked with 2 and 5 are drawn. There are ten equally likely outcomes in this sample space, which is

$$S = \{(1,2),(1,3),(1,4),(1,5),(2,3),\\ (2,4),(2,5),(3,4),(3,5),(4,5)\}.$$

(a) Both numbers in the outcome pair are even. This event is {(2, 4)}, which is called a simple event since it consists of only one outcome.

(b) One number in the pair is even and the other number is odd. This event is

$$\{(1,2),(1,4),(2,3),(2,5),(3,4),(4,5)\}.$$

(c) Each slip of paper has a different number written on it, so it is not possible to draw two slips marked with the same number. This event is $\emptyset$, which is called an impossible event since it contains no outcomes.

17. $S = \{HH, THH, HTH, TTHH, THTH, HTTH,\\ TTTH, TTHT, THTT, HTTT, TTTT\}$

$n(S) = 11$. The outcomes are not equally likely.

(a) The coin is tossed four times. This event is written {*TTHH*, *THTH*, *HTTH*, *TTTH*, *TTHT*, *THTT*, *HTTT*, *TTTT*}.

(b) Exactly two heads are tossed. This event is written {*HH*, *THH*, *HTH*, *TTHH*, *THTH*, *HTTH*}.

(c) No heads are tossed. This event is written {*TTTT*}.

For Exercises 19–24, use the sample space

$$S = \{1,2,3,4,5,6\}.$$

19. "Getting a 2" is the event $E = \{2\}$, so $n(E) = 1$ and $n(S) = 6$.

If all the outcomes in a sample space S are equally likely, then the probability of an event E is

$$P(E) = \frac{n(E)}{n(S)}.$$

In this problem,

$$P(E) = \frac{n(E)}{n(S)} = \frac{1}{6}.$$

21. "Getting a number less than 5" is the event $E = \{1,2,3,4\}$, so $n(E) = 4$.

$$P(E) = \frac{4}{6} = \frac{2}{3}.$$

23. "Getting a 3 or a 4" is the event $E = \{3,4\}$, so $n(E) = 2$.

$$P(E) = \frac{2}{6} = \frac{1}{3}.$$

For Exercises 25–34, the sample space contains all 52 cards in the deck, so $n(S) = 52$.

25. Let E be the event "a 9 is drawn." There are four 9's in the deck, so $n(E) = 4$.

$$P(9) = P(E) = \frac{n(E)}{n(S)} = \frac{4}{52} = \frac{1}{13}$$

27. Let F be the event "a black 9 is drawn." There are two black 9's in the deck, so $n(F) = 2$.

$$P(\text{black } 9) = P(F) = \frac{n(F)}{n(S)} = \frac{2}{52} = \frac{1}{26}$$

29. Let G be the event "a 9 of hearts is drawn." There is only one 9 of hearts in a deck of 52 cards, so $n(G) = 1$.

$$P(9 \text{ of hearts}) = P(G) = \frac{n(G)}{n(S)} = \frac{1}{52}$$

31. Let H be the event "a 2 or a queen is drawn." There are four 2's and four queens in the deck, so $n(H) = 8$.

$$P(2 \text{ or queen}) = P(H) = \frac{n(H)}{n(S)} = \frac{8}{52} = \frac{2}{13}$$

33. Let E be the event "a red card or a ten is drawn." There are 26 red cards and 4 tens in the deck. But 2 tens are red cards and are counted twice. Use the result from the previous section.

$$\begin{aligned} n(E) &= n(\text{red cards}) + n(\text{tens}) - n(\text{red tens}) \\ &= 26 + 4 - 2 \\ &= 28 \end{aligned}$$

Now calculate the probability of E.

$$\begin{aligned} P(\text{red cards or ten}) &= \frac{n(E)}{n(S)} \\ &= \frac{28}{52} \\ &= \frac{7}{13} \end{aligned}$$

For Exercises 35–40, the sample space consists of all the marbles in the jar. There are $3 + 4 + 5 + 8 = 20$ marbles, so $n(S) = 20$.

35. 3 of the marbles are white, so

$$P(\text{white}) = \frac{3}{20}.$$

37. 5 of the marbles are yellow, so

$$P(\text{yellow}) = \frac{5}{20} = \frac{1}{4}.$$

39. $3 + 4 + 5 = 12$ of the marbles are not black, so

$$P(\text{not black}) = \frac{12}{20} = \frac{3}{5}.$$

41. It is possible to establish an exact probability for this event, so it is not an empirical probability.

43. It is not possible to establish an exact probability for this event, so this an empirical probability.

45. It is not possible to establish an exact probability for this event, so this is an empirical probability.

47. The gambler's claim is a mathematical fact, so this is not an empirical probability.

49. The outcomes are not equally likely.

51. E: worker is female

F: worker has worked less than 5 yr

G: worker contributes to a voluntary retirement plan

(a) E' occurs when E does not, so E' is the event "worker is male."

(b) $E \cap F$ occurs when both E and F occur, so $E \cap F$ is the event "worker is female and has worked less than 5 yr."

(c) $E \cup G'$ is the event "worker is female or does not contribute to a voluntary retirement plan."

(d) F' occurs when F does not, so F' is the event "worker has worked 5 yr or more."

(e) $F \cup G$ occurs when F or G occurs or both, so $F \cup G$ is the event "worker has worked less than 5 yr or contributes to a voluntary retirement plan."

(f) $F' \cap G'$ occurs when F does not and G does not, so $F' \cap G'$ is the event "worker has worked 5 yr or more and does not contribute to a voluntary retirement plan."

53. **(a)** From the solution to Exercise 42 in Section 7.2 we know that 80 investors made all three types of investments, so the probability that a randomly chosen professor invested in stocks and bonds and certificates of deposit is $\frac{80}{150} = \frac{8}{15}$.

(b) From the solution to Exercise 42 in Section 7.2 we find that $80 - 65$ or 15 professors invested only in bonds, so the probability that a randomly chosen professor invested only in bonds is $\frac{15}{150} = \frac{1}{10}$.

55. E: person smokes

F: person has a family history of heart disease

G: person is overweight

(a) G': "person is not overweight."

(b) $F \cap G$: "person has a family history of heart disease and is overweight."

(c) $E \cup G'$: "person smokes or is not overweight."

57. **(a)** $P(\text{heart disease}) = \frac{615{,}651}{2{,}424{,}059} = 0.2540$

(b)
$$\begin{aligned} P(\text{cancer or heart disease}) &= \frac{1{,}175{,}838}{2{,}424{,}059} \\ &= 0.4851 \end{aligned}$$

(c) P(not accident and not diabetes mellitus)

$$= \frac{2{,}424{,}059 - 117{,}075 - 70{,}905}{2{,}424{,}059}$$
$$= \frac{2{,}236{,}079}{2{,}424{,}059}$$
$$= 0.9225$$

59. $P(\text{served } 20-29 \text{ years}) = \frac{17}{100} = 0.17$

61. **(a)** $P(\text{III Corps}) = \frac{22{,}083}{70{,}076} \approx 0.3151$

(b) $P(\text{lost in battle}) = \frac{22{,}557}{70{,}076} \approx 0.3219$

(c) $P(\text{I Corps lost in battle}) = \frac{7661}{20{,}706} \approx 0.3700$

(d) P(I Corps not lost in battle)

$$= \frac{20{,}706 - 7661}{20{,}706} \approx 0.6300$$

P(II Corps not lost in battle)

$$= \frac{20{,}666 - 6603}{20{,}666} \approx 0.6805$$

P(III Corps not lost in battle)

$$= \frac{22{,}083 - 8007}{22{,}083} \approx 0.6374$$

P(Cavalry not lost in battle)

$$= \frac{6621 - 286}{6621} \approx 0.9568$$

The Cavalry had the highest probability of not being lost in battle.

(e) $P(\text{I Corps loss}) = \frac{7661}{20{,}706} \approx 0.3700$

$P(\text{II Corps loss}) = \frac{6603}{20{,}666} \approx 0.3195$

$P(\text{III Corps loss}) = \frac{8007}{22{,}083} \approx 0.3626$

$P(\text{Cavalry loss}) = \frac{286}{6621} \approx 0.0432$

I Corps had the highest probability of loss.

63. There were 342 in attendance.

(a) $P(\text{speaks Cantonese}) = \frac{150}{342} = \frac{25}{57}$

(b) $P(\text{does not speaks Cantonese}) = \frac{192}{342} = \frac{32}{57}$

(c) P(woman who did not light firecracker).

$$= \frac{72}{342} = \frac{4}{19}$$

7.4 Basic Concepts of Probability

Your Turn 1

Let A stand for the event "ace" and C stand for the event "club."

$$P(A \cup C) = P(A) + P(C) - P(A \cap C)$$
$$= \frac{4}{52} + \frac{13}{52} - \frac{1}{52}$$
$$= \frac{16}{52} = \frac{4}{13}$$

Your Turn 2

Let E stand for the event "eight" and B stand for the event "both dice show the same number." From Figure 18 we see that E contains the 5 events 6-2, 5-3, 4-4, 3-5, and 2-6. B contains the 6 events 1-1, 2-2, 3-3, 4-4, 5-5, and 6-6. Only the even 4-4 belongs to both E and B. The sample space contains 36 equally likely events.

$$P(E \cup B) = P(E) + P(B) - P(E \cap B)$$
$$= \frac{5}{36} + \frac{6}{36} - \frac{1}{36}$$
$$= \frac{10}{36} = \frac{5}{18}$$

Your Turn 3

The complement of the event "sum < 11" is "sum $= 11$ or sum $= 12$." Figure 18 shows that "sum$= 11$ or sum $= 12$" contains the 3 events 5-6, 6-5, and 6-6.

$$P(\text{sum} < 11) = 1 - P(\text{sum} = 11 \text{ or sum} = 12)$$
$$= 1 - \frac{3}{36}$$
$$= \frac{33}{36} = \frac{11}{12}$$

Your Turn 4

Let E be the event "snow tomorrow." Since $P(E) = \frac{3}{10}$, $P(E') = 1 - P(E) = \frac{7}{10}$. The odds in favor of snow tomorrow are

$$\frac{P(E)}{P(E')} = \frac{3/10}{7/10} = \frac{3}{7}.$$

We can write these odds as 3 to 7 or 3:7.

Your Turn 5

Let E be the event "package delivered on time." The odds in favor of E are 17 to 3, so $P(E') = \frac{3}{17+3} = \frac{3}{20}$. The probability that the package will not be delivered on time is $\frac{3}{20}$.

Your Turn 6

If the odds against the horse winning are 7 to 3,

$P(\text{loses}) = \frac{7}{7+3} = \frac{7}{10}$, so $P(\text{wins}) = 1 - \frac{7}{10} = \frac{3}{10}$.

7.4 Exercises

3. A person can own a dog and own an MP3 player at the same time. No, these events are not mutually exclusive.

5. A person can be retired and be over 70 years old at the same time. No, these events are not mutually exclusive.

7. A person cannot be one of the ten tallest people in the United States and be under 4 feet tall at the same time. Yes, these events are mutually exclusive.

9. When two dice are rolled, there are 36 equally likely outcomes.

(a) Of the 36 ordered pairs, there is only one for which the sum is 2, namely $\{(1,1)\}$. Thus,

$$P(\text{sum is } 2) = \frac{1}{36}.$$

(b) $\{(1,3),(2,2),(3,1)\}$ comprise the ways of getting a sum of 4. Thus,

$$P(\text{sum is } 4) = \frac{3}{36} = \frac{1}{12}.$$

(c) $\{(1,4),(2,3),(3,2),(4,1)\}$ comprise the ways of getting a sum of 5. Thus,

$$P(\text{sum is } 5) = \frac{4}{36} = \frac{1}{9}.$$

(d) $\{(1,5),(2,4),(3,3),(4,2),(5,1)\}$ comprise the ways of getting a sum of 6. Thus,

$$P(\text{sum is } 6) = \frac{5}{36}.$$

11. Again, when two dice are rolled there are 36 equally likely outcomes.

(a) Here, the event is the union of four mutually exclusive events, namely, the sum is 9, the sum is 10, the sum is 11, and the sum is 12. Hence,

$$\begin{aligned} P(\text{sum is 9 or more}) &= P(\text{sum is } 9) + P(\text{sum is } 10) \\ &\quad + P(\text{sum is } 11) + (\text{sum is } 12) \\ &= \frac{4}{36} + \frac{3}{36} + \frac{2}{36} + \frac{1}{36} \\ &= \frac{10}{36} = \frac{5}{18}. \end{aligned}$$

(b) $P(\text{sum is less than } 7)$

$$\begin{aligned} &= P(2) + P(3) + P(4) + P(5) + P(6) \\ &= \frac{1}{36} + \frac{2}{36} + \frac{3}{36} + \frac{4}{36} + \frac{5}{36} \\ &= \frac{15}{36} \\ &= \frac{5}{12} \end{aligned}$$

(c) $P(\text{sum is between 5 and 8})$

$$\begin{aligned} &= P(\text{sum is } 6) + P(\text{sum is } 7) \\ &= \frac{5}{36} + \frac{6}{36} \\ &= \frac{11}{36} \end{aligned}$$

13. $P(\text{first die is 3 or sum is 8})$

$$\begin{aligned} &= P(\text{first die is } 3) + P(\text{sum is } 8) \\ &\quad - P(\text{first die is 3 and sum is 8}) \\ &= \frac{6}{36} + \frac{5}{36} - \frac{1}{36} \\ &= \frac{10}{36} \\ &= \frac{5}{18} \end{aligned}$$

15. **(a)** The events E, "9 is drawn," and F, "10 is drawn," are mutually exclusive, so $P(E \cap F) = 0$. Using the union rule,

$$P(9 \text{ or } 10) = P(9) + P(10) = \frac{4}{52} + \frac{4}{52} = \frac{8}{52} = \frac{2}{13}.$$

(b)

$$P(\text{red or } 3) = P(\text{red}) + P(3) - P(\text{red and } 3) = \frac{26}{52} + \frac{4}{52} - \frac{2}{52} = \frac{28}{52} = \frac{7}{13}$$

(c) Since these events are mutually exclusive,

$$P(9 \text{ or black } 10) = P(9) + P(\text{black } 10) = \frac{4}{52} + \frac{2}{52} = \frac{6}{52} = \frac{3}{26}.$$

(d) $P(\text{heart or black}) = \frac{13}{52} + \frac{26}{52} = \frac{39}{52} = \frac{3}{4}.$

(e) $P(\text{face card or diamond})$

$$= P(\text{face card}) + P(\text{diamond}) - P(\text{face card and diamond}) = \frac{12}{52} + \frac{13}{52} - \frac{3}{52} = \frac{22}{52} = \frac{11}{26}$$

17. (a) Since these events are mutually exclusive,

$$P(\text{brother or uncle}) = P(\text{brother}) + P(\text{uncle}) = \frac{2}{13} + \frac{3}{13} = \frac{5}{13}.$$

(b) Since these events are mutually exclusive,

$$P(\text{brother or cousin}) = P(\text{brother}) + P(\text{cousin}) = \frac{2}{13} + \frac{5}{13} = \frac{7}{13}.$$

(c) Since these events are mutually exclusive,

$$P(\text{brother or mother}) = P(\text{brother}) + P(\text{mother}) = \frac{2}{13} + \frac{1}{13} = \frac{3}{13}.$$

19. (a) There are 5 possible numbers on the first slip drawn, and for each of these, 4 possible numbers on the second, so the sample space contains $5 \cdot 4 = 20$ ordered pairs. Two of these ordered pairs have a sum of 9: (4, 5) and (5, 4). Thus,

$$P(\text{sum is } 9) = \frac{2}{20} = \frac{1}{10}.$$

(b) The outcomes for which the sum is 5 or less are (1, 2), (1, 3), (1, 4), (2, 1), (2, 3), (3, 1), (3, 2), and (4,1). Thus,

$$P(\text{sum is 5 or less}) = \frac{8}{20} = \frac{2}{5}.$$

(c) Let A be the event "the first number is 2" and B the event "the sum is 6." Use the union rule.

$$P(A \cup B) = P(A) + P(B) - P(A \cap B) = \frac{4}{20} + \frac{4}{20} - \frac{1}{20} = \frac{7}{20}$$

21. Since $P(E \cap F) = 0.16$, the overlapping region $E \cap F$ is assigned the probability 0.16 in the diagram. Since $P(E) = 0.26$ and $P(E \cap F) = 0.16$, the region in E but not F is given the label 0.10. Similarly, the remaining regions are labeled.

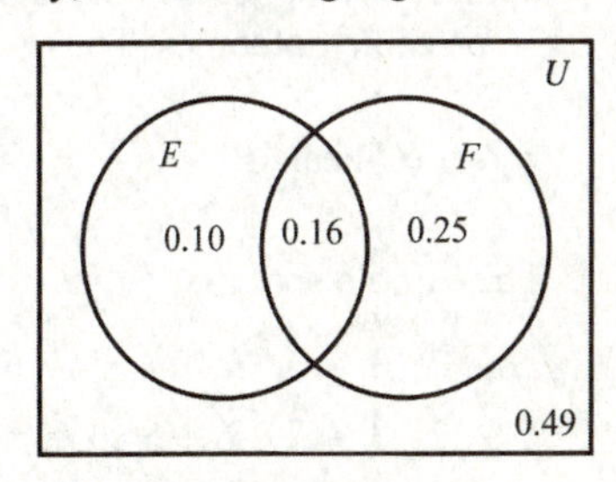

(a)
$$P(E \cup F) = 0.10 + 0.16 + 0.25 = 0.51$$

Consequently, the part of U outside $E \cup F$ receives the label

$$1 - 0.51 = 0.49.$$

(b)
$$P(E' \cap F) = P(\text{in } F \text{ but not in } E) = 0.25$$

(c) The region $E \cap F'$ is that part of E which is not in F. Thus,

$$P(E \cap F') = 0.10.$$

(d)
$$P(E' \cup F') = P(E') + P(F') - P(E' \cap F') = 0.74 + 0.59 - 0.49 = 0.84$$

23. **(a)** The sample space is

3 – 1 3 – 1 3 – 5 3 – 5 3 – 9 3 – 9
3 – 1 3 – 1 3 – 5 3 – 5 3 – 9 3 – 9
4 – 1 4 – 1 4 – 5 4 – 5 4 – 9 4 – 9
4 – 1 4 – 1 4 – 5 4 – 5 4 – 9 4 – 9
8 – 1 8 – 1 8 – 5 8 – 5 8 – 9 8 – 9
8 – 1 8 – 1 8 – 5 8 – 5 8 – 9 8 – 9

where the first number in each pair is the number that appears on A and the second the number that appears on B. B beats A in 20 of 36 possible outcomes. Thus,

$$P(B \text{ beats } A) = \frac{20}{36} = \frac{5}{9}.$$

(b) The sample space is

1 – 2 1 – 2 1 – 6 1 – 6 1 – 7 1 – 7
1 – 2 1 – 2 1 – 6 1 – 6 1 – 7 1 – 7
5 – 2 5 – 2 5 – 6 5 – 6 5 – 7 5 – 7
5 – 2 5 – 2 5 – 6 5 – 6 5 – 7 5 – 7
9 – 2 9 – 2 9 – 6 9 – 6 9 – 7 9 – 7
9 – 2 9 – 2 9 – 6 9 – 6 9 – 7 9 – 7

where the first number in each pair is the number that appears on B and the second the number that appears on C. C beats B in 20 of 36 possible outcomes. Thus,

$$P(C \text{ beats } B) = \frac{20}{36} = \frac{5}{9}.$$

(c) The sample space is

3 – 2 3 – 2 3 – 6 3 – 6 3 – 7 3 – 7
3 – 2 3 – 2 3 – 6 3 – 6 3 – 7 3 – 7
4 – 2 4 – 2 4 – 6 4 – 6 4 – 7 4 – 7
4 – 2 4 – 2 4 – 6 4 – 6 4 – 7 4 – 7
8 – 2 8 – 2 8 – 6 8 – 6 8 – 7 8 – 7
8 – 2 8 – 2 8 – 6 8 – 6 8 – 7 8 – 7

where the first number in each pair is the number that appears on A and the second the number that appears on C. A beats C in 20 of 36 possible outcomes. Thus,

$$P(A \text{ beats } C) = \frac{20}{36} = \frac{5}{9}.$$

27. Let E be the event "a 3 is rolled."

$$P(E) = \frac{1}{6} \text{ and } P(E') = \frac{5}{6}.$$

The odds in favor of rolling a 3 are

$$\frac{P(E)}{P(E')} = \frac{\frac{1}{6}}{\frac{5}{6}} = \frac{1}{5},$$

which is written "1 to 5."

29. Let E be the event "a 2, 3, 4, or 5 is rolled." Here $P(E) = \frac{4}{6} = \frac{2}{3}$ and $P(E') = \frac{1}{3}$. The odds in favor of E are

$$\frac{P(E)}{P(E')} = \frac{\frac{2}{3}}{\frac{1}{3}} = \frac{2}{1},$$

which is written "2 to 1."

31. **(a)** Yellow: There are 3 ways to win and 15 ways to lose. The odds in favor of drawing yellow are 3 to 15, or 1 to 5.

(b) Blue: There are 11 ways to win and 7 ways to lose; the odds in favor of drawing blue are 11 to 7.

(c) White: There are 4 ways to win and 14 ways to lose; the odds in favor of drawing white are 4 to 14, or 2 to 7.

(d) Not white: Since the odds in favor of white are 2 to 7, the odds in favor of not white are 7 to 2.

35. Each of the probabilities is between 0 and 1 and the sum of all the probabilities is

$$0.09 + 0.32 + 0.21 + 0.25 + 0.13 = 1,$$

so this assignment is possible.

$$0.92 + 0.03 + 0 + 0.02 + 0.03 = 1.$$

37. The sum of the probabilities

$$\frac{1}{3} + \frac{1}{4} + \frac{1}{6} + \frac{1}{8} + \frac{1}{10} = \frac{117}{120} < 1,$$

so this assignment is not possible.

39. This assignment is not possible because one of the probabilities is -0.08, which is not between 0 and 1. A probability cannot be negative.

41. The answers that are given are theoretical. Using the Monte Carlo method with at least 50 repetitions on a graphing calculator should give values close to these.

(a) 0.2778

(b) 0.4167

43. The answers that are given are theoretical. Using the Monte Carlo method with at least 100 repetitions should give values close to these.

(a) 0.0463

(b) 0.2963

47. Let C be the event "the calculator has a good case," and let B be the event "the calculator has good batteries."

$$\begin{aligned} &P(C \cap B) \\ &= 1 - P[(C \cap B)'] \\ &= 1 - P(C' \cup B') \\ &= 1 - [P(C') + P(B') - P(C' \cap B')] \\ &= 1 - (0.08 + 0.11 - 0.03) \\ &= 0.84 \end{aligned}$$

Thus, the probability that the calculator has a good case and good batteries is 0.84.

49. **(a)** $P(\$500 \text{ or more}) = 1 - P(\text{less than } \$500)$
$= 1 - (0.21 + 0.17)$
$= 1 - 0.38$
$= 0.62$

(b) $P(\text{less than } \$1000) = 0.21 + 0.17 + 0.16$
$= 0.54$

(c) $P(\$500 \text{ to } \$2999) = 0.16 + 0.15 + 0.12$
$= 0.43$

(d) $P(\$3000 \text{ or more}) = 0.08 + 0.07 + 0.04$
$= 0.19$

51. **(a)** The probability of Female and 16 to 24 years old is the entry in the first row in the Female column, which is 0.061.

(b) 16 to 54 years old includes the first two rows of the table, so adding the totals for these two rows we find that the probability is $0.127 + 0.634 = 0.761$.

(c) For Male or 25 to 54, we add the totals of the second row and the Male column and then subtract the value for Male and 25 to 54 in order not to count it twice. Thus the probability is $0.634 + 0.531 - 0.343 = 0.822$.

(d) For Female or 16 to 24 we add the totals of the first row and the Female column and then subtract the value for Female and 16 to 24 in order not to count it twice. Thus the probability is $0.127 + 0.469 - 0.061 = 0.535$.

53. $P(C) = 0.039, P(M \cap C) = 0.035,$
$P(M \cup C) = 0.491$

Place the given information in a Venn diagram by starting with 0.035 in the intersection of the regions for M and C.

Since $P(C) = 0.039, 0.039 - 0.035 = 0.004$

goes inside region C, but outside the intersection of C and M. Thus,

$$P(C \cap M') = 0.004.$$

Since

$$P(M \cup C) = 0.491, 0.491 - 0.035 - 0.004 = 0.452$$

goes inside region M, but outside the intersection of C and M. Thus, $P(M \cap C') = 0.452$. The labeled regions have probability

$$0.452 + 0.035 + 0.004 = 0.491.$$

Since the entire region of the Venn diagram must have probability 1, the region outside M and C, or $M' \cap C'$, has probability

$$1 - 0.491 = 0.509.$$

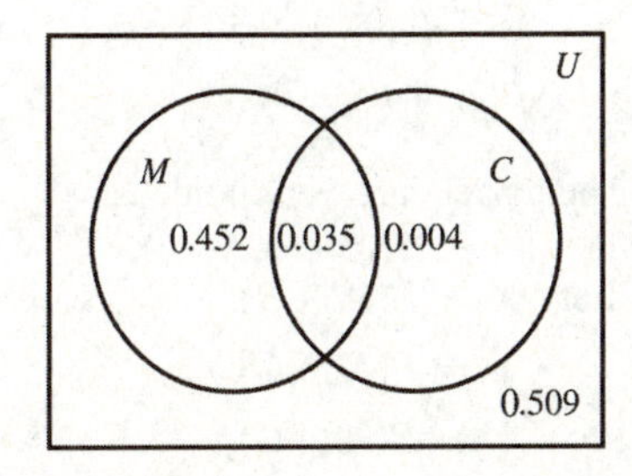

(a) $P(C') = 1 - P(C)$
$= 1 - 0.039$
$= 0.961$

(b) $P(M) = 0.452 + 0.035$
$= 0.487$

(c) $P(M') = 1 - P(M)$
$= 1 - 0.487$
$= 0.513$

(d) $P(M' \cap C') = 0.509$

(e) $P(C \cap M') = 0.004$

(f) $P(C \cup M')$
$= P(C) + P(M') - P(C \cap M')$
$= 0.039 + 0.513 - 0.004$
$= 0.548$

55. **(a)** Now red is no longer dominant, and RW or WR results in pink, so

$$P(\text{red}) = P(RR) = \frac{1}{4}.$$

(b) $P(\text{pink}) = P(RW) + P(WR)$

$$= \frac{1}{4} + \frac{1}{4} = \frac{1}{2}$$

(c) $P(\text{white}) = P(WW) = \frac{1}{4}$

57. Let L be the event "visit results in lab work" and R be the event "visit results in referral to specialist." We are given the probability a visit results in neither is 35%, so $P((L \cup R)') = 0.35$. Since $P(L \cup R) = 1 - P((L \cup R)')$, we have

$$P(L \cup R) = 1 - 0.35 = 0.65.$$

We are also given $P(L) = 0.40$ and $P(R) = 0.30$. Using the union rule for probability,

$$\begin{aligned} P(L \cup R) &= P(L) + P(R) - P(L \cap R) \\ 0.65 &= 0.40 + 0.30 - P(L \cap R) \\ 0.65 &= 0.50 - P(L \cap R) \\ P(L \cap R) &= 0.05 \end{aligned}$$

The correct answer choice is **a**.

59. Let $x = P(A \cap B)$,
$y = P(B \cap C)$,
$z = P(A \cap C)$,
and $w = P((A \cup B \cup C)')$.

If an employee must choose exactly two or none of the supplementary coverages A, B, and C, then $P(A \cap B \cap C) = 0$ and the probabilities of the region representing a single choice of coverages A, B, or C are also 0. We can represent the choices and probabilities with the following Venn diagram.

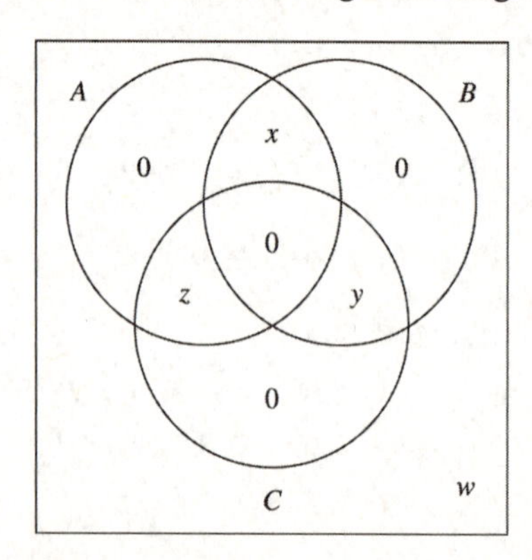

The information given leads to the following system of equations.

$$\begin{aligned} x + y + z + w &= 1 \\ x \quad + z \quad &= \frac{1}{4} \\ x + y \quad &= \frac{1}{3} \\ y + z \quad &= \frac{5}{12} \end{aligned}$$

Using a graphing calculator or computer program, the solution to the system is $x = 1/2$, $y = 1/4$, $z = 1/6$, $w = 1/2$. Since the probability that a randomly chosen employee will choose no supplementary coverage is w, the correct answer choice is **c**.

61. Since 55 of the workers were women, $130 - 55 = 75$ were men. Since 3 of the women earned more than \$40,000, $55 - 3 = 52$ of them earned \$40,000 or less. Since 62 of the men earned \$40,000 or less, $75 - 62 = 13$ earned more than \$40,000. These data for the 130 workers can be summarized in the following table.

	Men	Women
\$40,000 or less	62	52
Over \$40,000	13	3

(a) P(a woman earning \$40,000 or less)

$$= \frac{52}{130} = 0.4$$

(b) P(a man earning more than \$40,000)

$$= \frac{13}{130} = 0.1$$

(c) P(a man or is earning more than \$40,000)

$$= \frac{62 + 13 + 3}{130}$$
$$= \frac{78}{130} = 0.6$$

(d) P(a woman or is earning \$40,000 or less)

$$= \frac{52 + 3 + 62}{130}$$
$$= \frac{117}{130} = 0.9$$

63. Let A be the set of refugees who came to escape abject poverty and B be the set of refugees who came to escape political oppression. Then $P(A) = 0.80$, $P(B) = 0.90$, and $P(A \cap B) = 0.70$.

$$P(A \cup B) = P(A) + P(B) - P(A \cap B)$$
$$= 0.80 + 0.90 - 0.70 = 1$$
$$P(A' \cap B') = 1 - P(A \cap B)$$
$$= 1 - 1 = 0$$

The probability that a refugee in the camp was neither poor nor seeking political asylum is 0.

65. The odds of winning are 3 to 2; this means there are 3 ways to win and 2 ways to lose, out of a total of $2 + 3 = 5$ ways altogether. Hence, the probability of losing is $\frac{2}{5}$.

67. (a) P(somewhat or extremely intolerant of Facists)
$= P$(somewhat intolerant of Facists)
$+ P$(extremely intolerant of Facists)
$$= \frac{27.1}{100} + \frac{59.5}{100} = \frac{86.6}{100} = 0.866$$

(b) P(completely tolerant of Communists)
$= P$(no intolerance at all of Communists)
$$= \frac{47.8}{100} = 0.478$$

69. (a) There are $67 + 25 = 92$ possible judging combinations with Sasha Cohen finishing in first place. The probability of this outcome is, therefore, $92/220 = 23/55$.

(b) There are 67 possible judging combinations with Irina Slutskaya finishing in second place. The probability of this outcome is, therefore, 67/220.

(c) There are $92 + 67 = 159$ possible judging combinations with Shizuka Arahawa finishing in third place. The probability of this outcome is, therefore, 159/220.

71. The probabilities are as follows:

$$\frac{1}{1 + 195{,}199{,}999} = \frac{1}{195{,}200{,}000} = 0.0000000051$$

$$\frac{1}{1 + 835{,}499} = \frac{1}{835{,}500} = 0.0000012$$

$$\frac{1}{1 + 157.6} = \frac{1}{158.6} = 0.0063$$

$$\frac{1}{1 + 59.32} = \frac{1}{60.32} = 0.0166$$

7.5 Conditional Probability; Independent Events

Your Turn 1

Reduce the sample space to B' and then find $n(A \cap B')$ and $n(B')$.

$$P(A|B') = \frac{P(A \cap B')}{P(B')} = \frac{n(A \cap B')}{n(B')} = \frac{30}{55} = \frac{6}{11}$$

Your Turn 2

$$P(E \cup F) = P(E) + P(F) - P(E \cap F)$$
$$0.80 = 0.56 + 0.64 - P(E \cap F)$$
$$P(E \cap F) = 0.56 + 0.64 - 0.80$$
$$P(E \cap F) = 0.40$$
$$P(E \mid F) = \frac{P(E \cap F)}{P(F)}$$
$$= \frac{0.40}{0.64} = 0.625$$

Your Turn 3

Since at least one coin is a tail, the sample space is reduced to $\{ht, th, tt\}$. Two of these equally likely outcomes have exactly one head, so $P(\text{one head}|\text{at least one tail}) = \frac{2}{3}$.

Your Turn 4

Let C represent "lives on campus" and A represent "has a car on campus." Using the given information, $P(A|C) = \frac{1}{4}$ and $P(C) = \frac{4}{5}$. By the product rule,

$$P(A \cap C) = P(A|C) \cdot P(C)$$
$$= \frac{1}{4} \cdot \frac{4}{5} = \frac{1}{5}.$$

Your Turn 5

Using Figure 23, we follow the A branch and then the U branch. Multiplying along the tree we find the probability of the composite branch, which represents $A \cap U$ or the event that A is in charge of the campaign and it produces unsatisfactory results.

$$P(A \cap U) = \frac{2}{3} \cdot \frac{1}{4} = \frac{1}{6}$$

Your Turn 6

There are two paths that result in one NY plant and one Chicago plant: C first and NY second, and NY first and C second. From the tree, $P(\text{C, NY}) = \frac{1}{2} \cdot \frac{1}{5} = \frac{1}{10}$, and $P(\text{NY, C}) = \frac{1}{6} \cdot \frac{3}{5} = \frac{1}{10}$. So

P(one NY plant and one Chicago plant)
$$= P(\text{C, NY}) + P(\text{NY, C}) = \frac{1}{10} + \frac{1}{10} = \frac{1}{5}.$$

Your Turn 7

Successive rolls of a die are independent events, so

$$P(\text{two fives in a row}) = P(\text{five}) \cdot P(\text{five}) = \frac{1}{6} \cdot \frac{1}{6} = \frac{1}{36}.$$

Your Turn 8

Let M represent the event "you do your math homework" and H represent the event "you do your history assignment."

$$P(M) = 0.8$$
$$P(H) = 0.7$$
$$P(M \cup H) = 0.9$$

Now solve for $P(M \cap H)$.

$$P(M \cup H) = P(M) + P(H) - P(M \cap H)$$
$$0.9 = 0.8 + 0.7 - P(M \cap H)$$
$$P(M \cap H) = 0.8 + 0.7 - 0.9$$
$$P(M \cap H) = 0.6$$

However, $P(M)P(H) = (0.8)(0.7) = 0.56$, so $P(M \cap H) \neq P(M)P(H)$. Since M and H do not satisfy the product rule for independent events, they are not independent.

7.5 Exercises

1. Let A be the event "the number is 2" and B be the event "the number is odd."

The problem seeks the conditional probability $P(A|B)$. Use the definition

$$P(A|B) = \frac{P(A \cap B)}{P(B)}.$$

Here, $P(A \cap B) = 0$ and $P(B) = \frac{1}{2}$. Thus,

$$P(A|B) = \frac{0}{\frac{1}{2}} = 0.$$

3. Let A be the event "the number is even" and B be the event "the number is 6." Then

$$P(A|B) = \frac{P(A \cap B)}{P(B)} = \frac{\frac{1}{6}}{\frac{1}{6}} = 1.$$

5. $P(\text{sum of } 8|\text{greater than } 7)$

$$= \frac{P(8 \cap \text{greater than } 7)}{P(\text{greater than } 7)}$$
$$= \frac{n(8 \cap \text{greater than } 7)}{n(\text{greater than } 7)}$$
$$= \frac{5}{15} = \frac{1}{3}$$

7. The event of getting a double given that 9 was rolled is impossible; hence,

$$P(\text{double}|\text{sum of } 9) = 0.$$

9. Use a reduced sample space. After the first card drawn is a heart, there remain 51 cards, of which 12 are hearts. Thus,

$$P(\text{heart on 2nd}|\text{heart on 1st}) = \frac{12}{51} = \frac{4}{17}.$$

11. Use a reduced sample space. After the first card drawn is a jack, there remain 51 cards, of which 11 are face cards. Thus,

$$P(\text{face card on 2nd}|\text{ jack on lst}) = \frac{11}{51}.$$

13. $P(\text{a jack and a } 10)$

$$= P(\text{jack followed by } 10) + P(10 \text{ followed by jack})$$
$$= \frac{4}{52} \cdot \frac{4}{51} + \frac{4}{52} \cdot \frac{4}{51}$$
$$= \frac{16}{2652} + \frac{16}{2652}$$
$$= \frac{32}{2652} = \frac{8}{663}$$

15. $P(\text{two black cards})$

$$= P(\text{black on 1st}) \cdot P(\text{black on 2nd}|\text{black on 1st})$$
$$= \frac{26}{52} \cdot \frac{25}{51}$$
$$= \frac{650}{2652} = \frac{25}{102}$$

19. Examine a table of all possible outcomes of rolling a red die and rolling a green die (such as Figure 18 in Section 7.4). There are 9 outcomes of the 36 total outcomes that correspond to rolling "red die comes up even and green die comes up even"—in other words, corresponding to $A \cap B$. Therefore,

$$P(A \cap B) = \frac{9}{36} = \frac{1}{4}.$$

We also know that $P(A) = 1/2$ and $P(B) = 1/2$. Since

$$P(A \cap B) = \frac{1}{4} = \frac{1}{2} \cdot \frac{1}{2} = P(A) \cdot P(B),$$

the events A and B are independent.

21. Notice that $P(F|E) \neq P(F)$: the knowledge that a person lives in Dallas affects the probability that the person lives in Dallas or Houston. Therefore, the events are dependent.

23. **(a)** The events that correspond to "sum is 7" are (2, 5), (3, 4), (4, 3), and (5, 2), where the first number is the number on the first slip of paper and the second number is the number on the second. Of these, only (3, 4) corresponds to "first is 3," so

$$P(\text{first is } 3|\text{sum is } 7) = \frac{1}{4}.$$

(b) The events that correspond to "sum is 8" are (3, 5) and (5, 3). Of these, only (3, 5) corresponds to "first is 3," so

$$P(\text{first is } 3|\text{sum is } 8) = \frac{1}{2}.$$

25. **(a)** Many answers are possible; for example, let B be the event that the first die is a 5. Then

$$P(A \cap B) = P(\text{sum is 7 and first is 5}) = \frac{1}{36}$$

$$P(A) \cdot P(B) = P(\text{sum is 7}) \cdot P(\text{first is 5}) = \frac{6}{36} \cdot \frac{1}{6} = \frac{1}{36}$$

so, $P(A \cap B) = P(A) \cdot P(B)$.

(b) Many answers are possible; for example, let B be the event that at least one die is a 5.

$$P(A \cap B) = P(\text{sum is 7 and at least one is a 5}) = \frac{2}{36}$$

$$P(A) \cdot P(B) = P(\text{sum is 7}) \cdot P(\text{at least one is a 5}) = \frac{6}{36} \cdot \frac{11}{6}$$

so, $P(A \cap B) \neq P(A) \cdot P(B)$.

29. Since A and B are independent events,

$$P(A \cap B) = P(A) \cdot P(B) = \frac{1}{4} \cdot \frac{1}{5} = \frac{1}{20}.$$

Thus,

$$P(A \cup B) = P(A) + P(B) - P(A \cap B) = \frac{1}{4} + \frac{1}{5} - \frac{1}{20} = \frac{2}{5}.$$

31. At the first booth, there are three possibilities: shaker 1 has heads and shaker 2 has heads; shaker 1 has tails and shaker 2 has heads; shaker 1 has heads and shaker 2 has tails. We restrict ourselves to the condition that at least one head has appeared. These three possibilities are equally likely so the probability of two heads is $\frac{1}{3}$. At the second booth we are given the condition of one head in one shaker. The probability that the second shaker has one head is $\frac{1}{2}$. Therefore, you stand the best chance at the second booth.

33. No, these events are not independent.

35. Assume that each box is equally likely to be drawn from and that within each box each marble is equally likely to be drawn. If Laura does not redistribute the marbles, then the probability of winning the Porsche is $\frac{1}{2}$, since the event of a pink marble being drawn is equivalent to the event of choosing the first of the two boxes.

If however, Laura puts 49 of the pink marbles into the second box with the 50 blue marbles, the probability of a pink marble being drawn increases to $\frac{74}{99}$. The probability of the first box being chosen is $\frac{1}{2}$, and the probability of drawing a pink marble from this box is 1. The probability of the second box being chosen is $\frac{1}{2}$, and the probability of drawing a pink marble from this box is $\frac{49}{99}$. Thus, the probability of drawing a pink marble is $\frac{1}{2} \cdot 1 + \frac{1}{2} \cdot \frac{49}{99} = \frac{74}{99}$. Therefore Laura increases her chances of winning by redistributing some marbles.

37. The probability that a customer cashing a check will fail to make a deposit is

$$P(D'|C) = \frac{n(D' \cap C)}{n(C)} = \frac{30}{90} = \frac{1}{3}.$$

39. The probability that a customer making a deposit will not cash a check is

$$P(C'|D) = \frac{n(C' \cap D)}{n(D)} = \frac{20}{80} = \frac{1}{4}.$$

41. **(a)** Since the separate flights are independent, the probability of 4 flights in a row is

$$(0.773)(0.773)(0.773)(0.773) \approx 0.3570$$

43. Let W be the event "withdraw cash from ATM" and C be the event "check account balance at ATM."

$$\begin{aligned} P(C \cup W) &= P(C) + P(W) - P(C \cap W) \\ 0.96 &= 0.32 + 0.92 - P(C \cap W) \\ -0.28 &= -P(C \cap W) \\ P(C \cap W) &= 0.28 \end{aligned}$$

$$\begin{aligned} P(W|C) &= \frac{P(C \cap W)}{P(C)} \\ &= \frac{0.28}{0.32} \\ &\approx 0.875 \end{aligned}$$

The probability that she uses an ATM to get cash given that she checked her account balance is 0.875.

Use the following tree diagram for Exercise 45.

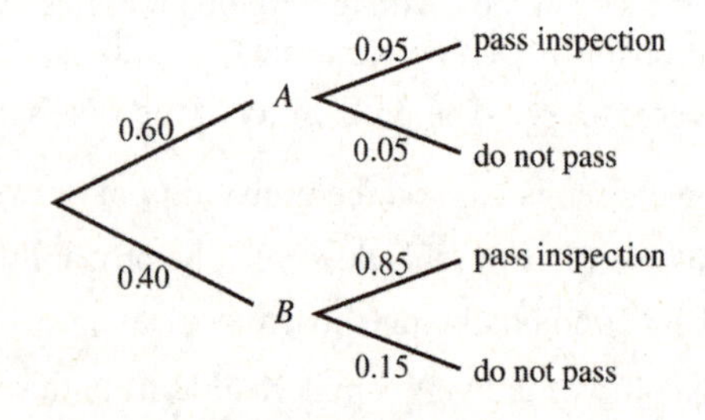

45. Since 40% of the production comes off B, $P(B) = 0.40$. Also, $P(\text{pass}|B) = 0.85$, so $P(\text{not pass}|B) = 0.15$. Therefore,

$$\begin{aligned} P(\text{not pass} \cap B) &= P(B) \cdot P(\text{not pass}|B) \\ &= 0.40(0.15) \\ &= 0.06 \end{aligned}$$

47. The sample space is

$$\{RW, WR, RR, WW\}.$$

The event "red" is $\{RW, WR, RR\}$, and the event "mixed" is $\{RW, WR\}$.

$$\begin{aligned} P(\text{mixed}|\text{red}) &= \frac{n(\text{mixed and red})}{n(\text{red})} \\ &= \frac{2}{3}. \end{aligned}$$

Use the following tree diagram for Exercises 49 through 53.

1st child	2nd child	3rd child	Branch	Probability
1/2 B	1/2 B	1/2 B	1	1/8
		1/2 G	2	1/8
	1/2 G	1/2 B	3	1/8
		1/2 G	4	1/8
1/2 G	1/2 B	1/2 B	5	1/8
		1/2 G	6	1/8
	1/2 G	1/2 B	7	1/8
		1/2 G	8	1/8

49. $P(\text{all girls}|\text{first is a girl})$

$$\begin{aligned} &= \frac{P(\text{all girls and first is a girl})}{P(\text{first is a girl})} \\ &= \frac{n(\text{all girls and first is a girl})}{n(\text{first is a girl})} \\ &= \frac{1}{4} \end{aligned}$$

51. $P(\text{all girls}|\text{second is a girl})$

$$\begin{aligned} &= \frac{P(\text{all girls and second is a girl})}{P(\text{second is a girl})} \\ &= \frac{n(\text{all girls and second is a girl})}{n(\text{second is a girl})} \\ &= \frac{1}{4} \end{aligned}$$

53. $P(\text{all girls}|\text{at least 1 girl})$

$$\begin{aligned} &= \frac{P(\text{all girls and at least 1 girl})}{P(\text{at least 1 girl})} \\ &= \frac{n(\text{all girls and at least 1 girl})}{n(\text{at least 1 girl})} \\ &= \frac{1}{7} \end{aligned}$$

55. $P(C) = 0.039$, the total of the C row.

57. $$\begin{aligned} P(M \cup C) &= P(M) + P(C) - P(M \cap C) \\ &= 0.487 + 0.039 - 0.035 \\ &= 0.491 \end{aligned}$$

59. $$\begin{aligned} P(C|M) &= \frac{P(C \cap M)}{P(M)} \\ &= \frac{0.035}{0.487} \\ &\approx 0.072 \end{aligned}$$

61. By the definition of independent events, C and M are independent if

$$P(C|M) = P(C).$$

From Exercises 55 and 59,

$$P(C) = 0.039$$

and $$P(C|M) = 0.072.$$

Since $P(C|M) \neq P(C)$, events C and M are not independent, so we say that they are dependent. This means that red-green color blindness does not occur equally among men and women.

63. First draw a tree diagram.

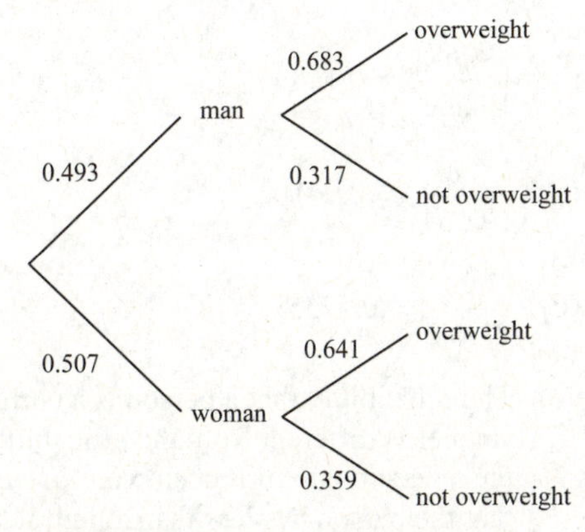

(a) $$\begin{aligned} P(\text{overweight man}) &= (0.493)(0.683) \\ &= 0.3367 \end{aligned}$$

(b) $$\begin{aligned} P(\text{overweight}) &= 0.3367 + (0.507)(0.641) \\ &= 0.3367 + 0.3250 \\ &= 0.6617 \end{aligned}$$

(c) The two events "man" and "overweight" are independent if

$$P(\text{overweight man}) = P(\text{overweight})P(\text{man}).$$
$$P(\text{overweight})P(\text{man}) = (0.6617)(0.493) = 0.3262.$$

Since $0.3367 \neq 0.3262$, the events are not independent

65. $$P(C|F) = \frac{n(C \cap F)}{n(F)} = \frac{7}{229}$$

67. $$\begin{aligned} P(B'|H') &= P((A \cup C \cup D) \mid F) \\ &= \frac{n((A \cup C \cup D) \cap F)}{n(F)} \\ &= \frac{34 + 7 + 150}{229} \\ &= \frac{191}{229} \end{aligned}$$

69. Let H be the event "patient has high blood pressure,"
N be the event "patient has normal blood pressure,"
L be the event "patient has low blood pressure,"
R be the event "patient has a regular heartbeat,"
and I be the event "patient has an irregular heartbeat.

We wish to determine $P(R \cap L)$.

Statement (i) tells us $P(H) = 0.14$ and statement (ii) tells us $P(L) = 0.22$. Therefore,

$$\begin{aligned} P(H) + P(N) + P(L) &= 1 \\ 0.14 + P(N) + 0.22 &= 1 \\ P(N) &= 0.64. \end{aligned}$$

Statement (iii) tells us $P(I) = 0.15$. This and statement (iv) lead to

$$P(I \cap H) = \frac{1}{3}P(I) = \frac{1}{3}(0.15) = 0.05.$$

Statement (v) tells us

$$P(N \cap I) = \frac{1}{8}P(N) = \frac{1}{3}(0.64) = 0.08.$$

Make a table and fill in the data just found.

	H	N	L	Totals
R	–	–	–	–
I	0.05	0.08	–	0.15
Totals	0.14	0.64	0.22	1.00

To determine $P(R \cap L)$, find $P(I \cap L)$.

$$\begin{aligned} P(I) &= P(I \cap H) + P(I \cap N) + P(I \cap L) \\ 0.15 &= 0.05 + 0.08 + P(I \cap L) \\ 0.15 &= 0.13 + P(I \cap L) \end{aligned}$$
$$P(I \cap L) = 0.02$$

Now calculate $P(R \cap L)$.

$$P(L) = P(R \cap L) + P(I \cap L)$$
$$0.22 = P(R \cap L) + 0.02$$
$$P(I \cap L) = 0.20$$

The correct answer choice is **e**.

71. (a) The total number of males is $2(5844) + 6342 = 18{,}030$. The total number of infants is $2(17{,}798) = 35{,}596$. So among infants who are part of a twin pair, the proportion of males is $\frac{18{,}030}{35{,}596} = 0.5065$.

To answer parts (b) through (g) we make the assumption that the event "twin comes from an identical pair" is independent of the event "twin is male." We can then construct the following tree diagram.

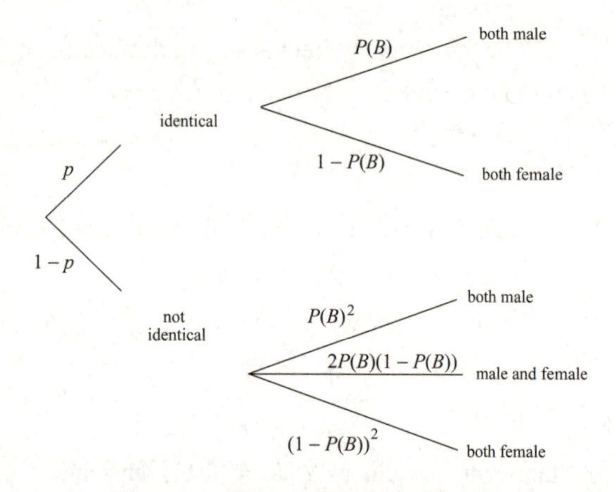

(b) The event that the pair of twins is male happens along two compound branches. Multiplying the probabilities along these branches and adding the products we get

$$pP(B) + (1 - p)(P(B))^2.$$

(c) Using the values from (a) and (b) together with the fraction of mixed twins, we solve this equation:

$$\frac{5844}{17{,}798} = p(0.506518) + (1 - p)(0.506518)^2$$
$$0.328352 = 0.506518p + 0.256560 - 0.256560p$$
$$0.071792 = 0.249958p$$
$$p = \frac{0.071792}{0.249958}$$
$$p = 0.2872$$

So our estimate for p is 0.2872. Note that if you use fewer places in the value for $P(B)$ you may get a slightly different answer.

(d) Multiplying along the two branches that result in two female twins and adding the products gives

$$p(1 - P(B)) + (1 - p)(1 - P(B))^2.$$

(e) The equation to solve will now be the following:

$$\frac{5612}{17{,}798} = p(1 - 0.506518) + (1 - p)(1 - 0.506518)^2$$

The answer will be the same as in part (c)

(f) Now only the "not identical" branch is involved, and the expression is

$$2(1 - p)P(B)(1 - P(B)).$$

(g) The equation to solve will now be the following:

$$\frac{6342}{17{,}798} = 2(1 - p)(0.506518)(1 - 0.506518)$$

Again the answer will be the same as in part (c). Note that because of our independence assumption these three estimates for p must agree.

73. (a) $\frac{39.66}{192.64} = 0.2059$

(b) $\frac{28.68}{192.64} = 0.1489$

(c) $\frac{7.90}{192.64} = 0.0410$

(d) $\frac{7.90}{28.68} = 0.2755$

(e) The probability that a person is a current smoker is different from the probability that the a person is a current smoker *given* that that the person has less than a high school diploma. Thus knowing that a person has less than a high school diploma changes our estimate of the probability that the person is a smoker, so these events are not independent. Alternatively, we could note that the product of the answers to (a) and (b), which is

$P(\text{smoker}) \cdot P(\text{less than HS diploma})$

is equal to $(0.2059)(0.1489) = 0.0307$, while according to (c), P(smoker *and* less than HS diploma) $= 0.0410$. Since these values are different, the two events "current smoker" and "less than a high school diploma" are not independent.

75. **(a)** In this exercise, it is easier to work with complementary events. Let E be the event "at least one of the faults erupts." Then the complementary event E' is "none of the faults erupts," and we can use $P(E) = 1 - P(E')$. Consider the event E': "none of the faults erupts." This means "the first fault does not erupt and the second fault does not erupt and . . . and the seventh fault does not erupt." Letting F_i denote the event "the ith fault erupts," we wish to find

$$P(E') = P(F_1' \cap F_2' \cap F_3' \cap F_4' \cap F_5' \cap F_6' \cap F_7').$$

Since we are assuming the events are independent, we have

$$\begin{aligned} P(E') &= P(F_1' \cap F_2' \cap F_3' \cap F_4' \cap F_5' \cap F_6' \cap F_7') \\ &= P(F_1') \cdot P(F_2') \cdot P(F_3') \cdot P(F_4') \cdot P(F_5') \cdot P(F_6') \cdot P(F_7') \end{aligned}$$

Now use $P(F_i') = 1 - P(F_i)$ and perform the calculations.

$$\begin{aligned} P(E') &= P(F_1') \cdot P(F_2') \cdot P(F_3') \cdot P(F_4') \cdot P(F_5') \cdot P(F_6') \cdot P(F_7') \\ &= (1 - 0.27) \cdot (1 - 0.21) \cdot \ldots \cdot (1 - 0.03) \\ &= (0.73)(0.79)(0.89)(0.90)(0.96)(0.97)(0.97) \\ &\approx 0.42 \end{aligned}$$

Therefore,

$$\begin{aligned} P(E) &= 1 - P(E') \\ &\approx 1 - 0.42 \approx 0.58. \end{aligned}$$

77. **(a)** $P(\text{second class}) = \dfrac{357}{1316} \approx 0.2713$

(b) $P(\text{surviving}) = \dfrac{499}{1316} \approx 0.3792$

(c) $P(\text{surviving} \mid \text{first class}) = \dfrac{203}{325} \approx 0.6246$

(d) $P(\text{surviving} \mid \text{child and third class}) = \dfrac{27}{79}$

≈ 0.3418

(e) $P(\text{woman} \mid \text{first class and survived}) = \dfrac{140}{203}$

≈ 0.6897

(f) $P(\text{third class} \mid \text{man and survived}) = \dfrac{75}{146}$

≈ 0.5137

(g) $P(\text{survived} \mid \text{man}) = \dfrac{146}{805} \approx 0.1814$

$P(\text{survived} \mid \text{man and third class}) = \dfrac{75}{462}$

≈ 0.1623

No, the events are not independent.

79. First draw the tree diagram.

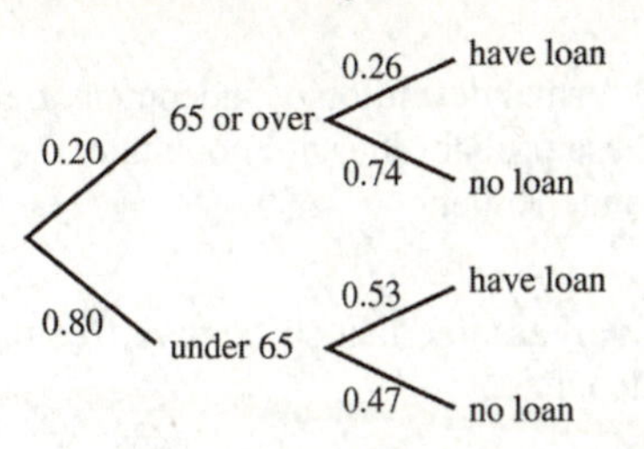

(a) $P(\text{person is 65 or over and has a loan})$

$= P(\text{65 or over}) \cdot P(\text{has loan} \mid \text{65 or over})$

$= 0.20(0.26) = 0.052$

(b)

$$\begin{aligned} P(\text{person has a loan}) &= P(\text{65 or over and has loan}) \\ &\quad + P(\text{under 65 and has loan}) \\ &= 0.20(0.26) + 0.80(0.53) \\ &= 0.052 + 0.424 \\ &= 0.476 \end{aligned}$$

81.

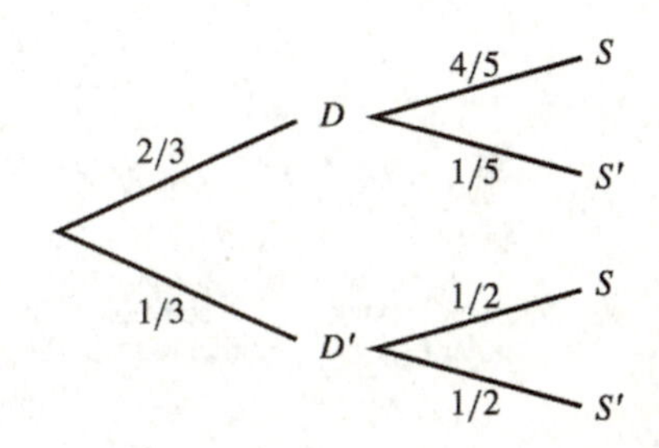

From the tree diagram, we see that the probability that a person

(a) drinks diet soft drinks is

$$\frac{2}{3}\left(\frac{4}{5}\right) + \frac{1}{3}\left(\frac{1}{2}\right) = \frac{8}{15} + \frac{1}{6} = \frac{21}{30} = \frac{7}{10};$$

(b) diets, but does not drink diet soft drinks is

$$\frac{2}{3}\left(\frac{1}{5}\right) = \frac{2}{15}.$$

83. Let F_i be the event "the ith burner fails." The event "all four burners fail" is equivalent to the event "the first burner fails and the second burner fails and the third burner fails and the fourth burner fails"—that is, the event $F_1 \cap F_2 \cap F_3 \cap F_4$. We are told that the burners are independent. Therefore

$$\begin{aligned} P(F_1 \cap F_2 \cap F_3 \cap F_4) &= P(F_1) \cdot P(F_2) \cdot P(F_3) \cdot P(F_4) \\ &= (0.001)(0.001)(0.001)(0.001) \\ &= 0.000000000001 = 10^{-12}. \end{aligned}$$

85. $P(\text{luxury car}) = 0.04$ and $P(\text{luxury car} \mid \text{CPA}) = 0.17$

Use the formal definition of independent events. Since these probabilities are not equal, the events are not independent.

87. (a) We will assume that successive free throws are independent.

$$P(0) = 0.4$$
$$P(1) = (0.6)(0.4) = 0.24$$
$$P(2) = (0.6)(0.6) = 0.36$$

(b) Let p be her season free throw percentage.

$$P(0) = 1 - p$$
$$P(2) = p^2$$

Solve:

$$p^2 = 1 - p$$
$$p^2 + p - 1 = 0$$
$$p = \frac{-1 + \sqrt{5}}{2} \approx 0.618$$

89. (a)

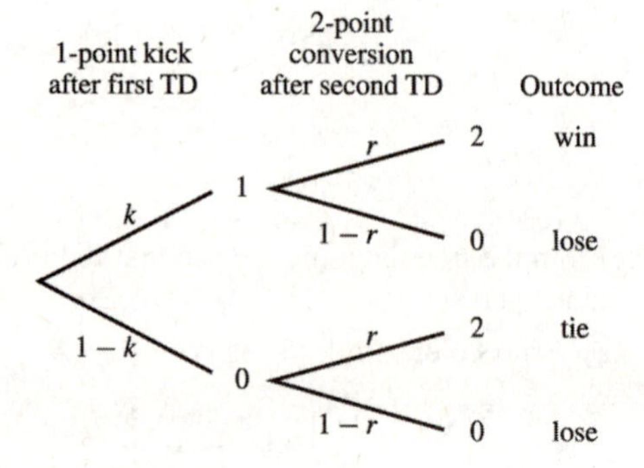

From the tree diagram,

$$P(\text{win}) = kr$$
$$P(\text{tie}) = (1 - k)r$$
$$P(\text{lose}) = k(1 - r) + (1 - k)(1 - r)$$
$$= k - kr + 1 - r - k + kr$$
$$= 1 - r$$

(b)

2-point conversion after first TD
Points after second TD
Outcome
r
1 − r
2
0
k
1 − k
r
1 − r
1
0
2
0
win
tie
tie
lose

From the tree diagram,

$$P(\text{win}) = rk$$
$$P(\text{tie}) = r(1 - k) + (1 - r)r$$
$$= r - rk + r - r^2$$
$$= 2r - rk - r^2$$
$$= r(2 - k - r)$$
$$P(\text{lose}) = (1 - r)(1 - r)$$
$$= (1 - r)^2$$

(c) $P(\text{win})$ is the same under both strategies.

(d) If $r < 1$, $(1 - r) > (1 - r)^2$. The probability of losing is smaller for the 2-point first strategy.

7.6 Bayes' Theorem

Your Turn 1

Let E be the event "passes math exam" and F be the event "attended review session."

$$P(E|F) = 0.8$$
$$P(E|F') = 0.65$$
$$P(F) = 0.6 \text{ so } P(F') = 0.4$$

Now apply Bayes' Theorem.

$$\begin{aligned} P(F|E) &= \frac{P(F)P(E|F)}{P(F)P(E|F) + P(F')P(E|F')} \\ &= \frac{(0.6)(0.8)}{(0.6)(0.8) + (0.4)(0.65)} \\ &= 0.6486 \end{aligned}$$

So the probability that the student attended the review session given that the student passed the exam is 0.6486.

Your Turn 2

$$\begin{aligned} \text{Let } F_1 &= \text{in English I} \\ F_2 &= \text{in English II} \\ F_3 &= \text{in English III} \\ E &= \text{received help from writing center} \end{aligned}$$

Express the given information in terms of these variables.

$$\begin{aligned} P(F_1) &= 0.12 \quad P(E \mid F_1) = 0.80 \\ P(F_2) &= 0.68 \quad P(E \mid F_2) = 0.40 \\ P(F_3) &= 0.20 \quad P(E \mid F_3) = 0.11 \end{aligned}$$

$$\begin{aligned} P(F_1|E) &= \frac{P(F_1)P(E|F_1)}{P(F_1)P(E|F_1) + P(F_2)P(E|F_2) + P(F_3)P(E|F_3)} \\ &= \frac{(0.12)(0.80)}{(0.12)(0.80) + (0.68)(0.40) + (0.20)(0.11)} \\ &= 0.2462 \end{aligned}$$

Given that a student received help from the writing center, the probability that the student is in English I is 0.2462.

7.6 Exercises

1. Use Bayes' theorem with two possibilities M and M'.

$$\begin{aligned} P(M|N) &= \frac{P(M) \cdot P(N|M)}{P(M) \cdot P(N|M) + P(M') \cdot P(N|M')} \\ &= \frac{0.4(0.3)}{0.4(0.3) + 0.6(0.4)} \\ &= \frac{0.12}{0.12 + 0.24} \\ &= \frac{0.12}{0.36} = \frac{12}{36} = \frac{1}{3} \end{aligned}$$

3. Using Bayes' theorem,

$$P(R_1|Q) = \frac{P(R_1)\cdot P(Q|R_1)}{P(R_1)\cdot P(Q|R_1) + P(R_2)\cdot P(Q|R_2) + P(R_3)\cdot P(Q|R_3)}$$
$$= \frac{0.15(0.40)}{(0.15)(0.40) + 0.55(0.20) + 0.30(0.70)}$$
$$= \frac{0.06}{0.38} = \frac{6}{38} = \frac{3}{19}.$$

5. Using Bayes' theorem,

$$P(R_3|Q) = \frac{P(R_3)\cdot P(Q|R_3)}{P(R_1)\cdot P(Q|R_1) + P(R_2)\cdot P(Q|R_2) + P(R_3)\cdot P(Q|R_3)}$$
$$= \frac{0.30(0.70)}{(0.15)(0.40) + 0.55(0.20) + 0.30(0.70)}$$
$$= \frac{0.21}{0.38} = \frac{21}{38}.$$

7. We first draw the tree diagram and determine the probabilities as indicated below.

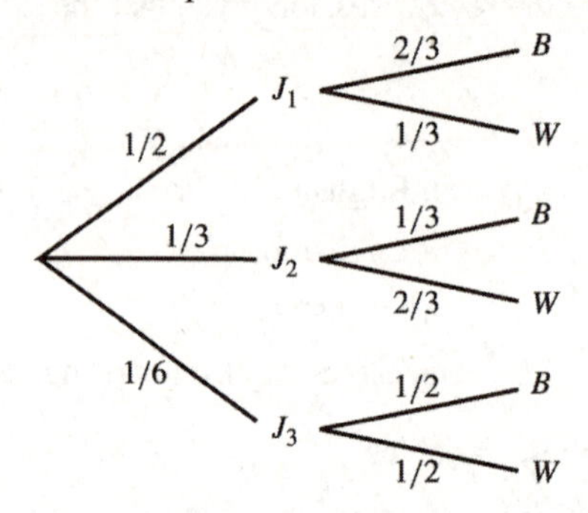

We want to determine the probability that if a white ball is drawn, it came from the second jar. This is $P(J_2\,|\,W)$. Use Bayes' theorem.

$$P(J_2|W) = \frac{P(J_2)\cdot P(W|J_2)}{P(J_2)\cdot P(W|J_2) + P(J_1)\cdot P(W|J_1) + P(J_3)\cdot P(W|J_3)} = \frac{\frac{1}{3}\cdot\frac{2}{3}}{\frac{1}{3}\cdot\frac{2}{3} + \frac{1}{2}\cdot\frac{1}{3} + \frac{1}{6}\cdot\frac{1}{2}}$$
$$= \frac{\frac{2}{9}}{\frac{2}{9} + \frac{1}{6} + \frac{1}{12}} = \frac{\frac{2}{9}}{\frac{17}{36}} = \frac{8}{17}$$

9. Let G represent "good worker," B represent "bad worker," S represent "pass the test," and F represent "fail the test." The given information if $P(G) = 0.70, P(B) = P(G') = 0.30, P(S|G) = 0.85$ (and therefore $P(F|G) = 0.15$), and $P(S|B) = 0.35$ (and therefore $P(F|B) = 0.65$). If passing the test is made a requirement for employment, then the percent of the new hires that will turn out to be good workers is

$$P(G|S) = \frac{P(G)\cdot P(S|G)}{P(G)\cdot P(S|G) + P(B)\cdot P(S|B)}$$
$$= \frac{0.70(0.85)}{0.70(0.85) + 0.30(0.35)}$$
$$= \frac{0.595}{0.700}$$
$$= 0.85.$$

85% of new hires become good workers.

11. Let Q represent "qualified" and A represent "approved by the manager." Set up the tree diagram.

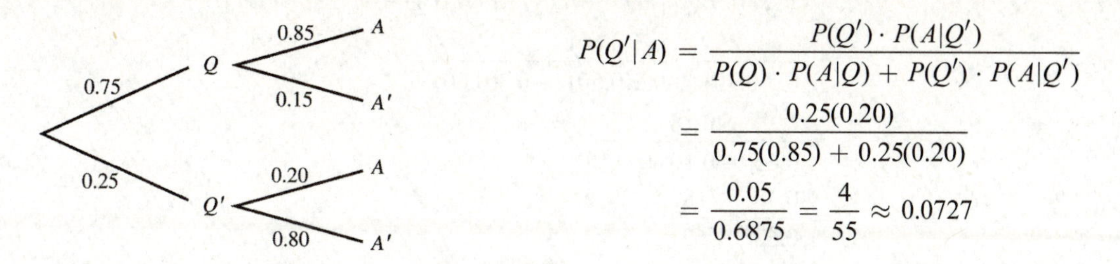

$$P(Q'|A) = \frac{P(Q') \cdot P(A|Q')}{P(Q) \cdot P(A|Q) + P(Q') \cdot P(A|Q')}$$
$$= \frac{0.25(0.20)}{0.75(0.85) + 0.25(0.20)}$$
$$= \frac{0.05}{0.6875} = \frac{4}{55} \approx 0.0727$$

13. Let D represent "damaged," A represent "from supplier A," and B represent "from supplier B." Set up the tree diagram.

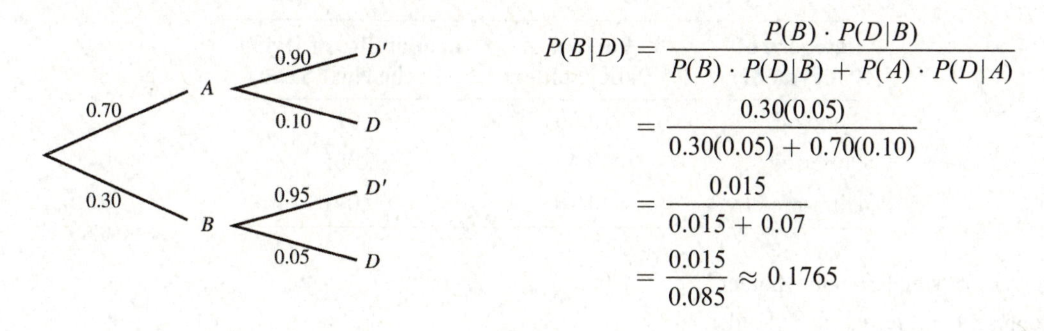

$$P(B|D) = \frac{P(B) \cdot P(D|B)}{P(B) \cdot P(D|B) + P(A) \cdot P(D|A)}$$
$$= \frac{0.30(0.05)}{0.30(0.05) + 0.70(0.10)}$$
$$= \frac{0.015}{0.015 + 0.07}$$
$$= \frac{0.015}{0.085} \approx 0.1765$$

15. Start with the tree diagram, where the first state refers to the companies and the second to a defective appliance.

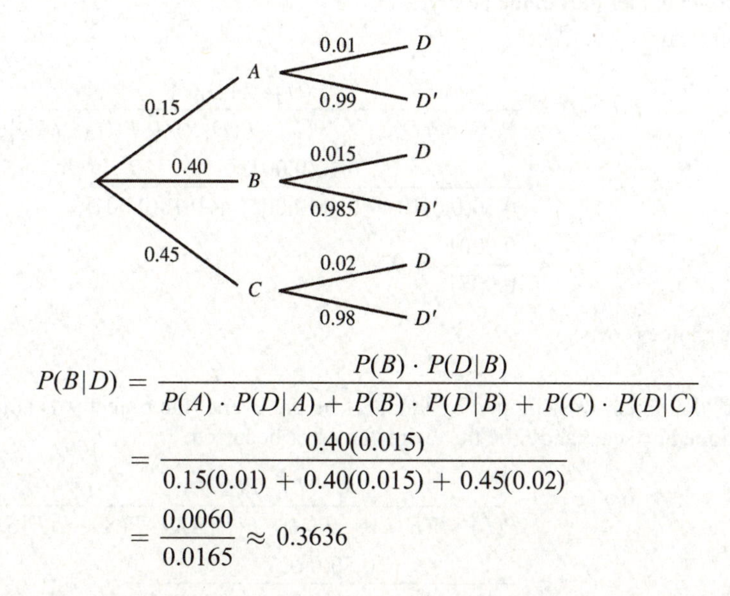

$$P(B|D) = \frac{P(B) \cdot P(D|B)}{P(A) \cdot P(D|A) + P(B) \cdot P(D|B) + P(C) \cdot P(D|C)}$$
$$= \frac{0.40(0.015)}{0.15(0.01) + 0.40(0.015) + 0.45(0.02)}$$
$$= \frac{0.0060}{0.0165} \approx 0.3636$$

17. Let H represent "high rating," F_1 represent "sponsors college game," F_2 represent "sponsors baseball game," and F_3 represent "sponsors pro football game."

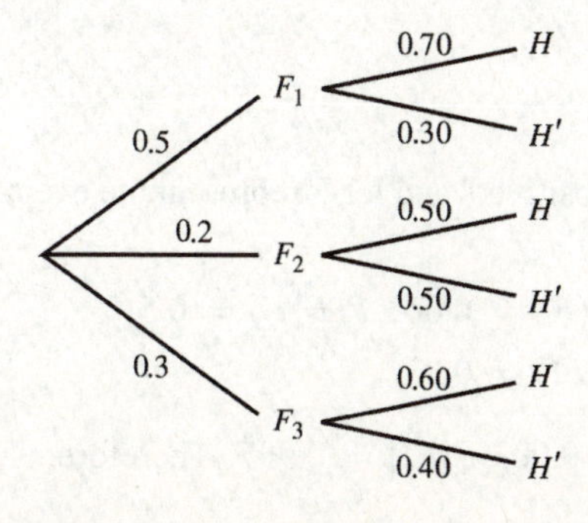

$$P(F_3|H) = \frac{P(F_3) \cdot P(H|F_3)}{P(F_1) \cdot P(H|F_1) + P(F_2) \cdot P(H|F_2) + P(F_3) \cdot P(H|F_3)}$$
$$= \frac{0.3(0.60)}{0.5(0.70) + 0.2(0.50) + 0.3(0.60)}$$
$$= \frac{0.18}{0.35 + 0.10 + 0.18}$$
$$= \frac{0.18}{0.63} = \frac{18}{63} = \frac{2}{7}$$

19. Using the given information, construct a table similar to the one in the previous exercise.

Category of Policyholder	Portion of Policyholders	Probability of Dying in the Next Year
Standard	0.50	0.010
Preferred	0.40	0.005
Ultra-preferred	0.10	0.001

Let S represent "standard policyholder,"
R represent "preferred policyholder,"
U represent "ultra-preferred policyholder,"
and D represent "policyholder dies in the next year."
We wish to find $P(U|D)$.

$$P(U|D) = \frac{P(U) \cdot P(D|U)}{P(S) \cdot P(D|S) + P(R) \cdot P(D|R) + P(U) \cdot P(D|U)}$$
$$= \frac{0.10(0.001)}{0.50(0.010) + 0.40(0.005) + 0.10(0.001)}$$
$$= \frac{0.0001}{0.0071} \approx 0.141$$

The correct answer choice is **d**.

21. Let L be the event "the object was shipped by land," A be the event "the object was shipped by air," S be the event "the object was shipped by sea," and E be the event "an error occurred."

$$P(L|E) = \frac{P(L) \cdot P(E|L)}{P(L) \cdot P(E|L) + P(A) \cdot P(E|A) + P(S) \cdot P(E|S)}$$
$$= \frac{0.50(0.02)}{0.50(0.02) + 0.40(0.04) + 0.10(0.14)}$$
$$= \frac{0.0100}{0.0400} = 0.25$$

The correct response is **c**.

23. Let E represent the event "hemoccult test is positive," and let F represent the event "has colorectal cancer." We are given

$$P(F) = 0.003, P(E|F) = 0.5,$$
$$\text{and } P(E|F') = 0.03$$

and we want to find $P(F|E)$. Since $P(F) = 0.003, P(F') = 0.997$. Therefore,

$$P(F|E) = \frac{P(F) \cdot P(E|F)}{P(F) \cdot P(E|F) + P(F') \cdot P(E|F')} = \frac{0.003 \cdot 0.5}{0.003 \cdot 0.5 + 0.997 \cdot 0.03} \approx 0.0478.$$

25. $P(T^+ \mid D^+) = 0.796$

$P(T^- \mid D^-) = 0.902$

$P(D^+) = 0.005$ so $P(D^-) = 0.995$

We can now fill in the complete tree.

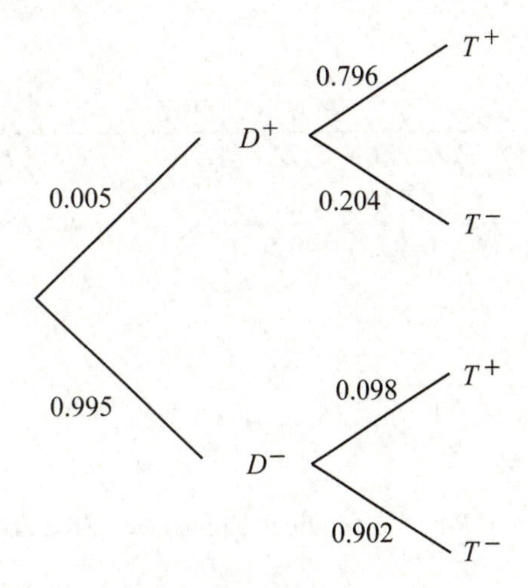

(a)
$$P(D^+ \mid T^+) = \frac{P(D^+)P(T^+ \mid D^+)}{P(D^+)P(T^+ \mid D^+) + P(D^-)P(T^+ \mid D^-)}$$
$$= \frac{(0.005)(0.796)}{(0.005)(0.796) + (0.995)(0.098)}$$
$$= 0.039$$

(b)
$$P(D^- \mid T^-) = \frac{P(D^-)P(T^- \mid D^-)}{P(D^-)P(T^- \mid D^-) + P(D^+)P(T^- \mid D^+)}$$
$$= \frac{(0.995)(0.902)}{(0.995)(0.902) + (0.005)(0.204)}$$
$$= 0.999$$

(c)
$$P(D^+ \mid T^-) = \frac{P(D^+)P(T^- \mid D^+)}{P(D^+)P(T^- \mid D^+) + P(D^-)P(T^- \mid D^-)}$$
$$= \frac{(0.005)(0.204)}{(0.005)(0.204) + (0.995)(0.902)}$$
$$= 0.001$$

Alternatively, since $P(D^- \mid T^-) + P(D^+ \mid T^-) = 1$, we can subtract the answer to (b) from 1.

(d) The right half of the tree stays the same, but $P(D^+)$ is now 0.015 and $P(D^-)$ is 0.985.

$$P(D^+ \mid T^+) = \frac{P(D^+)P(T^+ \mid D^+)}{P(D^+)P(T^+ \mid D^+) + P(D^-)P(T^+ \mid D^-)}$$
$$= \frac{(0.015)(0.796)}{(0.015)(0.796) + (0.985)(0.098)}$$
$$= 0.110$$

27. **(a)** Let H represent "heavy smoker," L be "light smoker," N be "nonsmoker," and D be "person died." Let $x = P(D|N)$, that is, let x be the probability that a nonsmoker died. Then $P(D|L) = 2x$ and $P(D|H) = 4x$. Create a table.

Level of Smoking	Probability of Level	Probability of Death for Level
H	0.2	$4x$
L	0.3	$2x$
N	0.5	x

We wish to find $P(H|D)$.

$$\begin{aligned} P(H|D) &= \frac{P(H) \cdot P(D|H)}{P(H) \cdot P(D|H) + P(L) \cdot P(D|L) + P(N) \cdot P(D|N)} \\ &= \frac{0.2(4x)}{0.2(4x) + 0.3(2x) + 0.5(x)} \\ &= \frac{0.8x}{1.9x} \\ &\approx 0.42 \end{aligned}$$

The correct answer choice is **d**.

29. Let H represent "person has the disease" and R be "test indicates presence of the disease." We wish to determine $P(H|R)$.

Construct a table as before.

Category of Person	Probability of Population	Probability of Presence of Disease
H	0.01	0.950
H'	0.99	0.005

$$\begin{aligned} P(H|R) &= \frac{P(H) \cdot P(R|H)}{P(H) \cdot P(R|H) + P(H') \cdot P(R|H')} \\ &= \frac{0.01(0.950)}{0.01(0.950) + 0.99(0.005)} \\ &= \frac{0.00950}{0.01445} \approx 0.657 \end{aligned}$$

The correct answer choice is **b**.

31. We start with a tree diagram based on the given information. Let W stand for "woman" and A stand for "abstains from alcohol."

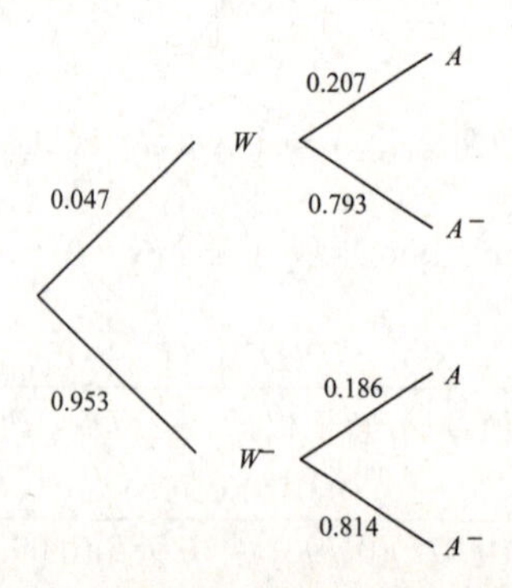

(a) $P(A) = P(A \mid W)P(W) + P(A \mid W^-)P(W^-)$
$= (0.047)(0.207) + (0.953)(0.186)$
$= 0.1870$

(b) $$P(W^- \mid A) = \frac{P(W^-)P(A \mid W^-)}{P(W^-)P(A \mid W^-) + P(W)P(A \mid W)}$$
$$= \frac{(0.953)(0.186)}{(0.953)(0.186) + (0.047)(0.207)}$$
$$= 0.9480$$

33. P(between 35 and 44 | never married) (for a randomly selected man)

$$= \frac{(0.186)(0.204)}{(0.186)(0.204) + (0.132)(0.901) + (0.186)(0.488) + (0.348)(0.118) + (0.148)(0.044)}$$
$$= 0.1285$$

35. P(between 45 and 64|never married) (for a randomly selected woman)

$$= \frac{(0.345)(0.092)}{(0.345)(0.092) + (0.121)(0.825) + (0.172)(0.366) + (0.178)(0.147) + (0.184)(0.040)}$$
$$= 0.1392$$

In Exercise 37, let S stand for "smokes" and S^- for "does not smoke."

37. $P(18 - 44|S)$

$$= \frac{P(18-44)P(S|18-44)}{P(18-44)P(S\mid 18-44) + P(45-64)P(S|45-65) + P(65-74)P(S|65-74) + P(>75)P(S|>75)}$$
$$= \frac{(0.49)(0.23)}{(0.49)(0.23) + (0.34)(0.22) + (0.09)(0.12) + (0.08)(0.06)}$$
$$= 0.5549$$

39.

Category	Proportion of Population	Probability of Being Picked Up
Has terrorist ties	$\frac{1}{1{,}000{,}000}$	0.99
Does not have terrorists ties	$\frac{999{,}999}{1{,}000{,}000}$	0.01

$$P(\text{Has terrorist ties}|\text{Picked up}) = \frac{\frac{1}{1{,}000{,}000}(0.99)}{\frac{1}{1{,}000{,}000}(0.99) + \frac{999{,}999}{1{,}000{,}000}(0.01)}$$
$$= \frac{\frac{1}{1{,}000{,}000}(0.99)}{\frac{1}{1{,}000{,}000}(0.99) + \frac{999{,}999}{1{,}000{,}000}(0.01)} \cdot \frac{1{,}000{,}000}{1{,}000{,}000}$$
$$= \frac{0.99}{10{,}000.98}$$
$$\approx 9.9 \times 10^{-5}$$

Chapter 7 Review Exercises

1. True

2. True

3. False: The union of a set with itself has the same number of elements as the set.

4. False: The intersection of a set with itself has the same number of elements as the set.

5. False: If the sets share elements, this procedure gives the wrong answer.

6. True

7. False: This procedure is correct only if the two events are mutually exclusive.

8. False: We can calculate this probability by assuming a sample space in which each card in the 52-card deck is equally likely to be drawn.

9. False: If two events A and B are mutually exclusive, then $P(A \cap B) = 0$ and this will not be equal to $P(A)P(B)$ if $P(A)$ and $P(B)$ are greater than 0.

10. True

11. False: In general these two probabilities are different. For example, for a draw from a 52-card deck, $P(\text{heart}|\text{queen}) = 1/4$ and $P(\text{queen}|\text{heart}) = 1/13$.

12. True

13. $9 \in \{8, 4, -3, -9, 6\}$

    Since 9 is not an element of the set, this statement is false.

15. $2 \notin \{0, 1, 2, 3, 4\}$

    Since 2 is an element of the set, this statement is false.

17. $\{3, 4, 5\} \subseteq \{2, 3, 4, 5, 6\}$

    Every element of {3, 4, 5} is an element of {2, 3, 4, 5, 6}, so this statement is true.

19. $\{3, 6, 9, 10\} \subseteq \{3, 9, 11, 13\}$

    10 is an element of {3, 6, 9, 10}, but 10 is not an element of {3, 9, 11, 13}. Therefore, {3, 6, 9, 10} is not a subset of {3, 9, 11, 13}. The statement is false.

21. $\{2, 8\} \not\subseteq \{2, 4, 6, 8\}$

    Since both 2 and 8 are elements of {2, 4, 6, 8}, {2, 8} is a subset of {2, 4, 6, 8}. This statement is false.

In Exercises 23–32

$$U = \{a, b, c, d, e, f, g, h\},$$
$$K = \{c, d, e, f, h\},$$
$$\text{and } R = \{a, c, d, g\}.$$

23. K has 5 elements, so it has $2^5 = 32$ subsets.

25. K' (the complement of K) is the set of all elements of U that do *not* belong to K.

    $$K' = \{a, b, g\}$$

27. $K \cap R$ (the intersection of K and R) is the set of all elements belonging to both set K and set R.

    $$K \cap R = \{c, d\}$$

29. $(K \cap R)' = \{a, b, e, f, g, h\}$ since these elements are in U but not in $K \cap R$. (See Exercise 27.)

31. $\emptyset' = U$

33. $A \cap C$ is the set of all female employees in the K.O. Brown Company who are in the accounting department.

35. $A \cup D$ is the set of all employees in the K.O. Brown Company who are in the accounting department *or* have MBA degrees or both.

37. $B' \cap C'$ is the set of all male employees who are not in the sales department.

39. $A \cup B'$ is the set of all elements which belong to A or do not belong to B, or both.

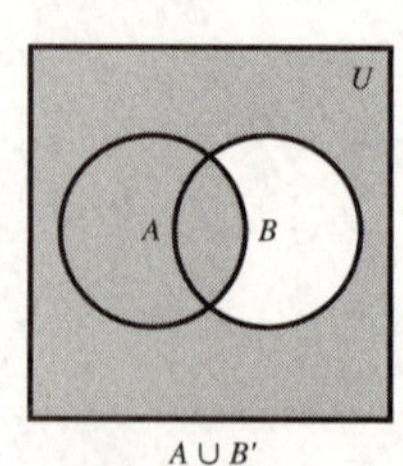

$A \cup B'$

41. $(A \cap B) \cup C$

First find $A \cap B$.

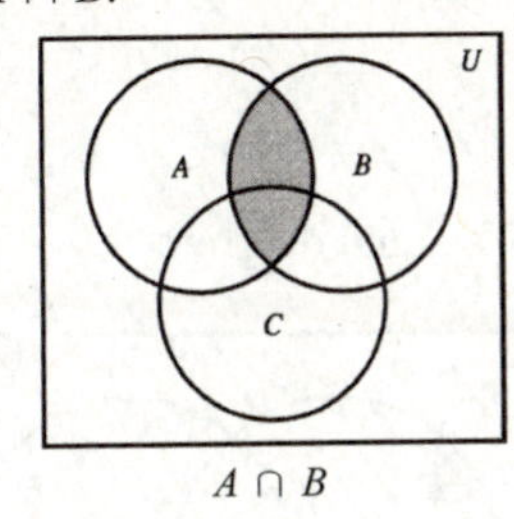

$A \cap B$

Now find the union of this region with C.

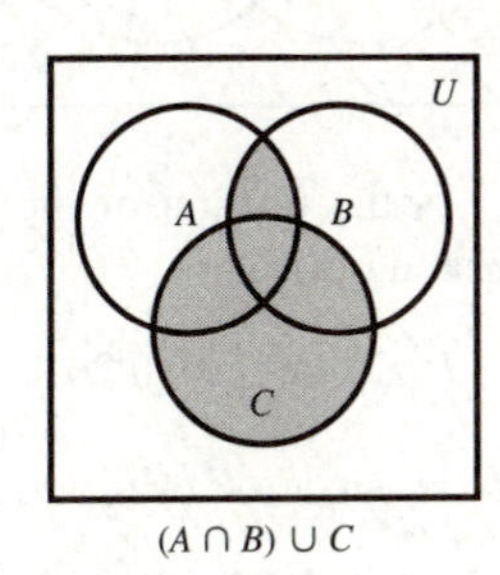

$(A \cap B) \cup C$

43. The sample space for rolling a die is

$$S = \{1, 2, 3, 4, 5, 6\}.$$

45. The sample space of the possible weights is

$$S = \{0, 0.5, 1, 1.5, 2, \ldots, 299.5, 300\}.$$

47. The sample space consists of all ordered pairs (a, b) where a can be 3, 5, 7, 9, or 11, and b is either R (red) or G (green). Thus,

$$S = \{(3, R), (3, G), (5, R), (5, G), (7, R), (7, G), (9, R), (9, G), (11, R), (11, G)\}.$$

49. The event F that the second ball is green is

$$F = \{(3, G), (5, G), (7, G), (9, G), (11, G)\}.$$

51. There are 13 hearts out of 52 cards in a deck. Thus,

$$P(\text{heart}) = \frac{13}{52} = \frac{1}{4}.$$

53. There are 3 face cards in each of the four suits.

$$P(\text{face card}) = \frac{12}{52}$$

$$P(\text{heart}) = \frac{13}{52}$$

$$P(\text{face card and heart}) = \frac{3}{52}$$

$$\begin{aligned} P(\text{face card or heart}) &= P(\text{face card}) + P(\text{heart}) \\ &\quad - P(\text{face card and heart}) \\ &= \frac{12}{52} + \frac{13}{52} - \frac{3}{52} \\ &= \frac{22}{52} = \frac{11}{26} \end{aligned}$$

55. There are 4 queens of which 2 are red, so

$$\begin{aligned} P(\text{red}\,|\,\text{queen}) &= \frac{n(\text{red and queen})}{n(\text{queen})} \\ &= \frac{2}{4} = \frac{1}{2}. \end{aligned}$$

57. There are 4 kings of which all 4 are face cards. Thus,

$$\begin{aligned} P(\text{face card}\,|\,\text{king}) &= \frac{n(\text{face card and king})}{n(\text{king})} \\ &= \frac{4}{4} = 1. \end{aligned}$$

63. If A and B are nonempty and independent, then

$$P(A \cap B) = P(A) \cdot P(B).$$

For mutually exclusive events, $P(A \cap B) = 0$, which would mean $P(A) = 0$ or $P(B) = 0$. So independent events with nonzero probabilities are not mutually exclusive. But independent events one of which has zero probability are mutually exclusive.

65. Let C be the event "a club is drawn." There are 13 clubs in the deck, so $n(C) = 13$, $P(C') = \frac{13}{52} = \frac{1}{4}$, and $P(C') = 1 - P(C) = \frac{3}{4}$. The odds in favor of drawing a club are

$$\frac{P(C)}{P(C')} = \frac{\frac{1}{4}}{\frac{3}{4}} = \frac{1}{3},$$

which is written "1 to 3."

67. Let R be the event "a red face card is drawn" and Q be the event "a queen is drawn." Use the union rule for probability to find $P(R \cup Q)$.

$$P(R \cup Q) = P(R) + P(Q) - P(R \cap Q)$$
$$= \frac{6}{52} + \frac{4}{52} - \frac{2}{52}$$
$$= \frac{8}{52} = \frac{2}{13}$$
$$P(R \cup Q)' = 1 - P(R \cup Q)$$
$$= 1 - \frac{2}{13} = \frac{11}{13}$$

The odds in favor of drawing a red face card or a queen are

$$\frac{P(R \cup Q)}{P(R \cup Q)'} = \frac{\frac{2}{13}}{\frac{11}{13}} = \frac{2}{11},$$

which is written "2 to 11."

69. The sum is 8 for each of the 5 outcomes 2-6, 3-5, 4-4, 5-3, and 6-2. There are 36 outcomes in all in the sample space.

$$P(\text{sum is } 8) = \frac{5}{36}$$

71. $P(\text{sum is at least } 10)$

$$= P(\text{sum is } 10) + P(\text{sum is } 11) + P(\text{sum is } 12)$$
$$= \frac{3}{36} + \frac{2}{36} + \frac{1}{36}$$
$$= \frac{6}{36} = \frac{1}{6}$$

73. The sum can be 9 or 11. $P(\text{sum is } 9) = \frac{4}{36}$ and $P(\text{sum is } 11) = \frac{2}{36}$.

$P(\text{sum is odd number greater than } 8)$

$$= \frac{4}{36} + \frac{2}{36}$$
$$= \frac{6}{36} = \frac{1}{6}$$

75. Consider the reduced sample space of the 11 outcomes in which at least one die is a four. Of these, 2 have a sum of 7, 3-4 and 4-3. Therefore,

$P(\text{sum is } 7 | \text{at least one die is a } 4)$

$$= \frac{2}{11}.$$

77. $P(E) = 0.51,\ P(F) = 0.37,\ P(E \cap F) = 0.22$

(a) $P(E \cup F) = P(E) + P(F) - P(E \cap F)$
$= 0.51 + 0.37 - 0.22$
$= 0.66$

(b) Draw a Venn diagram.

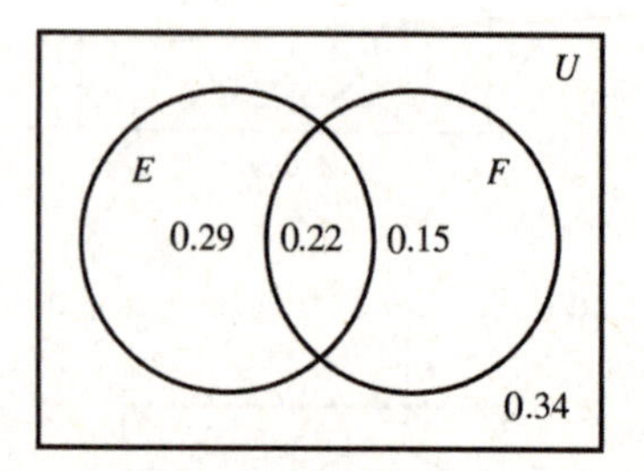

$E \cap F'$ is the portion of the diagram that is inside E and outside F.

$$P(E \cap F') = 0.29$$

(c) $E' \cup F$ is outside E or inside F, or both.

$$P(E' \cup F) = 0.22 + 0.15 + 0.34 = 0.71.$$

(d) $E' \cap F'$ is outside E and outside F.

$$P(E' \cap F') = 0.34$$

79. First make a tree diagram. Let A represent "box A" and K represent "black ball."

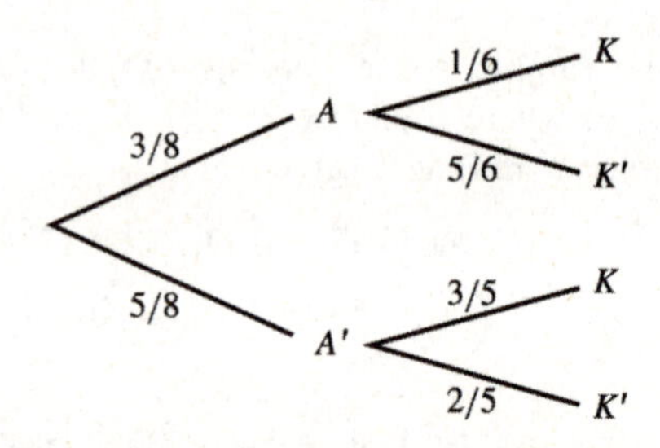

Use Bayes' theorem.

$$P(A|K) = \frac{P(A) \cdot P(K|A)}{P(A) \cdot P(K|A) + P(A') \cdot P(K|A')}$$
$$= \frac{\frac{3}{8} \cdot \frac{1}{6}}{\frac{3}{8} \cdot \frac{1}{6} + \frac{5}{8} \cdot \frac{3}{5}}$$
$$= \frac{\frac{1}{16}}{\frac{7}{16}} = \frac{1}{7}$$

81. First make a tree diagram letting C represent "a competent shop" and R represent "an appliance is repaired correctly."

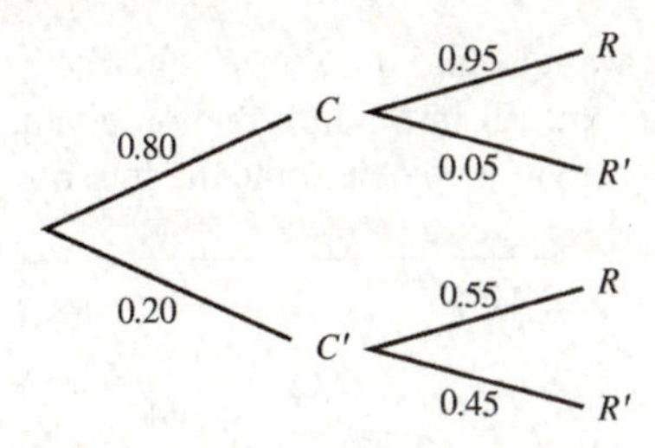

To obtain $P(C|R)$, use Bayes' theorem.

$$\begin{aligned} P(C|R) &= \frac{P(C)\cdot P(R|C)}{P(C)\cdot P(R|C) + P(C')\cdot P(R|C')} \\ &= \frac{0.80(0.95)}{0.80(0.95) + 0.20(0.55)} \\ &= \frac{0.76}{0.87} \approx 0.8736 \end{aligned}$$

83. Refer to the tree diagram for Exercise 81. Use Bayes' theorem.

$$\begin{aligned} P(C|R') &= \frac{P(C)\cdot P(R'|C)}{P(C)\cdot P(R'|C) + P(C')\cdot P(R'|C')} \\ &= \frac{0.80(0.05)}{0.80(0.05) + 0.20(0.45)} \\ &= \frac{0.04}{0.13} \approx 0.3077 \end{aligned}$$

85. To find $P(R)$, use

$$\begin{aligned} P(R) &= P(C)\cdot P(R|C) + P(C')\cdot P(R|C') \\ &= 0.80(0.95) + 0.20(0.55) = 0.87. \end{aligned}$$

87. **(a)** "A customer buys neither machine" may be written $(E \cup F)'$ or $E' \cap F'$.

(b) "A customer buys at least one of the machines" is written $E \cup F$.

89. Use Bayes' theorem to find the required probabilities.

(a) Let D be the event "item is defective" and E_k be the event "item came from supplier k," $k = 1, 2, 3, 4$.

$$\begin{aligned} P(D) &= P(E_1)\cdot P(D|E_1) + P(E_2)\cdot P(D|E_2) \\ &\quad + P(E_3)\cdot P(D|E_3) + P(E_4)\cdot P(D|E_4) \\ &= 0.17(0.01) + 0.39(0.02) + 0.35(0.05) \\ &\quad + 0.09(0.03) \\ &= 0.0297 \end{aligned}$$

(b) Find $P(E_4|D)$. Using Bayes' theorem, the numerator is

$$P(E_4)\cdot P(D|E_4) = 0.09(0.03) = 0.0027.$$

The denominator is $P(E_1)\cdot P(D|E_1) + P(E_2)\cdot P(D|E_2) + P(E_3)\cdot P(D|E_3) + P(E_4)\cdot P(D|E_4)$, which, from part (a), equals 0.0297.

Therefore,

$$P(E_4|D) = \frac{0.0027}{0.0297} \approx 0.0909.$$

(c) Find $P(E_2|D)$. Using Bayes' theorem with the same denominator as in part (a),

$$\begin{aligned} P(E_2|D) &= \frac{P(E_2)\cdot P(D|E_2)}{0.0418} \\ &= \frac{0.39(0.02)}{0.0297} \\ &= \frac{0.0078}{0.0297} \\ &\approx 0.2626. \end{aligned}$$

(d) Since $P(D) = 0.0297$ and $P(D|E_4) = 0.03$,

$$P(D) \neq P(D|E_4)$$

Therefore, the events are not independent.

91. Let E represent "customer insures exactly one car" and S represent "customer insures a sports car." Let x be the probability that a customer who insures exactly one car insures a sports car, or $P(S|E)$. Make a tree diagram.

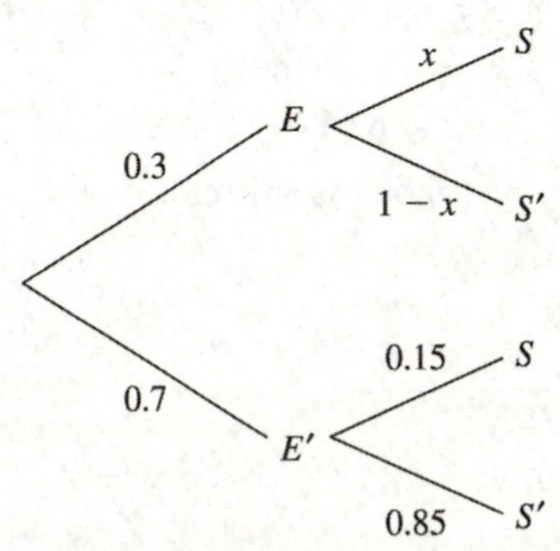

We are told that 20% of the customers insure a sports car, or $P(S) = 0.20$.

$$\begin{aligned} P(S) &= P(E)\cdot P(S|E) + P(E')\cdot P(S|E') \\ 0.20 &= 0.30(x) + 0.70(0.15) \\ 0.20 &= 0.3x + 0.105 \\ 0.3x &= 0.095 \\ x &\approx 0.316666667 \end{aligned}$$

Therefore, the probability that a customer insures a car other than a sports car is

$$\begin{aligned} P(S') &= 1 - P(S) \\ &\approx 1 - 0.316666667 \\ &= 0.683333333. \end{aligned}$$

Finally, the probability that a randomly selected customer insures exactly one car and that car is not a sports car is

$$\begin{aligned} P(E \cap S') &= P(E) \cdot P(S') \\ &\approx 0.3 \cdot 0.683333333 \\ &\approx 0.21. \end{aligned}$$

The correct answer choice is **b**.

93. Let C represent "the automobile owner purchases collision coverage" and D represent "the automobile owner purchases disability coverage." We want to find $P(C' \cap D') = P[(C \cup D)'] = 1 - P(C \cup D)$. We are given that $P(C) = 2 \cdot P(D)$ and that $P(C \cap D) = 0.15$. Let $x = P(D)$.

$$\begin{aligned} P(C \cap D) &= P(C) \cdot P(D) \\ 0.15 &= 2x \cdot x \\ 0.075 &= x^2 \\ x &= \sqrt{0.075} \\ x &\approx 0.2739 \end{aligned}$$

So $P(D) = x \approx 0.2739$ and $P(C) = 2x$.

$$\begin{aligned} P(C \cup D) &= P(C) + P(D) - P(C \cap D) \\ &= 2(0.2739) + 0.2739 - 0.15 \\ &= 0.6720 \end{aligned}$$

$$\begin{aligned} P(C' \cap D') &= 1 - P(C \cup D) \\ &= 1 - 0.6720 \\ &\approx 0.33 \end{aligned}$$

The correct choice is answer **b**.

95. (a)

	N_2	T_2
N_1	N_1N_2	N_1T_2
T_1	T_1N_2	T_1T_2

Since the four combinations are equally likely, each has probability $\frac{1}{2}$.

(b) $P(\text{two trait cells}) = P(T_1T_2) = \frac{1}{4}$

(c) $P(\text{one normal cell and one trait cell})$

$$= P(N_1T_2) + P(T_1N_2)$$
$$= \tfrac{1}{4} + \tfrac{1}{4} = \tfrac{1}{2}$$

(d) $P(\text{not a carrier and does not have disease})$

$$= P(N_1N_2) = \tfrac{1}{4}$$

97. We want to find $P(A' \cap B' \cap C' | A')$. Use a Venn diagram, fill in the information given, and use the diagram to help determine the missing values.

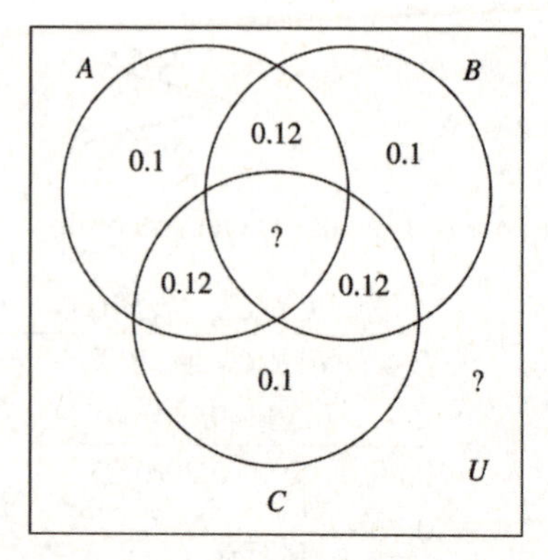

To determine $P(A \cap B \cap C)$, we are told that "The probability that a woman has all three risk factors, given that she has A and B, is 1/3." Therefore, $P(A \cap B \cap C | A \cap B) = 1/3$.

Let $x = P(A \cap B)$; then, using the diagram as a guide,

$$\begin{aligned} P(A \cap B \cap C) + P(A \cap B \cap C') &= P(A \cap B) \\ \frac{1}{3}x + 0.12 &= x \\ 0.12 &= \frac{2}{3}x \\ x &= 0.18 \end{aligned}$$

So, $P(A \cap B \cap C) = (1/3)(0.16) = 0.06$. By DeMorgan's laws, we have

$$A' \cap B' \cap C' = (A \cup B \cup C)'$$

so that

$$\begin{aligned} P(A' \cap B' \cap C') &= P[(A \cup B \cup C)'] \\ &= 1 - P(A \cup B \cup C) \\ &= 1 - [3(0.10) + 3(0.12) + 0.06] \\ &= 0.28. \end{aligned}$$

Therefore,

$$\begin{aligned} P(A' \cap B' \cap C' | A') &= \frac{P(A' \cap B' \cap C' \cap A')}{P(A')} \\ &= \frac{P(A' \cap B' \cap C')}{P(A')} \\ &= \frac{0.28}{0.6} \approx 0.467. \end{aligned}$$

The correct answer choice is **c**.

99. Let C be the set of viewers who watch situation comedies,
G be the set of viewers who watch game shows,
and M be the set of viewers who watch movies.

We are given the following information.

$n(C) = 20 \quad n(G) = 19 \quad n(M) = 27$
$n(M \cap G') = 19$
$n(C \cap G') = 15$
$n(C \cap M) = 10$
$n(C \cap G \cap M) = 3$
$n(C' \cap G' \cap M') = 7$

Start with $C \cap G \cap M$: $n(C \cap G \cap M) = 3$.

Since $n(C \cap M) = 10$, the number of people who watched comedies and movies but not game shows, or $n(C \cap G' \cap M)$, is $10 - 3 = 7$.

Since $n(M \cap G') = 19$, $n(C' \cap G' \cap M) = 19 - 7 = 12$.

Since $n(M) = 27$, $n(C' \cap G \cap M) = 27 - 3 - 7 - 12 = 5$.

Since $n(C \cap G') = 15$, $n(C \cap G' \cap M') = 15 - 7 = 8$.

Since $n(C) = 20$,
$n(C \cap G \cap M') = 20 - 8 - 3 - 7 = 2$.

Finally, since $n(G) = 19$, $n(C' \cap G \cap M') = 19 - 2 - 3 - 5 = 9$.

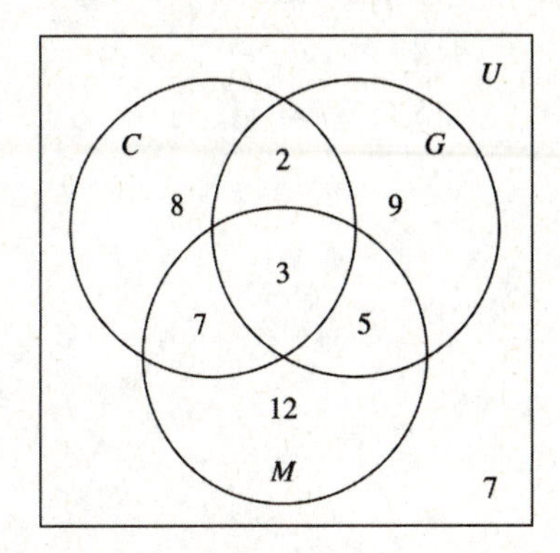

(a) $n(U) = 8 + 2 + 9 + 7 + 3 + 5 + 12 + 7 = 53$

(b) $n(C \cap G' \cap M) = 7$

(c) $n(C' \cap G' \cap M) = 12$

(d) $n(M') = n(U) - n(M) = 53 - 27 = 26$

101. Let C be the event "the culprit penny is chosen." Then

$$p(C|HHH) = \frac{P(C \cap HHH)}{P(HHH)}.$$

These heads will result two different ways. The culprit coin is chosen $\frac{1}{3}$ of the time and the probability of a head on any one flip is $\frac{3}{4}$: $P(C \cap HHH) = \frac{1}{3}\left(\frac{3}{4}\right)^3 \approx 0.1406$. If a fair (innocent) coin is chosen, the probability of a head on any one flip is $\frac{1}{2}$: $P(C'|HHH) = \frac{2}{3}\left(\frac{1}{2}\right)^3 \approx 0.0833$. Therefore,

$$\begin{aligned} P(C|HHH) &= \frac{P(C \cap HHH)}{P(HHH)} \\ &= \frac{P(C \cap HHH)}{P(C \cap HHH) + P(C' \cap HHH)} \\ &\approx \frac{0.1406}{0.1406 + 0.0833} \\ &\approx 0.6279 \end{aligned}$$

103. $P(\text{earthquake}) = \frac{9}{9+1} = \frac{9}{10} = 0.90$

105. Let W be the set of western states,
S be the set of small states, and
E be the set of early states.

We are given the following information.

$n(W) = 24$ $\quad n(S) = 22$ $\quad n(E) = 26$

$n(W' \cap S' \cap E') = 9$ $\quad n(W \cap S) = 14$

$n(S \cap E) = 11$ $\quad n(W \cap S \cap E) = 7$

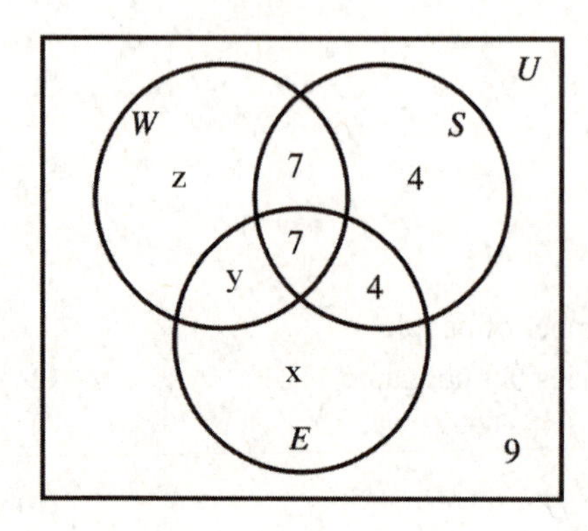

First, put 7 in $W \cap S \cap E$ and 9 in $W' \cap S' \cap E'$.
Complete $S \cap E$ with 4 for a total of 11.
Complete $W \cap S$ with 7 for a total of 14.
Complete S with 4 for a total of 22.

To complete the rest of the diagram requires solving some equations. Let the incomplete region of E be x, the incomplete region of $W \cap E$ be y, and the incomplete region of W be z. Then, using the given values and the fact that $n(U) = 50$,

$$\begin{aligned} x + y \quad &= 15 \\ y + z &= 10 \\ x + y + z &= 50 - 22 - 9 = 19. \end{aligned}$$

The solution to the system is $x = 9$, $y = 6$, and $z = 4$.

Complete $W \cap E$ with.
Complete E with 9 for a total of 26.
Complete W with 4 for a total of 24.
The completed diagram is as follows.

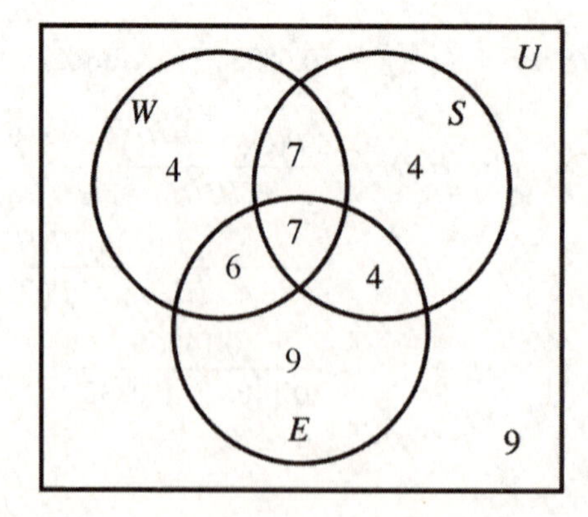

(a) $n(W \cap S' \cap E') = 4$

(b) $n(W' \cap S') = n((W \cup S)') = 18$

107. Let R be "a red side is facing up" and
RR be "the 2-sided red card is chosen."

If a red side is facing up, we want to find $P(RR|R)$ since the other possibility would be a green side is facing down.

$$P(RR|R) = \frac{P(RR)}{P(R)} = \frac{\frac{1}{3}}{\frac{1}{2}} = \frac{2}{3}$$

No, the bet is not a good bet.

109. Let G be the set of people who watched gymnastics,
B be the set of people who watched baseball,
and S be the set of people who watched soccer.

We want to find $P(G' \cap B' \cap S')$ or, by DeMorgan's laws, $P[(G \cup B \cup S)']$.

We are given the following information.

$P(G) = 0.28$ $\quad P(B) = 0.29$ $\quad P(S) = 0.19$ $\quad P(G \cap B) = 0.14$

$P(B \cap S) = 0.12$ $\quad P(G \cap S) = 0.10$ $\quad P(G \cap B \cap S) = 0.08$

Start with $P(G \cap B \cap S) = 0.08$ and work from the inside out.

Since $P(G \cap S) = 0.10$, $P(G \cap B' \cap S) = 0.02$.

Since $P(B \cap S) = 0.12$, $P(G' \cap B \cap S) = 0.04$.

Since $P(G \cap B) = 0.14$, $P(G \cap B \cap S') = 0.06$.

Since $P(S) = 0.19, P(G' \cap B' \cap S) = 0.19 - 0.14 = 0.05$.

Since $P(B) = 0.29, P(G' \cap B \cap S') = 0.29 - 0.18 = 0.11$.

Since $P(G) = 0.28, P(G \cap B' \cap S') = 0.28 - 0.16 = 0.12$.

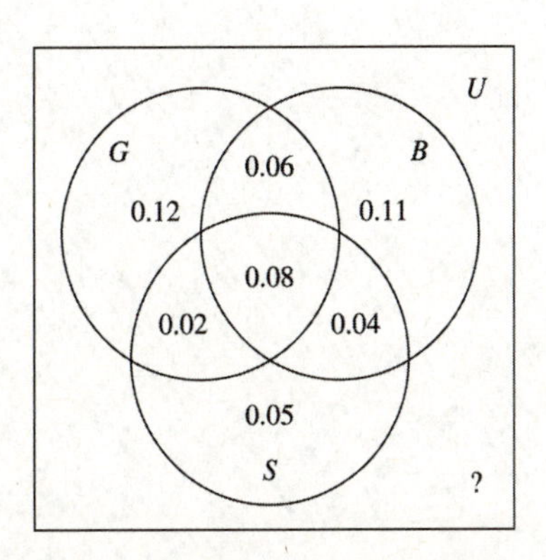

Therfore,

$$\begin{aligned} P(G' \cap B' \cap S) &= P[(G \cup B \cup S)'] \\ &= 1 - P(G \cup B \cup S) \\ &= 1 - (0.12 + 0.06 + 0.11 + 0.02 + 0.08 + 0.04 + 0.05) = 0.52. \end{aligned}$$

$$P(G' \cap B' \cap S) = 1 - (0.12 + 0.06 + 0.11 + 0.02 + 0.08 + 0.04 + 0.05) = 0.52.$$

The correct answer choice is **d**.

Chapter 8

COUNTING PRINCIPLES; FURTHER PROBABILITY TOPICS

8.1 The Multiplication Principle; Permutations

Your Turn 1

Each of the four digits can be one of the ten digits $0, 1, 2, \ldots, 10$, so there are $10 \cdot 10 \cdot 10 \cdot 10$ or 10,000 possible sequences. If no digit is repeated, there are 10 choices for the first place, 9 for the second, 8 for the third, and 7 for the fourth, so there are $10 \cdot 9 \cdot 8 \cdot 7 = 5040$ possible sequences.

Your Turn 2

Any of the 8 students could be first in line. Then there are 7 choices for the second spot, 6 for the third, and so on, with only one student remaining for the last spot in line. There are $8 \cdot 7 \cdot 6 \cdot 5 \cdot 4 \cdot 3 \cdot 2 \cdot 1 = 40{,}320$ different possible lineups.

Your Turn 3

The teacher has 8 ways to fill the first space (say the one on the left), 7 choices for the next book, and so on, leaving 4 choices for the last book on the right. So the number of possible arrangements is $8 \cdot 7 \cdot 6 \cdot 5 \cdot 4 = 6720.$

Your Turn 4

There are 6 letters. If we use 3 of the 6, the number of permutations is

$$P(6,3) = \frac{6!}{(6-3)!} = \frac{6 \cdot 5 \cdot 4 \cdot 3 \cdot 2 \cdot 1}{3 \cdot 2}$$
$$= 6 \cdot 5 \cdot 4 = 120.$$

Your Turn 5

If the panel sits in a row, there are 4! or 24 ways of arranging the four class groups. Within the groups, there are 2! ways of arranging the freshmen, 2! ways of arranging the sophomores, 2! ways of arranging the juniors, and 3! ways of arranging the seniors. Using the multiplication principle, the number of ways of seating the panel with the classes together is $24 \cdot 2! \cdot 2! \cdot 2! \cdot 3! = 24 \cdot 2 \cdot 2 \cdot 2 \cdot 6 = 1152.$

Your Turn 6

The word *Tennessee* contains 9 letters, consisting of 1 t, 4 e's, 2 n's and 2 s's. Thus the number of possible arrangements is

$$\frac{9!}{1!4!2!2!} = 3780.$$

Your Turn 7

The student has $4 + 5 + 3 + 2 = 14$ pairs of socks, so if the pairs were distinguishable there would be 14! possible selections for the next two weeks. But since the pairs of each color are identical, the number of distinguishable selections is

$$\frac{14!}{4!5!3!2!} = 2{,}522{,}520.$$

8.1 Exercises

1. $6! = 6 \cdot 5 \cdot 4 \cdot 3 \cdot 2 \cdot 1 = 720$

3. $15! = 15 \cdot 14 \cdot 13 \cdot 12 \cdot 11 \cdot 10 \cdot 9 \cdot 8 \cdot 7 \cdot 6 \cdot 5 \cdot 4 \cdot 3 \cdot 2 \cdot 1$
$\approx 1.308 \times 10^{12}$

5. $P(13,2) = \frac{13!}{(13-2)!} = \frac{13!}{11!}$
$= \frac{13 \cdot 12 \cdot 11!}{11!}$
$= 156$

7. $P(38,17) = \frac{38!}{(38-17)!} = \frac{38!}{21!}$
$\approx 1.024 \times 10^{25}$

9. $P(n,0) = \frac{n!}{(n-0)!} = \frac{n!}{n!} = 1$

11. $P(n,1) = \frac{n!}{(n-1)!} = \frac{n(n-1)!}{(n-1)!} = n$

13. By the multiplication principle, there will be $6 \cdot 3 \cdot 2 = 36$ different home types available.

15. There are 4 choices for the first name and 5 choices for the middle name, so, by the multiplication principle, there are $4 \cdot 5 = 20$ possible arrangements.

19. There is exactly one 3-letter subset of the letters A, B, and C, namely A, B, and C.

21. **(a)** initial

This word contains 3 i's, 1 n, 1 t, 1 a, and 1 ℓ. Use the formula for distinguishable permutations with $n = 7, n_1 = 3, \ n_2 = 1,$ $n_3 = 1, n_4 = 1,$ and $n_5 = 1.$

$$\frac{n!}{n_1!n_2!n_3!n_4!n_5!} = \frac{7!}{3!1!1!1!1!} = \frac{7 \cdot 6 \cdot 5 \cdot 4 \cdot 3!}{3!} = 840$$

There are 840 distinguishable permutations of the letters.

(b) little

Use the formula for distinguishable permutetions with $n = 6, n_1 = 2, \ n_2 = 1,$ $n_3 = 2,$ and $n_4 = 1.$

$$\frac{6!}{2!1!2!1!} = \frac{6!}{2!2!} = \frac{6 \cdot 5 \cdot 4 \cdot 3 \cdot 2 \cdot 1}{2 \cdot 1 \cdot 2 \cdot 1} = 180$$

There are 180 distinguishable permutations.

(c) decreed

Use the formula for distinguishable permutations with $n = 7, \ n_1 = 2,$ $n_2 = 3, \ n_3 = 1,$ and $n_4 = 1.$

$$\frac{7!}{2!3!1!1!} = \frac{7!}{2!3!} = \frac{7 \cdot 6 \cdot 5 \cdot 4 \cdot 3!}{2 \cdot 1 \cdot 3!} = 420$$

There are 420 distinguishable permutations.

23. **(a)** The 9 books can be arranged in

$P(9,9) = 9! = 362{,}880$ ways.

(b) The blue books can be arranged in 4! ways, the green books can be arranged in 3! ways, and the red books can be arranged in 2! ways. There are 3! ways to choose the order of the 3 groups of books. Therefore, using the multiplication principle, the number of possible arrangements is

$$4!3!2!3! = 24 \cdot 6 \cdot 2 \cdot 6 = 1728.$$

(c) Use the formula for distinguishable permutations with $n = 9, n_1 = 4, n_2 = 3,$ and $n_3 = 2$. The number of distinguishable arrangements is

$$\frac{9!}{4!3!2!} = \frac{9 \cdot 8 \cdot 7 \cdot 6 \cdot 5 \cdot 4!}{4! \cdot 6 \cdot 2} = 1260.$$

(d) There are 4 choices for the blue book, 3 for the green book, and 2 for the red book. The total number of arrangements is

$$4 \cdot 3 \cdot 2 = 24.$$

(e) From part (d) there are 24 ways to select a blue, red, and green book if the order does not matter. There 3! ways to choose the order. Using the multiplication principle, the number of possible ways is

$$24 \cdot 3! = 24 \cdot 6 = 144.$$

25. $10! = 10 \cdot 9!$

To find the value of 10!, multiply the value of 9! by 10.

27. **(a)** The number 13! has 2 factors of five so there must be 2 ending zeros in the answer.

(b) The number 27! has 6 factors of five (one each in 5, 10, 15, and 20 and two factors in 25), so there must be 6 ending zeros in the answer.

(c) The number 75! has $15 + 3 = 18$ factors of five (one each in $5, 10, \ldots, 75$ and two factors each in 25, 50, and 75), so there must be 18 ending zeros in the answer.

29. $P(4,4) = \dfrac{4!}{(4-4)!} = \dfrac{4!}{0!}$

If $0! = 0$, then $P(4,4)$ would be undefined.

31. **(a)** By the multiplication principle, since there are 7 pastas and 6 sauces, the number of different bowls is $7 \cdot 6 = 42$.

(b) If we exclude the two meat sauces there are 4 sauces left and the number of bowls is now $7 \cdot 4 = 28$.

33. **(a)** Since there are 11 slots, the 11 commercials can be arranged in $11! = 39{,}916{,}800$ ways.

(b) Use the multiplication principle. We can put either stores or restaurants first (2 choices). Then there are 6! orders for the stores and 5! orders for the restaurants, so the number of groupings is $2 \cdot 6! \cdot 5! = 172{,}800$.

(c) Since the number of restaurants is one more than the number of stores, a restaurant must come first. This eliminates the first choice in part (b), but we still can order the restaurants and the stores freely within each category, so the answer is $6! \cdot 5! = 86{,}400$.

35. If each species were to be assigned 3 initials, since there are 26 different letters in the alphabet, there could be $26^3 = 17{,}576$ different 3-letter designations. This would not be enough. If 4 initials were used, the biologist could represent $26^4 = 456{,}976$ different species, which is more than enough. Therefore, the biologist should use at least 4 initials.

37. The number of ways to seat the people is

$$\begin{aligned} P(6,6) &= \frac{6!}{0!} = \frac{6!}{1} \\ &= 6 \cdot 5 \cdot 4 \cdot 3 \cdot 2 \cdot 1 \\ &= 720. \end{aligned}$$

39. The number of possible batting orders is

$$\begin{aligned} P(19,9) &= \frac{19!}{(19-9)!} = \frac{19!}{10!} \\ &= 33{,}522{,}128{,}640 \\ &\approx 3.352 \times 10^{10}. \end{aligned}$$

41. **(a)** The number of ways 5 works can be arranged is

$$P(5,5) = 5! = 120.$$

(b) If one of the 2 overtures must be chosen first, followed by arrangements of the 4 remaining pieces, then

$$P(2,1) \cdot P(4,4) = 2 \cdot 24 = 48$$

is the number of ways the program can be arranged.

43. **(a)** There are 4 tasks to be performed in selecting 4 letters for the call letters. The first task may be done in 2 ways, the second in 25, the third in 24, and the fourth in 23. By the multiplication principle, there will be

$$2 \cdot 25 \cdot 24 \cdot 23 = 27{,}600$$

different call letter names possible.

(b) With repeats possible, there will be

$$2 \cdot 26 \cdot 26 \cdot 26 = 2 \cdot 26^3 \quad \text{or} \quad 35{,}152$$

call letter names possible.

(c) To start with W or K, make no repeats, and end in R, there will be

$$2 \cdot 24 \cdot 23 \cdot 1 = 1104$$

possible call letter names.

45. **(a)** Our number system has ten digits, which are 1 through 9 and 0.

There are 3 tasks to be performed in selecting 3 digits for the area code. The first task may be done in 8 ways, the second in 2, and the third in 10. By the multiplication principle, there will be

$$8 \cdot 2 \cdot 10 = 160$$

different area codes possible.

There are 7 tasks to be performed in selecting 7 digits for the telephone number. The first task may be done in 8 ways, and the other 6 tasks may each be done in 10 ways. By the multiplication principle, there will be

$$8 \times 10^6 = 8{,}000{,}000$$

different telephone numbers possible within each area code.

(b) Some numbers, such as 911, 800, and 900, are reserved for special purposes and are therefore unavailable for use as area codes.

47. **(a)** There were

$$26^3 \cdot 10^3 = 17{,}576{,}000$$

license plates possible that had 3 letters followed by 3 digits.

(b) There were

$$10^3 \cdot 26^3 = 17{,}576{,}000$$

new license plates possible when plates were also issued having 3 digits followed by 3 letters.

(c) There were

$$26 \cdot 10^3 \cdot 26^3 = 456{,}976{,}000$$

new license plates possible when plates were also issued having 1 letter followed by 3 digits and then 3 letters.

49. If there are no restrictions on the digits used, there would be

$$10^5 = 100{,}000$$

different 5-digit zip codes possible.

If the first digit is not allowed to be 0, there would be

$$9 \cdot 10^4 = 90{,}000$$

zip codes possible.

51. There are 3 possible identical shapes on each card.

There are 3 possible shapes for the identical shapes.

There are 3 possible colors.

There are 3 possible styles.

Therefore, the total number of cards is $3 \cdot 3 \cdot 3 \cdot 3 = 81$.

53. There are 3 possible answers for the first question and 2 possible answers for each of the 19 other questions. The number of possible objects is

$$3 \cdot 2^{19} = 1{,}572{,}864.$$

20 questions are not enough.

55. **(a)** Since the starting seat is not counted, the number of arrangements is

$$P(19,19) = 19! \approx 1.216451 \times 10^{17}.$$

(b) Since the starting bead is not counted and the necklace can be flipped, the number of arrangements is

$$\frac{P(14,14)}{2} = \frac{14!}{2} = 43{,}589{,}145{,}600.$$

8.2 Combinations

Your Turn 1

Use the combination formula.

$$C(10,4) = \frac{10!}{6!4!} = 210$$

Your Turn 2

Since the group of students contains either 3 or 4 students out of 15, it can be selected in $C(15,3) + C(15,4)$ ways.

$$\begin{aligned} C(15,3) + C(15,4) &= \frac{15!}{12!3!} + \frac{15!}{11!4!} \\ &= 455 + 1365 \\ &= 1820 \end{aligned}$$

Your Turn 3

(a) Use permutations.

$$P(10,4) = 10 \cdot 9 \cdot 8 \cdot 7 = 5040$$

(b) Use combinations.

$$C(15,3) = \frac{15!}{12!3!} = 455$$

(c) Use combinations.

$$C(8,2) = \frac{8!}{6!2!} = 28$$

(d) Use combinations and permutations. First pick 4 rooms; this is an unordered selection:

$$C(6,4) = \frac{6!}{2!4!} = 15$$

Now assign the patients to the rooms; this is an ordered selection:

$$P(4,4) = 4! = 24$$

The number of possible assignments is $15 \cdot 24 = 360$.

Your Turn 4

The committee is an unordered selection.

$$C(20,3) = \frac{20!}{17!3!} = 1140$$

If the selection includes assignment to one of the three offices we have an ordered selection.

$$P(20,3) = 20 \cdot 19 \cdot 18 = 6840$$

Your Turn 5

There are $C(4,2)$ ways to select 2 aces from the 4 aces in the deck and $C(48,3)$ ways to select the 3 remaining cards from the 48 non-aces. Now use the multiplication principle.

$$C(4,2) \cdot C(48,3) = 6 \cdot 17{,}296 = 103{,}776$$

8.2 Exercises

3. To evaluate $C(8,3)$, use the formula

$$C(n,r) = \frac{n!}{(n-r)!r!}$$

with $n = 8$ and $r = 3$.

$$\begin{aligned} C(8,3) &= \frac{8!}{(8-3)!3!} \\ &= \frac{8!}{5!3!} \\ &= \frac{8 \cdot 7 \cdot 6 \cdot 5!}{5! \cdot 3 \cdot 2 \cdot 1} = 56 \end{aligned}$$

5. To evaluate $C(44, 20)$, use the formula

$$C(n, r) = \frac{n!}{(n-r)!r!}$$

with $n = 44$ and $r = 20$.

$$\begin{aligned} C(44, 20) &= \frac{44!}{(44-20)!20!} \\ &= \frac{44!}{24!20!} \\ &= 1.761 \times 10^{12} \end{aligned}$$

7. $$\begin{aligned} C(n, 0) &= \frac{n!}{(n-0)!0!} \\ &= \frac{n!}{n! \cdot 1} \\ &= 1 \end{aligned}$$

9. $$\begin{aligned} C(n, 1) &= \frac{n!}{(n-1)!1!} \\ &= \frac{n(n-1)!}{(n-1)! \cdot 1} \\ &= n \end{aligned}$$

11. There are 13 clubs, from which 6 are to be chosen. The number of ways in which a hand of 6 clubs can be chosen is

$$C(13, 6) = \frac{13!}{7!6!} = 1716.$$

13. **(a)** There are

$$C(5, 2) = \frac{5!}{3!2!} = \frac{5 \cdot 4 \cdot 3!}{3! \cdot 2 \cdot 1} = 10$$

different 2-card combinations possible.

(b) The 10 possible hands are

$$\{1,2\}, \{2,3\}, \{3,4\}, \{4,5\}, \{1,3\},$$
$$\{2,4\}, \{3,5\}, \{1,4\}, \{2,5\}, \{1,5\}.$$

Of these, 7 contain a card numbered less than 3.

15. Choose 2 letters from {L, M, N}; order is important.

(a)

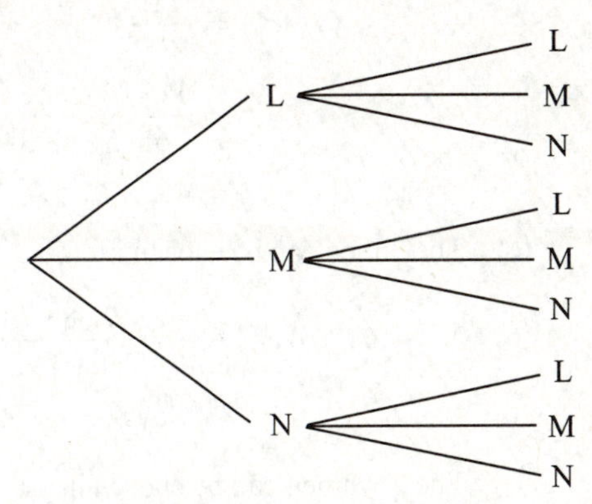

There are 9 ways to choose 2 letters if repetition is allowed.

(b)

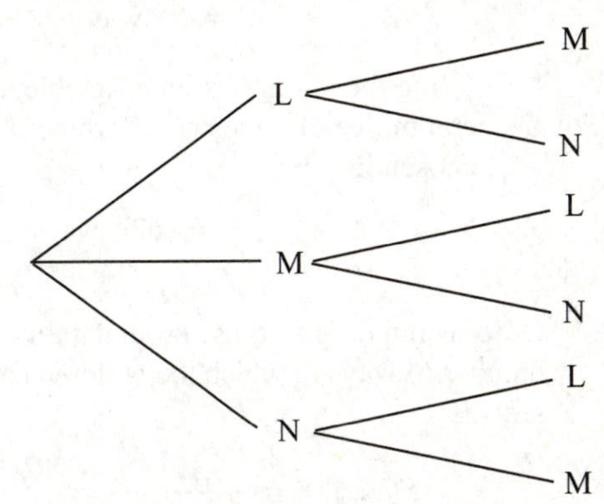

There are 6 ways to choose 2 letters if no repeats are allowed.

(c) The number of combinations of 3 elements taken 2 at a time is

$$C(3, 2) = \frac{3!}{1!2!} = 3.$$

This answer differs from both parts (a) and (b).

17. Order does not matter in choosing members of a committee, so use combinations rather than permutations.

(a) The number of committees whose members are all men is

$$\begin{aligned} C(9, 5) &= \frac{9!}{4!5!} \\ &= \frac{9 \cdot 8 \cdot 7 \cdot 6 \cdot 5!}{4 \cdot 3 \cdot 2 \cdot 1 \cdot 5!} \\ &= 126. \end{aligned}$$

(b) The number of committees whose members are all women is

$$C(11,5) = \frac{11!}{6!5!} = \frac{11 \cdot 10 \cdot 9 \cdot 8 \cdot 7 \cdot 6!}{6! \cdot 5 \cdot 4 \cdot 3 \cdot 2 \cdot 1} = 462.$$

(c) The 3 men can be chosen in

$$C(9,3) = \frac{9!}{6!3!} = \frac{9 \cdot 8 \cdot 7 \cdot 6!}{6! \cdot 3 \cdot 2 \cdot 1} = 84 \text{ ways.}$$

The 2 women can be chosen in

$$C(11,2) = \frac{11!}{9!2!} = \frac{11 \cdot 10 \cdot 9!}{9! \cdot 2 \cdot 1} = 55 \text{ ways.}$$

Using the multiplication principle, a committee of 3 men and 2 women can be chosen in

$$84 \cdot 55 = 4620 \text{ ways.}$$

19. Order is important, so use permutations. The number of ways in which the children can find seats is

$$P(12,11) = \frac{12!}{(12-11)!} = \frac{12!}{1!} = 12! = 479{,}001{,}600.$$

21. Since order does not matter, the answers are combinations.

(a) $C(16,2) = \dfrac{16!}{14!2!} = \dfrac{16 \cdot 15 \cdot 14!}{14! \cdot 2 \cdot 1} = 120$

120 samples of 2 marbles can be drawn.

(b) $C(16,4) = 1820$

1820 samples of 4 marbles can be drawn.

(c) Since there are 9 blue marbles in the bag, the number of samples containing 2 blue marbles is

$$C(9,2) = 36.$$

23. Since order does not matter, use combinations.

(a) $C(5,3) = \dfrac{5!}{2!3!} = \dfrac{5 \cdot 4 \cdot 3!}{2 \cdot 1 \cdot 3!} = 10$

There are 10 possible samples with all black jelly beans.

(b) There is only 1 red jelly bean, so there are no samples in which all 3 are red.

(c) $C(3,3) = 1$

There is 1 sample with all yellow.

(d) $C(5,2)C(1,1) = 10 \cdot 1 = 10$

There are 10 samples with 2 black and 1 red.

(e) $C(5,2)C(3,1) = 10 \cdot 3 = 30$

There are 30 samples with 2 black and 1 yellow.

(f) $C(3,2)C(5,1) = 3 \cdot 5 = 15$

There are 15 samples with 2 yellow and 1 black.

(g) There is only 1 red jelly bean, so there are no samples containing 2 red jelly beans.

25. Show that $C(n,r) = C(n,n-r)$.

Work with each side of the equation separately.

$$C(n,r) = \frac{n!}{r!(n-r)!}$$

$$C(n,n-r) = \frac{n!}{(n-r)![n-(n-r)]!} = \frac{n!}{(n-r)!r!}$$

Since both results are the same, we have shown that

$$C(n,r) = C(n,n-r).$$

27. Use combinations since order does not matter.

(a) First consider how many pairs of circles there are. This number is

$$C(6,2) = \frac{6!}{2!4!} = 15.$$

Each pair intersects in two points. The total number of intersection points is $2 \cdot 15 = 30$.

(b) The number of pairs of circles is

$$C(n,2) = \frac{n!}{(n-2)!2!} = \frac{n(n-1)(n-2)!}{(n-2)! \cdot 2} = \frac{1}{2}n(n-1).$$

Each pair intersects in two points. The total number of points is

$$2 \cdot \frac{1}{2}n(n-1) = n(n-1).$$

29. Since the assistants are assigned to different managers, this amounts to an ordered selection of 3 from 8.

$$P(8,3) = 8 \cdot 7 \cdot 6 = 336$$

31. Order is important in arranging a schedule, so use permutations.

(a) $P(6, 6) = \frac{6!}{0!} = 6! = 720$

She can arrange her schedule in 720 ways if she calls on all 6 prospects.

(b) $P(6, 4) = \frac{6!}{2!} = 360$

She can arrange her schedule in 360 ways if she calls on only 4 of the 6 prospects.

33. There are 2 types of meat and 6 types of extras. Order does not matter here, so use combinations.

(a) There are $C(2, 1)$ ways to choose one type of meat and $C(6, 3)$ ways to choose exactly three extras. By the multiplication principle, there are

$$C(2, 1)C(6, 3) = 2 \cdot 20 = 40$$

different ways to order a hamburger with exactly three extras.

(b) There are

$$C(6, 3) = 20$$

different ways to choose exactly three extras.

(c) "At least five extras" means "5 extras or 6 extras." There are $C(6,5)$ different ways to choose exactly 5 extras and $C(6, 6)$ ways to choose exactly 6 extras, so there are

$$C(6, 5) + C(6, 6) = 6 + 1 = 7$$

different ways to choose at least five extras.

35. Select 8 of the 16 smokers and 8 of the 22 non-smokers; order does not matter in the group, so use combinations. There are

$$C(16, 8)C(22, 8) = 4{,}115{,}439{,}900$$

different ways to select the study group.

37. Order does not matter in choosing a delegation, so use combinations. This committee has $5 + 4 = 9$ members.

(a) There are

$$\begin{aligned} C(9, 3) &= \frac{9!}{6!\,3!} \\ &= \frac{9 \cdot 8 \cdot 7 \cdot 6!}{6! \cdot 3 \cdot 2 \cdot 1} \\ &= 84 \text{ possible delegations.} \end{aligned}$$

(b) To have all Democrats, the number of possible delegations is

$$C(5, 3) = 10.$$

(c) To have 2 Democrats and 1 Republican, the number of possible delegations is

$$C(5, 2)C(4, 1) = 10 \cdot 4 = 40.$$

(d) We have previously calculated that there are 84 possible delegations, of which 10 consist of all Democrats. Those 10 delegations are the only ones with no Republicans, so the remaining $84 - 10 = 74$ delegations include at least one Republican.

39. Order does not matter in choosing the panel, so use combinations.

$$\begin{aligned} C(45, 3) &= \frac{45!}{42!\,3!} \\ &= \frac{45 \cdot 44 \cdot 43 \cdot 42!}{3 \cdot 2 \cdot 1 \cdot 42!} \\ &= 14{,}190 \end{aligned}$$

The publisher was wrong. There are 14,190 possible three judge panels.

41. Since the cards are chosen at random, that is, order does not matter, the answers are combinations.

(a) There are 4 queens and 48 cards that are not queens. The total number of hands is

$$C(4, 4)C(48, 1) = 1 \cdot 48 = 48.$$

(b) Since there are 12 face cards (3 in each suit), there are 40 nonface cards. The number of ways to choose no face cards (all 5 nonface cards) is

$$C(40, 5) = \frac{40!}{35!\,5!} = 658{,}008.$$

(c) If there are exactly 2 face cards, there will be 3 nonface cards. The number of ways in which the face cards can be chosen is $C(12,2)$, while the number of ways in which the nonface cards can be chosen is $C(40,3)$. Using the multiplication principle, the number of ways to get this result is

$$\begin{aligned} C(12, 2)C(40, 3) &= 66 \cdot 9880 \\ &= 652{,}080. \end{aligned}$$

(d) If there are at least 2 face cards, there must be either 2 face cards and 3 nonface cards, 3 face cards and 2 nonface cards, 4 face cards and 1 nonface card, or 5 face cards. Use the multiplication principle as in part (c) to find the number of ways to obtain each of these possibilities. Then add these numbers. The total number of ways to get at least 2 face cards is

$$\begin{aligned}&C(12,2)C(40,3)+C(12,3)C(40,2)\\&\qquad+C(12,4)C(40,1)+C(12,5)\\&=66\cdot 9880+220\cdot 780\\&\qquad+495\cdot 40+792\\&=652{,}080+171{,}600+19{,}800+792\\&=844{,}272.\end{aligned}$$

(e) The number of ways to choose 1 heart is $C(13, 1)$, the number of ways to choose 2 diamonds is $C(13, 2)$, and the number of ways to choose 2 clubs is $C(13, 2)$. Using the multiplication principle, the number of ways to get this result is

$$\begin{aligned}C(13,1)C(13,2)C(13,2)&=13\cdot 78\cdot 78\\&=79{,}092.\end{aligned}$$

43. Since order does not matter, use combinations.

2 good hitters: $C(5,2)C(4,1)=10\cdot 4=40$

2 good hitters: $C(5,3)C(4,0)=10\cdot 1=10$

The total number of ways is $40+10=50$.

45. Since order does not matter, use combinations.

(a) There are

$$C(20,5)=15{,}504$$

different ways to select 5 of the orchids.

(b) If 2 special orchids must be included in the show, that leaves 18 orchids from which the other 3 orchids for the show must be chosen. This can be done in

$$C(18,3)=816$$

different ways.

47. Since order is not important, use combinations. To pick 5 of the 6 winning numbers, we must also pick 1 of the 93 losing numbers. Therefore, the number of ways to pick 5 of the 6 winning numbers is

$$C(6,5)C(93,1)=6\cdot 93=558.$$

49. **(a)** The number of different committees possible is

$$\begin{aligned}&C(5,2)+C(5,3)+C(5,4)+C(5,5)\\&=10+10+5+1=26.\end{aligned}$$

(b) The total number of subsets is

$$2^5=32.$$

The number of different committees possible is

$$\begin{aligned}&2^5-C(5,1)-C(5,0)\\&=32-5-1\\&=26.\end{aligned}$$

51. **(a)** If the letters can be repeated, there are 26^6 choices of 6 letters, and there are 10 choices for the digit, giving

$26^6\cdot 10=3{,}089{,}157{,}760$ passwords.

(b) For nonrepeating letters we have $P(26,6)\cdot 10$ or 1,657,656,000 passwords.

53. **(a)** A pizza can have 3, 2, 1, or no toppings. The number of possibilities is

$$\begin{aligned}&C(17,3)+C(17,2)+C(17,1)+C(17,0)\\&=680+136+17+1\\&=834.\end{aligned}$$

There are also four speciality pizzas, so the number of different pizzas is $834+4=838$.

(b) The number of 4forAll Pizza possibilities if all four pizzas are different is

$$C(838,4)=20{,}400{,}978{,}015.$$

The number of 4forAll Pizza possibilities if there are three different pizzas (2 pizzas are the same and the other 2 are different) is

$$\begin{aligned}838\cdot C(837,2)&=838\cdot 349{,}866\\&=293{,}187{,}708.\end{aligned}$$

The number of 4forAll Pizza possibilities if there are two different pizzas (3 pizzas are the same or 2 pizzas and 2 pizzas are the same) is

$$\begin{aligned}&838\cdot 837+C(838,2)\\&=701{,}406+350{,}703\\&=1{,}052{,}109.\end{aligned}$$

The number of 4forAll Pizza possibilities if all four are the same is 838. The total number of 4forAll Pizza possibilities is

$$\begin{aligned}&20{,}400{,}978{,}015+293{,}187{,}708\\&\qquad+1{,}052{,}109+838\\&=20{,}695{,}218{,}670\end{aligned}$$

(c) Using the described method, there would be 837 vertical lines and 4 X's or 841 objects, so the total number is

$$C(841,4)=20{,}695{,}218{,}670.$$

55. **(a)** $C(9,3) = \dfrac{9!}{6!3!} = 84$

(b) $9 \cdot 9 \cdot 9 = 729$

(d) $C(17,3) = \dfrac{17!}{14!3!} = 680$

(e) First pick the two boneless buffalo wing flavors; there are $C(5,2) = 10$ ways of doing this. Then we still have 7 non-wing options, plus the 5 buffalo chicken wings available for the third item. So our total is $10 \cdot 12 = 120$.

57. **(a)** The number of ways the names can be arranged is

$$18! \approx 6.402 \times 10^{15}.$$

(b) 4 lines consist of a 3 syllable name repeated, followed by a 2 syllable name and then a 4 syllable name. Including order, the number of arrangements is

$$10 \cdot 4 \cdot 4 \cdot 9 \cdot 3 \cdot 3 \cdot 8 \cdot 2 \cdot 2 \cdot 7 \cdot 1 \cdot 1$$
$$= 2{,}903{,}040.$$

2 lines consist of a 3 syllable name repeated, followed by two more 3 syllable names. Including order, the number of arrangements is

$$6 \cdot 5 \cdot 4 \cdot 3 \cdot 2 \cdot 1 = 720.$$

The number of ways the similar 4 lines can be arranged among the 6 total lines is

$$C(6,4) = 15.$$

The number of arrangements that fit the pattern is

$$2{,}903{,}040 \cdot 720 \cdot 15 \approx 3.135 \times 10^{10}.$$

8.3 Probability Applications of Counting Principles

Your Turn 1

Using method 2,

$$P(1 \text{ NY}, 1 \text{ Chicago}) = \frac{C(3,1)C(1,1)}{C(6,2)} = \frac{1}{5}.$$

Your Turn 2

Since 8 of the nurses are men, $22 - 8 = 14$ of them are women. We choose 4 nurses, 2 men and 2 women.

$$P(2 \text{ men among 4 selected}) = \frac{C(8,2)C(14,2)}{C(22,4)} = \frac{2548}{7315} \approx 0.3483$$

Your Turn 3

The probability that the container will be shipped is the probability of selecting 3 working engines for testing when there are $12 - 4 = 8$ working engines in the container. This is

$$P(\text{all 3 work}) = \frac{C(8,3)}{C(12,3)} = \frac{56}{220} = \frac{14}{55}.$$

The probability that at least one defective engine is in the batch is

$$P(\text{at least one defective}) = 1 - \frac{14}{55} = \frac{41}{55} \approx 0.7455.$$

Your Turn 4

$$P(2 \text{ aces}, 2 \text{ kings}, 1 \text{ other}) = \frac{C(4,2)C(4,2)C(44,1)}{C(52,5)}$$
$$= \frac{6 \cdot 6 \cdot 44}{2{,}598{,}960}$$
$$\approx 0.0006095$$

Your Turn 5

If the slips are chosen without replacement, there are $P(7,3) = 7 \cdot 6 \cdot 5 = 210$ ordered selections. Only one of these spells "now" so $P(\text{now}) = \frac{1}{210}$. If the slips are chosen with replacement there are 7 choices for each and thus $7 \cdot 7 \cdot 7 = 343$ selections. Again, only one of these spells "now" so in this case $P(\text{now}) = \frac{1}{343}$.

Your Turn 6

Using method 1 we compute the number of ways to arrange 14 pieces of fruit which come in 4 kinds: 2 kiwis, 3 apricots, 4 pineapples and 5 coconuts. Assuming the pieces of each kind of fruit are indistinguishable (all kiwis look alike, and so on), the number of arrangements is $\frac{14!}{2!3!4!5!} = 2{,}522{,}520$. If we keep the four kinds together (all kiwis next to each other, and so on) there are $4! = 24$ arrangements of the kinds that keep them together. So

$$P(\text{all of same kind together}) = \frac{24}{2{,}522{,}520}$$
$$\approx 9.514 \times 10^{-6}.$$

8.3 Exercises

1. There are $C(11,3)$ samples of 3 apples.

$$C(11,3) = \frac{11 \cdot 10 \cdot 9}{3 \cdot 2 \cdot 1} = 165$$

There are $C(7,3)$ samples of 3 red apples.

$$C(7,3) = \frac{7 \cdot 6 \cdot 5}{3 \cdot 2 \cdot 1} = 35$$

Thus,

$$P(\text{all red apples}) = \frac{35}{165} = \frac{7}{33}.$$

3. There are $C(4,2)$ samples of 2 yellow apples.

$$C(4,2) = \frac{4 \cdot 3}{2 \cdot 1} = 6$$

There are $C(7,1) = 7$ samples of 1 red apple. Thus, there are $6 \cdot 7 = 42$ samples of 3 in which 2 are yellow and 1 red. Thus,

$$P(\text{2 yellow and 1 red apple}) = \frac{42}{165} = \frac{14}{55}.$$

For Exercises 5 through 9 the number of possible 5-member committees is $C(20,5) = 15{,}504$.

5. $P(\text{all men}) = \frac{C(9,5)}{C(20,5)}$

$$= \frac{126}{15{,}504} \approx 0.008127$$

7. $P(\text{3 men, 2 women}) = \frac{C(9,3)C(11,2)}{C(20,5)}$

$$= \frac{4620}{15{,}504} \approx 0.2980$$

9. $P(\text{at least 4 women})$

$$= P(\text{4 women}) + P(\text{5 women})$$

$$= \frac{C(11,4)C(9,1) + C(11,5)C(9,0)}{C(20,5)}$$

$$= \frac{3532}{15{,}504} \approx 0.2214$$

11. The number of 2-card hands is

$$C(52,2) = \frac{52 \cdot 51}{2 \cdot 1} = 1326.$$

13. There are $C(52,2) = 1326$ different 2-card hands. The number of 2-card hands with exactly one ace is

$$C(4,1)C(48,2) = 4 \cdot 48 = 192.$$

The number of 2-card hands with two aces is

$$C(4,2) = 6.$$

Thus there are 198 hands with at least one ace. Therefore,

$P(\text{the 2-card hand contains an ace})$

$$= \frac{198}{1326} = \frac{33}{221} \approx 0.149.$$

15. There are $C(52,2) = 1326$ different 2-card hands. There are $C(13,2) = 78$ ways to get a 2-card hand where both cards are of a single named suit, but there are 4 suits to choose from. Thus,

$P(\text{two cards of same suit})$

$$= \frac{4 \cdot C(13,2)}{C(52,2)} = \frac{312}{1326} = \frac{52}{221} \approx 0.235.$$

17. There are $C(52,2) = 1326$ different 2-card hands. There are 12 face cards in a deck, so there are 40 cards that are not face cards. Thus,

$P(\text{no face cards})$

$$= \frac{C(40,2)}{C(52,2)} = \frac{780}{1326} = \frac{130}{221} \approx 0.588.$$

19. There are 26 choices for each slip pulled out, and there are 5 slips pulled out, so there are

$$26^5 = 11{,}881{,}376$$

different "words" that can be formed from the letters. If the "word" must be "chuck," there is only one choice for each of the 5 letters (the first slip must contain a "c," the second an "h," and so on). Thus,

$P(\text{word is "chuck"})$

$$= \frac{1^5}{26^5} = \left(\frac{1}{26}\right)^5 \approx 8.417 \times 10^{-8}.$$

21. There are $26^5 = 11{,}881{,}376$ different "words" that can be formed. If the "word" is to have no repetition of letters, then there are 26 choices for the first letter, but only 25 choices for the second (since the letters must all be different), 24 choices for the third, and so on. Thus,

$P(\text{all different letters})$

$$= \frac{26 \cdot 25 \cdot 24 \cdot 23 \cdot 22}{26^5}$$

$$= \frac{1 \cdot 25 \cdot 24 \cdot 23 \cdot 22}{26^4}$$

$$= \frac{303{,}600}{456{,}976}$$

$$= \frac{18{,}975}{28{,}561} \approx 0.664.$$

25. $P(\text{at least 2 presidents have the same birthday})$
$= 1 - P(\text{no 2 presidents have the same birthday})$

The number of ways that 43 people can have the same or different birthdays is $(365)^{43}$. The number of ways that 43 people can have all different birthdays is the number of permutations of 365 things taken 43 at a time or $P(365, 43)$. Thus,

$$P(\text{at least 2 presidents have the same birthday}) = 1 - \frac{P(365, 43)}{365^{43}}.$$

(Be careful to realize that the symbol P is sometimes used to indicate permutations and sometimes used to indicate probability; in this solution, the symbol is used both ways.)

27. Since there are 435 members of the House of Representatives, and there are only 365 days in a year, it is a certain event that at least 2 people will have the same birthday. Thus,

$$P(\text{at least 2 members have the same birthday}) = 1.$$

29. $P(\text{matched pair})$

$$= P(\text{2 black or 2 brown or 2 blue})$$
$$= P(\text{2 black}) + P(\text{2 brown}) + 2(\text{2 blue})$$
$$= \frac{C(9,2)}{C(17,2)} + \frac{C(6,2)}{C(17,2)} + \frac{C(2,2)}{C(17,2)}$$
$$= \frac{36}{136} + \frac{15}{136} + \frac{1}{136}$$
$$= \frac{52}{136} = \frac{13}{34}$$

31. There are 6 letters so the number of possible spellings (counting duplicates) is $6! = 720$. Since the letter 1 is repeated 2 times and the letter t is repeated 2 times, the spelling "little" will occur $2!2! = 4$ times. The probability that "little" will be spelled is $\frac{4}{720} = \frac{1}{180}$.

33. Each of the 4 people can choose to get off at any one of the 7 floors, so there are 7^4 ways the four people can leave the elevator. The number of ways the people can leave at different floors is the number of permutations of 7 things (floors) taken 4 at a time or

$$P(7,4) = 7 \cdot 6 \cdot 5 \cdot 4 = 840.$$

The probability that no 2 passengers leave at the same floor is

$$\frac{P(7,4)}{7^4} = \frac{840}{2401} \approx 0.3499.$$

Thus, the probability that at least 2 passengers leave at the same floor is

$$1 - 0.3499 = 0.6501.$$

(Note the similarity of this problem and the "birthday problem.")

35. $P(\text{at least one \$100-bill})$

$$= P(\text{1 \$100-bill}) + P(\text{2 \$100-bills})$$
$$= \frac{C(2,1)C(4,1)}{C(6,2)} + \frac{C(2,2)C(4,0)}{C(6,2)}$$
$$= \frac{8}{15} + \frac{1}{15} = \frac{9}{15} = \frac{3}{5}$$

$$P(\text{no \$100-bill}) = \frac{C(2,0)\,C(4,2)}{C(6,2)} = \frac{6}{15} = \frac{2}{5}$$

It is more likely to get at least one \$100-bill.

37. There are $C(9,2)$ possible ways to choose 2 nondefective typewriters out of the $C(11,2)$ possible ways of choosing any 2. Thus,

$$P(\text{no defective}) = \frac{C(9,2)}{C(11,2)} = \frac{36}{55}.$$

39. There are $C(9,4)$ possible ways to choose 4 nondefective typewriters out of the $C(11,4)$ possible ways of choosing any 4. Thus,

$$P(\text{no defective}) = \frac{C(9,4)}{C(11,4)} = \frac{126}{330} = \frac{21}{55}.$$

41. There are $C(12,5) = 792$ ways to pick a sample of 5. It will be shipped if all 5 are good. There are $C(10,5) = 252$ ways to pick 5 good ones, so

$$P(\text{all good}) = \frac{252}{792} = \frac{7}{22} \approx 0.318.$$

43. $P(\text{not Scottsdale customer})$

$$= \frac{\left[\begin{array}{l}\text{number of choices of 4 out of the}\\ \text{5 non-Scottsdale customers}\end{array}\right]}{\text{number of choices of 4 out of the 6 customers}}$$
$$= \frac{C(5,4)}{C(6,4)} = \frac{5}{15} = \frac{1}{3}$$

45. There are 20 people in all, so the number of possible 5-person committees is $C(20, 5) = 15{,}504$. Thus, in parts (a)-(g), $n(S) = 15{,}504$.

(a) There are $C(10, 3)$ ways to choose the 3 men and $C(10, 2)$ ways to choose the 2 women. Thus,

$$P(\text{3 men and 2 women}) = \frac{C(10,3)C(10,2)}{C(20,5)} = \frac{120 \cdot 45}{15{,}504} = \frac{225}{646} \approx 0.348.$$

(b) There are $C(6, 3)$ ways to chose the 3 Miwoks and $C(9, 2)$ ways to choose the 2 Pomos. Thus,

$$P(\text{exactly 3 Miwoks and 2 Pomos}) = \frac{C(6,3)C(9,2)}{C(20,5)} = \frac{20 \cdot 36}{15{,}504} = \frac{15}{323} \approx 0.046.$$

(c) Choose 2 of the 6 Miwoks, 2 of the 5 Hoopas, and 1 of the 9 Pomos. Thus,

$$P(\text{2 Miwoks, 2 Hoopas, and a Pomo}) = \frac{C(6,2)C(5,2)C(9,1)}{C(20,5)} = \frac{15 \cdot 10 \cdot 9}{15{,}504} = \frac{225}{2584} \approx 0.087.$$

(d) There cannot be 2 Miwoks, 2 Hoopas, and 2 Pomos, since only 5 people are to be selected. Thus,

$P(\text{2 Miwoks, 2 Hoopas, and 2 Pomos}) = 0.$

(e) Since there are more women then men, there must be 3, 4, or 5 women.

$$P(\text{more women than men}) = \frac{C(10,3)C(10,2) + C(10,4)C(10,1) + C(10,5)C(10,0)}{C(20,5)} = \frac{7752}{15{,}504} = \frac{1}{2}$$

(f) Choose 3 of 5 Hoopas and any 2 of the 15 non-Hoopas.

$$P(\text{exactly 3 Hoopas}) = \frac{C(5,3)C(15,2)}{C(20,5)} = \frac{175}{2584} \approx 0.068$$

(g) There can be 2 to 5 Pomos, the rest chosen from the 11 non-Pomos.

$$P(\text{at least 2 Pomos}) = \frac{C(9,2)C(11,3) + C(9,3)C(11,2) + C(9,4)C(11,1) + C(9,5)C(11,0)}{C(20,5)} = \frac{503}{646} \approx 0.779$$

47. There are 21 books, so the number of selection of any 6 books is

$$C(21, 6) = 54{,}264.$$

(a) The probability that the selection consisted of 3 Hughes and 3 Morrison books is

$$\frac{C(9,3)C(7,3)}{C(21,6)} = \frac{85 \cdot 35}{54{,}264} = \frac{2940}{54{,}264} \approx 0.0542.$$

(b) A selection containing exactly 4 Baldwin books will contain 2 of the 16 books by the other authors, so the probability is

$$\frac{C(5,4)C(16,2)}{C(21,6)} = \frac{5 \cdot 120}{54{,}264} = \frac{600}{54{,}264} \approx 0.0111.$$

(c) The probability of a selection consisting of 2 Hughes, 3 Baldwin, and 1 Morrison book is

$$\frac{C(9,2)C(5,3)C(7,1)}{C(21,6)} = \frac{30 \cdot 10 \cdot 7}{54{,}264} = \frac{2520}{54{,}264} \approx 0.0464.$$

(d) A selection consisting of at least 4 Hughes books may contain 4, 5, or 6 Hughes books, with any remaining books by the other authors. Therefore, the probability is

$$\frac{\begin{pmatrix} C(9,4)C(12,2) + C(9,5)C(12,1) \\ + C(9,6)C(12,0) \end{pmatrix}}{C(21,6)} = \frac{126 \cdot 66 + 126 \cdot 12 + 84}{54{,}264} = \frac{8316 + 1512 + 84}{54{,}264} = \frac{9912}{54{,}264} \approx 0.1827.$$

(e) Since there are 9 Hughes books and 5 Baldwin books, there are 14 books written by males. The probability of a selection with exactly 4 books written by males is

$$\frac{C(14,2)C(7,2)}{C(21,6)} = \frac{1001 \cdot 21}{54,264}$$

$$= \frac{21,021}{54,264} \approx 0.3874.$$

(f) A selection with no more than 2 books written by Baldwin may contain 0, 1, or 2 books by Baldwin, with the remaining books by the other authors. Therefore, the probability is

$$\frac{C(5,0)C(16,6) + C(5,1)C(16,5) + C(5,2)C(16,4)}{C(21,6)}$$

$$= \frac{8008 + 5 \cdot 4368 + 10 \cdot 1820}{54,264}$$

$$= \frac{8008 + 21,840 + 18,200}{54,264}$$

$$= \frac{48,048}{54,264} \approx 0.8854.$$

49. A straight flush could start with an ace, 2, 3, 4, …, 7, 8, or 9. This gives 9 choices in each of 4 suits, so there are 36 choices in all. Thus,

$$P(\text{straight flush}) = \frac{36}{C(52,2)} = \frac{36}{2,598,960}$$

$$\approx 0.00001385$$

$$= 1.385 \times 10^{-5}.$$

51. A straight could start with an ace, 2, 3, 4, 5, 6, 7, 8, 9, or 10 as the low card, giving 40 choices. For each succeeding card, only the suit may be chosen. Thus, the number of straights is

$$40 \cdot 4^4 = 10,240.$$

But this also counts the straight flushes, of which there are 36 (see Exercise 49), and the 4 royal flushes. There are thus 10,200 straights that are not also flushes, so

$$P(\text{straight}) = \frac{10,200}{2,598,960} \approx 0.0039.$$

53. There are 13 different values of cards and 4 cards of each value. Choose 2 values out of the 13 for the values of the pairs. The number of ways to select the 2 values is $C(13, 2)$. The number of ways to select a pair for each value is $C(4, 2)$. There are $52 - 8 = 44$ cards that are neither of these 2 values, so the number of ways to select the fifth card is $C(44,1)$. Thus,

$$P(\text{two pairs}) = \frac{C(13,2)\,C(4,2)\,C(4,2)\,C(44,1)}{C(52,5)}$$

$$= \frac{123,552}{2,598,960} \approx 0.0475.$$

55. There are $C(52,13)$ different 13-card bridge hands. Since there are only 13 hearts, there is exactly one way to get a bridge hand containing only hearts. Thus,

$$P(\text{only hearts}) = \frac{1}{C(52,13)} \approx 1.575 \cdot 10^{-12}.$$

57. There are $C(4,2)$ ways to obtain 2 aces, $C(4,2)$ ways to obtain 2 kings, and $C(44,9)$ ways to obtain the remaining 9 cards. Thus,

$$P(\text{exactly 2 aces and exactly 2 kings})$$

$$= \frac{C(4,2)\,C(4,2)\,C(44,9)}{C(52,13)} \approx 0.0402.$$

For Exercises 59 through 65, use the fact that the number of 7-card selections is $C(52, 7)$.

59. Pick a kind for the pair: 13 choices
Choose 2 suits out of the 4 suits for this kind: $C(4, 2)$
Then pick 5 kinds out of the 12 remaining: $C(12, 5)$
For each of these 5 kinds, pick one of the 4 suits: 4^5
The product of these factors gives the numerator and the denominator is $C(52, 7)$.

$$\frac{13 \cdot C(4,2) \cdot C(12,5) \cdot 4^5}{C(52,7)} \approx 0.4728$$

61. Pick a kind for the three-of-a-kind: 13 choices
Choose 3 suits out of the 4 suits for this kind: $C(4, 3)$
There are now 4 cards remaining which must be 4 of the 12 kinds remaining: $C(12, 4)$
For each of these 4 kinds we pick one of the 4 suits: 4^4
The product of these factors gives the numerator and the denominator is $C(52, 7)$.

$$\frac{13 \cdot C(4,3) \cdot C(12,4) \cdot 4^4}{C(52,7)} \approx 0.0493$$

63. Pick a suit: 4
Now we either get exactly 5 of this suit and 2 of other suits: $C(13,5) \cdot C(39,2)$
...or exactly 6 of this suit and 1 of another suit: $C(13,6) \cdot C(39,1)$
...or all 7 of our chosen suit: $C(13,7)$. We now add these options over our usual denominator.

$$\frac{4 \cdot [C(13,5) \cdot C(39,2) + C(13,6) \cdot C(39,1) + C(13,7)]}{C(52,7)}$$
$$\approx 0.0306$$

65. We need at least 3 hearts out of 5 cards. This can happen in three ways:
3 hearts, 2 non-hearts: $C(11,3) \cdot C(39,2)$
4 hearts, 1 non-heart: $C(11,4) \cdot C(39,1)$
5 hearts: $C(11,5)$
Add these options over the usual denominator.

$$\frac{C(11,3) \cdot C(39,2) + C(11,4) \cdot C(39,1) + C(11,5)}{C(52,7)}$$
$$\approx 0.0640$$

67. To find the probability of picking 5 of the 6 lottery numbers correctly, we must recall that the total number of ways to pick the 6 lottery numbers is $C(99, 6) = 1{,}120{,}529{,}256$. To pick 5 of the 6 winning numbers, we must also pick 1 of the 93 losing numbers. Therefore, the number of ways of picking 5 of the 6 winning numbers is

$$C(6,5)\, C(93,1) = 558.$$

Thus, the probability of picking 5 of the 6 numbers correctly is

$$\frac{C(6,5)\, C(93,1)}{C(99,6)} \approx 4.980 \times 10^{-7}.$$

69. The probability of picking six numbers out of 49 is

$$P(6 \text{ out of } 49) = \frac{1}{C(49,6)} = \frac{1}{13{,}983{,}816}$$

The probability of picking of picking five numbers out of 52 is

$$P(5 \text{ out of } 52) = \frac{1}{C(52,5)} = \frac{1}{2{,}598{,}960}$$

The probability of winning the lottery when picking five out of 52 is higher.

71. **(a)** The number of ways to select 6 numbers between 1 and 49 is $C(49,6) = 13{,}983{,}816$. The number of ways to select 3 of the 6 numbers, while not selecting the bonus number is

$$C(6,3)\, C(42,3) = 20 \cdot 11{,}480 = 229{,}600.$$

The probability of winning fifth prize is

$$\frac{229{,}600}{13{,}983{,}816} \approx 0.01642.$$

(b) The number of ways to select 2 of the 6 numbers plus the bonus number is

$$C(6,2)\, C(1,1)\, C(42,3) = 15 \cdot 1 \cdot 11{,}480 = 172{,}200.$$

The probability of winning sixth prize is

$$\frac{172{,}200}{13{,}983{,}816} \approx 0.01231.$$

73. **(a)** There were 28 games played in the season, since the numbers in the "Won" column have a sum of 28 (and the numbers in the "Lost" column have a sum of 28).

(b) Assuming no ties, each of the 28 games had 2 possible outcomes; either Team A won and Team B lost, or else Team A lost and Team B won. By the multiplication principle, this means that there were

$$2^{28} = 268{,}435{,}456$$

different outcomes possible.

(c) Any one of the 8 teams could have been the one that won all of its games, any one of the remaining 7 teams could have been the one that won all but one of its games, and so on, until there is only one team left, and it is the one that lost all of its games. By the multiplication principle, this means that there were

$$8! = 8 \cdot 7 \cdot 6 \cdot 5 \cdot 4 \cdot 3 \cdot 2 \cdot 1 = 40{,}320$$

different "perfect progressions" possible.

(d) Thus,

$$P(\text{"perfect progression" in an 8-team league}) = \frac{8!}{2^{28}} \approx 0.0001502 = 1.502 \times 10^{-4}.$$

(e) If there are n teams in the league, then the "Won" column will begin with $n-1$, followed by $n-2$, then $n-3$, and so on down to 0. It can be shown that the sum of these n numbers is $\frac{n(n-1)}{2}$, so there are $2^{n(n-1)/2}$ different win/lose progressions possible. The n teams can be ordered in $n!$

different ways, so there are $n!$ different "perfect progressions" possible. Thus,

P("perfect progression" in an n-team league)

$$= \frac{n!}{2^{n(n-1)/2}}.$$

75. **(a)** There are only 4 ways to win in just 4 calls: the 2 diagonals, the center column, and the center row. There are $C(75,4)$ combinations of 4 numbers that can occur. The probability that a person will win bingo after just 4 numbers are called is

$\frac{4}{C(75,4)} \approx 3.291 \times 10^{-6}.$

(b) There is only 1 way to get an L. It can occur in as few as 9 calls. There are $C(75,9)$ combinations of 9 numbers that can occur in 9 calls , so the probability of an L in 9 calls is $\frac{1}{C(75,9)} \approx 7.962 \times 10^{-12}.$

(c) There is only 1 way to get an X-out. It can ocur in as few as 8 calls. There are $C(75,8)$ combinations of 8 numbers that can occur. The probability that an X-out occurs in 8 calls is $\frac{1}{C(75,8)} \approx 5.927 \times 10^{-11}.$

(d) Four columns contain a permutation of 15 numbers taken 5 at a time. One column contains a permutation of 15 numbers taken 4 at a time. The number of distinct cards is

$P(15,5)^4 \cdot P(15,4) \approx 5.524 \times 10^{26}.$

8.4 Binomial Probability

Your Turn 1

$$\begin{aligned} P(\text{exactly 2 of 6}) &= C(6,2)(0.59)^2(0.41)^4 \\ &= 15(0.3481)(0.0283) \\ &\approx 0.1475 \end{aligned}$$

Your Turn 2

$$\begin{aligned} P(\text{4 heads in 8 tosses}) &= C(8,4)\left(\frac{1}{2}\right)^4\left(\frac{1}{2}\right)^4 \\ &= 70\left(\frac{1}{2}\right)^8 \\ &\approx 0.2734 \end{aligned}$$

Your Turn 3

P(2 or 3 defective of 15)

$$\begin{aligned} &= C(15,2)(0.01)^2(0.99)^{13} + C(15,3)(0.01)^3(0.99)^{12} \\ &\approx 0.009214 + 0.000403 \\ &= 0.009617 \end{aligned}$$

Your Turn 4

P(at least one incorrect charge)

$$\begin{aligned} &= 1 - P(\text{no incorrect charges in 4}) \\ &= 1 - C(4,0)(0.29)^0(0.71)^4 \\ &\approx 1 - 0.2541 \\ &= 0.7459 \end{aligned}$$

Your Turn 5

P(at most 3 incorrect charges in 6)

$$\begin{aligned} &= P(0) + P(1) + P(2) + P(3) \\ &= C(6,0)(0.29)^0(0.71)^6 + C(6,1)(0.29)^1(0.71)^5 \\ &\quad + C(6,2)(0.29)^2(0.71)^4 + C(6,3)(0.29)^3(0.71)^3 \\ &\approx 0.9372 \end{aligned}$$

8.4 Exercises

1. This is a Bernoulli trial problem with $P(\text{success}) = P(\text{girl}) = \frac{1}{2}$. The probability of exactly x successes in n trials is

$$C(n,x)p^x(1-p)^{n-x},$$

where p is the probability of success in a single trial. We have $n = 5, x = 2,$ and $p = \frac{1}{2}$

Note that

$$1 - p = 1 - \frac{1}{2} = \frac{1}{2}.$$

P(exactly 2 girls and 3 boys)

$$\begin{aligned} &= C(5,2)\left(\frac{1}{2}\right)^2\left(\frac{1}{2}\right)^3 \\ &= \frac{10}{32} = \frac{5}{16} \approx 0.313 \end{aligned}$$

3. We have
$n = 5, x = 0, p = \frac{1}{2}$, and $1 - p = \frac{1}{2}$.

$$P(\text{no girls}) = C(5,0)\left(\frac{1}{2}\right)^0\left(\frac{1}{2}\right)^5 = \frac{1}{32} \approx 0.031$$

5. "At least 4 girls" means either 4 or 5 girls.

P(at least 4 girls)

$$= C(5,4)\left(\frac{1}{2}\right)^4\left(\frac{1}{2}\right)^1 + C(5,5)\left(\frac{1}{2}\right)^5\left(\frac{1}{2}\right)^0$$

$$= \frac{5}{32} + \frac{1}{32} = \frac{6}{32} = \frac{3}{16} \approx 0.188$$

7. P(no more than 3 boys)

$$= 1 - P(\text{at least 4 boys})$$
$$= 1 - P(\text{4 boys or 5 boys})$$
$$= 1 - [P(\text{4 boys}) + P(\text{5 boys})]$$
$$= 1 - \left(\frac{5}{32} + \frac{1}{32}\right)$$
$$= 1 - \frac{6}{32}$$
$$= 1 - \frac{3}{16} = \frac{13}{16} \approx 0.813$$

9. On one roll, $P(1) = \frac{1}{6}$. We have $n = 12$, $x = 12$, and $p = \frac{1}{6}$. Note that $1 - p = \frac{5}{6}$. Thus,

$$P(\text{exactly 12 ones}) = C(12,12)\left(\frac{1}{6}\right)^{12}\left(\frac{5}{6}\right)^0 \approx 4.594 \times 10^{-10}.$$

11.
$$P(\text{exactly 1 one}) = C(12,1)\left(\frac{1}{6}\right)^1\left(\frac{5}{6}\right)^{11} \approx 0.2692$$

13. "No more than 3 ones" means 0, 1, 2, or 3 ones. Thus,

P(no more than 3 ones)

$$= P(\text{0 ones}) + P(\text{1 one}) + P(\text{2 ones}) + P(\text{3 ones})$$

$$= C(12,0)\left(\frac{1}{6}\right)^0\left(\frac{5}{6}\right)^{12} + C(12,1)\left(\frac{1}{6}\right)^1\left(\frac{5}{6}\right)^{11}$$
$$+ C(12,2)\left(\frac{1}{6}\right)^2\left(\frac{5}{6}\right)^{10} + C(12,3)\left(\frac{1}{6}\right)^3\left(\frac{5}{6}\right)^9$$

≈ 0.8748.

For Exercises 15 and 17 we have
$n = 6, p = \frac{1}{4}$, and $1 - p = \frac{3}{4}$.

15.
$$P(\text{all heads}) = C(6,6)\left(\frac{1}{4}\right)^6\left(\frac{3}{4}\right)^0 = \frac{1}{4096} \approx 0.0002441$$

17. P(no more than 3 heads)

$$= P(\text{0 heads}) + P(\text{1 head}) + P(\text{2 heads}) + P(\text{3 heads})$$

$$= C(6,0)\left(\frac{1}{4}\right)^0\left(\frac{3}{4}\right)^6 + C(6,1)\left(\frac{1}{4}\right)^1\left(\frac{3}{4}\right)^5$$
$$+ C(6,2)\left(\frac{1}{4}\right)^2\left(\frac{3}{4}\right)^4 + C(6,3)\left(\frac{1}{4}\right)^3\left(\frac{3}{4}\right)^3$$

$$= \frac{3942}{4096} \approx 0.9624$$

21. $C(n,r) + C(n,r+1)$

$$= \frac{n!}{r!(n-r)!} + \frac{n!}{(r+1)![n-(r+1)]!}$$

$$= \frac{n!(r+1)}{r!(r+1)(n-r)!} + \frac{n!(n-r)}{(r+1)![n-(r+1)]!(n-r)}$$

$$= \frac{rn! + n!}{(r+1)!(n-r)!} + \frac{n(n!) - rn!}{(r+1)!(n-r)!}$$

$$= \frac{rn! + n! + n(n!) - rn!}{(r+1)!(n-r)!}$$

$$= \frac{n!(n+1)}{(r+1)!(n-r)!}$$

$$= \frac{(n+1)!}{(r+1)![(n+1)-(r+1)]!}$$

$$= C(n+1, r+1)$$

23. Since the potential callers are not likely to have birthdates that are distributed evenly throughout the twentieth century, the use of binomial probabilities is not applicable and thus, the probabilities that are computed are not correct.

For Exercises 25 and 27 we define a success to be the event that a customer is charged incorrectly. In this situation, $n = 15, p = \frac{1}{30}$ and $1 - p = \frac{29}{30}$.

25. P(0 incorrect charges)

$$= C(15,0)\left(\frac{1}{30}\right)^0\left(\frac{29}{30}\right)^{15} \approx 0.6014$$

27. $P(\text{at least 2 incorrect charges})$
$$= 1 - P(0 \text{ or } 1 \text{ incorrect charges})$$
$$= 1 - C(15,0)\left(\frac{1}{30}\right)^0\left(\frac{29}{30}\right)^{15} - C(15,1)\left(\frac{1}{30}\right)^1\left(\frac{29}{30}\right)^{14}$$
$$\approx 0.0876$$

For Exercises 29 and 31, we define a success to be the event that the family hardly ever pays off the balance. In this situation, $n = 20, p = 0.254$ and $1 - p = 0.746$.

29. $P(6) = C(20,6)(0.254)^6(0.746)^{14}$
$$\approx 0.1721$$

31.

$P(\text{at least } 4)$
$$= 1 - C(20,0)(0.254)^0(0.746)^{20} - C(20,1)(0.254)^1(0.746)^{19} - C(20,2)(0.254)^2(0.746)^{18} - C(20,3)(0.254)^3(0.746)^{17}$$
$$\approx 0.7868$$

33. We have $n = 6, x = 2, p = \frac{1}{5}$, and $1 - p = \frac{4}{5}$. Thus,
$$P(\text{exactly 2 correct}) = C(6,2)\left(\frac{1}{5}\right)^2\left(\frac{4}{5}\right)^4 \approx 0.2458.$$

35. We have

$P(\text{at least 4 correct})$
$$= P(4 \text{ correct}) + P(5 \text{ correct}) + P(6 \text{ correct})$$
$$= C(6,4)\left(\frac{1}{5}\right)^4\left(\frac{4}{5}\right)^2 + C(6,5)\left(\frac{1}{5}\right)^5\left(\frac{4}{5}\right)^1 + C(6,6)\left(\frac{1}{5}\right)^6\left(\frac{4}{5}\right)^0$$
$$\approx 0.0170.$$

37. $n = 20, p = 0.05, x = 0$
$$P(0 \text{ defective transistors}) = C(20,0)(0.05)^0(0.95)^{20} \approx 0.3585$$

39. Let success mean producing a defective item. Then we have $n = 75, p = 0.05$, and $1 - p = 0.95$.

(a) If there are exactly 5 defective items, then $x = 5$. Thus,

$P(\text{exactly 5 defective})$
$$= C(75,5)(0.05)^5(0.95)^{70} \approx 0.1488.$$

(b) If there are no defective items, then $x = 0$. Thus,

$P(\text{none defective})$
$$= C(75,0)(0.05)^0(0.95)^{75} \approx 0.0213.$$

(c) If there is at least 1 defective item, then we are interested in $x \geq 1$. We have

$P(\text{at least one defective})$
$$= 1 - P(x = 0) \approx 1 - 0.021 = 0.9787.$$

41. (a) Since 80% of the "good nuts" are good, 20% of the "good nuts" are bad. Let's let success represent "getting a bad nut." Then 0.2 is the probability of success in a single trial. The probability of 8 successes in 20 trials is
$$C(20,8)(0.2)^8(1 - 0.2)^{20-8} = C(20,8)(0.2)^8(0.8)^{12} \approx 0.0222$$

(b) Since 60% of the "blowouts" are good, 40% of the "blowouts" are bad. Let's let success represent "getting a bad nut." Then 0.4 is the probability of success in a single trial. The probability of 8 successes in 20 trials is
$$C(20,8)(0.4)^8(1 - 0.4)^{20-8} = C(20,8)(0.4)^8(0.6)^{12} \approx 0.1797$$

(c) The probability that the nuts are "blowouts" is
$$\frac{\left(\text{Probability of "Blowouts" having 8 bad nuts out of 20}\right)}{\left(\text{Probability of "Good Nuts" or "Blowouts" having 8 bad nuts of 20}\right)}$$
$$= \frac{0.3\left[C(20,8)(0.4)^8(0.6)^{12}\right]}{0.7\left[C(20,8)(0.2)^8(0.8)^{12}\right] + 0.3\left[C(20,8)(0.4)^8(0.6)^{12}\right]}$$
$$\approx 0.7766.$$

43. $n = 15, p = 0.85$

$$P(\text{all } 15) = C(15,15)(0.85)^{15}(0.15)^0 \approx 0.0874$$

45. $n = 15, p = 0.85$

$$\begin{aligned} P(\text{not all}) &= 1 - P(\text{all } 15) \\ &= 1 - C(15,15)(0.85)^{15}(0.15)^0 \\ &\approx 0.9126 \end{aligned}$$

47. $n = 100, p = 0.012, x = 2$

$$\begin{aligned} &P(\text{exactly 2 sets of twins}) \\ &= C(100,2)(0.012)^2(0.988)^{98} \\ &\approx 0.2183 \end{aligned}$$

49. We have $n = 10{,}000$,
$p = 2.5 \cdot 10^{-7} = 0.00000025$, and
$1 - p = 0.99999975$. Thus,

$$\begin{aligned} &P(\text{at least 1 mutation occurs}) \\ &= 1 - P(\text{none occurs}) \\ &= 1 - C(10{,}000, 0)p^0(1-p)^{10{,}000} \\ &= 1 - (0.99999975)^{10{,}000} \\ &\approx 0.0025. \end{aligned}$$

51. $n = 53, p = 0.042$

(a) The probability that exactly 5 men are color-blind is

$$P(5) = C(53,5)(0.042)^5(0.958)^{48} \approx 0.0478.$$

(b) The probability that no more than 5 men are color-blind is

$$\begin{aligned} &P(\text{no more than 5 men are color-blind}) \\ &= C(53,0)(0.042)^0(0.958)^{53} \\ &\quad + C(53,1)(0.042)^1(0.958)^{52} \\ &\quad + C(53,2)(0.042)^2(0.958)^{51} \\ &\quad + C(53,3)(0.042)^3(0.958)^{50} \\ &\quad + C(53,4)(0.042)^4(0.958)^{49} \\ &\quad + C(53,5)(0.042)^5(0.958)^{48} \\ &\approx 0.9767. \end{aligned}$$

(c) The probability that at least 1 man is color-blind is

$$\begin{aligned} &1 - P(\text{0 men are color-blind}) \\ &= 1 - C(53,0)(0.042)^0(0.958)^{53} \\ &\approx 0.8971. \end{aligned}$$

53. **(a)** Since the probability of a particular band matching is 1 in 4 or $\frac{1}{4}$, the probability that 5 bands match is $\left(\frac{1}{4}\right)^5 = \frac{1}{1024}$ or 1 chance in 1024.

(b) The probability that 20 bands match is $\left(\frac{1}{4}\right)^{20} \approx \frac{1}{1.1\times10^{12}}$ or about 1 chance in 1.1×10^{12}.

(c) If 20 bands are compared, the probability that 16 or more bands match is

$$\begin{aligned} &P(\text{at least } 16) \\ &= P(16) + P(17) + P(18) + P(19) + P(20) \\ &= C(20,16)\left(\frac{1}{4}\right)^{16}\left(\frac{3}{4}\right)^4 + C(20,17)\left(\frac{1}{4}\right)^{17}\left(\frac{3}{4}\right)^3 \\ &\quad + C(20,18)\left(\frac{1}{4}\right)^{18}\left(\frac{3}{4}\right)^2 + C(20,19)\left(\frac{1}{4}\right)^{19}\left(\frac{3}{4}\right)^1 \\ &\quad + C(20,20)\left(\frac{1}{4}\right)^{20}\left(\frac{3}{4}\right)^0 \end{aligned}$$

$$\begin{aligned} &= 4845\left(\frac{1}{4}\right)^{16}\left(\frac{3}{4}\right)^4 + 1140\left(\frac{1}{4}\right)^{17}\left(\frac{3}{4}\right)^3 \\ &\quad + 190\left(\frac{1}{4}\right)^{18}\left(\frac{3}{4}\right)^2 + 20\left(\frac{1}{4}\right)^{19}\left(\frac{3}{4}\right)^1 + \left(\frac{1}{4}\right)^{20} \cdot 1 \\ &= \left(\frac{1}{4}\right)^{16}\left[4845\left(\frac{81}{256}\right) + 1140\left(\frac{1}{4}\right)\left(\frac{27}{64}\right) + 190\left(\frac{1}{16}\right)\left(\frac{9}{16}\right) \right. \\ &\quad \left. + 20\left(\frac{1}{64}\right)\left(\frac{3}{4}\right) + \frac{1}{256}\right] \\ &= \left(\frac{1}{4}\right)^{16}\left(\frac{392{,}445 + 30{,}780 + 1710 + 60 + 1}{256}\right) \\ &= \frac{424{,}996}{4^{20}} = \frac{1}{\frac{4^{20}}{424{,}996}} \\ &\approx \frac{1}{2{,}587{,}110} \end{aligned}$$

or about 1 chance in 2.587×10^6.

55. $n = 4800, p = 0.001$

$P(\text{more than } 1)$

$= 1 - P(1) - P(0)$

$= 1 - C(4800,1)(0.001)^1(0.999)^{4799}$

$- C(4800,0)(0.001)^0(0.999)^{4800}$

≈ 0.9523

57. First, find the probability that one group of ten has at least 9 participants complete the study. $n = 10, P = 0.8,$

$$\begin{aligned} P(\text{at least 9 complete}) &= P(9) + P(10) \\ &= C(10,9)(0.8)^9(0.2)^1 \\ &\quad + C(10,10)(0.8)^{10}(0.2)^0 \\ &\approx 0.3758 \end{aligned}$$

The probability that 2 or more drop out in one group is $1 - 0.3758 = 0.6242$. Thus, the probability that at least 9 participants complete the study in one of the two groups, but not in both groups, is

$(0.3758)(0.6242) + (0.6242)(0.3758) \approx 0.469.$

The answer is e.

59. $n = 12, x = 7, p = 0.83$

$P(7) = C(12,7)(0.83)^7(0.17)^5 \approx 0.0305$

61. $n = 12, p = 0.83$

$$\begin{aligned} P(\text{at least 9}) &= P(9) + P(10) + P(11) + P(12) \\ &= C(12,9)(0.83)^9(0.17)^3 + C(12,10)(0.83)^{10}(0.17)^2 \\ &\quad + C(12,11)(0.83)^{11}(0.17)^1 + C(12,12)(0.83)^{12}(0.17)^0 \\ &\approx 0.8676 \end{aligned}$$

63. $n = 10, p = 0.322, 1 - p = 0.678$

(a) $P(2) = C(10,2)(0.322)^2(0.678)^8 \approx 0.2083$

(b) $$\begin{aligned} P(3 \text{ or fewer}) &= C(10,0)(0.322)^0(0.678)^{10} \\ &\quad + C(10,1)(0.322)^1(0.678)^9 \\ &\quad + C(10,2)(0.322)^2(0.678)^8 \\ &\quad + C(10,3)(0.322)^3(0.678)^7 \\ &\approx 0.5902 \end{aligned}$$

(c) If exactly 5 *do not* belong to a minority, then exactly $10 - 5 = 5$ *do* belong to a minority, and this probability is

$P(5) = C(10,5)(0.322)^5(0.678)^5 \approx 0.1250.$

(d) If 6 or more *do not* belong to a minority, then at most 4 *do* belong to a minority, and this probability is $P(\text{at most } 4)$

$$\begin{aligned} P(\text{at most 4}) &= C(10,0)(0.322)^0(0.678)^{10} \\ &\quad + C(10,1)(0.322)^1(0.678)^9 \\ &\quad + C(10,2)(0.322)^2(0.678)^8 \\ &\quad + C(10,3)(0.322)^3(0.678)^7 \\ &\quad + C(10,4)(0.322)^4(0.678)^6 \\ &\approx 0.8095 \end{aligned}$$

65. **(a)** Using the binomcdf function on a graphing calculator, we find

$$\begin{aligned} P(\text{at least 30}) &= 1 - P(29 \text{ or fewer}) \\ &= 1 - \text{binomcdf}(40, 0.74, 29) \\ &\approx 1 - 0.4740 \\ &= 0.5260 \end{aligned}$$

(b) Using the binomcdf function on a graphing calculator, we find

$$\begin{aligned} P(\text{at least 30}) &= 1 - P(29 \text{ or fewer}) \\ &= 1 - \text{binomcdf}(40, 0.83, 29) \\ &\approx 1 - 0.0657 \\ &= 0.9343 \end{aligned}$$

67. **(a)** Suppose the National League wins the series in four games. Then they must win all four games and $P = C(4,4)(0.5)^4(0.5)^0 = 0.0625$. Since the probability that the American League wins the series in four games is equally likely, the probability the series lasts four games is $2(0.0625) = 0.125$.

Suppose the National League wins the series in five games. Then they must win exactly three of the previous four games and $P = C(4,3)(0.5)^3(0.5)^1 \cdot (0.5) = 0.125$. Since the probability that the American League wins the series in five games is equally likely, the probability the series lasts five games is $2(0.125) = 0.25$. Suppose the National League wins the series in six games. Then they must win exactly three of the previous five games and

$P = C(5,3)(0.5)^3(0.5)^2 \cdot (0.5) = 0.15625.$

Since the probability that the American League wins the series in six games is equally likely, the probability the series lasts six games is $2(0.15625) = 0.3125$. Suppose the National League wins the series in seven games. Then they must win exactly three of

the previous six games and
$P = C(6,3)(0.5)^3(0.5)^3 \cdot (0.5) = 0.15625.$
Since the Probability that the American League wins the series in seven games is equally likely, the probability the series last seven games is $2(0.15625) = 0.3125$.

(b) Suppose the better team wins the series in four games. Then they must win all four games and $P = C(4,4)(0.73)^4(0.27)^0 \approx 0.2840$. Suppose the other team wins the series in four games. Then they must win all four games and
$P = C(4,4)(0.27)^4(0.73)^0 \approx 0.0053$. The probability the series lasts four games is the sum of two probabilities, 0.2893.

Suppose the better team wins the series in five games. Then they must win exactly three of the previous four games and
$P = C(4,3)(0.73)^3(0.27)^1 \cdot (0.73) \approx 0.3067.$
Suppose the other team wins the series in five games. Then they must win exactly three of the previous four games and
$C(4,3)(0.27)^3(0.73)^1 \cdot (0.27) \approx 0.0155$. The probability the series lasts five games is the sum of the two probabilities, 0.3222.

Suppose the better team wins the series in six games. Then they must win exactly three of the previous five games and
$P = C(5,3)(0.73)^3(0.27)^2 \cdot (0.73) \approx 0.2070.$
Suppose the other team wins the series in six games. Then they must win exactly three of the previous five games and
$P = C(5,3)(0.27)^3(0.73)^2 \cdot (0.27) \approx 0.0283.$
The probability the series lasts six games is the sum of the two probabilities, 0.2353.

Suppose the better team wins the series in seven games. Then they must win exactly three of the previous six games and
$P = C(6,3)(0.73)^3(0.27)^3 \cdot (0.73) \approx 0.1118.$
Suppose the other team wins the series in seven games. Then they must win exactly three of the previous six games and
$P = C(6,3)(0.27)^3(0.73)^3 \cdot (0.27) \approx 0.0413.$
The probability the series lasts seven games is the sum of the two probabilities, 0.1531.

8.5 Probability Distributions; Expected Value

Your Turn 1

$$P(x = 0) = \frac{C(3,0)C(9,2)}{C(12,2)} = \frac{6}{11}$$

$$P(x = 1) = \frac{C(3,1)C(9,1)}{C(12,2)} = \frac{9}{22}$$

$$P(x = 2) = \frac{C(3,2)C(9,0)}{C(12,2)} = \frac{1}{22}$$

The distribution is shown in the following table:

x	0	1	2
$P(x)$	6/11	9/22	1/22

Your Turn 2

Let the random variable x represent the number of tails.

$$P(x = 0) = C(3,0)\left(\frac{1}{2}\right)^0\left(\frac{1}{2}\right)^3 = \frac{1}{8}$$

$$P(x = 1) = C(3,1)\left(\frac{1}{2}\right)^1\left(\frac{1}{2}\right)^2 = \frac{3}{8}$$

$$P(x = 2) = C(3,2)\left(\frac{1}{2}\right)^2\left(\frac{1}{2}\right)^1 = \frac{3}{8}$$

$$P(x = 3) = C(3,3)\left(\frac{1}{2}\right)^3\left(\frac{1}{2}\right)^0 = \frac{1}{8}$$

The distribution is shown n the following table:

x	0	1	2	3
$P(x)$	1/8	3/8	3/8	1/8

Here is the histogram.

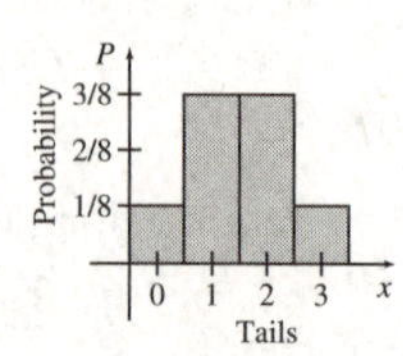

Your Turn 3

The expected payback is

$$955\left(\frac{1}{1000}\right) + 495\left(\frac{1}{1000}\right) + 245\left(\frac{1}{1000}\right) + (-5)\left(\frac{997}{1000}\right)$$
$$= \frac{-3250}{1000}$$
$$= -3.25 \quad \text{or} \quad -\$3.25.$$

Your Turn 4

Let the random variable x represents the number of male engineers in the sample of 5.

$$P(x=0) = C(5,0)(0.816)^0(0.184)^5 \approx 0.00021$$
$$P(x=1) = C(5,1)(0.816)^1(0.184)^4 \approx 0.00468$$
$$P(x=2) = C(5,2)(0.816)^2(0.184)^3 \approx 0.04148$$
$$P(x=3) = C(5,3)(0.816)^3(0.184)^2 \approx 0.18395$$
$$P(x=4) = C(5,4)(0.816)^4(0.184)^1 \approx 0.40790$$
$$P(x=5) = C(5,5)(0.816)^5(0.184)^0 \approx 0.36179$$
$$E(x) \approx (0)(0.0021) + (1)(0.00468) + (2)(0.04148) + (3)(0.18395) + (4)(0.40790) + (5)(0.36179) = 4.0803$$

In fact the exact value of the expectation can be computed quickly using the formula $E(x) = np$. For this example, $n = 5$ and $p = 0.816$ so $np = (5)(0.816) = 4.08$.

Your Turn 5

The expected number of girls in a family of a dozen children is $12\left(\frac{1}{2}\right) = 6$.

8.5 Exercises

1. Let x denote the number of heads observed. Then x can take on 0, 1, 2, 3, or 4 as values. The probabilities are as follows.

$$P(x=0) = C(4,0)\left(\frac{1}{2}\right)^0\left(\frac{1}{2}\right)^4 = \frac{1}{16}$$
$$P(x=1) = C(4,1)\left(\frac{1}{2}\right)^1\left(\frac{1}{2}\right)^3 = \frac{4}{16} = \frac{1}{4}$$
$$P(x=2) = C(4,2)\left(\frac{1}{2}\right)^2\left(\frac{1}{2}\right)^2 = \frac{6}{16} = \frac{3}{8}$$
$$P(x=3) = C(4,3)\left(\frac{1}{2}\right)^3\left(\frac{1}{2}\right)^1 = \frac{4}{16} = \frac{1}{4}$$
$$P(x=4) = C(4,4)\left(\frac{1}{2}\right)^4\left(\frac{1}{2}\right)^0 = \frac{1}{16}$$

Therefore, the probability distribution is as follows.

Number of Heads	0	1	2	3	4
Probability	$\frac{1}{16}$	$\frac{1}{4}$	$\frac{3}{8}$	$\frac{1}{4}$	$\frac{1}{16}$

3. Let x denote the number of aces drawn. Then x can take on values 0, 1, 2, or 3. The probabilities are as follows.

$$P(x=0) = C(3,0)\left(\frac{48}{52}\right)\left(\frac{47}{51}\right)\left(\frac{46}{50}\right) \approx 0.7826$$
$$P(x=1) = C(3,1)\left(\frac{4}{52}\right)\left(\frac{48}{51}\right)\left(\frac{47}{50}\right) \approx 0.2042$$
$$P(x=2) = C(3,2)\left(\frac{4}{52}\right)\left(\frac{3}{51}\right)\left(\frac{48}{50}\right) \approx 0.0130$$
$$P(x=3) = C(3,3)\left(\frac{4}{52}\right)\left(\frac{3}{51}\right)\left(\frac{2}{50}\right) \approx 0.0002$$

Therefore, the probability distribution is as follows.

Number of Aces	0	1	2	3
Probability	0.7826	0.2042	0.0130	0.0002

5. Use the probabilities that were calculated in Exercise 1. Draw a histogram with 5 rectangles, corresponding to $x = 0, x = 1, x = 2, x = 3$, and $x = 4$. $P(x \le 2)$ corresponds to

$$P(x=0) + P(x=1) + P(x=2),$$

so shade the first 3 rectangles in the histogram.

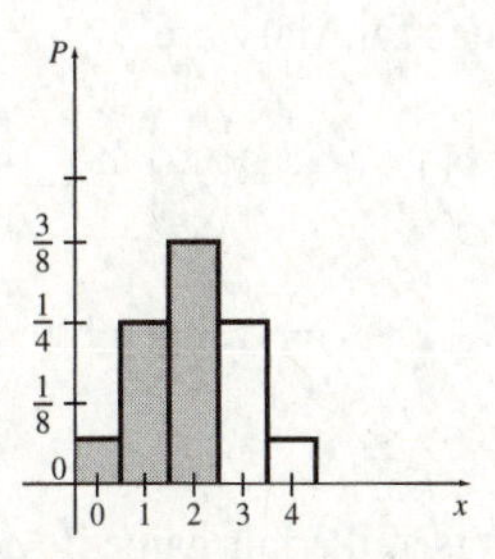

7. Use the probabilities that were calculated in Exercise 3. Draw a histogram with 4 rectangles, corresponding to $x = 0, x = 1, x = 2,$ and

$x = 3$. $P(\text{at least one ace}) = P(x \geq 1)$ corresponds to

$$P(x = 1) + P(x = 2) + P(x = 3),$$

so shade the last 3 rectangles.

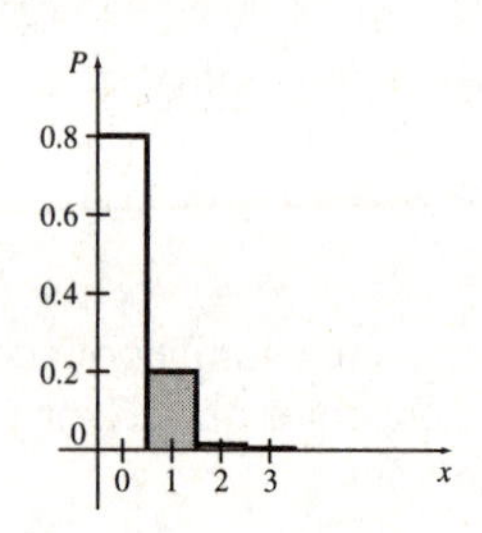

9. $E(x) = 2(0.1) + 3(0.4) + 4(0.3) + 5(0.2)$
$= 3.6$

11. $E(z) = 9(0.14) + 12(0.22) + 15(0.38) + 18(0.19) + 21(0.07)$
$= 14.49$

13. It is possible (but not necessary) to begin by writing the histogram's data as a probability distribution, which would look as follows.

x	1	2	3	4
$P(x)$	0.2	0.3	0.1	0.4

The expected value of x is

$$E(x) = 1(0.2) + 2(0.3) + 3(0.1) + 4(0.4) = 2.7.$$

15. The expected value of x is

$$E(x) = 6(0.1) + 12(0.2) + 18(0.4) + 24(0.2) + 30(0.1) = 18.$$

17. Using the data from Example 5, the expected winnings for Mary are

$$E(x) = -1.2\left(\frac{1}{4}\right) + 1.2\left(\frac{1}{4}\right) + 1.2\left(\frac{1}{4}\right) + (-1.2)\left(\frac{1}{4}\right) = 0.$$

Yes, it is still a fair game if Mary tosses and Donna calls.

19. **(a)**

Number of Yellow Marbles	Probability
0	$\frac{C(3,0)\,C(4,3)}{C(7,3)} = \frac{4}{35}$
1	$\frac{C(3,1)\,C(4,2)}{C(7,3)} = \frac{18}{35}$
2	$\frac{C(3,2)\,C(4,1)}{C(7,3)} = \frac{12}{35}$
3	$\frac{C(3,3)\,C(4,0)}{C(7,3)} = \frac{1}{35}$

Draw a histogram with four rectangles corresponding to $x = 0, 1, 2,$ and 3.

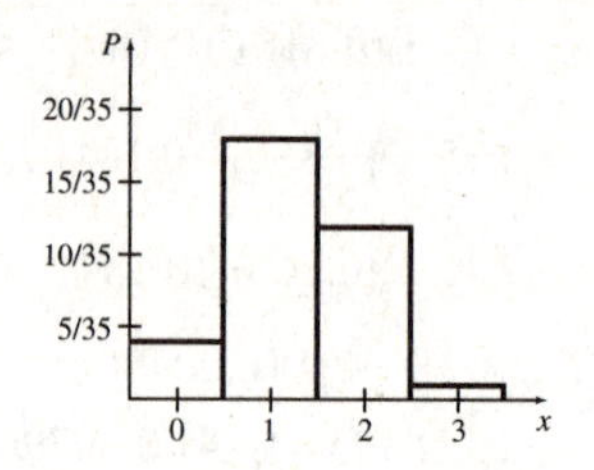

(b) Expected number of yellow marbles

$$= 0\left(\frac{4}{35}\right) + 1\left(\frac{18}{35}\right) + 2\left(\frac{12}{35}\right) + 3\left(\frac{1}{35}\right)$$
$$= \frac{45}{35} = \frac{9}{7} \approx 1.286$$

21. **(a)** Let x be the number of times 1 is rolled. Since the probability of getting a 1 on any single roll is $\frac{1}{6}$, the probability of any other outcome is $\frac{5}{6}$. Use combinations since the order of outcomes is not important.

$$P(x = 0) = C(4,0)\left(\frac{1}{6}\right)^0\left(\frac{5}{6}\right)^4 = \frac{625}{1296}$$

$$P(x = 1) = C(4,1)\left(\frac{1}{6}\right)^1\left(\frac{5}{6}\right)^3 = \frac{125}{324}$$

$$P(x = 2) = C(4,2)\left(\frac{1}{6}\right)^2\left(\frac{5}{6}\right)^2 = \frac{25}{216}$$

$$P(x = 3) = C(4,3)\left(\frac{1}{6}\right)^3\left(\frac{5}{6}\right)^1 = \frac{5}{324}$$

$$P(x = 4) = C(4,4)\left(\frac{1}{6}\right)^4\left(\frac{5}{6}\right)^0 = \frac{1}{1296}$$

x	0	1	2	3	4
$P(x)$	$\frac{625}{1296}$	$\frac{125}{324}$	$\frac{25}{216}$	$\frac{5}{324}$	$\frac{1}{1296}$

(b) $$E(x) = 0\left(\frac{625}{1296}\right) + 1\left(\frac{125}{324}\right) + 2\left(\frac{25}{216}\right) + 3\left(\frac{5}{324}\right) + 4\left(\frac{1}{1296}\right) = \frac{2}{3}$$

23. Set up the probability distribution.

Number of Women	0	1	2
Probability	$\frac{C(3,0)\,C(5,2)}{C(8,2)}$	$\frac{C(3,1)\,C(5,1)}{C(8,2)}$	$\frac{C(3,2)\,C(5,0)}{C(8,2)}$
Simplified	$\frac{5}{14}$	$\frac{15}{28}$	$\frac{3}{28}$

$$E(x) = 0\left(\frac{5}{14}\right) + 1\left(\frac{15}{28}\right) + 2\left(\frac{3}{28}\right) = \frac{21}{28} = \frac{3}{4} = 0.75$$

25. Set up the probability distribution as in Exercise 20.

Number of Women	Probability	Simplified
0	$\frac{C(13,0)\,C(39,2)}{C(52,2)}$	$\frac{741}{1326}$
1	$\frac{C(13,1)\,C(39,1)}{C(52,2)}$	$\frac{507}{1326}$
2	$\frac{C(13,2)\,C(39,0)}{C(52,2)}$	$\frac{78}{1326}$

$$E(x) = 0\left(\frac{741}{1326}\right) + 1\left(\frac{507}{1326}\right) + 2\left(\frac{78}{1326}\right) = \frac{663}{1326} = \frac{1}{2}$$

29. **(a)** First list the possible sums, 5, 6, 7, 8, and 9, and find the probabilities for each. The total possible number of results are $4 \cdot 3 = 12$. There are two ways to draw a sum of 5 (2 then 3, and 3 then 2). The probability of 5 is $\frac{2}{12} = \frac{1}{6}$. There are two ways to draw a sum of 6 (2 then 4, and 4 then 2). The probability of 6 is $\frac{2}{12} = \frac{1}{6}$. There are four ways to draw a sum of 7 (2 then 5, 3 then 4, 4 then 3, and 5 then 2). The probability of 7 is $\frac{4}{12} = \frac{1}{3}$. There are two ways to draw a sum of 8 (3 then 5, and 5 then 3). The probability of 8 is $\frac{2}{12} = \frac{1}{6}$. There are two ways to draw a sum of 9 (4 then 5, and 5 then 4).The probability of 9 is $\frac{2}{12} = \frac{1}{6}$. The distribution is as follows.

Sum	5	6	7	8	9
Probability	$\frac{1}{6}$	$\frac{1}{6}$	$\frac{1}{3}$	$\frac{1}{6}$	$\frac{1}{6}$

(b)

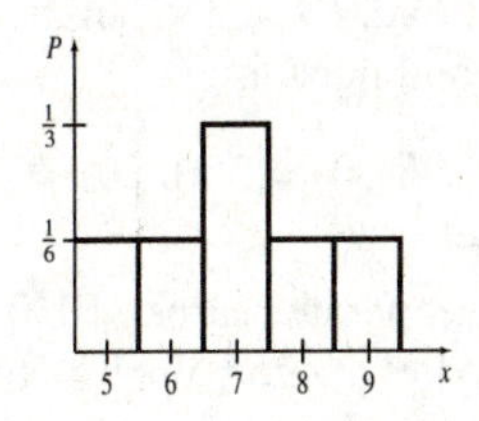

(c) The probability that the sum is even is $\frac{1}{6} + \frac{1}{6} = \frac{1}{3}$. Thus the odds are 1 to 2.

(d) $$E(x) = \frac{1}{6}(5) + \frac{1}{6}(6) + \frac{1}{3}(7) + \frac{1}{6}(8) + \frac{1}{6}(9) = 7$$

31. We first compute the amount of money the company can expect to pay out for each kind of policy. The sum of these amounts will be the total amount the company can expect to pay out. For a single \$100,000 policy, we have the following probability distribution.

	Pay	Don't Pay
Outcome	\$100,000	\$100,000
Probability	0.0012	0.9998

$$E(\text{payoff}) = 100{,}000(0.0012) + 0(0.9998) = \$120$$

For all 100 such policies, the company can expect to pay out

$$100(120) = \$12{,}000.$$

For a single \$50,000 policy,

$$E(\text{payoff}) = 50{,}000(0.0012) + 0(0.9998) = \$60.$$

For all 500 such policies, the company can expect to pay out

$$500(60) = \$30,000.$$

Similarly, for all 1000 policies of \$10,000, the company can expect to pay out

$$1000(12) = \$12,000.$$

Thus, the total amount the company can expect to pay out is

$$\$12,000 + \$30,000 + \$12,000 = \$54,000.$$

33. **(a)** Expected number of good nuts in 50 "blow outs" is

$$E(x) = 50(0.60) = 30.$$

(b) Since 80% of the "good nuts" are good, 20% are bad. Expected number of bad nuts in 50 "good nuts" is

$$E(x) = 50(0.20) = 10.$$

35. The tour operator earns \$1050 if 1 or more tourists do not show up. The tour operator earns \$950 if all tourists show up. The probability that all tourists show up is $(0.98)^{21} \approx 0.6543$. The expected revenue is
$1050(0.3457) + 950(0.6543) = 984.57$

The answer is e.

37. **(a)** Expected cost of Amoxicillin:

$$E(x) = 0.75(\$59.30) + 0.25(\$96.15) = \$68.51$$

Expected cost of Cefaclor:

$$E(x) = 0.90(\$69.15) + 0.10(\$106.00) = \$72.84$$

(b) Amoxicillin should be used to minimize total expected cost.

39. $E(x) = 250(0.74) = 185$

We would expect 38 low-birth-weight babies to graduate from high school.

41. **(a)** Using binomial probability, $n = 48, x = 0,$ $p = 0.0976.$

$$P(0) = C(48,0)(0.0976)^0(0.9024)^{48} \approx 0.007230$$

(b) Using combinations, the probability is

$$\frac{C(74,48)}{C(82,48)} \approx 5.094 \times 10^{-4}.$$

(c) Using binomial probability, $n = 6, x = 5,$ $p = 0.1.$

$$P(0) = C(6,5)(0.1)^5(0.9)^1 + (0.1)^6$$
$$= 5.5 \times 10^{-5}$$

(d) Using binomial probability, $n = 6, p = 0.1.$

$P(\text{at least } 2)$

$$= 1 - C(6,0)(0.1)^0(0.9)^6 - C(6,1)(0.1)^1(0.9)^5$$
$$\approx 0.1143$$

43. **(a)** We define a success to be a cat sitting in the chair with Kimberly. For this situation, $n = 4$; $x = 0, 1, 2, 3,$ or 4; $p = 0.3$; and $1 - p = 0.7.$

Number of Cats	Probability
0	$C(4,0)(0.3)^0(0.7)^4 = 0.2401$
1	$C(4,1)(0.3)^1(0.7)^3 = 0.4116$
2	$C(4,2)(0.3)^2(0.7)^2 = 0.2646$
3	$C(4,3)(0.3)^3(0.7)^1 = 0.0756$
4	$C(4,4)(0.3)^4(0.7)^0 = 0.0081$

(b) Expected number of cats

$$= 0(0.2401) + 1(0.4116) + 2(0.2646)$$
$$+ 3(0.0756) + 4(0.0081)$$
$$= 1.2$$

(c) Expected number of cats

$$= np = 4(0.3) = 1.2$$

45. Below is the probability distribution of x, which stands for the person's payback.

x	\$398	\$78	−\$2
$P(x)$	$\frac{1}{500} = 0.002$	$\frac{3}{500} = 0.006$	$\frac{497}{500} = 0.994$

The expected value of the person's winnings is

$$E(x) = 398(0.002) + 78(0.006) + (-2)(0.994)$$
$$\approx -\$0.72 \text{ or } -72¢.$$

Since the expected value of the payback is not 0, this is not a fair game.

47. There are 13 possible outcomes for each suit. That would make $13^4 = 28{,}561$ total possible outcomes. In one case, you win \$5000 (minus the \$1 cost to play the game). In the other 28,560, cases, you lose your dollar.

$$E(x) = 4999\left(\frac{1}{28{,}561}\right) + (-1)\left(\frac{28{,}560}{28{,}561}\right) = -82¢$$

49. There are $18 + 20 = 38$ possible outcomes. In 18 cases you win a dollar and in 20 you lose a dollar; hence,

$$E(x) = 1\left(\frac{18}{38}\right) + (-1)\left(\frac{20}{38}\right) = -\frac{1}{19}, \text{ or about } -5.3¢.$$

51. You have one chance in a thousand of winning \$500 on a \$1 bet for a net return of \$499. In the 999 other outcomes, you lose your dollar.

$$E(x) = 499\left(\frac{1}{1000}\right) + (-1)\left(\frac{999}{1000}\right) = -\frac{500}{1000} = -50¢$$

53. Let x represent the payback. The probability distribution is as follows.

x	$P(x)$
100,000	$\frac{1}{2{,}000{,}000}$
40,000	$\frac{2}{2{,}000{,}000}$
10,000	$\frac{2}{2{,}000{,}000}$
0	$\frac{1{,}999{,}995}{2{,}000{,}000}$

The expected value is

$$E(x) = 100{,}000\left(\frac{1}{2{,}000{,}000}\right) + 40{,}000\left(\frac{2}{2{,}000{,}000}\right) + 10{,}000\left(\frac{2}{2{,}000{,}000}\right) + 0\left(\frac{1{,}999{,}995}{2{,}000{,}000}\right)$$
$$= 0.05 + 0.04 + 0.01 + 0$$
$$= \$0.10 = 10¢.$$

Since the expected payback is 10¢, if entering the context costs 100¢, then it would be worth it to enter. The expected net return is $-\$0.90$.

55. **(a)** The possible scores are 0, 2, 3, 4, 5, 6. Each score has a probability of $\frac{1}{6}$.

$$E(x) = 0\left(\frac{1}{6}\right) + 2\left(\frac{1}{6}\right) + 3\left(\frac{1}{6}\right) + 4\left(\frac{1}{6}\right) + 5\left(\frac{1}{6}\right) + 6\left(\frac{1}{6}\right) = \frac{1}{6}(20) = \frac{10}{3}$$

(b) The possible scores are

0 which has a probability of $\frac{11}{36}$,

4 which has a probability of $\frac{1}{36}$,

5 which has a probability of $\frac{2}{36}$,

6 which has a probability of $\frac{3}{36}$,

7 which has a probability of $\frac{4}{36}$,

8 which has a probability of $\frac{5}{36}$,

9 which has a probability of $\frac{4}{36}$,

10 which has a probability of $\frac{3}{36}$,

11 which has a probability of $\frac{2}{36}$,

12 which has a probability of $\frac{1}{36}$.

$$E(x) = 0\left(\frac{11}{36}\right) + 4\left(\frac{1}{36}\right) + 5\left(\frac{2}{36}\right) + 6\left(\frac{3}{36}\right) + 7\left(\frac{4}{36}\right) + 8\left(\frac{5}{36}\right) + 9\left(\frac{4}{36}\right) + 10\left(\frac{3}{36}\right) + 11\left(\frac{2}{36}\right) + 12\left(\frac{1}{36}\right)$$
$$= \frac{4}{36} + \frac{10}{36} + \frac{18}{36} + \frac{28}{36} + \frac{40}{36} + \frac{36}{36} + \frac{30}{36} + \frac{22}{36} + \frac{12}{36}$$
$$= \frac{200}{36} = \frac{50}{9}$$

(c) If a single die does not result in a score of zero, the possible scores are 2, 3, 4, 5, 6 with each of these having a probability of $\frac{1}{5}$.

$$E(x) = 2\left(\frac{1}{5}\right) + 3\left(\frac{1}{5}\right) + 4\left(\frac{1}{5}\right) + 5\left(\frac{1}{5}\right) + 6\left(\frac{1}{5}\right) = \frac{1}{5}(20) = 4$$

Thus, if a player rolls n dice the expected average score is

$$n \cdot E(x) = n \cdot 4 = 4n.$$

(d) If a player rolls n dice, a nonzero score will occur whenever each die rolls a number other than 1. For each die there are 5 possibilities so the possible scoring ways for n dice is 5^n. When rolling one die there are 6 possibilities so the possible outcomes for n dice is 6^n. The probability of rolling a scoring set of dice is $\frac{5^n}{6^n}$; thus the expected value of the player's score when rolling n dice is $E(x) = \frac{5^n(4n)}{6^n}$.

57. Let x represent the number of hits. Since $p = 0.342, 1 - p = 0.658$.

$$P(0) = C(4,0)(0.342)^0(0.658)^4 = 0.1875$$

$$P(1) = C(4,1)(0.342)^1(0.658)^3 = 0.3897$$

$$P(2) = C(4,2)(0.342)^2(0.658)^2 = 0.3038$$

$$P(3) = C(4,3)(0.342)^3(0.658)^1 = 0.1053$$

$$P(4) = C(4,4)(0.342)^4(0.658)^0 = 0.0137$$

The distribution is shown in the following table.

x	0	1	2	3	4
$P(x)$	0.1875	0.3897	0.3038	0.1053	0.0137

The expected number of hits is $np = (4)(0.342) = 1.368$.

Chapter 8 Review Exercises

1. True

2. True

3. True

4. True

5. False: The probability of at least two occurrences is the probability of two or more occurrences.

6. True

7. True

8. False: Binomial probability applies to trials with exactly two outcomes.

9. True

10. False: For example, the random variable that assigns 0 to a head and 1 to a tail has expected value 1/2 for a fair coin.

11. True

12. False: The expected value of a fair game is 0.

13. 6 shuttle vans can line up at the airport in

$$P(6,6) = 6! = 720$$

different ways.

15. 3 oranges can be taken from a bag of 12 in

$$C(12,3) = \frac{12!}{9!3!} = \frac{12 \cdot 11 \cdot 10}{3 \cdot 2 \cdot 1} = 220$$

different ways.

17. **(a)** The sample will include 1 of the 2 rotten oranges and 2 of the 10 good oranges. Using the multiplication principle, this can be done in

$$C(2,1)C(10,2) = 2 \cdot 45 = 90 \text{ ways.}$$

(b) The sample will include both of the 2 rotten oranges and 1 of the 10 good oranges. This can be done in

$$C(2,2)C(10,1) = 1 \cdot 10 = 10 \text{ ways.}$$

(c) The sample will include 0 of the 2 rotten oranges and 3 of the 10 good oranges. This can be done in

$$C(2,0)C(10,3) = 1 \cdot 120 = 120 \text{ ways.}$$

(d) If the sample contains at most 2 rotten oranges, it must contain 0, 1, or 2 rotten oranges. Adding the results from parts (a), (b), and (c), this can be done in

$$90 + 10 + 120 = 220 \text{ ways.}$$

19. **(a)** $P(5,5) = 5! = 120$

(b) $P(4,4) = 4! = 24$

21. **(a)** The order within each list is not important. Use combinations and the multiplication principle. The choice of three items from column A can be made in $C(8, 3)$ ways, and the choice of two from column B can be made in $C(6, 2)$ ways. Thus, the number of possible dinners is

$$C(8,3)C(6,2) = 56 \cdot 15 = 840.$$

(b) There are

$$C(8,0) + C(8,1) + C(8,2) + C(8,3)$$

ways to pick up to 3 items from column A. Likewise, there are

$$C(6,0) + C(6,1) + C(6,2)$$

ways to pick up to 2 items from column B. We use the multiplication principle to obtain

$$\begin{aligned}&\left[C(8,0) + C(8,1) + C(8,2) + C(8,3)\right]\\&\qquad \cdot \left[C(6,0) + C(6,1) + C(6,2)\right]\\&= (1 + 8 + 28 + 56)(1 + 6 + 15)\\&= 93(22) = 2046.\end{aligned}$$

Since we are assuming that the diner will order at least one item, subtract 1 to exclude the dinner that would contain no items. Thus, the number of possible dinners is 2045.

25. There are $C(13, 3)$ ways to choose the 3 balls and $C(4, 3)$ ways to get all black balls. Thus,

$$P(\text{all black}) = \frac{C(4,3)}{C(13,3)} = \frac{4}{286} = \frac{2}{143} \approx 0.0140.$$

27. There are $C(4, 2)$ ways to get 2 black balls and $C(7, 1)$ ways to get 1 green ball. Thus,

$$P(\text{2 black and 1 green}) = \frac{C(4,2)C(7,1)}{C(11,3)} = \frac{(6.7)}{286} = \frac{42}{286} = \frac{21}{143} \approx 0.1469.$$

29. There are $C(2, 1)$ ways to get 1 blue ball and $C(11,2)$ ways to get 2 nonblue balls. Thus,

$$P(\text{exactly 1 blue}) = \frac{C(2,1)C(11,2)}{C(13,3)} = \frac{2 \cdot 55}{286} = \frac{110}{286} = \frac{5}{13} \approx 0.3846.$$

31. This is a Bernoulli trial problem with $P(\text{success}) = P(\text{girl}) = \frac{1}{2}$. Here, $n = 6, p = \frac{1}{2}$, and $x = 3$.

$$P(\text{exactly 3 girls}) = C(6,3)\left(\frac{1}{2}\right)^3\left(\frac{1}{2}\right)^3 = \frac{20}{64} = \frac{5}{16} \approx 0.313$$

33. $P(\text{at least 4 girls})$

$$\begin{aligned}&= P(\text{4 girls}) + P(\text{5 girls}) + P(\text{6 girls})\\&= C(6,4)\left(\frac{1}{2}\right)^4\left(\frac{1}{2}\right)^2 + C(6,5)\left(\frac{1}{2}\right)^5\left(\frac{1}{2}\right)^1\\&\qquad + C(6,6)\left(\frac{1}{2}\right)^6\left(\frac{1}{2}\right)^0\\&= \frac{22}{64} = \frac{11}{32} \approx 0.344\end{aligned}$$

35. $P(\text{both red})$

$$= \frac{C(26,2)}{C(52,2)} = \frac{325}{1326} = \frac{25}{102} \approx 0.245$$

37. $P(\text{at least 1 card is a spade})$

$$\begin{aligned}&= 1 - P(\text{neither is a spade})\\&= 1 - \frac{C(39,2)}{C(52,2)} = 1 - \frac{741}{1326}\\&= \frac{585}{1326} = \frac{15}{34} \approx 0.441\end{aligned}$$

39. There are 12 face cards and 40 nonface cards in an ordinary deck.

$$\begin{aligned}&P(\text{at least 1 face card})\\&= P(\text{1 face card}) + P(\text{2 face cards})\\&= \frac{C(12,1)C(40,1)}{C(52,2)} + \frac{C(12,2)}{C(52,2)}\\&= \frac{480}{1326} + \frac{66}{1326}\\&= \frac{546}{1326} \approx 0.4118\end{aligned}$$

41. This is a Bernoulli trial problem.

(a) $P(\text{success}) = P(\text{head}) = \frac{1}{2}$. Hence, $n = 3$ and $p = \frac{1}{2}$.

Number of Heads	Probability
0	$C(3,0)\left(\frac{1}{2}\right)^0\left(\frac{1}{2}\right)^3 = 0.125$
1	$C(3,1)\left(\frac{1}{2}\right)^1\left(\frac{1}{2}\right)^2 = 0.375$
2	$C(3,2)\left(\frac{1}{2}\right)^2\left(\frac{1}{2}\right)^1 = 0.375$
3	$C(3,3)\left(\frac{1}{2}\right)^3\left(\frac{1}{2}\right)^0 = 0.125$

(b)

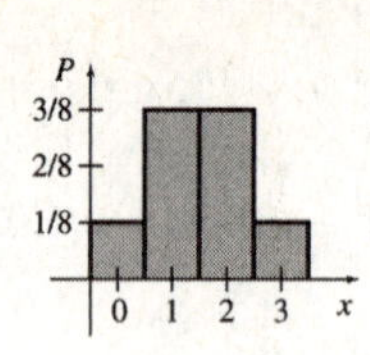

(c) $E(x) = 0(0.125) + 1(0.375) + 2(0.375) + 3(0.125)$

$= 1.5$

43. The probability that corresponds to the shaded region of the histogram is the total of the shaded areas, that is,

$$1(0.3) + 1(0.2) + 1(0.1) = 0.6.$$

45. The probability of rolling a 6 is $\frac{1}{6}$, and your net winnings would be \$2. The probability of rolling a 5 is $\frac{1}{6}$, and your net winnings would be \$1. The probability of rolling something else is $\frac{4}{6}$, and your net winnings would be $-\$2$. Let x represent your winnings. The expected value is

$$E(x) = 2\left(\frac{1}{6}\right) + 1\left(\frac{1}{6}\right) + (-2)\left(\frac{4}{6}\right) = -\frac{5}{6} \approx -\$0.833 \text{ or } -83.3¢.$$

This is not a fair game since the expected value is not 0.

47. (a)

Number of Aces	Probability
0	$\frac{C(4,0)C(48,3)}{C(52,3)} = \frac{17{,}296}{22{,}100}$
1	$\frac{C(4,1)C(48,2)}{C(52,3)} = \frac{4512}{22{,}100}$
2	$\frac{C(4,2)C(48,1)}{C(52,3)} = \frac{288}{22{,}100}$
3	$\frac{C(4,3)C(48,0)}{C(52,3)} = \frac{4}{22{,}100}$

$$E(x) = 0\left(\frac{17{,}296}{22{,}100}\right) + 1\left(\frac{4512}{22{,}100}\right) + 2\left(\frac{288}{22{,}100}\right) + 3\left(\frac{4}{22{,}100}\right)$$

$$= \frac{5100}{22{,}100} = \frac{51}{221} = \frac{3}{13} \approx 0.231$$

(b)

Number of Clubs	Probability
0	$\frac{C(13,0)C(39,3)}{C(52,3)} = \frac{9139}{22{,}100}$
1	$\frac{C(13,1)C(39,2)}{C(52,3)} = \frac{9633}{22{,}100}$
2	$\frac{C(13,2)C(39,1)}{C(52,3)} = \frac{3042}{22{,}100}$
3	$\frac{C(13,3)C(39,0)}{C(52,3)} = \frac{286}{22{,}100}$

$$E(x) = 0\left(\frac{9139}{22{,}100}\right) + 1\left(\frac{9633}{22{,}100}\right) + 2\left(\frac{3042}{22{,}100}\right) + 3\left(\frac{286}{22{,}100}\right)$$

$$= \frac{16{,}575}{22{,}100} = \frac{3}{4} = 0.75$$

49. We define a success to be the event that a student flips heads and is on the committee. In this situation, $n = 6$; $x = 1, 2, 3, 4$, or 5; $p = \frac{1}{2}$; and $1 - p = \frac{1}{2}$.

$$P(x = 1,2,3,4, \text{ or } 5)$$
$$= 1 - P(x = 6) - P(x = 0)$$
$$= 1 - C(6,6)\left(\frac{1}{2}\right)^6\left(\frac{1}{2}\right)^0 - C(6,0)\left(\frac{1}{2}\right)^0\left(\frac{1}{2}\right)^6$$
$$= 1 - \frac{1}{64} - \frac{1}{64} = \frac{62}{64} = \frac{31}{32}$$

51. (a) Given a set with n elements, the number of subsets of size

0 is $C(n,0) = 1$,

1 is $C(n,1) = n$,

2 is $C(n,2) = \frac{n(n-1)}{2}$, and

n is $C(n,n) = 1$.

(b) The total number of subsets is

$C(n, 0) + C(n, 1) + C(n, 2) + \cdots + C(n, n).$

(d) Let $n = 4$.

$$C(4,0) + C(4,1) + C(4,2) + C(4,3) + C(4,4)$$
$$= 1 + 4 + 6 + 4 + 1 = 16 = 2^4 = 2^n$$

Let $n = 5$.

$$C(5,0) + C(5,1) + C(5,2) + C(5,3) + C(5,4) + C(5,5)$$
$$= 1 + 5 + 10 + 10 + 5 + 1 = 32 = 2^5 = 2^n$$

(e) The sum of the elements in row n of Pascal's triangle is 2^n.

53. Use the multiplication principle.

$$3 \cdot 8 \cdot 2 = 48$$

55. $n = 12, x = 0, p = \frac{1}{6}$

$$P(0) = C(12,0)\left(\frac{1}{6}\right)^0\left(\frac{5}{6}\right)^{12} \approx 0.1122$$

57. $n = 12, x = 10, p = \frac{1}{6}$

$$P(10) = C(12,10)\left(\frac{1}{6}\right)^{10}\left(\frac{5}{6}\right)^2$$
$$\approx 7.580 \times 10^{-7}$$

59. $n = 12, p = \frac{1}{6}$

$$\begin{aligned} P(\text{at least 2}) &= 1 - P(\text{at most 1}) \\ &= 1 - P(0) - P(1) \\ &= 1 - C(12,0)\left(\frac{1}{6}\right)^0\left(\frac{5}{6}\right)^{12} \\ &\quad - C(12,1)\left(\frac{1}{6}\right)^1\left(\frac{5}{6}\right)^{11} \\ &\approx 0.6187 \end{aligned}$$

61. The expected value is $\frac{1}{6}(12) = 2$.

63. Observe that for $a + b = 7$,

$$P(a)P(b) = \left(\frac{1}{2^{a+1}}\right)\left(\frac{1}{2^{b+1}}\right) = \frac{1}{2^{a+b+2}} = \frac{1}{2^9}.$$

The probability that exactly seven claims will be received during a given two-week period is

$$P(0)P(7) + P(1)P(6) + P(2)P(5) + P(3)P(4) + P(4)P(3) + P(5)P(2) + P(6)P(1) + P(7)P(0)$$
$$= 8\left(\frac{1}{2^9}\right) = \frac{1}{64}.$$

The answer is d.

65. Denote by S the event that a product is successful.

Denote by U the event that a product is unsuccessful.

Denote by Q the event of passing quality control.

We must calculate the conditional probabilities $P(S|Q)$ and $P(U|Q)$ using Bayes' Theorem in order to calculate the expected net profit (in millions).

$$E = 40P(S|Q) - 15P(U|Q).$$
$$P(S) = P(U) = 0.5$$
$$P(Q|S) = 0.8, P(Q|U) = 0.25$$

$$\begin{aligned} P(S|Q) &= \frac{P(S) \cdot P(Q|S)}{P(S) \cdot P(Q|S) + P(U) \cdot P(Q|U)} \\ &= \frac{0.5(0.8)}{0.5(0.8) + 0.5(0.25)} \\ &= \frac{0.4}{0.4 + 0.125} = 0.762 \end{aligned}$$

$$\begin{aligned} P(U|Q) &= \frac{P(U) \cdot P(Q|U)}{P(U) \cdot P(Q|U) + P(S) \cdot P(Q|S)} \\ &= \frac{0.125}{0.525} = 0.238 \end{aligned}$$

Therefore,

$$\begin{aligned} E &= 40P(S|Q) - 15P(U|Q) \\ &= 40(0.762) - 15(0.238) \\ &\approx 27. \end{aligned}$$

So the expected net profit is \$27 million, or the correct answer is e.

67. Let $I(x)$ represent the airline's net income if x people show up.

$$\begin{aligned} I(0) &= 0 \\ I(1) &= 400 \\ I(2) &= 2(400) = 800 \\ I(3) &= 3(400) = 1200 \\ I(4) &= 3(400) - 400 = 800 \\ I(5) &= 3(400) - 2(400) = 400 \\ I(6) &= 3(400) - 3(400) = 0 \end{aligned}$$

Let $P(x)$ represent the probability that x people will show up. Use the binomial probability formula to find the values of $P(x)$.

$$P(0) = C(6,0)(0.6)^0(0.4)^6 = 0.0041$$
$$P(1) = C(6,1)(0.6)^1(0.4)^5 = 0.0369$$
$$P(2) = C(6,2)(0.6)^2(0.4)^4 = 0.1382$$
$$P(3) = C(6,3)(0.6)^3(0.4)^3 = 0.2765$$
$$P(4) = C(6,4)(0.6)^4(0.4)^2 = 0.3110$$
$$P(5) = C(6,5)(0.6)^5(0.4)^1 = 0.1866$$
$$P(6) = C(6,6)(0.6)^6(0.4)^0 = 0.0467$$

(a) $E(I) = 0(0.0041) + 400(0.0369) + 800(0.1382) + 1200(0.2765) + 800(0.3110) + 400(0.1866) + 0(0.0467) = \$780.56$

(b) $n = 3$

x	0	1	2	3
Income	0	100	200	300
$P(x)$	0.064	0.288	0.432	0.216

$E(I) = 0(0.064) + 400(0.288) + 800(0.432) + 1200(0.216) = \$720$

On the basis of all the calculations, the table given in the exercise is completed as follows.

x	Income	$P(x)$
0	0	0.004
1	400	0.037
2	800	0.038
3	1200	0.276
4	800	0.311
5	400	0.187
6	0	0.047

$n = 4$

x	1	1	2	3	4
Income	0	400	800	1200	800
$P(x)$	0.0256	0.1536	0.3456	0.3456	0.1296

$E(I) = 0(0.0256) + 400(0.1536) + 800(0.3456) + 1200(0.3456) + 800(0.1296) = \$856.32$

$n = 5$

x	Income	$P(x)$
0	0	0.01024
1	400	0.0768
2	800	0.2304
3	1200	0.3456
4	800	0.2592
5	400	0.07776

$E(I) = 0(0.01024) + 400(0.0768) + 800(0.2304) + 1200(0.3456) + 800(0.2596) + 400(0.07776) = \$868.22$

Since $E(I)$ is greatest when $n = 5$, the airlines should book 5 reservations to maximize revenue.

69. $C(40,5)\left(\frac{1}{8}\right)^5\left(\frac{7}{8}\right)^{35} \approx 0.1875$

71. This is a set of binomial trials with $n = 5, p = 0.48$, and $1 - p = 0.52$.

(a) $P(0 \text{ women}) = C(5,0)(0.48)^0(0.52)^5 \approx 0.0380$

$P(1 \text{ women}) = C(5,1)(0.48)^1(0.52)^4 \approx 0.1755$

$P(2 \text{ women}) = C(5,2)(0.48)^2(0.52)^3 \approx 0.3240$

$P(3 \text{ women}) = C(5,3)(0.48)^3(0.52)^2 \approx 0.2990$

$P(4 \text{ women}) = C(5,4)(0.48)^4(0.52)^1 \approx 0.1380$

$P(5 \text{ women}) = C(5,5)(0.48)^5(0.52)^0 \approx 0.0255$

The distribution is shown in the following table.

Number of women	Probability
0	0.0380
1	0.1755
2	0.3240
3	0.2990
4	0.1380
5	0.0255

(b)

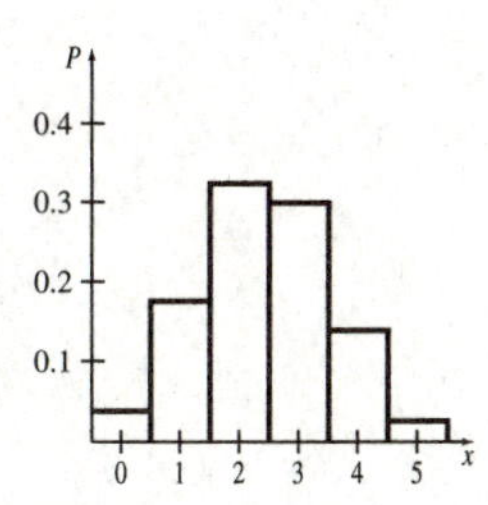

(c) Expected number of women $= np = 5(0.48) = 2.4.$

73. (a)

Number Who Did Not Do Homework	Probability
0	$\frac{C(3,0)C(7,5)}{C(10,5)} = \frac{21}{252} = \frac{1}{12}$
1	$\frac{C(3,1)C(7,4)}{C(10,5)} = \frac{105}{252} = \frac{5}{12}$
2	$\frac{C(3,2)C(7,3)}{C(10,5)} = \frac{105}{252} = \frac{5}{12}$
3	$\frac{C(3,3)C(7,2)}{C(10,5)} = \frac{21}{252} = \frac{1}{12}$

(b) Draw a histogram with four rectangles.

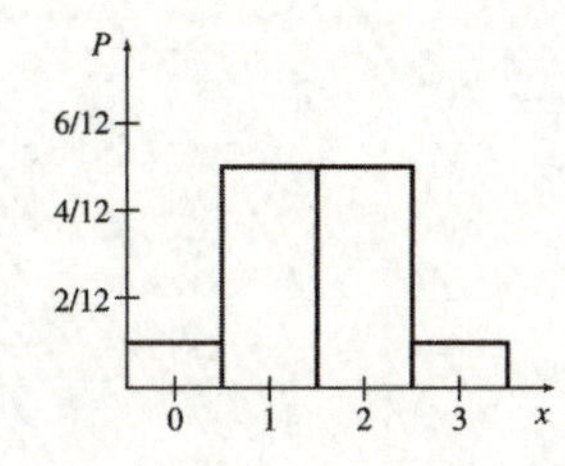

(c) Expected number who did not do homework

$$= 0\left(\frac{1}{12}\right) + 1\left(\frac{5}{12}\right) + 2\left(\frac{5}{12}\right) + 3\left(\frac{1}{12}\right)$$

$$= \frac{18}{12} = \frac{3}{2}$$

75. It costs $2(0.44 + 0.04) = 0.96$ to play the game.

x	\$1999.18	–\$0.96
$P(x)$	$\frac{1}{8000}$	$\frac{7999}{8000}$

$$E(x) = \$1999.18\left(\frac{1}{8000}\right) - \$0.96\left(\frac{7999}{8000}\right)$$

$$= -\$0.71$$

77. (a) The probability of the outcome 000 is 0.001, so the expected number of occurrences of this outcome in 30 years of play is $(30)(365)(0.001) = 10.95.$

(b) In 7 years of play the expected number of wins for 000 is $(7)(365)(0.001) \approx 2.56.$

79. (a) (i) When 5 socks are selected, we could get 1 matching pair and 3 odd socks or 2 matching pairs and 1 odd sock.

First consider 1 matching pair and 3 odd socks. The number of ways this could be done is

$$C(10,1)\left[C(18,3) - C(9,1)C(16,1)\right] = 6720.$$

$C(10,1)$ gives the number of ways for 1 pair, while $\left[C(18,3) - C(9,1)C(16,1)\right]$ gives the number of ways for choosing the remaining 3 socks from the 18 socks left. We must subtract the number of ways the last 3 socks could contain a pair from the 9 pairs remaining.

Next consider 2 matching pairs and 1 odd sock. The number of ways this could be done is

$$C(10,2)C(16,1) = 720.$$

$C(10,2)$ gives the number of ways for choosing 2 pairs, while $C(16,1)$ gives the number of ways for choosing the 1 odd sock.

The total number of ways is

$$6720 + 720 = 7440.$$

Then

$$P(\text{matching pair}) = \frac{7440}{C(20,5)} \approx 0.4799.$$

(ii) When 6 socks are selected, we could get 3 matching pairs and no odd socks or 2 matching pairs and 2 odd socks or 1 matching pair and 4 odd socks. The number of ways of obtaining 3 matching pairs is $C(10, 3) = 120$. The number of ways of obtaining 2 matching pairs and 2 odd socks is

$$C(10,2)\left[C(16,2) - C(8,1)\right] = 5040.$$

The 2 odd socks must come from the 16 socks remaining but cannot be one of the 8 remaining pairs.

The number of ways of obtaining 1 matching pair and 4 odd socks is

$$C(10,1)\left[C(18,4) - C(9,2) - C(9\,1)[C(16,2) - 8]\right]$$
$$= 20{,}160.$$

The 4 odd socks must come from the 18 socks remaining but cannot be 2 pairs and cannot be 1 pair and 2 odd socks.

The total number of ways is

$120 + 5040 + 20{,}160 - 25{,}320.$

Thus,

$$P(\text{matching pair}) = \frac{25{,}320}{C(20,6)} \approx 0.6533.$$

(c) Suppose 6 socks are lost at random. The worst case is they are 6 odd socks. The best case is they are 3 matching pairs.

First find the number of ways of selecting 6 odd socks. This is

$$C(10,6)C(2,1)C(2,1)C(2,1)C(2,1)C(2,1)C(2,1)$$
$$= 13{,}440.$$

The $C(10,6)$ gives the number of ways of choosing 6 different socks from the 10 pairs. But with each pair, $C(2,1)$ gives the number of ways of selecting 1 sock. Then

$$P(6\,\text{odd socks}) = \frac{13{,}440}{C(20,6)}$$
$$\approx 0.3467.$$

Next find the number of ways of selecting three matching pairs. This is $C(10,3) = 120$. Then

$$P(3\,\text{matching pairs}) = \frac{120}{C(20,6)}$$
$$\approx 0.003096.$$

Chapter 9

STATISTICS

9.1 Frequency Distributions; Measures of Central Tendency

Your Turn 1

$$\bar{x} = \frac{(12 + 17 + 21 + 25 + 27 + 38 + 49)}{7}$$

$$= \frac{189}{7} = 27$$

Your Turn 2

Interval	Midpoint, x	Frequency, f	Product, xf
0–6	3	2	6
7–13	10	4	40
14–20	17	7	119
21–27	24	10	240
28–34	31	3	93
35–41	38	1	38
Totals:		$n = 27$	536

$$\bar{x} = \frac{\sum xf}{n} = \frac{536}{27} \approx 19.85$$

Your Turn 3

The data are given in order: 12,17,21,25,27,38,49. The middle number is 25 so this is the median.

9.1 Exercises

1. (a)-(b) Since 0–24 is to be the first interval and there are 25 numbers between 0 and 24 inclusive, we will let all six intervals be of size 25. The other five intervals are 25–49, 50–74, 75–99, 100–124, and 125–149. Making a tally of how many data values lie in each interval leads to the following frequency distribution.

Interval	Frequency
0–24	4
25–49	8
50–74	5
75–99	10
100–124	4
125–149	5

(c) Draw the histogram. It consists of 6 bars of equal width having heights as determined by the frequency of each interval. See the histogram in part (d).

(d) To construct the frequency polygon, join consecutive midpoints of the tops of the histogram bars with line segments.

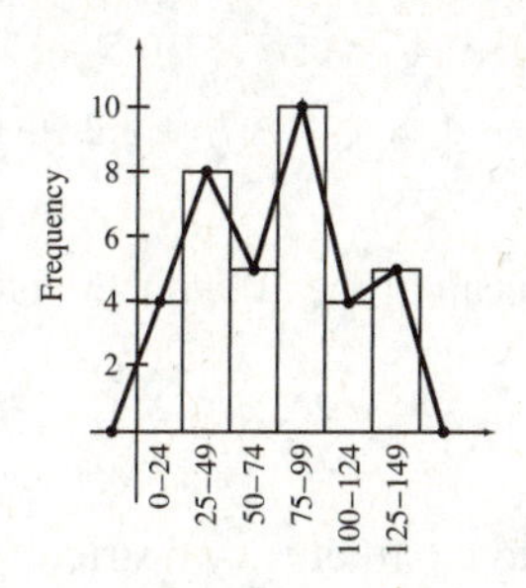

3. (a)-(b) There are eight intervals starting with 0–19. Making a tally of how many data values lie in each interval leads to the following frequency distribution.

Interval	Frequency
0–19	4
20–39	5
40–59	4
60–79	5
80–99	9
100–119	3
120–139	4
140–159	2

(c) Draw the histogram. It consists of 8 rectangles of equal width having heights as determined by the frequency of each interval. See the histogram in part (d).

(d) To construct the frequency polygon, join consecutive midpoints of the tops of the histogram bars with line segments.

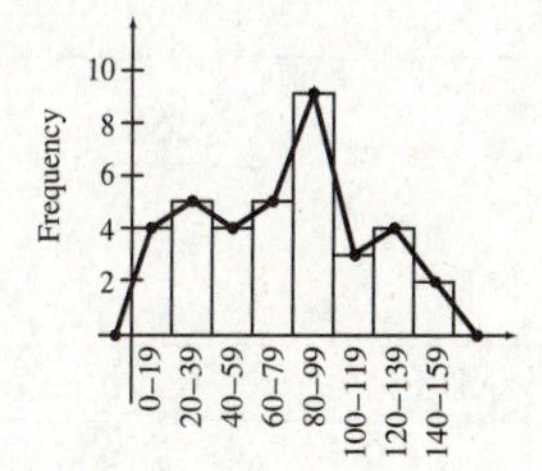

7. $\bar{x} = \frac{\sum x}{n}$

$$= \frac{8 + 10 + 16 + 21 + 25}{5}$$

$$= \frac{80}{5} = 16$$

9. $\sum x = 30{,}200 + 23{,}700 + 33{,}320$
$+ 29{,}410 + 24{,}600 + 27{,}750$
$+ 27{,}300 + 32{,}680$
$= 228{,}960$

The mean of the 8 numbers is

$$\overline{x} = \frac{\sum x}{n} = \frac{228{,}960}{8} = 28{,}620.$$

11. $\sum x = 9.4 + 11.3 + 10.5 + 7.4 + 9.1$
$+ 8.4 + 9.7 + 5.2 + 1.1 + 4.7$
$= 76.8$

The mean of the 10 numbers is

$$\overline{x} = \frac{\sum x}{n} = \frac{76.8}{10} = 7.68.$$

13. Add to the frequency distribution a new column, "Value × Frequency."

Value	Frequency	Value × Frequency
4	6	$4 \cdot 6 = 24$
6	1	$6 \cdot 1 = 6$
9	3	$9 \cdot 3 = 27$
15	2	$15 \cdot 2 = 30$
Totals:	12	87

The mean is

$$\overline{x} = \frac{\sum xf}{n} = \frac{87}{12} = 7.25.$$

15. **(a)**

Interval	Midpoint, x	Frequency, f	Product, xf
0–24	12	4	48
25–49	37	8	296
50–74	62	5	310
75–99	87	10	870
100–124	112	4	448
125–149	137	5	685
Totals:		$n = 36$	2657

$$\overline{x} = \frac{\sum xf}{n} = \frac{2657}{36} \approx 73.81$$

(b)

Interval	Midpoint, x	Frequency, f	Product, xf
0–19	9.5	4	38
20–39	29.5	5	147.5
40–59	49.5	4	198
60–79	69.5	5	347.5
80–99	89.5	9	805.5
100–119	109.5	3	328.5
120–139	129.5	4	518
140–159	149.5	2	299
Totals:		$n = 36$	2682

$$\overline{x} = \frac{\sum xf}{n} = \frac{2682}{36} = 74.5$$

17. 27, 35, 39, 42, 47, 51, 54

The numbers are already arranged in numerical order, from smallest to largest. The median is the middle number, 42.

19. 100, 114, 125, 135, 150, 172

The median is the mean of the two middle numbers, which is

$$\frac{125 + 135}{2} = \frac{260}{2} = 130.$$

21. Arrange the numbers in numerical order, from smallest to largest.

3.4, 9.1, 27.6, 28.4, 29.8, 32.1, 47.6, 59.8

There are eight numbers here; the median is the mean of the two middle numbers, which is

$$\frac{28.4 + 29.8}{2} = \frac{58.2}{2} = 29.1.$$

23. Using a graphing calculator, $\overline{x} \approx 73.861$ and the median is 80.5.

25. 4, 9, 8, 6, 9, 2, 1, 3

The mode is the number that occurs most often. Here, the mode is 9.

27. 55, 62, 62, 71, 62, 55, 73, 55, 71

The mode is the number that occurs most often. Here, there are two modes, 55 and 62, since they both appear three times.

29. 6.8, 6.3, 6.3, 6.9, 6.7, 6.4, 6.1, 6.0

The mode is 6.3.

33.

Interval	Midpoint, x	Frequency, f	Product, xf
0–24	12	4	48
25–49	37	8	296
50–74	62	5	310
75–99	87	10	870
100–124	112	4	448
125–149	137	5	685
Totals:		36	2657

The mean of this collection of grouped data is

$$\bar{x} = \frac{\sum xf}{n} = \frac{2657}{36} \approx 73.8.$$

The interval 75–99 contains the most data values, 10, so it is the modal class.

37. Find the mean of the numbers in the Production column.

$$\bar{x} = \frac{\sum x}{n} = \frac{20{,}959}{10} = 2095.9$$

The mean production of wheat is 2095.9 million bushels.

The middle two numbers in the Production column, when we list them in order, are 2103 and 2157. The average of these two numbers is the median.

$$\frac{2103 + 2157}{2} = \frac{4260}{2} = 2130$$

The median production of wheat is 2130 million bushels.

39. Find the mean for the grouped data. Note that the frequency is in thousands.

$$\bar{x} = \frac{\sum xf}{n} = \frac{664{,}840{,}000{,}000}{14{,}518{,}000} = 45{,}794$$

The estimated mean income for African American households in 2008 is $45,794.

41. **(a)** Find the mean of the numbers in the Complaints column.

$$\bar{x} = \frac{\sum x}{n} = \frac{3484}{10} = 348.4$$

The mean number of complaints is 348.4 complaints per airline.

The median is the mean of the two middle numbers.

$$\text{median} = \frac{350 + 149}{2} = 249.5$$

The median number of complaints is 249.5 complaints per airline.

(b) The averages found are not meaningful because not all airlines carry the same number of passengers.

(c) Find the mean of the numbers in the Complaints per 100,000 column.

$$\bar{x} = \frac{\sum x}{n} = \frac{11.87}{10} = 1.187$$

The mean number of complaints per 100,000 passengers boarding is 1.187.

The median is the mean of the two middle numbers once the values have been sorted.

$$\text{median} = \frac{0.87 + 1.56}{2} = 1.215$$

The median number of complaints per 100,000 passengers boarding is 1.215.

43. Find the mean.

$$\bar{x} = \frac{\sum x}{n} = \frac{16 + 12 + \ldots + 2}{13} = \frac{96}{13} \approx 7.38$$

The mean number of recognized blood types is 7.38.

Find the median.

The values are listed in order.

Since there are 13 values, the median is the seventh value, 7.

The median number of recognized blood types is 7.

The values 7, 5, and 4 each occur the greatest number of times, 2.

The modes are 7, 5, and 4.

45. **(a)** Find the mean of the numbers in the maximum temperature column.

$$\bar{x} = \frac{\sum x}{n} = \frac{666}{12} = 55.5$$

The mean of the maximum temperatures is 55.5°F. To find the median, list the 12 maximum temperatures from smallest to largest.

39, 39, 40, 44, 47, 50, 51, 60, 69, 70, 78, 79

The median is the mean of the two middle values.

$$\frac{50 + 51}{2} = 50.5°\text{F}$$

(b) Find the mean of the numbers in the minimum temperature column.

$$\bar{x} = \frac{\sum x}{n} = \frac{347}{12} \approx 28.9$$

The mean of the minimum temperatures is about 28.9°F.

To find the median, list the 12 minimum temperatures from smallest to largest.

16, 18, 20, 21, 24, 26, 31, 32, 37, 37, 42, 43

The median is the mean of the two middle values.

$$\frac{26 + 31}{2} = 28.5°\text{F}$$

47. **(a)** $\frac{5,700,000,000 \cdot 1 + 100 \cdot 80,000}{1 + 80,000}$

$= \frac{5,708,000,000}{80,001}$

$\approx 71,349$

The average worth of a citizen of Chukotka is $71,349.

49. **(a)** $\bar{x} = \frac{\Sigma x}{n} = \frac{206,333,389}{25} = 8,253,336$

The mean salary is $8,253,336.

Since the salaries are listed in order and there are 25 values, the median is the 13th value, which is $5,500,000.

The only value that occurs more than once is $5,500,000, which occurs twice. This is the mode.

(b) Most of the team earned below the mean salary, some well below. Either the mode or median makes a better description of the data.

9.2 Measures of Variation

Your Turn 1

The range is the difference of the largest and smallest values, or $35 - 7 = 28$.

For the variance, construct a table.

x	x^2
7	49
11	121
16	256
17	289
19	361
35	1225
Total: 105	2301

From the table we see that the mean is

$\bar{x} = \frac{105}{6} = 17.5.$

The variance is

$$s^2 = \frac{\Sigma x^2 - n\bar{x}^2}{n-1} = \frac{2301 - (6)(17.5)^2}{6-1} = 92.7$$

The standard deviation is $\sqrt{92.7} \approx 9.628$.

Your Turn 2

Using the grouped data we first find the mean:

$$\bar{x} = \frac{\Sigma xf}{n} = \frac{536}{27} = 19.852$$

Interval	x	x^2	f	fx^2
0–6	3	9	2	18
7–13	10	100	4	400
14–20	17	289	7	2023
21–27	24	576	10	5760
28–34	31	961	3	2883
35–41	38	1444	1	1444
Total:			27	12,528

The variance is

$$s^2 = \frac{\Sigma fx^2 - n\bar{x}^2}{n-1} = \frac{12,528 - (27)(19.852)^2}{27-1}$$
$$\approx 72.586$$

The standard deviation is $s = \sqrt{72.586} \approx 8.52$

9.2 Exercises

1. The standard deviation of a sample of numbers is the square root of the variance of the sample.

3. The range is the difference of the highest and lowest numbers in the list, or $85 - 52 = 33$.

To find the standard deviation, first find the mean.

$$\bar{x} = \frac{72 + 61 + 57 + 83 + 52 + 66 + 85}{7}$$
$$= \frac{476}{7} = 68$$

To prepare for calculating the standard deviation, construct a table.

x	x^2
72	5184
61	3721
57	3249
83	6889
52	2704
66	4356
85	7225
Total:	33,328

The variance is

$$s^2 = \frac{\sum x^2 - n\bar{x}^2}{n-1} = \frac{33{,}328 - 7(68)^2}{7-1} = \frac{33{,}328 - 32{,}368}{6} = 160$$

and the standard deviation is $s = \sqrt{160} \approx 12.6.$

5. The range is $287 - 241 = 46$. The mean is

$$\bar{x} = \frac{241 + 248 + 251 + 257 + 252 + 287}{6} = 256.$$

x	x^2
241	58,081
248	61,504
251	63,001
257	66,049
252	63,504
287	82,369
Total:	394,508

The standard deviation is

$$s = \sqrt{\frac{\sum x^2 - n\bar{x}^2}{n-1}} = \sqrt{\frac{394{,}508 - 6(256)^2}{5}} = \sqrt{258.4} \approx 16.1.$$

7. The range is $27 - 3 = 24$. The mean is

$$\bar{x} = \frac{\sum x}{n} = \frac{140}{10} = 14.$$

x	x^2
3	9
7	49
4	16
12	144
15	225
18	324
19	361
27	729
24	576
11	121
Total:	2554

The standard deviation is

$$s = \sqrt{\frac{\sum x^2 - n\bar{x}^2}{n-1}} = \sqrt{\frac{2554 - 10(14)^2}{9}} = \sqrt{66} \approx 8.1.$$

9. Using a graphing calculator, enter the 36 numbers into a list. Using the 1-Var Stats feature of a TI-84 Plus calculator, the standard deviation is found to be $Sx \approx 40.04793754$, or 40.05.

11. Expand the table to include columns for the midpoint x of each interval for xf, x^2, and fx^2.

Interval	f	x	xf	x^2	fx^2
30–39	4	12	48	144	576
40–49	8	37	296	1396	10,952
50–59	5	62	310	3844	19,220
60–69	10	87	870	7569	75,690
70–79	4	112	448	12,544	50,176
80–89	5	137	685	18,769	93,845
Total:	36		2657		250,459

The mean of the grouped data is

$$\bar{x} = \frac{\sum xf}{n} = \frac{2657}{36} \approx 73.8.$$

The standard deviation for the grouped data is

$$s = \sqrt{\frac{\sum fx^2 - n\bar{x}^2}{n-1}} = \sqrt{\frac{250{,}459 - 36(73.8)^2}{35}} \approx \sqrt{1554} \approx 39.4.$$

13. Use $k = 3$ in Chebyshev's theorem.

$$1 - \frac{1}{k^2} = 1 - \frac{1}{3^2} = \frac{8}{9},$$

So at least $\frac{8}{9}$ of the distribution is within 3 standard deviations of the mean.

15. Use $k = 5$ in Chebyshev's theorem.

$$1 - \frac{1}{k^2} = 1 - \frac{1}{5^2} = \frac{24}{25},$$

so at least $\frac{24}{25}$ of the distribution is within 5 standard deviations of the mean.

17. We have $36 = 60 - 3 \cdot 8 = \bar{x} - 3s$ and $84 = 60 + 3 \cdot 8 = \bar{x} + 3s$, so Chebyshev's theorem applies with $k = 3$. Hence, at least

$$1 - \frac{1}{k^2} = 1 - \frac{1}{9} = \frac{8}{9}$$

of the numbers lie between 36 and 84.

19. The answer here is the complement of the answer to Exercise 16. It was found there that at least 8/9 of the distribution of the numbers are between 36 and 84, so at most $1 - 8/9 = 1/9$ of the numbers are less than 36 or more than 84.

23.
$$\begin{aligned} s^2 &= \frac{\Sigma(x - \bar{x})^2}{n - 1} \\ &= \frac{\Sigma\left(x^2 - 2x\bar{x} + \bar{x}^2\right)}{n - 1} \\ &= \frac{\Sigma x^2 - 2\bar{x}\,\Sigma x + n\bar{x}^2}{n - 1} \\ &= \frac{\Sigma x^2 - 2\bar{x}\,(n\bar{x}) + n\bar{x}^2}{n - 1} \\ &= \frac{\Sigma x^2 - n\bar{x}^2}{n - 1} \end{aligned}$$

25. 15, 18, 19, 23, 25, 25, 28, 30, 34, 38

(a)
$$\begin{aligned} \bar{x} &= \frac{1}{10}(15 + 18 + 19 + 23 + 25 \\ &\quad + 25 + 28 + 30 + 34 + 38) \\ &= \frac{1}{10}(255) = 25.5 \end{aligned}$$

The mean life of the sample of Brand X batteries is 25.5 hr.

x	x^2
15	225
18	324
19	361
23	529
25	625
25	625
28	784
30	900
34	1156
38	1444
Total:	6973

$$\begin{aligned} s &= \sqrt{\frac{\Sigma x^2 - n\bar{x}^2}{n - 1}} \\ &= \sqrt{\frac{6973 - 10(25.5)^2}{9}} \\ &\approx \sqrt{52.28} \approx 7.2 \end{aligned}$$

The standard deviation of the Brand X lives is 7.2 hr.

(b) Forever Power has a smaller standard deviation (4.1 hr, as opposed to 7.2 hr for Brand X), which indicates a more uniform life.

(c) Forever Power has a higher mean (26.2 hr, as opposed to 25.5 hr for Brand X), which indicates a longer average life.

27.

Sample Number	(a) $\bar{x}$	(b) s
1	$\frac{1}{3}$	2.1
2	2	2.6
3	$-\frac{1}{3}$	1.5
4	0	2.6
5	$\frac{5}{3}$	2.5
6	$\frac{7}{3}$	0.6
7	1	1.0
8	$\frac{4}{3}$	2.1
9	$\frac{7}{3}$	0.6
10	$\frac{2}{3}$	1.2

(c) $\bar{X} = \frac{\Sigma\bar{x}}{n} \approx \frac{11.3}{10} = 1.13$

(d) $\bar{s} = \frac{\Sigma s}{n} = \frac{16.8}{10} = 1.68$

(e) The upper control limit for the sample means is

$$\begin{aligned} \bar{X} + k_1\bar{s} &= 1.13 + 1.954(1.68) \\ &\approx 4.41. \end{aligned}$$

The lower control limit for the sample means is

$$\begin{aligned} \bar{X} - k_1\bar{s} &= 1.13 - 1.954(1.68) \\ &\approx -2.15. \end{aligned}$$

(f) The upper control limit for the sample standard deviations is

$$k_2\bar{s} = 2.568(1.68) \approx 4.31.$$

The lower control limit for the sample standard deviations is

$$k_3\bar{s} = 0(1.68) = 0.$$

29. This exercise should be solved using a calculator with a standard deviation key. The answers are $\bar{x} = 1.8158$ mm and $s = 0.4451$ mm.

31. **(a)** This exercise should be solved using a calculator with a standard deviation key. The answers are $\bar{x} = 7.3571$ and $s = 0.1326$.

(b) $\bar{x} + 2s = 7.3571 + 2(0.1326) = 7.6223$

$\bar{x} - 2s = 7.3571 - 2(0.1326) = 7.0919$

All the data, or 100%, are within these two values, that is, within 2 standard deviations of the mean.

33. **(a)** Find the mean.

$$\bar{x} = \frac{\sum x}{n} = \frac{84 + 91 + \cdots + 164}{7}$$

$$= \frac{894}{7} = 127.71$$

The mean is 127.71 days.

Find the standard deviation with a graphing calculator or spreadsheet.

$$s = 30.16$$

The standard deviation is 30.16 days.

(b) $\bar{x} + 2s = 127.71 + 2(30.16) = 188.03$

$\bar{x} - 2s = 127.71 - 2(30.16) = 67.39$

All seven of these cancers have doubling times that are within two standard deviations of the mean.

35. **(a)** Using a graphing calculator, the standard deviation is approximately $9,267,188. In Section 9.2 we found the mean to be $8,253,336.

(b) $\bar{x} - 2s \approx -7,239,488$

$\bar{x} + 2s \approx 25,773,864$

Note that the lower limit here is effectively 0. Only the highest-paid player, Alex Rodriguez, has a salary more than 2 standard deviations from the mean. This is 1/25 or 4% of the team.

9.3 The Normal Distribution

Your Turn 1

(a) Using the T1-84 Plus graphing calculator, type `normal cdf (-1E99,-0.76,0,1).` Pressing ENTER gives an answer of 0.2236.

(b) Enter `1-normal cdf (-1E99,-1.36,0,1)` to get the answer, 0.9131.

(c) Enter `normal cdf (-1E99,1.33,0,1) - normal cdf (-1E99,-1.22,0,1)`. The answer is 0.7970.

Your Turn 2

(a) Using the Tl-84 Plus, `type invNorm (.025,0,1)`. Then pressing ENTER gives a z-score of –1.96.

(b) If 20.9% of the area is to the right, 79.1% is to the left. Enter `invNorm (.791,0,1)` to get a z-score of approximately 0.81.

Your Turn 3

$$z = \frac{x - \mu}{\sigma} = \frac{20 - 35}{20} = -0.75$$

Your Turn 4

For Example 4, $\mu = 1200$ and $\sigma = 150$. The calculator solution looks like this:

`normal cdf (-1EE99,1425,1200,150)-`
`normal cdf (-1EE99,1275,1200,150).`

Pressing ENTER now gives an answer of 0.2417. To solve the problem using the normal curve table, first convert the given mileage values to z-scores:

$$z = \frac{x - \mu}{\sigma} = \frac{1275 - 1200}{150} = 0.5$$

$$z = \frac{x - \mu}{\sigma} = \frac{1425 - 1200}{150} = 1.5$$

Now subtract the area to the left of 0.5 from the area to the left of 1.5: $0.9332 - 0.6915 = 0.2417$

9.3 Exercises

1. The peak in a normal curve occurs directly above *the mean.*

3. For normal distributions where $\mu \neq 0$ or $\sigma \neq 1$, z-scores are found by using the formula

$$z = \frac{x - \mu}{\sigma}.$$

5. Use the table, "Area Under a Normal Curve to the Left of z", in the Appendix. To find the percent of the area under a normal curve between the mean and 1.70 standard deviations from the mean, subtract the table entry for $z = 0$ (representing the mean) from the table entry for $z = 1.7$.

$$0.9554 - 0.5000 = 0.4554$$

Therefore, 45.54% of the area lies between μ and $\mu + 1.7\sigma$.

7. Subtract the table entry for $z = -2.31$ from the table entry for $z = 0$.

$$0.5000 - 0.0104 = 0.4896$$

48.96% of the area lies between μ and $\mu - 2.31\sigma$.

9. $P(0.32 \le z \le 3.18)$
$= P(z \le 3.18) - P(z \le 0.32)$
$=$ (area to the left of3.18)
$-$ (area to the left of 0.32)
$= 0.9993 - 0.6255$
$= 0.3738$ or 37.38%

11. $P(-1.83 \le z \le -0.91)$
$= P(z \le -0.91) - P(z \le -1.83)$
$= 0.1814 - 0.0336$
$= 0.1478$ or 14.78%

13. $P(-2.95 \le z \le 2.03)$
$= P(z \le 2.03) - P(z \le -2.95)$
$= 0.9788 - 0.0016$
$= 0.9772$ or 97.72%

15. 5% of the total area is to the left of z.

Use the table backwards. Look in the body of the table for an area of 0.05, and find the corresponding z using the left column and top column of the table.

The closest values to 0.05 in the body of the table are 0.0505, which corresponds to $z = -1.64$, and 0.0495, which corresponds to $z = -1.65$.

17. 10% of the total area is to the right of z.

If 10% of the area is to the right of z, then 90% of the area is to the left of z. The closest value to 0.90 in the body of the table is 0.8997, which corresponds to $z = 1.28$.

19. For any normal distribution, the value of $P(x \le \mu)$ is 0.5 since half of the distribution is less than the mean. Similarly, $P(x \ge \mu)$ is 0.5 since half of the distribution is greater than the mean.

21. According to Chebyshev's theorem, the probability that a number will lie within 3 standard deviations of the mean of a probability distribution is at least

$$1 - \frac{1}{3^2} = 1 - \frac{1}{9} = \frac{8}{9} \approx 0.8889.$$

Using the normal distribution, the probability that a number will lie within 3 standard deviations of the mean is 0.9974.

These values are not contradictory since "at least 0.8889" means 0.8889 or more. For the normal distribution, the value is more.

In Exercises 23–27, let x represent the life of a light bulb.

$$\mu = 500, \sigma = 100$$

23. Less than 500 hr

$$z = \frac{x - \mu}{\sigma} = \frac{500 - 500}{100} = 0, \text{ so}$$

$P(x < 500) = P(z < 0)$
$=$ area to the left of $z = 0$
$= 0.5000.$

Hence, $0.5000(10{,}000) = 5000$ bulbs can be expected to last less than 500 hr.

25. Between 350 and 550 hr

For $x = 350$,

$$z = \frac{350 - 500}{100} = -1.5,$$

and for $x = 550$,

$$z = \frac{550 - 500}{100} = 0.5.$$

Then

$P(350 < x < 550)$
$= P(-1.5 < z < 0.5)$
$=$ area between $z = -1.5$
and $z = 0.5$
$= 0.6915 - 0.0668 = 0.6247.$

Hence, $0.6247(10{,}000) = 6247$ bulbs should last between 350 and 550 hr.

27. More than 440 hr

For $x = 440$,

$$z = \frac{440 - 500}{100} = -0.6.$$

Then

$$\begin{aligned} P(x > 440) &= P(z > -0.6) \\ &= \text{area to the right of } z = -0.6 \\ &= 1 - 0.2743 \\ &= 0.7257. \end{aligned}$$

Hence, $0.7257(10,000) = 7257$ bulbs should last more than 440 hr.

29. Here, $\mu = 16.5, \sigma = 0.5$.

For $x = 16$,

$$z = \frac{16 - 16.5}{0.5} = -1.$$

$$P(x < 16) = P(z < -1) = 0.1587$$

The fraction of the boxes that are underweight is 0.1587.

31. Here, $\mu = 16.5, \sigma = 0.2$.

For $x = 16$,

$$z = \frac{16 - 16.5}{0.2} = -2.5.$$

$$P(x < 16) = P(z < -2.5) = 0.0062$$

The fraction of the boxes that are underweight is 0.0062.

In Exercises 33–37, let x represent the weight of a chicken.

33. More than 1700 g means $x > 1700$.

For $x = 1700$,

$$z = \frac{1700 - 1850}{150} = -1.0.$$

$$\begin{aligned} P(x > 1700) &= 1 - P(x \le 1700) \\ &= 1 - P(z \le -1.0) \\ &= 1 - 0.1587 \\ &= 0.8413 \end{aligned}$$

Thus, 84.13% of the chickens will weigh more than 1700 g.

35. Between 1750 and 1900 g means $1750 \le x \le 1900$.

For $x = 1750$,

$$z = \frac{1750 - 1850}{150} = -0.67.$$

For $x = 1900$,

$$z = \frac{1900 - 1850}{150} = 0.33.$$

$$\begin{aligned} &P(1750 \le x \le 1900) \\ &= P(-0.67 \le z \le 0.33) \\ &= P(z \le 0.33) - P(z \le -0.67) \\ &= 0.6293 - 0.2514 \\ &= 0.3779 \end{aligned}$$

Thus, 37.79% of the chickens will weigh between 1750 and 1900 g.

37. More than 2100 g or less than 1550 g

$$\begin{aligned} &P(x < 1550 \text{ or } x > 2100) \\ &= 1 - P(1550 \le x \le 2100). \end{aligned}$$

For $x = 1550$,

$$z = \frac{1550 - 1850}{150} = -2.00.$$

For $x = 2100$,

$$z = \frac{2100 - 1850}{150} = 1.67.$$

$$\begin{aligned} &P(x < 1550 \text{ or } x > 2100) \\ &= P(z \le -2.00) + [1 - P(z \le 1.67)] \\ &= 0.0228 + (1 - 0.9525) \\ &= 0.0228 + 0.0475 \\ &= 0.0703 \end{aligned}$$

Thus, 7.03% of the chickens will weigh more than 2100 g or less than 1550 g.

39. **(a)** In a normal distribution, 68.26% of the distribution is within one standard deviation of the mean. This leaves 31.74% for the two tails that are more than one standard deviation from the mean. Since the distribution is symmetrical, half of this tail area is on the left, so $\frac{0.3174}{2} = 0.1587$ or 15.87% of the population is more than one standard deviation below the mean.

(b) For a mean of \$169 million and a standard deviation of \$300, values more than one standard deviation below the mean would all be negative, so 0% of colleges and universities had expenditures in this range.

41. Let x represent the number of ounces of milk in a carton.

$$\mu = 32.2, \sigma = 1.2$$

Find the z-score for $x = 32$.

$$z = \frac{x - \mu}{\sigma} = \frac{32 - 32.2}{1.2} \approx -0.17$$

$$\begin{aligned} P(x < 32) &= P(z < -0.17) \\ &= \text{area to the left of } z = -0.17 \\ &= 0.4325 \end{aligned}$$

The probability that a filled carton will contain less than 32 oz is 0.4325.

43. Let x represent the weight of an egg (in ounces).

$$\mu = 1.5, \sigma = 0.4$$

Find the z-score for $x = 2.2$.

$$z = \frac{x - \mu}{\sigma} = \frac{2.2 - 1.5}{0.4} = 1.75,$$

so

$$\begin{aligned} &P(x > 2.2) \\ &= P(z > 1.75) \\ &= \text{area to the right of } z = 1.75 \\ &= 1 - (\text{area to the left of } z = 1.75) \\ &= 1 - 0.9599 \\ &= 0.0401. \end{aligned}$$

Thus, out of 5 dozen eggs, we expect $0.0401(60) = 2.406$ eggs, or about 2, to be graded extra large.

45. The Recommended Daily Allowance is

$$\begin{aligned} \mu + 2.5\sigma &= 1200 + 2.5(60) \\ &= 1350 \text{ units.} \end{aligned}$$

47. The Recommended Daily Allowance is

$$\begin{aligned} \mu + 2.5\sigma &= 1200 + 2.5(92) \\ &= 1430 \text{ units.} \end{aligned}$$

49. Let x represent a driving speed.

$$\mu = 52, \sigma = 8$$

At the 85th percentile, the area to the left is 0.8500, which corresponds to about $z = 1.04$. Find the x-value that corresponds to this z-score.

$$\begin{aligned} z &= \frac{x - \mu}{\sigma} \\ 1.04 &= \frac{x - 52}{8} \\ 8.32 &= x - 52 \\ 60.32 &= x \end{aligned}$$

The 85th percentile speed for this road is 60.32 mph.

51. $$\begin{aligned} P\left(x \geq \mu + \frac{3}{2}\sigma\right) &= P(z \geq 1.5) \\ &= 1 - P(z \leq 1.5) \\ &= 1 - 0.9332 = 0.0668 \end{aligned}$$

Thus, 6.68% of the student receive A's.

53. $$\begin{aligned} &P\left(\mu - \frac{1}{2}\sigma \leq x \leq \mu + \frac{1}{2}\sigma\right) \\ &= P(-0.5 \leq z \leq 0.5) \\ &= P(z \leq 0.5) - P(z \leq -0.5) \\ &= 0.6915 - 0.3085 \\ &= 0.383 \end{aligned}$$

Thus, 38.3% of the students receive C's

In Exercises 55 and 57, let x represents a student's test score.

55. Since the top 8% get A's, we want to find the number a for which

$$P(x \geq a) = 0.08,$$
$$\text{or} \quad P(x \leq a) = 0.92.$$

Read the table backwards to find the z-score for an area of 0.92, which is 1.41. Find the value of x that corresponds to $z = 1.41$.

$$\begin{aligned} z &= \frac{x - \mu}{\sigma} \\ 1.41 &= \frac{x - 76}{8} \\ 11.28 &= x - 76 \\ 87.28 &= x \end{aligned}$$

The bottom cutoff score for an A is 87.

57. 28% of the students will receive D's and F's, so to find the bottom cutoff score for a C we need to find the number c for which

$$P(x \leq c) = 0.28.$$

Read the table backwards to find the z-score for an area of 0.28, which is -0.58. Find the value of x that corresponds to $z = -0.58$.

$$-0.58 = \frac{x - 76}{8}$$
$$-4.64 = x - 76$$
$$71.36 = x$$

The bottom cutoff score for a C is 71.

59. **(a)** The area above the 55th percentile is equal to the area below the 45th percentile.

$$\begin{aligned} 2P(z > 0.55) &= 2[1 - P(z \leq 0.55)] \\ &= 2(1 - 0.7088) \\ &= 2(0.2912) \\ &= 0.5824 \end{aligned}$$

$0.58 = 58\%$

(b) The area above the 60th percentile is equal to the area below the 40th percentile.

$$\begin{aligned} 2P(z > 0.6) &= 2[1 - P(z \leq 0.6)] \\ &= 2(1 - 0.7257) \\ &= 2(0.2743) \\ &= 0.5486 \end{aligned}$$

$0.55 = 55\%$

The probability that the student will be above the 60th percentile or below the 40th percentile is 55%.

61. **(a)** $\mu = 93, \sigma = 16$

For $x = 130.5$,

$$z = \frac{130.5 - 93}{16} = 2.34.$$

Then,

$$\begin{aligned} P(x \geq 130.5) &= P(z \geq 2.34) \\ &= \text{area to the right of } 2.34 \\ &= 1 - 0.9904 = 0.0096 \end{aligned}$$

The probability is about 0.01 that a person from this time period would have a lead level of 130.5 ppm or higher. Yes, this provides evidence that Jackson suffered from lead poisoning during this time period.

(b) $\mu = 10, \sigma = 5$

For $x = 130.5$,

$$z = \frac{130.5 - 10}{5} = 24.1.$$

Then,

$$\begin{aligned} P(x \geq 130.5) &= P(z \geq 24.1) \\ &= \text{area to the right of } 24.1 \\ &\approx 0 \end{aligned}$$

The probability is essentially 0 by these standards.

From this we can conclude that Andrew Jackson had lead poisioning.

63.

	Reference	Models
(a) Head size:	$z = \frac{55 - 55.3}{2.0} = -0.15$ $P(x \geq 55) = P(z \geq -0.15) = 1 - 0.4404 = 0.5596$	$z = \frac{55 - 50.0}{2.4} = 2.08$ $P(x \geq 55) = P(z \geq 2.08) = 1 - 0.9812 = 0.0188$
(b) Neck size:	$z = \frac{23.9 - 32.7}{1.4} = -6.29$ $P(x \leq 23.9) = P(z \leq -6.29) \approx 0$	$z = \frac{23.9 - 31.0}{1.0} = -7.1$ $P(x \leq 23.9) = P(z \leq -7.1) \approx 0$
(c) Bust size:	$z = \frac{82.3 - 90.3}{5.5} = -1.45$ $P(x \geq 82.3) = P(z \geq -1.45) = 1 - 0.0735 = 0.9265$	$z = \frac{82.3 - 87.4}{3.0} = -1.70$ $P(x \geq 82.3) = P(z \geq -1.70) = 1 - 0.0446 = 0.9554$
(d) Wrist size:	$z = \frac{10.6 - 16.1}{0.8} = -6.88$ $P(x \leq 10.6) = P(z \leq -6.88) \approx 0$	$z = \frac{10.6 - 15.0}{0.6} = -7.33$ $P(x \leq 10.6) = P(z \leq -7.33) \approx 0$
(e) Waist size:	$z = \frac{40.7 - 69.8}{4.7} = -6.19$ $P(x \leq 40.7) = P(z \leq -6.19) \approx 0$	$z = \frac{40.7 - 65.7}{3.5} = -7.14$ $P(x \leq 40.7) = P(z \leq -7.14) \approx 0$

9.4 Normal Approximation to the Binomial Distribution

Your Turn 1

For this experiment, $n = 12$ and $p = \frac{1}{6}$.

$$\mu = np = (12)\left(\frac{1}{6}\right) = 2$$

$$\sigma = \sqrt{np(1-p)} = \sqrt{(12)\left(\frac{1}{6}\right)\left(\frac{5}{6}\right)} = \sqrt{\frac{5}{3}} \approx 1.291$$

Your Turn 2

First find the mean and standard deviation using $n = 120$ and $p = \frac{1}{5} = 0.2$.

$$\mu = np = (120)(0.2) = 24$$

$$\sigma = \sqrt{np(1-p)} = \sqrt{(120)(0.2)(0.8)} = \sqrt{19.2} \approx 4.3818$$

Let x represent the number of correct answers. To find the probability of at least 32 correct answers using the normal distribution we must find $P(x > 31.5)$. Since the corresponding z-score is $z = \frac{x - \mu}{\sigma} = \frac{31.5 - 24}{4.3818} \approx 1.7116$, we need to compute $P(z \geq 1.7116) = 1 - P(z \leq 1.7116)$. Using tables or a graphing calculator we find $P(z \leq 1.7116) \approx 0.9564$ so our answer is $1 - 0.9564 = 0.0436$. The probability of getting at least 32 correct by random guessing is about 0.0436; depending on how you round intermediate results you may get a slightly different answer.

9.4 Exercises

1. In order to find the mean and standard deviation of a binomial distribution, you must know the number of trials and the probability of a success on each trial.

3. Let x represent the number of heads tossed. For this experiment, $n = 16$, $x = 4$, and $p = \frac{1}{2}$.

(a) $P(x = 4) = C(16,4)\left(\frac{1}{2}\right)^4\left(1 - \frac{1}{2}\right)^4$

≈ 0.0278

(b) $\mu = np = 16\left(\frac{1}{2}\right) = 8$

$\sigma = \sqrt{np(1-p)}$

$= \sqrt{16\left(\frac{1}{2}\right)\left(\frac{1}{2}\right)}$

$= \sqrt{4} = 2$

For $x = 3.5$,

$$z = \frac{3.5 - 8}{2} = -2.25.$$

For $x = 4.5$,

$$z = \frac{4.5 - 8}{2} = -1.75.$$

$P(z < -1.75) - P(z < -2.25)$
$= 0.0401 - 0.0122 = 0.0279$

5. Let x represent the number of tails tossed. For this experiment, $n = 16$; $x = 13, 14, 15$, or 16; and $p = \frac{1}{2}$.

(a) $P(x = 13, 14, 15, \text{ or } 16)$

$$= C(16,13)\left(\frac{1}{2}\right)^{13}\left(1 - \frac{1}{2}\right)^3$$
$$+ C(16,14)\left(\frac{1}{2}\right)^{14}\left(1 - \frac{1}{2}\right)^2$$
$$+ C(16,15)\left(\frac{1}{2}\right)^{15}\left(1 - \frac{1}{2}\right)^1$$
$$+ C(16,16)\left(\frac{1}{2}\right)^{16}\left(1 - \frac{1}{2}\right)^0$$
$$\approx 0.00854 + 0.00183 + 0.00024$$
$$+ 0.00001$$
$$\approx 0.0106$$

(b) $\mu = np = 16\left(\frac{1}{2}\right) = 8$

$\sigma = \sqrt{np(1-p)}$

$= \sqrt{16\left(\frac{1}{2}\right)\left(\frac{1}{2}\right)}$

$= \sqrt{4} = 2$

For $x = 12.5$,

$$z = \frac{12.5 - 8}{2} = 2.25.$$

$P(z > 2.25) = 1 - P(z \le 2.25)$
$= 1 - 0.9878$
$= 0.0122$

In Exercises 7 and 9, let x represent the number of heads tossed. Since $n = 1000$ and $p = \frac{1}{2}$,

$$\mu = np = 1000\left(\frac{1}{2}\right) = 500$$

and

$$\sigma = \sqrt{np(1-p)}$$
$$= \sqrt{1000\left(\frac{1}{2}\right)\left(\frac{1}{2}\right)}$$
$$= \sqrt{250}$$
$$\approx 15.8.$$

7. To find P(exactly 500 heads), find the z-scores for $x = 499.5$ and $x = 500.5$.

For $x = 499.5$,

$$z = \frac{499.5 - 500}{15.8} \approx -0.03.$$

For $x = 500.5$,

$$z = \frac{500.5 - 500}{15.8} \approx 0.03.$$

Using the table,

$P(\text{exactly 500 heads}) = 0.5120 - 0.4880$
$= 0.0240.$

9. Since we want 475 heads or more, we need to find the area to the right of $x = 474.5$. This will be $1 -$ (the area to the left of $x = 474.5$). Find the z-score for $x = 474.5$.

$$z = \frac{474.5 - 500}{15.8} \approx -1.61$$

The area to the left of 474.5 is 0.0537, so

$P(\text{480 heads or more}) = 1 - 0.0537$
$= 0.9463.$

11. Let x represent the number of 5's tossed.

$n = 120, p = \frac{1}{6}$

$\mu = np = 120\left(\frac{1}{6}\right) = 20$

$$\sigma = \sqrt{np(1-p)} = \sqrt{120\left(\frac{1}{6}\right)\left(\frac{5}{6}\right)} \approx 4.08$$

Since we want the probability of getting exactly twenty 5's, we need to find the area between $x = 19.5$ and $x = 20.5$. Find the corresponding z-scores.

For $x = 19.5$,

$$z = \frac{19.5 - 20}{4.08} \approx -0.12.$$

For $x = 20.5$,

$$z = \frac{20.5 - 20}{4.08} \approx 0.12.$$

Using values from the table,

$$P(\text{exactly twenty 5's}) = 0.5478 - 0.4522 = 0.0956.$$

13. Let x represent the number of 3's tossed.

$$n = 120, p = \frac{1}{6}$$

$$\mu = 20, \sigma \approx 4.08$$

(These values for μ and σ are calculated in the solution for Exercise 11.)

Since

$$P(\text{more than fifteen 3's}) = 1 - P(\text{fifteen 3's or less}),$$

find the z-score for $x = 15.5$.

$$z = \frac{15.5 - 20}{4.08} \approx -1.10$$

From the table, $P(z < -1.10) = 0.1357$.

Thus,

$$P(\text{more than fifteen 3's}) = 1 - 0.1357 = 0.8643.$$

15. Let x represent the number of times the chosen number appears.

$n = 130$; $x = 26, 27, 28, \ldots, 130$; and $p = \frac{1}{6}$

$\mu = np = 130\left(\frac{1}{6}\right) = \frac{65}{3}$

$\sigma = \sqrt{np(1-p)} = \sqrt{130\left(\frac{1}{6}\right)\left(\frac{5}{6}\right)} = \frac{5}{6}\sqrt{26}$

For $x = 25.5$,

$$z = \frac{25.5 - \frac{65}{3}}{\frac{5}{6}\sqrt{26}} \approx 0.90.$$

$$P(z > 0.90) = 1 - P(z \le 0.90) = 1 - 0.8159 = 0.1841$$

17. **(a)** Let x represent the number of heaters that are defective.

$n = 10{,}000, p = 0.02$

$\mu = np = 10{,}000(0.02) = 200$

$\sigma = \sqrt{np(1-p)} = \sqrt{10{,}000(0.02)(0.98)} = 14$

To find $P(\text{fewer than } 170)$, find the z-score for $x = 169.5$.

$$z = \frac{169.5 - 200}{14} \approx -2.18$$

$P(\text{fewer than } 170) = 0.0146$

(b) Let x represent the number of heaters that are defective.

$n = 10{,}000, p = 0.02, \mu = np = 200$

$\sigma = \sqrt{np(1-p)} = \sqrt{10{,}000(0.02)(0.98)} = 14$

We want the area to the right of $x = 222.5$. For $x = 222.5$,

$$z = \frac{222.5 - 200}{14} \approx 1.61.$$

$$P(\text{more than 222 defects}) = P(x \ge 222.5) = P(z \ge 1.61) = 1 - P(z \le 1.61) = 1 - 0.9463 = 0.0537$$

19. Use a calculator or computer to complete this exercise. The answers are given.

(a) $P(\text{all 58 like it}) = 1.04 \times 10^{-9} \approx 0$

(b) $P(\text{exactly 28, 29, or 30 like it}) = 0.0018$

21. Let x be the number of nests escaping predation.

$n = 24, p = 0.3$

$\mu = np = 24(0.3) = 7.2$

$$\sigma = \sqrt{np(1-p)} = \sqrt{24(0.3)(0.7)}$$
$$= \sqrt{5.04} \approx 2.245$$

To find P(at least half escape predation), find the z-score for $x = 11.5$.

$$z = \frac{11.5 - 7.2}{2.245} \approx 1.92$$
$$p(z > 1.92) = 1 - 0.9726 = 0.0274$$

23. Let x represent the number of hospital patients struck by falling coconuts.

(a) $n = 20; x = 0$ or 1; and $p = 0.025$

$$p(x = 0 \text{ or } 1) = C(20,0)(0.025)^0(0.975)^{20}$$
$$+ (20,1)(0.025)^1(0.975)^{19}$$
$$\approx 0.60269 + 0.30907$$
$$\approx 0.9118$$

(b) $n = 2000;\ x = 0, 1, 2, \ldots,$ or $70;\ p = 0.025$

$$\mu = np = 2000(0.025) = 50$$
$$\sigma = \sqrt{np(1-p)} = \sqrt{2000(0.025)(0.975)}$$
$$= \sqrt{48.75}$$

To find P(70 or less), find the z-score for $x = 70.5$.

$$z = \frac{70.5 - 50}{\sqrt{48.75}} \approx 2.94$$

$$P(x < 2.94) = 0.9984$$

25. This exercise should be solved by calculator or computer methods. The answers, which may vary slightly, are

(a) 0.0001,

(b) 0.0002 and

(c) 0.0000.

27. Let x represent the number of motorcyclists injured between 3 p.m. and 6 p.m. $n = 200$ and $p = 0.241$. To use the normal approximation we compute μ and σ.

$$\mu = np = (200)(0.241) = 48.2$$
$$\sigma = \sqrt{np(1-p)}$$
$$= \sqrt{(200)(0.241)(0.759)}$$
$$\approx 6.0485$$

To find $P(x < 51)$, find the z-score for 50.5.

$$z = \frac{50.5 - 48.2}{6.0485} \approx 0.3803$$
$$P(z \le 0.3803) \approx 0.6480 \text{ (by table)}$$

The probability that at most 50 in the sample were injured between 3 p.m. and 6 p.m. is 0.6480.

29. Let x represent the number of ninth grade students who have tried cigarette smoking. $n = 500$ and $p = 0.463$. To use the normal approximation we compute μ and σ.

$$\mu = np = (500)(0.463) = 231.5$$
$$\sigma = \sqrt{np(1-p)} = \sqrt{(500)(0.463)(0.537)}$$
$$\approx 11.1497$$

To find $P(x < 251)$ find the z-score for 250.5.

$$z = \frac{250.5 - 231.5}{11.1497} \approx 1.7041$$
$$P(z \le 1.704) \approx 0.9554 \text{ (by table)}$$

The probability that at most half in the sample of 500 have tried cigarette smoking is approximately 0.9554.

31. **(a)** The numbers are too large for the calculator to handle.

(b) $n = 5{,}825{,}043, p = 0.5$

$$\mu = np = 5{,}825{,}043(0.5) = 2{,}912{,}522$$
$$\sigma = \sqrt{np(1-p)}$$
$$= \sqrt{5{,}825{,}043(0.5)(0.5)}$$
$$\approx 1206.8$$

$$z = \frac{2{,}912{,}253 - 2{,}912{,}522}{1206.8} = -0.22$$
$$z = \frac{2{,}912{,}790 - 2{,}912{,}522}{1206.8} = 0.22$$

$$\begin{aligned} &P(2{,}912{,}253 \le x \le 2{,}912{,}790) \\ &= P(-0.22 \le z \le 0.22) \\ &= P(z \le 0.22) - P(z \le -0.22) \\ &= 0.5871 - 0.4129 \\ &= 0.1742 \end{aligned}$$

33. Let x represent the number of questions.

$n = 100; x = 60, 61, 62, \ldots,$ or $100; p = \frac{1}{2}$

$$\mu = np = 180\left(\frac{1}{2}\right) = 50$$

$$\sigma = \sqrt{np(1-p)} = \sqrt{100\left(\frac{1}{2}\right)\left(\frac{1}{2}\right)} = \sqrt{25} = 5$$

To find P(60 or more correct), find the z-score for $x = 59.5$.

$$z = \frac{59.5 - 50}{5} = 1.90$$

$$\begin{aligned} P(z > 1.90) &= 1 - P(z \le 1.90) \\ &= 1 - 0.9713 \\ &= 0.0287 \end{aligned}$$

Chapter 9 Review Exercises

1. True

2. False: Any symmetrical distribution which peaks in the middle will have equal mean and mode.

3. False: Variance measures spread

4. True

5. False: A large variance indicates that the data are widely spread.

6. False: The mode is the most frequently occurring value.

7. False: This will be true only if $\mu(x) = 0$.

8. True

9. True

10. True

11. False: The expected value of a sample variance is the population variance.

12. True

15. **(a)** since 450-474 is to be the first interval, let all the intervals be of size 25. The largest data value is 566, so the last interval that will be needed is 550-574. The frequency distribution is as follows.

Interval	Frequency
450–474	5
475–499	6
500–524	5
525–549	2
550–574	2

(b) Draw the histogram. It consists of equal width having heights as determined by the frequency of each interval. See the histogram in part (c).

(c) Construct the frequency polygon by joining consecutive midpoints of the tops of the histogram bars with line segments.

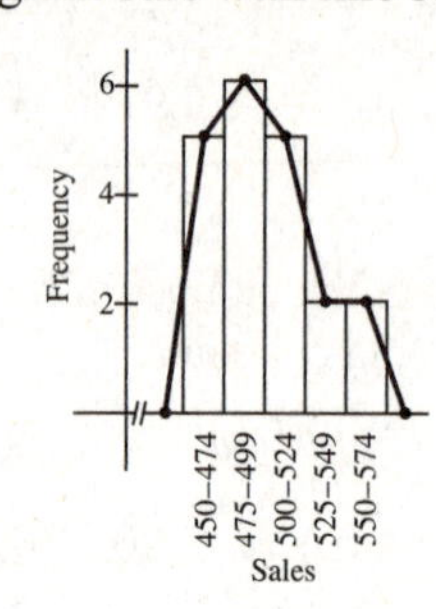

17.
$$\begin{aligned} \sum x &= 30 + 24 + 34 + 30 + 29 \\ &\quad + 28 + 30 + 29 \\ &= 234 \end{aligned}$$

The mean of the 8 numbers is

$$\bar{x} = \frac{\sum x}{n} = \frac{234}{8} = 29.25.$$

19.

Interval	Midpoint, x	Frequency, f	Product, xf
10–19	14.5	6	87
20–29	24.5	12	294
30–39	34.5	14	483
40–49	44.5	10	445
50–59	54.5	8	436
Totals:		50	1745

The mean of this collection of grouped data is

$$\bar{x} = \frac{\sum xf}{n} = \frac{1745}{50} = 34.9.$$

23. Arrange the numbers in numerical order, from smallest to largest.

35, 36, 36, 38, 38, 42, 44, 48

There are 8 numbers here; the median is the mean of the two middle numbers, which is

$$\frac{38 + 38}{2} = \frac{76}{2} = 38.$$

The mode is the number that occurs most often. Here, there are two modes, 36 and 38, since they both appear twice.

25. The modal class is the interval with the greatest frequency; in this case, the modal class is 30–39.

29. The range is $93 - 26 = 67$, the difference of the highest and lowest numbers in the distribution.

The mean is

$$\bar{x} = \frac{\sum x}{n} = \frac{520}{10} = 52.$$

Construct a table with the values of x and x^2.

x	x^2
26	676
43	1849
51	2601
29	841
37	1369
56	3136
29	841
82	6724
74	5476
93	8649
Total:	32,162

The standard deviation is:

$$s = \sqrt{\frac{\sum x^2 - n\bar{x}^2}{n - 1}}$$

$$= \sqrt{\frac{32{,}162 - 10(52)^2}{9}}$$

$$\approx \sqrt{569.1} \approx 23.9.$$

31. Recall that when working with grouped data, x represents the midpoint of each interval. Complete the following table.

Interval	f	x	xf	x^2	fx^2
10–19	6	14.5	87.0	210.25	1261.50
20–29	12	24.5	294.0	600.25	7203.00
30–39	14	34.5	483.0	1190.25	16,663.50
40–49	10	44.5	445.0	1980.25	19,802.50
50–59	8	54.5	436.0	2970.25	23,762.00
Totals:	50		1745		68,692.50

Use the formulas for grouped frequency distribution to find the mean and then the standard deviation. (The mean was also calculated in Exercise 19.)

$$\bar{x} = \frac{\sum xf}{n} = \frac{1745}{50} = 34.9$$

$$s = \sqrt{\frac{\sum fx^2 - n\bar{x}^2}{n - 1}}$$

$$= \sqrt{\frac{68{,}692.5 - 50(34.9)^2}{49}}$$

$$\approx 12.6$$

33. A skewed distribution has the largest frequency at one end rather than in the middle.

35. To the left of $z = 0.84$

Using the standard normal curve table,

$$P(z < 0.84) = 0.7995.$$

37. Between $z = 1.53$ and $z = 2.82$

$$P(1.53 \le z \le 2.82)$$
$$= P(z \le 2.82) - P(z \le 1.53)$$
$$= 0.9976 - 0.9370$$
$$= 0.0606$$

39. The normal distribution is not a good approximation of a binomial distribution that has a value of p close to 0 or 1 because the histogram of such a binomial distribution is skewed and therefore not close to the shape of a normal distribution.

41.

Number of Heads, x	Frequency, f	xf	fx^2
0	1	0	0
1	5	5	5
2	7	14	28
3	5	15	45
4	2	8	32
Totals:	20	42	110

(a) $\bar{x} = \dfrac{\sum xf}{n} = \dfrac{42}{20} = 2.1$

$$s = \sqrt{\frac{\sum fx^2 - n\bar{x}^2}{n - 1}}$$

$$= \sqrt{\frac{110 - 20(2.1)^2}{20 - 1}}$$

$$\approx 1.07$$

(b) For this binomial experiment,

$$\mu = np = 4\left(\frac{1}{2}\right) = 2,$$

and

$$\sigma = \sqrt{np(1-p)} = \sqrt{4\left(\frac{1}{2}\right)\left(\frac{1}{2}\right)} = \sqrt{1} = 1.$$

(c) The answer to parts (a) and (b) should be close.

43. **(a)** For Stock I,

$$\bar{x} = \frac{11 + (-1) + 14}{3} = 8,$$

so, the mean (average return) is 8%.

$$s = \sqrt{\frac{\sum x^2 - n\bar{x}^2}{n-1}} = \sqrt{\frac{318 - 3(8)^2}{2}} = \sqrt{63} \approx 7.9$$

so the standard deviation is 7.9%.

For Stock II,

$$\bar{x} = \frac{9 + 5 + 10}{3} = 8,$$

so the mean is also 8%.

$$s = \sqrt{\frac{\sum x^2 - n\bar{x}^2}{n-1}} = \sqrt{\frac{206 - 3(8)^2}{2}} = \sqrt{7} \approx 2.6,$$

so the standard deviation is 2.6%.

(b) Both stocks offer an average (mean) return of 8%. The smaller standard deviation for Stock II indicates a more stable return and thus greater security.

45. Let x represents the number of overstuffed frankfurters.

$$n = 500, p = 0.04, 1 - p = 0.96$$

We also need the following results.

$$\mu = np = 500(0.04) = 20$$

$$\sigma = \sqrt{np(1-p)} = \sqrt{500(0.04)(0.96)} = \sqrt{19.2} \approx 4.38$$

(a) P(twenty-five or fewer) or, equivalently, $P(x \leq 25)$

First, using the binomial probability formula:

$$\begin{aligned} P(x \leq 25) &= C(500,0)(0.04)^0(0.96)^{500} \\ &\quad + C(500,1)(0.04)^1(0.96)499 \\ &\quad + \cdots + C(500,25)(0.04)^{25}(0.96)^{475} \\ &= (1.4 \times 10^{-9}) + (2.8 \times 10^{-8}) \\ &\quad + \cdots + 0.0446 \\ &\approx 0.8924 \end{aligned}$$

(To evaluate the sum, use a calculator or computer program. For example, using a TI-83/84 Plus calculator, enter the following:

```
sum(seq((500nCrX)(0.04^X)
    (0.96^(500 – X)),X,0,25,1))
```

The displayed result is 0.8923644609.)

Second, using the normal approximation:

To find $P(x \leq 25)$, first find the z-score for 25.5.

$$z = \frac{25.5 - 20}{4.38} \approx 1.26$$

$$P(z < 1.26) = 0.8962$$

(b) P(exactly twenty-five) or $P(x = 25)$

Using the binomial probability formula:

$$P(x = 25) = C(500,25)(0.04)^{25}(0.96)^{475} \approx 0.0446$$

Using the normal approximation:

P(exactly twenty-five) corresponds to the area under the normal curve between $x = 24.5$ and $x = 25.5$. The correspond-ing z-scores are found as follows.

$$z = \frac{24.5 - 20}{4.38} \approx 1.03 \text{ and}$$

$$z = \frac{24.5 - 20}{4.38} \approx 1.26$$

$$\begin{aligned} P(x = 25) &= P(24.5 < x < 25.5) \\ &= P(1.03 < z < 1.26) \\ &= P(z < 1.26) - P(z < 1.03) \\ &= 0.8962 - 0.8485 \\ &= 0.0477 \end{aligned}$$

(c) P(at least 30), or equivalently, $P(x \geq 30)$

Using the binomial probability formula:

This is the complementary event to "less than 30," which requires fewer calculations. We can use the results from Part (a) to reduce the amount of work even more.

$$\begin{aligned} &P(x < 30) \\ &= P(x \leq 25) + P(x = 26) \\ &\quad + P(x = 27) + P(x = 28) \\ &\quad + P(x = 29) \\ &= 0.8924 + C(500,26)(0.04)^{26}(0.96)^{474} \\ &\quad + \cdots + C(500,29)(0.04)^{29}(0.96)^{471} \\ &\approx 0.9804 \end{aligned}$$

Therefore,

$$\begin{aligned} P(x \geq 30) &= 1 - P(x < 30) \\ &= 1 - 0.9804 \\ &= 0.0196. \end{aligned}$$

Using the normal approximation:

Again, use the complementary event. To find $P(x < 30)$, find the z-score for 29.5.

$$z = \frac{29.5 - 20}{4.38} \approx 2.17$$

$$P(z < 2.17) = 0.9850$$

Therefore,

$$\begin{aligned} P(x \geq 30) &= 1 - P(x < 30) \\ &= 1 - 0.9850 \\ &= 0.0150. \end{aligned}$$

47. The table below records the mean and standard deviation for diet A and for diet B.

	$\overline{x}$	s
Diet A	2.7	2.26
Diet B	1.3	0.95

(a) Diet A had the greater mean gain, since the mean for diet A is larger.

(b) Diet B had a more consistent gain, since diet B has a smaller standard deviation.

49. Let x represents the number of flies that are killed.

$$n = 1000;\ x = 0, 1, 2, \ldots, 986;\ p = 0.98$$

$$\mu = np = 1000(0.98) = 980$$

$$\sigma = \sqrt{np(1-p)} = \sqrt{1000(0.98)(0.02)} = \sqrt{19.6}$$

To find P(no more that 986), find the z-score for $x = 986.5$.

$$z = \frac{986.5 - 980}{\sqrt{19.6}} \approx 1.47$$

$$P(z < 1.47) = 0.9292$$

51. Again, let x represents the number of flies that are killed.

$$n = 1000;\ x = 973, 974, 975, \ldots, 993;\ p = 0.98$$

As in Exercise 49, $\mu = 980$ and $\sigma = \sqrt{19.6}$. To find P(between 973 and 993), find the z-scores for $x = 972.5$ and $x = 993.5$.

For $x = 972.5$,

$$z = \frac{972.5 - 980}{\sqrt{19.6}} \approx -1.69.$$

For $x = 993.5$,

$$z = \frac{993.5 - 980}{\sqrt{19.6}} \approx 3.05.$$

$$\begin{aligned} &P(-1.69 \leq z \leq 3.05) \\ &= P(z \leq 3.05) - P(z \leq -1.69) \\ &= 0.9989 - 0.0455 \\ &= 0.9534 \end{aligned}$$

53. No more that 40 min/day

$$\mu = 42,\ \sigma = 12$$

Find the z-score for $x = 40$.

$$z = \frac{40 - 42}{12} \approx -0.17$$

$$P(x \leq 40) = P(z \leq -0.17) = 0.4325$$

43.25% of the residents commute no more that 40 min/day.

55. Between 38 and 60 min/day

$$\mu = 42,\ \sigma = 12$$

Find the z-scores for $x = 38$ and $x = 60$.

For $x = 38$,

$$z = \frac{38 - 42}{12} \approx -0.33.$$

For $x = 60$,

$$z = \frac{60 - 42}{12} = 1.5.$$

$$
\begin{aligned}
P(38 \le x \le 60) &= P(-0.33 \le z \le 1.5) \\
&= P(z \le 1.5) - P(z \le -0.33) \\
&= 0.9332 - 0.3707 \\
&= 0.5625
\end{aligned}
$$

56.25% of the residents commute between 38 and 60 min/day.

57. **(a)** The mean is

$$\bar{x} = \frac{\sum x}{n} = \frac{1852}{13} \approx 142.46.$$

To find the median, arrange the values in order. The middle value is the 7th in the list, which is 140 (corresponding to Los Angeles), so this is the median. There are no repeated values so there is no mode.

(b) Find the standard deviation by using the 1-variable statistics available on a graphing calculator. Enter the data as a list in `L1` and then enter `1-Var Stats L1`. The standard deviation s is given as the value of Sx, which is approximately 48.67.

(c) $\bar{x} - s = 142.46 - 48.67 = 93.79$

$\bar{x} + s = 142.46 + 48.67 = 191.13$

Five of the data values, or 38.5%, fall within one standard deviation of the mean.

(d) $\bar{x} - 2s = 142.46 - (2)(48.67) = 45.12$

$\bar{x} + 2s = 142.46 + (2)(48.67) = 239.80$

All the data values, or 100%, fall within two standard deviations of the mean.

59. **(a)**

$$
\begin{aligned}
P(x \ge 35) &= P\left(z \ge \frac{35 - 28.0}{5.5}\right) \\
&= P(z \ge 1.27) \\
&= 1 - P(z < 1.27) \\
&= 1 - 0.8980 \\
&= 0.1020
\end{aligned}
$$

(b)

$$
\begin{aligned}
P(x \ge 35) &= P\left(z \ge \frac{35 - 32.2}{8.4}\right) \\
&= P(z \ge 0.33) \\
&= 1 - P(z < 0.33) \\
&= 1 - 0.6293 \\
&= 0.3707
\end{aligned}
$$

(d)

$$
\begin{aligned}
P(x \ge 1.4) &= P\left(z \ge \frac{1.4 - 1.64}{0.08}\right) \\
&= P(z \ge -3) \\
&= 1 - P(z < -3) \\
&= 1 - 0.0013 \\
&= 0.9987
\end{aligned}
$$

(e)

$$
\begin{aligned}
&P(z < -1.5) + P(z > 1.5) \\
&= 2P(z < -1.5) \\
&= 2(0.0668) \\
&= 0.1336
\end{aligned}
$$

Chapter 10

NONLINEAR FUNCTIONS

10.1 Properties of Functions

Your Turn 1

$$y = \frac{1}{\sqrt{x^2 - 4}}$$

Since the denominator cannot be zero and the radicand cannot be negative, the domain includes only those values of x satisfying $x^2 - 4 > 0$. Using the methods for solving a quadratic inequality, we get the domain $(-\infty, -2) \cup (2, \infty)$. Since the denominator can never be negative, y cannot be negative or zero. So, the range is $(0, \infty)$.

Your Turn 2

$f(x) = 2x^2 - 3x - 4$

(a) $f(x+h) = 2(x+h)^2 - 3(x+h) - 4$
$= 2(x^2 + 2xh + h^2) - 3(x+h) - 4$
$= 2x^2 + 4xh + 2h^2 - 3x - 3h - 4$

(b) $f(x) = -5$

$2x^2 - 3x - 4 = -5$

$2x^2 - 3x + 1 = 0$

$(2x - 1)(x - 1) = 0$

$x = \frac{1}{2}$ or $x = 1$

10.1 Exercises

1. The x-value of 82 corresponds to two y-values, 93 and 14. In a function, each value of x must correspond to exactly one value of y.

The rule is not a function.

3. Each x-value corresponds to exactly one y-value.

The rule is a function.

5. $y = x^3 + 2$

Each x-value corresponds to exactly one y-value.

The rule is a function.

7. $x = |y|$

Each value of x (except 0) corresponds to two y-values.

The rule is not a function.

9. $y = 2x + 3$

x	-2	-1	0	1	2	3
y	-1	1	3	5	7	9

Pairs: $(-2,-1)$, $(-1, 1)$, $(0, 3)$, $(1, 5)$, $(2, 7)$, $(3, 9)$

Range: $\{-1, 1, 3, 5, 7, 9\}$

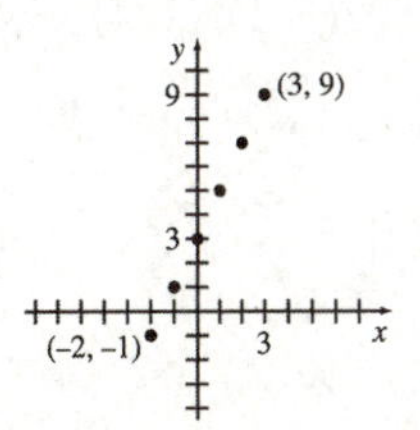

11. $2y - x = 5$

$2y = 5 + x$

$y = \frac{1}{2}x + \frac{5}{2}$

x	-2	-1	0	1	2	3
y	$\frac{3}{2}$	2	$\frac{5}{2}$	3	$\frac{7}{2}$	4

Pairs: $\left(-2, \frac{3}{2}\right)$, $(-1, 2)$, $\left(0, \frac{5}{2}\right)$, $(1, 3)$, $\left(2, \frac{7}{2}\right)$, $(3, 4)$.

Range: $\left\{\frac{3}{2}, 2, \frac{5}{2}, 3, \frac{7}{2}, 4\right\}$

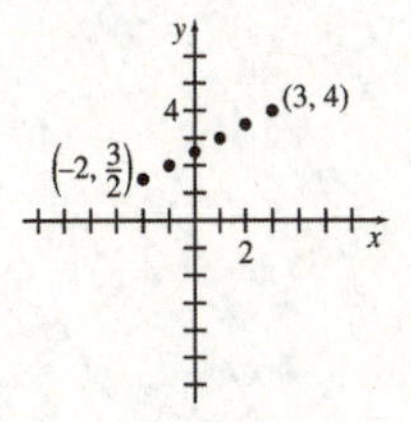

13. $y = x(x + 2)$

x	-2	-1	0	1	2	3
y	0	-1	0	3	8	15

Pairs: $(-2, 0), (-1, -1), (0, 0), (1, 3), (2, 8), (3, 15)$

Range: $\{-1, 0, 3, 8, 15\}$

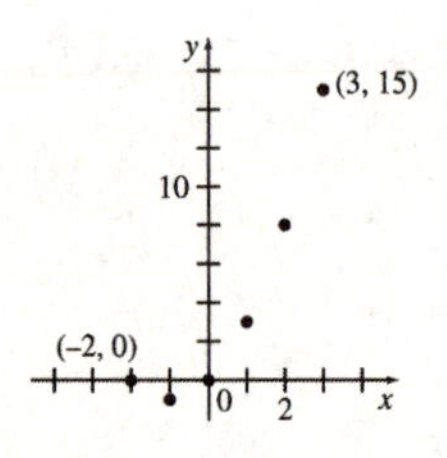

15. $y = x^2$

x	-2	-1	0	1	2	3
y	4	1	0	1	4	9

Pairs: $(-2, 4), (-1, 1), (0, 0), (1, 1), (2, 4), (3, 9)$

Range: $\{0, 1, 4, 9\}$

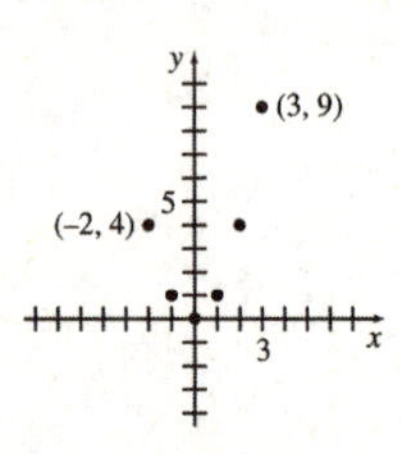

17. $f(x) = 2x$

x can take on any value, so the domain is the set of real numbers, $(-\infty, \infty)$.

19. $f(x) = x^4$

x can take on any value, so the domain is the set of real numbers, $(-\infty, \infty)$.

21. $f(x) = \sqrt{4 - x^2}$

For $f(x)$ to be a real number, $4 - x^2 \geq 0$.
Solve $4 - x^2 = 0$.

$$(2 - x)(2 + x) = 0$$
$$x = 2 \text{ or } x = -2$$

The numbers form the intervals $(-\infty, -2)$, $(-2, 2)$, and $(2, \infty)$.

Values in the interval $(-2, 2)$ satisfy the inequality; $x = 2$ and $x = -2$ also satisfy the inequality. The domain is $[-2, 2]$.

23. $f(x) = (x - 3)^{1/2} = \sqrt{x - 3}$

For $f(x)$ to be a real number,

$$x - 3 \geq 0$$
$$x \geq 3.$$

The domain is $[3, \infty)$.

25. $f(x) = \dfrac{2}{1 - x^2} = \dfrac{2}{(1 - x)(1 + x)}$

Since division by zero is not defined, $(1 - x) \cdot (1 + x) \neq 0$.

When $(1 - x)(1 + x) = 0$,

$$1 - x = 0 \quad \text{or} \quad 1 + x = 0$$
$$x = 1 \quad \text{or} \quad x = -1$$

Thus, x can be any real number except ± 1. The domain is

$$(-\infty, -1) \cup (-1, 1) \cup (1, \infty).$$

27. $f(x) = -\sqrt{\dfrac{2}{x^2 - 16}} = -\sqrt{\dfrac{2}{(x - 4)(x + 4)}}.$

$(x - 4) \cdot (x + 4) > 0$, since $(x - 4) \cdot (x + 4) < 0$ would produce a negative radicand and $(x - 4) \cdot (x + 4) = 0$ would lead to division by zero.

Solve $(x - 4) \cdot (x + 4) = 0$.

$$x - 4 = 0 \quad \text{or} \quad x + 4 = 0$$
$$x = 4 \quad \text{or} \quad x = -4$$

Use the values -4 and 4 to divide the number line into 3 intervals, $(-\infty, -4)$, $(-4, 4)$ and $(4, \infty)$. Only the values in the intervals $(-\infty, -4)$ and $(4, \infty)$ satisfy the inequality.

The domain is $(-\infty, -4) \cup (4, \infty)$.

29. $f(x) = \sqrt{x^2 - 4x - 5} = \sqrt{(x - 5)(x + 1)}$

See the method used in Exercise 21.

$$(x - 5)(x + 1) \geq 0$$

when $x \geq 5$ and when $x \leq -1$. The domain is $(-\infty, -1] \cup [5, \infty)$.

31. $f(x) = \dfrac{1}{\sqrt{3x^2 + 2x - 1}} = \dfrac{1}{\sqrt{(3x - 1)(x + 1)}}$

$(3 - 2)(x + 1) > 0$, since the radicand cannot be negative and the denominator of the function cannot be zero.

Solve $(3 - 1)(x + 1) = 0$.

$$3 - 1 = 0 \quad \text{or} \quad x + 1 = 0$$
$$x = \tfrac{1}{3} \quad \text{or} \quad x = -1$$

Use the values -1 and $\frac{1}{3}$ to divide the number line into 3 intervals, $(-\infty, -1)$, $(-1, 4)$ and $\left(\frac{1}{3}, \infty\right)$.

Only the values in the intervals $(-\infty, -1)$ and $\left(\frac{1}{3}, \infty\right)$ satisfy the inequality.

The domain is $(-\infty, -1) \cup \left(\frac{1}{3}, \infty\right)$.

33. By reading the graph, the domain is all numbers greater than or equal to -5 and less than 4. The range is all numbers greater than or equal to -2 and less than or equal to 6.

Domain: $[-5, 4)$; range: $[-2, 6]$

35. By reading the graph, x can take on any value, but y is less than or equal to 12.

Domain: $(-\infty, \infty)$; range: $(-\infty, 12]$

37. The domain is all real numbers between the end points of the curve, or $[-2; 4]$.

The range is all real numbers between the minimum and maximum values of the function or $[0, 4]$.

(a) $f(-2) = 0$

(b) $f(0) = 4$

(c) $f\left(\frac{1}{2}\right) = 3$

(d) From the graph, $f(x) = 1$ when $x = -1.5, 1.5,$ or 2.5.

39. The domain is all real numbers between the endpoints of the curve, or $[-2, 4]$.

The range is all real numbers between the minimum and maximum values of the function or $[-3, 2]$.

(a) $f(-2) = -3$

(b) $f(0) = -2$

(c) $f\left(\frac{1}{2}\right) = -1$

(d) From the graph, $f(x) = 1$ when $x = 2.5$.

41. $f(x) = 3x^2 - 4x + 1$

(a) $f(4) = 3(4)^2 - 4(4) + 1$
$= 48 - 16 + 1$
$= 33$

(b) $f\left(-\frac{1}{2}\right) = 3\left(-\frac{1}{2}\right)^2 - 4\left(-\frac{1}{2}\right) + 1$
$= \frac{3}{4} + 2 + 1$
$= \frac{15}{4}$

(c) $f(a) = 3(a)^2 - 4(a) + 1$
$= 3a^2 - 4a + 1$

(d) $f\left(\frac{2}{m}\right) = 3\left(\frac{2}{m}\right)^2 - 4\left(\frac{2}{m}\right) + 1$
$= \frac{12}{m^2} - \frac{8}{m} + 1$
or $\frac{12 - 8m + m^2}{m^2}$

(e)
$$f(x) = 1$$
$$3x^2 - 4x + 1 = 1$$
$$3x^2 - 4x = 0$$
$$x(3x - 4) = 0$$
$$x = 0 \text{ or } x = \frac{4}{3}$$

43. $f(x) = \begin{cases} \frac{2x+1}{x-4} & \text{if } x \neq 4 \\ 7 & \text{if } x = 4 \end{cases}$

(a) $f(4) = 7$

(b) $f\left(-\frac{1}{2}\right) = \dfrac{2\left(-\frac{1}{2}\right) + 1}{\left(-\frac{1}{2}\right) - 4} = \dfrac{0}{-\frac{9}{2}} = 0$

(c) $f(a) = \dfrac{2a + 1}{a - 4}$ if $a \neq 4$
$f(a) = 7$ if $a = 4$

(d) $f\left(\frac{2}{m}\right) = \dfrac{2\left(\frac{2}{m}\right) + 1}{\frac{2}{m} - 4} = \dfrac{\frac{4}{m} + 1}{\frac{2}{m} - 4}$
$= \dfrac{\frac{4+m}{m}}{\frac{2-4m}{m}} = \dfrac{4 + m}{2 - 4m}$ if $m \neq \frac{1}{2}$

$f\left(\frac{2}{m}\right) = 7$ if $m = \frac{1}{2}$

(e) $\dfrac{2x+1}{x-4} = 1$

$2x + 1 = x - 4$

$x = -5$

45. $f(x) = 6x^2 - 2$

$$\begin{aligned} f(t+1) &= 6(t+1)^2 - 2 \\ &= 6(t^2 + 2t + 1) - 2 \\ &= 6t^2 + 12t + 6 - 2 \\ &= 6t^2 + 12t + 4 \end{aligned}$$

47. $g(r + h)$

$$\begin{aligned} &= (r+h)^2 - 2(r+h) + 5 \\ &= r^2 + 2hr + h^2 - 2r - 2h + 5 \end{aligned}$$

49. $g\left(\dfrac{3}{q}\right) = \left(\dfrac{3}{q}\right)^2 - 2\left(\dfrac{3}{q}\right) + 5$

$$= \frac{9}{q^2} - \frac{6}{q} + 5$$

or $\dfrac{9 - 6q + 5q^2}{q^2}$

51. $f(x) = 2x + 1$

(a) $f(x+h) = 2(x+h) + 1$
$= 2x + 2h + 1$

(b) $f(x+h) - f(x)$

$$\begin{aligned} &= 2x + 2h + 1 - 2x - 1 \\ &= 2h \end{aligned}$$

(c) $\dfrac{f(x+h) - f(x)}{h}$

$$\begin{aligned} &= \frac{2h}{h} \\ &= 2 \end{aligned}$$

53. $f(x) = 2x^2 - 4x - 5$

(a) $f(x+h)$

$$\begin{aligned} &= 2(x+h)^2 - 4(x+h) - 5 \\ &= 2(x^2 + 2hx + h^2) - 4x - 4h - 5 \\ &= 2x^2 + 4hx + 2h^2 - 4x - 4h - 5 \end{aligned}$$

(b) $f(x+h) - f(x)$

$$\begin{aligned} &= 2x^2 + 4hx + 2h^2 - 4x - 4h - 5 \\ &\quad - (2x^2 - 4x - 5) \\ &= 2x^2 + 4hx + 2h^2 - 4x - 4h - 5 \\ &\quad - 2x^2 + 4x + 5 \\ &= 4hx + 2h^2 - 4h \end{aligned}$$

(c) $\dfrac{f(x+h) - f(x)}{h}$

$$\begin{aligned} &= \frac{4hx + 2h^2 - 4h}{h} \\ &= \frac{h(4x + 2h - 4)}{h} \\ &= 4x + 2h - 4 \end{aligned}$$

55. $f(x) = \dfrac{1}{x}$

(a) $f(x+h) = \dfrac{1}{x+h}$

(b) $f(x+h) - f(x)$

$$\begin{aligned} &= \frac{1}{x+h} - \frac{1}{x} \\ &= \left(\frac{x}{x}\right)\frac{1}{x+h} - \frac{1}{x}\left(\frac{x+h}{x+h}\right) \\ &= \frac{x - (x+h)}{x(x+h)} \\ &= \frac{-h}{x(x+h)} \end{aligned}$$

(c) $\dfrac{f(x+h) - f(x)}{h}$

$$\begin{aligned} &= \frac{1}{h}\left[\frac{-h}{x(x+h)}\right] \\ &= \frac{-1}{x(x+h)} \end{aligned}$$

57. A vertical line drawn anywhere through the graph will intersect the graph in only one place. The graph represents a function.

59. A vertical line drawn through the graph may intersect the graph in two places. The graph does not represent a function.

61. A vertical line drawn anywhere through the graph will intersect the graph in only one place. The graph represents a function.

63. $$\begin{aligned} f(x) &= 3x \\ f(-x) &= 3(-x) \\ &= -(3x) \\ &= -f(x) \end{aligned}$$

The function is odd.

65. $$\begin{aligned} f(x) &= 2x^2 \\ f(-x) &= 2(-x)^2 \\ &= 2x^2 \\ &= f(x) \end{aligned}$$

The function is even.

67. $$\begin{aligned} f(x) &= \frac{1}{x^2+4} \\ f(-x) &= \frac{1}{(-x)^2+4} \\ &= \frac{1}{x^2+4} \\ &= f(x) \end{aligned}$$

The function is even.

69. $$\begin{aligned} f(x) &= \frac{x}{x^2-9} \\ f(-x) &= \frac{-x}{(-x)^2-9} \\ &= -\frac{x}{x^2-9} \\ &= -f(x) \end{aligned}$$

The function is odd.

71. If x is a whole number of days, the cost of renting a saw in dollars is $S(x) = 28x + 8$. For x in whole days and a fraction of a day, substitute the next whole number for x in $28x + 8$, because a fraction of a day is charged as a whole day.

(a) $S\left(\frac{1}{2}\right) = S(1) = 28(1) + 8 = 36$

The cost is \$36.

(b) $S(1) = 28(1) + 8 = 36$

The cost is \$36.

(c) $$\begin{aligned} S\left(1\frac{1}{4}\right) &= S(2) = 28(2) + 8 \\ &= 56 + 8 = 64 \end{aligned}$$

The cost is \$64.

(d) $$\begin{aligned} S\left(3\frac{1}{2}\right) &= S(4) = 28(4) + 8 \\ &= 112 + 8 = 120 \end{aligned}$$

The cost is \$120.

(e) $S(4) = 28(4) + 8 = 112 + 8 = 120$

The cost is \$120.

(f) $$\begin{aligned} S\left(4\frac{1}{10}\right) &= S(5) = 28(5) + 8 \\ &= 140 + 8 = 148 \end{aligned}$$

The cost is \$148.

(g) $$\begin{aligned} S\left(4\frac{9}{10}\right) &= S(5) = 28(5) + 8 \\ &= 140 + 8 = 148 \end{aligned}$$

The cost is \$148.

(h) To continue the graph, continue the horizontal bars up and to the right.

(i) The independent variable is x, the number of full and partial days.

(j) The dependent variable is S, the cost of renting a saw.

(k) S is not a linear function. Its graph is not a continuous straight line. S is a step function.

73. (a)

$$\begin{aligned} f(250{,}000) &= 0.40(150{,}000) + 0.333(100{,}000) \\ &= 93{,}300 \end{aligned}$$

The maximum amount that an attorney can receive for a \$250,000 jury award is \$93,300.

(b) $$\begin{aligned} f(350{,}000) &= 0.40(150{,}000) \\ &\quad + 0.333(150{,}000) \\ &\quad + 0.30(50{,}000) \\ &= 124{,}950 \end{aligned}$$

The maximum amount that an attorney can receive for a \$350,000 jury award is \$124,500.

(c)

$$\begin{aligned} f(550{,}000) &= 0.40(150{,}000) + 0.333(150{,}000) \\ &\quad + 0.30(200{,}000) + 0.24(50{,}000) \\ &= 181{,}950 \end{aligned}$$

The maximum amount that an attorney can receive for a \$550,000 jury award is \$181,950.

(d)

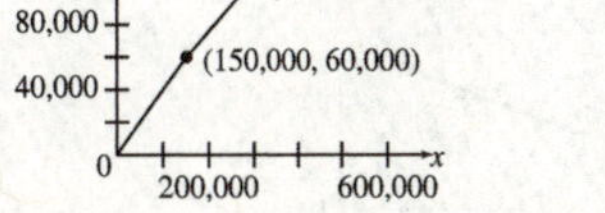

f(x)
160,000
120,000
80,000
40,000
0
200,000
600,000
x
(500,000, 169,950)
(300,000, 109,950)
(150,000, 60,000)

75. **(a)** The curve in the graph crosses the point with x-coordinate 17:37 and y-coordinate of approximately 140. So, at time 17 hours, 37 minutes the whale reaches a depth of about 140 m.

(b) The curve in the graph crosses the point with x-coordinate 17:39 and y-coordinate of approximately 240. So, at time 17 hours, 39 minutes the whale reaches a depth of about 250 m.

77. **(a)** (i) By the given function f, a muskrat weighing 800 g expends

$$f(800) = 0.01(800)^{0.88}$$
$$\approx 3.6, \text{or approximately}$$

3.6 kcal/km when swimming at the surface of the water.

(ii) A sea otter weighing 20,000 g expends

$$f(20{,}000) = 0.01(20{,}000)^{0.88}$$
$$\approx 61, \text{ or approximately}$$

61 kcal/km when swimming at the surface of the water.

(b) If z is the number of kilograms of an animal's weight, then $x = g(z) = 1000z$ is the number of grams since 1 kilogram equals 1000 grams.

(c)
$$\begin{aligned} f(g(z)) &= f(1000z) \\ &= 0.01(1000z)^{0.88} \\ &= 0.01(1000^{0.88})z^{0.88} \\ &\approx 4.4z^{0.88} \end{aligned}$$

79. **(a)** $P = 2l + 2w$

However, $lw = 500$, so $l = \dfrac{500}{w}$.

$$P(w) = 2\left(\frac{500}{w}\right) + 2w$$

$$P(w) = \frac{1000}{w} + 2w$$

(b) Since $l = \frac{500}{w}$, $w \neq 0$ but w could be any positive value. Therefore, the domain of P is $0 < w < \infty$, or $(0, \infty)$.

(c)

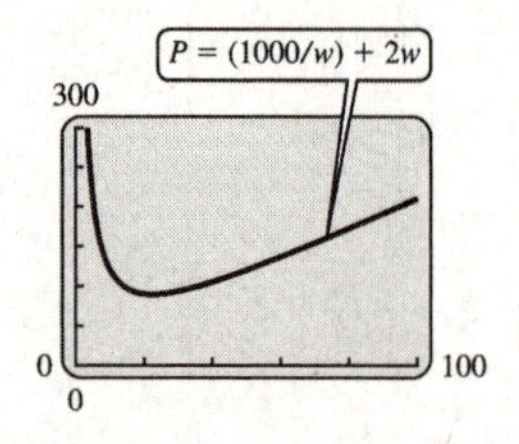

10.2 Quadratic Functions; Translation and Reflection

Your Turn 1

$f(x) = 2x^2 - 6x - 1$

(a)
$$\begin{aligned} y &= 2x^2 - 6x - 1 \\ &= 2(x^2 - 3x) - 1 \\ &= 2\left(x^2 - 3x + \frac{9}{4}\right) - 1 - 2\left(\frac{9}{4}\right) \\ &= 2\left(x - \frac{3}{2}\right)^2 - \frac{11}{2} \end{aligned}$$

(b) $y = 2(0)^2 - 6(0) - 1 = -1$

The y-intercept is -1.

(c)

$$\begin{aligned} 0 &= 2\left(x - \frac{3}{2}\right)^2 - \frac{11}{2} \\ \frac{11}{2} &= 2\left(x - \frac{3}{2}\right)^2 \\ \frac{11}{4} &= \left(x - \frac{3}{2}\right)^2 \\ \pm\frac{\sqrt{11}}{2} &= x - \frac{3}{2} \\ \frac{3 \pm \sqrt{11}}{2} &= x \end{aligned}$$

(d) Since $y = 2\left(x - \frac{3}{2}\right)^2 - \frac{11}{2}$ is in the form $y = a(x - h)^2 + k$, we get $h = \frac{3}{2}$ and $k = -\frac{11}{2}$. So the vertex is $\left(\frac{3}{2}, -\frac{11}{2}\right)$.

(e)

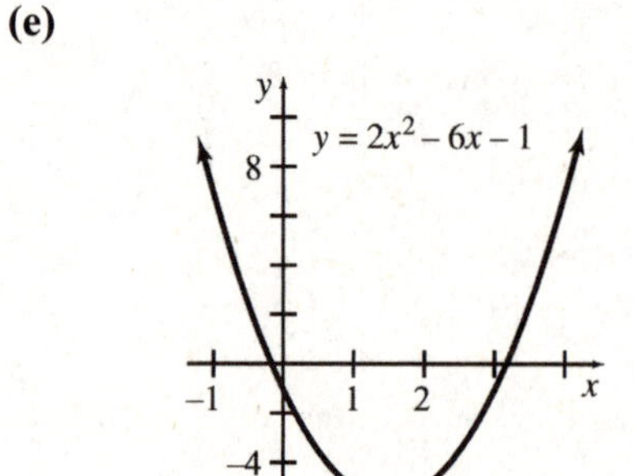

Your Turn 2

Let x be the number of \$40 decreases in the price.

$$\text{Price per person} = 1650 - 40x$$
$$\text{Number of people} = 900 + 80x$$

$$\begin{aligned} R(x) &= (1650 - 40x)(900 + 80x). \\ &= 1{,}485{,}000 + 96{,}000x - 3200x^2 \end{aligned}$$
$$x = \frac{-b}{2a} = \frac{-96{,}000}{2(-3200)} = \frac{-96{,}000}{-6400} = 15$$

The y-coordinate is

$$\begin{aligned} y &= 1{,}485{,}000 + 96{,}000(15) - 3200(15)^2 \\ &= 2{,}205{,}000 \end{aligned}$$

The maximum revenue is \$2,205,000 when $1650 - 40(15) = \$1050$ per person is charged.

Your Turn 3

$R(x) = -x^2 + 40x$ and $C(x) = 8x + 102$

(a)
$$\begin{aligned} R(x) &= C(x) \\ -x^2 + 40x &= 8x + 192 \\ 0 &= x^2 - 32x + 192 \\ 0 &= (x - 24)(x - 8) \end{aligned}$$

The two graphs cross when $x = 8$ or $x = 20$. The minimum break-even quantity is $x = 8$, so the deli owner must sell 8 lb of cream cheese to break even.

(b) The maximum revenue for $R(x) = -x^2 + 40x$ is at

$$x = \frac{-b}{2a} = \frac{-40}{2(-1)} = 20.$$

The maximum revenue is

$$R(20) = -20^2 + 40(20) = 400, \text{ or } \$400.$$

(c)
$$\begin{aligned} P(x) &= R(x) - C(x) \\ &= -x^2 + 40x - 8x - 192 \\ &= -x^2 + 32x - 192 \end{aligned}$$

The maximum point of $P(x)$ is at

$$x = \frac{-b}{2a} = \frac{-32}{2(-1)} = 16.$$

$$P(16) = -(16)^2 + 32(16) - 192 = 64$$

So, the maximum profit is \$64.

10.2 Exercises

3. The graph of $y = x^2 - 3$ is the graph of $y = x^2$ translated 3 units downward.

This is graph **D**.

5. The graph of $y = (x - 3)^2 + 2$ is the graph of $y = x^2$ translated 3 units to the right and 2 units upward.

This is graph **A**.

7. The graph of $y = -(3 - x)^2 + 2$ is the same as the graph of $y = -(x - 3)^2 + 2$. This is the graph of $y = x^2$ reflected in the x-axis, translated 3 units to the right, and translated 2 units upward.

This is graph **C**.

9.
$$\begin{aligned} y &= 3x^2 + 9x + 5 \\ &= 3(x^2 + 3x) + 5 \\ &= 3\left(x^2 + 3x + \frac{9}{4}\right) + 5 - 3\left(\frac{9}{4}\right) \\ &= 3\left(x + \frac{3}{2}\right)^2 - \frac{7}{4} \end{aligned}$$

The vertex is $\left(-\frac{3}{2}, -\frac{7}{4}\right)$.

11.
$$\begin{aligned} y &= -2x^2 + 8x - 9 \\ &= -2(x^2 - 4x) - 9 \\ &= -2(x^2 - 4x + 4) - 9 - (-2)(4) \\ &= -2(x - 2)^2 - 1 \end{aligned}$$

The vertex is $(2, -1)$.

13.
$$\begin{aligned} y &= x^2 + 5x + 6 \\ y &= (x + 3)(x + 2) \end{aligned}$$

Set $y = 0$ to find the x-intercepts.

$$\begin{aligned} 0 &= (x + 3)(x + 2) \\ x &= -3, x = -2 \end{aligned}$$

The x-intercepts are -3 and -2. Set $x = 0$ to find the y-intercept.

$$\begin{aligned} y &= 0^2 + 5(0) + 6 \\ y &= 6 \end{aligned}$$

The y-intercept is 6.

The x-coordinate of the vertex is

$$x = \frac{-b}{2a} = \frac{-5}{2} = -\frac{5}{2}.$$

Substitute to find the y-coordinate.

$$y = \left(-\frac{5}{2}\right)^2 + 5\left(-\frac{5}{2}\right) + 6 = \frac{25}{4} - \frac{25}{2} + 6 = -\frac{1}{4}$$

The vertex is $\left(-\frac{5}{2}, -\frac{1}{4}\right)$

The axis is $x = -\frac{5}{2}$, the vertical line through the vertex.

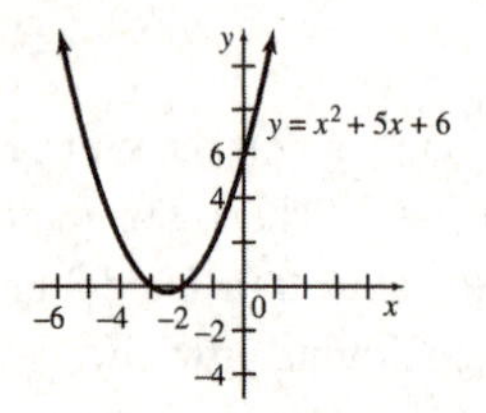

15. $y = -2x^2 - 12x - 16$

$= -2(x^2 + 6x + 8)$

$= -2(x + 4)(x + 2)$

Let $y = 0$.

$$0 = -2(x + 4)(x + 2)$$
$$x = -4, x = -2$$

-4 and -2 are the x-intercepts. Let $x = 0$.

$$y = -2(0)^2 + 12(0) - 16$$

-16 is the y-intercept.

Vertex: $x = \frac{-b}{2a} = \frac{12}{-4} = -3$

$$y = -2(-3)^2 - 12(-3) - 16$$
$$= -18 + 36 - 16 = 2$$

The vertex is $(-3, 2)$.

The axis is $x = -3$, the vertical line through the vertex.

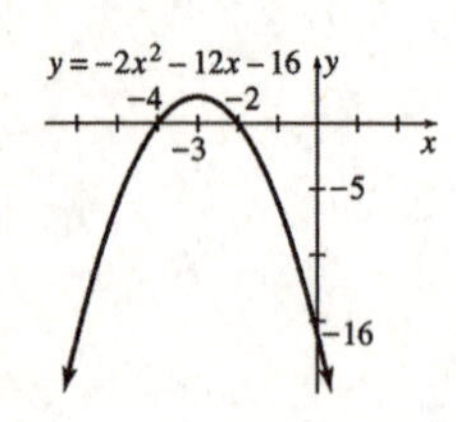

17. $y = 2x^2 + 8x - 8$

Let $y = 0$.

$$2x^2 + 8x - 8 = 0$$
$$x^2 + 4x - 4 = 0$$

$$x = \frac{-4 \pm \sqrt{4^2 - 4(1)(-4)}}{2(1)}$$
$$= \frac{-4 \pm \sqrt{32}}{2} = \frac{-4 \pm 4\sqrt{2}}{2}$$
$$= -2 \pm 2\sqrt{2}$$

The x-intercepts are $-2 \pm 2\sqrt{2} \approx 0.83$ or -4.83.

Let $x = 0$.

$$y = 2(0)^2 + 8(0) - 8 = -8$$

The y-intercept is -8.

The x-coordinate of the vertex is

$$x = \frac{-b}{2a} = -\frac{8}{4} = -2.$$

If $x = -2$,

$$y = 2(-2)^2 + 8(-2) - 8$$
$$= 8 - 16 - 8 = -16.$$

The vertex is $(-2, -16)$.

The axis is $x = -2$.

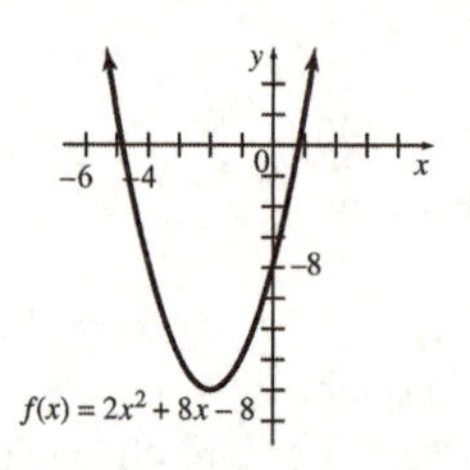

19. $f(x) = 2x^2 - 4x + 5$

Let $f(x) = 0$.

$$0 = 2x^2 - 4x + 5$$
$$x = \frac{-(-4) \pm \sqrt{(-4)^2 - 4(2)(5)}}{2(2)}$$
$$= \frac{4 \pm \sqrt{16 - 40}}{4}$$
$$= \frac{4 \pm \sqrt{-24}}{4}$$

Since the radicand is negative, there are no x-intercepts.

Let $x = 0$.

$$y = 2(0)^2 - 4(0) + 5$$
$$y = 5$$

5 is the y-intercept.

Vertex: $x = \dfrac{-b}{2a} = \dfrac{-(-4)}{2(2)} = \dfrac{4}{4} = 1$

$y = 2(1)^2 - 4(1) + 5 = 2 - 4 + 5 = 3$

The vertex is (1, 3).

The axis is $x = 1$.

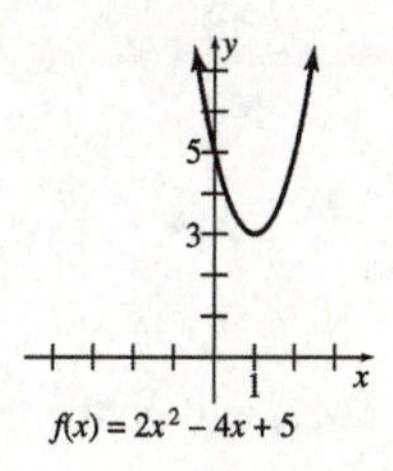

21. $f(x) = -2x^2 + 16x - 21$

Let $f(x) = 0$

Use the quadratic formula.

$$\begin{aligned} x &= \frac{-16 \pm \sqrt{16^2 - 4(-2)(-21)}}{2(-2)} \\ &= \frac{-16 \pm \sqrt{88}}{-4} \\ &= \frac{-16 \pm 2\sqrt{22}}{-4} \\ &= 4 \pm \frac{\sqrt{22}}{2} \end{aligned}$$

The x-intercepts are $4 + \frac{\sqrt{22}}{2} \approx 6.35$ and $4 - \frac{\sqrt{22}}{2} \approx 1.65$.

Let $x = 0$.

$$\begin{aligned} y &= -2(0)^2 + 16(0) - 21 \\ y &= -21 \end{aligned}$$

-21 is the y-intercept.

Vertex: $x = \dfrac{-b}{2a} = \dfrac{-16}{2(-2)} = \dfrac{-16}{-4} = 4$

$$\begin{aligned} y &= -2(4)^2 + 16(4) - 21 \\ &= -32 + 64 - 21 = 11 \end{aligned}$$

The vertex is (4, 11).

The axis is $x = 4$.

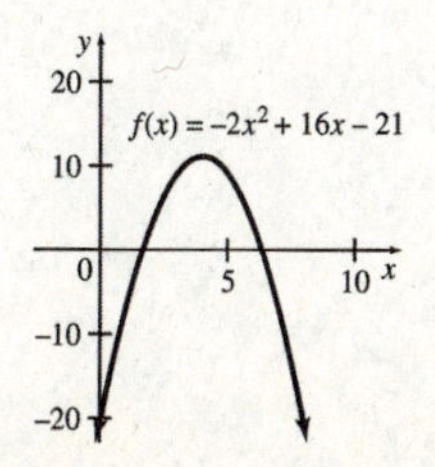

23. $y = \dfrac{1}{3}x^2 - \dfrac{8}{3}x + \dfrac{1}{3}$

Let $y = 0$.

$$0 = \frac{1}{3}x^2 - \frac{8}{3}x + \frac{1}{3}$$

Multiply by 3.

$$\begin{aligned} 0 &= x^2 - 8x + 1 \\ x &= \frac{-(-8) \pm \sqrt{(-8)^2 - 4(1)(1)}}{2(1)} \\ &= \frac{8 \pm \sqrt{64 - 4}}{2} = \frac{8 \pm \sqrt{60}}{2} \\ &= \frac{8 \pm 2\sqrt{15}}{2} = 4 \pm \sqrt{15} \end{aligned}$$

The x-intercepts are $4 + \sqrt{15} \approx 7.87$ and $4 - \sqrt{15} \approx 0.13$.

Let $x = 0$.

$$y = \frac{1}{3}(0)^2 - \frac{8}{3}(0) + \frac{1}{3}$$

$\frac{1}{3}$ is the y-intercept.

Vertex: $x = \dfrac{-b}{2a} = \dfrac{-\left(-\frac{8}{3}\right)}{2\left(\frac{1}{3}\right)} = \dfrac{\frac{8}{3}}{\frac{2}{3}} = 4$

$$\begin{aligned} y &= \frac{1}{3}(4)^2 - \frac{8}{3}(4) + \frac{1}{3} \\ &= \frac{16}{3} - \frac{32}{3} + \frac{1}{3} = -\frac{15}{3} = -5 \end{aligned}$$

The vertex is (4, –5).

The axis is $x = 4$.

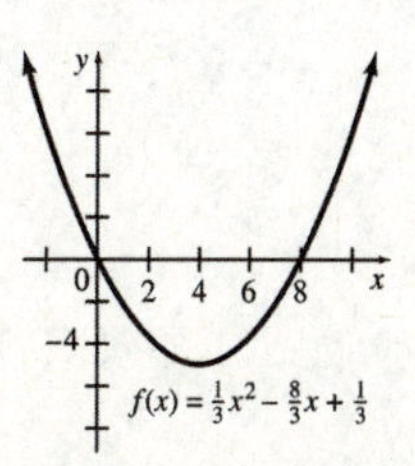

25. The graph of $y = \sqrt{x + 2} - 4$ is the graph of $y = \sqrt{x}$ translated 2 units to the left and 4 units downward.

This is graph **D**.

27. The graph of $y = \sqrt{-x+2} - 4$ is the graph of $y = \sqrt{-(x-2)} - 4$, which is the graph of $y = \sqrt{x}$ reflected in the y-axis, translated 2 units to the right, and translated 4 units downward.

This is graph **C**.

29. The graph of $y = -\sqrt{x+2} - 4$ is the graph of $y = \sqrt{x}$ reflected in the x-axis, translated 2 units to the left, and translated 4 units downward.
This is graph **E**.

31. The graph of $y = -f(x)$ is the graph of $y = f(x)$ reflected in the x-axis.

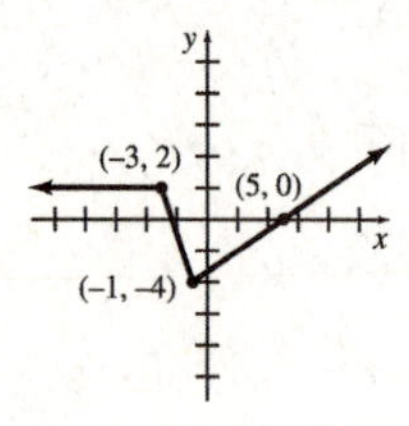

33. The graph of $y = f(-x)$ is the graph of $y = f(x)$ reflected in the y-axis.

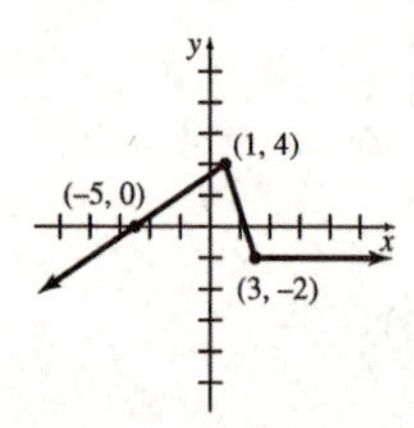

35. $f(x) = \sqrt{x-2} + 2$

Translate the graph of $f(x) = \sqrt{x}$ 2 units right and 2 units up.

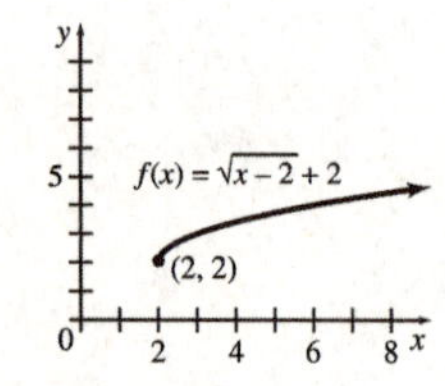

37. $f(x) = -\sqrt{2-x} - 2$
$= -\sqrt{-(x-2)} - 2$

Reflect the graph of $f(x)$ vertically and horizontally. Translate the graph 2 units right and 2 units down.

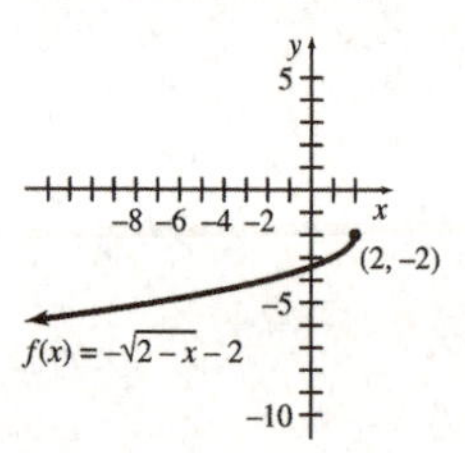

39. If $0 < a < 1$, the graph of $f(ax)$ will be the graph of $f(x)$ stretched horizontally.

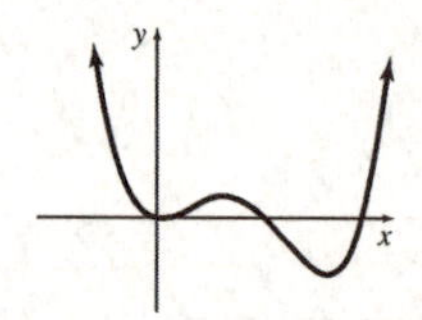

41. If $-1 < a < 0$, the graph of $f(ax)$ will be reflected horizontally, since a is negative. It will be stretched horizontally.

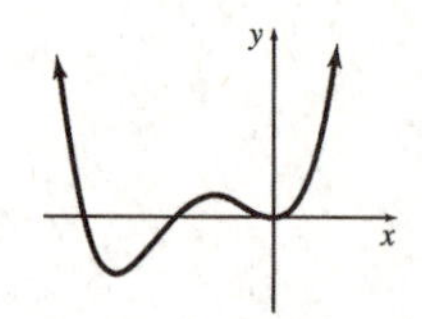

43. If $0 < a < 1$, the graph of $af(x)$ will be flatter than the graph of $f(x)$. Each y-value is only a fraction of the height of the original y-values.

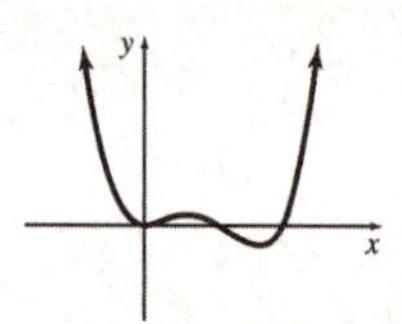

45. If $-1 < a < 0$, the graph will be reflected vertically, since a will be negative. Also, because a is a fraction, the graph will be flatter because each y-value will only be a fraction of its original height.

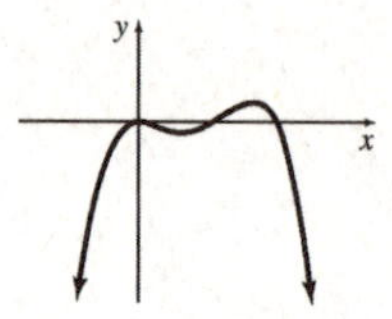

47. **(a)** Since the graph of $y = f(x)$ is reflected vertically to obtain the graph of $y = -f(x)$, the x-intercept is unchanged. The x-intercept of the graph of $y = f(x)$ is r.

(b) Since the graph of $y = f(x)$ is reflected horizontally to obtain the graph of $y = f(-x)$, the x-intercept of the graph of $y = f(-x)$ is $-r$.

(c) Since the graph of $y = f(x)$ is reflected both horizontally and vertically to obtain the graph of $y = -f(-x)$, the x-intercept of the graph of $y = -f(-x)$ is $-r$.

49. **(a)**

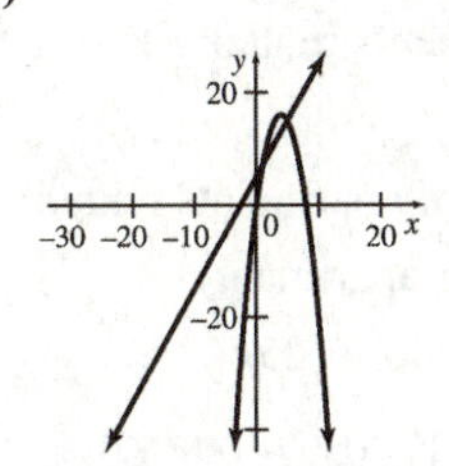

(b) Break-even quantities are values of $x =$ batches of widgets for which revenue and cost are equal.

Set $R(x) = C(x)$ and solve for x.

$$-x^2 + 8x = 2x + 5$$
$$x^2 - 6x + 5 = 0$$
$$(x - 5)(x - 1) = 0$$
$$x - 5 = 0 \text{ or } x - 1 = 0$$
$$x = 5 \text{ or } \quad x = 1$$

So, the break-even quantities are 1 and 5. The minimum break-even quantity is 1 batch of widgets.

(c) The maximum revenue occurs at the vertex of R. Since $R(x) = -x^2 + 8x$, then the x-coordinate of the vertex is

$$x = -\frac{b}{2a} = -\frac{8}{2(-1)} = 4.$$

So, the maximum revenue is

$$R(4) = -4^2 + 8(4) = 16, \text{ or } \$16{,}000.$$

(d) The maximum profit is the maximum difference $R(x) - C(x)$. Since

$$P(x) = R(x) - C(x)$$
$$= -x^2 + 8x - (2x + 5)$$
$$= -x^2 + 6x - 5$$

is a quadratic function, we can find the maximum profit by finding the vertex of P. This occurs at

$$x = -\frac{b}{2a} = \frac{-6}{2(-1)} = 3.$$

Therefore, the maximum profit is

$$P(3) = -(3)^2 + 6(3) - 5 = 4, \text{ or } \$4000.$$

51. **(a)**

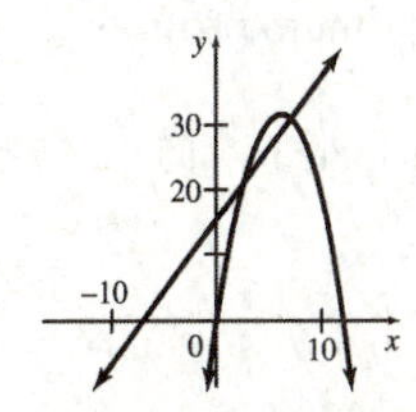

(b) Break-even quantities are values of $x =$ batches of widgets for which revenue equals cost.

Set $R(x) = C(x)$ and solve for x.

$$-\frac{4}{5}x^2 + 10x = 2x + 15$$
$$\frac{4}{5}x^2 - 8x + 15 = 0$$
$$4x^2 - 40x + 75 = 0$$
$$4x^2 - 10x - 30x + 75 = 0$$
$$2x(2x - 5) - 15(2x - 5) = 0$$
$$(2x - 5)(2x - 15) = 0$$
$$2x - 5 = 0 \text{ or } 2x - 15 = 0$$
$$x = 2.5 \quad \text{or} \quad x = 7.5$$

So, the break-even quantities are 2.5 and 7.5 with the minimum break-even quantity being 2.5 batches of widgets.

(c) The maximum revenue occurs at the vertex of R. Since $R(x) = -\frac{4}{5}x^2 + 10x$, then the x-coordinate of the vertex is

$$x = -\frac{b}{2a} = -\frac{10}{2\left(-\frac{4}{5}\right)} = 6.25.$$

So, the maximum revenue is

$$R(6.25) = 31.25, \text{ or } \$31{,}250.$$

(d) The maximum profit is the maximum difference $R(x) - C(x)$. Since

$$P(x) = R(x) - C(x)$$
$$= -\frac{4}{5}x^2 + 10x - (2x + 15)$$
$$= -\frac{4}{5}x^2 + 8x - 15$$

is a quadratic function, we can find the maximum profit by finding the vertex of P. This occurs at

$$x = -\frac{b}{2a} = -\frac{8}{2\left(-\frac{4}{5}\right)} = 5.$$

Therefore, the maximum profit is

$$P(5) = -\frac{4}{5}5^2 + 8(5) - 15 = 5, \text{ or } \$5000.$$

53. $R(x) = 8000 + 70x - x^2$

$$= -x^2 + 70x + 8000$$

The maximum revenue occurs at the vertex.

$$x = \frac{-b}{2a} = \frac{-70}{2(-1)} = 35$$
$$y = 8000 + 70(35) - (35)^2$$
$$= 8000 + 2450 - 1225$$
$$= 9225$$

The vertex is (35, 9225). The maximum revenue of \$9225 is realized when 35 seats are left unsold.

55. $p = 500 - x$

(a) The revenue is

$$R(x) = px$$
$$= (500 - x)(x)$$
$$= 500x - x^2.$$

(b)

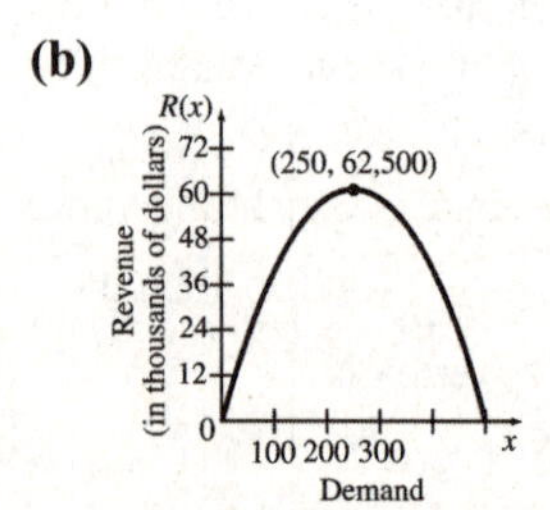

(c) From the graph, the vertex is halfway between $x = 0$ and $x = 500$, so $x = 250$ units corresponds to maximum revenue. Then the price is

$$p = 500 - x$$
$$= 500 - 250 = \$250.$$

Note that price, p, cannot be read directly from the graph of

$$R(x) = 500x - x^2.$$

(d) $R(x) = 500x - x^2$

$$= -x^2 + 500x$$

Find the vertex.

$$x = \frac{-b}{2a} = \frac{-500}{2(-1)} = 250$$
$$y = -(250)^2 + 500(250)$$
$$= 62,500$$

The vertex is (250, 62,500).

The maximum revenue is \$62,500.

57. Let $x =$ the number of \$25 increases.

(a) Rent per apartment: $800 + 25x$

(b) Number of apartments rented: $80 - x$

(c) Revenue:

$R(x) =$ (number of apartments rented) $\times$ (rent per apartment)

$$= (80 - x)(800 + 25x)$$
$$= -25x^2 + 1200x + 64,000$$

(d) Find the vertex:

$$x = \frac{-b}{2a} = \frac{-1200}{2(-25)} = 24$$
$$y = -25(24)^2 + 1200(24) + 64,000$$
$$= 78,400$$

The vertex is (24, 78,400). The maximum revenue occurs when $x = 24$.

(e) The maximum revenue is the y-coordinate of the vertex, or \$78,400.

59. $S(x) = 1 - 0.058x - 0.076x^2$

(a) $0.50 = 1 - 0.058x - 0.076x^2$

$$0.076x^2 + 0.058x - 0.50 = 0$$
$$76x^2 + 58x - 500 = 0$$
$$38x^2 + 29x - 250 = 0$$

$$x = \frac{-29 \pm \sqrt{(29)^2 - 4(38)(-250)}}{2(38)}$$
$$= \frac{-29 \pm \sqrt{38,841}}{76}$$

$$\frac{-29 - \sqrt{38,841}}{76} \approx -2.97$$

and $\frac{-29 + \sqrt{38,841}}{76} \approx 2.21$

We ignore the negative value.

The value $x = 2.2$ represents 2.2 decades or 22 years, and 22 years after 65 is 87.

The median length of life is 87 years.

(b) If nobody lives, $S(x) = 0$.

$$1 - 0.058x - 0.076x^2 = 0$$

$$76x^2 + 58x - 1000 = 0$$

$$38x^2 + 29x - 500 = 0$$

$$x = \frac{-29 \pm \sqrt{(29)^2 - 4(38)(-500)}}{2(38)}$$

$$= \frac{-29 \pm \sqrt{76{,}841}}{76}$$

$$\frac{-29 - \sqrt{76{,}841}}{76} \approx -4.03$$

and $$\frac{-29 + \sqrt{76{,}841}}{76} \approx 3.27$$

We ignore the negative value.

The value $x = 3.3$ represents 3.3 decades or 33 years, and 33 years after 65 is 98.

Virtually nobody lives beyond 98 years.

61. (a) The vertex of the quadratic function $y = 0.057x - 0.001x^2$ is at

$$x = -\frac{b}{2a} = -\frac{0.057}{2(-0.001)} = 28.5.$$

Since the coefficient of the leading term, -0.001, is negative, then the graph of the function opens downward, so a maximum is reached at 28.5 weeks of gestation.

(b) The maximum splenic artery resistance reached at the vertex is

$$y = 0.057(28.5) - 0.001(28.5)^2 \approx 0.81.$$

(c) The splenic artery resistance equals 0, when $y = 0$.

$0.057x - 0.001x^2 = 0$ Substitute in the expression in x for y

$x(0.057 - 0.001x) = 0$ Factor

$x = 0$ or $0.057 - 0.001x = 0$ Set each factor equal to 0.

$$x = \frac{0.057}{0.001} = 57$$

So, the splenic artery resistance equals 0 at 0 weeks or 57 weeks of gestation.

No, this is not reasonable because at $x = 0$ or 57 weeks, the fetus does not exist.

63. (a)

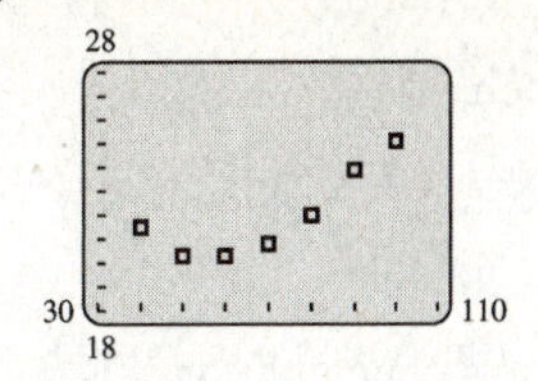

(b) Quadratic

(c) $y = 0.002726x^2 - 0.3113x + 29.33$

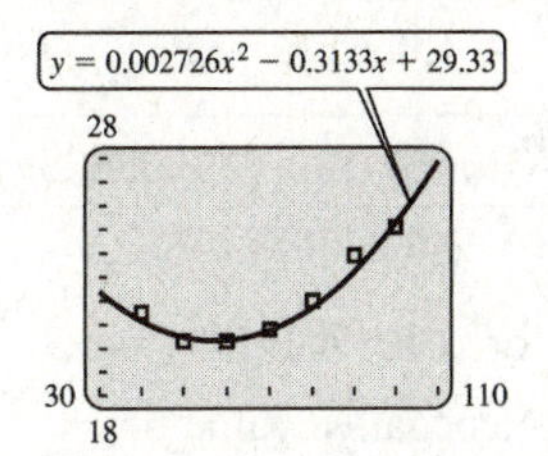

(d) Given that $(h, k) = (60, 20.3)$, the equation has the form

$$y = a(x - 60)^2 + 20.3.$$

Since (100, 25.1) is also on the curve.

$$25.1 = a(100 - 60)^2 + 20.3$$

$$4.8 = 1600a$$

$$a = 0.003$$

A quadratic function that models the data is $f(x) = 0.003(x - 60)^2 + 20.3$.

(e)

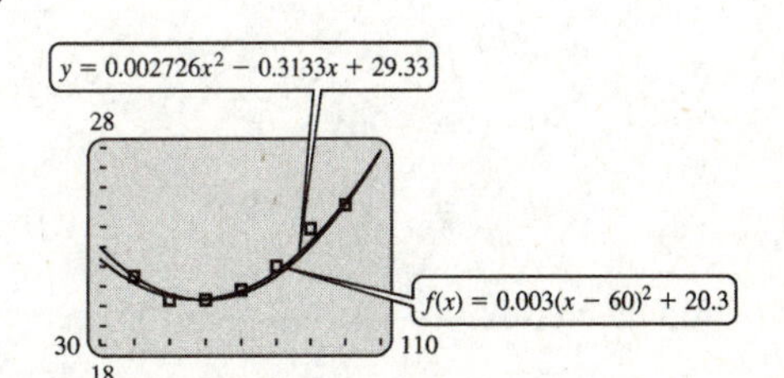

The two graphs are very close.

65. $f(x) = 60.0 - 2.28x + 0.0232x^2$

$$\frac{-b}{2a} = -\frac{-2.28}{2(0.0232)} \approx 49.1$$

The minimum value occurs when $x \approx 49.1$. The age at which the accident rate is a minimum is 49 years. The minimum rate is

$$f(49.1) = 60.0 - 2.28(49.1) + 0.0232(49.1)^2$$

$$= 60.0 - 111.948 + 55.930792$$

$$\approx 3.98.$$

67. $y = 0.056057x^2 + 1.06657x$

(a) If $x = 25$ mph,

$$y = 0.056057(25)^2 + 1.06657(25)$$
$$y \approx 61.7.$$

At 25 mph, the stopping distance is approximately 61.7 ft.

(b) $$0.056057x^2 + 1.06657x = 150$$
$$0.056057x^2 + 1.06657x - 150 = 0$$

$$x = \frac{-1.06657 \pm \sqrt{(1.06657)^2 - 4(0.056057)(-150)}}{2(0.056057)}$$

$x \approx 43.08$ or $x \approx -62.11$

We ignore the negative value.

To stop within 150 ft, the fastest speed you can drive is 43.08 mph or about 43 mph.

69. Let $x =$ the length of the lot and $y =$ the width of the lot.

The perimeter is given by

$$P = 2x + 2y$$
$$380 = 2x + 2y$$
$$190 = x + y$$
$$190 - x = y.$$

Area $= xy$ (quantity to be maximized)

$$A = x(190 - x)$$
$$= 190x - x^2$$
$$= -x^2 + 190x$$

Find the vertex: $\frac{-b}{2a} = \frac{-190}{-2} = 95$

$$y = -(95)^2 + 190(95)$$
$$= 9025$$

This is a parabola with vertex (95,9025) that opens downward. The maximum area is the value of A at the vertex, or 9025 sq ft.

71. Sketch the culvert on the xy-axes as a parabola that opens upward with vertex at (0, 0).

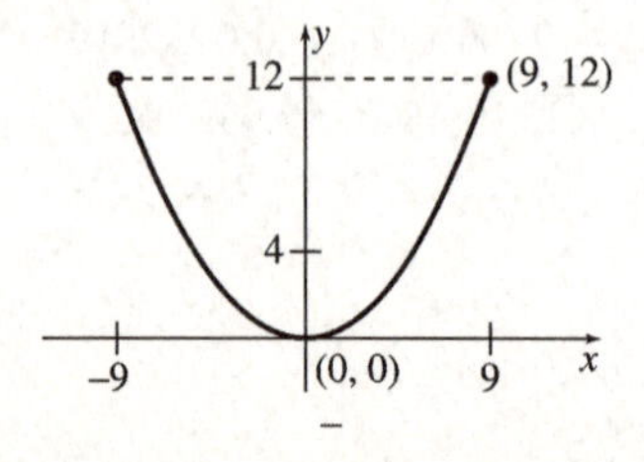

The equation is of the form $y = ax^2$, Since the culvert is 18 ft wide at 12 ft from its vertex, the points (9, 12) and (−9, 12) are on the parabola. Use (9, 12) as one point on the parabola.

$$12 = a(9)^2$$
$$12 = 81a$$
$$\frac{12}{81} = a$$
$$\frac{4}{27} = a$$

Thus, $y = \frac{4}{27}x^2$.

To find the width 8 feet from the top, find the points with

$$y\text{-value} = 12 - 8 = 4.$$

Thus,

$$4 = \frac{4}{27}x^2$$
$$108 = 4x^2$$
$$27 = x^2$$
$$x^2 = 27$$
$$x = \pm\sqrt{27}$$
$$x = \pm 3\sqrt{3}.$$

The width of the culvert is $2\left(3\sqrt{3}\right) = 6\sqrt{3}$ ft ≈ 10.39 ft.

10.3 Polynomial and Rational Functions

Your Turn 1

Graph $f(x) = 64 - x^6$.

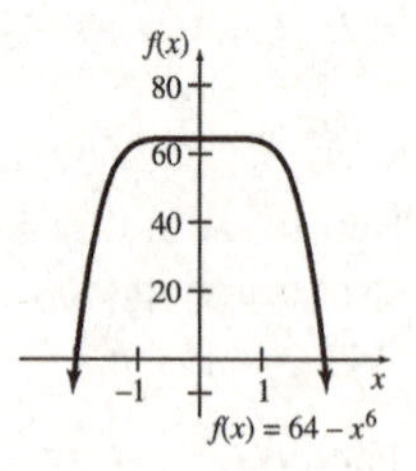

Your Turn 2

Graph $y = \frac{4x - 6}{x - 3}$.

The value $x = 3$ makes the denominator 0, so 3 is not in the domain and the line $x = 3$ is a vertical asymptote. Let x get larger and larger.

Then $\frac{4x-6}{x-3} \approx \frac{4x}{x} = 4$ as x gets larger and larger.
So, the line $y = 4$ is a horizontal asymptote.

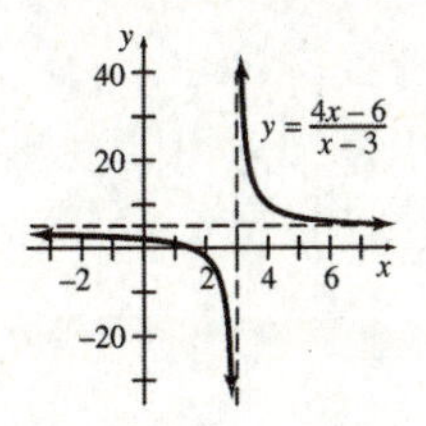

10.3 Exercises

3. The graph of $f(x) = (x - 2)^3 + 3$ is the graph of $y = x^3$ translated 2 units to the right and

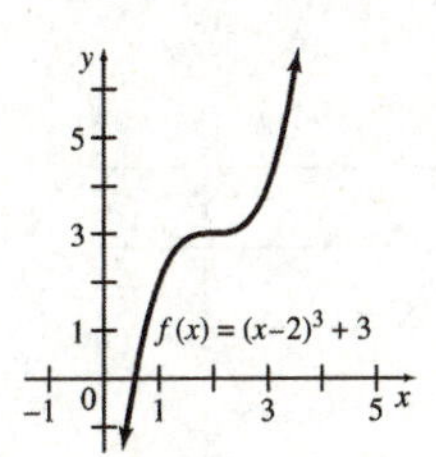

5. The graph of $f(x) = -(x + 3)^4 + 1$ is the graph of $y = x^4$ reflected horizontally, translated 3 units to the left, and translated 1 unit upward.

7. The graph of $y = x^3 - 7x - 9$ has the right end up, the left end down, at most two turning points, and a y-intercept of -9.

   This is graph **D**.

9. The graph of $y = -x^3 - 4x^2 + x + 6$ has the right end down, the left end up, at most two turning points, and a y-intercept of 6.

   This is graph **E**.

11. The graph of $y = x^4 - 5x^2 + 7$ has both ends up, at most three turning points, and a y-intercept of 7.

    This is graph **I**.

13. The graph of $y = -x^4 + 2x^3 + 10x + 15$ has both ends down, at most three turning points, and a y-intercept of 15.

    This is graph **G**.

15. The graph of $y = -x^5 + 4x^4 + x^3 - 16x^2 + 12x + 5$ has the right end down, the left end up, at most four turning points, and a y-intercept of 5.

    This is graph **A**.

17. The graph of $y = \frac{2x^2 + 3}{x^2 + 1}$ has no vertical asymptote, the line with equation $y = 2$ as a horizontal asymptote, and a y-intercept of 3.

    This is graph **D**.

19. The graph $y = \frac{-2x^2 - 3}{x^2 + 1}$ has no vertical asymptote, the line with equation $y = -2$ as a horizontal asymptote, and a y-intercept of -3.

    This is graph **E**.

21. The right end is up and the left end is up. There are three turning points.
    The degree is an even integer equal to 4 or more.
    The x^n term has a $+$ sign.

23. The right end is up and the left end is down. There are four turning points. The degree is an odd integer equal to 5 or more. The x^n term has a $+$ sign.

25. The right end is down and the left end is up. There are six turning points. The degree is an odd integer equal to 7 or more. The x^n term has a $-$ sign.

27. $y = \frac{-4}{x + 2}$

    The function is undefined for $x = -2$, so the line $x = -2$ is a vertical asymptote.

x	-102	-12	-7	-5	-3	-1	8	98
$x + 2$	-100	-10	-5	-3	-1	1	10	100
y	0.04	0.4	0.8	1.3	4	-4	-0.4	-0.04

The graph approaches $y = 0$, so the line $y = 0$ (the x-axis) is a horizontal asymptote.

Asymptotes: $y = 0$, $x = -2$

x-intercept:

none

y-intercept:

-2, the value when $x = 0$

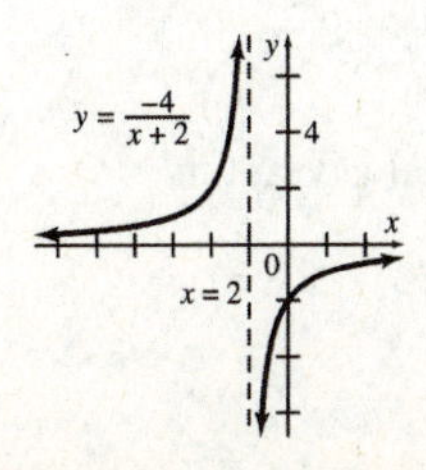

29. $y = \dfrac{2}{3 + 2x}$

$3 + 2x = 0$ when $2x = -3$ or $x = -\frac{3}{2}$, so the line $x = -\frac{3}{2}$ is a vertical asymptote.

x	-51.5	-6.5	-2	-1	3.5	48.5
$3 + 2x$	-100	-10	-1	1	10	100
y	-0.02	-0.2	-2	2	0.2	0.02

The graph approaches $y = 0$, so the line $y = 0$ (the x-axis) is a horizontal asymptote.

Asymptote: $y = 0$, $x = -\frac{3}{2}$

x-intercept:

none

y-intercept:

$\frac{2}{3}$, the value when $x = 0$

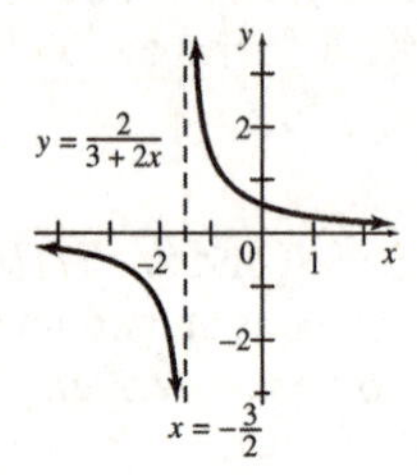

31. $y = \dfrac{2x}{x - 3}$

$x - 3 = 0$ when $x = 3$, so the line $x = 3$ is a vertical asymptote.

x	-97	-7	-1	1	2	2.5
$2x$	-194	-14	-2	2	4	5
$x - 3$	-100	-10	-4	-2	-1	-0.5
y	1.94	1.4	0.5	-1	-4	-10

x	3.5	4	5	7	11	103
$2x$	7	8	10	14	22	206
$x - 3$	0.5	1	2	4	8	100
y	14	8	5	3.5	2.75	2.06

As x gets larger,

$$\frac{2x}{x - 3} \approx \frac{2x}{x} = 2.$$

Thus, $y = 2$ is a horizontal asymptote.

Asymptotes: $y = 2$, $x = 3$

x-intercept:

0, the value when $y = 0$

y-intercept:

0, the value when $x = 0$

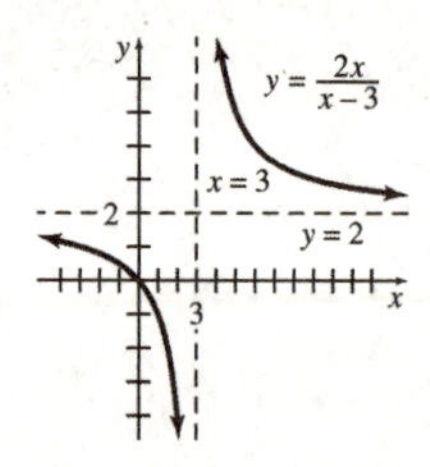

33. $y = \dfrac{x + 1}{x - 4}$

$x - 4 = 0$ when $x = 4$, so $x = 4$ is a vertical asymptote.

x	-96	-6	-1	0	3
$x + 1$	-95	-5	0	1	4
$x - 4$	-100	-10	-5	-4	-1
y	0.95	0.5	0	-0.25	-4

x	3.5	4.5	5	14	104
$x + 1$	4.5	5.5	6	15	105
$x - 4$	-0.5	0.5	1	10	100
y	-9	11	6	1.5	1.05

As x gets larger,

$$\frac{x + 1}{x - 4} \approx \frac{x}{x} = 1.$$

Thus, $y = 1$ is a horizontal asymptote.
Asymptotes: $y = 1$, $x = 4$

x-intercept: -1, the value when $y = 0$

y-intercept: $-\frac{1}{4}$, the value when $x = 0$

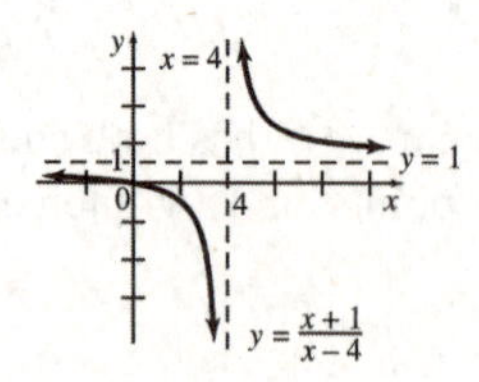

35. $y = \dfrac{3 - 2x}{4x + 20}$

$4x + 20 = 0$ when $4x = -20$ or $x = -5$, so the line $x = -5$ is a vertical asymptote.

x	-8	-7	-6	-4	-3	-2
$3 - 2x$	-26	-23	-20	-14	-11	-8
$4x + 20$	-12	-8	-4	4	8	12
y	2.17	2.88	5	-3.5	-1.38	-0.67

As x gets larger,

$$\frac{3-2x}{4x+20} \approx \frac{-2x}{4x} = -\frac{1}{2}.$$

Thus, the line $y = -\frac{1}{2}$ is a horizontal asymptote.

Asymptotes: $x = -5$, $y = -\frac{1}{2}$

x-intercept:

$\frac{3}{2}$, the value when $y = 0$

y-intercept:

$\frac{3}{20}$, the value when $x = 0$

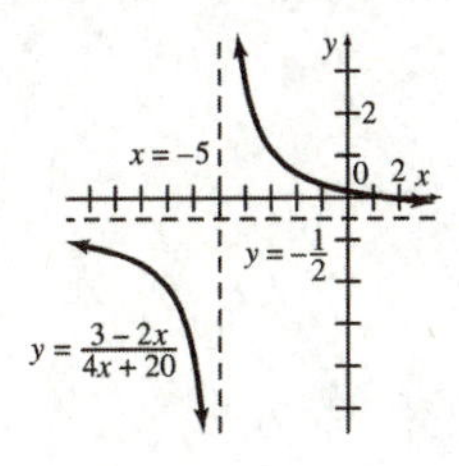

37. $y = \frac{-x-4}{3x+6}$

$3x + 6 = 0$ when $3x = -6$ or $x = -2$, so the line $x = -2$ is a vertical asymptote.

x	−5	−4	−3	−1	0	1
$-x-4$	1	0	−1	−3	−4	−5
$3x+6$	−9	−6	−3	3	6	9
y	−0.11	0	0.33	−1	−0.67	−0.56

As x gets larger,

$$\frac{-x-4}{3x+6} \approx \frac{-x}{3x} = -\frac{1}{3}.$$

The line $y = -\frac{1}{3}$ is a horizontal asymptote.

Asymptotes: $y = -\frac{1}{3}, x = -2$

x-intercept:

-4, the value when $y = 0$

y-intercept:

$-\frac{2}{3}$, the value when $x = 0$

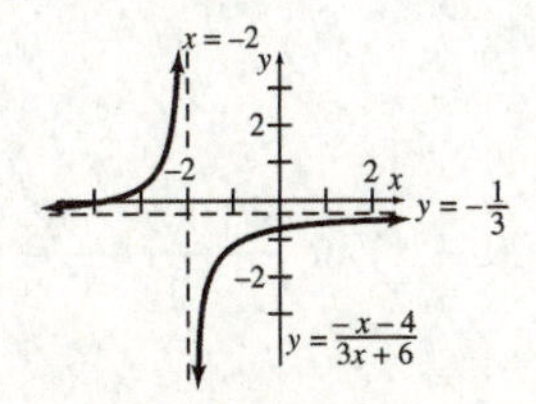

39. $$\begin{aligned} y &= \frac{x^2+7x+12}{x+4} \\ &= \frac{(x+3)(x+4)}{x+4} \\ &= x+3, x \neq -4 \end{aligned}$$

There are no asymptotes, but there is a hole at $x = -4$.

x-intercept: -3, the value when $y = 0$.

y-intercept: 3, the value when $x = 0$.

41. For a vertical asymptote at $x = 1$, put $x - 1$ in the denominator. For a horizontal asymptote at $y = 2$, the degree of the numerator must equal the degree of the denominator and the quotient of their leading terms must equal 2. So, $2x$ in the numerator would cause y to approach 2 as x gets larger.

So, one possible answer is $y = \frac{2x}{x-1}$.

43. $f(x) = (x-1)(x-2)(x+3)$,
$g(x) = x^3 + 2x^2 - x - 2$,
$h(x) = 3x^3 + 6x^2 - 3x - 6$

(a) $f(1) = (0)(-1)(4) = 0$

(b) $f(x)$ is zero when $x = 2$ and when $x = -3$.

(c) $$\begin{aligned} g(-1) &= (-1)^3 + 2(-1)^2 - (-1) - 2 \\ &= -1 + 2 + 1 - 2 = 0 \\ g(1) &= (1)^3 + 2(1)^2 - (1) - 2 \\ &= 1 + 2 - 1 - 2 = 0 \\ g(-2) &= (-2)^3 + 2(-2)^2 - (-2) - 2 \\ &= -8 + 8 + 2 - 2 = 0 \end{aligned}$$

(d) $g(x) = [x-(-1)](x-1)[x-(-2)]$
$g(x) = (x+1)(x-1)(x+2)$

(e) $$\begin{aligned} h(x) &= 3g(x) \\ &= 3(x+1)(x-1)(x+2) \end{aligned}$$

(f) If f is a polynomial and $f(a) = 0$ for some number a, then one factor of the polynomial is $x - a$.

45. $f(x) = \dfrac{1}{x^5 - 2x^3 - 3x^2 + 6}$

(a) Two vertical asymptotes appear, one at $x = -1.4$ and one at $x = 1.4$.

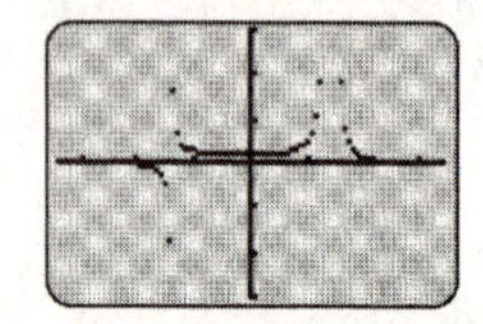

(b) Three vertical asymptotes appear, one at $x = -1.414$, one at $x = 1.414$, and one at $x = 1.442$.

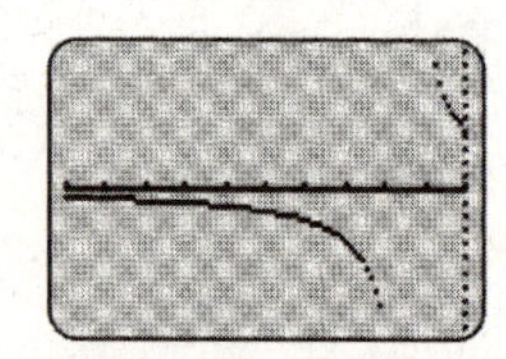

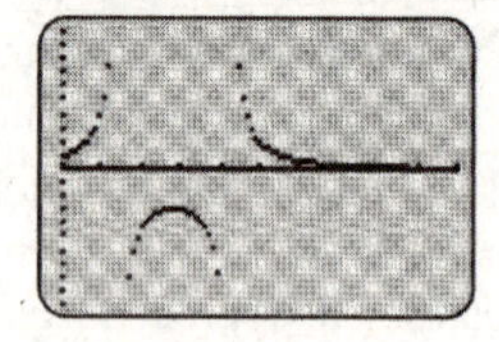

47. $\overline{C}(x) = \dfrac{220{,}000}{x + 475}$

(a) If $x = 25$,

$$\overline{C}(25) = \frac{220{,}000}{25 + 475} = \frac{220{,}000}{500} = \$440.$$

If $x = 50$,

$$\overline{C}(50) = \frac{220{,}000}{50 + 475} = \frac{220{,}000}{525} \approx \$419.$$

If $x = 100$,

$$\overline{C}(100) = \frac{220{,}000}{100 + 475} = \frac{220{,}000}{575} \approx \$383.$$

If $x = 200$,

$$\overline{C}(200) = \frac{220{,}000}{200 + 475} = \frac{220{,}000}{675} \approx \$326.$$

If $x = 300$,

$$\overline{C}(300) = \frac{220{,}000}{300 + 475} = \frac{220{,}000}{775} \approx \$284.$$

If $x = 400$,

$$\overline{C}(400) = \frac{220{,}000}{400 + 475} = \frac{220{,}000}{875} \approx \$251.$$

(b) A vertical asymptote occurs when the denominator is 0.

$$x + 475 = 0$$
$$x = -475$$

A horizontal asymptote occurs when $\overline{C}(x)$ approaches a value as x gets larger. In this case, $\overline{C}(x)$ approaches 0.

The asymptotes are $x = -475$ and $y = 0$.

(c) x-intercepts:

$0 = \dfrac{220{,}000}{x + 475}$; no such x, so no x-intercepts

y-intercepts:

$$\overline{C}(0) = \frac{220{,}000}{0 + 475} \approx 463.2$$

(d) Use the following ordered pairs: (25,440), (50,419), (100,383), (200,326), (300,284), (400,251).

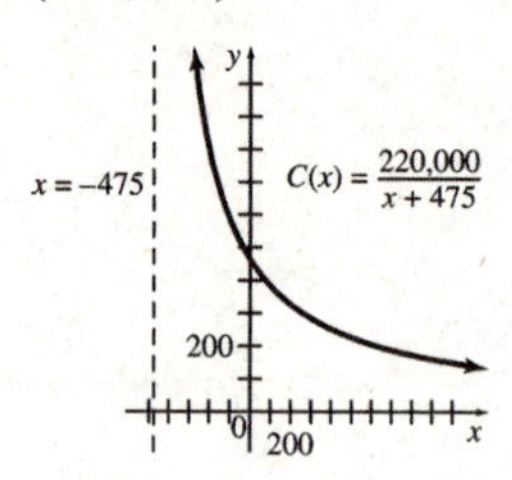

49. Quadratic functions with roots at $x = 0$ and $x = 100$ are of the form $f(x) = ax(100 - x)$.

$f_1(x)$ has a maximum of 100, which occurs at the vertex. The x-coordinate of the vertex lies between the two roots.

The vertex is (50, 100).

$$100 = a(50)(100 - 50)$$
$$100 = a(50)(50)$$
$$\frac{100}{2500} = a$$
$$\frac{1}{25} = a$$

$$f_1(x) = \frac{1}{25}x(100 - x) \text{ or } \frac{x(100 - x)}{25}$$

$f_2(x)$ has a maximum of 250, occurring at (50, 250).

$$250 = a(50)(100 - 50)$$
$$250 = a(50)(50)$$
$$\frac{250}{2500} = a$$
$$\frac{1}{10} = a$$

$$f_2(x) = \frac{1}{10}x(100 - x) \text{ or } \frac{x(100 - x)}{10}$$

$$f_1(x) \cdot f_2(x) = \left[\frac{x(100-x)}{25}\right] \cdot \left[\frac{x(100-x)}{10}\right]$$

$$= \frac{x^2(100-x)^2}{250}$$

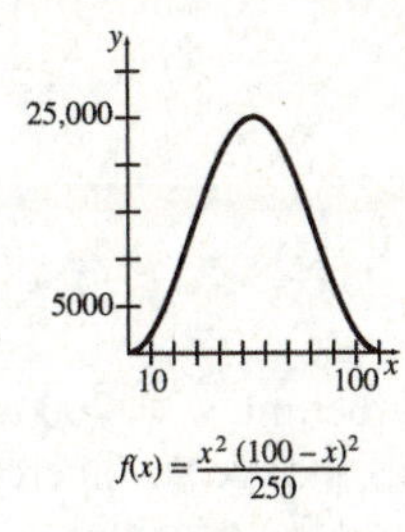

51. $y = \frac{6.7x}{100 - x}$,

Let x = percent of pollutant;
y = cost in thousands.

(a) $x = 50$

$$y = \frac{6.7(50)}{100 - 50} = 6.7$$

The cost is $6700.

$x = 70$

$$y = \frac{6.7(70)}{100 - 70} \approx 15.6$$

The cost is $15,600.

$x = 80$

$$y = \frac{6.7(80)}{100 - 80} = 26.8$$

The cost is $26,800.

$x = 90$

$$y = \frac{6.7(90)}{100 - 90} = 60.3$$

The cost is $60,300.

$x = 95$

$$y = \frac{6.7(95)}{100 - 95}$$

The cost is $127,300.

$x = 98$

$$y = \frac{6.7(98)}{100 - 98} = 328.3$$

The cost is $328,300.

$x = 99$

$$y = \frac{6.7(99)}{100 - 99} = 668.3$$

The cost is $663,300.

(b) No, because $x = 100$ makes the denominator zero, so $x = 100$ is a vertical asymptote.

(c)

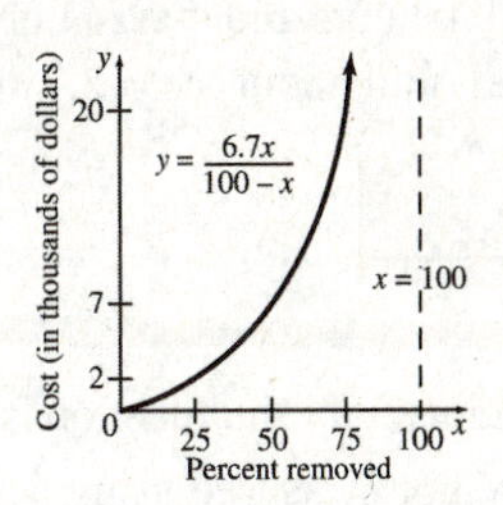

53. **(a)** $a = \frac{k}{d}$

$k = ad$

d	a	$k = ad$
36.000	9.37	337.32
36.125	9.34	337.4075
36.250	9.31	337.4875
36.375	9.27	337.19625
36.500	9.24	337.26
36.625	9.21	337.31625
36.750	9.18	337.365
36.875	9.15	337.40625
37.000	9.12	337.44

We find the average of the nine values of k by adding them and dividing by 9. This gives 337.35, or, rounding to the nearest integer, $k = 337$. Therefore,

$$a = \frac{337}{d}.$$

(b) When $d = 40.50$,

$$a = \frac{337}{40.50} \approx 8.32.$$

The strength for 40.50 diopter lenses is 8.32 mm of arc.

55. $A(x) = 0.003631x^3 - 0.03746x^2 + 0.1012x + 0.009$

(a)

x	0	1	2	3	4	5
$A(x)$	0.009	0.076	0.091	0.073	0.047	0.032

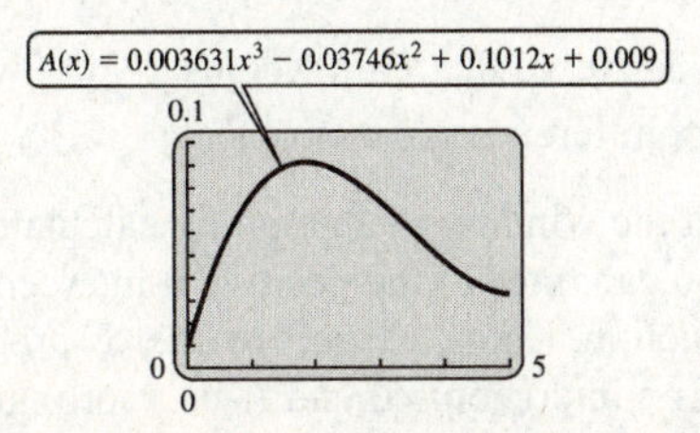

(b) The peak of the curve comes at about $x = 2$ hours.

(c) The curve rises to a y-value of 0.08 at about $x = 1.1$ hours and stays at or above that level until about $x = 2.7$ hours.

57. $f(x) = \dfrac{\lambda x}{1 + (ax)^b}$

(a) A reasonable domain for the function is $[0, \infty)$. Populations are not measured using negative numbers and they may get extremely large.

(b) If $\lambda = a = b = 1$, the function becomes

$$f(x) = \frac{x}{1 + x^2}.$$

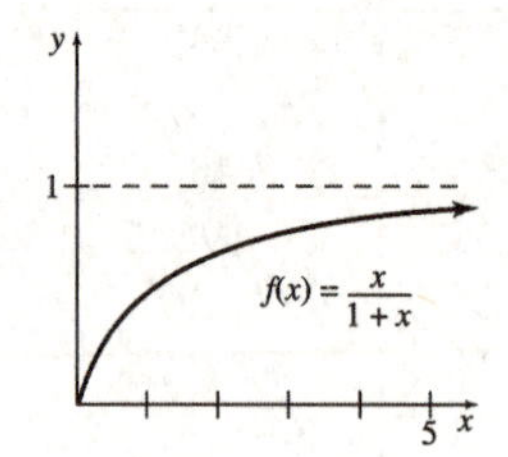

(c) If $\lambda = a = 1$ and $b = 2$, the function becomes

$$f(x) = \frac{x}{1 + x}.$$

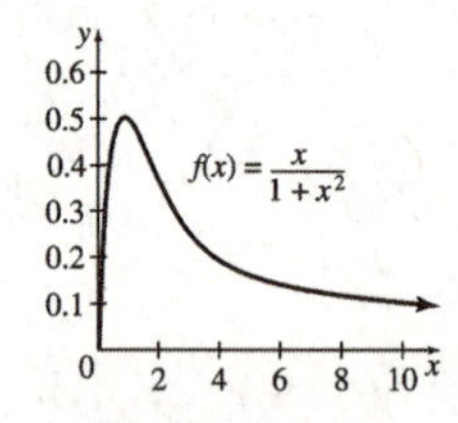

(d) As seen from the graphs, when b increases, the population of the next generation, $f(x)$, gets smaller when the current generation, x, is larger.

59. (a) When $c = 30, w = \frac{30^3}{100} - \frac{1500}{30} = 220,$ so the brain weights 220 g when its circumference measures 30 cm. When $c = 40, w = \frac{40^3}{100} - \frac{1500}{40} = 602.5,$ so the brain weighs 602.5 g when its circumference is 40 cm. When $c = 50,\ w = \frac{50^3}{100} - \frac{1500}{50}$ $= 1220,$ so the brain weighs 1220 g when its circumference is 50 cm.

(b) Set the window of a graphing calculator so you can trace to the positive x-intercept of the function. Using a "root" or "zero" program, this x-intercept is found to be approximately 19.68. Notice in the graph that positive c values less than 19.68 correspond to negative w values. Therefore, the answer is $c < 19.68$.

(c)

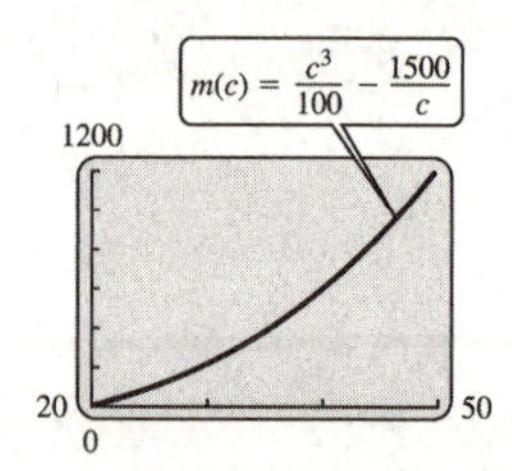

(d) One method is to graph the line $y = 700$ on the graph found in part (c) and use an "intercept" program to find the point of intersection of the two graphs. This point has the approximate coordinates (41.9, 700). Therefore, an infant has a brain weighing 700 g when the circumference measures 41.9 cm.

61. (a) Using a graphing calculator with the given data, for the 4.0-foot pendulum and for

$n = 1,$

$$4.0 = k(2.22)^1$$
$$k = 1.80;$$

$n = 2,$

$$4.0 = k(2.22)^2$$
$$k = 0.812;$$

$n = 3,$

$$4.0 = k(2.22)^3$$
$$k = 0.366.$$

(b)

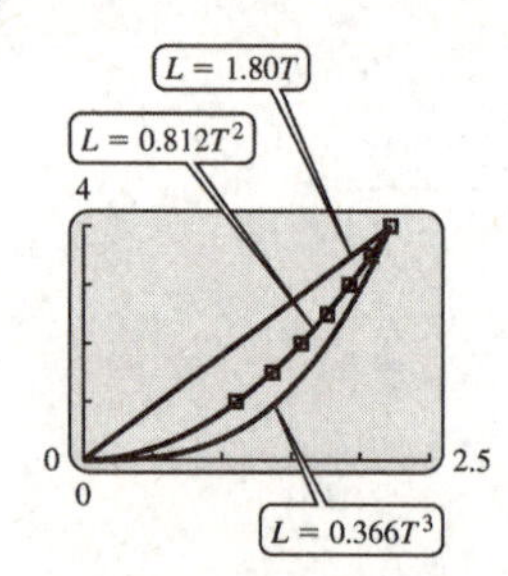

From the graphs, $L = 0.812T^2$ is the best fit.

(c) $5.0 = 0.812T^2$

$$T^2 = \frac{5.0}{0.812}$$

$$T = \sqrt{\frac{5.0}{0.812}}$$

$$T \approx 2.48$$

The period will be 2.48 seconds.

(d) $L = 0.812T^2$

If L doubles, T^2 doubles, and T increases by factor of $\sqrt{2}$.

(e) $L \approx 0.822T^2$ which is very close to $L = 0.812T^2$.

10.4 Exponential Functions

Your Turn 1

$$25^{x/2} = 125^{x+3}$$
$$(5^2)^{x/2} = (5^3)^{x+3}$$
$$5^x = 5^{3x+9}$$
$$x = 3x + 9$$
$$2x = -9$$
$$x = -\frac{9}{2}$$

Your Turn 2

$$A = P\left(1 + \frac{r}{m}\right)^{tm}$$
$$= 4400\left(1 + \frac{0.0325}{4}\right)^{5(4)}$$
$$= 5172.97$$

The interest amounts to $5172.97 - 4400 = \$772.97$.

Your Turn 3

$$A = Pe^{rt}$$
$$= 800e^{4(0.0315)}$$
$$= 907.43$$

The amount after 4 years will be \$907.43.

10.4 Exercises

1.

number of folds	1	2	3	4	5 ...	10 ...	50
layers of paper	2	4	8	16	32 ...	1024 ...	2^{50}

$$2^{50} = 1.125899907 \times 10^{15}$$

3. The graph of $y = 3^x$ is the graph of an exponential function $y = a^x$ with $a > 1$.

This is graph **E**.

5. The graph of $y = \left(\frac{1}{3}\right)^{1-x}$ is the graph of $y = (3^{-1})^{1-x}$ or $y = 3^{x-1}$. This is the graph of $y = 3^x$ translated 1 unit to the right.

This is graph **C**.

7. The graph of $y = 3(3)^x$ is the same as the graph of $y = 3^{x+1}$. This is the graph of $y = 3^x$ translated 1 unit to the left.

This is graph **F**.

9. The graph of $y = 2 - 3^{-x}$ is the same as the graph of $y = -3^{-x} + 2$. This is the graph of $y = 3^x$ reflected in the x-axis, reflected in the y-axis, and translated up 2 units.

This is graph **A**.

11. The graph of $y = 3^{x-1}$ is the graph of $y = 3^x$ translated 1 unit to the right.

This is graph **C**.

13.
$$2^x = 32$$
$$2^x = 2^5$$
$$x = 5$$

15.
$$3^x = \frac{1}{81}$$
$$3^x = \frac{1}{3^4}$$
$$3^x = 3^{-4}$$
$$x = -4$$

17.
$$4^x = 8^{x+1}$$
$$(2^2)^x = (2^3)^{x+1}$$
$$2^{2x} = 2^{3x+3}$$
$$2x = 3x + 3$$
$$-x = 3$$
$$x = -3$$

19.
$$\begin{aligned} 16^{x+3} &= 64^{2x-5} \\ (2^4)^{x+3} &= (2^6)^{2x-5} \\ 2^{4x+12} &= 2^{12x-30} \\ 4x+12 &= 12x-30 \\ 42 &= 8x \\ \frac{21}{4} &= x \end{aligned}$$

21.
$$\begin{aligned} e^{-x} &= (e^4)^{x+3} \\ e^{-x} &= e^{4x+12} \\ -x &= 4x+12 \\ -5x &= 12 \\ x &= -\frac{12}{5} \end{aligned}$$

23.
$$\begin{aligned} 5^{-|x|} &= \frac{1}{25} \\ 5^{-|x|} &= 5^{-2} \\ |x| &= 2 \\ x &= 2 \text{ or } x = -2 \end{aligned}$$

25.
$$\begin{aligned} 5^{x^2+x} &= 1 \\ 5^{x^2+x} &= 5^0 \\ x^2+x &= 0 \\ x(x+1) &= 0 \\ x = 0 \quad &\text{or} \quad x+1 = 0 \\ x = 0 \quad &\text{or} \quad x = -1 \end{aligned}$$

27.
$$\begin{aligned} 27^x &= 9^{x^2+x} \\ (3^3)^x &= (3^2)^{x^2+x} \\ 3^{3x} &= 3^{2x^2+2x} \\ 3x &= 2x^2+2x \\ 0 &= 2x^2-x \\ 0 &= x(2x-1) \\ x = 0 \quad &\text{or} \quad 2x-1 = 0 \\ x = 0 \quad &\text{or} \quad x = \frac{1}{2} \end{aligned}$$

29. Graph of $y = 5e^x + 2$

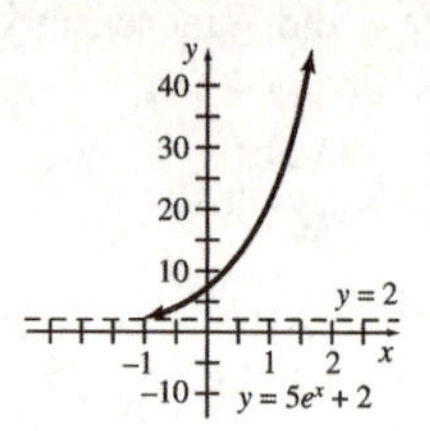

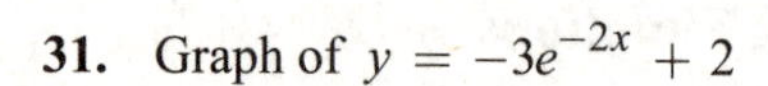

31. Graph of $y = -3e^{-2x} + 2$

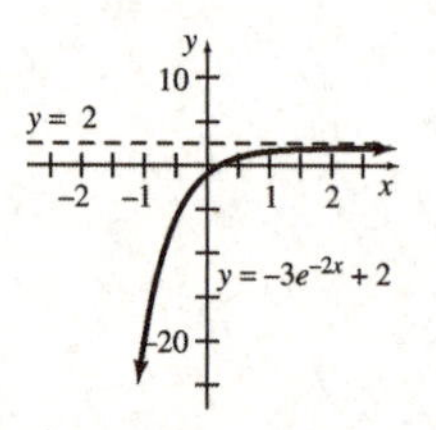

37. $A = P\left(1+\frac{r}{m}\right)^{tm}$, $P = 10{,}000$, $r = 0.04$, $t = 5$

(a) annually, $m = 1$

$$\begin{aligned} A &= 10{,}000\left(1+\frac{0.04}{1}\right)^{5(1)} \\ &= 10{,}000(1.04)^5 \\ &= \$12{,}166.53 \\ \text{Interest} &= \$12{,}166.53 - \$10.000 \\ &= \$21{,}66.53 \end{aligned}$$

(b) semiannually, $m = 2$

$$\begin{aligned} A &= 10{,}000\left(1+\frac{0.04}{2}\right)^{5(2)} \\ &= 10{,}000(1.02)^{10} \\ &= \$12{,}189.94 \\ \text{Interest} &= \$12{,}189.94 - \$10{,}000 \\ &= \$2189.94 \end{aligned}$$

(c) quarterly, $m = 4$

$$\begin{aligned} A &= 10{,}000\left(1+\frac{0.04}{4}\right)^{5(4)} \\ &= 10{,}000(1.01)^{20} \\ &= \$12{,}201.90 \\ \text{Interest} &= \$12{,}201.90 - \$10{,}000 \\ &= \$2201.90 \end{aligned}$$

(d) monthly, $m = 12$

$$A = 10{,}000\left(1 + \frac{0.04}{12}\right)^{5(12)}$$
$$= 10{,}000(1.00\overline{3})^{60}$$
$$= \$12{,}209.97$$
$$\text{Interest} = \$12{,}209.97 - \$10{,}000$$
$$= \$2209.97$$

(e) $A = 10{,}000e^{(0.04)(5)} = \$12{,}214.03$

$$\text{Interest} = \$12{,}214.03 - \$10{,}000$$
$$= \$2214.03$$

39. For 6% compounded annually for 2 years,

$$A = 18{,}000(1 + 0.06)^2$$
$$= 18{,}000(1.06)^2$$
$$= 20{,}224.80$$

For 5.9% compounded monthly for 2 years,

$$A = 18{,}000\left(1 + \frac{0.059}{12}\right)^{12(2)}$$
$$= 18{,}000\left(\frac{12.059}{12}\right)^{24}$$
$$= 20{,}248.54$$

The 5.9% investment is better. The additional interest is

$$\$20{,}248.54 - \$20{,}224.80 = \$23.74.$$

41. $A = Pe^{rt}$

(a) $r = 3\%$

$$A = 10e^{0.03(3)} = \$10.94$$

(b) $r = 4\%$

$$A = 10e^{0.04(3)} = \$11.27$$

(c) $r = 5\%$

$$A = 10e^{0.05(3)} = \$11.62$$

43. $1200 = 500\left(1 + \frac{r}{4}\right)^{(14)(4)}$

$$\frac{1200}{500} = \left(1 + \frac{r}{4}\right)^{56}$$
$$2.4 = \left(1 + \frac{r}{4}\right)^{56}$$
$$1 + \frac{r}{4} = (2.4)^{1/56}$$
$$4 + r = 4(2.4)^{1/56}$$
$$r = 4(2.4)^{1/56} - 4$$
$$r \approx 0.0630$$

The required interest rate is 6.30%.

45. $y = (0.92)^t$

(a)

t	y
0	$(0.92)^0 = 1$
1	$(0.92)^1 = 0.92$
2	$(0.92)^2 \approx 0.85$
3	$(0.92)^3 \approx 0.78$
4	$(0.92)^4 \approx 0.72$
5	$(0.92)^5 \approx 0.66$
6	$(0.92)^6 \approx 0.61$
7	$(0.92)^7 \approx 0.56$
8	$(0.92)^8 \approx 0.51$
9	$(0.92)^9 \approx 0.47$
10	$(0.92)^{10} \approx 0.43$

(b)

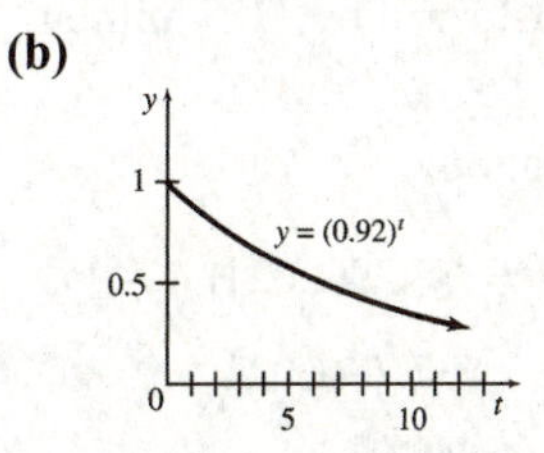

(c) Let $x =$ the cost of the house in 10 years.

Then, $0.43x = 165{,}000$

$$x \approx 383{,}721.$$

In 10 years, the house will cost about $384,000.

(d) Let $x =$ the cost of the book in 8 years.

Then, $0.51x = 50$

$$x \approx 98$$

In 8 years, the textbook will cost about $98.

47. $A(t) = 3100e^{0.0166t}$

(a) 1970: $t = 10$

$$\begin{aligned} A(20) &= 3100e^{(0.0166)(10)} \\ &= 3100e^{0.166} \\ &\approx 3659.78 \end{aligned}$$

The function gives a population of about 3660 million in 1970.

This is very close to the actual population of about 3686 million.

(b) 2000: $t = 40$

$$\begin{aligned} A(50) &= 3100e^{0.0166(40)} \\ &= 3100e^{0.664} \\ &\approx 6021.90 \end{aligned}$$

The function gives a population of 6022 million in 2000.

(c) 2015: $t = 55$

$$\begin{aligned} &= 3100e^{0.0166(55)} \\ &= 3100e^{0.913} \\ &= 7724.54 \end{aligned}$$

From the function, we estimate that the world population in 2015 will be 7725 million.

49. (a) Hispanic population:

$$\begin{aligned} h(t) &= 37.79(1.021)^t \\ h(5) &= 37.79(1.021)^5 \\ &\approx 41.93 \end{aligned}$$

The projected Hispanic population in 2005 is 41.93 million, which is slightly less than the actual value of 42.69 million.

(b) Asian population:

$$\begin{aligned} h(t) &= 11.14(1.023)^t \\ h(5) &= 11.14(1.023)^t \\ &\approx 12.48 \end{aligned}$$

The projected Asian population in 2005 is 12.48 million, which is very close to the actual value of 12.69 million.

(c) Annual Hispanic percent increase:

$$1.021 - 1 = 0.021 = 2.1\%$$

Annual Asian percent increase:

$$1.023 - 1 = 0.023 = 2.3\%$$

The Asian population is growing at a slightly faster rate.

(d) Black population:

$$\begin{aligned} b(t) &= 0.5116t + 35.43 \\ b(5) &= 0.5116(5) + 35.43 \\ &\approx 37.99 \end{aligned}$$

The projected Black population in 2005 is 37.99 million, which is extremely close to the actual value of 37.91 million.

(e) Hispanic population:

Double the actual 2005 value is

$$2(42.69) = 85.38 \text{ million.}$$

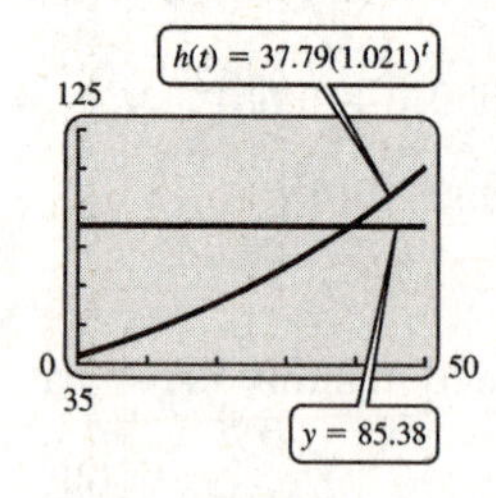

The doubling point is reached when $t \approx 39$, or in the year 2039.

Asian population:

Double the actual 2005 value is

$$2(12.69) = 25.38 \text{ million.}$$

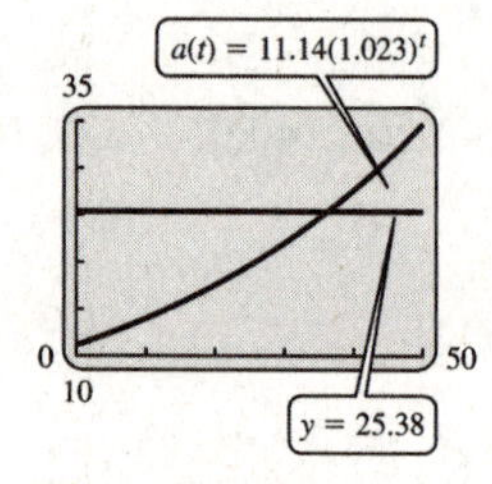

The doubling point is reached when $t \approx 36$, or in the 2036.

Black population:

Double the actual 2005 value is

$$2(37.91) = 7582 \text{ million.}$$

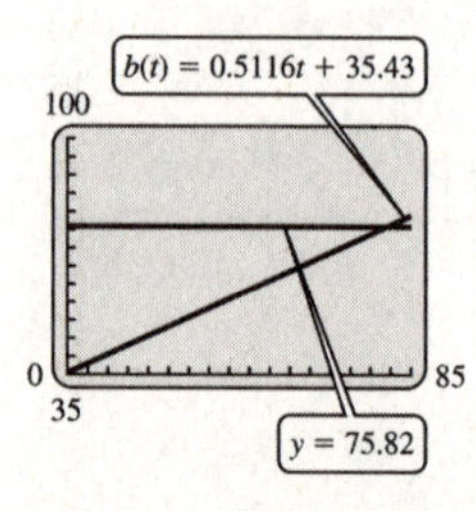

The doubling point is reached when $t \approx 79$, or in the year 2079.

51. (a)

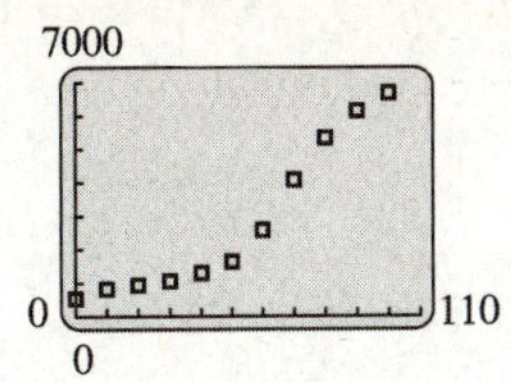

The emissions appear to grow exponentially.

(b) $f(x) = f_0 a^x$

$f_0 = 534$

Use the point (100, 6672) to find a.

$$6672 = 534a^{100}$$
$$a^{100} = \frac{6672}{534}$$
$$a = \sqrt[100]{\frac{6672}{534}}$$
$$\approx 1.026$$
$$f(x) = 534(1.026)^x$$

(c) $1.026 - 1 = 0.026 = 2.6\%$

(d) Double the 2000 value is $2(6672) = 13{,}344$.

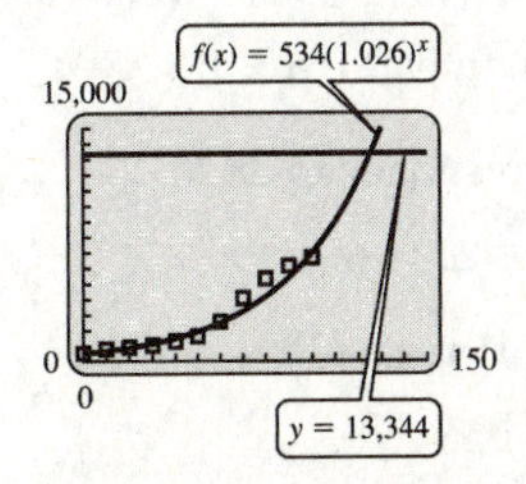

The doubling point is reached when $x \approx 125.4$. The first year in which emissions equal or exceed that threshold is 2026.

53. (a) When $x = 0$, $P = 1013$.

When $x = 10{,}000$, $P = 265$.

First we fit $P = ae^{kx}$.

$$1013 = ae^0$$
$$a = 1013$$
$$P = 1013e^{kx}$$
$$265 = 1013e^{k(10{,}000)}$$
$$\frac{265}{1013} = e^{10{,}000k}$$
$$10{,}000k = \ln\left(\frac{265}{1013}\right)$$
$$k = \frac{\ln\left(\frac{265}{1013}\right)}{10{,}000} \approx -1.34 \times 10^{-4}$$

Therefore $P = 1013e^{(-1.34\times10^{-4})x}$.

Next we fit $P = mx + b$.

We use the points $(0, 1013)$ and $(10{,}000, 265)$.

$$m = \frac{265 - 1013}{10{,}000 - 0} = -0.0748$$
$$b = 1013$$

Therefore $P = -0.0748x + 1013$.

Finally, we fit $P = \frac{1}{ax+b}$.

$$1013 = \frac{1}{a(0) + b}$$
$$b = \frac{1}{1013} \approx 9.87 \times 10^{-4}$$
$$P = \frac{1}{ax + \frac{1}{1013}}$$
$$265 = \frac{1}{10{,}000a + \frac{1}{1013}}$$
$$\frac{1}{265} = 10{,}000a + \frac{1}{1013}$$
$$10{,}000a = \frac{1}{265} - \frac{1}{1013}$$
$$a = \frac{\frac{1}{265} - \frac{1}{1013}}{10{,}000} \approx 2.79 \times 10^{-7}$$

Therefore,

$$P = \frac{1}{(2.79 \times 10^{-7})x + (9.87 \times 10^{-4})}.$$

(b)

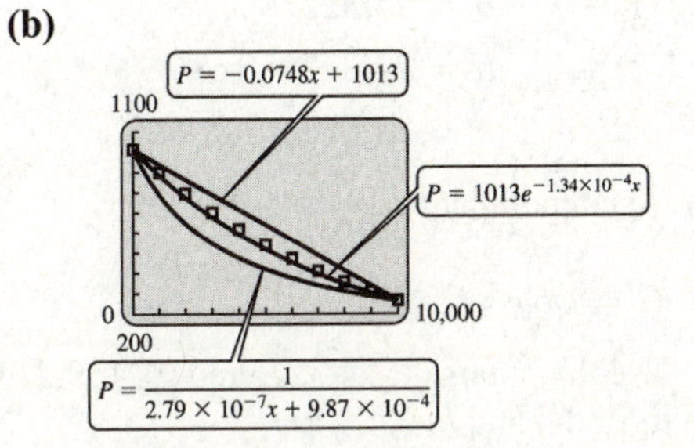

$P = 1013e^{(-1.34\times10^{-4})x}$ is the best fit.

(c) $P(1500) = 1013e^{-1.34\times10^{-4}(1500)} \approx 829$

$P(11{,}000) = 1013e^{-1.34\times10^{-4}(11{,}000)} \approx 232$

We predict that the pressure at 1500 meters will be 829 millibars, and at 11,000 meters will be 232 millibars.

(d) Using exponential regression, we obtain $P = 1038(0.99998661)^x$ which differs slightly from the function found in part (b) which can be rewritten as

$$P = 1013(0.99998660)^x.$$

55. (a) $C = mt + b$

Use the points (1, 24,322) and (9, 159,213) to find m.

$$m = \frac{159{,}213 - 24{,}322}{9 - 1} = \frac{134{,}891}{8} \approx 16{,}861$$

Use the point (1, 24,322) to find b.

$$\begin{aligned} y &= 16{,}861t + b \\ 24{,}322 &= 16{,}861(1) + b \\ b &= 7461 \\ C &= 16{,}861t + 7461 \end{aligned}$$

$$C = at^2 + b$$

Use the points (1, 24,322) and (9, 159,213) to find a and b.

$$24{,}322 = a(1)^2 + b \qquad 159{,}213 = a(9)^2 + b$$
$$24{,}322 = a + b \qquad 159{,}213 = 81a + b$$

$$\begin{aligned} 81a + b &= 159{,}213 \\ a + b &= 24{,}322 \\ \hline 80a &= 134{,}891 \\ a &= \frac{134{,}891}{80} \approx 1686 \end{aligned}$$

$$\begin{aligned} a + b &= 24{,}322 \\ b &= 24{,}322 - 1686 \\ b &= 22{,}636 \end{aligned}$$

$$C = 1686t^2 + 22{,}636$$

$$C = ab^t$$

Use the points (1, 24,322) and (9, 159,213) to find a and b.

$$24{,}322 = ab^1 = ab \qquad 159{,}213 = ab^9$$

$$\frac{159{,}213}{24{,}322} = \frac{ab^9}{ab} = b^8$$

$$b = \sqrt[8]{\frac{159{,}213}{24{,}322}} \approx 1.2647$$

$$\begin{aligned} ab &= 24{,}322 \\ (1.2647)a &= 24{,}322 \\ a &= \frac{24{,}322}{1.2647} \approx 19{,}231 \\ C &= 19{,}231(1.2647)^t \end{aligned}$$

(b)

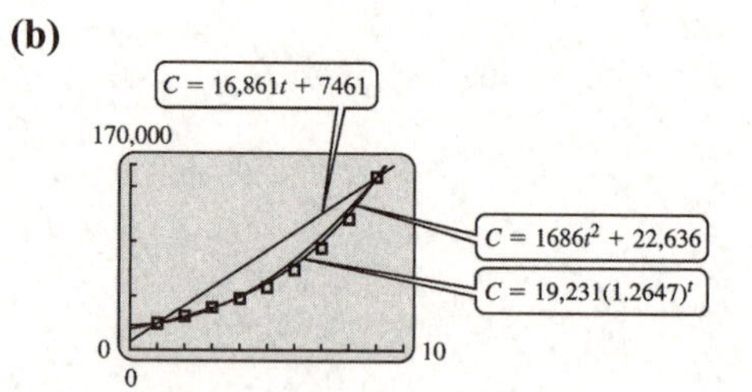

The function $C = 19{,}231(1.2647)^t$ is the best fit.

(c) The regression function is $C = 19{,}250(1.2579)^t$. This is very close to the function in part (b).

(d) $x = 10$ corresponds to 2010.

$$\begin{aligned} C &= 16{,}861t + 7461 \\ &= 16{,}861(10) + 7461 \\ &= 176{,}071 \\ &\approx 176{,}100 \end{aligned}$$

$$\begin{aligned} C &= 1686t^2 + 22{,}636 \\ &= 1686(10)^2 + 22{,}636 \\ &= 191{,}236 \\ &\approx 191{,}200 \end{aligned}$$

$$\begin{aligned} C &= 19{,}231(1.2647)^t \\ &= 19{,}231(1.2647)^{10} \\ &= 201{,}315.43 \\ &\approx 201{,}300 \end{aligned}$$

$$\begin{aligned} C &= 19{,}250(1.2579)^t \\ &= 19{,}250(1.2579)^{10} \\ &= 190{,}937.80 \\ &\approx 190{,}900 \end{aligned}$$

10.5 Logarithmic Functions

Your Turn 1

$5^{-2} = \frac{1}{25}$ means $\log_5\left(\frac{1}{25}\right) = -2$

Your Turn 2

$\log_3\left(\frac{1}{81}\right)$

We seek a number x such that

$$\begin{aligned} 3^x &= \frac{1}{81} \\ 3^x &= (3)^{-4} \\ x &= -4 \end{aligned}$$

Your Turn 3

$$\begin{aligned} \log_a\left(\frac{x^2}{y^3}\right) &= \log_a x^2 - \log_a y^3 \\ &= 2\log_a x - 3\log_a y \end{aligned}$$

Your Turn 4

$$\log_3 50 = \frac{\ln 50}{\ln 3} \approx 3.561$$

Your Turn 5

$$\begin{aligned} \log_2 x + \log_2 (x+2) &= 3 \\ \log_2 x(x+2) &= 3 \\ x(x+2) &= 2^3 \\ x^2 + 2x &= 8 \\ x^2 + 2x - 8 &= 0 \\ x = -4 \quad \text{or} \quad x &= 2 \end{aligned}$$

Since the logarithm of a negative number is undefined, the only solution is $x = 2$.

Your Turn 6

$$\begin{aligned} 2^{x+1} &= 3^x \\ \ln 2^{x+1} &= \ln 3^x \\ (x+1)\ln 2 &= x\ln 3 \\ x\ln 2 + \ln 2 &= x\ln 3 \\ x\ln 3 - x\ln 2 &= \ln 2 \\ x(\ln 3 - \ln 2) &= \ln 2 \\ x\ln\left(\tfrac{3}{2}\right) &= \ln 2 \\ x &= \frac{\ln 2}{\ln 1.5} \approx 1.7095 \end{aligned}$$

Your Turn 7

$$e^{0.025x} = (e^{0.025})^x \approx 1.0253^x$$

10.5 Exercises

1. $5^3 = 125$

Since $a^y = x$ means $y = \log_a x$, the equation in logarithmic form is

$$\log_5 125 = 3.$$

3. $3^4 = 81$

The equation in logarithmic form is

$$\log_3 81 = 4.$$

5. $3^{-2} = \frac{1}{9}$

The equation in logarithmic form is

$$\log_3 \frac{1}{9} = -2.$$

7. $\log_2 32 = 5$

Since $y = \log_a x$ means $a^y = x$, the equation in exponential form is

$$2^5 = 32.$$

9. $\ln\frac{1}{e} = -1$

The equation in exponential form is

$$e^{-1} = \frac{1}{e}.$$

11.

$$\begin{aligned} \log 100{,}000 &= 5 \\ \log_{10} 100{,}000 &= 5 \\ 10^5 &= 100{,}000 \end{aligned}$$

When no base is written, $\log_{10}$ is understood.

13. Let $\log_8 64 = x$.

$$\begin{aligned} \text{Then, } 8^x &= 64 \\ 8^x &= 8^2 \\ x &= 2. \end{aligned}$$

Thus, $\log_8 64 = 2$.

15. $\log_4 64 = x$

$$4^x = 64$$
$$4^x = 4^3$$
$$x = 3$$

17. $\log_2 \frac{1}{16} = x$

$$2^x = \frac{1}{16}$$
$$2^x = 2^{-4}$$
$$x = -4$$

19. $\log_2 \sqrt[3]{\frac{1}{4}} = x$

$$2^x = \left(\frac{1}{4}\right)^{1/3}$$
$$2^x = \left(\frac{1}{2^2}\right)^{1/3}$$
$$2^x = 2^{-2/3}$$
$$x = -\frac{2}{3}$$

21. $\ln e = x$

Recall that ln means $\log_e$.

$$e^x = e$$
$$x = 1$$

23. $\ln e^{5/3} = x$

$$e^x = e^{5/3}$$
$$x = \frac{5}{3}$$

25. The logarithm to the base 3 of 4 is written $\log_3 4$. The subscript denotes the base.

27. $\log_5 (3k) = \log_5 3 + \log_5 k$

29. $\log_3 \frac{3p}{5k}$

$$= \log_3 3p - \log_3 5k$$
$$= (\log_3 3 + \log_3 p) - (\log_3 5 + \log_3 k)$$
$$= 1 + \log_3 p - \log_3 5 - \log_3 k$$

31. $\ln \frac{3\sqrt{5}}{\sqrt[3]{6}}$

$$= \ln 3\sqrt{5} - \ln \sqrt[3]{6}$$
$$= \ln 3 \cdot 5^{1/2} - \ln 6^{1/3}$$
$$= \ln 3 + \ln 5^{1/2} - \ln 6^{1/3}$$
$$= \ln 3 + \frac{1}{2}\ln 5 - \frac{1}{3}\ln 6$$

33. $\log_b 32 = \log_b 2^5$

$$= 5\log_b 2$$
$$= 5a$$

35. $\log_b 72b = \log_b 72 + \log_b b$

$$= \log_b 72 + 1$$
$$= \log_b 2^3 \cdot 3^3 + 1$$
$$= \log_b 2^3 + \log_b 3^2 + 1$$
$$= 3 \log_b 2 + 2\log_b 3 + 1$$
$$= 3a + 2c + 1$$

37. $\log_5 30 = \frac{\ln 30}{\ln 5}$

$$\approx \frac{3.4012}{1.6094}$$
$$\approx 2.113$$

39. $\log_{1.2} 0.95 = \frac{\ln 0.95}{\ln 1.2}$

$$\approx -0.281$$

41. $\log_x 36 = -2$

$$x^{-2} = 36$$
$$(x^{-2})^{-1/2} = 36^{-1/2}$$
$$x = \frac{1}{6}$$

43. $\log_8 16 = z$

$$8^z = 16$$
$$(2^3)^z = 2^4$$
$$2^{3z} = 2^4$$
$$3z = 4$$
$$z = \frac{4}{3}$$

45. $$\begin{aligned} \log_r 5 &= \frac{1}{2} \\ r^{1/2} &= 5 \\ (r^{1/2})^2 &= 5^2 \\ r &= 25 \end{aligned}$$

47. $$\begin{aligned} \log_5 (9x - 4) &= 1 \\ 5^1 &= 9x - 4 \\ 9 &= 9x \\ 1 &= x \end{aligned}$$

49. $$\begin{aligned} \log_9 m - \log_9 (m - 4) &= -2 \\ \log_9 \frac{m}{m-4} &= -2 \\ 9^{-2} &= \frac{m}{m-4} \\ \frac{1}{81} &= \frac{m}{m-4} \\ m - 4 &= 81m \\ -4 &= 80m \\ -0.05 &= m \end{aligned}$$

This value is not possible since $\log_9 (-0.05)$ does not exist.

Thus, there is no solution to the original equation.

51. $$\begin{aligned} \log_3 (x-2) + \log_3 (x+6) &= 2 \\ \log_3 [(x-2)(x+6)] &= 2 \\ (x-2)(x+6) &= 3^2 \\ x^2 + 4x - 12 &= 9 \\ x^2 + 4x - 21 &= 0 \\ (x+7)(x-3) &= 0 \\ x = -7 \text{ or } x &= 3 \end{aligned}$$

$x = -7$ does not check in the original equation. The only solution is 3.

53. $$\begin{aligned} \log_2 (x^2 - 1) - \log_2 (x+1) &= 2 \\ \log_2 \frac{x^2-1}{x+1} &= 2 \\ 2^2 &= \frac{x^2-1}{x+1} \\ 4 &= \frac{(x-1)(x+1)}{x+1} \\ 4 &= x - 1 \\ x &= 5 \end{aligned}$$

55. $$\begin{aligned} \ln x + \ln 3x &= -1 \\ \ln 3x^2 &= -1 \\ 3x^2 &= e^{-1} \\ x^2 &= \frac{e^{-1}}{3} \\ x &= \sqrt{\frac{e^{-1}}{3}} = \frac{1}{\sqrt{3e}} \approx 0.3502 \end{aligned}$$

57. $$\begin{aligned} 2^x &= 6 \\ \ln 2^x &= \ln 6 \\ x \ln 2 &= \ln 6 \\ x &= \frac{\ln 6}{\ln 2} \approx 2.5850 \end{aligned}$$

59. $$\begin{aligned} e^{k-1} &= 6 \\ \ln e^{k-1} &= \ln 6 \\ (k-1) \ln e &= \ln 6 \\ k - 1 &= \frac{\ln 6}{\ln e} \\ k - 1 &= \frac{\ln 6}{1} \\ k &= 1 + \ln 6 \\ &\approx 2.7918 \end{aligned}$$

61. $$\begin{aligned} 3^{x+1} &= 5^x \\ \ln 3^{x+1} &= \ln 5^x \\ (x+1) \ln 3 &= x \ln 5 \\ x \ln 3 + \ln 3 &= x \ln 5 \\ x \ln 5 - x \ln 3 &= \ln 3 \\ x(\ln 5 - \ln 3) &= \ln 3 \\ x &= \frac{\ln 3}{\ln(5/3)} \approx 2.1507 \end{aligned}$$

63. $$\begin{aligned} 5(0.10)^x &= 4(0.12)^x \\ \ln[5(0.10)^x] &= \ln[4(0.12)^x] \\ \ln 5 + x \ln 0.10 &= \ln 4 + x \ln 0.12 \\ x(\ln 0.12 - \ln 0.10) &= \ln 5 - \ln 4 \\ x &= \frac{\ln 5 - \ln 4}{\ln 0.12 - \ln 0.10} \\ &= \frac{\ln 1.25}{\ln 1,2} \\ &\approx 1.2239 \end{aligned}$$

65. $10^{x+1} = e^{(\ln 20)(x+1)}$

67. $e^{3x} = (e^3)^x \approx 20.09^x$

69. $f(x) = \log(5 - x)$

$$5 - x > 0$$
$$-x > -5$$
$$x < 5$$

The domain of f is $x < 5$.

71. $\log A - \log B = 0$

$$\log \frac{A}{B} = 0$$
$$\frac{A}{B} = 10^0 = 1$$
$$A = B$$
$$A - B = 0$$

Thus, solving $\log A - \log B = 0$ is equivalent to solving $A - B = 0$

73. Let $m = \log_a \frac{x}{y}$, $n = \log_a x$, and $p = \log_a y$.

Then $a^m = \frac{x}{y}$, $a^n = x$, and $a^p = y$.

Substituting gives

$$a^m = \frac{x}{y} = \frac{a^n}{a^p} = a^{n-p}.$$

So $m = n - p$.

Therefore,

$$\log_a \frac{x}{y} = \log_a x - \log_a y.$$

75. From Example 8, the doubling time t in years when $m = 1$ is given by

$$t = \frac{\ln 2}{\ln(1 + r)}.$$

(a) Let $r = 0.03$.

$$t = \frac{\ln 2}{\ln(1.03)}$$
$$= 23.4 \text{ years}$$

(b) Let $r = 0.06$.

$$t = \frac{\ln 2}{\ln 1.06} = 11.9 \text{ years}$$

(c) Let $r = 0.08$.

$$t = \frac{\ln 2}{\ln 1.08}$$
$$= 9.0 \text{ years}$$

(d) Since $0.001 \le 0.03 \le 0.05$, for $r = 0.03$, we use the rule of 70.

$$\frac{70}{100r} = \frac{70}{100(0.03)} = 23.3 \text{ years}$$

Since $0.05 \le 0.06 \le 0.12$, for $r = 0.06$, we use the rule of 72.

$$\frac{72}{100r} = \frac{72}{100(0.06)} = 12 \text{ years}$$

For $r = 0.08$, we use the rule of 72.

$$\frac{72}{100(0.08)} = 9 \text{ years}$$

77.

$$A = Pe^{rt}$$
$$1200 = 500e^{r \cdot 14}$$
$$2.4 = e^{14r}$$
$$\ln(2.4) = \ln e^{14r}$$
$$\ln(2.4) = 14r$$
$$\frac{\ln(2.4)}{14} = r$$
$$0.0625 \approx r$$

The interest rate should be 6.25%.

79. After x years at Humongous Enterprises, your salary would be $45{,}000\ (1 + 0.04)^x$ or $45{,}000\ (1.04)^x$. After x years at Crabapple Inc., your salary would be $30{,}000\ (1 + 0.06)^x$ or $30{,}000\ (1.06)^x$.

First we find when the salaries would be equal.

$$45{,}000(1.04)^x = 30{,}000(1.06)^x$$
$$\frac{(1.04)^x}{(1.06)^x} = \frac{30{,}000}{45{,}000}$$
$$\left(\frac{1.04}{1.06}\right)^x = \frac{2}{3}$$
$$\log\left(\frac{1.04}{1.06}\right)^x = \log\left(\frac{2}{3}\right)$$
$$x \log\left(\frac{1.04}{1.06}\right) = \log\left(\frac{2}{3}\right)$$

$$x = \frac{\log\left(\frac{2}{3}\right)}{\log\left(\frac{1.04}{1.06}\right)}$$

$$x \approx 21.29$$

$$2013 + 21.29 = 2034.29$$

Therefore, on July 1, 2035, the job at Crabapple, Inc., will pay more.

81. **(a)** The total number of individuals in the community is 50 + 50, or 100.

Let $P_1 = \frac{50}{100} = 0.5,\ P_2 = 0.5.$

$$\begin{aligned} H &= -1[P_1 \ln P_1 + P_2 \ln P_2] \\ &= -1[0.5 \ln 0.5 + 0.5 \ln 0.5] \\ &\approx 0.693 \end{aligned}$$

(b) For 2 species, the maximum diversity is ln 2.

(c) Yes, $\ln 2 \approx 0.693$.

83. **(a)** 3 species, $\frac{1}{3}$ each:

$$P_1 = P_2 = P_3 = \frac{1}{3}$$

$$\begin{aligned} H &= -(P_1 \ln P_1 + P_2 \ln P_2 + P_3 \ln P_3) \\ &= -3\left(\frac{1}{3}\ln\frac{1}{3}\right) \\ &= -\ln\frac{1}{3} \\ &\approx 1.099 \end{aligned}$$

(b) 4 species, $\frac{1}{4}$ each:

$$P_1 = P_2 = P_3 = P_4 = \frac{1}{4}$$

$$\begin{aligned} H &= -(P_1 \ln P_1 + P_2 \ln P_2 + P_3 \ln P_3 + P_4 \ln P_4) \\ &= -4\left(\frac{1}{4}\ln\frac{1}{4}\right) \\ &= -\ln\frac{1}{4} \\ &\approx 1.386 \end{aligned}$$

(c) Notice that

$$-\ln\frac{1}{3} = \ln(3^{-1})^{-1} = \ln 3 \approx 1.099$$

and

$$-\ln\frac{1}{4} = \ln(4^{-1})^{-1} = \ln 4 \approx 1.386$$

by Property c of logarithms, so the populations are at a maximum index of diversity.

85. $C(t) = C_0 e^{-kt}$

When $t = 0$, $C(t) = 2$, and when $t = 3$, $C(t) = 1$.

$$\begin{aligned} 2 &= C_0 e^{-k(0)} \\ C_0 &= 2 \\ 1 &= 2e^{-3k} \\ \frac{1}{2} &= e^{-3k} \\ -3k &= \ln\frac{1}{2} = \ln 2^{-1} = -\ln 2 \\ k &= \frac{\ln 2}{3} \\ T &= \frac{1}{k}\ln\frac{C_2}{C_1} \\ T &= \frac{1}{\frac{\ln 2}{3}}\ln\frac{5C_1}{C_1} \\ T &= \frac{3\ln 5}{\ln 2} \approx 7.0 \end{aligned}$$

The drug should be given about every 7 hours.

87. **(a)** $h(t) = 37.79(1.021)^t$

Double the 2005 population is 2(42.69) = 85.38 million

$$\begin{aligned} 85.38 &= 37.79(1.021)^t \\ \frac{85.38}{37.79} &= (1.021)^t \\ \log_{1.021}\left(\frac{85.38}{37.79}\right) &= t \\ t &= \frac{\ln\left(\frac{85.38}{37.79}\right)}{\ln 1.021} \approx 39.22 \end{aligned}$$

The Hispanic population is estimated to double their 2005 population in 2039.

(b) $h(t) = 11.14(1.023)^t$

Double the 2005 population is 2(12.69) = 25.38 million

$$\begin{aligned} 25.38 &= 11.14(1.023)^t \\ \frac{25.38}{11.14} &= (1.023)^t \\ \log_{1.023}\left(\frac{25.38}{11.14}\right) &= t \\ t &= \frac{\ln\left(\frac{25.38}{11.14}\right)}{\ln 1.023} \approx 36.21 \end{aligned}$$

The Asian population is estimated to double their 2005 population in 2036.

89.
$$C = B\log_2\left(\frac{s}{n}+1\right)$$
$$\frac{C}{B} = \log_2\left(\frac{s}{n}+1\right)$$
$$2^{C/B} = \frac{s}{n}+1$$
$$\frac{s}{n} = 2^{C/B} - 1$$

91. Let I_1 be the intensity of the sound whose decibel rating is 85.

(a)
$$10\log\frac{I_1}{I_0} = 85$$
$$\log\frac{I_1}{I_0} = 8.5$$
$$\log I_1 - \log I_0 = 8.5$$
$$\log I_1 = 8.5 + \log I_0$$

Let I_2 be the intensity of the sound whose decibel rating is 75.
$$10\log\frac{I_2}{I_0} = 75$$
$$\log\frac{I_2}{I_0} = 7.5$$
$$\log I_2 - \log I_0 = 7.5$$
$$\log I_0 = \log I_2 - 7.5$$

Substitute for I_0 in the equation for $\log I_1$.
$$\begin{aligned}\log I_1 &= 8.5 + \log I_0\\ &= 8.5 + \log I_2 - 7.5\\ &= 1 + \log I_2\end{aligned}$$
$$\log I_1 - \log I_2 = 1$$
$$\log\frac{I_1}{I_2} = 1$$

Then $\frac{I_1}{I_2} = 10$, so $I_2 = \frac{1}{10} I_1$. This means the intensity of the sound that had a rating of 75 decibels is $\frac{1}{10}$ as intense as the sound that had a rating of 85 decibels.

93. $\text{pH} = -\log[\text{H}^+]$

(a) For pure water:
$$7 = -\log[\text{H}^+]$$
$$-7 = \log[\text{H}^+]$$
$$10^{-7} = [\text{H}^+]$$

For acid rain:
$$4 = -\log[\text{H}^+]$$
$$-4 = \log[\text{H}^+]$$
$$10^{-4} = [\text{H}^+]$$
$$\frac{10^{-4}}{10^{-7}} = 10^3 = 1000$$

The acid rain has a hydrogen ion concentration 1000 times greater than pure water.

(b) For laundry solution:
$$11 = -\log[\text{H}^+]$$
$$10^{-11} = [\text{H}^+]$$

For black coffee:
$$5 = -\log[\text{H}^+]$$
$$10^{-5} = [\text{H}^+]$$
$$\frac{10^{-5}}{10^{-11}} = 10^6 = 1{,}000{,}000$$

The coffee has a hydrogen ion concentration 1,000,000 times greater than the laundry mixture.

10.6 Applications: Growth and Decay; Mathematics of Finance

Your Turn 1

$$y = y_0e^{kt}$$
$$18 = 5e^{k(16)}$$
$$e^{16k} = \ln(18/5)$$
$$16k = \ln(18/5)$$
$$k = \frac{\ln(18/5)}{16} \approx 0.08$$
$$y = 5e^{0.08t}$$

Your Turn 2

Let $A(t) = \frac{1}{10}A_0$ and $k = -(\ln 2/5600)$

$$A(t) = A_0 e^{kt}$$
$$\frac{1}{10}A_0 = A_0 e^{-(\ln 2/5600)t}$$
$$\frac{1}{10} = e^{-(\ln 2/5600)t}$$
$$\ln\left(\frac{1}{10}\right) = \ln e^{-(\ln 2/5600)t}$$
$$\ln\left(\frac{1}{10}\right) = -(\ln 2/5600)t$$
$$t = -\frac{5600}{\ln 2}\ln\left(\frac{1}{10}\right)$$
$$\approx 18602.80$$

The age of the sample is about 18,600 years.

Your Turn 3

(a) 4.25% compounded monthly

$$\left(1+\frac{0.0425}{12}\right)^{12} - 1 = (1.0035417)^{12} - 1$$
$$\approx 0.0433$$

The effective rate is 4.33%.

(b) 3.75% compounded continuously

$$e^{0.0375} - 1 \approx 0.0382$$

The effective rate is 3.82%.

Your Turn 4

$A = 50,000$, $r = 0.315$, and $m = 4$.

$$50,000 = 30,000\left(1+\frac{0.0315}{4}\right)^{4t}$$
$$\frac{5}{3} = (1.007875)^{4t}$$
$$t = \frac{\ln(5/3)}{4\ln(1.007875)} \approx 16.28$$

We need to round up to the nearest quarter, so \$30,000 will grow to \$50,000 in 16.5 years.

Your Turn 5

$$A = Pe^{rt}$$
$$4500 = 3200e^{7r}$$
$$1.40625 = e^{7r}$$
$$\ln 1.40625 = \ln e^{7r}$$
$$7r = \ln 1.40625$$
$$r = \frac{\ln 1.40625}{7} \approx 0.0487$$

The interest rate needed is 4.87%.

10.6 Exercises

5. Assume that $y = y_0 e^{kt}$ represents the amount remaining of a radioactive substance decaying with a half-life of T. Since $y = y_0$ is the amount of the substance at time $t = 0$, then $y = \frac{y_0}{2}$ is the amount at time $t = T$. Therefore, $\frac{y_0}{2} = y_0 e^{kT}$, and solving for k yields

$$\frac{1}{2} = e^{kT}$$
$$\ln\left(\frac{1}{2}\right) = kT$$
$$k = \frac{\ln\left(\frac{1}{2}\right)}{T}$$
$$= \frac{\ln(2^{-1})}{T}$$
$$= -\frac{\ln 2}{T}.$$

7. $r = 4\%$ compounded quarterly,

$m = 4$

$$r_E = \left(1+\frac{r}{m}\right)^m - 1$$
$$= \left(1+\frac{0.04}{4}\right)^4 - 1$$
$$\approx 0.0406$$
$$\approx 4.06\%$$

9. $r = 8\%$ compounded continuously

$$\begin{aligned} r_E &= e^r - 1 \\ &= e^{0.08} - 1 \\ &\approx 0.0833 = 8.33\% \end{aligned}$$

11. $A = \$10{,}000$, $r = 6\%$, $m = 4$, $t = 8$

$$\begin{aligned} P &= A\left(1 + \frac{r}{m}\right)^{-tm} \\ &= 10{,}000\left(1 + \frac{0.06}{4}\right)^{-8(4)} \\ &\approx \$6209.93 \end{aligned}$$

13. $A = \$7300$, $r = 5\%$ compounded continuously, $t = 3$

$$\begin{aligned} A &= Pe^{rt} \\ P &= \frac{A}{e^{rt}} \\ &= \frac{7300}{e^{0.05(3)}} \approx \$6283.17 \end{aligned}$$

15. $r = 9\%$ compounded semiannually

$$\begin{aligned} r_E &= \left(1 + \frac{0.09}{2}\right)^2 - 1 \\ &\approx 0.0920 = 9.20\% \end{aligned}$$

17. $r = 6\%$ compounded monthly

$$\begin{aligned} r_E &= \left(1 + \frac{0.06}{12}\right)^{12} - 1 \\ &\approx 0.0617 \\ &\approx 6.17\% \end{aligned}$$

19. (a) $A = \$307{,}000, t = 3, r = 6\%, m = 2$

$$\begin{aligned} A &= P\left(1 + \frac{r}{m}\right)^{mt} \\ 307{,}000 &= P\left(1 + \frac{0.06}{2}\right)^{3(2)} \\ 307{,}000 &= P(1.03)^6 \\ \frac{307{,}000}{(1.03)^6} &= P \\ \$257{,}107.67 &= P \end{aligned}$$

(b)
$$\begin{aligned} \text{Interest} &= 307{,}000 - 257{,}107.67 \\ &= \$49{,}892.33 \end{aligned}$$

(c)
$$\begin{aligned} P &= \$200{,}000 \\ A &= 200{,}000(1.03)^6 \\ &= 238{,}810.46 \end{aligned}$$

The additional amount needed is

$$\begin{aligned} &307{,}000 - 238{,}810.46 \\ &= \$68{,}189.54. \end{aligned}$$

(d)
$$\begin{aligned} A &= Pe^{rt} \\ 307{,}000 &= 200{,}000e^{3r} \\ 1.535 &= e^{3r} \\ \ln 1.535 &= \ln e^{3r} \\ 3r &= \ln 1.535 \\ r &= \frac{\ln 1.535}{3} \approx 0.1428 \end{aligned}$$

The interest rate needed is 14.28%.

21. $P = \$60{,}000$

(a) $r = 8\%$ compounded quarterly:

$$\begin{aligned} A &= P\left(1 + \frac{r}{m}\right)^{tm} \\ &= 60{,}000\left(1 + \frac{0.08}{47}\right)^{5(4)} \\ &\approx \$89{,}156.84 \end{aligned}$$

$r = 7.75\%$ compounded continuously

$$\begin{aligned} A &= Pe^{rt} \\ &= 60{,}000e^{0.775(5)} \\ &\approx \$88{,}397.58 \end{aligned}$$

Linda will earn more money at 8% compoundded quarterly.

(b) She will earn \$759.26 more.

(c) $r = 8\%, m = 4$:

$$\begin{aligned} r_E &= \left(1 + \frac{r}{m}\right)^m - 1 \\ &= \left(1 + \frac{0.08}{4}\right)^4 - 1 \\ &\approx 0.0824 \\ &= 8.24\% \end{aligned}$$

$r = 7.75\%$ compounded continuously:

$$\begin{aligned} r_E &= e^r - 1 \\ &= e^{0.0775} - 1 \\ &\approx 0.0806 \\ &= 8.06\% \end{aligned}$$

(d) $A = \$80{,}000$

$$\begin{aligned} A &= Pe^{rt} \\ 80{,}000 &= 60{,}000e^{0.0775t} \\ \frac{4}{3} &= e^{0.0775t} \\ \ln\frac{4}{3} &= \ln e^{0.0775t} \\ \ln 4 - \ln 3 &= 0.0775t \\ \frac{\ln 4 - \ln 3}{0.0775} &= t \\ 3.71 &= t \end{aligned}$$

\$60,000 will grow to \$80,000 in about 3.71 years.

(e) $60{,}000\left(1 + \frac{0.08}{4}\right)^{4x} \geq 80{,}000$

$$\begin{aligned} (1.02)^{4x} &\geq \frac{80{,}000}{60{,}000} \\ (1.02)^{4x} &\geq \frac{4}{3} \\ \log(1.02)^{4x} &\geq \log\left(\frac{4}{3}\right) \\ 4x\log(1.02) &\geq \log\left(\frac{4}{3}\right) \\ x &\geq \frac{\log\left(\frac{4}{3}\right)}{4\log(1.02)} \approx 3.63 \end{aligned}$$

We need to round up to the nearest quarter, so it will take 3.75 years.

23. $S(x) = 1000 - 800e^{-x}$

(a) $$\begin{aligned} S(0) &= 1000 - 800e^{0} \\ &= 1000 - 800 \\ &= 200 \end{aligned}$$

(b) $$\begin{aligned} S(x) &= 500 \\ 500 &= 1000 - 800e^{-x} \\ -500 &= -800e^{-x} \\ \frac{5}{8} &= e^{-x} \\ \ln\frac{5}{8} &= \ln e^{-x} \\ -\ln\frac{5}{8} &= x \\ 0.47 &\approx x \end{aligned}$$

Sales reach 500 in about $\frac{1}{2}$ year.

(c) Since $800e^{-x}$ will never actually be zero, $S(x) = 1000 - 800e^{-x}$ will never be 1000.

(d) Graphing the function $y = S(x)$ on a graphing calculator will show that there is a horizontal asymptote at $y = 1000$. This indicates that the limit on sales is 1000 units.

25. (a) $P = P_0e^{kt}$

When $t = 1650$, $P = 500$.

When $t = 2010$, $P = 6756$.

$$\begin{aligned} 500 &= P_0e^{1650k} \\ 6756 &= P_0e^{2010k} \\ \frac{6756}{500} &= \frac{P_0e^{2010k}}{P_0e^{1650k}} \\ \frac{6756}{500} &= e^{360k} \\ 360k &= \ln\left(\frac{6765}{500}\right) \\ k &= \frac{\ln\left(\frac{6765}{500}\right)}{360} \approx 0.007232 \end{aligned}$$

Substitute this value into $500 = P_0e^{1650k}$ to find P_0.

$$\begin{aligned} 500 &= P_0e^{1650(0.007232)} \\ P_0 &= \frac{500}{e^{1650(0.007232)}} \\ P_0 &= 0.003286 \end{aligned}$$

Therefore, $P(t) = 0.003286e^{0.007232t}$.

(b) $$\begin{aligned} P(1) &= 0.003286e^{0.007232} \\ &\approx 0.0033 \text{ million, or } 3300. \end{aligned}$$

The exponential equation gives a world population of only 3300 in the year 1.

(c) No, the answer in part (b) is too small. Exponential growth does not accurately describe population growth for the world over a long period of time.

27. $y = y_0e^{kt}$

$y = 40{,}000$, $y_0 = 25{,}000$, $t = 10$

(a) $$\begin{aligned} 40{,}000 &= 25{,}000e^{k(10)} \\ 1.6 &= e^{10k} \\ \ln 1.6 &= 10k \\ 0.047 &= k \end{aligned}$$

The equation is

$$y = 25{,}000e^{0.047t}.$$

(b) $y = 25{,}000e^{0.047t}$

$= 25{,}000(e^{0.047})^t$

$= 25{,}000(1.048)^t$

(c) $y = 60{,}000$

$60{,}000 = 25{,}000e^{0.047t}$

$2.4 = e^{0.047t}$

$\ln 2.4 = 0.047t$

$18.6 = t$

There will be 60,000 bacteria in about 18.6 hours.

29. $f(t) = 500e^{0.1t}$

(a) $f(t) = 3000$

$3000 = 500e^{0.1t}$

$6 = e^{0.1t}$

$\ln 6 = 0.1t$

$17.9 \approx t$

It will take 17.9 days.

(b) If $t = 0$ corresponds to January 1, the date January 17 should be placed on the product. January 18 would be more than 17.9 days.

31. (a) From the graph, the risks of chromosomal abnormality per 1000 at ages 20, 35, 42, and 49 are 2, 5, 24, and 125, respectively.

(Note: It is difficult to read the graph accurately. If you read different values from the graph, your answers to parts (b)-(e) may differ from those given here.)

(b) $y = Ce^{kt}$

When $t = 20, y = 2,$ and when $t = 35,$ $y = 5.$

$2 = Ce^{20k}$

$5 = Ce^{35k}$

$\frac{5}{2} = \frac{Ce^{35k}}{Ce^{20k}}$

$2.5 = e^{15k}$

$15k = \ln 2.5$

$k = \frac{\ln 2.5}{15}$ $k = 0.061$

(c) $y = Ce^{kt}$

When $t = 42, y = 29,$ and when $t = 49,$ $y = 125.$

$24 = Ce^{42k}$

$125 = Ce^{49k}$

$\frac{125}{24} = \frac{Ce^{49k}}{Ce^{42k}}$

$\frac{125}{24} = e^{7k}$

$7k = \ln\left(\frac{125}{24}\right)$

$k = \frac{\ln\left(\frac{125}{24}\right)}{7}$ $k \approx 0.24$

(d) Since the values of k are different, we cannot assume the graph is of the form $y = Ce^{kt}$.

(e) The results are summarized in the following table.

n	Value of k for [20, 35]	Value of k for [42, 49]
2	0.00093	0.0017
3	2.3×10^{-5}	2.5×10^{-5}
4	6.3×10^{-7}	4.1×10^{-7}

The value of n should be somewhere between 3 and 4.

33. $\frac{1}{2}A_0 = A_0e^{-0.053t}$

$\frac{1}{2} = e^{-0.053t}$

$\ln\frac{1}{2} = -0.053t$

$-\ln 2 = -0.053t$

$t = \frac{\ln 2}{0.52}$

$t \approx 13$

The half-life of plutonium 241 is about 13 years.

35. (a) $A(t) = A_0\left(\frac{1}{2}\right)^{t/13}$

$A(100) = 4.0\left(\frac{1}{2}\right)^{100/13}$

$A(100) \approx 0.0193$

After 100 years, about 0.0193 gram will remain.

(b)

$$0.1 = 4.0\left(\frac{1}{2}\right)^{t/13}$$

$$\frac{0.1}{4.0} = \left(\frac{1}{2}\right)^{t/13}$$

$$\ln 0.025 = \frac{t}{13}\ln\left(\frac{1}{2}\right)$$

$$t = \frac{13 \ln 0.025}{\ln\left(\frac{1}{2}\right)}$$

$$t \approx 69.19$$

It will take 69 years.

37. **(a)** $y = y_0 e^{kt}$

When $t = 0$, $y = 500$, so $y_0 = 500$.

When $t = 3$, $y = 386$.

$$386 = 500e^{3k}$$

$$\frac{386}{500} = e^{3k}$$

$$e^{3k} = 0.772$$

$$3k = \ln 0.772$$

$$k = \frac{\ln 0.772}{3}$$

$$k \approx -0.0863$$

$$y = 500e^{-0.0863t}$$

(b) From part (a), we have

$$k = \frac{\ln\left(\frac{386}{500}\right)}{3}.$$

$$\begin{aligned} y &= 500e^{kt} \\ &= 500e^{[\ln(386/500)/3]t} \\ &= 500e^{\ln(386/500)\cdot(t/3)} \\ &= 500\,[e^{\ln(386/500)}]^{t/3} \\ &= 500(386/500)^{t/3} \\ &= 500(0.722)^{t/3} \end{aligned}$$

(c) $\frac{1}{2}y_0 = y_0 e^{-0.0863t}$

$$\ln\frac{1}{2} = -0.0863t$$

$$t = \frac{\ln\left(\frac{1}{2}\right)}{-0.0863}$$

$$t \approx 8.0$$

The half-life is about 8.0 days.

39. $y = 40e^{-0.004t}$

(a) $t = 180$

$$\begin{aligned} y &= 40e^{-0.004(180)} = 40e^{-0.72} \\ &\approx 19.5 \text{ watts} \end{aligned}$$

(b)

$$20 = 20e^{-0.0004t}$$

$$\frac{1}{2} = e^{-0.004t}$$

$$\ln\frac{1}{2} = -0.004t$$

$$\frac{\ln 2}{0.004} = t$$

$$173 \approx t$$

It will take about 173 days.

(c) The power will never be completely gone. The power will approach 0 watts but will never be exactly 0.

41. $P(t) = 100e^{-0.1t}$

(a) $P(4) = 100e^{-0.1(4)} \approx 67\%$

(b) $P(10) = 100e^{-0.1(10)} \approx 37\%$

(c)

$$10 = 100e^{-0.1t}$$

$$0.1 = e^{-0.1t}$$

$$\ln(0.1) = -0.1t$$

$$\frac{-\ln(0.1)}{0.1} = t$$

$$23 \approx t$$

It would take about 23 days.

(d)

$$1 = 100e^{-0.1t}$$

$$0.01 = e^{-0.1t}$$

$$\ln(0.01) = -0.1t$$

$$\frac{-\ln(0.01)}{0.1} = t$$

$$46 \approx t$$

It would take about 46 days.

43. $t = 9, T_0 = 18, C = 5, k = 0.6$

$$f(t) = T_0 + Ce^{-kt}$$

$$\begin{aligned} f(t) &= 18 + 5e^{-0.6(9)} \\ &= 18 + 5e^{-5.4} \\ &\approx 18.02 \end{aligned}$$

The temperature is about 18.02°.

45. $C = -14.6, k = 0.6, T_0 = 18°,$

$f(t) = 10°$

$f(t) = T_0 + Ce^{-kt}$

$$f(t) = 18 + (-14.6)e^{-0.6t}$$
$$-8 = -14.6e^{-0.6t}$$
$$0.5479 = e^{-0.6t}$$
$$\ln 0.5479 = -0.6t$$
$$\frac{-\ln 0.5479}{0.6} = t$$
$$1 \approx t$$

It would take about 1 hour for the pizza to thaw.

Chapter 10 Review Exercises

1. True

2. False; for example $f(x) = \frac{x}{x+1}$ is a rational function but not an exponential function.

3. True

4. True

5. False; an exponential function has the form $f(x) = a^x$.

6. False; the vertical asymptote is at $x = 6$.

7. True

8. False; the domain includes all numbers except $x = 2$ and $x = -2$.

9. False; the amount is $A = 2000\left(1 + \frac{0.04}{12}\right)^{24}$.

10. False; the logarithmic function $f(x) = \log_a x$ is not defined for $a = 1$.

11. False; $\ln(5 + 7) = \ln 12 \neq \ln 5 + \ln 7$

12. False; $(\ln 3)^4 \neq 4 \ln 3$ since $(\ln 3)^4$ means $(\ln 3)(\ln 3)(\ln 3)(\ln 3)$.

13. False; $\log_{10} 0$ is undefined since $10^x = 0$ has no solution.

14. True

15. False; $\ln(-2)$ is undefined.

16. False; $\frac{\ln 4}{\ln 8} = 0.6667$ and $\ln 4 - \ln 8 = \ln(1/2) \approx -0.6931$.

17. True

18. True

23. $y = (2x - 1)(x + 1)$

$= 2x^2 + x - 1$

x	-3	-2	-1	0	1	2	3
y	14	5	0	-1	1	9	20

Pairs: $(-3, 14), (-2, 5), (-1, 0), (0, -1), (1, 2), (2, 9), (3, 20)$

Range: $\{-1, 0, 2, 5, 9, 14, 20\}$

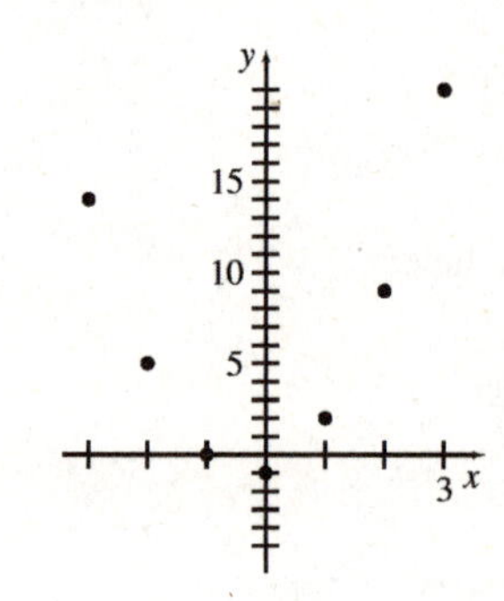

25. $f(x) = 5x^2 - 3$ and $g(x) = -x^2 + 4x + 1$

(a) $f(-2) = 5(-2)^2 - 3 = 17$

(b) $g(3) = -(3)^2 + 4(3) + 1 = 4$

(c) $f(-k) = 5(-k)^2 - 3 = 5k^2 - 3$

(d) $g(3m) = -(3m)^2 + 4(3m) + 1$
$= -9m^2 + 12m + 1$

(e) $f(x + h) = 5(x + h)^2 - 3$
$= 5(x^2 + 2xh + h^2) - 3$
$= 5x^2 + 10xh + 5h^2 - 3$

(f)

$$g(x+h) = -(x+h)^2 + 4(x+h) + 1$$
$$= -(x^2 + 2xh + h^2) + 4x + 4h + 1$$
$$= -x^2 - 2xh - h^2 + 4x + 4h + 1$$

(g) $\frac{f(x+h) - f(x)}{h}$

$$= \frac{5(x+h)^2 - 3 - (5x^2 - 3)}{h}$$
$$= \frac{5(x^2 + 2hx + h^2) - 3 - 5x^2 + 3}{h}$$
$$= \frac{5x^2 + 10hx + 5h^2 - 5x^2}{h}$$
$$= \frac{10hx + 5h^2}{h} = 10x + 5h$$

(h)

$$\frac{g(x+h) - g(x)}{h}$$
$$= \frac{-(x+h)^2 + 4(x+h) + 1 - (-x^2 + 4x + 1)}{h}$$
$$= \frac{-(x^2 + 2xh + h^2) + 4x + 4h + 1 + x^2 - 4x - 1}{h}$$
$$= \frac{-x^2 - 2xh - h^2 + 4h + x^2}{h}$$
$$= \frac{-2xh - h^2 + 4h}{h}$$
$$= -2x - h + 4$$

27. $y = \frac{3x - 4}{x}$

$x \neq 0$

Domain: $(-\infty, 0) \cup (0, \infty)$

29. $y = \ln(x + 7)$

$$x + 7 > 0$$
$$x > -7$$

Domain: $(-7, \infty)$.

31. $y = 2x^2 + 3x - 1$

The graph is a parabola.

Let $y = 0$.

$$0 = 2x^2 + 3x - 1$$
$$x = \frac{-3 \pm \sqrt{3^2 - 4(2)(-1)}}{2(2)}$$
$$= \frac{-3 \pm \sqrt{9 + 8}}{4}$$
$$= \frac{-3 \pm 17}{4}$$

The x-intercepts are $\frac{-3 + \sqrt{17}}{4} \approx 0.28$ and $\frac{-3 - \sqrt{17}}{4} \approx -1.48$.

Let $x = 0$.

$$y = 2(0)^2 + 3(0) - 1$$

-1 is the y-intercept.

Vertex: $x = \frac{-b}{2a} = \frac{-3}{2(2)} = -\frac{3}{4}$

$$y = 2\left(-\frac{3}{4}\right)^2 + 3\left(-\frac{3}{4}\right) - 1$$
$$= \frac{9}{8} - \frac{9}{4} - 1$$
$$= -\frac{17}{8}$$

The vertex is $\left(-\frac{3}{4}, -\frac{17}{8}\right)$.

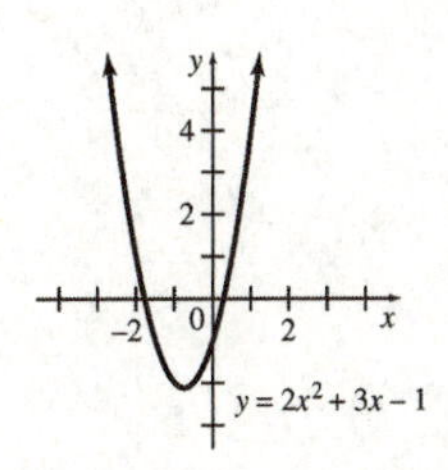

33. $y = -x^2 + 4x + 2$

Let $y = 0$.

$$0 = -x^2 + 4x + 2$$
$$x = \frac{-4 \pm \sqrt{4^2 - 4(-1)(2)}}{2(-1)}$$
$$= \frac{-4 \pm \sqrt{24}}{-2}$$
$$= 2 \pm \sqrt{6}$$

The x-intercepts are $2 + \sqrt{6} \approx 4.45$ and $2 - \sqrt{6} \approx -0.45$.

Let $x = 0$.

$$y = -0^2 + 4(0) + 2$$

2 is the y-intercept.

Vertex: $x = \frac{-b}{2a} = \frac{-4}{2(-1)} = \frac{-4}{-2} = 2$

$$y = -2^2 + 4(2) + 2 = 6$$

The vertex is (2, 6).

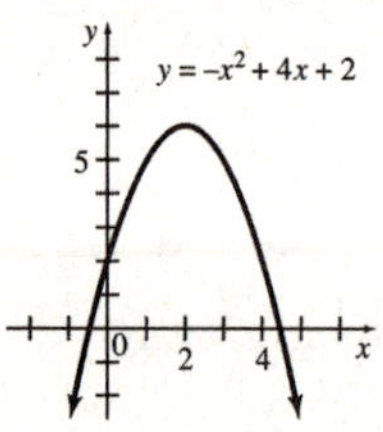

35. $f(x) = x^3 - 3$

Translate the graph of $f(x) = x^3$ 3 units down.

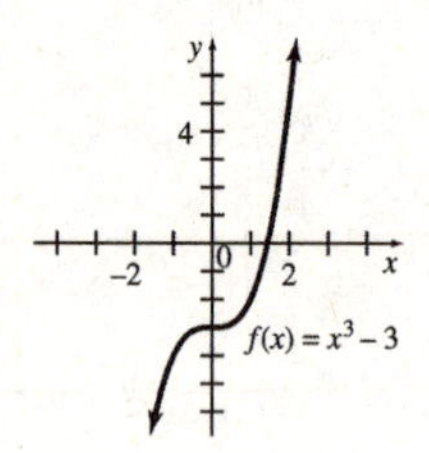

37. $y = -(x - 1)^4 + 4$

Translate the graph of $y = x^4$ 1 unit to the right and reflect vertically. Translate 4 units upward.

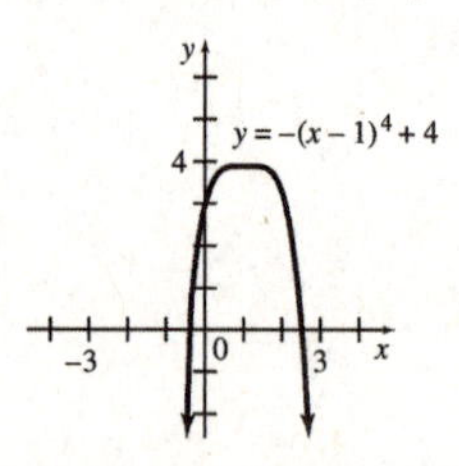

39. $f(x) = \frac{8}{x}$

Vertical asymptote: $x = 0$

Horizontal asymptote:

$\frac{8}{x}$ approaches zero as x gets larger.

$y = 0$ is an asymptote.

x	−4	−3	−2	−1	1	2	3	4
y	−2	−2.7	−4	−8	8	4	2.7	2

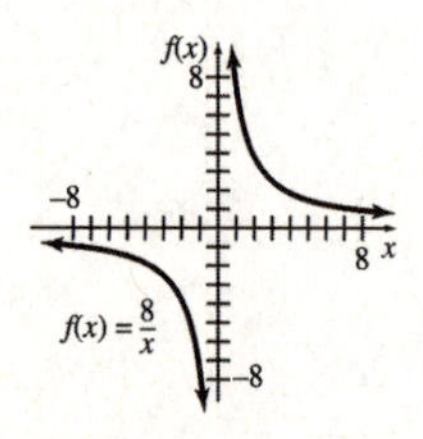

41. $f(x) = \frac{4x - 2}{3x + 1}$

Vertical asymptote:

$$3x + 1 = 0$$

$$x = -\frac{1}{3}$$

Horizontal asymptote:

As x gets larger,

$$\frac{4x - 2}{3x - 1} \approx \frac{4x}{3x} = \frac{4}{3}.$$

$y = \frac{4}{3}$ is an asymptote.

x	− 3	−2	−1	0	1	2	3
y	1.75	2	3	−2	0.5	0.86	1

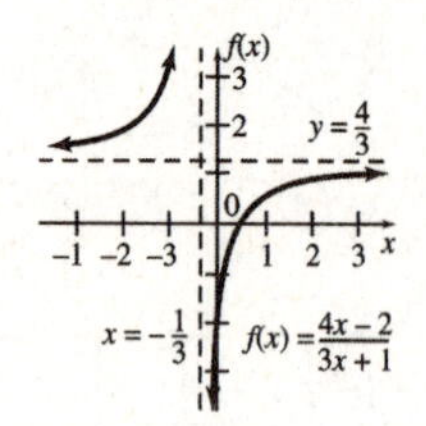

43. $y = 4^x$

x	−2	−1	0	1	2
y	$\frac{1}{16}$	$\frac{1}{4}$	1	4	16

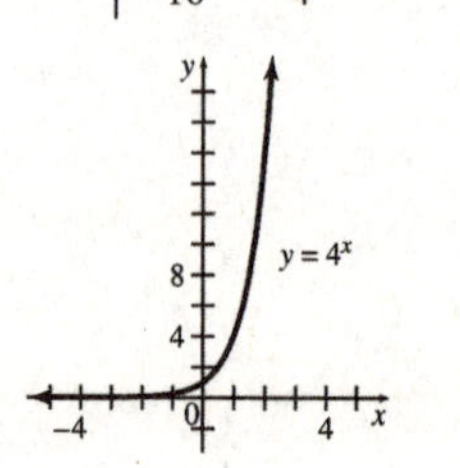

45. $y = \left(\frac{1}{5}\right)^{2x-3}$

x	0	1	2
y	125	5	$\frac{1}{5}$

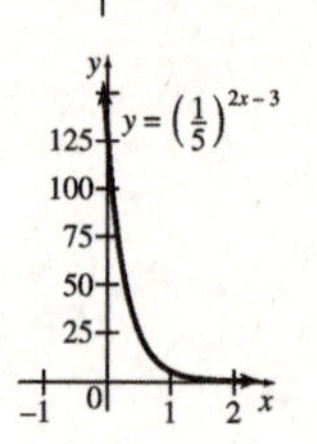

47. $y = \log_2(x-1)$

$2^y = x - 1$

$x = 1 + 2^y$

x	2	3	5	9
y	0	1	2	3

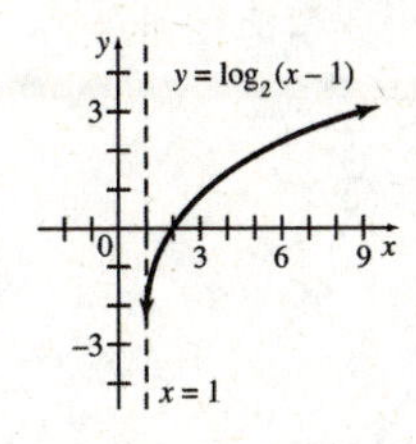

49. $y = -\ln(x+3)$

$-y = \ln(x+3)$

$e^{-y} = x + 3$

$e^{-y} - 3 = x$

x	−2.63	−2	−0.28	4.39
y	1	0	−1	−2

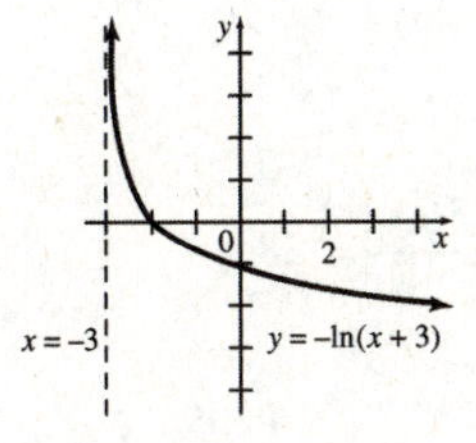

51. $2^{x+2} = \frac{1}{8}$

$2^{x+2} = \frac{1}{2^3}$

$2^{x+2} = 2^{-3}$

$x + 2 = -3$

$x = -5$

53. $9^{2y+3} = 27^y$

$(3^2)^{2y+3} = (3^3)y$

$3^{4y+6} = 3^{3y}$

$4y + 6 = 3y$

$y = -6$

55. $3^5 = 243$

The equation in logarithmic form is

$$\log_3 243 = 5.$$

57. $e^{0.8} = 2.22554$

The equation in logarithmic form is

$$\ln 2.22554 = 0.8.$$

59. $\log_2 32 = 5$

The equation in exponential form is

$$2^5 = 32.$$

61. $\ln 82.9 = 4.41763$

The equation in exponential form is

$$e^{4.41763} = 82.9.$$

63. $\log_3 81 = x$

$3^x = 81$

$3^x = 3^4$

$x = 4$

65. $\log_4 8 = x$

$4^x = 8$

$(2^2)^x = 2^3$

$2x = 3$

$x = \frac{3}{2}$

67. $\log_5 3k + \log_{5.} 7k^3$

$= \log_5 3k(7k^3)$

$= \log_5(21k^4)$

69. $4\log_3 y - 2\log_3 x$

$= \log_3 y^4 - \log_3 x^2$

$= \log_3\left(\frac{y^4}{x^2}\right)$

71. $6^p = 17$

$\ln 6^p = \ln 17$

$p\ln 6 = \ln 17$

$p = \frac{\ln 17}{\ln 6}$

≈ 1.581

73.
$$\begin{aligned} 2^{1-m} &= 7 \\ \ln 2^{1-m} &= \ln 7 \\ (1-m)\ln 2 &= \ln 7 \\ 1-m &= \frac{\ln 7}{\ln 2} \\ -m &= \frac{\ln 7}{\ln 2} - 1 \\ m &= 1 - \frac{\ln 7}{\ln 2} \\ &\approx -1.807 \end{aligned}$$

75.
$$\begin{aligned} e^{-5-2x} &= 5 \\ \ln e^{-5-2x} &= \ln 5 \\ -5-2x &= \ln 5 \\ -2x &= \ln 5 + 5 \\ x &= \frac{\ln 5 + 5}{-2} \\ &\approx -3.305 \end{aligned}$$

77.
$$\begin{aligned} \left(1+\frac{m}{3}\right)^5 &= 15 \\ \left[\left(1+\frac{m}{3}\right)^5\right]^{1/5} &= 15^{1/5} \\ 1+\frac{m}{3} &= 15^{1/5} \\ \frac{m}{3} &= 15^{1/5} - 1 \\ m &= 3(15^{1/5} - 1) \\ &\approx 2.156 \end{aligned}$$

79.
$$\begin{aligned} \log_k 64 &= 6 \\ k^6 &= 64 \\ k^6 &= 2^6 \\ k &= 2 \end{aligned}$$

81.
$$\begin{aligned} \log(4p+1) + \log p &= \log 3 \\ \log[p(4p+1)] &= \log 3 \\ \log(4p^2 + p) &= \log 3 \\ 4p^2 + p &= 3 \\ 4p^2 + p - 3 &= 0 \\ (4p-3)(p+1) &= 0 \end{aligned}$$

$$4p - 3 = 0 \quad \text{or} \quad p + 1 = 0$$
$$p = \frac{3}{4} \qquad\qquad p = -1$$

p cannot be negative, so $p = \frac{3}{4}$.

83. $f(x) = a^x; a > 0, a \neq 1$

(a) The domain is $(-\infty, \infty)$.

(b) The range is $(0, \infty)$.

(c) The y-intercept is 1.

(d) The x-axis, $y = 0$, is a horizontal asymptote.

(e) The function is increasing if $a > 1$.

(f) The function is decreasing if $0 < a < 1$.

87. $y = \dfrac{7x}{100 - x}$

(a) $y = \dfrac{7(80)}{100-80} = \dfrac{560}{20} = 28$

The cost is $28,000.

(b) $y = \dfrac{7(50)}{100-50} = \dfrac{350}{50} = 7$

The cost is $7000.

(c) $\dfrac{7(90)}{100-90} = \dfrac{630}{10} = 63$

The cost is $63,000.

(d) Plot the points (80, 28), (50, 7), and (90, 63).

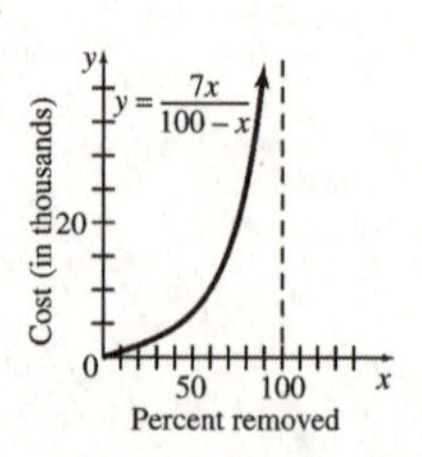

(e) No, because all of the pollutant would be removed when $x = 100$, at which point the denominator of the function would be zero.

89. $P = \$2781.36, r = 4.8\%, t = 6, m = 4$

$$A = P\left(1 + \frac{r}{m}\right)^{tm}$$

$$A = 2781.36\left(1 + \frac{0.048}{4}\right)^{(6)(4)}$$

$$= 2781.36(1.012)^{24}$$
$$= \$3703.31$$

$$\text{Interest} = \$3703.31 - \$2781.36$$
$$= \$921.95$$

91. $P = \$12{,}104, r = 6.2\%, t = 4$

$$A = Pe^{rt}$$
$$A = 12{,}104e^{0.062(4)}$$
$$= 12{,}104e^{0.248}$$
$$= \$15{,}510.79$$

93. $P = \$12{,}000, r = 0.05, t = 8$

$$A = 12{,}000e^{0.05(8)}$$
$$= 12{,}000e^{0.40}$$
$$= \$17{,}901.90$$

95. \$2100 deposited at 4% compounded quarterly.

$$A = P\left(1 + \frac{r}{m}\right)^{tm}$$

To double:

$$2(2100) = 2100\left(1 + \frac{0.04}{4}\right)^{t \cdot 4}$$
$$2 = 1.01^{4t}$$
$$\ln 2 = 4t \ln 1.01$$
$$t = \frac{\ln 2}{4 \ln 1.01}$$
$$\approx 17.4$$

Because interest is compounded quarterly, round the result up to the nearest quarter, which is 17.5 years or 70 quarters.

To triple:

$$3(2100) = 2100\left(1 + \frac{0.04}{4}\right)^{t \cdot 4}$$
$$3 = 1.01^{4t}$$
$$\ln 3 = 4t \ln 1.01$$
$$t = \frac{\ln 3}{4 \ln 1.01}$$
$$\approx 27.6$$

Because interest is compounded quarterly, round the result up to the nearest quarter, which is 27.75 years or 111 quarters.

97. $r = 6\%, m = 12$

$$r_E = \left(1 + \frac{r}{m}\right)^m - 1$$
$$= \left(1 + \frac{0.06}{12}\right)^{12} - 1$$
$$= 0.0617 = 6.17\%$$

99. $A = \$2000, r = 6\%, t = 5, m = 1$

$$P = A\left(1 + \frac{r}{m}\right)^{-tm}$$
$$= 2000\left(1 + \frac{0.06}{1}\right)^{-5(1)}$$
$$= 2000(1.06)^{-5}$$
$$= \$1494.52$$

101. $r = 7\%, t = 8, m = 2, P = 10{,}000$

$$A = P\left(1 + \frac{r}{m}\right)^{tm}$$
$$= 10{,}000\left(1 + \frac{0.07}{2}\right)^{8(2)}$$
$$= 10{,}000(1.035)^{16}$$
$$= \$17{,}339.86$$

103. $P = \$6000, A = \$8000, t = 3$

$$A = Pe^{rt}$$
$$8000 = 6000e^{3r}$$
$$\frac{4}{3} = e^{3r}$$
$$\ln 4 - \ln 3 = 3r$$
$$r = \frac{\ln 4 - \ln 3}{3}$$
$$r \approx 0.0959 \text{ or about } 9.59\%$$

105. **(a)** $n = 1000 - (p - 50)(10),\ p \geq 50$
$= 1000 - 10p + 500$
$= 1500 - 10p$

(b) $R = pn$
$R = p(1500 - 10p)$

(c) $p \geq 50$

Since n cannot be negative,

$$1500 - 10p \geq 0$$
$$-10p \geq -1500$$
$$p \leq 150.$$

Therefore, $50 \leq p \leq 150$.

(d) Since $n = 1500 - 10p$,

$$10p = 1500 - n$$
$$p = 150 - \frac{n}{10}.$$
$$R = pn$$
$$R = \left(150 - \frac{n}{10}\right)n$$

(e) Since she can sell at most 1000 tickets, $0 \leq n \leq 1000$.

(f) $R = -10p^2 + 1500p$

$$\frac{-b}{2a} = \frac{-1500}{2(-10)} = 75$$

The price producing maximum revenue is \$75.

(g) $R = -\frac{1}{10}n^2 + 150n$

$$\frac{-b}{2a} = \frac{-150}{2\left(-\frac{1}{10}\right)}$$
$$= 750$$

The number of tickets producing maximum revenue is 750.

(h) $R(p) = -10p^2 + 1500p$

$$R(75) = -10(75)^2 + 1500(75)$$
$$= -56,250 + 112,500$$
$$= 56,250$$

The maximum revenue is \$56,250.

(i)

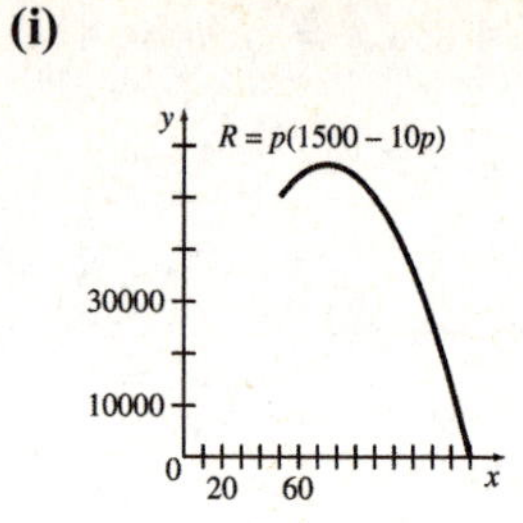

(j) The revenue starts at \$50,000 when the price is \$50, rises to a maximum of \$56,250 when the price is \$75, and falls to 0 when the price is \$150.

107. $C(x) = x^2 + 4x + 7$

(a)

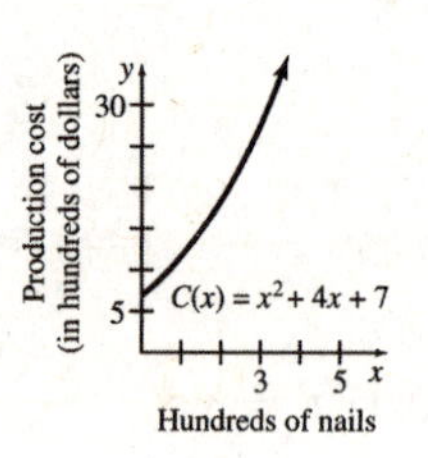

(b)

$$C(x + 1) - C(x)$$
$$= (x + 1)^2 + 4(x + 1) + 7 - (x^2 + 4x + 7)$$
$$= x^2 + 2x + 1 + 4x + 4 + 7 - x^2 - 4x - 7$$
$$= 2x + 5$$

(c) $A(x) = \frac{C(x)}{x} = \frac{x^2 + 4x + 7}{x}$

$$= x + 4 + \frac{7}{x}$$

(d) $A(x + 1) - A(x)$

$$= (x + 1) + 4 + \frac{7}{x + 1} - \left(x + 4 + \frac{7}{x}\right)$$
$$= x + 1 + 4 + \frac{7}{x + 1} - x - 4 - \frac{7}{x}$$
$$= 1 + \frac{7}{x + 1} - \frac{7}{x}$$
$$= 1 + \frac{7x - 7(x + 1)}{x(x + 1)}$$
$$= 1 + \frac{7x - 7x - 7}{x(x + 1)}$$
$$= 1 - \frac{7}{x(x + 1)}$$

109. $F(x) = -\frac{2}{3}x^2 + \frac{14}{3}x + 96$

The maximum fever occurs at the vertex of the parabola.

$$x = \frac{-b}{2a} = \frac{-\frac{14}{3}}{-\frac{4}{3}} = \frac{7}{2}$$

$$\begin{aligned} y &= -\frac{2}{3}\left(\frac{7}{2}\right)^2 + \frac{14}{3}\left(\frac{7}{2}\right) + 96 \\ &= -\frac{2}{3}\left(\frac{49}{4}\right) + \frac{49}{3} + 96 \\ &= -\frac{49}{6} + \frac{49}{3} + 96 \\ &= -\frac{49}{6} + \frac{98}{6} + \frac{576}{6} = \frac{625}{6} \approx 104.2 \end{aligned}$$

The maximum fever occurs on the third day. It is about 104.2°F.

111. **(a)**

300

0 13

0

(b) $y = 2.384t^2 - 42.55t + 269.2$

$y = -0.4931t^3 + 11.26t^2$
$\quad - 82.00t + 292.9$

$y = 213.8(0.9149)^t$

(c)

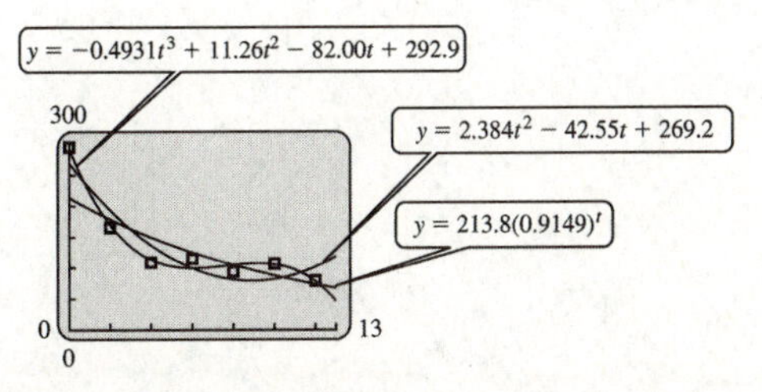

The cubic function seems to best capture the behavior of the data.

(d) $x = 20$ corresponds to 2015.

$$\begin{aligned} y &= 2.384(20)^2 - 42.55(20) \\ &\quad + 269.2 \approx 372 \end{aligned}$$

$$\begin{aligned} y &= -0.4931(20)^3 + 11.26(20)^2 \\ &\quad - 82.00(20) + 292.9 \\ &\approx -788 \end{aligned}$$

$$y = 213.8(0.9149)^{20} \approx 36$$

The only realistic value is given by the exponential function because the pattern of the data suggests that the number of cases decrease over time.

113. This function has a maximum value at $x \approx 187.9$. At $x \approx 187.9$, $y \approx 345$. The largest girth for which this formula gives a reasonable answer is 187.9 cm. The predicted mass of a polar bear with this girth is 345 kg.

115. $p(t) = \frac{1.79 \cdot 10^{11}}{(2026.87 - t)^{0.99}}$

(a) $p(2010) \approx 10.915$ billion

This is about 4.006 billion more than the estimate of 6.909 billion.

(b) $p(2020) \approx 26.56$ billion

$p(2025) \approx 96.32$ billion

117. Graph

$$y = c(t) = e^{-t} - e^{-2t}$$

on a graphing calculator and locate the maximum point. A calculator shows that the x-coordinate of the maximum point is about 0.69, and the y-coordinate is exactly 0.25. Thus, the maximum concentration of 0.25 occurs at about 0.69 minutes.

119. **(a)** $S = -3.404 + 103.2 \ln A$

(b) $S = 81.26A^{0.3011}$

(c)

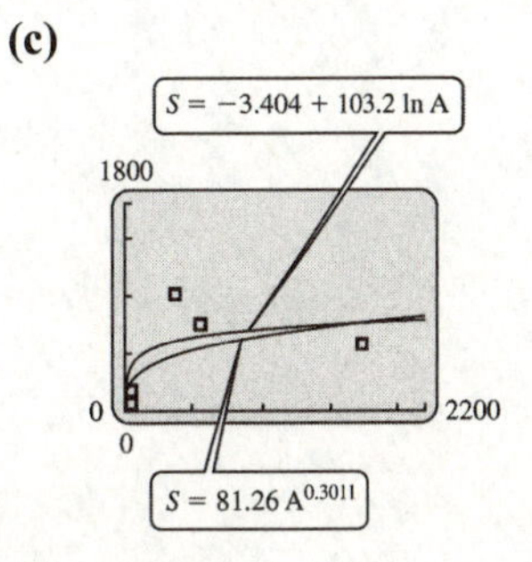

(d) $S \approx 234$

$S \approx 163$

Neither number is close to the actual number of 421.

121. $t = (1.26 \times 10^9)\dfrac{\ln\left[1 + 8.33\left(\frac{A}{K}\right)\right]}{\ln 2}$

(a) $A = 0, K > 0$

$$t = (1.26 \times 10^9)\frac{\ln[1 + 8.33(0)]}{\ln 2}$$
$$= (1.26 \times 10^9)(0) = 0 \text{ years}$$

(b) $t = (1.26 \times 10^9)\dfrac{\ln[1 + 8.33(0.212)]}{\ln 2}$

$$= (1.26 \times 10^9)\frac{\ln 2.76596}{\ln 2}$$
$$= 1{,}849{,}403{,}169$$

or about 1.85×10^9 years

(c) As r increases, t increases, but at a slower and slower rate. As r decreases, t decreases at a faster and faster rate

Chapter 11

THE DERIVATIVE

11.1 Limits

Your Turn 1

$f(x) = x^2 + 2$

x	0.9	0.99	0.999	0.9999	1	1.0001	1.001	1.01	1.1
$f(x)$	2.81	2.9801	2.998001	2.99980001	3	3.00020001	3.002001	3.0201	3.21

The table suggests that, as x get closer and closer to 1 from either side, $f(x)$ gets closer and closer to 3. So, $\lim_{x\to 1}(x^2 + 2) = 3$.

Your Turn 2

$$f(x) = \frac{x^2 - 4}{x - 2} = \frac{(x + 2)\cancel{(x - 2)}}{\cancel{(x - 2)}}$$
$$= x + 2, \text{ provided } x \neq 2$$

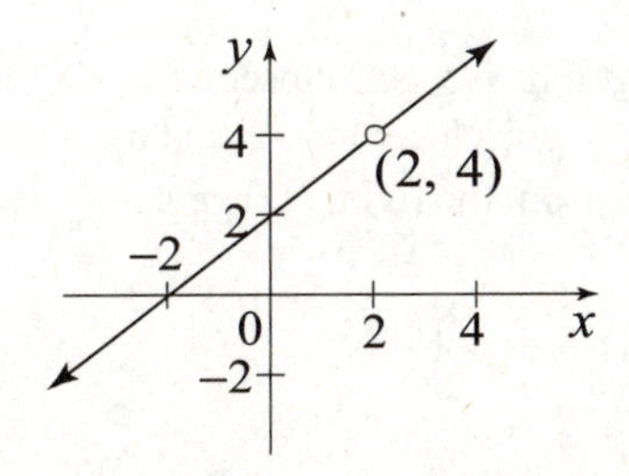

The graph of $y = \frac{x^2-4}{x-2}$ is the graph of $y = x + 2$, except there is a hole at (2, 4).

Looking at the graph, we see that as x is close to, but not equal to 2, $f(x)$ approaches 4.

$$\lim_{x\to 2}\frac{x^2 - 4}{x - 2} = 4$$

Your Turn 3

Find $\lim_{x\to 3} f(x)$ if $f(x) = \begin{cases} 2x - 1 & \text{if } x \neq 3 \\ 1 & \text{if } x = 3 \end{cases}$

The graph of f is shown in the next column.

$$\lim_{x\to 3} f(x) = 5$$

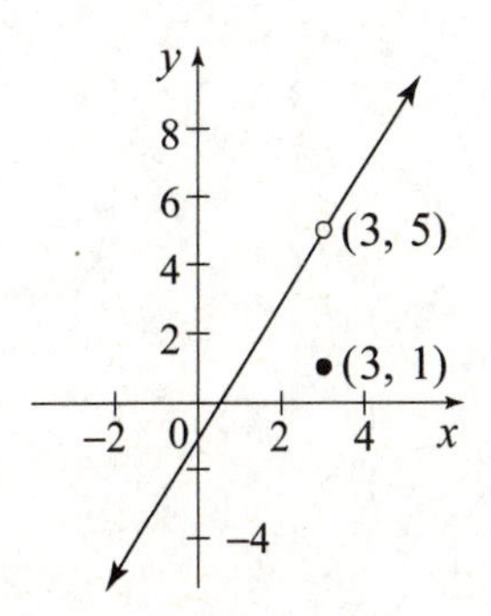

Your Turn 4

Find $\lim_{x\to 0}\frac{2x - 1}{x}$.

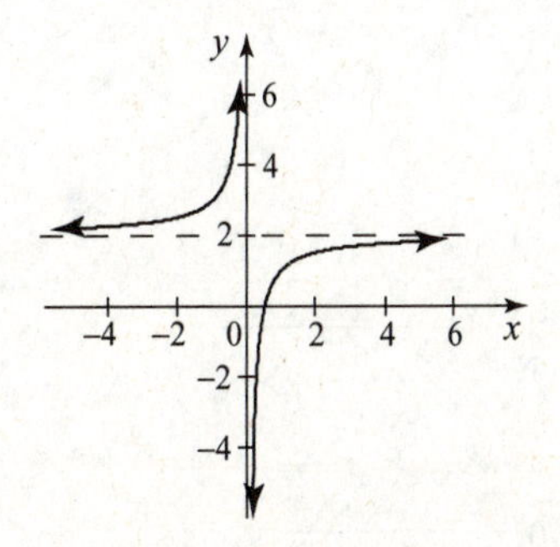

$$\lim_{x\to 0^-} f(x) = \infty$$
$$\lim_{x\to 0^+} f(x) = -\infty$$

Since there is no real number that $f(x)$ approaches as x approaches 0 from either side, nor does $f(x)$ approach either ∞ or $-\infty$, $\lim_{x\to 0}\frac{2x-1}{x}$ does not exist.

Your Turn 5

Let $\lim_{x\to 2} f(x) = 3$ and $\lim_{x\to 2} g(x) = 4$.

$$\begin{aligned}\lim_{x\to 2}[f(x)+g(x)]^2 &= \left[\lim_{x\to 2}[f(x)+g(x)]\right]^2\\ &= \left[\lim_{x\to 2} f(x) + \lim_{x\to 2} g(x)\right]^2\\ &= [3+4]^2\\ &= 7^2\\ &= 49\end{aligned}$$

Your Turn 6

$$\begin{aligned}\lim_{x\to -3}\frac{x^2-x-12}{x+3} &= \lim_{x\to -3}\frac{(x-4)\cancel{(x+3)}}{\cancel{x+3}} \quad (x \neq 3)\\ &= \lim_{x\to -3} x-4\\ &= (-3)-4\\ &= -7\end{aligned}$$

Your Turn 7

$$\begin{aligned}\lim_{x\to 1}\frac{\sqrt{x}-1}{x-1} &= \lim_{x\to 1}\frac{\sqrt{x}-1}{x-1}\cdot\frac{\sqrt{x}+1}{\sqrt{x}+1}\\ &= \lim_{x\to 1}\frac{(\sqrt{x})^2-1}{(x-1)(\sqrt{x}+1)}\\ &= \lim_{x\to 1}\frac{\cancel{x-1}}{\cancel{(x-1)}(\sqrt{x}+1)}\\ &= \lim_{x\to 1}\frac{1}{\sqrt{x}+1} = \frac{1}{1+1} = \frac{1}{2}\end{aligned}$$

Your Turn 8

$$\begin{aligned}\lim_{x\to\infty}\frac{2x^2+3x-4}{6x^2-5x+7} &= \lim_{x\to\infty}\frac{\frac{2x^2}{x^2}+\frac{3x}{x^2}-\frac{4}{x^2}}{\frac{6x^2}{x^2}-\frac{5x}{x^2}+\frac{7}{x^2}}\\ &= \lim_{x\to\infty}\frac{2+\frac{3}{x}-\frac{4}{x^2}}{6-\frac{5}{x}+\frac{7}{x^2}}\\ &= \frac{2+0-0}{6-0+0} = \frac{2}{6} = \frac{1}{3}\end{aligned}$$

11.1 Exercises

1. Since $\lim_{x\to 2^-} f(x)$ does not equal $\lim_{x\to 2^+} f(x)$, $\lim_{x\to 2} f(x)$ does not exist. The answer is c.

3. Since $\lim_{x\to 4^-} f(x) = \lim_{x\to 4^+} f(x) = 6$, $\lim_{x\to 4} f(x) = 6$. The answer is b.

5. **(a)** By reading the graph, as x gets closer to 3 from the left or right, $f(x)$ gets closer to 3.

$$\lim_{x\to 3} f(x) = 3$$

(b) By reading the graph, as x gets closer to 0 from the left or right, $f(x)$ gets closer to 1.

$$\lim_{x\to 0} f(x) = 1.$$

7. **(a)** By reading the graph, as x gets closer to 0 from the left or right, $f(x)$ gets closer to 0.

$$\lim_{x\to 0} f(x) = 0$$

(b) By reading the graph, as x gets closer to 2 from the left, $f(x)$ gets closer to -2, but as x gets closer to 2 from the right, $f(x)$ gets closer to 1.

$\lim_{x\to 2} f(x)$ does not exist.

9. **(a)** (i) By reading the graph, as x gets closer to -2 from the left, $f(x)$ gets closer to -1.

$$\lim_{x\to -2^-} f(x) = -1$$

(ii) By reading the graph, as x gets closer to -2 from the right, $f(x)$ gets closer to $-\frac{1}{2}$.

$$\lim_{x\to -2^+} f(x) = -\frac{1}{2}$$

(iii) Since $\lim_{x\to -2^-} f(x) = -1$ and $\lim_{x\to -2^+} f(x) = -\frac{1}{2}$, $\lim_{x\to -2} f(x)$ does not exist.

(iv) $f(-2)$ does not exist since there is no point on the graph with an x-coordinate of -2.

(b) (i) By reading the graph, as x gets closer to -1 from the left, $f(x)$ gets closer to $-\frac{1}{2}$.

$$\lim_{x\to -1^-} f(x) = -\frac{1}{2}$$

(ii) By reading the graph, as x gets closer to -1 from the right, $f(x)$ gets closer to $-\frac{1}{2}$.

$$\lim_{x\to -1^+} f(x) = -\frac{1}{2}$$

(iii) Since $\lim_{x\to -1^-} f(x) = -\frac{1}{2}$ and $\lim_{x\to -1^+} f(x) = -\frac{1}{2}$, $\lim_{x\to -1} f(x) = -\frac{1}{2}$.

(iv) $f(-1) = -\frac{1}{2}$ since $\left(-1, -\frac{1}{2}\right)$ is a point of the graph.

11. By reading the graph, as x moves further to the right, $f(x)$ gets closer to 3. Therefore, $\lim_{x\to\infty} f(x) = 3$.

13. $\lim_{x\to 2} F(x)$ in Exercise 6 exists because $\lim_{x\to 2^-} F(x) = 4$ and $\lim_{x\to 2^+} F(x) = 4$.
$\lim_{x\to -2} f(x)$ in Exercise 9 does not exist since $\lim_{x\to -2^-} f(x) = -1$, but $\lim_{x\to -2^+} f(x) = -\frac{1}{2}$.

15. From the table, as x approaches 1 from the left or the right, $f(x)$ approaches 4.

$$\lim_{x\to 1} f(x) = 4$$

17. $k(x) = \dfrac{x^3 - 2x - 4}{x - 2}$; find $\lim_{x\to 2} k(x)$.

x	1.9	1.99	1.999
$k(x)$	9.41	9.9401	9.9941

x	2.001	2.01	2.1
$k(x)$	10.006	10.0601	10.61

As x approaches 2 from the left or the right, $k(x)$ approaches 10.

$$\lim_{x\to 2} k(x) = 10$$

19. $h(x) = \dfrac{\sqrt{x} - 2}{x - 1}$; find $\lim_{x\to 1} h(x)$.

x	0.9	0.99	0.999
$h(x)$	10.51317	100.50126	1000.50013

x	1.001	1.01	1.1
$h(x)$	−999.50012	−99.50124	−9.51191

$$\lim_{x\to 1^-} = \infty$$
$$\lim_{x\to 1^+} = -\infty$$

Thus, $\lim_{x\to 1} h(x)$ does not exist.

21.
$$\lim_{x\to 4}[f(x) - g(x)] = \lim_{x\to 4} f(x) - \lim_{x\to 4} g(x) = 9 - 27 = -18$$

23.
$$\lim_{x\to 4} \frac{f(x)}{g(x)} = \frac{\lim_{x\to 4} f(x)}{\lim_{x\to 4} g(x)} = \frac{9}{27} = \frac{1}{3}$$

25.
$$\lim_{x\to 4} \sqrt{f(x)} = \lim_{x\to 4}[f(x)^{1/2}] = \left[\lim_{x\to 4} f(x)\right]^{1/2} = 9^{1/2} = 3$$

27.
$$\lim_{x\to 4} 2^{f(x)} = 2^{\lim_{x\to 4} f(x)} = 2^9 = 512$$

29.
$$\lim_{x\to 4} \frac{f(x) + g(x)}{2g(x)} = \frac{\lim_{x\to 4}[f(x) + g(x)]}{\lim_{x\to 4} 2g(x)} = \frac{\lim_{x\to 4} f(x) + \lim_{x\to 4} g(x)}{2\lim_{x\to 4} g(x)} = \frac{9 + 27}{2(27)} = \frac{36}{54} = \frac{2}{3}$$

31. $\lim_{x\to 3}\frac{x^2-9}{x-3}=\lim_{x\to 3}\frac{(x-3)(x+3)}{x-3}$
$=\lim_{x\to 3}(x+3)$
$=\lim_{x\to 3}x+\lim_{x\to 3}3$
$=3+3$
$=6$

33. $\lim_{x\to 1}\frac{5x^2-7x+2}{x^2-1}=\lim_{x\to 1}\frac{(5x-2)(x-1)}{(x+1)(x-1)}$
$=\lim_{x\to 1}\frac{5x-2}{x+1}$
$=\frac{5-2}{2}$
$=\frac{3}{2}$

35. $\lim_{x\to -2}\frac{x^2-x-6}{x+2}=\lim_{x\to -2}\frac{(x-3)(x+2)}{x+2}$
$=\lim_{x\to -2}(x-3)$
$=\lim_{x\to -2}x+\lim_{x\to -2}(-3)$
$=-2-3$
$=-5$

37. $\lim_{x\to 0}\frac{\frac{1}{x+3}-\frac{1}{3}}{x}$
$=\lim_{x\to 0}\left(\frac{1}{x+3}-\frac{1}{3}\right)\left(\frac{1}{x}\right)$
$=\lim_{x\to 0}\left[\frac{3}{3(x+3)}-\frac{x+3}{3(x+3)}\right]\left(\frac{1}{x}\right)$
$=\lim_{x\to 0}\frac{3-x-3}{3(x+3)(x)}$
$=\lim_{x\to 0}\frac{-x}{3(x+3)x}$
$=\lim_{x\to 0}\frac{-1}{3(x+3)}$
$=\frac{-1}{3(0+3)}$
$=-\frac{1}{9}$

39. $\lim_{x\to 25}\frac{\sqrt{x}-5}{x-25}$
$=\lim_{x\to 25}\frac{\sqrt{x}-5}{x-25}\cdot\frac{\sqrt{x}+5}{\sqrt{x}+5}$
$=\lim_{x\to 25}\frac{x-25}{(x-25)(\sqrt{x}+5)}$
$=\lim_{x\to 25}\frac{1}{\sqrt{x}+5}$
$=\frac{1}{\sqrt{25}+5}$
$=\frac{1}{10}$

41. $\lim_{h\to 0}\frac{(x+h)^2-x^2}{h}$
$=\lim_{h\to 0}\frac{x^2+2hx+h^2-x^2}{h}$
$=\lim_{h\to 0}\frac{2hx+h^2}{h}$
$=\lim_{h\to 0}\frac{h(2x+h)}{h}$
$=\lim_{h\to 0}(2x+h)$
$=2x+0=2x$

43. $\lim_{x\to\infty}\frac{3x}{7x-1}=\lim_{x\to\infty}\frac{\frac{3x}{x}}{\frac{7x}{x}-\frac{1}{x}}$
$=\lim_{x\to\infty}\frac{3}{7-\frac{1}{x}}$
$=\frac{3}{7-0}=\frac{3}{7}$

45. $\lim_{x\to -\infty}\frac{-3x^2+2x}{2x^2-2x+1}$
$=\lim_{x\to -\infty}\frac{\frac{3x^2}{x^2}+\frac{2x}{x^2}}{\frac{2x^2}{x^2}-\frac{2x}{x^2}+\frac{1}{x^2}}$
$=\lim_{x\to -\infty}\frac{3+\frac{2}{x}}{2-\frac{2}{x}+\frac{1}{x^2}}$
$=\frac{3-0}{2+0+0}=\frac{3}{2}$

47. $$\lim_{x\to\infty}\frac{3x^3+2x-1}{2x^4-3x^3-2}$$
$$=\lim_{x\to\infty}\frac{\frac{3x^3}{x^4}+\frac{2x}{x^4}-\frac{1}{x^4}}{\frac{2x^4}{x^4}-\frac{3x^3}{x^4}-\frac{2}{x^4}}$$
$$=\lim_{x\to\infty}\frac{\frac{3}{x}+\frac{2}{x^3}-\frac{1}{x^4}}{2-\frac{3}{x}-\frac{2}{x^4}}$$
$$=\frac{0+0-0}{2-0-0}=0$$

49. $$\lim_{x\to\infty}\frac{2x^3-x-3}{6x^2-x-1}$$
$$=\lim_{x\to\infty}\frac{\frac{2x^3}{x^2}-\frac{x}{x^2}-\frac{3}{x^2}}{\frac{6x^2}{x^2}-\frac{x}{x^2}-\frac{1}{x^2}}$$
$$=\lim_{x\to\infty}\frac{2x-\frac{1}{x}-\frac{3}{x^2}}{6-\frac{1}{x}-\frac{1}{x^2}}=\infty$$

The limit does not exist.

51. $$\lim_{x\to\infty}\frac{2x^2-7x^4}{9x^2+5x-6}=\lim_{x\to\infty}\frac{\frac{2x^2}{x^2}-\frac{7x^4}{x^2}}{\frac{9x^2}{x^2}+\frac{5x}{x^2}-\frac{6}{x^2}}$$
$$=\lim_{x\to\infty}\frac{2-7x^2}{9+\frac{5}{x}-\frac{6}{x^2}}$$

The denominator approaches 9, while the numerator becomes a negative number that is larger and larger in magnitude, so

$$\lim_{x\to\infty}\frac{2x^2-7x^4}{9x^2+5x-6}=-\infty \text{ (does not exist).}$$

53. $\lim_{x\to-1^-} f(x)=1$ and $\lim_{x\to-1^+} f(x)=1$.

Therefore $\lim_{x\to-1} f(x)=1$.

55. **(a)** $\lim_{x\to 3} f(x)=2$.

(b) $\lim_{x\to 5} f(x)$ does not exist since $\lim_{x\to 5^-} f(x)=2$ and $\lim_{x\to 5^+} f(x)=8$.

57. Find $\lim_{x\to 3} f(x)$, where $f(x)=\frac{x^2-9}{x-3}$.

x	2.9	2.99	2.999	3.001	3.01	3.1
$f(x)$	5.9	5.99	5.999	6.001	6.01	6.1

$$\lim_{x\to 3} f(x)=\lim_{x\to 3}\frac{x^2-9}{x-3}=6.$$

59. Find $\lim_{x\to 1} f(x)$, where $f(x)=\frac{5x^2-7x+2}{x^2-1}$.

x	0.9	0.99	0.999	1.001	1.01	1.1
$f(x)$	1.316	1.482	1.498	1.502	1.517	1.667

$$\lim_{x\to 1} f(x)=\lim_{x\to 1}\frac{5x^2-7x+2}{x^2-1}=1.5=\frac{3}{2}.$$

61. **(a)** $\lim_{x\to -2}\frac{3x}{(x+2)^3}$ does not exist since

$$\lim_{x\to -2^+}\frac{3x}{(x+2)^3}=-\infty$$

and $\lim_{x\to -2^-}\frac{3x}{(x+2)^3}=\infty$.

(b) Since $(x+2)^3=0$ when $x=-2$, $x=-2$ is the vertical asymptote of the graph of $F(x)$.

(c) The two answers are related. Since $x=-2$ is a vertical asymptote, we know that $\lim_{x\to -2} F(x)$ does not exist.

65. **(a)** $\lim_{x\to -\infty} e^x=0$ since, as the graph goes further to the left, e^x gets closer to 0.

(b) The graph of e^x has a horizontal asymptote at $y=0$ since $\lim_{x\to -\infty} e^x=0$.

67. **(a)** $\lim_{x\to 0^+}\ln x=-\infty$ (does not exist) since, as the graph gets closer to $x=0$, the value of $\ln x$ get smaller.

(b) The graph of $y=\ln x$ has a vertical asymptote at $x=0$ since $\lim_{x\to 0^+}\ln x=-\infty$.

71. $\lim_{x\to 1} \frac{x^4 + 4x^3 - 9x^2 + 7x - 3}{x - 1}$

(a)

x	1.01	1.001	1.0001	0.99	0.999	0.9999
$f(x)$	5.0908	5.009	5.0009	4.9108	4.991	4.9991

As $x \to 1^-$ and as $x \to 1^+$, we see that $f(x) \to 5$.

(b) Graph

$$y = \frac{x^4 + 4x^3 - 9x^2 + 7x - 3}{x - 1}$$

on a graphing calculator. One suitable choice for the viewing window is $[-6,6]$ by $[-10, 40]$ with Xscl = 1, Yscl = 10.

Because $x - 1 = 0$ when $x = 1$, we know that the function is undefined at this x-value. The graph does not show an asymptote at $x = 1$. This indicates that the rational expression that defines this function is not written in lowest terms, and that the graph should have an open circle to show a "hole" in the graph at $x = 1$. The graphing calculator doesn't show the hole, but if we try to find the value of the function at $x = 1$, we see that it is undefined. (Using the TABLE feature on a TI-84 Plus, we see that for $x = 1$, the y-value is listed as "ERROR.")

By viewing the function near $x = 1$, and using the ZOOM feature, we see that as x gets close to 1 from the left or the right, y gets close to 5, suggesting that

$$\lim_{x\to 1} \frac{x^4 + 4x^3 - 9x^2 + 7x - 3}{x - 1} = 5.$$

73. $\lim_{x\to -1} \frac{x^{1/3} + 1}{x + 1}$

(a)

x	-1.01	-1.001	-1.0001
$f(x)$	0.33223	0.33322	0.33332

x	-0.99	-0.999	-0.9999
$f(x)$	0.33445	0.33344	0.33334

We see that as $x \to -1^-$ and as $x \to -1^+$, $f(x) \to 0.3333$ or $\frac{1}{3}$.

(b) Graph $y = \frac{x^{1/3} + 1}{x + 1}$.

One suitable choice for the viewing window is $[-5, 5]$ by $[-2, 2]$

Because $x + 1 = 0$ when $x = -1$, we know that the function is undefined at this x-value. The graph does not show an asymptote at $x = -1$. This indicates that the rational expression that defined this function is not written lowest terms, and that the graph should have an open circle to show a "hole" in the graph at $x = -1$. The graphing calculator doesn't show the hole, but if we try to find the value of the function at $x = -1$, we see that it is undefined. (Using the TABLE feature on a TI-83, we see that for $x = -1$, the y-value is listed as "ERROR.")

By viewing the function near $x = -1$ and using the ZOOM feature, we see that as x gets close to -1 from the left or right, y gets close to 0.3333, suggesting that

$$\lim_{x\to -1} \frac{x^{1/3} + 1}{x + 1} = 0.3333 \text{ or } \frac{1}{3}.$$

75. $\lim_{x\to\infty} \frac{\sqrt{9x^2 + 5}}{2x}$

Graph the functions on a graphing calculator. A good choice for the viewing window is $[-10, 10]$ by $[-5, 5]$.

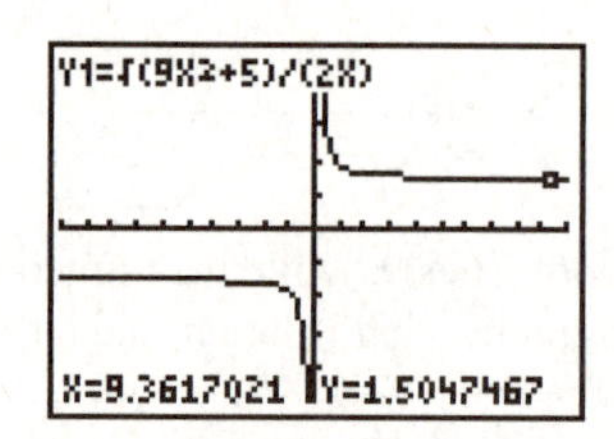

(a) The graph appears to have horizontal asymptotes at $y = \pm 1.5$. We see that as $x \to \infty$, $y \to 1.5$, so we determine that

$$\lim_{x\to\infty} \frac{\sqrt{9x^2 + 5}}{2x} = 1.5.$$

(b) As $x \to \infty$,

$$\frac{\sqrt{9x^2 + 5}}{2x} \to \frac{3\,|x|}{2x}.$$

Since $x > 0, |x| = x$, so

$$\frac{3\,|x|}{2x} = \frac{3x}{2x} = \frac{3}{2}.$$

Thus,

$$\lim_{x\to\infty} \frac{\sqrt{9x^2 + 5}}{2x} = \frac{3}{2} \text{ or } 1.5.$$

77. $\lim_{x\to-\infty} \frac{\sqrt{36x^2+2x+7}}{3x}$

Graph this function on a graphing calculator. A good choice for the viewing window is $[-10, 10]$ by $[-5, 5]$.

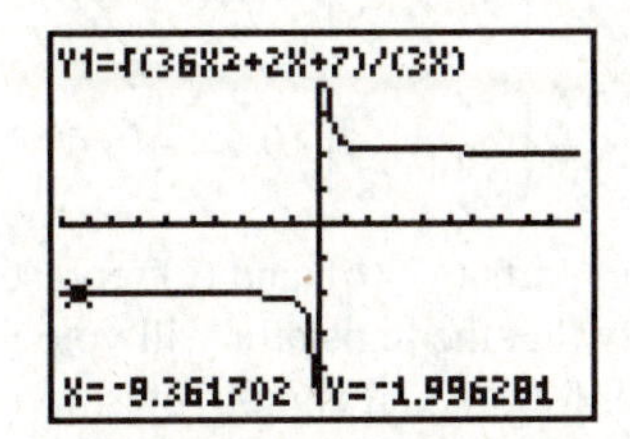

(a) The graph appears to have horizontal asymptotes at $y = \pm 2$. We see that as $x \to -\infty, y \to -2$, so we determine that

$$\lim_{x\to-\infty} \frac{\sqrt{36x^2+2x+7}}{3x} = -2.$$

(b) As $x \to -\infty$,

$$\frac{\sqrt{36x^2+2x+7}}{3x} \to \frac{6|x|}{3x}.$$

Since $x < 0$, $|x| = -x$, so

$$\frac{6|x|}{3x} = \frac{6(-x)}{3x} = -2.$$

Thus,

$$\lim_{x\to-\infty} \frac{\sqrt{36x^2+2x+7}}{3x} = -2.$$

79. $\lim_{x\to\infty} \frac{(1+5x^{1/3}+2x^{5/3})^3}{x^5}$

Graph this function on a graphing calculator. A good choice for the viewing window is $[-20, 20]$ by $[0, 20]$ with Xscl $= 5$, Yscl $= 5$.

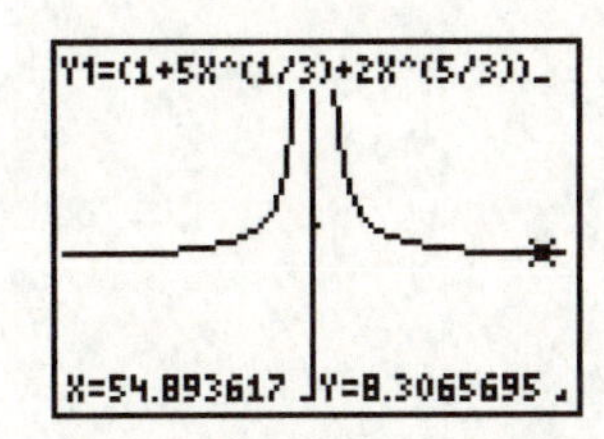

(a) The graph appears to have a horizontal asymptote at $y = 8$. We see that as $x \to \infty$, $y \to 8$, so we determine that

$$\lim_{x\to\infty} \frac{(1+5x^{1/3}+2x^{5/3})^3}{x^5} = 8.$$

(b) As $x \to \infty$,

$$\frac{(1+5x^{1/3}+2x^{5/3})^3}{x^5} \to \frac{8x^5}{x^5} = 8.$$

Thus, $\lim_{x\to\infty} \frac{(1+5x^{1/3}+2x^{5/3})^3}{x^5} = 8.$

83. (a) $\lim_{x\to 53} T(x) = 3$ cents because $T(x)$ is constant at 3 cents as x approaches 53 from the left or the right.

(b) $\lim_{x\to 09^-} T(x) = 7.25$ cents because as x approaches 09 from the left, $T(x)$ is constant at 7.25 cents.

(c) $\lim_{x\to 09^+} T(x) = 8.25$ cents because as x approaches 09 from the right, $T(x)$ is constant at 8.25 cents.

(d) $\lim_{x\to 09} T(x)$ .does not exist because

$$\lim_{x\to 09^-} T(x) \neq \lim_{x\to 09^+} T(x).$$

(e) $T(09) = 8.25$ cents since $(09, 8.25)$ is a point on the graph.

85. $C(x) = 15{,}000 + 6x$

$$\overline{C}(x) = \frac{C(x)}{x} = \frac{15{,}000+6x}{x} = \frac{15{,}000}{x} + 6$$

$$\lim_{x\to\infty} \overline{C}(x) = \lim_{x\to\infty} \frac{15{,}000}{x} + 6 = 0 + 6 = 6$$

This means that the average cost approaches \$6 as the number of tapes produced becomes very large.

87. $P(s) = \frac{63s}{s+8}$

$$\lim_{s\to\infty} \frac{63s}{s+8} = \lim_{s\to\infty} \frac{\frac{63s}{s}}{\frac{s}{s}+\frac{8}{s}}$$

$$= \lim_{s\to\infty} \frac{63}{1+\frac{8}{s}}$$

$$= \frac{63}{1+0}$$

$$= 63$$

The number of items of work a new employee produces gets closer and closer to 63 as the number of days of training increases.

89. $\lim_{n\to\infty}\left[\frac{R}{i-g}\left[1-\left(\frac{1+g}{1+i}\right)^n\right]\right]$

$= \frac{R}{i-g}\lim_{n\to\infty}\left[1-\left(\frac{1+g}{1+i}\right)^n\right]$

$= \frac{R}{i-g}\left[\lim_{n\to\infty}1-\lim_{n\to\infty}\left(\frac{1+g}{1+i}\right)^n\right]$

assuming $i > g$,

$= \frac{R}{i-g}[1-0]$

$= \frac{R}{i-g}$

91. (a) $D(t) = 155(1 - e^{-0.0133t})$

$D(20) = 155\left(1 - e^{-0.0133(20)}\right)$

$= 155(1 - e^{-0.266})$

≈ 36.2

The depth of the sediment layer deposited below the bottom of the lake in 1970 was 36.2 cm.

(b) $\lim_{t\to\infty} D(t) = \lim_{t\to\infty} 155(1 - e^{-0.0133t})$

$= 155 \lim_{t\to\infty}(1 - e^{-0.0133t})$

$= 155\left(\lim_{t\to\infty}1 - \lim_{t\to\infty}e^{-0.0133t}\right)$

$= 155(1) - 155\lim_{t\to\infty}e^{-0.0133t}$

$= 155 - 155(0) = 155$

Thus,

$$\lim_{t\to\infty} D(t) = 155.$$

Going back in time (t is years before 1990), the depth of the sediment approaches 155 cm.

93. (a) $p_2 = \frac{1}{2} + \left(0.7 - \frac{1}{2}\right)[1 - 2(0.2)]^2 = 0.572$

(b) $p_4 = \frac{1}{2} + \left(0.7 - \frac{1}{2}\right)[1 - 2(0.2)]^4 = 0.526$

(c) $p_8 = \frac{1}{2} + \left(0.7 - \frac{1}{2}\right)[1 - 2(0.2)]^8 = 0.503$

(d) $\lim_{n\to\infty} P_n = \lim_{n\to\infty}\left[\frac{1}{2} + \left(p_0 - \frac{1}{2}\right)(1-2p)^n\right]$

$= \frac{1}{2} + \lim_{n\to\infty}\left(p_0 - \frac{1}{2}\right)(1-2p)^n$

$= \frac{1}{2} + \left(p_0 - \frac{1}{2}\right)\lim_{n\to\infty}(1-2p)^n$

$= \frac{1}{2} + \left(p_0 - \frac{1}{2}\right)\cdot 0 = \frac{1}{2}$

The number in parts (a), (b), and (c) represent the probability that the legislator will vote yes on the second, fourth, and eighth votes. In (d), as the number of roll calls increases, the probability gets close to 0.5, but is never less than 0.5.

11.2 Continuity

Your Turn 1

$f(x) = \sqrt{5x+3}$

The Square root function is discontinuous wherever $5x + 3 < 0$. There is a discontinuity when

$5a + 3 < 0$, or $a < -\frac{3}{5}$.

Your Turn 2

$$f(x) = \begin{cases} 5x - 4 & \text{if } x < 0 \\ x^2 & \text{if } 0 \le x \le 3 \\ x + 6 & \text{if } x > 3 \end{cases}$$

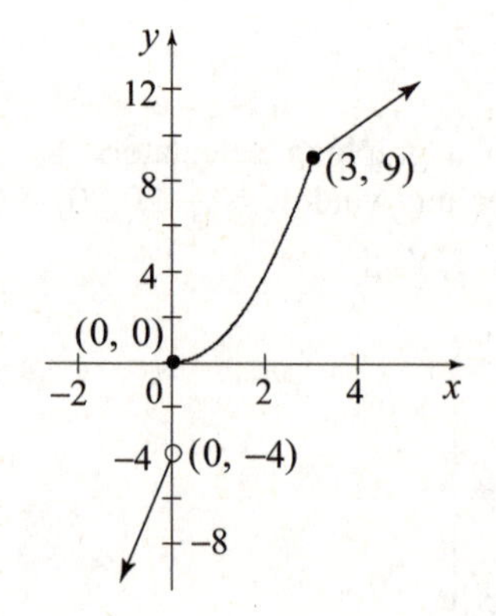

$\lim_{x\to 0^-} f(x) = -4$

$\lim_{x\to 0^+} f(x) = 0$

Because $\lim_{x\to 0^-} f(x) \ne \lim_{x\to 0^+} f(x)$, the limit doesn't exist, so f is discontinuous at $x = 0$.

11.2 Exercises

1. Discontinuous at $x = -1$

(a) $f(-1)$ does not exist.

(b) $\lim_{x \to -1^-} f(x) = \frac{1}{2}$

(c) $\lim_{x \to -1^+} f(x) = \frac{1}{2}$

(d) $\lim_{x \to -1} f(x) = \frac{1}{2}$ (since (a) and (b) have the same answers)

(e) $f(-1)$ does not exist.

3. Discontinuous at $x = 1$

(a) $f(1) = 2$

(b) $\lim_{x \to 1^-} f(x) = -2$

(c) $\lim_{x \to 1^+} f(x) = -2$

(d) $\lim_{x \to 1} f(x) = -2$ (since (a) and (b) have the same answers)

(e) $\lim_{x \to 1} f(x) \neq f(1)$

5. Discontinuous at $x = -5$ and $x = 0$

(a) $f(-5)$ does not exist. $f(0)$ does not exist.

(b) $\lim_{x \to -5^-} f(x) = \infty$ (limit does not exist)

$\lim_{x \to 0^-} f(x) = 0$

(c) $\lim_{x \to -5^+} f(x) = -\infty$ (limit does not exist)

$\lim_{x \to 0^+} f(x) = 0$

(d) $\lim_{x \to -5} f(x)$ does not exist, since the answers to (a) and (b) are different. $\lim_{x \to 0} f(x) = 0$, since the answers to (a) and (b) are the same.

(e) $f(-5)$ does not exist and $\lim_{x \to -5} f(x)$ does not exist. $f(0)$ does not exist.

7. $f(x) = \dfrac{5 + x}{x(x - 2)}$

$f(x)$ is discontinuous at $x = 0$ and $x = 2$ since the denominator equals 0 at these two values.

$\lim_{x \to 0} f(x)$ does not exist since $\lim_{x \to 0^-} f(x) = \infty$ and $\lim_{x \to 0^+} f(x) = -\infty$.

$\lim_{x \to 2} f(x)$ does not exist since $\lim_{x \to 2^-} f(x) = -\infty$ and $\lim_{x \to 2^+} f(x) = \infty$.

9. $f(x) = \dfrac{x^2 - 4}{x - 2}$

$f(x)$ is discontinuous at $x = 2$ since the denominator equals zero at that value.

Since for $x \neq 2$

$$\frac{x^2 - 4}{x - 2} = \frac{(x + 2)(x - 2)}{x - 2} = x + 2,$$

$\lim_{x \to 2} f(x) = 2 + 2 = 4.$

11. $p(x) = x^2 - 4x + 11$

Since $p(x)$ is a polynomial function, it is continuous everywhere and thus discontinuous nowhere.

13. $p(x) = \dfrac{|x + 2|}{x + 2}$

$p(x)$ is discontinuous at $x = -2$ since the denominator is zero at that value.

since $\lim_{x \to -2^-} p(x) = -1$ and $\lim_{x \to -2^+} p(x) = 1$, $\lim_{x \to -2} p(x)$ does not exist.

15. $k(x) = e^{\sqrt{x-1}}$

The function is undefined for $x < 1$, so the function is discontinuous for $a < 1$. The limit as x approaches any $a < 1$ does not exist because the function is undefined for $x < 1$.

17. As x approaches 0 from the left or the right, $\left|\frac{x}{x-1}\right|$ approaches 0 and $r(x) = \ln\left|\frac{x}{x-1}\right|$ goes to $-\infty$. So $\lim_{x \to 0} r(x)$ does not exist. As x approaches 1 from the left or the right, $\left|\frac{x}{x-1}\right|$ goes to ∞ and so does $r(x) = \ln\left|\frac{x}{x-1}\right|$. So $\lim_{x \to 1} r(x)$ does not exist.

19. $f(x) = \begin{cases} 1 & \text{if } x < 2 \\ x + 3 & \text{if } 2 \le x \le 4 \\ 7 & \text{if } x > 4 \end{cases}$

(a)

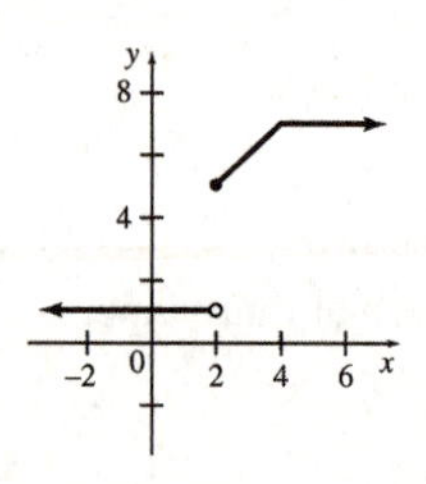

(b) $f(x)$ is discontinuous at $x = 2$.

(c) $\lim\limits_{x \to 2^-} f(x) = 1$

$\lim\limits_{x \to 2^+} f(x) = 5$

21. $g(x) = \begin{cases} 11 & \text{if } x < -1 \\ x^2 + 2 & \text{if } -1 \le x \le 3 \\ 11 & \text{if } x > 3 \end{cases}$

(a)

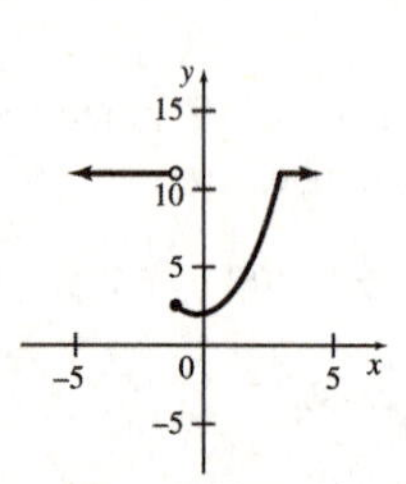

(b) $g(x)$ is discontinuous at $x = -1$.

(c) $\lim\limits_{x \to -1^-} g(x) = 11$

$\lim\limits_{x \to -1^+} g(x) = (-1)^2 + 2 = 3$

23. $h(x) = \begin{cases} 4x + 4 & \text{if } x \le 0 \\ x^2 - 4x + 4 & \text{if } x > 0 \end{cases}$

(a)

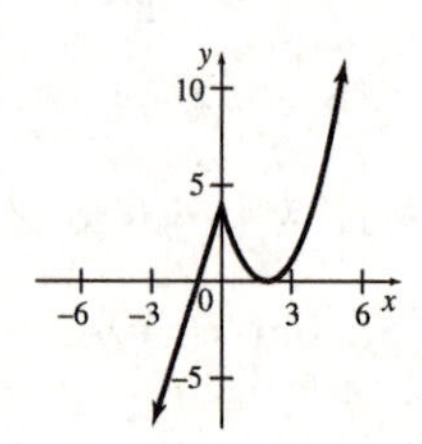

(b) There are no points of discontinuity.

25. Find k so that $kx^2 = x + k$ for $x = 2$.

$$k(2)^2 = 2 + k$$
$$4k = 2 + k$$
$$3k = 2$$
$$k = \frac{2}{3}$$

27. $\dfrac{2x^2 - x - 15}{x - 3} = \dfrac{(2x + 5)(x - 3)}{x - 3} = 2x + 5$

Find k so that $2x + 5 = kx - 1$ for $x = 3$.

$$2(3) + 5 = k(3) - 1$$
$$6 + 5 = 3k - 1$$
$$11 = 3k - 1$$
$$12 = 3k$$
$$4 = k$$

31. $f(x) = \dfrac{x^2 + x + 2}{x^3 - 0.9x^2 + 4.14x - 5.4} = \dfrac{P(x)}{Q(x)}$

(a) Graph

$$Y_1 = \frac{P(x)}{Q(x)} = \frac{x^2 + x + 2}{x^3 - 0.9x^2 + 4.14x - 5.4}$$

on a graphing calculator. A good choice for the viewing window is $[-3,3]$ by $[-10,10]$.

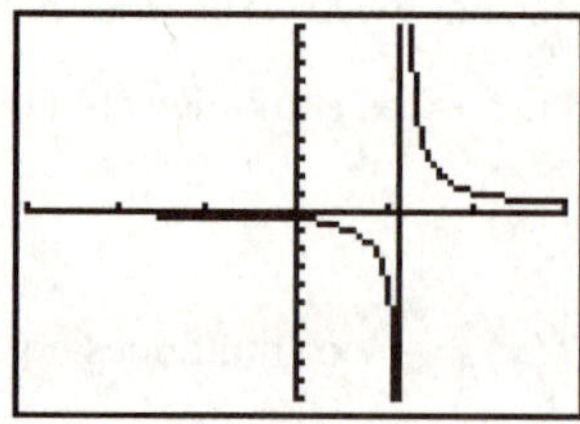

The graph has a vertical asymptote at $x = 1.2$, which indicates that f is discontinuous at $x = 1.2$.

(b) Graph

$$Y_2 = Q(x) = x^3 - .09x^2 + 4.14x - 5.4$$

using the same viewing window.

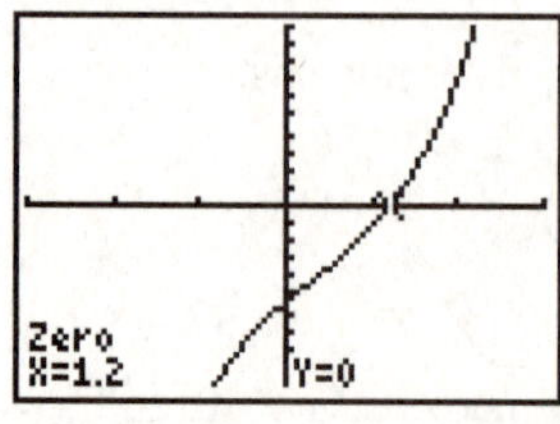

We see that this graph has one x-intercept, 1.2. This indicates that 1.2 is the only real solution of the equation $Q(x) = 0$.

This result verifies our answer from part (a) because a rational function of the form

$$f(x) = \frac{P(x)}{Q(x)}$$

will be discontinuous wherever $Q(x) = 0$.

33. $g(x) = \dfrac{x + 4}{x^2 + 2x - 8}$

$= \dfrac{x + 4}{(x - 2)(x + 4)}$

$= \dfrac{1}{x - 2}, x \neq -4$

If $g(x)$ is defined so that $g(-4) = \frac{1}{-4-2} = -\frac{1}{6}$, then the function becomes continuous at -4. It cannot be made continuous at 2. The correct answer is (a).

35. **(a)** $\lim_{x \to 6} P(x)$

As x approaches 6 from the left or the right, the value of $P(x)$ for the corresponding point on the graph approaches 500.

Thus, $\lim_{x \to 6} P(x) = \$500$.

(b) $\lim_{x \to 10^-} P(x) = \$1500$

because, as x approaches 10 from the left, $P(x)$ approaches \$1500.

(c) $\lim_{x \to 10^+} P(x) = \$1000$ because, as x approaches 10 from the right, $P(x)$ approaches \$1000.

(d) Since $\lim_{x \to 10^+} P(x) \neq \lim_{x \to 10^-} P(x)$,

$\lim_{x \to 10} P(x)$ does not exist.

(e) From the graph, the function is discontinuous at $x = 10$. This may be the result of a change of shifts.

(f) From the graph, the second shift will be as profitable as the first shift when 15 units are produced.

37. In dollars,

$$F(x) = \begin{cases} 1.25x & \text{if } 0 < x \leq 100 \\ 1.00x & \text{if } x > 100. \end{cases}$$

(a) $F(80) = 1.25(80) = \$100$

(b) $F(150) = 1.00(150) = \$150$

(c) $F(100) = 1.25(100) = \$125$

(d) F is discontinuous at $x = 100$.

39. $C(x)$ is a step function.

(a) $\lim_{x \to 3^-} C(x) = \$2.92$

(b) $\lim_{x \to 3^+} C(x) = \$3.76$

(c) $\lim_{x \to 3} C(x)$ does not exist.

(d) $C(3) = \$2.92$

(e) $\lim_{x \to 14^+} C(x) = \$10.56$

(f) $\lim_{x \to 14^-} C(x) = \$10.56$

(g) $\lim_{x \to 14} C(x) = \$10.56$

(h) $C(14) = \$10.56$

(i) $C(t)$ is discontinuous at 1, 2, 3, 4, 5, 6, 7, 8, 12, 16, 20, 24, 28, 32, 36, 40, 44, 48, 52, 56, and 60.

41.

$$W(t) \begin{cases} 48 + 3.64t + 0.6363t^2 + 0.00963t^3, & 1 \leq t \leq 28 \\ -1{,}004 + 65.8t, & 28 < y \leq 56 \end{cases}$$

(a) $W(25) = 48 + 3.64(25) + 0.6363(25)^2 + 0.00963(25)^3$

≈ 687.156

A male broiler at 25 days weighs about 687 grams.

(b) $W(t)$ is not a continuous function.

At $t = 28$

$\lim_{t \to 28^-} W(t)$

$= \lim_{t \to 28^-} 48 + 3.64t + 0.6363t^2 + 0.00963t^3$

$= 48 + 3.64(28) + 0.6363(28)^2 + 0.00963(28)^3$

≈ 860.18

and $\lim_{t \to 28^-} W(t) \neq \lim_{t \to 28^+} (-1004 + 65.8t)$

$= -1004 + 65.8(28)$

$= 838.4$

so $\lim_{t \to 28^-} W(t) \neq \lim_{t \to 28^+} W(t)$

Thus $W(t)$ is discontinuous.

(c)

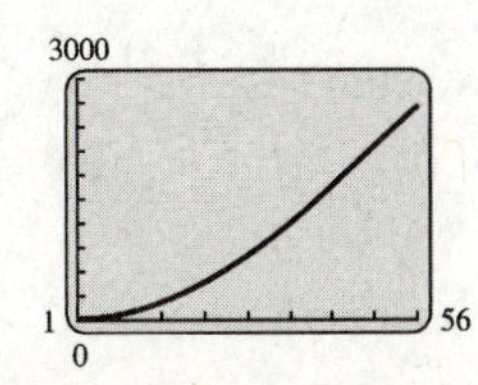

11.3 Rates of Change

Your Turn 1

$A(t) = 11.14(1.023)^t$

Average rate of change from $t = 0$ (2000) to $t = 10$ (2010) is

$$\frac{A(10) - A(0)}{10 - 0} = \frac{11.14(1.023)^{10} - 11.14(1.023)^0}{10}$$
$$= \frac{13.98 - 11.14}{10} = \frac{2.84}{10} \approx 0.0284,$$

or 0.0284 million.

The U.S. Asian population increased, on average, by 284,000 people per year.

Your Turn 2

Average rate of change in percent of households between 2007 and 2009 is

$$\frac{L(2009) - L(2007)}{9 - 7} = \frac{73.5 - 82.0}{2}$$
$$= \frac{-8.5}{2} = -4.25$$

On average, the percent of U.S. households with landline telephones decreased by 4.25% per year.

Your Turn 3

For $t = 2$, the instantaneous velocity is $\lim\limits_{h \to 0} \frac{s(2+h) - s(2)}{h}$ feet per second.

$$\begin{aligned} s(2 + h) &= 2(2 + h)^2 - 5(2 + h) + 40 \\ &= 2(4 + 4h + h^2) - 10 - 5h + 40 \\ &= 8 + 8h + h^2 - 10 - 5h + 40 \\ &= h^2 + 3h + 38 \end{aligned}$$

$$\begin{aligned} s(2) &= 2(2)^2 - 5(2) + 40 \\ &= 2(4) - 10 + 40 \\ &= 38 \end{aligned}$$

$$\begin{aligned} \lim_{h \to 0} \frac{s(2 + h) - s(2)}{h} &= \lim_{h \to 0} \frac{h^2 + 3h + 38 - 38}{h} \\ &= \lim_{h \to 0} \frac{h^2 + 3h}{h} \\ &= \lim_{h \to 0} \frac{h(h + 3)}{h} \\ &= \lim_{h \to 0} h + 3 = 3, \end{aligned}$$

or 3 feet per second.

Your Turn 4

$C(x) = x^2 - 2x + 12$

The instantaneous rate of change of cost when $x = 4$ is

$$\lim_{h \to 0} \frac{C(h + 4) - C(4)}{h}$$
$$= \lim_{h \to 0} \frac{[(h + 4)^2 - 2(h + 4) + 12] - [4^2 - 2(4) + 12]}{h}$$
$$= \lim_{h \to 0} \frac{h^2 + 8h + 16 - 2h - 8 + 12 - 16 + 8 - 12}{h}$$
$$= \lim_{h \to 0} \frac{h^2 + 6h}{h} = \lim_{h \to 0} \frac{h(h + 6)}{h}$$
$$= \lim_{h \to 0} h + 6 = 0 + 6 = 6$$

When $x = 4$, the cost increases at a rate of \$6 per unit.

Your Turn 5

$A(t) = 11.14(1.023)^t$, $t = 10$ corresponds to 2010

$$\lim_{h \to 0} \frac{11.14(1.023)^{10+h} - 11.14(1.023)^{10}}{h}$$

Use the TABLE feature on a TI-84 Plus calculator.

h	$\frac{11.14(1.023)^{10+h} - 11.14(1.023)^{10}}{h}$
1	0.32164
0.1	0.31836
0.01	0.31803
0.001	0.318
0.0001	0.318
0.00001	0.318

The limit seems to be approaching 0.318 million. The instantaneous rate of change in the U.S. Asian population is about 318,000 people per year in 2010.

11.3 Exercises

1. $y = x^2 + 2x = f(x)$ between $x = 1$ and $x = 3$

Average rate of change

$$= \frac{f(3) - f(1)}{3 - 1}$$
$$= \frac{15 - 3}{2}$$
$$= 6$$

3. $y = -3x^3 + 2x^2 - 4x + 1 = f(x)$ between $x = -2$ and $x = 1$

Average rate of change

$$= \frac{f(1) - f(-2)}{1 - (-2)}$$

$$= \frac{(-4) - (-41)}{1 - (-2)} = \frac{-45}{3} = -15$$

5. $y = \sqrt{x} = f(x)$ between $x = 1$ and $x = 4$

Average rate of change

$$= \frac{f(4) - f(1)}{4 - 1}$$

$$= \frac{2 - 1}{3}$$

$$= \frac{1}{3}$$

7. $y = e^x = f(x)$ between $x = -2$ and $x = 0$

Average rate of change

$$= \frac{f(0) - f(-2)}{0 - (-2)}$$

$$= \frac{1 - e^{-2}}{2}$$

$$\approx 0.4323$$

9.

$$\lim_{h \to 0} \frac{s(6 + h) - s(6)}{h}$$

$$= \lim_{h \to 0} \frac{(6 + h)^2 + 5(6 + h) + 2 - [6^2 + 5(6) + 2]}{h}$$

$$= \lim_{h \to 0} \frac{h^2 + 17h + 68 - 68}{h} = \lim_{h \to 0} \frac{h^2 + 17h}{h}$$

$$= \lim_{h \to 0} \frac{h(h + 17)}{h} = \lim_{h \to 0} (h + 17) = 17$$

The instantaneous velocity at $t = 6$ is 17.

11. $s(t) = 5t^2 - 2t - 7$

$$\lim_{h \to 0} \frac{s(2 + h) - s(2)}{h}$$

$$= \lim_{h \to 0} \frac{[5(2 + h)^2 - 2(2 + h) - 7] - [5(2)^2 - 2(2) - 7]}{h}$$

$$= \lim_{h \to 0} \frac{[20 + 20h + 5h^2 - 4 - 2h - 7] - [20 - 4 - 7]}{h}$$

$$= \lim_{h \to 0} \frac{9 + 18h + 5h^2 - 9}{h} = \lim_{h \to 0} \frac{18h + 5h^2}{h}$$

$$= \lim_{h \to 0} \frac{h(18 + 5h)}{h} = \lim_{h \to 0} (18 + 5h) = 18$$

The instantaneous velocity at $t = 2$ is 18.

13. $s(t) = t^3 + 2t + 9$

$$\lim_{x \to 0} \frac{s(1 + h) - s(1)}{h}$$

$$= \lim_{h \to 0} \frac{[(1 + h)^3 + 2(1 + h) + 9] - [(1)^3 + 2(1) + 9]}{h}$$

$$= \lim_{h \to 0} \frac{[1 + 3h + 3h^2 + h^3 + 2 + 2h + 9] - [1 + 2 + 9]}{h}$$

$$= \lim_{h \to 0} \frac{h^3 + 3h^2 + 5h + 12 - 12}{h}$$

$$= \lim_{h \to 0} \frac{h^3 + 3h^2 + 5h}{h} = \lim_{h \to 0} \frac{h(h^2 + 3h + 5)}{h}$$

$$= \lim_{h \to 0} (h^2 + 3h + 5) = 5$$

The instantaneous velocity at $t = 1$ is 5.

15. $f(x) = x^2 + 2x$ at $x = 0$

$$\lim_{h \to 0} \frac{f(0 + h) - f(0)}{h}$$

$$= \lim_{h \to 0} \frac{(0 + h)^2 + 2(0 + h) - [0^2 + 2(0)]}{h}$$

$$= \lim_{h \to 0} \frac{h^2 + 2h}{h}$$

$$= \lim_{h \to 0} \frac{h(h + 2)}{h}$$

$$= \lim_{h \to 0} h + 2 = 2$$

The instantaneous rate of change at $x = 0$ is 2.

17. $g(t) = 1 - t^2$ at $t = -1$

$$\lim_{h\to 0} \frac{g(-1+h) - g(-1)}{h}$$
$$= \lim_{h\to 0} \frac{1 - (-1+h)^2 - [1 - (-1)^2]}{h}$$
$$= \lim_{h\to 0} \frac{1 - (1 - 2h + h^2) - 1 + 1}{h}$$
$$= \lim_{h\to 0} \frac{2h - h^2}{h}$$
$$= \lim_{h\to 0} \frac{h(2-h)}{h}$$
$$= \lim_{h\to 0} (2 - h) = 2$$

The instantaneous rate of change at $t = -1$ is 2.

19. $f(x) = x^x$ at $x = 2$

h	
0.01	$\frac{f(2+0.01) - f(2)}{0.01} = \frac{2.01^{2.01} - 2^2}{0.01} = 6.84$
0.001	$\frac{f(2+0.001) - f(2)}{0.001} = \frac{2.001^{2.001} - 2^2}{0.001} = 6.779$
0.0001	$\frac{f(2+0.0001) - f(2)}{0.00001} = \frac{2.0001^{2.0001} - 2^2}{0.0001} = 6.773$
0.00001	$\frac{f(2+0.00001) - f(2)}{0.00001} = \frac{2.00001^{2.00001} - 2^2}{0.00001} = 6.7727$
0.000001	$\frac{f(2+0.000001) - f(2)}{0.000001} = \frac{2.000001^{2.000001} - 2^2}{0.000001} = 6.7726$

The instantaneous rate of change at $x = 2$ is 6.773.

21. $f(x) = x^{\ln x}$ at $x = 2$

h	
0.01	$\frac{f(2+0.01) - f(2)}{0.01} = \frac{2.01^{\ln 2.01} - 2^{\ln 2}}{0.01} = 1.1258$
0.001	$\frac{f(2+0.001) - f(2)}{0.001} = \frac{2.001^{\ln 2.001} - 2^{\ln 2}}{0.001} = 1.1212$

h	
0.0001	$\frac{f(2+0.0001) - f(2)}{0.0001} = \frac{2.0001^{\ln 2.0001} - 2^{\ln 2}}{0.0001} = 1.1207$
0.00001	$\frac{f(2+0.00001) - f(2)}{0.00001} = \frac{2.00001^{\ln 2.00001} - 2^{\ln 2}}{0.00001} = 1.1207$

The instantaneous rate of change at $x = 2$ is 1.121.

25. $P(x) = 2x^2 - 5x + 6$

(a) $P(4) = 18$
$P(2) = 4$

Average rate of change of profit

$$= \frac{P(4) - P(2)}{4 - 2}$$
$$= \frac{18 - 4}{2} = \frac{14}{2} = 7,$$

which is \$700 per item.

(b) $P(3) = 9$
$P(2) = 4$

Average rate of change of profit

$$= \frac{P(3) - P(2)}{3 - 2} = \frac{9 - 4}{1} = 5$$

which is \$500 per item.

(c) $\lim_{h \to 0} \frac{P(2+h) - P(2)}{h}$

$= \lim_{h \to 0} \frac{2(2+h)^2 - 5(2+h) + 6 - 4}{h}$

$= \lim_{h \to 0} \frac{8 + 8h + 2h^2 - 10 - 5h + 2}{h}$

$= \lim_{h \to 0} \frac{2h^2 + 3h}{h} = \lim_{h \to 0} \frac{h(2h+3)}{h}$

$= \lim_{h \to 0} (2h + 3) = 3,$

which is \$300 per item.

(d) $\lim_{h \to 0} \frac{P(4+h) - P(4)}{h}$

$= \lim_{h \to 0} \frac{2(4+h)^2 - 5(4+h) + 6 - 18}{h}$

$= \lim_{h \to 0} \frac{32 + 16h + 2h^2 - 20 - 5h - 12}{h}$

$= \lim_{h \to 0} \frac{2h^2 + 11h}{h}$

$= \lim_{h \to 0} \frac{h(2h+11)}{h}$

$= \lim_{h \to 0} 2h + 11 = 11,$

which is \$1100 per item.

27. $N(p) = 80 - 5p^2, 1 \le p \le 4$

(a) Average rate of change of demand is

$\frac{N(3) - N(2)}{3 - 2} = \frac{35 - 60}{1}$

$= -25$ boxes per dollar.

(b) Instantaneous rate of change when p is 2 is

$\lim_{h \to 0} \frac{N(2+h) - N(2)}{h}$

$= \lim_{h \to 0} \frac{80 - 5(2+h)^2 - [80 - 5(2)^2]}{h}$

$= \lim_{h \to 0} \frac{80 - 20 - 20h - 5h^2 - (80 - 20)}{h}$

$= \lim_{h \to 0} \frac{-5h^2 - 20h}{h} = -20$ boxes per dollar.

Around the \$2 point, a \$1 price increase (say, from \$1.50 to \$2.50) causes a drop in demand of about 20 boxes.

(c) Instantaneous rate of change when p is 3 is

$\lim_{h \to 0} \frac{80 - 5(3+h)^2 - [80 - 5(3)^2]}{h}$

$= \lim_{h \to 0} \frac{80 - 45 - 30h - 5h^2 - 80 + 45}{h}$

$= \lim_{h \to 0} \frac{-30h - 5h^2}{h}$

$= -30$ boxes per dollar.

(d) As the price increases, the demand decreases; this is an expected change.

29. $A(t) = 1000e^{0.05t}$

(a) Average rate of change in the total amount from $t = 0$ to $t = 5$:

$\frac{A(5) - A(0)}{5 - 0} = \frac{1000e^{0.05(5)} - 1000e^{0.05(0)}}{5}$

$= 56.8051,$

which is \$56.81 per year.

(b) Average rate of change in the total amount from $t = 5$ to $t = 10$:

$\frac{A(10) - A(5)}{10 - 5} = \frac{1000e^{0.05(10)} - 1000e^{0.05(5)}}{5}$

$= 72.93917,$

which is \$72.94 per year.

(c) Instantaneous rate of change for $t = 5$:

$\lim_{h \to 0} \frac{1000e^{0.05(5+h)} - 1000e^{0.05(5)}}{h}$

Use the TABLE feature on a TI-84 Plus calculator to estimate the limit.

h	$\frac{1000e^{0.05(5+h)} - 1000e^{0.05(5)}}{h}$
1	65.833
0.1	64.362
0.01	64.217
0.001	64.203
0.0001	64.201
0.00001	64.201

The limit seems to be approaching 64.201. So, the instantaneous rate of change for $t = 5$ is about \$64.20 per year.

31. Let $P(t)$ = the price per gallon of gasoline for the month t, where $t = 1$ represents January, $t = 2$ represents February, and so on.

(a) $P(1) = 329$ (cents)

$P(7) = 435$ (cents)

Average change in price from January to July:

$$\frac{P(7) - P(1)}{7 - 1} = \frac{435 - 329}{6} = \frac{106}{6} = 17.6667$$

On average, the price of gasoline increased about 17.7 cents per gallon per month.

(b) $P(7) = 435$ (cents), $P(12) = 195$ (cents)

Average change in price from July to December:

$$\frac{P(12) - P(7)}{12 - 7} = \frac{195 - 435}{5} = \frac{-240}{5} = -48$$

On average, the price of gasoline decreased about 48.0 cents per gallon per month.

(c) $P(1) = 329$ (cents), $P(12) = 195$ (cents)

Average change in price from January to December:

$$\frac{P(12) - P(1)}{12 - 1} = \frac{195 - 329}{11} = \frac{-134}{11} = -12.1818$$

On average, the price of gasoline decreased about 12.2 cents per gallon per month.

33. $p(t) = t^2 + t$

(a) $p(1) = 1^2 + 1 = 2$

$p(4) = 4^2 + 4 = 20$

Average rate of change

$$= \frac{p(4) - p(1)}{4 - 1} = \frac{20 - 2}{3} = 6$$

The average rate of change is 6% per day.

(b) $$\lim_{h \to 0} \frac{p(3 + h) - p(3)}{h}$$

$$= \lim_{h \to 0} \frac{(3 + h)^2 + (3 + h) - (3^2 + 3)}{h}$$

$$= \lim_{h \to 0} \frac{9 + 6h + h^2 + 3 + h - 12}{h}$$

$$= \lim_{h \to 0} \frac{h^2 + 7h}{h}$$

$$= \lim_{h \to 0} \frac{h(h + 7)}{h}$$

$$= \lim_{h \to 0} (h + 7) = 7$$

The instantaneous rate of change is 7% per day. The number of people newly infected on day 3 is about 7% of the total population.

35. (a) $P(1) = 3, P(2) = 5$

$$\text{Average rate of change} = \frac{P(2) - P(1)}{2 - 1} = \frac{5 - 3}{2 - 1} = \frac{2}{1} = 2$$

From 1 min to 2 min, the population of bacteria increases, on the average, 2 million per min.

(b) $P(2) = 5,\ P(3) = 4.2$

$$\text{Average rate of change} = \frac{P(3) - P(2)}{3 - 2} = \frac{4.2 - 5}{3 - 2} = \frac{-0.8}{1} = -0.8$$

From 2 min to 3 min, the population of bacteria decreases, on the average, 0.8 million or 800,000 per min.

(c) $P(3) = 4.2,\ P(4) = 2$

$$\text{Average rate of change} = \frac{P(4) - P(3)}{4 - 3} = \frac{2 - 4.2}{4 - 3} = \frac{-2.2}{1} = -2.2$$

From 3 min to 4 min, the population of bacteria decreases, on the average, 2.2 million per min.

(d) $P(4) = 2,\ P(5) = 1$

$$\text{Average rate of change} = \frac{P(5) - P(4)}{5 - 4} = \frac{1 - 2}{5 - 4} = \frac{-1}{1} = -1$$

From 4 min to 5 min, the population decreases, on the average, -1 million per min.

(e) The population increased up to 2 min after the bactericide was introduced, but decreased after 2 min.

(f) The rate of decrease of the population slows down at about 3 min. The graph becomes less and less steep after that point.

37. **(a)**

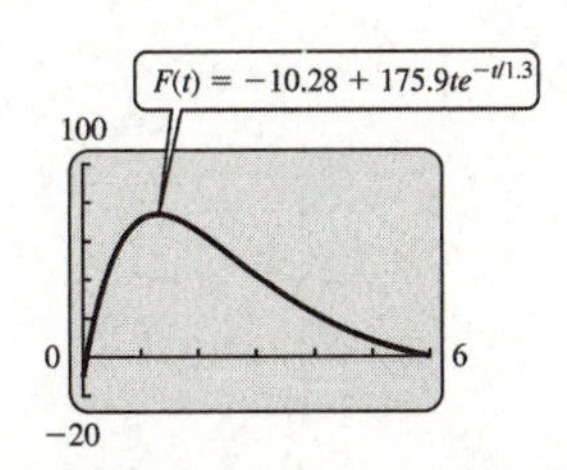

(b) The average rate of change during the first hour is

$$\frac{F(1) - F(0)}{1 - 0} \approx 81.51$$

kilojoules per hour per hour.

(c) Store $F(t)$ in a function menus of a graphing calculator. Store $\frac{Y_1(1+X)-Y_1(1)}{X}$ as Y_2 in the function menu, where Y_1 represents $F(t)$. Substitute small values for X in Y_2 perhaps with use of a table feature of the graphing calculator. As X is allowed to get smaller, Y_2 approaches 18.81 kilojoules per hour per hour.

(d) Through use of a MAX/feature program of a graphing calculator, the maximum point seen in part (a) is estimated to occur at approximately $t = 1.3$ hours.

39. Let $I(t)$ represent immigration (in thousands) in year t.

(a) $$\frac{I(1955) - I(1905)}{1955 - 1905} = \frac{238 - 1026}{50} = \frac{-788}{50} = -15.76$$

The average rate of change is $-15{,}760$ immigrants per year.

(b) $$\frac{I(2005) - I(1955)}{2005 - 1955} = \frac{1122 - 238}{50} = \frac{884}{50} = 17.68$$

The average rate of change is 17,680 immigrants per year.

(c) $$\frac{I(2005) - I(1905)}{2005 - 1905} = \frac{1122 - 1026}{100} = \frac{96}{100} = 0.96$$

The average rate of change is 960 immigrants per year.

(d) $$\frac{-15{,}760 + 17{,}680}{2} = \frac{1920}{2} = 960$$

They are equal. This will not be true for all time periods. (It is true for only time periods of equal length.)

(e) 2009 is 4 years after 2005.

$1{,}122{,}000 + 4(960/\text{year}) = 1{,}125{,}840$

The predicted number of immigrants in 2009 is about 1,126,000 immigrants. The predicted value is about 5,000 less than the actual number of 1,130,818.

41. **(a)** $$\frac{T(3000) - T(1000)}{3000 - 1000} = \frac{23 - 15}{2000} = \frac{8}{2000} = \frac{4}{1000}$$

From 1000 to 3000 ft, the temperature changes about 4° per 1000 ft; the temperature rises (on the average).

(b) $$\frac{T(5000) - T(1000)}{5000 - 1000} = \frac{22 - 15}{4000} = \frac{7}{4000} = \frac{1.75}{1000}$$

From 1000 to 5000 ft, the temperature changes about 1.75° per 1000 ft; the temperature rises (on the average).

(c) $$\frac{T(9000) - T(3000)}{9000 - 3000} = \frac{15 - 23}{6000} = \frac{-8}{6000} = \frac{-\frac{4}{3}}{1000}$$

From 3000 to 9000 ft, the temperature changes about $-\frac{4}{3}^{\circ}$ per 1000 ft; the temperature falls (on the average).

(d) $$\frac{T(9000) - T(1000)}{9000 - 1000} = \frac{15 - 15}{8000} = 0$$

From 1000 to 9000 ft, the temperature changes about 0° per 1000 ft; the temperature stays constant (on the average).

(e) The temperature is highest at 3000 ft and lowest at 1000 ft. If 7000 ft is changed to 10,000 ft, the lowest temperature would be at 10,000 ft.

(f) The temperature at 9000 ft is the same as 1000 ft.

43. **(a)** Average rate of change from 0.5 to 1:

$$\frac{f(1) - f(0.5)}{1 - 0.5} = \frac{55 - 30}{0.5} = 50 \text{ mph}$$

Average rate of change from 1 to 1.5:

$$\frac{f(1.5) - f(1)}{1.5 - 1} = \frac{80 - 55}{0.5} = 50 \text{ mph}$$

Estimate of instantaneous velocity is

$$\frac{50 + 50}{2} = 50 \text{ mph.}$$

(b) Average rate of change from 1.5 to 2:

$$\frac{f(2) - f(1.5)}{2 - 1.5} = \frac{104 - 80}{0.5} = 48 \text{ mph}$$

Average rate of change from 2 to 2.5

$$\frac{f(2.5) - f(2)}{2.5 - 2} = \frac{124 - 104}{0.5} = 40 \text{ mph}$$

Estimate of instantaneous velocity is

$$\frac{48 + 40}{2} = 44 \text{ mph.}$$

11.4 Definition of the Derivative

Your Turn 1

$f(x) = x^2 - x,\ x = -2$ and $x = 1$.

$$\text{Slope of secant line} = \frac{f(1) - f(-2)}{1 - (-2)} = \frac{0 - 6}{3} = -2$$

Use the point-slope form and the point $(1, f(1))$, or $(1, 0)$.

$$y - y_1 = m(x - x_1)$$
$$y - 0 = -2(x - 1)$$
$$y = -2x + 2$$

Your Turn 2

$$f(x) = x^2 - x$$

$$\begin{aligned} f'(x) &= \lim_{h \to 0} \frac{f(x+h) - f(x)}{h} \\ &= \lim_{h \to 0} \frac{[(x+h)^2 - (x+h)] - [x^2 - x]}{h} \\ &= \lim_{h \to 0} \frac{x^2 + 2xh + h^2 - x - h - x^2 + x}{h} \\ &= \lim_{h \to 0} \frac{2xh - h + h^2}{h} = \lim_{h \to 0} \frac{h(2x - 1 + h)}{h} \\ &= \lim_{h \to 0} (2x - 1 + h) = 2x - 1 + 0 \\ &= 2x - 1 \end{aligned}$$

$$f'(-2) = 2(-2) - 1 = -5$$

Your Turn 3

$$f(x) = x^3 - 1$$

$$\begin{aligned} f'(x) &= \lim_{h \to 0} \frac{f(x+h) - f(x)}{h} \\ &= \lim_{h \to 0} \frac{[(x+h)^3 - 1] - [x^3 - 1]}{h} \\ &= \lim_{h \to 0} \frac{x^3 + 3x^2h + 3xh^2 + h^3 - 1 - x^3 + 1}{h} \\ &= \lim_{h \to 0} \frac{3x^2h + 3xh^2 + h^3}{h} \\ &= \lim_{h \to 0} \frac{h(3x^2 + 3xh + h^2)}{h} \\ &= \lim_{h \to 0} (3x^2 + 3xh + h^2) \\ &= 3x^2 + 0 + 0 \\ &= 3x^2 \end{aligned}$$

$$f'(-1) = 3(-1)^2 = 3$$

Your Turn 4

$$f(x) = -\frac{2}{x}$$

$$\begin{aligned} f'(x) &= \lim_{h\to 0}\frac{f(x+h)-f(x)}{h} \\ &= \lim_{h\to 0}\frac{\left(-\frac{2}{x+h}\right)-\left(-\frac{2}{x}\right)}{h} \\ &= \lim_{h\to 0}\left[-\frac{2}{x+h}+\frac{2}{x}\right]\cdot\frac{1}{h} \\ &= \lim_{h\to 0}\frac{-2x+2x+2h}{x(x+h)}\cdot\frac{1}{h} \\ &= \lim_{h\to 0}\frac{2h}{x(x+h)}\cdot\frac{1}{h} = \lim_{h\to 0}\frac{2}{x(x+h)} \\ &= \frac{2}{x(x+0)} \\ &= \frac{2}{x^2} \end{aligned}$$

Your Turn 5

$$f(x) = 2\sqrt{x}$$

$$\begin{aligned} f'(x) &= \lim_{h\to 0}\frac{f(x+h)-f(x)}{h} \\ &= \lim_{h\to 0}\frac{2\sqrt{x+h}-2\sqrt{x}}{h} \\ &= \lim_{h\to 0}\frac{2\sqrt{x+h}-2\sqrt{x}}{h}\cdot\frac{2\sqrt{x+h}+2\sqrt{x}}{2\sqrt{x+h}+2\sqrt{x}} \\ &= \lim_{h\to 0}\frac{4(x+h)-4x}{h\left(2\sqrt{x+h}+2\sqrt{x}\right)} \\ &= \lim_{h\to 0}\frac{4x+4h-4x}{h\left(2\sqrt{x+h}+2\sqrt{x}\right)} \\ &= \lim_{h\to 0}\frac{4h}{h\left(2\sqrt{x+h}+2\sqrt{x}\right)} \\ &= \lim_{h\to 0}\frac{4}{2\sqrt{x+h}+2\sqrt{x}} \\ &= \frac{4}{2\sqrt{x}+2\sqrt{x}} = \frac{4}{4\sqrt{x}} \\ &= \frac{1}{\sqrt{x}} \end{aligned}$$

Your Turn 6

$$C(x) = 10x - 0.002x^2$$

$$\begin{aligned} C'(x) &= \lim_{h\to 0}\frac{C(x+h)-C(x)}{h} \\ &= \lim_{h\to 0}\frac{10(x+h)-0.002(x+h)^2-10x+0.002x^2}{h} \\ &= \lim_{h\to 0}\frac{10x+10h-0.002x^2-0.004xh-0.002h^2-10x+0.002x^2}{h} \\ &= \lim_{h\to 0}\frac{10h-0.004xh-0.002h^2}{h} \\ &= \lim_{h\to 0}\frac{h(10-0.004x-0.002h)}{h} \\ &= \lim_{h\to 0}(10-0.004x-0.002h) \\ &= 10-0.004x+0 \\ &= 10-0.004x \end{aligned}$$

$$C'(100) = 10 - 0.004(100) = 10 - 0.4 = 9.60$$

The rate of change of the cost when $x = 100$ is \$9.60.

Your Turn 7

$f(x) = 2\sqrt{x}$ at $x = 4$

From Your Turn 5, we have $f'(x) = \frac{1}{\sqrt{x}}$.

At $x = 4$: $f'(4) = \frac{1}{\sqrt{4}} = \frac{1}{2}$ and $f(4) = 2\sqrt{4} = 2(2) = 4$

Slope of the tangent line at $(4, f(4))$, or $(4, 4)$ is $\frac{1}{2}$.

$$\begin{aligned} y - y_1 &= m(x - x_1) \\ y - 4 &= \tfrac{1}{2}(x - 4) \\ y - 4 &= \tfrac{1}{2}x - 2 \\ y &= \tfrac{1}{2}x + 2 \end{aligned}$$

11.4 Exercises

1. **(a)** $f(x) = 5$ is a horizontal line and has slope 0; the derivative is 0.

(b) $f(x) = x$ has slope 1; the derivative is 1.

(c) $f(x) = -x$ has slope of -1, the derivative is -1.

(d) $x = 3$ is vertical and has undefined slope; the derivative does not exist.

(e) $y = mx + b$ has slope m; the derivative is m.

3. $f(x) = \frac{x^2-1}{x+2}$ is not differentiable when $x + 2 = 0$ or $x = -2$ because the function is undefined and a vertical asymptote occurs there.

5. Using the points $(5,3)$ and $(6,5)$, we have

$$\begin{aligned} m &= \frac{5-3}{6-5} = \frac{2}{1} \\ &= 2. \end{aligned}$$

7. Using the points $(-2,2)$ and $(2,3)$, we have

$$m = \frac{3-2}{2-(-2)} = \frac{1}{4}.$$

9. Using the points $(-3,-3)$ and $(0,-3)$, we have

$$m = \frac{-3-(-3)}{0-3} = \frac{0}{-3} = 0.$$

11. $f(x) = 3x - 7$

Step 1 $f(x + h)$
$$\begin{aligned} &= 3(x+h) - 7 \\ &= 3x + 3h - 7 \end{aligned}$$

Step 2 $f(x + h) - f(x)$
$$\begin{aligned} &= 3x + 3h - 7 - (3x - 7) \\ &= 3x + 3h - 7 - 3x + 7 \\ &= 3h \end{aligned}$$

Step 3 $\frac{f(x+h) - f(x)}{h} = \frac{3h}{h} = 3$

Step 4 $f'(x) = \lim_{h\to 0} \frac{f(x+h) - f(x)}{h}$
$$= \lim_{h\to 0} 3 = 3$$

$f'(-2) = 3,\ f'(0) = 3\ f'(3) = 3$

13. $f(x) = -4x^2 + 9x + 2$

Step 1 $f(x + h)$
$$\begin{aligned} &= -4(x+h)^2 + 9(x+h) + 2 \\ &= -4(x^2 + 2xh + h^2) + 9x + 9h + 2 \\ &= -4x^2 - 8xh - 4h^2 + 9x + 9h + 2 \end{aligned}$$

Step 2 $f(x + h) - f(x)$
$$\begin{aligned} &= -4x^2 - 8xh - 4h^2 + 9x + 9h + 2 \\ &\quad -(-4x^2 + 9x + 2) \\ &= -8xh - 4h^2 + 9h \\ &= h(-8x - 4h + 9) \end{aligned}$$

Step 3 $\frac{f(x+h) - f(x)}{h}$
$$\begin{aligned} &= \frac{h(-8x - 4h + 9)}{h} \\ &= -8x - 4h + 9 \end{aligned}$$

Step 4 $f'(x) = \lim_{h\to 0} \frac{f(x+h) - f(x)}{h}$
$$\begin{aligned} &= \lim_{h\to 0} (-8x - 4h + 9) \\ &= -8x + 9 \end{aligned}$$

$$\begin{aligned} f'(-2) &= -8(-2) + 9 = 25 \\ f'(0) &= -8(0) + 9 = 9 \\ f'(3) &= -8(3) + 9 = -15 \end{aligned}$$

15. $f(x) = \frac{12}{x}$

$$f(x+h) = \frac{12}{x+h}$$

$$\begin{aligned} f(x+h) - f(x) &= \frac{12}{x+h} - \frac{12}{x} \\ &= \frac{12x - 12(x+h)}{x(x+h)} \\ &= \frac{12x - 12x - 12h}{x(x+h)} \\ &= \frac{-12h}{x(x+h)} \end{aligned}$$

$$\begin{aligned} \frac{f(x+h) - f(x)}{h} &= \frac{-12h}{hx(x+h)} \\ &= \frac{-12}{x(x+h)} \\ &= \frac{-12}{x^2 + xh} \end{aligned}$$

$$\begin{aligned} f'(x) &= \lim_{h\to 0} \frac{f(x+h) - f(x)}{h} \\ &= \lim_{h\to 0} \frac{-12}{x^2 + xh} \\ &= \frac{-12}{x^2} \end{aligned}$$

$$f'(-2) = \frac{-12}{(-2)^2} = \frac{-12}{4} = -3$$

$f'(0) = \frac{-12}{0^2}$ which is undefined so $f'(0)$ does not exist.

$$\begin{aligned} f'(3) &= \frac{-12}{3^2} \\ &= \frac{-12}{9} = -\frac{4}{3} \end{aligned}$$

17. $f(x) = \sqrt{x}$

Steps 1-3 are combined.

$$\begin{aligned} &\frac{f(x+h) - f(x)}{h} \\ &= \frac{\sqrt{x+h} - \sqrt{x}}{h} \\ &= \frac{\sqrt{x+h} - \sqrt{x}}{h} \cdot \frac{\sqrt{x+h} + \sqrt{x}}{\sqrt{x+h} + \sqrt{x}} \\ &= \frac{x+h-x}{h(\sqrt{x+h} + \sqrt{x})} \\ &= \frac{1}{\sqrt{x+h} + \sqrt{x}} \end{aligned}$$

$$\begin{aligned} f'(x) &= \lim_{h\to 0} \frac{f(x+h) - f(x)}{h} \\ &= \lim_{h\to 0} \frac{1}{\sqrt{x+h} + \sqrt{x}} = \frac{1}{2\sqrt{x}} \end{aligned}$$

$f'(-2) = \frac{1}{2\sqrt{-2}}$ which is undefined so $f'(-2)$ does not exist.

$f'(0) = \frac{1}{2\sqrt{0}} = \frac{1}{0}$ which is undefined so $f'(0)$ does not exist.

$$f'(3) = \frac{1}{2\sqrt{3}}$$

19. $f(x) = 2x^3 + 5$

Steps 1-3 are combined.

$$\begin{aligned} &\frac{f(x+h) - f(x)}{h} \\ &= \frac{2(x+h)^3 + 5 - (2x^3 + 5)}{h} \\ &= \frac{2(x^3 + 3x^2h + 3xh^2 + h^3) + 5 - 2x^3 - 5}{h} \\ &= \frac{2x^3 + 6x^2h + 6xh^2 + 2h^3 + 5 - 2x^3 - 5}{h} \\ &= \frac{6x^2h + 6xh^2 + 2h^3}{h} \\ &= \frac{h(6x^2 + 6xh + 2h^2)}{h} \\ &= 6x^2 + 6xh + 2h^2 \end{aligned}$$

$$f'(x) = \lim_{h\to 0}(6x^2 + 6xh + 2h^2) = 6x^2$$

$$f'(-2) = 6(-2)^2 = 24$$

$$f'(0) = 6(0)^2 = 0$$

$$f'(3) = 6(3)^2 = 54$$

21. (a) $f(x) = x^2 + 2x;\ x = 3, x = 5$

$$\begin{aligned} \text{Slope of secant line} &= \frac{f(5) - f(3)}{5 - 3} \\ &= \frac{(5)^2 + 2(5) - [(3)^2 + 2(3)]}{2} \\ &= \frac{35 - 15}{2} \\ &= 10 \end{aligned}$$

Now use $m = 10$ and $(3, f(3)) = (3,15)$ in the point-slope form.

$$y - 15 = 10(x - 3)$$
$$y - 15 = 10x - 30$$
$$y = 10x - 30 + 15$$
$$y = 10x - 15$$

(b) $f(x) = x^2 + 2x;\ x = 3$

$$\frac{f(x+h) - f(x)}{h}$$
$$= \frac{[(x+h)^2 + 2(x+h)] - (x^2 + 2x)}{h}$$
$$= \frac{(x^2 + 2hx + h^2 + 2x + 2h) - (x^2 + 2x)}{h}$$
$$= \frac{2hx + h^2 + 2h}{h} = 2x + h + 2$$
$$f'(x) = \lim_{h\to 0}(2x + h + 2) = 2x + 2$$

$f'(3) = 2(3) + 2 = 8$ is the slope of the tangent line at $x = 3$.

Use $m = 8$ and $(3,15)$ in the point-slope form.

$$y - 15 = 8(x - 3)$$
$$y = 8x - 9$$

23. (a) $f(x) = \frac{5}{x};\ x = 2, x = 5$

$$\text{Slope of secant line} = \frac{f(5) - f(2)}{5 - 2}$$
$$= \frac{\frac{5}{5} - \frac{5}{2}}{3} = \frac{1 - \frac{5}{2}}{3}$$
$$= -\frac{1}{2}$$

Now use $m = -\frac{1}{2}$ and $(5, f(5)) = (5, 1)$ in the point-slope form.

$$y - 1 = -\frac{1}{2}[x - 5]$$
$$y - 1 = -\frac{1}{2}x + \frac{5}{2}$$
$$y = -\frac{1}{2}x + \frac{5}{2} + 1$$
$$y = -\frac{1}{2}x + \frac{7}{2}$$

(b) $f(x) = \frac{5}{x};\ x = 2$

$$\frac{f(x+h) - f(x)}{h} = \frac{\frac{5}{x+h} - \frac{5}{x}}{h}$$
$$= \frac{\frac{5x - 5(x+h)}{(x+h)x}}{h}$$
$$= \frac{5x - 5x - 5h}{h(x+h)(x)}$$
$$= \frac{-5h}{h(x+h)x}$$
$$= \frac{-5}{(x+h)x}$$

$$f'(x) = \lim_{h\to 0}\frac{-5}{(x+h)(x)} = -\frac{5}{x^2}$$

$f'(2) = \frac{-5}{2^2} = -\frac{5}{4}$ is the slope of the tangent line at $x = 2$.

Now use $m = -\frac{5}{4}$ and $\left(2, \frac{5}{2}\right)$ in the point-slope form.

$$y - \frac{5}{2} = -\frac{5}{4}(x - 2)$$
$$y - \frac{5}{2} = -\frac{5}{4}x + \frac{10}{4}$$
$$y = -\frac{5}{4}x + 5$$
$$5x + 4y = 20$$

25. (a) $f(x) = 4\sqrt{x};\ x = 9, x = 16$

$$\text{Slope of secant line} = \frac{f(16) - f(9)}{16 - 9}$$
$$= \frac{4\sqrt{16} - 4\sqrt{9}}{7}$$
$$= \frac{16 - 12}{7} = \frac{4}{7}$$

Now use $m = \frac{4}{7}$ and $(9, f(9)) = (9, 12)$ in the point-slope form.

$$y - 12 = \frac{4}{7}(x - 9)$$
$$y - 12 = \frac{4}{7}x - \frac{36}{7}$$
$$y = \frac{4}{7}x - \frac{36}{7} + 12$$
$$y = \frac{4}{7}x + \frac{48}{7}$$

(b) $f(x) = 4\sqrt{x};\ x = 9$

$$\frac{f(x+h)-f(x)}{h}$$
$$= \frac{4\sqrt{x+h}-4\sqrt{x}}{h}\cdot\frac{4\sqrt{x+h}+4\sqrt{x}}{4\sqrt{x+h}+4\sqrt{x}}$$
$$= \frac{16(x+h)-16x}{h(4\sqrt{x+h}+4\sqrt{x})}$$
$$f'(x) = \lim_{h\to 0}\frac{16(x+h)-16x}{h(4\sqrt{x+h}+4\sqrt{x})}$$
$$= \lim_{h\to 0}\frac{16h}{h(4\sqrt{x+h}+4\sqrt{x})}$$
$$= \lim_{h\to 0}\frac{4}{(\sqrt{x+h}+\sqrt{x})} = \frac{4}{2\sqrt{x}}$$
$$= \frac{2}{\sqrt{x}}$$

$f'(9) = \frac{2}{\sqrt{9}} = \frac{2}{3}$ is the slope of the tangent line at $x = 9$.

Use $m = \frac{2}{3}$ and $(9, 12)$ in the point-slope form.

$$y - 12 = \frac{2}{3}(x-9)$$
$$y = \frac{2}{3}x + 6$$
$$3y = 2x + 18$$

27. $f(x) = -4x^2 + 11x$

$$\frac{f(x+h)-f(x)}{h}$$
$$= \frac{-4(x+h)^2 + 11(x+h) - (-4x^2+11x)}{h}$$
$$= \frac{-8xh - 4h^2 + 11h}{h}$$
$$f'(x) = \lim_{h\to 0}(-8x - 4h + 11) = -8x + 11$$
$$f'(2) = -8(2) + 11 = -5$$
$$f'(16) = -8(16) + 11 = -117$$
$$f'(-3) = -8(-3) + 11 = 35$$

29. $f(x) = e^x$

$$\frac{f(x+h)-f(x)}{h} = \frac{e^{x+h}-e^x}{h}$$
$$f'(x) = \lim_{h\to 0}\frac{e^{x+h}-e^x}{h}$$

$f'(2) \approx 7.3891;\ f'(16) \approx 8{,}886{,}111;\ f'(-3) \approx 0.0498$

31. $f(x) = -\frac{2}{x}$

$$\frac{f(x+h)-f(x)}{h} = \frac{\frac{-2}{x+h} - \left(\frac{-2}{x}\right)}{h}$$
$$= \frac{\frac{-2x+2(x+h)}{(x+h)x}}{h}$$
$$= \frac{2h}{h(x+h)x} = \frac{2}{(x+h)x}$$
$$f'(x) = \lim_{h\to 0}\frac{2}{(x+h)x} = \frac{2}{x^2}$$
$$f'(2) = \frac{2}{2^2} = \frac{1}{2}$$
$$f'(16) = \frac{2}{16^2} = \frac{2}{256} = \frac{1}{128}.$$
$$f'(-3) = \frac{2}{(-3)^2} = \frac{2}{9}$$

33. $f(x) = \sqrt{x}$

$$\frac{f(x+h)-f(x)}{h}$$
$$= \frac{\sqrt{x+h}-\sqrt{x}}{h}\cdot\frac{\sqrt{x+h}+\sqrt{x}}{\sqrt{x+h}+\sqrt{x}}$$
$$= \frac{(x+h)-x}{h(\sqrt{x+h}+\sqrt{x})}$$
$$= \frac{h}{h(\sqrt{x+h}+\sqrt{x})} = \frac{1}{\sqrt{x+h}+\sqrt{x}}$$
$$f'(x) = \lim_{h\to 0}\frac{1}{\sqrt{x+h}+\sqrt{x}} = \frac{1}{2\sqrt{x}}$$
$$f'(2) = \frac{1}{2\sqrt{2}}$$
$$f'(16) = \frac{1}{2\sqrt{16}} = \frac{1}{8}$$

$f'(-3) = \frac{1}{2\sqrt{-3}}$ is not a real number, so $f'(-3)$ does not exist.

35. At $x = 0$, the graph of $f(x)$ has a sharp point. Therefore, there is no derivative for $x = 0$.

37. For $x = -3$ and $x = 0$, the tangent to the graph of $f(x)$ is vertical. For $x = -1$, there is a gap in the graph of $f(x)$. For $x = 2$, the function $f(x)$ does not exist. For $x = 3$ and $x = 5$, the graph of $f(x)$ has sharp points. Therefore, no derivative exists for $x = -3$, $x = -1$, $x = 0$, $x = 2$, $x = 3$, and $x = 5$.

39. (a) The rate of change of $f(x)$ is positive when $f(x)$ is increasing, that is, on $(a, 0)$ and (b, c).

(b) The rate of change of $f(x)$ is negative when $f(x)$ is decreasing, that is, on $(0, b,)$.

(c) The rate of change is zero when the tangent to the graph is horizontal, that is, at $x = 0$ and $x = b$.

41. The zeros of graph (b) correspond to the turning points of graph (a), the points where the derivative is zero. Graph (a) gives the distance, while graph (b) gives the velocity.

43. $f(x) = x^x, a = 3$

(a)

h	
0.01	$\frac{f(3+0.01)-f(3)}{0.01} = \frac{3.01^{3.01}-3^3}{0.01} = 57.3072$
0.001	$\frac{f(3+0.001)-f(3)}{0.001} = \frac{3.001^{3.001}-3^3}{0.001} = 56.7265$
0.00001	$\frac{f(3+0.00001)-f(3)}{0.00001} = \frac{3.00001^{3.00001}-3^3}{0.00001} = 56.6632$
0.000001	$\frac{f(3+0.000001)-f(3)}{0.000001} = \frac{3.000001^{3.000001}-3^3}{0.000001} = 56.6626$
0.0000001	$\frac{f(3+0.0000001)-f(3)}{0.0000001} = \frac{3.0000001^{3.0000001}-3^3}{0.0000001} = 56.6625$

It appears that $f'(3) \approx 56.66$.

(b) Graph the function on a graphing calculator and move the cursor to an x-value near $x = 3$. A good choice for the initial viewing window is [0, 4] by [0, 60] with Xscl = 1, Yscl = 10.

Now zoom in on the function several times. Each time you zoom in, the graph will look less like a curve and more like a straight line. Use the TRACE feature to select two points on the graph, and record their coordinates. Use these two points to compute the slope. The result will be close to the most accurate value found in part (a), which is 56.66.

Note: In this exercise, the method used in part (a) gives more accurate results than the method used in part (b).

45. $f(x) = x^{1/x},\ a = 3$

(a)

h	
0.01	$\frac{f(3+0.01)-f(3)}{0.01} = \frac{3.01^{1/3.01}-3^{1/3}}{0.01} = -0.0160$
0.001	$\frac{f(3+0.001)-f(3)}{0.001} = \frac{3.001^{1/3.001}-3^{1/3}}{0.001} = -0.0158$
0.0001	$\frac{f(3+0.0001)-f(3)}{0.0001} = \frac{3.0001^{1/3.0001}-3^{1/3}}{0.0001} = -0.0158$

It appears that $f'(3) = -0.0158$.

(b) Graph the function on a graphing calculator and move the cursor to an x-value near $x = 3$. A good choice for the initial viewing window is [0, 5] by [0, 3].

Follow the procedure outlined in the solution for Exercise 43, part (b). Note that near $x = 3$, the graph is very close to a horizontal line, so we expect that it slope will be close to 0. The final result will be close to the value found in part (a) of this exercise, which is -0.0158.

49. $D(p) = -2p^2 - 4p + 300$

D is demand; p is price.

(a) Given that $D'(p) = -4p - 4$, the rate of change of demand with respect to price is $-4p - 4$, the derivative of the function $D(p)$.

(b) $D'(10) = -4(10) - 4$
$= -44$

The demand is decreasing at the rate of about 44 items for each increase in price of \$1.

51. $R(x) = 20x - \dfrac{x^2}{500}$

(a) $R'(x) = 20 - \dfrac{1}{250}x$

At $y = 1000$,

$$R'(1000) = 20 - \frac{1}{250}(1000)$$
$$= \$16 \text{ per table.}$$

(b) The marginal revenue for the 100lst table is approximately $R'(1000)$. From (a), this is about \$16.

(c) The actual revenue is

$$R(1001) - R(1000) = 20(1001) - \frac{1001^2}{500}$$
$$-\left[20(1000) - \frac{1000^2}{500}\right]$$
$$= 18{,}015.998 - 18{,}000$$
$$= \$15.998 \text{ or } \$16.$$

(d) The marginal revenue gives a good approximation of the actual revenue from the sale of the 100lst table.

53. (a) $f(x) = 0.0000329x^3 - 0.00405x^2 + 0.0613x + 2.34$

$f(10) = 2.54$
$f(20) = 2.03$
$f(30) = 1.02$

(b)

$Y_1 = 0.0000329x^3 - 0.00405x^2 + 0.0613x + 2.34$

$nDeriv(Y_1, x, 0) \approx 0.061$
$nDeriv(Y_1, x, 10) \approx -0.019$
$nDeriv(Y_1, x, 20) \approx -0.079$
$nDeriv(Y_1, x, 30) \approx -0.120$
$nDeriv(Y_1, x, 35) \approx -0.133$

55. The derivative at (2, 4000) can be approximated by the slope of the line through (0, 2000) and (2, 4000).
The derivative is approximately

$$\frac{4000 - 2000}{2 - 0} = \frac{2000}{2} = 1000.$$

Thus the shellfish population is increasing at a rate of 1000 shellfish per unit time.
The derivative at about (10,10,300) can be approximated by the slope of the line through (10,10,300) and (13,12,000). The derivative is approximately

$$\frac{12{,}000 - 10{,}300}{13 - 10} = \frac{1700}{3} \approx 570.$$

The shellfish population is increasing at a rate of about 583 shellfish per unit time. The derivative at about (13, 11,250) can be approximated by the slope of the line through (13, 11,250) and (16, 12,000). The derivative is approximately

$$\frac{12{,}000 - 11{,}250}{16 - 13} = \frac{750}{3} \approx 250.$$

The shellfish population is increasing at a rate of 250 shellfish per unit time.

57. (a) Set $M(v) = 150$ and solve for v.

$$0.0312443v^2 - 101.39v + 82{,}264 = 150$$
$$0.0312443v^2 - 101.39v + 82{,}114 = 0$$

Solve using the quadratic formula.

Let D equal the discriminant.

$$\begin{aligned} D &= b^2 - 4ac \\ &= (-101.39)^2 - 4(0.0312443)(82{,}114) \\ &\approx 17.55 \end{aligned}$$

$$v = \frac{101.39 \pm \sqrt{D}}{2(0.0312443)}$$

$v \approx 1690$ meter per second or

$v \approx 1560$ meters per second.

Since the functions is defined only for $v \geq 1620$, the only solution is 1690 meters per second.

(b) Calculate $\lim\limits_{h \to 0} \dfrac{M(1700 + h) - M(1700)}{h}$

$$\begin{aligned} &M(1700 + h) \\ &= 0.0312443(1700 + h)^2 - 101.39(1700 + h) + 82{,}264 \\ &= 90{,}296.027 + 106.23062h + 0.0312443h^2 - 172{,}363 - 101.39h + 82{,}264 \\ &= 0.01312443h^2 + 4.84062h + 197.027 \end{aligned}$$

$M(1700) = 197.027$, so the derivative of $M(v)$ at $v = 1700$ is

$$\begin{aligned} &\lim_{h \to 0} \left(\frac{0.0312443h^2 + 4.84062h + 197.027 - 197.027}{h} \right) \\ &= \lim_{h \to 0} \left(\frac{0.0312443h^2 + 4.84062h}{h} \right) \\ &= \lim_{h \to 0} (0.0312443h + 4.84062) \\ &= 4.84062 \\ &\approx 4.84 \text{ days per meter per second} \end{aligned}$$

The increase in velocity for this cheese from 1700 m/s to 1701 m/s indicates that the approximate age of the cheese has increased by 4.84 days.

59. (a) The derivative does not exist at the two "corners" or "sharp points." The x-values of these points are 0.75 and 3.

(b) To find $T'(0.5)$, calculate the slope of the line segment with positive slope, since this portion of the graph includes the value $x = 0.5$ (marked as $\frac{1}{2}$ hr on the graph.) To find this slope, use the points (0,100) (the starting point) and (0.75,875) (the beginning point of the cleaning cycle).

$$\begin{aligned} m = T'(0.5) &= \frac{875 - 100}{0.75 - 0} \\ &= \frac{775}{0.75} \approx 1033 \end{aligned}$$

The oven temperature is increasing at 1033° per hour.

(c) To find $T'(2)$, find the slope of the line segment containing the value $x = 2$. This segment is horizontal, so

$$m = T'(2) = 0.$$

The oven temperature is not changing.

(d) To find $T'(3.5)$, calculate the slope of the line segment with negative slope, since this portion of the graph includes the value $x = 3.5$. To find this slope, use the points (3,875) (the end of the cleaning cycle) and (3.75,100) (the stopping point).

$$m = T'(3.5) = \frac{100 - 875}{3.75 - 3} = \frac{-775}{0.75} \approx -1033$$

The oven temperature is decreasing at 1033° per hour.

61. (a) At 40 oz the tangent looks horizontal; thus the derivative for a 40-ounce bat is about 0 mph per oz.

The slope of the graph at $x = 30$ can be estimated using the points (30, 79) and (32, 80).

$$\text{slope} = \frac{80 - 79}{32 - 30} = 0.5$$

Thus, the derivative for a 30 ounce bat is about 0.5 mph per oz .

(b) The optimal bat is 40 oz.

11.5 Graphical Differentiation

Your Turn 1

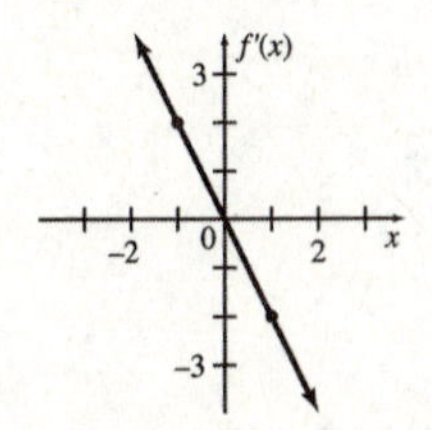

Your Turn 2

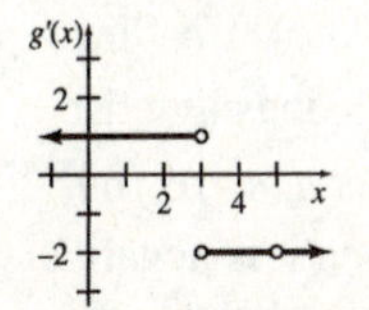

11.5 Exercises

3. Since the x-intercepts of the graph of f' occur whenever the graph of f has a horizontal tangent line, Y_1 is the derivative of Y_2. Notice that Y_1 has 2 x-intercepts; each occurs at an x-value where the tangent line to Y_2 is horizontal.

Note also that Y_1 is positive whenever Y_2 is increasing, and that Y_1 is negative whenever Y_2 is decreasing.

5. Since the x-intercepts of the graph of f' occur whenever the graph of f has a horizontal tangent line, Y_2 is the derivative of Y_1. Notice that Y_2 has 1 x-intercept which occurs at the x-value where the tangent line to Y_1 is horizontal. Also notice that the range on which Y_1 is increasing, Y_2 is positive and the range on which it is decreasing, Y_2 is negative.

7. To graph f', observe the intervals where the slopes of tangent lines are positive and where they are negative to determine where the derivative is positive and where it is negative. Also, whenever f has a horizontal tangent, f' will be 0, so the graph of f' will have an x-intercept. The x-values of the three turning point on the graph of f become the three x-intercepts of the graph of f.

Estimate the magnitude of the slope at several points by drawing tangents to the graph of f.

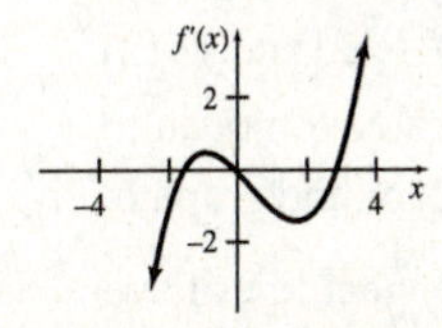

9. On the interval $(-\infty,-2)$, the graph of f is a horizontal line, so its slope is 0. Thus, on this interval, the graph of f' is $y = 0$ on $(-\infty,-2)$. On the interval $(-2, 0)$, the graph of f is a straight line, so its slope is constant. To find this slope, use the points $(-2, 2)$ and $(0,0)$.

$$m = \frac{2-0}{-2-0} = \frac{2}{-2} = -1$$

On the interval $(0, 1)$, the slope is also constant. To find this slope, use the points $(0, 0)$ and $(1, 1)$.

$$m = \frac{1-0}{1-0} = 1$$

On the interval $(1,\infty)$, the graph is again a horizontal line, so $m = 0$. The graph of f' will be made up of portions of the y-axis and the lines $y = -1$ and $y = 1$.

Because the graph of f has "sharp points" or "corners" at $x = -2$, $x = 0$, and $x = 1$, we know that $f'(-2)$, $f'(0)$, and $f'(1)$ do not exist. We show this on the graph of f' by using open circles at the endpoints of the portions of the graph.

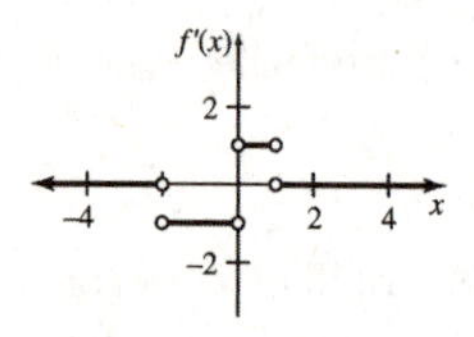

11. On the interval $(-\infty,-2)$, the graph of f is a straight line, so its slope is constant. To find this slope, use the points $(-4, 2)$ and $(-2, 0)$.

$$m = \frac{0-2}{-2-(-4)} = \frac{-2}{2} = -1$$

On the interval $(2,\infty)$, the slope of f is also constant. To find this slope, use the points $(2, 0)$ and $(3, 2)$.

$$m = \frac{2-0}{3-2} = \frac{2}{1} = 2$$

Thus, we have $f'(x) = -1$ on $(-\infty,-2)$ and $f'(x) = 2$ on $(2,\infty)$.

Because f is discontinuous at $x = -2$ and $x = 2$, we know that $f'(-2)$ and $f'(2)$ do not exist, which we indicate with open circles at $(-2,-1)$ and $(2, 2)$ on the graph of f'.

On the interval $(-2, 2)$ all tangent lines have positive slopes, so the graph of f' will be above the y-axis. Notice that the slope of f (and thus the y-value of f') decreases on $(-2, 0)$ and increases on $(0, 2)$ with a minimum value on this interval of about 1 at $x = 0$.

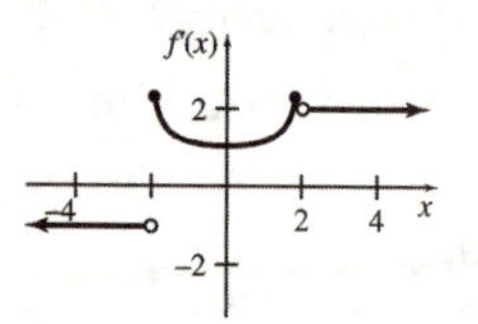

13. We observe that the slopes of tangent lines are positive on the interval $(-\infty,0)$ and negative on the interval $(0,\infty)$, so the value of f' will be positive on $(-\infty,0)$ and negative on $(0,\infty)$. Since f is undefined at $x = 0$, $f'(0)$ does not exist.

Notice that the graph of f becomes very flat when $|x| \to \infty$. The *value* of f approaches 0 and also the *slope* approaches 0. Thus, $y = 0$ (the x-axis) is a horizontal asymptote for both the graph of f and the graph of f'.

As $x \to 0^-$ and $x \to 0^+$, the graph of f gets very steep, so $|f'(x)| \to \infty$. Thus, $x = 0$ (the y-axis) is a vertical asymptote for both the graph of f and the graph of f'.

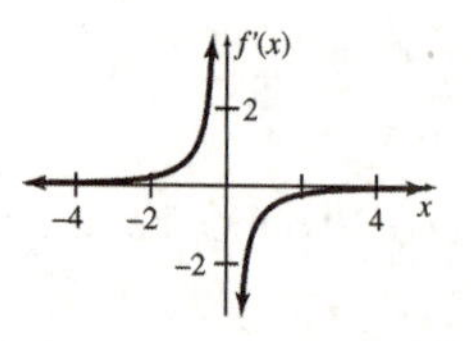

15. The slope of $f(x)$ is undefined at $x = -2, -1, 0, 1$, and 2, and the graph approaches vertical (unbounded slope) as x approaches those values. Accordingly, the graph of $f'(x)$ has vertical asymptotes at $x = -2, -1, 0, 1$, and 2. $f(x)$ has turning points (zero slope) at $x = -1.5, -0.5, 0.5$, and 1.5, so the graph of $f'(x)$ crosses the x-axis at those values. Elsewhere, the graph of $f'(x)$ is negative where $f(x)$ is decreasing and positive where $f(x)$ is increasing.

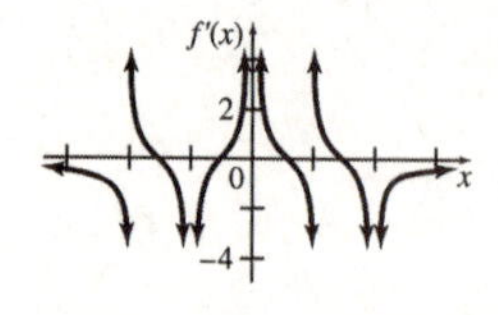

17. The graph of G decreases steadily with varying degrees of steepness. The steepness increases (that is, the slopes of the tangent lines becomes more negative) between $t = 12$ and $t = 16$. Since G is discontinuous at $t = 16$, $G'(16)$ doesn't exist. The graph continues to decrease after $t = 16$, but the slopes of the tangent lines become less negative as the curve gets flatter. So, the derivative values are increasing toward 0.

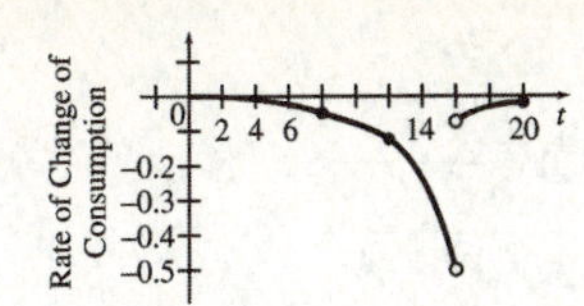

19. The graph rises steadily, with varying degrees of steepness. The graph is steepest around 1976 and nearly flat around 1950 and 1980. Accordingly, the rate of change is always positive, with a maximum value around 1976 and values near zero around 1950 and 1980.

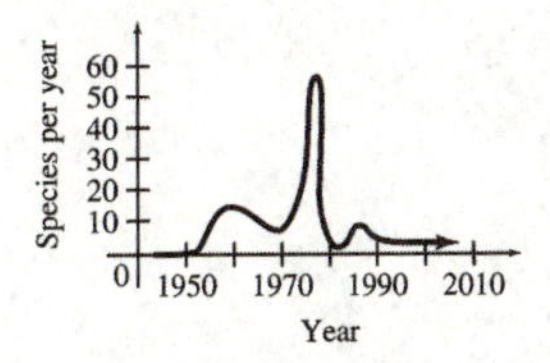

21. The growth rate of the function $y = f(x)$ is given by the derivative of this function $y' = f(x)$. We use the graph of f to sketch the graph of f'. First, notice as x increases, y increases throughout the domain of f, but at a slower and slower rate. The slope of f is positive but always decreasing, and approaches 0 as t gets large. Thus, y' will always be positive and decreasing. It will approach but never reach 0.

To plot point on the graph of f', we need to estimate the slope of f at several points. From the graph of f, we obtain the values given in the following table.

t	y'
2	1000
10	700
13	250

Use these points to sketch the graph.

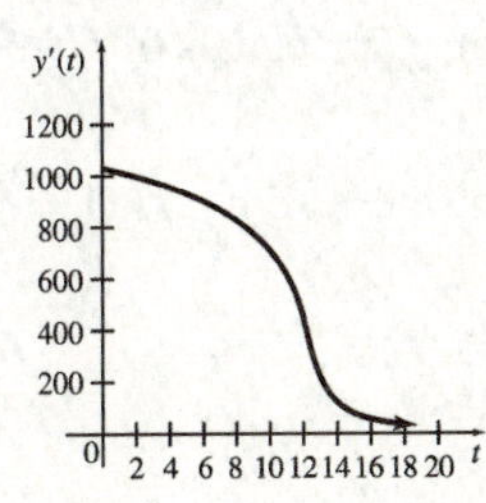

23.

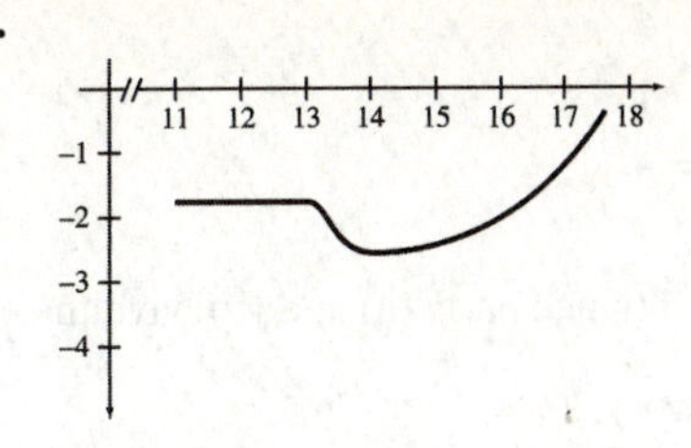

About 9 cm; about 2.6 cm less per year

Chapter 11 Review Exercises

1. True

2. True

3. True

4. False; for example, if $f(x) = \frac{x^2-4}{x+2}$, $\lim_{x \to -2} f(x) = -4$, but the graph of $f(x) = \frac{x^2-4}{x+2}$ has a hole at the point $(-2, -4)$.

5. True

6. False; for example, the rational function $f(x) = \frac{5}{x+1}$ is discontinuous at $x = -1$.

7. False; the derivative gives the instantaneous rate of change of a function.

8. True

9. True

10. True

11. False; the slope of the tangent line gives the instantaneous rate of change.

12. False; for example, the function $f(x) = |x|$ is continuous at $x = 0$, but $f'(0)$ does not exist. The graph of $f(x) = |x|$ has a "corner" at $x = 0$.

17. (a) $\lim_{x\to -3^-} = 4$

(b) $\lim_{x\to -3^+} = 4$

(c) $\lim_{x\to -3} = 4$ (since parts (a) and (b) have the same answer)

(d) $f(-3) = 4$, since $(-3, 4)$ is a point of the graph.

19. (a) $\lim_{x\to 4^-} f(x) = \infty$

(b) $\lim_{x\to 4^+} f(x) = -\infty$

(c) $\lim_{x\to 4} f(x)$ does not exist since limits in (a) and (b) do not exist.

(d) $f(4)$ does not exist since the graph has no point with an x-value of 4.

21. $\lim_{x\to -\infty} g(x) = \infty$ since the y-value gets very large as the x-value gets very small.

23. $\lim_{x\to 6} \frac{2x+7}{x+3} = \frac{2(6)+7}{6+3} = \frac{19}{9}$

25.
$$\begin{aligned}\lim_{x\to 4} \frac{x^2-16}{x-4} &= \lim_{x\to 4} \frac{(x-4)(x+4)}{x-4}\\ &= \lim_{x\to 4}(x+4)\\ &= 4+4\\ &= 8\end{aligned}$$

27.
$$\begin{aligned}\lim_{x\to -4} \frac{2x^2+3x-20}{x+4} &= \lim_{x\to -4} \frac{(2x-5)(x+4)}{x+4}\\ &= \lim_{x\to -4}(2x-5)\\ &= 2(-4)-5\\ &= -13\end{aligned}$$

29.
$$\begin{aligned}\lim_{x\to 9} \frac{\sqrt{x}-3}{x-9} &= \lim_{x\to 9} \frac{\sqrt{x}-3}{x-9}\cdot\frac{\sqrt{x}+3}{\sqrt{x}+3}\\ &= \lim_{x\to 9} \frac{x-9}{(x-9)(\sqrt{x}+3)}\\ &= \lim_{x\to 9} \frac{1}{\sqrt{x}+3}\\ &= \frac{1}{\sqrt{9}+3} = \frac{1}{6}\end{aligned}$$

31.
$$\begin{aligned}\lim_{x\to\infty} \frac{2x^2+5}{5x^2-1} &= \lim_{x\to\infty} \frac{\frac{2x^2}{x^2}+\frac{5}{x^2}}{\frac{5x^2}{x^2}-\frac{1}{x^2}}\\ &= \lim_{x\to\infty} \frac{2+\frac{5}{x^2}}{5-\frac{1}{x^2}}\\ &= \frac{2+0}{5-0}\\ &= \frac{2}{5}\end{aligned}$$

33.
$$\begin{aligned}&\lim_{x\to -\infty}\left(\frac{3}{8}+\frac{3}{x}-\frac{6}{x^2}\right)\\ &= \lim_{x\to -\infty}\frac{3}{8}+\lim_{x\to -\infty}\frac{3}{x}-\lim_{x\to -\infty}\frac{6}{x^2}\\ &= \frac{3}{8}+0-0\\ &= \frac{3}{8}\end{aligned}$$

35. As shown on the graph, $f(x)$ is discontinuous at x_2 and x_4.

37. $f(x)$ is discontinuous at $x = 0$ and $x = -\frac{1}{3}$ since that is where the denominator of $f(x)$ equals 0. $f(0)$ and $f\left(-\frac{1}{3}\right)$ do not exist.

$\lim_{x\to 0} f(x)$ does not exist since $\lim_{x\to 0^+} f(x) = -\infty$, but $\lim_{x\to 0^-} f(x) = \infty$. $\lim_{x\to -\frac{1}{3}} f(x)$ does not exist since $\lim_{x\to -\frac{1}{3}^-} = -\infty$, but $\lim_{x\to -\frac{1}{3}^+} f(x) = \infty$.

39. $f(x)$ is discontinuous at $x = -5$ since that is where the denominator of $f(x)$ equals 0. $f(-5)$ does not exist.

$\lim_{x\to -5} f(x)$ does not exist since $\lim_{x\to -5^-} f(x) = \infty$, but $\lim_{x\to -5^+} f(x) = -\infty$.

41. $f(x) = x^2 + 3x - 4$ is continuous everywhere since f is a polynomial function.

43. (a)

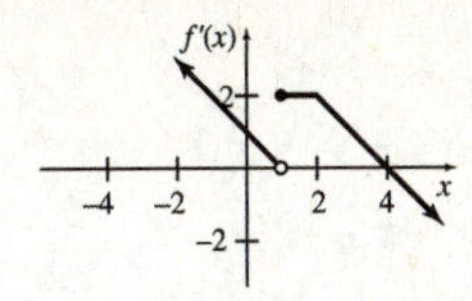

(b) The graph is discontinuous at $x = 1$.

(c) $\lim_{x \to 1^-} f(x) = 0$; $\lim_{x \to 1^+} f(x) = 2$

45. $f(x) = \dfrac{x^4 + 2x^3 + 2x^2 - 10x + 5}{x^2 - 1}$

(a) Find the values of $f(x)$ when x is close to 1.

x	y
1.1	2.6005
1.01	2.06
1.001	2.006
1.0001	2.0006
0.99	1.94
0.999	1.994
0.9999	1.9994

It appears that $\lim_{x \to 1} f(x) = 2$.

(b) Graph

$$y = \frac{x^4 + 2x^3 + 2x^2 - 10x + 5}{x^2 - 1}$$

on a graphing calculator. One suitable choice for the viewing window is $[-2, 6]$ by $[-10, 10]$. Because $x^2 - 1 = 0$ when $x = -1$ or $x = 1$, this function is discontinuous at these two x-values. The graph shows a vertical asymptote at $x = -1$ but not at $x = 1$. The graph should have an open circle to show a "hole" in the graph at $x = 1$. The graphing calculator doesn't show the hole, but trying to find the value of the function of $x = 1$ will show that this value is undefined.

By viewing the function near $x = 1$ and using the ZOOM feature, we see that as x gets close to 1 from the left or the right, y gets close to 2, suggesting that

$$\lim_{x \to 1} \frac{x^4 + 2x^3 + 2x^2 - 10x + 5}{x^2 - 1} = 2.$$

47. $y = 6x^3 + 2 = f(x)$; from $x = 1$ to $x = 4$

$$f(4) = 6(4)^3 + 2 = 386$$
$$f(1) = 6(1)^3 + 2 = 8$$

Average rate of change:

$$= \frac{386 - 8}{4 - 1} = \frac{378}{3} = 126$$

$$y' = 18x$$

Instantaneous rate of change at $x = 1$:

$$f'(1) = 18(1) = 18$$

49. $y = \dfrac{-6}{3x - 5} = f(x)$; from $x = 4$ to $x = 9$

$$f(9) = \frac{-6}{3(9) - 5} = \frac{-6}{22} = -\frac{3}{11}$$
$$f(4) = \frac{-6}{3(4) - 5} = -\frac{6}{7}$$

Average rate of change:

$$= \frac{\frac{-3}{11} - \left(-\frac{6}{7}\right)}{9 - 4} = \frac{\frac{-21+66}{77}}{5} = \frac{45}{5(77)} = \frac{9}{77}$$

$$y' = \frac{(3x - 5)(0) - (-6)(3)}{(3x - 5)^2} = \frac{18}{(3x - 5)^2}$$

Instantaneous rate of change at $x = 4$:

$$f'(4) = \frac{18}{(3 \cdot 4 - 5)^2} = \frac{18}{7^2} = \frac{18}{49}$$

51. (a) $f(x) = 3x^2 - 5x + 7$; $x = 2$, $x = 4$

Slope of secant line

$$= \frac{f(4) - f(2)}{4 - 2}$$
$$= \frac{[3(4)^2 - 5(4) + 7] - [3(2)^2 - 5(2) + 7]}{2}$$
$$= \frac{35 - 9}{2}$$
$$= 13$$

Now use $m = 13$ and $2, f(2) = (2,9)$ in the point-slope form.

$$y - 9 = 13(x - 2)$$
$$y - 9 = 13x - 26$$
$$y = 13x - 26 + 9$$
$$y = 13x - 17$$

(b) $f(x) = 3x^2 - 5x + 7;\ x = 2$

$$\frac{f(x+h) - f(x)}{h}$$
$$= \frac{[3(x+h)^2 - 5(x+h) + 7] - [3x^2 - 5x + 7]}{h}$$
$$= \frac{3x^2 + 6xh + 3h^2 - 5x - 5h + 7 - 3x^2 + 5x - 7}{h}$$
$$= \frac{6xh + 3h^2 - 5h}{h}$$
$$= 6x + 3h - 5$$

$$f'(x) = \lim_{h \to 0} 6x + 3h - 5$$
$$= 6x - 5$$
$$f'(2) = 6(2) - 5$$
$$= 7$$

Now use $m = 7$ and $(2, f(2)) = (2,9)$ in the point-slope form.

$$y - 9 = 7(x - 2)$$
$$y - 9 = 7x - 14$$
$$y = 7x - 14 + 9$$
$$y = 7x - 5$$

53. (a) $f(x) = \frac{12}{x-1};\ x = 3,\ x = 7$

$$\text{Slope of secant line} = \frac{f(7) - f(3)}{7 - 3}$$
$$= \frac{\frac{12}{7-1} - \frac{12}{3-1}}{4}$$
$$= \frac{2 - 6}{4}$$
$$= -1$$

Now use $m = -1$ and $(3, f(x)) = (3,6)$ in the point-slope form.

$$y - 6 = -1(x - 3)$$
$$y - 6 = -x + 3$$
$$y = -x + 3 + 6$$
$$y = -x + 9$$

(b) $f(x) = \frac{12}{x-1};\ x = 3$

$$\frac{f(x+h) - f(x)}{h} = \frac{\frac{12}{x+h-1} - \frac{12}{x-1}}{h}$$
$$= \frac{12(x-1) - 12(x+h-1)}{h(x-1)(x+h-1)}$$
$$= \frac{-12h}{h(x-1)(x+h-1)}$$
$$= -\frac{12}{(x-1)(x+h-1)}$$

$$f'(x) = \lim_{h \to 0} -\frac{12}{(x-1)(x+h-1)}$$
$$= -\frac{12}{(x-1)^2}$$
$$f'(3) = -\frac{12}{(3-1)^2}$$
$$= -3$$

Now use $m = -3$ and $(3, f(x)) = (3,6)$ in the point-slope form.

$$y - 6 = -3(x - 3)$$
$$y - 6 = -3x + 9$$
$$y = -3x + 9 + 6$$
$$y = -3x + 15$$

55. $y = 4x^2 + 3x - 2 = f(x)$

$$
\begin{aligned}
y' &= \lim_{h\to 0} \frac{f(x+h) - f(x)}{h} \\
&= \lim_{h\to 0} \frac{[4(x+h)^2 + 3(x+h) - 2] - [4x^2 + 3x - 2]}{h} \\
&= \lim_{h\to 0} \frac{4(x^2 + 2xh + h^2) + 3x + 3h - 2 - 4x^2 - 3x + 2}{h} \\
&= \lim_{h\to 0} \frac{4x^2 + 8xh + 4h^2 + 3x + 3h - 2 - 4x^2 - 3x + 2}{h} \\
&= \lim_{h\to 0} \frac{8xh + 4h^2 + 3h}{h} \\
&= \lim_{h\to 0} \frac{h(8x + 4h + 3)}{h} \\
&= \lim_{h\to 0} (8x + 4h + 3) \\
&= 8x + 3
\end{aligned}
$$

57. $f(x) = (\ln x)^x$, $x_0 = 3$

(a)

h	
0.01	$\frac{f(3+0.01) - f(3)}{0.01} = \frac{(\ln 3.01)^{3.01} - (\ln 3)^3}{0.01} = 1.3385$
0.001	$\frac{f(3+0.001) - f(3)}{0.001} = \frac{(\ln 3.001)^{3.001} - (\ln 3)^3}{0.001} = 1.3323$
0.0001	$\frac{f(3+0.0001) - f(3)}{0.0001} = \frac{(\ln 3.0001)^{3.0001} - (\ln 3)^3}{0.0001} = 1.3317$
0.00001	$\frac{f(3+0.00001) - f(3)}{0.00001} = \frac{(\ln 3.00001)^{3.00001} - (\ln 3)^3}{0.00001} = 1.3317$

It appears that $f'(3) \approx 1.332$.

(b) Using a graphing calculator will confirm this result.

59. On the interval $(-\infty, 0)$, the graph of f is a straight line, so its slope is constant. To find this slope, use the points $(-2, 2)$ and $(0, 0)$.

$$m = \frac{0-2}{0-(-2)} = \frac{-2}{2} = -1$$

Thus, the value of f' will be -1 on this interval.

The graph of f has a sharp point at 0, so $f'(0)$ does not exist. To show this, we use an open circle on the graph of f' at $(0, -1)$.

We also observe that the slope of f is positive but decreasing from $x = 0$ to about $x = 1$, and then negative from there on. As $x \to \infty, f(x) \to 0$ and also $f'(x) = 0$.

Use this information to complete the graph of f'.

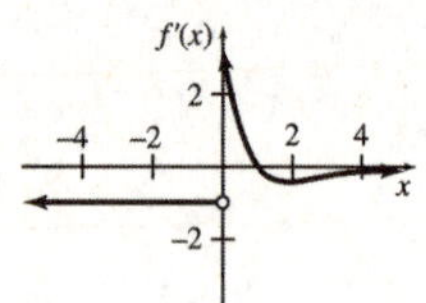

61.
$$\lim_{x\to\infty} \frac{cf(x) - dg(x)}{f(x) - g(x)}$$
$$= \frac{\lim_{x\to\infty} [cf(x) - dg(x)]}{\lim_{x\to\infty} [f(x) - g(x)]}$$
$$= \frac{\lim_{x\to\infty} [cf(x)] - \lim_{x\to\infty} [dg(x)]}{\lim_{x\to\infty} [f(x)] - \lim_{x\to\infty} [g(x)]}$$
$$= \frac{c \lim_{x\to\infty} [f(x)] - d \lim_{x\to\infty} [g(x)]}{\lim_{x\to\infty} [f(x)] - \lim_{x\to\infty} [g(x)]}$$
$$= \frac{c \cdot c - d \cdot d}{c - d} = \frac{(c+d)(c-d)}{c-d}$$
$$= c + d$$

The answer is (e).

63. $C(x) = \begin{cases} 1.50x & \text{for } 0 < x \leq 125 \\ 1.35x & \text{for } x > 125 \end{cases}$

(a) $C(100) = 1.50(100) = \$150$

(b) $C(125) = 1.50(125) = \$187.50$

(c) $C(140) = 1.35(140) = \$189$

(d)

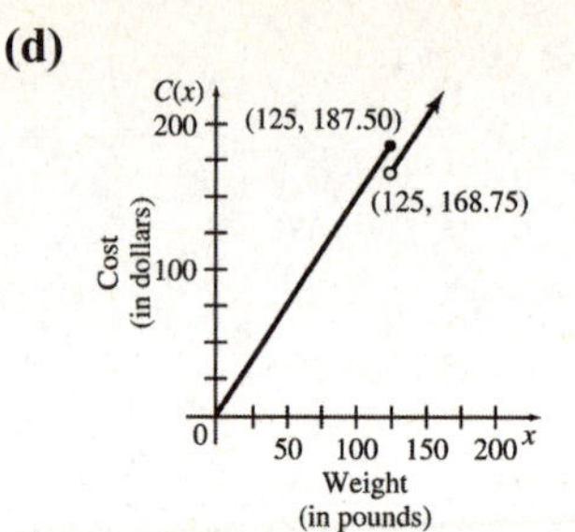

(e) By reading the graph, $C(x)$ is discontinuous at $x = \$125$.

The average cost per pound is given by $\overline{C}(x) = \frac{C(x)}{x}$.

$$\overline{C}(x) = \begin{cases} 1.50 & \text{for } 0 < x \leq 125 \\ 1.35 & \text{for } x > 125 \end{cases}$$

(f) $\overline{C}(100) = \$1.50$

(g) $\overline{C}(125) = \$1.50$

(h) $\overline{C}(140) = \$1.35$

The marginal cost is given by

$$C(x) = \begin{cases} 1.50 & \text{for } 0 < x \leq 125 \\ 1.35 & \text{for } x > 125. \end{cases}$$

(i) $C'(100) = 1.50$; the 101st pound will cost \$1.50.

(j) $C'(140) = 1.35$; the 141st pound will cost \$1.35.

65. (b) The value of x for which the average cost is smallest is $x = 7.5$. This can be found by drawing a line from the origin to any point of $C(x)$. At $x = 7.5$, you will get a line with the smallest slope.

(c) The marginal cost equals the average cost at the point where the average cost is smallest.

67.

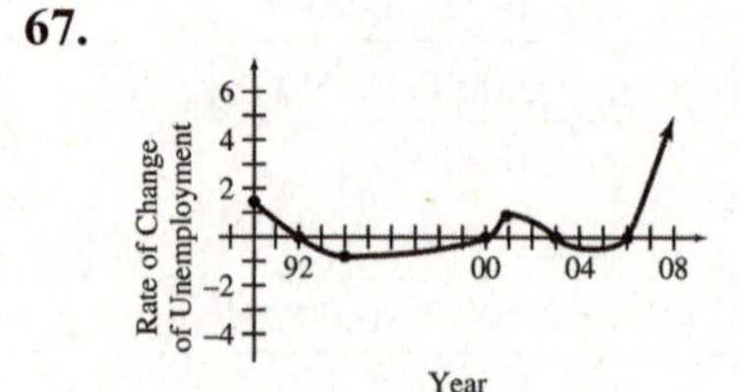

In 2008, the unemployment rate is about 5.4%; the rate of change of the unemployment rate is about 2% per year.

69. $V(t) = -t^2 + 6t - 4$

(a)

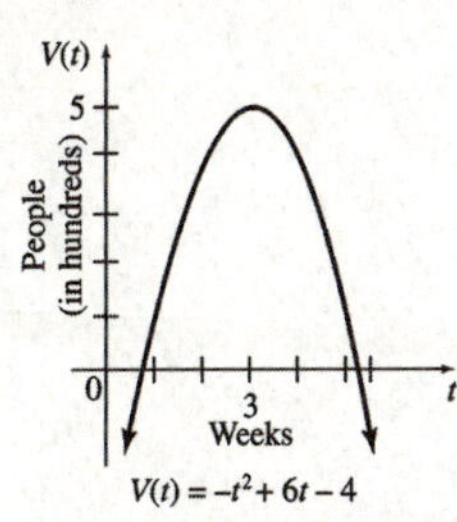

(b) The x-intercepts of the parabola are 0.8 and 5.2, so a reasonable domain would be [0.8, 5.2], which represents the time period from 0.8 to 5.2 weeks.

(c) The number of cases reaches a maximum at the vertex;

$$x = \frac{-b}{2a} = \frac{-6}{-2} = 3$$

$$V(3) = -3^2 + 6(3) - 4 = 5$$

The vertex of the parabola is $(3, 5)$. This represents a maximum at 3 weeks of 500 cases.

(d) The rate of change function is

$$V'(t) = -2t + 6.$$

(e) The rate of change in the number of cases at the maximum is

$$V'(3) = -2(3) + 6 = 0.$$

(f) The sign of the rate of change up to the maximum is + because the function is increasing. The sign of the rate of change after the maximum is – because the function is decreasing.

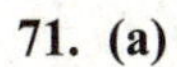

71. (a)

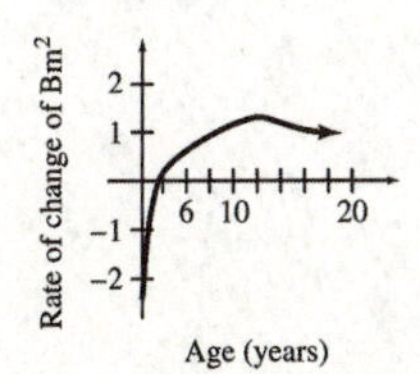

(b)

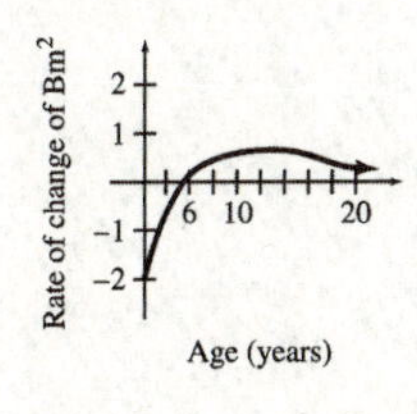

73. (a) The graph is discontinuous nowhere.

(b) The graph is not differentiable where the graph makes a sudden change, namely at $x = 50$, $x = 130$, $x = 230$, and $x = 770$.

(c)

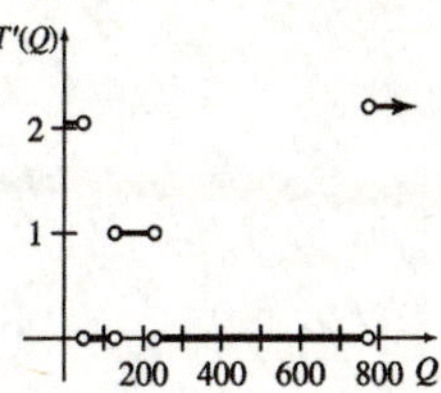

Chapter 12

CALCULATING THE DERIVATIVE

12.1 Techniques for Finding Derivatives

Your Turn 1

$$f(t) = \frac{1}{\sqrt{t}} = t^{-1/2}$$
$$f'(t) = -\tfrac{1}{2}t^{-1/2-1}$$
$$= -\tfrac{1}{2}t^{-3/2}$$
$$= -\frac{1}{2t^{3/2}} \text{ or } -\frac{1}{2\sqrt{t^3}}$$

Your Turn 2

$$y = 3\sqrt{x} = 3x^{1/2}$$
$$\frac{dy}{dx} = 3 \cdot \frac{1}{2}x^{-1/2}$$
$$= \frac{3}{2}x^{-1/2}$$
$$= \frac{3}{2\sqrt{x}}$$

Your Turn 3

$$h(t) = -3t^2 + 2\sqrt{t} + \frac{5}{t^4} - 7$$
$$= 3t^2 + 2t^{1/2} + 5t^{-4} - 7$$
$$h'(t) = -6t + t^{-1/2} - 20t^{-5}$$
$$= -6t + \frac{1}{\sqrt{t}} - \frac{20}{t^5}$$

Your Turn 4

Find the marginal cost of the cost function $C(x) = 5x^3 - 10x^2 + 75$ when $x = 100$.

$$C'(x) = 15x^2 - 20x$$
$$C'(100) = 15(100)^2 - 20(100)$$
$$= 150{,}000 - 2000$$
$$= 148{,}000$$

The marginal cost when $x = 100$ is \$148,000.

Your Turn 5

Find the marginal revenue of the demand function $p = 16 - 1.25q$ when $q = 5$.

$$R(q) = qp = q(16 - 1.25q) = 16q - 1.25q^2$$
$$R'(q) = 16 - 2.5q$$
$$R'(5) = 16 - 2.5(5) = 3.5$$

The marginal revenue for 5 units is \$3.50 per unit.

12.1 Exercises

1. $y = 12x^3 - 8x^2 + 7x + 5$

$$\frac{dy}{dx} = 12(3x^{3-1}) - 8(2x^{2-1}) + 7x^{1-1} + 0$$
$$= 36x^2 - 16x + 7$$

3. $y = 3x^4 - 6x^3 + \frac{x^2}{8} + 5$

$$\frac{dy}{dx} = 3(4x^{4-1}) - 6(3x^{3-1}) + \frac{1}{8}(2x^{2-1}) + 0$$
$$= 12x^3 - 18x^2 + \frac{1}{4}x$$

5. $y = 6x^{3.5} - 10x^{0.5}$

$$\frac{dy}{dx} = 6(3.5x^{3.5-1}) - 10(0.5x^{0.5-1})$$
$$= 21x^{2.5} - 5x^{-0.5} \text{ or } 21x^{2.5} - \frac{5}{x^{0.5}}$$

7. $y = 8\sqrt{x} + 6x^{3/4} = 8x^{1/2} + 6x^{3/4}$

$$\frac{dy}{dx} = 8\left(\frac{1}{2}x^{1/2-1}\right) + 6\left(\frac{3}{4}x^{3/4-4}\right)$$
$$= 4x^{-1/2} + \frac{9}{2}x^{-1/4} \text{ or } \frac{4}{x^{1/2}} + \frac{9}{2x^{1/4}}$$

9. $y = 10x^{-3} + 5x^{-4} - 8x$

$$\frac{dy}{dx} = 10(-3x^{-3-1}) + 5(-4x^{-4-1}) - 8x^{1-1}$$
$$= -30x^{-4} - 20x^{-5} - 8 \text{ or } \frac{-30}{x^4} - \frac{20}{x^5} - 8$$

11. $f(t) = \frac{7}{t} - \frac{5}{t^3}$

$= 7t^{-1} - 5t^{-3}$

$f'(t) = 7(-1t^{-1-1}) - 5(-3t^{-3-1})$

$= -7t^{-2} + 15t^{-4}$ or $\frac{-7}{t^2} + \frac{15}{t^4}$

13. $y = \frac{6}{x^4} - \frac{7}{x^3} + \frac{3}{x} + \sqrt{5}$

$= 6x^{-4} - 7x^{-3} + 3x^{-1} + \sqrt{5}$

$\frac{dy}{dx} = 6(-4x^{-4-1}) - 7(-3x^{-3-1})$

$+ 3(-1x^{-1-1}) + 0$

$= -24x^{-5} + 21x^{-4} - 3x^{-2}$

or $\frac{-24}{x^5} + \frac{21}{x^4} - \frac{3}{x^2}$

15. $p(x) = -10x^{-1/2} + 8x^{-3/2}$

$p'(x) = -10\left(-\frac{1}{2}x^{-3/2}\right) + 8\left(-\frac{3}{2}x^{-5/2}\right)$

$= 5x^{-3/2} - 12x^{-5/2}$ or $\frac{5}{x^{3/2}} - \frac{12}{x^{5/2}}$

17. $y = \frac{6}{4\sqrt{x}} = 6x^{-1/4}$

$\frac{dy}{dx} = 6\left(-\frac{1}{4}\right)x^{-5/4}$

$= -\frac{3}{2}x^{-5/4}$ or $\frac{-3}{2x^{5/4}}$

19. $f(x) = \frac{x^3 + 5}{x} = x^2 + 5x^{-1}$

$f'(x) = 2x^{2-1} + 5(-1x^{-1-1})$

$= 2x - 5x^{-2}$ or $2x - \frac{5}{x^2}$

21. $g(x) = (8x^2 - 4x)^2$

$= 64x^4 - 64x^3 + 16x^2$

$g'(x) = 64(4x^{4-1}) - 64(3x^{3-1}) + 16(2x^{2-1})$

$= 256x^3 - 192x^2 + 32x$

23. A quadratic function has degree 2. When the derivative is taken, the power will decrease by 1 and the derivative function will be linear, so the correct choice is (b).

27. $D_x\left[9x^{-1/2} + \frac{2}{x^{3/2}}\right]$

$= D_x[9x^{-1/2} + 2x^{-3/2}]$

$= 9\left(-\frac{1}{2}x^{-3/2}\right) + 2\left(-\frac{3}{2}x^{-5/2}\right)$

$= -\frac{9}{2}x^{-3/2} - 3x^{-5/2}$ or $\frac{-9}{2x^{3/2}} - \frac{3}{x^{5/2}}$

29. $f(x) = \frac{x^4}{6} - 3x$

$= \frac{1}{6}x^4 - 3x$

$f'(x) = \frac{1}{6}(4x^3) - 3$

$= \frac{2}{3}x^3 - 3$

$f'(-2) = \frac{2}{3}(-2)^3 - 3$

$= -\frac{16}{3} - 3 = -\frac{25}{3}$

31. $y = x^4 - 5x^3 + 2;\ x = 2$

$y' = 4x^3 - 15x^2$

$y'(2) = 4(2)^3 - 15(2)^2$

$= -28$

The slope of tangent line at $x = 2$ is -28.

Use $m = -28$ and $(x_1, y_1) = (2, -22)$ to obtain the equation.

$y - (-22) = -28(x - 2)$

$y = -28x + 34$

33. $y = -2x^{1/2} + x^{3/2}$

$y' = -2\left(\frac{1}{2}x^{-1/2}\right) + \frac{3}{2}x^{1/2}$

$= -x^{-1/2} + \frac{3}{2}x^{1/2}$

$= -\frac{1}{x^{1/2}} + \frac{3x^{1/2}}{2}$

$y'(9) = -\frac{1}{(9)^{1/2}} + \frac{3(9)^{1/2}}{2}$

$= -\frac{1}{3} + \frac{9}{2}$

$= \frac{25}{6}$

The slope of the tangent line at $x = 9$ is $\frac{25}{6}$.

35. $f(x) = 9x^2 - 8x + 4$

$f'(x) = 18x - 8$

Let $f'(x) = 0$ to find the point where the slope of the tangent line is zero.

$$18x - 8 = 0$$
$$18x = 8$$
$$x = \frac{8}{18} = \frac{4}{9}$$

Find the y-coordinate.

$$f(x) = 9x^2 - 8x + 4$$
$$f\left(\frac{4}{9}\right) = 9\left(\frac{4}{9}\right)^2 - 8\left(\frac{4}{9}\right) + 4$$
$$= 9\left(\frac{16}{81}\right) - \frac{32}{9} + 4$$
$$= \frac{16}{9} - \frac{32}{9} + \frac{36}{9} = \frac{20}{9}$$

The slope of the tangent line is zero at one point, $\left(\frac{4}{9}, \frac{20}{9}\right)$.

37. $f(x) = 2x^3 + 9x^2 - 60x + 4$

$f'(x) = 6x^2 + 18x - 60$

If the tangent line is horizontal, then its slope is zero and $f'(x) = 0$.

$$6x^2 + 18x - 60 = 0$$
$$6(x^2 + 3x - 10) = 0$$
$$6(x + 5)(x - 2) = 0$$
$$x = -5 \text{ or } x = 2$$

Thus, the tangent line is horizontal at $x = -5$ and $x = 2$.

39. $f(x) = x^3 - 4x^2 - 7x + 8$

$f'(x) = 3x^2 - 8x - 7$

If the tangent line is horizontal, then its slope is zero and $f'(x) = 0$.

$$3x^2 - 8x - 7 = 0$$
$$x = \frac{8 \pm \sqrt{64 + 84}}{6}$$
$$x = \frac{8 \pm \sqrt{148}}{6}$$
$$x = \frac{8 \pm 2\sqrt{37}}{6}$$
$$x = \frac{2(4 \pm \sqrt{37})}{6}$$
$$x = \frac{4 \pm \sqrt{37}}{3}$$

Thus, the tangent line is horizontal at $x = \frac{4 \pm \sqrt{37}}{3}$.

41. $f(x) = 6x^2 + 4x - 9$

$f'(x) = 12x + 4$

If the slope of the tangent line is -2, $f'(x) = -2$.

$$12x + 4 = -2$$
$$12x = -6$$
$$x = -\frac{1}{2}$$
$$f\left(-\frac{1}{2}\right) = -\frac{19}{2}$$

The slope of the tangent line is -2 at $\left(-\frac{1}{2}, -\frac{19}{2}\right)$.

43. $f(x) = x^3 + 6x^2 + 21x + 2$

$f'(x) = 3x^2 + 12x + 21$

If the slope of the tangent line is 9, $f'(x) = 9$.

$$3x^2 + 12x + 21 = 9$$
$$3x^2 + 12x + 12 = 0$$
$$3(x^2 + 4x + 4) = 0$$
$$3(x + 2)^2 = 0$$
$$x = -2$$
$$f(-2) = -24$$

The slope of the tangent line is 9 at $(-2, -24)$.

45. $f(x) = \frac{1}{2}g(x) + \frac{1}{4}h(x)$

$$f'(x) = \frac{1}{2}g'(x) + \frac{1}{4}h'(x)$$
$$f'(2) = \frac{1}{2}g'(2) + \frac{1}{4}h'(2)$$
$$= \frac{1}{2}(7) + \frac{1}{4}(14) = 7$$

49. $\frac{f(x)}{k} = \frac{1}{k} \cdot f(x)$

Use the rule for the derivative of a constant times a function.

$$\frac{d}{dx}\left[\frac{f(x)}{k}\right] = \frac{d}{dx}\left[\frac{1}{k} \cdot f(x)\right]$$
$$= \frac{1}{k} f'(x)$$
$$= \frac{f'(x)}{k}$$

51. The demand is given by $q = 5000 - 100p$. Solve for p.

$$p = \frac{5000 - q}{100}$$
$$R(q) = q\left(\frac{5000 - q}{100}\right)$$
$$= \frac{5000q - q^2}{100}$$
$$R'(q) = \frac{5000 - 2q}{100}$$

(a) $R'(1000) = \frac{5000 - 2(1000)}{100}$
$= 30$

(b) $R'(2500) = \frac{5000 - 2(2500)}{100}$
$= 0$

(c) $R'(3000) = \frac{5000 - 2(3000)}{100}$
$= -10$

53. $p(q) = \frac{1000}{q^2} + 1000$

If R is the revenue function, $R(q) = qp(q)$.

$$R(q) = q\left(\frac{1000}{q^2} + 1000\right)$$
$$R(q) = 1000q^{-1} + 1000q$$
$$R'(q) = -1000q^{-2} + 1000$$
$$R'(q) = 1000 - \frac{1000}{q^2}$$
$$R'(q) = 1000\left(1 - \frac{1}{q^2}\right)$$
$$R'(10) = 1000\left(1 - \frac{1}{10^2}\right)$$
$$R'(10) = 1000\left(\frac{99}{100}\right)$$
$$= 990$$

The marginal revenue is \$990.

55. $S(t) = 100 - 100t^{-1}$

$$S'(t) = -100(-1t^{-2})$$
$$= 100t^{-2}$$
$$= \frac{100}{t^2}$$

(a) $S'(1) = \frac{100}{(1)^2} = \frac{100}{1} = 100$

(b) $S'(10) = \frac{100}{(10)^2} = \frac{100}{100} = 1$

57. **(a)** 1982 when $t = 50$:

$$C(50) = 0.008446(50)^2 - 0.08924(50) + 1.254$$
$$= 17.907$$
$$\approx 18 \text{ cents}$$

2002 when $t = 70$:

$$C(70) = 0.008446(70)^2 - 0.08924(70) + 1.254$$
$$= 36.3926$$
$$\approx 36 \text{ cents}$$

(b) $C'(t) = 0.016892t - 0.08924$

1982 when $t = 50$:

$$C'(50) = 0.016892(50) - 0.08924$$
$$= 0.75536$$
$$\approx 0.755 \text{ cent per year}$$

2002 when $t = 70$:

$$C'(70) = 0.016892(70) - 0.08924$$
$$= 1.0932$$
$$\approx 1.09 \text{ cents per year}$$

(c) Using a graphing calculator, a cubic function that models the postage cost data is

$$C(t) = -0.0001549t^3 + 0.02699t^2 - 0.6484t + 3.212$$
$$C'(t) = -0.0004657t^2 + 0.05398t - 0.6484$$
$$C'(50) \approx 0.889 \text{ cent per year}$$
$$C'(70) \approx 0.853 \text{ cent per year}$$

59. $N(t) = 0.00437t^{3.2}$

$N'(t) = 0.013984t^{2.2}$

(a) $N'(5) \approx 0.4824$

(b) $N'(10) \approx 2.216$

61. $V(t) = -2159 + 1313t - 60.82t^2$

(a) $V(3) = -2159 + 1313(3) - 60.82(3)^2 = 1232.62 \text{ cm}^3$

(b) $V'(t) = 1313 - 121.64t$

$V'(3) = 1313 - 121.64(3) = 948.08 \text{ cm}^3/\text{yr}$

63. $v = 2.69l^{1.86}$

$$\frac{dv}{dl} = (1.86)2.69l^{1.86-1} \approx 5.00l^{0.86}$$

65. $t = 0.0588s^{1.125}$

(a) When $s = 1609$, $t \approx 238.1$ seconds, or 3 minutes, 58.1 seconds.

(b) $\frac{dt}{ds} = 0.0588(1.125s^{1.125-1}) = 0.06615s^{0.125}$

When $s = 100$, $\frac{dt}{ds} \approx 0.118$ sec/m.
At 100 meters, the fastest possible time increases by 0.118 seconds for each additional meter.

(c) Yes, they have been surpassed. In 2000, the world record in the mile stood at 3:43.13. (Ref:www.runnersworld.com)

67. $\text{BMI} = \frac{703w}{h^2}$

(a) $6'8'' = 80$ in.

$$\text{BMI} = \frac{703(250)}{80^2} \approx 27.5$$

(b) $\text{BMI} = \frac{703w}{80^2} = 24.9$ implies

$$w = \frac{24.9(80)^2}{703} \approx 227.$$

A 250-lb person needs to lose 23 pounds to get down to 227 lbs.

(c) If $f(h) = \frac{703(125)}{h^2} = 87{,}875h^{-2}$, then

$$f'(h) = 87{,}875(-2h^{-2-1}) = -175{,}750h^{-3} = -\frac{175{,}750}{h^3}$$

(d) $f'(65) = -\frac{175{,}750}{65^3} \approx -0.64$

For a 125-lb female with a height of 65 in. (5′5″), the BMI decreases by 0.64 for each additional inch of height.

(e) Sample Chart

ht/wt	140	160	180	200
60	27	31	35	39
65	23	27	30	33
70	20	23	26	29
75	17	20	22	25

69. $s(t) = 18t^2 - 13t + 8$

(a) $v(t) = s'(t) = 18(2t) - 13 + 0 = 36t - 13$

(b) $v(0) = 36(0) - 13 = -13$

$v(5) = 36(5) - 13 = 167$

$v(10) = 36(10) - 13 = 347$

71. $s(t) = -3t^3 + 4t^2 - 10t + 5$

(a) $v(t) = s'(t) = -3(3t^2) + 4(2t) - 10 + 0$
$= -9t^2 + 8t - 10$

(b) $v(0) = -9(0)^2 + 8(0) - 10 = -10$
$v(5) = -9(5)^2 + 8(5) - 10$
$= -225 + 40 - 10 = -195$
$v(10) = -9(10)^2 + 8(10) - 10$
$= -900 + 80 - 10 = -830$

73. $s(t) = -16t^2 + 64t$

(a) $v(t) = s'(t) = -16(2t) + 64 = -32t + 64$
$v(2) = -32(2) + 64 = -64 + 64 = 0$
$v(3) = -32(3) + 64 = -96 + 64 = -32$

The ball's velocity is 0 ft/sec after 2 seconds and -32 ft/sec after 3 seconds.

(b) As the ball travels upward, its speed decreases because of the force of gravity until, at maximum height, its speed is 0 ft/sec.

In part (a), we found that $v(2) = 0$.

It takes 2 seconds for the ball to reach its maximum height.

(c) $s(2) = -16(2)^2 + 64(2)$
$= -16(4) + 128$
$= -64 + 128$
$= 64$

It will go 64 ft high.

75. $y_1 = 4.13x + 14.63$
$y_2 = -0.033x^2 + 4.647x + 13.347$

(a) When $x = 5$, $y_1 \approx 35$ and $y_2 \approx 36$.

(b) $\frac{dy_1}{dx} = 4.13$

$\frac{dy_2}{dx} = 0.033(2x) + 4.647$
$= -0.066x + 4.647$

When $x = 5$, $\frac{dy_1}{dx} = 4.13$ and $\frac{dy_2}{dx} \approx 4.32$.
These values are fairly close and represent the rate of change of four years for a dog for one year of a human, for a dog that is actually 5 years old.

(c) With the first two points eliminated, the dog age increases in 2-year steps and the human age increases in 8-year steps, for a slope of 4. The equation has the form $y = 4x + b$.
A value of 16 for b makes the numbers come out right. $y = 4x + b$. For a dog of age $x = 5$ years or more, the equivaltent human age is given by $y = 4x + 16$.

12.2 Derivatives of Products and Quotients

Your Turn 1

$$y = (x^3 + 7)(4 - x^2)$$
$$\frac{dy}{dx} = (x^3 + 7)(-2x) + (4 - x^2)(3x^2)$$
$$= -2x^4 - 14x + 12x^2 - 3x^4$$
$$= -5x^4 + 12x^2 - 14x$$

Your Turn 2

$$f(x) = \frac{3x + 2}{5 - 2x}$$
$$f'(x) = \frac{(5 - 2x)(3) - (3x + 2)(-2)}{(5 - 2x)^2}$$
$$= \frac{19}{(5 - 2x)^2}$$

Your Turn 3

$$D_x\left[\frac{(5x-3)(2x+7)}{3x+7}\right]$$

$$= \frac{(3x+7)D_x[(5x-3)(2x+7)] - [(5x-3)(2x+7)]D_x(3x+7)}{(3x+7)^2}$$

$$= \frac{(3x+7)[(5x-3)(2)+(2x+7)(5)] - (10x^2+29x-21)(3)}{(3x+7)^2}$$

$$= \frac{(3x+7)(10x-6+10x+35) - (30x^2+87x-63)}{(3x+7)^2}$$

$$= \frac{60x^2+227x+203-30x^2-87x+63}{(3x+7)^2}$$

$$= \frac{30x^2+140x+266}{(3x+7)^2}$$

Your Turn 4

$$C(x) = \frac{4x+50}{x+2}$$

The marginal average cost is

$$\frac{d}{dx}\left[\frac{C(x)}{x}\right] = \frac{d}{dx}\left(\frac{4x+50}{x^2+2x}\right)$$

$$= \frac{(x^2+2x)(4)-(4x+50)(2x+2)}{(x^2+2x)^2}$$

$$= \frac{4x^2+8x-(8x^2+108x+100)}{(x^2+2x)^2}$$

$$= \frac{4x^2+8x-8x^2-108x-100}{(x^2+2x)^2}$$

$$= \frac{-4x^2-100x-100}{(x^2+2x)^2}$$

12.2 Exercises

1. $y = (3x^2+2)(2x-1)$

$$\frac{dy}{dx} = (3x^2+2)(2)+(2x-1)(6x)$$

$$= 6x^2+4+12x^2-6x$$

$$= 18x^2-6x+4$$

3. $y = (2x-5)^2$

$$= (2x-5)(2x-5)$$

$$\frac{dy}{dx} = (2x-5)(2)+(2x-5)(2)$$

$$= 4x-10+4x-10$$

$$= 8x-20$$

5. $k(t) = (t^2-1)^2 = (t^2-1)(t^2-1)$

$$k'(t) = (t^2-1)(2t)+(t^2-1)(2t)$$

$$= 2t^3-2t+2t^3-2t$$

$$= 4t^3-4t$$

7. $y = (x+1)(\sqrt{x}+2)$

$$= (x+1)(x^{1/2}+2)$$

$$\frac{dy}{dx} = (x+1)\left(\frac{1}{2}x^{-1/2}\right)+(x^{1/2}+2)(1)$$

$$= \frac{1}{2}x^{1/2}+\frac{1}{2}x^{-1/2}+x^{1/2}+2$$

$$= \frac{3}{2}x^{1/2}+\frac{1}{2}x^{-1/2}+2$$

$$\text{or } \frac{3x^{1/2}}{2}+\frac{1}{2x^{1/2}}+2$$

9. $p(y) = (y^{-1} + y^{-2})(2y^{-3} - 5y^{-4})$

$$\begin{aligned} p'(y) &= (y^{-1} + y^{-2})(-6y^{-4} + 20y^{-5}) \\ &\quad + (-y^{2} - 2y^{-3})(2y^{-3} - 5y^{-4}) \\ &= -6y^{-5} + 20y^{-6} - 6y^{-6} + 20y^{-7} \\ &\quad - 2y^{-5} + 5y^{-6} - 4y^{-6} + 10y^{-7} \\ &= -8y^{-5} + 15y^{-6} + 30y^{-7} \end{aligned}$$

11. $f(x) = \dfrac{6x + 1}{3x + 10}$

$$\begin{aligned} f'(x) &= \frac{(3x + 10)(6) - (6x + 1)(3)}{(3x + 10)^2} \\ &= \frac{18x + 60 - 18x - 3}{(3x + 10)^2} \\ &= \frac{57}{(3x + 10)^2} \end{aligned}$$

13. $y = \dfrac{5 - 3t}{4 + t}$

$$\begin{aligned} \frac{dy}{dx} &= \frac{(4 + t)(-3) - (5 - 3t)(1)}{(4 + t)^2} \\ &= \frac{-12 - 3t - 5 + 3t}{(4 + t)^2} \\ &= \frac{-17}{(4 + t)^2} \end{aligned}$$

15. $y = \dfrac{x^2 + x}{x - 1}$

$$\begin{aligned} \frac{dy}{dx} &= \frac{(x - 1)(2x + 1) - (x^2 + x)(1)}{(x - 1)^2} \\ &= \frac{2x^2 + x - 2x - 1 - x^2 - x}{(x - 1)^2} \\ &= \frac{x^2 - 2x - 1}{(x - 1)^2} \end{aligned}$$

17. $f(t) = \dfrac{4t^2 + 11}{t^2 + 3}$

$$\begin{aligned} f'(t) &= \frac{(t^2 + 3)(8t) - (4t^2 + 11)(2t)}{(t^2 + 3)^2} \\ &= \frac{8t^3 + 24t - 8t^3 - 22t}{(t^2 + 3)^2} \\ &= \frac{2t}{(t^2 + 3)^2} \end{aligned}$$

19.

$g(x) = \dfrac{x^2 - 4x + 2}{x^2 + 3}$

$$\begin{aligned} g'(x) &= \frac{(x^2 + 3)(2x - 4) - (x^2 - 4x + 2)(2x)}{(x^2 + 3)^2} \\ &= \frac{2x^3 - 4x^2 + 6x - 12 - 2x^3 + 8x^2 - 4x}{(x^2 + 3)^2} \\ &= \frac{4x^2 + 2x - 12}{(x^2 + 3)^2} \end{aligned}$$

21. $p(t) = \dfrac{\sqrt{t}}{t - 1}$

$$= \frac{t^{1/2}}{t - 1}$$

$$\begin{aligned} p'(t) &= \frac{(t - 1)\left(\frac{1}{2}t^{-1/2}\right) - t^{1/2}(1)}{(t - 1)^2} \\ &= \frac{\frac{1}{2}t^{1/2} - \frac{1}{2}t^{-1/2} - t^{1/2}}{(t - 1)^2} \\ &= \frac{-\frac{1}{2}t^{1/2} - \frac{1}{2}t^{-1/2}}{(t - 1)^2} \\ &= \frac{-\frac{\sqrt{t}}{2} - \frac{1}{2\sqrt{t}}}{(t - 1)^2} \text{ or } \frac{-t - 1}{2\sqrt{t}(t - 1)^2} \end{aligned}$$

23. $y = \dfrac{5x + 6}{\sqrt{x}} = \dfrac{5x + 6}{x^{1/2}} = 5x^{1/2} + 6x^{-1/2}$

$$\frac{dy}{dx} = \frac{5}{2}x^{-1/2} - 3x^{-3/2} \text{ or } \frac{5x - 6}{2x\sqrt{x}}$$

25. $h(z) = \dfrac{z^{2.2}}{z^{3.2} + 5}$

$$h'(z) = \frac{(z^{3.2} + 5)(2.2z^{1.2}) - z^{2.2}(3.2z^{2.2})}{(z^{3.2} + 5)^2} = \frac{2.2z^{4.4} + 11z^{1.2} - 3.2z^{4.4}}{(z^{3.2} + 5)^2} = \frac{-z^{4.4} + 11z^{1.2}}{(z^{3.2} + 5)^2}$$

27. $f(x) = \dfrac{(3x^2 + 1)(2x - 1)}{5x + 4}$

$$f'(x) = \frac{(5x + 4)[(3x^2 + 1)(2) + (6x)(2x - 1)] - (3x^2 + 1)(2x - 1)(5)}{(5x + 4)^2}$$

$$= \frac{(5x + 4)(18x^2 - 6x + 2) - (3x^2 + 1)(10x - 5)}{(5x + 4)^2}$$

$$= \frac{90x^3 - 30x^2 + 10x + 72x^2 - 24x + 8 - 30x^3 + 15x^2 - 10x + 5}{(5x + 4)^2}$$

$$= \frac{60x^3 + 57x^2 - 24x + 13}{(5x + 4)^2}$$

29. $h(x) = f(x)g(x)$

$h'(x) = f(x)g'(x) + g(x)f'(x)$

$h'(3) = f(3)g'(3) + g(3)f'(3) = 9(5) + 4(8) = 77$

31. In the first step, the two terms in the numerator are reversed. The correct work follows.

$$D_x\left(\frac{2x + 5}{x^2 - 1}\right)$$

$$= \frac{(x^2 - 1)(2) - (2x + 5)(2x)}{(x^2 - 1)^2}$$

$$= \frac{2x^2 - 2 - 4x^2 - 10x}{(x^2 - 1)^2}$$

$$= \frac{-2x^2 - 10x - 2}{(x^2 - 1)^2}$$

33. $f(x) = \dfrac{x}{x - 2}$, at (3,3)

$$m = f'(x) = \frac{(x - 2)(1) - x(1)}{(x - 2)^2} = -\frac{2}{(x - 2)^2}$$

At (3,3),

$$m = -\frac{2}{(3 - 2)^2} = -2,$$

Use the point-slope from.

$$y - 3 = -2(x - 3)$$
$$y = -2x + 9$$

35. **(a)** $f(x) = \dfrac{3x^3 + 6}{x^{2/3}}$

$$f'(x) = \frac{(x^{2/3})(9x^2) - (3x^3 + 6)(\frac{2}{3}x^{-1/3})}{(x^{2/3})^2} = \frac{9x^{8/3} - 2x^{8/3} - 4x^{-1/3}}{x^{4/3}} = \frac{7x^{8/3} - \frac{4}{x^{1/3}}}{x^{4/3}} = \frac{7x^3 - 4}{x^{5/3}}$$

(b) $f(x) = 3x^{7/3} + 6x^{-2/3}$

$$f'(x) = 3\left(\frac{7}{3}x^{4/3}\right) + 6\left(-\frac{2}{3}x^{-5/3}\right) = 7x^{4/3} - 4x^{-5/3}$$

(c) The derivatives are equivalent.

37. $f(x) = \dfrac{u(x)}{v(x)}$

$$f'(x) = \lim_{h \to 0} \frac{f(x+h) - f(x)}{h} = \lim_{h \to 0} \frac{\frac{u(x+h)}{v(x+h)} - \frac{u(x)}{v(x)}}{h} = \lim_{h \to 0} \frac{u(x+h)v(x) - u(x)v(x+h)}{hv(x+h)v(x)}$$

$$= \lim_{h \to 0} \frac{u(x+h)v(x) - u(x)v(x) + u(x)v(x) - u(x)v(x+h)}{hv(x+h)v(x)}$$

$$= \lim_{h \to 0} \frac{v(x)[u(x+h) - u(x)] - u(x)[v(x+h) - v(x)]}{hv(x+h)v(x)}$$

$$= \lim_{h \to 0} \frac{v(x)\frac{u(x+h)-u(x)}{h} - u(x)\frac{v(x+h)-v(x)}{h}}{v(x+h)v(x)} = \frac{v(x) \cdot u'(x) - u(x)v'(x)}{[v(x)]^2}$$

39. Graph the numerical derivative of $f(x) = (x^2 - 2)(x^2 - \sqrt{2})$ for x ranging from -2 to 2. The derivative crosses the x-axis at 0 and at approximately -1.307 and 1.307.

41. $C(x) = \dfrac{3x + 2}{x + 4}$

$$\overline{C}(x) = \frac{C(x)}{x} = \frac{3x + 2}{x^2 + 4x}$$

(a) $\overline{C}(10) = \dfrac{3(10) + 2}{10^2 + 4(10)} = \dfrac{32}{140} \approx 0.2286$ hundreds of dollars or \$22.86 per unit

(b) $\overline{C}(20) = \dfrac{3(20) + 2}{(20)^2 + 4(20)} = \dfrac{62}{480} \approx 0.1292$ hundreds of dollars or \$12.92 per unit

(c) $\overline{C}(x) = \dfrac{3x + 2}{x^2 + 4x}$ per unit

(d) $\overline{C}'(x) = \dfrac{(x^2 + 4x)(3) - (3x + 2)(2x + 4)}{(x^2 + 4x)^2} = \dfrac{3x^2 + 12x - 6x^2 - 12x - 4x - 8}{(x^2 + 4x)^2} = \dfrac{-3x^2 - 4x - 8}{(x^2 + 4x)^2}$

43. $M(d) = \dfrac{100d^2}{3d^2 + 10}$

(a) $M'(d) = \dfrac{(3d^2 + 10)(200d) - (100d^2)(6d)}{(3d^2 + 10)^2} = \dfrac{600d^3 + 2000d - 600d^3}{(3d^2 + 10)^2} = \dfrac{2000d}{(3d^2 + 10)^2}$

(b) $M'(2) = \dfrac{2000(2)}{[3(2)^2 + 10]^2} = \dfrac{4000}{484} \approx 8.3$

This means the new employee can assemble about 8.3 additional bicycles per day after 2 days of training.

$$M'(5) = \frac{2000(5)}{[3(5)^2 + 10]^2} = \frac{10{,}000}{7225} \approx 1.4$$

This means the new employee can assemble about 1.4 additional bicycles per day after 5 days of training.

45. $\bar{C}(x) = \dfrac{C(x)}{x}$

Let $u(x) = C(x)$, with $u'(x) = C'(x)$

Let $v(x) = x$ with $v'(x) = 1$. Then, by the quotient rule,

$$\bar{C}(x) = \frac{v(x) \cdot u'(x) - u(x) \cdot v'(x)}{[v(x)]^2} = \frac{x \cdot C'(x) - C(x) \cdot 1}{x^2} = \frac{xC'(x) - C(x)}{x^2}$$

47. Let $C(t)$ be the cost as a function of time and $q(t)$ be the quantity as a function of time.

Then $\bar{C}(t) = \frac{C(t)}{q(t)}$ is the revenue as a function of time. Let $t = t_1$ represent last month.

$$\begin{aligned}\bar{C}'(t) &= \frac{q(t)C'(t) - C(t)q'(t)}{[g(t)]^2}\\ \bar{C}'(t_1) &= \frac{q(t_1)C'(t_1) - C(t_1)q'(t_1)}{[g(t_1)]^2}\\ &= \frac{(12{,}500)(1200) - (27{,}000)(350)}{(12{,}500)^2}\\ &= 0.03552\end{aligned}$$

The average cost is increasing at a rate of \$0.03552 per gallon per month.

49. $f(x) = \dfrac{Kx}{A + x}$

(a) $f'(x) = \dfrac{(A + x)K - Kx(1)}{(A + x)^2}$

$f'(x) = \dfrac{AK}{(A + x)^2}$

(b) $f'(A) = \dfrac{AK}{(A + A)^2}$

$= \dfrac{AK}{4A^2} = \dfrac{K}{4A}$

51. $R(w) = \dfrac{30(w - 4)}{w - 1.5}$

(a) $R(5) = \dfrac{30(5 - 4)}{5 - 1.5}$

≈ 8.57 min

(b) $R(7) = \dfrac{30(7 - 4)}{7 - 1.5}$

≈ 16.36 min

(c)

$$\begin{aligned}R'(w) &= \frac{(w - 1.5)(30) - 30(w - 4)(1)}{(w - 1.5)^2}\\ &= \frac{30w - 45 - 30w + 120}{(w - 1.5)^2}\\ &= \frac{75}{(w - 1.5)^2}\end{aligned}$$

$$R'(5) = \frac{75}{(5 - 1.5)^2} \approx 6.12 \ \frac{\text{min}^2}{\text{kcal}}$$

$$R'(7) = \frac{75}{(7 - 1.5)^2} \approx 2.48 \ \frac{\text{min}^2}{\text{kcal}}$$

53.
$$f(t) = \frac{90t}{99t - 90}$$
$$f'(t) = \frac{(99t - 90)(90) - (90t)(99)}{(99t - 90)^2}$$
$$= \frac{-8100}{(99t - 90)^2}$$

(a)
$$f'(1) = \frac{-8100}{(99 - 90)^2} = \frac{-8100}{9^2}$$
$$= \frac{-8100}{81} = -100 \text{ facts/hr}$$

(b)
$$f'(10) = \frac{-8100}{[99(10) - 90]^2} = \frac{-8100}{(900)^2}$$
$$= \frac{-8100}{810{,}000}$$
$$= -\frac{1}{100} \text{ or } -0.01 \text{ facts/hr}$$

12.3 The Chain Rule

Your Turn 1

Let $f(x) = 2x - 1$ and $g(x) = \sqrt{3x + 5}$.

$$g(0) = \sqrt{3 \cdot 0 + 5} = \sqrt{5}$$
$$f[g(0)] = 2\sqrt{5} - 1$$

$$f(0) = 2 \cdot 0 - 1 = -1$$
$$g[f(0)] - \sqrt{3(-1) + 5} = \sqrt{2}$$

Your Turn 2

Let $f(x) = 2x - 3$ and $g(x) = x^2 + 1$.

$$g[f(x)] = g(2x - 3)$$
$$= (2x - 3)^2 + 1$$
$$= 4x^2 - 12x + 9 + 1$$
$$= 4x^2 - 12x + 10$$

Your Turn 3

Write $h(x) = (2x - 3)^3$ in the form $h(x) = f[g(x)]$.

One possible answer is $h(x) = f[g(x)]$ where $g(x) = 2x - 3$ and $f(x) = x^3$.

Your Turn 4

$$y = (5x^2 - 6x)^{-2}$$
$$\frac{dy}{dx} = -2(5x^2 - 6x)^{-3} \cdot (10x - 6)$$
$$= \frac{-2(10x - 6)}{(5x^2 - 6x)^3} = \frac{-20x + 12}{(5x^2 - 6x)^3}$$

Your Turn 5

$$D_x[(x^2 - 7)^{10}] = 10(x^2 - 7)^9(2x) = 20x(x^2 - 7)^9$$

Your Turn 6

$$y = x^2(5x - 1)^3$$
$$\frac{dy}{dx} = x^2 \cdot 3(5x - 1)^2(5) + (5x - 1)^3 \cdot 2x$$
$$= 15x^2(5x - 1)^2 + 2x(5x - 1)^3$$
$$= x(5x - 1)^2[15x + 2(5x - 1)]$$
$$= x(5x - 1)^2(25x - 2)$$

Your Turn 7

$$D_x\left[\frac{(4x - 1)^3}{x + 3}\right] = \frac{(x + 3)[3(4x - 1)^2(4)] - (4x - 1)^3(1)}{(x + 3)^2}$$
$$= \frac{12(x + 3)(4x - 1)^2 - (4x - 1)^3}{(x + 3)^2}$$
$$= \frac{(4x - 1)^2[12(x + 3) - (4x - 1)]}{(x + 3)^2}$$
$$= \frac{(4x - 1)^2(8x + 37)}{(x + 3)^2}$$

12.3 Exercises

In Exercises 1 through 6, $f(x) = 5x^2 - 2x$ and $g(x) = 8x + 3$.

1.
$$g(2) = 8(2) + 3 = 19$$
$$f[g(2)] = f[19]$$
$$= 5(19)^2 - 2(19)$$
$$= 1805 - 38 = 1767$$

3.
$$f(2) = 5(2)^2 - 2(2)$$
$$= 20 - 4 = 16$$
$$g[f(2)] = g[16]$$
$$= 8(16) + 3$$
$$= 128 + 3 = 131$$

5. $g(k) = 8k + 3$

$$\begin{aligned} f[g(k)] &= f[8k + 3] \\ &= 5(8k + 3)^2 - 2(8k + 3) \\ &= 5(64k^2 + 48k + 9) - 16k - 6 \\ &= 320k^2 + 224k + 39 \end{aligned}$$

7. $f(x) = \frac{x}{8} + 7; g(x) = 6x - 1$

$$\begin{aligned} f[g(x)] &= \frac{6x - 1}{8} + 7 \\ &= \frac{6x - 1}{8} + \frac{56}{8} \\ &= \frac{6x + 55}{8} \end{aligned}$$

$$\begin{aligned} g[f(x)] &= 6\left[\frac{x}{8} + 7\right] - 1 \\ &= \frac{6x}{8} + 42 - 1 \\ &= \frac{3x}{4} + 41 \\ &= \frac{3x}{4} + \frac{164}{4} \\ &= \frac{3x + 164}{4} \end{aligned}$$

9. $f(x) = \frac{1}{x}; g(x) = x^2$

$$f[g(x)] = \frac{1}{x^2}$$

$$\begin{aligned} g[f(x)] &= \left(\frac{1}{x}\right)^2 \\ &= \frac{1}{x^2} \end{aligned}$$

11. $f(x) = \sqrt{x + 2}; g(x) = 8x^2 - 6$

$$\begin{aligned} f[g(x)] &= \sqrt{(8x^2 - 6) + 2} \\ &= \sqrt{8x^2 - 4} \end{aligned}$$

$$\begin{aligned} g[f(x)] &= 8(\sqrt{x + 2})^2 - 6 \\ &= 8x + 16 - 6 \\ &= 8x + 10 \end{aligned}$$

13. $f(x) = \sqrt{x + 1}; g(x) = \frac{-1}{x}$

$$\begin{aligned} f[g(x)] &= \sqrt{\frac{-1}{x} + 1} \\ &= \sqrt{\frac{x - 1}{x}} \end{aligned}$$

$$g[f(x)] = \frac{-1}{\sqrt{x + 1}}$$

15. $y = (5 - x^2)^{3/5}$

If $f(x) = x^{3/5}$ and $g(x) = 5 - x^2$, then

$$y = f[g(x)] = (5 - x^2)^{3/5}.$$

17. $y = -\sqrt{13 + 7x}$

If $f(x) = -\sqrt{x}$ and

$g(x) = 13 + 7x,$

then $y = f[g(x)] = -\sqrt{13 + 7x}.$

19. $y = (x^2 + 5x)^{1/3} - 2(x^2 + 5x)^{2/3} + 7$

If $f(x) = x^{1/3} - 2x^{2/3} + 7$ and

$g(x) = x^2 + 5x,$

then

$$y = f[g(x)] = (x^2 + 5x)^{1/3} - 2(x^2 + 5x)^{2/3} + 7.$$

21. $y = (8x^4 - 5x^2 + 1)^4$

Let $f(x) = x^4$ and $g(x) = 8x^4 - 5x^2 + 1.$

Then $(8x^4 - 5x^2 + 1)^4 = f[g(x)].$

Use the alternate form of the chain rule.

$$\begin{aligned} \frac{dy}{dx} &= f'[g(x)] \cdot g'(x) \\ f'(x) &= 4x^3 \\ f'[g(x)] &= 4[g(x)]^3 = 4(8x^4 - 5x^2 + 1)^3 \\ g'(x) &= 32x^3 - 10x \\ \frac{dy}{dx} &= 4(8x^4 - 5x^2 + 1)^3(32x^3 - 10x) \end{aligned}$$

23. $k(x) = -2(12x^2 + 5)^{-6}$

Use the generalized power rule with $u = 12x^2 + 5$, $n = -6$, and $u' = 24x$.

$$\begin{aligned} k'(x) &= -2[-6(12x^2 + 5)^{-6-1} \cdot 24x] \\ &= -2[-144x(12x^2 + 5)^{-7}] \\ &= 288x(12x^2 + 5)^{-7} \end{aligned}$$

25. $s(t) = 45(3t^3 - 8)^{3/2}$

Use the generalized power rule with $u = 3t^3 - 8$, $n = \frac{3}{2}$, and $u' = 9t^2$.

$$\begin{aligned} s'(t) &= 45\left[\frac{3}{2}(3t^3 - 8)^{1/2} \cdot 9t^2\right] \\ &= 45\left[\frac{27}{2}t^2(3t^3 - 8)^{1/2}\right] \\ &= \frac{1215}{2}t^2(3t^3 - 8)^{1/2} \end{aligned}$$

27. $g(t) = -3\sqrt{7t^3 - 1}$
$= -3\sqrt{(7t^3 - 1)^{1/2}}$

Use generalized power rule with $u = 7t^3 - 1$, $n = \frac{1}{2}$, and $u' = 21t^2$.

$$\begin{aligned} g'(t) &= -3\left[\frac{1}{2}(7t^3 - 1)^{-1/2} \cdot 21t^2\right] \\ &= -3\left[\frac{21}{2}t^2(7t^3 - 1)^{-1/2}\right] \\ &= \frac{-63}{2}t^2 \cdot \frac{1}{(7t^3 - 1)^{1/2}} \\ &= \frac{-63t^2}{2\sqrt{7t^3 - 1}} \end{aligned}$$

29. $m(t) = -6t(5t^4 - 1)^4$

Use the product rule and the power rule.

$$\begin{aligned} m'(t) &= -6t[4(5t^4 - 1)^3 \cdot 20t^3] + (5t^4 - 1)^4(-6) \\ &= -480t^4(5t^4 - 1)^3 - 6(5t^4 - 1)^4 \\ &= -6(5t^4 - 1)^3[80t^4 + (5t^4 - 1)] \\ &= -6(5t^4 - 1)^3(85t^4 - 1) \end{aligned}$$

31. $y = (3x^4 + 1)^4(x^3 + 4)$

Use the product rule and the power rule.

$$\begin{aligned} \frac{dy}{dx} &= (3x^4 + 1)^4(3x^2) + (x^3 + 4)[4(3x^4 + 1)^3 \cdot 12x^3] \\ &= 3x^2(3x^4 + 1)^4 + 48x^3(x^3 + 4)(3x^4 + 1)^3 \\ &= 3x^2(3x^4 + 1)^3[3x^4 + 1 + 16x(x^3 + 4)] \\ &= 3x^2(3x^4 + 1)^3(3x^4 + 1 + 16x^4 + 64) \\ &= 3x^2(3x^4 + 1)^3(19x^4 + 64x + 1) \end{aligned}$$

33. $q(y) = 4y^2(y^2 + 1)^{5/4}$

Use the product rule and the power rule.

$$\begin{aligned} q'(y) &= 4y^2 \cdot \frac{5}{4}(y^2 + 1)^{1/4}(2y) + 8y(y^2 + 1)^{5/4} \\ &= 10y^3(y^2 + 1)^{1/4} + 8y(y^2 + 1)^{5/4} \\ &= 2y(y^2 + 1)^{1/4}[5y^2 + 4(y^2 + 1)^{4/4}] \\ &= 2y(y^2 + 1)^{1/4}(9y^2 + 4) \end{aligned}$$

35. $y = \frac{-5}{(2x^3 + 1)^2} = -5(2x^3 + 1)^{-2}$

$$\begin{aligned} \frac{dy}{dx} &= -5[-2(2x^3 + 1)^{-3} \cdot 6x^2] \\ &= -5[-12x^2(2x^3 + 1)^{-3}] \\ &= 60x^2(2x^3 + 1)^{-3} \\ &= \frac{60x^2}{(2x^3 + 1)^3} \end{aligned}$$

37. $r(t) = \frac{(5t - 6)^4}{3t^2 + 4}$

$$\begin{aligned} r'(t) &= \frac{(3t^2 + 4)[4(5t - 6)^3 \cdot 5] - (5t - 6)^4(6t)}{(3t^2 + 4)^2} \\ &= \frac{20(3t^2 + 4)(5t - 6)^3 - 6t(5t - 6)^4}{(3t^2 + 4)^2} \\ &= \frac{2(5t - 6)^3[10(3t^2 + 4) - 3t(5t - 6)]}{(3t^2 + 4)^2} \\ &= \frac{2(5t - 6)^3(30t^2 + 40 - 15t^2 + 18t)}{(3t^2 + 4)^2} \\ &= \frac{2(5t - 6)^3(15t^2 + 18t + 40)}{(3t^2 + 4)^2} \end{aligned}$$

39. $y = \frac{3x^2 - x}{(2x-1)^5}$

$$\frac{dy}{dx} = \frac{(2x-1)^5(6x-1) - (3x^2 - x)[5(2x-1)^4 \cdot 2]}{[(2x-1)^5]^2}$$

$$= \frac{(2x-1)^5(6x-1) - 10(3x^2 - x)(2x-1)^4}{(2x-1)^{10}}$$

$$= \frac{(2x-1)^4[(2x-1)(6x-1) - 10(3x^2 - x)]}{(2x-1)^{10}}$$

$$= \frac{12x^2 - 2x - 6x + 1 - 30x^2 + 10x}{(2x-1)^6}$$

$$= \frac{-18x^2 + 2x + 1}{(2x-1)^6}$$

43. **(a)** $D_x(f[g(x)])$ *at* $x = 1$

$$= f'[g(1)] \cdot g'(1)$$

$$= f'(2) \cdot \left(\frac{2}{7}\right)$$

$$= -7\left(\frac{2}{7}\right) = -2$$

(b) $D_x(f[g(x)])$ *at* $x = 2$

$$= f'[g(2)] \cdot g'(2)$$

$$= f'(3) \cdot \left(\frac{3}{7}\right)$$

$$= -8\left(\frac{3}{7}\right) = -\frac{24}{7}$$

45. $f(x) = \sqrt{x^2 + 16};\ x = 3$

$$f(x) = (x^2 + 16)^{1/2}$$

$$f'(x) = \frac{1}{2}(x^2 + 16)^{-1/2}(2x)$$

$$f'(x) = \frac{x}{\sqrt{x^2 + 16}}$$

$$f'(3) = \frac{3}{\sqrt{3^2 + 16}} = \frac{3}{5}$$

$$f(3) = \sqrt{3^2 + 16} = 5$$

We use $m = \frac{3}{5}$ and the point $P(3, 5)$in the point-slope form

$$y - 5 = \frac{3}{5}(x - 3)$$

$$y - 5 = \frac{3}{5}x - \frac{9}{5}$$

$$y = \frac{3}{5}x + \frac{16}{5}$$

47. $f(x) = x(x^2 - 4x + 5)^4;\ x = 2$

$$f'(x) = x \cdot 4(x^2 - 4x + 5)^3 \cdot (2x - 4) + 1 \cdot (x^2 - 4x + 5)^4$$

$$= (x^2 - 4x + 5)^3 \cdot [4x(2x - 4) + (x^2 - 4x + 5)]$$

$$= (x^2 - 4x + 5)^3(9x^2 - 20x + 5)$$

$$f'(2) = (1)^3(1) = 1$$

$$f(2) = 2(1)^4 = 2$$

We use $m = 1$ and the point $P(2, 2)$.

$$y - 2 = 1(x - 2)$$

$$y - 2 = x - 2$$

$$y = x$$

49. $f(x) = \sqrt{x^3 - 6x^2 + 9x + 1}$

$$f(x) = (x^3 - 6x^2 + 9x + 1)^{1/2}$$

$$f'(x) = \frac{1}{2}(x^3 - 6x^2 + 9x + 1)^{-1/2} \cdot (3x^2 - 12x + 9)$$

$$f'(x) = \frac{3(x^2 - 4x + 3)}{2\sqrt{x^3 - 6x^2 + 9x + 1}}$$

If the tangent line is horizontal, its slope is zero and $f'(x) = 0$.

$$\frac{3(x^2 - 4x + 3)}{2\sqrt{x^3 - 6x^2 + 9x + 1}} = 0$$

$$3(x^2 - 4x + 3) = 0$$

$$3(x - 1)(x - 3) = 0$$

$$x = 1 \text{ or } x = 3$$

The tangent line is horizontal $x = 1$ and $x = 3$.

53. $D(p) = \frac{-p^2}{100} + 500;\ p(c) = 2c - 10$

The demand in terms of the cost is

$$\begin{aligned} D(c) &= D[p(c)] \\ &= \frac{-(2c-10)^2}{100} + 500 \\ &= \frac{-4(c-5)^2}{100} + 500 \\ &= \frac{-c^2 + 10c - 25}{25} + 500 \\ &= \frac{-c^2 + 10c - 25 + 12{,}500}{25} \\ &= \frac{-c^2 + 10c + 12{,}475}{25}. \end{aligned}$$

55. $A = 1500\left(1 + \frac{r}{36{,}500}\right)^{1825}$

$\frac{dA}{dr}$ is the rate of change of A with respect to r.

$$\begin{aligned} \frac{dA}{dr} &= 1500(1825)\left(1 + \frac{r}{36{,}500}\right)^{1824}\left(\frac{1}{36{,}500}\right) \\ &= 75\left(1 + \frac{r}{36{,}500}\right)^{1824} \end{aligned}$$

(a) For $r = 6\%$,

$$\frac{dA}{dr} = 75\left(1 + \frac{6}{36{,}500}\right)^{1824} = \$101.22.$$

(b) For $r = 8\%$,

$$\frac{dA}{dr} = 75\left(1 + \frac{8}{36{,}500}\right)^{1824} = \$111.86.$$

(c) For $r = 9\%$,

$$\frac{dA}{dr} = 75\left(1 + \frac{9}{36{,}500}\right)^{1824} = \$117.59.$$

57. $V = \frac{60{,}000}{1 + 0.3t + 0.1t^2}$

The rate of change of the value is

$$\begin{aligned} V'(t) &= \frac{(1 + 0.3t + 0.1t^2)(0) - 60{,}000(0.3 + 0.2t)}{(1 + 0.3t + 0.1t^2)^2} \\ &= \frac{-60{,}000(0.3 + 0.2t)}{(1 + 0.3t + 0.1t^2)^2}. \end{aligned}$$

(a) 2 years after purchase, the rate of change in the value is

$$\begin{aligned} V'(2) &= \frac{-60{,}000[0.3 + 0.2(2)]}{[1 + 0.3(2) + 0.1(2)^2]^2} \\ &= \frac{-60{,}000(0.3 + 0.4)}{(1 + 0.6 + 0.4)^2} \\ &= \frac{-42{,}000}{4} \\ &= -\$10{,}500. \end{aligned}$$

(b) 4 years after purchase, the rate of change in the value is

$$\begin{aligned} V'(4) &= \frac{-60{,}000[0.3 + 0.2(4)]}{[1 + 0.3(4) + 0.1(4)^2]^2} \\ &= \frac{-66{,}000}{14.44} \\ &= -\$4570.64. \end{aligned}$$

59. $P(x) = 2x^2 + 1;\ x = f(a) = 3a + 2$

$$\begin{aligned} P[f(a)] &= 2(3a+2)^2 + 1 \\ &= 2(9a^2 + 12a + 4) + 1 \\ &= 18a^2 + 24a + 9 \end{aligned}$$

61. **(a)** $r(t) = 2t;\ A(r) = \pi r^2$

$$\begin{aligned} A[r(t)] &= \pi(2t)^2 \\ &= 4\pi t^2 \end{aligned}$$

$A = 4\pi t^2$ gives the area of the pollution in terms of the time since the pollutants were first emitted.

(b) $D_t A[r(t)] = 8\pi t$

$D_t A[r(4)] = 8\pi(4) = 32\pi$

At 12 P.M., the area of pollution is changing at the rate of 32π mi²/hr.

63. $C(t) = \frac{1}{2}(2t+1)^{-1/2}$

$$\begin{aligned} C'(t) &= \frac{1}{2}\left(-\frac{1}{2}\right)(2t+1)^{-3/2}(2) \\ &= -\frac{1}{2}(2t+1)^{-3/2} \end{aligned}$$

(a) $C'(0) = -\frac{1}{2}[2(0)+1]^{-3/2}$

$$\begin{aligned} &= -\frac{1}{2} \\ &= -0.5 \end{aligned}$$

(b) $C'(4) = -\frac{1}{2}[2(4)+1]^{-3/2}$
$= -\frac{1}{2}(9)^{-3/2}$
$= \frac{-1}{2} \cdot \frac{1}{(\sqrt{9})^3}$
$= -\frac{1}{54}$
≈ -0.02

(c) $C'(7.5) = -\frac{1}{2}[2(7.5)+1]^{-3/2}$
$= -\frac{1}{2}(16)^{-3/2}$
$= -\frac{1}{2}\left(\frac{1}{(\sqrt{16})^3}\right)$
$= -\frac{1}{128}$
≈ -0.008

(d) C is always decreasing because

$$C' = -\tfrac{1}{2}(2t+1)^{-3/2}$$

is always negative for $t \geq 0$.

(The amount of calcium in the bloodstream will continue to decrease over time.)

65. $V(r) = \frac{4}{3}\pi r^3,\ S(r) = 4\pi r^2,\ r(t) = 6 - \frac{3}{17}t$

(a) $r(t) = 0$ when $6 - \frac{3}{17}t = 0;$

$t = \frac{17(6)}{3} = 34$ min.

(b) $\frac{dV}{dr} = 4\pi r^2, \frac{dS}{dr} = 8\pi r, \frac{dr}{dt} = -\frac{3}{17}$

$\frac{dV}{dt} = \frac{dV}{dr} \cdot \frac{dr}{dt} = -\frac{12}{17}\pi r^2$
$= -\frac{12}{17}\pi\left(6 - \frac{3}{17}t\right)^2$

$\frac{dS}{dt} = \frac{dS}{dr} \cdot \frac{dr}{dt} = -\frac{24}{17}\pi r$
$= -\frac{24}{17}\pi\left(6 - \frac{3}{17}t\right)$

When $t = 17$,

$\frac{dV}{dt} = -\frac{12}{17}\pi\left[6 - \frac{3}{17}(17)\right]^2$
$= -\frac{108}{17}\pi$ mm^3/min

$\frac{dS}{dt} = -\frac{24}{17}\pi\left[6 - \frac{3}{17}(17)\right]$
$= -\frac{72}{17}\pi$ mm^2/min

At $t = 17$ minutes, the volume is decreasing by $\frac{108}{17}\pi$ mm^3 per minute and the surface area is decreasing by $\frac{72}{17}\pi$ mm^2 per minute.

67. (a) $y = ((x^3)^2)^2$
$\frac{dy}{dx} = 2((x^3)^2) \cdot \frac{d}{dx}((x^3)^2)$
$= 2x^6 \cdot 2(x^3) \cdot \frac{d}{dx}(x^3)$
$= 2x^6 \cdot 2x^3 \cdot 3x^2$
$= 12x^{11}$

(b) $y = ((x^3)^2)^2 = x^{12}$
$\frac{dy}{dx} = 12x^{12-1} = 12x^{11}$

12.4 Derivatives of Exponential Functions

Your Turn 1

(a) $y = 4^{3x}$
$\frac{dy}{dx} = (\ln 4) \cdot 4^{3x} \cdot 3 = 3(\ln 4)4^{3x}$

(b) $y = e^{7x^3+5}$
$\frac{dy}{dx} = e^{7x^3+5} \cdot 21x^2 = 21x^2e^{7x^3+5}$

Your Turn 2

$y = (x^2+1)^2e^{2x}$
$\frac{dy}{dx} = (x^2+1) \cdot e^{2x} \cdot 2 + e^{2x} \cdot 2(x^2+1) \cdot 2x$
$= 2(x^2+1)e^{2x}(x^2+2x+1)$
$= 2e^{2x}(x^2+1)(x+1)^2$

Your Turn 3

$$f(x) = \frac{100}{5 + 2e^{-0.01x}}$$

$$f'(x) = \frac{(5 + 2e^{-0.01x}) \cdot 0 - 100 \cdot 2e^{-0.01x} \cdot (-0.01)}{(5 + 2e^{-0.01x})^2}$$

$$= \frac{2e^{-0.01x}}{(5 + 2e^{-0.01x})^2}$$

Your Turn 4

$$Q(t) = 100e^{-0.421t}$$

$$\frac{dQ}{dt} = 100 \cdot e^{-0.421t}(-0.421)$$

$$= -42.1e^{-0.421t}$$

After 2 years ($t = 2$), the rate of change of the quantity present is

$$\frac{dQ}{dt} = -42.1e^{-0.421(2)}$$

$$= -42.1e^{-0.842}$$

$$\approx -18.1 \text{ grams per year}$$

12.4 Exercises

1. $y = e^{4x}$

Let $g(x) = 4x$,

with $g'(x) = 4$.

$$\frac{dy}{dx} = 4e^{4x}$$

3. $y = -8e^{3x}$

$$\frac{dy}{dx} = -8(3e^{3x}) = -24e^{3x}$$

5. $y = -16e^{2x+1}$

$g(x) = 2x + 1$

$g'(x) = 2$

$$\frac{dy}{dx} = -16(2e^{2x+1}) = -32e^{2x+1}$$

7. $y = e^{x^2}$

$g(x) = x^2$

$g'(x) = 2x$

$$\frac{dy}{dx} = 2xe^{x^2}$$

9. $y = 3e^{2x^2}$

$g(x) = 2x^2$

$g'(x) = 4x$

$$\frac{dy}{dx} = 3\left(4xe^{2x^2}\right)$$

$$= 12xe^{2x^2}$$

11. $y = 4e^{2x^2-4}$

$g(x) = 2x^2 - 4$

$g'(x) = 4x$

$$\frac{dy}{dx} = 4\left[(4x)e^{2x^2-4}\right]$$

$$= 16xe^{2x^2-4}$$

13. $y = xe^x$

Use the product rule.

$$\frac{dy}{dx} = xe^x + e^x \cdot 1 = e^x(x + 1)$$

15. $y = (x + 3)^2 e^{4x}$

Use the product rule.

$$\frac{dy}{dx} = (x + 3)^2(4)e^{4x} + e^{4x} \cdot 2(x + 3)$$

$$= 4(x + 3)^2 e^{4x} + 2(x + 3)e^{4x}$$

$$= 2(x + 3)e^{4x}[2(x + 3) + 1]$$

$$= 2(x + 3)(2x + 7)e^{4x}$$

17. $y = \dfrac{x^2}{e^x}$

Use the quotient rule.

$$\frac{dy}{dx} = \frac{e^x(2x) - x^2e^x}{(e^x)^2}$$

$$= \frac{xe^x(2 - x)}{e^{2x}}$$

$$= \frac{x(2 - x)}{e^x}$$

19. $y = \dfrac{e^x + e^{-x}}{x}$

$$\frac{dy}{dx} = \frac{x(e^x - e^{-x}) - (e^x + e^{-x})}{x^2}$$

21. $p = \dfrac{10{,}000}{9 + 4e^{-0.2t}}$

$$\frac{dp}{dt} = \frac{(9 + 4e^{-0.2t}) \cdot 0 - 10{,}000[0 + 4(-0.2)e^{-0.2t}]}{(9 + 4e^{-0.2t})^2}$$

$$= \frac{8000e^{-0.2t}}{(9 + 4e^{-0.2t})^2}$$

23. $f(z) = \left(2z + e^{-z^2}\right)^2$

$$f'(z) = 2\left(2z + e^{-z^2}\right)^1\left(2 - 2ze^{-z^2}\right)$$

$$= 4\left(2z + e^{-z^2}\right)\left(1 - ze^{-z^2}\right)$$

25. $y = 7^{3x+1}$

Let $g(x) = 3x + 1$, with $g'(x) = 3$. Then

$$\frac{dy}{dx} = (\ln 7)(7^{3x+1}) \cdot 3 = 3(\ln 7)7^{3x+1}$$

27. $y = 3 \cdot 4^{x^2+2}$

Let $g(x) = x^2 + 2$, with $g'(x) = 2x$. Then

$$\frac{dy}{dx} = 3(\ln 4)4^{x^2+2} \cdot 2x = 6x(\ln 4)4^{x^2+2}$$

29. $s = 2 \cdot 3^{\sqrt{t}}$

Let $g(t) = \sqrt{t}$, with $g'(t) = \frac{1}{2\sqrt{t}}$. Then

$$\frac{ds}{dt} = 2(\ln 3)3^{\sqrt{t}} \cdot \frac{1}{2\sqrt{t}}$$

$$= \frac{(\ln 3)3^{\sqrt{t}}}{\sqrt{t}}$$

31. $y = \dfrac{te^t + 2}{e^{2t} + 1}$

Use the quotient rule and product rule.

$$\frac{dy}{dt} = \frac{(e^{2t} + 1)(te^t + e^t \cdot 1) - (te^t + 2)(2e^{2t})}{(e^{2t} + 1)^2}$$

$$= \frac{(e^{2t} + 1)(te^t + e^t) - (te^t + 2)(2e^{2t})}{(e^{2t} + 1)^2}$$

$$= \frac{te^{3t} + e^{3t} + te^t + e^t - 2te^{3t} - 4e^{2t}}{(e^{2t} + 1)^2}$$

$$= \frac{-te^{3t} + e^{3t} + te^t + e^t - 4e^{2t}}{(e^{2t} + 1)^2}$$

$$= \frac{(1 - t)e^{3t} - 4e^{2t} + (1 + t)e^t}{(e^{2t} + 1)^2}$$

33. $f(x) = e^{x\sqrt{3x+2}}$

Let $g(x) = x\sqrt{3x + 2}$.

$$g'(x) = 1 \cdot \sqrt{3x + 2} + x\left(\frac{3}{2\sqrt{3x + 2}}\right)$$

$$= \sqrt{3x + 2} + \frac{3x}{2\sqrt{3x + 2}}$$

$$= \frac{2(3x + 2)}{2\sqrt{3x + 2}} + \frac{3x}{2\sqrt{3x + 2}}$$

$$= \frac{9x + 4}{2\sqrt{3x + 2}}$$

$$f'(x) = e^{x\sqrt{3x+2}} \cdot \left(\frac{9x + 4}{2\sqrt{3x + 2}}\right)$$

35. $y = y_o e^{kt}$

$$\frac{dy}{dx} = \frac{d}{dt}\left[y_o e^{kt}\right] = y_o k e^{kt}$$

$$= k(y_o e^{kt})$$

$$= ky$$

37. Graph the function $y = e^x$.

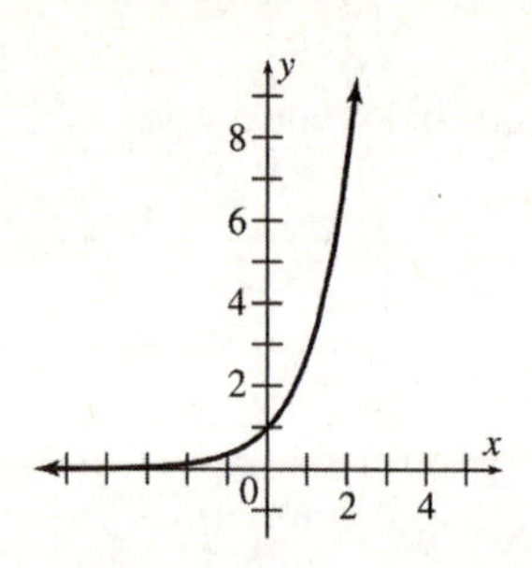

Sketch the lines tangent to the graph at $x = -1$, 0, 1, 2.

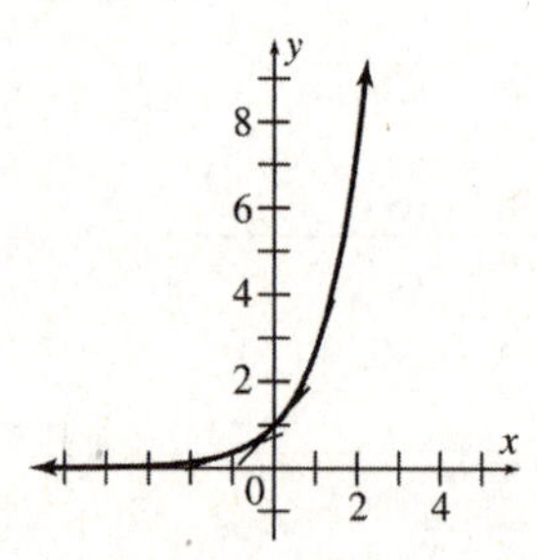

Estimate the slopes of the tangent lines at these points.

At $x = -1$ the slope is a little steeper than $\frac{1}{3}$ or approximately $0.\overline{3}$.

At $x = 0$ the slope is 1.

At $x = 1$ the slope is a little steeper than $\frac{5}{2}$ or 2.5.

At $x = 2$ the slope is a little steeper than $7\frac{1}{3}$ or $7.\overline{3}$.

Note that $e^{-1} \approx 0.36787944$, $e^0 = 1$, $e^1 = e \approx 2.7182812$, and $e^2 \approx 7.3890561$. The values are close enough to the slopes of the tangent lines to convince us that $\frac{de^x}{dx} = e^x$.

39. $C(x) = \sqrt{900 - 800 \cdot 1.1^{-x}}$

$$C(x) = [900 - 800(1.1^{-x})]^{1/2}$$

$$= \frac{1}{2}[900 - 800(1.1^{-x})]^{-1/2} \cdot [-800(\ln 1.1)(1.1^{-x})(-1)]$$

$$C'(x) = \frac{(400 \ln 1.1)(1.1^{-x})}{\sqrt{900 - 800(1.1^{-x})}}$$

(a) $C'(0) = \frac{400 \ln 1.1}{\sqrt{100}} \approx 3.81$

The marginal cost is \$3.81.

(b) $C'(20) = \frac{(400 \ln 1.1)(1.1^{-20})}{\sqrt{900 - 800(1.1^{-20})}} \approx 0.20$

The marginal cost is \$.20.

(c) As x becomes larger and larger, $C'(x)$ approaches zero.

41. $y = 100e^{-0.03045t}$

(a) For $t = 0$,

$$y = 100e^{-0.03045(0)} = 100e^0 = 100\%.$$

(b) For $t = 2$,

$$y = 100e^{-0.03045(2)} = 100e^{-0.0609} \approx 94\%.$$

(c) For $t = 4$,

$$y = 100e^{-0.03045(4)} \approx 89\%.$$

(d) For $t = 6$,

$$y' = 100e^{-0.03045(6)} \approx 83\%.$$

(e) $y' = 100(-0.03045)e^{-0.3045t} = -3.045e^{-0.03045t}$

For $t = 0$,

$$y' = -3.045e^{-0.03045(0)} = -3.045.$$

(f) For $t = 2$,

$$y' = -3.045e^{-0.03045(2)} \approx -2.865.$$

(g) The percent of these cars on the road is decreasing, but at a slower rate as they age.

43. **(a)** $G_0 = 2$, $m = 250$, $k = 0.0018$

$$G(t) = \frac{250}{1 + \left(\frac{250}{2} - 1\right)e^{-0.0018(250)t}} = \frac{250}{1 + 124e^{-0.45t}}$$

(b)

$$G'(t) = \frac{(1+124e^{-0.45t})(0) - 250(124e^{-0.45t})(-0.45)}{(1+124e^{-0.45t})^2}$$
$$= \frac{13{,}950e^{-0.45t}}{(1+124e^{-0.45t})^2}$$

1995 when $t = 5$:

$$G(5) = \frac{250}{1+124e^{-0.45(5)}} \approx 17.8$$
$$G'(5) = \frac{13{,}950e^{-0.45(5)}}{\left(1+124e^{-0.45(5)}\right)^2} \approx 7.4$$

The number of Internet users in 1995 is about 17.8 million, and the growth rate is about 7.4 million users per year .

(c) 2000 when $t = 10$:

$$G(10) = \frac{250}{1+124e^{-0.45(10)}} \approx 105.2$$
$$G'(10) = \frac{13{,}950e^{-0.45(10)}}{\left(1+124e^{-0.45(10)}\right)^2} \approx 27.4$$

The number of Internet users in 2000 is about 105.2 million, and the growth rate is about 27.4 million users per year.

(d) 2010 when $t = 20$:

$$G(20) = \frac{250}{1+124e^{-0.45(20)}} \approx 246.2$$
$$G'(20) = \frac{13{,}950e^{-0.45(20)}}{(1+124e^{-0.45(20)})^2} \approx 1.7$$

The number is Internet users in 2010 is about 246.2 million, and the growth rate is about 1.7 million users per year.

(e) The rate of growth increases for a while and then gradually decreases to 0.

45. $h(t) = 37.79(1.021)^t$

$$h'(t) = 37.79(\ln 1.021)(1.021)^t$$
$$= 0.789(1.021)^t$$

(a) For 2015, $t = 15$:

$$h(15) = 37.79(1.021)^{15} \approx 51.6$$

The Hispanic population in the United States will be about 51,600,000 in 2015.

(b) $h'(15) = 0.789(\ln 1.021)(1.021)^{15} \approx 1.07$

The Hispanic population in the United States will be increasing at the rate of 1,070,000 people per year at the end of the year 2015.

47. $G(t) = \dfrac{mG_0}{G_0 + (m - G_0)e^{kmt}}$, where $G_o = 400$; $m = 5200$; and $k = 0.0001$.

(a)
$$G(t) = \frac{(5200)(400)}{400 + (5200 - 400)e^{(-0.0001)(5200)t}}$$
$$= \frac{(400)(5200)}{400 + 4800e^{-0.52t}}$$
$$= \frac{5200}{1+12e^{-0.52t}}$$

(b)

$$G(t) = 5200(1+12e^{-0.52t}) - 1$$
$$G'(t) = -5200(1+12e^{-0.52t})^{-2}(-6.24e^{-0.52t})$$
$$= \frac{32{,}448e^{-0.52t}}{(1+12e^{-0.52t})^2}$$
$$G(1) = \frac{5200}{1+12e^{-0.52}} \approx 639$$
$$G'(1) = \frac{32{,}448e^{-0.52}}{(1+12e^{-0.52})^2} \approx 292$$

(c)
$$G(4) = \frac{5200}{1+12e^{-2.08}} \approx 2081$$
$$G'(4) = \frac{32{,}448e^{-2.08}}{(1+12e^{-2.08})^2} \approx 649$$

(d)
$$G(10) = \frac{5200}{1+12e^{-5.2}} \approx 4877$$
$$G'(10) = \frac{34{,}448e^{-5.2}}{(1+12e^{-5.2})^2} \approx 167$$

(e) It increases for a while and then gradually decreases to 0.

49. $V(t) = 1100[1023e^{-0.02415t} + 1]^{-4}$

(a)
$$V(240) = 1100[1023e^{-0.02415(240)} + 1]^{-4}$$
$$\approx 3.857\,\text{cm}^3$$

(b)
$$V = \frac{4}{3}\pi r^3, \text{ so } r(V) = \sqrt[3]{\frac{3V}{4\pi}}$$
$$r(3.857) = \sqrt[3]{\frac{3(3.857)}{4\pi}} \approx 0.973\,\text{cm}$$

(c)

$$V(t) = 1100[1023e^{-0.02415t} + 1]^{-4} = 0.5$$
$$[1023e^{-0.02415t} + 1]^{-4} = \frac{1}{2200}$$
$$(1023e^{-0.02415t} + 1)^4 = 2200$$
$$1023e^{-0.02415t} + 1 = 2200^{1/4}$$
$$1023e^{-0.02415t} = 2200^{1/4} - 1$$
$$e^{-0.02415t} = \frac{2200^{1/4} - 1}{1023}$$
$$-0.02415t = \ln\left(\frac{2200^{1/4} - 1}{1023}\right)$$
$$t = \frac{1}{-0.02415}\ln\left(\frac{2200^{1/4} - 1}{1023}\right)$$
$$\approx 214 \text{ months}$$

The tumor has been growing for almost 18 years.

(d) As t goes to infinity, $e^{-0.02415t}$ goes to zero, and $V(t) = 1100[1023e^{-0.02415t} + 1]^{-4}$ goes to 1100 cm^3, which corresponds to a sphere with a radius of $\sqrt[3]{\frac{3(1100)}{4\pi}} \approx 6.4$ cm. It makes sense that a tumor growing in a person's body reaches a maximum volume of this size.

(e) By the chain rule,

$$\frac{dV}{dt} = 1100(-4)[1023e^{-0.2415t} + 1]^{-5}$$
$$\cdot (1023)(e^{-0.2415t})(-0.02415)$$
$$= 108{,}703.98[1023e^{-0.02415t} + 1]^{-5}e^{-0.2415t}$$

At $t = 240$, $\frac{dV}{dt} \approx 0.282$.

At 240 months old, the tumor is increasing in volume at the instantaneous rate of 0.282 cm^3/month.

51. $p(x) = 0.001131e^{0.1268x}$

(a) $p(40) = 0.001131e^{0.1268(40)} \approx 0.180$

(b) When $p(x) = 1$,

$$0.001131e^{0.1268x} = 1$$
$$e^{0.1268x} = \frac{1}{0.001131}$$
$$0.1268x = \ln\frac{1}{0.001131}$$
$$x = \frac{1}{0.1268}\ln\frac{1}{0.001131}$$
$$\approx 54$$

This represents the year 2024.

(c)
$$p'(x) = 0.001131e^{0.1268x}(0.1268)$$
$$= 0.0001434108e^{0.1268x}$$
$$p'(40) = 0.0001434108e^{0.1268(40)}$$
$$\approx 0.023$$

The marginal increase in the proportion per year in 2010 is approximately 0.023.

53.
$$W_1(t) = 509.7(1 - 0.941e^{-0.00181t})$$
$$W_2(t) = 498.4(1 - 0.889e^{-0.00219t})^{1.25}$$

(a) Both W_1 and W_2 are strictly increasing functions, so they approach their maximum values as t approaches ∞.

$$\lim_{t\to\infty} W_1(t) = \lim_{t\to\infty} 509.7(1 - 0.941e^{-0.00181t})$$
$$= 509.7(1 - 0) = 509.7$$
$$\lim_{t\to\infty} W_2(t) = \lim_{t\to\infty} 498.4(1 - 0.889e^{-0.00219t})^{1.25}$$
$$= 498.4(1 - 0)^{1.25} = 498.4$$

So, the maximum values of W_1 and W_2 are 509.7 kg and 498.4 kg respectively.

(b)
$$0.9(509.7) = 509.7(1 - 0.941e^{-0.00181t})$$
$$0.9 = 1 - 0.941e^{-0.00181t}$$
$$\frac{0.1}{0.941} \approx e^{-0.00181t}$$
$$1239 \approx t$$
$$0.9(498.4) = 498.4(1 - 0.889e^{-0.00219t})^{1.25}$$
$$0.9 = (1 - 0.889e^{-0.00219t})^{1.25}$$
$$\frac{1 - 0.9^{0.8}}{0.889} = e^{-0.00219t}$$
$$1095 \approx t$$

Respectively, it will take the average beef cow about 1239 days or 1095 days to reach 90% of its maximum.

(c)

$$W_1'(t) = (509.7)(-0.941)(-0.00181)e^{-0.00181t}$$
$$\approx 0.868126e^{-0.00181t}$$
$$W_1'(750) \approx 0.868126e^{-0.00181(750)}$$
$$\approx 0.22 \text{ kg/day}$$
$$W_2'(t) = (498.4)(1.25)(1 - 0.889e^{-0.00219t})^{0.25}$$
$$\cdot (-0.889)(-0.00219)e^{-0.00219t}$$
$$\approx 1.21292e^{-0.00219t}(1 - 0.889e^{-0.00219t})^{0.25}$$
$$W_2'(750) \approx 1.12192e^{-0.00219(750)}$$
$$\cdot \left(1 - 0.889e^{-0.00219(750)}\right)^{0.25}$$
$$\approx 0.22 \text{ kg/day}$$

Both functions yield a rate of change of about 0.22 kg per day.

(d) Looking at the graph, the growth patterns of the two functions are very similar.

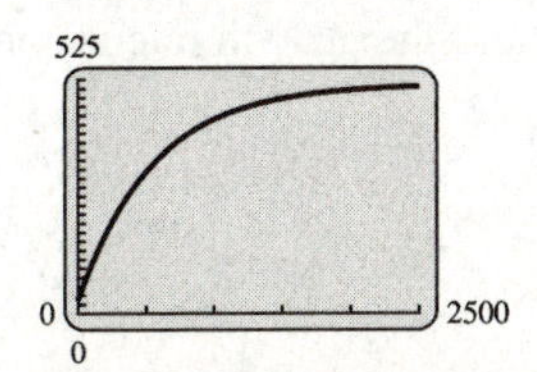

(e) The graphs of the rate of change of the two functions are also very similar.

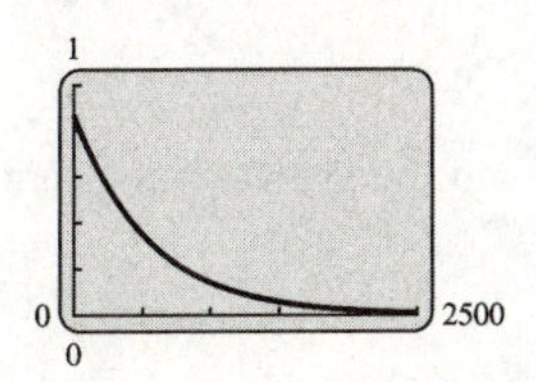

55. **(a)** $G_0 = 0.00369,\ m = 1,\ k = 3.5$

$$G(t) = \frac{1}{1 + \left(\frac{1}{0.00369} - 1\right)e^{-3.5(1)t}}$$
$$= \frac{1}{1 + 270e^{-3.5t}}$$

(b)

$$G'(t) = -(1 + 270e^{-3.5t})^{-2} \cdot 270e^{-3.5t}(-3.5)$$
$$= \frac{945e^{-3.5t}}{(1 + 270e^{-3.5t})^2}$$
$$G(1) = \frac{1}{1 + 270e^{-3.5(1)}} \approx 0.109$$
$$G'(1) = \frac{945e^{-3.5(1)}}{\left[1 + 270e^{-3.5(1)}\right]^2} \approx 0.341$$

The proportion is 0.109 and the rate of growth is 0.341 per century.

(c)
$$G(2) = \frac{1}{1 + 270e^{-3.5(2)}}$$
$$\approx 0.802$$
$$G'(2) = \frac{945e^{-3.5(2)}}{[1 + 270e^{-3.5(2)}]^2}$$
$$\approx 0.555$$

The proportion is 0.802 and the rate of growth is 0.555 per century.

(d)
$$G(3) = \frac{1}{1 + 270e^{-3.5(3)}}$$
$$\approx 0.993$$
$$G'(3) = \frac{945e^{-3.5(2)}}{[1 + 270e^{-3.5(2)}]^2}$$
$$\approx 0.0256$$

The proportion is 0.993 and the rate of growth is 0.0256 per century.

(e) The rate of growth increases for a while and then gradually decreases to 0.

57. $G_0 = 1.603,\ m = 6.8,\ k = 0.0440$

(a) $$G(t) = \frac{6.8}{1 + \left(\frac{6.8}{1.603} - 1\right)e^{-0.0440(6.8)t}}$$
$$= \frac{6.8}{1 + 3.242e^{-0.2992t}}$$

(b) $$G'(t) = \frac{(1 + 3.242e^{-0.2992t})(0) - 6.8(3.242e^{-0.2992t})(-0.2992)}{(1 + 3.242e^{-0.2992t})}$$
$$= \frac{6.59604e^{-0.2992t}}{(1 + 3.242e^{-0.2992t})^2}$$

For 2004, $t = 2$:

$$G(2) = \frac{6.8}{1 + 3.242e^{-0.2992(2)}} \approx 2.444$$
$$G'(2) = \frac{6.59604e^{-0.2992(2)}}{\left(1 + 3.242e^{-0.2992(2)}\right)^2} \approx 0.468$$

In 2002, about 2.444 million students enrolled in at least one online course and the growth rate is about 0.468 million students per year.

(c) For 2006, $t = 4$:

$$G(4) = \frac{6.8}{1 + 3.242e^{-0.2992(4)}} \approx 3.435$$
$$G'(4) = \frac{6.59604e^{-0.2992(4)}}{\left(1 + 3.242e^{-0.2992(4)}\right)^2} \approx 0.509$$

In 2006, about 3.435 million students enrolled in at least one online course and the growth rate is about 0.509 million students per year.

(d) For 2010, $t = 8$:

$$G(8) = \frac{6.8}{1 + 3.242e^{-0.2992(8)}} \approx 5.247$$
$$G'(8) = \frac{6.59604e^{-0.2992(8)}}{\left(1 + 3.242e^{-0.2992(8)}\right)^2} \approx 0.359$$

In 2010, about 5.247 million students enrolled in at least one online course and the growth rate is about 0.359 million students per year.

(e) The growth rate increases at first and then decreases.

59. $Q(t) = CV(1 - e^{-t/RC})$

(a) $$I_c = \frac{dQ}{dt} = CV\left[0 - e^{-t/RC}\left(-\frac{1}{RC}\right)\right]$$
$$= CV\left(\frac{1}{RC}\right)e^{-t/RC} = \frac{V}{R}e^{-t/RC}$$

(b) When $C = 10^{-5}$ farads, $R = 10^7$ ohms, and $V = 10$ volts, after 200 seconds

$$I_c = \frac{10}{10^7}e^{-200/(10^7 \cdot 10^{-5})}$$
$$\approx 1.35 \times 10^{-7}\text{amps}$$

61. $t(r) = 218 + 31(0.933)^n$

(a) $t(60) = 218 + 31(0.933)^{60}$
≈ 218.5 sec

(b) $t'(n) = (31 \ln 0.933)(0.933)^n$
$t'(60) = (31 \ln 0.933)(0.933)^{60}$
≈ -0.034

The record is decreasing by 0.034 seconds per year at the end of 2010.

(c) As $n \to \infty$, $(0.933)^n \to 0$ and

$t(n) \to 218$. If the estimate is correct, then this is the least amount of time that it will ever take a human to run a mile.

63. **(a)** $s(0) = 693.9 - 34.38\left(e^{0.01003(0)} + e^{-0.01003(0)}\right)$
$= 693.9 - 34.38(2)$
$= 625.14$

The height of the Gateway Arch at its center is 625.14ft.

(b) $m = s'(t) = -34.38(e^{0.01003x}(0.01003) + e^{-0.01003x}(-0.01003))$
$= -34.38(0.1003e^{0.01003x} - 0.01003e^{-0.01003x})$
$m = s'(0) = -34.38(0.01003(1) - 0.01003(1)) = 0$

The slope of the line tangent to the curve when $x = 0$ is 0, which makes sense because the tangent line has to be a horizontal line at the center of the arch(the highest point on the curve)

(c) $s'(150) = -34.38\left(0.01003e^{0.01003(150)} - 0.01003e^{-0.01003(150)}\right)$
$= -1.476$

The rate of change of the height of the arch $x = 150$ ft is -1.476 ft per foot from the center.

12.5 Derivatives of Logarithmic Functions

Your Turn 1

$$f(x) = \log_3 x$$
$$f'(x) = \frac{1}{(\ln 3)x}$$

Your Turn 2

(a) $y = \ln(2x^3 - 3)$

$$\frac{dy}{dx} = \frac{1}{2x^3 - 3} \cdot \frac{d}{dx}(2x^3 - 3)$$
$$= \frac{1}{2x^3 - 3}(6x^2)$$
$$= \frac{6x^2}{2x^3 - 3}$$

(b) $f(x) = \log_4(5x + 3x^3)$

$$f'(x) = \frac{1}{(\ln 4)(5x + 3x^3)} \cdot \frac{d}{dx}(5x + 3x^3)$$
$$= \frac{5 + 9x^2}{(\ln 4)(5x + 3x^3)}$$

Your Turn 3

(a) $y = \ln|2x + 6|$

$$\frac{dy}{dx} = \frac{1}{2x + 6} \cdot \frac{d}{dx}(2x + 6)$$
$$= \frac{2}{2x + 6} = \frac{2}{2(x + 3)} = \frac{1}{x + 3}$$

(b) $f(x) = x^2 \ln 3x$

$$f'(x) = x^2 \cdot \frac{1}{3x}(3) + \ln 3x(2x)$$
$$= x + 2x \ln 3x$$

(c) $s(t) = \dfrac{\ln(t^2 - 1)}{t + 1}$

$$s'(t) = \frac{(t + 1)\left(\frac{1}{t^2 - 1}\right)(2t) - \ln(t^2 - 1)(1)}{(t + 1)^2}$$
$$= \frac{\frac{2t}{t-1} - \ln(t^2 - 1)}{(t + 1)^2}$$
$$= \frac{2t - (t - 1)\ln(t^2 - 1)}{(t - 1)(t + 1)^2}$$

12.5 Exercises

1. $y = \ln(8x)$

$$\frac{dy}{dx} = \frac{d}{dx}(\ln 8x)$$
$$= \frac{d}{dx}(\ln 8 + \ln x)$$
$$= \frac{d}{dx}(\ln 8) + \frac{d}{dx}(\ln x)$$
$$= 0 + \frac{1}{x} = \frac{1}{x}$$

3. $y = \ln(8 - 3x)$

$$g(x) = 8 - 3x$$
$$g'(x) = -3$$
$$\frac{dy}{dx} = \frac{g'(x)}{g(x)} = \frac{-3}{8 - 3x} \text{ or } \frac{3}{3x - 8}$$

5. $y = \ln|4x^2 - 9x|$

$$g(x) = 4x^2 - 9x$$
$$g'(x) = 8x - 9$$
$$\frac{dy}{dx} = \frac{g'(x)}{g(x)} = \frac{8x - 9}{4x^2 - 9x}$$

7. $y = \ln\sqrt{x + 5}$

$$g(x) = \sqrt{x + 5} = (x + 5)^{1/2}$$
$$g'(x) = \frac{1}{2}(x + 5)^{-1/2}$$
$$\frac{dy}{dx} = \frac{\frac{1}{2}(x + 5)^{-1/2}}{(x + 5)^{1/2}} = \frac{1}{2(x + 5)}$$

9. $y = \ln(x^4 + 5x^2)^{3/2}$

$$= \frac{3}{2}\ln(x^4 + 5x^2)$$
$$\frac{dy}{dx} = \frac{3}{2}D_x[\ln(x^4 + 5x^2)]$$
$$g(x) = x^4 + 5x^2$$
$$g'(x) = 4x^3 + 10x$$
$$\frac{dy}{dx} = \frac{3}{2}\left(\frac{4x^3 + 10x}{x^4 + 5x^2}\right)$$
$$= \frac{3}{2}\left[\frac{2x(2x^2 + 5)}{x^2(x^2 + 5)}\right]$$
$$= \frac{3(2x^2 + 5)}{x(x^2 + 5)}$$

11. $y = -5x\ln(3x+2)$

Use the product rule.

$$\frac{dy}{dx} = -5x\left[\frac{d}{dx}\ln(3x+2)\right] + \ln(3x+2)\left[\frac{d}{dx}(-5x)\right]$$
$$= -5x\left(\frac{3}{3x+2}\right) + [\ln(3x+2)](-5)$$
$$= -\frac{15x}{3x+2} - 5\ln(3x+2)$$

13. $s = t^2\ln|t|$

$$\frac{ds}{dt} = t^2\cdot\frac{1}{t} + 2t\ln|t|$$
$$= t + 2t\ln|t|$$
$$= t(1 + 2\ln|t|)$$

15. $y = \dfrac{2\ln(x+3)}{x^2}$

Use the quotient rule.

$$\frac{dy}{dx} = \frac{x^2\left(\frac{2}{x+3}\right) - 2\ln(x+3)\cdot 2x}{(x^2)^2}$$
$$= \frac{\frac{2x^2}{x+3} - 4x\ln(x+3)}{x^4}$$
$$= \frac{2x^2 - 4x(x+3)\ln(x+3)}{x^4(x+3)}$$
$$= \frac{x[2x - 4(x+3)\ln(x+3)]}{x^4(x+3)}$$
$$= \frac{2x - 4(x+3)\ln(x+3)}{x^3(x+3)}$$

17. $y = \dfrac{\ln x}{4x+7}$

Use the quotient rule.

$$\frac{dy}{dx} = \frac{(4x+7)\left(\frac{1}{x}\right) - (\ln x)(4)}{(4x+7)^2}$$
$$= \frac{\frac{4x+7}{x} - 4\ln x}{(4x+7)^2}$$
$$= \frac{4x+7-4x\ln x}{x(4x+7)^2}$$

19. $y = \dfrac{3x^2}{\ln x}$

$$\frac{dy}{dx} = \frac{(\ln x)(6x) - 3x^2\left(\frac{1}{x}\right)}{(\ln x)^2}$$
$$= \frac{6x\ln x - 3x}{(\ln x)^2}$$

21. $y = (\ln|x+1|)^4$

$$\frac{dy}{dx} = 4(\ln|x+1|)^3\left(\frac{1}{x+1}\right)$$
$$= \frac{4(\ln|x+1|)^3}{x+1}$$

23. $y = \ln|\ln x|$

$$g(x) = \ln x$$
$$g'(x) = \frac{1}{x}$$
$$\frac{dy}{dx} = \frac{g'(x)}{g(x)}$$
$$= \frac{\frac{1}{x}}{\ln x}$$
$$= \frac{1}{x\ln x}$$

25. $y = e^{x^2}\ln x,\ x > 0$

$$\frac{dy}{dx} = e^{x^2}\left(\frac{1}{x}\right) + (\ln x)(2x)e^{x^2}$$
$$= \frac{e^{x^2}}{x} + 2xe^{x^2}\ln x$$

27. $y = \dfrac{e^x}{\ln x},\ x > 0$

Use the quotient rule.

$$\frac{dy}{dx} = \frac{(\ln x)e^x - e^x\left(\frac{1}{x}\right)}{(\ln x)^2}\cdot\frac{x}{x}$$
$$= \frac{xe^x\ln x - e^x}{x(\ln x)^2}$$

29. $g(z) = (e^{2z} + \ln z)^3$

$$g'(z) = 3(e^{2z} + \ln z)^2\left(e^{2z}\cdot 2 + \frac{1}{z}\right)$$
$$= 3(e^{2z} + \ln z)^2\left(\frac{2ze^{2z}+1}{z}\right)$$

31.
$$y = \log(6x)$$
$$g(x) = 6x \text{ and } g'(x) = 6.$$
$$\frac{dy}{dx} = \frac{1}{\ln 10}\left(\frac{6}{6x}\right)$$
$$= \frac{1}{x \ln 10}$$

33.
$$y = \log|1 - x|$$
$$g(x) = 1 - x \text{ and } g'(x) = -1.$$
$$\frac{dy}{dx} = \frac{1}{\ln 10} \cdot \frac{-1}{1 - x}$$
$$= -\frac{1}{(\ln 10)(1 - x)}$$
$$\text{or } \frac{1}{(\ln 10)(x - 1)}$$

35.
$$y = \log_5 \sqrt{5x + 2}$$
$$g(x) = \sqrt{5x + 2} \text{ and } g'(x) = \frac{5}{2\sqrt{5x + 2}}.$$
$$\frac{dy}{dx} = \frac{1}{\ln 5} \cdot \frac{\frac{5}{2\sqrt{5x+2}}}{\sqrt{5x + 2}}$$
$$= \frac{5}{2 \ln 5(5x + 2)}$$

37.
$$y = \log_3 (x^2 + 2x)^{3/2}$$
$$g(x) = (x^2 + 2x)^{3/2} \text{ and}$$
$$g'(x) = \frac{3}{2}(x^2 + 2x)^{1/2} \cdot (2x + 2)$$
$$= 3(x + 1)(x^2 + 2x)^{1/2}$$
$$\frac{dy}{dx} = \frac{1}{\ln 3} \cdot \frac{3(x + 1)(x^2 + 2x)^{1/2}}{(x^2 + 2x)^{3/2}}$$
$$= \frac{3(x + 1)}{(\ln 3)(x^2 + 2x)}$$

39.
$$w = \log_8(2^p - 1)$$
$$g(p) = 2^p - 1$$
$$g'(p) = (\ln 2)2^p$$
$$\frac{dw}{dp} = \frac{1}{\ln 8} \cdot \frac{(\ln 2)2^p}{(2^p - 1)}$$
$$= \frac{(\ln 2)2^p}{(\ln 8)(2^p - 1)}$$

41.
$$f(x) = e^{\sqrt{x}} \ln(\sqrt{x} + 5)$$
Use the product rule.
$$f'(x) = e^{\sqrt{x}}\left(\frac{\frac{1}{2\sqrt{x}}}{\sqrt{x} + 5}\right) + [\ln(\sqrt{x} + 5)]e^{\sqrt{x}}\left(\frac{1}{2\sqrt{x}}\right)$$
$$= \frac{e^{\sqrt{x}}}{2}\left[\frac{1}{\sqrt{x}(\sqrt{x} + 5)} + \frac{\ln(\sqrt{x} + 5)}{\sqrt{x}}\right]$$

43. $f(t) = \dfrac{\ln(t^2 + 1) + t}{\ln(t^2 + 1) + 1}$

Use the quotient rule.

$$u(t) = \ln(t^2 + 1) + t, u'(t) = \frac{2t}{t^2 + 1} + 1$$
$$v(t) = \ln(t^2 + 1) + 1, v'(t) = \frac{2t}{t^2 + 1}$$

$$f'(t) = \frac{[\ln(t^2 + 1) + 1]\left(\frac{2t}{t^2+1} + 1\right) - [\ln(t^2 + 1) + t]\left(\frac{2t}{t^2+1}\right)}{[\ln(t^2 + 1) + 1]^2}$$
$$= \frac{\frac{2t\ln(t^2+1)}{t^2+1} + \frac{2t}{t^2+1} + \ln(t^2 + 1) + 1 - \frac{2t\ln(t^2+1)}{t^2+1} - \frac{2t^2}{t^2+1}}{[\ln(t^2 + 1) + 1]^2}$$
$$= \frac{\frac{2t-2t^2}{t^2+1} + \ln(t^2 + 1) + 1}{[\ln(t^2 + 1) + 1]^2}$$
$$= \frac{2t - 2t^2 + (t^2 + 1)\ln(t^2 + 1) + t^2 + 1}{(t^2 + 1)[\ln(t^2 + 1) + 1]^2}$$
$$= \frac{-t^2 + 2t + 1 + (t^2 + 1)\ln(t^2 + 1)}{(t^2 + 1)[\ln(t^2 + 1) + 1]^2}$$

49. Use the derivative of $\ln x$.

$$\frac{d\ln[u(x)v(x)]}{dx} = \frac{1}{u(x)v(x)} \cdot \frac{d[u(x)v(x)]}{dx}$$
$$\frac{d\ln u(x)}{dx} = \frac{1}{u(x)} \cdot \frac{d[u(x)]}{dx}$$
$$\frac{d\ln v(x)}{dx} = \frac{1}{v(x)} \cdot \frac{d[v(x)]}{dx}$$

Then since $\ln[u(x)v(x)] = \ln u(x) + \ln v(x)$,

$$\frac{1}{u(x)v(x)} \cdot \frac{d[u(x)v(x)]}{dx} = \frac{1}{u(x)} \cdot \frac{d[u(x)]}{dx} + \frac{1}{v(x)} \cdot \frac{d[v(x)]}{dx}.$$

Multiply both sides of this equation by $u(x)v(x)$. Then $\dfrac{d[u(x)v(x)]}{dx} = v(x)\dfrac{d[u(x)]}{dx} + u(x)\dfrac{d[v(x)]}{dx}$.

This is the product rule.

51. Graph the function $y = \ln x$. Sketch lines tagent to the graph at $x = \frac{1}{2}, 1, 2, 3, 4$. Estimate the slopes of the tangent lines at these points.

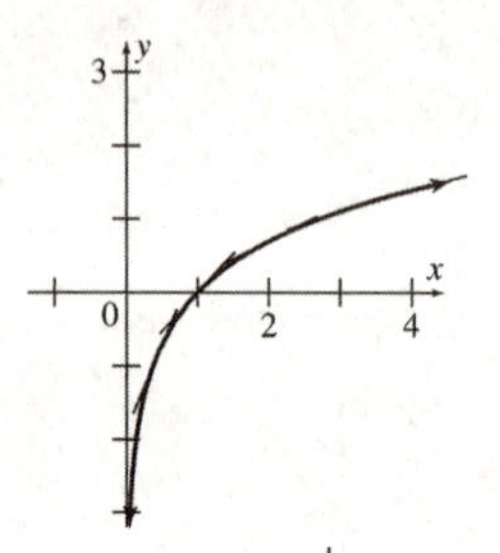

x	slope of tangent
$\frac{1}{2}$	2
1	1
2	$\frac{1}{2}$
3	$\frac{1}{3}$
4	$\frac{1}{4}$

The values of the slopes at x are $\frac{1}{x}$.

Thus we see that $\frac{d \ln x}{dx} = \frac{1}{x}$.

53. **(a)**

$$h(x) = u(x)^{v(x)}$$

$$\frac{d}{dx}\ln h(x) = \frac{d}{dx}\ln[u(x)^{v(x)}]$$

$$= \frac{d}{dx}[v(x)\ln u(x)]$$

$$= v(x)\frac{d}{dx}\ln u(x) + (\ln u(x))v'(x)$$

$$= v(x)\frac{u'(x)}{u(x)} + (\ln u(x))v'(x)$$

$$= \frac{v(x)u'(x)}{u(x)} + (\ln u(x))v'(x)$$

(b) Since $\frac{d}{dx}\ln h(x) = \frac{h'(x)}{h(x)}$,

$$h'(x) = h(x)\frac{d}{dx}\ln h(x)$$

$$= h(x)\left[\frac{v(x)u'(x)}{u(x)} + (\ln u(x))v'(x)\right]$$

$$= u(x)^{v(x)}\left[\frac{v(x)u'(x)}{u(x)} + (\ln u(x))v'(x)\right]$$

55.

$$h(x) = (x^2+1)^{5x}$$

$$u(x) = x^2 + 1, u'(x) = 2x$$

$$v(x) = 5x, v'(x) = 5$$

$$h'(x) = (x^2+1)^{5x}\left[\frac{5x(2x)}{x^2+1} + \ln(x^2+1)\cdot(5)\right]$$

$$= (x^2+1)^{5x}\left[\frac{10x^2}{x^2+1} + 5\ln(x^2+1)\right]$$

57. $p = 100 + \frac{50}{\ln q}, q > 1$

(a) $R = pq$

$$R = 100q + \frac{50q}{\ln q}$$

The marginal revenue is

$$\frac{dR}{dq} = 100 + \frac{(\ln q)(50) - 50q\left(\frac{1}{q}\right)}{(\ln q)^2}$$

$$= 100 + \frac{50(\ln q - 1)}{(\ln q)^2}.$$

(b) The revenue from one more unit is $\frac{dR}{dq}$ for $q = 8$.

$$100 + \frac{50(\ln 8 - 1)}{(\ln 8)^2} = \$112.48$$

(c) The manager can use the information from (b) to decide if it is reasonable to sell additional items.

59.

$$C(x) = 5\log_2 x + 10$$

$$\overline{C}(x) = \frac{C(x)}{x} = \frac{5\log_2 x + 10}{x}$$

$$\overline{C}'(x) = \frac{x\cdot 5\cdot \frac{1}{\ln 2}\cdot\frac{1}{x} - (5\log_2 x + 10)\cdot 1}{x^2}$$

$$= \frac{5 - (\ln 2)(5\log_2 x + 10)}{x^2 \ln 2}$$

(a) $\overline{C}'(10) = \frac{5 - (\ln 2)(5\log_2 10 + 10)}{(10^2)\ln 2}$

$$\approx -0.19396$$

(b) $\overline{C}'(20) = \frac{5 - (\ln 2)(5\log_2 20 + 10)}{(20^2)\ln 2}$

$$\approx -0.06099$$

61. $\ln\left(\frac{N(t)}{N_0}\right) = 9.8901e^{-e^{2.54197-0.2167t}}$

(a) $\frac{N(t)}{1000} = e^{9.8901e^{-e^{2.54197-0.2167t}}}$

$N(t) = 1000e^{9.8901e^{-e^{2.54197-0.2167t}}}$

(b) $N'(20) \approx 1{,}307{,}416$ bacteria/hour

Twenty hours into the experiment, the number of bacteria is increasing at a rate of 1,307,416 per hour

(c) $S(t) = \ln\left(\frac{N(t)}{N_0}\right)$

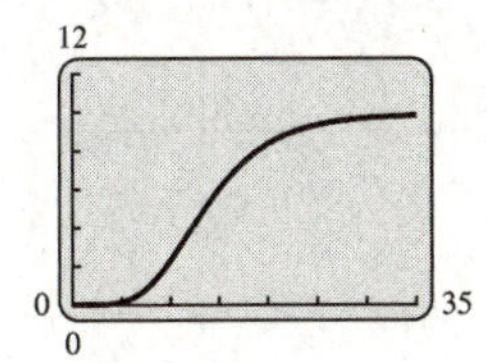

(d)

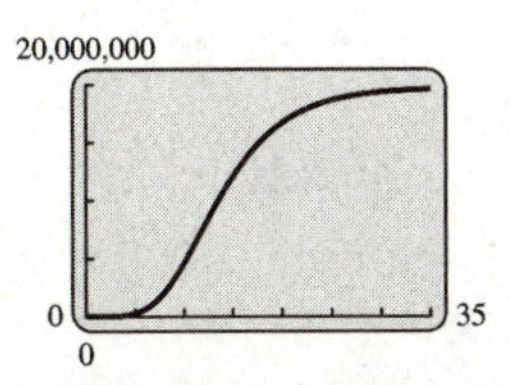

The two graphs have the same general shape, but $N(t)$ is scaled much larger.

(e) $\lim_{t\to\infty} S(t) = \lim_{t\to\infty} 9.8901e^{-e^{2.54197-0.2167t}}$

$= 9.8901$

$S(t) = \ln\left(\frac{N(t)}{N_0}\right)$

$N(t) = N_0 e^{S(t)}$

$\lim_{t\to\infty} N(t) = N_0 e^{\lim_{t\to\infty} S(t)} = 1000e^{9.8901}$

$\approx 19{,}734{,}033$ bacteria

63. $\log y = 1.54 - 0.008x - 0.658\log x$

(a) $y(x) = 10^{(1.54-0.008x-0.658\log x)}$

$= 10^{1.54}(10^{-0.008x})(10^{-0.658\log x})$

$= 10^{1.54}(10^{0.008})^{-x}(10^{\log x})^{-0.658}$

$\approx 34.7(1.0186)^{-x}x^{-0.658}$

(b) (i) $y(20) = 34.7(1.0186)^{-20}20^{-0.658}$

≈ 3.343

(ii) $y(40) = 34.7(1.0186)^{-40}40^{-0.658}$

≈ 1.466

(c)

$$\frac{dy}{dx} = -34.7(1.0186)^{-x}x^{-0.658}\left(\ln(1.0186) + \frac{0.658}{x}\right)$$

(i) $\frac{dy}{dx}(20) = -0.17160... \approx -0.172$

(ii) $\frac{dy}{dx}(40) = -0.051120... \approx -0.0511$

65. $P(t) = (t + 100)\ln(t + 2)$

$P'(t) = (t + 100)\left(\frac{1}{t+2}\right) + \ln(t + 2)$

$P'(2) = (102)\left(\frac{1}{4}\right) + \ln(4) \approx 26.9$

$P'(8) = (108)\left(\frac{1}{10}\right) + \ln(10) \approx 13.1$

67. $M = \frac{2}{3}\log\frac{E}{0.007}$

(a) $8.9 = \frac{2}{3}\log\frac{E}{0.007}$

$13.35 = \log\frac{E}{0.007}$

$10^{13.35} = \frac{E}{0.007}$

$E = 0.007(10^{13.35})$

$\approx 1.567 \times 10^{11}$ kWh

(b) $10{,}000{,}000 \times 247$ kWh/month

$= 2{,}470{,}000{,}000$ kWh/month

$$\frac{1.567 \times 10^{11}\text{ kWh}}{2{,}470{,}000{,}000\text{ kWh/month}} \approx 63.4 \text{ months}$$

(c) $M = \frac{2}{3}\log E - \frac{2}{3}\log 0.007$

$\frac{dM}{dE} = \frac{2}{3}\left(\frac{1}{(\ln 10)E}\right) = \frac{2}{(3\ln 10)E}$

When $E = 70{,}000$,

$\frac{dM}{dE} = \frac{2}{(3\ln 10)70{,}000}$

$\approx 4.14 \times 10^{-6}$

(d) $\frac{dM}{dE}$ varies inversely with E, so as E increases, $\frac{dM}{dE}$ decreases and approaches zero.

Chapter 12 Review Exercises

1. False; the derivative of π^3 is 0 because π^3 is a constant.

2. True

3. False; the derivative of a product $u(x) \cdot v(x)$ is $u(x) \cdot v'(x) + v(x) \cdot u'(x)$.

4. True

5. False; the chain rule is used to take the derivative of a composition of functions.

6. False; the derivative ce^x is ce^x for any constant c.

7. False; the derivative of 10^x is $(\ln 10)10^x$.

8. True

9. True

10. False; the derivative of $\log x$ is $\frac{1}{(\ln 10)x}$, whereas the derivative of $\ln x$ is $\frac{1}{x}$.

11.
$$y = 5x^3 - 7x^2 - 9x + \sqrt{5}$$
$$\frac{dy}{dx} = 5(3x^2) - 7(2x) - 9 + 0$$
$$= 15x^2 - 14x - 9$$

13.
$$y = 9x^{8/3}$$
$$\frac{dy}{dx} = 9\left(\frac{8}{3}x^{5/3}\right)$$
$$= 24x^{5/3}$$

15.
$$f(x) = 3x^{-4} + 6\sqrt{x}$$
$$= 3x^{-4} + 6x^{1/2}$$
$$f'(x) = 3(-4x^{-5}) + 6\left(\tfrac{1}{2}x^{-1/2}\right)$$
$$= -12x^{-5} + 3x^{-1/2} \text{ or } -\frac{12}{x^5} + \frac{3}{x^{1/2}}$$

17.
$$k(x) = \frac{3x}{4x + 7}$$
$$k'(x) = \frac{(4x + 7)(3) - (3x)(4)}{(4x + 7)^2}$$
$$= \frac{12x + 21 - 12x}{(4x + 7)^2}$$
$$= \frac{21}{(4x + 7)^2}$$

19.
$$y = \frac{x^2 - x + 1}{x - 1}$$
$$\frac{dy}{dx} = \frac{(x - 1)(2x - 1) - (x^2 - x + 1)(1)}{(x - 1)^2}$$
$$= \frac{2x^2 - 3x + 1 - x^2 + x - 1}{(x - 1)^2}$$
$$= \frac{x^2 - 2x}{(x - 1)^2}$$

21.
$$f(x) = (3x^2 - 2)^4$$
$$f'(x) = 4(3x^2 - 2)^3[3(2x)]$$
$$= 24x(3x^2 - 2)^3$$

23.
$$y = \sqrt{2t^7 - 5} = (2t^7 - 5)^{1/2}$$
$$\frac{dy}{dt} = \frac{1}{2}(2t^7 - 5)^{1/2}[2(7t^6)]$$
$$= 7t^6(2t^7 - 5)^{-1/2} \text{ or } \frac{7t^6}{(2t^7 - 5)^{1/2}}$$

25.
$$y = 3x(2x + 1)^3$$
$$\frac{dy}{dx} = 3x(3)(2x + 1)^2(2) + (2x + 1)^3(3)$$
$$= (18x)(2x + 1)^2 + 3(2x + 1)^3$$
$$= 3(2x + 1)^2[6x + (2x + 1)]$$
$$= 3(2x + 1)^2(8x + 1)$$

27.

$$r(t) = \frac{5t^2 - 7t}{(3t+1)^3}$$

$$r'(t) = \frac{(3t+1)^3(10t-7) - (5t^2 - 7t)(3)(3t+1)^2(3)}{[(3t+1)^3]^2}$$

$$= \frac{(3t+1)^3(10t-7) - 9(5t^2 - 7t)(3t+1)}{(3t+1)^6}$$

$$= \frac{(3t+1)(10t-7) - 9(5t^2 - 7t)}{(3t+1)^4}$$

$$= \frac{30t^2 - 11t - 7 - 45t^2 + 63t}{(3t+1)^4}$$

$$= \frac{-15t^2 + 52t - 7}{(3t+1)^4}$$

29. $p(t) = t^2(t^2+1)^{5/2}$

$$p'(t) = t^2 \cdot \frac{5}{2}(t^2+1)^{3/2} \cdot 2t + 2t(t^2+1)^{5/2}$$

$$= 5t^3(t^2+1)^{3/2} + 2t(t^2+1)^{5/2}$$

$$= t(t^2+1)^{3/2}[5t^2 + 2(t^2+1)^1]$$

$$= t(t^2+1)^{3/2}(7t^2+2)$$

31. $y = -6e^{2x}$

$$\frac{dy}{dx} = -6(2e^{2x}) = -12e^{2x}$$

33. $y = e^{-2x^3}$

$$g(x) = -2x^3$$

$$g'(x) = -6x^2$$

$$y' = -6x^2e^{-2x^3}$$

35. $y = 5x \cdot e^{2x}$

Use the product rule.

$$\frac{dy}{dx} = 5x(2e^{2x}) + e^{2x(5)}$$

$$= 10xe^{2x} + 5e^{2x}$$

$$= 5e^{2x}(2x+1)$$

37. $y = \ln(2 + x^2)$

$$g(x) = 2x + x^2$$

$$g'(x) = 2x$$

$$\frac{dy}{dx} = \frac{2x}{2+x^2}$$

39. $y = \dfrac{\ln|3x|}{x-3}$

$$\frac{dy}{dx} = \frac{(x-3)\left(\frac{1}{3x}\right)(3) - (\ln|3x|)(1)}{(x-3)^2}$$

$$= \frac{x - 3 - x\ln|3x|}{x(x-3)^2}$$

41. $y = \dfrac{xe^x}{\ln(x^2-1)}$

$$\frac{dy}{dx} = \frac{\ln(x^2-1)[xe^x + e^x] - xe^x\left(\frac{1}{x^2-1}\right)(2x)}{[\ln(x^2-1)]^2}$$

$$= \frac{e^x(x+1)\ln(x^2-1) - \frac{2x^2e^x}{x^2-1}}{[\ln(x^2-1)]^2} \cdot \frac{x^2-1}{x^2-1}$$

$$= \frac{e^x(x+1)(x^2-1)\ln(x^2-1) - 2x^2e^x}{(x^2-1)[\ln(x^2-1)]^2}$$

43. $s = (t^2 + e^t)^2$

$$s' = 2(t^2 + e^t)(2t + e^t)$$

45. $y = 3 \cdot 10^{-x^2}$

$$\frac{dy}{dx} = 3 \cdot (\ln 10)10^{-x^2}(-2x)$$

$$= -6x(\ln 10) \cdot 10^{-x^2}$$

47. $g(z) = \log_2(z^3 + z + 1)$

$$g'(z) = \frac{1}{\ln 2} \cdot \frac{3z^2+1}{z^3+z+1}$$

$$= \frac{3z^2+1}{(\ln 2)(z^3+z+1)}$$

49. $f(x) = e^{2x}\ln(xe^x + 1)$

Use the product rule.

$$f'(x) = e^{2x}\left(\frac{e^x + xe^x}{xe^x + 1}\right) + [\ln(xe^x+1)](2e^{2x})$$

$$= \frac{(1+x)e^{3x}}{xe^x + 1} + 2e^{2x}\ln(xe^x + 1)$$

51. **(a)** $D_x(f[g(x)])$ at $x = 2$

$$= f'[g(2)]g'(2)$$
$$= f'(1)\left(\frac{3}{10}\right)$$
$$= -5\left(\frac{3}{10}\right)$$
$$= -\frac{3}{2}$$

(b) $D_x(f[g(x)])$ at $x = 3$

$$= f'[g(3)]g'(3)$$
$$= f'(2)\left(\frac{4}{11}\right)$$
$$= -6\left(\frac{4}{11}\right)$$
$$= -\frac{24}{11}$$

53. $y = x^2 - 6x$; tangent at $x = 2$

$$\frac{dy}{dx} = 2x - 6$$

Slope $= y'(2) = 2(2) - 6 = -2$

Use $(2, -8)$ and point-slope form.

$$y - (-8) = -2(x - 2)$$
$$y + 8 = -2x + 4$$
$$y + 2x = -4$$
$$y = -2x - 4$$

55. $y = \frac{3}{x-1}$; tangent at $x = -1$

$$y = \frac{3}{x - 1} = 3(x - 1)^{-1}$$
$$\frac{dy}{dx} = 3(-1)(x - 1)^{-2}(1)$$
$$= -3(x - 1)^{-2}$$

Slope $= y'(-1) = -3(-1 - 1)^{-2} = -\frac{3}{4}$

Use $\left(-1, -\frac{3}{2}\right)$ and point-slope form.

$$y - \left(-\frac{3}{2}\right) = -\frac{3}{4}[x - (-1)]$$
$$y + \frac{3}{2} = -\frac{3}{4}(x + 1)$$
$$y + \frac{6}{4} = -\frac{3}{4}x - \frac{3}{4}$$
$$y = -\frac{3}{4}x - \frac{9}{4}$$

57. $y = \sqrt{6x - 2}$; tangent at $x = 3$

$$y = \sqrt{6x - 2} = (6x - 2)^{1/2}$$
$$\frac{dy}{dx} = \frac{1}{2}(6x - 2)^{-1/2}(6)$$
$$= 3(6x - 2)^{-1/2}$$

$$\text{slope} = y'(3) = 3(6 \cdot 3 - 2)^{-1/2}$$
$$= 3(16)^{-1/2}$$
$$= \frac{3}{16^{1/2}} = \frac{3}{4}$$

Use $(3, 4)$ and point-slope form.

$$y - 4 = \frac{3}{4}(x - 3)$$
$$y - \frac{16}{4} = \frac{3}{4}x - \frac{9}{4}$$
$$y = \frac{3}{4}x + \frac{7}{4}$$

59. $y = e^x$; $x = 0$

$$\frac{dy}{dx} = e^x$$

The value of $\frac{dy}{dx}$ when $x = 0$ is the slope

$m = e^0 = 1$.

When $x = 0$, $y = e^0 = 1$. Use $m = 1$ with $P(0, 1)$.

$$y - 1 = 1(x - 0)$$
$$y = x + 1$$

61. $y = \ln x; x = 1$

$\frac{dy}{dx} = \frac{1}{x}$

The value of $\frac{dy}{dx}$ when $x = 1$ is the slope $m = \frac{1}{1} = 1$.

When $x = 1$, $y = \ln 1 = 0$.

Use $m = 1$ with $P(1,0)$.

$$y - 0 = 1(x - 1)$$
$$y = x - 1$$

63. The slope of the graph of $y = x + k$ is 1. First, we find the point on the graph of $f(x) = \sqrt{2x-1}$ at which the slope is also 1.

$$f(x) = (2x-1)^{1/2}$$
$$f'(x) = \frac{1}{2}(2x-1)^{-1/2}(2)$$
$$f'(x) = \frac{1}{\sqrt{2x-1}}$$

The slope is 1 when

$$\frac{1}{\sqrt{2x-1}} = 1$$
$$1 = \sqrt{2x-1}$$
$$1 = 2x - 1$$
$$2x = 2$$
$$x = 1,$$

and

$$f(1) = 1.$$

Therefore, at $P(1,1)$ on the graph of $f(x) = \sqrt{2x-1}$, the slope is 1. An equation of the tangent line is

$$y - 1 = 1(x - 1)$$
$$y - 1 = x - 1$$
$$y = x + 0.$$

Any tangent line intersects the curve in exactly one point.

From this we see that if $k = 0$, there is one point of intersection.

The graph of f is below the line $y = x + 0$. Therefore, if $k > 0$, the graph of $y = x + k$ will not intersect the graph.

Consider the point $Q\left(\frac{1}{2}, 0\right)$ on the graph. We find an equation of the line through Q with slope 1.

$$y - 0 = 1\left(x - \frac{1}{2}\right)$$
$$y = x - \frac{1}{2}$$

The line with a slope of 1 through $Q\left(\frac{1}{2}, 0\right)$ will intersect the graph in two points. One is Q and the other is some point on the graph to the right of Q.

The graph of $y = x + 0$ intersects the graph in one point, while the graph of $y = x - \frac{1}{2}$ intersects it in two points. If we use a value of k in $y = x + k$ with $-\frac{1}{2} < k < 0$, we will have a line with a y-intercept between $-\frac{1}{2}$ and a 0 and a slope of 1 which will intersect the graph in two points.

If k, the y-intercept, is less than $-\frac{1}{2}$, the graph of $y = x + k$ will be below point Q and will intersect the graph of f in exactly one point.

To summarize, the graph of $y = x + k$ will intersect the graph of $f(x) = \sqrt{2x-1}$ in

(1) no points if $k > 0$;

(2) exactly one point if $k = 0$ or if $k < -\frac{1}{2}$;

(3) exactly two points if $-\frac{1}{2} \le k < 0$

65. Using the result $\widehat{fg} = \hat{f} + \hat{g}$, the total amount of tuition collected goes up by approximately $2\% + 3\% = 5\%$.

Let $T =$ tuition per person before the increase and S the number of students before the increase. Then the new tuition is $1.03T$ and the new numbers of students is $1.02S$, so the total amount of tuition collected is $(1.03T)(1.02S) = 1.0506TS$, which is an increase of 5.06%.

67. $C(x) = \sqrt{x+1}$

$$\overline{C}(x) = \frac{C(x)}{x} = \frac{\sqrt{x+1}}{x} = \frac{(x+1)^{1/2}}{x}$$

$$\overline{C}'(x) = \frac{x[\frac{1}{2}(x+1)^{-1/2}] - (x+1)^{1/2}(1)}{x^2}$$

$$= \frac{\frac{1}{2}x(x+1)^{-1/2} - (x+1)^{1/2}}{x^2}$$

$$= \frac{x(x+1)^{-1/2} - 2(x+1)^{1/2}}{2x^2}$$

$$= \frac{(x+1)^{-1/2}[x - 2(x+1)]}{2x^2}$$

$$= \frac{(x+1)^{-1/2}(-x-2)}{2x^2}$$

$$= \frac{-x-2}{2x^2(x+1)^{1/2}}$$

69. $C(x) = (x^2+3)^3$

$$\overline{C}(x) = \frac{C(x)}{x} = \frac{(x^2+3)^3}{x}$$

$$\overline{C}'(x) = \frac{x[3(x^2+3)^2(2x)] - (x^2+3)^3(1)}{x^2}$$

$$= \frac{6x^2(x^2+3)^2 - (x^2+3)^3}{x^2}$$

$$= \frac{(x^2+3)^2[6x^2 - (x^2+3)]}{x^2}$$

$$= \frac{(x^2+3)^2(5x^2-3)}{x^2}$$

71. $C(x) = 10 - e^{-x}$

$$\overline{C}(x) = \frac{C(x)}{x}$$

$$\overline{C}(x) = \frac{10 - e^{-x}}{x}$$

$$\overline{C}'(x) = \frac{x(e^{-x}) - (10 - e^{-x}) \cdot 1}{x^2}$$

$$= \frac{e^{-x}(x+1) - 10}{x^2}$$

73. $S(x) = 1000 + 60\sqrt{x} + 12x$

$$= 1000 + 60x^{1/2} + 12x$$

$$\frac{dS}{dx} = 60\left(\frac{1}{2}x^{-1/2}\right) + 12$$

$$= 30x^{-1/2} + 12 = \frac{30}{\sqrt{x}} + 12$$

(a) $\frac{dS}{dx}(9) = \frac{30}{\sqrt{9}} + 12 = \frac{30}{3} + 12 = 22$

Sales will increase by \$22 million when \$1000 more is spent on research.

(b) $\frac{dS}{dx}(16) = \frac{30}{\sqrt{16}} + 12 = \frac{30}{4} + 12 = 19.5$

Sales will increase by \$19.5 million when \$1000 more is spent on research.

(c) $\frac{dS}{dx}(25) = \frac{30}{\sqrt{25}} + 12 = \frac{30}{5} + 12 = 18$

Sales will increase by \$18 million when \$1000 more is spent on research.

(d) As more money is spent on research, the increase in sales is decreasing.

75. $T(x) = \frac{1000 + 60x}{4x + 5}$

$$T'(x) = \frac{(4x+5)(60) - (1000+60x)(4)}{(4x+5)^2}$$

$$= \frac{240x + 300 - 4000 - 240x}{(4x+5)^2}$$

$$= \frac{-3700}{(4x+5)^2}$$

(a) $T'(9) = \frac{-3700}{[4(9)+5]^2} = \frac{-3700}{1681} \approx -2.201$

Costs will decrease \$2201 for the next \$100 spent on training.

(b) $T'(19) = \frac{-3700}{[4(19)+5]^2}$

$$= \frac{-3700}{6561} \approx -0.564$$

Costs will decrease \$564 for the next \$100 spent on training.

(c) Costs will always decrease because

$$T'(x) = \frac{-3700}{(4x+5)^2}$$

will always be negative.

77. $A(r) = 1000e^{12r/100}$

$$A'(r) = 1000e^{12r/100} \cdot \frac{12}{100}$$
$$= 120e^{12r/100}$$
$$A'(5) = 120e^{0.6} \approx 218.65$$

The balance increases by approximately \$218.65 for every 1% increase in the interest rate when the rate is 5%.

79. $p(t) = -0.00132t^4 + 0.0665t^3 - 1.127t^2 + 11.581t + 105.655$

$$p'(t) = -0.00528t^3 + 0.1.995t^2 - 2.254t + 11.581$$

(a) For 1995, $t = 15$:

$$p'(15) = -0.00528(15)^3 + 0.1.995(15)^2 - 2.254(15) + 11.581 \approx 4.839$$

In 1995 the volume of mail was increasing by about 4,839,000,000 pieces per year.

(b) For 2005, $t = 25$:

$$p'(25) = -0.00528(25)^3 + 0.1.995(25)^2 - 2.254(25) + 11.581 \approx -2.582$$

In 1995 the volume of mail was decreasing by about 2,582,000,000 pieces per year.

81. **(a)** Using the regression feature on a graphing calculator, a cubic function that models the data is

$$y = (1.799 \times 10^{-5})t^3 - (3.177 \times 10^{-4})t^2 - 0.06866t + 2.504.$$

Using the regression feature on a graphing calculator, a quartic function that models the data is

$$y = (-1.112 \times 10^{-6})t^4 + (2.920 \times 10^{-4})t^3 - 0.02238t^2 + 0.6483t - 4.278.$$

(b) Using the cubic function, $\frac{dy}{dx}$ at $x = 105$ is about 0.59 dollar per year. Using the quartic function, $\frac{dy}{dx}$ at $x = 105$ is about 0.46 dollar per year.

83. $G(t) = \dfrac{mG_0}{G_0 + (m - G_0)e^{-kmt}}$, where

$m = 30{,}000$, $G_o = 2000$, and $k = 5.10^{-6}$.

(a)

$$G(t) = \frac{(30{,}000)(2000)}{2000 + (30{,}000 - 2000)e^{-5.10^{-6}}(30{,}000)t}$$
$$= \frac{30{,}000}{1 + 14e^{-0.15t}}$$

(b)

$$G(t) = 30{,}000(1 + 14e^{-0.15t})^{-1}$$
$$G(6) = 30{,}000(1 + 14e^{-0.90})^{-1} \approx 4483$$
$$G'(t) = -30{,}000(1 + 14e^{-0.15t})^{-2}(-2.1e^{-0.15t})$$
$$= \frac{63{,}000e^{-0.15t}}{(1 + 14e^{-0.15t})^2}$$
$$G'(6) = \frac{63{,}000e^{-0.90}}{(1 + 14e^{-0.90})^2} \approx 572$$

The population is 4483, and the rate of growth is 572.

85. $M(t) = 3583e^{-e^{-0.020(t-66)}}$

(a) $M(250) = 3583e^{-e^{-0.020(250-66)}} \approx 3493.76$ grams,

or about 3.5 kilograms

(b) As $t \to \infty$, $-e^{-0.020(t-66)} \to 0$,

$e^{-e^{-0.020(t-66)}} \to 1$, and $M(t) \to 3583$ grams or about 3.6 kilograms.

(c) 50% of 3583 is 1791.5.

$$1791.5 = 3583e^{-e^{-0.020(t-66)}}$$
$$\ln\left(\frac{1791.5}{3583}\right) = -e^{-0.020(t-66)}$$
$$\ln\left(\ln\frac{3583}{1791.5}\right) = -0.020(t - 66)$$
$$t = -\frac{1}{0.020}\ln\left(\ln\frac{3583}{1791.5}\right) + 66$$
$$\approx 84 \text{ days}$$

(d)

$D_t M(t)$

$$= 3583e^{-e^{-0.020(t-66)}} D_t\left(-e^{-0.020(t-66)}\right)$$

$$= 3583e^{-e^{-0.020(t-56)}}\left(-e^{-0.020(t-66)}\right)(-0.020)$$

$$= 71.66e^{-e^{-0.020(t-66)}}\left(e^{-0.020(t-66)}\right)$$

When $t = 250$, $D_t M(t) \approx 1.76$ g/day.

(e)

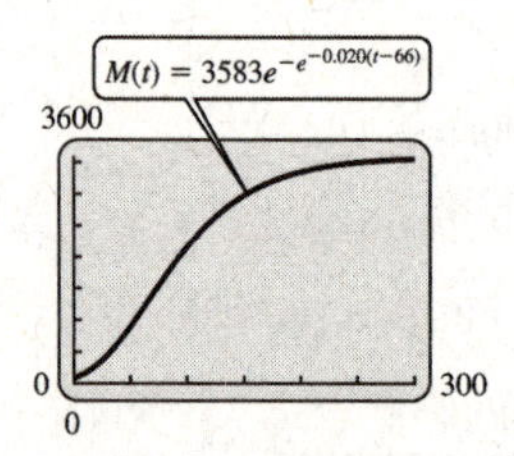

Growth is initially rapid, then tapers off.

(f)

Day	*Weight*	*Rate*
50	904	24.90
100	2159	21.87
150	2974	11.08
200	3346	4.59
250	3494	1.76
300	3550	0.66

87. $C(t) = 19{,}231(1.2647)^t$

$C'(t) = 19{,}231(\ln 1.2647)(1.2647)^t$

(a) For 2025, $t = 5$:

$$C'(5) = 19{,}231(\ln 1.2647)(1.2647)^5 \approx 14{,}612$$

In 2005 the rate of change in the energy capacity is 14,612 megawatts per year.

(b) For 2010, $t = 10$:

$$C'(10) = 19{,}231(\ln 1.2647)(1.2647)^{10} \approx 47{,}276$$

In 2010 the rate of change in the energy capacity is 47,276 megawatts per year.

(c) For 2015, $t = 15$:

$$C'(15) = 19{,}231(\ln 1.2647)(1.2647)^{15} \approx 152{,}960$$

In 2015 the rate of change in the energy capacity is 152,960 megawatts per year.

89.

$$p(x) = 1.757(1.0249)^{x-1930}$$
$$p'(x) = 1.757(\ln 1.0249)(1.0249)^{x-1930}$$
$$p'(2000) = 1.757(\ln 1.0249)(1.0249)^{2000-1930} \approx 0.242$$

The production of corn is increasing at a rate of 0.242 billion bushels per year in 2000.

91.

$$f(x) = k(x-49)^6 + .8$$
$$f'(x) = k \cdot 6(x-49)^5 = (3.8 \times 10^{-9})(6)(x-49)^5 = (2.28 \times 10^{-8})(x-49)^5$$

(a) $f'(20) = (2.28 \times 10^{-8})(20-49)^5 \approx -0.4677$

fatalities per 1000 licensed drivers per 100 million miles per year.

At the age of 20, each extra year results in a decrease of 0.4677 fatalities per 1000 licensed drivers per 100 million miles.

(b) $f'(60) = (2.28 \times 10^{-8})(60-49)^5 \approx 0.003672$

fatalities per 1000 licensed drivers per 100 million miles per year.

At the age of 60, each extra year results in an increase of 0.003672 fatalities per 1000 licensed drivers per 100 million miles.

Chapter 13

GRAPHS AND THE DERIVATIVE

13.1 Increasing and Decreasing Functions

Your Turn 1

Scanning across the x-values from left to right we see that the function is increasing on $(-1, 2)$ and $(4, \infty)$ and decreasing on $(-\infty, -1)$ and $(2, 4)$.

Your Turn 2

$$f(x) = -x^3 - 2x^2 + 15x + 10$$
$$f'(x) = -3x^2 - 4x + 15$$

Set $f'(x)$ equal to 0 and solve for x.

$$-3x^2 - 4x + 15 = 0$$
$$3x^2 + 4x - 15 = 0$$
$$(3x - 5)(x + 3) = 0$$
$$x = \frac{5}{3} \quad \text{or} \quad x = -3$$

The only critical numbers are -3 and $5/3$. These points determine three intervals: $(-\infty, -3)$, $(-3, 5/3)$, and $(5/3, \infty)$. Determine the sign of f' in each interval by picking a test point and evaluating $f'(x)$.

$$f'(-4) = -(3(-4) - 5)((-4) + 3) = -(-17)(-1) < 0$$
$$f'(0) = -(3(0) - 5)((0) + 3) = -(-5)(3) > 0$$
$$f'(2) = -(3(2) - 5)((2) + 3) = -(1)(5) < 0$$

The arrows in each interval in the figure below indicate where f is increasing or decreasing.

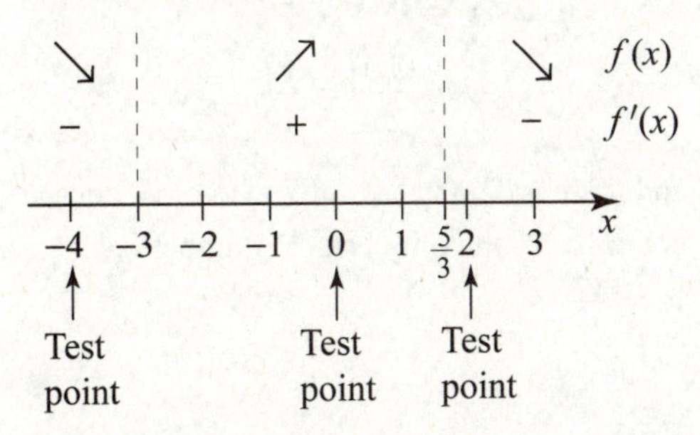

The function $f(x) = -x^3 - 2x^2 + 15x + 10$ is increasing on $(-3, 5/3)$ and decreasing on $(-\infty, -3)$ and $(5/3, \infty)$.

Since $f(-3) = -26$ and $f\left(\frac{5}{3}\right) = \frac{670}{27} \approx 24.8$, the graph goes through $(-3, -26)$ and $\left(\frac{5}{3}, \frac{670}{27}\right)$.

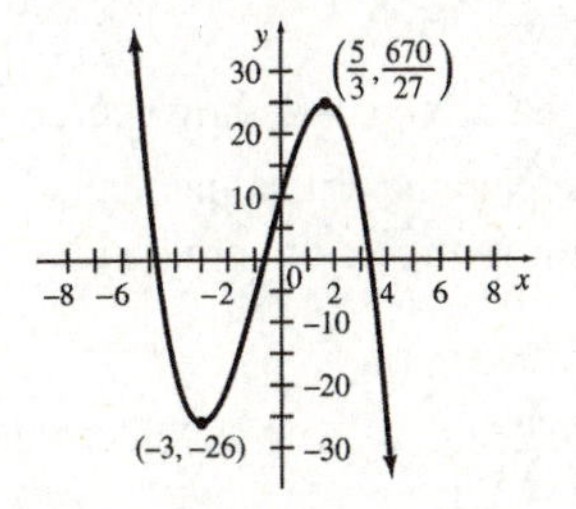

Your Turn 3

$$f(x) = (2x + 4)^{2/5}$$
$$f'(x) = \frac{2}{5}(2x + 4)^{-3/5}(2)$$
$$= \frac{4}{5(2x + 4)^{3/5}}$$

Find the critical numbers. $f'(x)$ is never 0 so we find any values where $f'(x)$ fails to be defined by setting the denominator equal to 0.

$$5(2x + 4)^{3/5} = 0$$
$$(2x + 4)^{3/5} = 0$$
$$2x + 4 = 0^{5/3} = 0$$
$$2x = -4$$
$$x = -2$$

Since $f'(-2)$ does not exist but $f(-2)$ is defined, $x = -2$ is a critical number.

Choosing -3 and -1 as test points, we find that $f'(-3) < 0$ and $f'(-1) > 0$. Thus the function f is decreasing on $(-\infty, -2)$ and increasing on $(-2, \infty)$. Using the fact that $f(-2) = 0$ and $f(0) = 4^{2/5} \approx 1.74$ we can sketch the following graph. Note that the graph is not smooth at $x = -2$, where the derivative is not defined.

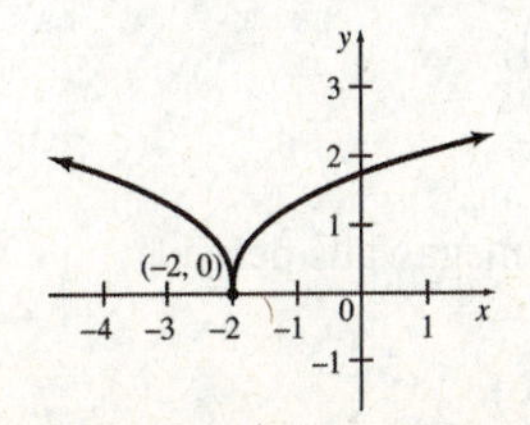

Your Turn 4

$$f(x) = \frac{-2x}{x+2}; \text{ the domain of } f \text{ excludes } -2$$

$$\begin{aligned} f'(x) &= \frac{(x+2)(-2)-(-2x)(1)}{(x+2)^2} \\ &= \frac{-2x-4+2x}{(x+2)^2} \\ &= -\frac{4}{(x+2)^2} \end{aligned}$$

$f'(x)$ is never 0 and fails to exist at $x = -2$; however, f is undefined at $x = -2$ so there are no critical numbers. Apply the first derivative test for numbers on either side of $x = -2$.

$$f'(-3) = -\frac{4}{(-3+2)^2} < 0$$

$$f'(-1) = -\frac{4}{(-1+2)^2} < 0$$

Since $f'(x)$ is negative wherever it is defined, the function is never increasing and is decreasing on $(-\infty, -2)$ and $(-2, \infty)$.

Next find any asymptotes. Since $x = -2$ makes the denominator of f equal to 0, the line $x = -2$ is a vertical asymptote.

$$\lim_{x\to\infty} \frac{-2x}{x+2} = \lim_{x\to\infty} \frac{-2}{1+\frac{2}{x}} = -2$$

The line $y = -2$ is a horizontal asymptote.

The x-intercept is (0, 0). The graph is shown below.

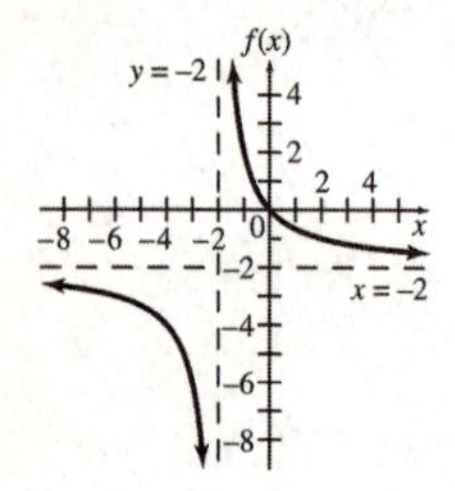

13.1 Exercises

1. By reading the graph, f is
   (a) increasing on $(1, \infty)$ and
   (b) decreasing on $(-\infty, 1)$.

3. By reading the graph, g is
   (a) increasing on $(-\infty, -2)$ and
   (b) decreasing on $(-2, \infty)$.

5. By reading the graph, h is
   (a) increasing on $(-\infty, -4)$ and $(-2, \infty)$ and
   (b) decreasing on $(-4, -2)$.

7. By reading the graph, f is
   (a) increasing on $(-7, -4)$ and $(-2, \infty)$ and
   (b) decreasing on $(-\infty, -7)$ and $(-4, -2)$.

9. (a) Since the graph of $f'(x)$ is positive for $x < -1$ and $x > 3$, the intervals where $f(x)$ is increasing are $(-\infty, -1)$ and $(3, \infty)$.
   (b) Since the graph of $f'(x)$ is negative for $-1 < x < 3$, the interval where $f(x)$ is decreasing is $(-1, 3)$.

11. (a) Since the graph of $f'(x)$ is positive for $x < -8, -6 < x < -2.5$ and $x > -1.5$, the intervals where $f(x)$ is increasing are $(-\infty, -8), (-6, -2.5)$, and $(-1.5, \infty)$.
    (b) Since the graph of $f'(x)$ is negative for $-8 < x < -6$ and $-2.5 < x < -1.5$, the intervals where $f(x)$ is decreasing are $(-8, -6)$ and $(-2.5, -1.5)$.

13. $y = 2.3 + 3.4x - 1.2x^2$

    (a) $y' = 3.4 - 2.4x$

    y' is zero when

    $$3.4 - 2.4x = 0$$

    $$x = \frac{3.4}{2.4} = \frac{17}{12}$$

    and there are no values of x where y' does not exist, so the only critical number is $x = \frac{17}{12}$.

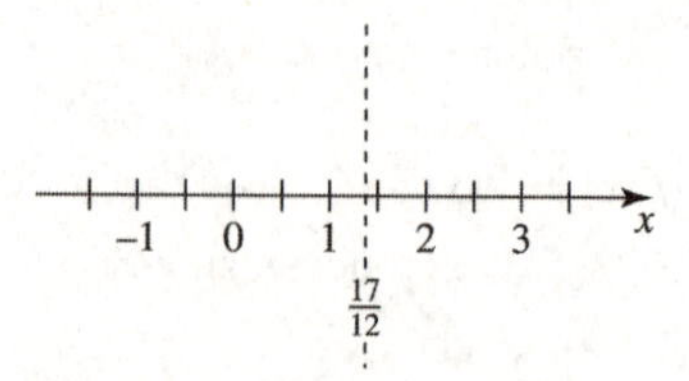

Test a point in each interval.

When
$x = 0, y' = 3.4 - 2.4(0) = 3.4 > 0.$

When
$x = 2, y' = 3.4 - 2.4(2) = -1.4 < 0.$

(b) The function is increasing on $\left(-\infty, \frac{17}{12}\right)$.

(c) The function is decreasing on $\left(\frac{17}{12}, \infty\right)$.

15. $f(x) = \frac{2}{3}x^3 - x^2 - 24x - 4$

(a) $f'(x) = 2x^2 - 2x - 24$
$= 2(x^2 - x - 12)$
$= 2(x + 3)(x - 4)$

$f'(x)$ is zero when $x = -3$ or $x = 4$, so the critical numbers are -3 and 4.

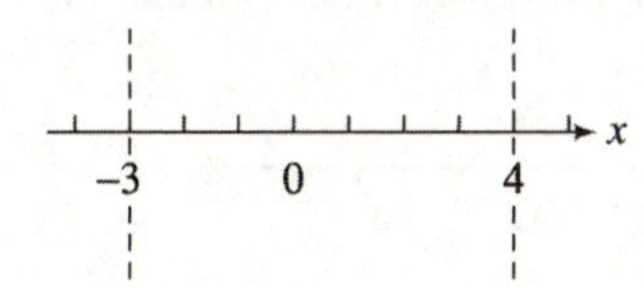

Test a point in each interval.

$f'(-4) = 16 > 0$
$f'(0) = -24 < 0$
$f'(5) = 16 > 0$

(b) f is increasing on $(-\infty, -3)$ and $(4, \infty)$.

(c) f is decreasing on $(-3, 4)$.

17. $f(x) = 4x^3 - 15x^2 - 72x + 5$

(a) $f'(x) = 12x^2 - 30x - 72$
$= 6(2x^2 - 5x - 12)$
$= 6(2x + 3)(x - 4)$

$f'(x)$ is zero when $x = -\frac{3}{2}$ or $x = 4$, so the critical numbers are $-\frac{3}{2}$ and 4.

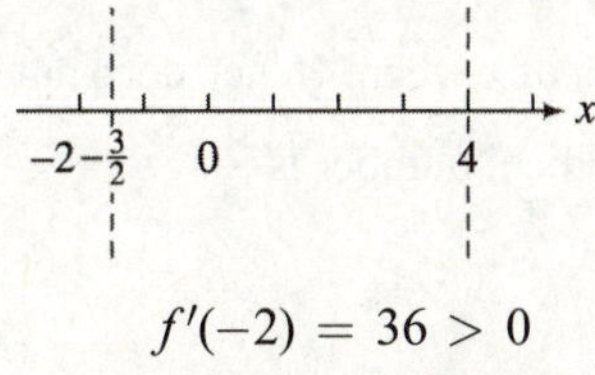

$f'(-2) = 36 > 0$
$f'(0) = -72 < 0$
$f'(5) = 78 > 0$

(b) f is increasing on $\left(-\infty, -\frac{3}{2}\right)$ and $(4, \infty)$.

(c) f is decreasing on $\left(-\frac{3}{2}, 4\right)$.

19. $f(x) = x^4 + 4x^3 + 4x^2 + 1$

(a) $f'(x) = 4x^3 + 12x^2 + 8x$
$= 4x(x^2 + 3x + 2)$
$= 4x(x + 2)(x + 1)$

$f'(x)$ is zero when $x = 0$, $x = -2$, or $x = -1$, so the critical numbers are 0, -2, and -1.

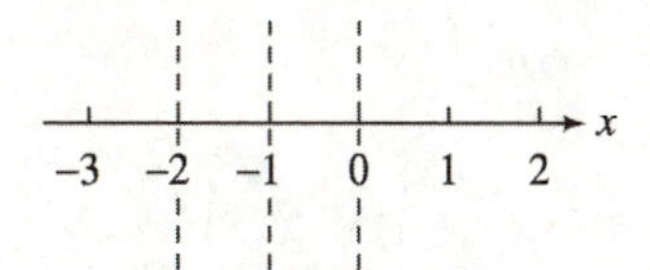

Test a point in each interval.

$f'(-3) = -12(-1)(-2) = -24 < 0$
$f'(-1.5) = -6(.5)(-.5) = 1.5 > 0$
$f'(-.5) = -2(1.5)(.5) = -1.5 < 0$
$f'(1) = 4(3)(2) = 24 > 0$

(b) f is increasing on $(-2, -1)$ and $(0, \infty)$.

(c) f is decreasing on $(-\infty, -2)$ and $(-1, 0)$.

21. $y = -3x + 6$

(a) $y' = -3 < 0$

There are no critical numbers since y' is never 0 and always exists.

(b) Since y' is always negative, the function is increasing on no interval.

(c) y' is always negative, so the function is decreasing everywhere, or on the interval $(-\infty, \infty)$.

23. $f(x) = \frac{x + 2}{x + 1}$

(a) $f'(x) = \frac{(x + 1)(1) - (x + 2)(1)}{(x + 1)^2}$
$= \frac{-1}{(x + 1)^2}$

The derivative is never 0, but it fails to exist at $x = -1$. Since -1 is not in the domain of f, however, -1 is not a critical number.

$$f'(-2) = -1 < 0$$
$$f'(0) = -1 < 0$$

(b) f is increasing on no interval.

(c) f is decreasing everywhere that it is defined, on $(-\infty, -1)$ and on $(-1, \infty)$.

25. $y = \sqrt{x^2 + 1}$
$= (x^2 + 1)^{1/2}$

(a) $y' = \frac{1}{2}(x^2 + 1)^{-1/2}(2x)$

$$= x(x^2 + 1)^{-1/2}$$

$$= \frac{x}{\sqrt{x^2 + 1}}$$

$y' = 0$ when $x = 0$.

Since y does not fail to exist for any x, and since $y' = 0$ when $x = 0$, 0 is the only critical number.

$$y'(1) = \frac{1}{\sqrt{2}} > 0$$

$$y'(-1) = \frac{-1}{\sqrt{2}} < 0$$

(b) y is increasing on $(0, \infty)$.

(c) y is decreasing on $(-\infty, 0)$.

27. $f(x) = x^{2/3}$

(a) $f'(x) = \frac{2}{3}x^{-1/3} = \frac{2}{3x^{1/3}}$

$f'(x)$ is never zero, but fails to exist when $x = 0$, so 0 is the only critical number.

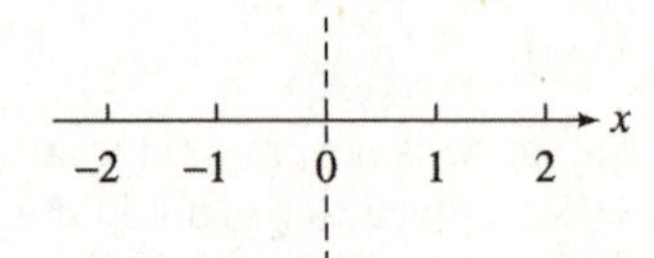

$$f'(-1) = -\frac{2}{3} < 0$$

$$f'(1) = \frac{2}{3} > 0$$

(b) f is increasing on $(0, \infty)$.

(c) f is decreasing on $(-\infty, 0)$.

29. $y = x - 4\ln(3x - 9)$

(a) $y' = 1 - \frac{12}{3x - 9} = 1 - \frac{4}{x - 3}$

$$= \frac{x - 7}{x - 3}$$

y' is zero when $x = 7$. The derivative does not exist at $x = 3$, but note that the domain of f is $(3, \infty)$.

Thus, the only critical number is 7.

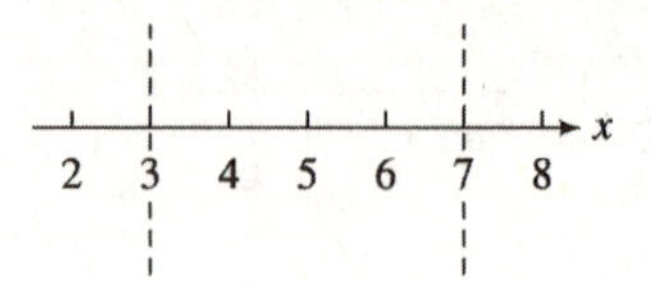

Choose a value in the intervals $(3, 7)$ and $(7, \infty)$.

$$f'(4) = -3 < 0$$

$$f(8) = \frac{1}{5} > 0$$

(b) The function is increasing on $(7, \infty)$.

(c) The function is decreasing on $(3, 7)$.

31. $f(x) = xe^{-3x}$

(a) $f'(x) = e^{-3x} + x(-3e^{-3x})$

$$= (1 - 3x)e^{-3x}$$

$$= \frac{1 - 3x}{e^{3x}}$$

$f'(x)$ is zero when $x = \frac{1}{3}$ and three are no values of x where $f'(x)$ does not exist, so the critical number is $\frac{1}{3}$.

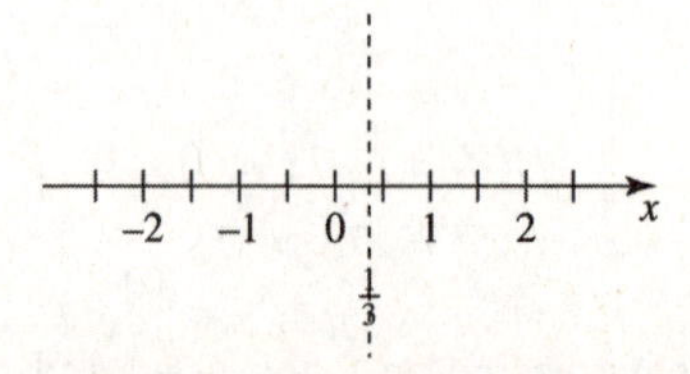

Test a point in each interval.

$$f'(0) = \frac{1-3(0)}{e^{3(0)}} = 1 > 0$$

$$f'(1) = \frac{1-3(1)}{e^{3(1)}} = -\frac{2}{e^3} < 0$$

(b) The function is increasing on $\left(-\infty, \frac{1}{3}\right)$.

(c) The function is decreasing on $\left(\frac{1}{3}, \infty\right)$.

33. $f(x) = x^2 2^{-x}$

(a) $f'(x) = x^2[\ln 2(2^{-x})(-1)] + (2^{-x})2x$

$= 2^{-x}(-x^2 \ln 2 + 2x)$

$= \dfrac{x(2 - x\ln 2)}{2^x}$

$f'(x)$ **is zero when** $x = 0$ **or** $x = \frac{2}{\ln 2}$ and there are no values of x where $f'(x)$ does not exist. The critical numbers are 0 and $\frac{2}{\ln 2}$.

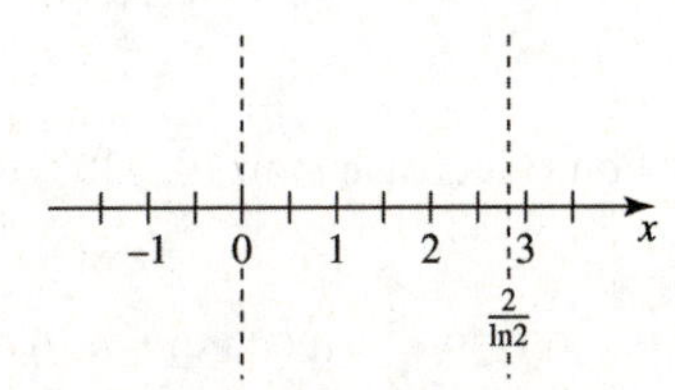

Test a point in each interval.

$$f'(-1) = \frac{(-1)(2-(-1)\ln 2)}{2^{-1}} = -2(2+\ln 2) < 0$$

$$f'(1) = \frac{(1)(2-(1)\ln 2)}{2^1} = \frac{2-\ln 2}{2} > 0$$

$$f'(3) = \frac{(3)(2-(3)\ln 2)}{2^3} = \frac{3(2-3\ln 2)}{8} < 0$$

(b) The function is increasing on $\left(0, \frac{2}{\ln 2}\right)$.

(c) The function is decreasing on $(-\infty, 0)$ and $\left(\frac{2}{\ln 2}, \infty\right)$.

35. $y = x^{2/3} - x^{5/3}$

(a) $y' = \frac{2}{3}x^{-1/3} - \frac{5}{3}x^{2/3} = \dfrac{2-5x}{3x^{1/3}}$

$y' = 0$ when $x = \frac{2}{5}$. The derivative does not exist at $x = 0$. So the critical numbers are 0 and $\frac{2}{5}$.

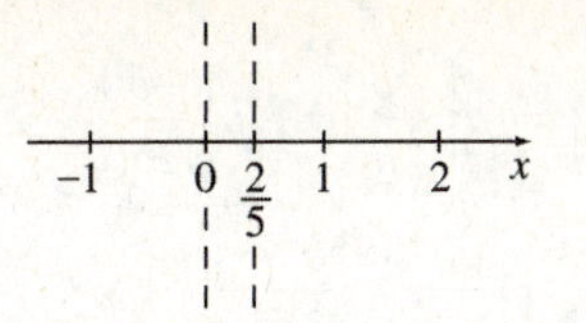

Test a point in each interval.

$$y'(-1) = \frac{7}{-3} < 0$$

$$y'\left(\frac{1}{5}\right) = \frac{1}{3\left(\frac{1}{5}\right)^{1/3}} = \frac{5^{1/3}}{2} > 0$$

$$y'(1) = \frac{-3}{3} = -1 < 0$$

(b) y is increasing on $\left(0, \frac{2}{5}\right)$.

(c) y is decreasing on $(-\infty, 0)$ and $\left(\frac{2}{5}, \infty\right)$.

39. $f(x) = ax^2 + bx + c, a < 0$

$f'(x) = 2ax + b$

Let $f'(x) = 0$ to find the critical number.

$$2ax + b = 0$$
$$2ax = -b$$
$$x = \frac{-b}{2a}$$

Choose a value in the interval $\left(-\infty, -\frac{-b}{2a}\right)$.

Since $a < 0$,

$$\frac{-b}{2a} - \frac{-1}{2a} = \frac{-b+1}{2a} < \frac{-b}{2a}.$$

$$f'\left(\frac{-b+1}{2a}\right) = 2a\left(\frac{-b+1}{2a}\right) + b$$
$$= 1 < 0$$

Choose a value in the interval $\left(\frac{-b}{2a}, \infty\right)$.

Since $a < 0$,

$$\frac{-b}{2a} - \frac{-1}{2a} = \frac{-b-1}{2a} < \frac{-b}{2a}.$$

$$f'\left(\frac{-b-1}{2a}\right) = 2a\left(\frac{-b-1}{2a}\right) + b$$
$$= -1 > 0$$

f is increasing on $\left(-\infty, \frac{-b}{2a}\right)$ and decreasing on $\left(\frac{-b}{2a}, \infty\right)$.

This tells us that the curve opens downward and $x = \frac{-b}{2a}$ is the x-coordinate of the vertex.

$$f\left(\frac{-b}{2a}\right) = a\left(\frac{-b}{2a}\right)^2 + b\left(\frac{-b}{2a}\right) + c$$
$$= \frac{ab^2}{4a^2} - \frac{b^2}{2a} + c$$
$$= \frac{b^2}{4a} - \frac{2b^2}{4a} + \frac{4bc}{4a}$$
$$= \frac{4ac - b^2}{4a}$$

The vertex is $\left(\frac{-b}{2a}, \frac{4ac-b^2}{4a}\right)$ or $\left(-\frac{b^2}{2a}, \frac{4ac-b^2}{4a}\right)$.

41. $f(x) = \ln x$

$f'(x) = \frac{1}{x}$

$f'(x)$ is undefined at $x = 0$. $f'(x)$ never equals zero. Note that $f(x)$ has a domain of $(0, \infty)$. Pick a value in the interval $(0, \infty)$.

$$f'(2) = \frac{1}{2} > 0$$

$f(x)$ is increasing on $(0, \infty)$.

$f(x)$ is never decreasing.

Since $f(x)$ never equals zero, the tangent line is horizontal nowhere.

43. $f(x) = e^{0.001x} - \ln x$

$f'(x) = 0.001e^{0.001x} - \frac{1}{x}$

Note that $f(x)$ is only defined for $x > 0$. Use a graphing calculator to plot $f'(x)$ for $x > 0$.

(a) $f'(x) > 0$ on about $(567, \infty)$, so $f(x)$ is increasing on about $(567, \infty)$.

(b) $f'(x) < 0$ on about $(0, 567)$, so $f(x)$ is decreasing on about $(0, 567)$.

45. $H(r) = \frac{300}{1 + 0.03r^2} = 300(1 + 0.03r^2)^{-1}$

$H'(r) = 300[-1(1 + 0.03r^2)^{-2}(0.06r)]$

$= \frac{-18r}{(1 + 0.03r^2)^2}$

Since r is a mortgage rate (in percent), it is always positive. Thus, $H'(r)$ is always negative.

(a) H is increasing on nowhere.

(b) H is decreasing on $(0, \infty)$.

47. $C(x) = 0.32x^2 - 0.00004x^3$

$R(x) = 0.848x^2 - 0.0002x^3$

$P(x) = R(x) = C(x)$

$= (0.848x^2 - 0.0002x^3)$

$- (0.32x^2 - 0.00004x^3)$

$= 0.528x^2 - 0.00016x^3$

$P'(x) = 1.056x - 0.00048x^2$

$1.056x - 0.00048x^2 = 0$

$x(1.056 - 0.00048x) = 0$

$x = 0$ or $x = 2200$

Choose $x = 1000$ and $x = 3000$ as test points.

$P'(1000) = 1.056(1000) - 0.00048(1000)^2 = 576$

$P'(3000) = 1.056(3000) - 0.00048(3000)^2$

$= -1152$

The function is increasing on $(0, 2200)$.

49. $A(t) = 0.0000329t^3 - 0.00450t^2 + 0.0613t + 2.34$

$A'(t) = 0.0000987t^2 - 0.009t + 0.0613$

Set $A'(t)$ equal to 0 and solve for t.

$0.0000987t^2 - 0.009t + 0.0613 = 0$

$$t = \frac{0.009 \pm \sqrt{(0.009)^2 - 4(0.0000987)(0.0613)}}{(2)(0.0000987)}$$

$t \approx 83.8$ or $t = 7.4$

Since $0 \leq t \leq 50, t \approx 7.4$.

Check the sign of A' on either side of this critical number.

$A'(7) \approx 0.00314 > 0$

$A'(8) \approx -0.00438 < 0$

(a) The projected year-end assets function is increasing on the interval $(0, 7.4)$, or from 2000 to about the middle of 2007.

(b) The projected year-end assets function is decreasing on the interval $(7.4, 50)$, or from about the middle of 2007 to 2050.

51. **(a)** These curves are graphs of functions since they all pass the vertical line test.

(b) The graph for particulates increases from April to July; it decreases from July to November; it is constant from January to April and November to December.

(c) All graphs are constant from January to April and November to December. When the temperature is low, as it is during these months, air pollution is greatly reduced.

53. $A(x) = 0.003631x^3 - 0.03746x^2 + 0.1012x + 0.009$

$A'(x) = 0.010893x^2 - 0.07492x + 0.1012$

Solve for $A'(x) = 0$.

$x \approx 1.85$ or $x \approx 5.03$

Choose $x = 1$ and $x = 4$ as test points.

$$A'(1) = 0.010893(1)^2 - 0.07492(1) + 0.1012 = 0.037173$$

$$A'(4) = 0.010893(4)^2 - 0.07492(4) + 0.1012 = -0.024192$$

(a) The function is increasing on $(0, 1.85)$.

(b) The function is decreasing on $(1.85, 5)$.

55. $K(t) = \frac{5t}{t^2+1}$

$$K'(t) = \frac{5(t^2+1) - 2t(5t)}{(t^2+1)^2} = \frac{5t^2 + 5 - 10t^2}{(t^2+1)^2} = \frac{5 - 5t^2}{(t^2+1)^2}$$

$K'(t) = 0$ when

$$\frac{5-5t^2}{(t^2+1)^2} = 0$$
$$5 - 5t^2 = 0$$
$$5t^2 = 5$$
$$t = \pm 1.$$

Since t is the time after a drug is administered, the function applies only for $[0, \infty)$, so we discard $t = -1$. Then 1 divides the domain into two intervals.

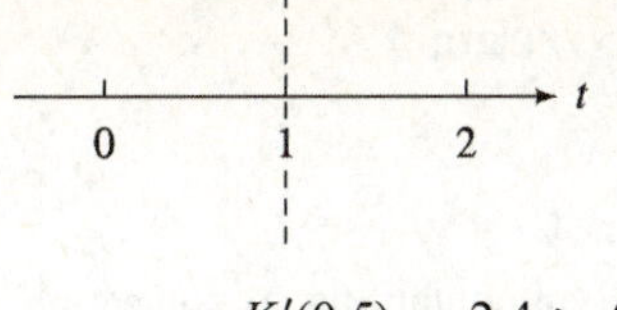

$$K'(0.5) = 2.4 > 0$$
$$K'(2) = -0.6 < 0$$

(a) K is increasing on $(0, 1)$.

(b) K is decreasing on $(1, \infty)$.

57. **(a)** $F(t) = -10.28 + 175.9te^{-t/1.3}$

$$F'(t) = (175.9)(e^{-t/1.3}) + (175.9.9t)\left(-\frac{1}{1.3}e^{-t/1.3}\right) = (175.9)(e^{-t/1.3})\left(1 - \frac{t}{1.3}\right) \approx 175.9e^{-t/1.3}(1 - 0.769t)$$

(b) $F'(t)$ is equal to 0 at $t = 1.3$. Therefore, 1.3 is a critical number. Since the domain is $(0, \infty)$, test values in the intervals from $(0, 1.3)$ and $(1.3, \infty)$.

$F'(1) \approx 18.83 > 0$ and $F'(2) \approx -20.32 < 0$

$F'(t)$ is increasing on $(0, 1.3)$ and decreasing on $(1.3, \infty)$.

59. $f(x) = \frac{1}{\sqrt{2\pi}}e^{-x^2/2}$

$$f'(x) = \frac{1}{\sqrt{2\pi}}e^{-x^2/2}(-x) = \frac{-x}{\sqrt{2\pi}}e^{-x^2/2}$$

$f'(x) = 0$ when $x = 0$.

Choose a value from each of the intervals $(-\infty, 0)$ and $(0, \infty)$.

$$f'(-1) = \frac{1}{\sqrt{2\pi}}e^{-1/2} > 0$$
$$f'(1) = \frac{-1}{\sqrt{2\pi}}e^{-1/2} < 0$$

The function is increasing on $(-\infty, 0)$ and decreasing on $(0, \infty)$.

61. **(a)** $(2500, 5750)$

(b) $(5750, 6000)$

(c) $(2800, 4800)$

(d) $(2500, 2800)$ and $(4800, 6000)$

13.2 Relative Extrema

Your Turn 1

There is an open interval containing x_1 for which $f(x) \geq f(x_1)$.

There is an open interval containing x_2 for which $f(x) \leq f(x_2)$.

There is an open interval containing x_3 for which $f(x) \geq f(x_3)$.

There are relative minima of $f(x_1)$ at $x = x_1$ and $f(x_3)$ at $x = x_3$ and a relative maximum of $f(x_2)$ at $x = x_2$.

Your Turn 2

$$f(x) = -x^3 - 2x^2 + 15x + 10$$
$$f'(x) = -3x^2 - 4x + 15$$

Set $f'(x)$ equal to 0 and solve for x to find the critical numbers.

$$-3x^2 - 4x + 15 = 0$$
$$-(3x^2 + 4x - 15) = 0$$
$$-(3x - 5)(x + 3) = 0$$

$$3x - 5 = 0 \text{ or } x + 3 = 0$$
$$x = 5/3 \quad \text{or } x = -3$$

The critical numbers are -3 and $5/3$. They divide the domain of f into the three intervals $(-\infty, -3)$, $(-3, 5/3)$, and $(5/3, \infty)$. Pick test points in each interval to determine the sign of f' on that interval. For example, choose $x = -4$, $x = 0$, and $x = 2$.

$$f'(-4) = -17$$
$$f'(0) = 15$$
$$f'(2) = -5$$

Thus the derivative is negative on $(-\infty, -3)$, positive on $(-3, 5/3)$, and negative on $(5/3, \infty)$. The graph below shows this information and indicates where f is increasing or decreasing.

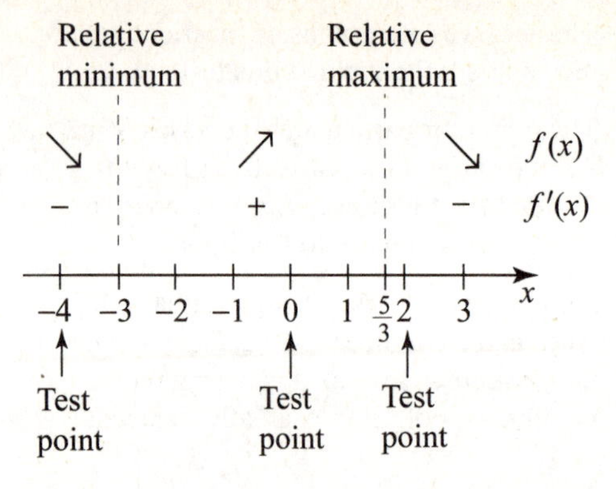

By the first derivative test, we know that the function has a relative minimum of $f(-3) = -26$ at $x = -3$ and a relative maximum of $f(5/3) = 670/27 \approx 24.8$ at $x = 5/3$.

Your Turn 3

$$f(x) = x^{2/3} - x^{5/3}$$
$$f'(x) = \frac{2}{3}x^{-1/3} - \frac{5}{3}x^{2/3}$$
$$= \frac{1}{3}\left(\frac{2 - 5x}{x^{1/3}}\right)$$

$x = 0$ is a critical number because $f(x)$ is defined at $x = 0$ but $f'(x)$ is not. To find any other critical numbers, assume $x \neq 0$ and set $f'(x) = 0$ and solve for x.

$$\frac{1}{3}\left(\frac{2 - 5x}{x^{1/3}}\right) = 0$$
$$2 - 5x = 0$$
$$x = \frac{2}{5}$$

Use the critical numbers to divide the line into three intervals, $(-\infty, 0)$, $(0, 2/5)$, and $(2/5, \infty)$. Choose test points in each interval; for example, choose -1; $-1/5$, and 1.

$$f'(-1) = -2.33$$
$$f'(1/5) = 0.57$$
$$f'(1) = -1$$

Thus the derivative is negative on $(-\infty, -1)$, positive on $(-1, 2/5)$, and negative on $(2/5, \infty)$. The graph below shows this information and indicates where f is increasing or decreasing.

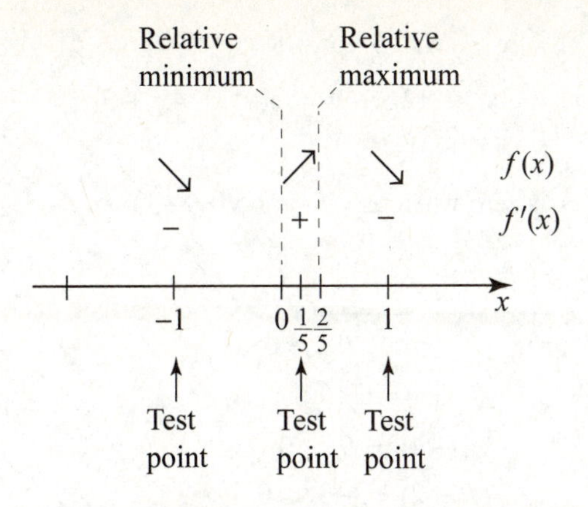

The first derivative test identifies $f(2/5)$ as a relative maximum, and there is a relative minimum at $x = 0$. Thus the function has a relative maximum of

$$f\left(\frac{2}{5}\right) = \frac{3}{5}\left(\frac{2}{5}\right)^{2/3} \approx 0.3257 \text{ at } x = \frac{2}{5}$$

and a relative minimum of $f(0) = 0$ at $x = 0$.

Your Turn 4

$$f(x) = x^2e^x$$

Using the product rule:

$$\begin{aligned} f'(x) &= x^2e^x + 2xe^x \\ &= e^x(x^2 + 2x) \end{aligned}$$

Find the critical values:

$$\begin{aligned} e^x(x^2 + 2x) &= 0 \\ x^2 + 2x &= 0 \\ x(x + 2) &= 0 \end{aligned}$$

$$x + 2 = 0 \quad \text{or} \quad x = 0$$
$$x = -2 \quad \text{or} \quad x = 0$$

There are two critical numbers, dividing the line into the intervals $(-\infty, -2)$, $(-2, 0)$ and $(0, \infty)$. Choose a test point in each interval, for example -3, -1, and 1.

$$\begin{aligned} f'(-3) &= 0.149 \\ f'(-1) &= -0.368 \\ f'(1) &= 8.155 \end{aligned}$$

Thus the derivative is positive on $(-\infty, -2)$, negative on $(-2, 0)$, and positive on $(0, \infty)$. The graph below shows this information and indicates where f is increasing or decreasing.

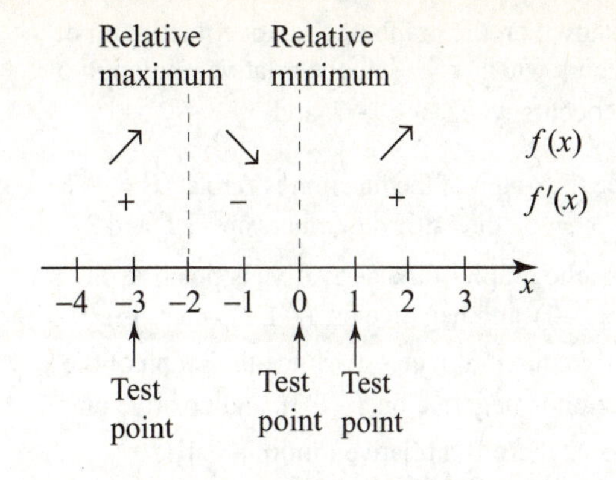

The function has a relative maximum of $4e^{-2} \approx 0.5413$ at $x = -2$ and a relative minimum of $f(0) = 0$ at $x = 0$.

Your Turn 5

$$\begin{aligned} C(q) &= 100 + 10q \\ p = D(q) &= 50 - 2q \\ P(q) &= R(q) - C(q) \\ &= qD(q) - C(q) \\ &= q(50 - 2q) - (100 + 10q) \\ &= 50q - 2q^2 - 100 + 10q \\ &= -2q^2 + 40q - 100 \\ P'(q) &= -4q + 40 \\ &= -4(q - 10) \end{aligned}$$

P' is 0 only when $q = 10$, and this is the only critical number. $P'(5) = 20$ and $P(20) = -40$ so P is increasing as q approaches 10 from the left and decreasing to the right of $q = 10$. The value of P at $q = 10$ is $P(10) = -2(100) + 40(1) - 100 = 100$.

Thus the maximum weekly profit is \$100 when the demand is 10 items per week. The company should charge $D(10) = 50 - 2(10) = 30$, or \$30 per item.

13.2 Exercises

1. As shown on the graph, the relative minimum of -4 occurs when $x = 1$.

3. As shown on the graph, the relative maximum of 3 occurs when $x = -2$.

5. As shown on the graph, the relative maximum of 3 occurs when $x = -4$ and the relative minimum of 1 occurs when $x = -2$.

7. As shown on the graph, the relative maximum of 3 occurs when $x = -4$; the relative minimum of -2 occurs when $x = -7$ and $x = -2$.

9. Since the graph of the function is zero at $x = -1$ and $x = 3$, the critical numbers are -1 and 3.

Since the graph of the derivative is positive on $(-\infty, -1)$ and negative on $(-1, 3)$, there is a relative maximum at -1. Since the graph of the function is negative on $(-1, 3)$ and positive on $(3, \infty)$, there is a relative minimum at 3.

11. Since the graph of the derivative is zero at $x = -8$, $x = -6$, $x = -2.5$ and $x = -1.5$, the critical numbers are $-8, -6, -2.5$, and -1.5.

Since the graph of the derivative is positive on $(-\infty, -8)$ and negative on $(-8, -6)$, there is a relative maximum at -8. Since the graph of the derivative is negative on $(-8, -6)$ and positive on $(-6, -2.5)$, there is a relative minimum at -6. Since the graph of the derivative is positive on $(-6, -2.5)$ and negative on $(-2.5, -1.5)$, there is a relative maximum at -2.5. Since the graph of the derivative is negative on $(-2.5, -1.5)$ and positive on $(-1.5, \infty)$, there is a relative minimum at -1.5.

13. $f(x) = x^2 - 10x + 33$

$f'(x) = 2x - 10$

$f'(x)$ is zero when $x = 5$.

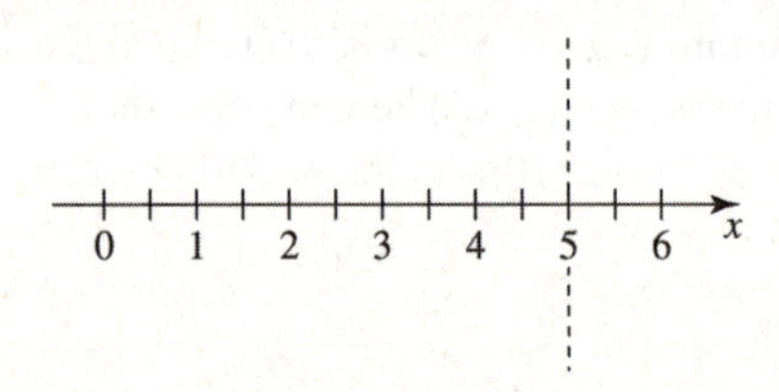

$$f'(0) = -10 < 0$$
$$f'(6) = 2 > 0$$

f is decreasing on $(-\infty, 5)$ and increasing on $(5, \infty)$. Thus, a relative minimum occurs at $x = 5$.

$$f(5) = 8$$

Relative minimum of 8 at 5

15. $f(x) = x^3 + 6x^2 + 9x - 8$

$f'(x) = 3x^2 + 12x + 9 = 3(x^2 + 4x + 3)$
$= 3(x + 3)(x + 1)$

$f'(x)$ is zero when $x = -1$ or $x = -3$.

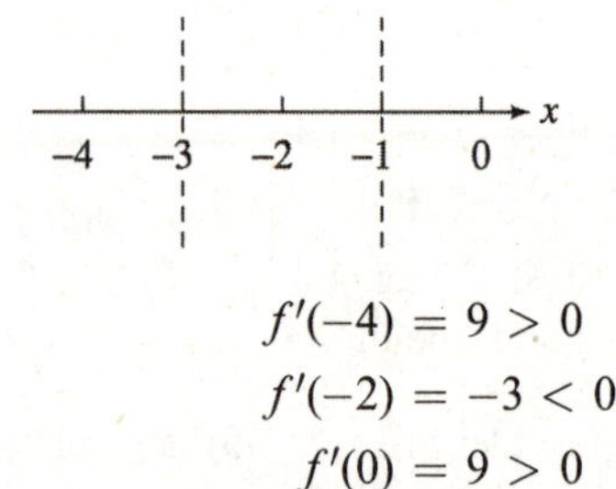

$$f'(-4) = 9 > 0$$
$$f'(-2) = -3 < 0$$
$$f'(0) = 9 > 0$$

Thus, f is increasing on $(-\infty, -3)$, decreasing on $(-3, -1)$, and increasing on $(-1, \infty)$.

f has a relative maximum at -3 and a relative minimum at -1.

$$f(-3) = -8$$
$$f(-1) = -12$$

Relative maximum of -8 at -3; relative minimum of -12 at -1

17. $f(x) = -\frac{4}{3}x^3 - \frac{21}{2}x^2 - 5x + 8$

$f'(x) = -4x^2 - 21x - 5$
$= (-4x - 1)(x + 5)$

$f'(x)$ is zero when $x = -5$, or $x = -\frac{1}{4}$.

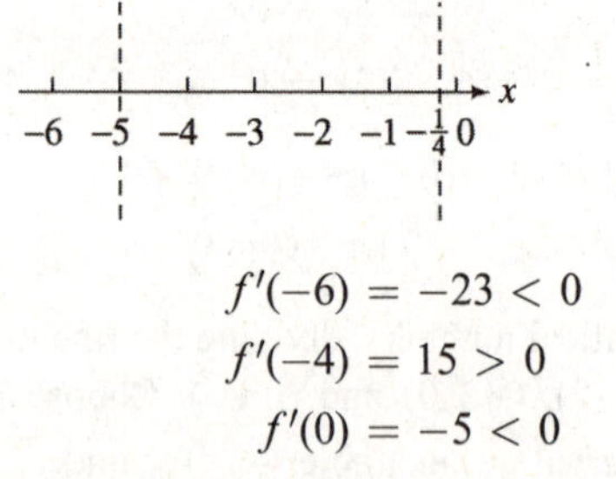

$$f'(-6) = -23 < 0$$
$$f'(-4) = 15 > 0$$
$$f'(0) = -5 < 0$$

f is decreasing on $(-\infty, -5)$, increasing on $\left(-5, -\frac{1}{4}\right)$, and decreasing on $\left(-\frac{1}{4}, \infty\right)$. f has a relative minimum at -5 and a relative maximum at $-\frac{1}{4}$.

$$f(-5) = -\frac{377}{6}$$
$$f\left(-\frac{1}{4}\right) = \frac{827}{96}$$

Relative maximum of $\frac{827}{96}$ at $-\frac{1}{4}$; relative minimum of $-\frac{377}{6}$ at -5.

19. $f(x) = x^4 - 18x^2 - 4$

$$f'(x) = 4x^3 - 36x$$
$$= 4x(x^2 - 9)$$
$$= 4x(x + 3)(x - 3)$$

$f'(x)$ is zero when $x = 0$ or $x = -3$ or $x = 3$.

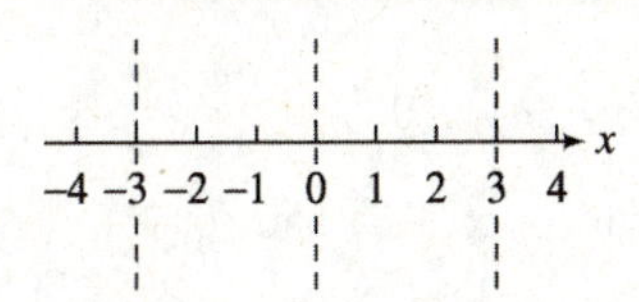

$$f'(-4) = 4(-4)^3 - 36(-4) = -112 < 0$$
$$f'(-1) = -4 + 36 = 32 > 0$$
$$f'(1) = 4 - 36 = -32 < 0$$
$$f'(4) = 4(4)^3 - 36(4) = 112 > 0$$

f is decreasing on $(-\infty, -3)$ and $(0, 3)$; f is increasing on $(-3, 0)$ and $(3, \infty)$.

$$f(-3) = -85$$
$$f(0) = -4$$
$$f(3) = -85$$

Relative maximum of -4 at 0; relative minimum of -85 at 3 and -3

21. $f(x) = 3 - (8 + 3x)^{2/3}$

$$f'(x) = -\frac{2}{3}(8 + 3x)^{-1/3}(3)$$
$$= -\frac{2}{(8 + 3x)^{1/3}}$$

Critical number:

$$8 + 3x = 0$$
$$x = -\frac{8}{3}$$

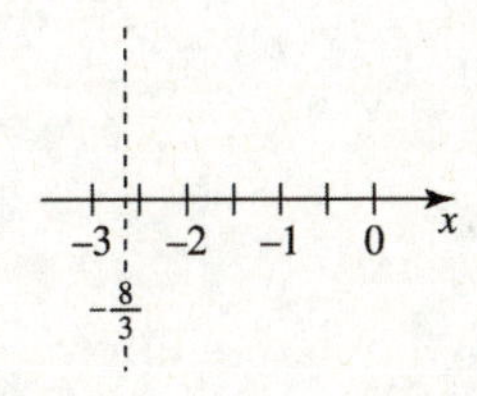

$$f'(-3) = 2 > 0$$
$$f'(0) = -1 < 0$$

f is increasing on $\left(-\infty, -\frac{8}{3}\right)$ and decreasing on $\left(-\frac{8}{3}, \infty\right)$.

$$f\left(-\frac{8}{3}\right) = 3$$

Relative maximum of 3 at $-\frac{8}{3}$.

23. $f(x) = 2x + 3x^{2/3}$

$$f'(x) = 2 + 2x^{-1/3}$$
$$= 2 + \frac{2}{\sqrt[3]{x}}$$

Find the critical numbers.

$f'(x) = 0$ when

$$2 + \frac{2}{\sqrt[3]{x}} = 0$$
$$\frac{2}{\sqrt[3]{x}} = -2$$
$$x = (-1)^3$$
$$x = -1.$$

$f'(x)$ does not exist when

$$\sqrt[3]{x} = 0$$
$$x = 0.$$

$$f'(-2) = 2 + \frac{2}{\sqrt[3]{-2}} \approx 0.41 > 0$$
$$f'\left(-\frac{1}{2}\right) = 2 + \frac{2}{\sqrt[3]{-\frac{1}{2}}}$$
$$= 2 + \frac{2\sqrt[3]{2}}{-1} \approx -0.52 < 0$$
$$f'(1) = 2 + \frac{2}{\sqrt[3]{1}} = 4 > 0$$

f is increasing on $(-\infty, -1)$ and $(0, \infty)$.

f is decreasing on $(-1, 0)$.

$$f(-1) = 2(-1) + 3(-1)^{2/3} = 1$$
$$f(0) = 0$$

Relative maximum of 1 at -1; relative minimum of 0 at 0.

25. $f(x) = x - \frac{1}{x}$

$f'(x) = 1 + \frac{1}{x^2}$ is never zero, but fails to exist at $x = 0$

Since $f(x)$ also fails to exist at $x = 0$, there are no critical numbers and no relative extrema.

27. $f(x) = \frac{x^2 - 2x + 1}{x - 3}$

$$f'(x) = \frac{(x-3)(2x-2) - (x^2 - 2x + 1)(1)}{(x-3)^2}$$

$$= \frac{x^2 - 6x + 5}{(x-3)^2}$$

Find the critical numbers:

$$x^2 - 6x + 5 = 0$$
$$(x-5)(x-1) = 0$$
$$x = 5 \text{ or } x = 1$$

Note that $f(x)$ and $f'(x)$ do not exist at $x = 3$, so the only critical numbers are 1 and 5.

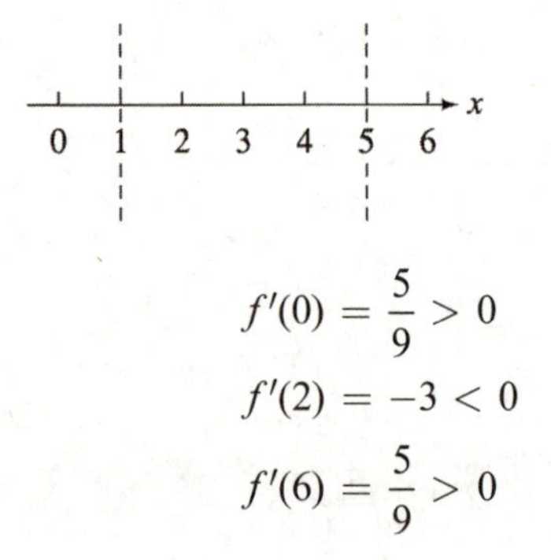

$$f'(0) = \frac{5}{9} > 0$$
$$f'(2) = -3 < 0$$
$$f'(6) = \frac{5}{9} > 0$$

$f(x)$ is increasing on $(-\infty, 1)$ and $(5, \infty)$.

$f(x)$ is decreasing on $(1, 5)$.

$$f(1) = 0$$
$$f(5) = 8$$

Relative maximum of 0 at 1; relative minimum of 8 at 5

29. $f(x) = x^2e^x - 3$

$$f'(x) = x^2e^x + 2xe^x$$
$$= xe^x(x + 2)$$

$f'(x)$ is zero at $x = 0$ and $x = -2$.

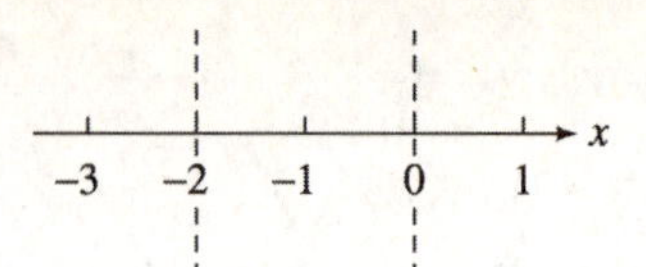

$$f'(-3) = 3e^{-3} = \frac{3}{e^3} > 0$$
$$f'(-1) = -e^{-1} = \frac{-1}{e} < 0$$
$$f'(1) = 3e^1 > 0$$

f is increasing on $(-\infty, -2)$ and $(0, \infty)$.

f is decreasing on $(-2, 0)$.

$$f(0) = 0 \cdot e^0 - 3 = -3$$
$$f(-2) = (-2)^2e^{-2} - 3$$
$$= \frac{4}{e^2} - 3$$
$$\approx -2.46$$

Relative minimum of -3 at 0; relative maximum of -2.46 at -2.

31. $f(x) = 2x + \ln x$

$$f'(x) = 2 + \frac{1}{x} = \frac{2x + 1}{x}$$

$f'(x)$ is zero at $x = -\frac{1}{2}$. The domain of $f(x)$ is $(0, \infty)$. Therefore $f'(x)$ is never zero in the domain of $f(x)$.

$f'(1) = 3 > 0$. Since $f(x)$ is always increasing, f has no relative extrema.

33. $f(x) = \frac{2^x}{x}$

$$f'(x) = \frac{(x)\ln 2(2^x) - 2^x(1)}{x^2}$$
$$= \frac{2^x(x \ln 2 - 1)}{x^2}$$

Find the critical numbers:

$$x\ln 2 - 1 = 0 \quad \text{or} \quad x^2 = 0$$
$$x = \frac{1}{\ln 2} \qquad\qquad x = 0$$

Since f is defined for $x = 0$, 0 is not a critical number. $x = \frac{1}{\ln 2} \approx 1.44$ is the only critical number.

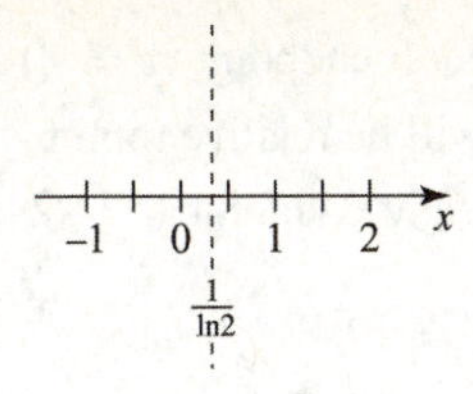

$$f'(1) \approx -0.6137 < 0$$
$$f'(2) \approx 0.3863 > 0$$

f is decreasing on $\left(0, \frac{1}{\ln 2}\right)$ and increasing on $\left(\frac{1}{\ln 2}, \infty\right)$.

$$f\left(\frac{1}{\ln 2}\right) = \frac{2^{1/\ln 2}}{\frac{1}{\ln 2}}$$
$$= \ln 2\left(e^{\ln 2}\right)^{1/\ln 2} = e \ln 2$$

Relative minimum of $e \ln 2$ at $\frac{1}{\ln 2}$.

35. $y = -2x^2 + 12x - 5$

$$y' = -4x + 12$$
$$= -4(x - 3)$$

The vertex occurs when $y' = 0$ or when

$$x - 3 = 0$$
$$x = 0$$

When $x = 3$,

$$y = -2(3)^2 + 12(3) - 5 = 13$$

The vertex is $(3, 13)$.

37. $f(x) = x^5 - x^4 + 4x^3 - 30x^2 + 5x + 6$

$$f'(x) = 5x^4 - 4x^3 + 12x^2 - 60x + 5$$

Graph f' on a graphing calculator. A suitable choice for the viewing window is $[-4, 4]$ by $[-50, 50]$, Yscl $= 10$.

Use the calculator to estimate the x-intercepts of this graph. These numbers are the solutions of the equation $f'(x) = 0$ and thus the critical numbers for f. Rounded to three decimal places, these x-values are 0.085 and 2.161.

Examine the graph of f' near $x = 0.085$ and $x = 2.161$. Observe that $f'(x) > 0$ to the left of $x = 0.085$ and $f'(x) < 0$ to the right of $x = 0.085$. Also observe that $f'(x) < 0$ to the left of $x = 2.161$ and $f'(x) > 0$ to the right of $x = 2.161$. The first derivative test allows us to conclude that f has a relative maximum at $x = 0.085$ and a relative minimum at $x = 2.161$.

$$f(0.085) \approx 6.211$$
$$f(2.161) \approx -57.607$$

Relative maximum of 6.211 at 0.085; relative minimum of -57.607 at 2.161.

39. $f(x) = 2|x + 1| + 4|x - 5| - 20$

Graph this function in the window $[-10, 10]$ by $[-15, 30]$, Yscl $= 5$.

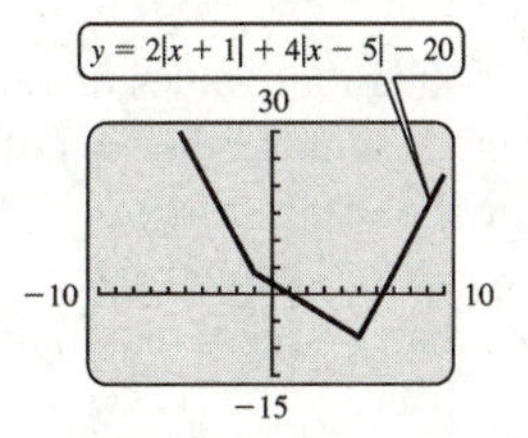

The graph shows that f has no relative maxima, but there is a relative minimum at $x = 5$.

(Note that the graph has a sharp point at $(5, -8)$, indicating that $f'(-5)$ does not exist.)

41. $C(q) = 80 + 18q$; $p = 70 - 2q$

$$P(q) = R(q) - C(q) = pq - C(q)$$
$$= (70 - 2q)q - (80 + 18q)$$
$$= -2q^2 + 52q - 80$$

(a) Since the graph of P is a parabola that opens downward, we know that its vertex is a maximum point. To find the q-value of this point, we find the critical number.

$$P'(q) = -4q + 52$$
$$P'(q) = 0 \text{ when}$$
$$-4q + 52 = 0$$
$$4q = 52$$
$$q = 13$$

The number of units that produce maximum profit is 13.

(b) If $q = 13$,

$$p = 70 - 2(13) = 44$$

The price that produces maximum profit is \$44.

(c) $P(13) = -2(13)^2 + 52(13) - 80 = 258$

The maximum profit is \$258.

43. $C(q) = 100 + 20qe^{-0.01q}$; $p = 40e^{-0.01q}$

$$P(q) = R(q) - C(q) = pq - C(q)$$
$$= (40e^{-0.01q})q - (100 + 20qe^{-0.01q})$$
$$= 20qe^{-0.01q} - 100$$

(a) $P'(q) = 20e^{-0.01q} + 20qe^{-0.01q}(-0.01)$
$$= (20 - 0.2q)e^{-0.01q}$$

Solve $P'(q) = 0$.

$$(20 - 0.2q)e^{-0.01q} = 0$$
$$20 - 0.2q = 0$$
$$q = 100$$

Since $e^{-0.01q} > 0$ for all values of q, the sign of $P'(q)$ is the same as the sign of $20 - 0.2q$. For $q < 100$, $P'(q) > 0$; for $q > 100$, $P'(q) < 0$. Therefore, the number of units that produces maximum profit is 100.

(b) If $q = 100$,

$$p = 40e^{-0.01(100)}$$
$$= 40e^{-1}$$
$$\approx 14.72$$

The price per unit that produces maximum profit is \$14.72.

(c) $P(100) = 20(100)e^{-0.01(100)} - 100$
$$= 2000e^{-1} - 100$$
$$\approx 635.76$$

The maximum profit is \$635.76.

45. $P(t) = -0.005846t^3 + 0.1614t^2 - 0.4910t + 20.47$

$P'(t) = -0.017538 + 0.3228t - 0.4190$

Use a graphing calculator to find the critical numbers for $0 \leq t \leq 24$.

The critical numbers are $t \approx 1.673$ and $t \approx 16.733$. These divide the domain of P into three intervals, and we pick a test point in each: Pick $t = 1$, $t = 10$, and $t = 20$.

$$P'(1) \approx -0.186$$
$$P'(10) \approx 0.983$$
$$P'(20) \approx -1.050$$

P is decreasing on $(0, 1.673)$ and $(16.733, 24)$ and increasing on $(1.673, 16.733)$. There will be relative maxima at the left endpoint ($t = 0$) and at $t = 16.733$; there will be relative minima at $t = 1.673$ and at the right endpoint ($t = 24$).

$$P(0) = 20.47$$
$$P(1.673) \approx 20.07$$
$$P(16.733) \approx 30.06$$
$$P(24) \approx 20.84$$

We convert the time values to hours and minutes by multiplying the decimal parts by 60: $(0.673)(60) \approx 40$ and $(0.733)(60) \approx 44$.

Recalling that the function P gives the power in thousands of megawatts, we have the final result:

Relative maximum of 20,470 megawatts at midnight ($t = 0$)

Relative minimum of 20,070 megawatts at 1:40 AM.

Relative maximum of 30,060 megawatts at 4:44 PM.

Relative minimum of 20,840 megawatts at midnight ($t = 24$).

47. $p = D(q) = 200e^{-0.1q}$

$$R(q) = pq$$
$$= 200qe^{-0.1q}$$
$$R'(q) = 200qe^{-0.1q}(-0.1) + 200e^{-0.1q}$$
$$= 20e^{-0.1q}(10 - q)$$

$R'(q) = 0$ when $q = 10$, the only critical number. Use the first derivative test to verify that $q = 10$ gives the maximum revenue.

$$R'(9) = 20e^{-0.9} > 0$$
$$R'(11) = -20e^{-1.1} < 0$$

The maximum revenue results when $q = 10$

$p = D(10) = \frac{200}{e} \approx 73.58$, or when telephones are sold at \$73.58.

49. $C(x) = 0.002x^3 = 9x + 6912$

$$\overline{C}(x) = \frac{C(x)}{x} = 0.002x^2 + 9 + \frac{6912}{x}$$

$$\overline{C}'(x) = 0.004x - \frac{6912}{x^2}$$

$\overline{C}'(x) = 0$ when

$$0.004x - \frac{6912}{x^2} = 0$$
$$0.004x^3 = 6912$$
$$x^3 = 1{,}728{,}000$$
$$x = 120$$

A product level of 120 units will produce the minimum average cost per unit.

51. $a(t) = 0.008t^3 - 0.288t^2 + 2.304t + 7$

$a'(t) = 0.024t^2 - 0.576t + 2.304$

Set $a' = 0$ and solve for t.

$$0.024t^2 - 0.576t + 2.304 = 0$$
$$0.024(t^2 - 24t + 96) = 0$$
$$t^2 - 24t + 96 = 0$$
$$t \approx 5.07 \text{ or } t \approx 18.93$$

$t = 5.07 = 5$ hours $+ 0.07 \cdot 60$ minutes corresponds to 5:04 P.M.

$t = 18.93 = 18$ hours $+ 0.93 \cdot 60$ minutes corresponds to 6:56 A.M.

53. $M(t) = 369(0.93)^t(t)^{0.36}$

$$M'(t) = (369)(0.93)^t \ln(0.93)(t^{0.36}) + 369(0.93)^t(0.36)(t)^{-0.64}$$
$$= (369t^{0.36})(0.93^t \ln 0.93) + \frac{132.84(0.93)^t}{t^{0.64}}$$

$M'(t) = 0$ when $t \approx 4.96$.

Verify that $t \approx 4.96$ gives a maximum.

$$M'(4) > 0$$
$$M'(5) < 0$$

Find $M(4.96)$

$$M(4.96) = 369(0.93)^{4.96}(4.96)^{0.36} \approx 485.22$$

The female moose reaches a maximum weight of about 458.22 kilograms at about 4.96 years.

55. $D(x) = -x^4 + 8x^3 + 80x^2$

$$D'(x) = -4x^3 + 24x^2 + 160x$$
$$= -4x(x^2 - 6x - 40)$$
$$= -4x(x + 4)(x - 10)$$

$D'(x) = 0$ when $x = 0$, $x = -4$, or $x = 10$.

Disregard the nonpositive values.

Verify that $x = 10$ gives a maximum.

$$D'(9) = 468 > 0$$
$$D'(11) = -660 < 0$$

The speaker should aim for a degree of discrepancy of 10.

57. $s(t) = -16t^2 + 40t + 3$

$s'(t) = -32t + 40$

(a) when $s'(t) = 0$,

$$-32t + 40 = 0$$
$$32t = 40$$
$$t = \frac{40}{32} = \frac{5}{4}$$

Verify that $t = 5/4$ gives a maximum.

$$s'(1) = 8$$
$$s'(2) = -24$$

Now find the height when $t = 5/4$.

$$s\left(\frac{5}{4}\right) = -16\left(\frac{5}{4}\right)^2 + 40\left(\frac{5}{4}\right) + 3$$
$$= -25 + 50 + 3$$
$$= 28$$

The maximum height of the cork is 28 feet.

(b) The cork remains in the air as long as $s(t) > 0$. Use the quadratic formula to solve $s(t) = 0$.

$$-16t^2 + 40t + 3 = 0$$
$$t = \frac{-40 \pm \sqrt{40^2 - 4(-16)(3)}}{2(-16)}$$
$$= \frac{-5 \pm \sqrt{28}}{-4}$$
$$= \frac{5 \pm 2\sqrt{7}}{4}$$
$$\approx -0.073, 2.573$$

Only the positive solution is relevant, so the cork stays in the air for about 2.57 seconds.

13.3 Higher Derivatives, Concavity, and the Second Derivative Test

Your Turn 1

$$f(x) = 5x^4 - 4x^3 + 3x$$
$$f'(x) = 20x^3 - 12x^2 + 3$$
$$f''(x) = 60x^2 - 24x$$
$$f''(1) = 60(1^2) - 24(1) = 36$$

Your Turn 2

(a) Use the chain rule.

$$f(x) = (x^3 + 1)^2$$
$$f'(x) = (2)(x^3 + 1)(3x^2) = 6x^2(x^3 + 1) = 6x^5 + 6x^2$$
$$f''(x) = 30x^4 + 12x$$

(b) Use the product rule.

$$f(x) = xe^x$$
$$f'(x) = xe^x + (1)e^x = xe^x + e^x$$

Now note that we have already found the derivative of the first term of $f'(x)$, which is the original function $f(x)$.

Thus

$$f''(x) = (xe^x + e^x) + e^x = 2e^x + xe^x$$

(c) Use the quotient rule.

$$h(x) = \frac{\ln x}{x}$$
$$h'(x) = \frac{(x)\left(\frac{1}{x}\right) - (\ln x)(1)}{x^2} = \frac{1 - \ln x}{x^2}$$
$$h''(x) = \frac{(x^2)\left(-\frac{1}{x}\right) - (1 - \ln x)(2x)}{x^4} = \frac{-x - 2x + 2x\ln x}{x^4} = \frac{-3 + 2\ln x}{x^3}$$

Your Turn 3

$$s(t) = t^3 - 3t^2 - 24t + 10$$
$$v(t) = s'(t) = 3t^2 - 6t - 24$$
$$a(t) = v'(t) = s''(t) = 6t - 6$$

First find when v changes sign.

$$v(t) = 3t^2 - 6t - 24 = 0$$
$$3(t^2 - 2t - 8) = 0$$
$$(t - 4)(t + 2) = 0$$
$$t = 4 \text{ or } t = -2$$

Only the positive value of t is relevant. Check the velocity at times before and after 4, say at $t = 2$ and $t = 6$.

$$v(2) = -24 < 0$$
$$v(6) = 48 > 0$$

Thus the car backs up for the first four seconds and then goes forward.

Now find where a changes sign.

$$a(t) = 6t - 6 = 0$$
$$6t - 6 = 0$$
$$t = 1$$

Check the acceleration at times before and after 1, say at $t = 1/2$ and $t = 2$.

$$a(1/2) = -3 < 0$$
$$a(2) = 6 > 0$$

Now we construct the following graph showing the signs of v and a.

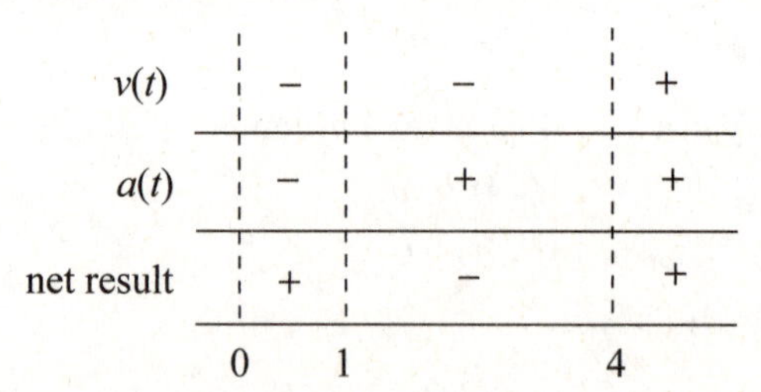

The car speeds up (in the backward direction) for $0 < t < 1$; it slows down (still in the backward direction) for $1 < t < 4$; it speeds up (in the forward direction) for $t > 4$. (All times are in seconds.)

Your Turn 4

$$f(x) = x^5 - 30x^3$$
$$f'(x) = 5x^4 - 90x^2$$
$$f''(x) = 20x^3 - 180x$$

Factor $f''(x)$ and create a number line.

$$\begin{aligned} f''(x) &= 20x^3 - 180x \\ &= 20(x^3 - 9x) \\ &= 20(x)(x-3)(x+3) \end{aligned}$$

The zeros of $f''(x)$ are $-3, 0,$ and 3, and they divide the domain of f into four intervals:

$$(-\infty, -3), (-3, 0), (0, 3), (3, \infty)$$

Choose a test point in each interval and find the sign of $f''(x)$ at the test points. For example, choose $-4, -1, 1,$ and 4.

$$\begin{aligned} f''(-4) &= -560 \\ f''(-1) &= 160 \\ f''(1) &= -160 \\ f''(4) &= 560 \end{aligned}$$

Here is the corresponding number line.

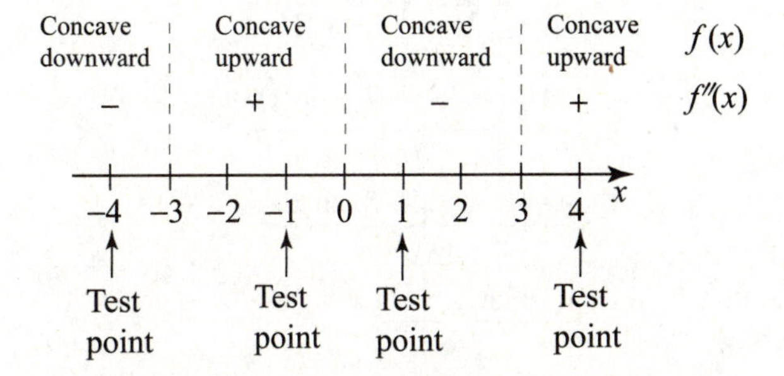

The figure shows that f is concave downward on $(-\infty, -3)$ and $(0, 3)$ and concave upward on $(-3, 0)$ and $(3, \infty)$. Inflection points occur where $f''(x)$ changes sign, so we evaluate f at these points:

$$\begin{aligned} f(-3) &= 567 \\ f(0) &= 0 \\ f(3) &= -567 \end{aligned}$$

Thus the inflection points are $(-3, 567)$, $(0, 0)$ and $(3, -567)$.

Your Turn 5

$$\begin{aligned} f(x) &= -2x^3 + 3x^2 + 72x \\ f'(x) &= -6x^2 + 6x + 72 \end{aligned}$$

Solve the equation $f'(x) = 0$.

$$\begin{aligned} -6x^2 + 6x + 72 &= 0 \\ x^2 - x - 12 &= 0 \\ (x+3)(x-4) &= 0 \\ x = -3 \text{ or } x &= 4 \end{aligned}$$

Now use the second derivative test.

$$\begin{aligned} f''(x) &= -12x + 6 \\ f''(-3) &= 42 \\ f''(4) &= -42 \end{aligned}$$

Since $f''(-3) > 0$, $x = -3$ leads to a relative minimum; since $f''(4) < 0$, $x = 4$ leads to a relative maximum. Thus we have a relative minimum of $f(-3) = -135$ at $x = -3$ and a relative maximum of $f(4) = 208$ at $x = 4$.

13.3 Exercises

1.
$$\begin{aligned} f(x) &= 5x^3 - 7x^2 + 4x + 3 \\ f'(x) &= 15x^2 - 14x + 4 \\ f''(x) &= 30x - 14 \\ f''(0) &= 30(0) - 14 = -14 \\ f''(2) &= 30(2) - 14 = 46 \end{aligned}$$

3.
$$\begin{aligned} f(x) &= 4x^4 - 3x^3 - 2x^2 + 6 \\ f'(x) &= 16x^3 - 9x^2 - 4x \\ f''(x) &= 48x^2 - 18x - 4 \\ f''(0) &= 48(0)^2 - 18(0) - 4 = -4 \\ f''(2) &= 48(2)^2 - 18(2) - 4 = 152 \end{aligned}$$

5.
$$\begin{aligned} f(x) &= 3x^2 - 4x + 8 \\ f'(x) &= 6x - 4 \\ f''(x) &= 6 \\ f''(0) &= 6 \\ f''(2) &= 6 \end{aligned}$$

7.
$$\begin{aligned} f(x) &= \frac{x^2}{1+x} \\ f'(x) &= \frac{(1+x)(2x) - x^2(1)}{(1+x)^2} \\ &= \frac{2x + x^2}{(1+x)^2} \end{aligned}$$

$$f''(x) = \frac{(1+x)^2(2+2x) - (2x+x^2)(2)(1+x)}{(1+x)^4}$$
$$= \frac{(1+x)(2+2x) - (2x+x^2)(2)}{(1+x)^3}$$
$$= \frac{2}{(1+x)^3}$$
$$f''(0) = 2$$
$$f''(2) = \frac{2}{27}$$

9. $f(x) = \sqrt{x^2+4} = (x^2+4)^{1/2}$

$$f'(x) = \frac{1}{2}(x^2+4)^{-1/2} \cdot 2x$$
$$= \frac{x}{(x^2+4)^{1/2}}$$
$$f''(x) = \frac{(x^2+4)^{1/2}(1) - x\left[\frac{1}{2}(x^2+4)^{-1/2}\right]2x}{x^2+4}$$
$$= \frac{(x^2+4)^{1/2} - \frac{x^2}{(x^2+4)^{1/2}}}{x^2+4}$$
$$= \frac{(x^2+4) - x^2}{(x^2+4)^{3/2}}$$
$$= \frac{4}{(x^2+4)^{3/2}}$$
$$f''(0) = \frac{4}{(0^2+4)^{3/2}}$$
$$= \frac{4}{4^{3/2}} = \frac{4}{8} = \frac{1}{2}$$
$$f''(2) = \frac{4}{(2^2+4)^{3/2}}$$
$$= \frac{4}{8^{3/2}} = \frac{4}{16\sqrt{2}} = \frac{1}{4\sqrt{2}}$$

11. $f(x) = 32x^{3/4}$

$$f'(x) = 24x^{-1/4}$$
$$f''(x) = -6x^{-5/4} = -\frac{6}{x^{5/4}}$$

$f''(0)$ does not exist.

$$f''(2) = -\frac{6}{2^{5/4}}$$
$$= -\frac{3}{2^{1/4}}$$

13. $f(x) = 5e^{-x^2}$

$$f'(x) = 5e^{-x^2}(-2x) = -10xe^{-x^2}$$
$$f''(x) = -10xe^{-x^2}(-2x) + e^{-x^2}(-10)$$
$$= 20x^2e^{-x^2} - 10e^{-x^2}$$
$$f''(0) = 20(0^2)e^{-0^2} - 10e$$
$$= 0 - 10 = -10$$
$$f''(2) = 20(2^2)e^{-(2^2)} - 10e^{-(2^2)}$$
$$= 80e^{-4} - 10e^{-4} = 70e^{-4}$$
$$\approx 1.282$$

15. $f(x) = \frac{\ln x}{4x}$

$$f'(x) = \frac{4x\left(\frac{1}{x}\right) - (\ln x)(4)}{(4x)^2}$$
$$= \frac{4 - 4\ln x}{16x^2} = \frac{1 - \ln x}{4x^2}$$
$$f''(x) = \frac{4x^2\left(-\frac{1}{x}\right) - (1 - \ln x)8x}{16x^4}$$
$$= \frac{-4x - 8x + 8x\ln x}{16x^4}$$
$$= \frac{-12x + 8x\ln x}{16x^4}$$
$$= \frac{4x(-3 + 2\ln x)}{16x^4}$$
$$= \frac{-3 + 2\ln x}{4x^3}$$

$f''(0)$ does not exist because ln 0 is undefined.

$$f''(2) = \frac{-3 + 2\ln 2}{4(2)^3} = \frac{-3 + 2\ln 2}{32} \approx 0.050$$

17. $f(x) = 7x^4 + 6x^3 + 5x^2 + 4x + 3$

$$f'(x) = 28x^3 + 18x^2 + 10x + 4$$
$$f''(x) = 84x^2 + 36x + 10$$
$$f'''(x) = 168x + 36$$
$$f^{(4)}(x) = 168$$

19. $f(x) = 5x^5 - 3x^4 + 2x^3 + 7x^2 + 4$

$f'(x) = 25x^4 - 12x^3 + 6x^2 + 14x$

$f''(x) = 100x^3 - 36x^2 + 12x + 14$

$f'''(x) = 300x^2 - 72x + 12$

$f^{(4)}(x) = 600x - 72$

21. $f(x) = \dfrac{x-1}{x+2}$

$f'(x) = \dfrac{(x+2)-(x-1)}{(x+2)^2} = \dfrac{3}{(x+2)^2}$

$f''(x) = \dfrac{-3(2)(x+2)}{(x+2)^4} = \dfrac{-6}{(x+2)^3}$

$f'''(x) = \dfrac{(-6)(-3)(x+2)^2}{(x+2)^6}$

$= 18(x+2)^{-4}$ or $\dfrac{18}{(x+2)^4}$

$f^{(4)}(x) = \dfrac{-18(4)(x+2)^3}{(x+2)^8}$

$= -72(x+2)^{-5}$ or $\dfrac{-72}{(x+2)^5}$

23. $f(x) = \dfrac{3x}{x-2}$

$f'(x) = \dfrac{(x-2)(3) - 3x(1)}{(x-2)^2} = \dfrac{-6}{(x-2)^2}$

$f''(x) = \dfrac{-6(-2)(x-2)}{(x-2)^4} = \dfrac{12}{(x-2)^3}$

$f'''(x) = \dfrac{-12(3)(x-2)^2}{(x-2)^6} = -36(x-2)^{-4}$

or $\dfrac{-36}{(x-2)^4}$

$f^{(4)}(x) = \dfrac{-36(-4)(x-2)^3}{(x-2)^8} = 144(x-2)^{-5}$

or $\dfrac{144}{(x-2)^5}$

25. $f(x) = \ln x$

(a) $f'(x) = \dfrac{1}{x} = x^{-1}$

$f''(x) = -x^{-2} = \dfrac{-1}{x^2}$

$f'''(x) = 2x^{-3} = \dfrac{2}{x^3}$

$f^{(4)}(x) = -6x^{-4} = \dfrac{-6}{x^4}$

$f^{(5)}(x) = 24x^{-5} = \dfrac{24}{x^5}$

(b) $f^{(n)}(x) = \dfrac{(-1)^{n-1}(n-1)!}{x^n}$

27. Concave upward on $(2, \infty)$

Concave downward on $(-\infty, 2)$

Inflection point at $(2, 3)$

29. Concave upward on $(-\infty, -1)$ and $(8, \infty)$

Concave downward on $(-1, 8)$

Inflection points at $(-1, 7)$ and $(8, 6)$

31. Concave upward on $(2, \infty)$

Concave downward on $(-\infty, 2)$

No points inflection

33. $f(x) = x^2 + 10x - 9$

$f'(x) = 2x + 10$

$f''(x) = 2 > 0$ for all x.

Always concave upward

No inflection points

35. $f(x) = -2x^3 + 9x^2 + 168x - 3$

$$f'(x) = -6x^2 + 18x + 168$$
$$f''(x) = -12x + 18$$
$$f''(x) = -12x + 18 > 0 \text{ when}$$
$$-6(2x - 3) > 0$$
$$2x - 3 < 0$$
$$x < \frac{3}{2}.$$

Concave upward on $\left(-\infty, \frac{3}{2}\right)$

$$f''(x) = -12x + 18 < 0 \text{ when}$$
$$-6(2x - 3) < 0$$
$$2x - 3 > 0$$
$$x > \frac{3}{2}.$$

Concave downward on $\left(\frac{3}{2}, \infty\right)$

$$f''(x) = -12x + 18 = 0 \text{ when}$$
$$-6(2x + 3) = 0$$
$$2x + 3 = 0$$
$$x = \frac{3}{2}.$$
$$f\left(\frac{3}{2}\right) = \frac{525}{2}$$

Inflection point at $\left(\frac{3}{2}, \frac{525}{2}\right)$

37. $f(x) = \dfrac{3}{x - 5}$

$$f'(x) = \frac{-3}{(x - 5)^2}$$
$$f''(x) = \frac{-3(-2)(x - 5)}{(x - 5)^4} = \frac{6}{(x - 5)^3}$$

$$f''(x) = \frac{6}{(x - 5)^3} > 0 \text{ when}$$
$$(x - 5)^3 > 0$$
$$x - 5 > 0$$
$$x > 5.$$

Concave upward on $(5, \infty)$

$$f''(x) = \frac{6}{(x - 5)^3} < 0 \text{ when}$$
$$(x - 5)^3 < 0$$
$$x - 5 < 0$$
$$x < 5.$$

Concave downward on $(-\infty, 5)$

$f''(x) \neq 0$ for any value for x; it does not exist when $x = 5$. There is a change of concavity there, but no inflection point since $f(5)$ does not exist.

39. $f(x) = x(x + 5)^2$

$$f'(x) = x(2)(x + 5) + (x + 5)^2$$
$$= (x + 5)(2x + x + 5)$$
$$= (x + 5)(3x + 5)$$
$$f''(x) = (x + 5)(3) + (3x + 5)$$
$$= 3x + 15 + 3x + 5 = 6x + 20$$
$$f''(x) = 6x + 20 > 0 \text{ when}$$
$$2(3x + 10) > 0$$
$$3x > -10$$
$$x > -\frac{10}{3}.$$

Concave upward on $\left(-\frac{10}{3}, \infty\right)$

$$f''(x) = 6x + 20 < 0 \text{ when}$$
$$2(3x + 10) < 0$$
$$3x < -10$$
$$x < -\frac{10}{3}.$$

Concave downward on $\left(-\infty, -\frac{10}{3}\right)$

$$f\left(-\frac{10}{3}\right) = -\frac{10}{3}\left(-\frac{10}{3} + 5\right)^2$$
$$= \frac{-10}{3}\left(\frac{-10 + 15}{3}\right)^2$$
$$= -\frac{10}{3} \cdot \frac{25}{9} = -\frac{250}{27}$$

Inflection point at $\left(-\frac{10}{3}, -\frac{250}{27}\right)$

41. $f(x) = 18x - 18e^{-x}$

$$f'(x) = 18 - 18e^{-x}(-1) = 18 + 18e^{-x}$$
$$f''(x) = 18e^{-x}(-1) = -18e^{-x}$$
$$f''(x) = -18e^{-x} < 0 \text{ for all } x$$

$f(x)$ is never concave upward and always concave downward. There are no points of inflection since $-18e^{-x}$ is never equal to 0.

43. $f(x) = x^{8/3} - 4x^{5/3}$

$f'(x) = \frac{8}{3}x^{5/3} - \frac{20}{3}x^{2/3}$

$f''(x) = \frac{40}{9}x^{2/3} - \frac{40}{9}x^{-1/3} = \frac{40(x-1)}{9x^{1/3}}$

$f''(x) = 0$ when $x = 1$

$f''(x)$ fails to exist when $x = 0$

Note that both $f(x)$ and $f'(x)$ exist at $x = 0$. Check the sign of $f''(x)$ in the three intervals determined by $x = 0$ and $x = 1$ using test points.

$$f''(-1) = \frac{40(-2)}{9(-1)} = \frac{80}{9} > 0$$

$$f''\left(\frac{1}{8}\right) = \frac{40\left(-\frac{7}{8}\right)}{9\left(\frac{1}{2}\right)} = -\frac{70}{9} < 0$$

$$f''(8) = \frac{40(7)}{9(2)} = \frac{140}{9} > 0$$

Concave upward on $(-\infty, 0)$ and $(1, \infty)$; concave downward on $(0, 1)$

$$f(0) = (0)^{8/3} - 4(0)^{5/3} = 0$$

$$f(1) = (1)^{8/3} - 4(1)^{5/3} = -3$$

Inflection points at $(0, 0)$ and $(1, -3)$

45. $f(x) = \ln(x^2 + 1)$

$f'(x) = \frac{2x}{x^2 + 1}$

$f''(x) = \frac{(x^2 + 1)(2) - (2x)(2x)}{(x^2 + 1)^2}$

$= \frac{-2x^2 + 2}{(x^2 + 1)^2}$

$f''(x) = \frac{-2x^2 + 2}{(x^2 + 1)^2} > 0$ when

$-2x^2 + 2 > 0$

$-2x^2 > -2$

$x^2 < 1$

$-1 < x < 1$

Concave upward on $(-1, 1)$

$f''(x) = \frac{-2x^2 + 2}{(x^2 + 1)^2} < 0$ when

$-2x^2 + 2 < 0$

$-2x^2 < -2$

$x^2 > 1$

$x > 1$ or $x < -1$

Concave downward on $(-\infty, -1)$ and $(1, \infty)$

$$f(1) = \ln[(1)^2 + 1] = \ln 2$$

$$f(-1) = \ln[(-1)^2 + 1] = \ln 2$$

Inflection points at $(-1, \ln 2)$ and $(1, \ln 2)$

47. $f(x) = x^2 \log|x|$

$f'(x) = 2x\log|x| + x^2\left(\frac{1}{x\ln 10}\right)$

$= 2x\log|x| + \frac{x}{\ln 10}$

$f''(x) = 2\log|x| + 2x\left(\frac{1}{x\ln 10}\right) + \frac{1}{\ln 10}$

$= 2\log|x| + \frac{3}{\ln 10}$

$f''(x) > 0$ when

$$2\log|x| + \frac{3}{\ln 10} > 0$$

$$2\log|x| > -\frac{3}{\ln 10}$$

$$\log|x| > -\frac{3}{2\ln 10}$$

$$\frac{\ln|x|}{\ln 10} > -\frac{3}{2\ln 10}$$

$$\ln|x| > -\frac{3}{2}$$

$$|x| > e^{-3/2}$$

$$x > e^{-3/2} \text{ or } x < -e^{-3/2}$$

Concave upward on $(-\infty, -e^{-3/2})$ and $(e^{-3/2}, \infty)$

$f''(x) < 0$ when

$$2 \log |x| + \frac{3}{\ln 10} < 0$$

$$2 \log |x| < -\frac{3}{\ln 10}$$

$$\log |x| < -\frac{3}{2 \ln 10}$$

$$\frac{\ln |x|}{\ln 10} < -\frac{3}{2 \ln 10}$$

$$\ln |x| < -\frac{3}{2}$$

$$|x| < e^{-3/2}$$

$$-e^{-3/2} < x < e^{-3/2}$$

Note that $f(x)$ is not defined at $x = 0$.

Concave downward on $(-e^{-3/2}, 0)$ and $(0, e^{-3/2})$.

$$f(-e^{-3/2}) = (-e^{-3/2})^2 \log|-e^{-3/2}|$$

$$= e^{-3} \log e^{-3/2} = -\frac{3e^{-3}}{2 \ln 10}$$

$$f(e^{-3/2}) = (e^{-3/2})^2 \log|e^{-3/2}|$$

$$= e^{-3} \log e^{-3/2} = -\frac{3e^{-3}}{2 \ln 10}$$

Inflection points at $\left(-e^{-3/2}, -\frac{3e^{-3}}{2 \ln 10}\right)$ and $\left(e^{-3/2}, -\frac{3e^{-3}}{2 \ln 10}\right)$

49. Since the graph of $f'(x)$ is increasing on $(-\infty, 0)$ and $(4, \infty)$, the function is concave upward on $(-\infty, 0)$ and $(4, \infty)$. Since the graph of $f'(x)$ is decreasing on $(0, 4)$, the function is concave downward on $(0, 4)$. The inflection points are at 0 and 4.

51. Since the graph of $f'(x)$ is increasing on $(-7, 3)$ and $(12, \infty)$, the function is concave upward on $(-7, 3)$ and $(12, \infty)$. Since the graph of $f'(x)$ is decreasing on $(-\infty, -7)$ and $(3, 12)$, the function is concave downward on $(-\infty, -7)$ and $(3, 12)$. The inflection points are at $-7, 3,$ and 12.

53. Choose $f(x) = x^k$, where $1 < k < 2$.

If $k = \frac{4}{3}$, then

$$f'(x) = \frac{4}{3}x^{1/3} \qquad f''(x) = \frac{4}{9}x^{-2/3} = \frac{4}{9x^{2/3}}$$

Critical number: 0

Since $f'(x)$ is negative when $x < 0$ and positive when $x > 0$, $f(x) = x^{4/3}$ has a relative minimum at $x = 0$.

If $k = \frac{5}{3}$, then

$$f(x) = \frac{5}{3}x^{2/3} \qquad f''(x) = \frac{10}{9}x^{-1/3} = \frac{10}{9x^{1/3}}$$

$f''(x)$ is never 0, and does not exist when $x = 0$; so, the only candidate for an inflection point is at $x = 0$.

Since $f''(x)$ is negative when $x < 0$ and positive when $x > 0$, $f(x) = x^{5/3}$ has an inflection point at $x = 0$.

55. **(a)** The slope of the tangent line to $f(x) = e^x$ as $x \to -\infty$ is close to 0 since the tangent line is almost horizontal, and a horizontal line has a slope of 0.

(b) The slope of the tangent line to $f(x) = e^x$ as $x \to 0$ is close to 1 since the first derivative represents the slope of the tangent line, $f'(x) = e^x$, and $e^0 = 1$.

57.

$$f(x) = -x^2 - 10x - 25$$

$$f'(x) = -2x - 10$$

$$= -2(x + 5) = 0$$

Critical number: -5

$f''(x) = -2 < 0$ for all x.

The curve is concave downward, which means a relative maximum occurs at $x = -5$.

59. $f(x) = 3x^3 - 3x^2 + 1$

$f'(x) = 9x^2 - 6x$

$= 3x(3x - 2) = 0$

Critical numbers: 0 and $\frac{2}{3}$

$f''(x) = 18x - 6$

$f''(0) = -6 < 0,$ which means that a relative maximum occurs at $x = 0$.

$f''\left(\frac{2}{3}\right) = 6 > 0,$ which means that a relative minimum occurs at $x = \frac{2}{3}$.

61. $f(x) = (x + 3)^4$

$f'(x) = 4(x + 3)^3 = 0$

Critical number: $x = -3$

$$f''(x) = 12(x + 3)^2$$
$$f''(-3) = 12(-3 + 3)^2 = 0$$

The second derivative test fails.
Use the first derivative test.

−5 −4 −3 −2 −1 x

$$f'(-4) = 4(-4 + 3)^2$$
$$= 4(-1)^3 = -4 < 0$$

This indicates that f is decreasing on $(-\infty, -3)$.

$$f'(0) = 4(0 + 3)^3$$
$$= 4(3)^3 = 108 > 0$$

This indicates that f is increasing on $(-3, \infty)$. A relative minimum occurs at -3.

63. $f(x) = x^{7/3} + x^{4/3}$

$f'(x) = \frac{7}{3}x^{4/3} + \frac{4}{3}x^{1/3}$

$f'(x) = 0$ when

$$\frac{7}{3}x^{4/3} + \frac{4}{3}x^{1/3} = 0$$

$$\frac{x^{1/3}}{3}(7x + 4) = 0$$

$$x = 0 \text{ or } x = -\frac{4}{7}.$$

Critical numbers: $-\frac{4}{7}, 0$

$$f''(x) = \frac{28}{9}x^{1/3} + \frac{4}{9}x^{-2/3}$$

$$f''\left(-\frac{4}{7}\right) = \frac{28}{9}\left(-\frac{4}{7}\right)^{-1/3} + \frac{4}{9}\left(-\frac{4}{7}\right)^{-2/3} \approx -1.9363$$

Relative maximum occurs at $-\frac{4}{7}$.

$f''(0)$ does not exist, so the second derivative test fails.

Use the first derivative test.

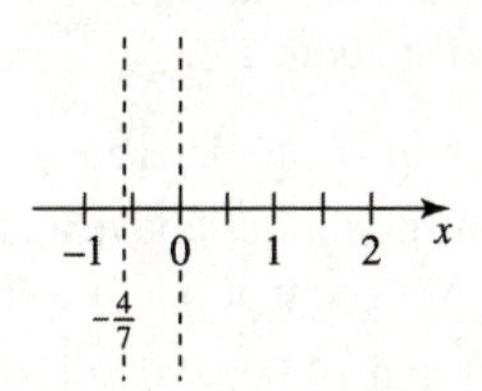

$$f'\left(-\frac{1}{2}\right) = \frac{7}{3}\left(-\frac{1}{2}\right)^{4/3} + \frac{4}{3}\left(-\frac{1}{2}\right)^{1/3} \approx -0.1323$$

This indicates that f is decreasing on $\left(-\frac{4}{7}, 0\right)$.

$$f'(1) = \frac{7}{3}(1)^{4/3} + \frac{4}{3}(1)^{1/3} = \frac{11}{3}$$

This indicates that f is increasing on $(0, \infty)$.
Relative minimum occurs at 0.

65. $f'(x) = x^3 - 6x^2 + 7x + 4$

$f''(x) = 3x^2 - 12x + 7$

Graph f' and f'' in the window $[-5, 5]$ by $[-5, 15]$, Xscl $= 0.5$.

Graph of f':

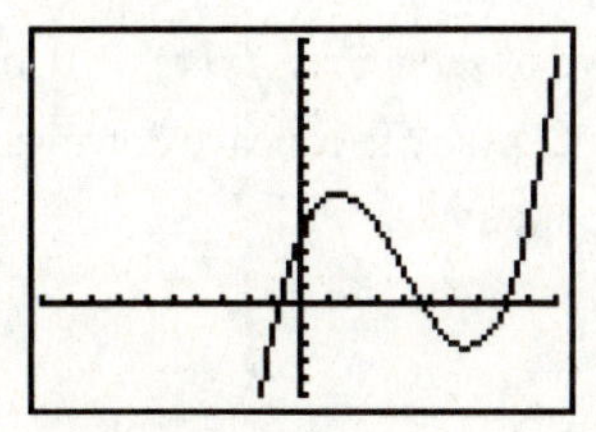

Graph of f'':

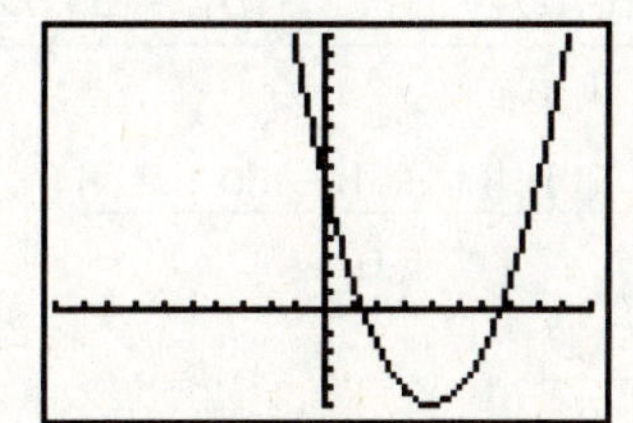

(a) f has relative extrema where $f'(x) = 0.$ Use the graph to approximate the x-intercepts of the graph of f'. These numbers are the solutions of the equation $f'(x) = 0.$ We find that the critical numbers of f are about -0.4, 2.4, and 4.0.

By either looking at the graph of f' and applying the first derivative test or by looking at the graph of f'' and applying the second derivative test, we see that f has relative minima at about -0.4 and 4.0 and a relative maximum at about 2.4.

(b) Examine the graph of f' to determine the intervals where the graph lies above and below the x-axis. We see that $f'(x) > 0$ on about $(-0.4, 2.4)$ and $(4.0, \infty)$, indicating that f is increasing on the same intervals. We also see that $f'(x) < 0$ on about $(-\infty, -0.4)$ and $(2.4, 4.0)$, indicating that f is decreasing on the same intervals.

(c) Examine the graph of f''. We see that this graph has two x-intercepts, so there are two x-values where $f''(x) = 0.$ These x-values are about 0.7 and 3.3. Because the sign of f'' changes at these two values, we see that the x-values of the inflection points of the graph of f are about 0.7 and 3.3.

(d) We observe from the graph of f'' that $f''(x) > 0$ on about $(-\infty, 0.7)$ and $(3.3, \infty)$, so f is concave upward on the same intervals.

Likewise, we observe that $f''(x) < 0$ on about $(0.7, 3.3)$, so f is concave downward on the same interval.

67. $f'(x) = \dfrac{1 - x^2}{(x^2 + 1)^2}$

$$f''(x) = \frac{(x^2 + 1)^2(-2x) - (1 - x^2)(2)(x^2 + 1)2x}{(x^2 + 1)^4}$$

$$= \frac{-2x(x^2 + 1)^2[(x^2 + 1) + 2(1 - x^2)]}{(x^2 + 1)^4}$$

$$= \frac{-2x(3 - x^2)}{(x^2 + 1)^3}$$

Graph f' and f'' in the window $[-3, 3]$ by $[-1.5, 1, 5]$, Xscl $= 0.2$.

Graph of f':

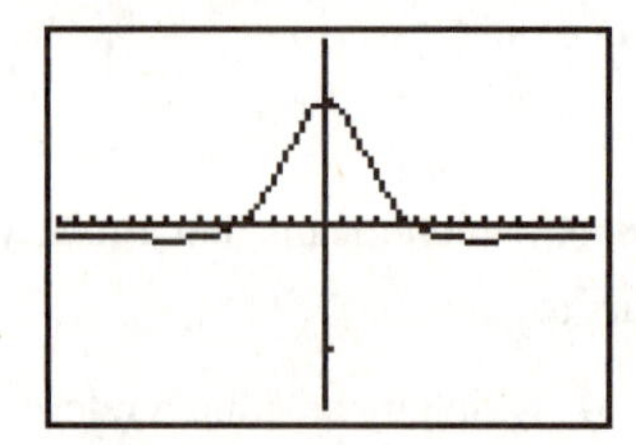

Graph of f'':

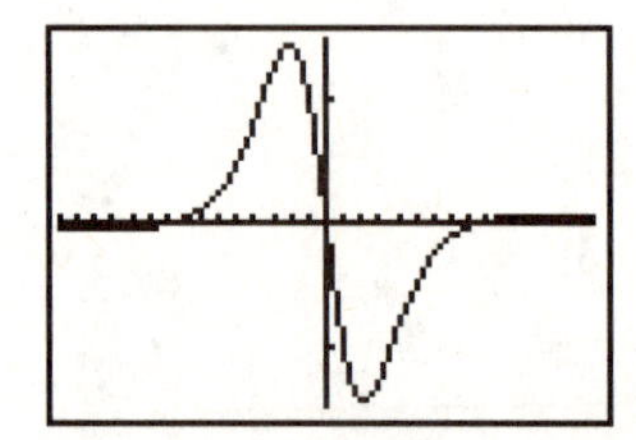

(a) The critical numbers of f are the x-intercepts of the graph of f'. (Note that there are no values where f' does not exist.) We see from the graph that these x-values are -1 and 1.

By either looking at the graph of f' and applying the first derivative test or by looking at the graph of f'' and applying the second derivative test, we see that f has a relative minimum at -1 and a relative maximum at 1.

(b) Examine the graph of f' to determine the intervals where the graph lies above and below the x-axis. We see that $f'(x) > 0$ on $(-1, 1)$, indicating that f is increasing on the same interval. We also see that $f'(x) < 0$ on $(-\infty, -1)$ and $(1, \infty)$, indicating that f is decreasing on the same intervals.

(c) Examine the graph of f''. We see that the graph has three x-intercepts, so there are three values where $f''(x) = 0.$ These x-values are about -1.7, 0, and about 1.7. Because the sign of f'' and thus the concavity of f changes at these three values, we see that the x-values of the inflection points of the graph of f are about -1.7, 0, and about 1.7.

(d) We observe from the graph of f'' that $f''(x) > 0$ on about $(-1.7, 0)$ and $(1.7, \infty)$, so f is concave upward on the same intervals.

Likewise, we observe that $f''(x) < 0$ on about $(-\infty, -1.7)$ and $(0, 1.7)$, so f is concave downward on the same intervals.

69. There are many examples. The easiest is $f(x) = \sqrt{x}$. This graph is increasing and concave downward.

$$f'(x) = \frac{1}{2}x^{-1/2} = \frac{1}{2\sqrt{x}}$$

$f'(0)$ does not exist, while $f'(x) > 0$ for all $x > 0$. (Note that the domain of f is $[0, \infty)$.)

As x increases, the value of $f'(x)$ decreases, but remains positive. It approaches zero, but never becomes zero or negative.

71. $A(t) = 0.0000329t^3 - 0.00450t^2 + 0.0613t + 2.34$

$$A'(t) = 0.0000987t^2 - 0.009t + 0.0613$$
$$A''(t) = 0.0001974x - 0.009$$

Assets will decrease most rapidly when A' has a minimum. A has a single inflection point when $A''(t) = 0$, which occurs when

$$0.0001974t - 0.009 = 0$$
$$0.0001974t = 0.009$$
$$t = \frac{0.009}{0.0001974} \approx 45.6.$$

This t-value corresponds to a minimum of A' and the minimum rate of decrease is $A'(45.6) \approx -0.144$. Since $t = 0$ represents the year 2000, the maximum rate of decrease of Social Security assets will occur in around the middle of 2045.

73. $R(x) = \frac{4}{27}(-x^3 + 66x^2 + 1050x - 400)$
$0 \le x \le 25$

$$R'(x) = \frac{4}{27}(-3x^2 + 132x + 1050)$$
$$R''(x) = \frac{4}{27}(-6x + 132)$$

A point of diminishing returns occurs at a point of inflection, or where $R''(x) = 0$.

$$\frac{4}{27}(-6x + 132) = 0$$
$$-6x + 132 = 0$$
$$6x = 132$$
$$x = 22$$

Test $R''(x)$ to determine whether concavity changes at $x = 22$.

$$R''(20) = \frac{4}{27}(-6 \cdot 20 + 132) = \frac{16}{9} > 0$$
$$R''(24) = \frac{4}{27}(-6 \cdot 24 + 132) = -\frac{16}{9} < 0$$

$R(x)$ is concave upward on $(0, 22)$ and concave downward on $(22, 25)$.

$$R(22) = \frac{4}{27}[-(22)^3 + 66(22)^2 + 1060(22) - 400]$$
$$\approx 6517.9$$

The point of diminishing returns is $(22, 6517.9)$.

75. $R(x) = -0.6x^3 + 3.7x^2 + 5x,\ 0 \le x \le 6$

$$R'(x) = -1.8x^2 + 7.4x + 5$$
$$R''(x) = -3.6x + 7.4$$

A point of diminishing returns occurs at a point of inflection or where $R''(x) = 0$.

$$-3.6x + 7.4 = 0$$
$$-3.6x = -7.4$$
$$x = \frac{-7.4}{-3.6x} \approx 2.06$$

Test $R''(x)$ to determine whether concavity changes at $x = 2.05$.

$$R''(2) = -3.6(2) + 7.4$$
$$= -7.2 + 7.4 = 0.2 > 0$$
$$R''(3) = -3.6(3) + 7.4$$
$$= -10.8 + 7.4 = -3.4 < 0$$

$R(x)$ is concave upward on $(0, 2.06)$ and concave downward on $(2.06, 6)$.

$$R(2.06) = -0.6(2.06)^3 + 3.7(2.06)^2 + 5(2.06)$$
$$\approx 20.8$$

The point of diminishing returns is $(2.06, 20.8)$.

77. Let $D(q)$ represent the demand function. The revenue function, $R(q)$, is $R(q) = qD(q)$. The marginal revenue is given by

$$R'(q) = qD'(q) + D(q)(1) = qD'(q) + D(q).$$

$$R''(q) = qD''(q) + D'(q)(1) + D'(q) = qD''(q) + 2D'(q)$$

gives the rate of decline of marginal revenue. $D'(q)$ gives the rate of decline of price. If marginal revenue declines more quickly than price,

$$qD''(q) + 2D'(q) - D'(q) < 0$$

or $qD''(q) + D'(q) < 0.$

79. (a) $R(t) = t^2(t - 18) + 96t + 1000,$

$0 < t < 8$

$$= t^3 - 18t^2 + 96t + 1000$$

$$R'(t) = 3t^2 - 36t + 96$$

Set $R'(t) = 0.$

$$3t^2 - 36t + 96 = 0$$

$$t^2 - 12t + 32 = 0$$

$$(t - 8)(t - 4) = 0$$

$$t = 8 \text{ or } t = 4$$

8 is not in the domain of $R(t)$.

$$R''(t) = 6t - 36$$

$R''(4) = -12 < 0$ implies that $R(t)$ is maximized at $t = 4$, so the population is maximized at 4 hours.

(b) $R(4) = 16(-14) + 96(4) + 1000$

$$= -224 + 384 + 1000$$

$$= 1160$$

The maximum population is 1160 million.

81. $K(x) = \dfrac{3x}{x^2 + 4}$

(a) $K'(x) = \dfrac{3(x^2 + 4) - (2x)(3x)}{(x^2 + 4)^2}$

$$= \frac{-3x^2 + 12}{(x^2 + 4)^2} = 0$$

$$-3x^2 + 12 = 0$$

$$x^2 = 4$$

$$x = 2 \text{ or } x = -2$$

For this application, the domain of K is $[0, \infty)$, so the only critical number is 2.

$$K''(x) = \frac{(x^2 + 4)^2(-6x) - (-3x^2 + 12)(2)(x^2 + 4)(2x)}{(x^2 + 4)^4}$$

$$= \frac{-6x(x^2 + 4) - 4x(-3x^2 + 12)}{(x^2 + 4)^3}$$

$$= \frac{6x^3 - 72x}{(x^2 + 4)^3}$$

$K''(2) = \frac{-96}{512} = -\frac{3}{16} < 0$ implies that $K(x)$ is maximized at $x = 2$.

Thus, the concentration is a maximum after 2 hours.

(b) $K(2) = \dfrac{3(2)}{(2)^2 + 4} = \dfrac{3}{4}$

The maximum concentration is $\frac{3}{4}$%.

83. $G(t) = \dfrac{10{,}000}{1 + 49e^{-0.1t}}$

$$G'(t) = \frac{(1 + 49e^{-0.1t})(0) - (10{,}000)(-4.9e^{-0.1t})}{(1 + 49e^{-0.1t})^2}$$

$$= \frac{49{,}000e^{-0.1t}}{(1 + 49e^{-0.1t})^2}$$

To find $G''(t)$, apply the quotient rule to find the derivative of $G'(t)$.

The numerator of $G''(t)$ will be

$$(1 + 49e^{-0.1t})^2(-4900e^{-0.1t})$$
$$-(49{,}000e^{-0.1t})(2)(1 + 49e^{-0.1t})(-4.9e^{-0.1t})$$
$$= (1 + 49e^{-0.1t})(-4900e^{-0.1t})$$
$$\cdot [(1 + 49e^{-0.1t}) - 20(4.9e^{-0.1t})]$$
$$= (-4900e^{-0.1t})[1 + 49e^{-0.1t} - 98e^{-0.1t}]$$
$$= (-4900e^{-0.1t})(1 - 49e^{-0.1t}).$$

Thus,

$$G''(t) = \frac{(-4900e^{-0.1t})(1 - 49e^{-0.1t})}{(1 + 49e^{-0.1t})^4}.$$

$G''(t) = 0$ when $-4900e^{-0.1t} = 0$ or $1 - 49e^{-0.1t} = 0.$

$-4900e^{-0.1t} < 0$, and thus never equals zero.

$$1 - 49e^{-0.1t} = 0$$
$$1 = 49e^{-0.1t}$$
$$\frac{1}{49} = e^{-0.1t}$$
$$\ln\left(\frac{1}{49}\right) = -0.1t$$
$$\ln 1 - \ln 49 = 0.1t$$
$$t = 10 \ln 49$$
$$t \approx 38.9182$$

The point of inflection is $(38.9182, 5000)$.

85. $L(t) = Be^{-ce^{-kt}}$

$$L'(t) = Be^{-ce^{-kt}}(-ce^{-kt})'$$
$$= Be^{-ce^{-kt}}[-ce^{-kt}(-kt)']$$
$$= Bcke^{-ce^{-kt}-kt}$$
$$L''(t) = Bcke^{-ce^{-kt}-kt}(-ce^{-kt} - kt)'$$
$$= Bcke^{-ce^{-kt}-kt}[-ce^{-kt}(-kt)' - k]$$
$$= Bcke^{-ce^{-kt}-kt}(cke^{-kt} - k)$$
$$= Bck^2e^{-ce^{-kt}-kt}(ce^{-kt} - 1)$$

$L''(t) = 0$ when $ce^{-kt} - 1 = 0$

$$ce^{-kt} - 1 = 0$$
$$\frac{c}{e^{kt}} = 1$$
$$e^{kt} = c$$
$$kt = \ln c$$
$$t = \frac{\ln c}{k}$$

Letting $c = 7.267963$ and $k = 0.670840$

$$t = \frac{\ln 7.267963}{0.670840} \approx 2.96 \text{ years}$$

Verify that there is a point of inflection at $t = \frac{\ln c}{k} \approx 2.96$. For

$$L''(t) = Bck^2e^{-ce^{-kt}-kt}(ce^{-kt} - 1),$$

we only need to test the factor $ce^{-kt} - 1$ on the intervals determined by $t \approx 2.96$ since the other factors are always positive.

$L''(1)$ has the same sign as

$$7.267963e^{-0.670840(1)} - 1 \approx 2.72 > 0.$$

$L''(3)$ has the same sign as

$$7.267963e^{-0.670840(3)} - 1 \approx -0.029 < 0.$$

Therefore L, is concave up on $\left(0, \frac{\ln c}{k} \approx 2.96\right)$ and concave down on $\left(\frac{\ln c}{k}, \infty\right)$, so there is a point of inflection at $t = \frac{\ln c}{k} \approx 2.96$ years.

This signifies the time when the rate of growth begins to slow down since L changes from concave up to concave down at this inflection point.

87. $v'(x) = -35.98 + 12.09x - 0.4450x^2$

$$v'(x) = 12.09 - 0.89x$$
$$v''(x) = -0.89$$

Since $-0.89 < 0$, the function is always concave down.

89. Since the rate of violent crimes is decreasing but at a slower rate than in previous years, we know that $f'(t) < 0$ but $f''(t) > 0$. Note that since $f'(t) < 0$, f is decreasing, and since $f''(t) > 0$, the graph of f is concave upward.

91. $s(t) = -16t^2$

$$v(t) = s'(t) = -32t$$

(a) $v(3) = -32(3) = -96$ ft/sec

(b) $v(5) = -32(5) = -160$ ft/sec

(c) $v(8) = -32(8) = -256$ ft/sec

(d) $a(t) = v'(t) = s''(t)$
$= -32 \text{ ft/sec}^2$

93. $s(t) = 256t - 16t^2$

$$v(t) = s'(t) = 256 - 32t$$
$$a(t) = v'(t) = s''(t) = -32$$

To find when the maximum height occurs, set $s'(t) = 0$.

$$256 - 32t = 0$$
$$t = 8$$

Find the maximum height.

$$s(8) = 256(8) - 16(8^2)$$
$$= 1024$$

The maximum height of the ball is 1024 ft. The ball hits the ground when $s = 0$.

$$256t - 16t^2 = 0$$
$$16t(16 - t) = 0$$
$$t = 0 \quad \text{(initial moment)}$$
$$t = 16 \quad \text{(final moment)}$$

The ball hits the ground 16 seconds after being thrown.

95. The car was moving most rapidly when $t \approx 6$, because acceleration was positive on $(0, 6)$ and negative after $t = 6$, so velocity was a maximum at $t = 6$.

13.4 Curve Sketching

Your Turn 1

$$f(x) = -x^3 + 3x^2 + 9x - 10$$
$$f'(x) = -3x^2 + 6x + 9$$
$$f''(x) = -6x + 6$$

Use the first derivative to find intervals where the function is increasing or decreasing.

$$-3x^2 + 6x + 9 = 0$$
$$-3(x^2 - 2x - 3) = 0$$
$$(x + 1)(x - 3) = 0$$
$$x = -1 \text{ or } x = 3$$

These critical numbers divide the line into three intervals, $(-\infty, -1)$, $(-1, 3)$ and $(3, \infty)$. Evaluate the derivative at a test point in each region.

$$f'(-2) = -15$$
$$f'(0) = 9$$
$$f'(4) = -4$$

This shows that f is decreasing on $(-\infty, -1)$, increasing on $(-1, 3)$, and decreasing on $(3, \infty)$. By the first derivative test, f has a relative minimum of $f(-1) = -15$ at $x = -1$ and a relative maximum of $f(3) = 17$ at $x = 3$.

Use the second derivative to find the intervals where the function is concave upward or downward.

$$-6x + 6 = 0$$
$$6x = x$$
$$x = 1$$

The value where the second derivative is 0 divides the line into two intervals, $(-\infty, 1)$ and $(1, \infty)$. Evaluate $f''(x)$ at a point in each interval.

$$f''(0) = 6$$
$$f''(2) = -6$$

This shows that f is concave upward on $(-\infty, 1)$ and concave downward on $(1, \infty)$. The graph has an inflection point at $(1, f(1))$, or $(1, 1)$.

This information is summarized in the following table.

Interval	$(-\infty, -1)$	$(-1, 1)$	$(1, 3)$	$(3, \infty)$
Sign of f'	$-$	$+$	$+$	$-$
Sign of f''	$+$	$+$	$-$	$-$
f increasing or decreasing	Decreasing	Increasing	Increasing	Decreasing
Concavity of f	Upward	Upward	Downward	Downward
Shape of graph				

Now sketch the graph using this information.

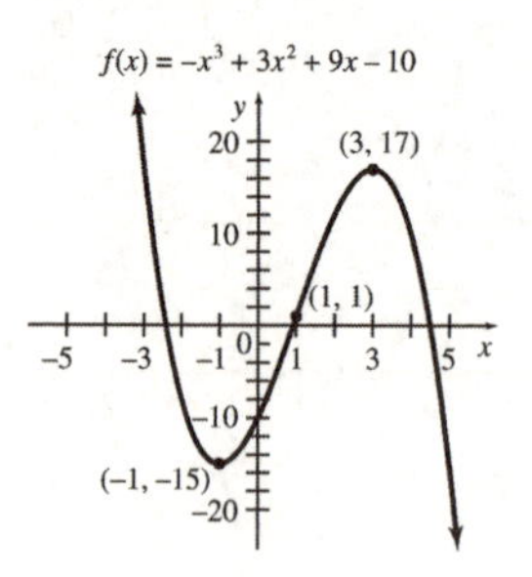

Your Turn 2

$$f(x) = 4x + \frac{1}{x} = \frac{4x^2 + 1}{x}$$

Because $x = 0$ makes the denominator 0 but not the numerator, the line $x = 0$ (that is, the y-axis) is a vertical asymptote. Neither $\lim_{x \to -\infty} f(x)$ nor $\lim_{x \to \infty} f(x)$ exists, so there is no horizontal asymptote. But since the term $1/x$ gets very small as $|x|$ gets large, the graph of f approaches the line $y = 4x$ as $|x|$ becomes larger and larger, so this line is an oblique asymptote.

$f(-x) = -f(x)$, so the graph of the left side can be found by rotating the right side around the origin by 180°.

Find any critical numbers.

$$f'(x) = 4 - \frac{1}{x^2}$$
$$4 - \frac{1}{x^2} = 0$$
$$x^2 = \frac{1}{4}$$
$$x = \frac{1}{2} \text{ or } -\frac{1}{2}$$

The critical numbers are $-1/2$ and $1/2$, which together with the location of the vertical asymptote divide the line into four regions, $(-\infty, -1/2)$, $(-1/2, 0)$, $(0, 1/2)$, and $(1/2, \infty)$. Evaluate the derivative at a test point in each region.

$$f'(-1) = 3$$
$$f'\left(-\frac{1}{4}\right) = -12$$
$$f'\left(\frac{1}{4}\right) = -12$$
$$f'(1) = 3$$

This shows that f has a relative maximum of $f(-1/2) = -4$ at $x = -1/2$ and a relative minimum of $f(1/2) = 4$ at $0\,x = 1/2$.

Now find the intervals where the graph is concave upward or concave downward.

$$f''(x) = \frac{2}{x^3}$$

The second derivative is never 0, but concavity might change where the second derivative is undefined, at $x = 0$. In fact,

$$f''(x) < 0 \text{ for } x < 0,$$
$$f''(x) > 0 \text{ for } x > 0,$$

so the graph is concave downward to the left of the origin and concave upward to the right of the origin. This provides enough information to sketch the graph.

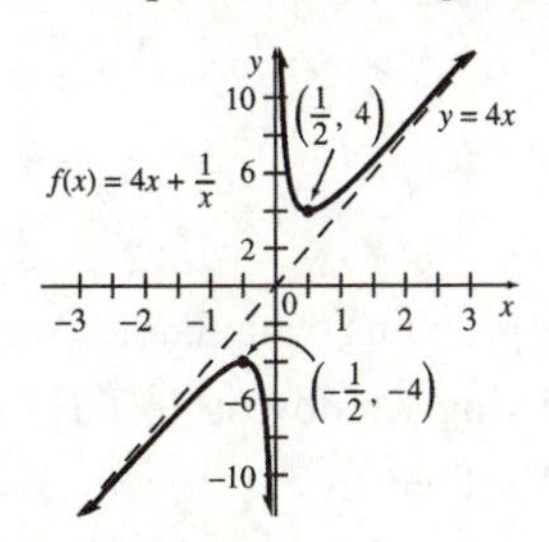

Your Turn 3

$$f(x) = \frac{4x^2}{x^2 + 4}$$

The denominator of f is never 0, so f has no vertical asymptotes. However,

$$\lim_{x\to-\infty} \frac{4x^2}{x^2 + 4} = 4 \text{ and } \lim_{x\to\infty} \frac{4x^2}{x^2 + 4} = 4,$$

so the line $y = 4$ is a horizontal asymptote. Note that $f(-x) = f(x)$, so the graph of f is symmetrical around the y-axis,

Find any critical numbers.

$$f'(x) = \frac{(x^2 + 4)(8x) - (4x^2)(2x)}{(x^2 + 4)^2}$$
$$= \frac{8x^3 + 32x - 8x^3}{(x^2 + 4)^2}$$
$$= \frac{32x}{(x^2 + 4)^2}$$

The derivative is 0 only when $x = 0$, and the denominator of the derivative is never 0. Therefore the critical number divides the line into just two regions, $(-\infty, 0)$ and $(0, \infty)$. Evaluate the derivative at a test point in each region.

$$f'(-1) = -1.28$$
$$f'(1) = 1.28$$

Thus f is decreasing as x approaches 0 from the left, and increasing to the right of $x = 0$, so f has a relative minimum of $f(0) = 0$ at $x = 0$.

Now find the intervals where the graph is concave upward or concave downward.

$$f'(x) = \frac{32x}{(x^2 + 4)^2}$$
$$f''(x) = \frac{(x^2 + 4)^2(32) - (32x)(2)(x^2 + 4)(2x)}{(x^2 + 4)^4}$$
$$= \frac{(x^2 + 4)[(x^2 + 4)(32) - 128x^2]}{(x^2 + 4)^4}$$
$$= \frac{128 - 96x^2}{(x^2 + 4)^3}$$

The denominator is never 0, but the numerator is 0 when

$$128 - 96x^2 = 0$$
$$96x^2 = 128$$
$$3x^2 = 4$$
$$x^2 = \frac{4}{3}$$
$$x = \pm\sqrt{\frac{4}{3}} \approx \pm 1.155$$

The zeros of the second derivative divide the line into three regions,

$$\left(-\infty, -\sqrt{\frac{4}{3}}\right), \left(-\sqrt{\frac{4}{3}}, \sqrt{\frac{4}{3}}\right) \text{ and } \left(\sqrt{\frac{4}{3}}, \infty\right).$$

Evaluate $f''(x)$ at a point in each interval.

$$f''(-2) = -0.5$$
$$f''(0) = 2$$
$$f''(2) = -0.5$$

According to this information, the graph is concave downward on the interval $\left(-\infty, -\sqrt{\frac{4}{3}}\right)$,

concave upward on the interval $\left(-\sqrt{\frac{4}{3}}, \sqrt{\frac{4}{3}}\right)$,

and concave downward on the interval $\left(\sqrt{\frac{4}{3}}, \infty\right)$.

Thus there will be two inflection points. Find the corresponding values of f.

$$f\left(-\sqrt{\frac{4}{3}}\right) = 1$$

$$f\left(\sqrt{\frac{4}{3}}\right) = 1$$

The inflection points are at $\left(-\sqrt{\frac{4}{3}}, 1\right)$ and $\left(\sqrt{\frac{4}{3}}, 1\right)$.

This provides enough information to sketch the graph.

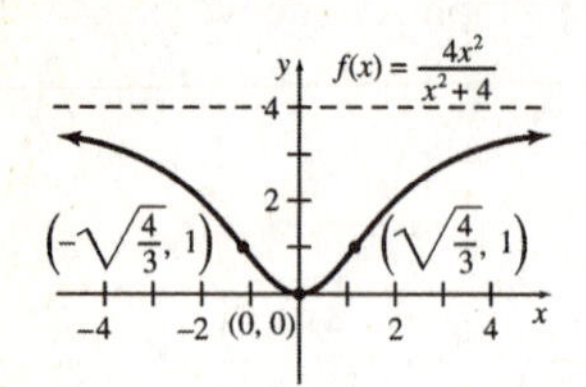

Your Turn 4

$f(x) = (x + 2)e^{-x}$

Since $\lim_{x\to\infty} (x + 2)e^{-x} = 0$, the line $y = 0$ (the x-axis) is a horizontal asymptote for the graph. Neither $f(-x) = -f(x)$ nor $f(-x) = f(x)$ is true, so the graph has no symmetry.

Find any critical numbers.

$$f'(x) = (x + 2)(-e^{-x}) + (1)(e^{-x})$$
$$= -(1 + x)e^{-x}$$

The derivative is 0 at only one point, where $1 + x = 0$ or $x = -1$. This critical number divides the line into two regions, $(-\infty, -1)$ and $(-1, \infty)$. Evaluate the derivative at a test point in each region.

$$f'(-2) \approx 7.389$$
$$f'(0) = -1$$

Thus f is increasing on the interval $(-\infty, -1)$ and decreasing on the interval $(-1, \infty)$, and has a relative maximum of $f(-1) \approx 2.718$ at $x = -1$.

Now find the intervals where the graph is concave upward or concave downward.

$$f'(x) = -(1 + x)e^{-x}$$
$$f''(x) = (1 + x)e^{-x} + (-1)e^{-x}$$
$$= xe^{-x}$$

The second derivative is 0 only at $x = 0$. Evaluate $f''(x)$ at points on either side of $x = 0$.

$$f''(-1) \approx -2.718$$
$$f''(1) \approx 0.368$$

The graph is concave downward to the left of $x = 0$ and concave upward to the right of $x = 0$. There is an inflection point at $(0, f(0))$, or $(0, 2)$.

This provides enough information to sketch the graph.

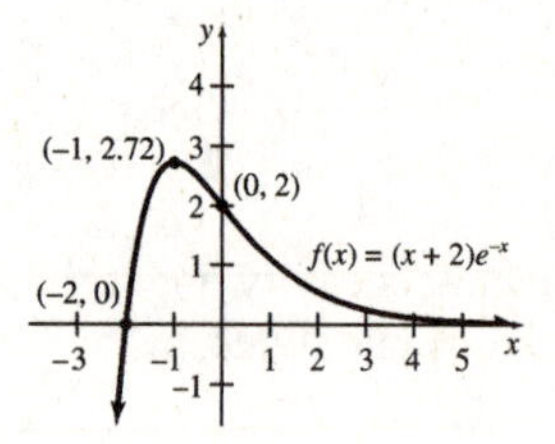

13.4 Exercises

1. Graph $y = x \ln |x|$ on a graphing calculator. A suitable choice for the viewing window is $[-1, 1]$ by $[-1, 1]$, Xscl = 0.1, Yscl = 0.1.

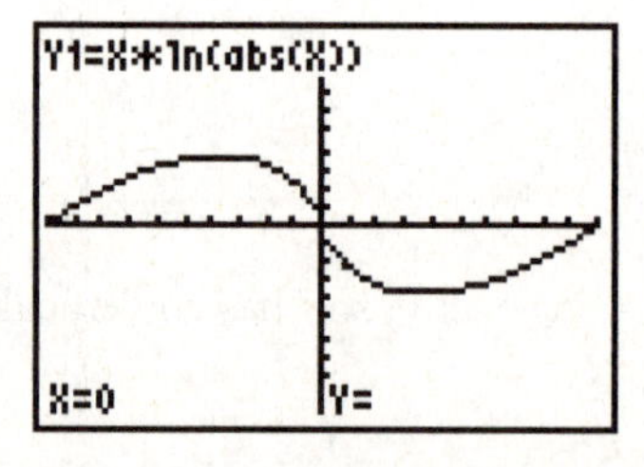

The calculator shows no y-value when $x = 0$ because 0 is not in the domain of this function. However, we see from the graph that

$$\lim_{x\to 0^-} x \ln |x| = 0$$

and

$$\lim_{x\to 0^+} x \ln |x| = 0.$$

Thus,

$$\lim_{x\to 0} x \ln |x| = 0.$$

3. $f(x) = -2x^3 - 9x^2 + 108x - 10$

Domain is $(-\infty, \infty)$.

$$f(-x) = -2(-x)^3 - 9(-x)^2 + 108(-x) - 10$$
$$= 2x^3 - 9x^2 - 108x - 10$$

No symmetry

$$f'(x) = -6x^2 - 18x + 108$$
$$= -6(x^2 + 3x - 18)$$
$$= -6(x + 6)(x - 3)$$

$f'(x) = 0$ when $x = -6$ or $x = 3$.

Critical numbers: -6 and 3

Critical points: $(-6, -550)$ and $(3, 179)$

$$f''(x) = -12x - 18$$
$$f''(-6) = 54 > 0$$
$$f''(3) = -54 < 0$$

Relative maximum at 3, relative minimum at -6

Increasing on $(-6, 3)$

Decreasing on $(-\infty, -6)$ and $(3, \infty)$

$$f''(x) = -12x - 18 = 0$$
$$-6(2x + 3) = 0$$
$$x = -\frac{3}{2}$$

Point of inflection at $(-1.5, -185.5)$

Concave upward on $(-\infty, -1.5)$

Concave downward on $(-1.5, \infty)$

y-intercept:

$$y = -2(0)^3 - 9(0)^2 + 108(0) - 10 = -10$$

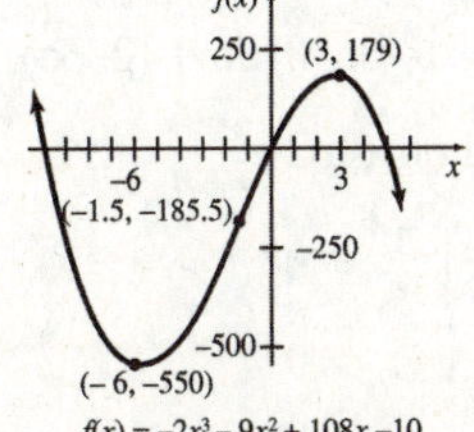

5. $f(x) = -3x^3 + 6x^2 - 4x - 1$

Domain is $(-\infty, \infty)$.

$$f(-x) = -3(-x)^3 + 6(-x)^2 - 4(-x) - 1$$
$$= 3x^3 + 6x^2 + 4x - 1$$

No symmetry

$$f'(x) = -9x^2 + 12x - 4$$
$$= -(3x - 2)^2$$
$$(3x - 2)^2 = 0$$
$$x = \frac{2}{3}$$

Critical number: $\frac{2}{3}$

$$f\left(\frac{2}{3}\right) = -3\left(\frac{2}{3}\right)^3 + 6\left(\frac{2}{3}\right)^2 - 4\left(\frac{2}{3}\right) - 1 = -\frac{17}{9}$$

Critical point: $\left(\frac{2}{3}, -\frac{17}{9}\right)$

$$f'(0) = -9(0)^2 + 12(0) - 4 = -4 < 0$$
$$f'(1) = -9(1)^2 + 12(1) - 4 = -1 < 0$$

No relative extremum at $\left(\frac{2}{3}, -\frac{17}{9}\right)$

Decreasing on $(-\infty, \infty)$

$$f''(x) = -18x + 12$$
$$= -6(3x - 2)$$
$$3x - 2 = 0$$
$$x = \frac{2}{3}$$

Point of inflection at $\left(\frac{2}{3}, -\frac{17}{9}\right)$

$$f''(0) = -18(0) + 12 = 12 > 0$$
$$f''(1) = -18(1) + 12 = -6 < 0$$

Concave upward on $\left(-\infty, \frac{2}{3}\right)$

Concave upward on $\left(\frac{2}{3}, \infty\right)$

Point of inflection at $\left(\frac{2}{3}, -\frac{17}{9}\right)$

y-intercept:

$$y = -3(0)^3 + 6(0)^2 - 4(0) - 1 = -1$$

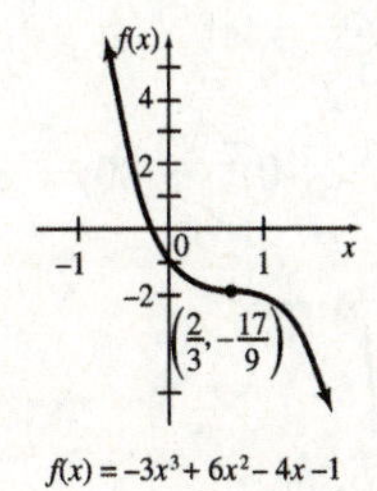

7. $f(x) = x^4 - 24x^2 + 80$

Domain is $(-\infty, \infty)$.

$$f(-x) = (-x)^4 - 24(-x)^2 + 80$$
$$= x^4 - 24x^2 + 80 = f(x)$$

The graph is symmetric about the *y*-axis.

$f'(x) = 4x^3 - 48x$

$$4x^3 - 48x = 0$$
$$4x(x^2 - 12) = 0$$
$$4x(x - 2\sqrt{3})(x + 2\sqrt{3}) = 0$$

Critical numbers: $-2\sqrt{3}$, 0, and $2\sqrt{3}$

Critical points: $(-2\sqrt{3}, -64)$, $(0, 80)$, and $(2\sqrt{3}, -64)$

$f''(x) = 12x^2 - 48$

$$f''(-2\sqrt{3}) = 12(-2\sqrt{3})^2 - 48 = 96 > 0$$
$$f''(0) = 12(0)^2 - 48 = -48 < 0$$
$$f''(2\sqrt{3}) = 12(2\sqrt{3})^2 - 48 = 96 > 0$$

Relative maximum at 0, relative minima at $-2\sqrt{3}$ and $2\sqrt{3}$

Increasing on $(-2\sqrt{3}, 0)$ and $(2\sqrt{3}, \infty)$

Decreasing on $(-\infty, -2\sqrt{3})$ and $(0, 2\sqrt{3})$

$$12x^2 - 48 = 0$$
$$12(x^2 - 4) = 0$$
$$x = \pm 2$$

Points of inflection at $(-2, 0)$ and $(2, 0)$

Concave upward on $(-\infty, -2)$ and $(2, \infty)$

Concave downward on $(-2, 2)$

x-intercepts: $0 = x^4 - 24x^2 + 80$

Let $u = x^2$.

$$u^2 - 24u + 80 = 0$$
$$(u - 4)(u - 20) = 0$$
$$u = 4 \quad \text{or} \quad u = 20$$
$$x = \pm 2 \quad \text{or} \quad x = \pm 2\sqrt{5}$$

y-intercept: $y = (0)^4 - 24(0)^2 + 80 = 80$

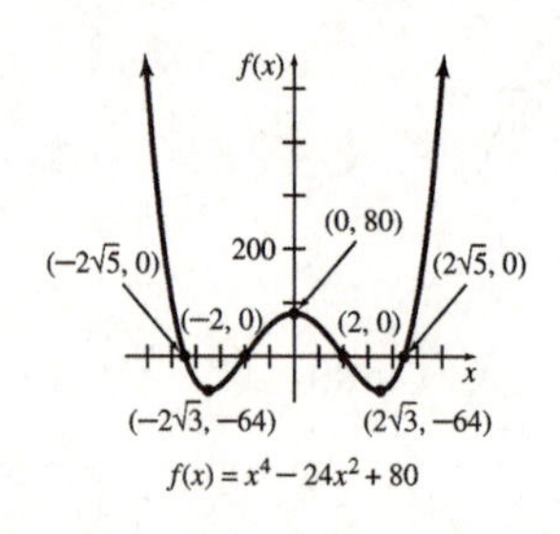

9. $f(x) = x^4 - 4x^3$

Domain is $(-\infty, \infty)$.

$$f(-x) = (-x)^4 - 4(-x)^3$$
$$= x^4 + 4x^3 \neq f(x) \text{ or } -f(x)$$

The graph is not symmetric about the y-axis or the origin.

$f'(x) = 4x^3 - 12x^2$

$4x^3 - 12x^2 = 0$

$4x^2(x - 3) = 0$

Critical numbers: 0 and 3

Critical points: $(0, 0)$ and $(3, -27)$

$f''(x) = 12x^2 - 24x$

$f''(0) = 12(0)^2 - 24(0) = 0$

$f''(3) = 12(3)^2 - 24(3) = 36 > 0$

Second derivative test fails for 0. Use first derivative test.

$$f'(-1) = 4(-1)^3 - 12(-1)^2 = -16 < 0$$
$$f'(1) = 4(1)^3 - 12(1)^2 = -8 < 0$$

Neither a relative minimum nor maximum at 0

Relative minimum at 3

Increasing on $(3, \infty)$

Decreasing on $(-\infty, 3)$

$$12x^2 - 24x = 0$$
$$12x(x - 2) = 0$$
$$x = 0 \text{ or } x = 2$$

Points of inflection at $(0, 0)$ and $(2, -16)$

Concave upward on $(-\infty, 0)$ and $(2, \infty)$

Concave downward on $(0, 2)$

x-intercepts: $x^4 - 4x^3 = 0$

$$x^3(x - 4) = 0$$
$$x = 0 \text{ or } x = 4$$

y-intercepts: $y = (0)^4 - 4(0)^3 = 0$

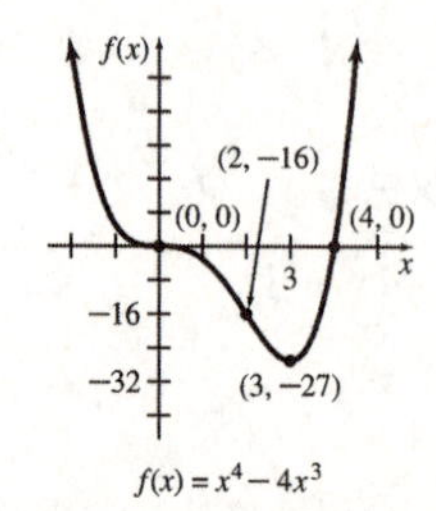

11. $f(x) = 2x + \dfrac{10}{x}$

$= 2x + 10x^{-1}$

Since $f(x)$ does not exist when $x = 0$, the domain is $(-\infty, 0) \cup (0, \infty)$.

$$f(-x) = 2(-x) + 10(-x)^{-1}$$
$$= -(2x + 10x^{-1})$$
$$= -f(x)$$

The graph is symmetric about the origin.

$$f'(x) = 2 - 10x^{-2}$$

$$2 - \frac{10}{x^2} = 0$$

$$\frac{2(x^2 - 5)}{x^2} = 0$$

$$x = \pm\sqrt{5}$$

Critical numbers: $-\sqrt{5}$ and $\sqrt{5}$

Critical points: $(-\sqrt{5}, -4\sqrt{5})$ and $(\sqrt{5}, 4\sqrt{5})$

Test a point in the intervals $(-\infty, -\sqrt{5})$, $(-\sqrt{5}, 0)$, $(0, \sqrt{5})$, and $(\sqrt{5}, \infty)$.

$$f'(-3) = 2 - 10(-3)^{-2} = \frac{8}{9} > 0$$

$$f'(-1) = 2 - 10(-1)^{-2} = -8 < 0$$

$$f'(1) = 2 - 10(1)^{-2} = -8 < 0$$

$$f'(3) = 2 - 10(3)^{-2} = \frac{8}{9} > 0$$

Relative maximum at $-\sqrt{5}$

Relative minimum at $\sqrt{5}$

Increasing on $(-\infty, -\sqrt{5})$ and $(\sqrt{5}, \infty)$

Decreasing on $(-\sqrt{5}, 0)$ and $(0, \sqrt{5})$

(Recall that $f(x)$ does not exist at $x = 0$.)

$$f''(x) = 20x^{-3} = \frac{20}{x^3}$$

$f''(x) = \dfrac{20}{x^3}$ is never equal to zero.

There are no inflection points.

Test a point in the intervals $(-\infty, 0)$ and $(0, \infty)$.

$$f''(-1) = \frac{20}{(-1)^3} = -20 < 0$$

$$f''(1) = \frac{20}{(1)^3} = 20 > 0$$

Concave upward on $(0, \infty)$

Concave downward on $(-\infty, 0)$

$f(x)$ is never zero, so there are no x-intercepts.

$f(x)$ does not exist for $x = 0$, so there is no y-intercept.

Vertical asymptote at $x = 0$

$y = 2x$ is an oblique asymptote.

$$f(-x) = 2(-x) + 10(-x)^{-1}$$
$$= -(2x + 10x^{-1})$$
$$= -f(x)$$

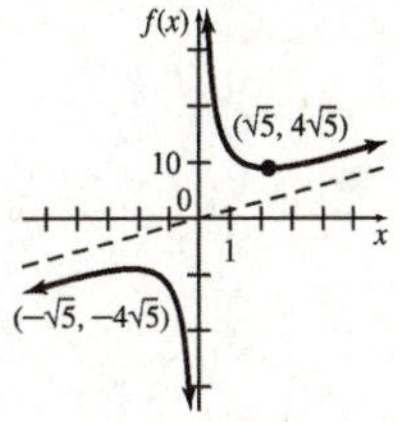

13. $f(x) = \dfrac{-x + 4}{x + 2}$

Since $f(x)$ does not exist when $x = -2$, the domain is $(-\infty, -2) \cup (-2, \infty)$.

$$f(-x) = \frac{-(-x) + 4}{(-x) + 2} = \frac{x + 4}{-x + 2}$$

The graph is not symmetric about the y-axis or the origin.

$$f'(-x) = \frac{(x + 2)(-1) - (-x + 4)(1)}{(x + 2)^2}$$
$$= \frac{-6}{(x + 2)^2}$$

$f'(x) < 0$ and is never zero. $f'(x)$ fails to exist for $x = -2$.

No critical numbers; no relative extrema

Decreasing on $(-\infty, -2)$ and $(-2, \infty)$

$$f''(x) = \frac{12}{(x + 2)^3}$$

$f''(x)$ fails to exist for $x = -2$.

No points of inflection

Test a point in the intervals $(-\infty, -2)$ and $(-2, \infty)$.

$f''(-3) = -12 < 0$

$f''(-1) = 12 > 0$

Concave upward on $(-2, \infty)$

Concave downward on $(-\infty, -2)$

x-intercept: $\frac{-x+4}{x+2} = 0$

$x = 4$

y-intercept: $y = \frac{-0+4}{0+2} = 2$

Vertical asymptote at $x = -2$

Horizontal asymptote at $y = -1$

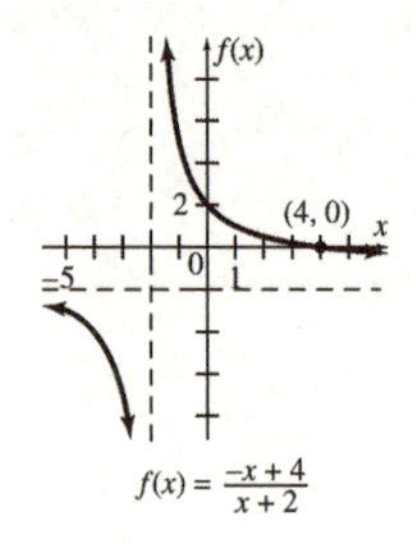

15. $f(x) = \frac{1}{x^2 + 4x + 3}$

$= \frac{1}{(x+3)(x+1)}$

Since $f(x)$ does not exist when $x = -3$ and $x = -1$, the domain is $(-\infty, -3) \cup (-3, -1) \cup (-1, \infty)$.

$$f(-x) = \frac{1}{(-x)^2 + 4(-x) + 3} = \frac{1}{x^2 - 4x + 3}$$

The graph is not symmetric about the y-axis or the origin.

$$f'(x) = \frac{0 - (2x+4)}{(x^2 + 4x + 3)^2} = \frac{-2(x+2)}{[(x+3)(x+1)]^2}$$

Critical number: -2

Test a point in the intervals $(-\infty, -3)$, $(-3, -2)$, $(-2, -1)$, and $(-1, \infty)$.

$$f'(-4) = \frac{-2(-4+2)}{[(-4+3)(-4+1)]^2} = \frac{4}{9} > 0$$

$$f'\left(-\frac{5}{2}\right) = \frac{-2\left(-\frac{5}{2} + 2\right)}{\left[\left(-\frac{5}{2} + 3\right)\left(-\frac{5}{2} + 1\right)\right]^2} = \frac{16}{9} > 0$$

$$f'\left(-\frac{3}{2}\right) = \frac{-2\left(-\frac{3}{2} + 2\right)}{\left[\left(-\frac{3}{2} + 3\right)\left(-\frac{3}{2} + 1\right)\right]^2} = -\frac{16}{9} < 0$$

$$f'(0) = \frac{-2(0+2)}{[(0+3)(0+1)]^2} = -\frac{4}{9} < 0$$

$$f(-2) = \frac{1}{(-2+3)(-2+1)} = -1$$

Relative maximum at $(-2, -1)$

Increasing on $(-\infty, -3)$ and $(-3, -2)$

Decreasing on $(-2, -1)$ and $(-1, \infty)$

$$f''(x) = \frac{(x^2 + 4x + 3)^2(-2) - (-2x - 4)(2)(x^2 + 4x + 3)(2x + 4)}{(x^2 + 4x + 3)^4}$$

$$= \frac{-2(x^2 + 4x + 3)[(x^2 + 4x + 3) + (-2x - 4)(2x + 4)]}{(x^2 + 4x + 3)^4}$$

$$= \frac{-2(x^2 + 4x + 3 - 4x^2 - 16x - 16)}{(x^2 + 4x + 3)^3}$$

$$= \frac{-2(-3x^2 - 12x - 13)}{(x^2 + 4x + 3)^3}$$

$$= \frac{2(3x^2 + 12x + 13)}{[(x+3)(x+1)]^3}$$

Since $3x^2 + 12x + 13 = 0$ has no real solutions, there are no x-values where $f''(x) = 0$. $f''(x)$ does not exist where $x = -3$ and $x = -1$. Since $f(x)$ does not exist at these x-values, there are no points of inflection.

Test a point in the intervals $(-\infty, -3)$, $(-3, -1)$, and $(-1, \infty)$.

$$f''(-4) = \frac{2[3(-4)^2 + 12(-4) + 13]}{[(-4+3)(-4+1)]^3} = \frac{26}{27} > 0$$

$$f''(-2) = \frac{2[3(-2)^2 + 12(-2) + 13]}{[(-2+3)(-2+1)]^3} = -2 < 0$$

$$f''(0) = \frac{2[3(0)^2 + 12(0) + 13]}{[(0+3)(0+1)]^3} = \frac{26}{27} > 0$$

Concave upward on $(-\infty, -3)$ and $(-1, \infty)$

Concave downward on $(-3, -1)$

$f(x)$ is never zero, so there are no x-intercepts.

y-intercept: $y = \dfrac{1}{(0+3)(0+1)} = \dfrac{1}{3}$

Vertical asymptotes where $f(x)$ is undefined at $x = -3$ and $x = -1$.

Horizontal asymptote at $y = 0$

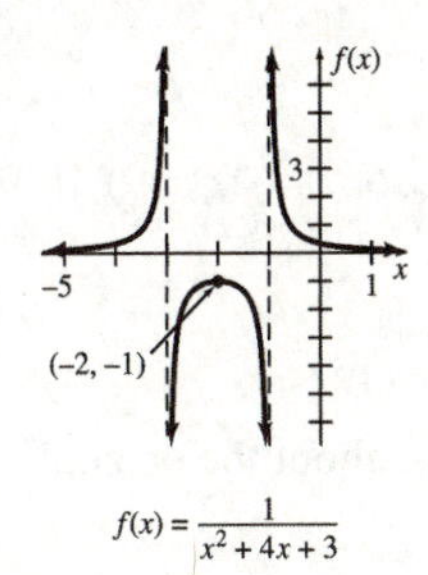

17. $f(x) = \dfrac{x}{x^2+1}$

Domain is $(-\infty, \infty)$

$$f(-x) = \frac{-x}{(-x)^2+1} = -\frac{x}{x^2+1} = -f(x)$$

The graph is symmetric about the origin.

$$f'(x) = \frac{(x^2+1)(1) - x(2x)}{(x^2+1)^2}$$
$$= \frac{1-x^2}{(x^2+1)^2}$$
$$1 - x^2 = 0$$

Critical numbers: 1 and -1

Critical points: $\left(1, \frac{1}{2}\right)$ and $\left(-1, -\frac{1}{2}\right)$

$$f''(x) = \frac{(x^2+1)^2(-2x) - (1-x^2)(2)(x^2+1)(2x)}{(x^2+1)^4}$$
$$= \frac{-2x^3 - 2x - 4x + 4x^3}{(x^2+1)^3} = \frac{2x^3 - 6x}{(x^2+1)^3}$$

$f''(1) = -\dfrac{1}{2} < 0$

$f''(-1) = \dfrac{1}{2} > 0$

Relative maximum at 1

Relative minimum at -1

Increasing on $(-1, 1)$

Decreasing on $(-\infty, -1)$ and $(1, \infty)$

$$f''(x) = \frac{2x^3 - 6x}{(x^2+1)^3} = 0$$
$$2x^3 - 6x = 0$$
$$2x(x^2 - 3) = 0$$
$$x = 0, x = \pm\sqrt{3}$$

Inflection points at $(0, 0)$, $\left(\sqrt{3}, \frac{\sqrt{3}}{4}\right)$ and $\left(-\sqrt{3}, -\frac{\sqrt{3}}{4}\right)$

Concave upward on $(-\sqrt{3}, 0)$ and $(\sqrt{3}, \infty)$

Concave downward on $(-\infty, -\sqrt{3})$ and $(0, \sqrt{3})$

x-intercept: $0 = \dfrac{x}{x^2+1}$

$0 = x$

y-intercept: $y = \dfrac{0}{0^2+1} = 0$

Horizontal asymptote at $y = 0$

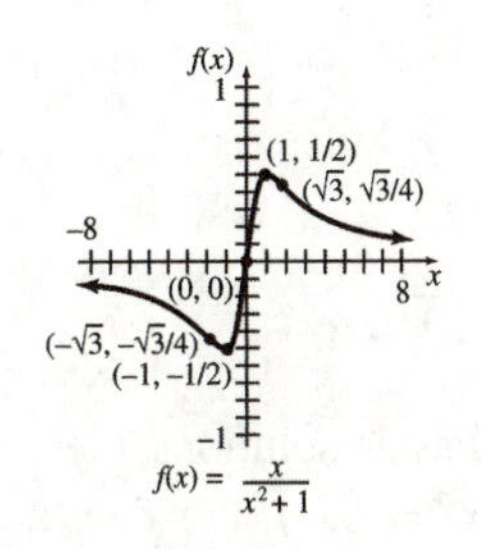

19. $f(x) = \dfrac{1}{x^2-9}$

$$= \frac{1}{(x+3)(x-3)}$$

Since $f(x)$ does not exist when $x = -3$ and $x = 3$, the domain is $(-\infty, -3) \cup (-3, 3) \cup (3, \infty)$.

$$f(-x) = \frac{1}{(-x)^2 - 9} = \frac{1}{x^2-9} = f(x)$$

The graph is symmetric about the y-axis.

$$f'(x) = \frac{-2x}{(x^2-9)^2}$$

Critical number: 0

Critical point: $\left(0, -\frac{1}{9}\right)$

Test a point in the intervals $(-\infty, -3)$, $(-3, 0)$, $(0, 3)$, and $(3, \infty)$.

$$f'(-4) = \frac{-2(-4)}{[(-4)^2 - 9]^2} = \frac{8}{49} > 0$$

$$f'(-1) = \frac{-2(-4)}{[(-1)^2 - 9]^2} = \frac{1}{32} > 0$$

$$f'(1) = \frac{-2(1)}{[(1)^2 - 9]^2} = -\frac{1}{32} < 0$$

$$f'(4) = \frac{-2(4)}{[(4)^2 - 9]^2} = -\frac{8}{49} < 0$$

Relative maximum at $\left(0, -\frac{1}{9}\right)$

Increasing on $(-\infty, -3)$ and $(-3, 0)$

Decreasing on $(0, 3)$ and $(3, \infty)$

$$f''(x) = \frac{(x^2 - 9)^2(-2) - (-2x)(2)(x^2 - 9)(2x)}{(x^2 - 9)^4}$$

$$= \frac{-2(x^2 - 9)[(x^2 - 9) + (-2x)(2x)]}{(x^2 + 4)^4}$$

$$= \frac{-2(x^2 - 9 - 4x^2)}{(x^2 - 9)^3}$$

$$= \frac{-2(-3x^2 - 9)}{(x^2 - 9)^3}$$

$$= \frac{6(x^2 + 3)}{[(x + 3)(x - 3)]^3}$$

Since $x^2 + 3 = 0$ has no solutions, there are no x-values where $f''(x) = 0$. $f''(x)$ does not exist where $x = -3$ and $x = 3$. Since $f(x)$ does not exist at these x-values, there are no points of inflection.

Test a point in the intervals $(-\infty, -3)$, $(-3, 3)$, and $(3, \infty)$.

$$f''(-4) = \frac{6[(-4)^2 + 3]}{[(-4 + 3)(-4 - 3)]^3} = \frac{114}{343} > 0$$

$$f''(0) = \frac{6[(0)^2 + 3]}{[(0 + 3)(0 - 3)]^3} = -\frac{2}{81} < 0$$

$$f''(4) = \frac{6[(4)^2 + 3]}{[(4 + 3)(4 - 3)]^3} = \frac{114}{343} > 0$$

Concave upward on $(-\infty, -3)$ and $(3, \infty)$

Concave downward on $(-3, 3)$

$f(x)$ is never zero, so there are no x-intercepts.

y-intercept: $y = \frac{1}{0^2 - 9} = -\frac{1}{9}$

Vertical asymptotes where $f(x)$ is undefined at $x = -3$ and $x = 3$.

Horizontal asymptote at $y = 0$

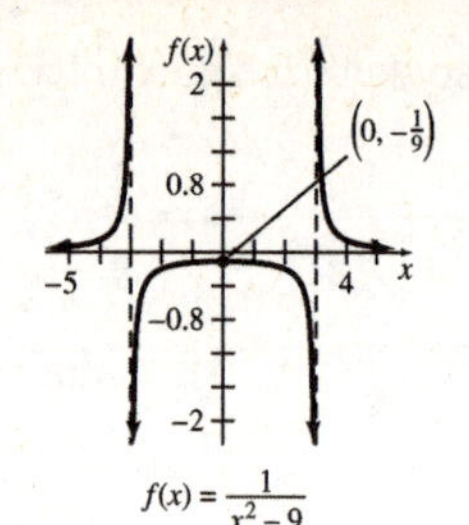

21. $f(x) = x \ln |x|$

The domain of this function is $(-\infty, 0) \cup (0, \infty)$.

$$f(-x) = -x \ln|-x|$$
$$= -x \ln|x| = -f(x)$$

The graph is symmetric about the origin.

$$f'(x) = x \cdot \frac{1}{x} + \ln |x|$$
$$= 1 + \ln|x|$$

$f'(x) = 0$ when

$$0 = 1 + \ln |x|$$
$$-1 = \ln |x|$$
$$e^{-1} = |x|$$
$$x = \pm\frac{1}{e} \approx \pm 0.37.$$

Critical numbers: $\pm\frac{1}{e} \approx \pm 0.37$.

$$f'(-1) = 1 + \ln |-1| = 1 > 0$$
$$f'(-0.1) = 1 + \ln |-0.1| \approx -1.3 < 0$$
$$f'(0.1) = 1 + \ln |0.1| \approx -1.3 < 0$$
$$f'(1) = 1 + \ln |1| = 1 > 0$$

$$f\left(\frac{1}{e}\right) = \frac{1}{e} \ln\left|\frac{1}{e}\right| = -\frac{1}{e}$$

$$f\left(-\frac{1}{e}\right) = -\frac{1}{e} \ln\left|-\frac{1}{e}\right| = \frac{1}{e}$$

Relative maximum of $\left(-\frac{1}{e}, \frac{1}{e}\right)$; relative minimum of $\left(\frac{1}{e}, -\frac{1}{e}\right)$.

Increasing on $\left(-\infty, -\frac{1}{e}\right)$ and $\left(\frac{1}{e}, \infty\right)$ and decreasing on $\left(-\frac{1}{e}, 0\right)$ and $\left(0, \frac{1}{e}\right)$.

$$f''(x) = \frac{1}{x}$$

$$f''(-1) = \frac{1}{-1} = -1 < 0$$

$$f''(1) = \frac{1}{1} = 1 > 0$$

Concave downward on $(-\infty, 0)$;

Concave upward on $(0, \infty)$.

There is no y-intercept.

x-intercept: $0 = x\ln|x|$

$$x = 0 \quad \text{or} \quad \ln|x| = 0$$
$$|x| = e^0 = 1$$
$$x = \pm 1$$

Since 0 is not in the domain, the only x-intercepts are -1 and 1.

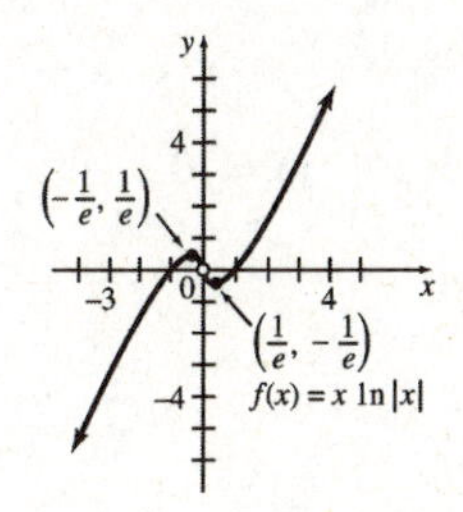

23. $f(x) = \dfrac{\ln x}{x}$

Note that the domain of this function is $(0, \infty)$.

$f(-x) = \dfrac{\ln(-x)}{-x}$ does not exist when $x \geq 0$,

no symmetry.

$$f'(x) = \frac{x\left(\frac{1}{x}\right) - \ln x(1)}{x^2}$$
$$= \frac{1 - \ln x}{x^2}$$

Critical numbers:

$$1 - \ln x = 0$$
$$1 = \ln x$$
$$e^1 = x$$

$$f(e) = \frac{\ln e}{e} = \frac{1}{e}$$

Critical points: $\left(e, \frac{1}{e}\right)$

$$f'(1) = \frac{1 - \ln 1}{1^2} = \frac{1}{1} = 1 > 0$$
$$f'(3) = \frac{1 - \ln 3}{3^2} = -0.01 < 0$$

There is a relative maximum at $\left(e, \frac{1}{e}\right)$.

The function is increasing on $(0, e)$ and decreasing on (e, ∞).

$$f''(x) = \frac{x^2\left(-\frac{1}{x}\right) - (1 - \ln x)2x}{x^4}$$
$$= \frac{-x - 2x(1 - \ln x)}{x^4}$$
$$= \frac{-x[1 + 2(1 - \ln x)]}{x^4}$$
$$= \frac{-(1 + 2 - 2\ln x)}{x^3}$$
$$= \frac{-3 + 2\ln x}{x^3}$$

$$f''(x) = 0 \text{ when } -3 + 2\ln x = 0$$
$$2\ln x = 3$$
$$\ln x = \frac{3}{2} = 1.5$$
$$x = e^{1.5} \approx 4.48.$$

$$f''(1) = \frac{-3 + 2\ln 1}{1^3} = -3 < 0$$
$$f''(5) = \frac{-3 + 2\ln 5}{5^3} \approx 0.0018 > 0$$

Inflection point at $\left(e^{1.5}, \frac{1.5}{e^{1.5}}\right) \approx (4.48, 0.33)$

Concave downward on $(0, e^{1.5})$; concave upward on $(e^{1.5}, \infty)$

$$f(e^{1.5}) = \frac{\ln e^{1.5}}{e^{1.5}} = \frac{1.5}{e^{1.5}}$$
$$= \frac{3}{2e^{1.5}} \approx 0.33$$

Since $x \neq 0$, there is no y-intercept.

x-intercept: $f(x) = 0$ when $\ln x = 0$

$$x = e^0 = 1$$

Vertical asymptote at $x = 0$

Horizontal asymptote at $y = 0$

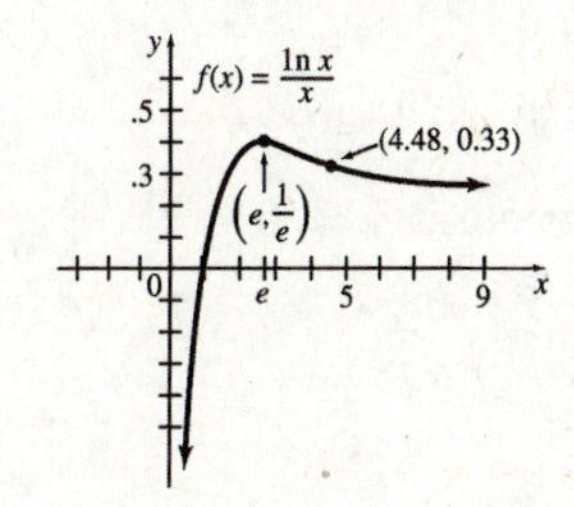

25. $f(x) = xe^{-x}$

Domain is $(-\infty, \infty)$.

$f(-x) = -xe^{x}$

The graph has no symmetry.

$$f'(x) = -xe^{-x} + e^{-x}$$
$$= e^{-x}(1 - x)$$

$f'(x) = 0$ when $e^{-x}(1 - x) = 0$

$$x = 1$$

Critical numbers: 1

Critical points: $\left(1, \frac{1}{e}\right)$

$$f'(0) = e^{-0}(1 - 0) = 1 > 0$$
$$f'(2) = e^{-2}(1 - 2) = \frac{-1}{e^2} < 0$$

Relative maximum at $\left(1, \frac{1}{e}\right)$

Increasing on $(-\infty, 1)$; decreasing on $(1, \infty)$

$$f''(x) = e^{-x}(-1) + (1 - x)(-e^{-x})$$
$$= -e^{-x}(1 + 1 - x)$$
$$= -e^{-x}(2 - x)$$

$f'' = 0$ when $-e^{-x}(2 - x) = 0$

$$x = 2.$$

$$f''(0) = -e^{-0}(2 - 0) = -2 < 0$$
$$f''(3) = -e^{-3}(2 - 3) = \frac{1}{e^3} > 0$$

Inflection point at $\left(2, \frac{2}{e^2}\right)$

Concave downward on $(-\infty, 2)$, concave upward on $(2, \infty)$

x-intercept: $0 = xe^{-x}$

$$x = 0$$

y-intercept: $y = 0 \cdot e^{-0} = 0$

Horizontal asymptote at $y = 0$

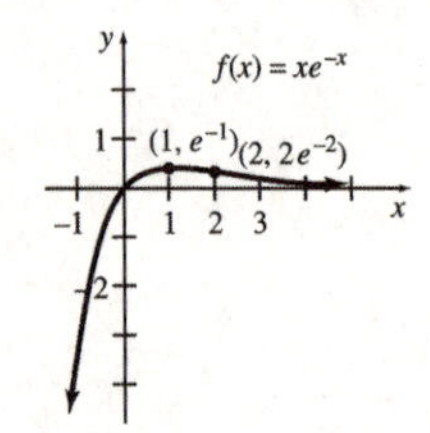

27. $f(x) = (x - 1)e^{-x}$

Domain is $(-\infty, \infty)$

$f(-x) = (-x - 1)e^{x}$

The graph has no symmetry.

$$f'(x) = -(x - 1)e^{-x} + e^{-x}(1)$$
$$= e^{-x}[-(x - 1) + 1]$$
$$= e^{-x}(2 - x)$$

$f'(x) = 0$ when $e^{-x}(2 - x) = 0$

$$x = 2.$$

Critical number: 2

Critical point: $\left(2, \frac{1}{e^2}\right)$

$$f''(x) = -e^{-x} + (2 - x)(-e^{-x})$$
$$= -e^{-x}[1 + (2 - x)]$$
$$= -e^{-x}(3 - x)$$

$$f''(2) = -e^{-2}(3 - 2) = \frac{-1}{e^2} < 0$$

Relative maximum at $\left(2, \frac{1}{e^2}\right)$

$$f'(0) = e^{-0}(2 - 0) = 2 > 0$$
$$f'(3) = e^{-3}(2 - 3) = \frac{-1}{e^3} < 0$$

Increasing on $(-\infty, 2)$; decreasing on $(2, \infty)$.

$f''(x) = 0$ when $-e^{-x}(3 - x) = 0$

$$x = 3.$$

$$f''(0) = -e^{-0}(3 - 0) = -3 < 0$$
$$f''(4) = -e^{-4}(3 - 4) = \frac{1}{e^4} > 0$$

Inflection point at $\left(3, \frac{2}{e^3}\right)$

Concave downward on $(-\infty, 3)$; concave upward on $(3, \infty)$

$$f(3) = (3 - 1)e^{-3} = \frac{2}{e^3}$$

y-intercept: $y = (0 - 1)e^{-0}$

$$= (-1)(1) = -1$$

x-intercept: $0 = (x - 1)e^{-x}$

$$x - 1 = 0$$
$$x = 1$$

Horizontal asymptote at $y = 0$

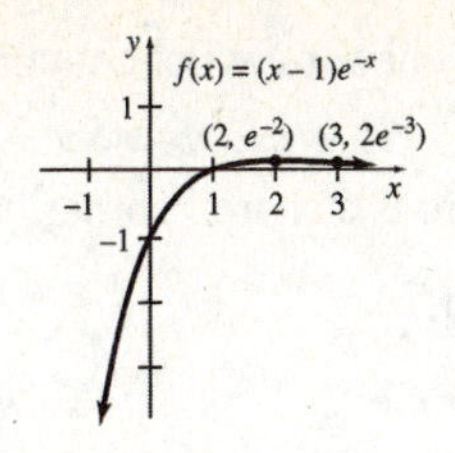

29. $f(x) = x^{2/3} - x^{5/3}$

Domain is $(-\infty, \infty)$.

$f(-x) = x^{2/3} + x^{5/3}$

The graph has no symmetry.

$$f'(x) = \frac{2}{3}x^{-1/3} - \frac{5}{3}x^{2/3} = \frac{2-5x}{3x^{1/3}}$$

$f'(x) = 0$ when $2 - 5x = 0$

Critical number: $x = \frac{2}{5}$

$$f\left(\frac{2}{5}\right) = \left(\frac{2}{5}\right)^{2/3} - \left(\frac{2}{5}\right)^{5/3} = \frac{3 \cdot 2^{2/3}}{5^{5/3}} \approx 0.326$$

Critical point: $(0.4, 0.326)$

$$f''(x) = \frac{3x^{1/3}(-5) - (2-5x)(3)\left(\frac{1}{3}\right)x^{-2/3}}{(3x^{1/3})^2} = \frac{-15x^{1/3} - (2-5x)x^{-2/3}}{9x^{2/3}} = \frac{-15x - (2-5x)}{9x^{4/3}} = \frac{-10x-2}{9x^{4/3}}$$

$$f''\left(\frac{2}{5}\right) = \frac{-10\left(\frac{2}{5}\right) - 2}{9\left(\frac{2}{5}\right)^{4/3}} \approx -2.262 < 0$$

Relative maximum at $\left(\frac{2}{5}, \frac{3 \cdot 2^{2/3}}{5^{5/3}}\right) \approx (0.4, 0.326)$

$f'(x)$ does not exist when $x = 0$

Since $f''(0)$ is undefined, use the first derivative test.

$$f'(-1) = \frac{2-5(-1)}{3(-1)^{1/3}} = \frac{7}{-3} < 0$$

$$f'\left(\frac{1}{8}\right) = \frac{2-5\left(\frac{1}{8}\right)}{3\left(\frac{1}{8}\right)^{1/3}} = \frac{11}{12} > 0$$

$$f'(1) = \frac{2-5}{3 \cdot 1^{1/3}} = -1 < 0$$

Relative minimum at $(0, 0)$

f increases on $\left(0, \frac{2}{5}\right)$.

f decreases on $(-\infty, 0)$ and $\left(\frac{2}{5}, \infty\right)$.

$f''(x) = 0$ when $-10x - 2 = 0$

$$x = -\frac{1}{5}$$

$f''(x)$ undefined when $9x^{4/3} = 0$

$$x = 0$$

$$f''(-1) = \frac{-10(-1)-2}{9(-1)^{4/3}} = \frac{8}{9} > 0$$

$$f''\left(-\frac{1}{8}\right) = \frac{-10\left(-\frac{1}{8}\right) - 2}{9\left(-\frac{1}{8}\right)^{4/3}} = -\frac{4}{3} < 0$$

$$f''(1) = \frac{-10(1)-2}{9(1)^{4/3}} = -\frac{4}{3} < 0$$

Concave upward on $\left(-\infty, -\frac{1}{5}\right)$

Concave upward on $\left(-\frac{1}{5}, \infty\right)$

Inflection point at $\left(-\frac{1}{5}, \frac{6}{5^{5/3}}\right) \approx (-0.2, 0.410)$

y-intercept: $y = 0^{2/3} - 0^{5/3} = 0$

x-intercept: $0 = x^{2/3} - x^{5/3} = x^{2/3}(1-x)$

$x = 0$ or $x = 1$

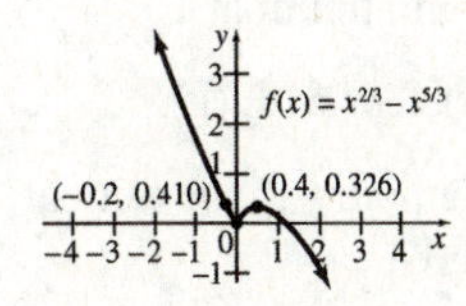

31. For Exercises 3, 7, and 9, the relative maxima or minima are outside the vertical window of $-10 \le y \le 10$.

For Exercise 11, the default window shows only a small portion of the graph.

For Exercise 15, the default window does not allow the graph to properly display the vertical asymptotes.

33. For Exercises 17, 19, 23, 25, and 27, the y-coordinate of the relative minimum, relative maximum, or inflection points is so small, it may be hard to distinguish.

For Exercises 35–39 other graphs are possible.

35. (a) indicates a smooth, continuous curve except where there is a vertical asymptote.

(b) indicates that the function decreases on both sides of the asymptote, so there are no relative extrema.

(c) gives the horizontal asymptote $y = 2$.

(d) and (e) indicate that concavity does not change left of the asymptote, but that the right portion of the graph changes concavity at $x = 2$ and $x = 4$.

There are inflection points at 2 and 4.

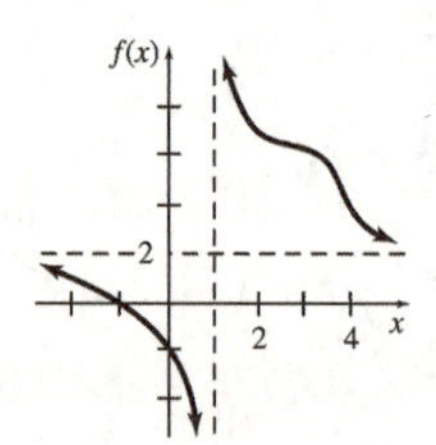

37. (a) indicates that there can be no asymptotes, sharp "corners", holes, or jumps. The graph must be one smooth curve.

(b) and (c) indicate relative maxima at -3 and 4 and a relative minimum at 1.

(d) and (e) are consistent with (g).

(f) indicates turning points at the critical numbers -3 and 4.

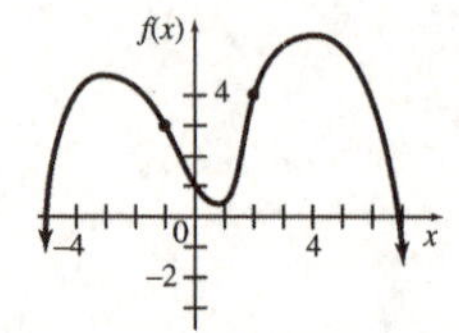

39. (a) indicates that the curve may not contain breaks.

(b) indicates that there is a sharp "corner" at 4.

(c) gives a point at $(1, 5)$.

(d) shows critical numbers.

(e) and (f) indicate (combined with (c) and (d)) a relative maximum at $(1, 5)$, and (combined with (b)) a relative minimum at 4.

(g) is consistent with (b).

(h) indicates the curve is concave upward on $(2, 3)$.

(i) indicates the curve is concave downward on $(-\infty, 2)$, $(3, 4)$ and $(4, \infty)$.

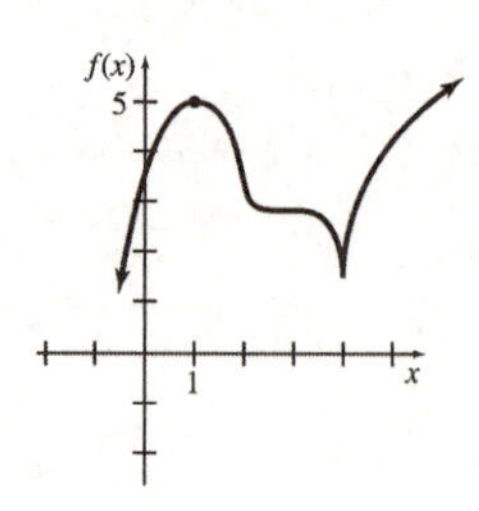

Chapter 13 Review Exercises

1. True

2. False: The function is increasing on this interval.

3. False: The function could have neither a minimum nor a maximum; consider $f(x) = x^3$ at $c = 0$.

4. True

5. False: Consider $f(x) = x^{3/2}$ at $c = 0$.

6. True

7. False: The function is concave upward.

8. False: Consider $f(x) = x^4$ at $c = 0$.

9. False: Consider $f(x) = x^4$, which has a relative minimum at $c = 0$.

10. False: Polynomials are rational functions, and nonconstant polynomials have neither vertical nor horizontal asymptotes.

11. True

12. False: Consider $f(x) = x^2$ on $(-1, 1)$ with $c = 0$.

17. $f(x) = x^2 + 9x + 8$

$f'(x) = 2x + 9$

$f'(x) = 0$ when $x = -\frac{9}{2}$ and f' exists everywhere.

Critical number: $-\frac{9}{2}$

Test an x-value in the intervals $\left(-\infty, -\frac{9}{2}\right)$ and $\left(-\frac{9}{2}, -\infty\right)$.

$f'(-5) = -1 < 0$

$f'(-4) = 1 > 0$

f is increasing on $\left(-\frac{9}{2}, \infty\right)$ and decreasing on $\left(-\infty, -\frac{9}{2}\right)$.

19. $f(x) = -x^3 + 2x^2 + 15x + 16$

$$\begin{aligned} f'(x) &= -3x^2 + 4x + 15 \\ &= -(3x^2 - 4x - 15) \\ &= -(3x + 5)(x - 3) \end{aligned}$$

$f'(x) = 0$ when $x = -\frac{5}{3}$ or $x = 3$ and f' exists everywhere.

Critical numbers: $-\frac{5}{3}$ and 3

Test an x-value in the intervals $\left(-\infty, -\frac{5}{3}\right)$, $\left(-\frac{5}{3}, 3\right)$, and $(3, \infty)$.

$f'(-2) = -5 < 0$

$f'(0) = 15 > 0$

$f'(4) = -17 < 0$

f is increasing on $\left(-\frac{5}{3}, 3\right)$ and decreasing on $\left(-\infty, -\frac{5}{3}\right)$ and $(3, \infty)$.

21. $f(x) = \dfrac{16}{9 - 3x}$

$f'(x) = \dfrac{16(-1)(-3)}{(9 - 3x)^2} = \dfrac{48}{(9 - 3x)^2}$

$f'(x) > 0$ for all x $(x \neq 3)$, and f is not defined for $x = 3$.

f is increasing on $(-\infty, 3)$ and $(3, \infty)$ and never decreasing.

23. $f(x) = \ln|x^2 - 1|$

$f'(x) = \dfrac{2x}{x^2 - 1}$

f is not defined for $x = -1$ and $x = 1$.

$f'(x) = 0$ when $x = 0$.

Test an x-value in the intervals $(-\infty, -1)$, $(-1, 0)$, $(0, 1)$, and $(1, \infty)$.

$f'(-2) = -\dfrac{4}{3} < 0$

$f'\left(-\dfrac{1}{2}\right) = \dfrac{4}{3} > 0$

$f'\left(\dfrac{1}{2}\right) = -\dfrac{4}{3} < 0$

$f'(2) = \dfrac{4}{3} > 0$

f is increasing on $(-1, 0)$ and $(1, \infty)$ and decreasing on $(-\infty, -1)$ and $(0, 1)$.

25. $f(x) = -x^2 + 4x - 8$

$f'(x) = -2x + 4 = 0$

Critical number: $x = 2$

$f''(x) = -2 < 0$ for all x, so $f(2)$ is a relative maximum.

$f(2) = -4$

Relative maximum of -4 at 2

27. $f(x) = 2x^2 - 8x + 1$

$f'(x) = 4x - 8 = 0$

Critical number: $x = 2$

$f''(x) = 4 > 0$ for all x, so $f(2)$ is a relative minimum.

$f(2) = -7$

Relative minimum of -7 at 2

29. $f(x) = 2x^3 + 3x^2 - 36x + 20$

$$\begin{aligned} f'(x) = 6x^2 + 6x - 36 &= 0 \\ 6(x^2 + x - 6) &= 0 \\ (x + 3)(x - 2) &= 0 \end{aligned}$$

Critical numbers: -3 and 2

$$f''(x) = 12x + 6$$
$$f''(-3) = -30 < 0, \text{ so a maximum occurs at } x = -3.$$
$$f''(2) = 30 > 0, \text{ so a minimum occurs at } x = 2.$$
$$f(-3) = 101$$
$$f(2) = -24$$

Relative maximum of 101 at -3

Relative minimum of -24 at 2

31. $$f(x) = \frac{xe^x}{x-1}$$
$$f'(x) = \frac{(x-1)(xe^x + e^x) - xe^x(1)}{(x-1)^2}$$
$$= \frac{x^2e^x - xe^x - xe^x - e^x - xe^x}{(x-1)^2}$$
$$= \frac{x^2e^x - xe^x - e^x}{(x-1)^2}$$
$$= \frac{e^x(x^2 - x - 1)}{(x-1)^2}$$

$f'(x)$ is undefined at $x = 1$, but 1 is not in the domain of $f(x)$.

$f'(x) = 0$ when $x^2 - x - 1 = 0$

$$x = \frac{1 \pm \sqrt{1 - 4(1)(-1)}}{2}$$
$$= \frac{1 \pm \sqrt{5}}{2}$$
$$\frac{1+\sqrt{5}}{2} \approx 1.618 \text{ or } \frac{1-\sqrt{5}}{2} = -0.618$$

Critical numbers are -0.618 and 1.618.

$$f'(1.4) = \frac{e^{1.4}(1.4^2 - 1.4 - 1)}{(1.4-1)^2} \approx -11.15 < 0$$
$$f'(2) = \frac{e^2(2^2 - 2 - 1)}{(2-1)^2} = e^2 \approx 7.39 > 0$$
$$f'(-1) = \frac{e^{-1}[(-1)^2 - (-1) - 1]}{(-1-1)^2} \approx 0.09 > 0$$
$$f'(0) = \frac{e^0(0^2 - 0 - 1)}{(0-1)^2} = -1 < 0$$

There is a relative maximum at $(-0.618, 0.206)$ and a relative minimum at $(1.618, 13.203)$.

33. $$f(x) = 3x^4 - 5x^2 - 11x$$
$$f'(x) = 12x^3 - 10x - 11$$
$$f''(x) = 36x^2 - 10$$
$$f''(1) = 36(1)^2 - 10 = 26$$
$$f''(-3) = 36(-3)^2 - 10 = 314$$

35. $$f(x) = \frac{4x+2}{3x-6}$$
$$f'(x) = \frac{(3x-6)(4) - (4x+2)(3)}{(3x-6)^2}$$
$$= \frac{12x - 24 - 12x - 6}{(3x-6)^2} = \frac{-30}{(3x-6)^2}$$
$$= -30(3x-6)^{-2}$$
$$f''(x) = -30(-2)(3x-6)^{-3}(3)$$
$$= 180(3x-6)^{-3} \text{ or } \frac{180}{(3x-6)^3}$$
$$f''(1) = 180[3(1) - 6]^{-3} = -\frac{20}{3}$$
$$f''(-3) = 180[3(-3) - 6]^{-3} = -\frac{4}{75}$$

37. $$f(t) = \sqrt{t^2+1} = (t^2+1)^{1/2}$$
$$f'(t) = \frac{1}{2}(t^2+1)^{-1/2}(2t) = t(t^2+1)^{-1/2}$$
$$f''(t) = (t^2+1)^{-1/2}(1) + t\left[\left(-\frac{1}{2}\right)(t^2+1)^{-3/2}(2t)\right]$$
$$= (t^2+1)^{-1/2} - t^2(t^2+1)^{-3/2}$$
$$= \frac{1}{(t^2+1)^{1/2}} - \frac{t^2}{(t^2+1)^{3/2}} = \frac{t^2+1-t^2}{(t^2+1)^{3/2}}$$
$$= (t^2+1)^{-3/2} \text{ or } \frac{1}{(t^2+1)^{3/2}}$$
$$f''(1) = \frac{1}{(1+1)^{3/2}} = \frac{1}{2^{3/2}} \approx 0.354$$
$$f''(-3) = \frac{1}{(9+1)^{3/2}} = \frac{1}{10^{3/2}} \approx 0.032$$

39. $f(x) = -2x^3 - \frac{1}{2}x^2 + x - 3$

Domain is $(-\infty, \infty)$

The graph has no symmetry.

$$f'(x) = -6x^2 - x + 1 = 0$$
$$(3x - 1)(2x + 1) = 0$$

Critical numbers: $\frac{1}{3}$ and $-\frac{1}{2}$

Critical points: $\left(\frac{1}{3}, -2.80\right)$ and $\left(-\frac{1}{2}, -3.375\right)$

$$f''(x) = -12x - 1$$
$$f''\left(\frac{1}{3}\right) = -5 < 0$$
$$f''\left(-\frac{1}{2}\right) = 5 > 0$$

Relative maximum at $\frac{1}{3}$

Relative minimum at $-\frac{1}{2}$

Increasing on $\left(-\frac{1}{2}, \frac{1}{3}\right)$

Decreasing on $\left(-\infty, -\frac{1}{2}\right)$ and $\left(\frac{1}{3}, \infty\right)$

$$f''(x) = -12x - 1 = 0$$
$$x = -\frac{1}{12}$$

Point of inflection at $\left(-\frac{1}{12}, -3.09\right)$

Concave upward on $\left(-\infty, -\frac{1}{12}\right)$

Concave downward on $\left(-\frac{1}{12}, \infty\right)$

y-intercept:

$$y = -2(0)^3 - \frac{1}{2}(0)^2 + (0) - 3 = -3$$

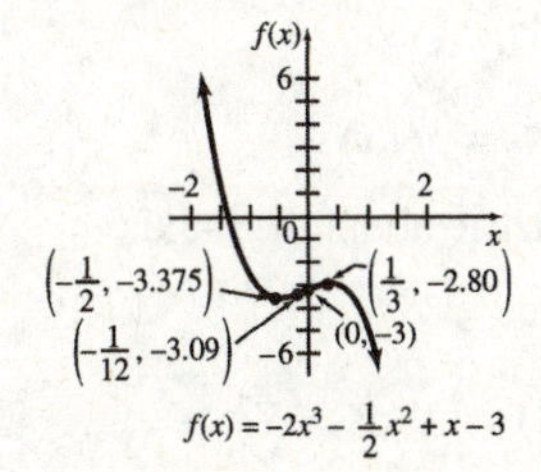

41. $f(x) = x^4 - \frac{4}{3}x^3 - 4x^2 + 1$

Domain is $(-\infty, \infty)$

The graph has no symmetry.

$$f'(x) = 4x^3 - 4x^2 - 8x = 0$$
$$4x(x^2 - x - 2) = 0$$
$$4x(x - 2)(x + 1) = 0$$

Critical numbers: 0, 2, and -1

Critical points: $(0, 1)$, $\left(2, -\frac{29}{3}\right)$ and $\left(-1, -\frac{2}{3}\right)$

$$f''(x) = 12x^2 - 8x - 8$$
$$= 4(3x^2 - 2x - 2)$$
$$f''(-1) = 12 > 0$$
$$f''(0) = -8 < 0$$
$$f''(2) = 24 > 0$$

Relative maximum at 0

Relative minima at -1 and 2

Increasing on $(-1, 0)$ and $(2, \infty)$

Decreasing on $(-\infty, -1)$ and $(0, 2)$

$$f''(x) = 4(3x^2 - 2x - 2) = 0$$
$$x = \frac{2 \pm \sqrt{4 - (-24)}}{6}$$
$$= \frac{1 \pm \sqrt{7}}{3}$$

Points of inflection at $\left(\frac{1+\sqrt{7}}{3}, -5.12\right)$ and $\left(\frac{1-\sqrt{7}}{3}, 0.11\right)$

Concave upward on $\left(-\infty, \frac{1-\sqrt{7}}{3}\right)$ and $\left(\frac{1+\sqrt{7}}{3}, \infty\right)$

Concave downward on $\left(\frac{1-\sqrt{7}}{3}, \frac{1+\sqrt{7}}{3}\right)$

y-intercept:

$$y = (0)^4 - \frac{4}{3}(0)^3 - 4(0)^2 + 1 = 1$$

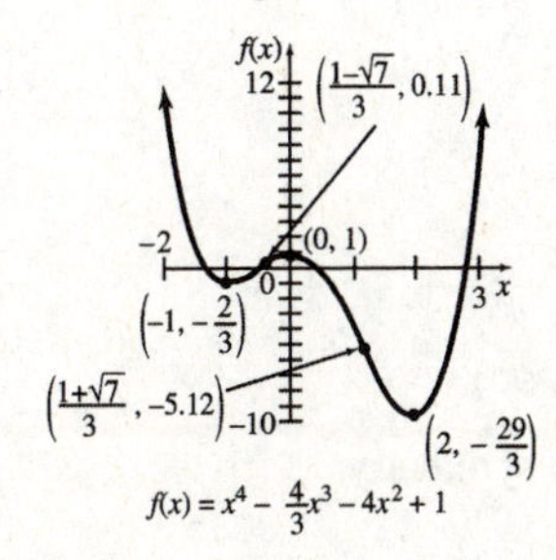

43. $f(x) = \frac{x - 1}{2x + 1}$

Domain is $\left(-\infty, -\frac{1}{2}\right) \cup \left(-\frac{1}{2}, \infty\right)$

The graph has no symmetry.

$$f'(x) = \frac{(2x + 1)(1) - (x - 1)(2)}{(2x + 1)^2}$$
$$= \frac{3}{(2x + 1)^2}$$

f' is never zero.

$f'\left(-\frac{1}{2}\right)$ does not exist, but $-\frac{1}{2}$ is not a critical number because $-\frac{1}{2}$ is not in the domain of f.

Thus, there are no critical numbers, so $f(x)$ has no relative extrema.

Increasing on $\left(-\infty, \frac{1}{2}\right)$ and $\left(\frac{1}{2}, \infty\right)$

$$f''(x) = \frac{-12}{(2x+1)^3}$$

$$f''(0) = -12 < 0$$

$$f''(-1) = 12 > 0$$

No inflection points

Concave upward on $\left(-\infty, -\frac{1}{2}\right)$

Concave downward on $\left(-\frac{1}{2}, \infty\right)$

x-intercept: $\frac{x-1}{2x+1} = 0$

$$x = 1$$

y-intercept: $y = \frac{0-1}{2(0)+1} = -1$

Vertical asymptote at $x = -\frac{1}{2}$

Horizontal asymptote at $y = \frac{1}{2}$

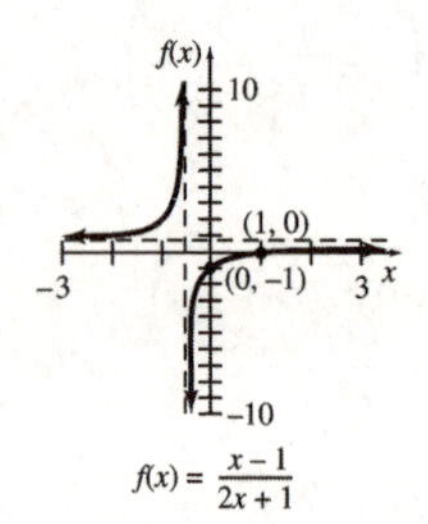

45. $f(x) = -4x^3 - x^2 + 4x + 5$

Domain is $(-\infty, \infty)$

The graph has no symmetry.

$$f'(x) = -12x^2 - 2x + 4$$

$$= -2(6x^2 + x - 2) = 0$$

$$(3x+2)(2x-1) = 0$$

Critical numbers: $-\frac{2}{3}$ and $\frac{1}{2}$

Critical points: $\left(-\frac{2}{3}, 3.07\right)$ and $\left(\frac{1}{2}, 6.25\right)$

$$f''(x) = -24x - 2$$

$$= -2(12x + 1)$$

$$f''\left(-\frac{2}{3}\right) = 14 > 0$$

$$f''\left(\frac{1}{2}\right) = -14 < 0$$

Relative maximum at $\frac{1}{2}$

Relative minimum at $-\frac{2}{3}$

Increasing on $\left(-\frac{2}{3}, \frac{1}{2}\right)$

Decreasing on $\left(-\infty, -\frac{2}{3}\right)$ and $\left(\frac{1}{2}, \infty\right)$

$$f''(x) = -2(12x + 1) = 0$$

$$x = -\frac{1}{12}$$

Point of inflection at $\left(-\frac{1}{12}, 4.66\right)$

Concave upward on $\left(-\infty, -\frac{1}{12}\right)$

Concave downward on $\left(-\frac{1}{12}, \infty\right)$

y-intercept:

$$y = -4(0)^3 - (0)^2 + 4(0) + 5 = 5$$

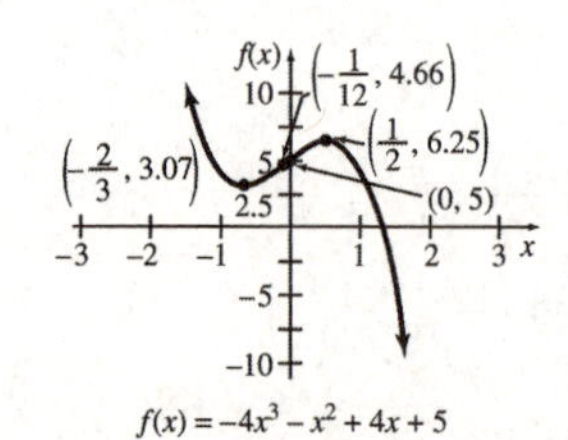

47. $f(x) = x^4 + 2x^2$

Domain is $(-\infty, \infty)$

$$f(-x) = (-x)^4 - 2(-x)^2$$

$$= x^4 + 2x^2 = f(x)$$

The graph is symmetric about the y-axis.

$$f'(x) = 4x^3 + 4x$$

$$= 4x(x^2 + 1) = 0$$

Critical number: 0

Critical point: (0, 0)

$$f''(x) = 12x^2 + 4 = 4(3x^2 + 1)$$

$$f''(0) = 4 > 0$$

Relative minimum at 0

Increasing on $(0, \infty)$

Decreasing on $(-\infty, 0)$

$f''(x) = 4(3x^2 + 1) \neq 0$ for any x

No points of inflection

$f''(-1) = 16 > 0$

$f''(1) = 16 > 0$

Concave upward on $(-\infty, \infty)$

x-intercept: 0; y-intercept: 0

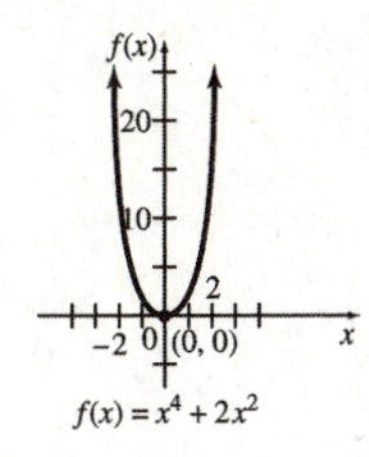

49. $f(x) = \dfrac{x^2 + 4}{x}$

Domain is $(-\infty, 0) \cup (0, \infty)$

$$f(-x) = \frac{(-x)^2 + 4}{-x}$$

$$= \frac{x^2 + 4}{-x} = -f(x)$$

The graph is symmetric about the origin.

$$f'(x) = \frac{x(2x) - (x^2 + 4)}{x^2}$$

$$= \frac{x^2 - 4}{x^2} = 0$$

Critical numbers: -2 and 2

Critical points: $(-2, -4)$ and $(2, 4)$

$$f''(x) = \frac{8}{x^3}$$

$f''(-2) = -1 < 0$

$f''(2) = 1 > 0$

Relative maximum at -2

Relative minimum at 2

Increasing on $(-\infty, -2)$ and $(2, \infty)$

Decreasing on $(-2, 0)$ and $(0, 2)$

$f''(x) = \frac{8}{x^3} > 0$ for all x.

No inflection points

Concave upward on $(0, \infty)$

Concave downward on $(-\infty, 0)$

No x- or y-intercepts

Vertical asymptote at $x = 0$

Oblique asymptote at $y = x$

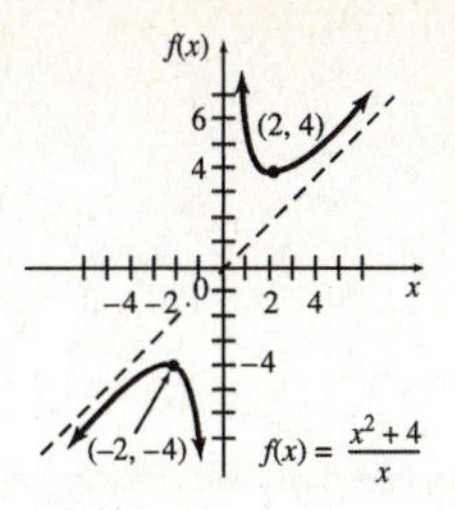

51. $f(x) = \dfrac{2x}{3 - x}$

Domain is $(-\infty, 3) \cup (3, \infty)$

The graph has no symmetry.

$$f'(x) = \frac{(3 - x)(2) - (2x)(-1)}{(3 - x)^2}$$

$$= \frac{6}{(3 - x)^2}$$

$f'(x)$ is never zero. $f'(3)$ does not exist, but since 3 is not in the domain of f, it is not a critical number. No critical numbers, so no relative extrema

$f'(0) = \frac{2}{3} > 0$

$f'(4) = 6 > 0$

Increasing on $(-\infty, 3)$ and $(3, \infty)$

$$f''(x) = \frac{12}{(3 - x)^3}$$

$f''(x)$ is never zero. $f''(3)$ does not exist, but since 3 is not in the domain of f, there is no inflection point at $x = 3$.

$f''(0) = \frac{12}{27} > 0$

$f''(4) = -12 < 0$

Concave upward on $(-\infty, 3)$

Concave downward on $(3, \infty)$

x-intercept: 0; y-intercept: 0

Vertical asymptote at $x = 3$

Horizontal asymptote at $y = -2$

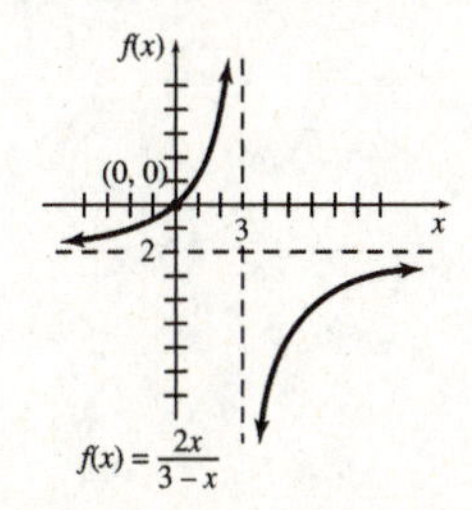

53. $f(x) = xe^{2x}$

Domain is $(-\infty, \infty)$.

$f(-x) = -xe^{-2x}$

The graph has no symmetry.

$$f'(x) = (1)(e^{2x}) + (x)(2e^{2x})$$
$$= e^{2x}(2x + 1)$$

$f'(x) = 0$ when $x = -\frac{1}{2}$.

Critical number: $-\frac{1}{2}$

Critical point: $\left(-\frac{1}{2}, -\frac{1}{2e}\right)$

$$f'(-1) = e^{2(-1)}[2(-1) + 1] = -e^{-2} < 0$$
$$f'(0) = e^{2(0)}[2(0) + 1] = 1 > 0$$

No relative maximum

Relative minimum at $\left(-\frac{1}{2}, -\frac{1}{2e}\right)$

Decreasing on $\left(-\infty, -\frac{1}{2}\right)$ and increasing on $\left(-\frac{1}{2}, \infty\right)$

$$f''(x) = 2e^{2x}(2x + 1) + e^{2x}(2)$$
$$= 4e^{2x}(x + 1)$$

$f''(x) = 0$ when $x = -1$.

$$f''(-2) = 4e^{2(-2)}[(-2) + 1] = -4e^{-4} < 0$$
$$f''(0) = 4e^{2(0)}[(0) + 1] = 4 > 0$$

Inflection point at $(-1, -e^{-2})$

Concave upward on $(-1, \infty)$

Concave downward on $(-\infty, -1)$

x-intercept: $xe^{2x} = 0$
$$x = 0$$

y-intercept: $y = (0)e^{2(0)} = 0$

Since $\lim\limits_{x\to-\infty} xe^{2x} = 0$, there is a horizontal asymptote at $y = 0$.

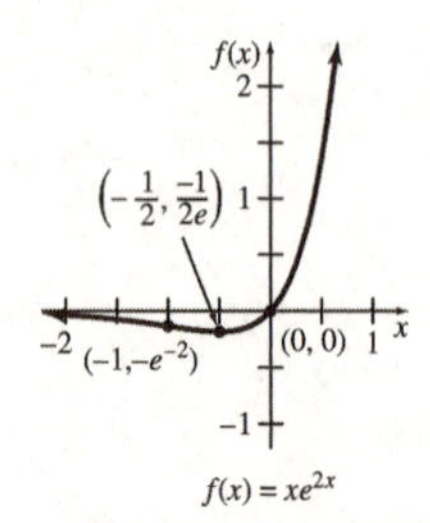

55. $f(x) = \ln(x^2 + 4)$

Domain is $(-\infty, \infty)$.

$f(-x) = \ln[(-x)^2 + 4] = \ln(x^2 + 4) = f(x)$

The graph is symmetric about the y-axis.

$$f'(x) = \frac{2x}{x^2 + 4}$$

$f'(x) = 0$ when $x = 0$.

Critical number: 0

Critical point: $(0, \ln 4)$

$$f'(-1) = \frac{2(-1)}{(-1)^2 + 4} = -\frac{2}{5} < 0$$
$$f'(1) = \frac{2(1)}{(1)^2 + 4} = \frac{2}{5} > 0$$

No relative maximum

Relative minimum at $(0, \ln 4)$

Increasing on $(0, \infty)$

Decreasing on $(-\infty, 0)$

$$f''(x) = \frac{(x^2 + 4)(2) - (2x)(2x)}{(x^2 + 4)^2}$$
$$= \frac{-2(x^2 - 4)}{(x^2 + 4)^2}$$

$f''(x) = 0$ when

$$x^2 - 4 = 0$$
$$x = \pm 2$$

$$f''(-3) = \frac{-2[(-3)^2 - 4]}{[(-3)^2 + 4]^2} = -\frac{10}{169} < 0$$
$$f''(0) = \frac{-2[(0)^2 - 4]}{[(0)^2 + 4]^2} = \frac{1}{2} > 0$$
$$f''(3) = \frac{-2[(3)^2 - 4]}{[(3)^2 + 4]^2} = -\frac{10}{169} < 0$$

Inflection points at $(-2, \ln 8)$ and $(2, \ln 8)$

Concave upward on $(-2, 2)$

Concave downward on $(-\infty, -2)$ and $(2, \infty)$

Since $f(x)$ never equals zero, there are no x-intercepts.

y-intercept: $y = \ln[(0)^2 + 4] = \ln 4$

No horizontal or vertical asymptotes.

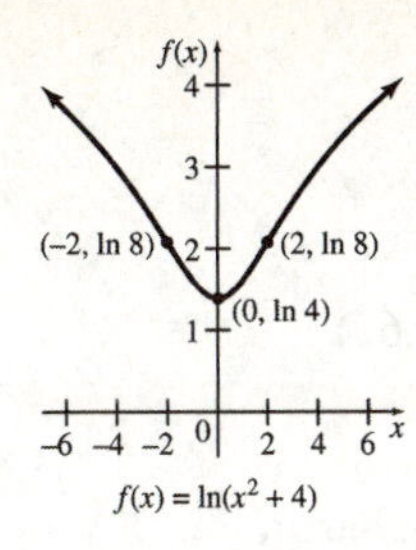

57. $f(x) = 4x^{1/3} + x^{4/3}$

Domain is $(-\infty, \infty)$.

$f(-x) = 4(-x)^{1/3} + (-x)^{4/3} = -4x^{1/3} + x^{4/3}$

The graph is not symmetric about the y-axis or origin.

$f'(x) = \frac{4}{3}x^{-2/3} + \frac{4}{3}x^{1/3}$

$f'(x) = 0$ when

$\frac{4}{3}x^{-2/3} + \frac{4}{3}x^{1/3} = 0$

$\frac{4}{3}x^{-2/3}(1 + x) = 0$

$x = -1$

$f'(x)$ is not defined when $x = 0$

Critical numbers: -1 and 0

Critical points: $(-1, -3)$ and $(0, 0)$

$f'(-8) = \frac{4}{3}(-8)^{-2/3} + \frac{4}{3}(-8)^{1/3} = -\frac{7}{3} < 0$

$f'\left(-\frac{1}{8}\right) = \frac{4}{3}\left(-\frac{1}{8}\right)^{-2/3} + \frac{4}{3}\left(-\frac{1}{8}\right)^{1/3} = \frac{14}{3} > 0$

$f'(1) = \frac{4}{3}(1)^{-2/3} + \frac{4}{3}(1)^{1/3} = \frac{8}{3} > 0$

No relative maximum

Relative minimum at $(-1, -3)$

Increasing on $(-1, \infty)$

Decreasing on $(-\infty, -1)$

$f''(x) = -\frac{8}{9}x^{-5/3} + \frac{4}{9}x^{-2/3}$

$f''(x) = 0$ when

$-\frac{8}{9}x^{-5/3} + \frac{4}{9}x^{-2/3} = 0$

$\frac{4}{9}x^{-5/3}(-2 + x) = 0$

$x = 2$

$f''(x)$ is not defined when $x = 0$

$f''(-1) = -\frac{8}{9}(-1)^{-5/3} + \frac{4}{9}(-1)^{-2/3} = \frac{4}{3} > 0$

$f''(1) = -\frac{8}{9}(1)^{-5/3} + \frac{4}{9}(1)^{-2/3} = -\frac{4}{9} < 0$

$f''(8) = -\frac{8}{9}(8)^{-5/3} + \frac{4}{9}(8)^{-2/3} = \frac{1}{12} > 0$

Inflection points at $(0, 0)$ and $(2, 6 \cdot 2^{1/3})$

Concave upward on $(-\infty, 0)$ and $(2, \infty)$

Concave downward on $(0, 2)$

x-intercept: $4x^{1/3} + x^{4/3} = 0$

$x^{1/3}(4 + x) = 0$

$x = 0$ or $x = -4$

y-intercept: $y = 4(0)^{1/3} + (0)^{4/3} = 0$

No horizontal or vertical asymptotes

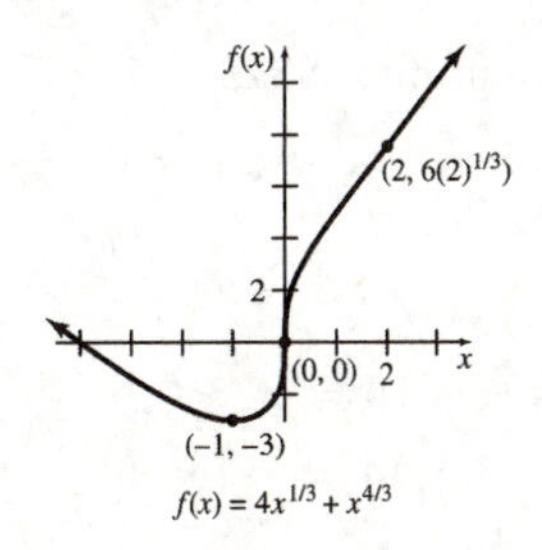

59.

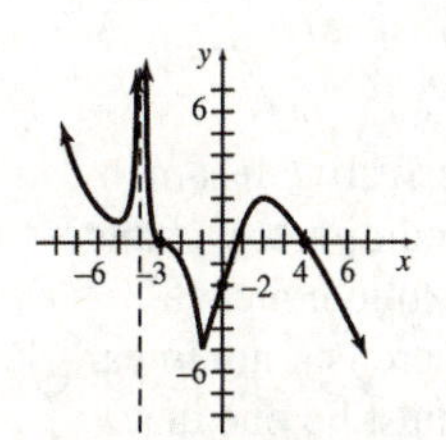

Other graphs are possible.

61. **(a)-(b)** If the price of the stock is falling faster and faster, $P(t)$ would be decreasing, so $P'(t)$ would be negative. $P(t)$ would be concave downward, so $P''(t)$ would also be negative.

63. **(a)** Profit = Income − Cost

$$\begin{aligned} P(q) &= qp - C(q) \\ &= q(-q^2 - 3q + 299) \\ &\quad - (-10q^2 + 250q) \\ &= -q^3 - 3q^2 + 299q \\ &\quad + 10q^2 - 250q \\ &= -q^3 + 7q^2 + 49q \end{aligned}$$

(b) $P'(q) = -3q^2 + 14q + 49$
$= (-3q - 7)(q - 7)$
$= -(3q + 7)(q - 7)$

$q = \frac{7}{3}$ (nonsensical) or $q = 7$

$P''(q) = -6q + 14$
$P''(7) = -6 \cdot 7 + 14 = -28 < 0$
(indicates a maximum)

7 brushes would produce the maximum profit.

(c) $p = -7^2 - 3(7) + 299$
$= -49 - 21 + 299 = 229$

$229 is the price that produces the maximum profit.

(d) $P(7) = -7^3 + 7(7^2) + 49(7) = 343$

The maximum profit is $343.

(e) $P''(q) = 0$ when $-6q + 14 = 0$

$$q = \frac{7}{3}$$

$P''(2) = -6(2) + 14 = 2 > 0$
$P''(3) = -6 \cdot 3 + 14 = -4 < 0$

The point of diminishing returns is $q = \frac{7}{3}$ (between 2 and 3 brushes).

65. **(a)** Since the second derivative has many sign changes, the graph continually changes from concave upward to concave downward. Since there is a nonlinear decline, the graph must be one that declines, levels off, declines, levels off, etc. Therefore, the first derivative has many critical numbers where the first derivative is zero.

(b) The curve is always decreasing except at frequent points of inflection.

67. Sketch the curve for $l_1(v) = 0.08e^{0.33v}$

$l_1'(v) = 0.0264e^{0.33v}$
$e^{0.33v} \neq 0$

$l_1(v)$ has no critical points.

$l_1''(v) = 0.008712e^{0.33v}$
$e^{0.33v} \neq 0$

$l_1(v)$ has no inflection points.

Sketch the curve for $l_2 = -0.87v^2 + 28.17v - 211.41$

$l_2'(v) = -1.74v + 28.17$
$-1.74v + 28.17 = 0$
$v \approx 16.19$

Critical point: (16.19, 16.62)

$l_2''(v) = -1.74$

$l_2(v)$ has no inflection points.

$l_2(v)$ has a relative maximum at (16.19, 16.62).

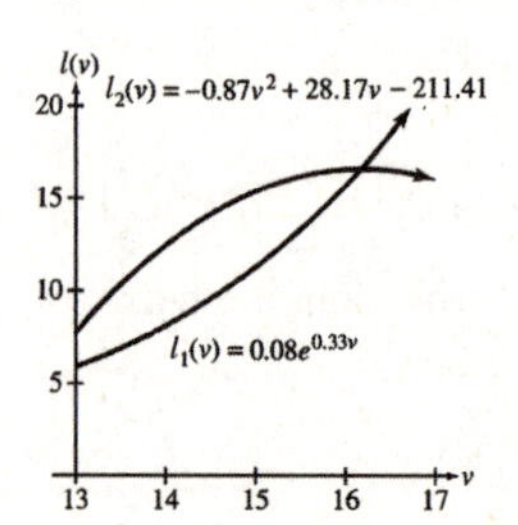

69. $y = 34.7(1.0186)^{-x}(x^{-0.658})$

In Chapter 12, the function was originally defined as
$\log y = 1.54 - 0.008x - 0.658 \log x$

so, $0 < x < \infty$, and $0 < y < \infty$.

The function will have a vertical asymptote at $x = 0$ and a horizontal asymptote at $y = 0$.

$$\frac{dy}{dx} = -34.7(1.0186)^{-x}x^{-0.658}\left[\ln(1.0186) + \frac{0.658}{x}\right]$$

For every value in the domain, $\frac{dy}{dx} < 0$, so y has no critical points and is decreasing on $(0, \infty)$.

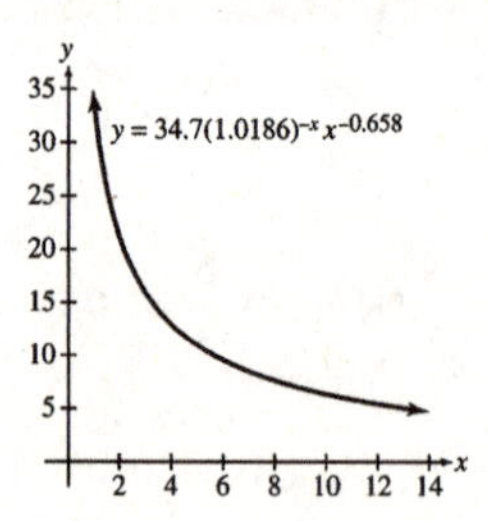

71. $y(t) = A^{c^t}$

$y'(t) = (\ln A)A^{c^t} \cdot \frac{d}{dt}c^t$
$= (\ln A)(\ln c)c^t A^{c^t}$

$y''(t) = (\ln A)(\ln c) \cdot [(\ln c)c^t A^{c^t}$
$+ c^t(\ln A)(\ln c)c^t A^{c^t}]$
$= (\ln A)(\ln c)^2 c^t A^{c^t}[1 + (\ln A)c^t]$

$$y''(t) = 0 \text{ when } 1 + \ln(A)c^t = 0$$

$$c^t = -\frac{1}{\ln A}$$

$$t \ln c = \ln\left(-\frac{1}{\ln A}\right)$$

$$t = -\frac{\ln(-\ln A)}{\ln c}$$

$$= -\frac{\ln[-\ln(0.3982 \cdot 10^{-291})]}{\ln 0.4152}$$

By properties of logarithms,

$$\begin{aligned} -\ln(0.3982 \cdot 10^{-291}) &= -[\ln(0.3982) + \ln(10^{-291})] \\ &= -[\ln(0.3982) - 291 \ln(10)] \\ &= -\ln(0.3982) + 291 \ln(10) \end{aligned}$$

So,

$$t = -\frac{\ln[-\ln(0.3982) + 291 \ln(10)]}{\ln(0.4152)}$$

$$\approx 7.405$$

At about 7.405 years the rate of learning to pass the test begins to slow down.

73. **(a)** The U.S. total inventory was at a relative maximum in 1965, 1973, 1976, 1983, 1986, and 1988.

(b) The U.S. total inventory was at its largest relative maximum from 1965 to 1967. During this period, the Soviet total inventory was concave upward. This means that the total inventory was increasing at an increasingly rapid rate.

Chapter 14

APPLICATIONS OF THE DERIVATIVE

14.1 Absolute Extrema

Your Turn 1

To find the absolute extrema of $f(x) = 3x^{2/3} - 3x^{5/3}$ on $[0, 8]$, first find the critical numbers in the interval $(0,8)$.

$$\begin{aligned} f'(x) &= 2x^{-1/3} - 5x^{2/3} \\ &= \frac{2}{x^{1/3}} - 5x^{2/3} \\ &= \frac{2}{x^{1/3}} - \frac{5x}{x^{1/3}} \\ &= \frac{1}{x^{1/3}}(2 - 5x) \end{aligned}$$

$f'(x)$ is 0 when $2 - 5x = 0$, that is, when $x = \frac{5}{2} = 0.4$. Now evaluate the function f at the critical value 0.4 and at the endpoints 0 and 8.

x	$f(x)$
0	0
0.4	0.977
8	−84

The absolute maximum of approximately 0.977 occurs when $x = 2/5$; the absolute minimum of -84 occurs when $x = 8$.

Your Turn 2

The domain of the function

$$f(x) = -x^4 - 4x^3 + 8x^2 + 20$$

is the open interval $(-\infty, \infty)$. As x approaches $+\infty$ and $-\infty$ the dominant term is $-x^4$, and this approaches $-\infty$. Thus the function has no absolute minimum. To find the absolute maximum, we evaluate f at the critical points.

$$\begin{aligned} f'(x) &= -4x^3 - 12x^2 + 16x \\ &= -4x(x^2 + 3x - 4) \\ &= -4x(x + 4)(x - 1) \end{aligned}$$

Setting $f'(x)$ equal to 0, we have $x = 0, x = -4$, or $x = 1$.

x	$f(x)$
−4	148
0	20
1	23

The absolute maximum is 148, at $x = -4$. As noted above, there is no absolute minimum.

14.1 Exercises

1. As shown on the graph, the absolute maximum occurs at x_3; there is no absolute minimum. (There is no functional value that is less than all others.)

3. As shown on the graph, there are no absolute extrema.

5. As shown on the graph, the absolute minimum occurs at x_1; there is no absolute maximum.

7. As shown on the graph, the absolute maximum occurs at x_1; the absolute minimum occurs at x_2.

11. $f(x) = x^3 - 6x^2 + 9x - 8; [0,5]$

Find critical numbers:

$$\begin{aligned} f'(x) = 3x^2 - 12x + 9 &= 0 \\ x^2 - 4x + 3 &= 0 \\ (x - 3)(x - 1) &= 0 \\ x = 1 \text{ or } x &= 3 \end{aligned}$$

x	$f(x)$	
0	−8	Absolute minimum
1	−4	
3	−8	Absolute minimum
5	12	Absolute maximum

13. $f(x) = \frac{1}{3}x^3 + \frac{3}{2}x^2 - 4x + 1; [-5, 2]$

Find critical numbers:

$$f'(x) = x^2 + 3x - 4 = 0$$
$$(x + 4)(x - 1) = 0$$
$$x = -4 \text{ or } x = 1$$

x	$f(x)$	
-4	$\frac{59}{3} \approx 19.67$	Absolute maximum
1	$-\frac{7}{6} \approx -1.17$	Absolute minimum
-5	$\frac{101}{6} \approx 16.83$	
2	$\frac{5}{3} \approx 1.67$	

15. $f(x) = x^4 - 18x^2 + 1; [-4, 4]$

$$f'(x) = 4x^3 - 36x = 0$$
$$4x(x^2 - 9) = 0$$
$$4x(x + 3)(x - 3) = 0$$
$$x = 0 \text{ or } x = -3 \text{ or } x = 3$$

x	$f(x)$	
-4	-31	
-3	-80	Absolute minimum
0	1	Absolute maximum
3	-80	Absolute minimum
4	-31	

17. $f(x) = \frac{1 - x}{3 + x}; [0, 3]$

$$f'(x) = \frac{-4}{(3 + x)^2}$$

No critical numbers

x	$f(x)$	
0	$\frac{1}{3}$	Absolute maximum
3	$-\frac{1}{3}$	Absolute minimum

19. $f(x) = \frac{x - 1}{x^2 + 1}; [1, 5]$

$$f'(x) = \frac{-x^2 + 2x + 1}{(x^2 + 1)^2}$$

$f'(x) = 0$ when

$$-x^2 + 2x + 1 = 0$$
$$x = 1 \pm \sqrt{2},$$

but $1 - \sqrt{2}$ is not in $[1, 5]$.

x	$f(x)$	
1	0	Absolute minimum
5	$\frac{2}{13} \approx 0.15$	
$1 + \sqrt{2}$	$\frac{\sqrt{2} - 1}{2} \approx 0.21$	Absolute maximum

21. $f(x) = (x^2 - 4)^{1/3}; [-2, 3]$

$$f'(x) = \frac{1}{3}(x^2 - 4)^{-2/3}(2x)$$
$$= \frac{2x}{3(x^2 - 4)^{2/3}}$$

$$f'(x) = 0 \text{ when } 2x = 0$$
$$x = 0$$

$f'(x)$ is undefined at $x = -2$ and $x = 2$, but $f(x)$ is defined there, so -2 and 2 are also critical numbers:

x	$f(x)$	
-2	0	
0	$(-4)^{1/3} \approx -1.587$	Absolute minimum
2	0	
3	$5^{1/3} \approx 1.710$	Absolute maximum

23. $f(x) = 5x^{2/3} + 2x^{5/3}; [-2, 1]$

$$f'(x) = \frac{10}{3}x^{-1/3} + \frac{10}{3}x^{2/3}$$
$$= \frac{10}{3x^{1/3}} + \frac{10x^{2/3}}{3}$$
$$= \frac{10x + 10}{3x^{1/3}}$$
$$= \frac{10(x + 1)}{3\sqrt[3]{x}}$$

$$f'(x) = 0 \text{ when } 10(x + 1) = 0$$
$$x + 1 = 0$$
$$x = -1.$$

$f'(x)$ is undefined at $x = 0$, but $f(x)$ is defined at $x = 0$, so 0 is also a critical number.

x	$f(x)$	
-2	1.587	
-1	3	
0	0	Absolute minimum
1	7	Absolute maximum

25. $f(x) = x^2 - 8 \ln x; [1, 4]$

$f'(x) = 2x - \frac{8}{x}$

$$f'(x) = 0 \text{ when } 2x - \frac{8}{x} = 0$$
$$2x = \frac{8}{x}$$
$$2x^2 = 8$$
$$x^2 = 4$$
$$x = -2 \quad \text{or} \quad x = 2$$

but $x = -2$ is not in the given interval.

Although $f'(x)$ fails to exist at $x = 0$, 0 is not in the specified domain for $f(x)$, so 0 is not a critical number.

x	$f(x)$	
1	1	
2	-1.545	Absolute minimum
4	4.910	Absolute maximum

27. $f(x) = x + e^{-3x}; [-1, 3]$

$f'(x) = 1 - 3e^{-3x}$

$$f'(x) = 0 \text{ when } 1 - 3e^{-3x} = 0$$
$$-3e^{-3x} = -1$$
$$e^{-3x} = \frac{1}{3}$$
$$-3x = \ln \frac{1}{3}$$
$$x = \frac{\ln 3}{3}$$

x	$f(x)$	
-1	19.09	Absolute maximum
$\frac{\ln 3}{3}$	0.6995	Absolute minimum
3	3.000	

29. $f(x) = \frac{-5x^4 + 2x^3 + 3x^2 + 9}{x^4 - x^3 + x^2 + 7}; [-1, 1]$

The indicated domain tells us the x-values to use for the viewing window, but we must experiment to find a suitable range for the y-values. In order to show the absolute extrema on $[-1, 1]$, we find that a suitable window is $[-1, 1]$ by $[0, 1.5]$ with Xscl = 0.1, Yscl = 0.1.

From the graph, we see that on $[-1, 1]$, f has an absolute maximum of 1.356 at about 0.6085 and an absolute minimum of 0.5 at -1.

31. $f(x) = 2x + \frac{8}{x^2} + 1, \ x > 0$

$$f'(x) = 2 - \frac{16}{x^3}$$
$$= \frac{2x^3 - 16}{x^3}$$
$$= \frac{2(x-2)(x^2 + 2x + 4)}{x^3}$$

Since the specified domain is $(0, \infty)$, a critical number is $x = 2$.

x	$f(x)$
2	7

There is an absolute minimum of 7 at $x = 2$; there is no absolute maximum, as can be seen by looking at the graph of f.

33. $f(x) = -3x^4 + 8x^3 + 18x^2 + 2$

$$f'(x) = -12x^3 + 24x^2 + 36x$$
$$= -12x(x^2 - 2x - 3)$$
$$= -12x(x-3)(x+1)$$

Critical numbers are 0, 3, and -1.

x	$f(x)$
-1	9
0	2
3	137

There is an absolute maximum of 137 at $x = 3$; there is no absolute minimum, as can be seen by looking at the graph of f.

35. $f(x) = \frac{x-1}{x^2+2x+6}$

$$f'(x) = \frac{(x^2+2x+6)(1)-(x-1)(2x+2)}{(x^2+2x+6)^2}$$
$$= \frac{x^2+2x+6-2x^2+2}{(x^2+2x+6)^2}$$
$$= \frac{-x^2+2x+8}{(x^2+2x+6)^2}$$
$$= \frac{-(x^2-2x-8)}{(x^2+2x+6)^2}$$
$$= \frac{-(x-4)(x+2)}{(x^2+2x+6)^2}$$

Critical numbers are 4 and -2.

x	$f(x)$
-2	$-\frac{1}{2}$
4	0.1

There is an absolute maximum of 0.1 at $x = 4$ and an absolute minimum of 0.5 at $x = -2$. This can be verified by looking at the graph of f.

37. $f(x) = \frac{\ln x}{x^3}$

$$f'(x) = \frac{x^3 \cdot \frac{1}{x} - 3x^2 \ln x}{x^6}$$
$$= \frac{x^2 - 3x^2 \ln x}{x^6}$$
$$= \frac{x^2(1-3\ln x)}{x^6}$$
$$= \frac{1-3\ln x}{x^4}$$

$f'(x) = 0$ when $x = e^{1/3}$, and $f'(x)$ does not exist when $x \le 0$. The only critical number is $e^{1/3}$.

x	$f(x)$
$e^{1/3}$	$\frac{1}{3}e^{-1} \approx 0.1226$

There is an absolute maximum of 0.1226 at $x = e^{1/3}$. There is no absolute minimum, as can be seen by looking at the graph of f.

39. $f(x) = 2x - 3x^{2/3}$

$$f'(x) = 2 - 2x^{-1/3} = 2 - \frac{2}{\sqrt[3]{x}} = \frac{2\sqrt[3]{x}-2}{\sqrt[3]{x}}$$

$f'(x) = 0$ when $2\sqrt[3]{x} - 2 = 0$
$$2\sqrt[3]{x} = 2$$
$$\sqrt[3]{x} = 1$$
$$x = 1$$

$f'(x)$ is undefined at $x = 0$, but $f(x)$ is defined at $x = 0$. So the critical numbers are 0 and 1.

(a) On $[-1, 0.5]$

x	$f(x)$
-1	-5
0	0
1	-1
0.5	-0.88988

Absolute minimum of -5 at $x = -1$; absolute maximum of 0 at $x = 0$

(b) On $[0.5, 2]$

x	$f(x)$
0.5	-0.88988
1	-1
2	-0.7622

Absolute maximum of about -0.76 at $x = 2$; absolute minimum of -1 at $x = 1$.

41. (a) Looking at the graph we see that there are relative maxima of 8496 in 2001, 7556 in 2004, 6985 in 2006, and 6700 in 2008. There are relative minima of 7127 in 2000, 7465 in 2003, 6748 in 2005, 5933 in 2007, and 5943 in 2009.

(b) The absolute maximum is 8496 (in 2001) and the absolute minimum is 5933 (in 2007).

43. $P(x) = -x^3 + 9x^2 + 120x - 400, x \ge 5$

$$P'(x) = -3x^2 + 18x + 120$$
$$= -3(x^2 - 6x - 40)$$
$$= -3(x-10)(x+4) = 0$$
$$x = 10 \quad \text{or} \quad x = -4$$

-4 is not relevant since $x \ge 5$, so the only critical number is 10.

The graph of $P'(x)$ is a parabola that opens downward, so $P'(x) > 0$ on the interval $[5, 10)$

and $P'(x) < 0$ on the interval $(10, \infty)$. Thus, $P(x)$ is a maximum at $x = 10$.

Since x is measured in hundreds thousands, 10 hundred thousand or 1,000,000 tires must be sold to maximize profit.

Also,

$$P(10) = -(10)^3 + 9(10)^2 + 120(10) - 400 = 700.$$

The maximum profit is \$700 thousand or \$700,000.

45. $C(x) = x^3 + 37x + 250$

(a) $1 \le x \le 10$

$$\overline{C}(x) = \frac{C(x)}{x} = \frac{x^3 + 37x + 250}{x} = x^2 + 37 + \frac{250}{x}$$

$$\overline{C}'(x) = 2x - \frac{250}{x^2} = \frac{2x^3 - 250}{x^2} = 0 \text{ when}$$

$$2x^3 = 250$$
$$x^3 = 125$$
$$x = 5.$$

Test for relative minimum.

$\overline{C}'(4) = -7.625 < 0$

$\overline{C}'(6) \approx 5.0556 > 0$

$\overline{C}(5) = 112$

$\overline{C}(1) = 1 + 37 + 250 = 288$

$\overline{C}(10) = 100 + 37 + 25 = 162$

The minimum for $1 \le x \le 10$ is 112.

(b) $10 \le x \le 20$

There are no critical values in this interval. Check the endpoints.

$\overline{C}(10) = 162$

$\overline{C}(20) = 400 + 37 + 12.5 = 449.5$

The minimum for $10 \le x \le 20$ is 162.

47. The value $x = 11.5$ minimizes $\frac{f(x)}{x}$ because this is the point where the line from the origin to the curve is tangent to the curve.

A production level of 11.5 units results in the minimum cost per unit.

49. The value $x = 100$ maximizes $\frac{f(x)}{x}$ because this is the point where the line from the origin to the curve is tangent to the curve.

A production level of 100 units results in the maximum profit per item produced.

51. $f(x) = \dfrac{x^2 + 36}{2x}, 1 \le x \le 12$

$$f'(x) = \frac{2x(2x) - (x^2 + 36)(2)}{(2x)^2} = \frac{4x^2 - 2x^2 - 72}{4x^2} = \frac{2x^2 - 72}{4x^2} = \frac{2(x^2 - 36)}{4x^2} = \frac{(x + 6)(x - 6)}{2x^2}$$

$f'(x) = 0$ when $x = 6$ and when $x = -6$. Only 6 is in the interval $1 \le x \le 12$.

Test for relative maximum or minimum.

$$f'(5) = \frac{(11)(-1)}{50} < 0$$

$$f'(7) = \frac{(13)(1)}{98} > 0$$

The minimum occurs at $x = 6$, or at 6 months. Since $f(6) = 6$, $f(1) = 18.5$, and $f(12) = 7.5$, the minimum percent is 6%.

53. Since we are only interested in the length during weeks 22 through 28, the domain of the function for this problem is [22, 28]. We now look for any critical numbers in this interval. We find

$$L'(t) = 0.788 - 0.02t$$

There is a critical number at $t = \frac{0.788}{0.02} = 39.4$. which is not in the interval. Thus, the maximum value will occur at one of the endpoints.

t	$L(t)$
22	5.4
28	7.2

The maximum length is about 7.2 millimeters.

55. $M(x) = -\frac{1}{45}x^2 + 2x - 20, 30 \le x \le 65$

$M'(x) = -\frac{1}{45}(2x) + 2 = -\frac{2x}{45} + 2$

When $M'(x) = 0$,

$$-\frac{2x}{45} + 2 = 0$$
$$2 = \frac{2x}{45}$$
$$45 = x.$$

x	$M(x)$
30	20
45	25
65	$\frac{145}{9} \approx 16.1$

The absolute maximum miles per gallon is 25 at 45 mph and the absolute minimum miles per gallon is about 16.1 at 65 mph.

57. Total area $A(x) = \pi\left(\frac{x}{2\pi}\right)^2 + \left(\frac{12 - x}{4}\right)^2$

$$= \frac{x^2}{4\pi} + \frac{(12 - x)^2}{16}$$

$$A'(x) = \frac{x}{2\pi} - \frac{12 - x}{8} = 0$$
$$\frac{4x - \pi(12 - x)}{8\pi} = 0$$
$$x = \frac{12\pi}{4 + \pi} \approx 5.28$$

x	Area
0	9
5.28	5.04
12	11.46

The total area is minimized when the piece used to form the circle is $\frac{12\pi}{4+\pi}$ feet, or about 5.28 feet long.

59. For the solution to Exercise 57, the piece of length x used to form the circle is $\frac{12\pi}{4+\pi}$ feet. The circle can be inscribed inside the square if the side of the square equals the diameter of the circle (that is, twice the radius).

side of the square $= 2$ (radius)

$$\frac{12 - x}{4} = 2\left(\frac{x}{2\pi}\right)$$
$$\frac{12 - x}{4} = \frac{x}{\pi}$$
$$4x = 12\pi - \pi x$$
$$x(4 + \pi) = 12\pi$$
$$x = \frac{12\pi}{4 + \pi}$$

Therefore, the circle formed by piece of length $x = \frac{12\pi}{4+\pi}$ can be inscribed inside the square.

14.2 Applications of Extrema

Your Turn 1

Assign a variable to the quantity to be minimized:

$$M = x^2y$$

Use the given condition to express M in terms of one variable, say x:

$$x + 3y = 30$$
$$y = \frac{30 - x}{3}$$
$$M = x^2\left(\frac{30 - x}{3}\right) = 10x^2 - \frac{1}{3}x^3$$

Find the domain of M:

Since x and y must be nonnegative, we have $x \ge 0$ and $\frac{30-x}{3} \ge 0$ which requires $x \le 30$. Thus the domain of M is $[0, 30]$.

Find the critical numbers:

$$\frac{dM}{dx} = 20x - x^2 = x(20 - x)$$

$\frac{dM}{dx}$ is 0 when $x = 0$ (already found as an endpoint) and when $x = 20$.

Evaluate M at the critical numbers and endpoints:

x	M
0	0
20	$\frac{4000}{3}$
30	0

The maximum of M occurs at $x = 20$.

The corresponding value of y is $\frac{30-20}{3}$ or $\frac{10}{3}$.

Your Turn 2

The first steps of the solution follow Example 2. As in Example 2, $300 - x$ will be the distance the professor must run along the trail. The first change in the model occurs when we compute the professor's total time. Since his speed through the woods is now 40 m/min rather than 70 m/min, his total time $T(x)$ is now

$$T(x) = \frac{300 - x}{160} + \frac{\sqrt{800^2 + x^2}}{40}.$$

As before, the domain of T is $[0, 300]$. The derivative of T is

$$T'(x) = -\frac{1}{160} + \frac{1}{40}\left(\frac{1}{2}\right)(800^2 + x^2)^{-1/2}(2x)$$

Now find the critical numbers by setting the derivative of T equal to 0.

$$-\frac{1}{160} + \frac{1}{40}\left(\frac{1}{2}\right)(800^2 + x^2)^{-1/2}(2x) = 0$$

$$\frac{x}{40\sqrt{800^2 + x^2}} = \frac{1}{160}$$

$$4x = \sqrt{800^2 + x^2}$$

$$16x^2 = 800^2 + x^2$$

$$15x^2 = 800^2$$

$$x = \frac{800}{\sqrt{15}}$$

$$x \approx 206.56$$

Calculate the total time at this critical number and at the endpoints of the domain.

x	T
0	21.875
206.56	21.24
300	21.36

The minimum travel time occurs for $x = 206.56$. The professor runs $300 - 206.56$ meters or about 93 meters along the path and then heads into the woods.

Your Turn 3

We follow the procedure of Example 3, but now our volume function is

$$V(x) = x(8 - 2x)^2 = 4(16x - 8x^2 + x^3).$$

For nonnegative side lengths we require $x \geq 0$ and $8 - 2x \geq 0$ or $x \leq 4$; the domain of V is thus $[0, 4]$.

Set the derivative of V equal to 0 and solve for x.

$$V'(x) = 4(16 - 16x + 3x^2) = 4(3x - 4)(x - 4)$$

This derivative has two positive roots, $x = 4$ and $x = 4/3$. We have already identified one of these as an endpoint of the domain and the other lies within the domain. Evaluating V at 0, 4/3 and 4 gives the following table.

x	V
0	0
4/3	$\frac{1024}{27}$
4	0

The maximum volume is 1024/27 m^3 and occurs when $x = 4/3$ m.

Note that we can solve this problem efficiently using a scaling argument. Since the proportions of the largest-volume box should not depend on the linear scale we adopt (that is, on the units in which we measure length), we can just note that our piece of metal is 8/12 or the 2/3 size of the one in Example 3. Our minimizing length x will be 2/3 of the value we found in Example 3, or $(2/3)(2) = 4/3$ m.

Your Turn 4

Follow the procedure of Example 4 with a volume of 500 cm^3 instead of 1000 cm^3.

$$V = \pi r^2 h = 500$$

so

$$h = \frac{500}{\pi r^2}$$

and

$$S = 2\pi r^2 + 2\pi r \frac{500}{\pi r^2}$$

$$= 2\pi r^2 + \frac{1000}{r}.$$

Excluding a radius of 0 gives a domain for S of $(0, \infty)$. Now find the critical numbers of S.

$$S' = 4\pi r - \frac{1000}{r^2}$$

$$4\pi r = \frac{1000}{r^2}$$

$$r^3 = \frac{250}{\pi}$$

$$r = \left(\frac{250}{\pi}\right)^{1/3} \approx 4.301$$

We can verify that S' is negative to the left of 4.3 and positive to the right (for example, $S'(4) \approx -12.2$ and $S'(5) \approx 22.8$). Thus the function S is decreasing as we move toward 4.3 from left and increasing as we move past 4.3 to the right, so there is a relative minimum at 4.3. Since there is only one critical number, the critical point theorem tells us that it corresponds to an absolute minimum of the area function.

Then $h \approx \frac{500}{\pi(4.301)^2} \approx 8.604$. The minimum surface area is obtained with a radius of 4.3 cm and a height of 8.6 cm.

As with the box in Your Turn 3, we could obtain this answer directly from Example 4 using a scaling argument. Changing the volume by a factor of 1/2 (from 1000 to 500) scales all linear measures by a factor of $(1/2)^{1/3}$, so we can find the new r and h by dividing the values found in Example 4 by $(1/2)^{1/3}$:

$$r = \frac{5.419}{2^{1/3}} \approx 4.301$$

$$h = \frac{10.84}{2^{1/3}} \approx 8.604$$

14.2 Exercises

1. $x + y = 180, P = xy$

(a) $y = 180 - x$

(b) $P = xy = x(180 - x)$

(c) Since $y = 180 - x$ and x and y are nonnegative numbers, $x \geq 0$ and $180 - x \geq 0$ or $x \leq 180$. The domain of P is $[0, 180]$.

(d) $P'(x) = 180 - 2x$

$$180 - 2x = 0$$
$$2(90 - x) = 0$$
$$x = 90$$

(e)

x	P
0	0
90	8100
180	0

(f) From the chart, the maximum value of P is 8100; this occurs when $x = 90$ and $y = 90$.

3. $x + y = 90$

Minimize x^2y.

(a) $y = 90 - x$

(b) Let $P = x^2y = x^2(90 - x)$
$$= 90x^2 - x^3.$$

(c) Since $y = 90 - x$ and x and y are nonnegative numbers, the domain of P is $[0, 90]$.

(d) $P' = 180x - 3x^2$

$$180x - 3x^2 = 0$$
$$3x(60 - x) = 0$$
$$x = 0 \text{ or } x = 60$$

(e)

x	P
0	0
60	108,000
90	0

(f) The maximum value of x^2y occurs when $x = 60$ and $y = 30$. The maximum value is 108,000.

5. $C(x) = \frac{1}{2}x^3 + 2x^2 - 3x + 35$

The average cost function is

$$A(x) = \bar{C}(x) = \frac{C(x)}{x}$$
$$= \frac{\frac{1}{2}x^3 + 2x^2 - 3x + 35}{x}$$
$$= \frac{1}{2}x^2 + 2x - 3 + \frac{35}{x}$$
$$\text{or } \frac{1}{2}x^2 + 2x - 3 + 35x^{-1}.$$

Then $A'(x) = x + 2 - 35x^{-2}$

$$\text{or } x + 2 - \frac{35}{x^2}.$$

Graph $y = A'(x)$ on a graphing calculator. A suitable choice for the viewing window is $[0,10]$ by $[-10,10]$. (Negative values of x are not meaningful in this application.) Using the calculator, we see that the graph has an x-intercept at $x \approx 2.722$. Thus, 2.722 is a critical number.

Now graph $y = A(x)$ and use this graph to confirm that a minimum occurs at $x \approx 2.722$.

Thus, the average cost is smallest at $x \approx 2.722$. At this value of x, $A \approx 19.007$.

7. $p(x) = 160 - \frac{x}{10}$

(a) Revenue from sale of x thousand candy bars:

$$R(x) = 1000xp$$
$$= 1000x\left(160 - \frac{x}{10}\right)$$
$$= 160{,}000x - 100x^2$$

(b) $R'(x) = 160{,}000 - 200x$

$$160{,}000 - 200x = 0$$
$$800 = x$$

The maximum revenue occurs when 800 thousand bars are sold.

(c) $R(800) = 160{,}000(800) - 100(800)^2$
$$= 64{,}000{,}000$$

The maximum revenue is 64,000,000 cents or \$640,000.

9. Let $x =$ the width

and $y =$ the length.

(a) The perimeter is

$$P = 2x + y$$
$$= 1400,$$

so

$$y = 1400 - 2x.$$

(b) Area $= xy = x(1400 - 2x)$
$$A(x) = 1400x - 2x^2$$

(c) $A' = 1400 - 4x$
$$1400 - 4x = 0$$
$$1400 = 4x$$
$$350 = x$$

$A'' = -4$, which implies that $x = 350$ m leads to the maximum area.

(d) If $x = 350$,

$$y = 1400 - 2(350) = 700.$$

The maximum area is $(350)(700)$
$= 245{,}000 \text{ m}^2$.

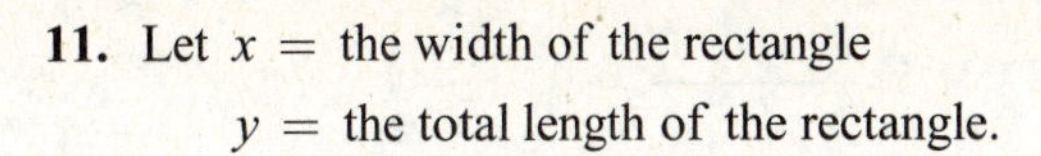

11. Let $x =$ the width of the rectangle
$y =$ the total length of the rectangle.

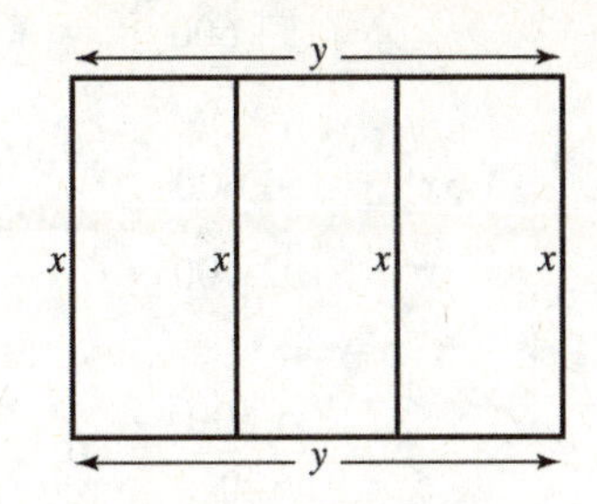

An equation for the fencing is

$$3600 = 4x + 2y$$
$$2y = 3600 - 4x$$
$$y = 1800 - 2x.$$

Area $= xy = x(1800 - 2x)$
$$A(x) = 1800x - 2x^2$$

$A' = 1800 - 4x$
$$1800 - 4x = 0$$
$$1800 = 4x$$
$$450 = x$$

$A'' = -4$, which implies that $x = 450$ is the location of a maximum.

If $x = 450$, $y = 1800 - 2(450) = 900$.

The maximum area is

$$(450)(900) = 405{,}000 \text{ m}^2.$$

13. Let $x =$ length at \$1.50 per meter
$y =$ width at \$3 per meter.

$$xy = 25{,}600$$
$$y = \frac{25{,}600}{x}$$

$$\text{Perimeter} = x + 2y = x + \frac{51{,}200}{x}$$

$$\text{Cost} = C(x) = x(1.5) + \frac{51{,}200}{x}(3)$$
$$= 1.5x + \frac{153{,}600}{x}$$

Minimize cost:

$$C'(x) = 1.5 - \frac{153{,}600}{x^2}$$

$$1.5 - \frac{153{,}600}{x^2} = 0$$
$$1.5 = \frac{153{,}600}{x^2}$$
$$1.5x^2 = 153{,}600$$
$$x^2 = 102{,}400$$
$$x = 320$$
$$y = \frac{25{,}600}{320} = 80$$

320 m at $1.50 per meter will cost $480. 160 m at $3 per meter will cost $480. The total cost will be $960.

15. Let $x =$ the number of refunds.
Then $90 + x =$ the number of passengers.

(a) Revenue $= R(x) = (90 + x)(1600 - 10x)$
$= 144{,}000 + 700x - 10x^2$

Assume that the number of refunds is nonnegative and that the number of refunds is limited to 160 so that the revenue will be nonnegative. Thus the domain of R is $[0,160]$. Now set the derivative of R equal to 0 and solve.

$$R' = 700 - 20x = 0$$
$$x = 35$$

Check the value of R at this critical number and at the endpoints of the domain:

x	R
0	144,000
35	156,250
160	0

Thus the maximum revenue is obtained with 35 refunds, which happens when there are 125 passengers.

(b) The maximum revenue is $156,250.

17. Let $x =$ the number of days to wait.

$\frac{12{,}000}{100} = 120 =$ the number of 100-lb groups collected already.

Then $7.5 - 0.15x =$ the price per 100 lb;
$4x =$ the number of 100-lb groups collected per day;
$120 + 4x =$ total number of 100-lb groups collected.

Revenue $= R(x)$
$= (7.5 - 0.15x)(120 + 4x)$
$= 900 + 12x - 0.6x^2$

$$R'(x) = 12 - 1.2x = 0$$
$$x = 10$$

$R''(x) = -1.2 < 0$ so $R(x)$ is maximized at $x = 10$.

The scouts should wait 10 days at which time their income will be maximized at

$$R(10) = 900 + 12(10) - 0.6(10)^2 = \$960.$$

19. Let $x =$ a side of the base
$h =$ the height of the box.

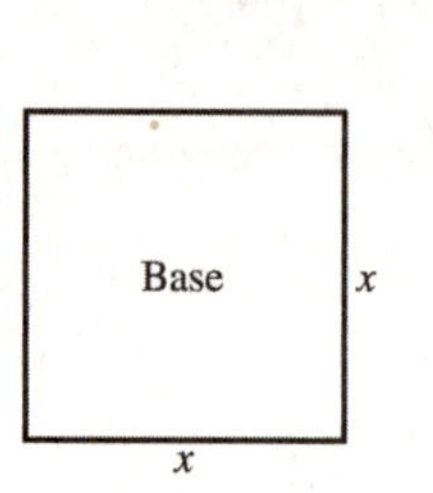

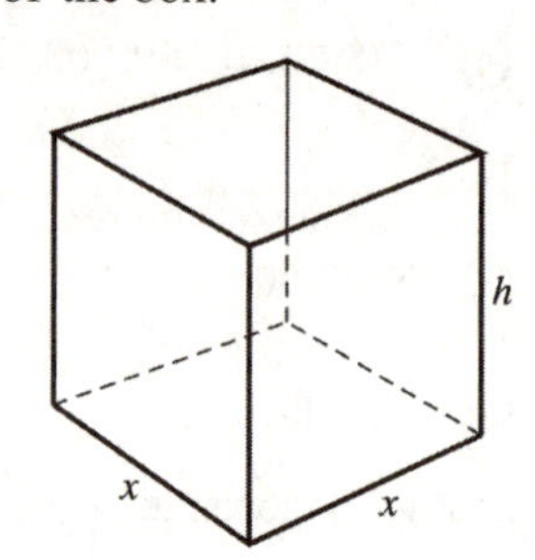

An equation for the volume of the box is

$$V = x^2h,$$
$$\text{so } 32 = x^2h$$
$$h = \frac{32}{x^2}.$$

The box is open at the top so the area of the surface material $m(x)$ in square inches is the area of the base plus the area of the four sides.

$$m(x) = x^2 + 4xh$$
$$= x^2 + 4x\left(\frac{32}{x^2}\right)$$
$$= x^2 + \frac{128}{x}$$
$$m'(x) = 2x - \frac{128}{x^2}$$

$$\frac{2x^3 - 128}{x^2} = 0$$
$$2x^3 - 128 = 0$$
$$2(x^3 - 64) = 0$$
$$x = 4$$

$m'(x) = 2 + \frac{256}{x^3} > 0$ since $x > 0$.

So, $x = 4$ minimizes the surface material. If $x = 4$.

$$h = \frac{32}{x^2} = \frac{32}{16} = 2.$$

The dimensions that will minimize the surface material are 4 in. by 4 in. by 2 in.

21. Let $x =$ the length of the side of the cutout square.

Then $3 - 2x =$ the width of the box and $8 - 2x =$ the length of the box.

$$V(x) = x(3 - 2x)(8 - 2x)$$
$$= 4x^3 - 22x^2 + 24x$$

The domain of V is $\left(0, \frac{3}{2}\right)$.

Maximize the volume.

$$V'(x) = 12x^2 - 44x + 24$$
$$12x^2 - 44x + 24 = 0$$
$$4(3x^2 - 11x + 6) = 0$$
$$4(3x - 2)(x - 3) = 0$$
$$x = \frac{2}{3} \quad \text{or} \quad x = 3$$

3 is not in the domain of V.

$$V''(x) = 24x - 44$$
$$V''\left(\frac{2}{3}\right) = -28 < 0$$

This implies that V is maximized when $x = \frac{2}{3}$. The box will have maximum volume when $x = \frac{2}{3}$ ft or 8 in.

23. Let $x =$ the length of a side of the top and bottom.

Then $x^2 =$ the area of the top and bottom

and $(3)(2x^2) =$ the cost for the top and bottom

Let $y =$ depth of box.

Then $xy =$ the area of one side,

$4xy =$ the total area of the sides.

and $(1.50)(4xy) =$ The cost of the sides

The total cost is

$$C(x) = (3)(2x^2) + (1.50)(4xy) = 6x^2 + 6xy.$$

The volume is

$$V = 16{,}000 = x^2y.$$
$$y = \frac{16{,}000}{x^2}$$

$$C(x) = 6x^2 + 6x\left(\frac{16{,}000}{x^2}\right) = 6x^2 + \frac{96{,}000}{x}$$

$$C'(x) = 12x - \frac{96{,}000}{x^2} = 0$$
$$x^3 = 8000$$
$$x = 20$$

$C''(x) = 12 + \frac{192{,}000}{x^3} > 0$ at $x = 20$, which implies that $C(x)$ is minimized when $x = 20$.

$$y = \frac{16{,}000}{(20)^2} = 40$$

So the dimensions of the box are x by x by y, or 20 cm by 20 cm by 40 cm.

$$C(20) = 6(20)^2 + \frac{96{,}000}{20} = 7200$$

The minimum total cost is \$7200.

25. (a) $S = 2\pi r^2 + 2\pi rh,\ V = \pi r^2 h$

$$S = 2\pi r^2 + \frac{2V}{r}$$

Treat V as a constant.

$$S' = 4\pi r - \frac{2V}{r^2}$$

$$4\pi r - \frac{2V}{r^2} = 0$$
$$\frac{4\pi r^3 - 2V}{r^2} = 0$$
$$4\pi r^3 - 2V = 0$$
$$2\pi r^3 - V = 0$$
$$2\pi r^3 = V$$
$$2\pi r^3 = \pi r^2 h$$
$$2r = h$$

27. From Example 4, we know that the surface area of the can is given by

$$S = 2\pi r^2 + \frac{2000}{r}.$$

Aluminum costs $3¢/\text{cm}^2$, so the cost of the aluminum to make the can is

$$0.03\left(2\pi r^2 + \frac{2000}{r}\right) = 0.06\pi r^2 + \frac{60}{r}.$$

The perimeter (or circumference) of the circular top is $2\pi r$. Since there is a 2¢/cm charge to seal the top and bottom, the sealing cost is

$$0.02(2)(2\pi r) = 0.08\pi r.$$

Thus, the total cost is given by the function

$$C(r) = 0.06\pi r^2 + \frac{60}{r} + 0.08\pi r$$
$$= 0.06\pi r^2 + 60r^{-1} + 0.08\pi r.$$

Then

$$C'(r) = 0.12\pi r - 60r^{-2} + 0.08\pi$$
$$= 0.12\pi r - \frac{60}{r^2} + 0.08\pi.$$

Graph

$$y = 0.12\pi x - \frac{60}{x^2} + 0.08\pi$$

on a graphing calculator. Since r must be positive in this application, our window should not include negative values of x. A suitable choice for the viewing window is [0,10] by [−10, 10]. From the graph, we find that $C'(x) = 0$ when $x \approx 5.206$. Thus, the cost is minimized when the radius is about 5.206 cm.

We can find the corresponding height by using the equation

$$h = \frac{1000}{\pi r^2}$$

from Example 4.

If $r = 5.206$.

$$h = \frac{1000}{\pi(5.206)^2} \approx 11.75.$$

To minimize cost, the can should have radius 5.206 cm and height 11.75 cm.

29. In Exercise 27 we found that the cost of the aluminum to make the can is $0.06\pi r^2 + \frac{60}{r}$, the cost to seal the top and bottom is $0.08\pi r$, and the cost to seal the vertical seam is $\frac{10}{\pi r^2}$. Thus, the total cost is now given by the function

$$C(r) = 0.06\pi r^2 + \frac{60}{r} + 0.08\pi r + \frac{10}{\pi r^2}$$
$$\text{or } 0.06\pi r^2 + 60r^{-1} + 0.08\pi r + \frac{10}{\pi}r^{-2}.$$

Then

$$C'(r) = 0.12\pi r - 60r^{-2} + 0.80\pi - \frac{20}{\pi}r^{-3}$$
$$\text{or } 0.12\pi r - \frac{60}{r^2} + 0.08\pi - \frac{20}{\pi r^3}.$$

Graph

$$y = 0.12\pi r - \frac{60}{r^2} + 0.08\pi - \frac{20}{\pi r^3}$$

on a graphing calculator. A suitable choice for the viewing window is [0, 10] by [−10, 10]. From the graph, we find that $C'(x) = 0$ when $x \approx 5.242$. Thus, the cost is minimized when the radius is about 5.242 cm. To find the corresponding height, use the equation

$$h = \frac{1000}{\pi r^2}$$

from Example 4.

If $r = 5.242$,

$$h = \frac{1000}{\pi(5.242)^2} \approx 11.58.$$

To minimize cost, the can should have radius 5.242 cm and height 11.58 cm.

31. Distance on shore: $9 - x$ miles

Cost on shore: $400 per mile

Distance underwater: $\sqrt{x^2 + 36}$

Cost underwater: $500 per mile

Find the distance from A, that is, $(9 - x)$, to minimize cost, $C(x)$.

$$C(x) = (9 - x)(400) + \left(\sqrt{x^2 + 36}\right)(500)$$
$$= 3600 - 400x + 500(x^2 + 36)^{1/2}$$
$$C'(x) = -400 + 500\left(\frac{1}{2}\right)(x^2 + 36)^{-1/2}(2x)$$
$$= -400 + \frac{500x}{\sqrt{x^2 + 36}}$$

If $C'(x) = 0$,

$$\frac{500x}{\sqrt{x^2 + 36}} = 400$$
$$\frac{5x}{4} = \sqrt{x^2 + 36}$$
$$\frac{25}{16}x^2 = x^2 + 36$$

$$\frac{9}{16}x^2 = 36$$

$$x = \frac{6 \cdot 4}{3} = 8.$$

(Discard the negative solution.)
Then the distance should be

$$9 - x = 9 - 8$$
$$= 1 \text{ mile from point } A.$$

33. $p(t) = \dfrac{20t^3 - t^4}{1000}$, $[0, 20]$

(a) $p'(t) = \dfrac{3}{50}t^2 - \dfrac{1}{250}t^3$

$$= \frac{1}{50}t^2\left[3 - \frac{1}{5}t\right]$$

Critical numbers:

$$\frac{1}{50}t^2 = 0 \text{ or } 3 - \frac{1}{5}t = 0$$
$$t = 0 \text{ or } \quad t = 15$$

t	$p(t)$
0	0
15	16.875
20	0

The number of people infected reaches a maximum in 15 days.

(b) $P(15) = 16.875\%$

35. $H(S) = f(S) - S$

$f(S) = 12S^{0.25}$

$H(S) = 12S^{0.25} - S$

$H'(S) = 3S^{-0.75} - 1$

$H'(S) = 0$ when

$$3S^{-0.75} - 1 = 0$$
$$S^{-0.75} = \frac{1}{3}$$
$$\frac{1}{S^{0.75}} = \frac{1}{3}$$
$$S^{3/4} = 3$$
$$S = 3^{4/3}$$
$$S = 4.327.$$

The number of creatures needed to sustain the population is $S_0 = 4.327$ thousand.

$H''(S) = \frac{-2.25}{S^{1.75}} < 0$ when $S = 4.327$, so $H(S)$ is maximized.

$$H(4.327) = 12(4.327)^{0.25} - 4.327$$
$$\approx 12.98$$

The maximum sustainable harvest is 12.98 thousand.

37. $N(t) = 20\left(\dfrac{t}{12} - \ln\left(\dfrac{t}{12}\right)\right) + 30;$

$1 \le t \le 15$

$$N'(t) = 20\left[\frac{1}{12} - \frac{12}{t}\left(\frac{1}{12}\right)\right]$$
$$= 20\left(\frac{1}{12} - \frac{1}{t}\right)$$
$$= \frac{20(t - 12)}{12t}$$

$N'(t) = 0$ when

$$t - 12 = 0$$
$$t = 12.$$

$N''(t)$ does not exist at $t = 0$, but 0 is not in the domain of N.
Thus, 12 is the only critical number.
To find the absolute extrema on $[1, 15]$, evaluate N at the critical number and at the endpoints.

t	$N(t)$
1	81.365
12	50
15	50.537

Use this table to answer the questions in (a)-(d).

(a) The number of bacteria will be a minimum at $t = 12$, which represents 12 days.

(b) The minimum number of bacteria is given by $N(12) = 50$, which represents 50 bacteria per ml.

(c) The number of bacteria will be a maximum at $t = 1$, which represents 1 day.

(d) The maximum number of bacteria is give by $N(1) = 81.365$, which represents 81.365 bacteria per ml.

39. $r = 0.1$, $P = 100$

$$f(S) = Se^{r(1-S/P)}$$

$$f'(S) = -\frac{1}{1000} \cdot Se^{0.1(1-S/100)} + e^{0.1(1-S/100)}$$

$$f'(S_0) = -0.001S_0e^{0.1(1-S_0/100)} + e^{0.1(1-S_0/100)}$$

Graph

$$Y_1 = -0.001xe^{0.1(1-x/100)} + e^{0.1(1-x/100)}$$

and

$$Y_2 = 1$$

on the same screen. A suitable choice for the viewing window is $[0, 60]$ by $[0.5, 1.5]$ with Xscl $= 10$ and Yscl $= 0.5$. By zooming or using the "intersect" option, we find the graphs intersect when $x \approx 49.37$.

The maximum sustainable harvest is 49.37.

41. Let $x =$ distance from P to A.

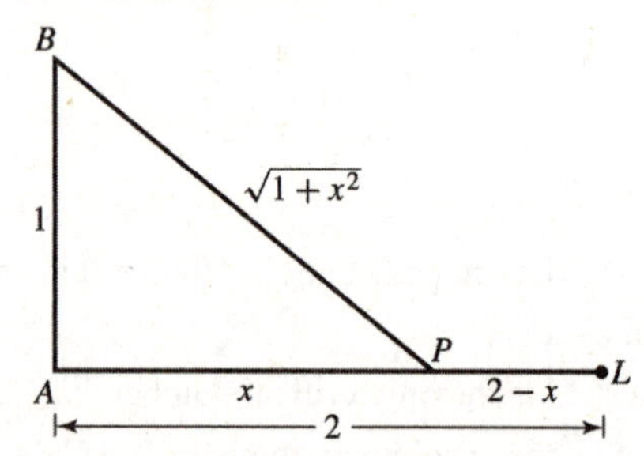

Energy used over land: 1 unit per mile

Energy used over water: $\frac{4}{3}$ units per mile

Distance over land: $(2 - x)$ mi

Distance over water: $\sqrt{1 + x^2}$ mi

Find the location of P to minimize energy used.

$$E(x) = 1(2 - x) + \frac{4}{3}\sqrt{1 + x^2}, \text{ where } 0 \le x \le 2.$$

$$E'(x) = -1 + \frac{4}{3}\left(\frac{1}{2}\right)(1 + x^2)^{-1/2}(2x)$$

If $E'(x) = 0$,

$$\frac{4}{3}x(1 + x^2)^{-1/2} = 1$$

$$\frac{4x}{3(1 + x^2)^{1/2}} = 1$$

$$\frac{4}{3}x = (1 + x^2)^{1/2}$$

$$\frac{16}{9}x^2 = 1 + x^2$$

$$\frac{7}{9}x^2 = 1$$

$$x^2 = \frac{9}{7}$$

$$x = \frac{3}{\sqrt{7}} = \frac{3\sqrt{7}}{7}.$$

x	$E(x)$
0	3.3333
1.134	2.8819
2	2.9814

The absolute minimum occurs at $x \approx 1.134$.

Point P is $\frac{3\sqrt{7}}{7} \approx 1.134$ mi from Point A.

43. (a) $f(S) = aSe^{-bS}$

$$\begin{aligned} f(S) &= Se^{r(1-S/P)} \\ &= Se^{r-rS/P} \\ &= Se^re^{-rS/P} \\ &= e^rSe^{-(r/P)S} \end{aligned}$$

Comparing the two terms, replace a with e^r and b with r/P.

(b) Shepherd:

$$f(S) = \frac{aS}{1 + (S/b)^c}$$

$$\begin{aligned} f'(S) &= \frac{[1 + (S/b)^c](a) - (aS)[c(S/b)^{c-1}(1/b)]}{[1 + (S/b)^c]^2} \\ &= \frac{a + a(S/b)^c - (acS/b)(S/b)^{c-1}}{[1 + (S/b)^c]^2} \\ &= \frac{a + a(S/b)^c - ac(S/b)^c}{[1 + (S/b)^c]^2} \\ &= \frac{a[1 + (1 - c)(S/b)^c]}{[1 + (S/b)^c]^2} \end{aligned}$$

Ricker:

$$\begin{aligned} f(S) &= aSe^{-bS} \\ f'(S) &= ae^{-bS} + aSe^{-bS}(-b) \\ &= ae^{-bS}(1 - bS) \end{aligned}$$

Berverton-Holt:

$$f(S) = \frac{aS}{1 + (S/b)}$$

$$\begin{aligned} f'(S) &= \frac{[1 + (S/b)](a) - aS(1/b)}{[1 + (S/b)]^2} \\ &= \frac{a + a(S/b) - a(S/b)}{[1 + (S/b)]^2} = \frac{a}{[1 + (S/b)]^2} \end{aligned}$$

(c) Shepherd:

$$f'(0) = \frac{a[1 + (1 - c)(0/b)^c]}{[1 + (0/b)^c]^2} = a$$

Ricker:

$$f'(0) = ae^{-b(0)}[1 - b(0)] = a$$

Beverton-Holt:

$$f'(0) = \frac{a}{[1 + (0/b)]^2} = a$$

The constant a represents the slope of the graph of $f(S)$ at $S = 0$.

(d) First find the critical numbers by solving $f'(S) = 0$.

Shepherd:

$$f'(S) = 0$$
$$a[1 + (1 - c)(S/b)^c] = 0$$
$$(1 - c)(S/b)^c = -1$$
$$(c - 1)(S/b)^c = 1$$

Substitute $b = 248.72$ and $c = 3.24$ and solve for S.

$$(3.24 - 1)(S/248.72)^{3.24} = 1$$
$$\left(\frac{S}{248.72}\right)^{3.24} = \frac{1}{2.24}$$
$$\frac{S}{248.72} = \left(\frac{1}{2.24}\right)^{1/3.24}$$
$$S = 248.72\left(\frac{1}{2.24}\right)^{1/3.24} \approx 193.914$$

Using the Shepherd model, next year's population is maximized when this year's population is about 194,000 tons. This can be verified by examing the graph of $f(S)$.

(e) First find the critical numbers by solving $f'(S) = 0$.

Ricker:

$$f'(S) = 0$$
$$ae^{-bS}(1 - bS) = 0$$
$$1 - bS = 0$$
$$S = \frac{1}{b}$$

Substitute $b = 0.0039$ and solve for S.

$$S = \frac{1}{0.0039}$$
$$S \approx 256.410$$

Using the Ricker model, next year's population is maximized when this year's population is about 256,000 tons. This can be verified by examining the graph of $f(S)$.

45. Let $8 - x =$ the distance the hunter will travel on the river.

Then $\sqrt{9 + x^2} =$ the distance he will travel on land.

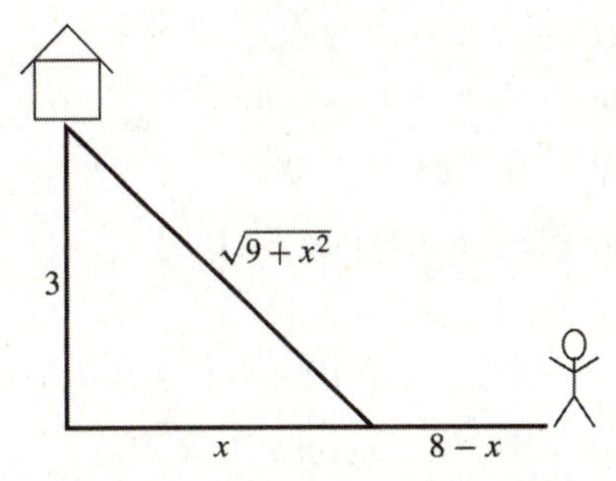

The rate on the river is 5 mph, the rate on land is 2 mph. Using $t = \frac{d}{r}$,

$\frac{8 - x}{5} =$ the time on the river,

$\frac{\sqrt{9 - x^2}}{2} =$ the time on land.

The total time is

$$T(x) = \frac{8 - x}{5} + \frac{\sqrt{9 + x^2}}{2}$$
$$= \frac{8}{5} - \frac{1}{5}x + \frac{1}{2}(9 + x^2)^{1/2}.$$
$$T' = -\frac{1}{5} + \frac{1}{4} \cdot 2x(9 + x^2)^{-1/2}$$

$$-\frac{1}{5} + \frac{x}{2(9 + x^2)^{1/2}} = 0$$
$$\frac{1}{5} = \frac{x}{2(9 + x^2)^{1/2}}$$
$$2(9 + x^2)^{1/2} = 5x$$
$$4(9 + x^2) = 25x^2$$
$$36 + 4x^2 = 25x^2$$
$$36 = 21x^2$$
$$\frac{6}{\sqrt{21}} = x$$
$$\frac{6\sqrt{21}}{21} = \frac{2\sqrt{21}}{7} = x$$

x	$T(x)$
0	3.1
$\frac{2\sqrt{21}}{7}$	2.98
8	4.27

Since the maximum time is 2.98 hr, the hunter should travel $8 - \frac{2\sqrt{21}}{7} = \frac{56-2\sqrt{21}}{7}$ or about 6.7 miles along the river.

47. Let $x =$ width.
Then $x =$ height
and $108 - 4x =$ length.

(since length plus girth $= 108$)

$$V(x) = l \cdot w \cdot h$$
$$= (108 - 4x)x \cdot x$$
$$= 108x^2 - 4x^3$$
$$V'(x) = 216x - 12x^2$$

Set $V'(x) = 0$, and solve for x.

$$216x - 12x^2 = 0$$
$$12x(18 - x) = 0$$
$$x = 0 \text{ or } x = 18$$

0 is not in the domain, so the only critical number is 18.

$$\text{Width} = 18$$
$$\text{Height} = 18$$
$$\text{Length} = 108 - 4(18) = 36$$

The dimensions of the box with maximum volume are 36 inches by 18 inches by 18 inches.

14.3 Further Business Applications: Economic Lot Size; Economic Order Quantity; Elasticity of Demand

Your Turn 1

Use Equation (3) with $k = 3$, $M = 18{,}000$ and $f = 750$.

$$q = \sqrt{\frac{2fM}{k}}$$
$$= \sqrt{\frac{2(750)(18{,}000)}{3}}$$
$$= \sqrt{(900)(10{,}000)}$$
$$= 3000$$

To minimize production costs there should be 3000 cans per batch, requiring $18{,}000/3{,}000 = 6$ batches per year.

Your Turn 2

Use Equation (3) with $k = 2, M = 320$ and $f = 30$.

$$q = \sqrt{\frac{2(30)(320)}{2}}$$
$$= \sqrt{9600}$$
$$\approx 97.98$$

Since q is very close to 98 we expect a q of 98 to minimize costs, but we will check both 97 and 98 using $T(q) = \frac{fM}{q} + \frac{kq}{2}$. $T(97) \approx 195.969$ and $T(98) \approx 195.959$, so the company should order 98 units in each batch. The time between orders will be about $12\left(\frac{98}{320}\right) = 3.675$ months.

Your Turn 3

$$q = 24{,}000 - 3p^2$$
$$\frac{dq}{dp} = -6p$$
$$E = -\frac{p}{q}\frac{dq}{dp} = \frac{6p^2}{24{,}000 - 3p^2}$$

When $p = 50$,

$$E = \frac{6(50^2)}{24{,}000 - 3(50^2)}$$
$$\approx 0.909,$$

which corresponds to inelastic demand.

Your Turn 4

$$q = 200e^{-0.4p}$$
$$\frac{dq}{dp} = -80e^{-0.4p}$$
$$E = -\frac{p}{q}\frac{dq}{dp} = \frac{80pe^{-0.4p}}{200e^{-0.4p}} = 0.4p$$

For $p = 100$ we have $E = 0.4(100) = 40$, which corresponds to elastic demand.

Your Turn 5

$$q = 3600 - 3p^2$$
$$\frac{dq}{dp} = -6p$$
$$E = -\frac{p}{q}\frac{dq}{dp} = \frac{6p^2}{3600 - 3p^2} = \frac{2p^2}{1200 - p^2}$$

The demand has unit elasticity when $E = 1$.

$$\frac{2p^2}{1200 - p^2} = 1$$
$$2p^2 = 1200 - p^2$$
$$3p^2 = 1200$$
$$p^2 = 400$$
$$p = 20$$

Testing value for p smaller and larger than 20 (say 15 and 25) we find

$$E(15) < 1$$
$$E(25) > 1$$

Thus demand is inelastic when $p < 20$ and elastic when $p > 20$.

Revenue is maximized at the price corresponding to unit elasticity, which is $p = \$20$. At this price the revenue is

$$pq = (20)[3600 - 3(20^2)] = (20)(2400) = 48{,}000$$

or $48,000.

14.3 Exercises

1. When $q < \sqrt{\frac{2fM}{k}}, T'(q) < -\frac{k}{2} + \frac{k}{2} = 0$; and when $q > \sqrt{\frac{2fM}{k}}, T'(q) > -\frac{k}{2} + \frac{k}{2} = 0$. Since the function $T(q)$ is decreasing before $q = \sqrt{\frac{2fM}{k}}$ and increasing after $q = \sqrt{\frac{2fM}{k}}$, there must be a relative minimum at $q = \sqrt{\frac{2fM}{k}}$. By the critical point theorem, there is an absolute minimum there.

3. The economic order quantity formula assumes that M, the total units needed per year, is known. Thus, c is the correct answer.

5. The demand function $q(p)$ is positive and increasing, so $\frac{dq}{dp}$ is positive. Since p_0 and q_0 are also positive, the elasticity $E = -\frac{p_0}{q_0} \cdot \frac{dq}{dp}$ is negative.

7. $q = m - np$ for $0 \le p \le \frac{m}{n}$

$$\frac{dq}{dp} = -n$$
$$E = -\frac{p}{q} \cdot \frac{dq}{dp}$$
$$E = -\frac{p}{m - np}(-n)$$
$$E = \frac{pn}{m - np} = 1$$
$$pn = m - np$$
$$2np = m$$
$$p = \frac{m}{2n}$$

Thus, $E = 1$ when $p = \frac{m}{2n}$, or at the midpoint of the demand curve on the interval $0 \le p \le \frac{m}{n}$.

9. Use equation (3) with $k = 1$, $M = 100{,}000$, and $f = 500$.

$$q = \sqrt{\frac{2fM}{k}} = \sqrt{\frac{2(500)(100{,}000)}{1}} = \sqrt{100{,}000{,}000} = 10{,}000$$

10,000 lamps should be made in each batch to minimize production costs.

11. From Exercise 9, $M = 100{,}000$, and $q = 10{,}000$. The number of batches per year is

$$\frac{M}{q} = \frac{100{,}000}{10{,}000} = 10.$$

13. Here $k = 0.50$, $M = 100{,}000$, and $f = 60$. We have

$$q = \sqrt{\frac{2fM}{k}} = \sqrt{\frac{2(60)(100{,}000)}{0.50}}$$
$$= \sqrt{24{,}000{,}000} \approx 4898.98$$

$T(4898) = 2449.489792$ and $T(4899) = 2449.489743$, so ordering 4899 copies per order minimizes the annual costs.

15. Using maximum inventory size,

$$T(q) = \frac{fM}{q} + gM + kq;\ (0, \infty)$$
$$T'(q) = \frac{-fM}{q^2} + k$$

Set the derivative equal to 0.

$$\frac{-fM}{q^2} + k = 0$$
$$k = \frac{fM}{q^2}$$
$$q^2k = fM$$
$$q^2 = \frac{fM}{k}$$
$$q = \sqrt{\frac{fM}{k}}$$

Since $\lim_{q\to 0} T(q) = \infty$,

$\lim_{q\to\infty} T(q) = \infty$, and

$q = \sqrt{\frac{fM}{k}}$ is the only critical value in $(0, \infty)$,

$q = \sqrt{\frac{fM}{k}}$ is the number of unit that should be ordered or manufactured to minimize total costs.

17. Assuming an annual cost, k_1, for storing a single unit, plus an annual cost per unit, k_2, that must be paid for each unit up to the maximum number of units stored, we have

$$T(q) = \frac{fM}{q} + gM + \frac{k_1q}{2} + k_2q;\ (0,\infty)$$
$$T'(q) = \frac{-fM}{q^2} + \frac{k_1}{2} + k_2$$

Set the derivative equal to 0.

$$\frac{-fM}{q^2} + \frac{k_1}{2} + k_2 = 0$$
$$\frac{k_1 + 2k_2}{2} = \frac{fM}{q^2}$$
$$\frac{q^2(k_1 + 2k_2)}{2} = fM$$
$$q^2 = \frac{2fM}{k_1 + 2k_2}$$
$$q = \sqrt{\frac{2fM}{k_1 + 2k_2}}$$

Since $\lim_{q\to 0} T(q) = \infty$, $\lim_{q\to\infty} T(q) = \infty$, and

$q = \sqrt{\frac{2fM}{k_1+2k_2}}$ is the only critical value in $(0,\infty)$,

$q = \sqrt{\frac{2fM}{k_1+2k_2}}$ is the number of units that should be ordered or manufactured to minimize the total cost in this case.

19. $q = 50 - \frac{p}{4}$

(a) $\frac{dq}{dp} = -\frac{1}{4}$

$$E = -\frac{p}{q} \cdot \frac{dq}{dp}$$
$$= -\frac{p}{50 - \frac{p}{4}}\left(-\frac{1}{4}\right)$$
$$= -\frac{p}{\frac{200-p}{4}}\left(-\frac{1}{4}\right)$$
$$= \frac{p}{200 - p}$$

(b) $R = pq$

$$\frac{dR}{dp} = q(1 - E)$$

When R is maximum, $q(1 - E) = 0$.

Since $q = 0$ means no revenue, set $1 - E = 0$, so $E = 1$.

From (a),

$$\frac{p}{200-p} = 1$$
$$p = 200 - p$$
$$p = 100.$$
$$q = 50 - \frac{p}{4}$$
$$= 50 - \frac{100}{4}$$
$$= 25$$

Total revenue is maximized if $q = 25$.

21. (a) $q = 37{,}500 - 5p^2$

$$\frac{dq}{dp} = -10p$$
$$E = \frac{-p}{q} \cdot \frac{dq}{dp}$$
$$= \frac{-p}{37{,}500 - 5p^2}(-10p)$$
$$= -\frac{10p^2}{37{,}500 - 5p^2}$$
$$= \frac{2p^2}{7500 - p^2}$$

(b) $R = pq$

$$\frac{dR}{dp} = q(1 - E)$$

When R is maximum, $q(1 - E) = 0$. Since $q = 0$ means no revenue, set $1 - E = 0$.

$$E = 1$$

From (a),

$$\frac{2p^2}{7500 - p^2} = 1$$
$$2p^2 = 7500 - p^2$$
$$3p^2 = 7500$$
$$p^2 = 2500$$
$$p = \pm 50.$$

Since p must be positive, $p = 50$.

$$q = 37{,}500 - 5p^2$$
$$= 37{,}500 - 5(50)^2$$
$$= 37{,}500 - 5(2500)$$
$$= 37{,}500 - 12{,}500$$
$$= 25{,}000.$$

23. $p = 400e^{-0.2q}$

In order to find the derivative $\frac{dq}{dp}$, we first need to solve for q in the equation $p = 400e^{-0.2q}$.

(a) $$\frac{p}{400} = e^{-0.2q}$$
$$\ln\left(\frac{p}{400}\right) = \ln(e^{-0.2q}) = -0.2q$$
$$q = \frac{\ln \frac{p}{400}}{-0.2} = -5\ln\left(\frac{p}{400}\right)$$

Now

$$\frac{dq}{dp} = -5\frac{1}{\frac{p}{400}} \cdot \frac{1}{400} = \frac{-5}{p}, \text{ and}$$
$$E = -\frac{p}{q} \cdot \frac{dq}{dp} = -\frac{p}{q} \cdot \frac{-5}{p} = \frac{5}{q}.$$

(b) $R = pq$

$$\frac{dR}{dp} = q(1 - E)$$

When R is maximum, $q(1 - E) = 0$. Since $q = 0$ means no revenue, set $1 - E = 0$.

$$E = 1$$

From part (a),

$$\frac{5}{q} = 1$$
$$5 = q$$

25. $q = 400 - 0.2p^2$

$$\frac{dq}{dp} = 0 - 0.4p$$
$$E = -\frac{p}{q} \cdot \frac{dq}{dp}$$
$$E = -\frac{P}{400 - 0.2p^2}(-0.4p)$$
$$= \frac{0.4p^2}{400 - 0.2p^2}$$

(a) If $p = \$20$,

$$E = \frac{(0.4)(20)^2}{400 - 0.2(20)^2}$$
$$= 0.5.$$

Since $E < 1$, demand is inelastic. This indicates that total revenue increases as price increases.

(b) If $p = \$40$,

$$E = \frac{(0.4)(40)^2}{400 - 0.2(40)^2} = 8.$$

Since $E > 1$, demand is elastic. This indicates that total revenue decreases as price increases.

27. $q = 2{,}431{,}129p^{-0.06}$

$$\frac{dq}{dp} = (-0.06)[2{,}431{,}129p^{-1.06}]$$

$$E = -\frac{p}{q}\frac{dq}{dp}$$

$$= -\frac{p}{2{,}431{,}129p^{-0.06}}(-0.06)[2{,}431{,}129p^{-1.06}]$$

$$= 0.06$$

At any price (including \$40/barrel) the elasticity is 0.06 and the demand is inelastic.

29. $q = 3{,}751{,}000p^{-2.826}$

$$\frac{dq}{dp} = (-2.826)[3{,}751{,}000p^{-3.826}]$$

$$E = -\frac{p}{q}\frac{dq}{dp}$$

$$= -\frac{p}{3{,}751{,}000p^{-2.826}}(-2.826)$$

$$[3{,}751{,}000p^{-3.826}]$$

$$= 2.826$$

At any price the elasticity is 2.826 and the demand is elastic.

31. (a) $q = 55.2 - 0.022p$

$$\frac{dq}{dp} = -0.022$$

$$E = -\frac{p}{q}\cdot\frac{dq}{dp}$$

$$= \frac{-p}{55.2 - 0.022p}\cdot(-0.022)$$

$$= \frac{0.022p}{55.2 - 0.022p}$$

When $p = \$166.10$,

$$E = \frac{3.6542}{55.2 - 3.6542} \approx 0.071.$$

(b) Since $E < 1$, the demand for airfare is inelastic at this price.

(c) $R = pq$

$$\frac{dR}{dp} = q(1 - E)$$

When R is maximum, $q(1 - E) = 0$.

Since $q = 0$ means no revenue, set $1 - E = 0$.

$$E = 1$$

From (a),

$$\frac{0.022p}{55.2 - 0.022p} = 1$$

$$0.022p = 55.2 - 0.022p$$

$$0.044p = 55.2$$

$$p \approx 1255$$

Total revenue is maximized if $p \approx \$1255$.

33.

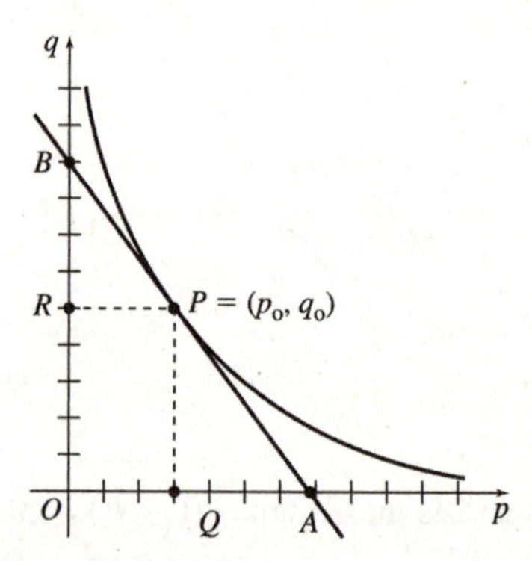

In the figure, we label P_0 as P. The slope of the tangent line is

$$\frac{dq}{dp} = -\frac{BR}{RP} = \frac{OB - OR}{-RP} = \frac{OB - q_0}{-p_0}$$

or

$$-p_0\frac{dq}{dp} = OB - q_0$$

$$-\frac{p_0}{q_0}\cdot\frac{dq}{dp} = \frac{OB}{q_0} - 1 = \frac{OB}{OR} - 1$$

Because triangles AOB and PRB are similar,

$$-\frac{p_0}{q_0}\cdot\frac{dq}{dp} = \frac{AB}{AP} - 1$$

$$= \frac{AB - AP}{AP} = \frac{PB}{PA}$$

But $E = -\frac{p_0}{q_0}\cdot\frac{dq}{dp}$ so the ratio $\frac{PB}{PA}$ equals the elasticity E.

14.4 Implicit Differentiation

Your Turn 1

$$x^2 + y^2 = xy$$
$$\frac{d}{dx}(x^2 + y^2) = \frac{d}{dx}xy$$
$$2x + 2y\frac{dy}{dx} = x\frac{dy}{dx} + y$$
$$(2y - x)\frac{dy}{dx} = y - 2x$$
$$\frac{dy}{dx} = \frac{y - 2x}{2y - x}$$

Your Turn 2

$$xe^y + x^2 = \ln y$$
$$\frac{d}{dx}(xe^y + x^2) = \frac{d}{dx}\ln y$$
$$\frac{d}{dx}xe^y + \frac{d}{dx}x^2 = \frac{d}{dx}\ln y$$
$$e^y + xe^y\frac{dy}{dx} + 2x = \frac{1}{y}\frac{dy}{dx}$$
$$\left(xe^y - \frac{1}{y}\right)\frac{dy}{dx} = -2x - e^y$$
$$(xye^y - 1)\frac{dy}{dx} = -2xy - ye^y$$
$$\frac{dy}{dx} = \frac{-2xy - ye^y}{xye^y - 1} = \frac{ye^y - 2xy}{1 - xye^y}$$

Your Turn 3

$$y^4 - x^4 - y^2 + x^2 = 0$$
$$\frac{d}{dx}(y^4 - x^4 - y^2 + x^2) = \frac{d}{dx}(0)$$
$$4y^3\frac{dy}{dx} - 4x^3 - 2y\frac{dy}{dx} + 2x = 0$$
$$(4y^3 - 2y)\frac{dy}{dx} = 4x^3 - 2x$$

At the point $(1, 1)$, $4y^3 - 2y \neq 0$ so we can divide both sides of the equation above by this factor.

$$\frac{dy}{dx} = \frac{4x^3 - 2x}{4y^3 - 2y}$$

At $(1, 1)$,

$$\frac{dy}{dx} = \frac{4x^3 - 2x}{4y^3 - 2y} = \frac{4 - 2}{4 - 2} = 1$$

so the slope of the tangent line is $m = 1$. Use the point-slope form of the equation of a line.

$$y - y_1 = m(x - x_1)$$
$$y - 1 = 1(x - 1)$$
$$y = x$$

The equation of the tangent line at the point $(1, 1)$ is $y = x$.

Your Turn 4

$$p = \frac{100{,}000}{q^2 + 100q} = 100{,}000(q^2 + 100q)^{-1}$$
$$\frac{d}{dp}p = \frac{d}{dp}[100{,}000(q^2 + 100q)^{-1}]$$
$$1 = -100{,}000\frac{2q + 100}{(q^2 + 100q)^2}\frac{dq}{dp}$$
$$\frac{dq}{dp} = -\frac{(q^2 + 100q)^2}{100{,}000(2q + 100)}$$

Substitute $q = 200$ in this expression for the derivative.

$$\frac{dq}{dp} = -\frac{(q^2 + 100q)^2}{100{,}000(2q + 100)}$$
$$= -\frac{[200^2 + 100(200)]^2}{(100{,}000)[2(200) + 100]}$$
$$= -72$$

When the price is 200, the rate of change of demand with respect to price is -72 units per unit change in price.

14.4 Exercises

1. $6x^2 + 5y^2 = 36$

$$\frac{d}{dx}(6x^2 + 5y^2) = \frac{d}{dx}(36)$$
$$\frac{d}{dx}(6x^2) + \frac{d}{dx}(5y^2) = \frac{d}{dx}(36)$$
$$12x + 5 \cdot 2y\frac{dy}{dx} = 0$$
$$10y\frac{dy}{dx} = -12x$$
$$\frac{dy}{dx} = -\frac{6x}{5y}$$

3. $8x^2 - 10xy + 3y^2 = 26$

$$\frac{d}{dx}(8x^2 - 10xy + 3y^2) = \frac{d}{dx}(26)$$

$$16x - \frac{d}{dx}(10xy) + \frac{d}{dx}(3y^2) = 0$$

$$16x - 10x\frac{dy}{dx} - y\frac{d}{dx}(10x) + 6y\frac{dy}{dx} = 0$$

$$16x - 10x\frac{dy}{dx} - 10y + 6y\frac{dy}{dx} = 0$$

$$(-10x + 6y)\frac{dy}{dx} = -16x + 10y$$

$$\frac{dy}{dx} = \frac{-16x + 10y}{-10x + 6y}$$

$$\frac{dy}{dx} = \frac{8x - 5y}{5x - 3y}$$

5. $5x^3 = 3y^2 + 4y$

$$\frac{d}{dx}(5x^3) = \frac{d}{dx}(3y^2 + 4y)$$

$$15x^2 = \frac{d}{dx}(3y^2) + \frac{d}{dx}(4y)$$

$$15x^2 = 6y\frac{dy}{dx} + 4\frac{dy}{dx}$$

$$\frac{15x^2}{6y + 4} = \frac{dy}{dx}$$

7. $3x^2 = \frac{2 - y}{2 + y}$

$$\frac{d}{dx}(3x^2) = \frac{d}{dx}\left(\frac{2 - y}{2 + y}\right)$$

$$6x = \frac{(2 + y)\frac{d}{dx}(2 - y) - (2 - y)\frac{d}{dx}(2 + y)}{(2 + y)^2}$$

$$6x = \frac{(2 + y)\left(-\frac{dy}{dx}\right) - (2 - y)\frac{dy}{dx}}{(2 + y)^2}$$

$$6x = \frac{-4\frac{dy}{dx}}{(2 + y)^2}$$

$$6x(2 + y)^2 = -4\frac{dy}{dx}$$

$$-\frac{3x(2 + y)^2}{2} = \frac{dy}{dx}$$

9. $2\sqrt{x} + 4\sqrt{y} = 5y$

$$\frac{d}{dx}(2x^{1/2} + 4y^{1/2}) = \frac{d}{dx}(5y)$$

$$x^{-1/2} + 2y^{-1/2}\frac{dy}{dx} = 5\frac{dy}{dx}$$

$$(2y^{-1/2} - 5)\frac{dy}{dx} = -x^{-1/2}$$

$$\frac{dy}{dx} = \frac{x^{-1/2}}{5 - 2y^{-1/2}}\left(\frac{x^{1/2}y^{1/2}}{x^{1/2}y^{1/2}}\right)$$

$$= \frac{y^{1/2}}{x^{1/2}(5y^{1/2} - 2)}$$

$$= \frac{\sqrt{y}}{\sqrt{x}(5\sqrt{y} - 2)}$$

11. $x^4y^3 + 4x^{3/2} = 6y^{3/2} + 5$

$$\frac{d}{dx}(x^4y^3 + 4x^{3/2}) = \frac{d}{dx}(6y^{3/2} + 5)$$

$$\frac{d}{dx}(x^4y^3) + \frac{d}{dx}(4x^{3/2}) = \frac{d}{dx}(6y^{3/2}) + \frac{d}{dx}(5)$$

$$4x^3y^3 + x^4 \cdot 3y^2\frac{dy}{dx} + 6x^{1/2} = 9y^{1/2}\frac{dy}{dx} + 0$$

$$4x^3y^3 + 6x^{1/2} = 9y^{1/2}\frac{dy}{dx} - 3x^4y^2\frac{dy}{dx}$$

$$4x^3y^3 + 6x^{1/2} = (9y^{1/2} - 3x^4y^2)\frac{dy}{dx}$$

$$\frac{4x^3y^3 + 6x^{1/2}}{9y^{1/2} - 3x^4y^2} = \frac{dy}{dx}$$

13. $e^{x^2y} = 5x + 4y + 2$

$$\frac{d}{dx}(e^{x^2y}) = \frac{d}{dx}(5x + 4y + 2)$$

$$e^{x^2y}\frac{d}{dx}(x^2y) = \frac{d}{dx}(5x) + \frac{d}{dx}(4y) + \frac{d}{dx}(2)$$

$$e^{x^2y}\left(2xy + x^2\frac{dy}{dx}\right) = 5 + 4\frac{dy}{dx} + 0$$

$$2xye^{x^2y} + x^2e^{x^2y}\frac{dy}{dx} = 5 + 4\frac{dy}{dx}$$

$$x^2e^{x^2y}\frac{dy}{dx} - 4\frac{dy}{dx} = 5 - 2xye^{x^2y}$$

$$(x^2e^{x^2y} - 4)\frac{dy}{dx} = 5 - 2xye^{x^2y}$$

$$\frac{dy}{dx} = \frac{5 - 2xye^{x^2y}}{x^2e^{x^2y} - 4}$$

15. $x + \ln y = x^2y^3$

$$\frac{d}{dx}(x + \ln y) = \frac{d}{dx}(x^2y^3)$$
$$1 + \frac{1}{y}\frac{dy}{dx} = 2xy^3 + 3x^2y^2\frac{dy}{dx}$$
$$\frac{1}{y}\frac{dy}{dx} - 3x^2y^2\frac{dy}{dx} = 2xy^3 - 1$$
$$\left(\frac{1}{y} - 3x^2y^2\right)\frac{dy}{dx} = 2xy^3 - 1$$
$$\frac{dy}{dx} = \frac{2xy^3 - 1}{\frac{1}{y} - 3x^2y^2}$$
$$= \frac{y(2xy^3 - 1)}{1 - 3x^2y^3}$$

17. $x^2 + y^2 = 25$; tangent at $(-3, 4)$

$$\frac{d}{dx}(x^2 + y^2) = \frac{d}{dx}(25)$$
$$2x + 2y\frac{dy}{dx} = 0$$
$$2y\frac{dy}{dx} = -2x$$
$$\frac{dy}{dx} = -\frac{x}{y}$$

$$m = -\frac{x}{y} = -\frac{-3}{4} = \frac{3}{4}$$
$$y - y_1 = m(x - x_1)$$
$$y - 4 = \frac{3}{4}[x - (-3)]$$
$$4y - 16 = 3x + 9$$
$$4y = 3x + 25$$
$$y = \frac{3}{4}x + \frac{25}{4}$$

19. $x^2y^2 = 1$; tangent at $(-1, 1)$

$$\frac{d}{dx}(x^2y^2) = \frac{d}{dx}(1)$$
$$x^2\frac{d}{dx}(y^2) + y^2\frac{d}{dx}(x^2) = 0$$
$$x^2(2y)\frac{dy}{dx} + y^2(2x) = 0$$
$$2x^2y\frac{dy}{dx} = -2xy^2$$
$$\frac{dy}{dx} = \frac{-2xy^2}{2x^2y} = -\frac{y}{x}$$

$$m = -\frac{y}{x} = -\frac{1}{-1} = 1$$
$$y - 1 = 1[x - (-1)]$$
$$y = x + 1 + 1$$
$$y = x + 2$$

21. $2y^2 - \sqrt{x} = 4$; tangent at $(16, 2)$

$$\frac{d}{dx}(2y^2 - \sqrt{x}) = \frac{d}{dx}(4)$$
$$4y\frac{dy}{dx} - \frac{1}{2}x^{-1/2} = 0$$
$$4y\frac{dy}{dx} = \frac{1}{2x^{1/2}}$$
$$\frac{dy}{dx} = \frac{1}{8yx^{1/2}}$$

$$m = \frac{1}{8yx^{1/2}} = \frac{1}{8(2)(16)^{1/2}}$$
$$= \frac{1}{8(2)(4)} = \frac{1}{64}$$

$$y - 2 = -\frac{1}{4}(x - 4)$$
$$y = -\frac{1}{4}x + 3$$
$$y - 2 = \frac{1}{64}(x - 16)$$
$$64y - 128 = x - 16$$
$$64y = x + 112$$
$$y = \frac{x}{64} + \frac{7}{4}$$

23. $e^{x^2+y^2} = xe^{5y} - y^2e^{5x/2}$; tangent at (2,1)

$$\frac{d}{dx}(e^{x^2+y^2}) = \frac{d}{dx}(xe^{5y} - y^2e^{5x/2})$$

$$e^{x^2+y^2} \cdot \frac{d}{dx}(x^2 + y^2) = e^{5y} + x\frac{d}{dx}(e^{5y}) - \left[2y\frac{dy}{dx}e^{5x/2} + y^2e^{5x/2}\frac{d}{dx}\left(\frac{5x}{2}\right)\right]$$

$$e^{x^2+y^2}\left(2x + 2y\frac{dy}{dx}\right) = e^{5y} + x \cdot 5e^{5y}\frac{dy}{dx} - 2ye^{5x/2}\frac{dy}{dx} - \frac{5}{2}y^2e^{5x/2}$$

$$\left(2ye^{x^2+y^2} - 5xe^{5y} + 2ye^{5x/2}\right)\frac{dy}{dx} = -2xe^{x^2+y^2} + e^{5y} - \frac{5}{2}y^2e^{5x/2}$$

$$\frac{dy}{dx} = \frac{-2xe^{x^2+y^2} + e^{5y} - \frac{5}{2}y^2e^{5x/2}}{2ye^{x^2+y^2} - 5xe^{5y} + 2ye^{5x/2}}$$

$$m = \frac{-4e^5 + e^5 - \frac{5}{2}e^5}{2e^5 - 10e^5 + 2e^5} = \frac{-\frac{11}{2}e^5}{-6e^5} = \frac{11}{12}$$

$$y - 1 = \frac{11}{12}(x - 2)$$

$$y = \frac{11}{12}x - \frac{5}{6}$$

25. $\ln(x + y) = x^3y^2 + \ln(x^2 + 2) - 4$; tangent at (1, 2)

$$\frac{d}{dx}[\ln(x + y)] = \frac{d}{dx}[x^3y^2 + \ln(x^2 + 2) - 4]$$

$$\frac{1}{x + y} \cdot \frac{d}{dx}(x + y) = 3x^2y^2 + x^3 \cdot 2y\frac{dy}{dx} + \frac{1}{x^2 + 2} \cdot \frac{d}{dx}(x^2 + 2) - \frac{d}{dx}(4)$$

$$\left(\frac{1}{x + y} - 2x^3y\right)\frac{dy}{dx} = 3x^2y^2 + \frac{2x}{x^2 + 2} - \frac{1}{x + y}$$

$$\frac{dy}{dx} = \frac{3x^2y^2 + \frac{2x}{x^2+2} - \frac{1}{x+y}}{\frac{1}{x+y} - 2x^3y}$$

$$m = \frac{3 \cdot 1 \cdot 4 + \frac{2 \cdot 1}{3} - \frac{1}{3}}{\frac{1}{3} - 2 \cdot 1 \cdot 2} = \frac{\frac{37}{3}}{\frac{-11}{3}} = -\frac{37}{11}$$

$$y - 2 = -\frac{37}{11}(x - 1)$$

$$y = -\frac{37}{11}x + \frac{59}{11}$$

27. $y^3 + xy - y = 8x^4; x = 1$

First, find the y-value of the point.

$$y^3 + (1)y - y = 8(1)^4$$
$$y^3 = 8$$
$$y = 2$$

The point is (1, 2).

Find $\frac{dy}{dx}$.

$$3y^2\frac{dy}{dx} + x\frac{dy}{dx} + y - \frac{dy}{dx} = 32x^3$$
$$(3y^2 + x - 1)\frac{dy}{dx} = 32x^3 - y$$
$$\frac{dy}{dx} = \frac{32x^3 - y}{3y^2 + x - 1}$$

At (1, 2),

$$\frac{dy}{dx} = \frac{32(1)^3 - 2}{3(2)^2 + 1 - 1} = \frac{30}{12} = \frac{5}{2}.$$

$$y - 2 = \frac{5}{2}(x - 1)$$
$$y - 2 = \frac{5}{2}x - \frac{5}{2}$$
$$y = \frac{5}{2}x - \frac{1}{2}$$

29. $y^3 + xy^2 + 1 = x + 2y^2; x = 2$

Find the y-value of the point.

$$y^3 + 2y^2 + 1 = 2 + 2y^2$$
$$y^3 + 1 = 2$$
$$y^3 = 1$$
$$y = 1$$

The point is (2, 1).

Find $\frac{dy}{dx}$.

$$3y^2\frac{dy}{dx} + x2y\frac{dy}{dx} + y^2 = 1 + 4y\frac{dy}{dx}$$
$$3y^2\frac{dy}{dx} + 2xy\frac{dy}{dx} - 4y\frac{dy}{dx} = 1 - y^2$$
$$(3y^2 + 2xy - 4y)\frac{dy}{dx} = 1 - y^2$$
$$\frac{dy}{dx} = \frac{1 - y^2}{3y^2 + 2xy - 4y}$$

At (2, 1),

$$\frac{dy}{dx} = \frac{1 - 1^2}{3(1)^2 + 2(2)(1) - 4(1)} = 0.$$
$$y - 0 = 0(x - 2)$$
$$y = 1$$

31. $2y^3(x - 3) + x\sqrt{y} = 3; x = 3$

Find the y-value of the point.

$$2y^3(3 - 3) + 3\sqrt{y} = 3$$
$$3\sqrt{y} = 3$$
$$\sqrt{y} = 1$$
$$y = 1$$

The point is (3, 1)

Find $\frac{dy}{dx}$.

$$2y^3(1) + 6y^2(x - 3)\frac{dy}{dx} + x\left(\frac{1}{2}\right)y^{-1/2}\frac{dy}{dx} + \sqrt{y} = 0$$
$$6y^2(x - 3)\frac{dy}{dx} + \frac{x}{2\sqrt{y}}\frac{dy}{dx} = -2y^3 - \sqrt{y}$$
$$\left[6y^2(x - 3) + \frac{x}{2\sqrt{y}}\right]\frac{dy}{dx} = -2y^3 - \sqrt{y}$$
$$\frac{dy}{dx} = \frac{-2y^3 - \sqrt{y}}{6y^2(x - 3) + \frac{x}{2\sqrt{y}}}$$
$$= \frac{-4y^{7/2} - 2y}{12y^{5/2}(x - 3) + x}$$

At (3, 1),

$$\frac{dy}{dx} = \frac{-4(1) - 2}{12(1)(3 - 3) + 3} = \frac{-6}{3} = -2.$$
$$y - 1 = -2(x - 3)$$
$$y - 1 = -2x + 6$$
$$y = -2x + 7$$

33. $x^{2/3} + y^{2/3} = 2; (1, 1)$

Find $\frac{dy}{dx}$.

$$\frac{2}{3}x^{-1/3} + \frac{2}{3}y^{-1/3}\frac{dy}{dx} = 0$$

$$\frac{2}{3}y^{-1/3}\frac{dy}{dx} = -\frac{2}{3}x^{-1/3}$$

$$\frac{dy}{dx} = \frac{-\frac{2}{3}x^{-1/3}}{\frac{2}{3}y^{-1/3}}$$

$$= -\frac{y^{1/3}}{x^{1/3}}$$

At (1, 1)

$$\frac{dy}{dx} = -\frac{1^{1/3}}{1^{1/3}} = -1$$

$$y - 1 = -1(x - 1)$$

$$y - 1 = -x + 1$$

$$y = -x + 2$$

35. $y^2(x^2 + y^2) = 20x^2$; (1, 2)

Find $\frac{dy}{dx}$.

$$2y(x^2 + y^2)\frac{dy}{dx} + y^2\left(2x + 2y\frac{dy}{dx}\right) = 40x$$

$$2x^2y\frac{dy}{dx} + 2y^3\frac{dy}{dx} + 2xy^2 + 2y^3\frac{dy}{dx} = 40x$$

$$2x^2y\frac{dy}{dx} + 4y^3\frac{dy}{dx} = -2xy^2 + 40x$$

$$(2x^2y + 4y^3)\left(\frac{dy}{dx}\right) = -2xy^2 + 40x$$

$$\frac{dy}{dx} = \frac{-2xy^2 + 40x}{2x^2y + 4y^3}$$

At (1, 2),

$$\frac{dy}{dx} = \frac{-2(1)(2)^2 + 40(1)}{2(1)^2(2) + 4(2)^3}$$

$$= \frac{32}{36} = \frac{8}{9}$$

$$y - 2 = \frac{8}{9}(x - 1)$$

$$y - 2 = \frac{8}{9}x - \frac{8}{9}$$

$$y = \frac{8}{9}x + \frac{10}{9}$$

37. $x^2 + y^2 = 100$

(a) Lines are tangent at points where $x = 6$. By substituting $x = 6$ in the equation, we find that the points are $(6, 8)$ and $(6, -8)$.

$$\frac{d}{dx}(x^2 + y^2) = \frac{d}{dx}(100)$$

$$2x + 2y\frac{dy}{dx} = 0$$

$$2y\frac{dy}{dx} = -2x$$

$$dy = -\frac{x}{y}$$

$$m_1 = -\frac{x}{y} = -\frac{6}{8} = -\frac{3}{4}$$

$$m_2 = -\frac{x}{y} = -\frac{6}{-8} = \frac{3}{4}$$

First tangent:

$$y - 8 = -\frac{3}{4}(x - 6)$$

$$y = -\frac{3}{4}x + \frac{25}{2}$$

Second tangent:

$$y - (-8) = \frac{3}{4}(x - 6)$$

$$y + 8 = \frac{3}{4}x - \frac{18}{4}$$

$$y = \frac{3}{4}x - \frac{25}{2}$$

(b)

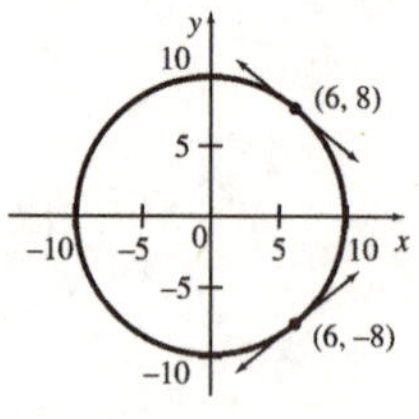

39. (a) $\sqrt{u} + \sqrt{2v + 1} = 5$

$$\frac{du}{dv}(\sqrt{u} + \sqrt{2v + 1}) = \frac{du}{dv}(5)$$

$$\frac{1}{2}u^{-1/2}\frac{du}{dv} + \frac{1}{2}(2v + 1)^{-1/2}(2) = 0$$

$$\frac{1}{2}u^{-1/2}\frac{du}{dv} = -\frac{1}{(2v + 1)^{1/2}}$$

$$\frac{du}{dv} = -\frac{2u^{1/2}}{(2v + 1)^{1/2}}$$

(b) $\sqrt{u} + \sqrt{2v+1} = 5$

$$\frac{dv}{du}(\sqrt{u} + \sqrt{2v+1}) = \frac{dv}{du}(5)$$

$$\frac{1}{2}u^{-1/2} + \frac{1}{2}(2v+1)^{-1/2}(2)\frac{dv}{du} = 0$$

$$(2v+1)^{-1/2}\frac{dv}{du} = -\frac{1}{2}u^{-1/2}$$

$$\frac{dv}{du} = -\frac{(2v+1)^{1/2}}{2u^{1/2}}$$

The derivatives are reciprocals.

41. $x^2 + y^2 + 1 = 0$

$$\frac{d}{dx}(x^2 + y^2) = \frac{d}{dx}(-1)$$

$$2x + 2y\frac{dy}{dx} = 0$$

$$\frac{dy}{dx} = \frac{-2x}{2y} = -\frac{x}{y}$$

If x and y are real numbers, x^2 and y^2 are nonnegative; 1 plus a nonnegative number cannot equal zero, so there is no function $y = f(x)$ that satisfies $x^2 + y^2 + 1 = 0$.

43. $C^2 = x^2 + 100\sqrt{x} + 50$

(a) $2C\frac{dC}{dx} = 2x + \frac{1}{2}(100)x^{-1/2}$

$$\frac{dC}{dx} = \frac{2x + 50x^{-1/2}}{2C}$$

$$\frac{dC}{dx} = \frac{x + 25x^{-1/2}}{C} \cdot \frac{x^{1/2}}{x^{1/2}}$$

$$\frac{dC}{dx} = \frac{x^{3/2} + 25}{Cx^{1/2}}$$

When $x = 5$, the approximate increase in cost of an additional unit is

$$\frac{(5)^{3/2} + 25}{(5^2 + 100\sqrt{5} + 50)^{1/2}(5)^{1/2}} = \frac{36.18}{(17.28)\sqrt{5}} \approx 0.94.$$

(b) $900(x-5)^2 + 25R^2 = 22,500$

$$R^2 = 900 - 36(x-5)^2$$

$$2R\frac{dR}{dx} = -72(x-5)$$

$$\frac{dR}{dx} = \frac{-36(x-5)}{R} = \frac{180 - 36x}{R}$$

When $x = 5$, the approximate change in revenue for a unit increase in sales is

$$\frac{180 - 36(5)}{R} = \frac{0}{R} = 0.$$

45. (a) $\ln q = D - 0.44 \ln p$

$$\frac{1}{q}\frac{dq}{dp} = -\frac{0.44}{p}$$

$$\frac{dp}{dq} = -0.44\frac{q}{p}$$

$$E = -\frac{p}{q}\frac{dq}{dp}$$

$$= -\frac{p}{q}\left(-0.44\frac{q}{p}\right)$$

$$= 0.44$$

E is less than 1, so the demand is inelastic.

(b) Solving for q first:

$$\ln q = D - 0.44 \ln p$$

$$e^{\ln q} = e^{D - 0.44 \ln p}$$

$$q = e^D p^{-0.44}$$

$$\frac{dq}{dp} = e^D(-0.678)p^{-1.44}$$

$$E = -\frac{p}{q}\frac{dq}{dp}$$

$$= \left(-\frac{p}{e^D p^{-0.44}}\right)(e^D(-0.44)p^{-1.44})$$

$$= 0.44$$

This is the same answer as found in part (a).

47. $b - a = (b+a)^3$

$$\frac{d}{db}(b - a) = \frac{d}{db}[(b+a)^3]$$

$$1 - \frac{da}{db} = 3(b+a)^2\frac{d}{db}(b+a)$$

$$1 - \frac{da}{db} = 3(b+a)^2\left(1 + \frac{da}{db}\right)$$

$$1 - \frac{da}{db} = 3(b+a)^2 + 3(b+a)^2\frac{da}{db}$$

$$-\frac{da}{db} - 3(b+a)^2\frac{da}{db} = 3(b+a)^2 - 1$$

$$[-1 - 3(b+a)^2]\frac{da}{db} = 3(b+a)^2 - 1$$

$$\frac{da}{db} = \frac{3(b+a)^2 - 1}{-1 - 3(b+a)^2}$$

$$\frac{da}{db} = 0$$

$$3(b+a)^2 - 1 = 0$$

$$b + a = \frac{1}{\sqrt{3}}$$

Since $b - a = (b+a)^3 = \left(\frac{1}{\sqrt{3}}\right)^3 = \frac{1}{3\sqrt{3}}.$

$$b + a = \frac{1}{\sqrt{3}}$$

$$-(b - a) = -\frac{1}{3\sqrt{3}}$$

$$2a = \frac{2}{3\sqrt{3}}$$

$$a = \frac{1}{3\sqrt{3}}$$

49. $s^3 - 4st + 2t^3 - 5t = 0$

$$3s^2\frac{ds}{dt} - \left(4t\frac{ds}{dt} + 4s\right) + 6t^2 - 5 = 0$$

$$3s^2\frac{ds}{dt} - 4t\frac{ds}{dt} - 4s + 6t^2 - 5 = 0$$

$$\frac{ds}{dt}(3s^2 - 4t) = 4s - 6t^2 + 5$$

$$\frac{ds}{dt} = \frac{4s - 6t^2 + 5}{3s^2 - 4t}$$

14.5 Related Rates

Your Turn 1

$x^3 + 2xy + y^2 = 1,$ where both x and y are functions of t. Given $x = 1, y = -2,$ and $dx/dy = 6,$ find dy/dt.

$$\frac{d}{dt}\left(x^3 + 2xy + y^2\right) = \frac{d}{dt}(1)$$

$$3x^2\frac{dx}{dt} + 2x\frac{dy}{dt} + 2y\frac{dx}{dt} + 2y\frac{dy}{dt} = 0$$

$$52x\frac{dy}{dt} + 2y\frac{dy}{dt} = -\left(3x^2\frac{dx}{dt} + 2y\frac{dx}{dt}\right)$$

$$\frac{dy}{dt} = \frac{dx}{dt}\left(-\frac{3x^2 + 2y}{2x + 2y}\right)$$

Now substitute the given values.

$$\frac{dy}{dt} = \frac{dx}{dt}\left(-\frac{3x^2 + 2y}{2x + 2y}\right)$$

$$= 6\left(-\frac{3(1^2) + 2(-2)}{2(1) + 2(-2)}\right)$$

$$= 6\left(-\frac{-1}{-2}\right)$$

$$= -3$$

Your Turn 2

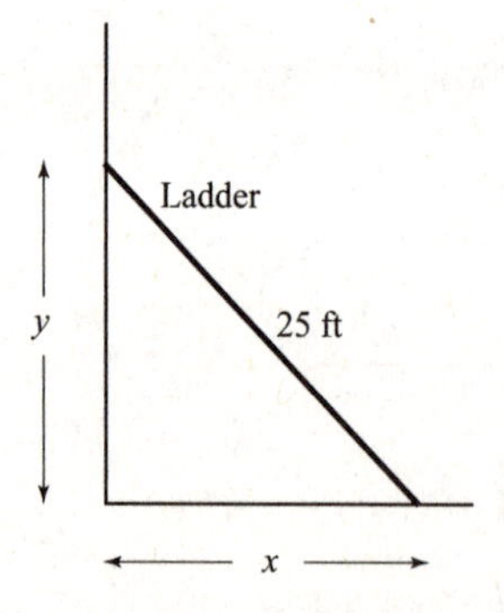

$x^2 + y^2 = 25^2,$ where both x and y are functions of t. We are interested in what happens when $x = 7$ ft. At this time, $y = \sqrt{25^2 - 7^2} = 24,$ and since the bottom of the ladder is slipping away from the building at 3 ft/min, $dx/dt = 3.$

$$\frac{d}{dt}\left(x^2 + y^2\right) = \frac{d}{dt}(25^2)$$

$$2x\frac{dx}{dt} + 2y\frac{dy}{dt} = 0$$

Now substitute the known values.

$$2(7)(3) + 2(24)\frac{dy}{dt} = 0$$
$$\frac{dy}{dt} = \frac{-2(7)(3)}{2(24)}$$
$$= -\frac{7}{8}$$

The latter is sliding *down* the side of the building at 7/8 ft/min.

Your Turn 3

In Example 4, differentiating the formula for the volume of a cone gives the following result:

$$\frac{dV}{dt} = \frac{1}{3}\pi\left[r^2\frac{dh}{dt} + (h)(2r)\frac{dr}{dt}\right]$$

For this problem,

$$\frac{dV}{dt} = -10,\ \frac{dr}{dt} = -0.4,\ r = 4, \text{ and } h = 20.$$

Substitute these values above and solve for dh/dt.

$$\frac{dV}{dt} = \frac{1}{3}\pi\left[r^2\frac{dh}{dt} + (h)(2r)\frac{dr}{dt}\right]$$
$$-10 = \frac{1}{3}\pi\left[(4^2)\frac{dh}{dt} + (20)(2)(4)(-0.4)\right]$$
$$-\frac{30}{\pi} + 64 = 16\frac{dh}{dt}$$
$$\frac{dh}{dt} = \frac{-\frac{30}{\pi} + 64}{16}$$
$$\frac{dh}{dt} \approx 3.4$$

The length increases at a rate of 3.4 cm per hour.

Your Turn 4

The revenue equation is

$$R = qp = q\left(2000 - \frac{q^2}{100}\right) = 2000q - \frac{q^3}{100}.$$

Differentiate with respect to t.

$$\frac{dR}{dt} = 2000\frac{dq}{dt} - \frac{3q^2}{100}\frac{dq}{dt}$$
$$= \left(2000 - \frac{3q^2}{100}\right)\frac{dq}{dt}$$

Substitute the know values for q and dq/dt, which are the same as in Example 5: $q = 200$, $dq/dt = 50$.

$$\frac{dR}{dt} = \left(2000 - \frac{3q^2}{100}\right)\frac{dq}{dt}$$
$$= \left(2000 - \frac{3(200^2)}{100}\right)(50)$$
$$= (800)(50) = 40{,}000$$

Revenue is increasing at the rate of \$40,000 per day.

14.5 Exercises

1. $y^2 - 8x^3 = -55;\ \frac{dx}{dt} = -4, x = 2, y = 3$

$$2y\frac{dy}{dt} - 24x^2\frac{dx}{dt} = 0$$
$$y\frac{dy}{dt} = 12x^2\frac{dx}{dt}$$
$$3\frac{dy}{dt} = 48(-4)$$
$$\frac{dy}{dt} = -64$$

3. $2xy - 5x + 3y^3 = -51;\ \frac{dx}{dt} = -6,$

$x = 3, y = -2$

$$2x\frac{dy}{dt} + 2y\frac{dx}{dt} - 5\frac{dx}{dt} + 9y^2\frac{dx}{dt} = 0$$
$$(2x + 9y^2)\frac{dy}{dt} + (2y - 5)\frac{dx}{dt} = 0$$
$$(2x + 9y^2)\frac{dy}{dt} = (5 - 2y)\frac{dx}{dt}$$
$$\frac{dy}{dt} = \frac{5 - 2y}{2x + 9y^2}\cdot\frac{dx}{dt}$$
$$= \frac{5 - 2(-2)}{2(3) + 9(-2)^2}\cdot(-6)$$
$$= \frac{9}{42}\cdot(-6) = \frac{-54}{42}$$
$$= -\frac{9}{7}$$

5. $\dfrac{x^2 + y}{x - y} = 9;\ \dfrac{dx}{dt} = 2, x = 4, y = 2$

$$\frac{(x - y)\left(2x\frac{dx}{dt} + \frac{dy}{dt}\right) - (x^2 + y)\left(\frac{dx}{dt} - \frac{dy}{dt}\right)}{(x - y)^2} = 0$$

$$\frac{2x(x - y)\frac{dx}{dt} + (x - y)\frac{dy}{dt} - (x^2 + y)\frac{dx}{dt} + (x^2 + y)\frac{dy}{dt}}{(x - y)^2} = 0$$

$$[2x(x - y) - (x^2 + y)]\frac{dx}{dt} + [(x - y) + (x^2 + y)]\frac{dy}{dt} = 0$$

$$\frac{dy}{dt} = \frac{[(x^2 + y) - 2x(x - y)]\frac{dx}{dt}}{(x - y) + (x^2 + y)}$$

$$\frac{dy}{dt} = \frac{(-x^2 + y + 2xy)\frac{dx}{dt}}{x + x^2}$$

$$= \frac{[-(4)^2 + 2 + 2(4)(2)](2)}{4 + 4^2}$$

$$= \frac{4}{20} = \frac{1}{5}$$

7. $xe^y = 3 + \ln x; \frac{dx}{dt} = 6, x = 2, y = 0$

$$e^y \frac{dx}{dt} + xe^y \frac{dy}{dt} = 0 + \frac{1}{x}\frac{dx}{dt}$$
$$xe^y \frac{dy}{dt} = \left(\frac{1}{x} - e^y\right)\frac{dx}{dt}$$
$$\frac{dy}{dt} = \frac{\left(\frac{1}{x} - e^y\right)\frac{dx}{dt}}{xe^y}$$
$$= \frac{(1 - xe^y)\frac{dx}{dt}}{x^2e^y}$$
$$= \frac{[1 - (2)e^0](6)}{2^2e^0}$$
$$= \frac{-6}{4} = -\frac{3}{2}$$

9. $C = 0.2x^2 + 10{,}000; x = 80, \frac{dx}{dt} = 12$

$$\frac{dC}{dt} = 0.2(2x)\frac{dx}{dt} = 0.2(160)(12) = 384$$

The cost is changing at a rate of \$384 per month.

11. $R = 50x - 0.4x^2; C = 5x + 15;$

$x = 40; \frac{dx}{dt} = 10$

(a) $\frac{dR}{dt} = 50\frac{dx}{dt} - 0.8x\frac{dx}{dt}$

$= 50(10) - 0.8(40)(10)$

$= 500 - 320$

$= 180$

Revenue is increasing at a rate of \$180 per day.

(b) $\frac{dC}{dt} = 5\frac{dx}{dt} = 5(10) = 50$

Cost is increasing at a rate of \$50 per day.

(c) Profit = Revenue − Cost

$P = R - C$

$$\frac{dP}{dt} = \frac{dR}{dt} - \frac{dC}{dt} = 180 - 50 = 130$$

Profit is increasing at a rate of \$130 per day.

13. $pq = 8000; p = 3.50, \frac{dp}{dt} = 0.15$

$$pq = 8000$$
$$p\frac{dq}{dt} + q\frac{dp}{dt} = 0$$
$$\frac{dq}{dt} = \frac{-q\frac{dp}{dt}}{p}$$
$$= \frac{-\left(\frac{8000}{3.50}\right)(0.15)}{3.50}$$
$$\approx -98$$

Demand is decreasing at a rate of approximately 98 units per unit time.

15. $V = k(R^2 - r^2); k = 555.6, R = 0.02$ mm,

$\frac{dR}{dt} = 0.003$ mm per minute; r is constant.

$$V = k(R^2 - r^2)$$
$$V = 555.6(R^2 - r^2)$$
$$\frac{dV}{dt} = 555.6\left(2R\frac{dR}{dt} - 0\right)$$
$$= 555.6(2)(0.02)(0.003)$$
$$= 0.067 \text{ mm/min}$$

17. $b = 0.22m^{0.87}$

$$\frac{db}{dt} = 0.22(0.87)m^{-0.13}\frac{dm}{dt}$$
$$= 0.1914m^{-0.13}\frac{dm}{dt}$$
$$\frac{dm}{dt} = \frac{m^{0.13}}{0.1914}\frac{db}{dt}$$
$$= \frac{25^{0.13}}{0.1914}(0.25)$$
$$\approx 1.9849$$

The rate of change of the total weight is about 1.9849 g/day.

19. $r = 140.2m^{0.75}$

(a) $\frac{dr}{dt} = 140.2(0.75)m^{-0.25}\frac{dm}{dt}$

$= 105.15m^{-0.25}\frac{dm}{dt}$

(b) $\frac{dr}{dt} = 105.15(250)^{-0.25}(2)$

≈ 52.89

The rate of change of the average daily metabolic rate is about 52.89 kcal/day^2.

21. $C = \frac{1}{10}(T - 60)^2 + 100$

$$\frac{dC}{dt} = \frac{1}{5}(T - 60)\frac{dT}{dt}$$

If $T = 76°$ and $\frac{dT}{dt} = 8$,

$$\frac{dC}{dt} = \frac{1}{5}(76 - 60)(8) = \frac{1}{5}(16)(8)$$
$$= 25.6.$$

The crime rate is rising at the rate of 25.6 crimes/month.

23. $x =$ The distance of the base of the ladder from the base of the building
$y =$ The distance up the side of the building to the top of the ladder

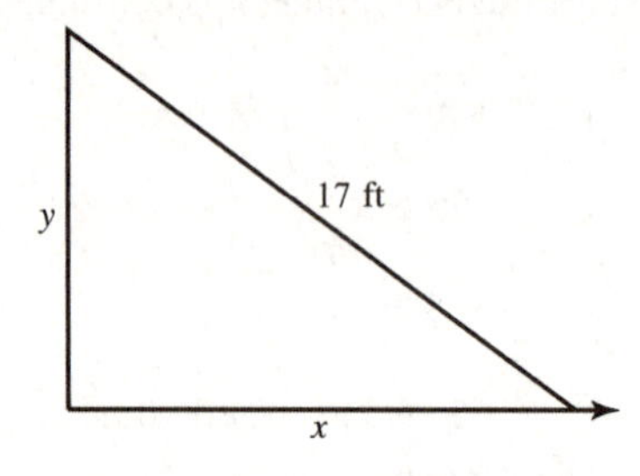

Find $\frac{dy}{dt}$ when $x = 8$ ft and $\frac{dx}{dt} = 9$ ft/min.

Since $y = \sqrt{17^2 - x^2}$, when $x = 8$, $y = 15$.

By the Pythagorean theorem,

$$x^2 + y^2 = 17^2.$$

$$\frac{d}{dx}(x^2 + y^2) = \frac{d}{dt}(17^2)$$
$$2x\frac{dx}{dt} + 2y\frac{dy}{dt} = 0$$
$$2y\frac{dy}{dt} = -2x\frac{dx}{dt}$$
$$\frac{dy}{dt} = \frac{-2x}{2y} \cdot \frac{dx}{dt} = -\frac{x}{y} \cdot \frac{dx}{dt}$$
$$= -\frac{8}{15}(9)$$
$$= -\frac{24}{5}$$

The ladder is sliding down the building at the rate of $\frac{24}{5}$ ft/min.

25. Let $r =$ the raius of the circle formed by the ripple.

Find $\frac{dA}{dt}$ when $r = 4$ ft and $\frac{dr}{dt} = 2$ ft/min.

$$A = \pi r^2$$
$$\frac{dA}{dt} = 2\pi r\frac{dr}{dt}$$
$$= 2\pi(4)(2) = 16\pi$$

The area is changing at the rate of 16π ft^2/min.

27. $V = x^3$, $x = 3$ cm, and $\frac{dV}{dt} = 2$ cm^3/min

$$\frac{dV}{dt} = 3x^2\frac{dx}{dt}$$
$$\frac{dx}{dt} = \frac{1}{3x^2}\frac{dV}{dt}$$
$$= \frac{1}{3 \cdot 3^2}(2) = \frac{2}{27} \text{ cm/min}$$

29. Let $y =$ the length of the man's shadow;
$x =$ the distane of the man from the lamp post;
$h =$ the height of the lamp post.

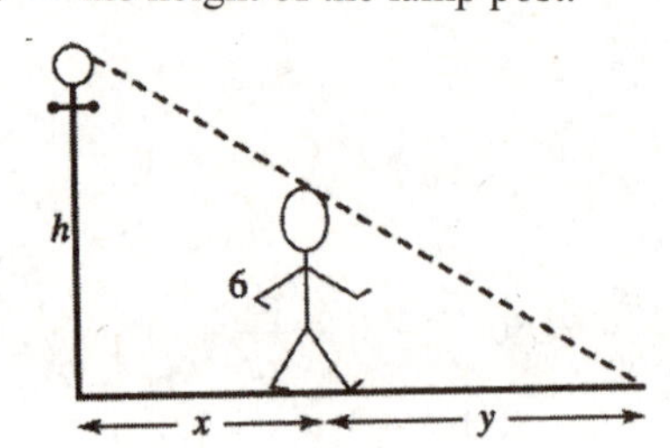

$$\frac{dx}{dt} = 50 \text{ ft/min}$$

Find $\frac{dy}{dt}$ when $x = 25$ ft.

Now $\frac{h}{x+y} = \frac{6}{y}$, by similar triangles.

When $x = 8$, $y = 10$,

$$\frac{h}{18} = \frac{6}{10}$$
$$h = 10.8.$$
$$\frac{10.8}{x + y} = \frac{6}{y},$$
$$10.8y = 6x + 6y$$
$$4.8y = 6x$$
$$y = 1.25x$$

$$\frac{dy}{dt} = 1.25\frac{dx}{dt}$$
$$= 1.25(50)$$
$$\frac{dy}{dt} = 62.5$$

The length of the shadow is increasing at the rate of 62.5 ft/min.

31. Let $x =$ the distance from the docks

$s =$ the length of the rope.

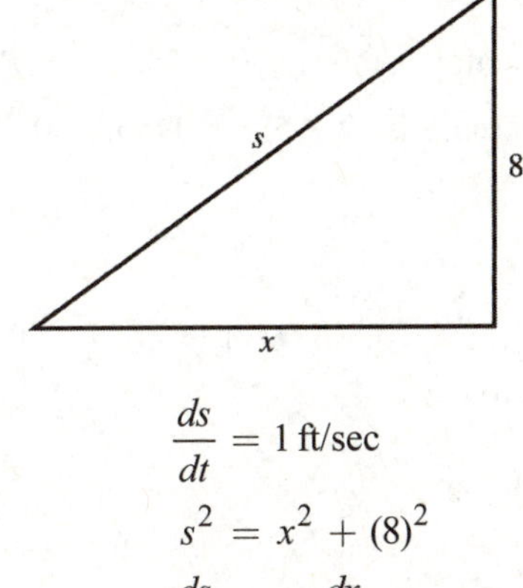

$$\frac{ds}{dt} = 1 \text{ ft/sec}$$
$$s^2 = x^2 + (8)^2$$
$$2s\frac{ds}{dt} = 2x\frac{dx}{dt} + 0$$
$$s\frac{ds}{dt} = x\frac{dx}{dt}$$

If $x = 8$,

$$s = \sqrt{(8)^2 + (8)^2} = \sqrt{128} = 8\sqrt{2}.$$

Then,

$$8\sqrt{2}(1) = 8\frac{dx}{dt}$$
$$\frac{dx}{dt} = \sqrt{2} \approx 1.41$$

The boat is approaching the dock at $\sqrt{2} \approx 1.41$ ft/sec.

14.6 Differentials: Linear Approximation

Your Turn 1

$y = 300x^{-2/3}$, $x = 8$, $dx = 0.05$

$$\frac{dy}{dx} = \left(-\frac{2}{3}\right)(300)x^{-5/3} = -200x^{-5/3}$$
$$dy = -200x^{-5/3}dx$$
$$= (-200)(8^{-5/3})(0.05)$$
$$= (-200)\left(\frac{1}{32}\right)\left(\frac{1}{20}\right)$$
$$= -\frac{5}{16}$$

Your Turn 2

Use the approximation formula for $f(x) = \sqrt{x}$ developed in Example 2:

$$f(x + \Delta x) \approx \sqrt{x} + \frac{1}{2\sqrt{x}}dx$$

For this problem, $x = 100$ and $dx = -1$.

$$f(99) = f(100 - 1)$$
$$\approx \sqrt{100} + \frac{1}{2\sqrt{100}}(-1)$$
$$= 10 - \frac{1}{20}$$
$$= 9.95$$

Your Turn 3

Use the approximation derived in Example 5:

$$dV = 4\pi r^2 dr$$

For this problem, $r = 1.25$ and $dr = \Delta r = \pm 0.025$.

$$dV = 4\pi(1.25)^2(\pm 0.025)$$
$$\approx 0.491$$
$$\approx 0.5$$

The maximum error in the volume is about 0.5 mm^3.

14.6 Exercises

1. $y = 2x^3 - 5x;\ x = -2,\ \Delta x = 0.1$

$dy = (6x^2 - 5)dx$

$\Delta y \approx (6x^2 - 5)\Delta x \approx [6(-2)^2 - 5](0.1) \approx 1.9$

3. $y = x^3 - 2x^2 + 3,\ x = 1,\ \Delta x = -0.1$

$dy = (3x^2 - 4x)dx$

$$\begin{aligned}\Delta y &\approx (3x^2 - 4x)\Delta x \\ &= [3(1^2) - 4(1)](-0.1) \\ &= 0.1\end{aligned}$$

5. $y = \sqrt{3x+2},\ x = 4,\ \Delta x = 0.15$

$$dy = 3\left(\frac{1}{2}(3x+2)^{-1/2}\right)dx$$

$$\Delta y \approx \frac{3}{2\sqrt{3x+2}}\Delta x \approx \frac{3}{2(3.74)}(0.15) \approx 0.060$$

7. $y = \dfrac{2x-5}{x+1};\ x = 2,\ \Delta x = -0.03$

$$\begin{aligned}dy &= \frac{(x+1)(2) - (2x-5)(1)}{(x+1)^2}dx \\ &= \frac{7}{(x+1)^2}dx\end{aligned}$$

$$\begin{aligned}\Delta y &\approx \frac{7}{(x+1)^2}\Delta x \\ &= \frac{7}{(2+1)^2}(-0.03) \\ &= -0.023\end{aligned}$$

9. $\sqrt{145}$

We know $\sqrt{144} = 12$, so $f(x) = \sqrt{x}$, $x = 144$, $dx = 1$.

$$\frac{dy}{dx} = \frac{1}{2}x^{-1/2}$$

$$dy = \frac{1}{2\sqrt{x}}dx$$

$$dy = \frac{1}{2\sqrt{144}}(1) = \frac{1}{24}$$

$$\begin{aligned}\sqrt{145} &\approx f(x) + dy = 12 + \frac{1}{24} \\ &\approx 12.0417\end{aligned}$$

By calculator, $\sqrt{145} \approx 12.0416$.

The difference is $|12.0417 - 12.0416| = 0.0001$.

11. $\sqrt{0.99}$

We know $\sqrt{1} = 1$, so $f(x) = \sqrt{x}$, $x = 1$, $dx = -0.01$.

$$\frac{dy}{dx} = \frac{1}{2}x^{-1/2}$$

$$dy = \frac{1}{2\sqrt{x}}dx$$

$$dy = \frac{1}{2\sqrt{1}}(-0.01) = -0.005$$

$$\begin{aligned}\sqrt{0.99} &\approx f(x) + dy = 1 - 0.005 \\ &= 0.995\end{aligned}$$

By calculator, $\sqrt{0.99} \approx 0.9950$.

The difference is $|0.995 - 0.9950| = 0$.

13. $e^{0.01}$

We know $e^0 = 1$, so $f(x) = e^x$, $x = 0$, $dx = 0.01$.

$$\frac{dy}{dx} = e^x$$

$$dy = e^x dx$$

$$dy = e^0(0.01) = 0.01$$

$$e^{0.01} \approx f(x) + dy = 1 + 0.01 = 1.01$$

By calculator, $e^{0.01} \approx 1.0101$.

The difference is $|1.01 - 1.0101| = 0.0001$.

15. $\ln 1.05$

We know $\ln 1 = 0$, so $f(x) = \ln x$, $x = 1$, $dx = 0.05$.

$$\frac{dy}{dx} = \frac{1}{x}$$

$$dy = \frac{1}{x}dx$$

$$dy = \frac{1}{1}(0.05) = 0.05$$

$$\ln 1.05 \approx f(x) + dy = 0 + 0.05 = 0.05$$

By calculator, $\ln 1.05 \approx 0.0488$.

The difference is $|0.05 - 0.0488| = 0.0012$.

17. Let $D =$ the demand in thousands of pounds;
$x =$ the price in dollars.

$$D(q) = -3q^3 - 2q^2 + 1500$$

(a) $q = 2, \Delta q = 0.10$

$$dD = (-9q^2 - 4q)dq$$
$$\Delta D \approx (-9q^2 - 4q)\,\Delta q$$
$$\approx [-9(4) - 4(2)](0.10)$$
$$\approx -4.4 \text{ thousand pounds}$$

(b) $q = 6, \Delta q = 0.15$

$$\Delta D \approx [-9(36) - 4(6)](0.15)$$
$$\approx -52.2 \text{ thousand pounds}$$

19. $R(x) = 12{,}000 \ln(0.01x + 1)$
$x = 100,\ \Delta x = 1$

$$dR = \frac{12{,}000}{0.01x + 1}(0.01)dx$$
$$\Delta R \approx \frac{120}{0.01x + 1}\Delta x$$
$$\approx \frac{120}{0.01(100) + 1}(1)$$
$$\approx \$60$$

21. If a cube is given a coating 0.1 in. thick, each edge increases in length by twice that amount, or 0.2 in. because there is a face at both ends of the edge.

$$V = x^3, x = 4, \Delta x = 0.2$$
$$dV = 3x^2dx$$
$$\Delta V \approx 3x^2\Delta x$$
$$= 3(4^2)(0.2)$$
$$= 9.6$$

For 1000 cubes $9.6(1000) = 9600 \text{ in.}^3$ of coating should be ordered.

23. (a) $A(x) = y = 0.003631x^3 - 0.03746x^2 + 0.1012x + 0.009$

Let $x = 1, dx = 0.2$.

$$\frac{dy}{dx} = 0.010893x^2 - 0.07492x + 0.1012$$
$$dy = (0.010893x^2 - 0.07492x + 0.1012)dx$$
$$\Delta y \approx (0.010893x^2 - 0.07492x + 0.1012)\Delta x$$
$$\approx (0.010893 \cdot 1^2 - 0.07492 \cdot 1 + 0.1012) \cdot 0.2$$
$$\approx 0.007435$$

The alcohol concentration increases by about 0.74 percent.

(b)

$$\Delta y \approx (0.010893 \cdot 3^2 - 0.07492 \cdot 3 + 0.1012) \cdot 0.2$$
$$\approx -0.005105$$

The alcohol concentration decreases by about 0.51 percent.

25. $P(x) = \dfrac{25x}{8 + x^2}$

$$dP = \frac{(8 + x^2)(25) - 25x(2x)}{(8 + x^2)^2}dx$$
$$= \frac{(8 + x^2)(25) - 25x(2x)}{(8 + x^2)^2}\Delta x$$

(a) $x = 2,\ \Delta x = 0.5$

$$dP = \frac{[(8 + 4)(25) - (25)(2)(4)](0.5)}{(8 + 4)^2}$$
$$= 0.347 \text{ million}$$

(b) $x = 3,\ \Delta x = 0.25$

$$dP = \frac{[(8 + 9)(25) - 25(3)(6)]0.25}{(8 + 9)^2}$$
$$\approx -0.022 \text{ million}$$

27. r changes from 14 mm to 16 mm, so $\Delta r = 2$.

$$V = \frac{4}{3}\pi r^3$$
$$dV = \frac{4}{3}(3)\pi r^2\,dr$$
$$\Delta V \approx 4\pi r^2\,\Delta r$$
$$= 4\pi(14)^2(2)$$
$$= 1568\pi \text{ mm}^3$$

29. r increases from 20 mm to 22 mm, so $\Delta r = 2$.

$$A = \pi r^2$$
$$dA = 2\pi r dr$$
$$\Delta A \approx 2\pi r\,\Delta r$$
$$= 2\pi(20)(2)$$
$$= 80\pi \text{ mm}^2$$

31. $W(t) = -3.5 + 197.5e^{-e^{-0.01394(t-108.4)}}$

(a) $dW = 197.5e^{-e^{-0.01394(t-108.4)}}(-1)e^{-0.01394(t-108.4)}(-0.01394)dt$

$= 2.75315e^{-e^{-0.01394(t-108.4)}}e^{-0.01394(t-108.4)}dt$

We are given $t = 80$ and $dt = 90 - 80 = 10$.

$$dW \approx 9.258$$

The pig will gain about 9.3 kg.

(b) The actual weight gain is calculated as

$$\begin{aligned} W(90) - W(80) &\approx 50.736 - 41.202 \\ &= 9.534 \end{aligned}$$

or about 9.5 kg.

33. $r = 3$ cm, $\Delta r = -0.2$ cm

$$\begin{aligned} V &= \frac{4}{3}\pi r^3 \\ dV &= 4\pi r^2 dr \\ \Delta V &\approx 4\pi r^2 \Delta r = 4\pi(9)(-0.2) = -7.2\pi \text{ cm}^3 \end{aligned}$$

35. $V = \frac{1}{3}\pi r^2 h;\ h = 13,\ dh = 0.2$

$$\begin{aligned} V &= \frac{1}{3}\pi\left(\frac{h}{15}\right)^2 h = \frac{\pi}{775}h^3 \\ dV &= \frac{\pi}{775}\cdot 3h^2 dh = \frac{\pi}{225}h^2 dh \\ \Delta V &\approx \frac{\pi}{225}h^2\Delta h \approx \frac{\pi}{225}(13^2)(0.2) \approx 0.472 \text{ cm}^3 \end{aligned}$$

37. $A = x^2;\ x = 4,\ dA = 0.01$

$$\begin{aligned} dA &= 2x\,dx \\ \Delta A &\approx 2x\,\Delta x \\ \Delta x &\approx \frac{\Delta A}{2x} \approx \frac{0.01}{2(4)} \approx 0.00125 \text{ cm} \end{aligned}$$

39. $V = \frac{4}{3}\pi r^3;\ r = 5.81,\ \Delta r = \pm 0.003$

$$\begin{aligned} dV &= \frac{4}{3}\pi(3r^2)dr \\ \Delta V &\approx \frac{4}{3}\pi(3r^2)\Delta r \\ &= 4\pi(5.81)^2(\pm 0.003) = \pm 0.405\pi \approx \pm 1.273 \text{ in.}^3 \end{aligned}$$

41. $h = 7.284$ in., $r = 1.09 \pm 0.007$ in.

$$\begin{aligned} V &= \frac{1}{3}\pi r^2 h \\ dV &= \frac{2}{3}\pi rh\,dr \\ \Delta V &\approx \frac{2}{3}\pi rh\,\Delta r \\ &= \frac{2}{3}\pi(1.09)(7.284)(0.007) \\ &= \pm 0.116 \text{ in.}^3 \end{aligned}$$

Chapter 14 Review Exercises

1. False: The absolute maximum might occur at the endpoint of the interval of interest.

2. True

3. False: It could have either. For example $f(x) = 1/(1 - x^2)$ has an absolute minimum of 1 on $(-1, 1)$.

4. True

5. True

6. True

7. True

8. True

9. True

10. True

11. $f(x) = -x^3 + 6x^2 + 1; [-1, 6]$

$f'(x) = -3x^2 + 12x = 0$ when $x = 0, 4$.

$$\begin{aligned} f(-1) &= 8 \\ f(0) &= 1 \\ f(4) &= 33 \\ f(6) &= 1 \end{aligned}$$

Absolute maximum of 33 at 4; absolute minimum of 1 at 0 and 6.

13. $f(x) = x^3 + 2x^2 - 15x + 3; [-4, 2]$

$f'(x) = 3x^2 + 4x - 15 = 0$ when

$$(3x - 5)(x + 3) = 0$$

$$x = \frac{5}{3} \quad \text{or} \quad x = -3.$$

$$\begin{aligned} f(-4) &= 31 \\ f(-3) &= 39 \\ f\left(\frac{5}{3}\right) &= -\frac{319}{27} \\ f(2) &= -11 \end{aligned}$$

Absolute maximum of 39 at -3; absolute minimum of $-\frac{319}{27}$ at $\frac{5}{3}$.

17. (a) $f(x) = \dfrac{2\ln x}{x^2}; [1, 4]$

$$\begin{aligned} f'(x) &= \frac{x^2\left(\frac{2}{x}\right) - (2\ln x)(2x)}{x^4} \\ &= \frac{2x - 4x\ln x}{x^4} \\ &= \frac{2 - 4\ln x}{x^3} \end{aligned}$$

$f'(x) = 0$ when

$$\begin{aligned} 2 - 4\ln x &= 0 \\ 2 &= 4\ln x \\ 0.5 &= \ln x \\ e^{0.5} &= x \\ x &\approx 1.6487. \end{aligned}$$

x	$f(x)$
1	0
$e^{0.5}$	0.36788
4	0.17329

Maximum is 0.37; minimum is 0.

(b) $[2, 5]$

Note that the critical number of f is not in the domain, so we only test the endpoints.

x	$f(x)$
2	0.34657
5	0.12876

Maximum is 0.35, minimum is 0.13.

21. $x^2 - 4y^2 = 3x^3y^4$

$$\frac{d}{dx}(x^2 - 4y^2) = \frac{d}{dx}(3x^3y^4)$$
$$2x - 8y\frac{dy}{dx} = 9x^2y^4 + 3x^3 \cdot 4y^3\frac{dy}{dx}$$
$$(-8y - 3x^3 \cdot 4y^3)\frac{dy}{dx} = 9x^2y^4 - 2x$$
$$\frac{dy}{dx} = \frac{2x - 9x^2y^4}{8y + 12x^3y^3}$$

23. $2\sqrt{y-1} = 9x^{2/3} + y$

$$\frac{d}{dx}[2(y-1)^{1/2}] = \frac{d}{dx}(9x^{2/3} + y)$$
$$2 \cdot \frac{1}{2} \cdot (y-1)^{-1/2}\frac{dy}{dx} = 6x^{-1/3} + \frac{dy}{dx}$$
$$[(y-1)^{-1/2} - 1]\frac{dy}{dx} = 6x^{-1/3}$$
$$\frac{1 - \sqrt{y-1}}{\sqrt{y-1}} \cdot \frac{dy}{dx} = \frac{6}{x^{1/3}}$$
$$\frac{dy}{dx} = \frac{6\sqrt{y-1}}{x^{1/3}(1 - \sqrt{y-1})}$$

25. $\frac{6 + 5x}{2 - 3y} = \frac{1}{5x}$

$$5x(6 + 5x) = 2 - 3y$$
$$30x + 25x^2 = 2 - 3y$$
$$\frac{d}{dx}(30x + 25x^2) = \frac{d}{dx}(2 - 3y)$$
$$30 + 50x = -3\frac{dy}{dx}$$
$$-\frac{30 + 50x}{3} = \frac{dy}{dx}$$

27. $\ln(xy + 1) = 2xy^3 + 4$

$$\frac{d}{dx}[\ln(xy + 1)] = \frac{d}{dx}(2xy^3 + 4)$$
$$\frac{1}{xy + 1} \cdot \frac{d}{dx}(xy + 1) = 2y^3 + 2x \cdot 3y^2\frac{dy}{dx} + \frac{d}{dx}(4)$$
$$\frac{1}{xy + 1}\left(y + x\frac{dy}{dx} + \frac{d}{dx}(1)\right) = 2y^3 + 6xy^2\frac{dy}{dx}$$
$$\frac{y}{xy + 1} + \frac{x}{xy + 1} \cdot \frac{dy}{dx} = 2y^3 + 6xy^2\frac{dy}{dx}$$
$$\left(\frac{x}{xy + 1} - 6xy^2\right)\frac{dy}{dx} = 2y^3 - \frac{y}{xy + 1}$$
$$\frac{dy}{dx} = \frac{2y^3 - \frac{y}{xy + 1}}{\frac{x}{xy + 1} - 6xy^2}$$
$$= \frac{2y^3(xy + 1) - y}{x - 6xy^2(xy + 1)}$$
$$= \frac{2xy^4 + 2y^3 - y}{x - 6x^2y^3 - 6xy^2}$$

29. $\sqrt{2y} - 4xy = -22$, tangent line at $(3, 2)$.

$$\frac{d}{dx}\left(\sqrt{2y} - 4xy\right) = \frac{d}{dx}(-22)$$
$$\frac{1}{2}(2)(2y)^{-1/2}\frac{dy}{dx} - \left(4y + 4x\frac{dy}{dx}\right) = 0$$
$$((2y)^{-1/2} - 4x)\frac{dy}{dx} = 4y$$
$$\frac{dy}{dx} = \frac{4y}{\frac{1}{\sqrt{2y}} - 4x}$$

To find the slope m of the tangent line, substitute 3 for x and 2 for y.

$$m = \frac{4y}{\frac{1}{2\sqrt{2y}} - 4x} = \frac{4(2)}{\frac{1}{\sqrt{2(2)}} - 4(3)}$$
$$= \frac{8}{\frac{1}{2} - 12} = \frac{16}{1 - 24} = -\frac{16}{23}$$

The equation of the tangent line is

$$y - y_1 = m(x - x_1)$$
$$y - 2 = -\frac{16}{23}(x - 3)$$
$$y - 2 - \frac{48}{23} = -\frac{16}{23}x$$
$$y = -\frac{16}{23}x + \frac{94}{23}.$$

We can also write this equation as $16x + 23y = 94$.

33. $y = 8x^3 - 7x^2,\ \frac{dx}{dt} = 4,\ x = 2$

$$\begin{aligned}\frac{dy}{dt} &= \frac{d}{dt}(8x^3 - 7x^2)\\ &= 24x^2\frac{dx}{dt} - 14x\frac{dx}{dt}\\ &= 24(2)^2(4) - 14(2)(4)\\ &= 272\end{aligned}$$

35. $y = \frac{1+\sqrt{x}}{1-\sqrt{x}},\ \frac{dx}{dt} = -4,\ x = 4$

$$\frac{dy}{dt} = \frac{d}{dt}\left[\frac{1+\sqrt{x}}{1-\sqrt{x}}\right]$$

$$= \frac{\left[\begin{aligned}&\left(1-\sqrt{x}\right)\left(\tfrac{1}{2}x^{-1/2}\tfrac{dx}{dt}\right)\\ &-1\left(1+\sqrt{x}\right)\left(-\tfrac{1}{2}\right)\left(x^{-1/2}\tfrac{dx}{dt}\right)\end{aligned}\right]}{\left(1-\sqrt{x}\right)^2}$$

$$= \frac{\left[\begin{aligned}&(1-2)\left(\tfrac{1}{2\cdot 2}\right)(-4)\\ &-(1+2)\left(\tfrac{-1}{2\cdot 2}\right)(-4)\end{aligned}\right]}{(1-2)^2}$$

$$= \frac{1-3}{1} = -2$$

37. $y = xe^{3x};\ \frac{dx}{dt} = -2, x = 1$

$$\begin{aligned}\frac{dy}{dt} &= \frac{d}{dt}(xe^{3x})\\ &= \frac{dx}{dt}\cdot e^{3x} + x\cdot\frac{d}{dt}(e^{3x})\\ &= \frac{dx}{dt}\cdot e^{3x} + xe^{3x}\cdot 3\frac{dx}{dt}\\ &= (1+3x)e^{3x}\frac{dx}{dt}\\ &= (1+3\cdot 1)e^{3(1)}(-2) = -8e^3\end{aligned}$$

41. $y = \frac{3x-7}{2x+1}; x = 2, \Delta x = 0.003$

$$dy = \frac{(3)(2x+1) - (2)(3x-7)}{(2x+1)^2}dx$$

$$dy = \frac{17}{(2x+1)^2}dx$$

$$\begin{aligned}\Delta y &\approx \frac{17}{(2x+1)^2}\Delta x\\ &= \frac{17}{(2[2]+1)^2}(0.003)\\ &= 0.00204\end{aligned}$$

43. $-12x + x^3 + y + y^2 = 4$

$$\begin{aligned}\frac{dy}{dx}(-12x + x^3 + y + y^2) &= \frac{d}{dx}(4)\\ -12 + 3x^2 + \frac{dy}{dx} + 2y\frac{dy}{dx} &= 0\\ (1+2y)\frac{dy}{dx} &= 12 - 3x^2\\ \frac{dy}{dx} &= \frac{12-3x^2}{1+2y}\end{aligned}$$

(a) If $\frac{dy}{dx} = 0$,

$$\begin{aligned}12 - 3x^2 &= 0\\ 12 &= 3x^2\\ \pm 2 &= x.\end{aligned}$$

$x = 2$:

$$\begin{aligned}-24 + 8 + y + y^2 &= 4\\ y + y^2 &= 20\\ y^2 + y - 20 &= 0\\ (y+5)(y-4) &= 0\\ y = -5 \text{ or } y &= 4\end{aligned}$$

$(2, -5)$ and $(2, 4)$ are critical points.

$x = -2$:

$$\begin{aligned}24 - 8 + y + y^2 &= 4\\ y + y^2 &= -12\\ y^2 + y + 12 &= 0\\ y &= \frac{-1 \pm \sqrt{1^2 - 48}}{2}\end{aligned}$$

This leads to imaginary roots.

$x = -2$ does not produce critical points.

(b)

x	y_1	y_2
1.9	−4.99	3.99
2	−5	4
2.1	−4.99	3.99

The point $(2, -5)$ is a relative minimum.

The point $(2, 4)$ is a relative maximum.

(c) There is no absolute maximum or minimum for x or y.

45. (a)

$$P(x) = -x^3 + 10x^2 - 12x$$
$$P'(x) = -3x^2 + 20x - 12 = 0$$
$$3x^2 - 20x + 12 = 0$$
$$(3x - 2)(x - 6) = 0$$
$$3x - 2 = 0 \quad \text{or} \quad x - 6 = 0$$
$$x = \frac{2}{3} \quad \text{or} \quad x = 6$$
$$P''(x) = -6x + 20$$
$$P''\left(\frac{2}{3}\right) = 16,$$

which implies that $x = \frac{2}{3}$ is the location of the minimum.

$$P''(6) = -16,$$

which implies that $x = 6$ is the location of the maximum. Thus, 600 boxes will produce a maximum profit.

(b) Maximum profit $= P(6)$

$$= -(6)^3 + 10(6)^2 - 12(6) = 72$$

The maximum profit is \$720.

47. Volume of cylinder $= \pi r^2 h$

Surface area of cylinder open at one end $= 2\pi rh + \pi r^2$.

$$V = \pi r^2 h = 27\pi$$
$$h = \frac{27\pi}{\pi r^2} = \frac{27}{r^2}$$
$$A = 2\pi r\left(\frac{27}{r^2}\right) + \pi r^2$$
$$= 54\pi r^{-1} + \pi r^2$$
$$A' = -54\pi r^{-2} + 2\pi r$$

If $A' = 0$,

$$2\pi r = \frac{54\pi}{r^2}$$
$$r^3 = 27$$
$$r = 3.$$

If $r = 3$,

$$A'' = 108\pi r^{-3} + 2\pi > 0,$$

so the value at $r = 3$ is a minimum.

For the maximum cost, the radius of the bottom should be 3 inches.

49. Here $k = 0.15, M = 20{,}000,$ and $f = 12$. We have

$$q = \sqrt{\frac{2fM}{k}} = \sqrt{\frac{2(12)20{,}000}{0.15}}$$
$$= \sqrt{3{,}200{,}000} \approx 1789$$

Ordering 1789 rolls each time minimizes annual cost.

51. Use equation (3) from Section 6.3 with $k = 1$, $M = 128{,}000$, and $f = 10$.

$$q = \sqrt{\frac{2fM}{k}} = \sqrt{\frac{2(10)(128{,}000)}{1}}$$
$$= \sqrt{2{,}560{,}000} = 1600$$

The number of lots that should be produced annually is

$$\frac{M}{q} = \frac{128{,}000}{1600} = 80.$$

53.

$$\ln q = D - 0.447 \ln p$$
$$\frac{1}{q}\frac{dq}{dp} = -\frac{0.47}{p}$$
$$\frac{dp}{dq} = -0.47\frac{q}{p}$$
$$E = -\frac{p}{q}\frac{dq}{dp}$$
$$= -\frac{p}{q}\left(-0.47\frac{q}{p}\right)$$
$$= 0.47$$

E is less than 1, so the demand is inelastic.

55. $A = \pi r^2$; $\frac{dr}{dt} = 4$ ft/min, $r = 7$ ft

$$\frac{dA}{dt} = 2\pi r \frac{dr}{dt}$$

$$\frac{dA}{dt} = 2\pi(7)(4) = 56\pi$$

The rate of change of the area is 56π ft^2/min.

57. (a)

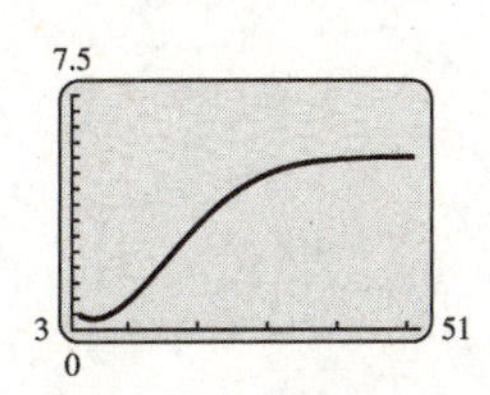

(b) We use a graphing calculator to graph

$$M'(t) = -0.4321173 + 0.1129024t - 0.0061518t^2 + 0.0001260t^3 - 0.0000008925t^4$$

on [3,51] by [0,0.3]. We find the maximum value of $M'(t)$ on this graph at about 15.41, or on about the 15th day.

59. Let x = the distance from the base of the ladder to the building;

y = the height on the building at the top of the ladder.

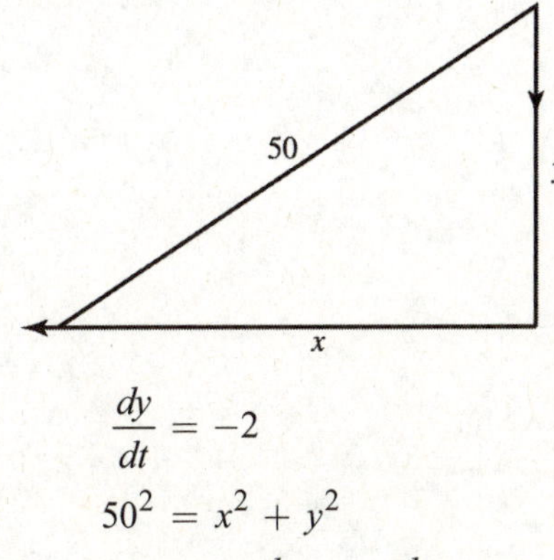

$$\frac{dy}{dt} = -2$$

$$50^2 = x^2 + y^2$$

$$0 = 2x\frac{dx}{dt} + 2y\frac{dy}{dt}$$

$$\frac{dx}{dt} = -\frac{y}{x}\frac{dy}{dt}$$

When $x = 30, y = \sqrt{2500 - (30)^2} = 40.$

So

$$\frac{dx}{dt} = \frac{-40}{30}(-2) = \frac{80}{30} = \frac{8}{3}$$

The base of the ladder is slipping away from the building at a rate of $\frac{8}{3}$ ft/min.

61. Let x = one-half of the width of the triangular cross section;

h = the height of the water;

V = the volume of the water.

$\frac{dV}{dt} = 3.5$ ft^3/min.

Find $\frac{dV}{dt}$ when $h = \frac{1}{3}$.

$$V = \begin{pmatrix}\text{Area of} \\ \text{triangular} \\ \text{side}\end{pmatrix}(\text{length})$$

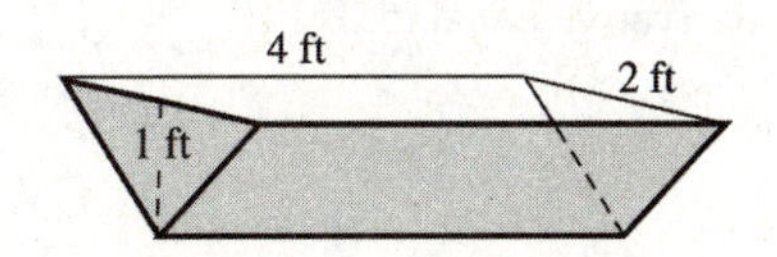

Area of triangular cross section

$$= \frac{1}{2}(\text{base})(\text{altitude})$$

$$= \frac{1}{2}(2x)(h) = xh$$

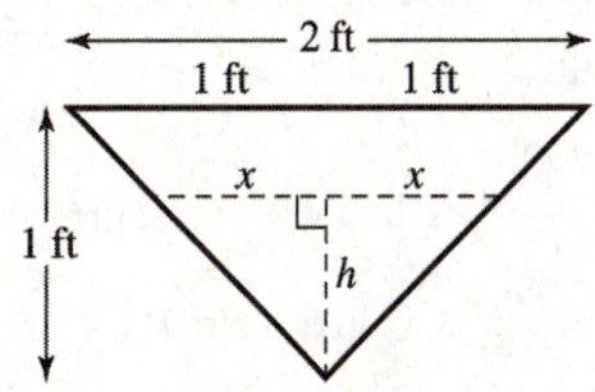

By similar triangles, $\frac{2x}{h} = \frac{2}{1}$, so $x = h$.

$$V = (xh)(4)$$

$$= h^2 \cdot 4$$

$$= 4h^2$$

$$\frac{dV}{dt} = 8h\frac{dh}{dt}$$

$$\frac{1}{8h} \cdot \frac{dV}{dt} = \frac{dh}{dt}$$

$$\frac{1}{8\left(\frac{1}{3}\right)}(3.5) = \frac{dh}{dt}$$

$$\frac{dh}{dt} = \frac{21}{16} = 1.3125$$

The depth of water is changing at the rate of 1.3125 ft/min.

63. $A = s^2;\ s = 9.2,\ \Delta s = \pm 0.04$

$$\begin{aligned} ds &= 2s\,ds \\ \Delta A &\approx 2s\Delta s \\ &= 2(9.2)(\pm 0.04) \\ &= \pm 0.736 \text{ in.}^2 \end{aligned}$$

65. We need to minimize y. Note that $x > 0$.

$$\frac{dy}{dx} = \frac{x}{8} - \frac{2}{x}$$

Set the derivative equal to 0.

$$\begin{aligned} \frac{x}{8} - \frac{2}{x} &= 0 \\ \frac{x}{8} &= \frac{2}{x} \\ x^2 &= 16 \\ x &= 4 \end{aligned}$$

Since $\lim_{x\to 0} y = \infty$, $\lim_{x\to\infty} y = \infty$, and $x = 4$ is the only critical value in $(0, \infty)$, $x = 4$ produces a minimum value.

$$\begin{aligned} y &= \frac{4^2}{16} - 2\ln 4 + \frac{1}{4} + 2\ln 6 \\ &= 1.25 + 2(\ln 6 - \ln 4) \\ &= 1.25 + 2\ln 1.5 \end{aligned}$$

The y coordinate of the Southern most point of the second boat's path is$1.25 + 2\ln 1.5$.

67. Distance on shore: $40 - x$ feet

Speed on shore: 5 feet per second

Distance in water: $\sqrt{x^2 + 40^2}$ feet

Speed in water: 3 feet for second

The total travel time t is

$$t = t_1 + t_2 = \frac{d_1}{v_1} + \frac{d_2}{v_2}.$$

$$\begin{aligned} t(x) &= \frac{40 - x}{5} + \frac{\sqrt{x^2 + 40^2}}{3} \\ &= 8 - \frac{x}{5} + \frac{\sqrt{x^2 + 1600}}{3} \\ t'(x) &= -\frac{1}{5} + \frac{1}{3} \cdot \frac{1}{2}(x^2 + 1600)^{-1/2}(2x) \\ &= -\frac{1}{5} + \frac{x}{3\sqrt{x^2 + 1600}} \end{aligned}$$

Minimize the travel time $t(x)$. If $t'(x) = 0$:

$$\begin{aligned} \frac{x}{3\sqrt{x^2 + 1600}} &= \frac{1}{5} \\ 5x &= 3\sqrt{x^2 + 1600} \\ \frac{5x}{3} &= \sqrt{x^2 + 1600} \\ \frac{25}{9}x^2 &= x^2 + 1600 \\ x^2 &= \frac{1600 \cdot 9}{16} \\ x &= \frac{40 \cdot 3}{4} = 30 \end{aligned}$$

(Discard the negative solution.)

To minimize the time, he should walk $40 - x = 40 - 30 = 10$ ft along the shore before paddling toward the desired destination. The minimum travel time is

$$\frac{40 - 30}{5} + \frac{\sqrt{30^2 + 40^2}}{3} \approx 18.67 \text{ seconds.}$$

Chapter 15

INTEGRATION

15.1 Antiderivatives

Your Turn 1

Find an antiderivative for the function $f(x) = 8x^7$.

Since the derivative of x^n is nx^{n-1}, the derivative of x^8 is $8x^7$. Thus x^8 is an antiderivative of $8x^7$. The general antiderivative is $x^8 + C$.

Your Turn 2

Find $\int \frac{1}{t^4}\,dt$.

Use the power rule with $n = -4$.

$$\int \frac{1}{t^4}\,dt = \int t^{-4}\,dt$$
$$= \frac{t^{-4+1}}{-4+1} + C$$
$$= \frac{t^{-3}}{-3} + C$$
$$= -\frac{1}{3t^3} + C$$

Your Turn 3

Find $\int (6x^2 + 8x - 9)\,dx$.

Use the sum or difference rule and the constant multiple rule.

$$\int (6x^2 + 8x - 9)\,dx = \int 6x^2\,dx + \int 8x\,dx - \int 9\,dx$$
$$= 6\int x^2\,dx + 8\int x\,dx - 9\int dx$$

Now use the power rule on each term.

$$6\int x^2\,dx + 8\int x\,dx - 9\int dx$$
$$= 6\left(\frac{x^3}{3}\right) + 8\left(\frac{x^2}{2}\right) - 9x + C$$
$$= 2x^3 + 4x^2 - 9x + C$$

Your Turn 4

$$\int \frac{x^3 - 2}{\sqrt{x}}\,dx = \int \left(\frac{x^3}{\sqrt{x}} - \frac{2}{\sqrt{x}}\right)$$
$$= \int \frac{x^3}{x^{1/2}}\,dx - \int \frac{2}{x^{1/2}}\,dx$$
$$= \int x^{5/2}\,dx - 2\int x^{-1/2}\,dx$$
$$= \frac{2}{7}x^{7/2} - 2\left(\frac{2}{1}x^{1/2}\right) + C$$
$$= \frac{2}{7}x^{7/2} - 4x^{1/2} + C$$

Your Turn 5

$$\int \left(\frac{3}{x} + e^{-3x}\right)dx = \int \frac{3}{x}\,dx + \int e^{-3x}\,dx$$
$$= 3\int \frac{1}{x}\,dx + \int e^{-3x}\,dx$$
$$= 3\ln|x| - \frac{1}{3}e^{-3x} + C$$

Your Turn 6

Suppose an object is thrown down from the top of the 2717-ft tall Burj Khalifa with an initial velocity of -20 ft/sec. Find when it hits the ground and how fast it is traveling when it hits the ground.

In Example 11 (b) we derived the formulas

$$v(t) = -32t - 20$$
$$s(t) = -16t^2 - 20t + 1100$$

for the velocity v and distance above the ground s for an object thrown down from the Willis Tower. The only change required for the new problem is to change 1100 to the height of the Burj Khalifa, 2717, so we have

$$s(t) = -16t^2 - 20t + 2717.$$

To find when the object hits the ground we solve

$$s(t) = -16t^2 - 20t + 2717 = 0.$$

Use the quadratic formula.

$$t = \frac{20 \pm \sqrt{20^2 - 4(-16)(2717)}}{2(-16)}$$

$$t \approx -13.62 \quad \text{or} \quad t \approx 12.42$$

Only the positive root is relevant. To find the speed on impact, substitute $t = 12.42$ into the formula for v.

$$\begin{aligned} v(12.42) &= -32(12.42) - 20 \\ &\approx -417 \end{aligned}$$

The object hits the ground after 12.42 sec, traveling downward at 417 ft/sec.

Your Turn 7

$$f'(x) = 3x^{1/2} + 4$$

$$\begin{aligned} f(x) &= \int f'(x)\,dx \\ &= \int (3x^{1/2} + 4)\,dx \\ &= 3\int x^{1/2}\,dx + 4\int dx \\ &= 3\left(\frac{2}{3}x^{3/2}\right) + 4x + C \\ &= 2x^{3/2} + 4x + C \end{aligned}$$

Since the graph of f is to go through the point $(1, -2)$, $f(1) = -2$.

$$\begin{aligned} 2(1)^{3/2} + 4(1) + C &= -2 \\ 2 + 4 + C &= -2 \\ C &= -8 \end{aligned}$$

Thus,

$$f(x) = 2x^{3/2} + 4x - 8.$$

15.1 Exercises

1. If $F(x)$ and $G(x)$ are both antiderivatives of $f(x)$, then there is a constant C such that

$$F(x) - G(x) = C.$$

The two functions can differ only by a constant.

5.
$$\begin{aligned} \int 6\,dk &= 6\int 1\,dk \\ &= 6\int k^0\,dk \\ &= 6 \cdot \frac{1}{1}k^{0+1} + C \\ &= 6k + C \end{aligned}$$

7.
$$\begin{aligned} &\int (2z + 3)\,dz \\ &= 2\int z\,dz + 3\int z^0\,dz \\ &= 2 \cdot \frac{1}{1+1}z^{1+1} + 3 \cdot \frac{1}{0+1}z^{0+1} + C \\ &= z^2 + 3z + C \end{aligned}$$

9.
$$\begin{aligned} &\int (6t^2 - 8t + 7)\,dt \\ &= 6\int t^2\,dt - 8\int t\,dt + 7t\int t^0\,dt \\ &= \frac{6t^3}{3} - \frac{8t^2}{2} + 7t + C \\ &= 2t^3 - 4t^2 + 7t + C \end{aligned}$$

11.
$$\begin{aligned} &\int (4z^3 + 3z^2 + 2z - 6)\,dz \\ &= 4\int z^3\,dz + 3\int z^2\,dz + 2\int z\,dz \\ &\quad -6\int z^0\,dz \\ &= \frac{4z^4}{4} + \frac{3z^3}{3} + \frac{2z^2}{2} - 6z + C \\ &= z^4 + z^3 + z^2 - 6z + C \end{aligned}$$

13.
$$\begin{aligned} \int (5\sqrt{z} + \sqrt{2})\,dz &= 5\int z^{1/2}\,dz + \sqrt{2}\int dz \\ &= \frac{5z^{3/2}}{\frac{3}{2}} + \sqrt{2}z + C \\ &= 5\left(\frac{2}{3}\right)z^{3/2} + \sqrt{2}z + C \\ &= \frac{10z^{3/2}}{3} + \sqrt{2}z + C \end{aligned}$$

15.
$$\begin{aligned} \int 5x(x^2 - 8)\,dx &= \int (5x^3 - 40x)\,dx \\ &= \frac{5x^4}{4} - \frac{40x^2}{2} + C \\ &= \frac{5x^4}{4} - 20x^2 + C \end{aligned}$$

17. $\int (4\sqrt{v} - 3v^{3/2})\,dv$

$$= 4\int v^{1/2}dv - 3\int v^{3/2}dv$$

$$= \frac{4v^{3/2}}{\frac{3}{2}} - \frac{3v^{5/2}}{\frac{5}{2}} + C$$

$$= \frac{8v^{3/2}}{3} - \frac{6v^{5/2}}{5} + C$$

19. $\int (10u^{3/2} - 14u^{5/2})\,du$

$$= 10\int u^{3/2}du - 14\int u^{5/2}\,du$$

$$= \frac{10u^{5/2}}{\frac{5}{2}} - \frac{14u^{7/2}}{\frac{7}{2}} + C$$

$$= 10\left(\frac{2}{5}\right)u^{5/2} - 14\left(\frac{2}{7}\right)u^{7/2} + C$$

$$= 4u^{5/2} - 4u^{7/2} + C$$

21. $\int \left(\frac{7}{z^2}\right)dz = \int 7z^{-2}dz$

$$= 7\int z^{-2}dz$$

$$= 7\left(\frac{z^{-2+1}}{-2+1}\right) + C$$

$$= \frac{7z^{-1}}{-1} + C$$

$$= -\frac{7}{z} + C$$

23. $\int \left(\frac{\pi^3}{y^3} - \frac{\sqrt{\pi}}{\sqrt{y}}\right)dy$

$$= \int \pi^3 y^{-3}dy - \int \sqrt{\pi}\, y^{-1/2}dy$$

$$= \pi^3\int y^{-3}dy - \sqrt{\pi}\int y^{-1/2}dy$$

$$= \pi^3\left(\frac{y^{-2}}{-2}\right) - \sqrt{\pi}\left(\frac{y^{-1/2}}{\frac{1}{2}}\right) + C$$

$$= -\frac{\pi^3}{2y^2} - 2\sqrt{\pi y} + C$$

25. $\int (-9t^{-2.5} - 2t^{-1})\,dt$

$$= -9\int t^{-2.5}dt - 2\int t^{-1}dt$$

$$= \frac{-9t^{-1.5}}{-1.5} - 2\int \frac{dt}{t}$$

$$= 6t^{-1.5} - 2\ln|t| + C$$

27. $\int \frac{1}{3x^2}dx = \int \frac{1}{3}x^{-2}dx$

$$= \frac{1}{3}\int x^{-2}dx$$

$$= \frac{1}{3}\left(\frac{x^{-1}}{-1}\right) + C$$

$$= -\frac{1}{3}x^{-1} + C$$

$$= -\frac{1}{3x} + C$$

29. $\int 3e^{-0.2x}dx = 3\int e^{-0.2x}dx$

$$= 3\left(\frac{1}{-0.2}\right)e^{-0.2x} + C$$

$$= \frac{3(e^{-0.2x})}{-0.2} + C$$

$$= -15e^{-0.2x} + C$$

31. $\int \left(-\frac{3}{x} + 4e^{-0.4x} + e^{0.1}\right)dx$

$$= -3\int \frac{dx}{x} + 4\int e^{-0.4x}dx + e^{0.1}\int dx$$

$$= -3\ln|x| + \frac{4e^{-0.4x}}{-0.4} + e^{0.1}x + C$$

$$= -3\ln|x| - 10e^{-0.4x} + e^{0.1}x + C$$

33. $\int \left(\frac{1+2t^3}{4t}\right)dt = \int \left(\frac{1}{4t} + \frac{t^2}{2}\right)dt$

$$= \frac{1}{4}\int \frac{1}{t}\,dt + \frac{1}{2}\int t^2 dt$$

$$= \frac{1}{4}\ln|t| + \frac{1}{2}\left(\frac{t^3}{3}\right) + C$$

$$= \frac{1}{4}\ln|t| + \frac{t^3}{6} + C$$

35. $\int (e^{2u} + 4u)\,du = \frac{e^{2u}}{2} + \frac{4u^2}{2} + C$

$= \frac{e^{2u}}{2} + 2u^2 + C$

37. $\int (x+1)^2 dx = \int (x^2 + 2x + 1)\,dx$

$= \frac{x^3}{3} + \frac{2x^2}{2} + x + C$

$= \frac{x^3}{3} + x^2 + x + C$

39. $\int \frac{\sqrt{x}+1}{\sqrt[3]{x}}\,dx = \int \left(\frac{\sqrt{x}}{\sqrt[3]{x}} + \frac{1}{\sqrt[3]{x}}\right) dx$

$= \int \left(x^{(1/2-1/3)} + x^{-1/3}\right) dx$

$= \int x^{1/6} dx + \int x^{-1/3} dx$

$= \frac{x^{7/6}}{\frac{7}{6}} + \frac{x^{2/3}}{\frac{2}{3}} + C$

$= \frac{6x^{7/6}}{7} + \frac{3x^{2/3}}{2} + C$

41. $\int 10^x dx = \frac{10^x}{\ln 10} + C$

43. Find $f(x)$ such that $f'(x) = x^{2/3}$, and $\left(1, \frac{3}{5}\right)$ is on the curve.

$\int x^{2/3} dx = \frac{x^{5/3}}{\frac{5}{3}} + C$

$f(x) = \frac{3x^{5/3}}{5} + C$

Since $\left(1, \frac{3}{5}\right)$ is on the curve,

$f(1) = \frac{3}{5}.$

$f(1) = \frac{3(1)^{5/3}}{5} + C = \frac{3}{5}$

$\frac{3}{5} + C = \frac{3}{5}$

$C = 0.$

Thus,

$f(x) = \frac{3x^{5/3}}{5}.$

45. $C'(x) = 4x - 5$; fixed cost is \$8.

$C(x) = \int (4x - 5)\,dx$

$= \frac{4x^2}{2} - 5x + k$

$= 2x^2 - 5x + k$

$C(0) = 2(0)^2 - 5(0) + k = k$

Since $C(0) = 8, k = 8.$

Thus,

$C(x) = 2x^2 - 5x + 8.$

47. $C'(x) = 0.03e^{0.01x}$; fixed cost \$8.

$C(x) = \int 0.03e^{0.01x} dx$

$= 0.03 \int e^{0.01x} dx$

$= 0.03\left(\frac{1}{0.01} e^{0.01x}\right) + k$

$= 3e^{0.01x} + k$

$C(0) = 3e^{0.01(0)} + k = 3(1) + k$

$= 3 + k$

Since $C(0) = 8, 3 + k = 8$, and $k = 5$.

Thus,

$C(x) = 3e^{0.01x} + 5.$

49. $C'(x) = x^{2/3} + 2$; 8 units cost \$58.

$C(x) = \int (x^{2/3} + 2)dx$

$= \frac{3x^{5/3}}{5} + 2x + k$

$C(8) = \frac{3(8)^{5/3}}{5} + 2(8) + k$

$= \frac{3(32)}{5} + 16 + k$

Since $C(8) = 58$,

$58 - 16 - \frac{96}{5} = k$

$\frac{114}{5} = k.$

Thus,

$C(x) = \frac{3x^{5/3}}{5} + 2x + \frac{114}{5}.$

51. $C'(x) = 5x - \frac{1}{x}$; 10 units cost \$94.20, so

$C(10) = 94.20.$

$$C(x) = \int \left(5x - \frac{1}{x}\right) dx = \frac{5x^2}{2} - \ln|x| + k$$

$$C(10) = \frac{5(10)^2}{2} - \ln(10) + k$$
$$= 250 - 2.30 + k.$$

Since $C(10) = 94.20,$

$$94.20 = 247.70 + k$$
$$-153.50 = k.$$

Thus, $C(x) = \frac{5x^2}{2} - \ln|x| - 153.50.$

53. $R'(x) = 175 - 0.02x - 0.03x^2$

$$R = \int (175 - 0.02x - 0.03x^2)\,dx$$
$$= 175x - 0.01x^2 - 0.01x^3 + C.$$

If $x = 0$, then $R = 0$ (no items sold means no revenue), and

$$0 = 175(0) - 0.01(0)^2 - 0.01(0)^3 + C$$
$$0 = C.$$

Thus, $R = 175x - 0.01x^2 - 0.01x^3$ gives the revenue function. Now, recall that $R = xp$, where p is the demand function. Then

$$175x - 0.01x^2 - 0.01x^3 = xp$$
$$175 - 0.01x - 0.01x^2 = p,$$

the demand function.

55. $R'(x) = 500 - 0.15\sqrt{x}$

$$R = \int (500 - 0.015\sqrt{x})\,dx$$
$$= 500x - 0.1x^{3/2} + C.$$

If $x = 0$, $R = 0$ (no items sold means no revenue), and

$$0 = 500(0) - 0.1(0)^{3/2} + C$$
$$0 = C.$$

Thus, $R = 500x - 0.1x^{3/2}$ gives the revenue function. Now, recall that $R = xp$, where p is the demand function. Then

$$500x - 0.1x^{3/2} = xp$$
$$500 - 0.1\sqrt{x} = p, \text{ the demand function.}$$

57. $f'(t) = 7.50t - 16.8$

(a) $f(t) = \int (7.50t - 16.8)\,dt$

$$= \frac{7.50}{2}t^2 - 16.8t + C$$
$$= 3.75t^2 + 16.8t + C$$

In 2005($t = 5$), $f(t) = 9.8$ and

$$9.8 = 3.75(5)^2 - 16.8(5) + C$$
$$9.8 = 9.75 + C$$
$$0.05 = C$$

Thus, $f(t) = 3.75t^2 - 16.8t + 0.05.$

(b) In 2009, $t = 9$ and

$$f(9) = 3.75(9)^2 - 16.8(9) + 0.05 = 152.6$$

The function derived in (a) predicts 152.6 billion monthly text messages, quite close to the actual value of 152.7 billion messages.

59. **(a)** $P'(x) = 50x^3 + 30x^2$; profit is -40 when no cheese is sold.

$$P(x) = \int (50x^3 + 30x^2)\,dx$$
$$= \frac{25x^4}{2} + 10x^3 + k$$

$$P(0) = \frac{25(0)^4}{2} + 10(0)^3 + k$$

Since

$$P(0) = -40,$$
$$-40 = k.$$

Thus,

$$P(x) = \frac{25x^4}{2} + 10x^3 - 40.$$

(b) $P(2) = \frac{25(2)^4}{2} + 10(2)^3 - 40 = 240$

The profit from selling 200 lbs of Brie cheese is \$240.

61.

$$\int \frac{g(x)}{x}\,dx = \int \frac{a - bx}{x}\,dx$$
$$= \int \left(\frac{a}{x} - b\right) dx$$
$$= a\int \frac{dx}{x} - b\int dx$$
$$= a\ln|x| - bx + C$$

Since x represents a positive quantity, the absolute value sign can be dropped.

$$\int \frac{g(x)}{x}\,dx = a \ln x - bx + C$$

63. $N'(t) = Ae^{kt}$

(a) $N(t) = \frac{A}{k}e^{kt} + C$

$A = 50,\ N(t) = 300$ when $t = 0$.

$$N(0) = \frac{50}{k}e^{0} + C = 300$$
$$N'(5) = 250$$

Therefore,

$$N'(5) = 50e^{5k} = 250$$
$$e^{5k} = 5$$
$$5k = \ln 5$$
$$k = \frac{\ln 5}{5}.$$

$$N(0) = \frac{50}{\frac{\ln 5}{5}} + C = 300$$
$$\frac{250}{\ln 5} + C = 300$$
$$C = 300 - \frac{250}{\ln 5} \approx 144.67$$

$$N(t) = \frac{50}{\frac{\ln 5}{5}}e^{(\ln 5/5)t} + 144.67$$
$$= 155.3337e^{0.321888t} + 144.67$$
$$\approx 155.3e^{0.3219} + 144.7$$

(b) $N(12) = 155.3337e^{0.321888(12)} + 144.67$
≈ 7537

There are 7537 cells present after 12 days.

65. $B'(t) = 0.06048t^2 - 1.292t + 15.86$

(a) $B(t) = \int (0.06048t^2 - 1.292t + 15.86)\,dt$

$$= \frac{0.06048}{3}t^3 - \frac{1.292}{2}t^2 + 15.86t + C$$
$$= 0.02016t^3 - 0.6460t^2 + 15.86t + C$$

In 1970, when $t = 0$, 839,700 or about 839.7 thousand degrees were conferred, so $B(0) = 839.7$ and thus $C = 839.7$ and the formula for B is

$$B(t) = 0.02016t^3 - 0.6460t^2 + 15.86t + 839.7.$$

(b) To project the number of bachelor's degrees conferred in 2015 we set t equal to 45 and evaluate $B(45)$.

$$B(45) = 0.02016(45)^3 - 0.6460(45)^2 + 15.86(45) + 839.7$$
$$\approx 2082$$

The formula predicts that 2082 thousand or about 2,082,000 bachelor's degrees will be conferred in 2015.

67. $a(t) = 5t^2 + 4$

$$v(t) = \int (5t^2 + 4)\,dt$$
$$= \frac{5t^3}{3} + 4t + C$$
$$v(0) = \frac{5(0)^3}{3} + 4(0) + C$$

Since $v(0) = 6, C = 6$.

$$v(t) = \frac{5t^3}{3} + 4t + 6$$

69. $a(t) = -32$

$$v(t) = \int -32\,dt = -32t + C_1$$
$$v(0) = -32(0) + C_1$$

Since $v(0) = 0, C_1 = 0$.

$$v(t) = -32t$$
$$s(t) = \int -32t\,dt$$
$$= \frac{-32t^2}{2} + C_2$$
$$= -16t^2 + C_2$$

At $t = 0$, the plane is at 6400 ft.

That is, $s(0) = 6400$.

$$s(0) = -16(0)^2 + C_2$$
$$6400 = 0 + C_2$$
$$C_2 = 6400$$
$$s(t) = -16t^2 + 6400$$

When the object hits the ground, $s(t) = 0$.

$$-16t^2 + 6400 = 0$$
$$-16t^2 = -6400$$
$$t^2 = 400$$
$$t = \pm 20$$

Discard -20 since time must be positive.

The object hits the ground in 20 sec.

71. $a(t) = \frac{15}{2}\sqrt{t} = 3e^{-t}$

$$v(t) = \int \left(\frac{15}{2}\sqrt{t} + 3e^{-t}\right) dt$$
$$= \int \left(\frac{15}{2}t^{1/2} + 3e^{-t}\right) dt$$
$$= \frac{15}{2}\left(\frac{t^{3/2}}{\frac{3}{2}}\right) + 3\left(\frac{1}{-1}e^{-t}\right) + C_1$$
$$= 5t^{3/2} - 3e^{-t} + C_1$$

$$v(0) = 5(0)^{3/2} - 3e^{-0} + C_1 = -3 + C_1$$

Since $v(0) = -3, C_1 = 0.$

$$v(t) = 5t^{3/2} - 3e^{-t}$$
$$s(t) = \int (5t^{3/2} - 3e^{-t})\, dt$$
$$= 5\left(\frac{t^{5/2}}{\frac{5}{2}}\right) - 3\left(-\frac{1}{1}e^{-t}\right) + C_2$$
$$= 2t^{5/2} + 3e^{-t} + C_2$$
$$s(0) = 2(0)^{5/2} + 3e^{-0} + C_2 = 3 + C_2$$

Since $s(0) = 4, C_2 = 1.$

Thus,

$$s(t) = 2t^{5/2} + 3e^{-t} + 1.$$

73. First find $v(t)$ by integrating $a(t)$:

$$v(t) = \int (-32)dt = -32t + k.$$

When $t = 5, v(t) = 0$:

$$0 = -32(5) + k$$
$$160 = k$$

and

$$v(t) = -32t + 160.$$

Now integrate $v(t)$ to find $h(t)$.

$$h(t) = \int (-32t + 160)dt = -16t^2 + 160t + C$$

Since $h(t) = 412$ when $t = 5$, we can substitute these values into the equation for $h(t)$ to get $C = 12$ and

$$h(t) = -16t^2 + 160t + 12.$$

Therefore, from the equation given in Exercise 72, the initial velocity v_0 is 160 ft/sec and the initial height of the rocket h_0 is 12 ft.

15.2 Substitution

Your Turn 1

Find $\int 8x(4x^2 + 8)^6 dx.$

Let $u = 4x^2 + 8.$

Then $du = 8x\,dx.$

Now substitute.

$$\int 8x(4x^2 + 8)^6 dx = \int (4x^2 + 8)^6 (8x\,dx)$$
$$= \int u^6 du$$
$$= \frac{1}{7}u^7 + C$$

Now replace u with $4x^2 + 8.$

$$\int 8x(4x^2 + 8)^6 dx = \frac{1}{7}u^7 + C$$
$$= \frac{1}{7}(4x^2 + 8)^7 + C$$

Your Turn 2

Find $\int x^3\sqrt{3x^4 + 10}\, dx.$

Let $u = 3x^4 + 10.$

Then $du = 12x^3\, dx.$

Multiply the integral by $\frac{12}{12}$ to introduce the factor of 12 needed for du, and then substitute.

$$\int x^3\sqrt{3x^4+10}\,dx = \frac{1}{12}\int 12x^3\sqrt{3x^4+10}\,dx$$
$$= \frac{1}{12}\int \sqrt{3x^4+10}\,(12x^3\,dx)$$
$$= \frac{1}{12}\int u^{1/2}du$$
$$= \frac{1}{12}\left(\frac{2}{3}u^{3/2}+C\right)$$
$$= \frac{1}{18}u^{3/2}+C$$

(Note that (1/12) C is just a different constant, which we can also call C.) Now replace u with $3x^4+10$.

$$\int x^3\sqrt{3x^4+10}\,dx = \frac{1}{18}u^{3/2}+C$$
$$= \frac{1}{18}(3x^4+10)^{3/2}+C$$

Your Turn 3

Find $\int \frac{x+1}{(4x^2+8x)^3}dx.$

Let $u = 4x^2+8x.$
Then $du = (8x+8)dx = 8(x+1)\,dx$

Multiply the integral by $\frac{8}{8}$ to introduce the factor of 8 needed for du, and then substitute.

$$\int \frac{x+1}{(4x^2+8x)^3}dx = \frac{1}{8}\int \frac{8(x+1)}{(4x^2+8x)^3}dx$$
$$= \frac{1}{8}\int \frac{1}{u^3}du$$
$$= \frac{1}{8}\int u^{-3}du$$
$$= \frac{1}{8}\left(-\frac{1}{2}u^{-2}+C\right)$$
$$= -\frac{1}{16u^2}+C$$

Now replace u with $4x^2+8x$.

$$\int \frac{x+1}{(4x^2+8x)^3}dx = -\frac{1}{16u^2}+C$$
$$= -\frac{1}{16(4x^2+8x)^2}$$

Your Turn 4

Find $\int \frac{x+3}{x^2+6x}dx.$

Let $u = x^2+6x.$
Then $du = 2x+6\,dx = 2(x+3)\,dx$

Multiply the integral by $\frac{2}{2}$ to introduce the factor of 2 needed for du, and then substitute.

$$\int \frac{x+3}{x^2+6x}dx = \frac{1}{2}\int \frac{2(x+3)}{x^2+6x}dx$$
$$= \frac{1}{2}\int \frac{1}{u}du$$
$$= \frac{1}{2}(\ln|u|+C)$$
$$= \frac{1}{2}\ln|u|+C$$

Now replace u with x^2+6x.

$$\int \frac{x+3}{x^2+6x}dx = \frac{1}{2}\ln|u|+C$$
$$= \frac{1}{2}|x^2+6x|+C$$

Your Turn 5

Find $\int x^3e^{x^4}\,dx.$

Let $u = x^4.$

Then $du = 4x^3\,dx.$

Multiply the integral by $\frac{4}{4}$ to introduce the factor of 4 needed for du, and then substitute.

$$\int x^3e^{x^4}\,dx = \frac{1}{4}\int 4x^3e^{x^4}\,dx$$
$$= \frac{1}{4}\int e^{x^4}(4x^3\,dx)$$
$$= \frac{1}{4}\int e^u\,du$$
$$= \frac{1}{4}e^u+C$$

Now replace u with x^4.

$$\int x^3e^{x^4}\,dx = \frac{1}{4}e^u+C = \frac{1}{4}e^{x^4}+C$$

Your Turn 6

Find $\int x\sqrt{3+x}\,dx.$

Let $u = 3 + x.$

Then $du = dx$ and $x = u - 3.$

Now substitute.

$$\begin{aligned}\int x\sqrt{3+x}\,dx &= \int (u-3)\sqrt{u}\,du\\ &= \int u\sqrt{u}\,du - 3\int \sqrt{u}\,du\\ &= \int u^{3/2}\,du - 3\int u^{1/2}\,du\\ &= \frac{2}{5}u^{5/2} - 3\left(\frac{2}{3}u^{3/2}\right) + C\\ &= \frac{2}{5}u^{5/2} - 2u^{3/2} + C\end{aligned}$$

Now replace u with $3 + x.$

$$\begin{aligned}\int x\sqrt{3+x}\,dx &= \frac{2}{5}u^{5/2} - 2u^{3/2} + C\\ &= \frac{2}{5}(3+x)^{5/2} - 2(3+x)^{3/2} + C\end{aligned}$$

15.2 Exercises

3. $\int 4(2x+3)^4\,dx = 2\int 2(2x+3)^4\,dx$

Let $u = 2x + 3,$ so that $du = 2\,dx.$

$$\begin{aligned}&= 2\int u^4 du\\ &= \frac{2\cdot u^5}{5} + C\\ &= \frac{2(2x+3)^5}{5} + C\end{aligned}$$

5. $\int \frac{2\,dm}{(2m+1)^3} = \int 2(2m+1)^{-3}\,dm$

Let $u = 2m + 1,$ so that $du = 2\,dm.$

$$\begin{aligned}&= \int u^{-3} du\\ &= \frac{u^{-2}}{-2} + C = \frac{-(2m+1)^{-2}}{2} + C\end{aligned}$$

7. $\int \frac{2x+2}{(x^2+2x-4)^4}dx$

$$= \int (2x+2)(x^2+2x-4)^{-4}dx$$

Let $w = x^2 + 2x - 4,$ so that $dw = (2x+2)dx.$

$$\begin{aligned}&= \int w^{-4} dw\\ &= \frac{w^{-3}}{-3} + C\\ &= -\frac{(x^2+2x-4)^{-3}}{3} + C\\ &= -\frac{1}{3(x^2+2x-4)^3} + C\end{aligned}$$

9. $\int z\sqrt{4z^2-5}\,dz = \int z(4z^2-5)^{1/2}dz$

$$= \frac{1}{8}\int 8z(4z^2-5)^{1/2}dz$$

Let $u = 4z^2 - 5,$ so that $du = 8z\,dz.$

$$\begin{aligned}&= \frac{1}{8}\int u^{1/2} du\\ &= \frac{1}{8}\cdot\frac{u^{3/2}}{\frac{3}{2}} + C\\ &= \frac{1}{8}\cdot\left(\frac{2}{3}\right)u^{3/2} + C\\ &= \frac{(4z^2-5)^{3/2}}{12} + C\end{aligned}$$

11. $\int 3x^2 e^{2x^3}\,dx = \frac{1}{2}\int 2\cdot 3x^2 e^{2x^3}\,dx$

Let $u = 2x^3,$ so that $du = 6x^2 dx.$

$$\begin{aligned}&= \frac{1}{2}\int e^u du\\ &= \frac{1}{2}e^u + C\\ &= \frac{e^{2x^3}}{2} + C\end{aligned}$$

13. $\int (1-t)e^{2t-t^2}dt$

$$= \frac{1}{2}\int 2(1-t)e^{2t-t^2}dt$$

Let $u = 2t - t^2$, so that $du = (2-2t)dt$.

$$= \frac{1}{2}\int e^u du$$

$$= \frac{e^u}{2} + C = \frac{e^{2t-t^2}}{2} + C$$

15. $\int \frac{e^{1/z}}{z^2}dz = -\int e^{1/z} \cdot \frac{-1}{z^2}dz$

Let $u = \frac{1}{z}$, so that $du = \frac{-1}{z^2}dx$.

$$\int \frac{e^{1/z}}{z^2}dz = -\int e^u du$$

$$= -e^u + C$$

$$= -e^{1/z} + C$$

17. $\int \frac{t}{t^2+2}dt$

Let $t^2 + 2 = u$, so that $2t\, dt = du$.

$$= \frac{1}{2}\int \frac{du}{u}$$

$$= \frac{1}{2}\ln|u| + C$$

$$= \frac{\ln(t^2+2)}{2} + C$$

19. $\int \frac{x^3+2x}{x^4+4x^2+7}dx$

Let $u = x^4 + 4x^2 + 7$.

Then $du = (4x^3 + 8x)dx = 4(x^3 + 2x)dx$.

$$\int \frac{x^3+2x}{x^4+4x^2+7}dx = \frac{1}{4}\int \frac{(4x^3+2x)}{x^4+4x^2+7}dx$$

$$= \frac{1}{4}\int \frac{1}{u}du = \frac{1}{4}\ln|u| + C$$

$$= \frac{1}{4}\ln(x^4+4x^2+7) + C$$

Since $x^4 + 4x^2 + 7 > 0$ for all x, we can write this answer as $\frac{1}{4}\ln(x^4 + 4x^2 + 7) + C$.

21. $\int \frac{2x+1}{(x^2+x)^3}dx$

$$= \int (2x+1)(x^2+x)^{-3}dx$$

Let $u = x^2 + x$, so that $du = (2x+1)dx$.

$$= \int u^{-3}du = \frac{u^{-2}}{-2} + C$$

$$= \frac{-1}{2u^2} + C = \frac{-1}{2(x^2+x)^2} + C$$

23. $\int p(p+1)^5 dp$

Let $u = p + 1$, so that $du = dp$; also, $p = u - 1$.

$$= \int (u-1)u^5 du$$

$$= \int (u^6 - u^5)du$$

$$= \frac{u^7}{7} - \frac{u^6}{6} + C$$

$$= \frac{(p+1)^7}{7} - \frac{(p+1)^6}{6} + C$$

25. $\int \frac{u}{\sqrt{u-1}}du$

$$= \int u(u-1)^{-1/2}du$$

Let $w = u - 1$, so that $dw = du$ and $u = w + 1$.

$$= \int (w+1)w^{-1/2}dw$$

$$= \int (w^{1/2} + w^{-1/2})\,dw$$

$$= \frac{w^{3/2}}{\frac{3}{2}} + \frac{w^{1/2}}{\frac{1}{2}} + C$$

$$= \frac{2(u-1)^{3/2}}{3} + 2(u-1)^{1/2} + C$$

27. $\int \left(\sqrt{x^2+12x}\right)(x+6)\,dx$

$$= \int (x^2+12x)^{1/2}(x+6)\,dx$$

Let $x^2 + 12x = u$, so that

$$(2x + 12)\,dx = du$$
$$2(x + 6)\,dx = du.$$

$$= \frac{1}{2}\int u^{1/2}du = \frac{1}{2}\left(\frac{2}{3}\right)u^{3/2} + C$$

$$= \frac{(x^2 + 12x)^{3/2}}{3} + C$$

29. $\displaystyle\int \frac{3(1 + 3\ln x)^2}{x}dx$

Let $u = 1 + 3\ln x$, so that $du = \frac{3}{x}dx.$

$$= \frac{1}{3}\int \frac{3(1 + 3\ln x)^2}{x}dx$$

$$= \frac{1}{3}\int u^2 du$$

$$= \frac{1}{3}\cdot\frac{u^3}{3} + C$$

$$= \frac{(1 + 3\ln x)^3}{9} + C$$

31. $\displaystyle\int \frac{e^{2x}}{e^{2x} + 5}dx$

Let $u = e^{2x} + 5$, so that $du = 2e^{2x}\,dx.$

$$= \frac{1}{2}\int \frac{du}{u}$$

$$= \frac{1}{2}\ln|u| + C$$

$$= \frac{1}{2}\ln|e^{2x} + 5| + C$$

$$= \frac{1}{2}\ln(e^{2x} + 5) + C$$

33. $\displaystyle\int \frac{\log x}{x}dx$

Let $u = \log x$, so that $du = \frac{1}{(\ln 10)x}dx.$

$$\int \frac{\log x}{x}dx = (\ln 10)\int \frac{\log x}{(\ln 10)x}dx = (\ln 10)\int u\,du$$

$$= (\ln 10)\left(\frac{u^2}{2}\right) + C$$

$$= \frac{(\ln 10)(\log x)^2}{2} + C$$

35. $\displaystyle\int x8^{3x^2+1}dx$

Let $u = 3x^2 + 1$, so that $du = 6x\,dx.$

$$= \frac{1}{6}\int 6x \cdot 8^{3x^2+1}dx$$

$$= \frac{1}{6}\int 8^u du$$

$$= \frac{1}{6}\left(\frac{8^u}{\ln 8}\right) + C$$

$$= \frac{8^{3x^2+1}}{6\ln 8} + C$$

39. **(a)** $R'(x) = 4x(x^2 + 27{,}000)^{-2/3}$

$$R(x) = \int 4x(x^2 + 27{,}000)^{-2/3}dx$$

$$= 2\int 2x(x^2 + 27{,}000)^{-2/3}dx$$

Let $u = x^2 + 27{,}000$, so that $du = 2x\,dx.$

$$R = 2\int u^{-2/3}du$$

$$= 2\cdot 3u^{1/3} + C$$

$$= 6(x^2 + 27{,}000)^{1/3} + C$$

$$R(125) = 6(125^2 + 27{,}000)^{1/3} + C$$

Since $R(125) = 29.591,$

$$6(125^2 + 27{,}000)^{1/3} + C = 29.591$$
$$C = -180$$

Thus,

$$R(x) = 6(x^2 + 27{,}000)^{1/3} - 180.$$

(b) $R(x) = 6(x^2 + 27{,}000)^{1/3} - 180 \geq 40$

$$6(x^2 + 27{,}000)^{1/3} \geq 220$$
$$(x^2 + 27{,}000)^{1/3} \geq 36.6667$$
$$x^2 + 27{,}000 \geq 49{,}296.43$$
$$x^2 \geq 22{,}296.43$$
$$x \geq 149.4$$

For a revenue of at least \$40,000, 150 players must be sold.

41. $C'(x) = \dfrac{60x}{5x^2 + e}$

(a) Let $u = 5x^2 + e$, so that $du = 10x\,dx$.

$$\begin{aligned} C(x) &= \int C'(x)\,dx \\ &= \int \frac{60x}{5x^2 + e}\,dx \\ &= 6\int \frac{du}{u} = 6\ln|u| + C \\ &= 6\ln(5x^2 + e) + C \end{aligned}$$

Since $C(0) = 10, C = 4$.

Therefore,

$$\begin{aligned} C(x) &= 6\ln|5x^2 + e| + 4 \\ &= 6\ln(5x^2 + e) + 4. \end{aligned}$$

(b) $C(5) = 6\ln(5 \cdot 5^2 + e) + 4 \approx 33.099$

Since this represents $33,099 dollars which is greater than $20,000, a new source of investment income should be sought.

43. $f'(t) = 4.0674 \cdot 10^{-4} t(t - 1970)^{0.4}$

(a) Let $u = t - 1970$. To get the t outside the parentheses in terms of u, solve $u = t - 1970$ for t to get $t = u + 1970$. Then $dt = du$ and we can substitute as follows.

$$\begin{aligned} f(t) = \int f'(t)dt &= \int 4.0674 \cdot 10^{-4} t(t - 1970)^{0.4}\,dt \\ &= \int 4.0674 \cdot 10^{-4}(u + 1970)(u)^{0.4}\,du \\ &= 4.0674 \cdot 10^{-4}\int (u + 1970)(u)^{0.4}\,du \\ &= 4.0674 \cdot 10^{-4}\int (u^{1.4} + 1970u^{0.4})du \\ &= 4.0674 \cdot 10^{-4}\left(\frac{u^{2.4}}{2.4} + \frac{1970u^{1.4}}{1.4}\right) + C \\ &= 4.0674 \cdot 10^{-4}\left[\frac{(t - 1970)^{2.4}}{2.4} + \frac{1970(t - 1970)^{1.4}}{1.4}\right] + C \end{aligned}$$

Since $f(1970) = 61.298, C = 61.298$.

Therefore, $f(t) = 4.0674 \cdot 10^{-4}\left[\dfrac{(t - 1970)^{2.4}}{2.4} + \dfrac{1970(t - 1970)^{1.4}}{1.4}\right] + 61.298.$

(b) $f(2015) = 4.0674 \cdot 10^{-4}\left[\dfrac{(2015 - 1970)^{2.4}}{2.4} + \dfrac{1970(2015 - 1970)^{1.4}}{1.4}\right] + 61.298 \approx 180.9.$

In the year 2015, there will be about 181,000 local transit vehicles.

15.3 Area and the Definite Integral

Your Turn 1

Approximate $\int_1^5 4x\,dx$ using four rectangles.

Find the area of the shaded region:

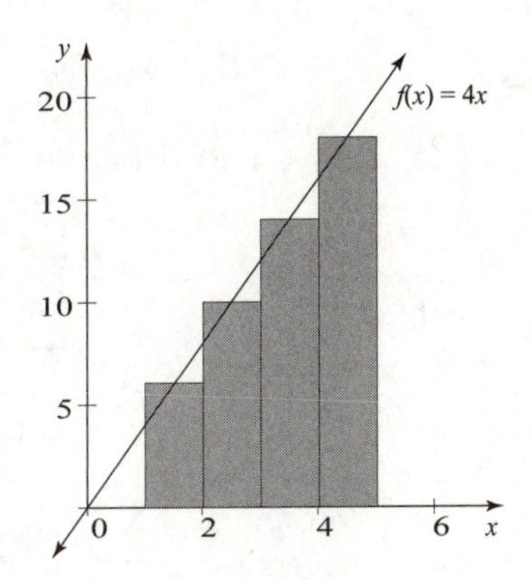

Build a table giving the heights of the rectangles, which are the values of $f(x) = 4x$ at the midpoint of each interval.

i	x_i	$f(x_i)$
1	1.5	6.0
2	2.5	10.0
3	3.5	14.0
4	4.5	18.0

For each interval, $\Delta x = 1$. The sum of the areas of the rectangles is

$$\sum_{i=1}^{4} f(x_i)\Delta x_i = f(1.5)\Delta x + f(2.5)\Delta x + f(3.5)\Delta x + f(4.5)\Delta x$$

$$= 1(6) + 1(10) + 1(14) + 1(18) = 48.$$

Thus our approximation to the integral is 48. In this case the approximation is exact.

Your Turn 2

A driver has the following velocities at various times:

Time (hr)	0	0.5	1	1.5	2
Velocity (mph)	0	50	56	40	48

Approximate the total distance traveled during the 2-hour period.

Using left endpoints:

$$\text{distance} = 0(0.5) + 50(0.5) + 56(0.5) + 40(0.5) = 73 \text{ miles}$$

Using right endpoints:

$$\text{distance} = 50(0.5) + 56(0.5) + 40(0.5) + 48(0.5) = 97 \text{ miles}$$

Averaging these two estimates:

$$\text{distance} = \frac{73 + 97}{2} = 85 \text{ miles}$$

15.3 Exercises

3. $f(x) = 2x + 5, x_1 = 0, x_2 = 2, x_3 = 4,$
$x_4 = 6$, and $\Delta x = 2$

(a) $\sum_{i=1}^{4} f(x_i)\Delta x$

$$= f(x_1)\Delta x + f(x_2)\Delta x + f(x_3)\Delta x + f(x_4)\Delta x$$
$$= f(0)(2) + f(2)(2) + f(4)(2) + f(6)(2)$$
$$= [2(0) + 5](2) + [2(2) + 5](2) + [2(4) + 5](2) + [2(6) + 5](2)$$
$$= 10 + 9(2) + 13(2) + 17(2) = 88$$

(b)

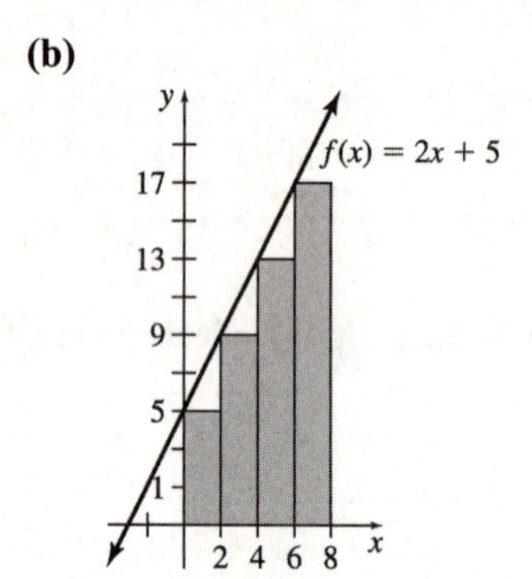

The sum of these rectangles approximates

$$\int_0^8 (2x + 5)\, dx.$$

5. $f(x) = 2x + 5$ from $x = 2$ to $x = 4$

For $n = 4$ rectangles:

$$\Delta x = \frac{4 - 2}{4} = 0.5$$

(a) Using the left endpoints:

i	x_i	$f(x_i)$
1	2	9
2	2.5	10
3	3	11
4	3.5	12

$$A = \sum_{1}^{4} f(x_i)\Delta x$$
$$= 9(0.5) + 10(0.5) + 11(0.5) + 12(0.5)$$
$$= 21$$

(b) Using the right endpoints:

i	x_i	$f(x_i)$
1	2.5	10
2	3	11
3	3.5	12
4	4	13

$$A = 10(0.5) + 11(0.5) + 12(0.5) + 13(0.5) = 23$$

(c) Average $= \dfrac{21 + 23}{2} = \dfrac{44}{2} = 22$

(d) Using the midpoints:

i	x_i	$f(x_i)$
1	2.25	9.5
2	2.75	10.5
3	3.25	11.5
4	3.75	12.5

$$A = \sum_{1}^{4} f(x_i)\Delta x = 9.5(0.5) + 10.5(0.5)$$
$$+ 11.5(0.5) + 12.5(0.5) = 22$$

7. $f(x) = -x^2 + 4$ from $x = -2$ to $x = 2$

For $n = 4$ rectangles:

$$\Delta x = \frac{2 - (-2)}{4} = 1$$

(a) Using the left endpoints:

i	x_i	$f(x_i)$
1	-2	$-(-2)^2 + 4 = 0$
2	-1	$-(-1)^2 + 4 = 3$
3	0	$-(0)^2 + 4 = 4$
4	1	$-(1)^2 + 4 = 3$

$$A = \sum_{i=1}^{4} f(x_i)\Delta x$$
$$= (0)(1) + (3)(1) + (4)(1) + (3)(1)$$
$$= 10$$

(b) Using the right endpoints:

i	x_i	$f(x_i)$
1	-1	3
2	0	4
3	1	3
4	2	0

Area $= 1(3) + 1(4) + 1(3) + 1(0) = 10$

(c) Average $= \dfrac{10 + 10}{2} = 10$

(d) Using the midpoints:

i	x_i	$f(x_i)$
1	$-\frac{3}{2}$	$\frac{7}{4}$
2	$-\frac{1}{2}$	$\frac{15}{4}$
3	$\frac{1}{2}$	$\frac{15}{4}$
4	$\frac{3}{2}$	$\frac{7}{4}$

$$\begin{aligned} A &= \sum_{i=1}^{4} f(x_i)\Delta x \\ &= \frac{7}{4}(1) + \frac{15}{4}(1) + \frac{15}{4}(1) + \frac{7}{4}(1) \\ &= 11 \end{aligned}$$

9. $f(x) = e^x + 1$ from $x = -2$ to $x = 2$

For $n = 4$ rectangles:

$$\Delta x = \frac{2 - (-2)}{4} = 1$$

(a) Using the left endpoints:

i	x_i	$f(x_i)$
1	-2	$e^{-2} + 1$
2	-1	$e^{-1} + 1$
3	0	$e^0 + 1 = 2$
4	1	$e^1 + 1$

$$\begin{aligned} A &= \sum_{i=1}^{4} f(x_i)\Delta x = \sum_{i=1}^{4} f(x_i)(1) = \sum_{i=1}^{4} f(x_i) \\ &= (e^{-2} + 1) + (e^{-1} + 1) + 2 + e^1 + 1 \\ &\approx 8.2215 \approx 8.22 \end{aligned}$$

(b) Using the right endpoints:

i	x_i	$f(x_i)$
1	-1	$e^{-1} + 1$
2	0	2
3	1	$e + 1$
4	2	$e^2 + 1$

$$\begin{aligned} \text{Area} &= 1(e^{-1} + 1) + 1(2) + 1(e + 1) + 1(e^2 + 1) \\ &\approx 15.4752 \approx 15.48 \end{aligned}$$

(c) $$\begin{aligned} \text{Average} &= \frac{8.2215 + 15.4752}{2} \\ &= 11.84835 \\ &\approx 11.85 \end{aligned}$$

(d) Using the midpoints:

i	x_i	$f(x_i)$
1	$-\frac{3}{2}$	$e^{-3/2} + 1$
2	$-\frac{1}{2}$	$e^{-1/2} + 1$
3	$\frac{1}{2}$	$e^{1/2} + 1$
4	$\frac{3}{2}$	$e^{3/2} + 1$

$$\begin{aligned} A &= \sum_{i=1}^{4} f(x_i)\Delta x \\ &= (e^{-3/2} + 1)(1) + (e^{-1/2} + 1)(1) \\ &\quad + (e^{1/2} + 1)(1) + (e^{3/2} + 1)(1) \\ &\approx 10.9601 \approx 10.96 \end{aligned}$$

11. $f(x) = \frac{2}{x}$ from $x = 1$ to $x = 9$

For $n = 4$ rectangles:

$$\Delta x = \frac{9 - 1}{4} = 2$$

(a) Using the left endpoints:

i	x_i	$f(x_i)$
1	1	$\frac{2}{1} = 2$
2	3	$\frac{2}{3}$
3	5	$\frac{2}{5} = 0.4$
4	7	$\frac{2}{7}$

$$\begin{aligned} A &= \sum_{i=1}^{4} f(x_i)\Delta x \\ &= (2)(2) + \frac{2}{3}(2) + (0.4)(2) + \left(\frac{2}{7}\right)(2) \\ &\approx 6.7048 \approx 6.70 \end{aligned}$$

(b) Using the right endpoints:

i	x_i	$f(x_i)$
1	3	$\frac{2}{3}$
2	5	$\frac{2}{5}$
3	7	$\frac{2}{7}$
4	9	$\frac{2}{9}$

$$\text{Area} = 2\left(\frac{2}{3}\right) + 2\left(\frac{2}{5}\right) + 2\left(\frac{2}{7}\right) + 2\left(\frac{2}{9}\right)$$
$$= \frac{4}{3} + \frac{4}{5} + \frac{4}{7} + \frac{4}{9} \approx 3.1492 \approx 3.15$$

(c) $\text{Average} = \dfrac{6.7 + 3.15}{2} = 4.93$

(d) Using the midpoints:

i	x_i	$f(x_i)$
1	2	1
2	4	$\frac{1}{2}$
3	6	$\frac{1}{3}$
4	8	$\frac{1}{4}$

$$A = \sum_{i=1}^{4} f(x_i)\Delta x$$
$$= 1(2) + \frac{1}{2}(2) + \frac{1}{3}(2) + \frac{1}{4}(2)$$
$$\approx 4.1667 \approx 4.17$$

13. **(a)** $\text{Width} = \dfrac{4-0}{4} = 1;\ f(x) = \dfrac{x}{2}$

$$\text{Area} = 1 \cdot f\left(\frac{1}{2}\right) + 1 \cdot f\left(\frac{3}{2}\right) + 1 \cdot f\left(\frac{5}{2}\right) + 1 \cdot f\left(\frac{7}{2}\right)$$
$$= \frac{1}{4} + \frac{3}{4} + \frac{5}{4} + \frac{7}{4} = \frac{16}{4} = 4$$

(b)

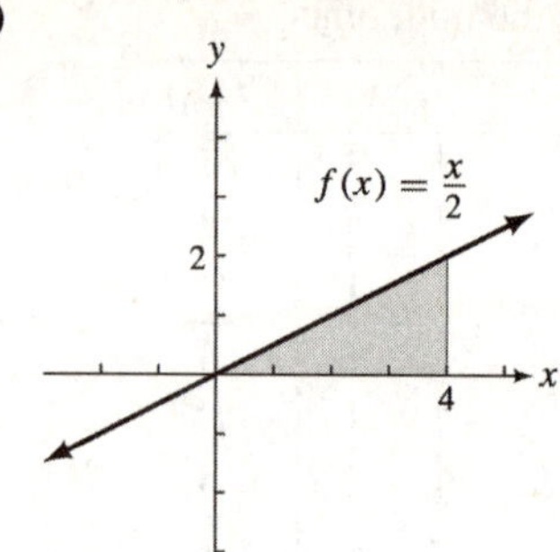

$$\int_0^4 f(x)dx = \int_0^4 \frac{x}{2}dx = \frac{1}{2}(\text{base})(\text{height})$$
$$= \frac{1}{2}(4)(2) = 4$$

15. **(a)** Area of triangle is $\frac{1}{2}$ ·base · height.

The base is 4; the height is 2.

$$\int_0^4 f(x)\,dx = \frac{1}{2} \cdot 4 \cdot 2 = 4$$

(b) The larger triangle has an area of $\frac{1}{2} \cdot 3 \cdot 3 = \frac{9}{2}$. The smaller triangle has an area of $\frac{1}{2} \cdot 1 \cdot 1 = \frac{1}{2}$. The sum is $\frac{9}{2} + \frac{1}{2} = \frac{10}{2} = 5$.

17. $\displaystyle\int_{-4}^{0} \sqrt{16 - x^2}\,dx$

Graph $y = \sqrt{16 - x^2}$.

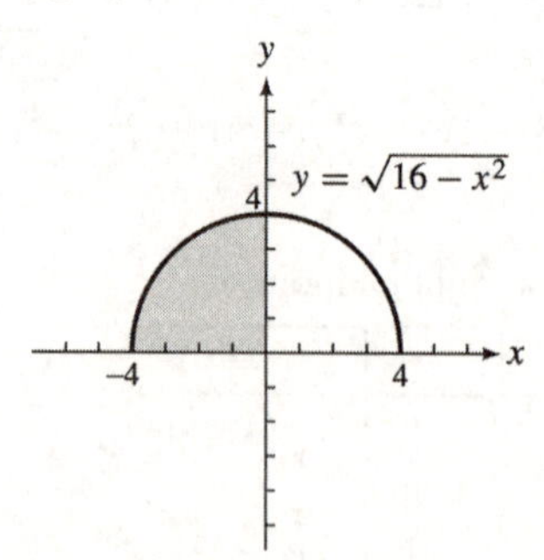

$\displaystyle\int_{-4}^{0} \sqrt{16 - x^2}\,dx$ is the area of the portion of the circle in the second quadrant, which is one-fourth of a circle. The circle has radius 4.

$$\text{Area} = \frac{1}{4}\pi r^2 = \frac{1}{4}\pi(4)^2 = 4\pi$$

19. $\int_2^5 (1 + 2x)\,dx$

Graph $y = 1 + 2x$.

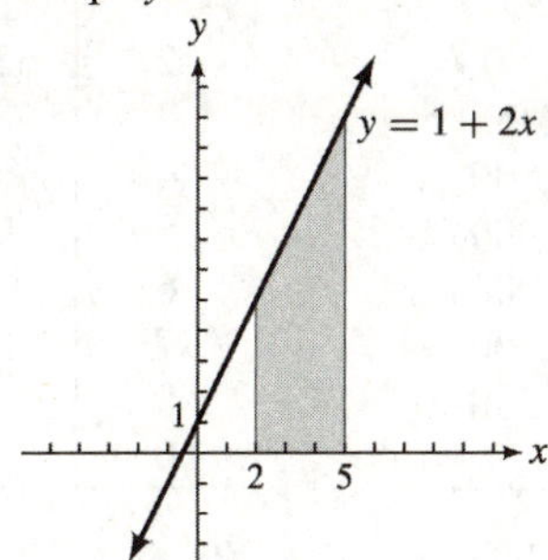

$\int_2^5 (1 + 2x)\,dx$ is the area of the trapezoid with $B = 11, b = 5$, and $h = 3$. The formula for the area is

$$A = \frac{1}{2}(B + b)h,$$

so we have

$$A = \frac{1}{2}(11 + 5)(3) = 24.$$

21. **(a)** With $n = 10, \Delta x = \frac{1-0}{10} = 0.1$, and $x_1 = 0 + 0.1 = 0.1$, use the command seq $(X^2, X, 0.1, 1, 0.1) \to L1$. The resulting screen is:

```
seq(X²,X,.1,1,.1
)→L₁
{.01 .04 .09 .1…
```

(b) Since $\sum_{i=1}^{n} f(x_i)\Delta x = \Delta x\left(\sum_{i=1}^{n} f(x_i)\right)$, use the command 0.1∗sum (L1) to approximate $\int_0^1 x^2 dx$. The resulting screen is:

```
.1*sum(L₁)
                .385
```

$$\int_0^1 x^2 dx \approx 0.385$$

(c) With $n = 100, \Delta x = \frac{1-0}{100} = 0.01$ and $x_1 = 0 + 0.01 = 0.01$, use the command seq $(X^2, X, 0.01, 1, 0.01) \to L1$. The resulting screen is:

Use the command 0.01∗sum(L1) to approximate $\int_0^1 x^2 dx$. The resulting screen is:

$$\int_0^1 x^2 dx \approx 0.33835$$

(d) With $n = 500, \Delta x = \frac{1-0}{500} = 0.002$, and $x_1 = 0 + 0.002 = 0.002$, use the command seq$(X^2, X, 0.002, 1, 0.002) \to L1$. The resulting screen is:

```
seq(X²,X,.002,1,
.002)→L₁
{4E-6 1.6E-5 3.…
```

Use the command 0.002∗sum(L1) to approximate $\int_0^1 x^2 dx$. The resulting screen is:

$$\int_0^1 x^2 dx \approx 0.334334$$

(e) As n gets larger the approximation for $\int_0^1 x^2 dx$ seems to be approaching 0.333333 or $\frac{1}{3}$. We estimate $\int_0^1 x^2 dx = \frac{1}{3}$.

25. Left endpoints:

Read values of the function on the graph every three years from 1997 to 2006. These values give us the heights of four rectangles. The width of each rectangle is $\Delta x = 3$. We estimate the area under the curve as follows:

$$A = \sum_{i=1}^{4} f(x_i)\Delta x$$
$$= 34(3) + 57(3) + 115(3) + 264(3)$$
$$= 1410$$

Right endpoints:

Read values of the function on the graph every three years from 2000 to 2009. We estimate the area under the curve as follows:

$$A = \sum_{i=1}^{4} f(x_i)\Delta x$$
$$= 57(3) + 115(3) + 264(3) + 697(3)$$
$$= 3399$$

Average: $\dfrac{1410 + 3399}{2} = 2404.5$ trillion BTUs

We estimate the total wind energy consumption over the 12-year period from 1997 to 2009 as 2404.5 trillion BTUs.

27. First read approximate data values from the graph. These readings are just estimates, and you may get different answers if your estimated readings differ from these. Month 1 represents mid-February.

Cows	
Month	Cases
1	3000
2	165,000
3	267,000
4	54,000
5	44,000
6	21,000
7	16,500
8	11,500
9	1000

Pigs	
Month	Cases
1	2000
2	62,000
3	68,000
4	3000
5	1000
6	9000
7	1000
8	0
9	0

(a) Left endpoints:

Add up the values corresponding to months 1 through 8 in the Cows table The total is 582,000 cases.

Right endpoints:

Add up the values corresponding to months 2 through 9. The total is 580,000 cases.

The average of these two values is 581,000 cases.

(b) Left endpoints:

Add up the values corresponding to months 1 through 8 in the Pigs table. The total is 146,000 cases.

Right endpoints:

Add up the values corresponding to months 2 through 9. The total is 144,000 cases.

The average of these two values is 145,000 cases.

29. Read the value of the function for every 5 sec from $x = 2.5$ to $x = 12.5$. These are the midpoints of rectangle with width $\Delta x = 5$. Then read the function for $x = 17$, which is the midpoint of a rectangle with width $\Delta x = 4$.

$$\sum_{i=1}^{4} f(x_i)\Delta x \approx 36(5) + 63(5) + 84(5) + 95(4) \approx 1295$$

$$\frac{1295}{3600}(5280) \approx 1900$$

The Porsche 928 traveled about 1900 ft.

31. Left endpoints:

Read values of the function from the table for every number of seconds from 2.0 to 19.3. These values give the heights of 10 rectangles. The width of each rectangle varies. We estimate the area under the curve as

$$\sum_{i=1}^{10} f(x_i)\Delta x$$

$$= 30(2.9 - 2.0) + 40(4.1 - 2.9) + 50(5.3 - 4.1) + 60(6.9 - 5.3) + 70(8.7 - 6.9) + 80(10.7 - 8.7)$$
$$+ 90(13.2 - 10.7) + 100(16.1 - 13.2) + 110(19.3 - 16.1) + 120(23.4 - 19.3)$$
$$= 1876$$

$$\frac{5280}{3600}(1876) \approx 2751$$

Right endpoints:

Read values of the function from the table for every number of seconds from 2.0 to 23.4. These values give the heights of 11 rectangles. The width of each rectangle varies. We estimate the area under the curve as

$$\sum_{i=1}^{11} f(x_i)\Delta x$$

$$= 30(2.0 - 0) + 40(2.9 - 2.0) + 50(4.1 - 2.9) + 60(5.3 - 4.1) + 70(6.9 - 5.3) + 80(8.7 - 6.9)$$
$$+ 90(10.7 - 8.7) + 100(13.2 - 10.7) + 110(16.1 - 13.2) + 120(19.3 - 16.1) + 130(23.4 - 19.3)$$
$$= 2150$$

$$\frac{5280}{3600}(2150) \approx 3153$$

Average: $\dfrac{2751 + 3153}{2} = \dfrac{5904}{2} = 2952$ ft

The distance traveled by the Mercedes-Benz S550 is about 2952 ft.

33. **(a)** Read values of the function on the plain glass graph every 2 hr from 6 to 6.These are at midpoints of the widths $\Delta x = 2$ and represent the heights of the rectangles.

$$f(x_i)\Delta x = 132(2) + 215(2) + 150(2) + 44(2) + 34(2) + 26(2) + 12(2) \approx 1226$$

The total heat gain was about 1230 BTUs per square foot.

(b) Read values on the ShadeScreen graph every 2 hr from 6 to 6.

$$\sum f(x_i)\Delta x = 38(2) + 25(2) + 16(2) + 12(2) + 10(2) + 10(2) + 5(2) \approx 232$$

The total heat gain was about 230 BTUs per square foot.

35. **(a)** Then area of a trapezoid is

$$A = \frac{1}{2}h(b_1 + b_2) = \frac{1}{2}(6)(1 + 2) = 9.$$

Car A has traveled 9 ft.

(b) Car A is furthest ahead of car B at 2 sec. Notice that from $t = 0$ to $t = 2$, $v(t)$ is larger for car A than for car B. For $t > 2$, $v(t)$ is larger for car B than for car A.

(c) As seen in part (a), car A drove 9 ft in 2 sec. The distance of car B can be calculated as follows:

$$\frac{2-0}{4} = \frac{1}{2} = \text{width}$$

$$\begin{aligned}\text{Distance} &= \frac{1}{2}\cdot v(0.25) + \frac{1}{2}v(0.75) + \frac{1}{2}v(1.25) + \frac{1}{2}v(1.75)\\ &= \frac{1}{2}(0.2) + \frac{1}{2}(1) + \frac{1}{2}(2.6) + \frac{1}{2}(5)\\ &= 4.4\end{aligned}$$

$$9 - 4.4 = 4.6$$

The furthest car A can get ahead of car B is about 4.6 ft.

(d) At $t = 3$, car A travels $\frac{1}{2}(6)(2 + 3) = 15$ ft and car B travels approximately 13 ft.

At $t = 3.5$, car A travels $\frac{1}{2}(6)(2.5 + 3.5) = 18$ ft and car B travels approximately 18.25 ft. Therefore, car B catches up with car A between 3 and 3.5 sec.

37. Using the left endpoints:

$$\begin{aligned}\text{Distance} &= v_0(1) + v_1(1) + v_2(1)\\ &= 10 + 6.5 + 6 = 22.5 \text{ ft}\end{aligned}$$

Using the right endpoints:

$$\begin{aligned}\text{Distance} &= v_1(1) + v_2(1) + v_3(1)\\ &= 6.5 + 6 + 5.5 = 18 \text{ ft}\end{aligned}$$

39. **(a)** Read values from the graph for every hour from 1 A.M. through 11 P.M. The values give the heights of 23 rectangles. The width of each rectangle is $\Delta x = 1$. We estimate the area under the curve as

$$\begin{aligned}A &= \sum_{i=1}^{23} f(x_i)\Delta x\\ &= 500(1) + 550(1) + 800(1) + 1600(1) + 4000(1) + 7000(1)\\ &\quad + 7000(1) + 5900(1) + 4500(1) + 3500(1) + 3100(1)\\ &\quad + 3100(1) + 3500(1) + 3800(1) + 4100(1) + 4800(1)\\ &\quad + 4750(1) + 4000(1) + 2500(1) + 2250(1) + 1800(1)\\ &\quad + 1500(1) + 1050(1)\\ &= 75{,}600 \text{ vehicles}\end{aligned}$$

(b) Read values from the graph for every hour from 1 A.M. through 11 P.M. The values give the heights of 23 rectangles. The width of each rectangle is $\Delta x = 1$. We estimate the area under the curve as

$$\begin{aligned}A &= \sum_{i=1}^{23} f(x_i)\Delta x\\ &= 500(1) + 400(1) + 400(1) + 700(1) + 1500(1)\\ &\quad + 3000(1) + 4100(1) + 3900(1) + 3200(1) + 3600(1)\\ &\quad + 4000(1) + 4000(1) + 4300(1) + 5200(1)\\ &\quad + 6000(1) + 6500(1) + 6400(1) + 6000(1) + 4700(1)\\ &\quad + 3100(1) + 2600(1) + 1900(1) + 1300(1)\\ &= 77{,}300 \text{ vehicles}\end{aligned}$$

15.4 The Fundamental Theorem of Calculus

Your Turn 1

Find $\int_1^3 3x^2\,dx$.

The indefinite integral is $\int 3x^2 dx = x^3 + C$.

By the Fundamental Theorem,

$$\int_1^3 3x^2 dx = x^3\Big|_1^3$$
$$= 3^3 - 1^3$$
$$= 27 - 1$$
$$= 26$$

Your Turn 2

Find $\int_3^5 (2x^3 - 3x + 4)\,dx$.

$$\int_3^5 (2x^3 - 3x + 4)\,dx$$
$$= 2\int_3^5 x^3\,dx - 3\int_3^5 x\,dx + 4\int_3^5 dx$$
$$= \frac{2}{4}x^4\Big|_3^5 - \frac{3}{2}x^2\Big|_3^5 + 4x\Big|_3^5$$
$$= \frac{1}{2}(5^4 - 3^4) - \frac{3}{2}(5^2 - 3^2) + 4(5-3)$$
$$= 256$$

Your Turn 3

Find $\int_1^3 \frac{2}{y}\,dy$.

$$\int_1^3 \frac{2}{y}\,dy = 2\int_1^3 \frac{1}{y}\,dy$$
$$= 2\ln|y|\,\Big|_1^3$$
$$= 2(\ln 3 - \ln 1)$$
$$= 2\ln 3 \text{ or } \ln 3^2 = \ln 9$$

Your Turn 4

Evaluate $\int_0^4 2x\sqrt{16 - x^2}\,dx$.

Using Method 1:

Let $u = 16 - x^2$.

Then $du = -2x$.

If $x = 4$, then $u = 16 - 4^2 = 0$.

If $x = 0$, then $u = 16 - 0^2 = 16$.

Now substitute.

Your Turn 5

Find the area between the graph of the function $f(x) = x^2 - 9$ and the x-axis from $x = 0$ to $x = 6$. Here is a graph of the function and the area to be found.

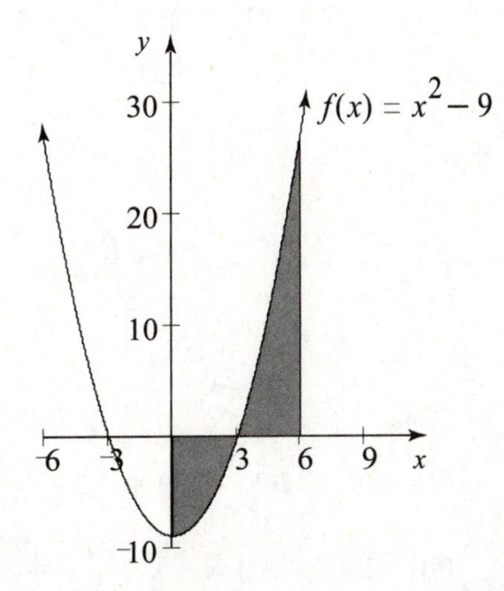

Since the curve is below the x-axis on the interval $(0, 3)$, the definite integral will count this areas as negative. The total positive area is thus

$$\left|\int_0^3 (x^2 - 9)\,dx\right| + \int_3^6 (x^2 - 9)\,dx$$
$$= \left|\left(\frac{1}{3}x^3 - 9x\right)\Big|_0^3\right| + \left(\frac{1}{3}x^3 - 9x\right)\Big|_3^6$$
$$|9 - 27| + (27 - 54 - (9 - 27))$$
$$= 18 + 18 + 18 = 54.$$

15.4 Exercises

1. $\displaystyle\int_{-2}^{4}(-3)\,dp = -3\int_{-2}^{4}dp = -3\cdot p\Big|_{-2}^{4}$

$= -3[4-(-2)]$

$= -18$

3. $\displaystyle\int_{-1}^{2}(5t-3)dt = 5\int_{-1}^{2}t\,dt - 3\int_{-1}^{2}dt$

$= \frac{5}{2}t^2\Big|_{-1}^{2} - 3t\Big|_{-1}^{2}$

$= \frac{5}{2}[2^2-(-1)^2] - 3[2-(-1)]$

$= \frac{5}{2}(4-1) - 3(2+1)$

$= \frac{15}{2} - 9$

$= \frac{15}{2} - \frac{18}{2} = -\frac{3}{2}$

5. $\displaystyle\int_{0}^{2}(5x^2-4x+2)\,dx$

$= 5\int_0^2 x^2dx - 4\int_0^2 x\,dx + 2\int_0^2 dx$

$= \frac{5x^3}{3}\Big|_0^2 - 2x^2\Big|_0^2 + 2x\Big|_0^2$

$= \frac{5}{3}(2^3-0^3) - 2(2^2-0^2) + 2(2-0)$

$= \frac{5}{3}(8) - 2(4) + 2(2) = \frac{40-24+12}{3}$

$= \frac{28}{3}$

7. $\displaystyle\int_0^2 3\sqrt{4u+1}\,du$

Let $4u+1=x$, so that $4\,du = dx$.

When $u=0$, $x=4(0)+1=1$.

When $u=2$, $x=4(2)+1=9$.

$\displaystyle\int_0^2 3\sqrt{4u+1}\,du$

$= \frac{3}{4}\int_0^2\sqrt{4u+1}\,(4\,du)$

$= \frac{3}{4}\int_1^9 x^{1/2}dx$

$= \frac{3}{4}\cdot\frac{x^{3/2}}{3/2}\Big|_1^9$

$= \frac{3}{4}\cdot\frac{2}{3}(9^{3/2}-1^{3/2})$

$= \frac{1}{2}(27-1) = \frac{26}{2} = 13$

9. $\displaystyle\int_0^4 2(t^{1/2}-t)\,dt = 2\int_0^4 t^{1/2}dt - 2\int_0^4 t\,dt$

$= 2\cdot\frac{t^{3/2}}{\frac{3}{2}}\Big|_0^4 - 2\cdot\frac{t^2}{2}\Big|_0^4$

$= \frac{4}{3}(4^{3/2}-0^{3/2}) - (4^2-0^2)$

$= \frac{32}{3} - 16 = -\frac{16}{3}$

11. $\displaystyle\int_1^4(5y\sqrt{y}+3\sqrt{y})dy$

$= 5\int_1^4 y^{3/2}dy + 3\int_1^4 y^{1/2}dy$

$= 5\left(\frac{y^{5/2}}{\frac{5}{2}}\right)\Big|_1^4 + 3\left(\frac{y^{3/2}}{\frac{3}{2}}\right)\Big|_1^4$

$= 2y^{5/2}\Big|_1^4 + 2y^{3/2}\Big|_1^4$

$= 2(4^{5/2}-1) + 2(4^{3/2}-1)$

$= 2(32-1) + 2(8-1)$

$= 62 + 14$

$= 76$

13. $\displaystyle\int_4^6 \frac{2}{(2x-7)^2}\,dx$

Let $u = 2x - 7$, so that $du = 2dx$.

When $x = 6$, $u = 2 \cdot 6 - 7 = 5$.

When $x = 4$, $u = 2 \cdot 4 - 7 = 1$.

$$\int_4^6 \frac{2}{(2x-7)^2}\,dx = \int_1^5 u^{-2}du$$
$$= \frac{u^{-1}}{-1}\bigg|_1^5$$
$$= -u^{-1}\Big|_1^5$$
$$= -\left(\tfrac{1}{5} - 1\right)$$
$$= -\left(-\tfrac{4}{5}\right)$$
$$= \frac{4}{5}$$

15. $\displaystyle\int_1^5 (6n^{-2} - n^{-3})\,dn$

$$= 6\int_1^5 n^{-2}dn - \int_1^5 n^{-3}dn$$
$$= 6 \cdot \frac{n^{-1}}{-1}\bigg|_1^5 - \frac{n^{-2}}{-2}\bigg|_1^5$$
$$= \frac{-6}{n}\bigg|_1^5 + \frac{1}{2n^2}\bigg|_1^5$$
$$= \frac{-6}{5} - \left(\frac{-6}{1}\right) + \left[\frac{1}{2(25)} - \frac{1}{2(1)}\right]$$
$$= \frac{-6}{5} - \frac{6}{1} + \frac{1}{50} - \frac{1}{2}$$
$$= \frac{108}{25}$$

17. $\displaystyle\int_{-3}^{-2}\left(2e^{-0.1y} + \frac{3}{y}\right)dy$

$$= 2\int_{-3}^{-2} e^{-0.1y}dy + \int_{-3}^{-2}\frac{3}{y}dy$$
$$= 2 \cdot \frac{e^{-0.1y}}{-0.1}\bigg|_{-3}^{-2} + 3\ln|y|\Big|_{-3}^{-2}$$
$$= -20e^{-0.1y}\Big|_{-3}^{-2} + 3\ln|y|\Big|_{-3}^{-2}$$
$$= 20e^{0.3} - 20e^{0.2} + 3\ln 2 - 3\ln 3$$
$$\approx 1.353$$

19. $\displaystyle\int_1^2\left(e^{4u} - \frac{1}{(u+1)^2}\right)du$

$$= \int_1^2 e^{4u}du - \int_1^2 \frac{1}{(u+1)^2}du$$
$$= \frac{e^{4u}}{4}\bigg|_1^2 - \frac{-1}{u+1}\bigg|_1^2$$
$$= \frac{e^8}{4} - \frac{e^4}{4} + \frac{1}{2+1} - \frac{1}{1+1}$$
$$= \frac{e^8}{4} - \frac{e^4}{4} - \frac{1}{6}$$
$$\approx 731.4$$

21. $\displaystyle\int_{-1}^0 y(2y^2 - 3)^5\,dy$

Let $u = 2y^3 - 3$, so that

$du = 4y\,dy$ and $\frac{1}{4}du = y\,dy$.

When $y = -1$, $u = 2(-1)^2 - 3 = -1$.

When $y = 0$, $u = 2(0)^2 - 3 = -3$.

$$\frac{1}{4}\int_{-1}^{-3} u^5du = \frac{1}{4} \cdot \frac{u^6}{6}\bigg|_{-1}^{-3}$$
$$= \frac{1}{24}u^6\Big|_{-1}^{-3}$$
$$= \frac{1}{24}(-3)^6 - \frac{1}{24}(-1)^6$$
$$= \frac{729}{24} - \frac{1}{24}$$
$$= \frac{728}{24} = \frac{91}{3}$$

23. $\displaystyle\int_1^{64}\frac{\sqrt{z} - 2}{\sqrt[3]{z}}\,dz$

$$= \int_1^{64}\left(\frac{z^{1/2}}{z^{1/2}} - 2z^{-1/3}\right)dz$$
$$= \int_1^{64} z^{1/6}dz - 2\int_1^{64} z^{-1/3}dz$$
$$= \frac{z^{7/6}}{\frac{7}{6}}\bigg|_1^{64} - 2\frac{z^{2/3}}{\frac{2}{3}}\bigg|_1^{64}$$
$$= \frac{6z^{7/6}}{7}\bigg|_1^{64} - 3z^{2/3}\Big|_1^{64}$$

$$= \frac{6(64)^{7/6}}{7} - \frac{6(1)^{7/6}}{7}$$
$$-3(64^{2/3} - 1^{2/3})$$
$$= \frac{6(128)}{7} - \frac{6}{7} - 3(16-1)$$
$$= \frac{768 - 6 - 315}{7} = \frac{447}{7} \approx 63.86$$

25. $\int_1^2 \frac{\ln x}{x}\,dx$

Let $u = \ln x$, so that

$$du = \frac{1}{x}dx.$$

When $x = 1,\ u = \ln 1 = 0.$

When $x = 2,\ u = \ln 2.$

$$\int_0^{\ln 2} u\,du = \left.\frac{u^2}{2}\right|_0^{\ln 2}$$
$$= \frac{(\ln 2)^2}{2} - 0$$
$$= \frac{(\ln 2)^2}{2}$$
$$\approx 0.2402$$

27. $\int_0^8 x^{1/3}\sqrt{x^{4/3} + 9}\,dx$

Let $u = x^{4/3} + 9$, so that

$du = \frac{4}{3}x^{1/3}dx$ and $\frac{3}{4}du = x^{1/3}dx.$

When $x = 0, u = 0^{4/3} + 9 = 9.$

When $x = 8, u = 8^{4/3} + 9 = 25.$

$$\frac{3}{4}\int_9^{25} \sqrt{u}du = \frac{3}{4}\int_9^{25} u^{1/2}du$$
$$= \frac{3}{4}\cdot\left.\frac{u^{3/2}}{\frac{3}{2}}\right|_9^{25}$$
$$= \left.\frac{1}{2}u^{3/2}\right|_9^{25}$$
$$= \frac{1}{2}(25)^{3/2} - \frac{1}{2}(9)^{3/2}$$
$$= \frac{125}{2} - \frac{27}{2} = 49$$

29. $\int_0^1 \frac{e^{2t}}{(3 + e^{2t})^2}\,dt$

Let $u = 3 + e^{2t}$, so that $du = 2e^{2t}\,dt.$

When $x = 1,\ u = 3 + e^{2\cdot 1} = 3 + e^2.$

When $x = 0,\ u = 3 + e^{2\cdot 0} = 4.$

$$\int_0^1 \frac{e^{2t}}{(3 + e^{2t})^2}\,dt = \frac{1}{2}\int_4^{3+e^2} u^{-2}du$$
$$= \left.\frac{1}{2}\cdot\frac{u^{-1}}{-1}\right|_4^{3+e^2} = \left.\frac{-1}{2u}\right|_4^{3+e^2}$$
$$= \frac{1}{8} - \frac{1}{2(3 + e^2)}$$
$$\approx 0.07687$$

31. $f(x) = 2x - 14; [6, 10]$

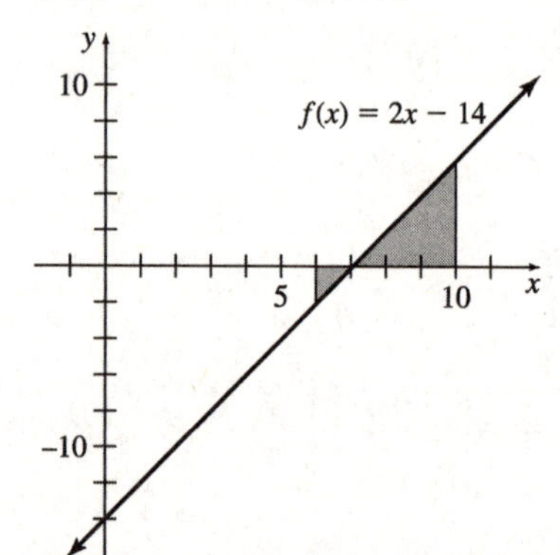

The graph crosses the x-axis at

$$0 = 2x - 14$$
$$2x = 14$$
$$x = 7.$$

This location is in the interval. The area of the region is

$$\left|\int_6^7 (2x - 14)\,dx\right| + \int_7^{10} (2x - 14)\,dx$$
$$= \left|(x^2 - 14x)\right|_6^7\Big| + (x^2 - 14x)\Big|_7^{10}$$
$$= |(7^2 - 98) - (6^2 - 84)|$$
$$+ (10^2 - 140) - (7^2 - 98)$$
$$= |-1| + (-40) - (-49)$$
$$= 10.$$

33. $f(x) = 2 - 2x^2; [0, 5]$

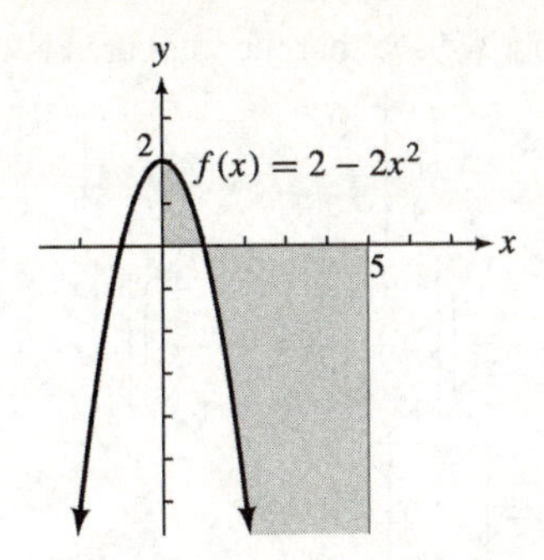

Find the points where the graph crosses the x-axis by solving $2 - 2x^2 = 0$.

$$\begin{aligned} 2 - 2x^2 &= 0 \\ 2x^2 &= 2 \\ x^2 &= 1 \\ x &= \pm 1. \end{aligned}$$

The only solution in the interval $[0, 5]$ is 1. The total area is

$$\int_0^1 (2 - 2x^2)\,dx + \left|\int_2^5 (2 - 2x^2)\,dx\right|$$

$$= \left(2x - \frac{2x^3}{3}\right)\Bigg|_0^1 + \left|\left(2x - \frac{2x^3}{3}\right)\Bigg|_1^5\right|$$

$$= 2 - \frac{2}{3} + \left|10 - \frac{2(5^3)}{3} - 2 + \frac{2}{3}\right|$$

$$= \frac{4}{3} + \left|\frac{-224}{3}\right|$$

$$= \frac{228}{3}$$

$$= 76.$$

35. $f(x) = x^3; [-1, 3]$

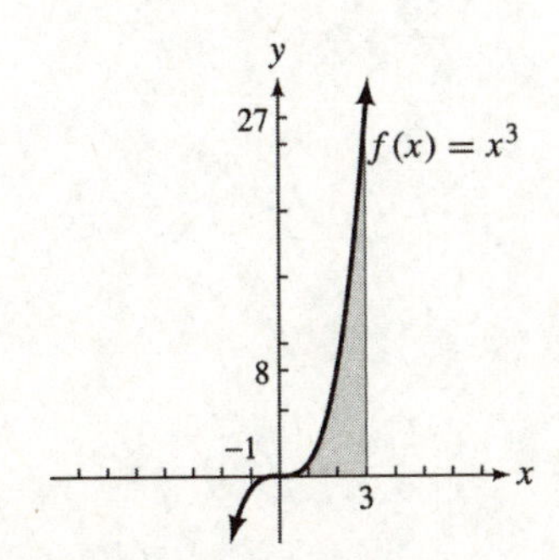

The solution

$$\begin{aligned} x^3 &= 0 \\ x &= 0 \end{aligned}$$

indicates that the graph crosses the x-axis at 0 in the given interval $[-1, 3]$.

The total area is

$$\left|\int_{-1}^0 x^3 dx\right| + \int_0^3 x^3 dx$$

$$= \left|\frac{x^4}{4}\right|_{-1}^0 + \left|\frac{x^4}{4}\right|_0^3$$

$$= \left|\left(0 - \frac{1}{4}\right)\right| + \left(\frac{3^4}{4} - 0\right)$$

$$= \frac{1}{4} + \frac{81}{4} = \frac{82}{4}$$

$$= \frac{41}{2}.$$

37. $f(x) = e^x - 1; [-1, 2]$

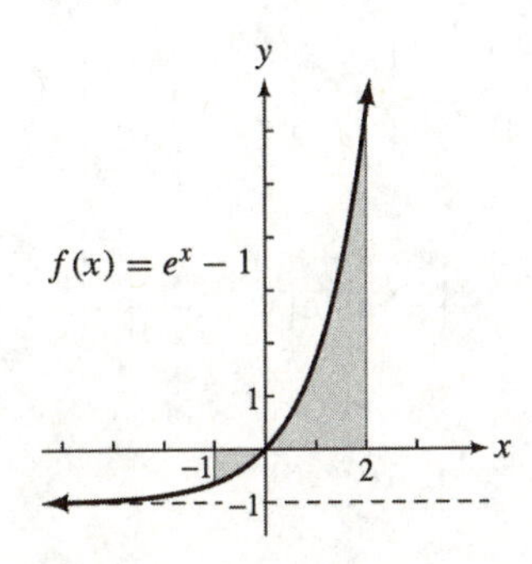

Solve

$$\begin{aligned} e^x - 1 &= 0. \\ e^x &= 1 \\ x \ln e &= \ln 1 \\ x &= 0 \end{aligned}$$

The graph crosses the x-axis at 0 in the given interval $[-1, 2]$.

The total area is

$$\left|\int_{-1}^0 (e^x - 1)\,dx\right| + \int_0^2 (e^x - 1)\,dx$$

$$= \left|(e^x - x)\right|_{-1}^0 + (e^x - x)\Big|_0^2$$

$$= |(1 - 0) - (e^{-1} + 1)| + (e^2 - 2) - (1 - 0)$$

$$= |1 - e^{-1} - 1| + e^2 - 2 - 1$$

$$= \frac{1}{e} + e^2 - 3$$

$$\approx 4.757.$$

39. $f(x) = \frac{1}{x} - \frac{1}{e}$; $[1, e^2]$

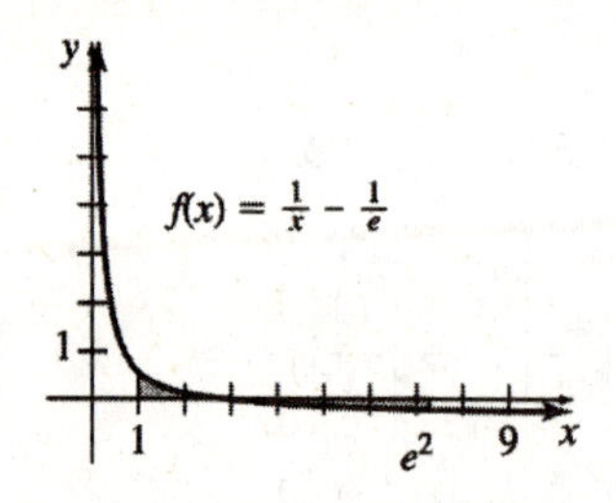

The graph crosses the x-axis at

$$0 = \frac{1}{x} - \frac{1}{e}$$
$$\frac{1}{x} = \frac{1}{e}$$
$$x = e.$$

This location is in the interval. The area of the region is

$$\int_1^e \left(\frac{1}{x} - \frac{1}{e}\right) dx + \left|\int_e^{e^2} \left(\frac{1}{x} - \frac{1}{e}\right) dx\right|$$
$$= \left|\ln|x| - \frac{x}{e}\right|_1^e + \left|\left(\ln|x| - \frac{x}{e}\right)\Big|_e^{e^2}\right|$$
$$= 0 - \left(-\frac{1}{e}\right) + |(2 - e) - 0|$$
$$= \frac{1}{e} + |2 - e|$$
$$= e - 2 + \frac{1}{e}.$$

41. $y = 4 - x^2$; $[0, 3]$

From the graph, we see that the total area is

$$\int_0^2 (4 - x^2)\, dx + \left|\int_2^3 (4 - x^2)\, dx\right|$$
$$= \left(4x - \frac{x^3}{3}\right)\Big|_0^2 + \left|\left(4x - \frac{x^3}{3}\right)\Big|_2^3\right|$$
$$= \left[\left(8 - \frac{8}{3}\right) - 0\right]$$
$$+ \left|\left[(12 - 9) - \left(8 - \frac{8}{3}\right)\right]\right|$$
$$= \frac{16}{3} + \left|3 - \frac{16}{3}\right|$$
$$= \frac{16}{3} + \frac{7}{3}$$
$$= \frac{23}{3}$$

43. $y = e^x - e$; $[0, 2]$

From the graph, we see that total area is

$$\left|\int_0^1 (e^x - e)\, dx\right| + \int_1^2 (e^x - e)\, dx$$
$$= \left|(e^x - xe)\Big|_0^1\right| + (e^x - xe)\Big|_1^2$$
$$= |(e^1 - e) - (e^0 + 0)| + (e^2 - 2e) - (e^1 - e)$$
$$= |-1| + e^2 - 2e$$
$$= 1 + e^2 - 2e \approx 2.952.$$

45. $\int_a^c f(x)\,dx = \int_a^b f(x)\,dx + \int_b^c f(x)\,dx$

47. $\int_0^{16} f(x)\,dx = \int_0^2 f(x)\,dx + \int_2^5 f(x)\,dx$

$+ \int_5^8 f(x)\,dx + \int_8^{16} f(x)\,dx$

$= \frac{1}{2} \cdot 2(1 + 3) + \frac{\pi(3^2)}{4}$

$- \frac{\pi(3^2)}{4} - \frac{1}{2}(3)(8)$

$= 4 + \frac{9}{4}\pi - \frac{9}{4}\pi - 12 = -8$

49. Prove: $\int_a^b f(x)\,dx$

$= \int_a^c f(x)\,dx + \int_c^b f(x)\,dx.$

Let $F(x)$ be an antiderivative of $f(x)$.

$\int_a^c f(x)\,dx + \int_c^b f(x)\,dx$

$= F(x)\Big|_a^c + F(x)\Big|_c^b$

$= [F(c) - F(a)] + [F(b) - F(c)]$

$= F(c) - F(a) + F(b) - F(c)$

$= F(b) - F(a)$

$= \int_a^b f(x)\,dx$

51. $\int_{-1}^4 f(x)\,dx$

$= \int_{-1}^0 (2x + 3) + dx \int_0^4 \left(-\frac{x}{4} - 3\right) dx$

$= (x^2 + 3x)\Big|_{-1}^0 + \left(-\frac{x^2}{8} - 3x\right)\Bigg|_0^4$

$= -(1 - 3) + (-2 - 12)$

$= 2 - 14$

$= -12$

53. **(a)** $g(t) = t^4$ and $c = 1$, use substitution.

$f(x) = \int_c^x g(t)dt$

$= \int_1^x t^4 dt$

$= \frac{t^5}{5}\Bigg|_1^x$

$= \frac{x^5}{5} - \frac{(1)^5}{5}$

$= \frac{x^5}{5} - \frac{1}{5}$

(b) $f'(x) = \frac{d}{dx}(f(x))$

$= \frac{d}{dx}\left(\frac{x^5}{5} - \frac{1}{5}\right)$

$= \frac{1}{5} \cdot \frac{d}{dx}(x^5) - \frac{d}{dx}\left(\frac{1}{5}\right)$

$= \frac{1}{5} \cdot 5x^4 - 0 = x^4$

Since $g(t) = t^4$, then $g(x) = x^4$ and we see $f'(x) = g(x)$.

(c) Let $g(t) = e^{t^2}$ and $c = 0$, then

$f(x) = \int_0^x e^{t^2}\,dt.$

$f(1) = \int_0^1 e^{t^2}\,dt$ and $f(1.01) = \int_0^{1.01} e^{t^2}\,dt.$

Use the fnInt command in the Math menu of your calculator to find $\int_0^1 e^{x^2}\,dx$ and $\int_0^{1.01} e^{x^2}\,dx$. The resulting screens are:

```
fnInt(e^(X²),X,0
,1)
          1.462651746
```

$$f(1) \approx 1.46265$$
$$f(1.01) \approx 1.49011$$

Use $\frac{f(1+h)-f(1)}{h}$ to approximate $f'(1)$ with $h = 0.01$

$$\frac{f(1+h)-f(1)}{h} = \frac{f(1.01)-f(1)}{0.01} \approx \frac{1.49011 - 1.46265}{0.01} = 2.746$$

So $f'(1) \approx 2.746$, and $g(1) = e^{1^2} = e \approx 2.718$.

55. $P'(t) = (3t+3)(t^2+2t+2)^{1/3}$

(a) $\int_0^3 3(t+1)(t^2+2t+2)^{1/3}dt$

Let $u = t^2 + 2t + 2$, so that $du = (2t+2)\,dt$ and $\frac{1}{2}du = (t+1)\,dt$.

When $t = 0$, $u = 0^2 + 2 \cdot 0 + 2 = 2$.

When $t = 3, u = 3^2 + 2 \cdot 3 + 2 = 17$.

$$\frac{3}{2}\int_2^{17} u^{1/3}du = \frac{3}{2} \cdot \frac{u^{4/3}}{\frac{4}{3}}\Bigg|_2^{17} = \frac{9}{8}u^{4/3}\Bigg|_2^{17} = \frac{9}{8}(17)^{4/3} - \frac{9}{8}(2)^{4/3} \approx 46.341$$

Total profits for the first 3 yr were $\frac{9000}{8}(17^{4/3} - 2^{4/3}) \approx \$46,341$.

(b) $\int_3^4 3(t+1)(t^2+2t+2)^{1/3}dt$

Let $u = t^2 + 2t + 2$, so that $du = (2t+2)\,dt = 2(t+1)\,dt$ and $\frac{3}{2}du = 3(t+1)\,dt$.

When $t = 3, u = 3^2 + 2 \cdot 3 + 2 = 17$.

When $t = 4, u = 4^2 + 2 \cdot 4 + 2 = 26$.

$$\frac{3}{2}\int_{17}^{26} u^{1/3}du = \frac{9}{8}u^{4/3}\Bigg|_{17}^{26} = \frac{9}{8}(26)^{4/3} - \frac{9}{8}(17)^{4/3} \approx 37.477$$

Profit in the fourth year was $\frac{9000}{8}(26^{4/3} - 17^{4/3}) \approx \$37,477$.

(c) $\lim_{x\to\infty} P'(t)$

$$= \lim_{x\to\infty}(3t+3)(t^2+2t+2)^{1/3} = \infty$$

The annual profit is slowly increasing without bound.

57. $P'(t) = 140t^{5/2}$

$$\int_0^4 140t^{5/2}dt = 140 \cdot \frac{t^{7/2}}{\frac{7}{2}}\Bigg|_0^4 = 40t^{7/2}\Big|_0^4 = 5120$$

Since 5120 is above the total level of acceptable pollution (4850), the factory cannot operate for 4 years without killing all the fish in the lake.

59. Growth rate is $0.6 + \frac{4}{(t+1)^3}$ ft/yr.

(a) Total growth in the second year is

$$\int_1^2 \left[0.6 + \frac{4}{(t+1)^3}\right]dt = \left[0.6t + \frac{4}{-2(t+1)^2}\right]\Bigg|_1^2$$
$$= \left[0.6(2) - \frac{2}{(2+1)^2}\right] - \left[0.6(1) - \frac{2}{(1+1)^2}\right]$$
$$= \frac{44}{45} - \frac{1}{10} \approx 0.8778\text{ ft.}$$

(b) Total growth in the third year is

$$\int_2^3 \left[0.6 + \frac{4}{(t+1)^3}\right] dt$$
$$= \left[0.6t + \frac{4}{-2(t+1)^2}\right]\Bigg|_2^3$$
$$= \left[0.6(3) - \frac{2}{(3+1)^2}\right] - \left[0.6(2) - \frac{2}{(2+1)^2}\right]$$
$$= \frac{67}{40} - \frac{44}{45}$$
$$\approx 0.6972 \text{ ft.}$$

61. $R'(t) = \frac{5}{t+1} + \frac{2}{\sqrt{t+1}}$

(a) Total reaction from $t = 1$ to $t = 12$ is

$$\int_1^{12} \left(\frac{5}{t+1} + \frac{2}{\sqrt{t+1}}\right) dt$$
$$= \left[5\ln(t+1) + 4\sqrt{t+1}\right]\Big|_1^{12}$$
$$= (5\ln 13 + 4\sqrt{13}) - (5\ln 2 + 4\sqrt{2})$$
$$\approx 18.12.$$

(b) Total reaction from $t = 12$ to $t = 24$ is

$$\int_{12}^{24} \left(\frac{5}{t+1} + \frac{2}{\sqrt{t+1}}\right) dt$$
$$= \left[5\ln(t+1) + 4\sqrt{t+1}\right]\Big|_{12}^{24}$$
$$= (5\ln 25 + 4\sqrt{25}) - (5\ln 13 + 4\sqrt{13})$$
$$\approx 8.847.$$

63. **(b)** $\int_0^{60} n(x)\, dx$

(c) $\int_5^{10} \sqrt{5x+1}\, dx$

Let $u = 5x + 1$. Then $du = 5\, dx$.

When $x = 5$, $u = 26$; when $x = 10$, $u = 51$.

$$\frac{1}{5}\int_{26}^{51} u^{1/2} du$$
$$= \frac{1}{5} \cdot \frac{u^{3/2}}{\frac{3}{2}}\Bigg|_{26}^{51}$$
$$= \frac{2}{15}u^{3/2}\Big|_{26}^{51}$$
$$= \frac{2}{15}(51^{3/2} - 26^{3/2})$$
$$\approx 30.89 \text{ million}$$

65. $v = k(R^2 - r^2)$

(a)
$$Q(R) = \int_0^R 2\pi v r\, dr$$
$$= \int_0^R 2\pi k(R^2 - r^2) r\, dr$$
$$= 2\pi k \int_0^R (R^2 r - r^3)\, dr$$
$$= 2\pi k \left(\frac{R^2 r^2}{2} - \frac{r^4}{4}\right)\Bigg|_0^R$$
$$= 2\pi k \left(\frac{R^4}{2} - \frac{R^4}{4}\right)$$
$$= 2\pi k \left(\frac{R^4}{4}\right)$$
$$= \frac{\pi k R^4}{2}$$

(b)
$$Q(0.4) = \frac{\pi k (0.4)^4}{2}$$
$$= 0.04k \text{ mm/min}$$

67. $E(t) = 753t^{-0.1321}$

(a) Since t is the age of the beagle in years, to convert the formula to days, let $T = 365t$, or $t = \frac{T}{365}$.

$$E(T) = 753\left(\frac{T}{365}\right)^{-0.1321}$$
$$\approx 1642T^{-0.1321}$$

Now, replace T with t.

$$E(t) = 1642t^{-0.1321}$$

(b) The beagle's age in days after one year is 365 days and after 3 years she is 1095 days old.

$$\int_{365}^{1095} 1642t^{-0.1321}dt$$

$$= 1642\frac{1}{0.8679}t^{0.8679}\Big|_{365}^{1095}$$

$$\approx 1892(1{,}095^{0.8679} - 365^{0.8679})$$

$$\approx 505{,}155$$

The beagle's total energy requirements are about 505,000 kJ/$W^{0.67}$, where W represents weight.

69. **(a)** $f(x) = 40.2 + 3.50x - 0.897x^2$

$$\int_0^9 (40.2 + 3.50x - 0.897x^2)\,dx$$

$$= (40.2x + 1.74x^2 - 0.299x^3)\Big|_0^9$$

$$\approx 286$$

The integral represents the population aged 0 to 90, which is about 286 million.

(b) $$\int_{4.5}^{6.5} (40.2 + 3.50x - 0.897x^2)\,dx$$

$$= (40.2x + 1.74x^2 - 0.299x^3)\Big|_{4.5}^{6.5}$$

$$\approx 64$$

The number of baby boomers is about 64 million.

71. $c'(t) = ke^{rt}$

(a) $c'(t) = 1.2\,e^{0.04t}$

(b) The amount of oil that the company will sell in the next ten years is given by the integral

$$\int_0^{10} 1.2e^{0.04t}\,dt.$$

(c) $$\int_0^{10} 1.2e^{0.04t}\,dx = \frac{1.2e^{0.04t}}{0.04}\Big|_0^{10}$$

$$= 30e^{0.04t}\Big|_0^{10}$$

$$= 30e^{0.4} - 30$$

$$\approx 14.75$$

This represents about 14.75 billion barrels of oil.

(d) $$\int_0^T 1.2e^{0.04t}\,dt = 30e^{0.04t}\Big|_0^T$$

$$= 30e^{0.04T} - 30$$

Solve

$$20 = 30e^{0.04T} - 30.$$

$$50 = 30e^{0.04T}$$

$$\frac{5}{3} = e^{0.04T}$$

$$\ln\frac{5}{3} = 0.04T\ln e$$

$$T = \frac{\ln\frac{5}{3}}{0.04}$$

$$\approx 12.8$$

The oil will last about 12.8 years.

(e) $$\int_0^T 1.2e^{0.02t}\,dt = 60e^{0.02t}\Big|_0^T$$

$$= 60e^{0.02T} - 60$$

Solve

$$20 = 60e^{0.02T} - 60.$$

$$80 = 60e^{0.02T}$$

$$\frac{4}{3} = e^{0.02T}$$

$$\ln\frac{4}{3} = 0.02T\ln e$$

$$T = \frac{\ln\frac{4}{3}}{0.02} \approx 14.4$$

The oil will last about 14.4 years.

15.5 The Area Between Two Curves

Your Turn 1

Find the area bounded by $f(x) = 4 - x^2$, $g(x) = x + 2$, $x = -2$, and $x = 1$. A sketch such as the one below shows that the two graphs intersect at the points $(-2, 0)$ and $(1, 3)$.

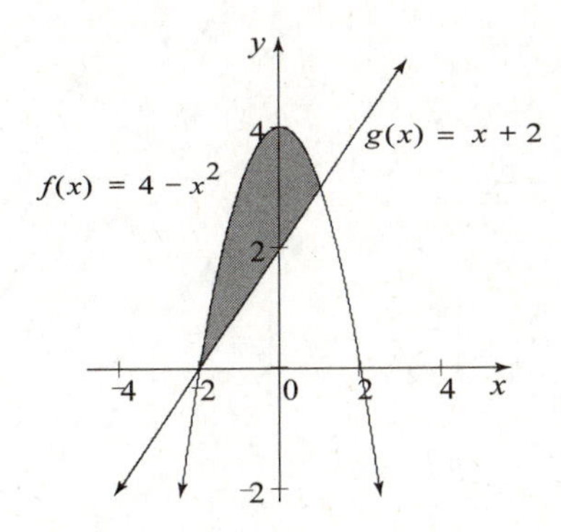

Over the interval $[-2, 1]$, $f(x) \geq g(x)$, so the area will be given by $\int_{-2}^{1} [f(x) - g(x)]dx$.

$$\int_{-2}^{1} [f(x) - g(x)]dx = \int_{-2}^{1} [4 - x^2 - (x + 2)]dx$$
$$= \int_{-2}^{1} (2 - x - x^2)dx$$
$$= \left(2x - \frac{1}{2}x^2 - \frac{1}{3}x^3\right)\Bigg|_{-2}^{1}$$
$$= \left(2 - \frac{1}{2} - \frac{1}{3}\right) - \left(-4 - 2 + \frac{8}{3}\right)$$
$$= \frac{9}{2}$$

Your Turn 2

Find the area between the curves $y = x^{1/4}$ and $y = x^2$. First find where these two curves intersect by setting the two righthand sides equal.

$$x^{1/4} = x^2$$
$$x = x^8$$
$$x^8 - x = 0$$
$$x(x^7 - 1) = 0$$
$$x = 0 \text{ or } x^7 - 1 = 0$$
$$x = 0 \text{ or } x = 1.$$

The corresponding y values are 0 and 1, respectively, so the curves intersect at $(0, 0)$ and $(1, 1)$, as shown in the graph below. Here $f(x) = x^{1/4}$ and $g(x) = x^2$.

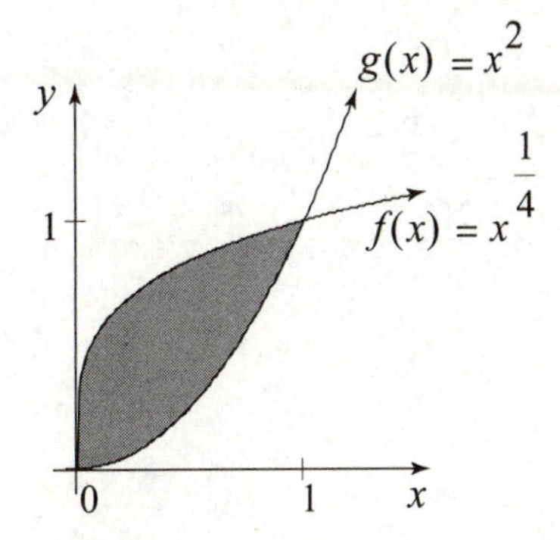

Over the interval $[0, 1]$, $f(x) \geq g(x)$, so the area is given by the integral $\int_0^1 [f(x) - g(x)]\,dx$.

$$\int_0^1 [f(x) - g(x)]\,dx = \int_0^1 (x^{1/4} - x^2)\,dx$$
$$= \left(\frac{4}{5}x^{5/2} - \frac{1}{3}x^3\right)\Bigg|_0^1$$
$$= \left(\frac{4}{5} - \frac{1}{3}\right) - (0 - 0) = \frac{7}{15}$$

Your Turn 3

Find the area enclosed by $y = x^2 - 3x$ and $y = 2x$ on $[0, 6]$. First find where the two graphs intersect.

$$x^2 - 3x = 2x$$
$$x^2 - 5x = 0$$
$$x(x - 5) = 0$$
$$x = 0 \quad \text{or} \quad x = 5.$$

The intersection points are $(0, 0)$ and $(5, 10)$, so we will need to use two integrals. On $(0, 5)$, $2x$ is the larger function and on $(5, 6)$, $x^2 - 3x$ is the larger function, as illustrated in the following graph.

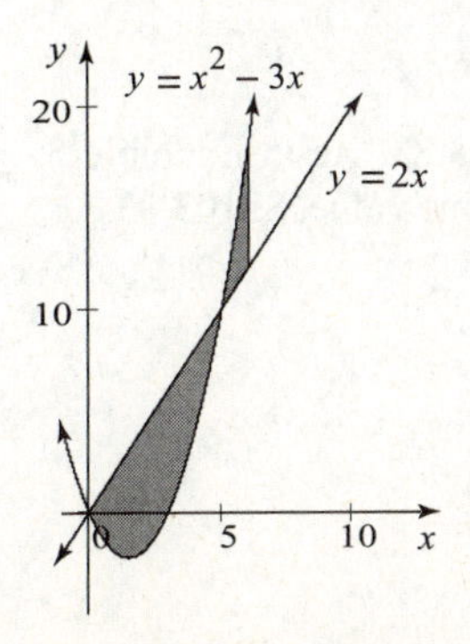

$$\text{Area} = \int_0^5 [2x - (x^2 - 3x)]dx + \int_5^6 [(x^2 - 3x) - 2x]dx$$

$$= \int_0^5 (5x - x^2)dx + \int_5^6 (x^2 - 5x)dx$$

$$= \left(\frac{5}{2}x^2 - \frac{1}{3}x^3\right)\Bigg|_0^5 + \left(\frac{1}{3}x^3 - \frac{5}{2}x^2\right)\Bigg|_5^6$$

$$= \left(\frac{125}{2} - \frac{125}{3} - 0\right) + \left(\frac{216}{3} - \frac{180}{2} - \frac{125}{3} + \frac{125}{2}\right)$$

$$= \frac{71}{3}$$

Your Turn 4

Find the consumers' surplus and the producers' surplus for oat bran when the price in dollars per ton is $D(q) = 600 - e^{q/3}$ when the demand is q tons, and the price in dollars per ton is $S(q) = e^{q/3} - 100$ when the demand is q tons.

First find the equilibrium quantity.

$$e^{q/3} - 100 = 600 - e^{q/3}$$
$$2e^{q/3} = 700$$
$$e^{q/3} = 350$$
$$\frac{q}{3} = \ln 350$$
$$q = 3\ln 350$$
$$q \approx 17.57380$$

The equilibrium price is

$$S(17.57380) = e^{17.57380/3} - 100$$
$$\approx 250.00$$

The consumers' surplus is given by the following integral:

$$\int_0^{17.57380} (600 - e^{q/3} - 250)\,dq$$

$$= \left(350q - 3e^{q/3}\right)\Bigg|_0^{17.57380}$$

$$= (350(17.57380) - 3e^{17.57380/3}) - (0 - 3)$$

$$\approx 5103.83$$

The consumers' surplus is \$5103.83. As in Example 5, the producers' surplus has the same value, \$5103.83.

15.5 Exercises

1. $x = -2, x = 1, y = 2x^2 + 5, y = 0$

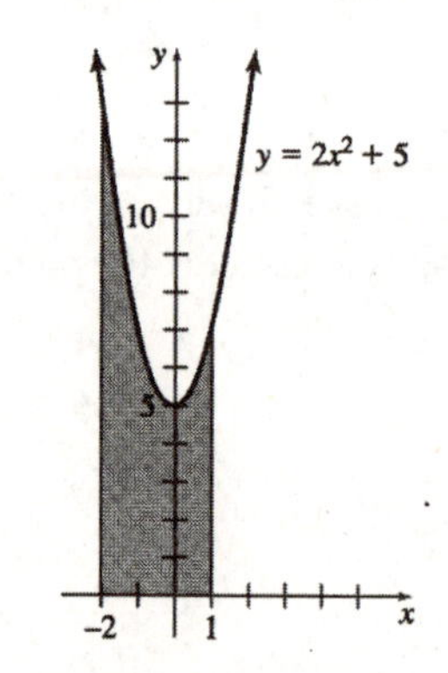

$$\int_{-2}^{1} [(2x^2 + 5) - 0] = \left(\frac{2x^3}{3} + 5x\right)\Bigg|_{-2}^{1}$$

$$= \left(\frac{2}{3} + 5\right) - \left(-\frac{16}{3} - 10\right)$$

$$= 21$$

3. $x = -3, x = 1, y = x^3 + 1, y = 0$

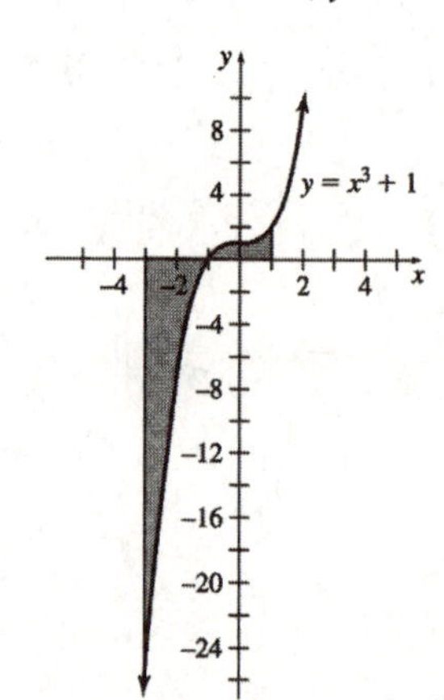

To find the points of intersection of the graphs, substitute for y.

$$x^3 + 1 = 0$$
$$x^3 = -1$$
$$x = -1$$

The region is composed of two separate regions because $y = x^3 + 1$ intersects $y = 0$ at $x = -1$.

Let $f(x) = x^3 + 1, g(x) = 0.$

In the interval $[-3, -1]$, $g(x) \geq f(x)$.

In the interval $[-1, 1]$, $f(x) \geq g(x)$.

$$\int_{-3}^{-1}[0-(x^3+1)dx]+\int_{-1}^{1}[(x^3+1)-0]dx$$

$$=\left(\frac{-x^4}{4}-x\right)\Bigg|_{-3}^{-1}+\left(\frac{x^4}{4}+x\right)\Bigg|_{-1}^{1}$$

$$=\left(-\frac{1}{4}+1\right)-\left(-\frac{81}{4}+3\right)+\left(\frac{1}{4}+1\right)-\left(\frac{1}{4}-1\right)$$

$$=20$$

5. $x=-2, x=1, y=2x, y=x^2-3$

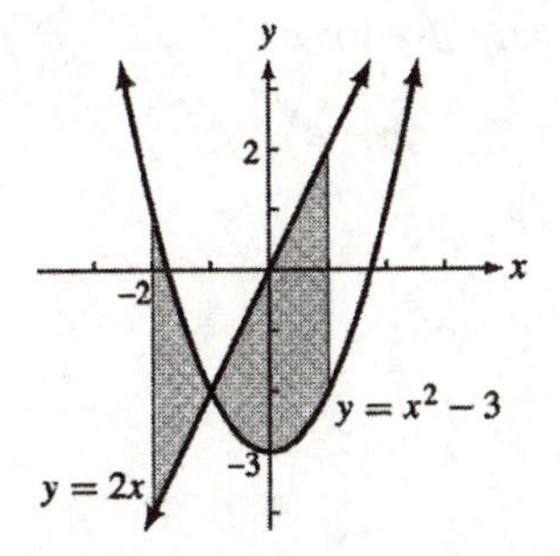

Find the points of intersection of the graphs of $y=2x$ and $y=x^2-3$ by substituting for y.

$$2x=x^2-3$$
$$0=x^2-2x-3$$
$$0=(x-3)(x+1)$$

The only intersection in $[-2,1]$ is at $x=-1$.

In the interval $[-2,-1]$, $(x^2-3)\geq 2x$.

In the interval $[-1,1]$, $2x\geq(x^2-3)$.

$$\int_{-2}^{-1}[(x^2-3)-(2x)]dx+\int_{-1}^{1}[(2x)-(x^2-3)]dx$$

$$=\int_{-2}^{-1}(x^2-3-2x)\,dx+\int_{-1}^{1}(2x-x^2+3)\,dx$$

$$=\left(\frac{x^3}{3}-3x-x^2\right)\Bigg|_{-2}^{-1}+\left(x^2-\frac{x^3}{3}+3x\right)\Bigg|_{-1}^{1}$$

$$=-\frac{1}{3}+3-1-\left(-\frac{8}{3}+6-4\right)+1-\frac{1}{3}+3$$
$$-\left(1+\frac{1}{3}-3\right)$$

$$=\frac{5}{3}+6=\frac{23}{3}$$

7. $y=x^2-30$
$y=10-3x$

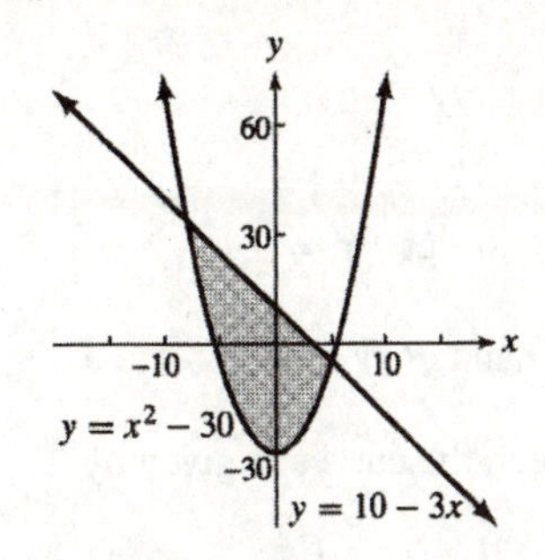

Find the points of intersection.

$$x^2-30=10-3x$$
$$x^2+3x-40=0$$
$$(x+8)(x-5)=0$$
$$x=-8 \text{ or } x=5$$

Let $f(x)=10-3x$ and $g(x)=x^2-30$.

The area between the curves is given by

$$\int_{-8}^{5}[f(x)-g(x)]\,dx$$

$$=\int_{-8}^{5}[(10-3x)-(x^2-30)]\,dx$$

$$=\int_{-8}^{5}(-x^2-3x+40)\,dx$$

$$=\left(\frac{-x^3}{3}-\frac{3x^3}{2}+40x\right)\Bigg|_{-8}^{5}$$

$$=\frac{-5^3}{3}-\frac{3(5)^2}{2}+40(5)$$
$$-\left[\frac{-(-8)^3}{3}-\frac{3(-8)^2}{2}+40(-8)\right]$$

$$=\frac{-125}{3}-\frac{75}{2}+200-\frac{512}{3}+\frac{192}{2}+320$$

$$\approx 366.1667.$$

9. $y=x^2, y=2x$

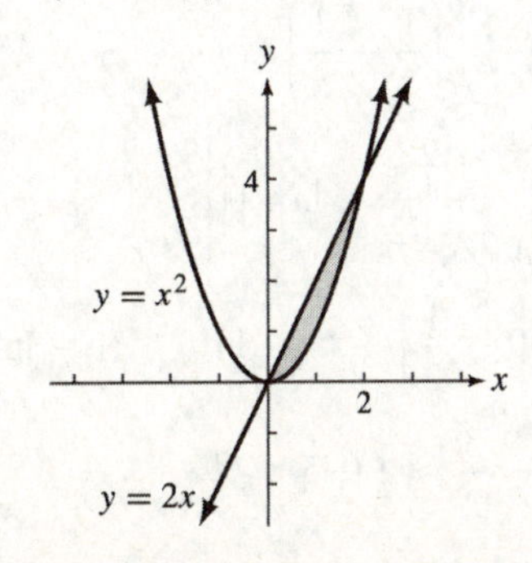

Find the points of intersection.

$$x^2 = 2x$$
$$x^2 - 2x = 0$$
$$x(x-2) = 0$$
$$x = 0 \quad \text{or} \quad x = 2$$

Let $f(x) = 2x$ and $g(x) = x^2$.

The area between the curves is given by

$$\int_0^2 [f(x) - g(x)]\,dx = \int_0^2 (2x - x^2)\,dx$$
$$= \left(\frac{2x^2}{2} - \frac{x^3}{3}\right)\Bigg|_0^2$$
$$= 4 - \frac{8}{2} = \frac{4}{3}.$$

11. $x = 1, x = 6, y = \frac{1}{x}, y = \frac{1}{2}$

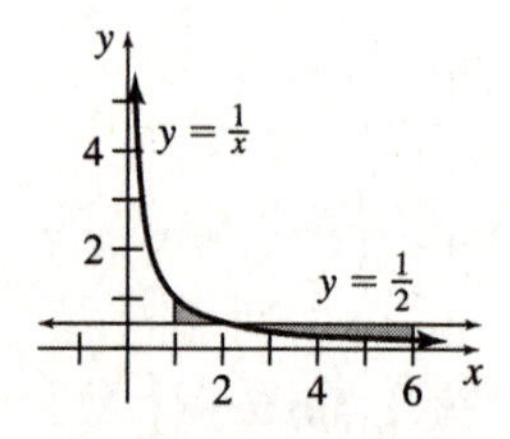

To find the points of intersection of the graphs, substitute for y.

$$\frac{1}{x} = \frac{1}{2}$$
$$x = 2$$

The region is composed of two separate regions because $y = \frac{1}{x}$ intersects $y = \frac{1}{2}$ at $x = 2$.

Let $f(x) = \frac{1}{x}, g(x) = \frac{1}{2}$.

In the interval $[1, 2]$, $f(x) \ge g(x)$.

In the interval $[2, 6]$, $g(x) \ge f(x)$.

$$\int_1^2 \left(\frac{1}{x} - \frac{1}{2}\right) dx + \int_2^6 \left(\frac{1}{2} - \frac{1}{x}\right) dx$$
$$= \left(\ln|x| - \frac{x}{2}\right)\Bigg|_1^2 + \left(\frac{x}{2} - \ln|x|\right)\Bigg|_2^6$$
$$= (\ln 2 - 1) - \left(0 - \frac{1}{2}\right) + (3 - \ln 6) - (1 - \ln 2)$$
$$= 2\ln 2 - \ln 6 + \frac{3}{2} \approx 1.095$$

13. $x = -1, x = 1, y = e^x, y = 3 - e^x$

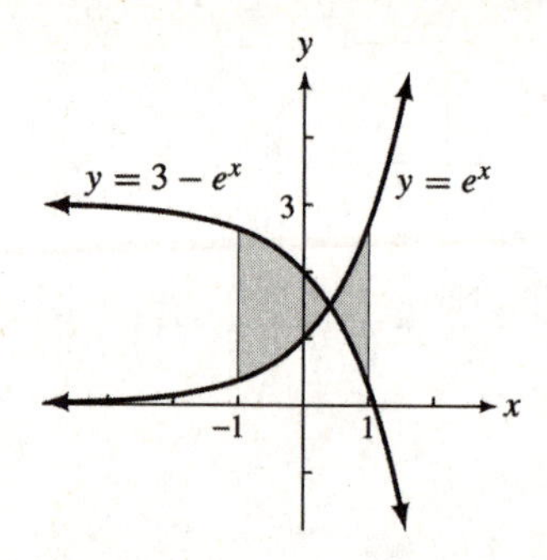

To find the point of intersection, set $e^x = 3 - e^x$ and solve for x.

$$e^x = 3 - e^x$$
$$2e^x = 3$$
$$e^x = \frac{3}{2}$$
$$\ln e^x = \ln\frac{3}{2}$$
$$x \ln e = \ln\frac{3}{2}$$
$$x = \ln\frac{3}{2}$$

The area of the region between the curves from $x = -1$ to $x = 1$ is

$$\int_{-1}^{\ln 3/2} [(3 - e^x) - e^x]\,dx$$
$$+ \int_{\ln 3/2}^{1} [e^x - (3 - e^x)]\,dx$$
$$= \int_{-1}^{\ln 3/2} (3 - 2e^x)\,dx + \int_{\ln 3/2}^{1} (2e^x - 3)\,dx$$
$$= (3x - 2e^x)\Big|_{-1}^{\ln 3/2} + (2e^x - 3x)\Big|_{\ln 3/2}^{1}$$
$$= \left[\left(3\ln\frac{3}{2} - 2e^{\ln 3/2}\right) - [3(-1) - 2e^{-1}]\right]$$
$$+ \left[2e^1 - 3(1) - \left(2e^{\ln 3/2} - 3\ln\frac{3}{2}\right)\right]$$
$$= \left[\left(3\ln\frac{3}{2} - 3\right) - \left(-3 - \frac{2}{e}\right)\right]$$
$$+ \left[2e - 3 - \left(3 - 3\ln\frac{3}{2}\right)\right]$$
$$= 6\ln\frac{3}{2} + \frac{2}{e} + 2e - 6 \approx 2.605.$$

15. $x = -1, x = 2, y = 2e^{2x}, y = e^{2x} + 1$

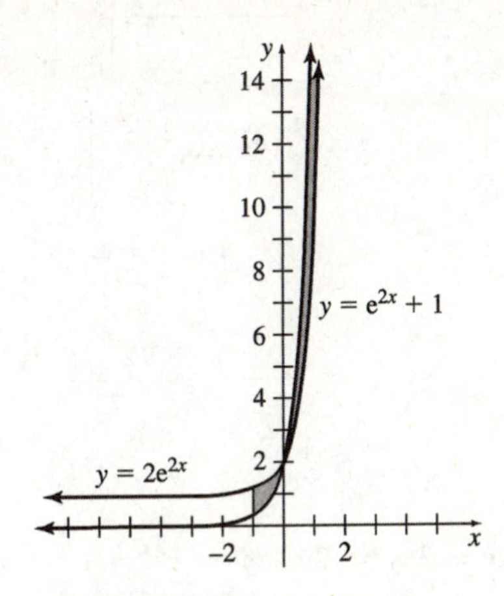

To find the points of intersection of the graphs, substitute for y.

$$2e^{2x} = e^{2x} + 1$$
$$e^{2x} = 1$$
$$2x = 0$$
$$x = 0$$

The region is composed of two separate regions because $y = 2e^{2x}$ intersects $y = e^{2x} + 1$ at $x = 0$.

Let $f(x) = 2e^{2x}, g(x) = e^{2x} + 1$.

In the interval $[-1, 0]$, $g(x) \geq f(x)$.

In the interval $[0, 2]$, $f(x) \geq g(x)$.

$$\int_{-1}^{0} (e^{2x} + 1 - 2e^{2x})dx$$
$$+ \int_{0}^{2} [2e^{2x} - (e^{2x} + 1)]\, dx$$
$$= \left(-\frac{e^{2x}}{2} + x\right)\Bigg|_{-1}^{0} + \left(\frac{e^{2x}}{2} - x\right)\Bigg|_{0}^{2}$$
$$= \left(-\frac{1}{2} + 0\right) - \left(-\frac{e^{-2}}{2} - 1\right)$$
$$+ \left(\frac{e^4}{2} - 2\right) - \left(\frac{1}{2} - 0\right)$$
$$= \frac{e^{-2} + e^4}{2} - 2 \approx 25.37$$

17. $y = x^3 - x^2 + x + 1, y = 2x^2 - x + 1$

Find the points of intersection.

$$x^3 - x^2 + x + 1 = 2x^2 - x + 1$$
$$x^3 - 3x^2 + 2x = 0$$
$$x(x^2 - 3x + 2) = 0$$
$$x(x - 2)(x - 1) = 0$$

The points of intersection are at $x = 0$, $x = 1$, and $x = 2$.

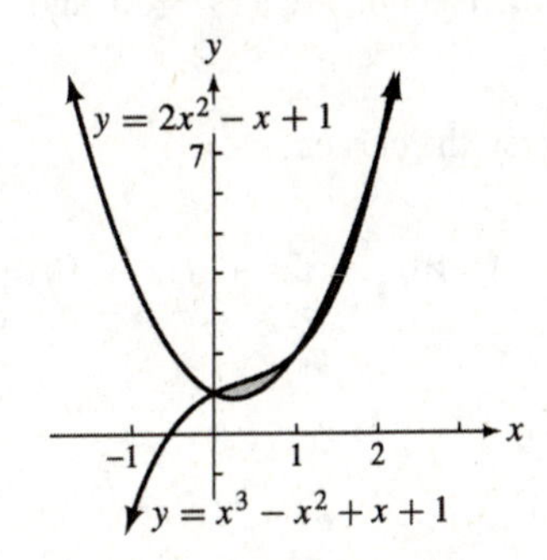

Area between the curves is

$$\int_{0}^{1} [(x^3 - x^2 + x + 1) - (2x^2 - x + 1)]\, dx$$
$$+ \int_{0}^{2} [(2x^2 - x + 1) - (x^3 - x^2 + x + 1)]\, dx$$
$$= \int_{0}^{1} (x^3 - 3x^2 + 2x)\, dx + \int_{1}^{2} (-x^3 + 3x^2 - 2x)]\, dx$$
$$= \left(\frac{x^4}{4} - x^3 + x^2\right)\Bigg|_{0}^{1} + \left(\frac{-x^4}{4} + x^3 - x^2\right)\Bigg|_{1}^{2}$$
$$= \left[\left(\frac{1}{4} - 1 + 1\right) - (0)\right]$$
$$+ \left[(-4 + 8 - 4) - \left(-\frac{1}{4} + 1 - 1\right)\right]$$
$$= \frac{1}{4} + \frac{1}{4}$$
$$= \frac{1}{2}.$$

19. $y = x^4 + \ln(x + 10),$

$y = x^3 + \ln(x + 10)$

Find the points of intersection.

$$\begin{aligned} x^4 + \ln(x+10) &= x^3 + \ln(x+10) \\ x^4 - x^3 &= 0 \\ x^3(x-1) &= 0 \\ x = 0 \text{ or } x &= 1 \end{aligned}$$

The points of intersection are at $x = 0$ and $x = 1$.

The area between the curves is

$$\int_0^1 [(x^3 + \ln(x+10)) - (x^4 + \ln(x+10))]\, dx$$

$$= \int_0^1 (x^3 - x^4)\, dx$$

$$= \left(\frac{x^4}{4} - \frac{x^5}{5}\right)\Bigg|_0^1$$

$$= \left(\frac{1}{4} - \frac{1}{5}\right) - (0) = \frac{1}{20}.$$

21. $y = x^{4/3}, y = 2x^{1/3}$

Find the points of intersection.

$$\begin{aligned} x^{4/3} &= 2x^{1/3} \\ x^{4/3} - 2x^{1/3} &= 0 \\ x^{1/3}(x - 2) &= 0 \\ x = 0 \text{ or } x &= 2 \end{aligned}$$

The points of intersection are at $x = 0$ and $x = 2$.

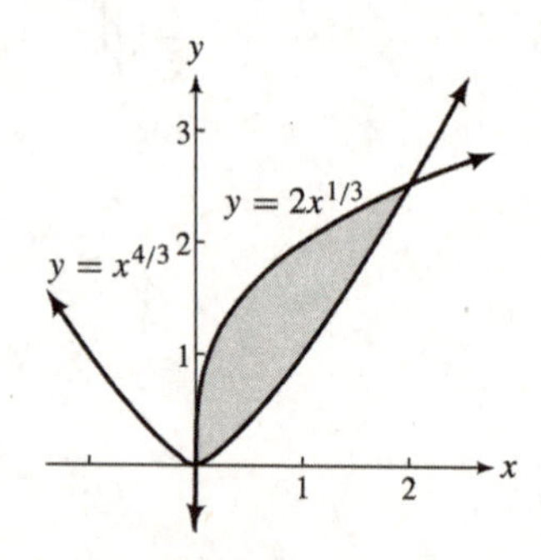

The area between the curves is

$$\int_0^2 (2x^{1/3} - x^{4/3})\, dx = 2\frac{x^{4/3}}{\frac{4}{3}} - \frac{x^{7/3}}{\frac{7}{3}}\Bigg|_0^2$$

$$= \frac{3}{2}x^{4/3} - \frac{3}{7}x^{7/3}\Bigg|_0^2$$

$$= \left[\frac{3}{2}(2)^{4/3} - \frac{3}{7}(2)^{7/3}\right] - 0$$

$$= \frac{3(2^{4/3})}{2} - \frac{3(2^{7/3})}{7}$$

$$\approx 1.62.$$

23. $x = 0, x = 3, y = 2e^{3x}, y = e^{3x} + e^6$

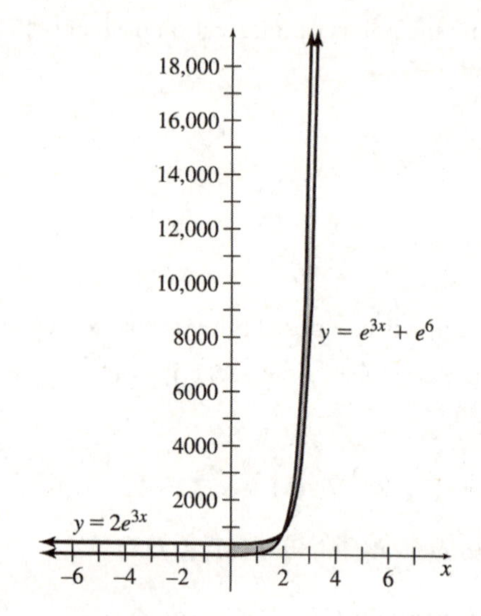

To find the points of intersection of the graphs, substitute for y.

$$\begin{aligned} 2e^{3x} &= e^{3x} + e^6 \\ e^{3x} &= e^6 \\ 3x &= 6 \\ x &= 2 \end{aligned}$$

The region is composed of two separate regions because $y = 2e^{3x}$ intersects $y = e^{3x} + e^6$ at $x = 2$.

Let $f(x) = 2e^{3x}, g(x) = e^{3x} + e^6.$

In the interval $[0, 2], g(x) \geq f(x).$

In the interval $[2, 3], f(x) \geq g(x).$

$$\int_0^2 (e^{3x} + e^6 - 2e^{3x})dx + \int_2^3 [2e^{3x} - (e^{3x} + e^6)]dx$$

$$= \left(-\frac{e^{3x}}{3} + e^6 x\right)\Big|_0^2 + \left(\frac{e^{3x}}{3} - e^6 x\right)\Big|_2^3$$

$$= \left(-\frac{e^6}{3} + 2e^6\right) - \left(-\frac{1}{3} + 0\right)$$

$$+ \left(\frac{e^9}{3} - 3e^6\right) - \left(\frac{e^6}{3} - 2e^6\right)$$

$$= \frac{e^9 + e^6 + 1}{3}$$

$$\approx 2836$$

25. Graph $y_1 = e^x$ and $y_2 = -x^2 - 2x$ on your graphing calculator. Use the intersect command to find the two intersection points. The resulting screens are:

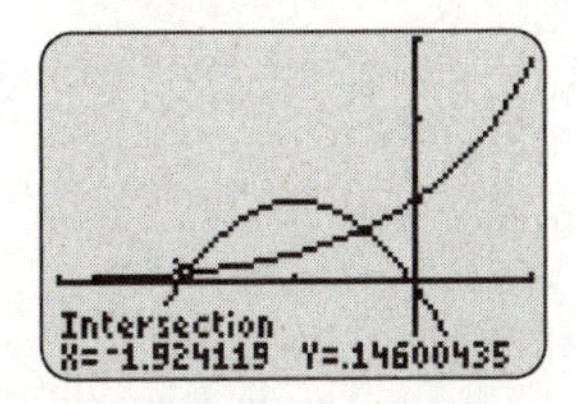

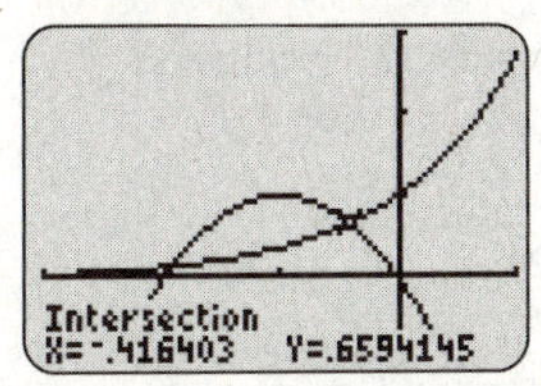

These screens show that $e^x = -x^2 - 2x$ when $x \approx -1.9241$ and $x \approx -0.4164$.

In the interval $[-1.9241, -0.4164]$,

$$e^x < -x^2 - 2x.$$

The area between the curves is given by

$$\int_{-1.9241}^{-0.4164} [(-x^2 - 2x) - e^x]dx.$$

Use the fnInt command to approximate this definite integral.

The resulting screen is:

```
fnInt((-X²-2X)-e
^(X),X,-1.9241,-
.4164)
        .6649863525
```

The last screen shows that the area is approximately 0.6650.

27. (a) It is profitable to use the machine until $S'(x) = C'(x)$.

$$150 - x^2 = x^2 + \frac{11}{4}x$$

$$2x^2 + \frac{11}{4}x - 150 = 0$$

$$8x^2 + 11x - 600 = 0$$

$$x = \frac{-11 \pm \sqrt{121 - 4(8)(-600)}}{16}$$

$$= \frac{-11 \pm 139}{16}$$

$$x = 8 \text{ or } x = -9.375$$

It will be profitable to use this machine for 8 years. Reject the negative solution.

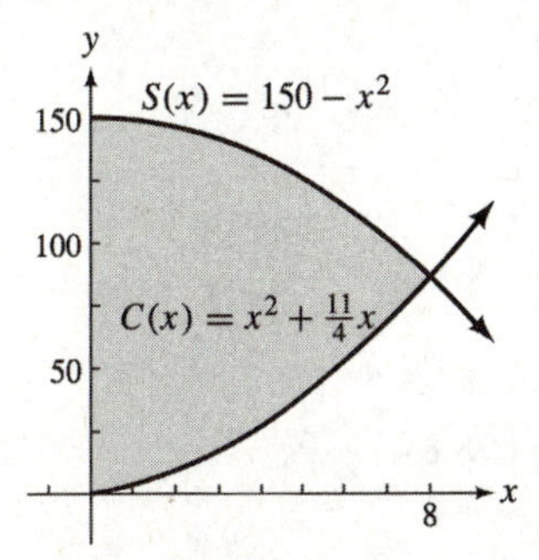

(b) Since $150 - x^2 > x^2 + \frac{11}{4}x$, in the interval $[0, 8]$, the net total saving in the first year are

$$\int_0^1 \left[(150 - x^2) - \left(x^2 + \frac{11}{4}x\right)\right] dx$$

$$= \int_0^1 \left(-2x^2 - \frac{11}{4}x + 150\right) dx$$

$$= \left(\frac{-2x^3}{3} - \frac{11x^2}{8} + 150x\right)\Big|_0^1$$

$$= -\frac{2}{3} - \frac{11}{8} + 150 \approx \$148.$$

(c) The net total savings over the entire period of use are

$$\int_0^8 \left[(150 - x^2) - \left(x^2 + \frac{11}{4}x\right)\right] dx$$

$$= \left(\frac{-2x^3}{3} - \frac{11x^2}{8} + 150x\right)\Big|_0^8$$

$$= \frac{-2(8^3)}{3} - \frac{11(8^2)}{8} + 150(8)$$

$$= \frac{-1024}{3} - \frac{704}{8} + 1200 \approx \$771.$$

29. **(a)** $E'(x) = e^{0.1x}$ and $I'(x) = 98.8 - e^{0.1x}$

To find the point of intersection, where profit will be maximized, set the functions equal to each other and solve for x.

$$\begin{aligned} e^{0.1x} &= 98.8 - e^{0.1x} \\ 2e^{0.1x} &= 98.8 \\ e^{0.1x} &= 49.4 \\ 0.1x &= \ln 49.4 \\ x &= \frac{\ln 49.4}{0.1} \approx 39 \end{aligned}$$

The optimum number of days for the job to last is 39.

(b) The total income for 39 days is

$$\begin{aligned} &\int_0^{39} (98.8 - e^{0.1x})\,dx \\ &= \left(98.8x - \frac{e^{0.1x}}{0.1}\right)\Bigg|_0^{39} \\ &= \left(98.8x - 10e^{0.1x}\right)\Big|_0^{39} \\ &= [98.8(39) - 10e^{3.9}] - (0 - 10) \\ &= \$3369.18. \end{aligned}$$

(c) The total expenditure for 39 days is

$$\begin{aligned} \int_0^{39} e^{0.1x}dx &= \frac{e^{0.1x}}{0.1}\Bigg|_0^{39} \\ &= 10e^{0.1x}\Big|_0^{39} \\ &= 10e^{3.9} - 10 \\ &= \$484.02. \end{aligned}$$

(d) Profit $=$ Income $-$ Expense
$= 3369.18 - 484.02 = \$2885.16$

31. $S(q) = q^{5/2} + 2q^{3/2} + 50;\ q = 16$ is the equilibrium quantity.

$$\text{Producers surplus} = \int_0^{q_0} [p_0 - S(q)]dq,$$

where p_0 is the equilibrium price and q_0 is equilibrium supply.

$$\begin{aligned} p_0 &= S(16) = (16)^{5/2} + 2(16)^{3/2} + 50 \\ &= 1202 \end{aligned}$$

Therefore, the producers' surplus is

$$\begin{aligned} &\int_0^{16} [1202 - (q^{5/2} + 2q^{3/2} + 50)]dq \\ &= \int_0^{16} (1152 - q^{5/2} - 2q^{3/2})\,dq \\ &= \left(1152q - \frac{2}{7}q^{7/2} - \frac{4}{5}q^{5/2}\right)\Bigg|_0^{16} \\ &= 1152(16) - \frac{2}{7}(16)^{7/2} - \frac{4}{5}(16)^{5/2} \\ &= 18{,}432 - \frac{32{,}768}{7} - \frac{4096}{5} \\ &= 12{,}931.66. \end{aligned}$$

The producers' surplus is 12,931.66.

33. $D(q) = \dfrac{200}{(3q+1)^2};\ q = 3$ is the equilibrium quantity.

$$\text{Consumers' surplus} = \int_0^{q_0} |D(q) - p_0|\,dq$$

$$p_0 = D(3) = 2$$

Therefore, the consumers' surplus is

$$\begin{aligned} &\int_0^3 \left[\frac{200}{(3q+1)^2} - 2\right] dq \\ &= \int_0^3 \frac{200}{(3q+1)^2}\,dq - \int_0^3 2\,dq. \end{aligned}$$

Let $u = 3q + 1$, so that

$$du = 3\,dq \text{ and } \frac{1}{3}du = dq.$$

$$\begin{aligned} &\int_0^3 \frac{200}{(3q+1)^2}\,dq - \int_0^3 2\,dq \\ &= \frac{1}{3}\int_1^{10} \frac{200}{u^2}\,du - \int_0^3 2\,dq \\ &= \frac{200}{3}\int_1^{10} u^{-2}du - \int_0^3 2\,dq \\ &= \frac{200}{3}\cdot\frac{u^{-1}}{-1}\Bigg|_1^{10} - 2q\Bigg|_0^3 \\ &= -\frac{200}{3u}\Bigg|_1^{10} - 6 \\ &= -\frac{200}{30} + \frac{200}{3} - 6 \\ &= 54 \end{aligned}$$

35. $S(q) = q^2 + 10q$

$D(q) = 900 - 20q - q^2$

(a) The graphs of the supply and demand functions are parabolas with vertices at $(-5, -25)$ and $(-10, 1900)$, respectively.

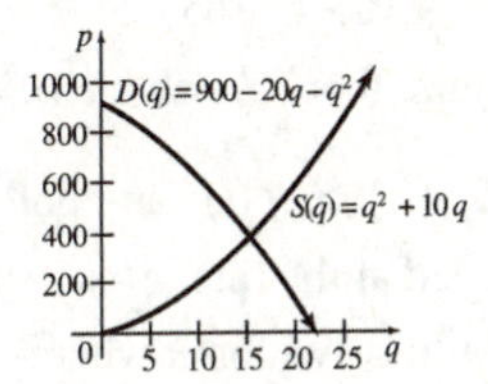

(b) The graphs intersect at the point where the y-coordinates are equal.

$$q^2 + 10q = 900 - 20q - q^2$$
$$2q^2 + 30q - 900 = 0$$
$$q^2 + 15q - 450 = 0$$
$$(q + 30)(q - 15) = 0$$
$$q = -30 \quad \text{or} \quad q = 15$$

Disregard the negative solution.

The supply and demand functions are in equilibrium when $q = 15$.

$$S(15) = 15^2 + 10(15) = 375$$

The point is $(15, 375)$.

(c) Find the consumers' surplus.

$$\int_0^{q_0} [D(q) - p_0)]\, dq$$
$$p_0 = D(15) = 375$$

$$\int_0^{15} [(900 - 20q - q^2) - 375]\, dq$$
$$= \int_0^{15} (525 - 20q - q^2)\, dq$$
$$= \left(525q - 10q^2 - \frac{1}{3}q^3\right)\Bigg|_0^{15}$$
$$= \left[525(15) - 10(15)^2 - \frac{1}{3}(15)^3\right] - 0 = 4500$$

The consumer's surplus is \$4500.

(d) Find the producers' surplus.

$$\int_0^{q_0} [p_0 - S(q)]\, dq$$
$$p_0 = S(15) = 375$$

$$\int_0^{15} [375 - (q^2 + 10q)]\, dq$$
$$= \int_0^{15} (375 - q^2 - 10q)\, dq$$
$$= \left(375q - \frac{1}{3}q^3 - 5q^2\right)\Bigg|_0^{15}$$
$$= \left[375(15) - \frac{1}{3}(15)^3 - 5(15)^2\right] - 0$$
$$= 3375$$

The producer's surplus is \$3375.

37. (a) $S(q) = q^2 + 10q$; $S(q) = 264$ is the price the government set.

$$264 = q^2 + 10q$$
$$0 = q^2 + 10q - 264$$
$$0 = (q - 12)(q + 22)$$
$$q = 12 \quad \text{or} \quad q = -22$$

Only 12 is a meaningful solution here. Thus, 12 units of oil will be produced.

(b) The consumers' surplus is given by

$$\int_0^{12} (900 - 20q - q^2 - 264)\, dq$$
$$= \int_0^{12} (636 - 20q - q^2)\, dq$$
$$= \left(636q - 10q^2 - \frac{1}{3}q^3\right)\Bigg|_0^{12}$$
$$= 636(12) - 10(12)^2 - \frac{1}{3}(12)^3 - 0$$
$$= 5616$$

Here the consumer' surplus is \$5616. In this case, the consumers' surplus is $5616 - 4500$ $=$ \$1116 larger.

(c) The producers' surplus is given by

$$\int_0^{12}[264-(q^2+10q)]\,dq$$
$$=\int_0^{12}(264-q^2-10q)\,dq$$
$$=\left(264q-\frac{1}{3}q^3-5q^2\right)\Big|_0^{12}$$
$$=264(12)-\frac{1}{3}(12)^3-5(12)^2-0$$
$$=1872$$

Here the producers' surplus is \$1872. In this case, the producers' surplus is 3375 − 1872 = \$1503 smaller.

(d) For the equilibrium price, the total consumers' and producers' surplus is

$$4500+3375=\$7875$$

For the government price, the total consumers' and producers' surplus is

$$5616+1872=\$7488.$$

The difference is

$$7875-7488=\$387.$$

39. (a) The pollution level in the lake is changing at the rate $f(t)-g(t)$ at any time t. We find the amount of pollution by integrating.

$$\int_0^{12}[f(t)-g(t)]\,dt$$
$$=\int_0^{12}[10(1-e^{-0.5t})-0.4t]\,dt$$
$$=\left(10t-10\cdot\frac{1}{-0.5}e^{-0.5t}-0.4\cdot\frac{1}{2}t^2\right)\Big|_0^{12}$$
$$=(20e^{-0.5t}+10t-0.2t^2)\Big|_0^{12}$$
$$=[20e^{-0.5(12)}+10(12)-0.2(12)^2]$$
$$-[20e^{-0.5(0)}+10(0)-0.2(0)^2]$$
$$=(20e^{-6}+91.2)-(20)$$
$$=20e^{-6}+71.2\approx 71.25$$

After 12 hours, there are about 71.25 gallons.

(b) The graphs of the functions intersect at about 25.00. So the rate that pollution enters the lake equals the rate the pollution is removed at about 25 hours.

(c) $$\int_0^{25}[f(t)-g(t)]\,dt$$
$$=(20e^{-0.5t}+10t-0.2t^2)\Big|_0^{25}$$
$$=[20e^{-0.5(25)}+10(25)-0.2(25)^2)]-20$$
$$=20e^{-12.5}+105\approx 105$$

After 25 hours, there are about 105 gallons.

(d) For $t>25$, $g(t)>f(t)$, and pollution is being removed at the rate $g(t)-f(t)$. So, we want to solve for c, where

$$\int_0^{c}[f(t)-g(t)]\,dt=0.$$

Alternatively, we could solve for c in

$$\int_{25}^{c}[g(t)-f(t)]\,dt=105.$$

One way to do this with a graphing calculator is to graph the function

$$y=\int_0^{x}[f(t)-g(t)]\,dt$$

and determine the values of x for which $y=0$. The first window shows how the function can be defined.

A suitable window for the graph is [0, 50] by [0, 110].

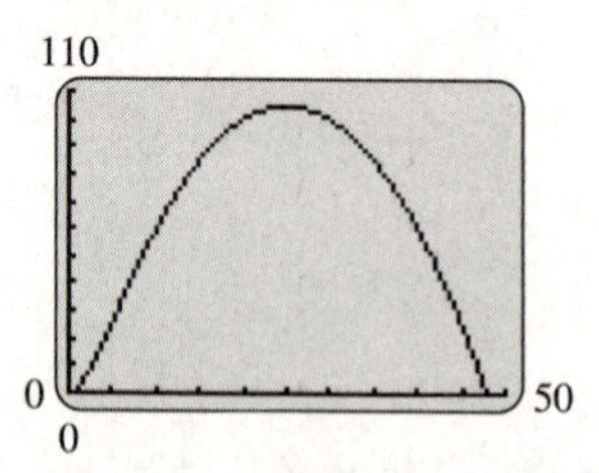

Use the calculator's features to approximate where the graph intersects the x-axis. These are at 0 and about 47.91. Therefore, the pollution will be removed from the lake after about 47.91 hours.

41. $I(x) = 0.9x^2 + 0.1x$

(a) $I(0.1) = 0.9(0.1)^2 + 0.1(0.1)$
$= 0.019$

The lower 10% of income producers earn 1.9% of total income of the population.

(b) $I(0.4) = 0.9(0.4)^2 + 0.1(0.4) = 0.184$

The lower 40% of income producers earn 18.4% of total income of the population.

(c) The graph of $I(x) = x$ is a straight line through the points (0, 0) and (1, 1). The graph of $I(x) = 0.9x^2 + 0.1x$ is a parabola with vertex $\left(-\frac{1}{18}, -\frac{1}{360}\right)$. Restrict the domain to $0 \le x \le 1$.

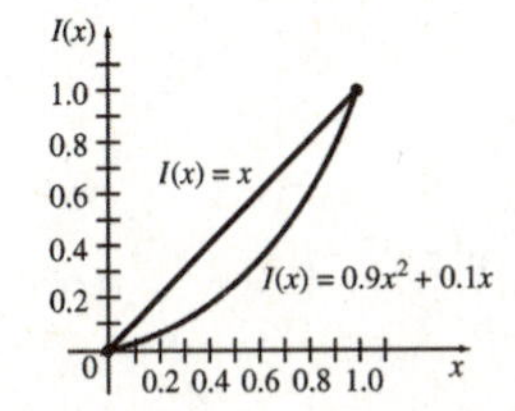

(d) To find the points of intersection, solve

$$x = 0.9x^2 + 0.1x.$$

$$0.9x^2 - 0.9x = 0$$
$$0.9x(x - 1) = 0$$
$$x = 0 \quad \text{or} \quad x = 1$$

The area between the curves is given by

$$\int_0^1 [x - (0.9x^2 + 0.1x)]\,dx$$
$$= \int_0^1 (0.9x - 0.9x^2)\,dx$$
$$= \left(\frac{0.9x^2}{2} - \frac{0.9x^3}{3}\right)\Bigg|_0^1$$
$$= \frac{0.9}{2} - \frac{0.9}{3} = 0.15.$$

(e) Income is distributed less equally in 2008 than in 1968.

15.6 Numerical Integration

Your Turn 1

Use the trapezoidal rule with $n = 4$ to approximate

$$\int_1^3 \sqrt{x^2 + 3}\,dx.$$

Here $f(x) = \sqrt{x^2 + 3}$, $a = 1, b = 3$, and $n = 4$. The subintervals have length $(3 - 1)/4 = 1/2$. The following table summarizes the information required.

i	x_i	$f(x_i)$
0	1	$f(1) = 2$
1	3/2	$f(3/2) \approx 2.29129$
2	2	$f(2) \approx 2.64575$
3	5/2	$f(5/2) \approx 3.04138$
4	3	$f(3) \approx 3.46410$

The trapezoidal rule gives

$$\int_1^3 \sqrt{x^2 + 3}\,dx$$
$$\approx \frac{3-1}{2}\left[\frac{1}{2}(2) + 2.29129 + 2.64757 + 3.04138 + \frac{1}{2}(3.46410)\right]$$
$$\approx 5.3552.$$

Your Turn 2

Use Simpson's rule with $n = 4$ to approximate

$$\int_1^3 \sqrt{x^2 + 3}\,dx.$$

Here $f(x) = \sqrt{x^2 + 3}$, $a = 1, b = 3$, and $n = 4$. The subintervals have length $(3 - 1)/4 = 1/2$. The following table summarizes the information required; it is the same as the table used in Your Turn 1.

i	x_i	$f(x_i)$
0	1	$f(1) = 2$
1	3/2	$f(3/2) \approx 2.29129$
2	2	$f(2) \approx 2.64575$
3	5/2	$f(5/2) \approx 3.04138$
4	3	$f(3) \approx 3.46410$

For Simpson's rule, the factor in front is $(b - a)/3n = (3 - 1)/12 = 1/6$. Simpson's rule thus gives

$$\int_1^3 \sqrt{x^2 + 3}\, dx$$
$$\approx \frac{1}{6}[2 + 4(2.29129) + 2(2.64757) + 4(3.04138) + 3.46410]$$
$$\approx 5.3477.$$

15.6 Exercises

1. $\int_0^2 (3x^2 + 2)\, dx$

$n = 4, b = 2, a = 0, f(x) = 3x^2 + 2$

i	x_i	$f(x_i)$
0	0	2
1	$\frac{1}{2}$	2.75
2	1	5
3	$\frac{3}{2}$	8.75
4	2	14

(a) Trapezoidal rule:

$$\int_0^2 (3x^2 + 2)\, dx$$
$$\approx \frac{2 - 0}{4}\left[\frac{1}{2}(2) + 2.75 + 5 + 8.75 + \frac{1}{2}(14)\right]$$
$$= 0.5(24.5)$$
$$= 12.25$$

(b) Simpson's rule:

$$\int_0^2 (3x^2 + 2)\, dx$$
$$\approx \frac{2 - 0}{3(4)}[2 + 4(2.75) + 2(5) + 4(8.75) + 14]$$
$$= \frac{2}{12}(72)$$
$$= 12$$

(c) Exact value:

$$\int_0^2 (3x^2 + 2)\, dx = (x^3 + 2x)\Big|_0^2$$
$$= (8 + 4) - 0$$
$$= 12$$

3. $\int_{-1}^3 \frac{3}{5 - x}\, dx$

$n = 4, b = 3, a = -1, f(x) = \frac{3}{5 - x}$

i	x_i	$f(x_i)$
0	−1	0.5
1	0	0.6
2	1	0.75
3	2	1
4	3	1.5

(a) Trapezoidal rule:

$$\int_{-1}^3 \frac{3}{5 - x}\, dx$$
$$\approx \frac{3 - (-1)}{4}\left[\frac{1}{2}(0.5) + 0.6 + 0.75 + 1 + \frac{1}{2}(1.5)\right]$$
$$= 1(3.35)$$
$$= 3.35$$

(b) Simpson's rule:

$$\int_{-1}^3 \frac{3}{5 - x}\, dx$$
$$\approx \frac{3 - (-1)}{3(4)}[0.5 + 4(0.6) + 2(0.75) + 4(1) + 1.5]$$
$$= \frac{1}{3}\left(\frac{99}{10}\right)$$
$$= \frac{33}{10} \approx 3.3$$

(c) Exact value:

$$\int_{-1}^3 \frac{3}{5 - x}\, dx = -3 \ln|5 - x|\Big|_{-1}^3$$
$$= -3(\ln|2| - \ln|6|)$$
$$= 3 \ln 3 \approx 3.296$$

5. $\int_{-1}^{2} (2x^3 + 1)\,dx$

$n = 4, b = 2, a = -1, f(x) = 2x^3 + 1$

i	x_i	$f(x)$
0	-1	-1
1	$-\frac{1}{4}$	$\frac{31}{32}$
2	$\frac{1}{2}$	$\frac{5}{4}$
3	$\frac{5}{4}$	$\frac{157}{32}$
4	2	17

(a) Trapezoidal rule:

$$\int_{-1}^{2} (2x^3 + 1)\,dx \approx \frac{2-(-1)}{4}\left[\frac{1}{2}(-1) + \frac{31}{32} + \frac{5}{4} + \frac{157}{32} + \frac{1}{2}(17)\right]$$

$$= 0.75(15.125)$$

$$\approx 11.34$$

(b) Simpson's rule:

$$\int_{-1}^{2} (2x^3 + 1)\,dx \approx \frac{2-(-1)}{3(4)}\left[-1 + 4\left(\frac{31}{32}\right) + 2\left(\frac{5}{4}\right) + 4\left(\frac{157}{32}\right) + 17\right]$$

$$= \frac{1}{4}(42) = 10.5$$

(c) Exact value:

$$\int_{-1}^{2} (2x^3 + 1)\,dx = \left(\frac{x^4}{2} + x\right)\Bigg|_{-1}^{2}$$

$$= (8 + 2) - \left(\frac{1}{2} - 1\right)$$

$$= \frac{21}{2}$$

$$= 10.5$$

7. $\int_{1}^{5} \frac{1}{x^2}\,dx$

$n = 4, b = 5, a = 1, f(x) = \frac{1}{x^2}$

i	x_i	$f(x_i)$
0	1	1
1	2	0.25
2	3	0.1111
3	4	0.0625
4	5	0.04

(a) Trapezoidal rule:

$$\int_{1}^{5} \frac{1}{x^2}\,dx \approx \frac{5-1}{4}\left[\frac{1}{2}(1) + 0.25 + 0.1111 + 0.0625 + \frac{1}{2}(0.04)\right]$$

$$\approx 0.9436$$

(b) Simpson's rule:

$$\int_{1}^{5} \frac{1}{x^2}\,dx \approx \frac{5-1}{12}[1 + 4(0.25) + 2(0.1111) + 4(0.0625) + 0.04)]$$

$$\approx 0.8374$$

(c) Exact value:

$$\int_{1}^{5} x^{-2}dx = -x^{-1}\Big|_{1}^{5}$$

$$= -\frac{1}{5} + 1$$

$$= \frac{4}{5} = 0.8$$

9. $\int_0^1 4xe^{-x^2}\,dx$

$n = 4, b = 1, a = 0, f(x) = 4xe^{-x^2}$

i	x_i	$f(x_i)$
0	0	0
1	$\frac{1}{4}$	$e^{-1/16}$
2	$\frac{1}{2}$	$2e^{-1/4}$
3	$\frac{3}{4}$	$3e^{-9/16}$
4	1	$4e^{-1}$

(a) Trapezoidal rule:

$$\int_0^1 4xe^{-x^2}\,dx$$
$$\approx \frac{1-0}{4}\left[\frac{1}{2}(0) + e^{-1/16} + 2e^{-1/4} + 3e^{-9/16} + \frac{1}{2}(4e^{-1})\right]$$
$$= \frac{1}{4}(e^{-1/16} + 2e^{-1/4} + 3e^{-9/16} + 2e^{-1})$$
$$\approx 1.236$$

(b) Simpson's rule:

$$\int_0^1 4xe^{-x^2}\,dx$$
$$\approx \frac{1-0}{3(4)}[0 + 4(e^{-1/16}) + 2(2e^{-1/4}) + 4(3e^{-9/16}) + 4e^{-1}]$$
$$= \frac{1}{12}(4e^{-1/16} + 4e^{-1/4} + 12e^{-9/16} + 4e^{-1})$$
$$\approx 1.265$$

(c) Exact value:

$$\int_0^1 4xe^{-x^2}\,dx = -2e^{-x^2}\Big|_0^1$$
$$= (-2e^{-1}) - (-2)$$
$$= 2 - 2e^{-1} \approx 1.264$$

11. $y = \sqrt{4 - x^2}$

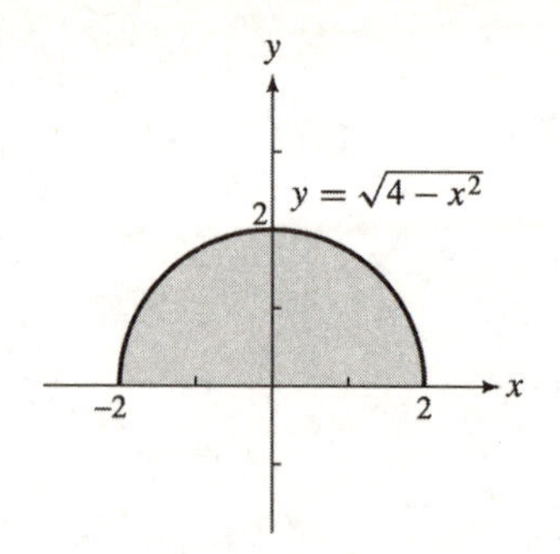

$n = 8, b = 2, a = -2, f(x) = \sqrt{4 - x^2}$

i	x_i	y
0	−2.0	0
1	−1.5	1.32289
2	−1.0	1.73205
3	−0.5	1.93649
4	0	2
5	0.5	1.93649
6	1.0	1.73205
7	1.5	1.32289
8	2.0	0

(a) Trapezoidal rule:

$$\int_{-2}^2 \sqrt{4 - x^2}\,dx$$
$$\approx \frac{2-(-2)}{8} \cdot \left[\frac{1}{2}(0) + 1.32289 + 1.73205 + \cdots + \frac{1}{2}(0)\right]$$
$$\approx 5.991$$

(b) Simpson's rule:

$$\int_{-2}^2 \sqrt{4 - x^2}\,dx$$
$$\approx \frac{2-(-2)}{3(8)} \cdot [0 + 4(1.32289) + 2(1.73205) + 4(1.93649) + 2(2) + 4(1.93649) + 2(1.73205) + 4(1.32289) + 0]$$
$$\approx 6.167$$

(c) Area of semicircle $= \frac{1}{2}\pi r^2 = \frac{1}{2}\pi(2)^2$

$$\approx 6.283$$

Simpson's rule is more accurate.

13. Since $f(x) > 0$ and $f''(x) > 0$ for all x between a and b, we know the graph of $f(x)$ on the interval from a to b is concave upward. Thus, the trapezoid that approximates the area will have an area greater than the actual area Thus,

$$T > \int_a^b f(x)\,dx.$$

The correct choice is (b).

15. **(a)**
$$\int_0^1 x^4 dx = \left(\frac{1}{5}\right)x^5\Big|_0^1 = \frac{1}{5} = 0.2$$

(b) $n = 4, b = 1, a = 0, f(x) = x^4$

$$\int_0^1 x^4 dx \approx \frac{1-0}{4}\left[\frac{1}{2}(0) + \frac{1}{256} + \frac{1}{16} + \frac{81}{256} + \frac{1}{2}(1)\right] = \frac{1}{4}\left(\frac{226}{256}\right) \approx 0.220703$$

$n = 8, b = 1, a = 0, f(x) = x^4$

$$\int_0^1 x^4 dx \approx \frac{1-0}{8}\left[\frac{1}{2}(0) + \frac{1}{4096} + \frac{1}{256} + \frac{81}{4096} + \frac{1}{16} + \frac{625}{4096} + \frac{81}{256} + \frac{2401}{4096} + \frac{1}{2}(1)\right] = \frac{1}{8}\left(\frac{6724}{4096}\right) \approx 0.20520$$

$n = 16, b = 1, a = 0, f(x) = x^4$

$$\int_0^1 x^4 dx \approx \frac{1-0}{16}\left[\frac{1}{2}(0) + \frac{1}{65,536} + \frac{1}{4096} + \frac{81}{65,536} + \frac{1}{256} + \frac{625}{65,536} + \frac{81}{4096} + \frac{2401}{65,536} + \frac{1}{16} + \frac{6561}{65,536} + \frac{625}{4096} + \frac{14,641}{65,536} + \frac{81}{256} + \frac{28,561}{65,536} + \frac{2401}{4096} + \frac{50,625}{65,536} + \frac{1}{2}(1)\right]$$
$$\approx \frac{1}{16}\left(\frac{211,080}{65,536}\right) \approx 0.201302$$

$n = 32, b = 1, a = 0, f(x) = x^4$

$$\int_0^1 x^4\,dx$$
$$\approx \frac{1-0}{32}\left[\frac{1}{2}(0) + \frac{1}{1,048,576} + \frac{1}{65,536} + \frac{81}{1,048,576} + \frac{1}{4096} + \frac{625}{1,048,576} + \frac{81}{65,536} + \frac{2401}{1,048,576} + \frac{1}{256} + \frac{6561}{1,048,576} + \frac{625}{65,536} + \frac{14,641}{1,048,576} + \frac{81}{4096} + \frac{28,561}{1,048,576} + \frac{2401}{65,536} + \frac{50,625}{1,048,576} + \frac{1}{16} + \frac{83,521}{1,048,576} + \frac{6561}{65,536} + \frac{130,321}{1,048,576} + \frac{625}{4096} + \frac{194,481}{1,048,576} + \frac{14,641}{65,536} + \frac{279,841}{1,048,576} + \frac{81}{256} + \frac{390,625}{1,048,576} + \frac{28,561}{65,536} + \frac{531,441}{1,048,576} + \frac{2401}{4096} + \frac{707,281}{1,048,576} + \frac{50,625}{65,536} + \frac{923,521}{1,048,576} + \frac{1}{2}(1)\right]$$
$$\approx \frac{1}{32}\left(\frac{6,721,808}{1,048,576}\right) \approx 0.200325$$

To find error for each value of n, subtract as indicated.

$n = 4$: $(0.220703 - 0.2) = 0.020703$

$n = 8$: $(0.205200 - 0.2) = 0.005200$

$n = 16$: $(0.201302 - 0.2) = 0.001302$

$n = 32$: $(0.200325 - 0.2) = 0.000325$

(c) $p = 1$

$$4^1(0.020703) = 4(0.020703) = 0.082812$$

$$8^1(0.005200) = 8(0.005200) = 0.0416$$

Since these are not the same, try $p = 2$.

$p = 2$:

$$4^2(0.020703) = 16(0.020703) = 0.331248$$

$$8^2(0.005200) = 64(0.005200) = 0.3328$$

$$16^2(0.001302) = 256(0.001302) = 0.333312$$

$$32^2(0.000325) = 1024(0.000325) = 0.3328$$

Since these values are all approximately the same, the correct choice is $p = 2$.

17. (a) $$\int_0^1 x^4 dx = \frac{1}{5}x^5\Big|_0^1 = \frac{1}{5} = 0.2$$

(b) $n = 4, b = 1, a = 0, f(x) = x^4$

$$\int_0^1 x^4 dx \approx \frac{1-0}{3(4)}\left[0 + 4\left(\frac{1}{256}\right) + 2\left(\frac{1}{16}\right) + 4\left(\frac{81}{256}\right) + 1\right] = \frac{1}{12}\left(\frac{77}{32}\right) \approx 0.2005208$$

$n = 8, b = 1, a = 0, f(x) = x^4$

$$\int_0^1 x^4\, dx \approx \frac{1-0}{3(8)}\left[0 + 4\left(\frac{1}{4096}\right) + 2\left(\frac{1}{256}\right) + 4\left(\frac{81}{4096}\right) + 2\left(\frac{1}{16}\right) + 4\left(\frac{625}{4096}\right) + 2\left(\frac{18}{256}\right) + 4\left(\frac{2401}{4096}\right) + 1\right] = \frac{1}{24}\left(\frac{4916}{1024}\right) \approx 0.2000326$$

$n = 16, b = 1, a = 0, f(x) = x^4$

$$\int_0^1 x^4 dx \approx \frac{1-0}{3(16)}\left[0 + 4\left(\frac{1}{65,536}\right) + 2\left(\frac{1}{4096}\right) + 4\left(\frac{81}{65,536}\right) + 2\left(\frac{1}{256}\right) + 4\left(\frac{625}{65,536}\right) + 2\left(\frac{81}{4096}\right) + 4\left(\frac{2401}{65,536}\right) + 2\left(\frac{1}{16}\right) + 4\left(\frac{6561}{65,536}\right) + 2\left(\frac{625}{4096}\right) + 4\left(\frac{14,641}{65,536}\right) + 2\left(\frac{81}{256}\right) + 4\left(\frac{28,561}{65,536}\right) + 2\left(\frac{2401}{4096}\right) + 4\left(\frac{50,625}{65,536}\right) + 1\right]$$

$$= \frac{1}{48}\left(\frac{157,288}{16,384}\right) \approx 0.2000020$$

$n = 32, b = 1, a = 0, f(x) = x^4$

$$\int_0^1 x^4\, dx \approx \frac{1-0}{3(32)}\left[0 + 4\left(\frac{1}{1,048,576}\right) + 2\left(\frac{1}{65,536}\right) + 4\left(\frac{81}{1,048,576}\right) + 2\left(\frac{1}{4096}\right) + 4\left(\frac{625}{1,048,576}\right) + 2\left(\frac{625}{65,536}\right) + 4\left(\frac{14,641}{1,048,576}\right) + 2\left(\frac{81}{4096}\right) + 4\left(\frac{28,561}{1,048,576}\right) + 2\left(\frac{2401}{65,536}\right) + 4\left(\frac{50,625}{1,048,576}\right) + 2\left(\frac{1}{16}\right) + 4\left(\frac{83,521}{1,048,576}\right) + 2\left(\frac{6561}{65,536}\right)\right.$$

$$+ 4\left(\frac{130,321}{1,048,576}\right) + 2\left(\frac{625}{4096}\right) + 4\left(\frac{194,481}{1,048,576}\right)$$
$$+ 2\left(\frac{14,641}{65,536}\right) + 4\left(\frac{279,841}{1,048,576}\right) + 2\left(\frac{81}{256}\right)$$
$$+ 4\left(\frac{390,625}{1,048,576}\right) + 2\left(\frac{28,561}{65,536}\right) + 4\left(\frac{531,441}{1,048,576}\right)$$
$$+ 2\left(\frac{2401}{4096}\right) + 4\left(\frac{707,281}{1,048,576}\right) + 2\left(\frac{50,625}{65,536}\right)$$
$$\left. + 4\left(\frac{923,521}{1,048,576}\right) + 1\right]$$
$$= \frac{1}{96}\left(\frac{50,033,168}{262,144}\right) \approx 0.2000001$$

To find error for each value of n, subtract as indicated.

$n = 4$: $(0.2005208 - 0.2) = 0.0005208$
$n = 8$: $(0.2000326 - 0.2) = 0.0000326$
$n = 16$: $(0.2000020 - 0.2) = 0.0000020$
$n = 32$: $(0.2000001 - 0.2) = 0.0000001$

(c) $p = 1$:

$$4^1(0.0005208) = 4(0.0005208) = 0.0020832$$
$$8^1(0.0000326) = 8(0.0000326) = 0.0002608$$

Try $p = 2$:

$$4^2(0.0005208) = 16(0.0005208) = 0.0083328$$
$$8^2(0.0000326) = 64(0.0000326) = 0.0020864$$

Try $p = 3$:

$$4^3(0.0005208) = 64(0.0005208) = 0.0333312$$
$$8^3(0.0000326) = 512(0.0000326) = 0.0166912$$

Try $p = 4$:

$$4^4(0.0005208) = 256(0.0005208) = 0.1333248$$
$$8^4(0.0000326) = 4096(0.0000326) = 0.1335296$$
$$16^4(0.0000020) = 65536(0.0000020) = 0.131072$$
$$32^4(0.0000001) = 1048576(0.0000001) = 0.1048576$$

These are the closest values we can get; thus, $p = 4$.

19. Midpoint rule:

$n = 4, b = 5, a = 1, f(x) = \frac{1}{x^2}, \Delta x = 1$

i	x_i	$f(x_i)$
0	$\frac{3}{2}$	$\frac{4}{9}$
2	$\frac{5}{2}$	$\frac{4}{25}$
3	$\frac{7}{2}$	$\frac{4}{49}$
4	$\frac{9}{2}$	$\frac{4}{81}$

$$\int_1^5 \frac{1}{x^2}dx \approx \sum_{i=1}^{4} f(x_i)\Delta x$$
$$= \frac{4}{9}(1) + \frac{4}{25}(1) + \frac{4}{49}(1) + \frac{4}{81}(1)$$
$$\approx 0.7355$$

Simpson's rule:

$m = 8, b = 5, a = 1, f(x) = \frac{1}{x^2}$

i	x_i	$f(x_i)$
0	1	1
1	$\frac{3}{2}$	$\frac{4}{9}$
2	2	$\frac{1}{4}$
3	$\frac{5}{2}$	$\frac{4}{25}$
4	3	$\frac{1}{9}$
5	$\frac{7}{2}$	$\frac{4}{49}$
6	4	$\frac{1}{16}$
7	$\frac{9}{2}$	$\frac{4}{81}$
8	5	$\frac{1}{25}$

$$\int_1^5 \frac{1}{x^2}\,dx$$

$$\approx \frac{5-1}{3(8)}\left[1+4\left(\frac{4}{9}\right)+2\left(\frac{1}{4}\right)+4\left(\frac{4}{25}\right)\right.$$

$$+2\left(\frac{1}{9}\right)+4\left(\frac{4}{49}\right)+2\left(\frac{1}{16}\right)$$

$$\left.+4\left(\frac{4}{81}\right)+\frac{1}{25}\right]$$

$$\approx \frac{1}{6}(4.82906)$$

$$\approx 0.8048$$

From #7 part a, $T \approx 0.9436$, when $n = 4$.

To verify the formula evaluate $\frac{2M+T}{3}$.

$$\frac{2M+T}{3} \approx \frac{2(0.7355)+0.9436}{3}$$

$$\approx 0.8048$$

21. **(a)**

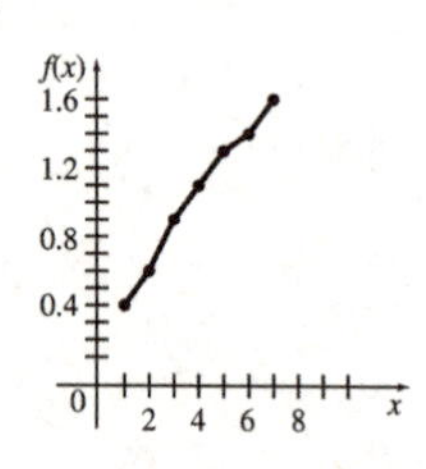

(b) $A = \frac{7-1}{6}\left[\frac{1}{2}(0.4)+0.6+0.9+1.1\right.$

$$\left.+1.3+1.4+\frac{1}{2}(1.6)\right]$$

$$= 6.3$$

(c) $A = \frac{7-1}{3(6)}[0.4+4(0.6)+2(0.9)$

$$+4(1.1)+2(1.3)+4(1.4)+1.6]$$

$$\approx 6.27$$

23. $y = e^{-t^2} + \frac{1}{t+1}$

The total reaction is

$$\int_1^9 \left(e^{-t^2}+\frac{1}{t+1}\right)dt.$$

$n = 8, b = 9, a = 1, f(t) = e^{-t^2} + \frac{1}{t+1}$

i	x_i	$f(x_i)$
0	1	0.8679
1	2	0.3516
2	3	0.2501
3	4	0.2000
4	5	0.1667
5	6	0.1429
6	7	0.1250
7	8	0.1111
8	9	0.1000

(a) Trapezoidal rule:

$$\int_1^9 \left(e^{-t^2}+\frac{1}{t+1}\right)dt$$

$$\approx \frac{9-1}{8}\left[\frac{1}{2}(0.8679)+0.3516+0.2501\right.$$

$$\left.+\cdots+\frac{1}{2}(0.1000)\right]$$

$$\approx 1.831$$

(b) Simpson's rule:

$$\int_1^9 \left(e^{-t^2}+\frac{1}{t+1}\right)dt$$

$$\approx \frac{9-1}{3(8)}[0.8679+4(0.3516)+2(0.2501)$$

$$+4(0.2000)+2(0.1667)+4(0.1429)$$

$$+2(0.1250)+4(0.1111)+0.1000]$$

$$= \frac{1}{3}(5.2739)$$

$$\approx 1.758$$

25. Note that heights may differ depending on the readings of the graph. Thus, answers may vary.
$n = 10,\ b = 20, a = 0$

i	x_i	$f(x_i)$
0	1	0
1	2	5
2	4	3
3	6	2
4	8	1.5
5	10	1.2
6	12	1
7	14	0.5
8	16	0.3
9	18	0.2
10	20	0.2

Area under curve for Formulation A

$$= \frac{20-0}{10}\left[\frac{1}{2}(0) + 5 + 3 + 2 + 1.5 + 1.2 + 1 + 0.5 + 0.3 + 0.2 + \frac{1}{2}(0.2)\right]$$

$= 2(14.8)$

≈ 30 mcg(h)/ml

This represents the total amount of drug available to the patient for each ml of blood.

27. As in Exercise 25, readings on the graph may vary, so answers may vary. The area both under the curve for Formulation A and above the minimum effective concentration line is on the interval $\left[\frac{1}{2}, 6\right]$.

Area under curve for Formulation A on $\left[\frac{1}{2}, 1\right]$, with $n = 1$

$$= \frac{1-\frac{1}{2}}{1}\left[\frac{1}{2}(2+6)\right]$$

$$= \frac{1}{2}(4) = 2$$

Area under curve for Formulation A on [1, 6], with $n = 5$

$$= \frac{6-1}{5}\left[\frac{1}{2}(6) + 5 + 4 + 3 + 2.4 + \frac{1}{2}(2)\right]$$

$= 18.4$

Area under minimum effective concentration line $\left[\frac{1}{2}, 6\right]$

$$= 5.5(2) = 11.0$$

Area under the curve for Formulation A and above minimum effective concentration line

$$= 2 + 18.4 - 110$$

$$\approx 9 \text{ mcg(h)/ml}$$

This represents the total erective amount of drug available to the patient for each ml of blood.

29. $y = b_0 w^{b_1} e^{-b_2 w}$

(a) if $t = 7w$ then $w = \frac{t}{7}$.

$$y = b_0\left(\frac{t}{7}\right)^{b_1} e^{-b_2 t/7}$$

(b) Replacing the constants with the given values, we have

$$y = 5.955\left(\frac{t}{7}\right)^{0.233} e^{-0.027t/7} dt$$

In 25 weeks, there are 175 days.

$$\int_0^{175} 5.955\left(\frac{t}{7}\right)^{0.233} e^{-0.027t/7} dt$$

$n = 10, b = 175, a = 0,$

$$f(t) = 5.955\left(\frac{t}{7}\right)^{0.233} e^{-0.027t/7}$$

i	t_i	$f(t_i)$
0	0	0
1	1.75	6.89
2	35	7.57
3	52.5	7.78
4	70	7.77
5	87.5	7.65
6	105	7.46
7	122.5	7.23
8	140	6.97
9	157.5	6.70
10	175	6.42

Trapezoidal rule:

$$\int_0^{175} 5.955\left(\frac{t}{7}\right)^{0.233} e^{-0.027t/7} dt$$

$$\approx \frac{175-0}{10}\left[\frac{1}{2}(0) + 6.89 + 7.57 + 7.78 + 7.77 + 7.65 + 7.46 + 7.23 + 6.97 + 6.70 + \frac{1}{2}(6.42)\right]$$

$= 17.5(69.23)$

$= 1211.525$

The total milk consumed is about 1212 kg.

Simpson's rule:

$$\int_0^{175} 5.955\left(\frac{t}{7}\right)^{0.233} e^{-0.027t/7} dt$$

$$\approx \frac{175-0}{3(10)}[0 + 4(6.89) + 2(7.57) + 4(7.78) + 2(7.77) + 4(7.65) + 2(7.46) + 4(7.23) + 2(6.97) + 4(6.70) + 6.42]$$

The total milk consumed is about 1231 kg.

(c) Replacing the constants with the given values, we have

$$y = 8.409\left(\frac{t}{7}\right)^{0.143} e^{-0.037t/7}.$$

In 25 weeks, there are 175 days.

$$\int_0^{175} 8.409\left(\frac{t}{7}\right)^{0.143} e^{-0.037t/7}\,dt$$

$n = 10, b = 175, a = 0,$

$$f(t) = 8.409\left(\frac{t}{7}\right)^{0.143} e^{-0.037t/7}$$

i	t_i	$f(t_i)$
0	0	0
1	17.5	8.74
2	35	8.80
3	52.5	8.50
4	70	8.07
5	87.5	7.60
6	105	7.11
7	122.5	6.63
8	140	6.16
9	157.5	5.71
10	175	5.28

Trapezoidal rule:

$$\int_0^{175} 8.409\left(\frac{t}{7}\right)^{0.143} e^{-0.037t/7}\,dt$$

$$\approx \frac{175-0}{10}\left[\frac{1}{2}(0) + 8.74 + 8.80 + 8.50 + 8.07 + 7.60 + 7.11 + 6.63 + 6.16 + 5.71 + \frac{1}{2}(5.28)\right]$$

$$= 17.5(69.96)$$

$$= 1224.30$$

The total milk consumed is about 1224 kg.

Simpson's rule:

$$\int_0^{175} 8.409\left(\frac{t}{7}\right)^{0.143} e^{-0.037t/7}\,dt$$

$$\approx \frac{175-0}{3(10)}[0 + 4(8.74) + 2(8.80) + 4(8.50) + 2(8.07) + 4(7.60) + 2(7.11) + 4(6.63) + 2(6.16) + 4(5.71) + 5.28]$$

$$= \frac{35}{6}(214.28)$$

$$= 1249.97$$

The total milk consumed is about 1250 kg.

31. (a)

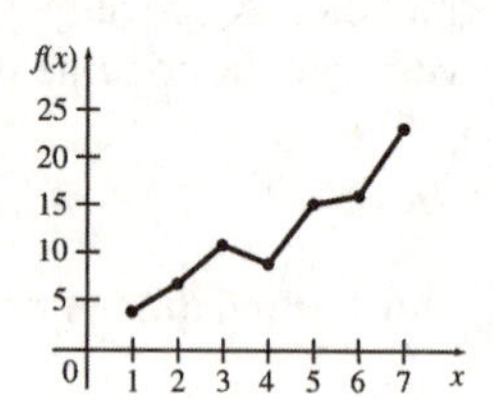

(b) $$\frac{7-1}{6}\,\frac{1}{2}$$

$$\left[(4) + 7 + 11 + 9 + 15 + 16 + \frac{1}{2}(23)\right]$$

$$= 71.5$$

(c) $$\frac{7-1}{3(6)}[4 + 4(7) + 2(11) + 4(9) + 2(15) + 4(16) + 23] = 69.0$$

33. We need to evaluate

$$\int_{12}^{36} (105e^{0.01x} + 32)\,dx.$$

Using a calculator program for Simpson's rule with $n = 20$, we obtain 3979.242 as the value of this integral. This indicates that the total revenue between the twelfth and thirty-sixth months is about 3979.

35. Use a calculator program for Simpson's rule with $n = 20$ to evaluate each of the integrals in this exercise.

(a) $$\int_{-1}^{1}\left(\frac{1}{\sqrt{2\pi}}e^{-x^2/2}\right)dx \approx 0.6827$$

The probability that a normal random variable is within 1 standard deviation of the mean is about 0.6827.

(b) $\int_{-2}^{2}\left(\frac{1}{\sqrt{2\pi}}e^{-x^2/2}\right)dx \approx 0.9545$

The probability that a normal random variable is within 2 standard deviation of the mean is about 0.9545.

(c) $\int_{-3}^{3}\left(\frac{1}{\sqrt{2\pi}}e^{-x^2/2}\right)dx \approx 0.9973$

The probability that a normal random variable is within 3 standard deviations of the mean is about 0.9973.

Chapter 15 Review Exercises

1. True

2. False: The statement is false for $n = -1$.

3. False: For example, if $f(x) = 1$ the first expression is equal to $x^2/2 + C$ and the second is equal to $x^2 + C$.

4. True

5. True

6. False: The derivative gives the instantaneous rate of change.

7. False: If the function is positive over the interval of integration the definite integral gives the exact area.

8. True

9. True

10. False: The definite integral may be positive, negative, or zero.

11. True

12. False: Sometimes true, but not in general.

13. False: The trapezoidal rule allows any number of intervals.

14. True

19. $\int(2x+3)\,dx = \frac{2x^2}{2} + 3x + C$
$= x^2 + 3x + C$

21. $\int(x^2 - 3x + 2)\,dx$
$= \frac{x^3}{3} - \frac{3x^2}{2} + 2x + C$

23. $\int 3\sqrt{x}dx = 3\int x^{1/2}\,dx$
$= \frac{3x^{3/2}}{\frac{3}{2}} + C$
$= 2x^{3/2} + C$

25. $\int(x^{1/2} + 3x^{-2/3})\,dx$
$= \frac{x^{3/2}}{\frac{3}{2}} + \frac{3x^{1/3}}{\frac{1}{3}} + C$
$= \frac{2x^{3/2}}{3} + 9x^{1/3} + C$

27. $\int\frac{-4}{x^3}\,dx = \int -4x^{-3}\,dx$
$= \frac{-4x^{-2}}{-2} + C$
$= 2x^{-2} + C$

29. $\int -3e^{2x}dx = \frac{-3e^{2x}}{2} + C$

31. $\int xe^{3x^2}\,dx = \frac{1}{6}\int 6xe^{3x^2}\,dx$

Let $u = 3x^2$, so that $du = 6x\,dx$.

$= \frac{1}{6}\int e^u du$
$= \frac{1}{6}e^u + C$
$= \frac{e^{3x^2}}{6} + C$

33. $\displaystyle\int \frac{3x}{x^2-1}\,dx = 3\left(\frac{1}{2}\right)\int \frac{2x\,dx}{x^2-1}$

Let $u = x^2 - 1$, so that $du = 2x\,dx$.

$$= \frac{3}{2}\int \frac{du}{u}$$
$$= \frac{3}{2}\ln|u| + C$$
$$= \frac{3\ln|x^2-1|}{2} + C$$

35. $\displaystyle\int \frac{x^2\,dx}{(x^3+5)^4} = \frac{1}{3}\int \frac{3x^2\,dx}{(x^3+5)^4}$

Let $u = x^3 + 5$, so that

$du = 3x^2dx.$

$$= \frac{1}{3}\int \frac{du}{u^4}$$
$$= \frac{1}{3}\int u^{-4}\,du$$
$$= \frac{1}{3}\left(\frac{u^{-3}}{-3}\right) + C$$
$$= \frac{-(x^3+5)^{-3}}{9} + C$$

37. $\displaystyle\int \frac{x^3}{e^{3x^4}}\,dx = \int x^3e^{-3x^4}$

$$= -\frac{1}{12}\int -12x^3e^{-3x^4}\,dx$$

Let $u = -3x^4$, so that $du = -12x^3\,dx$.

$$= -\frac{1}{12}\int e^u\,du$$
$$= -\frac{1}{12}\int e^u + C$$
$$= \frac{-e^{-3x^4}}{12} + C$$

39. $\displaystyle\int \frac{(3\ln x + 2)^4}{x}\,dx$

Let $u = 3\ln x + 2$ so that

$$du = \frac{3}{x}\,dx.$$

$$\int \frac{(3\ln x+2)^4}{x}\,dx = \frac{1}{3}\int \frac{3(3\ln x+2)^4}{x}\,dx$$
$$= \frac{1}{3}\int u^4\,du$$
$$= \frac{1}{3}\cdot\frac{u^5}{5} + C$$
$$= \frac{(3\ln x+2)^5}{15} + C$$

41. $f(x) = 3x + 1,\ x_1 = -1,\ x_2 = 0,\ x_3 = 1,$ $x_4 = 2,\ x_5 = 3$

$f(x_1) = -2,\ f(x_2) = 1,\ f(x_3) = 4,$ $f(x_4) = 7,\ f(x_5) = 10$

$$\sum_{i=1}^{5} f(x_i)$$
$$= f(1) + f(2) + f(3) + f(4) + f(5)$$
$$= -2 + 1 + 4 + 7 + 10$$
$$= 20$$

43. $f(x) = 2x + 3$, from $x = 0$ to $x = 4$

$$\Delta x = \frac{4-0}{4} = 1$$

i	x_i	$f(x_i)$
1	0	3
2	1	5
3	2	7
4	3	9

$$A = \sum_{i=1}^{4} f(x_i)\Delta x$$
$$= 3(1) + 5(1) + 7(1) + 9(1)$$
$$= 24$$

45. **(a)** Since $s(t)$ represents the odometer reading, the distance traveled between $t = 0$ and $t = T$ will be $s(T) - s(0)$.

(b) $\int_0^T v(t)\,dt = s(T) - s(0)$ is equivalent to the Fundamental Theorem of Calculus with $a = 0$, and $b = T$ because $s(t)$ is an antiderivative of $v(t)$.

47.
$$\int_1^2 (3x^2 + 5)\,dx = \left(\frac{3x^3}{3} + 5x\right)\Bigg|_1^2$$
$$= (2^3 + 10) - (1 + 5)$$
$$= 18 - 6$$
$$= 12$$

49.
$$\int_1^5 (3x^{-1} + x^{-3})\,dx = \left(3\ln|x| + \frac{x^{-2}}{-2}\right)\Bigg|_1^5$$
$$= \left(3\ln 5 - \frac{1}{50}\right) - \left(3\ln 1 - \frac{1}{2}\right)$$
$$= 3\ln 5 + \frac{12}{25} \approx 5.308$$

51. $\int_0^1 x\sqrt{5x^2 + 4}\,dx$

Let $u = 5x^2 + 4$, so that

$$du = 10x\,dx \text{ and } \frac{1}{10}du = x\,dx.$$

When $x = 0, u = 5(0^2) + 4 = 4$.

When $x = 1, u = 5(1^2) + 4 = 9$.

$$= \frac{1}{10}\int_4^9 \sqrt{u}\,du = \frac{1}{10}\int_4^9 u^{1/2}du$$
$$= \frac{1}{10}\cdot\frac{u^{3/2}}{3/2}\Bigg|_4^9 = \frac{1}{15}u^{3/2}\Bigg|_4^9$$
$$= \frac{1}{15}(9)^{3/2} - \frac{1}{15}(4)^{3/2}$$
$$= \frac{27}{15} - \frac{8}{15}$$
$$= \frac{19}{15}$$

53.
$$\int_0^2 3e^{-2x}dx = \frac{-3e^{-2x}}{2}\Bigg|_0^2$$
$$= \frac{-3e^{-4}}{2} + \frac{3}{2}$$
$$= \frac{3(1 - e^{-4})}{2} \approx 1.473$$

55. $\int_0^{1/2} x\sqrt{1 - 16x^4}\,dx$

Let $u = 4x^2$. Then $du = 8x\,dx$.

When $x = 0, u = 0$, and when $x = \frac{1}{2}, u = 1$.

Thus,

$$\int_0^{1/2} x\sqrt{1 - 16x^4}\,dx = \frac{1}{8}\int_0^1 \sqrt{1 - u^2}\,du.$$

Note that this integral represents the area of right upper quarter of a circle centered at the origin with a radius of 1.

Area of circle $= \pi r^2 = \pi(1^2) = \pi$

$$\int_0^1 \sqrt{1 - u^2}\,du = \frac{\pi}{4}$$
$$\frac{1}{8}\int_0^1 \sqrt{1 - u^2}\,du = \frac{1}{8}\cdot\frac{\pi}{4} = \frac{\pi}{32}$$

57. $\int_1^{e^5} \frac{\sqrt{25 - (\ln x)^2}}{x}\,dx$

Let $u = \ln x$. Then $du = \frac{1}{x}dx$.

When $x = e^5$, $u = \ln(e^5) = 5$.

When $x = 1, u = \ln(1) = 0$.

Thus,

$$\int_1^{e^5} \frac{\sqrt{25 - (\ln x)^2}}{x}\,dx = \int_0^5 \sqrt{25 - u^2}\,du.$$

Note that this integral represents the area of a right upper quarter of a circle centered at the origin with a radius of 5.

Area of circle $= \pi r^2 = \pi(5)^2 = 25\pi$

$$\int_0^5 \sqrt{25 - u^2}\,du = \frac{25\pi}{4}$$

59. $f(x) = \sqrt{4x-3};\ [1, 3]$

$$\begin{aligned}\text{Area} &= \int_1^3 \sqrt{4x-3}\,dx \\ &= \int_1^3 (4x-3)^{1/2}dx \\ &= \frac{2}{3}\cdot\frac{1}{4}\cdot(4x-3)^{3/2}\Big|_1^3 \\ &= \frac{1}{6}(9)^{3/2} - \frac{1}{6}(1)^{3/2} \\ &= \frac{1}{6}(26) \\ &= \frac{13}{3}\end{aligned}$$

61. $f(x) = xe^{x^2};\ [0, 2]$

$$\begin{aligned}\text{Area} &= \int_0^2 xe^{x^2}dx \\ &= \frac{e^{x^2}}{2}\Big|_0^2 \\ &= \frac{e^4}{2} - \frac{1}{2} \\ &= \frac{e^4-1}{2} \\ &\approx 26.80\end{aligned}$$

63. $f(x) = 5 - x^2,\ g(x) = x^2 - 3$

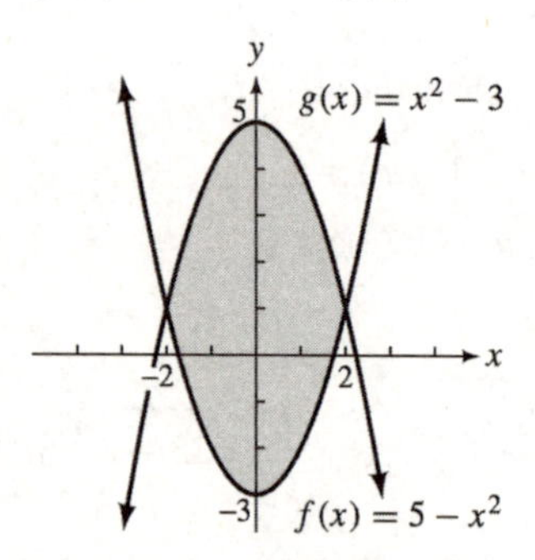

Points of intersection:

$$\begin{aligned}5 - x^2 &= x^2 - 3 \\ 2x^2 - 8 &= 0 \\ 2(x^2-4) &= 0 \\ x &= \pm 2\end{aligned}$$

Since $f(x) \ge g(x)$ in $[-2, 2]$, the area between the graphs is

$$\begin{aligned}\int_{-2}^{2}[f(x)-g(x)]dx &= \int_{-2}^{2}[(5-x^2)-(x^2-3)]dx \\ &= \int_{-2}^{2}(-2x^2+8)dx \\ &= \left(\frac{-2x^3}{3}+8x\right)\Big|_{-2}^{2} \\ &= -\frac{2}{3}(8) + 16 + \frac{2}{3}(-8) - 8(-2) \\ &= \frac{-32}{3} + 32 = \frac{64}{3}.\end{aligned}$$

65. $f(x) = x^2 - 4x,\ g(x) = x + 6,$
$x = -2,\ x = 4$

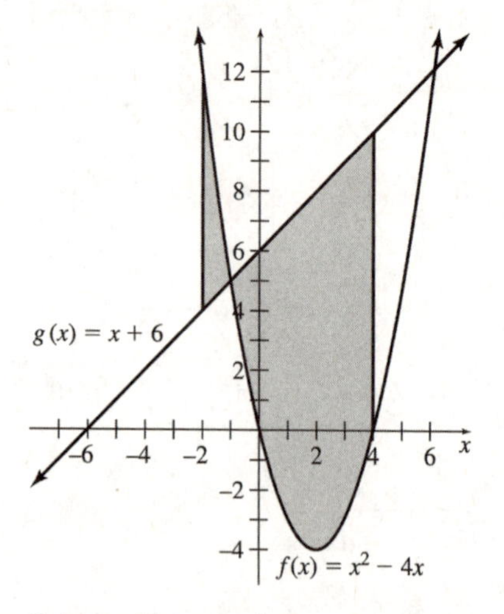

Points of intersection:

$$\begin{aligned}x^2 - 4x &= x + 6 \\ x^2 - 5x - 6 &= 0 \\ (x+1)(x-6) &= 0 \\ x &= -1 \text{ or } x = 6\end{aligned}$$

Thus, the area is

$$\int_{-2}^{-1}[x^2 - 4x - (x+6)]dx + \int_{-1}^{4}[x + 6 - (x^2 - 4x)]dx$$

$$= \left(\frac{x^3}{3} - \frac{5x^2}{2} - 6x\right)\Big|_{-2}^{-1} + \left(-\frac{x^3}{3} + \frac{5x^2}{2} + 6x\right)\Big|_{-1}^{4}$$

$$= \left(\frac{19}{6} + \frac{2}{3}\right) + \left(\frac{128}{3} + \frac{19}{6}\right) = \frac{149}{3}$$

67. $\int_1^3 \frac{\ln x}{x}\,dx$

Trapezoidal Rule:

$n = 4,\ b = 3,\ a = 1,\ f(x) = \frac{\ln x}{x}$

i	x_1	$f(x_i)$
0	1	0
1	1.5	0.27031
2	2	0.34657
3	2.5	0.36652
4	3	0.3662

$$\int_1^3 \frac{\ln x}{x}\,dx \approx \frac{3-1}{4}\left[\frac{1}{2}(0) + 0.27031 + 0.34657 + 0.36652 + \frac{1}{2}(0.3662)\right]$$
$$= 0.5833$$

Exact Value:

$$\int_1^3 \frac{\ln x}{x}\,dx$$
$$= \frac{1}{2}(\ln x)^2\Big|_1^3 = \frac{1}{2}(\ln 3)^2 - \frac{1}{2}(\ln 1)^2$$
$$\approx 0.6035$$

69. $\int_0^1 e^x\sqrt{e^x + 4}\,dx$

Trapezoidal Rule:

$n = 4, b = 1, a = 0, f(x) = e^x\sqrt{e^x + 4}$

i	x_i	$f(x_i)$
0	0	2.236
1	0.25	2.952
2	0.5	3.919
3	0.75	5.236
4	1	7.046

$$\int_0^1 e^x\sqrt{e^x + 4}\,dx$$
$$= \frac{1-0}{4}\left[\frac{1}{2}(2.236) + 2.952 + 3.919 + 5.236 + \frac{1}{2}(7.046)\right]$$
$$\approx 4.187$$

Exact value:

$$\int_0^1 e^x\sqrt{e^x + 4}\,dx = \int_0^1 e^x(e^x + 4)^{1/2}dx$$
$$= \frac{2}{3}(e^x + 4)^{3/2}\Big|_0^1$$
$$= \frac{2}{3}(e + 4)^{3/2} - \frac{2}{3}(5)^{3/2}$$
$$\approx 4.155$$

71. $\int_1^3 \frac{\ln x}{x}\,dx$

Simpson's rule:

$n = 4, b = 3, a = 1, f(x) = \frac{\ln x}{x}$

i	x_i	$f(x_i)$
0	1	0
1	1.5	0.27031
2	2	0.34657
3	2.5	0.36652
4	3	0.3662

$$\int_1^3 \frac{\ln x}{x}\,dx$$
$$\approx \frac{3-1}{3(4)}[0 + 4(0.27031) + 2(0.34657) + 4(0.36652) + 0.3662]$$
$$\approx 0.6011$$

This answer is close to the value of 0.6035 obtained from the exact integral in Exercise 67.

73. $\int_0^1 e^x\sqrt{e^x + 4}\,dx$

Simpson's rule:

$n = 4, b = 1, a = 0, f(x) = e^x\sqrt{e^x + 4}$

i	x_i	$f(x_i)$
0	0	2.236
1	0.25	2.952
2	0.5	3.919
3	0.75	5.236
4	1	7.046

$$\int_0^1 e^x\sqrt{e^x + 4}\,dx$$

$$= \frac{1-0}{3(4)}[2.236 + 4(2.952) + 2(3.919)$$
$$+ 4(5.236) + 7.046$$
$$\approx 4.156$$

This answer is close to the answer of 4.155 obtained from the exact integral in Exercise 69.

75. **(a)** $$\int_1^5 \left[\sqrt{x-1} - \left(\frac{x-1}{2}\right)\right] dx$$

$$= \int_1^5 \left(\sqrt{x-1} - \frac{x}{2} + \frac{1}{2}\,dx\right)$$

$$= \left(\frac{2}{3}(x-1)^{3/2} - \frac{x^2}{4} + \frac{x}{2}\right)\Bigg|_1^5$$

$$= \left(\frac{16}{3} - \frac{25}{4} + \frac{5}{2}\right) - \left(0 - \frac{1}{4} + \frac{1}{2}\right)$$

$$= \frac{16}{2} - 6 + 2 = \frac{4}{3}$$

(b) $n = 4, b = 5, a = 1,$

$$f(x) = \sqrt{x-1} - \frac{x}{2} + \frac{1}{2}$$

i	x_i	$f(x_i)$
0	1	0
1	2	0.5
2	3	0.41421
3	4	0.23205
4	5	0

$$\int_1^5 \left(\sqrt{x-1} - \frac{x}{2} + \frac{1}{2}\right) dx$$

$$= \left(\frac{5-1}{4}\right)\left[\frac{1}{2}(0) + 0.5 + 0.41421\right.$$
$$\left. + 0.23205 + \frac{1}{2}(0)\right]$$
$$= 1.146$$

(c) $$\int_1^5 \left(\sqrt{x-1} - \frac{x}{2} + \frac{1}{2}\right) dx$$

$$= \left(\frac{5-1}{3(4)}\right)[0 + 4(0.5) + 2(0.41421)$$
$$+ 4(0.23205) + 0]$$
$$= \left(\frac{1}{3}\right)(3.75662)$$
$$= 1.252$$

77. $$\int_{-2}^2 [x(x-1)(x+1)(x-2)(x+2)]^2 dx$$

(a) Trapezoidal Rule:

$n = 4, b = -2, a = 2,$

$$f(x) = [x(x-1)(x+1)(x-2)(x+2)]^2$$

i	x_i	$f(x_i)$
0	−2	0
1	−1	0
2	0	0
3	1	0
4	2	0

$$\int_{-2}^2 [x(x-1)(x+1)(x-2)(x+2)]^2 dx$$

$$\approx \frac{2-(-2)}{4}\left[\frac{1}{2}(0) + 0 + 0 + 0 + \frac{1}{2}(0)\right]$$
$$= 0$$

(b) Simpson's Rule:

$n = 4, b = 2, a = 2,$

$$f(x) = [x(x-1)(x+1)(x-2)(x+2)]^2$$

i	x_i	$f(x_i)$
0	−2	0
1	−1	0
2	0	0
3	1	0
4	2	0

$$\int_{-2}^2 [x(x-1)(x+1)(x-2)(x+2)]^2 dx$$

$$\approx \frac{2-(-2)}{3(4)}[0 + 4(0) + 2(0) + 4(0) + 0]$$
$$= 0$$

79. $C'(x) = 3\sqrt{2x-1}$; 13 units cost \$270.

$$C(x) = \int 3(2x-1)^{1/2}dx$$
$$= \frac{3}{2}\int 2(2x-1)^{1/2}dx$$

Let $u = 2x - 1$, so that

$$du = 2dx.$$
$$= \frac{3}{2}\int u^{1/2}du$$
$$= \frac{3}{2}\left(\frac{u^{3/2}}{3/2}\right) + C$$
$$= (2x-1)^{3/2} + C$$
$$C(13) = [2(13)-1]^{3/2} + C$$

Since $C(13) = 270$,

$$270 = 25^{3/2} + C$$
$$270 = 125 + C$$
$$C = 145.$$

Thus,

$$C(x) = (2x-1)^{3/2} + 145.$$

81. Read values for the rate of investment income accumulation for every 2 years from year 1 to year 9. These are the heights of rectangles with width $\Delta x = 2$.

Total accumulated income

$$= 11,000(2) + 9000(2) + 12,000(2) + 10,000(2) + 6000(2) \approx \$96,000$$

83. $S'(x) = 3\sqrt{2x+1} + 3$

$$S(x) = \int_0^4 (3\sqrt{2x+1} + 3)\,dx$$
$$= [(2x+1)^{3/2} + 3x]\Big|_0^4$$
$$= (27 + 12) - (1 + 0) = 38$$

Total sales = \$38,000.

85. $S(q) = q2 + 5q + 100$

$D(q) = 350 - q^2$

$S(q) = D(q)$ at the equilibrium point.

$$q^2 + 5q + 100 = 350 - q^2$$
$$2q^2 + 5q - 250 = 0$$
$$(-2q + 25)(q - 10) = 0$$
$$q = -\frac{25}{2} \quad \text{or} \quad q = 10$$

Since the number of units produced would not be negative, the equilibrium point occurs when $q = 10$.

Equilibrium supply

$$= (10)^2 + 5(10) + 100 = 250$$

Equilibrium demand

$$= 350 - (10)^2 = 250$$

(a) Producers' surplus

$$= \int_0^{10} \left[250 - (q^2 + 5q + 100)\right] dq$$
$$= \int_0^{10} (-q^2 - 5q + 150)\,dq$$
$$= \left(\frac{-q^3}{3} - \frac{5q^2}{2} + 150q\right)\Bigg|_0^{10}$$
$$= \frac{-1000}{3} - \frac{500}{2} + 1500$$
$$= \frac{\$2750}{3} \approx \$916.67$$

(b) Consumers' surplus

$$= \int_0^{10} [(350 - q^2) - 250]dq$$
$$= \int_0^{10} (100 - q^2)\,dq$$
$$= \left(100q - \frac{q^3}{3}\right)\Bigg|_0^{10}$$
$$= 1000 - \frac{1000}{3} = \frac{\$2000}{3} \approx \$666.67$$

87. **(a)** Total amount $\approx \frac{1}{2}(2.131) + 2.118 + 2.097 + 2.073 + 1.983 + 1.890 + 1.862 + 1.848 + 1.812 + \frac{1}{2}(1.938)$

$$\approx 17.718$$

The estimate is 17.718 billion barrels.

(b) The left endpoint sum is 17.814 and the right endpoint sum is 17.621. Their average is $\frac{17.814 + 17.621}{2} \approx 17.718.$

(d) The line of best fit has the equation $y = -0.03545x + 2.13475.$

$$\int_0^9 (-0.03545x + 2.13475)\,dx \approx 17.777$$

The integral yields an estimate of 17.777 billion barrels.

89. $f(t) = 100 - t\sqrt{0.4t^2 + 1}$

The total number of additional spiders in the first ten months is

$$\int_0^{10} (100 - t\sqrt{0.4t^2 + 1})\,dt,$$

where t is the time in months.

$$= \int_0^{10} 100dt - \int_0^{10} t\sqrt{0.4t^2 + 1}\,dt.$$

Let $u = 0.4t^2 + 1$, so that

$du = 0.8t\,dt$ and $\frac{1}{0.8}du = t\,dt.$

When $t = 10, u = 41.$

When $t = 0, u = 1.$

$$= \int_0^{10} 100\,dt - \frac{1}{0.8}\int_1^{41} u^{1/2}du$$

$$= 100t\Big|_0^{10} - \frac{5}{4}\cdot\frac{u^{3/2}}{\frac{3}{2}}\Bigg|_1^{41} = 1000 - \frac{5}{6}u^{3/2}\Bigg|_1^{41}$$

≈ 782

The total number of additional spiders in the first 10 months is about 782.

91. (a) The total area is the area of the triangle on [0, 12] with height 0.024 plus the area of the rectangle on [12, 17.6] with height 0.024.

$$A = \frac{1}{2}(12 - 0)(0.024) + (17.6 - 12)(0.024)$$
$$= 0.144 + 0.1344 = 0.2784$$

(b) On [0, 12] we define the function $f(x)$ with slope $\frac{0.024-0}{12-0} = 0.002$ and y-intercept 0.

$$f(x) = 0.002x$$

On [12, 17.6], define $g(x)$ as the constant value.

$$g(x) = 0.024.$$

The area is the sum of the integrals of these two functions.

$$A = \int_0^{12} 0.002x dx + \int_{12}^{17.6} 0.024dx$$
$$= 0.001x^2\Big|_0^{12} + 0.024x\Big|_{12}^{17.6}$$
$$= 0.001(12^2 - 0^2) + 0.024(17.6 - 12)$$
$$= 0.144 + 0.1344 = 0.2784$$

93. (a) $\int_0^{321} 1.87t^{1.49}e^{-0.189(\ln t)^2}\,dt$

Trapezoidal rule:

$n = 8, b = 321, a = 1,$

$$f(t) = 1.87t^{1.49}e^{-0.189(\ln t)^2}$$

i	x_i	$f(t_i)$
0	1	1.87
1	41	34.9086
2	81	33.9149
3	121	30.7147
4	161	27.5809
5	161	24.8344
6	201	22.4794
7	281	20.4622
8	321	18.7255

$$\text{Total amount} \approx \frac{321-1}{8}\left[\frac{1}{2}(1.87) + 34.9086 + 33.9149 + 30.7147 + 27.5809 + 24.8344 + 22.4794 + 20.4622 + \frac{1}{2}(18.7255)\right]$$

$$\approx 8208$$

The total milk production from $t = 1$ to $t = 321$ is approximately 8208 kg.

(b) Simpson's rule:

$n = 8, b = 321, a = 1, \ f(t) = 1.87t^{1.49}e^{-0.189(\ln t)^2}$

i	t_i	$f(t_i)$
0	1	1.87
1	41	34.9086
2	81	33.9149
3	121	30.7147
4	161	27.5809
5	201	24.8344
6	241	22.4794
7	281	20.4622
8	321	18.7255

$$\text{Total amount} \approx \frac{321-1}{8(3)}\left[1.87 + 4(34.9086) + 2(33.9149) + 4(30.7147) + 2(27.5809) + 4(24.8344) + 2(22.4794) + 4(20.4622) + 18.7255\right]$$

$$\approx 8430$$

The total milk production from $t = 1$ to $t = 321$ is approximately 8430 kg.

(c) Numerical evaluation gives $\int_0^{321} 1.87t^{1.49}e^{-0.189(\ln t)^2}\,dt \approx 8558$, or 8558 kg

95. $v(t) = t^2 - 2t$

$$s(t) = \int_0^t (t^2 - 2t)\,dt$$

$$s(t) = \frac{t^3}{3} - t^2 + s_0$$

If $t = 3, s = 8.$

$$8 = 9 - 9 + s_0$$
$$8 = s_0$$

Thus $s(t) = \dfrac{t^3}{3} - t^2 + 8.$

Chapter 16

FURTHER TECHNIQUES AND APPLICATIONS OF INTEGRATION

16.1 Integration by Parts

Your Turn 1

$$\int xe^{-2x}dx$$

Let $u = x$ and $dv = e^{-2x}dx.$

Then $du = dx$ and $v = \int e^{-2x}dx$

$$= -\frac{1}{2}e^{-2x}.$$

$$\int u\,dv = uv - \int v\,du$$

$$\int xe^{-2x}dx = -\frac{1}{2}xe^{-2x} - \int\left(-\frac{1}{2}e^{-2x}\right)dx$$

$$= -\frac{1}{2}xe^{-2x} + \frac{1}{2}\int e^{-2x}dx$$

$$= -\frac{1}{2}xe^{-2x} + \frac{1}{2}\left(-\frac{1}{2}\right)e^{-2x} + C$$

$$= -\frac{1}{2}xe^{-2x} - \frac{1}{4}e^{-2x} + C$$

$$= -\frac{1}{4}e^{-2x}(2x+1) + C$$

Your Turn 2

$$\int \ln 2x\,dx$$

Let $u = \ln 2x$ and $dv = dx.$

Then $du = \frac{1}{2x}(2) = \frac{1}{x}$ and $v = \int dx = x.$

$$\int u\,dv = uv - \int v\,du$$

$$\int \ln 2x\,dx = (\ln 2x)(x) - \int x\cdot\frac{1}{x}dx$$

$$= x\ln 2x - \int dx$$

$$= x\ln 2x - x + C$$

Your Turn 3

$$\int (3x^2+4)e^{2x}dx$$

Choose $3x^2 + 4$ as the part to be differentiated and e^{2x} as the part to be integrated.

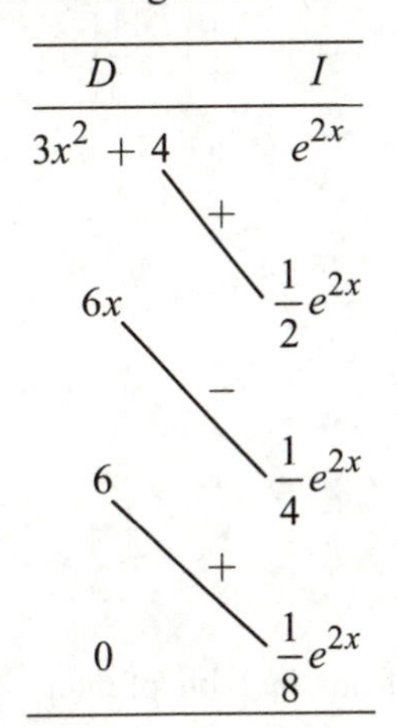

$$\int (3x^2+4)e^{2x}dx$$

$$= (3x^2+4)\left(\frac{1}{2}e^{2x}\right) - (6x)\left(\frac{1}{4}e^{2x}\right) + \left(\frac{1}{8}e^{2x}\right) + C$$

$$= \frac{1}{8}e^{2x}[4(3x^2+4) - 2(6x) + 6] + C$$

$$= \frac{1}{8}e^{2x}(12x^2 + 16 - 12x + 6) + C$$

$$= \frac{1}{4}e^{2x}(6x^2 - 6x + 11) + C$$

Your Turn 4

$$\int_1^e x^2\ln x\,dx$$

Let $u = \ln x$ and $dv = x^2dx.$

Then $du = \frac{1}{x}dx$ and $v = \int x^2dx = \frac{x^3}{3}.$

$$\int u\,dv = uv - \int v\,du$$

$$\int x^2 \ln x\, dx = \frac{x^3}{3}\ln x - \int\left(\frac{x^3}{3}\cdot\frac{1}{x}\right)dx$$
$$= \frac{x^3}{3}\ln x - \frac{1}{3}\int x^2 dx$$
$$= \frac{x^3}{3}\ln x - \frac{1}{3}\left(\frac{x^3}{3}\right) + C$$
$$= \frac{x^3}{3}\left(\ln x - \frac{1}{3}\right) + C$$
$$\int_1^e x^2 \ln x\, dx = \frac{x^3}{3}\left(\ln x - \frac{1}{3}\right)\Bigg|_1^e$$
$$= \frac{e^3}{3}\left(1 - \frac{1}{3}\right) - \frac{1}{3}\left(0 - \frac{1}{3}\right)$$
$$= \frac{2}{9}e^3 + \frac{1}{9}$$
$$= \frac{2e^3 + 1}{9}$$

Your Turn 5

$$\int \frac{1}{x\sqrt{4 + x^2}}dx$$

Use formula 10 from the table of integrals with $a = 2$.

$$\int \frac{1}{x\sqrt{4 + x^2}}dx = \int \frac{1}{x\sqrt{2^2 + x^2}}dx$$
$$= -\frac{1}{2}\ln\left|\frac{2 + \sqrt{4 + x^2}}{x}\right| + C$$

16.1 Exercises

1. $\int xe^x dx$

Let $dv = e^x dx$ and $u = x$.

Then $v = \int e^x dx$ and $du = dx$.

$v = e^x$

Use the formula

$$\int u\, dv = uv - \int v\, du.$$

$$\int xe^x dx = xe^x - \int e^x dx = xe^x - e^x + C$$

3. $\int (4x - 12)e^{-8x}dx$

Let $dv = e^{-8x}\, dx$ and $u = 4x - 12$

Then $v = \int e^{-8x}\, dx$ and $du = 4\,dx$.

$$v = \frac{e^{-8x}}{-8}$$

$$\int (4x - 12)e^{-8x}dx$$
$$= (4x - 12)\left(\frac{e^{-8x}}{-8}\right) - \int\left(\frac{e^{-8x}}{-8}\right)\cdot 4\,dx$$
$$= -\frac{4x}{8}e^{-8x} + \frac{12}{8}e^{-8x} - \left(-\frac{4}{8}\cdot\frac{e^{-8x}}{-8}\right) + C$$
$$= -\frac{x}{2}e^{-8x} + \frac{3}{2}e^{-8x} - \frac{1}{16}e^{-8x} + C$$
$$= \left(-\frac{x}{2} + \frac{23}{16}\right)e^{-8x} + C$$

5. $\int x \ln dx$

Let $dv = x\,dx$ and $u = \ln x$.

Then $v = \frac{x^2}{2}$ and $du = \frac{1}{x}\,dx$.

$$\int x \ln dx = \frac{x^2}{2}\ln x - \int \frac{x}{2}dx$$
$$= \frac{x^2 \ln x}{2} - \frac{x^2}{4} + C$$

7. $\int_0^1 \frac{2x + 1}{e^x}dx$

$$= \int_0^1 (2x + 1)e^{-x}\, dx$$

Let $dv = e^{-x}\, dx$ and $u = 2x + 1$.

Then $v = \int e^{-x}\, dx$ and $du = 2\,dx$.

$$v = -e^{-x}$$

$$\int \frac{2x + 1}{e^x}dx$$
$$= -(2x + 1)e^{-x} + \int 2e^{-x}dx$$
$$= -(2x + 1)e^{-x} - 2e^{-x}$$

$$\int_0^1 \frac{2x+1}{e^x}dx$$

$$= [-(2x+1)e^{-x} - 2e^{-x}]\Big|_0^1$$

$$= [-(3)e^{-1} - 2e^{-1}] - (-1 - 2)$$

$$= -5e^{-1} + 3$$

$$\approx 1.161$$

9. $\int \ln 3x\, dx$

Let $dv = dx$ and $u = \ln 3x$.

Then $v = x$ and $du = \frac{1}{x}dx$.

$$\int \ln 3x\, dx = x \ln 3x - \int dx$$

$$= x\ln 3x - x$$

$$\int \ln 3x\, dx = (x\ln 3x - x)\Big|_1^9$$

$$= (9\ln 27 - 9) - (\ln 3 - 1)$$

$$= 9\ln 3^3 - 9 - \ln 3 + 1$$

$$= 27\ln 3 - \ln 3 - 8$$

$$= 26\ln 3 - 8 \approx 20.56$$

11. The area is $\int_2^4 (x-2)e^x dx.$

Let $dv = e^x dx$ and $u = x - 2$.
Then $v = e^x$ and $du = dx$.

$$\int (x-2)e^x\, dx = (x-2)e^x - \int e^x dx$$

$$\int_1^4 (x-2)e^x\, dx = [(x-2)e^x - e^x]\Big|_2^4$$

$$= (2e^4 - e^4) - (0 - e^2)$$

$$= e^4 + e^2 \approx 61.99$$

13. $\int x^2 e^{2x}\, dx$

Let $u = x^2$ and $dv = e^{2x}\, dx$.

Use column integration.

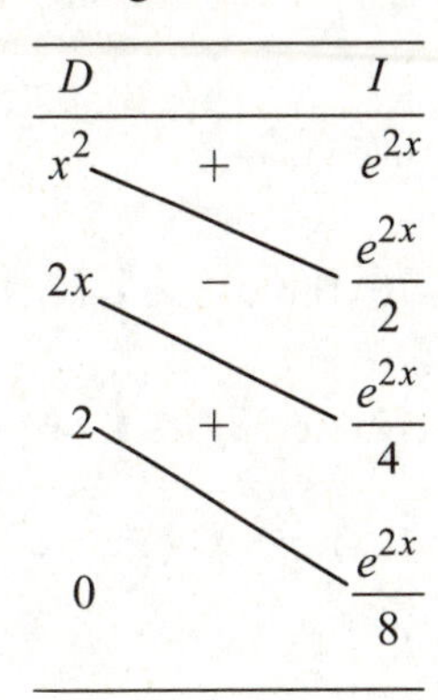

$$\int x^2e^{2x}\, dx = x^2\left(\frac{e^{2x}}{2}\right) - 2x\left(\frac{e^{2x}}{4}\right) + \frac{2e^{2x}}{8} + C$$

$$= \frac{x^2e^{2x}}{2} - \frac{xe^{2x}}{2} + \frac{e^{2x}}{4} + C$$

15. $\int x^2\sqrt{x+4}\, dx$

Let $u = x^2$ and $dv = (x+4)^{1/2}$. Use column integration.

D		I
x^2	+	$(x+4)^{1/2}$
$2x$	−	$\frac{2}{3}(x+4)^{3/2}$
2	+	$\left(\frac{2}{3}\right)\left(\frac{2}{5}\right)(x+4)^{5/2}$
0		$\left(\frac{2}{3}\right)\left(\frac{2}{5}\right)\left(\frac{2}{7}\right)(x+4)^{7/2}$

$$\int x^2\sqrt{x+4}\, dx$$

$$= x^2(x+4)^{3/2}\left(\frac{2}{3}\right) - 2x(x+4)^{5/2}\left(\frac{2}{3}\right)\left(\frac{2}{5}\right)$$

$$+ 2(x+4)^{7/2}\left(\frac{2}{3}\right)\left(\frac{2}{5}\right)\left(\frac{2}{7}\right) + C$$

$$= \frac{2}{3}x^2(x+4)^{3/2} - \frac{8}{15}x(x+4)^{5/2}$$

$$+ \frac{16}{105}(x+4)^{7/2} + C$$

17. $\int (8x + 10)\ln(5x)\,dx$

Let $dv = (8x + 10)\,dx$ and $u = \ln(5x)$.

Then $v = 4x^2 + 10x$ and $du = \frac{1}{x}\,dx$.

$$\int (8x + 10)\ln(5x)\,dx$$
$$= (4x^2 + 10x)\ln(5x) - \int (4x^2 + 10x)\left(\frac{1}{x}\right)dx$$
$$= (4x^2 + 10x)\ln(5x) - \int (4x + 10)\,dx$$
$$= (4x^2 + 10x)\ln(5x) - 2x^2 - 10x + C$$

19. $\int_1^2 (1 - x^2)e^{2x}\,dx$

Let $u = 1 - x^2$ and $dv = e^{2x}\,dx$.

Use column integration.

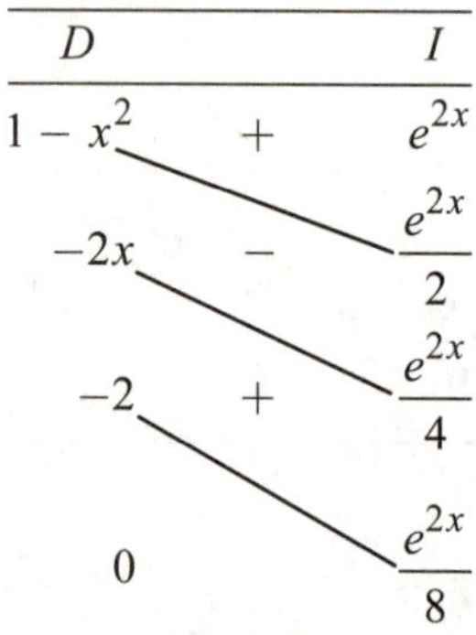

D		I
$1 - x^2$	+	e^{2x}
$-2x$	−	$\frac{e^{2x}}{2}$
-2	+	$\frac{e^{2x}}{4}$
0		$\frac{e^{2x}}{8}$

$$\int (1 - x^2)e^{2x}dx$$
$$= \frac{(1 - x^2)e^{2x}}{2} - \frac{(-2x)e^{2x}}{4} + \frac{(-2)e^{2x}}{8}$$
$$= \frac{(1 - x^2)e^{2x}}{2} + \frac{xe^{2x}}{2} - \frac{e^{2x}}{4}$$
$$= \frac{e^{2x}}{2}\left(1 - x^2 + x - \frac{1}{2}\right) = \frac{e^{2x}}{2}\left(\frac{1}{2} - x^2 + x\right)$$

$$\int_1^2 (1 - x^2)e^{2x}\,dx = \frac{e^{2x}}{2}\left(\frac{1}{2} - x^2 + x\right)\Bigg|_1^2$$
$$= \frac{e^4}{2}\left(-\frac{3}{2}\right) - \frac{e^2}{2}\left(\frac{1}{2}\right)$$
$$= -\frac{e^2}{4}(3e^2 + 1) \approx -42.80$$

21. $\int_0^1 \frac{x^3dx}{\sqrt{3 + x^2}} = \int_0^1 x^3(3x + x^2)^{-1/2}\,dx$

Let $dv = x(3 + x^2)^{-1/2}\,dx$ and $u = x^2$.

Then $v = \frac{2(3+x^2)^{1/2}}{2}$

$v = (3 + x^2)^{1/2}$ and $du = 2x\,dx$.

$$\int \frac{x^3dx}{\sqrt{3 + x^2}}$$
$$= x^2(3 + x^2)^{1/2} - \int 2x(3 + x^2)^{1/2}\,dx$$
$$= x^2(3 + x^2)^{1/2} - \frac{2}{3}(3 + x^2)^{3/2}$$

$$\int_0^1 \frac{x^3dx}{\sqrt{3 + x^2}}$$
$$= \left[x^2(3 + x^2)^{1/2} - \frac{2}{3}(3 + x^2)^{3/2}\right]\Bigg|_0^1$$
$$= 4^{1/2} - \frac{2}{3}(4^{3/2}) - 0 + \frac{2}{3}(3^{3/2})$$
$$= 2 - \frac{2}{3}(8) + \frac{2}{3}(3^{3/2})$$
$$= -\frac{10}{3} + 2\sqrt{3}$$
$$\approx 0.1308$$

23. $\int \frac{16}{\sqrt{x^2 + 16}}\,dx$

Use formula 5 from the table of integrals with $a = 4$.

$$\int \frac{16}{\sqrt{x^2 + 16}}\,dx = 16\int \frac{1}{\sqrt{x^2 + 4^2}}\,dx$$
$$= 16\ln\left|x + \sqrt{x^2 + 16}\right| + C$$

25. $\int \frac{3}{x\sqrt{121 - x^2}}\,dx = 3\int \frac{dx}{x\sqrt{11^2 - x^2}}$

If $a = 11$, this integral matches formula 9 in the table.

$$= 3\left(-\frac{1}{11}\ln\left|\frac{11 + \sqrt{121 - x^2}}{x}\right|\right) + C$$
$$= -\frac{3}{11}\ln\left|\frac{11 + \sqrt{121 - x^2}}{x}\right| + C$$

27. $\int \frac{-6}{x(4x+6)^2}\,dx$

Use formula 14 from the table of integrals with $a = 4$ and $b = 6$.

$$\int \frac{-6}{x(4x+6)^2}\,dx$$
$$= -6\int \frac{1}{x(4x+6)^2}\,dx$$
$$= -6\left[\frac{1}{6(4x+6)} + \frac{1}{6^2}\ln\left|\frac{x}{4x+6}\right|\right] + C$$
$$= \frac{-1}{(4x+6)} - \frac{1}{6}\ln\left|\frac{x}{4x+6}\right| + C$$

31. First find the indefinite integral using integration by parts.

$$\int u\,dv = uv - \int v\,du$$

Now substitute the given values.

$$\int_0^1 u\,dv = uv\Big|_0^1 - \int_0^1 v\,du$$
$$= [u(1)v(1) - u(0)v(0)] - 4$$
$$= (3)(-4) - (2)(1) - 4 = -18$$

33. $\int r\,ds = rs - \int s\,dr$

$$\int_0^2 r\,ds = rs\Big|_0^2 - \int_0^2 s\,dr$$
$$10 = r(2)s(2) - r(0)s(0) - 5$$
$$15 = r(s)s(2)$$

35. $\int x^n \cdot \ln|x|\,dx,\ n \neq -1$

Let $u = \ln|x|$ and $dv = x^n\,dx$.

Use column integration.

D		I
$\ln\lvert x\rvert$	+	x^n
$\frac{1}{x}$	−	$\frac{1}{n+1}x^{n+1}$

$$\int x^n \cdot \ln|x|\,dx$$
$$= \frac{1}{n+1}x^{n+1}\ln|x| - \int\left[\frac{1}{x}\cdot\frac{1}{n+1}x^{n+1}\right]dx$$
$$= \frac{1}{n+1}x^{n+1}\ln|x| - \int \frac{1}{n+1}x^n\,dx$$
$$= \frac{1}{n+1}x^{n+1}\ln|x| - \frac{1}{(n+1)^2}x^{n+1} + C$$
$$= x^{n+1}\left[\frac{\ln|x|}{n+1} - \frac{1}{(n+1)^2}\right] + C$$

37. $\int x\sqrt{x+1}\,dx$

(a) Let $u = x$ and $dv = \sqrt{x+1}\,dx$.
Use column integration.

D		I
x	+	$\sqrt{x+1}$
1	−	$\left(\frac{2}{3}\right)(x+1)^{3/2}$
0		$\left(\frac{4}{15}\right)(x+1)^{5/2}$

$$\int x\sqrt{x+1}\,dx$$
$$= \left(\frac{2}{3}\right)x(x+1)^{3/2} - \left(\frac{4}{15}\right)(x+1)^{5/2} + C$$

(b) Let $u = x + 1$; then $u - 1 = x$ and $du = dx$.

$$\int x\sqrt{x+1}\,dx$$
$$= \int (u-1)u^{1/2}du = \int (u^{3/2} - u^{1/2})\,du$$
$$= \frac{2}{5}u^{5/2} - \frac{2}{3}u^{3/2} + C$$
$$= \frac{2}{5}(x+1)^{5/2} - \frac{2}{3}(x+1)^{3/2} + C$$

(c) Both results factor as $\frac{2}{15}(x+1)^{3/2}(3x-2) + C$, so they are equivalent.

39. $R = \int_0^{12} (x+1)\ln(x+1)\,dx$

Let $u = \ln(x+1)$ and $dv = (x+1)\,dx$.

Then $du = \frac{1}{x+1}dx$ and $v = \frac{1}{2}(x+1)^2$.

$$\int (x+1)\ln(x+1)\,dx$$
$$= \frac{1}{2}(x+1)^2\ln(x+1) - \int\left[\frac{1}{2}(x+1)^2 \cdot \frac{1}{x+1}\right]dx$$
$$= \frac{1}{2}(x+1)^2\ln(x+1) - \int \frac{1}{2}(x+1)\,dx$$
$$= \frac{1}{2}(x+1)^2\ln(x+1) - \frac{1}{4}(x+1)^2 + C$$

$$\int_0^{12}(x+1)\ln(x+1)\,dx$$
$$= \left[\frac{1}{2}(x+1)^2\ln(x+1) - \frac{1}{4}(x+1)^2\right]\Bigg|_0^{12}$$
$$= \frac{169}{2}\ln 13 - 42 \approx \$174.74$$

41. The total accumulated growth of the microbe population during the first 2 days is given by

$$\int_0^2 27t\,e^{3t}\,dt.$$

Let $dv = e^{3t}\,dt$ and $u = 27t.$

Then $v = \frac{e^{3t}}{3}$ and $du = 27dt.$

$$\int 27te^{3t}dt = 27t \cdot \frac{e^{3t}}{3} - \int \frac{e^{3t}}{3} \cdot 27dt$$
$$= 9te^{3t} - 3e^{3t}$$

$$\int_0^2 27te^{3t}dt = (9te^{3t} - 3e^{3t})\Big|_0^2$$
$$= (18e^6 - 3e^6) - (0 - 3)$$
$$= 15e^6 + 3 \approx 6054$$

43. $\int_0^6 (-10.28 + 175.9te^{-t/1.3})dt$

$$= -10.28t + 175.9\int te^{-t/1.3}dt$$

Evaluate this integral using integration by parts.

Let $u = t$ and $dv = e^{-t/1.3}dt.$

Then $du = dt$ and $v = -1.3e^{-t/1.3}.$

$$\int te^{-t/1.3}dt$$
$$= (t)(-1.3e^{-t/1.3}) - \int (-1.3e^{-t/1.3})dt$$
$$= -1.3te^{-t/1.3} - 1.69e^{-t/1.3} + C$$

Substitute this expression in the earlier expression.

$$-10.28t + 175.9(-1.3te^{-t/1.3} - 1.69e^{-t/1.3})\Big|_0^6$$
$$= -10.28t - 228.67te^{-t/1.3} - 297.271e^{-t/1.3}\Big|_0^6$$
$$= (-61.68 - 1669.291e^{-6/1.3}) - (-297.271)$$
$$\approx 219.07$$

The total thermic energy is about 219 kJ.

16.2 Volume and Average Value

Your Turn 1

Find the volume of the solid of revolution formed by rotating about the x-axis the region bounded by $y = x^2 + 1$, $y = 0$, $x = -1$, and $x = 1$.

The region and the solid are shown below.

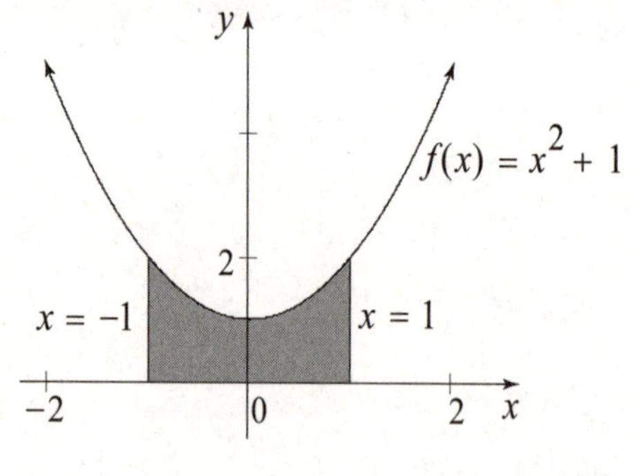

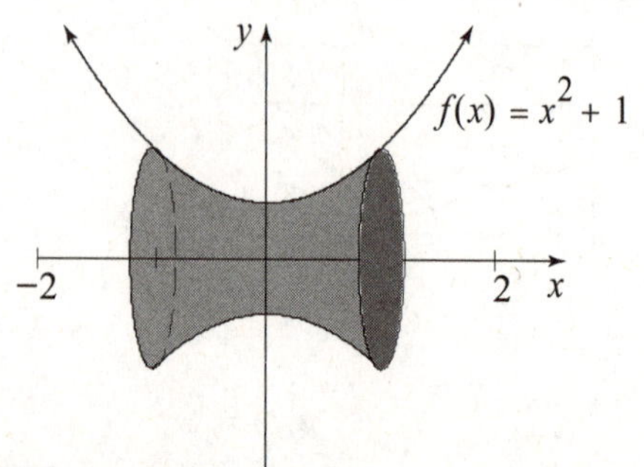

$$\begin{aligned}
V &= \int_{-1}^{1} \pi(x^2+1)^2\,dx \\
&= \pi\int_{-1}^{1} (x^4+2x^2+1)\,dx \\
&= \pi\left(\frac{x^5}{5}+\frac{2x^3}{3}+x\right)\Bigg|_{-1}^{1} \\
&= \pi\left[\left(\frac{1}{5}+\frac{2}{3}+1\right)-\left(-\frac{1}{5}-\frac{2}{3}-1\right)\right] \\
&= \pi\left(\frac{56}{15}\right)=\frac{56\pi}{15}
\end{aligned}$$

Your Turn 2

Find the average value of the function $f(x) = x + \sqrt{x}$ on the interval $[1, 4]$.

Use the formula for average value with $a = 1$ and $b = 4$.

$$\begin{aligned}
\frac{1}{4-1}\int_1^4 (x+\sqrt{x})\,dx &= \frac{1}{2}\left(\frac{x^2}{2}+\frac{2}{3}x^{3/2}\right)\Bigg|_1^4 \\
&= \frac{1}{3}\left(8+\frac{16}{3}-\frac{1}{2}-\frac{2}{3}\right) \\
&= \frac{1}{3}\left(\frac{73}{6}\right)=\frac{73}{18}
\end{aligned}$$

16.2 Exercises

1. $f(x) = x, y = 0, x = 0, x = 3$

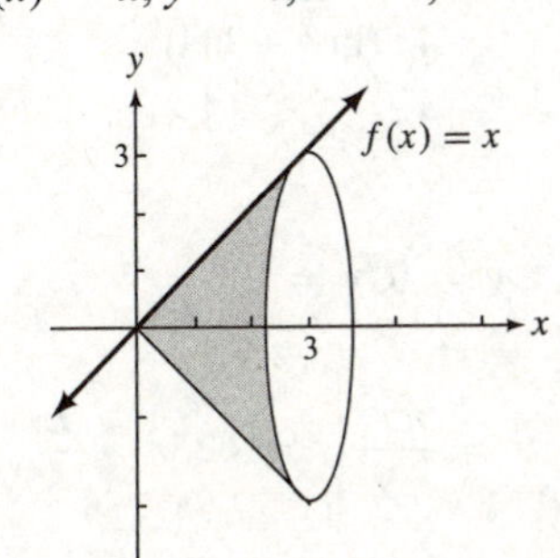

$$V = \pi\int_0^3 x^2\,dx = \frac{\pi x^3}{2}\Bigg|_0^3 = \frac{\pi(27)}{3} - 0 = 9\pi$$

3. $f(x) = 2x + 1, y = 0, x = 0, x = 4$

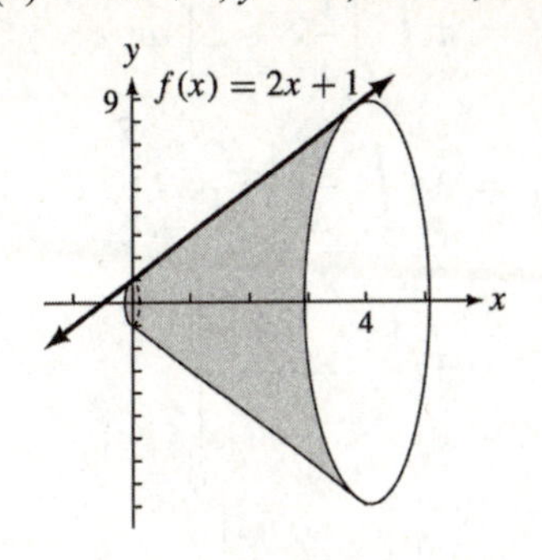

$$V = \pi\int_0^4 (2x+1)^2\,dx$$

Let $u = 2x + 1$. Then $du = 2\,dx$.

If $x = 4$, $u = 9$. If $x = 0$, $u = 1$.

$$\begin{aligned}
V &= \frac{1}{2}\pi\int_0^4 2(2x+1)^2\,dx \\
&= \frac{1}{2}\pi\int_1^9 u^2\,du \\
&= \frac{\pi}{2}\left(\frac{u^3}{3}\right)\Bigg|_1^9 \\
&= \frac{\pi}{2}\left(\frac{729}{3}-\frac{1}{3}\right) \\
&= \frac{728\pi}{6} \\
&= \frac{364\pi}{3}
\end{aligned}$$

5. $f(x) = \frac{1}{3}x + 2, y = 0, x = 1, x = 3$

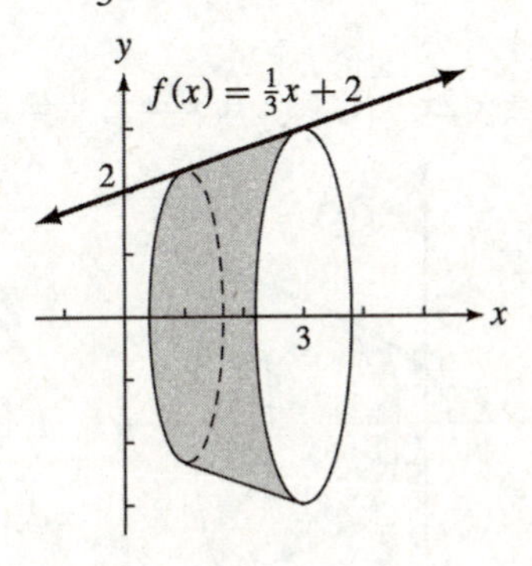

$$V = \pi\int_1^3\left(\frac{1}{3}x+2\right)^2 dx$$
$$= 3\pi\int_1^3\frac{1}{3}\left(\frac{1}{3}x+2\right)^2 dx$$
$$= 3\pi\left.\frac{\left(\frac{1}{3}x+2\right)^3}{3}\right|_1^3$$
$$= \pi\left.\left(\frac{1}{3}x+2\right)^3\right|_1^3$$
$$= 27\pi - \frac{343\pi}{27} = \frac{386\pi}{27}$$

7. $f(x) = \sqrt{x}, y = 0, x = 1, x = 4$

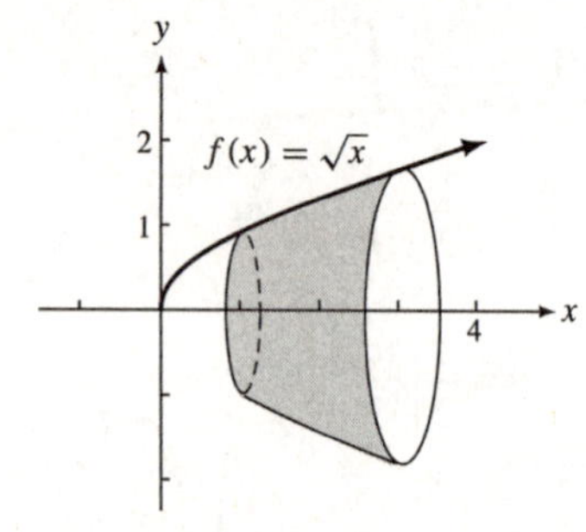

$$V = \pi\int_1^4(\sqrt{x})^2 dx = \pi\int_1^4 x\,dx$$
$$= \left.\frac{\pi x^2}{2}\right|_1^4$$
$$= 8\pi - \frac{\pi}{2} = \frac{15\pi}{2}$$

9. $f(x) = \sqrt{2x+1}, y = 0, x = 1, x = 4$

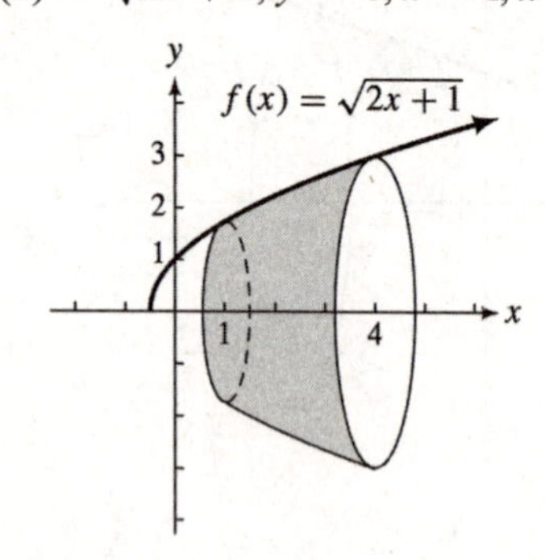

$$V = \pi\int_1^4(\sqrt{2x+1})^2 dx$$
$$= \pi\int_1^4(2x+1)\,dx$$
$$= \pi\left.\left(\frac{2x^2}{2}+x\right)\right|_1^4$$
$$= \pi[(16+4)-2]$$
$$= 18\pi$$

11. $f(x) = e^x; y = 0, x = 0; x = 2$

$$V = \pi\int_0^2 e^{2x}\,dx = \left.\frac{\pi e^{2x}}{2}\right|_0^2$$
$$= \frac{\pi e^4}{2} - \frac{\pi}{2}$$
$$= \frac{\pi}{2}(e^4-1)$$
$$\approx 84.19$$

13. $f(x) = \frac{2}{\sqrt{x}}, y = 0, x = 1, x = 3$

$$V = \pi\int_1^3\left(\frac{2}{\sqrt{x}}\right)^2 dx$$
$$= \pi\int_1^3\frac{4}{x}dx$$
$$= 4\pi\ln|x|\Big|_1^3$$
$$= 4\pi(\ln 3 - \ln 1)$$
$$= 4\pi\ln 3 \approx 13.81$$

15. $f(x) = x^2, y = 0, x = 1, x = 5$

$$V = \pi\int_1^5 x^4\,dx = \left.\frac{\pi x^5}{5}\right|_1^5 = 625\pi - \frac{\pi}{5} = \frac{3124\pi}{5}$$

17. $f(x) = 1 - x^2, y = 0$

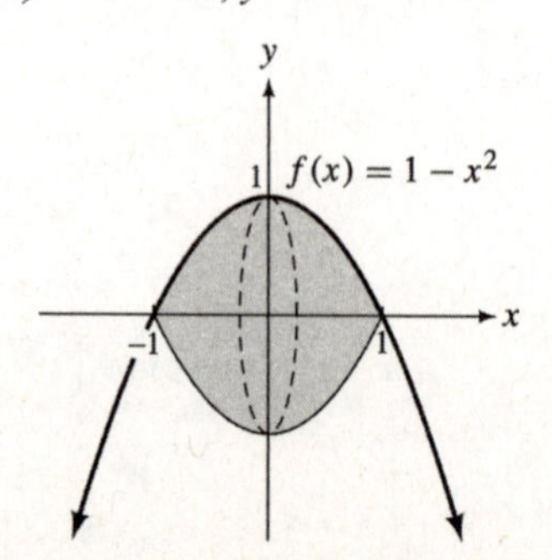

Since $f(x) = 1 - x^2$ intersects $y = 0$ where

$$\begin{aligned} 1 - x^2 &= 0 \\ x &= \pm 1, \\ a = -1 &\text{ and } b = 1. \end{aligned}$$

$$\begin{aligned} V &= \pi \int_{-1}^{1} (1 - x^2)^2 dx \\ &= \pi \int_{-1}^{1} (1 - 2x^2 + x^4)\, dx \\ &= \pi \left(x - \frac{2x^3}{3} + \frac{x^5}{5} \right) \Bigg|_{-1}^{1} \\ &= \pi \left(1 - \frac{2}{3} + \frac{1}{5} \right) - \pi \left(-1 + \frac{2}{3} - \frac{1}{5} \right) \\ &= 2\pi - \frac{4\pi}{3} + \frac{2\pi}{5} \\ &= \frac{16\pi}{15} \end{aligned}$$

19. $f(x) = \sqrt{1 - x^2}$

$r = \sqrt{1} = 1$

$$\begin{aligned} V &= \pi \int_{-1}^{1} (\sqrt{1 - x^2})^2 dx \\ &= \pi \int_{-1}^{1} (1 - x^2)\, dx \\ &= \pi \left(x - \frac{x^3}{3} \right) \Bigg|_{-1}^{1} \\ &= \pi \left(1 - \frac{1}{3} \right) - \pi \left(-1 + \frac{1}{3} \right) \\ &= 2\pi - \frac{2}{3}\pi = \frac{4\pi}{3} \end{aligned}$$

21. $f(x) = \sqrt{r^2 - x^2}$

$$\begin{aligned} V &= \pi \int_{-r}^{r} (\sqrt{r^2 - x^2})^2 dx \\ &= \pi \int_{-r}^{r} (r^2 - x^2)\, dx \\ &= \pi \left(r^2 x - \frac{x^3}{3} \right) \Bigg|_{-r}^{r} \\ &= \pi \left(r^3 - \frac{r^3}{3} \right) - \pi \left(-r^3 + \frac{r^3}{3} \right) \\ &= 2r^3\pi - \left(\frac{2r^3\pi}{3} \right) \\ &= \frac{4\pi r^3}{3} \end{aligned}$$

23. $f(x) = r, x = 0, x = h$

Graph $f(x) = r$; then show the solid of revolution formed by rotating about the x-axis the region bounded by $f(x), x = 0, x = h$.

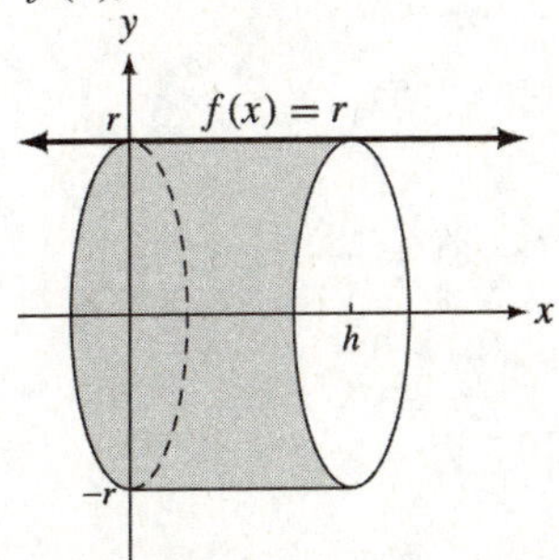

$$\begin{aligned} \int_0^h \pi r^2 dx &= \pi r^2 x \Bigg|_0^h \\ &= \pi r^2 h - 0 \\ &= \pi r^2 h \end{aligned}$$

25. $f(x) = x^2 - 4; [0, 5]$

Average value

$$= \frac{1}{5-0}\int_0^5 (x^2 - 4)dx$$
$$= \frac{1}{5}\left(\frac{x^3}{3} - 4x\right)\Big|_0^5$$
$$= \frac{1}{5}\left[\left(\frac{125}{3} - 20\right) - 0\right]$$
$$= \frac{13}{3} \approx 4.333$$

27. $f(x) = \sqrt{x+1}; [3, 8]$

Average value

$$= \frac{1}{8-3}\int_3^8 \sqrt{x+1}\,dx$$
$$= \frac{1}{5}\int_3^8 (x+1)^{1/2}dx$$
$$= \frac{1}{5}\cdot\frac{2}{3}(x+1)^{3/2}\Big|_3^8$$
$$= \frac{2}{15}(9^{3/2} - 4^{3/2})$$
$$= \frac{2}{15}(27 - 8) = \frac{38}{15} \approx 2.533$$

29. $f(x) = e^{x/7}; [0, 7]$

Average value

$$= \frac{1}{7-0}\int_0^7 e^{x/7}dx$$
$$= \frac{1}{7}\cdot 7e^{x/7}\Big|_0^7$$
$$= e^{x/7}\Big|_0^7 = e^1 - e^0$$
$$e - 1 \approx 1.718$$

31. $f(x) = x^2e^{2x}; [0, 2]$

Average value $= \dfrac{1}{2-0}\displaystyle\int_0^2 x^2e^{2x}\,dx$

Let $u = x^2$ and $dv = e^{2x}\,dx$.

Use column integration.

D		I
x^2	$+$	e^{2x}
$2x$	$-$	$\frac{1}{2}e^{2x}$
2	$+$	$\frac{1}{4}e^{2x}$
0		$\frac{1}{8}e^{2x}$

$$\frac{1}{2-0}\int_0^2 x^2e^{2x}\,dx$$
$$= \frac{1}{2}\left[(x^2)\left(\frac{1}{2}\right)e^{2x} - (2x)\left(\frac{1}{4}\right)e^{2x} + 2\left(\frac{1}{8}\right)e^{2x}\right]\Big|_0^2$$
$$= \frac{1}{2}\left(2e^4 - e^4 + \frac{1}{4}e^4 - \frac{1}{4}\right)$$
$$= \frac{5e^4 - 1}{8} \approx 34.00$$

33. $f(x) = e^{-x^2}, y = 0, x = -1, x = 1$

$$V = \pi\int_{-1}^1 (e^{-x^2})^2dx = \pi\int_{-2}^2 e^{-2x^2}dx$$

Using an integration feature on a graphing calculator to evaluate the integral, we obtain 3.758249634 ≈ 3.758.

35. Use the formula for average value with $a = 0$ and $b = 6$.

$$\frac{1}{6-0}\int_0^6 (37 + 6e^{-0.03t})dt$$
$$= \frac{1}{6}\left(37t + \frac{6}{-0.03}e^{-0.03t}\right)\Big|_0^6$$
$$= \frac{1}{6}(37t - 200e^{-0.03t})\Big|_0^6$$
$$= \frac{1}{6}[(222 - 200e^{-0.18}) - (0 - 200)]$$
$$= \frac{1}{6}(422 - 200e^{-0.18})$$
$$\approx 42.49$$

The average price is \$42.49.

37. Use the formula for average value with $a = 0$ and $b = 6$. The average price is

$$\frac{1}{30-0}\int_0^{30}(600 - 20\sqrt{30t})\,dt$$

$$= \frac{1}{30}\left(600t - 20\sqrt{30}\cdot\frac{2}{3}t^{3/2}\right)\Bigg|_0^{30}$$

$$= \frac{1}{30}\left(600t - \frac{40\sqrt{30}}{3}t^{3/2}\right)\Bigg|_0^{30}$$

$$= \frac{1}{30}(18{,}000 - 12{,}000)$$

$$= 200 \text{ cases}$$

39. $R(t) = te^{-0.1t}$

"During the nth hour" corresponds to the interval $(n-1, n)$.

The average intensity during nth hour is

$$\frac{1}{n-(n-1)}\int_{n-1}^{n} te^{-0.1t}dt = \int_{n-1}^{n} te^{-0.1t}dt$$

Let $u = t$ and $dv = e^{-0.1t}dt$.

D		I
t	$+$	$e^{-0.1t}$
1	$-$	$-10e^{-0.1t}$
0		$100e^{-0.1t}$

$$\int_{n-1}^{n} te^{-0.1t}\,dt$$

$$= (-10te^{-0.1t} - 100e^{-0.1t})\Big|_{n-1}^{n}$$

(a) Second hour, $n = 2$
Average intensity

$$= -10e^{-0.2}(12) + 10e^{-0.1}(11)$$

$$= 110e^{-0.1} - 120e^{-0.2} \approx 1.284$$

(b) Twelfth hour, $n = 12$
Average intensity

$$= -10e^{-1.2}(12+10) + 10e^{-1.1}(11+10)$$

$$= 210e^{-1.1} - 220e^{-1.2} \approx 3.640$$

(c) Twenty-fourth hour, $n = 24$
Average intensity

$$= -10e^{-2.4}(24+10) + 10e^{-2.3}(23+10)$$

$$= 330e^{-2.3} - 340e^{-2.4} \approx 2.241$$

41. For each part below, use
Average value

$$= \frac{1}{b-a}\int_a^b 45\ln(t+1)dt$$

$$= \frac{45}{b-a}\int_a^b \ln(t+1)dt.$$

Evaluate the integral using integration by parts.
Let $u = \ln(t+1)$ and $dv = dt$.

Then $du = \frac{1}{t+1}dt$ and $v = t$.

$$\int \ln(t+1)\,dt$$

$$= t\ln(t+1) - \int\frac{t}{t+1}dt$$

$$= t\ln(t+1) - \int\left(1 - \frac{1}{t+1}\right)dt$$

$$= t\ln(t+1) - t + \ln(t+1) + C$$

$$= (t+1)\ln(t+1) - t + C$$

Therefore
Average value

$$= \frac{1}{b-a}\int_a^b 45\ln(t+1)\,dt$$

$$= \frac{45}{b-a}[(t+1)\ln(t+1) - t]\Big|_a^b.$$

(a) The average number of items produced daily after 5 days is

$$\frac{45}{5-0}[(t+1)\ln(t+1) - t]\Big|_0^5$$

$$= 9[(6\ln 6 - 5) - (\ln 1 - 0)]$$

$$= 9(6\ln 6 - 5) \approx 51.76.$$

(b) The average number of items produced daily after 9 days is

$$\frac{45}{9-0}[(t+1)\ln(t+1) - t]\Big|_0^9$$

$$= 5(10\ln 10 - 9) \approx 70.13.$$

(c) The average number of items produced daily after 30 days is

$$\frac{45}{30-0}[(t+1)\ln(t+1) - t]\Big|_0^{30}$$

$$= \frac{3}{2}(31\ln 31 - 30) \approx 114.7.$$

43. From Exercise 22, the volume of an ellipsoid with horizontal axis of length $2a$ and vertical axis of length $2b$ is

$$V = \frac{4ab^2\pi}{3}.$$

For the Earth, $a = 6{,}356{,}752.3142$ and $b = 6{,}378{,}137$.

$$V = \frac{4(6{,}356{,}752.3142)(6{,}378{,}137)^2\pi}{3}$$
$$\approx 1.083 \times 10^{21}$$

The volume of the Earth is about 1.083×10^{21} cubic meters (m^3).

16.3 Continuous Money Flow

Your Turn 1

$f(t) = 810e^{kt}$

Use $f(1) = 797.94$ to find k.

$$f(1) = 810e^{k(1)}$$
$$797.94 = 810e^{k}$$
$$e^k = \frac{797.94}{810} \approx 0.9851$$
$$k \approx \ln 0.9851 \approx -0.015$$

$f(t) = 810e^{-0.015t}$

$$\text{Total income} = \int_0^2 810e^{-0.015t}\,dt$$
$$= -\frac{810}{0.015}e^{-0.015t}\Big|_0^2$$
$$= -54{,}000e^{-0.015t}\Big|_0^2$$
$$= -54{,}000(e^{-0.03} - 1)$$
$$\approx 1595.94$$

The total income is \$1595.94

Your Turn 2

Find the present value of an income given by $f(t) = 50{,}000t$ over the next 5 years if the interest rate is 3.5%.

$f(t) = 50{,}000t,\ 0 \le t \le 5$

$$P = \int_0^5 50{,}000te^{-0.035t}\,dt = 50{,}000\int_0^5 te^{-0.035t}\,dt$$

Use integration by parts.

Let $u = t$ and $dv = e^{-0.035t}dt$.

Then $du = dt$ and $v = -\frac{1}{0.035}e^{-0.035t}$
$= -28.57143e^{-0.035t}$.

$$\int_0^5 te^{-0.035t}\,dt$$
$$= -28.57143te^{-0.035t} - \frac{28.57143}{0.035}e^{-0.035t} + C$$
$$= (-28.57143t - 816.32657)e^{-0.035t} + C$$

$$P = 50{,}000\int_0^5 te^{-0.035t}\,dt$$
$$= 50{,}000(-28.57143t - 816.32657)e^{-0.035t}\Big|_0^5$$
$$= 50{,}000[(-28.57143(5) - 816.32657)e^{-0.035(5)} - (0 - 816.32657)]$$
$$= 50{,}000(-805.19351 + 816.32657)$$
$$\approx 556{,}653$$

The present value of the income is \$556,653.

Your Turn 3

Find the accumulated amount of money flow for an income given by $f(t) = 50{,}000t$ over the next 5 years if the interest rate is 3.5%.

$$A = e^{0.035(5)}\int_0^5 50{,}000te^{-0.035t}\,dt$$

The integral was computed in Your Turn 2. Using this value, the accumulated value is

$$e^{0.175}(556{,}653) \approx 663{,}111.$$

The accumulated amount of money flow is \$663,111.

Your Turn 4

Find the present value at the end of 8 years of the continuous flow of money given by

$$f(t) = 200t^2 + 100t + 50$$

at 5% compounded continuously.

$$P = \int_0^8 (200t^2 + 100t + 50)e^{-0.05t}\,dt$$

Using integration by parts as in Example 5, you can verify that

$$\int (200t^2 + 100t + 50)e^{-0.05t}dt$$
$$= -1000e^{-0.05t}(4t^2 + 162t + 3241) + C.$$

Thus $P = -1000e^{-0.05t}(4t^2 + 162t + 3241)\Big|_0^8$

$$= -1000(e^{-0.4}(4793) - 3241)$$
$$\approx 28{,}156.02$$

The present value at the end of 8 years of this flow of money is \$28,156.02.

16.3 Exercises

1. $f(t) = 1000$

(a) $P = \int_0^{10} 1000e^{-0.08t}\,dt$

$$= \frac{1000}{-0.08}e^{-0.08t}\Big|_0^{10}$$
$$= -12{,}500(e^{-0.8} - e^0)$$
$$= -12{,}500(e^{-0.8} - 1)$$
$$\approx 6883.387949$$

(We will use this value for P in part (b). Store it in your calculator without rounding.)

The present value is \$6883.39.

(b) $A = e^{0.08(10)} \int_0^{10} 1000e^{-0.08t}\,dt$

$$= e^{0.8}P$$
$$\approx 15{,}319.26161$$

The accumulated value is \$15,319.26.

3. $f(t) = 500$

(a) $P = \int_0^{10} 500e^{-0.08t}dt$

$$= \frac{500}{-0.08}e^{-0.08t}\Big|_0^{10}$$
$$= -6250(e^{-0.8} - e^0)$$
$$\approx 3441.693974$$

The present value is \$3441.69.

(b) $A = e^{0.08(10)} \int_0^{10} 500e^{-0.08t}dt$

$$= e^{0.8}P \approx 7659.630803$$

The accumulated value is \$7659.63.

5. $f(t) = 400e^{0.03t}$

(a) $P = \int_0^{10} 400e^{0.03t}e^{-0.08t}dt$

$$= 400\int_0^{10} e^{-0.05t}dt = \frac{400}{-0.05}e^{-0.05t}\Big|_0^{10}$$
$$= -8000(e^{-0.5} - e^0)$$
$$\approx 3147.754722$$

The present value is \$3147.75.

(b) $A = e^{0.08(10)} \int_0^{10} 400e^{0.03t}e^{-0.08t}\,dt$

$$= e^{0.8}P \approx 7005.456967$$

The accumulated value is \$7005.46.

7. $f(t) = 5000e^{-0.01t}$

(a) $P = \int_0^{10} 5000e^{-0.01t}e^{-0.08t}dt$

$$= 5000\int_0^{10} e^{-0.09t}dt$$
$$= \frac{5000}{-0.09}e^{-0.09t}\Big|_0^{10}$$
$$= -\frac{5000}{0.09}(e^{-0.9} - e^0)$$
$$\approx 32{,}968.35224$$

The present value is \$32,968.35.

(b) $A = e^{0.08(10)} \int_0^{10} 5000e^{-0.01t}e^{-0.08t}dt$

$$= e^{0.8}P$$
$$\approx 73{,}372.41725$$

The accumulated value is \$73,372.42.

9. $f(t) = 25t$

(a) $P = \int_0^{10} 25te^{-0.08t}\,dt$

$= 25\int_0^{10} te^{-0.08t}\,dt$

Find the antiderivative using integration by parts.

Let $u = t$ and $dv = e^{-0.08t}\,dt.$

Then $du = dt$ and $v = \frac{1}{-0.08}e^{-0.08t}$

$= -12.5e^{-0.08t}.$

$\int te^{-0.08t}\,dt$

$= t(-12.5e^{-0.08t}) - \int(-12.5e^{-0.08t})dt$

$= -12.5te^{-0.08t} + 12.5\int e^{-0.08t}\,dt$

$= -12.5te^{-0.08t} + \frac{12.5}{-0.08}e^{-0.08t} + C$

$= -(12.5t + 156.25)e^{-0.08t} + C$

Therefore

$P = 25[-(12.5t + 156.25)e^{-0.08t}]\Big|_0^{10}$

$= [-25(12.5t + 156.25)e^{-0.08t}]\Big|_0^{10}$

$= (-7031.25e^{-0.8}) - (-3906.25e^0)$

$\approx 746.9057211.$

The present value is \$746.91.

(b) $A = e^{0.08(10)}\int_0^{10} 25te^{-0.08t}\,dt$

$= e^{0.8}P$

≈ 1662.269252

The accumulated value is \$1662.27.

11. $f(t) = 0.01t + 100$

(a) $P = \int_0^{10}(0.01t + 100)e^{-0.08t}\,dt$

$= \int_0^{10} 0.01te^{-0.08t}\,dt$

$+ \int_0^{10} 100e^{-0.08t}\,dt$

$= 0.01\int_0^{10} te^{-0.08t}\,dt$

$+ 100\int_0^{10} e^{-0.08t}\,dt$

From Exercise 9, we know that

$\int te^{-0.08t}\,dt = -(12.5t + 156.25)e^{-0.08t} + C$

From Exercise 1, we know that

$\int e^{-0.08t}\,dt = -12.5e^{-0.08t} + C$

Substitute the given expressions and simplify.

$P = \{0.01[-(12.5t + 156.25)e^{-0.08t}]$

$+100(-12.5e^{-0.08t})\}\Big|_0^{10}$

$= [-(0.125t + 1251.5625)e^{-0.08t}]\Big|_0^{10}$

$= (-1252.8125e^{-0.8}) - (-1251.5625e^0)$

≈ 688.6375571

The Present value is \$688.64.

(b) $A = e^{0.08(10)}\int_0^{10}(0.01t + 100)e^{-0.08t}\,dt$

$= e^{0.8}P$

≈ 1532.591068

The accumulated value is \$1532.59.

13. $f(t) = 1000t - 100t^2$

(a) $P = \int_0^{10} (1000t - 100t^2)e^{-0.08t}\,dt$

$= 1000\int_0^{10} te^{-0.08t}\,dt$

$- 100\int_0^{10} t^2e^{-0.08t}\,dt$

From Exercise 9, we know that

$$\int te^{-0.08t}\,dt = -(12.5t + 156.25)e^{-0.08t} + C$$

Evaluate the antiderivative $\int t^2e^{-0.08t}\,dt$ using column integration. (Note that $\frac{1}{-0.08} = -12.5$.)

D		I
t^2	+	$e^{-0.08t}$
$2t$	−	$-12.5\,e^{-0.08t}$
2	+	$156.25\,e^{-0.08t}$
0		$-1953.125\,e^{-0.08t}$

Thus,

$\int_0^{10} t^2e^{-0.08t}\,dt$

$= (t^2)(-12.5e^{-0.08t})$

$- (2t)(156.25e^{-0.08t})$

$+ (2)(-1953.125e^{-0.08t}) + C$

$= -(12.5t^2 + 312.5t$

$+ 3906.25)e^{-0.08t} + C.$

Therefore:

$P = \{1000[-(12.5t + 156.25)e^{-0.08t}]$

$- 100[-(12.5t^2 + 312.5t$

$+ 3906.25)e^{-0.08t}]\}\Big|_0^{10}$

Collect like terms and simplify.

$P = [(1250t^2 + 18{,}750t + 234{,}375)e^{-0.08t}]\Big|_0^{10}$

$= (546{,}875e^{-0.8}) - (234{,}375e^0)$

$\approx 11{,}351.77725$

The principal value is \$11,351.78.

(b) $A = e^{0.08(10)}\int_0^{10} (1000t - 100t^2)e^{-0.08t}\,dt$

$= e^{0.8}P$

$\approx 25{,}263.84488$

The accumulated value is \$25,263.84.

15. $A = e^{0.04(3)}\int_0^3 20{,}000e^{-0.04t}\,dt$

$= e^{0.12}\left(\frac{20{,}000}{-0.04}e^{-0.04t}\right)\Bigg|_0^3$

$= e^{0.12}\left(\frac{20{,}000}{-0.04}e^{-0.12} + \frac{20{,}000}{0.04}\right)$

$\approx \$63{,}748.43$

17. **(a)** Present value

$= \int_0^8 5000e^{-0.01t}e^{-0.08t}\,dt$

$= \int_0^8 5000e^{-0.09t}\,dt$

$= \left(\frac{5000}{-0.09}e^{-0.09t}\right)\Bigg|_0^8$

$= \frac{5000e^{-0.72}}{-0.09} + \frac{5000}{0.09}$

$= \$28{,}513.76$

(b) Final amount

$= e^{0.08(8)}\int_0^8 5000e^{-0.01t}e^{-0.08t}\,dt$

$\approx e^{0.64}(28{,}513.76)$

$\approx \$54{,}075.81$

19. $P = \int_0^5 (1500 - 60t^2)e^{-0.05t}\,dt$

$= \int_0^5 1500e^{-0.05t}\,dt - \int_0^5 60t^2e^{-0.05t}\,dt$

$= 1500\int_0^5 e^{-0.05t}\,dt - 60\int_0^5 t^2e^{-0.05t}\,dt$

Find the second integral by column integration.

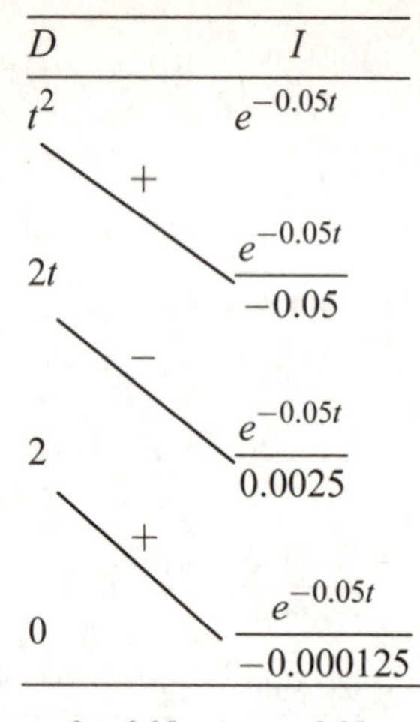

$$\int t^2 e^{-0.05t}\,dt = \frac{t^2 e^{-0.05t}}{-0.05} - \frac{2te^{-0.05t}}{0.0025} + \frac{2e^{-0.05}}{-0.000125} + C$$

$$= -e^{-0.05t}\left(\frac{t^2}{0.05} + \frac{2t}{0.0025} + \frac{2}{0.000125}\right) + C$$

Now add the first integral to this result.

$$1500\int_0^5 e^{-0.05t}\,dt - 60\int_0^5 t^2 e^{-0.05t}\,dt$$

$$= \frac{1500}{-0.05}e^{-0.05t}\Big|_0^5$$

$$+ 60e^{-0.05t}\left(\frac{t^2}{0.05} + \frac{2t}{0.0025} + \frac{2}{0.000125}\right)\Bigg|_0^5$$

$$= \frac{1500}{-0.05}(e^{-0.25} - 1)$$

$$+ 60\left[e^{-0.25}\left(\frac{25}{0.05} + \frac{10}{0.0025} + \frac{2}{0.000125}\right) - \frac{2}{0.000125}\right]$$

$$\approx 6636.977 - 2075.037$$

$$= \$4560.94$$

16.4 Improper Integrals

Your Turn 1

(a) $$\int_8^\infty \frac{1}{x^{1/3}}dx = \lim_{b\to\infty}\int_8^b \frac{1}{x^{1/3}}dx$$

$$= \lim_{b\to\infty}\left(\frac{3}{2}x^{2/3}\right)\Bigg|_8^b$$

$$= \lim_{b\to\infty}\left(\frac{3}{2}b^{2/3} - \frac{3}{2}(4)\right)$$

As $b \to \infty$, $b^{2/3} \to \infty$, so the limit above does not exist and the integral diverges.

(b) $$\int_8^\infty \frac{1}{x^{4/3}}dx$$

$$= \lim_{b\to\infty}\int_8^b \frac{1}{x^{4/3}}dx$$

$$= \lim_{b\to\infty}\left(-3x^{-1/3}\right)\Big|_8^b$$

$$= \lim_{b\to\infty}\left[-3b^{-1/3} - \left(-\frac{3}{8^{1/3}}\right)\right]$$

As $b \to \infty$, $b^{-1/3} = \frac{1}{b^{1/3}} \to 0$, so the integral is convergent and its value is $\frac{3}{8^{1/3}} = \frac{3}{2}$.

Your Turn 2

$$\int_0^\infty 5e^{-2x}\,dx = \lim_{b\to\infty}\int_0^b 5e^{-2x}\,dx$$

$$= \lim_{b\to\infty}\left(-\frac{5}{2}e^{-2x}\right)\Bigg|_0^b$$

$$= \lim_{b\to\infty}\left[-\frac{5}{2}e^{-2b} - \left(-\frac{5}{2}e^{-2(0)}\right)\right]$$

As $b \to \infty$, $e^{-2b} = \frac{1}{e^b} \to 0$, so the integral is convergent and its value is $\frac{5}{2}e^0 = \frac{5}{2}$.

16.4 Exercises

1. $$\int_3^\infty \frac{1}{x^2}dx = \lim_{b\to\infty}\int_3^b x^{-2}dx$$

$$= \lim_{b\to\infty} -x^{-1}\Big|_3^b$$

$$= \lim_{b\to\infty}\left(-\frac{1}{b} + \frac{1}{3}\right)$$

$$= \lim_{b\to\infty}\left(-\frac{1}{b}\right) + \lim_{b\to\infty}\frac{1}{3}$$

As $b \to \infty$, $-\frac{1}{b} \to 0$. The integral is convergent.

$$\int_3^\infty \frac{1}{x^2}dx = 0 + \frac{1}{3} = \frac{1}{3}$$

3. $$\int_4^{\infty} \frac{2}{\sqrt{x}}dx = \lim_{b\to\infty}\int_4^b 2x^{-1/2}dx$$
$$= \lim_{b\to\infty} 4x^{1/2}\Big|_4^b$$
$$= \lim_{b\to\infty} (4\sqrt{b} - 4\sqrt{4})$$
$$= \lim_{b\to\infty} 4\sqrt{b} - 8$$

As $b \to \infty, 4\sqrt{b} \to \infty$. The integral diverges.

5. $$\int_{-\infty}^{-1} \frac{2}{x^3}dx = \int_{-\infty}^{-1} 2x^{-3}dx$$
$$= \lim_{a\to-\infty}\int_a^{-1} 2x^{-3}dx$$
$$= \lim_{a\to-\infty}\left(\frac{2x^{-2}}{-2}\right)\Bigg|_a^{-1}$$
$$= \lim_{a\to-\infty}\left(-1 + \frac{1}{a^2}\right)$$

As $a \to -\infty, \frac{1}{a^2} \to 0$. The integral is convergent.

$$\int_{-\infty}^{-1} \frac{2}{x^3}dx = -1 + 0 = -1$$

7. $$\int_1^{\infty} \frac{1}{x^{1.0001}}dx$$
$$= \int_1^{\infty} x^{-1.0001}dx$$
$$= \lim_{b\to\infty}\int_1^b x^{-1.0001}dx$$
$$= \lim_{b\to\infty}\left(\frac{x^{-0.0001}}{-0.0001}\right)\Bigg|_1^b$$
$$= \lim_{b\to\infty}\left(-\frac{1}{(0.0001)b^{0.0001}} + \frac{1}{0.0001}\right)$$

As $b \to \infty, -\frac{1}{0.0001\,b^{0.0001}} \to 0$.

The integral is convergent.

$$\int_1^{\infty} \frac{1}{x^{1.0001}}dx = 0 + \frac{1}{0.0001} = 10{,}000$$

9. $$\int_{-\infty}^{-10} x^{-2}dx = \lim_{a\to-\infty}\int_a^{-10} x^{-2}dx$$
$$= \lim_{a\to-\infty}(-x^{-1})\Big|_a^{-10}$$
$$= \lim_{a\to-\infty}\left(\frac{1}{10} + \frac{1}{a}\right)$$
$$= \frac{1}{10} + 0 = \frac{1}{10}$$

The integral is convergent and its value is $\frac{1}{10}$.

11. $$\int_{-\infty}^{-1} x^{-8/3}dx = \lim_{a\to-\infty}\int_a^{-1} x^{-8/3}dx$$
$$= \lim_{a\to-\infty}\left(-\frac{3}{5}x^{-5/3}\right)\Bigg|_a^{-1}$$
$$= \lim_{a\to-\infty}\left(\frac{3}{5} + \frac{3}{5a^{5/3}}\right)$$
$$= \frac{3}{5} + 0 = \frac{3}{5}$$

The integral is convergent, and its value is $\frac{3}{5}$.

13. $$\int_0^{\infty} 8e^{-8x}dx = \lim_{b\to\infty}\int_0^b 8e^{-8x}dx$$
$$= \lim_{b\to\infty}\left(\frac{8e^{-8x}}{-8}\right)\Bigg|_0^b$$
$$= \lim_{b\to\infty}(-e^{-8b} + 1)$$
$$= \lim_{b\to\infty}\left(-\frac{1}{e^{8b}} + 1\right)$$
$$= 0 + 1 = 1$$

The integral is convergent, and its value is 1.

15. $$\int_{-\infty}^{0} 1000e^x dx = \lim_{a\to-\infty}\int_a^0 1000e^x dx$$
$$= \lim_{a\to-\infty}(1000e^x)\Big|_a^0$$
$$= \lim_{a\to-\infty}(1000 - 1000e^a)$$

As $a \to \infty, -1000e^a \to 0$. The integral is convergent.

$$\int_{-\infty}^{0} 1000e^x dx = 1000 - 0 = 1000$$

17. $\int_{-\infty}^{-1} \ln|x|dx = \lim_{a\to-\infty}\int_a^{-1} \ln|x|dx$

Let $u = \ln|x|$ and $dv = dx$.

Then $du = \frac{1}{x}dx$ and $v = x$.

$$\int \ln|x|dx = x\ln|x| - \int \frac{x}{x}dx$$
$$= x\ln|x| - x + C$$

$$\int_{-\infty}^{-1} \ln|x|\,dx = \lim_{a\to-\infty} (x\ln|x| - x)\Big|_a^{-1}$$
$$= \lim_{a\to-\infty} (-\ln 1 + 1 - a\ln|a| + a)$$
$$= \lim_{a\to-\infty} (1 + a - a\ln|a|)$$

The integral is divergent, since as $a \to -\infty$.

$(a - a\ln|a|) = -a(-1 + \ln|a|) \to \infty$.

19. $\int_0^{\infty} \frac{dx}{(x+1)^2}$

$$= \lim_{b\to\infty}\int_0^b \frac{dx}{(x+1)^2} \quad \textit{Use substitution}$$
$$= \lim_{b\to\infty} -(x+1)^{-1}\Big|_0^b$$
$$= \lim_{b\to\infty}\left(\frac{-1}{b+1} + 1\right)$$

As $b \to \infty$, $-\frac{1}{b+1} \to 0$. The integral is convergent.

$$\int_0^{\infty} \frac{dx}{(x+1)^2} = 0 + 1 = 1$$

21. $\int_{-\infty}^{-1} \frac{2x-1}{x^2-x}dx$

$$= \lim_{a\to-\infty}\int_0^{-1} \frac{2x-1}{x^2-x}dx \quad \textit{Use substitution}$$
$$= \lim_{a\to-\infty} \ln|x^2 - x|\Big|_a^{-1}$$
$$= \lim_{a\to-\infty} (\ln 2 - \ln|a^2 - a|)$$

As $a \to -\infty$, $\ln|a^2 - a| \to \infty$. The integral is divergent.

23. $\int_2^{\infty} \frac{1}{x\ln x}dx$

$$= \lim_{b\to\infty}\int_2^b \frac{1}{x\ln x}dx \quad \textit{Use substitution}$$
$$= \lim_{b\to\infty}\left[\ln(\ln x)\Big|_2^b\right]$$
$$= \lim_{b\to\infty} [\ln(\ln b) - \ln(\ln 2)]$$

As $b \to \infty$, $\ln(\ln b) \to \infty$. The integral is divergent.

25. $\int_0^{\infty} xe^{4x}dx = \lim_{b\to\infty}\int_0^b xe^{4x}dx$

Let $dv = e^{4x}dx$ and $u = x$.

Then $v = \frac{1}{4}e^{4x}dx$ and $du = dx$.

$$\int xe^{4x}dx = \frac{x}{4}e^{4x} - \int \frac{1}{4}e^{4x}dx$$
$$= \frac{x}{4}e^{4x} - \frac{1}{16}e^{4x} + C$$
$$= \frac{1}{16}(4x-1)e^{4x} + C$$

$$\int_0^{\infty} xe^{4x}dx$$
$$= \lim_{b\to\infty}\left[\frac{1}{16}(4x-1)e^{4x}\right]\Big|_0^b$$
$$= \lim_{b\to\infty}\left[\frac{1}{16}(4b-1)e^{4b} - \frac{1}{16}(-1)(1)\right]$$
$$= \lim_{b\to\infty}\left[\frac{1}{16}(4b-1)e^{4b} + \frac{1}{16}\right]$$

As $b \to \infty$, $\frac{1}{16}(4b-1)e^{4b} \to \infty$. The integral is divergent.

27. $\int_{-\infty}^{\infty} x^3e^{-x^4}dx$

$$= \int_{-\infty}^{0} x^3e^{-x^4}dx + \int_0^{\infty} x^3e^{-x^4}dx$$

We evaluate each of two improper integrals on the right.

$$\int_{-\infty}^{0} x^3 e^{-x^4} dx = \lim_{b \to -\infty} \int_{b}^{0} x^3 e^{-x^4} dx \quad \textit{Use substitution}$$

$$= \lim_{b \to -\infty} \left[-\frac{1}{4} e^{-x^4} \Big|_{b}^{0} \right]$$

$$= \lim_{b \to -\infty} \left[-\frac{1}{4} + \frac{1}{4e^{b^4}} \right]$$

As $b \to -\infty$, $\frac{1}{4e^{b^4}} \to 0$. The integral is convergent.

$$\int_{-\infty}^{0} x^3 e^{-x^4} dx = -\frac{1}{4} + 0 = -\frac{1}{4}$$

$$\int_{0}^{\infty} x^3 e^{-x^4} dx = \lim_{b \to \infty} \int_{0}^{b} x^3 e^{-x^4} dx \quad \textit{Use substitution}$$

$$= \lim_{b \to \infty} \left[-\frac{1}{4} e^{-x^4} \Big|_{0}^{b} \right]$$

$$= \lim_{b \to \infty} \left[\frac{1}{4e^{b^4}} + \frac{1}{4} \right]$$

As $b \to -\infty, -\frac{1}{4e^{b^4}} \to 0$. The integral is convergent.

$$\int_{0}^{\infty} x^3 e^{-x^4} dx = 0 + \frac{1}{4} = \frac{1}{4}$$

Since each of the improper integrals converges, the original improper integral converges.

$$\int_{-\infty}^{\infty} x^3 e^{-x^4} dx = -\frac{1}{4} + \frac{1}{4} = 0$$

29. $$\int_{-\infty}^{\infty} \frac{x}{x^2+1} dx$$

$$= \int_{-\infty}^{0} \frac{x}{x^2+1} dx + \int_{0}^{\infty} \frac{x}{x^2+1} dx$$

We evaluate the first improper integrals on the right.

$$\int_{-\infty}^{0} \frac{x}{x^2+1} dx$$

$$= \lim_{b \to -\infty} \int_{b}^{0} \frac{x}{x^2+1} dx \quad \textit{Use substitution}$$

$$= \lim_{b \to -\infty} \left[\frac{1}{2} \ln(x^2+1) \Big|_{b}^{0} \right]$$

$$= \lim_{b \to -\infty} \left[0 - \frac{1}{2} \ln(b^2+1) \right]$$

As $b \to -\infty$, $\ln(b^2+1) \to \infty$. The integral is divergent. Since one of the two improper integrals on the right diverges, the original improper integral diverges.

31. $f(x) = \frac{1}{x-1}$ for $(-\infty, 0]$

$$\int_{-\infty}^{0} \frac{1}{x-1} dx = \lim_{a \to -\infty} \int_{a}^{0} \frac{dx}{x-1} \ln|x-1|$$

$$= \lim_{a \to -\infty} \left(\ln|x-1| \Big|_{a}^{0} \right)$$

$$= \lim_{a \to -\infty} (\ln|-1| - \ln|a-1|)$$

But $\lim_{a \to -\infty} (\ln|a-1|) = \infty$.

The integral is divergent, so the area cannot be found.

33. $f(x) = \frac{1}{(x-1)^2}$ for $(-\infty, 0]$

$$\int_{-8}^{0} \frac{1}{(x-1)^2}$$

$$= \lim_{a \to -\infty} \int_{a}^{0} \frac{1}{(x-1)^2} \quad \textit{Use substitution}$$

$$= \lim_{a \to -\infty} -(x-1)^{-1} \Big|_{a}^{0}$$

$$= \lim_{a \to -\infty} \left(-\frac{1}{-1} + \frac{1}{a-1} \right)$$

As $a \to -\infty$, $\frac{1}{a-1} \to 0$. The integral is convergent.

$$= 1 + 0 = 1$$

Therefore, the area is 1.

35. $$\int_{-\infty}^{\infty} x e^{-x^2} dx$$

Let $u = -x^2$, so that $du = -2x\,dx$.

$$= \lim_{a \to -\infty} \left(-\frac{1}{2} \int_{a}^{0} -2x e^{-x^2} dx \right) + \lim_{b \to \infty} \left(-\frac{1}{2} \int_{0}^{b} -2x e^{-x^2} dx \right)$$

$$= \lim_{a\to-\infty}\left(-\frac{1}{2}e^{-x^2}\right)\Bigg|_a^0 + \lim_{b\to\infty}\left(-\frac{1}{2}e^{-x^2}\right)\Bigg|_0^b$$

$$= \lim_{a\to-\infty}\left(-\frac{1}{2}+\frac{1}{2e^{-a^2}}\right) + \lim_{b\to\infty}\left(-\frac{1}{2e^{b^2}}+\frac{1}{2}\right)$$

$$= -\frac{1}{2}+\frac{1}{2}=0$$

37. $\int_1^\infty \frac{1}{x^p}dx$

Case 1a $p < 1$:

$$\int_1^\infty \frac{1}{x^p}dx = \int_1^\infty x^{-p}dx = \lim_{a\to\infty}\int_1^a x^{-p}dx = \lim_{a\to\infty}\left[\frac{x^{-p+1}}{(-p+1)}\Bigg|_1^a\right]$$

$$= \lim_{a\to\infty}\left[\frac{1}{(-p+1)}(a^{-p+1}-1)\right] = \lim_{a\to\infty}\left[\frac{1}{(-p+1)}a^{1-p}-\frac{1}{(-p+1)}\right]$$

Since $p < 1$, $1-p$ is positive and, as $a \to \infty$, $a^{1-p} \to \infty$. The integral diverges.

Case 1b $p = 1$:

$$\int_1^\infty \frac{1}{x^p}dx = \int_1^\infty \frac{1}{x}dx = \lim_{a\to\infty}\int_1^a \frac{1}{x}dx = \lim_{a\to\infty}\left(\ln|x|\Big|_1^a\right) = \lim_{a\to\infty}(\ln|a|-\ln 1) = \lim_{a\to\infty}\ln|a|$$

As $a \to \infty$, $\ln|a| \to \infty$. The integral diverges.

Therefore, $\int_1^\infty \frac{1}{x^p}$ diverges when $p \le 1$.

Case 2 $p > 1$:

$$\int_1^\infty \frac{1}{x^p}dx = \lim_{x\to\infty}\int_1^a x^{-p}dx = \lim_{a\to\infty}\left(\frac{x^{-p+1}}{-p+1}\Bigg|_1^a\right) = \lim_{a\to\infty}\left[\frac{a^{-p+1}}{(-p+1)}-\frac{1}{(-p+1)}\right]$$

Since $p > 1$, $-p+1 < 0$; thus as $a \to \infty$, $\frac{a^{-p+1}}{(-p+1)} \to 0$.

Hence,

$$\lim_{a\to\infty}\left[\frac{a^{-p+1}}{(-p+1)}-\frac{1}{(-p+1)}\right] = 0-\frac{1}{(-p+1)} = \frac{-1}{-p+1} = \frac{1}{p-1}.$$

The integral converges.

39. **(a)** Use the *fnInt* feature on a graphing utility to obtain

$$\int_1^{20}\frac{1}{\sqrt{1+x^2}}dx \approx 2.808;$$

$$\int_1^{50}\frac{1}{\sqrt{1+x^2}}dx \approx 3.724;$$

$$\int_1^{100}\frac{1}{\sqrt{1+x^2}}dx \approx 4.417;$$

$$\int_1^{1000}\frac{1}{\sqrt{1+x^2}}dx \approx 6.720;$$

$$\int_1^{10,000}\frac{1}{\sqrt{1+x^2}}dx \approx 9.022.$$

(b) Since the values of the integrals in part a do not appear to be approaching some fixed finite number but get bigger, the integral $\int_1^\infty \frac{1}{\sqrt{1+x^2}}dx$ appears to be divergent.

(c) Use the *fnInt* feature on a graphing utility to obtain

$$\int_1^{20} \frac{1}{\sqrt{1+x^4}}\,dx \approx 0.8770;$$

$$\int_1^{50} \frac{1}{\sqrt{1+x^4}}\,dx \approx 0.9070;$$

$$\int_1^{100} \frac{1}{\sqrt{1+x^4}}\,dx \approx 0.9170;$$

$$\int_1^{1000} \frac{1}{\sqrt{1+x^4}}\,dx \approx 0.9260;$$

$$\int_1^{10,000} \frac{1}{\sqrt{1+x^4}}\,dx \approx 0.9269.$$

(d) Since the values of the integrals in part c appear to be approaching some fixed finite number, the integral

$$\int_1^{\infty} \frac{1}{\sqrt{1+x^4}}\,dx$$

appears to be convergent.

(e) For large x, we may consider $1 + x^2 \approx x^2$ and $1 + x^4 \approx x^4$.

Thus,

$$\frac{1}{\sqrt{1+x^2}} \approx \frac{1}{\sqrt{x^2}} = \frac{1}{x} \text{ and}$$

$$\frac{1}{\sqrt{1+x^4}} \approx \frac{1}{\sqrt{x^4}} = \frac{1}{x^2}.$$

In Example 1(a) we showed that $\int_1^{\infty} \frac{1}{x}\,dx$ diverges. Thus, we might guess that $\int_1^{\infty} \frac{1}{\sqrt{1+x^2}}\,dx$ diverges as well. In Exercise 1, we saw that $\int_2^{\infty} \frac{1}{x^2}\,dx$ converges. Thus, we might guess that $\int_1^{\infty} \frac{1}{\sqrt{1+x^4}}\,dx$ converges as well.

41. (a) Use the *fnInt* feature on a graphing utility to obtain

$$\int_0^{10} e^{-.00001x}\,dx \approx 9.9995;$$

$$\int_0^{50} e^{-.00001x}\,dx \approx 49.9875;$$

$$\int_0^{100} e^{-.00001x}\,dx \approx 99.9500;$$

$$\int_0^{1000} e^{-.00001x}\,dx \approx 995.0166.$$

(b) Since the values of the integrals in part a do not appear to be approaching some fixed finite number, the integral $\int_0^{\infty} e^{-0.00001x}\,dx$ appears to be divergent.

(c)
$$\int_0^{\infty} e^{-0.00001x}\,dx$$
$$= \lim_{b\to\infty} \int_0^b e^{-0.00001x}\,dx$$
$$= \lim_{b\to\infty} \left[\frac{e^{-0.00001x}}{-0.00001}\Bigg|_0^b\right]$$
$$= \lim_{b\to\infty} \left[-\frac{1}{0.00001e^{0.00001b}} + \frac{1}{0.00001}\right]$$
$$= 0 + 100,000 = 100,000$$

43.
$$\int_0^{\infty} 1,000,000e^{-0.05t}\,dt$$
$$= \lim_{b\to\infty} \int_0^b 1,000,000e^{-0.05t}\,dt$$
$$= \lim_{b\to\infty} \left(\frac{1,000,000}{-0.05}e^{-0.05t}\right)\Bigg|_0^b$$
$$= -20,000,000\left[\lim_{b\to\infty}(e^{-0.05b}) - e^0\right]$$

As $b \to \infty, e^{-0.05b} = \frac{1}{e^{0.05b}} \to 0$. The integral converges.

$$\int_0^{\infty} 1,000,000e^{-0.05t}\,dt$$
$$= -20,000,000(0 - 1)$$
$$= 20,000,000$$

The capital value is \$20,000,000.

45.
$$\int_0^{\infty} 1200e^{0.03t}e^{-0.07t}\,dt$$
$$= \lim_{b\to\infty} \int_0^b 1200e^{-0.04t}\,dt$$
$$= \lim_{b\to\infty} \left(\frac{1200}{-0.04}e^{-0.04t}\right)\Bigg|_0^b$$
$$= -30,000\left[\lim_{b\to\infty}(e^{-0.04b}) - e^0\right]$$

As $b \to \infty, e^{-0.04b} = \frac{1}{e^{0.04b}} \to 0$. The integral converges.

$$\int_0^\infty 1200e^{0.03t}e^{-0.07t}dt = -30{,}000(0-1)$$
$$= 30{,}000$$

The capital value is \$30,000.

47. $$\int_0^\infty 3000e^{-0.1t}dt$$
$$= \lim_{b\to\infty}\int_0^b 3000e^{-0.1t}dt$$
$$= \lim_{b\to\infty}\left.\frac{3000e^{-0.1b}}{-0.1}\right|_0^b$$
$$= \lim_{b\to\infty}\left(\frac{3000e^{-0.1b}}{-0.1}+\frac{3000}{0.1}\right)$$
$$= 0 + 30{,}000 = \$30{,}000$$

49. $$S = N\int_0^\infty \frac{a(1-e^{-kt})}{k}e^{-bt}dt$$
$$= \frac{Na}{k}\lim_{c\to\infty}\int_0^c (1-e^{-kt})(e^{-bt})dt$$
$$= \frac{Na}{k}\lim_{c\to\infty}\int_0^c (e^{-bt}-e^{-(b+k)t})dt$$
$$= \frac{Na}{k}\lim_{c\to\infty}\left[-\frac{1}{b}e^{-bt}+\frac{1}{b+k}e^{-(b+k)t}\right]\Bigg|_0^c$$
$$= \frac{Na}{k}\lim_{c\to\infty}\left[\left(-\frac{1}{b}e^{-bc}+\frac{1}{b+k}e^{-(b+k)c}\right) - \left(-\frac{1}{b}e^0+\frac{1}{b+k}e^0\right)\right]$$
$$= \frac{Na}{k}\left(0+0+\frac{1}{b}-\frac{1}{b+k}\right)$$
$$= \frac{Na}{k}\cdot\frac{(b+k)-b}{b(b+k)}$$
$$= \frac{Nak}{kb(b+k)}$$
$$= \frac{Na}{b(b+k)}$$

51. $$\int_0^\infty 50e^{-0.06t}dt = 50\lim_{b\to\infty}\int_0^b e^{-0.06t}dt$$
$$= 50\lim_{b\to\infty}\left.\frac{e^{-0.06t}}{-0.06}\right|_0^b$$
$$= \frac{50}{-0.06}\lim_{b\to\infty}(e^{-0.06b}-e^0)$$
$$= -\frac{50}{0.06}(0-1) = \frac{50}{0.06}$$
$$\approx 833.3$$

16.5 Solutions of Elementary and Separable Differential Equations

Your Turn 1

Find all solutions of $\frac{dy}{dx} = 12x^5 + \sqrt{x} + e^{5x}$.

$$y = \int (12x^5 + \sqrt{x} + e^{5x})dx$$
$$= 2x^6 + \frac{2}{3}x^{3/2} + \frac{1}{5}e^{5x} + C$$

Your Turn 2

Find the particular solution of $\frac{dy}{dx} - 12x^3 = 6x^2$, $y(2) = 60$.

First solve for dy/dx.

$$\frac{dy}{dx} = 12x^3 + 6x^2$$
$$y = 3x^4 + 2x^3 + C$$

Use the initial condition to find the value of C.

$$y(2) = 60$$
$$60 = 3(2)^4 + 2(2)^3 + C$$
$$60 = 48 + 16 + C$$
$$60 = 64 + C$$
$$-4 = C$$

The particular solution is $y = 3x^4 + 2x^3 - 4$.

Your Turn 3

Find the general solution of $\frac{dy}{dx} = \frac{x^2+1}{xy^2}$.

Separate the variables.

$$y^2\,dy = \frac{x^2+1}{x}dx$$
$$y^2\,dy = \left(x + \frac{1}{x}\right)dx$$

Now integrate.

$$y^2\,dy = \left(x + \frac{1}{x}\right)dx$$
$$\int y^2\,dy = \int \left(x + \frac{1}{x}\right)dx$$
$$\int y^2\,dy = \int x\,dx + \int \frac{1}{x}dx$$
$$\frac{1}{3}y^3 = \frac{1}{2}x^2 + \ln|x| + C$$
$$y^3 = \frac{3}{2}x^2 + 3\ln|x| + C$$
$$y = \left(\frac{3}{2}x^2 + 3\ln|x| + C\right)^{1/3}$$

Your Turn 4

Find the goat population in 5 years if the reserve can support 6000 goats, the growth rate is 15%, and there are currently 1200 goats in the area.

The general solution will be $y = N - Me^{-kt}$ as in Example 5, where N is the maximum population, k is the growth rate constant, and M is a constant to be determined using the initial population. For this problem, $N = 6000$ and $k = 20 = 15\% = 0.15$. Solve for M.

$$1200 = 6000 - Me^{(-0.15)(0)}$$
$$M = 6000 - 1200$$
$$M = 4800$$

The model is $y = 6000 - 4800e^{-0.15t}$.

Now find $y(5)$.

$$y(5) = 1200 - 4800e^{-0.15(5)}$$
$$= 6000 - 4800e^{-0.75} \approx 3733$$

The goat population in 5 years will be 3733.

16.5 Exercises

1. $\frac{dy}{dx} = -4x + 6x^2$

$$y = \int(-4x + 6x^2)\,dx$$
$$= -2x^2 + 2x^3 + C$$

3. $4x^3 - 2\frac{dy}{dx} = 0$

Solve for $\frac{dy}{dx}$.

$$\frac{dy}{dx} = 2x^3$$
$$y = 2\int x^3 dx$$
$$= 2\left(\frac{x^4}{4}\right) + C$$
$$= \frac{x^4}{2} + C$$

5. $y\frac{dy}{dx} = x^2$

Separate the variables and take antiderivatives.

$$\int y\,dy = \int x^2\,dx$$
$$\frac{y^2}{2} = \frac{x^3}{3} + K$$
$$y^2 = \frac{2}{3}x^3 + 2K$$
$$y^2 = \frac{2}{3}x^3 + C$$

7. $\frac{dy}{dx} = 2xy$

$$\int \frac{dy}{y} = \int 2x\,dx$$
$$\ln|y| = \frac{2x^3}{2} + C$$
$$\ln|y| = x^2 + C$$
$$e^{\ln|y|} = e^{x^2} + C$$

$$y = \pm e^{x^2} + C$$
$$y = \pm e^{x^2} \cdot e^C$$
$$y = ke^{x^2}$$

9.
$$\frac{dy}{dx} = 3x^2y - 2xy$$
$$\frac{dy}{dx} = y(3x^2 - 2x)$$
$$\int \frac{dy}{y} = \int (3x^2 - 2x)\,dx$$
$$\ln|y| = \frac{3x^3}{3} - \frac{2x^2}{2} + C$$
$$e^{\ln|y|} = e^{x^3 - x^2 + C}$$
$$y = \pm\left(e^{x^3 - x^2}\right)e^C$$
$$y = ke^{x^3 - x^2}$$

11.
$$\frac{dy}{dx} = \frac{y}{x}, x > 0$$
$$\int \frac{dy}{dx} = \int \frac{dx}{x}$$
$$\ln|y| = \ln x + C_1$$
$$e^{\ln|y|} = e^{\ln x + C_1}$$
$$y = \pm e^{\ln x} \cdot e^{C_1}$$
$$y = Ce^{\ln x}$$
$$y = Cx$$

13.
$$\frac{dy}{dx} = \frac{y^2 + 6}{2y}$$
$$\frac{2y}{y^2 + 6}dy = dx$$
$$\int \frac{2y}{y^2 + 6}dy = \int dx$$
$$\ln|(y^2 + 6)| = x + C$$

Since $y^2 + 6$ is always greater than 0 we can write this as $\ln(y^2 + 6) = x + C$.

15.
$$\frac{dy}{dx} = y^2e^{2x}$$
$$\int y^{-2}dy = \int e^{2x}dx$$
$$-y^{-1} = \frac{1}{2}e^{2x} + C$$
$$-\frac{1}{y} = \frac{1}{2}e^{2x} + C$$
$$y = \frac{-1}{\frac{1}{2}e^{2x} + C}$$

17.
$$\frac{dy}{dx} + 3x^2 = 2x$$
$$\frac{dy}{dx} = 2x = 3x^2$$
$$y = \frac{2x^2}{2} - \frac{3x^3}{3} + C$$
$$y = x^2 - x^3 + C$$

Since $y = 5$ when $x = 0$,

$$5 = 0 - 0 + C$$
$$C = 5.$$

Thus,

$$y = x^2 - x^3 + 5.$$

19.
$$2\frac{dy}{dx} = 4xe^{-x}$$
$$\frac{dy}{dx} = 2xe^{-x}$$

Use the table of integrals or integrate by parts.

$$y = 2(-x - 1)e^{-x} + C$$

Since $y = 42$ when $x = 0$,

$$42 = 2(0 - 1)(1) + C$$
$$42 = -2 + C$$
$$C = 44$$

Thus,

$$y = -2xe^{-x} - 2e^{-x} + 44.$$

21. $\frac{dy}{dx} = \frac{x^3}{y}$; $y = 5$ when $x = 0$.

$$\int y\,dy = \int x^3\,dx$$

$$\frac{y^2}{2} = \frac{x^4}{4} + C$$

$$y^2 = \frac{1}{2}x^4 + 2C$$

$$y^2 = \frac{1}{2}x^4 + k$$

Since $y = 5$ when $x = 0$,

$$25 = 0 + k$$

$$k = 25.$$

So $y^2 = \frac{1}{2}x^4 + 25$.

23. $(2x + 3)y = \frac{dy}{dx}$; $y = 1$ when $x = 0$.

$$\int (2x + 3)dx = \int \frac{dy}{y}$$

$$\frac{2x^2}{2} + 3x + C = \ln|y|$$

$$e^{x^2+3x+C} = e^{\ln|y|}$$

$$y = (e^{x^2+3x})(\pm e^C)$$

$$y = ke^{x^2+3x}$$

Since $y = 1$ when $x = 0$.

$$1 = ke^{0+0}$$

$$k = 1.$$

So $y = e^{x^2+3x}$.

25. $\frac{dy}{dx} = \frac{2x + 1}{y - 3}$; $y = 4$ when $x = 0$.

$$\int (y - 3)\,dy = \int (2x + 1)dx$$

$$\frac{y^2}{2} - 3y = \frac{2x^2}{2} + x + C$$

Since $y = 4$ when $x = 0$,

$$\frac{16}{2} - 12 = 0 + 0 + C$$

$$C = -4.$$

So,

$$\frac{y^2}{2} - 3y = x^2 + x - 4.$$

27. $\frac{dy}{dx} = \frac{y^2}{x}$; $y = 3$ when $x = e$.

$$\int y^{-2}dy = \int \frac{dx}{x}$$

$$-y^{-1} = \ln|x| + C$$

$$-\frac{1}{y} = \ln|x| + C$$

$$y = \frac{-1}{\ln|x| + C}$$

Since $y = 3$ when $x = e$,

$$3 = \frac{-1}{\ln e + C}$$

$$3 = \frac{-1}{1 + C}$$

$$3 + 3C = -1$$

$$3C = -4$$

$$C = -\frac{4}{3}.$$

So $\quad y = \frac{-1}{\ln|x| - \frac{4}{3}} = \frac{-3}{3\ln|x| - 4}.$

29. $\frac{dy}{dx} = (y - 1)^2 e^{x-1}$; $y = 2$ when $x = 1$.

$$\frac{dy}{(y - 1)^2} = e^{x-1}\,dx$$

$$\int (y - 1)^{-2}dy = \int e^{x-1}dx$$

$$\frac{(y - 1)^{-1}}{-1} = e^{x-1} + C$$

$$-\frac{1}{y - 1} = e^{x-1} + C$$

$$-(y - 1) = \frac{1}{e^{x-1} + C}$$

$$-y + 1 = \frac{1}{e^{x-1} + C}$$

$$1 - \frac{1}{e^{x-1} + C} = y$$

$$y = \frac{e^{x-1} + C}{e^{x-1} + C} - \frac{1}{e^{x-1} + C}$$

$$y = \frac{e^{x-1} + C - 1}{e^{x-1} + C}$$

$y = 2$, when $x = 1$.

$$2 = \frac{e^0 + C - 1}{e^0 + C}$$

$$2 = \frac{C}{1 + C}$$

$$2 + 2C = C$$

$$C = -2$$

$$y = \frac{e^{x-1} - 3}{e^{x-1} - 2}.$$

31. $\dfrac{dy}{dx} = \dfrac{k}{N}(N - y)y$

(a) $\dfrac{N\,dy}{(N - y)y} = k\,dx$

Since $\dfrac{1}{y} + \dfrac{1}{N - y} = \dfrac{N}{(N - y)y}$,

$$\int \frac{dy}{y} + \int \frac{dy}{N - y} = k\,dx$$

$$\ln\left|\frac{y}{N - y}\right| = kx + C$$

$$\frac{y}{N - y} = Ce^{kx}.$$

For $0 < y < N, Ce^{kx} > 0$.

For $0 < N < y, Ce^{kx} < 0$.

Solve for y.

$$y = \frac{Ce^{kx}N}{1 + Ce^{kx}} = \frac{N}{1 + C^{-1}e^{-kx}}$$

Let $b = C^{-1} > 0$ for $0 < y < N$.

$$y = \frac{N}{1 + be^{-kx}}$$

Let $-b = C^{-1} < 0$ for $0 < N < y$.

$$y = \frac{N}{1 - be^{-kx}}$$

(b) For $0 < y < N$; $t = 0$, $y = y_0$.

$$y_0 = \frac{N}{1 + be^0} = \frac{N}{1 + b}$$

Solve for b.

$$b = \frac{N - y_0}{y_0}$$

(c) For $0 < N < y; t = 0, y = y_0$.

$$y_0 = \frac{N}{1 - be^0} = \frac{N}{1 - b}$$

Solve for b.

$$b = \frac{y_0 - N}{y_0}$$

33. **(a)** $0 < y_0 < N$ implies that $y_0 > 0, N > 0$, and $N - y_0 > 0$.

Therefore,

$$b = \frac{N - y_0}{y_0} > 0.$$

Also, $e^{-kx} > 0$ for all x, which implies that $1 + be^{-kx} > 1$.

(1) $y(x) = \dfrac{N}{1 + be^{-kx}} < N$ since $1 + be^{-kx} > 1$.

(2) $y(x) = \dfrac{N}{1 + be^{-kx}} > 0$ since $N > 0$ and

$1 + be^{-kx} > 0.$

Combining statements (1) and (2), we have

$$0 < \frac{N}{1 + be^{-kx}} = y(x) = \frac{N}{1 + be^{-kx}} < N$$

or $0 < y(x) < N$ for all x.

(b) $\displaystyle\lim_{x\to\infty} \frac{N}{1 + be^{-kx}} = \frac{N}{1 + b(0)} = N$

$\displaystyle\lim_{x\to-\infty} \frac{N}{1 + be^{-kx}} = 0$

Note that as $x \to -\infty, 1 + be^{-kx}$ becomes infinitely large.

Therefore, the horizontal asymptotes are $y = N$ and $y = 0$.

(c)
$$y'(x) = \frac{(1 + be^{-kx})(0) - N(-kbe^{-kx})}{(1 + be^{-kx})^2} = \frac{Nkbe^{-kx}}{(1 + be^{-kx})^2} > 0 \text{ for all } x.$$

Therefore, $y(x)$ is an increasing function.

(d) To find $y''(x)$, apply the quotient rule to find the derivation of $y'(x)$. The numerator of $y''(x)$, is

$$\begin{aligned} y''(x) &= (1 + be^{-kx})^2 (-Nk^2be^{-kx}) \\ &\quad -Nkbe^{-kx}[-2kbe^{-kx}(1 + be^{-kx})] \\ &= -Nk^2be^{-kx}(1 - be^{-kx})(1 + be^{-kx}), \end{aligned}$$

and the denominator is

$$[(1 + be^{-kt})^2]^2 = (1 + be^{-kx})^4.$$

Thus,

$$y''(x) = \frac{-Nk^2be^{-kx}(1 - be^{-kx})}{(1 + be^{-kx})^3}.$$

$y''(x) = 0$ when

$$\begin{aligned} k - kbe^{-kx} &= 0 \\ be^{-kx} &= 1 \\ e^{-kx} &= \frac{1}{b} \\ -kx &= \ln\left(\frac{1}{b}\right) \\ x &= \frac{\ln\left(\frac{1}{b}\right)}{k} \\ &= \frac{\ln\left(\frac{1}{b}\right)^{-1}}{k} = \frac{\ln b}{k}. \end{aligned}$$

When $x = \frac{\ln b}{k}$,

$$\begin{aligned} y &= \frac{N}{1 + be^{-k\left(\frac{\ln b}{k}\right)}} = \frac{N}{1 + be^{(-\ln b)}} \\ &= \frac{N}{1 + be^{\ln(1/b)}} = \frac{N}{1 + b\left(\frac{1}{b}\right)} = \frac{N}{2}. \end{aligned}$$

Therefore, $\left(\frac{\ln b}{k}, \frac{N}{2}\right)$ is a point of inflection.

(e) To locate the maximum of $\frac{dy}{dx}$, we must consider, from part (d),

$$\frac{d}{dx}\left(\frac{dy}{dx}\right) = \frac{-Nkbe^{-kx}(k - kbe^{-kx})}{(1 + be^{-kx})^3}.$$

Since $y''(x) > 0$ for $x < \dfrac{\ln b}{k}$ and

$y''(x) < 0$ for $x > \dfrac{\ln b}{k}$,

we know that $x = \frac{\ln b}{k}$ locates a relative maximum of $\frac{dy}{dx}$.

35. $\frac{dy}{dx} = \frac{100}{32 - 4x}$

$$y = 100\left(-\frac{1}{4}\right)\ln|32 - 4x| + C$$
$$y = -25\ln|32 - 4x| + C$$

Now, $y = 1000$ when $x = 0$.

$$1000 = -25\ln|32| + C$$
$$C = 1000 + 25\ln 32$$
$$C \approx 1086.64$$

Thus,

$$y = -25\ln|32 - 4x| + 1086.64.$$

(a) Let $x = 3$.

$$y = -25\ln|32 - 12| + 1086.64$$
$$\approx \$1011.75$$

(b) Let $x = 5$.

$$y = -25\ln|32 - 20| + 1086.64$$
$$\approx \$1024.52$$

(c) Advertising expenditures can never reach \$8000. If $x = 8$, the denominator becomes zero.

37. $\frac{dy}{dt} = -0.05y$

See Example 5.

$$\int\frac{dy}{y} = \int -0.05\,dt$$
$$\ln|y| = -0.05t + C$$
$$e^{\ln|y|} = e^{-0.05t + C}$$
$$e^{\ln|y|} = e^{-0.05t}\cdot e^{C}$$
$$|y| = e^{-0.05t}\cdot e^{C}$$
$$y = Me^{-0.05t}$$

Let $y = 1$ when $t = 0$.

Solve M:

$$1 = Me^{0}$$
$$M = 1.$$

So $y = e^{-0.05t}$.

If $y = 0.50$,

$$0.5 = e^{-0.05t}$$
$$t = \frac{-\ln 0.5}{0.05} \approx 13.9$$

It will take about 13.9 years for \$1 to lose half its value.

39. $E = -\frac{p}{q}\cdot\frac{dq}{dp}$ with $p > 0$ and $q > 0$

If $E = 2$,

$$2 = -\frac{p}{q}\cdot\frac{dq}{dp}$$
$$\frac{2}{p}dp = -\frac{1}{q}dq$$
$$\int\frac{2}{p}dp = -\int\frac{1}{q}dq$$
$$2\ln p = -\ln q + K$$
$$\ln p^2 + \ln q = K$$
$$\ln(p^2q) = K$$
$$p^2q = e^K$$
$$p^2q = C$$
$$q = \frac{C}{p^2}.$$

41. $\frac{dA}{dt} = Ai$

$$\frac{dA}{A} = i\,dt$$
$$\int\frac{dA}{A} = \int i\,dt$$
$$\ln A = it + C$$
$$e^{\ln A} = e^{it+C}$$
$$A = Me^{it}$$

When $t = 0$, $A = 5000$. Therefore, $M = 5000$. Find i so that $A = 20{,}000$ when $t = 24$.

$$\begin{aligned} 20{,}000 &= 5000e^{24i} \\ 4 &= e^{24i} \\ \ln 4 &= 24i \\ i &= \frac{\ln 4}{24} \\ &= \frac{2\ln 2}{24} \\ &= \frac{\ln 2}{12} \end{aligned}$$

The answer is d.

43. **(a)** $\dfrac{dI}{dW} = 0.088(2.4 - I)$

Separate the variables and take antiderivatives.

$$\int \frac{dI}{2.4 - I} = \int 0.088\,dW$$
$$-\ln|2.4 - I| = 0.088W + k$$

Solve for I.

$$\begin{aligned} \ln|2.4 - I| &= -0.088W - k \\ |2.4 - I| &= e^{-0.088W-k} = e^{-k}e^{-0.088W} \\ I - 2.4 &= Ce^{-0.088W}, \text{where } C = \pm e^{-k}. \\ I &= 2.4 + Ce^{-0.088W} \end{aligned}$$

Since $I(0) = 1$, then

$$\begin{aligned} 1 &= 2.4 + Ce^0 \\ C &= 1 - 2.4 = -1.4. \end{aligned}$$

Therefore, $I = 2.4 - 1.4e^{-0.088W}$.

(b) Note that as W gets larger and larger $e^{-0.088W}$ approaches 0, so

$$\begin{aligned} \lim_{W\to\infty} I &= \lim_{W\to\infty} (2.4 - 1.4e^{-0.088W}) \\ &= 2.4 - 1.4(0) = 2.4, \end{aligned}$$

so I approaches 2.4.

45. **(a)** $\dfrac{dw}{dt} = k(C - 17.5w)$

C being constant implies that the calorie intake per day is constant.

(b) pounds/day $= k$(calories/day)

$$\frac{\text{pounds/day}}{\text{calories/day}} = k$$

The units of k are pounds/calorie.

(c) Since 3500 calories is equivalent to 1 pound, $k = \frac{1}{3500}$ and

$$\frac{dw}{dt} = \frac{1}{3500}(C - 17.5w).$$

(d) $\dfrac{dw}{dt} = \dfrac{1}{3500}(C - 17.5w)$; $w = w_0$ when $t = 0$.

$$\frac{3500}{C - 17.5w}dw = dt$$
$$\frac{3500}{-17.5}\int \frac{-17.5}{C - 17.5w}dw = \int dt$$
$$\begin{aligned} -200\ln|C - 17.5w| &= t + k \\ \ln|C - 17.5w| &= 0.005t - 0.005k \\ |C - 17.5w| &= e^{-0.005t-0.005k} \\ |C - 17.5w| &= e^{-0.005t} \cdot e^{-0.005k} \\ C - 17.5w &= e^{-0.005M}e^{-0.005t} \\ -17.5w &= -C + e^{-0.005M}e^{-0.005t} \\ w &= \frac{C}{17.5} - \frac{e^{-0.005M}}{17.5}e^{-0.005t} \end{aligned}$$

(e) Since $w = w_0$ when $t = 0$,

$$\begin{aligned} w_0 &= \frac{C}{17.5} - \frac{e^{-0.005M}}{17.5}(1) \\ w_0 - \frac{C}{17.5} &= -\frac{e^{-0.005M}}{17.5} \\ \frac{e^{-0.005M}}{17.5} &= \frac{C}{17.5} - w_0. \end{aligned}$$

Therefore,

$$\begin{aligned} w &= \frac{C}{17.5} - \left(\frac{C}{17.5} - w_0\right)e^{-0.005t} \\ w &= \frac{C}{17.5} + \left(w_0 - \frac{C}{17.5}\right)e^{-0.005t}. \end{aligned}$$

47. **(a)**

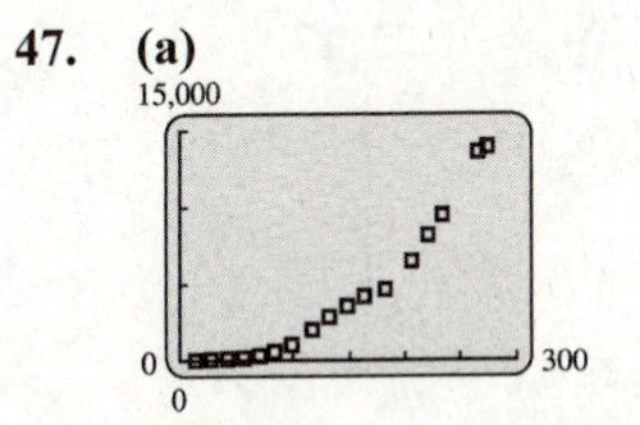

(b) $y = \dfrac{25{,}538}{1 + 110.28e^{-0.01819t}}$

(c)

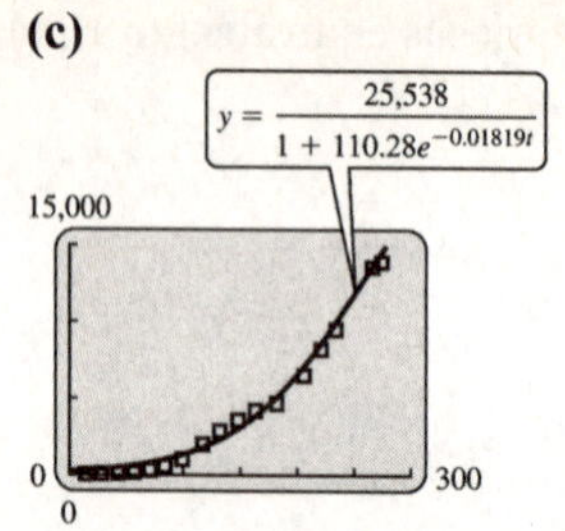

(d) As t gets very large, the value of the function in (b) approaches 25,538, so this is the limiting number of deaths predicted by the model.

49. $\dfrac{dy}{dt} = ky$

First separate the variables and integrate.

$$\frac{dy}{y} = k\,dt$$
$$\int \frac{dy}{y} = \int k\,dt$$
$$\ln|y| = kt + C.$$

Solve for y.

$$|y| = e^{kt+C_1} = e^{C_1}e^{kt}$$
$$y = Ce^{kt}, \text{ where } C = \pm e^{C_1}.$$
$$y(0) = 35.6, \text{ so } 35.6 = Ce^0 = C, \text{ and}$$
$$y = 35.6e^{kt}.$$

Since $y(50) = 102.6$, then $102.6 = 35.6e^{50k}$.

Solve for k.

$$e^{50k} = \frac{102.6}{35.6}$$
$$50k = \ln\left(\frac{102.3}{35.6}\right)$$
$$k = \frac{\ln\left(\frac{102.6}{35.6}\right)}{50} \approx 0.02117,$$
$$\text{so } y = 35.6e^{0.02117t}.$$

51. (a)

$$\frac{dy}{dt} = ky$$
$$\int \frac{dy}{y} = \int k\,dt$$
$$\ln|y| = kt + C$$
$$e^{\ln|y|} = e^{kt+C}$$
$$y = \pm(e^{kt})(e^{C})$$
$$y = Me^{kt}$$

If $y = 1$ when $t = 0$ and $y = 5$ when $t = 2$, we have the system of equations

$$1 = Me^{k(0)}$$
$$5 = Me^{2k}.$$
$$1 = M(1)$$
$$M = 1$$

Substitute.

$$5 = (1)e^{2k}$$
$$e^{2k} = 5$$
$$2k \ln e = \ln 5$$
$$k = \frac{\ln 5}{2}$$
$$\approx 0.8$$

(b) If $k = 0.8$ and $M = 1$,

$$y = e^{0.8t}.$$

When $t = 3$,

$$y = e^{0.8(3)}$$
$$= e^{2.4} \approx 11.$$

(c) When $t = 5$,

$$y = e^{0.8(5)}$$
$$= e^4 \approx 55.$$

(d) When $t = 10$,

$$y = e^{0.8(10)}$$
$$= e^8 \approx 3000.$$

53. $\dfrac{dy}{dx} = 7.5e^{-0.3y}$, $y = 0$ when $x = 0$.

$$e^{0.3y}\,dy = 7.5\,dx$$
$$\int e^{0.3y}\,dy = \int 7.5\,dx$$
$$\frac{e^{0.3y}}{0.3} = 7.5x + C$$
$$e^{0.3y} = 2.25x + C$$
$$1 = 0 + C = C$$
$$e^{0.3y} = 2.25x + 1$$
$$0.3y = \ln(2.25x + 1)$$
$$y = \frac{\ln(2.25x + 1)}{0.3}$$

When $x = 8$,

$$y = \frac{\ln[2.25(8) + 1]}{0.3} \approx 10 \text{ items}.$$

55. Let $t = 0$ be the time it started snowing. If h is the height of the snow and if the rate of snowfall is constant, $\frac{dh}{dt} = k_1$, where k_1 is a constant.

$$\frac{dh}{dt} = k_1 \text{ and } h = 0 \text{ when } t = 0.$$
$$dh = k_1\,dt$$
$$\int dh = \int k_1\,dt$$
$$h = k_1 t + C_1$$

Since $h = 0$ and $t = 0$, $0 = k_1(0) + C_1$.
Thus, $C_1 = 0$ and $h = k_1 t$.

Since the snowplow removes a constant volume of snow per hour and the volume is proportional to the height of the snow, the rate of travel of the snowplow is inversely proportional to the height of the snow.

$\frac{dx}{dt} = \frac{k^2}{h}$, where k_2 is a constant.

When $t = T, x = 0$.

When $t = T + 1, x = 2$.

When $t = T + 2, x = 3$.

Since $\frac{dy}{dt} = \frac{k^2}{h}$ and $h = k_1 t$,

$$\frac{dy}{dt} = \frac{k_2}{k_1 t}$$
$$\frac{dx}{dt} = \frac{k_2}{k_1} \cdot \frac{1}{t}.$$

Let $k_3 = \frac{k_2}{k_1}$. Then

$$\frac{dx}{dt} = k_3 \frac{1}{t}$$
$$dx = k_3 \frac{1}{t}\,dt$$
$$\int dx = \int k_3 \frac{1}{t}\,dt$$
$$x = k_3 \ln t + C_2.$$

Since $x = 0$, when $t = T$,

$$0 = k_3 \ln T + C_2$$
$$C_2 = -k_3 \ln T.$$

Thus,

$$x = k_3 \ln t - k_3 \ln T$$
$$x = k_3(\ln t - \ln T)$$
$$x = k_3 \ln\left(\frac{t}{T}\right).$$

Since $x = 2$, when $t = T + 1$,

$$2 = k_3 \ln\left(\frac{T+1}{T}\right). \qquad (1)$$

Since $x = 3$ when $t = T + 2$,

$$3 = k_3 \ln\left(\frac{T+2}{T}\right). \qquad (2)$$

We want to solve for T, so we divide equation (1) by equation (2).

$$\frac{2}{3} = \frac{k_3 \ln\left(\frac{T+1}{T}\right)}{k_3 \ln\left(\frac{T+2}{T}\right)}$$
$$\frac{2}{3} = \frac{\ln(T+1) - \ln T}{\ln(T+2) - \ln T}$$

$$2\ln(T+2) - 2\ln T = 3\ln(T+1) - 3\ln T$$
$$\ln(T+2)^2 - \ln T^2 - \ln(T+1)^3 + \ln T^3 = 0$$
$$\ln \frac{(T+2)^2 T^3}{T^2 (T+1)^3} = 0$$
$$\frac{T(T+2)^2}{(T+1)^3} = 1$$

$$T(T^2 + 4T + 4) = T^3 + 3T^2 + 3T + 1$$
$$T^3 + 4T^2 + 4T = T^3 + 3T^2 + 3T + 1$$
$$T^2 + T - 1 = 0$$
$$T = \frac{-1 \pm \sqrt{1+4}}{2}$$

$T = \frac{-1-\sqrt{5}}{2}$ is negative and is not a possible solution.

Thus, $T = \frac{-1+\sqrt{5}}{2} \approx 0.618$ hr.

0.618 hr $\approx$ 37 min and 5 sec

Now, 37 min and 5 sec before 8:00 A.M. is 7:22:55 A.M.

Thus, it started snowing at 7:22:55 A.M.

57. If $T = Ce^{-kt} + T_M$,

$$\lim_{t\to\infty} T = \lim_{t\to\infty} (Ce^{-kt} + T_M)$$
$$= \lim_{t\to\infty} (Ce^{-kt}) + T_M$$
$$= T_M$$

(The exponential term has limit 0 since $k > 0$.)

59. Use the formula from Exercise 57:

$$T = Ce^{-kt} + T_M.$$

(a) In this problem, $T_0 = 10, C = 88.6,$ and $k = 0.24.$

Therefore, $T = 88.6e^{-0.24t} + 10.$

(b)

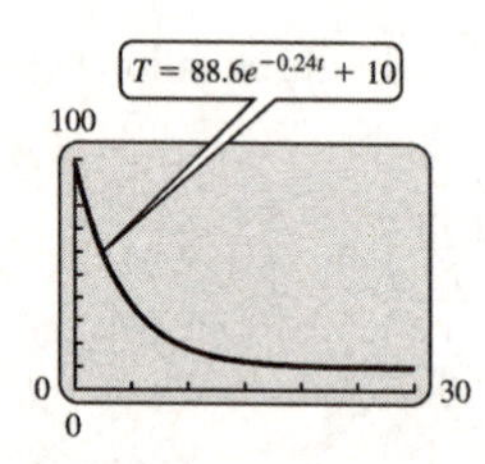

(c) The graph shows the most rapid decrease in the first few hours which is just after death.

(d) If $t = 4,$

$$T = 88.6e^{-0.24(4)} + 10$$
$$T \approx 43.9.$$

The temperature of the body will be 43.9°F after 4 hours.

(e)

$$40 = 88.6e^{-0.24t} + 10$$
$$88.6e^{-0.24t} = 30$$
$$e^{-0.24t} = \frac{30}{88.6}$$
$$-0.24t = \ln\left(\frac{30}{88.6}\right)$$
$$t = \frac{\ln\left(\frac{30}{88.6}\right)}{-0.24}$$
$$t \approx 4.5$$

The body will reach a temperature of 40°F in 4.5 hours.

Chapter 16 Review Exercises

1. False: This integral is best evaluated by substitution.

2. True

3. False: Using the substitution $u = x^2,$ $dv = xe^{-x^2}$ this integral requires only one integration by parts.

4. True

5. False: The integrand should be just $2x^2 + 3.$

6. False: The integrand should be $\pi(x^2 + 1).$

7. True

8. True

9. True

10. False: We must write the integral as $\lim_{a\to-\infty} \int_a^c xe^{-2x}dx + \lim_{b\to\infty} \int_c^b xe^{-2x}dx.$ The first of these integrals diverges so $\int_{-\infty}^{\infty} xe^{-2x}dx$ diverges.

11. True

12. False: $y = e^{2x}$ satisfies the given differential equation.

13. True

14. True

19. $\int \frac{3x}{\sqrt{x-2}}dx = \int 3x(x-2)^{-1/2}dx$

Let $u = 3x$ and $dv = (x-2)^{-1/2}dx.$

Then $du = 3\,dx$ and $v = 2(x-2)^{1/2}.$

$$\int \frac{3x}{\sqrt{x-2}}dx$$
$$= 6x(x-2)^{1/2} - 6\int (x-2)^{1/2}dx$$
$$= 6x(x-2)^{1/2} - \frac{6(x-2)^{3/2}}{\frac{3}{2}} + C$$
$$= 6x(x-2)^{1/2} - 4(x-2)^{3/2} + C$$

21. $\int (3x+6)e^{-3x}dx$

Let $u = 3x + 6$ and $dv = e^{-3x}dx.$

Then $du = 3dx$ and $v = \frac{1}{-3}e^{-3x}.$

$$\int (3x+6)e^{-3x}dx$$
$$= (3x+6)\left(-\frac{1}{3}e^{-3x}\right) - \int \left(-\frac{1}{3}e^{-3x}\right)3\,dx$$
$$= -(x+2)e^{-3x} + \int e^{-3x}dx$$
$$= -(x+2)e^{-3x} - \frac{1}{3}e^{-3x} + C$$

23. $\int (x-1)\ln|x|\,dx$

Let $\quad u = \ln|x| \quad$ and $\quad dv = (x-1)\,dx.$

Then $\quad du = \frac{1}{x}dx \quad$ and $\quad v = \frac{x^2}{2} - x.$

$$\int (x-1)\ln|x|\,dx$$
$$= \left(\frac{x^2}{2} - x\right)\ln|x| - \int \left(\frac{x}{2} - 1\right)dx$$
$$= \left(\frac{x^2}{2} - x\right)\ln|x| - \frac{x^2}{4} + x + C$$

25. $\int \frac{x}{\sqrt{16+8x^2}}dx$

Use substitution.

Let $u = 16 + 8x^2$. Then $du = 16x\,dx$.

$$\int \frac{x}{\sqrt{16+8x^2}}dx = \frac{1}{16}\int \frac{16x}{\sqrt{16+8x^2}}dx$$
$$= \frac{1}{16}\int \frac{1}{\sqrt{u}}du$$
$$= \frac{1}{16}\int u^{-1/2}du$$
$$= \frac{1}{16}(2)u^{1/2} + C$$
$$= \frac{1}{8}(16+8x^2)^{1/2} + C$$
$$= \frac{1}{8}\sqrt{16+8x^2} + C$$

27. $\int_0^1 x^2e^{x/2}dx$

Let $u = x^2$ and $dv = e^{x/2}\,dx.$

Use column integration.

D		I
x^2	$+$	$e^{x/2}$
$2x$	$-$	$2e^{x/2}$
2	$+$	$4e^{x/2}$
0		$8e^{x/2}$

$$\int_0^1 x^2e^{x/2}\,dx = (2x^2e^{x/2} - 8xe^{x/2} + 16e^{x/2})\Big|_0^1$$
$$= 2e^{1/2} - 8e^{1/2} + 16e^{1/2} - 16$$
$$= 10e^{1/2} - 16 \approx 0.4872$$

29. $A = \int_1^3 x^3(x^2-1)^{1/3}dx$

Let $\quad u = x^2$ and $dv = x(x^2-1)^{1/3}\,dx.$

Then $du = 2x\,dx$ and $v = \frac{3}{8}(x^2-1)^{4/3}.$

$$\int x^3(x^2-1)^{1/3}dx$$
$$= \frac{3x^2}{8}(x^2-1)^{4/3} - \frac{3}{4}\int x(x^2-1)^{4/3}dx$$
$$= \frac{3x^2}{8}(x^2-1)^{4/3} - \frac{3}{4}\left[\frac{1}{2}\cdot\frac{3}{7}(x^2-1)^{7/3}\right]$$
$$= \frac{3x^2}{8}(x^2-1)^{4/3} - \frac{9}{56}(x^2-1)^{7/3} + C$$
$$A = \left[\frac{3x^2}{8}(x^2-1)^{4/3} - \frac{9}{56}(x^2-1)^{7/3}\right]\Bigg|_0^3$$
$$= \frac{3}{8}(144) - \frac{9}{56}(128)$$
$$= 54 - \frac{144}{7} = \frac{234}{7} \approx 33.43$$

31. $f(x) = \sqrt{x-4};\ y = 0;\ x = 13$

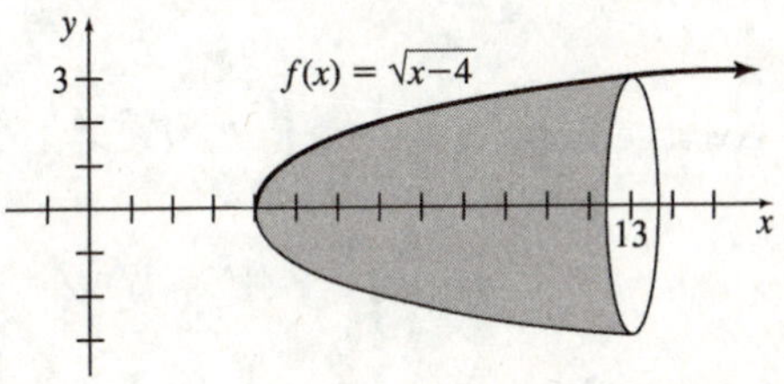

Since $f(x) = \sqrt{x-4}$ intersects $y = 0$ at $x = 4$, the integral has lower bound $a = 4$.

$$V = \pi \int_4^{13} (\sqrt{x - 4})^2 dx$$
$$= \pi \int_4^{13} (x - 4)dx$$
$$= \pi \left(\frac{x^2}{2} - 4x \right) \Bigg|_4^{13}$$
$$= \pi \left[\left(\frac{169}{2} - 52 \right) - (8 - 16) \right]$$
$$= \pi \left(\frac{65}{2} + 8 \right) = \frac{81}{2}\pi \approx 127.2$$

33. $f(x) = \dfrac{1}{\sqrt{x - 1}}, y = 0, x = 2, x = 4$

$$V = \pi \int_2^4 \left(\frac{1}{\sqrt{x - 1}} \right)^2 dx$$
$$= \pi \int_2^4 \frac{dx}{x - 1}$$
$$= \pi (\ln |x - 1|) \Big|_2^4$$
$$= \pi \ln 3 \approx 3.451$$

35. $f(x) = \dfrac{x^2}{4}, y = 0, x = 4$

Since $f(x) = \frac{x^2}{4}$ intersects $y = 0$ at $x = 0$, the integral has a lower bound, $a = 0$.

$$V = \pi \int_0^4 \left(\frac{x^2}{4} \right)^2 dx = \pi \int_0^4 \frac{x^4}{16}$$
$$= \frac{\pi}{16} \left(\frac{x^5}{5} \right) \Bigg|_0^4 = \frac{\pi}{16} \left(\frac{1024}{5} \right)$$
$$= \frac{64\pi}{5} \approx 40.21$$

39. $\text{Average value} = \dfrac{1}{2 - 0} \displaystyle\int_0^2 7x^2(x^3 + 1)^6 dx$

$$= \frac{7}{2} \int_0^2 x^2(x^3 + 1)^6 dx$$

Let $u = x^3 + 1$. Then $du = 3x^2 dx$.

$$\int x^2(x^3 + 1)^6 dx = \frac{1}{3} \int 3x^2(x^3 + 1)^6 dx$$
$$= \frac{1}{3} \int u^6 du$$
$$= \frac{1}{3} \cdot \frac{1}{7} u^7 + C$$
$$= \frac{1}{21}(x^3 + 1)^7 + C$$

$$\frac{7}{2} \int_0^2 x^2(x^3 + 1)^6 dx = \frac{7}{2} \cdot \frac{1}{21}(x^3 + 1)^7 \Big|_0^2$$
$$= \frac{1}{6}(9^7 - 1^7) = \frac{1}{6}(4{,}782{,}969 - 1)$$
$$= \frac{2{,}391{,}484}{3}$$

41. $\displaystyle\int_{-\infty}^{-5} x^{-2} dx = \lim_{a \to -\infty} \int_a^{-5} x^{-2} dx$

$$= \lim_{a \to -\infty} \left(\frac{x^{-1}}{-1} \right) \Bigg|_a^{-5}$$
$$= \lim_{a \to -\infty} \left(-\frac{1}{x} \right) \Bigg|_a^{-5}$$
$$= \frac{1}{5} + \lim_{a \to -\infty} \left(\frac{1}{a} \right)$$

As $a \to -\infty$, $\frac{1}{a} \to 0$. The integral converges.

$$\int_{-\infty}^{-5} x^{-2} dx = \frac{1}{5} + 0 = \frac{1}{5}$$

43. $\displaystyle\int_1^{\infty} 6e^{-x} dx = \lim_{b \to \infty} \int_1^b 6e^{-x} dx$

$$= \lim_{b \to \infty} - 6e^{-x} \Big|_1^b$$
$$= \lim_{b \to \infty} (-6e^{-b} + 6e^{-1})$$
$$= \lim_{b \to \infty} \left(\frac{-6}{e^b} + \frac{6}{e} \right)$$

As $b \to \infty, e^b \to \infty$, so $\frac{-6}{e^b} \to 0$. The integral converges.

$$\int_1^{\infty} 6e^{-x} dx = 0 + \frac{6}{e} = \frac{6}{e} \approx 2.207$$

45. $\int_4^{\infty} \ln(5x)dx = \lim_{b\to\infty} \int_4^b \ln(5x)dx$

Let $u = \ln(5x)$ and $dv = dx$.

Then $du = \frac{1}{x}dx$ and $v = x$.

$$\int \ln(5x)dx = x\ln(5x) - \int x \cdot \frac{1}{x}dx$$
$$= x\ln(5x) - \int dx$$
$$= x\ln(5x) - x + C$$

$$\lim_{b\to\infty} \int_4^b \ln(5x)dx$$
$$= \lim_{b\to\infty} [x\ln(5x) - x]\Big|_4^b$$
$$= \lim_{b\to\infty} [b\ln(5b) - b] - (4\ln 20 - 4)$$

As $b \to \infty$, $b\ln(5b) - b \to \infty$. The integral diverges.

47. $f(x) = 3e^{-x}$ for $[0, \infty)$

$$A = \int_0^{\infty} 3e^{-x}dx$$
$$= \lim_{b\to\infty} \int_0^b 3e^{-x}dx$$
$$= \lim_{b\to\infty} (-3e^{-x})\Big|_0^b$$
$$= \lim_{b\to\infty} \left(\frac{-3}{e^b} + 3\right)$$

As $b \to \infty$, $\frac{-3}{e^b} \to 0$.

$A = 0 + 3 = 3$

51. $\frac{dy}{dx} = 3x^2 + 6x$

$$dy = (3x^2 + 6x)dx$$
$$y = x^3 + 3x^2 + C$$

53. $\frac{dy}{dx} = 4e^{2x}$

$$dy = 4e^{2x}dx$$
$$y = 2e^{2x} + C$$

55. $\frac{dy}{dx} = \frac{3x + 1}{y}$

$$y\,dy = (3x + 1)\,dx$$
$$\frac{y^2}{2} = \frac{3x^2}{2} + x + C_1$$
$$y^2 = 3x^2 + 2x + C$$

57. $\frac{dy}{dx} = \frac{2y + 1}{x}$

$$\frac{dy}{2y + 1} = \frac{dy}{x}$$
$$\frac{1}{2}\left(\frac{2\,dy}{2y + 1}\right) = \frac{dx}{x}$$
$$\frac{1}{2}\ln|2y + 1| = \ln|x| + C_1$$
$$\ln|2y + 1|^{1/2} = \ln|x| + \ln k$$

Let $\ln k = C_1$

$$\ln|2y + 1|^{1/2} = \ln k|x|$$
$$|2y + 1|^{1/2} = k|x|$$
$$2y + 1 = k^2x^2$$
$$2y + 1 = Cx^2$$
$$2y = Cx^2 - 1$$
$$y = \frac{Cx^2 - 1}{2}$$

59. $\frac{dy}{dx} = x^2 - 6x$; $y = 3$ when $x = 0$.

$$dy = (x^2 - 6x)dx$$
$$y = \frac{x^3}{3} - 3x^2 + C$$

When $x = 0$, $y = 3$.

$$3 = 0 - 0 + C$$
$$C = 3$$
$$y = \frac{x^3}{3} - 3x^2 + 3$$

61. $\frac{dy}{dx} = (x+2)^3 e^y$; $y = 0$ when $x = 0$.

$$e^{-y}dy = (x+2)^3 dx$$
$$-e^{-y} = \frac{1}{4}(x+2)^4 + C$$
$$-1 = \frac{1}{4}(2)^4 + C$$
$$C = -5$$
$$e^{-y} = 5 - \frac{1}{4}(x+2)^4$$
$$y = -\ln\left[5 - \frac{1}{4}(x+2)^4\right]$$

Notice that $x < \sqrt[4]{20} - 2$.

63. $\frac{dy}{dx} = \frac{1-2x}{y+3}$; $y = 16$ when $x = 0$.

$$(y+3)dy = (1-2x)dx$$
$$\frac{y^2}{2} + 3y = x - x^2 + C$$
$$\frac{16^2}{2} + 3(16) = 0 + C$$
$$176 = C$$
$$\frac{y^2}{2} + 3y = x - x^2 + 176$$
$$y^2 + 6y = 2x - 2x^2 + 352$$

65. $R' = x(x-50)^{1/2}$

$$R = \int_{50}^{75} x(x-50)^{1/2}dx$$

Let $u = x$ and $dv = (x-50)^{1/2}$.

Then $du = dx$ and $v = \frac{2}{3}(x-50)^{3/2}$.

$$\int x(x-50)^{1/2}dx$$
$$= \frac{2}{3}x(x-50)^{3/2} - \frac{2}{3}\int (x-50)^{3/2}dx$$
$$= \frac{2}{3}x(x-50)^{3/2} - \frac{2}{3}\cdot\frac{2}{5}(x-50)^{5/2}$$

$$R = \left[\frac{2}{3}x(x-50)^{3/2} - \frac{4}{15}(x-50)^{5/2}\right]\Bigg|_{50}^{75}$$
$$= \frac{2}{3}(75)(25^{3/2}) - \frac{4}{15}(25^{5/2})$$
$$= 6250 - \frac{2500}{3}$$
$$= \frac{16{,}250}{3} \approx \$5416.67$$

67. $f(t) = 25{,}000$; 12 yr; 10%

$$P = \int_0^{12} 25{,}000e^{-0.10t}dt$$
$$= 25{,}000\left(\frac{e^{-0.10t}}{-0.10}\right)\Bigg|_0^{12}$$
$$\approx 250{,}000(-0.3012 + 1)$$
$$\approx \$174{,}701.45$$

69. $f(t) = 15t$; 18 mo; 8%

$$P = \int_0^{1.5} 15te^{-0.08t}dt$$
$$= 15\int_0^{1.5} te^{-0.08t}dt$$

Find the antiderivative using integration by parts.

Let $u = t$ and $dv = e^{-0.08t}dt$.

Then $du = dt$ and $v = \frac{1}{-0.08}e^{-0.08t}$
$= -12.5e^{-0.08t}$

$$\int te^{-0.08t}dt = -12.5te^{-0.08t} - \int(-12.5e^{-0.08t})dt$$
$$= -12.5te^{-0.08t} - 156.25e^{-0.08t} + C$$
$$P = 15\int_0^{1.5} te^{-0.08t}dt$$
$$= 15(-12.5te^{-0.08t} - 156.25e^{-0.08t})\Big|_0^{1.5}$$
$$= 15[(-18.75e^{-0.12} - 156.25e^{-0.12}) - (0 - 156.25)]$$
$$= 15(-175e^{-0.12} + 156.25)$$
$$\approx 15.58385362$$

The present value is \$15.58.

71. $f(t) = 500e^{-0.04t}$; 8 yr; 10% per yr

$$A = e^{0.1(8)}\int_0^8 500e^{-0.04t}\cdot e^{-0.1t}\,dt$$
$$= e^{0.8}\int_0^8 500e^{-0.14t}\,dt$$
$$= e^{0.8}\left(\frac{500}{-0.14}e^{-0.14t}\right)\bigg|_0^8$$
$$= e^{0.8}\left[\frac{500}{-0.14}(e^{-1.12} - 1)\right]$$
$$\approx 5354.971041$$

The accumulated value is \$5354.97.

73. $f(t) = 1000 + 200t$; 10 yr; 9% per yr

$$e^{(0.09)(10)}\int_0^{10}(1000 + 200t)e^{-0.09t}\,dt$$
$$= e^{0.9}\left[\frac{1000}{-0.09}e^{0.09t} + \frac{200}{(0.09)^2}(-0.09t - 1)e^{-0.09t}\right]\Bigg|_0^{10}$$
$$= e^{0.9}\left[\frac{1000}{-0.09}(e^{-0.9} - 1) + \frac{200}{(0.09)^2}(-1.9e^{-0.9} + 1)\right]$$
$$\approx \$30,035.17$$

75. $$e^{0.105(10)}\int_0^{10} 10{,}000e^{-0.105t}\,dt$$
$$= e^{1.05}\left(\frac{10{,}000e^{-0.105t}}{-0.105}\right)\bigg|_0^{10}$$
$$= \frac{10{,}000e^{1.05}}{-0.105}(e^{-1.05} - 1)$$
$$\approx -272{,}157.25(-0.65006)$$
$$\approx \$176{,}919.15$$

77. **(a)** $$\frac{dy}{dx} = 6e^{0.3x}$$
$$dy = 6e^{0.3x}dx$$
$$y = 20e^{0.3x} + C$$

When $x = 0,\ y = 0$

$$0 = 20e^0 + C$$
$$C = -20$$
$$y = 20e^{0.3x} - 20.$$

When $x = 6$,

$$y = 20e^{1.8} - 20$$
$$\approx 100.99.$$

Sales are \$10,099.

(b) When $x = 12$,

$$y = 20e^{3.6} - 20$$
$$\approx 711.96.$$

Sales are \$71,196.

79. $A = 300{,}000$ when $t = 0; r = 0.05,$ $D = -20{,}000$

(a) $$\frac{dA}{dt} = 0.05A - 20{,}000$$

(b) $$\frac{1}{0.05A - 20{,}000}dA = dt$$

$$\frac{1}{0.05}\ln|0.05A - 20{,}000| = t + C$$
$$\ln|0.05A - 20{,}000| = 0.05t + k$$
$$\ln|0.05(300{,}000) - 20{,}000| = k$$
$$k = \ln|-5000| = \ln 5000$$
$$\ln|0.05A - 20{,}000| = 0.0.5t + \ln 5000$$
$$|0.05A - 20{,}000| = 5000e^{0.05t}$$

Since $0.05A < 20{,}000$,

$$|0.05A - 20{,}000| = 20{,}000 - 0.05A$$
$$20{,}000 - 0.05A = 5000e^{0.05t}$$
$$A = \frac{1}{0.05 = 5}(20{,}000 - 5000e^{0.05t})$$
$$= 100{,}000(4 - e^{0.05t}).$$

When $t = 10$,

$$A = 100{,}000\left(4 - e^{0.05(10)}\right)$$
$$\approx \$235{,}127.87.$$

81. $$\int_0^5 0.5te^{-t}dt = 0.5\int_0^5 te^{-t}dt$$

Let $u = t$ and $dv = e^{-t}\,dt$.

Then $du = dt$ and $v = \frac{e^{-t}}{-1}$.

$$\int te^{-t}dt = \frac{te^{-t}}{-1} + \int e^{-t}dt$$
$$= -te^{-t} + \frac{e^{-t}}{-1}$$

$$0.5\int_0^5 te^{-t}dt = 0.5(-te^{-t} - e^{-t})\Big|_0^5$$
$$= 0.5(-5e^{-5} - e^{-5} + e^0)$$
$$\approx 0.4798$$

The total reaction over the first 5 hr is 0.4798.

83. $$\frac{dy}{dt} = \frac{-10}{1+5t}; \; y = 50 \text{ when } t = 0.$$

$$y = -2\ln(1+5t) + C$$
$$50 = -2\ln 1 + C$$
$$= C$$
$$y = 50 - 2\ln(1+5t)$$

(a) If $t = 24$,

$$y = 50 - 2\ln[1 + 5t(24)]$$
$$\approx 40 \text{ insects.}$$

(b) If $y = 0$,

$$50 = 2\ln(1+5t)$$
$$1 + 5t = e^{25}$$
$$t = \frac{e^{25}-1}{5}$$
$$\approx 1.44 \times 10^{10} \text{ hours}$$
$$\approx 6 \times 10^8 \text{ days} \approx 1.6 \text{ million years.}$$

85. $$y = \frac{N}{1+be^{-kt}}; \; y = y_i \text{ when } x = x_i,$$

$i = 1,2,3.$

t_1, t_2, t_3 are equally spaced:

$t_3 = 2t_2 - t_1$, so $t_1 + t_3 = 2t_2$,

or $t_2 = \frac{t_1+t_3}{2}$.

Show $N = \dfrac{\frac{1}{y_1} + \frac{1}{y_3} - \frac{2}{y_2}}{\frac{1}{y_1y_3} - \frac{1}{y_2^2}}$.

Let

$$A = \frac{1}{y_1} + \frac{1}{y_3} - \frac{2}{y_2}$$
$$= \frac{1+be^{-kt_1}}{N} + \frac{1+be^{-kt_3}}{N} - \frac{2\left(1+be^{-kt_2}\right)}{N}$$
$$= \frac{1}{N}\left(1 + be^{-kt_1} + 1 + be^{-kt_3} - 2 - 2be^{-kt_2}\right)$$
$$= \frac{b}{N}\left[e^{-kt_1} + e^{-kt_3} - 2e^{-kt_2}\right]$$

Let

$$B = \frac{1}{y_1y_3} - \frac{1}{y_2^2}$$
$$= \frac{\left(1+be^{-kt_1}\right)\left(1+be^{-kt_3}\right)}{N^2} - \frac{\left(1+be^{-kt_2}\right)^2}{N^2}$$
$$= \frac{1}{N^2}\Big[1 + be^{-kt_1} + e^{-kt_3} + b^2e^{-k(t_1+t_3)}$$
$$-1 - 2be^{-kt_2} - b^2e^{-2kt_2}\Big]$$
$$= \frac{b}{N^2}\Big[e^{-kt_1} + be^{-kt_3} + be^{-k(2t_2)}$$
$$-2e^{-kt_2} - be^{-2kt_2}\Big]$$
$$= \frac{b}{N^2}\left[e^{-kt_1} + e^{-kt_3} - 2e^{-kt_2}\right]$$

Clearly, $\frac{A}{B} = N$.

Hence,

$$N = \frac{\frac{1}{y_1} + \frac{1}{y_3} - \frac{2}{y_2}}{\frac{1}{y_1\,y_3} - \frac{1}{y_2^2}}.$$

87. From Exercise 86,

$$N = \frac{\frac{1}{40} + \frac{1}{204} - \frac{2}{106}}{\frac{1}{40(204)} - \frac{1}{(106)^2}} \approx 329.$$

(a) $y_0 = 40$,

$$b = \frac{N - y_0}{y_0} = \frac{329 - 40}{40}$$
$$\approx 7.23.$$

If 1920 corresponds to $t = 5$ decades, then

$$106 = \frac{329}{1 + 7.23e^{-5k}}$$
$$1 + 7.23e^{-5k} = \frac{329}{106}$$
$$e^{-5k} = \frac{1}{7.23}\left(\frac{329}{106} - 1\right)$$

$$\text{so } k = -\frac{1}{5}\ln\left[\frac{1}{7.23}\left(\frac{329}{106} - 1\right)\right]$$
$$\approx 0.247.$$

(b) $y = \dfrac{329}{1 + 7.23e^{-0.247t}}$

In 2010, $t = 14$. If $t = 14$,

$$y = \frac{329}{1 + 7.23e^{-3.458}} \approx 268 \text{ million}$$

The predicated population is 268 million which is less than the table value of 308.7 million.

(c) In 2030, $t = 16$, so

$$y = \frac{329}{1 + 7.23e^{-0.247(16)}} \approx 289 \text{ million.}$$

In 2050, $t = 18$, so

$$y = \frac{329}{1 + 7.23e^{-0.247(18)}} \approx 303 \text{ million.}$$

89. **(a)** $\dfrac{dx}{dt} = 1 - kx$

Separate the variables and integrate.

$$\int \frac{dx}{1 - kx} = \int dt$$

$$-\frac{1}{k} \ln |1 - kx| = t + C_1$$

Solve for x.

$$\ln|1 - kx| = -kt - kC_1$$

$$|1 - kx| = e^{-kC_1}e^{-kt}$$

$$1 - kx = Me^{-kt}, \text{ where } M = \pm e^{-kC_1}.$$

$$x = -\frac{1}{k}(Me^{-kt} - 1) = \frac{1}{k} + Ce^{-kt},$$

where $C = -\dfrac{M}{k}$.

(b) Since $k > 0$, then as t gets larger and larger, Ce^{-kt} approaches 0, so $\lim\limits_{x \to \infty} \left(\frac{1}{k} + Ce^{-kt} \right) = \frac{1}{k}$.

91. **(a)** $\overline{T} = \dfrac{1}{10 - 0} \displaystyle\int_0^{10} (160 - 0.05x^2)\, dx$

$$= \frac{1}{10}\left(160x - \frac{0.05x^3}{3}\right)\Bigg|_0^{10}$$

$$= \frac{1}{10}\left[160(10) - \frac{0.05}{3}(10)^3\right]$$

$$\approx \frac{1}{10}(1583.3) \approx 158.3°$$

(b) $\overline{T} = \dfrac{1}{40 - 10} \displaystyle\int_{10}^{40} (160 - 0.05x^2) dx$

$$= \frac{1}{30}\left(160x - \frac{0.05x^3}{3}\right)\Bigg|_{10}^{40}$$

$$= \frac{1}{30}\left[\left(160(40) - \frac{0.05(40)^3}{3}\right) - \left(160(10) - \frac{0.05(10)^3}{3}\right)\right]$$

$$\approx \frac{1}{30}(5333.33 - 1583.33)$$

$$= 125°$$

(c) $\overline{T} = \dfrac{1}{40 - 0} \displaystyle\int_0^{40} (160 - 0.05x^2) dx$

$$= \frac{1}{40}\left(160x - \frac{0.05x^3}{3}\right)\Bigg|_0^{40}$$

$$= \frac{1}{40}\left[\left((160)(40) - \frac{(0.05)(40)^3}{3}\right)\right.$$

$$\approx \frac{1}{40}(5{,}333.33) \approx 133.3°$$

93. $\dfrac{dT}{dt} = k(T - T_F)$; $T_F = 300$; $T = 40$ when $t = 0$. $T = 150$ when $t = 1$.

The solution to the differential equation is

$$T = Ce^{kt} + T_M$$

where C is a constant.

Here, $T = Ce^{kt} + 300$.

$$40 = C + 300$$

$$C = -260$$

$$T = 300 - 260e^{kt}$$

$$150 = 300 - 260e^{k}$$

$$e^k = \frac{15}{26}$$

$$k = \ln\left(\frac{15}{26}\right) \approx -0.55$$

$$T = 300 - 260e^{-0.55t}$$

$$250 = 300 - 260e^{-0.55t}$$

$$e^{-0.55t} = \frac{5}{26}$$

$$t = -\frac{1}{0.55} \ln\left(\frac{5}{26}\right) \approx 3 \text{ hr}$$

Chapter 17

MULTIVARIABLE CALCULUS

17.1 Functions of Several Variables

Your Turn 1

$$f(x, y) = 4x^2 + 2xy + \frac{3}{y}$$

$$f(2, 3) = 4(2)^2 + 2(2)(3) + \frac{3}{3}$$
$$= 16 + 12 + 1$$
$$= 29$$

Your Turn 2

$$f(x, y, z) = 4xz - 3x^2y + 2z^2$$
$$f(1, 2, 3) = 4(1)(3) - 3(1)^2(2) + 2(3)^2$$
$$= 12 - 6 + 18$$
$$= 24$$

Your Turn 3

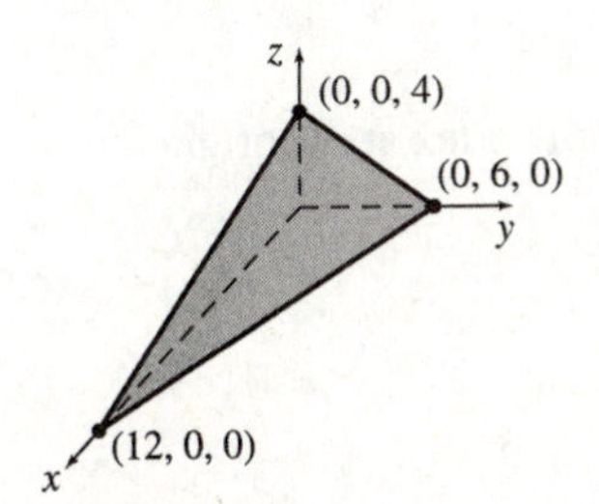

Your Turn 4

Use the Cobb-Douglas production function $z = x^{1/4}y^{3/4}$.

$$27 = x^{1/4}y^{3/4}$$
$$\frac{27}{x^{1/4}} = y^{3/4}$$
$$\left(\frac{27}{x^{1/4}}\right)^4 = \left(y^{3/4}\right)^4$$
$$y^3 = \frac{(27)^4}{x}$$
$$\left(y^3\right)^{1/3} = \left(\frac{(27)^4}{x}\right)^{1/3}$$
$$y = \frac{(27)^{4/3}}{x^{1/3}} = \frac{81}{x^{1/3}}$$

17.1 Exercises

1. $f(x, y) = 2x - 3y + 5$

   (a) $f(2, -1) = 2(2) - 3(-1) + 5 = 12$

   (b) $f(-4, 1) = 2(-4) - 3(1) + 5 = -6$

   (c) $f(-2, -3) = 2(-2) - 3(-3) + 5 = 10$

   (d) $f(0, 8) = 2(0) - 3(8) + 5 = -19$

3. $h(x, y) = \sqrt{x^2 + 2y^2}$

   (a) $h(5, 3) = \sqrt{25 + 2(9)} = \sqrt{43}$

   (b) $h(2, 4) = \sqrt{4 + 32} = 6$

   (c) $h(-1, -3) = \sqrt{1 + 18} = \sqrt{19}$

   (d) $h(-3, -1) = \sqrt{9 + 2} = \sqrt{11}$

5. $x + y + z = 9$

   If $x = 0$ and $y = 0$, $z = 9$.

   If $x = 0$ and $z = 0$, $y = 9$.

   If $y = 0$ and $z = 0$, $x = 9$.

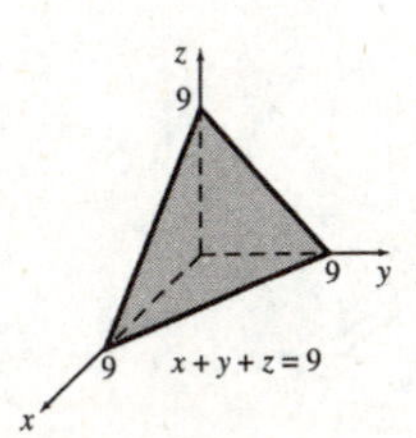

7. $2x + 3y + 4z = 12$

   If $x = 0$ and $y = 0$, $z = 3$.

   If $x = 0$ and $z = 0$, $y = 4$.

   If $y = 0$ and $z = 0$, $x = 6$.

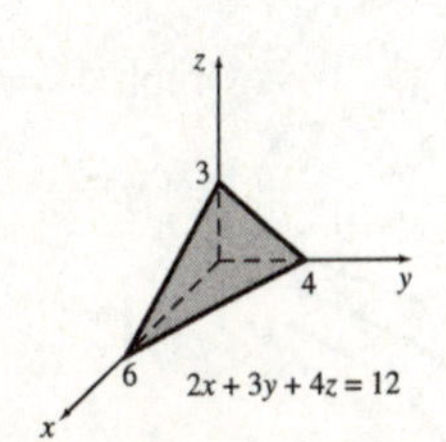

9. $x + y = 4$

If $x = 0$, $y = 4$.

If $y = 0$, $x = 4$.

There is no z-intercept.

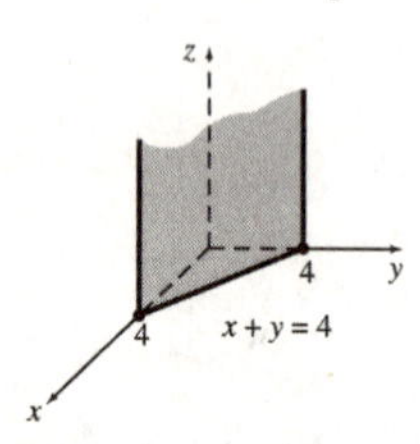

11. $x = 5$

The point $(5, 0, 0)$ is on the graph.

There are no y- or z-intercepts.

The plane is parallel to the yz-plane.

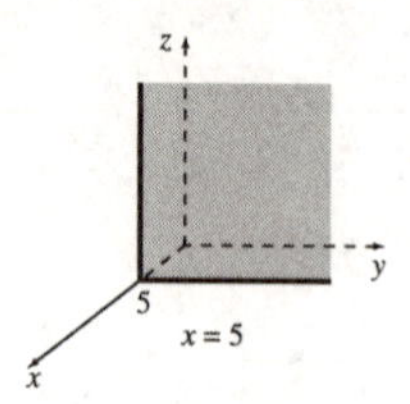

13. $3x + 2y + z = 24$

For $z = 0$, $3x + 2y = 24$. Graph the line $3x + 2y = 24$ in the xy-plane.

For $z = 2$, $3x + 2y = 22$. Graph the line $3x + 2y = 22$ in the plane $z = 2$.

For $z = 4$, $3x + 2y = 20$. Graph the line $3x + 2y = 20$ in the plane $z = 4$.

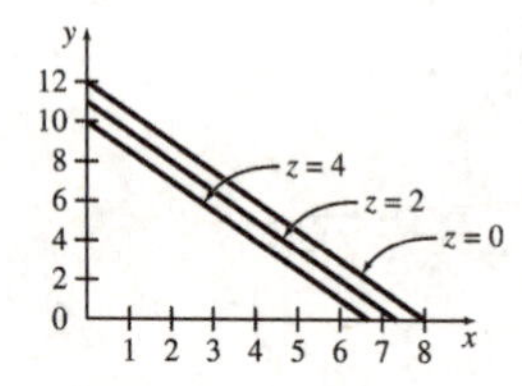

15. $y^2 - x = -z$

For $z = 0$, $x = y^2$. Graph $x = y^2$ in the xy-plane.

For $z = 2$, $x = y^2 + 2$. Graph $x = y^2 + 2$ in the plane $z = 2$.

For $z = 4$, $x = y^2 + 4$. Graph $x = y^2 + 4$ in the plane $z = 4$.

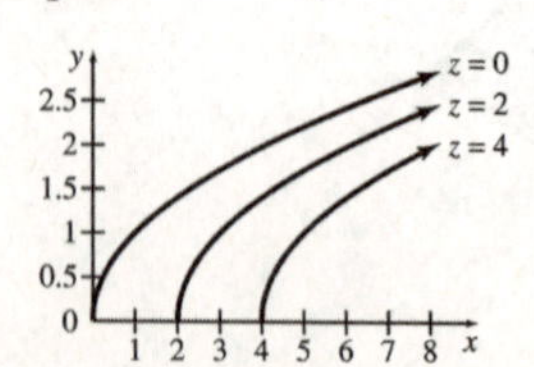

21. $z = x^2 + y^2$

The xz-trace is

$$z = x^2 + 0 = x^2.$$

The yz-trace is

$$z = 0 + y^2 = y^2.$$

Both are parabolas with vertices at the origin that open upward.

The xy-trace is

$$0 = x^2 + y^2.$$

This is a point, the origin.

The equation is represented by a paraboloid, as shown in (c).

23. $x^2 - y^2 = z$

The xz-trace is

$$x^2 = z,$$

which is a parabola with vertex at the origin that opens upward.

The yz-trace is

$$-y^2 = z,$$

which is a parabola with vertex at the origin that opens downward.

The xy-trace is

$$x^2 - y^2 = 0$$
$$x^2 = y^2$$
$$x = y \quad \text{or} \quad x = -y,$$

which are two lines that intersect at the origin.

The equation is represented by a hyperbolic paraboloid, as shown in (e).

25. $\frac{x^2}{16} + \frac{y^2}{25} + \frac{z^2}{4} = 1$

xz-trace:

$\frac{x^2}{16} + \frac{z^2}{4} = 1$, an ellipse

yz-trace:

$\frac{y^2}{25} + \frac{z^2}{4} = 1$, an ellipse

xy-trace:

$\frac{x^2}{16} + \frac{y^2}{25} = 1$, an ellipse

The graph is an ellipsoid, as shown in (b).

27. $f(x, y) = 4x^2 - 2y^2$

(a) $$\frac{f(x+h, y) - f(x, y)}{h}$$
$$= \frac{[4(x+h)^2 - 2y^2] - [4x^2 - 2y^2]}{h}$$
$$= \frac{4x^2 + 8xh + 4h^2 - 2y^2 - 4x^2 + 2y^2}{h}$$
$$= \frac{h(8x + 4h)}{h} = 8x + 4h$$

(b) $$\frac{f(x, y+h) - f(x, y)}{h}$$
$$= \frac{[4x^2 - 2(y+h)^2] - [4x^2 - 2y^2]}{h}$$
$$= \frac{4x^2 - 2y^2 - 4yh - 2h^2 - 4x^2 + 2y^2}{h}$$
$$= \frac{h(-4y - 2h)}{h}$$
$$= -4y - 2h$$

(c) $$\lim_{h\to 0} \frac{f(x+h, y) - f(x, y)}{h}$$
$$= \lim_{h\to 0} (8x + 4h)$$
$$= 8x + 4(0) = 8x$$

(d) $$\lim_{h\to 0} \frac{f(x, y+h) - f(x, y)}{h}$$
$$= \lim_{h\to 0} (-4y - 2h)$$
$$= -4y - 2(0) = -4y$$

29. $f(x, y) = xye^{x^2+y^2}$

(a) $$\lim_{h\to 0} \frac{f(1+h, 1) - f(1, 1)}{h}$$
$$= \lim_{h\to 0} \frac{(1+h)(1)e^{1+2h+h^2+1} - (1)(1)e^{1+1}}{h}$$
$$= \lim_{h\to 0} \frac{(1+h)e^{2+2h+h^2} - e^2}{h}$$
$$= e^2 \lim_{h\to 0} \frac{(1+h)e^{2h+h^2} - 1}{h}$$

The graphing calculator indicates that

$$\lim_{h\to 0} \frac{(1+h)e^{2h+h^2} - 1}{h} = 3, \text{ thus}$$
$$\lim_{h\to 0} \frac{f(1+h, 1) - f(1, 1)}{h} = 3e^2.$$

The slope of the tangent line in the direction of x at $(1, 1)$ is $3e^2$.

(b) $$\lim_{h\to 0} \frac{f(1, 1+h) - f(1, 1)}{h}$$
$$= \lim_{h\to 0} \frac{(1)(1+h)e^{1+1+2h+h^2} - (1)(1)e^{1+1}}{h}$$
$$= \lim_{h\to 0} \frac{(1+h)e^{2+2h+h^2} - e^2}{h}$$
$$= e^2 \lim_{h\to 0} \frac{(1+h)e^{2h+h^2} - 1}{h}$$

So, this limit reduces to the exact same limit as in part a. Therefore, since

$$\lim_{h\to 0} \frac{(1+h)e^{2h+h^2} - 1}{h} = 3,$$

then

$$\lim_{h\to 0} \frac{f(1, 1+h) - f(1, 1)}{h} = 3e^2.$$

The slope of the tangent line in the direction of y at $(1, 1)$ is $3e^2$.

31. $P(x, y) = 100\left[\frac{3}{5}x^{-2/5} + \frac{2}{5}y^{-2/5}\right]^{-5}$

(a) $P(32, 1)$
$$= 100\left[\frac{3}{5}(32)^{-2/5} + \frac{2}{5}(1)^{-2/5}\right]^{-5}$$
$$= 100\left[\frac{3}{5}\left(\frac{1}{4}\right) + \frac{2}{5}(1)\right]^{-5}$$
$$= 100\left(\frac{11}{20}\right)^{-5}$$
$$= 100\left(\frac{20}{11}\right)^{5}$$
$$\approx 1986.95$$

The production is approximately 1987 cameras.

(b) $P(1, 32)$

$$= 100\left[\frac{3}{5}(1)^{-2/5} + \frac{2}{5}(32)^{-2/5}\right]^{-5}$$

$$= 100\left[\frac{3}{5}(1) + \frac{2}{5}\left(\frac{1}{4}\right)\right]^{-5}$$

$$= 100\left(\frac{7}{10}\right)^{-5}$$

$$= 100\left(\frac{10}{7}\right)^{5}$$

$$\approx 595$$

The production is approximately 595 cameras.

(c) 32 work hours means that $x = 32$. 243 units of capital means that $y = 243$.

$P(32, 243)$

$$= 100\left[\frac{3}{5}(32)^{-2/5} + \frac{2}{5}(243)^{-2/5}\right]^{-5}$$

$$= 100\left[\frac{3}{5}\left(\frac{1}{4}\right) + \frac{2}{5}\left(\frac{1}{9}\right)\right]^{-5}$$

$$= 100\left(\frac{7}{36}\right)^{-5}$$

$$= 100\left(\frac{36}{7}\right)^{5}$$

$$\approx 359{,}767.81$$

The production is approximately 359,768 cameras.

33. $M = f(40, 0.06, 0.28)$

$$= \frac{(1 + 0.06)^{40}(1 - 0.28) + 0.28}{[1 + (1 - 0.28)(0.06)]^{40}}$$

$$= \frac{(1.06)^{40}(0.72) + 0.28}{[1 + (0.72)(0.06)]^{40}}$$

$$\approx 1.416$$

The multiplier is 1.416. Since $M > 1$, the IRA account grows faster.

35. $z = x^{0.6}y^{0.4}$ where $z = 500$

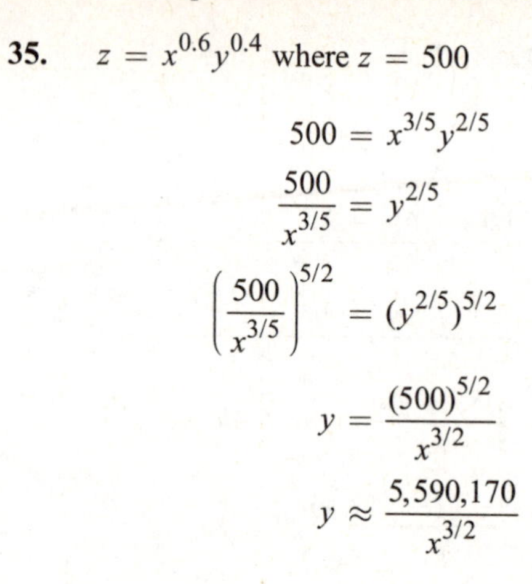

$$500 = x^{3/5}y^{2/5}$$

$$\frac{500}{x^{3/5}} = y^{2/5}$$

$$\left(\frac{500}{x^{3/5}}\right)^{5/2} = (y^{2/5})^{5/2}$$

$$y = \frac{(500)^{5/2}}{x^{3/2}}$$

$$y \approx \frac{5{,}590{,}170}{x^{3/2}}$$

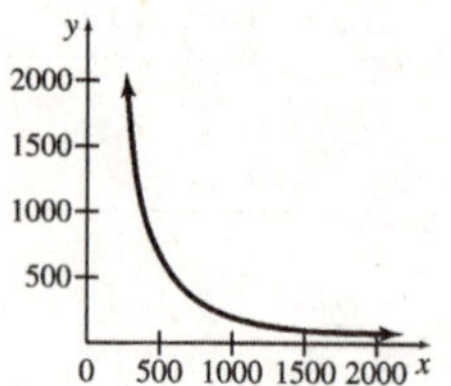

37. The cost function, C, is the sum of the products of the unit costs times the quantities x, y, and z. Therefore,

$$C(x, y, z) = 250x + 150y + 75z.$$

39. $A = 0.024265h^{0.3964}m^{0.5378}$

(a) $A = 0.024265(178)^{0.3964}(72)^{0.5378}$

$\approx 1.89 \text{ m}^2$

(b) $A = 0.024265(140)^{0.3964}(65)^{0.5378}$

$\approx 1.62 \text{ m}^2$

(c) $A = 0.024265(160)^{0.3964}(70)^{0.5378}$

$\approx 1.78 \text{ m}^2$

(d) Answers will vary.

41.

$$P(W, R, A) = 48 - 2.43W - 1.81R - 1.22A$$

(a) $P(5, 15, 0)$

$$= 48 - 2.43(5) - 1.81(15) - 1.22(0)$$

$$= 8.7$$

8.7% of fish will be intolerant to pollution.

(b) The maximum percentage will occur when the variable factors are a minimum, or when $W = 0, R = 0,$ and $A = 0.$

$$P(0, 0, 0) = 48 - 2.43(0) - 1.81(0) - 1.22(0) = 48$$

48% of fish will be intolerant to pollution.

(c) Any combination of values of $W, R,$ and A that result in $P = 0$ is a scenario that will drive the percentage of fish intolerant to pollution to zero.

If $R = 0$ and $A = 0$:

$$P(W, 0, 0) = 48 - 2.43W - 1.81(0) - 1.22(0) = 48 - 2.43W.$$

$$48 - 2.43W = 0$$
$$W = \frac{48}{2.43} \approx 19.75$$

So $W = 19.75,\ R = 0, A = 0$ is one scenario.

If $W = 10$ and $R = 10$:

$$P(10, 10, A) = 48 - 2.43(10) - 1.81(10) - 1.22A = 5.6 - 1.22A$$

$$5.6 - 1.22A = 0$$
$$A = \frac{5.6}{1.22} \approx 4.59$$

So $W = 10,\ R = 10,\ A = 4.59$ is another scenario.

(d) Since the coefficient of W is greater than the coefficients of R and A, a change in W will affect the value of P more than an equal change in R or A. Thus, the percentage of wetland (W) has the greatest influence on P.

43. $A(L,T,U,C) = 53.02 + 0.383L + 0.0015T + 0.0028U - 0.0003C$

(a) $A(266,\ 107{,}484,\ 31{,}697,\ 24{,}870)$

$$= 53.02 + 0.383(266) + 0.0015(107{,}484) + 0.0028(31{,}697) - 0.0003(24{,}870) \approx 397$$

The estimated number of accidents is 397.

45. $\ln(T) = 5.49 - 3.00\ln(F) + 0.18\ln(C)$

(a)
$$e^{\ln(T)} = e^{5.49-3.00\ln(F)+0.18\ln(C)}$$
$$T = e^{5.49}e^{-3.00\ln(F)}e^{0.18\ln(C)} = \frac{e^{5.49}e^{\ln(C^{0.18})}}{e^{\ln(F^3)}}$$
$$T \approx \frac{242.257C^{0.18}}{F^3}$$

(b) Replace F with 2 and C with 40 in the preceding formula.

$$T \approx \frac{242.257(40)^{0.18}}{(2)^3} \approx 58.82$$

T is about 58.8%. In other words, a tethered sow spends nearly 59% of the time doing repetitive behavior when she is fed 2 kg of food per day and neighboring sows spend 40% of the time doing repetitive behavior.

47. Let the area be given by $g(L,W,H)$.

Then,

$$g(L,W,H) = 2LW + 2WH + 2LH \text{ ft}^2.$$

17.2 Partial Derivatives

Your Turn 1

$$f(x, y) = 2x^2y^3 + 6x^5y^4$$
$$f_x(x, y) = 4xy^3 + 30x^4y^4$$
$$f_y(x, y) = 6x^2y^2 + 24x^5y^3$$

Your Turn 2

$$f(x, y) = e^{3x^2y}$$
$$f_x(x, y) = e^{3x^2y} \cdot \frac{\partial}{\partial x}(3x^2y) = 6xye^{3x^2y}$$
$$f_y(x, y) = e^{3x^2y} \cdot \frac{\partial}{\partial y}(3x^2y) = 3x^2e^{3x^2y}$$

Your Turn 3

$f(x, y) = xye^{x^2+y^3}$

Use the product rule.

$$f_x(x, y) = xy \cdot \frac{\partial}{\partial x}\left(e^{x^2+y^3}\right) + \left(e^{x^2+y^3}\right)\frac{\partial}{\partial x}(xy)$$
$$= xy\left(2xe^{x^2+y^3}\right) + \left(e^{x^2+y^3}\right)y$$
$$= (2x^2y + y)e^{x^2+y^3}$$
$$f_x(2, 1) = [2(2)^2(1) + 1]e^{2^2+1^3} = 9e^5$$
$$f_y(x, y) = xy \cdot \frac{\partial}{\partial y}\left(e^{x^2+y^3}\right) + \left(e^{x^2+y^3}\right)\frac{\partial}{\partial y}(xy)$$
$$= xy\left(3y^2e^{x^2+y^3}\right) + \left(e^{x^2+y^3}\right)x$$
$$= (3xy^3 + x)e^{x^2+y^3}$$
$$f_y(2, 1) = [3(2)(1)^3 + 2]e^{2^2+1^3} = 8e^5$$

Your Turn 4

$$f(x, y) = x^2e^{7y} + x^4y^5$$
$$f_x(x, y) = 2xe^{7y} + 4x^3y^5$$
$$f_y(x, y) = 7x^2e^{7y} + 5x^4y^4$$

$$f_{xx}(x, y) = \frac{\partial}{\partial x}f_x(x, y)$$
$$= \frac{\partial}{\partial x}(2xe^{7y} + 4x^3y^5)$$
$$= 2e^{7y} + 12x^2y^5$$
$$f_{yy}(x, y) = \frac{\partial}{\partial y}f_y(x, y)$$
$$= \frac{\partial}{\partial y}(7x^2e^{7y} + 5x^4y^4)$$
$$= 49x^2e^{7y} + 20x^4y^3$$

$$f_{xy}(x, y) = \frac{\partial}{\partial y}f_x(x, y)$$
$$= \frac{\partial}{\partial y}(2xe^{7y} + 4x^3y^5)$$
$$= 14xe^{7y} + 20x^3y^4$$
$$f_{yx}(x, y) = \frac{\partial}{\partial x}f_y(x, y)$$
$$= \frac{\partial}{\partial x}(7x^2e^{7y} + 5x^4y^4)$$
$$= 14xe^{7y} + 20x^3y^4$$

Note that, as in Example 6, $f_{xy}(x, y) = f_{yx}(x, y)$.

17.2 Exercises

1. $z = f(x, y) = 6x^2 - 4xy + 9y^2$

(a) $\frac{\partial z}{\partial x} = 12x - 4y$

(b) $\frac{\partial z}{\partial y} = -4x + 18y$

(c) $\frac{\partial f}{\partial x}(2, 3) = 12(2) - 4(3) = 12$

(d) $f_y(1, -2) = -4(1) + 18(-2)$
$= -40$

3.
$$f(x, y) = -4xy + 6y^3 + 5$$
$$f_x(x, y) = -4y$$
$$f_y(x, y) = -4x + 18y^2$$
$$f_x(2, -1) = -4(-1) = 4$$
$$f_y(-4, 3) = -4(-4) + 18(3)^2$$
$$= 16 + 18(9)$$
$$= 178$$

5.
$$f(x, y) = 5x^2y^3$$
$$f_x(x, y) = 10xy^3$$
$$f_y(x, y) = 15x^2y^2$$
$$f_x(2, -1) = 10(2)(-1)^3 = -20$$
$$f_y(-4, 3) = 15(-4)^2(3)^2 = 2160$$

7.
$$f(x, y) = e^{x+y}$$
$$f_x(x, y) = e^{x+y}$$
$$f_y(x, y) = e^{x+y}$$
$$f_x(2, -1) = e^{2-1}$$
$$= e^1 = e$$
$$f_y(-4, 3) = e^{-4+3}$$
$$= e^{-1}$$
$$= \frac{1}{e}$$

9.
$$f(x, y) = -6e^{4x-3y}$$
$$f_x(x, y) = -24e^{4x-3y}$$
$$f_y(x, y) = 18e^{4x-3y}$$
$$f_x(2, -1) = -24e^{4(2)-3(-1)} = -24e^{11}$$
$$f_y(-4, 3) = 18e^{4(-4)-3(3)} = 18e^{-25}$$

11. $f(x, y) = \dfrac{x^2 + y^3}{x^3 - y^2}$

$$f_x(x, y) = \frac{2x(x^3 - y^2) - 3x^2(x^2 + y^3)}{(x^3 - y^2)^2}$$
$$= \frac{2x^4 - 2xy^2 - 3x^4 - 3x^2y^3}{(x^3 - y^2)^2}$$
$$= \frac{-x^4 - 2xy^2 - 3x^2y^3}{(x^3 - y^2)^2}$$
$$f_y(x, y) = \frac{3y^2(x^3 - y^2) - (-2y)(x^2 + y^3)}{(x^3 - y^2)^2}$$
$$= \frac{3x^3y^2 - 3y^4 + 2x^2y + 2y^4}{(x^3 - y^2)^2}$$
$$= \frac{3x^3y^2 - y^4 + 2x^2y}{(x^3 - y^2)^2}$$
$$f_x(2, -1) = \frac{-2^4 - 2(2)(-1)^2 - 3(2^2)(-1)^3}{[2^3 - (-1)^2]^2}$$
$$= -\frac{8}{49}$$
$$f_y(-4, 3) = \frac{3(-4)^3(3)^2 - 3^4 + 2(-4)^2(3)}{[(-4)^3 - 3^2]^2}$$
$$= -\frac{1713}{5329}$$

13. $f(x, y) = \ln|1 + 5x^3y^2|$

$$f_x(x, y) = \frac{1}{1 + 5x^3y^2} \cdot 15x^2y^2 = \frac{15x^2y^2}{1 + 5x^3y^2}$$
$$f_y(x, y) = \frac{1}{1 + 5x^3y^2} \cdot 10x^3y = \frac{10x^3y}{1 + 5x^3y^2}$$
$$f_x(2, -1) = \frac{15(2)^2(-1)^2}{1 + 5(2)^3(-1)^2} = \frac{60}{41}$$
$$f_y(-4, 3) = \frac{10(-4)^3(3)}{1 + 5(-4)^3(3)^2} = \frac{1920}{2879}$$

15. $f(x, y) = xe^{x^2y}$

$$f_x(x, y) = e^{x^2y} \cdot 1 + x(2xy)(e^{x^2y})$$
$$= e^{x^2y}(1 + 2x^2y)$$
$$f_y(x, y) = x^3e^{x^2y}$$
$$f_x(2, -1) = e^{-4}(1 - 8) = -7e^{-4}$$
$$f_y(-4, 3) = -64e^{48}$$

17. $f(x, y) = \sqrt{x^4 + 3xy + y^4 + 10}$

$$f_x(x, y) = \frac{4x^3 + 3y}{2\sqrt{x^4 + 3xy + y^4 + 10}}$$
$$f_y(x, y) = \frac{3x + 4y^3}{2\sqrt{x^4 + 3xy + y^4 + 10}}$$
$$f_x(2, -1) = \frac{4(2)^3 + 3(-1)}{2\sqrt{2^4 + 3(2)(-1) + (-1)^4 + 10}}$$
$$= \frac{29}{2\sqrt{21}}$$
$$f_y(-4, 3) = \frac{3(-4) + 4(3)^3}{2\sqrt{(-4)^4 + 3(-4)(3) + 3^4 + 10}}$$
$$= \frac{48}{\sqrt{311}}$$

19.

$$f(x, y) = \frac{3x^2y}{e^{xy} + 2}$$
$$f_x(x, y) = \frac{6xy(e^{xy} + 2) - ye^{xy}(3x^2y)}{(e^{xy} + 2)^2}$$
$$= \frac{6xy(e^{xy} + 2) - 3x^2y^2e^{xy}}{(e^{xy} + 2)^2}$$
$$f_y(x, y) = \frac{3x^2(e^{xy} + 2) - xe^{xy}(3x^2y)}{(e^{xy} + 2)^2}$$
$$= \frac{3x^2(e^{xy} + 2) - 3x^3ye^{xy}}{(e^{xy} + 2)^2}$$
$$f_x(2, -1) = \frac{6(2)(-1)(e^{2(-1)} + 2) - 3(2)^2(-1)^2e^{2(-1)}}{(e^{2(-1)} + 2)^2}$$
$$= \frac{-12e^{-2} - 24 - 12e^{-2}}{(e^{-2} + 2)^2}$$
$$= \frac{-24(e^{-2} + 1)}{(e^{-2} + 2)^2}$$
$$f_y(-4, 3) = \frac{3(-4)^2\left(e^{(-4)(3)} + 2\right) - 3(-4)^3(3)e^{(-4)(3)}}{(e^{(-4)(3)} + 2)^2}$$
$$= \frac{48e^{-12} + 96 + 576e^{-12}}{(e^{-12} + 2)^2}$$
$$= \frac{624e^{-12} + 96}{(e^{-12} + 2)^2}$$

21.
$$f(x, y) = 4x^2y^2 - 16x^2 + 4y$$
$$f_x(x, y) = 8xy^2 - 32x$$
$$f_y(x, y) = 8x^2y + 4$$
$$f_{xx}(x, y) = 8y^2 - 32$$
$$f_{yy}(x, y) = 8x^2$$
$$f_{xy}(x, y) = f_{yx}(x, y) = 16xy$$

23.
$$R(x, y) = 4x^2 - 5xy^3 + 12y^2x^2$$
$$R_x(x, y) = 8x - 5y^3 + 24y^2x$$
$$R_y(x, y) = -15xy^2 + 24yx^2$$
$$R_{xx}(x, y) = 8 + 24y^2$$
$$R_{yy}(x, y) = -30xy + 24x^2$$
$$R_{xy}(x, y) = -15y^2 + 48xy$$
$$= R_{yx}(x, y)$$

25.
$$r(x, y) = \frac{6y}{x + y}$$
$$r_x(x, y) = \frac{(x + y)(0) - 6y(1)}{(x + y)^2}$$
$$= -6y(x + y)^{-2}$$
$$r_y(x, y) = \frac{(x + y)(6) - 6y(1)}{(x + y)^2}$$
$$= 6x(x + y)^{-2}$$
$$r_{xx}(x, y) = -6y(-2)(x + y)^{-3}(-1)$$
$$= \frac{12y}{(x + y)^3}$$
$$r_{yy}(x, y) = 6x(-2)(x + y)^{-3}(1)$$
$$= -\frac{12x}{(x + y)^3}$$
$$r_{xy}(x, y) = r_{yx}(x, y)$$
$$= -6y(-2)(x + y)^{-3}(1) + (x + y)^{-2}(-6)$$
$$= \frac{12y - 6(x + y)}{(x + y)^3}$$
$$= \frac{6y - 6x}{(x + y)^3}$$

27.
$$z = 9ye^x$$
$$z_x = 9ye^x$$
$$z_y = 9e^x$$
$$z_{xx} = 9ye^x$$
$$z_{yy} = 0$$
$$z_{xy} = z_{yx} = 9e^x$$

29.
$$r = \ln|x + y|$$
$$r_x = \frac{1}{x + y}$$
$$r_y = \frac{1}{x + y}$$
$$r_{xx} = \frac{-1}{(x + y)^2}$$
$$r_{yy} = \frac{-1}{(x + y)^2}$$
$$r_{xy} = r_{yx} = \frac{-1}{(x + y)^2}$$

31.
$$z = x\ln|xy|$$
$$z_x = \ln|xy| + 1$$
$$z_y = \frac{x}{y}$$
$$z_{xx} = \frac{1}{x}$$
$$z_{yy} = -xy^{-2} = \frac{-x}{y^2}$$
$$z_{xy} = z_{yx} = \frac{1}{y}$$

33. $f(x, y) = 6x^2 + 6y^2 + 6xy + 36x - 5$

First, $f_x = 12x + 6y + 36$ and
$f_y = 12y + 6x$.

We must solve the system

$$12x + 6y + 36 = 0$$
$$12y + 6x = 0.$$

Multiply both sides of the first equation by -2 and add.

$$\begin{array}{r} -24x - 12y - 72 = 0 \\ 6x + 12y \qquad = 0 \\ \hline -18x \qquad - 72 = 0 \\ x = -4 \end{array}$$

Substitute into either equation to get $y = 2$.
The solution is $x = -4$, $y = 2$.

35. $f(x,y) = 9xy - x^3 - y^3 - 6$

First, $f_x = 9y - 3x^2$ and $f_y = 9x - 3y^2$.

We must solve the system

$$9y - 3x^2 = 0$$
$$9x - 3y^2 = 0.$$

From the first equation, $y = \frac{1}{3}x^2$.

Substitute into the second equation to get

$$9x - 3\left(\frac{1}{3}x^2\right)^2 = 0$$
$$9x - 3\left(\frac{1}{9}x^4\right) = 0$$
$$9x - \frac{1}{3}x^4 = 0.$$

Multiply by 3 to get

$$27x - x^4 = 0.$$

Now factor.

$$x(27 - x^3) = 0$$

Set each factor equal to 0.

$$x = 0 \quad \text{or} \quad 27 - x^3 = 0$$
$$x = 3$$

Substitute into $y = \frac{x^2}{3}$.

$$y = 0 \text{ or } y = 3$$

The solutions are $x = 0, y = 0$ and $x = 3, y = 3$.

37. $f(x,y,z) = x^4 + 2yz^2 + z^4$

$f_x(x,y,z) = 4x^3$

$f_y(x,y,z) = 2z^2$

$f_z(x,y,z) = 4yz + 4z^3$

$f_{yz}(x,y,z) = 4z$

39. $f(x, y, z) = \dfrac{6x - 5y}{4z + 5}$

$f_x(x, y, z) = \dfrac{6}{4z + 5}$

$f_y(x, y, z) = \dfrac{-5}{4z + 5}$

$f_z(x,y,z) = \dfrac{-4(6x - 5y)}{(4z + 5)^2}$

$f_{yz}(x, y, z) = \dfrac{20}{(4z + 5)^2}$

41. $f(x, y, z) = \ln|x^2 - 5xz^2 + y^4|$

$f_x(x, y, z) = \dfrac{2x - 5z^2}{x^2 - 5xz^2 + y^4}$

$f_y(x, y, z) = \dfrac{4y^3}{x^2 - 5xz^2 + y^4}$

$f_z(x,y,z) = \dfrac{-10xz}{x^2 - 5xz^2 + y^4}$

$f_{yz}(x,y,z) = \dfrac{4y^3(10zx)}{(x^2 - 5xz^2 + y^4)^2}$

$= \dfrac{40xy^3z}{(x^2 - 5xz^2 + y^4)^2}$

43. $f(x, y) = \left(x + \dfrac{y}{2}\right)^{x+y/2}$

(a) $f_x(1, 2) = \lim\limits_{h\to 0} \dfrac{f(1 + h, 2) - f(1, 2)}{h}$

We will use a small value for h. Let $h = 0.00001$.

$$f_x(1, 2) \approx \frac{f(1.00001, 2) - f(1,2)}{0.00001}$$
$$\approx \frac{\left(1.00001 + \frac{2}{2}\right)^{1.00001+2/2} - \left(1 + \frac{2}{2}\right)^{1+2/2}}{0.00001}$$
$$\approx \frac{2.00001^{2.00001} - 2^2}{0.00001}$$
$$\approx 6.773$$

(b) $f_y(1, 2) = \lim_{h \to 0} \dfrac{f(1, 2+h) - f(1, 2)}{h}$

Again, let $h = 0.00001$.

$$f_y(1, 2) \approx \frac{f(1, 200001) - f(1, 2)}{0.00001}$$
$$\approx \frac{\left(1 + \frac{2.00001}{2}\right)^{1+2.00001/2} - \left(1 + \frac{2}{2}\right)^{1+2/2}}{0.00001}$$
$$\approx \frac{2.000005^{2.000005} - 2^2}{0.00001}$$
$$\approx 3.386$$

45. $M(x, y) = 45x^2 + 40y^2 - 20xy + 50$

(a) $M_y(x, y) = 80y - 20x$

$M_y(4, 2) = 80(2) - 20(4) = 80$

(b) $M_x(x, y) = 90x - 20y$

$M_x(3, 6) = 90(3) - 20(6) = 150$

(c) $\dfrac{\partial M}{\partial x}(2, 5) = 90(2) - 20(5) = 80$

(d) $\dfrac{\partial M}{\partial y}(6, 7) = 80(7) - 20(6) = 440$

47. $f(p, i) = 99p - 0.5pi - 0.0025p^2$

(a) $f(19{,}400, 8)$

$= 99(19{,}400) - 0.5(19{,}400)(8)$

$\quad - 0.0025(19{,}400)^2$

$= \$902{,}100$

The weekly sales are $902,100.

(b) $f_p(p, i) = 99 - 0.5i - 0.005p$, which represents the rate of change in weekly sales revenue per unit change in price when the interest rate remains constant.

$f_i(p, i) = -0.5p$, which represents the rate of change in weekly sales revenue per unit change in interest rate when the list price remains constant.

(c) $p = 19{,}400$ remains constant and i changes by 1 unit from 8 to 9.

$$f_i(p, i) = f_i(19{,}400, 8) = -0.5(19{,}400) = -9700$$

Therefore, sales revenue declines by $9700.

49. $f(x, y) = \left(\frac{1}{4}x^{-1/4} + \frac{3}{4}y^{-1/4}\right)^{-4}$

(a)
$$f(16,81) = \left[\frac{1}{4}(16)^{-1/4} + \frac{3}{4}(81)^{-1/4}\right]^{-4}$$
$$= \left(\frac{1}{4}\cdot\frac{1}{2} + \frac{3}{4}\cdot\frac{1}{3}\right)^{-4}$$
$$= \left(\frac{3}{8}\right)^{-4} \approx 50.56790123$$

50.57 hundred units are produced.

(b)

$$f_x(x, y) = -4\left(\frac{1}{4}x^{-1/4} + \frac{3}{4}y^{-1/4}\right)^{-5}\left[\frac{1}{4}\left(-\frac{1}{4}\right)x^{-5/4}\right]$$
$$= \frac{1}{4}x^{-5/4}\left(\frac{1}{4}x^{-1/4} + \frac{3}{4}y^{-1/4}\right)^{-5}$$

$$f_x(16,81) = \frac{1}{4}(16)^{-5/4}\left[\frac{1}{4}(16)^{-1/4} + \frac{3}{4}(81)^{-1/4}\right]^{-5}$$
$$= \frac{1}{4}\left(\frac{1}{32}\right)\left(\frac{3}{8}\right)^{-5} = \frac{256}{243}$$
$$\approx 1.053497942$$

$f_x(16,81) = 1.053$ hundred units and is the rate at which production is changing when labor changes by one unit (from 16 to 17) and capital remains constant.

$$f_y(x, y) = -4\left(\frac{1}{4}x^{-1/4} + \frac{3}{4}y^{-1/4}\right)^{-5}\left[\frac{3}{4}\left(-\frac{1}{4}\right)y^{-5/4}\right]$$
$$= \frac{3}{4}y^{-5/4}\left(\frac{1}{4}x^{-1/4} + \frac{3}{4}y^{-1/4}\right)^{-5}$$

$$f_y(16,81) = \frac{3}{4}(81)^{-5/4}\left[\frac{1}{4}(16)^{-1/4} + \frac{3}{4}(81)^{-1/4}\right]^{-5}$$
$$= \frac{3}{4}\left(\frac{1}{243}\right)\left(\frac{3}{8}\right)^{-5} = \frac{8192}{19{,}683}$$
$$\approx 0.4161967180$$

$f_y(16,81) = 0.4162$ hundred units and is the rate at which production is changing when capital changes by one unit (from 81 to 82) and labor remains constant.

(c) Using the value of $f_x(16, 81)$ found in (b), production would increase by approximately 105 units.

51. $z = x^{0.4}y^{0.6}$

The marginal productivity of labor is

$$\frac{\partial z}{\partial x} = 0.4x^{-0.6}y^{0.6} + x^{0.4} \cdot 0$$
$$= 0.4x^{-0.6}y^{0.6}.$$

The marginal productivity of capital is

$$\frac{\partial z}{\partial y} = x^{0.4}(0.6y^{-0.4}) + y^{0.6} \cdot 0$$
$$= 0.6x^{0.4}y^{-0.4}.$$

53. $f(w, v) = 25.92w^{0.68} + \frac{3.62w^{0.75}}{v}$

(a) $f(300,10) = 25.92(300)^{0.68} + \frac{3.62(300)^{0.75}}{10}$
$$\approx 1279.46$$

The value is about 1279 kcal/hr.

(b) $f_w(w,v) = 25.92(0.68)w^{-.32}$
$$+ \frac{3.62(0.75)w^{-0.25}}{v}$$
$$= \frac{17.6256}{w^{0.32}} + \frac{2.715}{w^{0.25}v}$$

$$f_w(300,10) = \frac{17.6256}{(300)^{0.32}} + \frac{2.715}{(300)^{0.25}(10)}$$
$$\approx 2.906$$

The value is about 2.906 kcal/hr/g. This means the instantaneous rate of change of energy usage for a 300 kg animal traveling at 10 kilometers per hour to walk or run 1 kilometer is about 2.906 kcal/hr/g.

55. $A = 0.024265h^{0.3964}m^{0.5378}$

(a) $A_m = (0.024265)(0.5378)h^{0.3964}m^{(0.5378-1)}$
$$= 0.013050h^{0.3964}m^{-0.4622}$$

When the mass m increases from 72 to 73 while the height h remains at 180 cm, the approximate change in body surface area is

$$0.013050(180)^{0.3964}(72)^{-0.4622} \approx 0.0142$$

or about 0.0142 m^2.

(b) $A_h = (0.024265)(0.3964)h^{(0.3964-1)}m^{0.5378}$
$$= 0.0096186h^{-0.6036}m^{0.5378}$$

When the height h increases from 160 to 161 while the mass m remains at 70, the approximate change in body surface area is

$$0.0096186(160)^{-0.6036}(70)^{0.5378}$$
$$\approx 0.00442$$

or about 0.00442 m^2.

57. $f(n,c) = \frac{1}{8}n^2 - \frac{1}{5}c + \frac{1937}{8}$

(a) $f(4,1200) = \frac{1}{8}(4) - \frac{1}{5}(1200) + \frac{1937}{8}$
$$= 2 - 240 + \frac{1937}{8} = 4.125$$

The client could expect to lose 4.125 lb.

(b) $\frac{\partial f}{\partial n} = \frac{1}{8}(2n) - \frac{1}{5}(0) + 0 = \frac{1}{4}n,$

which represents the rate of change of weight loss per unit change in number of workouts.

(c) $f_n(3,1100) = \frac{1}{4}(3) = \frac{3}{4}$ lb

represents an additional weight loss by adding the fourth workout.

59. $R(x,t) = x^2(a - x)t^2e^{-t} = (ax^2 - x^3)t^2e^{-t}$

(a) $\frac{\partial R}{\partial x} = (2ax - 3x^2)t^2e^{-t}$

(b) $\frac{\partial R}{\partial t} = x^2(a - x) \cdot [t^2 \cdot (-e^{-t}) + e^{-t} \cdot 2t]$
$$= x^2(a - x)(-t^2 + 2t)e^{-t}$$

(c) $\frac{\partial^2 R}{\partial x^2} = (2a - 6x)t^2e^{-t}$

(d) $\frac{\partial^2 R}{\partial x \partial t} = (2ax - 3x^2)(-t^2 + 2t)e^{-t}$

(e) $\frac{\partial R}{\partial x}$ gives the rate of change of the reaction per unit of change in the amount of drug administered.

$\frac{\partial R}{\partial t}$ gives the rate of change of the reaction for a 1-hour change in the time after the drug is administered.

61. $W(V,T)$

$$= 91.4 - \frac{(10.45 + 6.69\sqrt{V} - 0.447V)(91.4 - T)}{22}$$

(a)

$W(20,10)$

$$= 91.4 - \frac{(10.45 + 6.69\sqrt{20} - 0.447(20))(91.4 - 10)}{22}$$

$$\approx -24.9$$

The wind chill is $-24.9°F$ when the wind speed is 20 mph and the temperature is 10°F.

(b) Solve

$$-25 = 91.4 - \frac{(10.45 + 6.69\sqrt{V} - 0.447V)(91.4 - 5)}{22}$$

for V.

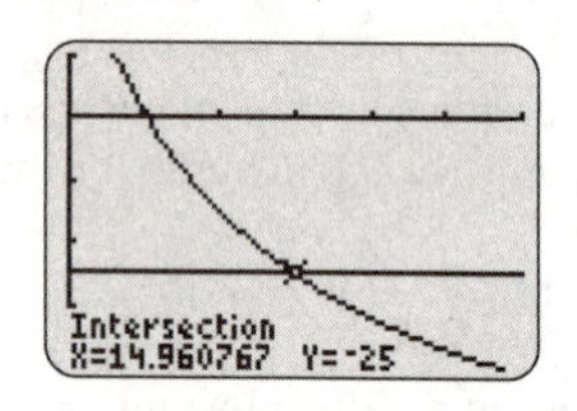

The wind speed is approximately 15 mph.

(c) $W_V = -\frac{1}{22}\left(\frac{6.69}{2\sqrt{V}} - 0.447\right)(91.4 - T)$

$$W_T = -\frac{1}{22}(10.45 + 6.69\sqrt{V} - 0.447V)(-1)$$

$$= \frac{1}{22}(10.45 + 6.69\sqrt{V} - 0.447V)$$

$$W_V(20, 10) = \frac{1}{22}\left(\frac{6.69}{2\sqrt{20}} - 0.447\right)(91.4 - 10)$$

$$\approx -1.114$$

When the temperature is held fixed at 10° F, the wind chill decreases approximately 1.1 degrees when the wind velocity increases by 1 mph.

$W_T(20,10) =$

$$\frac{1}{22}[10.45 + 6.69\sqrt{20} - 0.447(20)] \approx 1.429$$

When the wind velocity is held fixed at 20 mph, the wind chill increases approximately 1.429° F when the temperature increases from 10° F to 11° F.

(d) A sample table is

$T\backslash V$	5	10	15	20
30	27	16	9	4
20	16	3	−5	−11
10	6	−9	−18	−25
0	−5	−21	−32	−39

63. The rate of change in lung capacity with respect to age can be found by comparing the change in two lung capacity measurements to the difference in the respective ages when the height is held constant. So for a woman 58 inches tall, at age 20 the measured lung capacity is 1900 ml, and at age 25 the measured lung capacity is 1850 ml. So the rate of change in lung capacity with respect to age is

$$\frac{1900 - 1850}{20 - 25} = \frac{50}{-5}$$

$$= -10 \text{ ml per year.}$$

The rate of change in lung capacity with respect to height can be found by comparing the change in two lung capacity measurements to the difference in the respective heights when the age is held constant. So for a 20-year old woman the measured lung capacity for a woman 58 inches tall is 1900 ml and the measured lung capacity for a woman 60 inches tall is 2100 ml. So the rate of change in lung capacity with respect to height is

$$\frac{1900 - 2100}{58 - 60} = \frac{-200}{-2}$$

$$= 100 \text{ ml per in.}$$

The two rates of change remain constant throughout the table.

65. $F = \frac{mgR^2}{r^2} = mgR^2r^{-2}$

(a) $F_m = \frac{gR^2}{r^2}$ is the approximate rate of change in gravitational force per unit change in mass while distance is held constant.

$F_r = \frac{-2mgR^2}{r^3}$ is the approximate rate of change in gravitational force per unit change in distance while mass is held constant.

(b) $F_m = \frac{gR^2}{r^2}$, where all quantities are positive. Therefore, $F_m > 0$.

$F_r = \frac{-2mgR^2}{r^3}$, where m, g, R^2, and r^3 are positive.

Therefore, $F_r < 0$.

These results are reasonable since gravitational force increases when mass increases (m is in the numerator) and gravitational force decreases when distance increases (r is in the denominator).

67. $T = (s,w) = 105 + 265\log_2\left(\frac{2s}{w}\right)$

(a) $T(3,0.5) = 105 + 265\log_2\left[\frac{2(3)}{0.5}\right]$

$= 105 + 265\log_2 12$

≈ 1055

(b) $T(s,w) = 105 + 265\frac{\ln\left(\frac{2s}{w}\right)}{\ln 2}$

$= 105 + \frac{265}{\ln 2}[\ln(2s) - \ln(w)]$

$T_s(s,w) = \frac{265}{\ln 2}\left(\frac{1}{s}\right)$

$T_w(s,w) = -\frac{265}{\ln 2}\left(\frac{1}{w}\right)$

$T_s(3,0.5) = \frac{265}{3\ln 2} \approx 127.4$ msec/ft

If the distance the object is being moved increases from 3 feet to 4 feet, while keeping w fixed at 0.5 foot, the time to move the object increases by approximately 127.4 msec.

$T_w(3,0.5) = -\frac{265}{0.5\ln 2}$

≈ -764.5 msec/ft

It the width of the target area is increased by 1 foot, while keeping the distance fixed at 3 feet, the movement time decreases by approximately 764.5 msec.

17.3 Maxima and Minima

Your Turn 1

$f(x,y) = 4x^3 + 3xy + 4y^3$

$f_x(x,y) = 12x^2 + 3y,\ f_y(x,y) = 3x + 12y^2$

$$12x^2 + 3y = 0$$
$$3x + 12y^2 = 0$$

Solve this system by substitution. From the first equation,

$$y = -4x^2.$$

Substituting for y in the second equation gives

$$3x + 12\left(-4x^2\right)^2 = 0.$$

$$x + 64x^4 = 0$$
$$x(1 + 64x^3) = 0$$

$$x = 0 \text{ or } 1 + 64x^3 = 0$$
$$x = 0 \text{ or } x = -\frac{1}{4}$$

If $x = 0,\ y = -4(0)^2 = 0$.

If $x = -\frac{1}{4},\ y = -4\left(-\frac{1}{4}\right)^2 = -\frac{1}{4}$.

So the critical points are

$$(0,0) \text{ and } \left(-\frac{1}{4},-\frac{1}{4}\right).$$

Your Turn 2

Use the information from Your Turn 1:

$$f(x,y) = 4x^3 + 3xy + 4y^3$$
$$f_x(x,y) = 12x^2 + 3y$$
$$f_y(x,y) = 3x + 12y^2$$

The critical points are $(0, 0)$ and $(-1/4, -1/4)$.

Now compute the second partial derivatives.

$f_{xx}(x,y) = 24x,\ f_{yy}(x,y) = 24y,\ f_{xy}(x,y) = 3$

$$D(a,b) = f_{xx}(a,b)\cdot f_{yy}(a,b) - [f_{xy}(a,b)]^2$$
$$= (24a)(24b) - 9$$
$$= 576ab - 9$$

At the critical point $(0, 0)$,

$D(0,0) = 576(0)(0) - 9 = -9 < 0$, so this critical point is a saddle point.

At the critical point $(-1/4, -1/4)$,

$$D\left(-\frac{1}{4}, -\frac{1}{4}\right) = 576\left(-\frac{1}{4}\right)\left(-\frac{1}{4}\right) - 9$$
$$= 36 - 9$$
$$= 24$$

and

$$f_{xx}\left(-\frac{1}{4}, -\frac{1}{4}\right) = 24\left(-\frac{1}{4}\right) = -\frac{1}{6}.$$

Since $D > 0$ and $f_{xx} < 0$, there is a relative maximum at $(-1/4, -1/4)$.

17.3 Exercises

1. $f(x, y) = xy + y - 2x$

$f_x(x, y) = y - 2, f_y(x, y) = x + 1$

If $f_x(x, y) = 0, y = 2.$

If $f_y(x, y) = 0, x = -1.$

Therefore, $(-1, 2)$ is the critical point.

$$f_{xx}(x, y) = 0$$
$$f_{yy}(x, y) = 0$$
$$f_{xy}(x, y) = 1$$

For $(-1, 2)$,

$$D = 0 \cdot 0 - 1^2 = -1 < 0.$$

A saddle point is at $(-1, 2)$.

3. $f(x, y) = 3x^2 - 4xy + 2y^2 + 6x - 10$
$f_x(x, y) = 6x - 4y + 6$
$f_y(x, y) = -4x + 4y$

Solve the system $f_x(x, y) = 0, f_y(x, y) = 0.$

$$\begin{array}{r} 6x - 4y + 6 = 0 \\ \underline{-4x + 4y \quad\;\; = 0} \\ 2x \qquad + 6 = 0 \\ x = -3 \\ -4(-3) + 4y = 0 \\ y = -3 \end{array}$$

Therefore, $(-3, -3)$ is a critical point.

$$f_{xx}(x, y) = 6$$
$$f_{yy}(x, y) = 4$$
$$f_{xy}(x, y) = -4$$
$$D = 6 \cdot 4 - (-4)^2 = 8 > 0$$

Since $f_{xx}(x, y) = 6 > 0$, there is a relative minimum at $(-3, -3)$.

5. $f(x, y) = x^2 - xy + y^2 + 2x + 2y + 6$
$f_x(x, y) = 2x - y + 2,$
$f_y(x, y) = -x + 2y + 2$

Solve the system $f_x(x, y) = 0, f_y(x, y) = 0.$

$$\begin{array}{r} 2x - \;\; y + 2 = 0 \\ \underline{-x + 2y + 2 = 0} \\ 2x - \;\; y + 2 = 0 \\ \underline{-2x + 4y + 4 = 0} \\ 3y + 6 = 0 \\ y = -2 \\ -x + 2(-2) + 2 = 0 \\ x = -2 \end{array}$$

$(-2, -2)$ is the critical point.

$$f_{xx}(x, y) = 2$$
$$f_{yy}(x, y) = 2$$
$$f_{xy}(x, y) = -1$$

For $(-2, -2)$,

$$D = (2)(2) - (-1)^2 = 3 > 0.$$

Since $f_{xx}(x, y) > 0$, a relative minimum is at $(-2, -2)$.

7. $f(x, y) = x^2 + 3xy + 3y^2 - 6x + 3y$
$f_x(x, y) = 2x + 3y - 6,$
$f_y(x, y) = 3x + 6y + 3$

Solve the system $f_x(x, y) = 0, f_y(x, y) = 0.$

$$\begin{array}{r} 2x + 3y - 6 = 0 \\ 3x + 6y + 3 = 0 \\ -4x - 6y + 12 = 0 \\ \underline{3x + 6y + \;\; 3 = 0} \\ -x + 15 = 0 \\ x = 15 \end{array}$$

$$3(15) + 6y + 3 = 0$$
$$6y = -48$$
$$y = -8$$

$(15,-8)$ is the critical point.

$$f_{xx}(x,y) = 2$$
$$f_{yy}(x,y) = 6$$
$$f_{xy}(x,y) = 3$$

For $(15,-8)$,

$$D = 2 \cdot 6 - 9 = 3 > 0.$$

Since $f_{xx}(x,y) > 0$, a relative minimum is at $(15,-8)$.

9. $f(x,y) = 4xy - 10x^2 - 4y^2 + 8x + 8y + 9$

$$f_x(x,y) = 4y - 20x + 8$$
$$f_y(x,y) = 4x - 8y + 8$$

$$4y - 20x + 8 = 0$$
$$4x - 8y + 8 = 0$$

$$\begin{array}{r} 4y - 20x + 8 = 0 \\ -4y + 2x + 4 = 0 \\ \hline -18x + 12 = 0 \end{array}$$

$$x = \frac{2}{3}$$

$$4y - 20\left(\frac{2}{3}\right) + 8 = 0$$

The critical point is $\left(\frac{2}{3}, \frac{4}{3}\right)$.

$$f_{xx}(x,y) = -20$$
$$f_{yy}(x,y) = -8$$
$$f_{xy}(x,y) = 4$$

For $\left(\frac{2}{3}, \frac{4}{3}\right)$,

$$D = (-20)(-8) - 16 = 144 > 0.$$

Since $f_{xx}(x,y) < 0$, a relative maximum is at $\left(\frac{2}{3}, \frac{4}{3}\right)$.

11. $f(x,y) = x^2 + xy - 2x - 2y + 2$

$$f_x(x,y) = 2x + y - 2$$
$$f_y(x,y) = x - 2$$

$$2x + y - 2 = 0$$
$$x \qquad - 2 = 0$$
$$x = 2$$
$$2(2) + y - 2 = 0$$
$$y = -2$$

The critical point is $(2,-2)$.

$$f_{xx}(x,y) = 2$$
$$f_{yy}(x,y) = 0$$
$$f_{xy}(x,y) = 1$$

For $(2,-2)$,

$$D = 2 \cdot 0 - 1^2 = -1 < 0.$$

A saddle point is at $(2,-2)$.

13. $f(x,y) = 3x^2 + 2y^3 - 18xy + 42$

$$f_x(x,y) = 6x - 18y$$
$$f_y(x,y) = 6y^2 - 18x$$

If $f_x(x,y) = 0, 6x - 18y = 0$, or $x = 3y$. Substitute $3y$ for x in $f_y(x,y) = 0$ and solve for y.

$$6y^2 - 18(3y) = 0$$
$$6y(y - 9) = 0$$

$$y = 0 \quad \text{or} \quad y = 9$$

Then $\quad x = 0 \quad \text{or} \quad x = 27.$

Therefore, $(0, 0)$ and $(27, 9)$ are critical points.

$$f_{xx}(x,y) = 6$$
$$f_{yy}(x,y) = 12y$$
$$f_{xy}(x,y) = -18$$

For $(0, 0)$,

$$D = 6 \cdot 12(0) - (-18)^2 = -324 < 0.$$

There is a saddle point at $(0, 0)$.

For $(27, 9)$,

$$D = 6 \cdot 12(9) - (-18)^2 = 324 > 0.$$

Since $f_{xx}(x,y) = 6 > 0$, there is a relative minimum at $(27, 9)$.

15. $f(x, y) = x^2 + 4y^3 - 6xy - 1$

$f_x(x, y) = 2x - 6y,\ f_y(x, y) = 12y^2 - 6x$

Solve $f_x(x, y) = 0$ for x.

$$2x + 6 = 0$$
$$x = 3y$$

Substitute for x in $12y^2 - 6x = 0$.

$$12y^2 - 6(3y) = 0$$
$$6y(2y - 3) = 0$$
$$y = 0 \text{ or } y = \frac{3}{2}$$

Then $$x = 0 \text{ or } x = \frac{9}{2}.$$

The critical points are (0, 0) and $\left(\frac{9}{2}, \frac{3}{2}\right)$.

$$f_{xx}(x, y) = 2$$
$$f_{yy}(x, y) = 24y$$
$$f_{xy}(x, y) = -6$$

For (0, 0),

$$D = 2 \cdot 24(0) - (-6)^2$$
$$= -36 < 0.$$

A saddle point is at (0, 0).

For $\left(\frac{9}{2}, \frac{3}{2}\right)$,

$$D = 2 \cdot 24\left(\frac{3}{2}\right) - (-6)^2$$
$$= 36 > 0.$$

Since $f_{xx}(x, y) > 0$, a relative minimum is at $\left(\frac{9}{2}, \frac{3}{2}\right)$.

17. $f(x, y) = e^{x(y+1)}$

$f_x(x, y) = (y + 1)e^{x(y+1)}$

$f_y(x, y) = xe^{x(y+1)}$

If $$f_x(x, y) = 0$$
$$(y + 1)e^{x(y+1)} = 0$$
$$y + 1 = 0$$
$$y = -1.$$

If $$f_y(x, y) = 0$$
$$xe^{x(y+1)} = 0$$
$$x = 0.$$

Therefore, $(0, -1)$ is a critical point.

$$f_{xx}(x, y) = (y + 1)^2 e^{x(y+1)}$$
$$f_{yy}(x, y) = x^2 e^{x(y+1)}$$
$$f_{xy}(x, y) = (y + 1)e^{x(y+1)} \cdot x + e^{x(y+1)} \cdot 1$$
$$= (xy + x + 1)e^{x(y+1)}$$

For $(0, -1)$,

$$f_{xx}(0, -1) = (0)^2 e^0 = 0$$
$$f_{yy}(0, -1) = (0)^2 e^0 = 0$$
$$f_{xy}(0, -1) = (0 + 0 + 1)e^0 = 1$$
$$D = 0 \cdot 0 - 1^2 = -1 < 0$$

There is a saddle point at $(0, -1)$.

21. $z = -3xy + x^3 - y^3 + \frac{1}{8}$

$f_x(x, y) = -3y + 3x^2,\ f_y(x, y) = -3x - 3y^2$

Solve the system $f_x = 0,\ f_y = 0$.

$$-3y + 3x^2 = 0$$
$$-3x - 3y^2 = 0$$
$$-y + x^2 = 0$$
$$-x - y^2 = 0$$

Solve the first equation for y, substitute into the second, and solve for x.

$$y = x^2$$
$$-x - x^4 = 0$$
$$x(1 + x^3) = 0$$
$$x = 0 \quad \text{or} \quad x = -1$$

Then $$y = 0 \quad \text{or} \quad y = 1.$$

The critical points are (0, 0) and $(-1, 1)$.

$$f_{xx}(x, y) = 6x$$
$$f_{yy}(x, y) = -6y$$
$$f_{xy}(x, y) = -3$$

For (0, 0),

$$D = 0 \cdot 0 - (-3)^2 = -9 < 0.$$

A saddle point is at (0, 0).

For $(-1, 1)$,

$$D = -6(-6) - (-3)^2 = 27 > 0.$$

$$f_{xx}(x,y) = 6(-1) = -6 < 0.$$

$$f(-1,1) = -3(-1)(1) + (-1)^3 - 1^3 + \frac{1}{8} = \frac{9}{8}$$

A relative maximum of $\frac{9}{8}$ is at $(-1, 1)$.

The equation matches graph (a).

23. $z = y^4 - 2y^2 + x^2 - \frac{17}{16}$

$f_x(x,y) = 2x,\ f_y(x,y) = 4y^3 - 4y$

Solve the system $f_x = 0, f_y = 0$.

$$2x = 0 \quad (1)$$
$$4y^3 - 4y = 0 \quad (2)$$
$$4y(y^2 - 1) = 0$$
$$4y(y+1)(y-1) = 0$$

Equation (1) gives $x = 0$ and equation (2) gives $y = 0,\ y = -1,$ or $y = 1$.

The critical points are $(0, 0)$, $(0, -1)$ and $(0, 1)$.

$$f_{xx}(x,y) = 2,$$
$$f_{yy}(x,y) = 12y^2 - 4,$$
$$f_{xy}(x,y) = 0$$

For $(0, 0)$,

$$D = 2(12 \cdot 0^2 - 4) - 0 = -8 < 0.$$

A saddle point is at $(0, 0)$.

For $(0, -1)$,

$$D = 2[12(-1)^2 - 4] - 0 = 16 > 0.$$

$$f_{xx}(x,y) = 2 > 0$$

$$f(0,-1) = (-1)^4 - 2(-1)^2 + 0^2 - \frac{17}{16} = -2\tfrac{1}{16}$$

A relative minimum of $-2\frac{1}{16}$ is at $(0, -1)$.

For $(0, 1)$,

$$D = 2(12 \cdot 1^2 - 4) - 0 = 16 > 0$$

$$f_{xx}(x,y) = 2 > 0$$

$$f(0,1) = 1^4 - 2 \cdot 1^2 + 0^2 - \frac{17}{16} = -\frac{33}{16}$$

A relative minimum of $-\frac{33}{16}$ is at $(0, 1)$.

The equation matches graph (b).

25. $z = -x^4 + y^4 + 2x^2 - 2y^2 + \frac{1}{16}$

$f_x(x,y) = -4x^3 + 4x,\ f_y(x,y) = 4y^3 - 4y$

Solve $f_x(x,y) = 0,\ f_y(x,y) = 0.$

$$-4x^3 + 4x = 0 \quad (1)$$
$$4y^3 - 4y = 0 \quad (2)$$
$$-4x(x^2 - 1) = 0 \quad (1)$$
$$-4x(x+1)(x-1) = 0$$
$$4y(y^2 - 1) = 0 \quad (2)$$
$$4y(y+1)(y-1) = 0$$

Equation (1) gives $x = 0, -1,$ or 1.

Equation(2) gives $y = 0, -1,$ or 1.

Critical points are $(0, 0)$, $(0, -1)$, $(0, 1)$, $(-1, 0)$, $(-1, -1)$, $(-1, 1)$, $(1, 0)$, $(1, -1)$, $(1, 1)$.

$$f_{xx}(x,y) = -12x^2 + 4,$$
$$f_{yy}(x,y) = 12y^2 - 4$$
$$f_{xy}(x,y) = 0$$

For $(0, 0)$,

$$D = 4(-4) - 0 = -16 < 0.$$

For $(0, -1)$,

$$D = 4(8) - 0 = 32 > 0,$$

and $f_{xx}(x,y) = 4 > 0$.

$$f(0,-1) = -\frac{15}{16}$$

For $(0, 1)$,

$$D = 4(8) - 0 = 32 > 0,$$

and $f_{xx}(x,y) = 4 > 0$.

$$f(0,1) = -\frac{15}{16}$$

For $(-1, 0)$,

$$D = -8(-4) - 0 = 32 > 0,$$

and $f_{xx}(x,y) = -8 < 0$.

$$f(-1,0) = \frac{17}{16}$$

For $(-1,-1)$,

$$D = -8(8) - 0 = -64 < 0.$$

For $(-1, 1)$,

$$D = -8(8) - 0 = -64 < 0.$$

For $(1, 0)$,

$$D = -8(-4) = 32 > 0,$$

and $f_{xx}(x, y) = -8 < 0$.

$$f(1,0) = 1\tfrac{1}{16}$$

For $(1, -1)$,

$$D = -8(8) - 0 = -64 < 0.$$

For $(1, 1)$,

$$D = -8(8) - 0 = -64 < 0.$$

Saddle points are at $(0, 0)$, $(-1, -1)$, $(-1, 1)$, $(1, -1)$, and $(1, 1)$.

Relative maximum of $\frac{17}{16}$ is at $(-1, 0)$ and $(1, 0)$.

Relative minimum of $-\frac{15}{16}$ is at $(0, -1)$ and $(0, 1)$.

The equation matches graph (e).

27. $f(x, y) = 1 - x^4 - y^4$

$f_x(x, y) = -4x^3$, $f_y(x, y) = -4y^3$

The system

$f_x(x, y) = -4x^3 = 0$, $f_y(x, y) = -4y^3 = 0$

gives the critical point $(0, 0)$.

$$f_{xx}(x, y) = -12x^2$$
$$f_{yy}(x, y) = -12y^3$$
$$f_{xy}(x, y) = 0$$

For $(0, 0)$,

$$D = 0 \cdot 0 - 0^2 = 0.$$

Therefore, the test gives no information. Examine a graph of the function drawn by using level curves.

If $f(x, y) = 1$, then $x^4 + y^4 = 0$. The level curve is the point $(0, 0, 1)$.

If $f(x, y) = 0$, then $x^4 + y^4 = 1$. The level curve is the circle with center $(0, 0, 0)$ and radius 1.

If $f(x, y) = -15$, then $x^4 + y^4 = 16$. The level curve is the curve with center $(0,0,-15)$ and radius 2.

The xz-trace is

$$z = 1 - x^4.$$

This curve has a maximum at $(0, 0, 1)$ and opens downward.

The yz-trace is

$$z = 1 - y^4.$$

This curve also has a maximum at $(0, 0, 1)$ and opens downward.

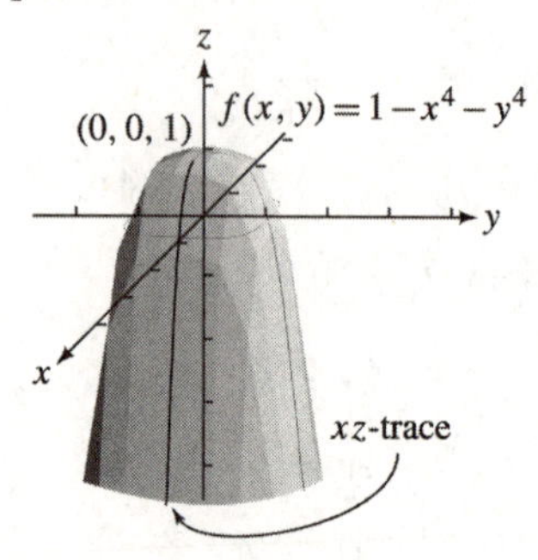

If $f(x, y) > 1$, then $x^4 + y^4 < 0$, which is impossible, so the function does not exist. Thus, the function has a relative maximum of 1 at $(0, 0)$.

31. $f(x, y) = x^2(y + 1)^2 + k(x + 1)^2y^2$

(a)
$$f_x(x, y) = 2x + 2ky^2(x + 1)$$
$$f_y(x, y) = 2x^2(y + 1) + 2k(x + 1)^2y$$
$$f_x(0,0) = 2(0) + 2k(0)^2(0 + 1) = 0$$
$$f_y(0,0) = 2(0)^2(0 + 1) + 2k(0 + 1)^2(0) = 0$$

Thus, $(0, 0)$ is a critical point for all values of k.

(b)
$$f_{xx}(x, y) = 2 + 2ky^2$$
$$f_{yy}(x, y) = 2x^2 + 2k(x + 1)^2$$
$$f_{xy}(x, y) = 4ky(x + 1)$$
$$f_{xx}(0,0) = 2 + 2k(0)^2 = 2$$
$$f_{yy}(0,0) = 2(0)^2 = 2k(0 + 1)^2 = 2k$$
$$f_{xy}(0,0) = 4k(0)(0 + 1) = 0$$
$$D = 2 \cdot 2k - 0^2 = 4k$$

$(0, 0)$ is a relative minimum when $4k > 0$, hence when $k > 0$. When $k = 0$, $D = 0$ so the test for relative extrema gives no information. But if $k = 0$, $f(xy) = x^2(y + 1)^2$, which is always greater than or equal to $f(0,0) = 0$. So (0, 0) is a relative minimum for $k \geq 0$.

35. $L(x,y) = \frac{3}{2}x^2 + y^2 - 2x - 2y - 2xy + 68,$

where x is the number of skilled hours and y is the number of semiskilled hours.

$$L_x(x,y) = 3x - 2 - 2y,$$
$$L_y(x,y) = 2y - 2 - 2x$$

$$\begin{aligned} 3x - 2 - 2y &= 0 \\ -2x - 2 + 2y &= 0 \\ \hline x - 4 \qquad &= 0 \\ x &= 4 \end{aligned}$$

$$\begin{aligned} -2(4) - 2 + 2y &= 0 \\ 2y &= 10 \\ y &= 5 \end{aligned}$$

Let $L_{xx}(x,y) = 3,\ L_{yy}(x,y) = 2,$ $L_{xy}(x,y) = -2.$

$D = 3(2) - (-2)^2 = 2 > 0$ and $L_{xx}(x,y) > 0.$

Relative minimum at (4, 5) is

$$\begin{aligned} L(4,5) &= \frac{3}{2}(4)^2 + (5)^2 - 2(4) - 2(5) \\ &\quad - 2(4)(5) + 68 \\ &= 59. \end{aligned}$$

So \$59 is a minimum cost, when $x = 4$ and $y = 5$.

37. $$\begin{aligned} R(x,y) &= 15 + 169x + 182y \\ &\quad - 5x^2 - 7y^2 - 7xy \end{aligned}$$

$$R_x(x,y) = 169 - 10x - 7y$$
$$R_y(x,y) = 182 - 14y - 7x$$

Solve the system $R_x = 0, R_y = 0.$

$$\begin{aligned} -10x - 7y + 169 &= 0 \\ -7x - 14y + 182 &= 0 \end{aligned}$$

$$\begin{aligned} 20x + 14y - 338 &= 0 \\ -7x - 14y + 182 &= 0 \\ \hline 13x \qquad - 156 &= 0 \\ x &= 12 \end{aligned}$$

$$\begin{aligned} -10(12) - 7y + 169 &= 0 \\ -7y &= -49 \\ y &= 7 \end{aligned}$$

(12, 7) is a critical point.

$$R_{xx} = -10$$
$$R_{yy} = -14$$
$$R_{xy} = -7$$
$$D = (-10)(-14) - (-7)^2 = 91 > 0$$

Since $R_{xx} = -10 < 0,$ there is a relative maximum at (12, 7).

$$\begin{aligned} R(12,7) &= 15 + 169(12) + 182(7) - 5(12)^2 \\ &\quad - 7(7)^2 - 7(12)(7) \\ &= 1666 \text{ (hundred dollars)} \end{aligned}$$

12 spas and 7 solar heaters should be sold to produce a maximum revenue of \$166,600.

39. $T(x,y) = x^4 + 16y^4 - 32xy + 40$

$$T_x(x,y) = 4x^3 - 32y$$
$$T_y(x,y) = 64y^3 - 32x$$
$$T_x(x,y) = 0$$

$$\begin{aligned} 4x^3 - 32y &= 0 \\ 4x^3 &= 32y \\ \frac{1}{8}x^3 &= y \end{aligned}$$

$$T_y(x,y) = 0$$

$$\begin{aligned} 64y^3 - 32x &= 0 \\ 64y^3 &= 32x \\ 2y^3 &= x \end{aligned}$$

Use substitution to solve the system of equations

$$\frac{1}{8}x^3 = y$$
$$2y^3 = x$$

$$\begin{aligned} y &= \frac{1}{8}(2y^3)^3 \\ y &= \frac{1}{8}(8)y^9 \\ y &= y^9 \\ y^9 - y &= 0 \\ y(y^8 - 1) &= 0 \end{aligned}$$

$$\begin{aligned} y = 0 \quad &\text{or} \quad y^8 - 1 = 0 \\ y = 0 \quad &\text{or} \quad y^8 = 1 \\ y = 0 \quad &\text{or} \quad y = 1 \end{aligned}$$

If $y = 0$, $x = 2(0)^3 = 0$.

If $y = 1$, $x = 2(1)^3 = 2$.

The critical points are (0, 0) and (2, 1).

$T_{xx}(x, y) = 12x^2$

$T_{yy}(x, y) = 192y^2$

$T_{xy}(x, y) = -32$

$T_{xx}(0,0) = 0$

$T_{yy}(0,0) = 0$

$T_{xy}(0,0) = -32$

$$D = 0 \cdot 0 - (-32)^2 = -1024$$

Since $D < 0$, there is a saddle point at (0, 0).

$T_{xx}(2,1) = 48$

$T_{yy}(2,1) = 192$

$T_{xy}(2,1) = -32$

$$D = 48 \cdot 192 - (-32)^2 = 8192$$

Since $D > 0$ and $T_{xx} > 0$, there is a relative minimum at (2, 1).

$$\begin{aligned} T(2,1) &= 2^4 + 16(1)^4 - 32(2)(1) + 40 \\ &= 16 + 16 - 64 + 40 \\ &= 8 \end{aligned}$$

Spend \$2000 on quality control and \$1000 on consulting, for a minimum time of 8 hours.

41. Using the procedure suggested, take the natural logarithm of the number-of-transistors data.

Time in years since 1985, t	Natural log of number of transistors in millionS w		
0	ln(0.275)	≈	−1.291
4	ln(1.2)	≈	0.182
8	ln(3.1)	≈	1.131
12	ln(7.5)	≈	2.015
14	ln(9.5)	≈	2.251
15	ln(42)	≈	3.738
20	ln(291)	≈	5.673
22	ln(820)	≈	6.709
24	ln(1900)	≈	7.550

(a) A linear fit of the data in the table above gives

$$w = r + st$$

$$w = -1.722 + 0.3653t$$

where w is the natural logarithm of the number of transistors in millions and t is the number of years since 1985. To convert this back to an exponential fit for the original data, compute

$$a = e^r = e^{-1.722} = 0.1787$$

and

$$b = e^s = e^{0.3652} = 1.441.$$

Thus an exponential fit to the original data has the form $y = a \cdot b^t$ with $a = 0.1787$ and $b = 1.441$:

$$y = 0.1787(1.441)^t$$

(b) Same as (a).

(c) Same as (a).

17.4 Lagrange Multipliers

Your Turn 1

Find the minimum value of $f(x, y) = x^2 + 2x + 9y^2 + 3y + 6xy$ subject to the constraint $2x + 3y = 12$.

Step 1 Rewrite the constraint as $g(x, y) = 2x + 3y - 12$.

Step 2 Form the Lagrange function.

$$\begin{aligned} F(x, y, \lambda) &= f(x, y) - \lambda \cdot g(x, y) \\ &= x^2 + 2x + 9y^2 + 3y \\ &\quad + 6xy - \lambda(2x + 3y - 12) \\ &= x^2 + 9y^2 + 6xy - 2x\lambda \\ &\quad - 3y\lambda + 2x + 3y + 12\lambda \end{aligned}$$

Step 3 Find the first partial derivatives of F.

$$F_x(x, y, \lambda) = 2x + 6y - 2\lambda + 2$$
$$F_y(x, y, \lambda) = 6x + 18y - 3\lambda + 3$$
$$F_\lambda(x, y, \lambda) = -2x - 3y + 12$$

Step 4 Form the system of equations

$$F_x(x, y, \lambda) = 0, F_y(x, y, \lambda) = 0, F_\lambda(x, y, \lambda) = 0.$$

$$\begin{aligned} 2x + 6y - 2\lambda + 2 &= 0 \\ 6x + 18y - 3\lambda + 3 &= 0 \\ -2x - 3y + 12 &= 0 \end{aligned}$$

Step 5 Solve the system of equations from Step 4.

First solve the first two equations for λ and then set these results equal to each other (remove the common factors of 2 from the first equation and 3 from the second equation).

$$x + 3y + 1 = \lambda$$
$$2x + 6y + 1 = \lambda$$

$$x + 3y = 2x + 6y$$
$$x + 3y = 0$$
$$x = -3y$$

Now substitute for x in the last equation.

$$-2(-3y) - 3y + 12 = 0$$
$$6y - 3y = -12$$
$$3y = -12$$
$$y = -4$$

Since $x = -3y$ we have $x = 12$ and $y = -4$. So a candidate for a minimum value of the function f will be $f(12,-4) = 12$.

To see if this is really a minimum we can evaluate f at a few nearby points that satisfy the constraint $2x + 3y = 12$. For example, let $x = 11$. Then $2(11) + 3y = 12$, so $y = -10/3$. $f(11,-10/3) = 13$, which is larger than our candidate. We conclude that the minimum value of f subject to the given constraint is 12.

Your Turn 2

As in Example 3, let $x, y,$ and z be the dimensions of the box. If the front and top are missing, the surface area is $2xy + xz + yz$ so the constraint is $2xy + xz + yz = 6$. As in Example 3, the volume is xyz. The problem is thus to maximize $f(x, y, z) = xyz$ subject to $2xy + xz + yz = 6$.

Step 1 Rewrite the constraint as

$$g(x, y, z) = 2xy + xz + yz - 6.$$

Step 2 Form the Lagrange function.

$$\begin{aligned} F(x, y, z, \lambda) &= f(x, y, z) - \lambda \cdot g(x, y, z) \\ &= xyz - \lambda\left(2xy + xz + yz - 6\right) \\ &= xyz - 2xy\lambda - xz\lambda - yz\lambda + 6\lambda \end{aligned}$$

Step 3 Find the first partial derivatives of F.

$$F_x(x, y, z, \lambda) = yz - 2y\lambda - z\lambda$$
$$F_y(x, y, z, \lambda) = xz - 2x\lambda - z\lambda$$
$$F_z(x, y, z, \lambda) = xy - x\lambda - y\lambda$$
$$F_\lambda(x, y, z, \lambda) = -2xy - xz - yz + 6$$

Step 4 Form the system of equations

$$F_x(x, y, z, \lambda) = 0,$$
$$F_y(x, y, z, \lambda) = 0,$$
$$F_z(x, y, z, \lambda) = 0,$$
$$F_\lambda(x, y, z, \lambda) = 0.$$

$$yz - 2y\lambda - z\lambda = 0$$
$$xz - 2x\lambda - z\lambda = 0$$
$$xy - x\lambda - y\lambda = 0$$
$$-2xy - xz - yz + 6 = 0$$

Step 5 Solve the system of equations from Step 4. First solve each of the first three equations for λ. This gives three expressions for λ.

$$\lambda = \frac{yz}{2y + z}, \quad \lambda = \frac{xz}{2x + z}, \quad \lambda = \frac{xy}{x + y}$$

Set the second and third expressions equal.

$$\frac{xz}{2x + z} = \frac{xy}{x + y}$$
$$x^2z + xyz = 2x^2y + xyz$$
$$x^2z = x^2(2y)$$

Since x is not zero, this shows that $z = 2y$.

Exactly the same calculation using the first and third expressions will show that $z = 2x$. Thus $x = y$. Now use the fourth equation:

$$-2xy - xz - yz + 6 = 0$$
$$-2x^2 - x(2x) - x(2x) + 6 = 0$$
$$6x^2 = 6$$
$$x = 1 \text{ (since } x \text{ is positive)}$$

Now we know that $x = 1$, $y = 1$, and $z = 2$. The box should be 2 ft. wide and 1 ft. high and long.

17.4 Exercises

1. Maximize $f(x,y) = 4xy$, subject to $x + y = 16$.

1. $g(x, y) = x + y - 16$
2. $F(x, y, \lambda) = 4xy - \lambda(x + y - 16)$.
3. $F_x(x, y, \lambda) = 4y - \lambda$
   $F_y(x, y, \lambda) = 4x - \lambda$
   $F_\lambda(x, y, \lambda) = -(x + y - 16)$

4. $4y - \lambda = 0 \quad (1)$
$4x - \lambda = 0 \quad (2)$
$x + y - 16 = 0 \quad (3)$

5. Equations (1) and (2) give $\lambda = 4y$ and $\lambda = 4x$. Thus,

$$4y = 4x$$
$$y = x.$$

Substituting into equation (3),

$$x + (x) - 16 = 0$$
$$x = 8.$$

So $y = 8.$

Maximum is $f(8,8) = 4(8)(8) = 256.$

3. Maximize $f(x, y) = xy^2$, subject to $x + 2y = 15$.

1. $g(x, y) = x + 2y - 15$

2. $F(x, y, \lambda) = xy^2 - \lambda(x + 2y - 15)$

3. $F_x(x, y, \lambda) = y^2 - \lambda$
$F_y(x, y, \lambda) = 2xy - 2\lambda$
$F_\lambda(x, y, \lambda) = -(x + 2y - 15)$

4. $y^2 - \lambda = 0 \quad (1)$
$2xy - 2\lambda = 0 \quad (2)$
$x + 2y - 15 = 0 \quad (3)$

5. Equations (1) and (2) give $\lambda = y^2$ and $\lambda = xy$. Thus,

$$y^2 = xy$$
$$y(y - x) = 0$$
$$y = 0 \quad \text{or} \quad y = x$$

Substituting $y = 0$ into equation (3),

$$x + 2(0) - 15 = 0$$
$$x = 15.$$

Substituting $y = x$ into equation (3)

$$x + 2(x) - 15 = 0$$
$$x = 5.$$

So $y = x = 5.$

Thus,

$$f(15,0) = 15(0)^2 = 0, \quad \text{and}$$
$$f(5,5) = 5(5)^2 = 125.$$

Since $f(5,5) > f(15,0)$, $f(5,5) = 125$ is a maximum.

5. Minimize $f(x, y) = x^2 + 2y^2 - xy$, subject to $x + y = 8$.

1. $g(x, y) = x + y - 8$

2. $F(x, y, \lambda)$
$= x^2 + 2y^2 - xy - \lambda(x + y - 8)$

3. $F_x(x, y, \lambda) = 2x - y - \lambda$
$F_y(x, y, \lambda) = 4y - x - \lambda$
$F_\lambda(x, y, \lambda) = -(x + y - 8)$

4. $2x - y - \lambda = 0$
$4y - x - \lambda = 0$
$x + y - 8 = 0$

5. Subtracting the second equation from the first equation to eliminate λ gives the new system of equations

$$x + y = 8$$
$$3x - 5y = 0.$$

Solve this system.

$$5x + 5y = 40$$
$$\underline{3x - 5y = 0}$$
$$8x = 40$$
$$x = 5$$

But $x + y = 8$, so $y = 3$.

Thus, $f(5,3) = 25 + 18 - 15 = 28$ is a minimum.

7. Maximize $f(x, y) = x^2 - 10y^2$, subject to $x - y = 18$.

1. $g(x, y) = x - y - 18$

2. $F(x, y, \lambda)$
$= x^2 - 10y^2 - \lambda(x - y - 18)$

3. $F_x(x, y, \lambda) = 2x - \lambda$
$F_y(x, y, \lambda) = -20y - \lambda$
$F_\lambda(x, y, \lambda) = -(x - y - 18)$

4. $2x - \lambda = 0$
$-20 + \lambda = 0$
$x - y - 18 = 0$

5. Adding the first two equations to eliminate λ gives

$$2x - 20y = 0$$
$$x = 10y.$$

Substituting $x = 10y$ in the third equation gives

$$10y - y = 18$$
$$y = 2$$
$$x = 20.$$

Thus,

$$f(20,2) = 20^2 - 10(2)^2$$
$$= 400 - 40 = 360.$$

$f(20,2) = 360$ is a maximum.

9. Maximize $f(x,y,z) = xyz^2$,

subject to $x + y + z = 6$.

1. $g(x,y,z) = x + y + z - 6$

2. $F(x,y,\lambda)$

   $= xyz^2 - \lambda(x + y + z - 6)$

3. $F_x(x,y,z,\lambda) = yz^2 - \lambda$

   $F_y(x,y,z,\lambda) = xz^2 - \lambda$

   $F_z(x,y,z,\lambda) = 2zxy - \lambda$

   $F_\lambda(x,y,z,\lambda) = -(x + y + z - 6)$

4. Setting F_x, F_y, F_z and F_λ equal to zero yields

$$yz^2 - \lambda = 0 \quad (1)$$
$$xz^2 - \lambda = 0 \quad (2)$$
$$2xyz - \lambda = 0 \quad (3)$$
$$x + y + z - 6 = 0. \quad (4)$$

5. $\lambda = yz^2$, $\lambda = xz^2$, and $\lambda = 2xyz$

$$yz^2 = xz^2$$
$$z^2(y - x) = 0$$
$$x = y \quad \text{or} \quad z = 0$$
$$yz^2 = 2xyz$$
$$2xyz - yz^2 = 0$$
$$yz(2x - z) = 0$$
$$y = 0 \text{ or } z = 0 \text{ or } z = 2x$$

In a similar way, the third equation

$$xz^2 = 2xyz$$

implies that $x = 0$ or $z = 0$ or $z = 2y$.

By the nature of the function to be maximized, $f(x,y,z) = xyz^2$, a nonzero maximum can come only from those points with nonzero coordinates.

Therefore, assume $y = x$ and $z = 2y = 2x$.

If $y = x$ and $z = 2x$ are substituted into equation (4), then

$$x + x + 2x - 6 = 0$$
$$x = \frac{3}{2}.$$

Thus, $y = \frac{3}{2}$ and $z = 3$, and

$$f\left(\frac{3}{2},\frac{3}{2},3\right) = \frac{3}{2}\cdot\frac{3}{2}\cdot 9$$
$$= \frac{81}{4} > 0.$$

So, $f\left(\frac{3}{2},\frac{3}{2},3\right) = \frac{81}{4} = 20.25$ is a maximum.

11. The problem can be restated as

Maximize $f(x,y) = 3xy^2$,

subject to $x + y = 24, x > 0, y > 0$.

1. $g(x,y) = x + y - 24$

2. $F(x,y,\lambda) = 3xy^2 - \lambda(x + y - 24)$

3. $F_x(x,y,\lambda) = 3y^2 - \lambda$

   $F_y(x,y,\lambda) = 6xy - \lambda$

   $F_\lambda(x,y,\lambda) = -(x + y - 24)$

4.
$$3y^2 - \lambda = 0 \quad (1)$$
$$6xy - \lambda = 0 \quad (2)$$
$$x + y - 24 = 0 \quad (3)$$

5. Equations (1) and (2) give $\lambda = 3y^2$ and $\lambda = 6xy$. Thus,

$$3y^2 = 6xy$$
$$3y^2 - 6xy = 0$$
$$3y(y - 2x) = 0$$
$$y = 0 \quad \text{or} \quad y = 2x.$$

Substituting $y = 0$ into equation (3),

$$x + (0) - 24 = 0$$
$$x = 24.$$

Substituting $y = 2x$ into equation (3),

$$x + (2x) - 24 = 0$$
$$3x - 24 = 0$$
$$x = 8.$$

So $y = 2x = 16.$

Thus,

$$f(24,0) = 3(24)(0)^2 = 0, \text{ and}$$
$$f(8,16) = 3(8)(16)^2 = 6144.$$

Since $f(8,16) > f(24,0), x = 8$ and $y = 16$ will maximize $f(x,y) = 3xy^2$.

13. Let x, y, and z be three number such that

$$x + y + z = 90$$

and $f(x,y,z) = xyz.$

1. $g(x,y,z) = x + y + z - 90$

2. $F(x,y,z)$
$= xyz - \lambda(x + y + z - 90)$

3. $F_x(x,y,z,\lambda) = yz - \lambda$
$F_y(x,y,z,\lambda) = xz - \lambda$
$F_\lambda(x,y,z,\lambda) = xy - \lambda$
$F_\lambda(x,y,z,\lambda) = -(x + y + z - 90)$

4.
$$yz - \lambda = 0 \quad (1)$$
$$xz - \lambda = 0 \quad (2)$$
$$xy - \lambda = 0 \quad (3)$$
$$x + y + z - 90 = 0 \quad (4)$$

5. $\lambda = yz, \lambda = xz,$ and $\lambda = xy$

$$yz = xz$$
$$yz - xz = 0$$
$$(y - x)z = 0$$
$$y - x = 0 \text{ or } z = 0$$
$$xz - xy = 0$$
$$x(z - y) = 0$$
$$x = 0 \text{ or } z - y = 0$$

Since $x = 0$ or $z = 0$ would not maximize $f(x,y,z) = xyz,$ then $y - x = 0$ and $z - y = 0$ imply that $y = x = z.$

Substituting into equation (4) gives

$$x + x + x - 90 = 0$$
$$x = 30.$$

$x = y = z = 30$ will maximize $f(x,y,z) = xyz$. The numbers are 30, 30, and 30.

15. Find the maximum and minimum of $f(x,y) = x^3 + 2xy + 4y^2$ subject to $x + 2y = 12.$

1. $g(x,y) = x + 2y - 12$

2. $F(x,y) = x^3 + 2xy + 4y^2 - \lambda(x + 2y - 12)$

3. $F_x(x,y,\lambda) = 3x^2 + 2y - \lambda$
$F_y(x,y,\lambda) = 2x + 8y - 2\lambda$
$F_\lambda(x,y,\lambda) = -x - 2y + 12$

4. $3x^2 + 2y - \lambda = 0$
$2x + 8y - 2\lambda = 0$
$-x - 2y + 12 = 0$

5. Solve the second equation for λ and substitute into the first equation.

$$2x + 8y - 2\lambda = 0$$
$$\lambda = x + 4y$$

The first equation is now

$$3x^2 + 2y - (x + 4y) = 0$$

or

$$3x^2 - x - 2y = 0$$

Solve the last equation for $-2y$ and substitute into this new equation.

$$-x - 2y + 12 = 0$$
$$-2y = x - 12$$
$$3x^2 - x + (x - 12) = 0$$
$$3x^2 - 12 = 0$$

Now solve for x.

$$3x^2 - 12 = 0$$
$$x^2 = 4$$
$$x = 2 \text{ or } x = -2$$

Find the corresponding y.

When $x = 2,\ -2 - 2y + 12 = 0$ so $y = 5.$

When $x = -2,\ -(-2) - 2y + 12 = 0$ so $y = 7.$

Thus our candidates for the locations of maxima or minima of f subject to the given constraint are $(2,5)$ and $(-2,7)$.

$f(2, 5) = 128$ and $f(-2, 7) = 160$, so probably the maximum is 160 at $(-2, 7)$ and the minimum is 128 at $(2, 5)$.

Try some nearby points that satisfy the constraints to check.

When $x = -2.2$, $y = (12 + 2.2)/2 = 7.1$;
$f(-2.2, 7.1) = 159.752$.

When $x = 2.2$, $y = (12 - 2.2)/2 = 4.9$;
$f(2.2, 4.9) = 128.248$.

This confirms our answer: There is a maximum value of 160 at $(-2, 7)$ and a minimum value of 128 at $(2, 5)$.

19. Consider the constraint and solve for y in terms of x.

$$3x - y = 9$$
$$y = 3x - 9$$

Then

$$f(x, y) = 8x^2y = 8x^2(3x - 9) = 24x^3 - 72x^2$$

So, $f(x, y) = 24x^3 - 72x^2 = f(x)$. Notice that f is unbounded; more specifically,

$$\lim_{y\to\infty} f(x) = \infty$$

and $\lim_{y\to-\infty} f(x) = -\infty$.

Therefore f, subject to the given constraint, has neither an absolute maximum nor an absolute minimum.

21. Minimize $f(x, y) = x^2 + 2x + 9y^2 + 4y + 8xy$ subject to $x + y = 1$.

(a) 1. $g(x, y) = x + y = 1$

2. $F(x, y) = x^2 + 2x + 9y^2 + 4y + 8xy - \lambda(x + y - 1)$

3. $F_x(x, y, \lambda) = 2x + 8y - \lambda + 2$
$F_y(x, y, \lambda) = 18y + 8x - \lambda + 4$
$F_\lambda(x, y, \lambda) = -x - y + 1$

4. $2x + 8y - \lambda + 2 = 0$
$18y + 8x - \lambda + 4 = 0$
$-x - y + 1 = 0$

5. Solve the system in Step 4. Solve the first two equations for λ and eliminate λ.

$$2x + 8y + 2 = 18y + 8x + 4$$
$$6x + 10y + 2 = 0$$
$$3x + 5y + 1 = 0$$

Now combine this equation in x and y with the last equation to form a system.

$$3x + 5y + 1 = 0$$
$$-x - y + 1 = 0$$

The solution of this system is $(3, -2)$. $f(3, -2) = -5$. Test nearby points that satisfy the constraint to see if -5 is indeed a minimum.

When $x = 3.2$, $y = 1 - 3.2 = -2.2$.

When $x = 2.8$, $y = 1 - 2.8 = -1.8$.

$f(3.2, -2.2) = f(2.8, -1.8) = -4.92$, so subject to the given constraint f has a minimum of -5 at $(3, -2)$.

(d) $f_{xx}(x, y) = 2$, $f_{yy}(x, y) = 18$, $f_{xy}(x, y) = 8$.
$D(3, -2) = 18 \cdot 2 - 64 = -28$, so $(3, 2)$ is a saddle point of the function f.

23. Maximize $f(x, y) = xy^2$ subject to $x + 2y = 60$.

1. $g(x, y) = x + 2y - 60$

2. $F(x, y) = xy^2 - \lambda(x + 2y - 60)$

3. $F_x(x, y, \lambda) = y^2 - \lambda$
$F_y(x, y, \lambda) = 2xy - 2\lambda$
$F_\lambda(x, y, \lambda) = -x - 2y + 60$

4. $y^2 - \lambda = 0$
$2xy - 2\lambda = 0$
$-x - 2y + 60 = 0$

5. Solve the first two equations for λ and eliminate λ.

$$y^2 = \lambda$$
$$xy = \lambda$$
$$y^2 = xy$$

Either $y = 0$ or $x = y$. If $y = 0$, then $x = 60$; if $x = y$, then $-3x = 60$ and $x = y = 20$. So our candidates for a maximum of f are (60, 0) and (20, 20). But $f(60,0) = 0$ so the maximum must be $f(20,20) = 8000$.

We can confirm this by solving the constraint for x.

$$x = 60 - 2y$$
$$f(x,y) = xy^2 = (60 - 2y)y^2 = 60y^2 - 2y^3.$$

This is now a function of one variable. The second derivative with respect to y is $120 - 12y$ which is negative when $y = 20$, so the value of 8000 found above is the maximum utility, obtained by purchasing 20 units of x and 20 units of y.

25. Maximize $f(x,y) = x^4y^2$ subject to $2x + 4y = 60$.

1. $g(x,y) = 2x + 4y - 60$

2. $F(x,y) = x^4y^2 - \lambda(2x + 4y - 60)$

3. $F_x(x,y,\lambda) = 4x^3y^2 - 2\lambda$
   $F_y(x,y,\lambda) = 2x^4y - 4\lambda$
   $F_\lambda(x,y,\lambda) = -2x - 4y + 60$

4. $$4x^3y^2 - 2\lambda = 0$$
   $$2x^4y - 4\lambda = 0$$
   $$-2x - 4y + 60 = 0$$

5. Solve the first two equations for 2λ and eliminate 2λ.

$$4x^3y^2 = 2\lambda$$
$$x^4y = 2\lambda$$
$$4x^3y^2 = x^4y$$

If either x or y equals 0, the utility will have a minimum value of 0. So we can assume $xy \neq 0$ and divide by x^3y to find that $4y = x$. Substitute for x in the last equation.

$$-2x - x + 60 = 0$$
$$x = 20,\ y = 20/4 = 5$$
$$f(20,5) = 4{,}000{,}000.$$

The maximum utility is 4,000,000, obtained by purchasing 20 units of x and 5 units of y.

27. Let x be the width and y be the length of a field such that the cost in dollars to enclose the field is

$$6x + 6y + 4x + 4y = 1200$$
$$10x + 10y = 1200.$$

The area is

$$f(x,y) = xy.$$

1. $g(x,y) = 10x + 10y - 1200$

2. $F(x,y) = xy - \lambda(10x + 10y - 1200)$

3. $F_x(x,y,\lambda) = y - 10\lambda$
   $F_y(x,y,\lambda) = x - 10\lambda$
   $F_\lambda(x,y,\lambda) = -(10x + 10y - 1200)$

4. $$y - 10\lambda = 0$$
   $$x - 10\lambda = 0$$
   $$10x + 10y - 1200 = 0$$

5. $10\lambda = y$ and $10\lambda = x$

$$y = x$$

Substituting into the third equation gives

$$10x + 10x - 1200 = 0$$
$$20x - 1200 = 0$$
$$x = 60$$
$$y = 60.$$

These dimensions, 60 feet by 60 feet, will maximize the area.

29. Maximize $C(x,y) = 2x^2 + 6y^2 + 4xy + 10$, subject to $x + y = 10$.

1. $g(x,y) = x + y - 10$

2. $F(x,y)$
   $= 2x^2 + 6y^2 + 4xy + 10$
   $- \lambda(x + y - 10)$

3. $F_x(x,y,\lambda) = 4x + 4y - \lambda$
   $F_y(x,y,\lambda) = 12y + 4x - \lambda$
   $F_\lambda(x,y,\lambda) = -(x + y - 10)$

4. $4x + 4y - \lambda = 0$
$12y + 4x - \lambda = 0$
$x + y - 10 = 0$

5. $\lambda = 4x + 4y$ and $\lambda = 12y + 4x$.

$$4x + 4y = 12y + 4x$$
$$8y = 0$$
$$y = 0$$

Since $x + y = 10$, $x = 10$.

10 large kits and no small kits will maximize the cost.

31. Maximize $f(x, y) = 3x^{1/3}y^{2/3}$, subject to $80x + 150y = 40{,}000$.

1. $g(x, y) = 80x + 150y - 40{,}000$

2. $F(x,y)$
$= 3x^{1/3}y^{2/3} - \lambda(80x + 150y - 40{,}000)$

3,4. $F_x(x, y, \lambda) = x^{-2/3}y^{2/3} - 80\lambda = 0$
$F_y(x, y, \lambda) = 2x^{1/3}y^{-1/3} - 150\lambda = 0$
$F_\lambda(x, y, \lambda) = -(80x + 150y - 40{,}000)$
$= 0$

5. $\dfrac{x^{-2/3}y^{2/3}}{80} = \dfrac{2x^{1/3}y^{-1/3}}{150}$

$$\frac{15y}{16} = x$$

Substitute into the third equation.

$$80\left(\frac{15y}{16}\right) + 150y - 40{,}000 = 0$$
$$y = 178\text{(rounded)}$$
$$x = \frac{15(178)}{16}$$
$$\approx 167$$

Use about 167 units of labor and 178 units of capital to maximize production.

33. Let x and y be the dimensions of the field such that $2x + 2y = 500$, and the area is $f(x, y) = xy$.

1. $g(x, y) = 2x + 2y - 500$

2. $F(x, y) = xy - \lambda(2x + 2y - 500)$

3. $F_x(x, y, \lambda) = y - 2\lambda$
$F_y(x, y, \lambda) = x - 2\lambda$
$F_\lambda(x, y, \lambda) = -(2x + 2y - 500)$

4. $y - 2\lambda = 0$
$x - 2\lambda = 0$
$2x + 2y - 500 = 0$

5. $2\lambda = y$ and $2\lambda = x$, so $x = y$.

$$2x + 2x - 500 = 0$$
$$4x - 500 = 0$$
$$x = 125$$

Thus, $y = 125$.

Dimensions of 125 m by 125 m will maximize the area.

35. Let x be the radius r of the circular base and y the height h of the can, such that the volume is $\pi x^2 y = 250\pi$.

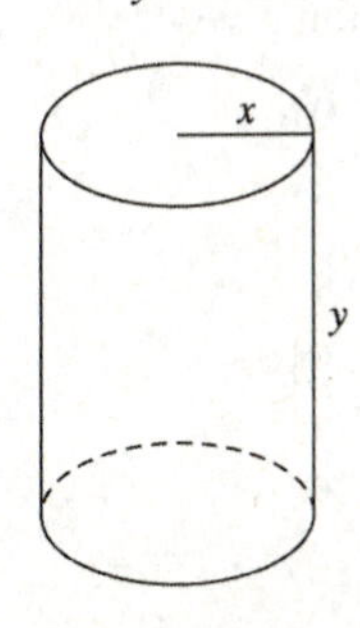

The surface area is

$$f(x, y) = 2\pi xy + 2\pi x^2.$$

1. $g(x, y) = \pi x^2 y + 250\pi$

2. $F(x, y) = 2\pi xy + 2\pi x^2$
$- \lambda(\pi x^2 y - 250\pi)$

3. $F_x(x, y, \lambda) = 2\pi y + 4\pi x - \lambda(2\pi xy)$
$F_y(x, y, \lambda) = 2\pi x - \lambda(\pi x^2)$
$F_\lambda(x, y, \lambda) = -(\pi x^2 y - 250\pi)$

4. $2\pi y + 4\pi x - \lambda(2\pi xy) = 0$
$2\pi x - \lambda \pi x^2 = 0$
$\pi x^2 y - 250\pi = 0$

Simplifying these equations gives

$$y + 2x - 1\lambda xy = 0$$
$$2x - 1\lambda x^2 = 0$$
$$x^2 y - 250 = 0.$$

5. From the second equation,

$$x(2 - \lambda x) = 0$$

$$x = 0 \quad \text{or} \quad \lambda = \frac{2}{x}.$$

If $x = 0$, the volume will be 0, which is not possible.

Substituting $x = \frac{2}{\lambda}$ into the first equation gives

$$y + 2\left(\frac{2}{\lambda}\right) - \lambda\left(\frac{2}{\lambda}\right)y = 0$$

$$y + \frac{4}{\lambda} - 2y = 0$$

$$\frac{4}{\lambda} = y$$

$$\lambda = \frac{4}{\lambda}.$$

Since $\lambda = \frac{2}{x}, y = 2x$.

Substituting into third equation gives

$$x^2(2x) - 250 = 0$$

$$2x^3 - 250 = 0$$

$$x = 5$$

$$y = 10.$$

Since $g(1,250) = 0$ and

$$f(1,250) = 502\pi > f(5,10) = 150\pi,$$

a can with radius of 5 inches and height of 10 inches will have a minimum surface area.

37. Let x, y, and z be the dimensions of the box such that the surface area is

$$xy + 2yz + 2xz = 500$$

and the volume is

$$f(x,y,z) = xyz.$$

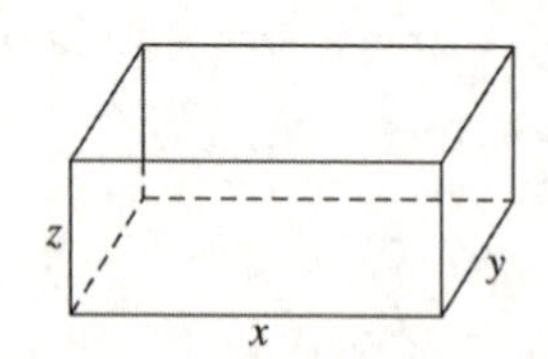

1. $g(x,y,z) - 500 = 0$

2. $F(x,y,z)$

$$= xyz - \lambda(xy + 2yz + 2xz - 500)$$

3,4.

$$F_x(x,y,z,\lambda) = yz - \lambda(y + 2z) = 0 \quad (1)$$

$$F_y(x,y,z,\lambda) = xz - \lambda(x + 2z) = 0 \quad (2)$$

$$Fz(x,y,z,\lambda) = xy - \lambda(2y + 2x) = 0 \quad (3)$$

$$F_\lambda(x,y,z,\lambda) = -(xy + 2xz + 2yz - 500) = 0 \quad (4)$$

Multiplying equation (1) by x, equation (2) by y, and equation (3) by z gives

$$xyz - \lambda x(y + 2z) = 0$$

$$xyz - \lambda y(x + 2z) = 0$$

$$xyz - \lambda z(2y + 2z) = 0.$$

5. Subtracting the first equation from the second equation gives

$$\lambda x(y + 2z) - \lambda y(x + 2z) = 0$$

$$2\lambda xz - 2\lambda yz = 0$$

$$\lambda z(x - y) = 0,$$

so

$$x = y.$$

Subtracting the third equation from the second equation gives

$$\lambda z(2y + 2x) - \lambda y(x + 2z) = 0$$

$$2\lambda xz - \lambda xy = 0$$

$$\lambda x(2z - y) = 0,$$

so

$$z = \frac{y}{2}.$$

Substituting into the fourth equation gives

$$y^2 + 2y\left(\frac{y}{2}\right) + 2y\left(\frac{y}{2}\right) - 500 = 0$$

$$3y^2 = 500$$

$$y = \sqrt{\frac{500}{3}}$$

$$\approx 12.9099$$

$$x \approx 12.9099$$

$$z \approx \frac{12.9099}{2}$$

$$\approx 6.4549.$$

The dimensions are 12.91 m by 12.91 m by 6.455 m.

39. Let x, y, and z be the dimensions of the box. The surface area is

$$2xy + 2xz + 2yz.$$

We must minimize

$$f(x, y, z) = 2xy + 2xz + 2yz$$

subject to $xyz = 125$.

1. $g(x, y, z) = xyz - 125$

2. $F(x, y, z)$
$$= 2xy + 2xz + 2yz - \lambda(xyz - 125)$$

3. $$F_x(x, y, z, \lambda) = 2y + 2z - \lambda yz$$
$$F_y(x, y, z, \lambda) = 2x + 2z - \lambda xz$$
$$F_z(x, y, z, \lambda) = 2x + 2y - \lambda xy$$
$$F_\lambda(x, y, z, \lambda) = -(xyz - 125)$$

4. $$2y + 2z - \lambda yz = 0 \quad (1)$$
$$2x + 2z - \lambda xz = 0 \quad (2)$$
$$2x + 2y - \lambda xy = 0 \quad (3)$$
$$xyz - 125 = 0 \quad (4)$$

5. Equations (1) and (2) give

$$\frac{2y + 2z}{yz} = \lambda \text{ and } \frac{2x + 2z}{xz} = \lambda.$$

Thus,

$$\frac{2y + 2z}{yz} = \frac{2x + 2z}{xz}$$
$$2xyz + 2xz^2 = 2xyz + 2yz^2$$
$$2xz^2 - 2yz^2 = 0$$

$$2z^2 = 0 \qquad \text{or} \quad x - y = 0$$
$$z = 0 \text{ (impossible)} \quad \text{or} \qquad x = y.$$

Equations (2) and (3) give

$$\frac{2x + 2z}{xz} = \lambda \text{ and } \frac{2x + 2y}{xy} = \lambda.$$

$$\frac{2x + 2z}{xz} = \frac{2x + 2y}{xy}$$

Thus,

$$2x^2y + 2xyz = 2x^2z + 2xyz$$
$$2x^2y - 2x^2z = 0$$
$$2x^2(y - z) = 0$$

$$2x^2 = 0 \qquad \text{or} \quad y - z = 0$$
$$x = 0 \text{ (impossible)} \quad \text{or} \qquad y = z.$$

Therefore, $x = y = z$. Substituting into equation (4) gives

$$x^3 - 125 = 0$$
$$x^3 = 125$$
$$x = 5.$$

Thus,

$$y = 5 \text{ and } z = 5.$$

The dimensions that will minimize the surface area are 5 m by 5 m by 5 m.

41. (a) The surface area of the box is

$$SA = xy + 2xz + 2yz.$$

Let the sides with area xz be those made from the free material (along with the bottom). This gives the constraint $2xz + xy = 4$. The volume of the box is xyz, thus it will take $\frac{400}{xyz}$ trips to transport the material at a cost of $\frac{400}{xyz}(.10) = \frac{40}{xyz}$.

The total cost also includes the cost of the material for the ends of the box: $(2yz)(20) = 40yz$.

Thus, the total cost is

$$f(x, y, z) = \frac{40}{xyz} + 40yz.$$

(b) Using a spreadsheet, $x = 2$ yards, $y = 1$ yard, $z = \frac{1}{2}$ yard.

17.5 Total Differentials and Approximations

Your Turn 1

$$f(x, y) = 3x^2y^4 + 6\sqrt{x^2 - 7y^2}$$
$$f_x(x, y) = 6xy^4 + \frac{6x}{\sqrt{x^2 - 7y^2}}$$
$$f_y(x, y) = 12x^2y^3 - \frac{42y}{\sqrt{x^2 - 7y^2}}$$

(a) $dz = f_x(x,y)\,dx + f_y(x,y)\,dy$

$$= \left(6xy^4 + \frac{6x}{\sqrt{x^2 - 7y^2}}\right)dx + \left(12x^2y^3 - \frac{42y}{\sqrt{x^2 - 7y^2}}\right)dy$$

(b) Evaluate dz for $x = 4$, $y = 1$, $dx = 0.02$, $dy = -0.03$.

$$dz = \left(6(4)(1)^4 + \frac{6(4)}{\sqrt{(4)^2 - 7(1)^2}}\right)(0.02) + \left(12(4)^2(1)^3 - \frac{42(1)}{\sqrt{(4)^2 - 7(1)^2}}\right)(-0.03)$$

$$= \left(24 + \frac{24}{3}\right)(0.02) + \left(192 - \frac{42}{3}\right)(-0.03)$$

$$= (32)(0.02) - 178(0.03)$$

$$= -4.7$$

Your Turn 2

As in Example 2, we let $f(x,y) = \sqrt{x^2 + y^2}$. As we found in Example 2,

$$dz = \left(\frac{x}{\sqrt{x^2 + y^2}}\right)dx + \left(\frac{y}{\sqrt{x^2 + y^2}}\right)dy$$

To approximate $\sqrt{5.03^2 + 11.99^2}$ we let $x = 5$, $y = 12$, $dx = 0.03$, $dy = -0.01$.

Then $f(x + dx,\ y + dy) \approx f(x,y) + dz$

$$\sqrt{5.03^2 + 11.99^2} \approx \sqrt{5^2 + 12^2} + \left(\frac{5}{\sqrt{5^2 + 12^2}}\right)(0.03) + \left(\frac{12}{\sqrt{5^2 + 12^2}}\right)(-0.01)$$

$$= 13 + \frac{0.15}{13} - \frac{0.12}{13}$$

$$\approx 13.0023$$

Your Turn 3

As calculated in Example 3, $\frac{dV}{V} = 2\frac{dr}{r} + \frac{dh}{h}$. Thus the maximum percent error in the volume if the errors in the radius and length are at most 4% and 2% is $\frac{dV}{V} = 2(0.04) + (0.02) = 0.10$ or 10%.

17.5 Exercises

1. $z = f(x,y) = 2x^2 + 4xy + y^2$

$x = 5, dx = 0.03, y = -1, dy = -0.02$

$$f_x(x,y) = 4x + 4y$$
$$f_y(x,y) = 4x + 2y$$
$$dz = (4x + 4y)dx + (4x + 2y)dy$$
$$= [4(5) + 4(-1)(0.03) + [4(5) + 2(-1)](-0.02)$$
$$= 0.48 - 0.36 = 0.12$$

3. $z = \frac{y^2 + 3x}{y^2 - x}$, $x = 4$, $y = -4$,

$dx = 0.01$, $dy = 0.03$

$$dz = \frac{(y^2 - x)\cdot 3 - (y^2 + 3x)\cdot(-1)}{(y^2 - x)^2}dx + \frac{(y^2 - x)\cdot 2y - (y^2 + 3x)\cdot 2y}{(y^2 - x)^2}dx$$

$$= \frac{4y^2}{(y^2 - x)^2}dx - \frac{8xy}{(y^2 - x)^2}dy$$

$$= \frac{4(-4)^2}{[(-4)^2 - 4]^2}(0.01) - \frac{8(4)(-4)}{[(-4)^2 - 4]^2}(0.03)$$

$$\approx 0.0311$$

5. $w = \frac{5x^2 + y^2}{z + 1}$

$x = -2$, $y = 1$, $z = 1$

$dx = 0.02$, $dy = -0.03$, $dz = 0.02$

$$f_x(x,y) = \frac{(z + 1)10x - (5x^2 + y^2)(0)}{(z + 1)^2} = \frac{10x}{z + 1}$$

$$f_y(x,y) = \frac{(z + 1)(2y) - (5x^2 + y^2)(0)}{(z + 1)^2} = \frac{2y}{z + 1}$$

$$f_z(x,y) = \frac{(z + 1)(0) - (5x^2 + y^2)(1)}{(z + 1)^2} = \frac{-5x^2 - y^2}{(z + 1)^2}$$

$$dw = \frac{10x}{z + 1}dx + \frac{2y}{z + 1}dy + \frac{-5x^2 - y^2}{(z + 1)^2}dz$$

Substitute the given values.

$$dw = \frac{-20}{2}(0.02) + \frac{2}{2}(-0.03) + \frac{[-5(4) - 1](0.02)}{(2)^2}$$
$$= -0.2 - 0.03 - \frac{21}{4}(0.02)$$
$$= -0.335$$

7. Let $z = f(x,y) = \sqrt{x^2 + y^2}$.

Then

$$dz = f_x(x,y)dx + f_y(x,y)dy$$
$$= \frac{1}{2}(x^2 + y^2)^{-1/2}(2x)dx + \frac{1}{2}(x^2 + y^2)^{-1/2}(2y)dy$$
$$= \frac{xdx + ydy}{\sqrt{x^2 + y^2}}.$$

To approximate $\sqrt{8.05^2 + 5.97^2}$, we let $x = 8$, $dx = 0.05$, $y = 6$ and $dy = -0.03$.

$$dz = \frac{8(0.05) + 6(-0.03)}{\sqrt{8^2 + 6^2}}$$
$$= \frac{4}{5}(0.05) + \frac{3}{5}(-0.03)$$
$$= 0.04 - 0.018 = 0.022$$

$$f(8.05, 5.97) = f(8,6) + \Delta z$$
$$\approx f(8,6) + dz$$
$$= \sqrt{8^2 + 6^2} + 0.222$$
$$= 10.022$$

Thus, $\sqrt{8.05^2 + 5.97^2} \approx 10.022$.

Using a calculator, $\sqrt{8.05^2 + 5.97^2} \approx 10.0221$.

The absolute value of the difference of the two results is $|10.022 - 10.0221| = 0.0001$.

9. Let $z = f(x,y) = (x^2 + y^2)^{1/3}$.

Then

$$dz = f_x(x,y)dx + f_y(x,y)dy$$

$$dz = \frac{1}{3}(x^2 + y^2)^{-2/3}(2x)dx + \frac{1}{3}(x^2 + y^2)^{-2/3}(2y)dy$$
$$= \frac{2x}{3(x^2 + y^2)^{2/3}}dx + \frac{2y}{3(x^2 + y^2)^{2/3}}dy$$

To approximate $(1.92^2 + 2.1^2)^{1/3}$, we let $x = 2$, $dx = -0.08$, $y = 2$, and $dy = 0.1$.

$$dz = \frac{2(2)}{3[(2)^2 + (2)^2]^{2/3}}(-0.08) + \frac{2(2)}{3[(2)^2 + (2)^2]^{2/3}}(0.1)$$
$$= \frac{4}{12}(-0.08) + \frac{4}{12}(0.1)$$
$$= 0.00\overline{6}$$

$$f(1.92, 2.1) = f(2,2) + \Delta z$$
$$\approx f(2,2) + dz$$
$$= 2 + 0.00\overline{6}$$
$$f(1.92, 2.1) \approx 2.0067$$

Using a calculator, $(1.92^2 + 2.1^2)^{1/3} \approx 2.0080$.

The absolute value of the difference of the two results is $|2.0067 - 2.0080| = 0.0013$.

11. Let $z = f(x,y) = xe^y$.

Then

$$dz = f_x(x,y)dx + f_y(x,y)dy$$
$$= e^y dx + xe^y dy.$$

To approximate $1.03e^{0.04}$, we let $x = 1$, $dx = 0.03$, $y = 0$, and $dy = 0.04$.

$$dz = e^0(0.03) + 1 \cdot e^0(0.04)$$
$$= 0.07$$

$$f(1.03, 0.04) = f(1,0) + \Delta z$$
$$\approx f(1,0) + dz$$
$$= 1 \cdot e^0 + 0.07$$
$$= 1.07$$

Thus, $1.03e^{0.04} \approx 1.07$.

Using a calculator, $1.03e^{0.04} \approx 1.0720$.

The absolute value of the difference of the two results is $|1.07 - 1.0720| = 0.0020$.

13. Let $z = f(x, y) = x \ln y$.

Then

$$dz = f_x(x, y)\,dx + f_y(x, y)\,dy$$
$$= \ln y\,dx + \frac{x}{y}dy$$

To approximate $0.99 \ln 0.98$, we let $x = 1$, $dx = -0.01, y = 1$, and $dy = -0.02$.

$$dz = \ln(1) \cdot (-0.01) + \frac{1}{1}(-0.02)$$
$$= -0.02$$

$$f(0.99, 0.98) = f(1,1) + \Delta z$$
$$\approx f(1,1) + dz$$
$$= 1 \cdot \ln(1) - 0.02$$
$$\approx -0.02$$

Thus, $0.99 \ln 0.98 \approx -0.02$.

Using a calculator, $0.99 \ln 0.98 \approx -0.0200$.

The absolute value of the difference of the two results is $|-0.02 - (-0.0200)| = 0$.

15. The volume of the can is

$$V = \pi r^2 h,$$

With

$r = 2.5\,\text{cm}, h = 14\,\text{cm}, dr = 0.08, dh = 0.16$.

$$dV = 2\pi rh dr + \pi r^2 dh$$
$$= 2\pi(2.5)(14)(0.08) + \pi(2.5)^2(0.16)$$
$$\approx 20.73$$

Approximately 20.73 cm^3 of aluminum are needed.

17. The volume of the box is

$$V = LWH$$

with $L = 10, W = 9$, and $H = 18$.

Since 0.1 inch is applied to each side and each dimension has a side at each end,

$$dL = dW = dH = 2(0.1) = 0.2$$
$$dV = WH\,dL + LH\,dW + LW\,dH.$$

Substitute.

$$dV = (9)(18)(0.2) + (10)(18)(0.2) + (10)(9)(0.2)$$
$$= 86.4$$

Approximately 86.4 in^3 are needed.

19. $z = x^{0.65}y^{0.35}$

$x = 50, y = 29,$

$dx = 52 - 50 = 2$

$dy = 27 - 29 = -2$

$$f_x(x, y) = y^{0.35}(0.65)(x^{-0.35}) = 0.65\left(\frac{y}{x}\right)^{0.35}$$

$$f_y(x, y) = (x^{0.65})(0.35)(y^{-0.65}) = 0.35\left(\frac{x}{y}\right)^{0.65}$$

$$dz = 0.65\left(\frac{y}{x}\right)^{0.35} dx + 0.35\left(\frac{x}{y}\right)^{0.65} dy$$

Substitute.

$$dz = 0.65\left(\frac{29}{50}\right)^{0.35}(2) + 0.35\left(\frac{50}{29}\right)^{0.65}(-2)$$
$$= 0.07694 \text{ unit}$$

21. The volume of the bone is

$$V = \pi r^2 h,$$

with $h = 7, r = 1.4, dr = 0.09, dh = 2(0.09) = 0.18$

$$dV = 2\pi rh\,dr + \pi r^2 dh$$
$$= 2\pi(1.4)(7)(0.09) + \pi(1.4)^2(0.18)$$
$$= 6.65$$

6.65 cm^3 of preservative are used.

23. $C = \frac{b}{a - v} = b(a - v)^{-1}$

$a = 160,$

$b = 200, v = 125$

$da = 145 - 160 = -15$

$db = 190 - 200 = -10$

$dv = 130 - 125 = 5$

$$dC = -b(a - v)^{-2}da + \frac{1}{a - v}db + b(a - v)^{-2}dv$$
$$= \frac{-b}{(a - v)^2}da + \frac{1}{a - v}db + \frac{b}{(a - v)^2}dv$$
$$= \frac{-200}{(160 - 125)^2}(-15) + \frac{1}{160 - 125}(-10) + \frac{200}{(160 - 125)^2}(5)$$
$$\approx 2.98 \text{ liters}$$

25. $C(t,g) = 0.6(0.96)^{(210t/1500)-1} + \dfrac{gt}{126t - 900}\left[1 - (0.96)^{(210t/1500)-1}\right]$

(a)

$$C(180,8) = 0.6(0.96)^{(210(180)/1500)-1} + \frac{(8)(180)}{126(180) - 900}\left[1 - (0.96)^{(210(180)/1500)-1}\right] \approx 0.2649$$

(b) $C_t(t,g)$

$$= 0.6(\ln 0.96)\left(\frac{210}{1500}\right)(0.96)^{(210t/1500)-1} + \frac{g(126t - 900) - 126(gt)}{(126t - 900)^2} \times [1 - (0.96)^{(210t/1500)-1}] - \frac{gt}{126t - 900}(\ln 0.96)\left(\frac{210}{1500}\right)(0.96)^{(210t/1500)-1}$$

$$C_g(t,g) = \frac{t}{126t - 900}\left[1 - (0.96)^{(210t/1500)-1}\right]$$

$$C(180 - 10, 8 + 1) \approx C(180,8) + C_t(180,8) \cdot (-10) + C_g(180,8) \cdot (1)$$
$$\approx 0.2649 + (-0.00115)(-10) + 0.00519(1)$$
$$\approx 0.2816$$

$$C(170,9) \approx 0.2817$$

The approximation is very good.

27. $P(A,B,D) = \dfrac{1}{1 + e^{3.68-0.016A-0.77B-0.12D}}$

(a) Since bird pecking is present, $B = 1$.

$$P(150,1,20) = \frac{1}{1 + e^{3.68-0.016(150)-0.77(1)-0.12(20)}} = \frac{1}{1 + e^{-1.89}} \approx 0.8688$$

The probability is about 87%.

(b) Since bird pecking is not present, $B = 0$.

$$P(150,0,20) = \frac{1}{1 + e^{3.68-0.016(150)-0.77(0)-0.12(20)}} = \frac{1}{1 + e^{-1.12}} \approx 0.7540$$

The probability is about 75%.

(c) Let $B = 0$. To simplify the notation, let $X = 3.68 - 0.016A - 0.12D$. Then

$$P(A,0,D) = \frac{1}{1 + e^{3.68-0.016A-0.12D}} = \frac{1}{1 + e^X}.$$

Some other values that we will need are

$$dA = 160 - 150 = 10$$
$$dD = 25 - 20 = 4$$
$$X(150,20) = 3.68 - 0.016(150) - 0.12(20) = -1.12$$
$$X_A = \frac{\partial X}{\partial A} = -0.016$$
$$X_D = \frac{\partial X}{\partial D} = -0.12.$$

$$P_A(A,0,D) = \frac{X_A e^X}{(1 + e^X)^2} = \frac{0.016e^X}{(1 + e^X)^2}$$

$$P_D(A,0,D) = \frac{X_D e^X}{(1 + e^X)^2} = \frac{0.12e^X}{(1 + e^X)^2}$$

$$dP = P_A(A,0,D)\,dA + P_D(A,0,D)\,dD = \frac{0.016e^X}{(1 + e^X)^2}dA + \frac{0.12e^X}{(1 + e^X)^2}dD$$

Substituting the given and calculated values,

$$dP = \frac{0.016e^{-1.12}}{(1 + e^{-1.12})^2}(10) + \frac{0.12e^{-1.12}}{(1 + e^{-1.12})^2}(5) = (0.016.10 + 0.12 \cdot 5)\frac{e^{-1.12}}{(1 + e^{-1.12})^2} \approx 0.76 \cdot 0.1855 \approx 0.14.$$

Therefore,

$$P(160,0,25) = P(150,0,20) + \Delta P \approx P(150,0,20) + dP = 0.75 + 0.14 = 0.89.$$

The probability is about 89%.
Using a calculator, $P(160,0,25) \approx 0.8676$, or about 87%.

29. The area is $A = \frac{1}{2}bh$ with $b = 15.8$ cm, $h = 37.5$ cm, $db = 1.1$ cm, and $dh = 0.8$ cm.

$$dA = \frac{1}{2}b\,dh + \frac{1}{2}h\,db = \frac{1}{2}(15.8)(0.8) + \frac{1}{2}(37.5)(1.1) = 26.945$$

The maximum possible error is 26.945 cm^2.

31. Let $z = f(L,W,H) = LWH$

Then

$$dz = f_L(L,W,H)dL + f_W(L,W,H)dW + f_H(L,W,H)dH = WHdL + LHdW + LWdH.$$

A maximum 1% error in each measurement means that the maximum values of dL, dW, and dH are given by $dL = 0.01L$, $dW = 0.01W$, and $dH = 0.01H$. Therefore,

$$dz = WH(0.01L) + LH(0.01W) + LW(0.01H) = 0.01LWH + 0.01LWH + 0.01LWH = 0.03LWH.$$

Thus, an estimate of the maximum error in calculating the volume is 3%.

33. The volume of a cone is $V = \frac{\pi}{3}r^2h.$

$$\frac{dV}{V} = \frac{\frac{2\pi}{3}rh}{\frac{\pi}{3}r^2h}dr + \frac{\frac{\pi}{3}r^2}{\frac{\pi}{3}r^2h}dh = \frac{2}{r}dr + \frac{1}{h}dh$$

When $r = 1$ and $h = 4$, a 1% change in radius changes the volume by 2%, and a 1% change in height changes the volume by $\frac{1}{4}$%. So the change produced by changing the radius is 8 times the change produced by changing the height.

17.6 Double Integrals

Your Turn 1

$$\int_1^3 (6x^2y^2 + 4xy + 8x^3 + 10y^4 + 3)\,dy$$

$$= (2x^2y^3 + 2xy^2 + 8x^3y + 2y^5 + 3y)\Big|_{y=1}^{y=3}$$

$$= (2x^2(3)^3 + 2x(3)^2 + 8x^3(3) + 2(3)^5 + 3(3)) - (2x^2(1)^3 + 2x(1)^2 + 8x^3(1) + 2(1)^5 + 3(1))$$

$$= 52x^2 + 16x + 16x^3 + 484 + 6$$

$$= 16x^3 + 52x^2 + 16x + 490$$

Your Turn 2

$$\int_0^2\left[\int_1^3 (6x^2y^2 + 4xy + 8x^3 + 10y^4 + 3)\,dy\right]dx$$

Use the result from Your Turn 1 to evaluate the inner integral.

$$\int_0^2 (16x^3 + 52x^2 + 16x + 490)\,dx$$

$$= \left(4x^4 + \frac{52}{3}x^3 + 8x^2 + 490x\right)\Bigg|_0^2$$

$$= 4(16) + \frac{52}{3}(8) + 8(4) + 490(2)$$

$$= \frac{3644}{3}$$

Integrating in the other order:

$$\int_1^3\left[\int_0^2 (6x^2y^2 + 4xy + 8x^3 + 10y^4 + 3)\,dx\right]dy$$

$$= \int_1^3\left[(2x^3y^2 + 2x^2y + 2x^4 + 10xy^4 + 3x)\Big|_{x=0}^{x=2}\right]dy$$

$$= \int_1^3 (16y^2 + 8y + 32 + 20y^4 + 6)\,dy$$

$$= \int_1^3 (20y^4 + 16y^2 + 8y + 38)\,dy$$

$$= \left(4y^5 + \frac{16}{3}y^3 + 4y^2 + 38y\right)\Bigg|_1^3$$

$$(972 + 144 + 36 + 114) - \left(4 + \frac{16}{3} + 4 + 38\right)$$

$$= \frac{3644}{3}$$

Your Turn 3

Let the region R be defined by $0 \le x \le 5$ and $1 \le y \le 6$.

$$\iint_R \frac{1}{\sqrt{x + y + 3}}\,dx\,dy$$

$$= \int_1^6\int_0^5 \frac{1}{\sqrt{x + y + 3}}\,dx\,dy$$

$$= \int_1^6 \left(2\sqrt{x + y + 3}\right)\Big|_{x=0}^{x=5} dy$$

$$= 2\int_1^6 \left(\sqrt{y + 8} - \sqrt{y + 3}\right)dy$$

$$= 2\frac{2}{3}[(y+8)^{3/2} - (y+3)^{3/2}]\Big|_1^6$$

$$= \frac{4}{3}[(14^{3/2} - 9^{3/2}) - (9^{3/2} - 4^{3/2})]$$

$$= \frac{4}{3}(14\sqrt{14} - 46)$$

$$= \frac{56\sqrt{14} - 184}{3}$$

Your Turn 4

The function $4 - x^3 - y^3$ is positive over the region $0 \le x \le 1$ and $0 \le y \le 1$, so the volume under the surface $z = 4 - x^3 - y^3$ over this region is

$$\iint_R (4 - x^3 - y^3)dx\,dy$$

$$= \int_0^1 \int_0^1 (4 - x^3 - y^3)\,dx\,dy.$$

$$\int_0^1 \int_0^1 (4 - x^3 - y^3)\,dx\,dy$$

$$= \int_0^1 \left(4x - \frac{x^4}{4} - xy^3\right)\Bigg|_{x=0}^{x=1} dy$$

$$= \int_0^1 \left(\frac{15}{4} - y^3\right)dy$$

$$= \left(\frac{15}{4}y - \frac{y^4}{4}\right)\Bigg|_0^1$$

$$= \frac{15}{4} - \frac{1}{4} = \frac{7}{2}$$

Your Turn 5

Find $\iint_R \left(x^3 + 4y\right)dy\,dx$ over the region bounded by

$y = 4x$ and $y = x^3$ for $0 \le x \le 2$.

The region is shown in the figure below.

Note that throughout the region, $4x \ge x^3$.

$$\iint_R (x^3 + 4y)dy\,dx$$

$$= \int_0^2 \int_{x^3}^{4x} (x^3 + 4y)\,dy\,dx$$

$$= \int_0^2 (x^3y + 2y^2)\Big|_{y=x^3}^{y=4x} dx$$

$$= \int_0^2 [(4x^4 + 32x^2) - (x^6 + 2x^6)]\,dx$$

$$= \int_0^2 (4x^4 + 32x^2 - 3x^6)\,dx$$

$$= \left(\frac{4}{5}x^5 + \frac{32}{3}x^3 - \frac{3}{7}x^7\right)\Bigg|_0^2$$

$$= \frac{128}{5} + \frac{256}{3} - \frac{384}{7}$$

$$= \frac{5888}{105}$$

17.6 Exercises

1. $$\int_0^5 (x^4y + y)dx = \left(\frac{x^5y}{5} - xy\right)\Bigg|_0^5$$

$$= (625y + 5y) - 0 = 630y$$

3. $$\int_4^5 x\sqrt{x^2 + 3y}dy$$

$$= \int_4^5 x(x^2 + 3y)^{1/2}dy$$

$$= \frac{2x}{9}[(x^2 + 3y)^{3/2}\Big|_4^5$$

$$= \frac{2x}{9}[(x^2 + 15)^{3/2} - (x^2 + 12)^{3/2}]$$

5. $$\int_4^9 \frac{3 + 5y}{\sqrt{x}}dx = (3 + 5y)\int_4^9 x^{-1/2}dx$$

$$= (3 + 5y)2x^{1/2}\Big|_4^9$$

$$= (3 + 5y)2\left[\sqrt{9} - \sqrt{4}\right]$$

$$= 6 + 10y$$

7. $$\int_2^6 e^{2x+3y}\,dx = \frac{1}{2}e^{2x+3y}\Big|_2^6 = \frac{1}{2}(e^{12+3y} - e^{4+3y})$$

9. $$\int_0^3 ye^{4x+y^2}\,dx$$

Let $u = 4x + y^2$; then $du = 2y\,dy$.

If $y = 0$ then $u = 4x$.

If $y = 3$ then $u = 4x + 9$.

$$\int_{4x}^{4x+9} e^u \cdot \frac{1}{2}du = \frac{1}{2}e^u\Big|_{4x}^{4x+9} = \frac{1}{2}(e^{4x+9} - e^{4x})$$

11. $$\int_1^2\int_0^5 (x^4y + y)\,dx\,dy$$

From Exercise 1

$$\int_0^5 (x^4y + y)\,dx = 630y.$$

Therefore,

$$\int_1^2\left[\int_0^5 (x^4y + y)\,dx\right]dy = \int_1^2 630y\,dy = 315y^2\Big|_1^2 = 315(4 - 1) = 945.$$

13. $$\int_0^1\left[\int_3^6 x\sqrt{x^2 + 3y}\,dx\right]dy$$

From Exercise 4,

$$\int_3^6 x\sqrt{x^2 + 3y}\,dx = \frac{1}{3}[(36 + 3y)^{3/2} - (9 + 3y)^3].$$

$$\int_0^1\left[\int_3^6 x\sqrt{x^2 + 3y}\,dx\right]dy = \int_0^1 \frac{1}{3}[(36 + 3y)^{3/2} - (9 + 3y)^{3/2}]\,dy$$

Let $u = 36 + 3y$. Then $du = 3\,dy$.

When $y = 0, u = 36$.

When $y = 1, u = 39$.

Let $z = 9 + 3y$. Then $dz = 3\,dy$.

When $y = 0, z = 9$.

When $y = 1, z = 12$.

$$\frac{1}{9}\left[\int_{36}^{39} u^{3/2}du - \int_9^{12} z^{3/2}dz\right]$$
$$= \frac{1}{9}\cdot\frac{2}{5}[(39)^{5/2} - (36)^{5/2} - (12)^{5/2} + (9)^{5/2}]$$
$$= \frac{2}{45}[(39)^{5/2} - (12)^{5/2} - 6^5 + 3^5]$$
$$= \frac{2}{45}(39^{5/2} - 12^{5/2} - 7533)$$

15. $$\int_1^2\left[\int_4^9 \frac{3 + 5y}{\sqrt{x}}\,dx\right]dy$$

From Exercise 5,

$$\int_4^9 \frac{3 + 5y}{\sqrt{x}}\,dx = 6 + 10y.$$

$$\int_1^2\left[\int_4^9 \frac{3 + 5y}{\sqrt{x}}\,dx\right]dy = \int_1^2 (6 + 10y)dy = 6y\Big|_1^2 + 5y^2\Big|_1^2 = 6(2 - 1) + 5(4 - 1) = 6 + 15 = 21$$

17. $$\int_1^3\int_1^3 \frac{dy\,dx}{xy} = \int_1^3\left[\int_1^3 \frac{1}{xy}\,dy\right]dx = \int_1^3\left(\frac{1}{x}\ln|y|\right)\Big|_1^3 dx = \int_1^3 \frac{\ln 3}{x}\,dx = (\ln 3)\ln|x|\Big|_1^3 = (\ln 3)(\ln 3 - 0) = (\ln 3)^2$$

19. $\int_2^4 \int_3^5 \left(\frac{x}{y} + \frac{y}{3}\right) dx\, dy$

$$= \int_2^4 \left(\frac{x^2}{2y} + \frac{yx}{3}\right)\Bigg|_3^5 dy$$

$$= \int_2^4 \left[\frac{25}{2y} + \frac{5y}{3} - \left(\frac{9}{2y} + \frac{3y}{3}\right)\right] dy$$

$$= \int_2^4 \left(\frac{16}{2y} + \frac{2y}{3}\right) dy$$

$$= \left(8 \ln|y| + \frac{y^2}{3}\right)\Bigg|_2^4$$

$$= 8(\ln 4 - \ln 2) + \frac{16}{3} - \frac{4}{3}$$

$$= 8 \ln \frac{4}{2} + \frac{12}{3}$$

$$= 8 \ln 2 + 4$$

21. $\iint_R (3x^2 + 4y) dx\, dy;$

$0 \le x \le 3, 1 \le y \le 4$

$$\iint_R (3x^2 + 4y) dx\, dy$$

$$= \int_1^4 \int_0^3 (3x^2 + 4y) dx\, dy$$

$$= \int_1^4 (x^3 + 4xy)\Big|_0^3 dy$$

$$= \int_1^4 (27 + 12y) dy$$

$$= (27y + 6y^2)\Big|_1^4$$

$$= (108 + 96) - (27 + 6) = 171$$

23. $\iint_R \sqrt{x + y}\, dy\, dx; 1 \le x \le 3, 0 \le y \le 1$

$$\iint_R \sqrt{x + y}\, dy\, dx$$

$$= \int_1^3 \int_0^1 (x + y)^{1/2} dy\, dx$$

$$= \int_1^3 \left[\frac{2}{3}(x + y)^{3/2}\right]\Bigg|_0^1 dx$$

$$= \int_1^3 \frac{2}{3}[(x + 1)^{3/2} - x^{3/2}]\, dx$$

$$= \frac{2}{3} \cdot \frac{2}{5}\left[(x + 1)^{5/2} - x^{5/2}\right]\Bigg|_1^3$$

$$= \frac{4}{15}(4^{5/2} - 3^{5/2} - 2^{5/2} + 1^{5/2})$$

$$= \frac{4}{15}(32 - 3^{5/2} - 2^{5/2} + 1)$$

$$= \frac{4}{15}(33 - 3^{5/2} - 2^{5/2})$$

25. $\iint_R \frac{3}{(x + y)^2} dy\, dx; 2 \le x \le 4, 1 \le y \le 6$

$$\iint_R \frac{3}{(x + y)^2} dy\, dx$$

$$= -3 \int_2^4 \int_1^6 (x + y)^{-2} dy\, dx$$

$$= -3 \int_2^4 (x + y)^{-1}\Big|_1^6 dx$$

$$= -3 \int_2^4 \left(\frac{1}{x + 6} - \frac{1}{x + 1}\right) dx$$

$$= -3(\ln|x + 6| - \ln|x + 1|)\Big|_2^4$$

$$= -3\left(\ln\left|\frac{x + 6}{x + 1}\right|\right)\Bigg|_2^4$$

$$= -3\left(\ln 2 - \ln \frac{8}{3}\right)$$

$$= -3 \ln \frac{2}{\frac{8}{3}}$$

$$= -3 \ln \frac{3}{4} \text{ or } 3 \ln \frac{4}{3}$$

27. $\iint_R ye^{(x+y^2)}dx\,dy;\ 2 \le x \le 3, 0 \le y \le 2$

$$\iint_R ye^{(x+y^2)}dx\,dy$$
$$= \int_0^2 \int_2^3 ye^{x+y^2}dx\,dy$$
$$= \int_0^2 ye^{x+y^2}\Big|_2^3 dy$$
$$= \int_0^2 (ye^{3+y^2} - ye^{2+y^2})\,dy$$
$$= e^3\int_0^2 ye^{y^2}dy - e^2\int_0^2 ye^{y^2}dy$$
$$= \frac{e^3}{2}(e^{y^2})\Big|_0^2 - \frac{e^2}{2}(e^{y^2})\Big|_0^2$$
$$= \frac{e^3}{2}(e^4 - e^0) - \frac{e^2}{2}(e^4 - e^0)$$
$$= \frac{1}{2}(e^7 - e^6 - e^3 + e^2)$$

29. $z = 8x + 4y + 10; -1 \le x \le 1, 0 \le y \le 3$

$$V = \int_{-1}^1 \int_0^3 (8x + 4y + 10)\,dy\,dx$$
$$= \int_{-1}^1 (8xy + 2y^2 + 10y)\Big|_0^3 dx$$
$$= \int_{-1}^1 (24x + 18 + 30 - 0)\,dx$$
$$= \int_{-1}^1 (24x + 48)\,dx$$
$$= (12x^2 + 48x)\Big|_{-1}^1$$
$$= (12 + 48) - (12 - 48) = 96$$

31. $z = x^2; 0 \le x \le 2, 0 \le y \le 5$

$$V = \int_0^2 \int_0^5 x^2 dy\,dx$$
$$= \int_0^2 x^2y\Big|_0^5 dx$$
$$= \int_0^2 5x^2 dx$$
$$= \frac{5}{3}x^3\Big|_0^2$$
$$= \frac{40}{3}$$

33. $z = x\sqrt{x^2 + y}; 0 \le x \le 1, 0 \le y \le 1$

$$V = \int_0^1 \int_0^1 x\sqrt{x^2 + y}\,dx\,dy$$

Let $u = x^2 + y$. Then $du = 2x\,dx$.

When $x = 0, u = y$.

When $x = 1, u = 1 + y$.

$$= \int_0^1 \left[\int_y^{1+y} u^{1/2}du\right] dy$$
$$= \int_0^1 \frac{1}{2}\left(\frac{2}{3}u^{3/2}\right)\Big|_y^{1+y} dy$$
$$= \int_0^1 \frac{1}{3}[(1 + y)^{3/2} - y^{3/2}]\,dy$$
$$= \frac{1}{3} \cdot \frac{2}{5}[(1 + y)^{5/2} - y^{5/2}]\Big|_0^1$$
$$= \frac{2}{15}(2^{5/2} - 1 - 1)$$
$$= \frac{2}{15}(2^{5/2} - 2)$$

35. $z = \dfrac{xy}{(x^2+y^2)^2}; 1 \le x \le 2, 1 \le y \le 4$

$$V = \int_1^2 \int_1^4 \frac{xy}{(x^2+y^2)^2}\,dy\,dx$$

$$= \int_1^2 \left[\int_1^4 xy(x^2+y^2)^{-2} dy \right] dx$$

$$= \int_1^2 \left[\int_1^4 \frac{1}{2}x(x^2+y^2)^{-2}(2y)dy \right] dx$$

$$= \int_1^2 \left[-\frac{1}{2}x(x^2+y^2)^{-1} \right]\Bigg|_1^4 dx$$

$$= \int_1^2 \left[-\frac{1}{2}x(x^2+16)^{-1} + \frac{1}{2}x(x^2+1)^{-1} \right] dx$$

$$= -\frac{1}{2}\int_1^2 \frac{1}{2}(x^2+16)^{-1}(2x)\,dx$$

$$+ \frac{1}{2}\int_1^2 \frac{1}{2}(x^2+1)^{-1}(2x)\,dx$$

$$= -\frac{1}{2}\cdot\frac{1}{2}\ln\left|x^2+16\right|\Bigg|_1^2$$

$$+ \frac{1}{2}\cdot\frac{1}{2}\ln\left|x^2+1\right|\Bigg|_1^2$$

$$= -\frac{1}{4}\cdot\ln 20 + \frac{1}{4}\ln 17$$

$$+ \frac{1}{4}\ln 5 - \frac{1}{4}\ln 2$$

$$= \frac{1}{4}(-\ln 20 + \ln 17 + \ln 5 - \ln 2)$$

$$= \frac{1}{4}\ln\frac{(17)(5)}{(20)(2)}$$

$$= \frac{1}{4}\ln\frac{17}{8}$$

37. $\iint_R xe^{xy}dx\,dy; 0 \le x \le 2; 0 \le y \le 1$

$$\iint_R xe^{xy}\,dx\,dy$$

$$= \int_0^2 \int_0^1 xe^{xy}\,dy\,dx$$

$$= \int_0^2 \frac{x}{x}e^{xy}\Bigg|_0^1 dx$$

$$= \int_0^2 (e^x - e^0)\,dx$$

$$= (e^x - x)\Big|_0^2$$

$$= e^2 - 2 - e^0 + 0$$

$$= e^2 - 3$$

39. $\displaystyle\int_2^4 \int_2^{x^2} (x^2+y^2)dy\,dx$

$$= \int_2^4 \left(x^2y + \frac{y^3}{3} \right)\Bigg|_2^{x^2} dx$$

$$= \int_2^4 \left(x^4 + \frac{x^6}{3} - 2x^2 - \frac{8}{3} \right) dx$$

$$= \left(\frac{x^5}{5} + \frac{x^7}{21} - \frac{2}{3}x^3 - \frac{8}{3}x \right)\Bigg|_2^4$$

$$= \frac{1024}{5} + \frac{16,384}{21} - \frac{2}{3}(64) - \frac{8}{3}(4)$$

$$- \left(\frac{32}{5} + \frac{128}{21} - \frac{16}{3} - \frac{16}{3} \right)$$

$$= \frac{1024}{5} - \frac{32}{5} + \frac{16,384 - 128}{21}$$

$$- \frac{128}{3} - \frac{32}{3} - \left(\frac{-32}{3} \right)$$

$$= \frac{992}{5} + \frac{16,256}{21} - \frac{128}{3}$$

$$= \frac{20,832}{105} + \frac{81,280}{105} - \frac{4480}{105}$$

$$= \frac{97,632}{105}$$

41. $\int_0^4 \int_0^x \sqrt{xy}\, dy\, dx$

$= \int_0^4 \int_0^x (xy)^{1/2}\, dy\, dx$

$= \int_0^4 \left[\frac{2(xy)^{3/2}}{3x} \right]\Bigg|_0^x dx$

$= \frac{2}{3} \int_0^4 \left[\frac{\left(\sqrt{x^2}\right)^3}{x} - \frac{0}{x} \right] dx$

$= \frac{2}{3} \int_0^4 x^2 dx = \frac{2}{3} \cdot \frac{x^3}{3}\Bigg|_0^4 = \frac{2}{9}(64)$

$= \frac{128}{9}$

43. $\int_2^6 \int_{2y}^{4y} \frac{1}{x}\, dx\, dy$

$= \int_2^6 (\ln |x|)\Bigg|_{2y}^{4y} dy$

$= \int_2^6 (\ln |4y| - \ln |2y|)\, dy$

$= \int_2^6 \ln \left| \frac{4y}{2y} \right| dy$

$= \int_2^6 \ln 2\, dy$

$= (\ln 2) y\Big|_2^6$

$= (\ln 2)(6 - 2) = 4 \ln 2$

Note: We can write 4 ln 2 as $\ln 2^4$, or ln 16.

45. $\int_0^4 \int_1^{e^x} \frac{x}{y}\, dy\, dx$

$= \int_0^4 (x \ln |y|)\Bigg|_1^{e^x} dx$

$= \int_0^4 (x \ln e^x - x \ln 1)\, dx$

$= \int_0^4 x^2\, dx = \frac{x^3}{3}\Bigg|_0^4 = \frac{64}{3}$

47. $\iint_R (5x + 8y)\, dy\, dx;\ 1 \le x \le 3,$

$0 \le y \le x - 1$

$\iint_R (5x + 8y)\, dy\, dx$

$= \int_1^3 \int_0^{x-1} (5x + 8y)\, dy\, dx$

$= \int_1^3 (5xy + 4y^2)\Bigg|_0^{x-1} dx$

$= \int_1^3 [5x(x - 1) + 4(x - 1)^2 - 0]\, dx$

$= \int_1^3 (9x^2 - 13x + 4)\, dx$

$= \left(3x^3 - \frac{13}{2}x^2 + 4x\right)\Bigg|_1^3$

$= \left(81 - \frac{117}{2} + 12\right) - \left(3 - \frac{13}{2} + 4\right)$

$= 34$

49. $\iint_R (4 - 4x^2)\, dy\, dx;\ 0 \le x \le 1,$

$0 \le y \le 2 - 2x$

$\iint_R (4 - 4x^2)\, dy\, dx$

$= \int_0^2 \int_0^{2-2x} 4(1 - x^2)\, dy\, dx$

$= \int_0^1 [4(1 - x^2)y]\Bigg|_0^{2(1-x)} dx$

$= \int_0^1 4(1 - x^2)(2)(1 - x)\, dx$

$= 8 \int_0^1 (1 - x - x^2 + x^3)\, dx$

$= 8\left(x - \frac{x^2}{2} - \frac{x^3}{3} + \frac{x^4}{4}\right)\Bigg|_0^1$

$= 8\left(1 - \frac{1}{2} - \frac{1}{3} + \frac{1}{4}\right)$

$= 8\left(\frac{1}{2} - \frac{1}{12}\right)$

$= 8 \cdot \frac{5}{12} = \frac{10}{3}$

51. $\iint_R e^{x/y^2}\,dx\,dy; 1 \le y \le 2, 0 \le x \le y^2$

$$\iint_R e^{x/y^2}\,dx\,dy$$

$$= \int_1^2 \int_0^{y^2} e^{x/y^2}\,dx\,dy$$

$$= \int_1^2 [y^2 e^{x/y^2}]\Big|_0^{y^2}\,dy$$

$$= \int_1^2 (y^2 e^{y^2/y^2} - y^2 e^0)\,dy$$

$$= \int_1^2 (ey^2 - y^2)\,dy$$

$$= (e-1)\frac{y^3}{3}\Big|_1^2$$

$$= (e-1)\left(\frac{8}{3} - \frac{1}{3}\right)$$

$$= \frac{7(e-1)}{3}$$

53. $\iint_R x^3 y\,dy\,dx$; R bounded by $y = x^2$, $y = 2x$

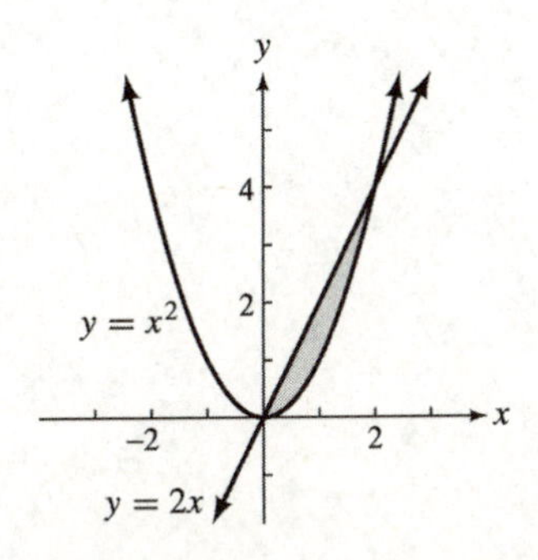

The points of intersection can be determined by solving the following system for x.

$$y = x^2$$
$$y = 2x$$

$$x^2 = 2x$$
$$x(x-2) = 0$$
$$x = 0 \quad \text{or} \quad x = 2$$

Therefore,

$$\iint_R x^3 y\,dx\,dy$$

$$= \int_0^2 \int_{x^2}^{2x} x^3 y\,dy\,dx = \int_0^2 \left(x^3 \frac{y^2}{2}\right)\Big|_{x^2}^{2x} dx$$

$$= \int_0^2 \left[x^3 \frac{(4x^2)}{2} - x^3 \frac{(x^4)}{2}\right] dx$$

$$= \int_0^2 \left(2x^5 - \frac{x^7}{2}\right) dx$$

$$= \left(\frac{1}{3}x^6 - \frac{1}{16}x^8\right)\Big|_0^2$$

$$= \frac{1}{3}\cdot 2^6 - \frac{1}{16}\cdot 2^8$$

$$= \frac{64}{3} - 16$$

$$= \frac{16}{3}.$$

55. $\iint_R \frac{dy\,dx}{y}$; R bounded by $y = x$, $y = \frac{1}{x}$, $x = 2$.

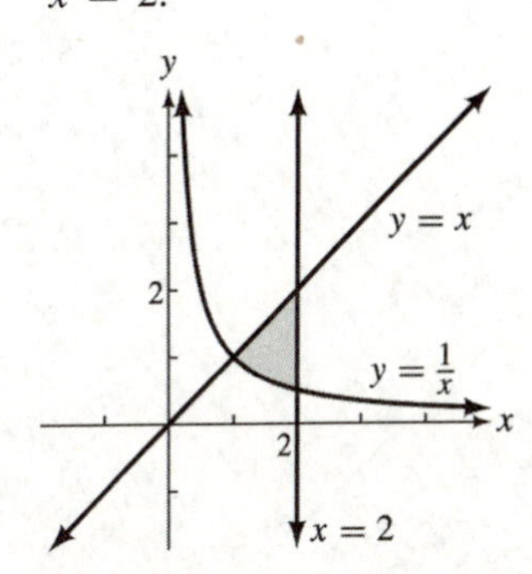

The graphs of $y = x$ and $y = \frac{1}{x}$ intersect at (1, 1).

$$\int_1^2 \int_{1/x}^{x} \frac{dy}{y}\,dx = \int_1^2 \ln y\Big|_{1/x}^{x} dx$$

$$= \int_1^2 \left(\ln x - \ln\frac{1}{x}\right) dx$$

$$= \int_1^2 2\ln x\,dx$$

$$= 2(x\ln x - x)\Big|_1^2$$

$$= 2[(2\ln 2 - 2) - (\ln 1 - 1)]$$

$$= 4\ln 2 - 2$$

57. $\int_0^{\ln 2} \int_{e^y}^2 \frac{1}{\ln x}\, dx\, dy$

Changing the order of integration,

$$\int_0^{\ln 2} \int_{e^y}^2 \frac{1}{\ln x}\, dx\, dy$$
$$= \int_1^2 \int_0^{\ln x} \frac{1}{\ln x}\, dy\, dx$$
$$= \int_1^2 \left[\frac{1}{\ln x} y \Big|_0^{\ln x} \right] dx$$
$$= \int_1^2 (1 - 0)\, dx$$
$$= x\Big|_1^2$$
$$= 2 - 1 = 1$$

61. $f(x, y) = 6xy + 2x; 2 \le x \le 5, 1 \le y \le 3$

The area of region R is

$A = (5 - 2)(3 - 1) = 6.$

The average value of f over R is

$$\frac{1}{A} \iint_R f(x, y)\, dy\, dx$$
$$= \frac{1}{6} \int_2^5 \int_1^3 (6xy + 2x)\, dy\, dx$$
$$= \frac{1}{6} \int_2^5 (3xy^2 + 2xy)\Big|_1^3\, dx$$
$$= \frac{1}{6} \int_2^5 [(27x + 6x) - (3x + 2x)]\, dx$$
$$= \frac{1}{6} \int_2^5 28x\, dx$$
$$= \frac{1}{6} 14x^2 \Big|_2^5$$
$$= \frac{7}{3}(25 - 4) = 49.$$

63. $f(x, y) = e^{-5y+3x};\ 0 \le x \le 2, 0 \le y \le 2$

The area of region R is

$(2 - 0)(2 - 0) = 4.$

The average value of f over R is

$$\frac{1}{4}\int_0^2\int_0^2 e^{-5y+3x}\,dy\,dx = \frac{1}{4}\int_0^2 -\frac{1}{5}e^{-5y+3x}\Big|_0^2\,dx = \frac{1}{4}\int_0^2 -\frac{1}{5}[e^{3x-10} - e^{3x}]\,dx$$

$$= -\frac{1}{20}\left[\frac{1}{3}e^{3x-10} - \frac{1}{3}e^{3x}\right]\Bigg|_0^2 = -\frac{1}{60}[e^{-4} - e^{6} - e^{-10} + 1] = \frac{e^{6} + e^{-10} - e^{-4} - 1}{60}.$$

65. The plane that intersects the axes has the equation

$z = 6 - 2x - 2y.$

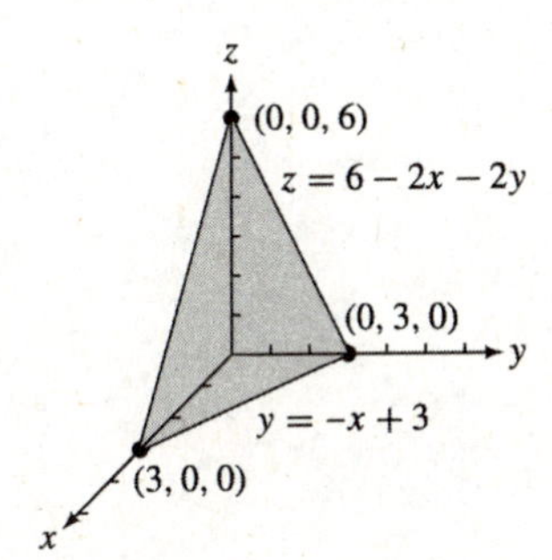

$$V = \iint_R f(x, y)\,dA$$

$$= \int_0^3\int_0^{-x+3} (6 - 2x - 2y)\,dy\,dx$$

$$= \int_0^3 (6y - 2xy - y^2)\Big|_0^{-x+3}\,dx$$

$$= \int_0^3 [-6x + 18 - 2x(-x + 3) - (3 - x)^2]\,dx$$

$$= \int_0^3 (-6x + 18 + 2x^2 - 6x - 9 + 6x - x^2)\,dx$$

$$= \int_0^3 (x^2 - 6x + 9)\,dx = \left(\frac{x^3}{3} - 3x^2 + 9x\right)\Bigg|_0^3$$

$$= (9 - 27 + 27) - 0 = 9$$

The volume is 9 in^3.

67. $P(x, y) = 500x^{0.2}y^{0.8}, 10 \le x \le 50,$
$20 \le y \le 40$

$A = 40 \cdot 20 = 800$

Average production:

$$\frac{1}{800}\int_{10}^{50}\int_{20}^{40} 500x^{0.2}y^{0.8}\,dy\,dx$$

$$= \frac{5}{8}\int_{10}^{50} \frac{x^{0.2}y^{1.8}}{1.8}\Bigg|_{20}^{40} dx = \frac{25}{72}\int_{10}^{50} x^{0.2}(40^{1.8} - 20^{1.8})\,dx$$

$$= \frac{25(40^{1.8} - 20^{1.8})}{72} \cdot \frac{x^{1.2}}{1.2}\Bigg|_{10}^{50} = \frac{125}{432}(40^{1.8} - 20^{1.8})(50^{1.2} - 10^{1.2})$$

$\approx 14{,}753$ units

69. $R = q_1p_1 + q_2p_2$ where
$q_1 = 300 - 2p_1,$
$q_2 = 500 - 1.2p_2, 25 \le p_1 \le 50,$
and $50 \le p_2 \le 75.$

$A = 25 \cdot 25 = 625$

$R = (300 - 2p_1)p_1 + (500 - 1.2p_2)p_2$

$R = 300p_1 - 2p_1^2 + 500p_2 - 1.2p_2^2$

Average Revenue:

$$\frac{1}{625}\int_{25}^{50}\int_{50}^{75}(300p_1 - 2p_1^2 + 500p_2 - 1.2p_2^2)dp_2dp_1$$

$$= \frac{1}{625}\int_{25}^{50}(300p_1p_2 - 2p_1^2p_2 + 250p_2^2 - 0.4p_2^3)\Big|_{50}^{75} dp_1$$

$$= \frac{1}{625}\int_{25}^{50}\begin{bmatrix} 22{,}500p_1 - 150p_1^2 + 1{,}406{,}250 - 168{,}750) \\ -(15{,}000p_1 - 100p_1^2 + 625{,}000 - 50{,}000) \end{bmatrix} dp_1$$

$$= \frac{1}{625}\int_{25}^{50}(662{,}500 + 7500p_1 - 50p_1^2)dp_1$$

$$= \frac{1}{625}\left(662{,}500p_1 + 3750p_1^2 - \frac{50p_1^3}{3}\right)\Bigg|_{25}^{50}$$

$$= \frac{1}{625}\left(33{,}125{,}000 + 9{,}375{,}000 - \frac{6{,}250{,}000}{3} - 16{,}562{,}500 - 2{,}343{,}750 + \frac{781{,}250}{3}\right)$$

$\approx \$34{,}833$

71. $P(x,y) = 36xy - x^3 - 8y^3$

Areas $= (8-0)(4-0)$

$= 32$

The average profit is

$$\frac{1}{32}\iint_R (36xy - x^3 - 8y^3)\,dy\,dx$$

$$= \frac{1}{32}\int_0^8\int_0^4 (36xy - x^3 - 8y^3)\,dy\,dx$$

$$= \frac{1}{32}\int_0^8 \left(\frac{36xy^2}{2} - x^3y - \frac{8y^4}{4}\right)\Bigg|_0^4 dx$$

$$= \frac{1}{32}\int_0^8 \left[\frac{36x(4-0)^2}{2} - x^3(4-0) - \frac{8(4-0)^4}{4}\right] dx$$

$$= \frac{1}{32}\int_0^8 (288x - 4x^3 - 512)dx$$

$$= \frac{1}{32}\left(\frac{288x^2}{2} - \frac{4x^4}{4} - 512x\right)\Bigg|_0^8$$

$$= \frac{1}{32}\left[\frac{288(8-0)^2}{2} - \frac{4(8-0)^4}{4} - 512(8-0)\right]$$

$$= \frac{1}{32}(9216 - 4096 - 4096)$$

$= \$32{,}000$

Chapter 17 Review Exercises

1. True

2. True

3. True

4. True

5. False: $f(x+h,y) = 3(x+h)^2 + 2(x+h)y + y^2$

6. False: (a, b) could be a saddle point.

7. False: No; near a saddle point the function takes on values both larger and smaller than its value at the saddle point.

8. True

9. False: We need to test values of the function at nearby points that satisfy the constraints to tell if the point found represents a maximum or minimum.

10. False: When dx and dy are interchanged, the limits on the first integral must be exchanged with the limits on the second integral.

11. True

12. False: The two integrals are over different regions, and neither region is a simple region of the sort that we deal with in this chapter.

17.
$$f(x,y) = -4x^2 + 6xy - 3$$
$$f(-1,2) = -4(-1)^2 + 6(-1)(2) - 3 = -19$$
$$f(6,-3) = -4(6)^2 + 6(6)(-3) - 3$$
$$= -4(36) + (-108) - 3$$
$$= -255$$

19.
$$f(x,y) = \frac{x-2y}{x+5y}$$
$$f(-1,2) = \frac{(-1)-2(2)}{(-1)+5(2)} = \frac{-5}{9} = -\frac{5}{9}$$
$$f(6,-3) = \frac{(6)-2(-3)}{(6)+5(-3)} = \frac{12}{-9} = -\frac{4}{3}$$

21. The plane $x + y + z = 4$ intersects the axes at (4, 0, 0), (0, 4, 0), and (0, 0, 4).

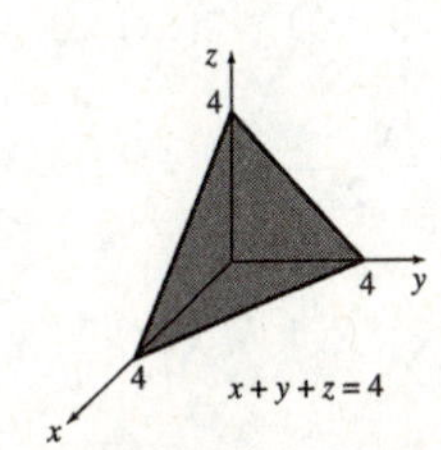

23. The plane $5x + 2y = 10$ intersects the x- and y-axes at (2, 0, 0) and (0, 5, 0). Note that there is no z-intercept since $x = y = 0$ is not a solution of the equation of the plane.

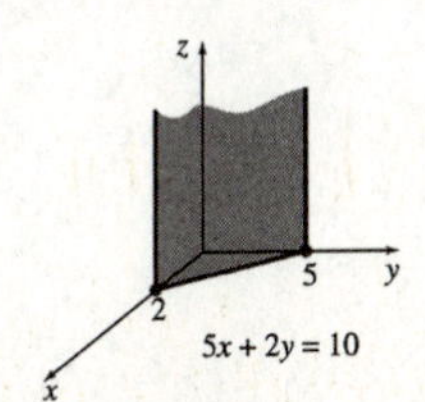

25. $x = 3$

The plane is parallel to the yz-plane. It intersects the x-axis at (3, 0, 0).

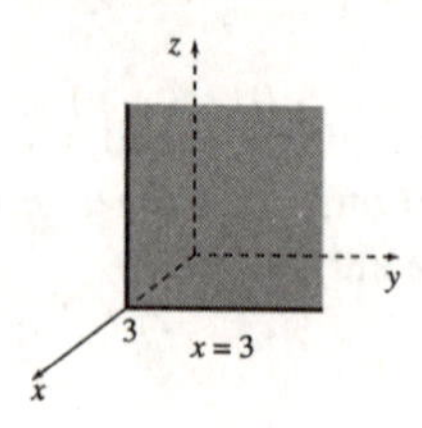

27. $z = f(x,y) = 3x^3 + 4x^2y - 2y^2$

(a) $\frac{\partial z}{\partial x} = 9x^2 + 8xy$

(b) $\frac{\partial z}{\partial y} = 4x^2 - 4y$

$\left(\frac{\partial z}{\partial y}\right)(-1,4) = 4(-1)^2 - 4(4) = -12$

(c) $f_{xy}(x,y) = 8x$

$f_{xy}(x,y)(2,-1) = 8(2) = 16$

29. $f(x,y) = 6x^2y^3 - 4y$

$f_x(x,y) = 12xy^3$

$f_y(x,y) = 18x^2y^2 - 4$

31. $f(x,y) = \sqrt{4x^2 + y^2}$

$$f_x(x,y) = \frac{1}{2}(4x^2 + y^2)^{-1/2}(8x) = \frac{4x}{(4x^2 + y^2)^{-1/2}}$$

$$f_y(x,y) = \frac{1}{2}(4x^2 + y^2)^{-1/2}(2y) = \frac{y}{(4x^2 + y^2)^{1/2}}$$

33. $f(x,y) = x^3e^{3y}$

$f_x(x,y) = 3x^2e^{3y}$

$f_y(x,y) = 3x^3e^{3y}$

35. $f(x,y) = \ln|2x^2 + y^2|$

$$f_x(x,y) = \frac{1}{2x^2 + y^2} \cdot 4x = \frac{4x}{2x^2 + y^2}$$

$$f_y(x,y) = \frac{1}{2x^2 + y^2} \cdot 2y = \frac{2y}{2x^2 + y^2}$$

37. $f(x,y) = 5x^3y - 6xy^2$

$f_x(x,y) = 15x^2y - 6y^2$

$f_{xx}(x,y) = 30xy$

$f_{xy}(x,y) = 15x^2 - 12y$

39. $f(x,y) = \frac{3x}{2x - y}$

$$f_x(x,y) = \frac{(2x - y) \cdot 3 - 3x \cdot 2}{(2x - y)^2} = \frac{-3y}{(2x - y)^2}$$

$$f_{xx}(x,y) = \frac{(2x - y)^2 \cdot 0 - (-3y) \cdot 2(2x - y) \cdot 2}{(2x - y)^4} = \frac{12y}{(2x - y)^3}$$

$$f_{xy}(x,y) = \frac{\left[(2x - y)^2 \cdot (-3) - (-3y) \cdot 2(2x - y) \cdot (-1)\right]}{(2x - y)^4} = \frac{-6x - 3y}{(2x - y)^3}$$

41. $f(x,y) = 4x^2e^{2y}$

$f_x(x,y) = 8xe^{2y}$

$f_{xx}(x,y) = 8e^{2y}$

$f_{xy}(x,y) = 16xe^{2y}$

43. $f(x,y) = \ln|2 - x^2y|$

$$f_x(x,y) = \frac{1}{2 - x^2y} \cdot (-2xy) = \frac{2xy}{x^2y - 2}$$

$$f_{xx}(x,y) = \frac{(x^2y - 2)2y - 2xy(2xy)}{(x^2y - 2)^2} = \frac{2y[(x^2y - 2) - 2x^2y]}{(x^2y - 2)^2} = \frac{2y(-x^2y - 2)}{(x^2y - 2)^2} = \frac{-2x^2y^2 - 4y}{(2 - x^2y)^2}$$

$$f_{xy}(x,y) = \frac{2x(x^2y - 2) - x^2(2xy)}{(x^2y - 2)^2} = \frac{2x[(x^2y - 2) - x^2y]}{(x^2y - 2)^2} = \frac{2x(-2)}{(x^2y - 2)^2} = \frac{-4x}{(2 - x^2y)^2}$$

45. $z = 2x^2 - 3y^2 + 12y$

$$z_x(x,y) = 4x$$
$$z_y(x,y) = -6y + 12$$

If $z_x(x,y) = 0, x = 0$. If $z_y(x,y) = 0, y = 2$.

Therefore, (0, 2) is a critical point.

$$z_{xx}(x,y) = 4$$
$$z_{yy}(x,y) = -6$$
$$z_{xy}(x,y) = 0$$
$$D = 4(-6) - 0^2 = -24 < 0$$

There is a saddle point at (0, 2).

47. $f(x,y) = x^2 + 3xy - 7x + 5y^2 - 16y$

$$f_x(x,y) = 2x + 3y - 7$$
$$f_y(x,y) = 3x + 10y - 16$$

Solve the system $f_x(x,y) = 0, f_y(x,y) = 0$.

$$2x + 3y - 7 = 0$$
$$3x + 10y - 16 = 0$$

$$-6x - 9y + 21 = 0$$
$$6x + 20y - 32 = 0$$
$$11y - 11 = 0$$
$$y = 1$$

$$2x + 3(1) - 7 = 0$$
$$2x = 4$$
$$x = 2$$

Therefore, (2, 1) is a critical point.

$$f_{xx}(x,y) = 2$$
$$f_{yy}(x,y) = 10$$
$$f_{xy}(x,y) = 3$$
$$D = 2 \cdot 10 - 3^2 = 11 > 0$$

Since $f_{xx} = 2 > 0$, there is a relative minimum at (2,1).

49. $z = \frac{1}{2}x^2 + \frac{1}{2}y^2 + 2xy - 5x - 7y + 10$

$$z_x(x,y) = x + 2y - 5$$
$$z_y(x,y) = y + 2x - 7$$

Setting $z_x = z_y = 0$ and solving yields

$$x + 2y = 5$$
$$2x + y = 7$$

$$-2x - 4y = -10$$
$$2x + y = 7$$
$$-3y = -3$$
$$y = 1, x = 3.$$

$z_{xx}(x,y) = 1, z_{yy}(x,y) = 1, z_{xy}(x,y) = 2$

For (3, 1),

$$D = 1 \cdot 1 - 4 = -3 < 0.$$

Therefore, z has a saddle point at (3, 1).

51. $z = x^3 + y^2 + 2xy - 4x - 3y - 2$

$$z_x(x,y) = 3x^2 + 2y - 4$$
$$z_y(x,y) = 2y + 2x - 3$$

Setting $z_x(x,y) = z_y(x,y) = 0$ yields

$$3x^2 + 2y - 4 = 0 \quad (1)$$
$$2y + 2x - 3 = 0. \quad (2)$$

Solving for $2y$ in equation (2) gives $2y = -2x + 3$.

Substitute into equation (1).

$$3x^2 + (-2x) + 3 - 4 = 0$$
$$3x^2 - 2x - 1 = 0$$
$$(3x + 1)(x - 1) = 0$$

$$x = -\frac{1}{3} \quad \text{or} \quad x = 1$$
$$y = \frac{11}{6} \quad \text{or} \quad y = \frac{1}{2}$$

$z_{xx}(x,y) = 6x,\ z_{yy}(x,y) = 2,\ z_{xy}(x,y) = 2$

For $\left(-\frac{1}{3}, \frac{11}{6}\right)$,

$$D = 6\left(-\frac{1}{3}\right)(2) - 4$$
$$= -4 - 4 = -8 < 0,$$

so z has a saddle point at $\left(-\frac{1}{3}, \frac{11}{6}\right)$.

$$D = 6(1)(2) - 4 = 8 > 0.$$

$z_{xx}\left(1, \frac{1}{2}\right) = 6 > 0$, so z has a relative minimum at $\left(1, \frac{1}{2}\right)$.

55. $f(x,y) = x^2 + y^2; x = y - 6.$

1. $g(x,y) = x - y + 6$
2. $F(x,y,\lambda) = x^2 + y^2 - \lambda(x - y + 6)$
3. $F_x(x,y,\lambda) = 2x - \lambda$
   $F_y(x,y,\lambda) = 2y + \lambda$
   $F_\lambda(x,y,\lambda) = -(x - y + 6)$
4. $$2x - \lambda = 0 \quad (1)$$
   $$2y + \lambda = 0 \quad (2)$$
   $$x - y + 6 = 0 \quad (3)$$
5. Equations (1) and (2) give $\lambda = 2x$, and $\lambda = -2y$. Thus,

$$2x = -2y$$
$$x = -y.$$

Substituting into equation (3),

$$(-y) - y + 6 = 0$$
$$y = 3.$$

So $\quad x = -3.$

And $\quad f(-3,3) = (-3)^2 + (3)^2 = 18.$

$$F_{xx}(x,y,\lambda) = 2$$
$$F_{yy}(x,y,\lambda) = 2$$
$$F_{xy}(x,y,\lambda) = 0$$

$$D = 2 \cdot 2 - 0^2 = 4 > 0$$

Since $F_{xx} > 0$, there is a relative minimum of 18 at $(-3,3)$.

57. Maximize $f(x,y) = xy^2$, subject to $x + y = 75$.

1. $g(x,y) = x + y - 75$
2. $F(x,y,\lambda) = xy^2 - \lambda(x + y - 75)$
3. $F_x(x,y,\lambda) = y^2 - \lambda$
   $F_y(x,y,\lambda) = 2xy - \lambda$
   $F_\lambda(x,y,\lambda) = -(x + y - 75)$
4. $$y^2 - \lambda = 0 \quad (1)$$
   $$2xy - \lambda = 0 \quad (2)$$
   $$x + y - 75 = 0 \quad (3)$$
5. Equations (1) and (2) give $\lambda = y^2$ and $\lambda = 2xy$. Thus,

$$y^2 = 2xy$$
$$y(y - 2x) = 0$$
$$y = 0 \quad \text{or} \quad y = 2x$$

Substituting $y = 0$ into equation (3),

$$x + (0) - 75 = 0$$
$$x = 75.$$

Substituting $y = 2x$ into equation (3),

$$x + (2x) - 75 = 0$$
$$x = 25.$$

So $\quad y = 2x = 50.$

Thus,

$f(75,0) = 75(0)^2 = 0$, and

$f(25,50) = 25(50)^2 = 62,500.$

Since $f(25,50) > f(75,0), x = 25$ and $y = 50$ will maximize $f(x,y) = xy^2$.

59. $z = f(x,y) = 6x^2 - 7y^2 + 4xy$
$x = 3, y = -1, dx = 0.03, dy = 0.01$

$$f_x(x,y) = 12x + 4y$$
$$f_y(x,y) = -14y + 4x$$
$$\begin{aligned} dz &= (12x + 4y)dx + (-14y + 4x)dy \\ &= [12(3) + 4(-1)](0.03) \\ &\quad + [-14(-1) + 4(3)](0.01) \\ &= 0.96 + 0.26 = 1.22 \end{aligned}$$

61. Let $z = f(x,y) = \sqrt{x^2 + y^2}$.

Then

$$dz = f_x(x,y)dx + f_y(x,y)\,dy.$$
$$\begin{aligned} dz &= \frac{1}{2}(x^2 + y^2)^{-1/2}(2x)\,dx \\ &\quad + \frac{1}{2}(x^2 + y^2)(2y)\,dx \\ &= \frac{x}{\sqrt{x^2 + y^2}}dx + \frac{y}{\sqrt{x^2 + y^2}}dy \end{aligned}$$

To approximate $\sqrt{5.1^2 + 12.05^2}$, we let $x = 5$, $dx = 0.1$, $y = 12$, and $dy = 0.05$.

Then,

$$\begin{aligned} dz &= \frac{5}{\sqrt{5^2 + 12^2}}(0.1) + \frac{12}{\sqrt{5^2 + 12^2}}(0.05) \\ &= \frac{5}{13}(0.1) + \frac{12}{13}(0.05) \approx 0.0846. \end{aligned}$$

Therefore,

$$\begin{aligned} f(5.1,12.05) &= f(5,12) + \Delta z \\ &\approx f(5,12) + dz \\ &= \sqrt{5^2 + 12^2} + 0.0846 \end{aligned}$$
$$f(5.1,12.05) \approx 13.0846$$

Using a calculator, $\sqrt{5.1^2 + 12.05^2} \approx 13.0848$.
The absolute value of the difference of the two results is $|13.0846 - 13.0848| = 0.0002$.

63. $\displaystyle\int_1^4 \frac{4y - 3}{\sqrt{x}}dx$

$$\begin{aligned} &= (4y - 3)(2\sqrt{x})\Big|_1^4 \\ &= (4y - 3)(2 \cdot 2 - 2 \cdot 1) \\ &= 8y - 6 \end{aligned}$$

65. $\displaystyle\int_0^5 \frac{6x}{\sqrt{4x^2 + 2y^2}}dx$

Let $u = 4x^2 + 2y^2$; then $du = 8x\,dx$.

When $x = 0$, $u = 2y^2$.

When $x = 5$, $u = 100 + 2y^2$.

$$\begin{aligned} &= \frac{3}{4}\int_{2y^2}^{100+2y^2} u^{-1/2}du \\ &= \frac{3}{4}(2u^{1/2})\Big|_{2y^2}^{100+2y^2} \\ &= \frac{3}{4} \cdot 2[(100 + 2y^2)^{1/2} - (2y^2)^{1/2}] \\ &= \frac{3}{2}[(100 + 2y^2)^{1/2} - (2y^2)^{1/2}] \end{aligned}$$

67. $\displaystyle\int_0^2 \left[\int_0^4 (x^2y^2 + 5x)\,dx\right]dy$

$$\begin{aligned} &= \int_0^2 \left(\frac{1}{3}x^3y^2 + \frac{5}{2}x^2\right)\Bigg|_0^4 dx \\ &= \int_0^2 \left(\frac{64}{3}y^2 + 40\right)dy \\ &= \left(\frac{64y^3}{9} + 40y\right)\Bigg|_0^2 \\ &= \frac{64}{9}(8) + 40(2) \\ &= \frac{512}{9} + \frac{720}{9} \\ &= \frac{1232}{9} \end{aligned}$$

69. $$\int_3^4\left[\int_2^5 \sqrt{6x+3y}\,dx\right]dy$$

$$= \int_3^4 \frac{1}{9}(6x+3y)^{3/2}\Big|_2^5\,dx$$

$$= \int_3^4 \frac{1}{9}[(30+3y)^{3/2} - (12+3y)^{3/2}]dy$$

$$= \frac{1}{3}\cdot\frac{1}{9}\cdot\frac{2}{5}\cdot[(30+3y)^{5/2} - (12+3y)^{5/2}]\Big|_3^4$$

$$= \frac{2}{135}[(42)^{5/2} - (24)^{5/2} - (39)^{5/2} + (21)^{5/2}]$$

71. $$\int_2^4\int_2^4 \frac{dx\,dy}{y} = \int_2^4\left(\frac{1}{y}x\right)\Big|_2^4\,dy$$

$$= \int_2^4\left[\frac{1}{y}(4-2)\right]dy$$

$$= 2\ln|y|\Big|_2^4$$

$$= 2\ln\left|\frac{4}{2}\right|$$

$$= 2\ln 2 \text{ or } \ln 4$$

73. $$\iint_R (x^2+2y^2)dx\,dy;\ 0\le x\le 5,\ 0\le y\le 2$$

$$\iint_R (x^2+2y^2)\,dx\,dy$$

$$= \int_0^2\int_0^5 (x^2+2y^2)\,dx\,dy$$

$$= \int_0^2\left(\frac{1}{3}x^3 + 2xy^2\right)\Big|_0^5\,dy$$

$$= \int_0^2\left[\left(\frac{125}{3}+10y^2\right) - 0\right]dy$$

$$= \int_0^2\left(\frac{125}{3}+10y^2\right)dy$$

$$= \left(\frac{125}{3}y + \frac{10}{3}y^3\right)\Big|_0^2$$

$$= \frac{250}{3} + \frac{80}{3} = 110$$

75. $$\iint_R \sqrt{y+x}\,dx\,dy;\ 0\le x\le 7, 1\le y\le 9$$

$$\iint_R \sqrt{y+x}\,dx\,dy$$

$$= \int_0^7\int_1^9 \sqrt{y+x}\,dy\,dx$$

$$= \int_0^7\left[\frac{2}{3}(y+x)^{3/2}\right]\Big|_1^9\,dx$$

$$= \int_0^7 \frac{2}{3}[(9+x)^{3/2} - (1+x)^{3/2}]dx$$

$$= \frac{2}{3}\cdot\frac{2}{5}[(9+x)^{5/2} - (1+x)^{5/2}]\Big|_0^7$$

$$= \frac{4}{15}[(16)^{5/2} - (8)^{5/2} - (9)^{5/2} + (1)^{5/2}]$$

$$= \frac{4}{15}[4^5 - (2\sqrt{2})^5 - 3^5 + 1]$$

$$= \frac{4}{15}(1024 - 32(4\sqrt{2}) - 243 + 1)$$

$$= \frac{4}{15}(782 - 128\sqrt{2})$$

$$= \frac{4}{15}(782 - 8^{5/2})$$

77. $z = x + 8y + 4;\ 0\le x\le 3, 1\le y\le 2$

$$V = \int_0^3\int_1^2 (x+8y+4)\,dy\,dx$$

$$= \int_0^3 (xy + 4y^2 + 4y)\Big|_1^2\,dx$$

$$= \int_0^3 [(2x+16+8) - (x+4+4)]\,dx$$

$$= \int_0^3 (x+16)\,dx$$

$$= \left(\frac{1}{2}x^2 + 16x\right)\Big|_0^3$$

$$= \left(\frac{9}{2}+48\right) - 0 = \frac{105}{2}$$

79. $\int_0^1 \int_0^{2x} xy\,dy\,dx$

$= \int_0^1 \left(\frac{xy^2}{2} \right) \Big|_0^{2x} dx$

$= \int_0^1 \frac{x}{2}(4x^2 - 0)\,dx$

$= \int_0^1 2x^3 dx$

$= \left(\frac{1}{2}x^4 \right) \Big|_0^1 = \frac{1}{2}$

81. $\int_0^1 \int_{x^2}^{x} x^3 y\,dy\,dx$

$\int_0^1 \left(\frac{x^3}{2} y^2 \right) \Big|_{x^2}^{x} dx$

$= \int_0^1 \frac{x^3}{2}(x^2 - x^4)\,dx$

$= \frac{1}{2} \int_0^1 (x^5 - x^7)\,dx$

$= \frac{1}{2} \left(\frac{x^6}{6} - \frac{x^8}{8} \right) \Big|_0^1$

$= \frac{1}{2} \left(\frac{1}{6} - \frac{1}{8} \right) = \frac{1}{2} \cdot \frac{1}{24} = \frac{1}{48}$

83. $\int_0^2 \int_{x/2}^{1} \frac{1}{y^2 + 1}\,dy\,dx$

Change the order of integration.

$\int_0^2 \int_{x/2}^{1} \frac{1}{y^2 + 1}\,dy\,dx$

$= \int_0^1 \int_0^{2y} \frac{1}{y^2 + 1}\,dx\,dy$

$= \int_0^1 \frac{x}{y^2 + 1} \Big|_0^{2y} dy$

$= \int_0^1 \left[\frac{1}{y^2 + 1}(2y) - \frac{1}{y^2 + 1}(0) \right] dy$

$= \int_0^1 \frac{2y}{y^2 + 1}\,dy$

$= \ln(y^2 + 1) \Big|_0^1$

$= \ln 2 - \ln 1$

$= \ln 2 - 0 = \ln 2$

85. $\iint_R (2x + 3y)\,dx\,dy;\ 0 \le y \le 1,$

$y \le x \le 2 - y$

$\int_0^1 \int_y^{2-y} (2x + 3y)\,dx\,dy$

$= \int_0^1 (x^2 + 3xy) \Big|_y^{2-y} dy$

$= \int_0^1 [(2 - y)^2 - y^2 + 3y(2 - y - y)]\,dy$

$= \int_0^1 (4 - 4y + y^2 - y^2 + 6y - 6y^2)\,dy$

$= \int_0^1 (4 + 2y - 6y^2)\,dy$

$= (4y + y^2 - 2y^3) \Big|_0^1$

$= 4 + 1 - 2 = 3$

87. $C(x, y) = 4x^2 + 5y^2 - 4xy + \sqrt{x}$

(a) $C(10,5)$

$$= 4(10)^2 + 5(5)^2 - 4(10)(5) + \sqrt{10}$$
$$= 400 + 125 - 200 + \sqrt{10}$$
$$= 325 + \sqrt{10} \approx 328.16$$

The cost is about \$328.16.

(b) $C(15,10)$

$$= 4(15)^2 + 5(10)^2 - 4(15)(10) + \sqrt{15}$$
$$= 900 + 500 - 600 + \sqrt{15}$$
$$= 800 + \sqrt{15} \approx 803.87$$

The cost is about \$803.87.

(c) $C(20, 20)$

$$= 4(20)^2 + 5(20)^2 - 4(20)(20) + \sqrt{20}$$
$$= 1600 + 2000 - 1600 + \sqrt{20}$$
$$= 2000 + \sqrt{20} \approx 2004.47$$

The cost is about \$2004.47.

89. $z = x^{0.7}y^{0.3}$

(a) The marginal productivity of labor is

$$\frac{\partial z}{\partial x} = 0.7x^{0.7-1}y^{0.3} = \frac{0.7y^{0.3}}{x^{0.3}}.$$

(b) The marginal productivity of capital is

$$\frac{\partial z}{\partial y} = 0.3x^{0.7}y^{0.3-1} = \frac{0.3x^{0.7}}{y^{0.7}}.$$

91. Maximize $f(x, y) = xy^3$ subject to $2x + 4y = 80$.

1. $g(x, y) = 2x + 4y - 80$

2. $F(x, y) = xy^3 - \lambda(2x + 4y - 80)$

3. $F_x(x, y, \lambda) = y^3 - 2\lambda$
$F_y(x, y, \lambda) = 3xy^2 - 4\lambda$
$F_\lambda(x, y, \lambda) = -2x - 4y + 80$

4. $$y^3 - 2\lambda = 0$$
$$3xy^2 - 4\lambda = 0$$
$$-2x - 4y + 80 = 0$$

5. Use the first and second equations to express 4λ and eliminate 4λ.

$$2y^3 = 4\lambda$$
$$3xy^2 = 4\lambda$$
$$2y^3 = 3xy^2$$

If y equals 0, the utility will have a minimum value of 0. So we can assume $y \neq 0$ and divide by y^2 to find that $2y = 3x$. Substitute $6x$ for $4y$ in the last equation.

$$-2x - 6x + 80 = 0$$
$$x = 10, y = \frac{3}{2}x = 15.$$
$$f(10,15) = 33,750.$$

The maximum utility is 33,750, obtained by purchasing 10 units of x and 15 units of y.

93. $C(x, y) = 100\ln(x^2 + y) + e^{xy/20}$
$x = 15,\ y = 9,\ dx = 1,\ dy = -1$

$$dC = \left(\frac{200x}{x^2 + y} + \frac{y}{20}e^{xy/20}\right)dx + \left(\frac{100}{x^2 + y} + \frac{x}{20}e^{xy/20}\right)dy$$

$dC(15,9)$

$$= \left(\frac{200(15)}{15^2 + 9} + \frac{9}{20}e^{(15)(9)/20}\right)(1) + \left(\frac{100}{15^2 + 9} + \frac{15}{20}e^{(9)(15)/20}\right)(-1)$$
$$= \frac{1450}{117} - \frac{3}{10}e^{27/4}$$
$$= -243.82$$

Costs decrease by \$243.82.

95. $V = \frac{4}{3}\pi r^3, r = 2 \text{ ft},$

$$dr = 1 \text{ in} = \frac{1}{12} \text{ ft}$$

$$dV = 4\pi r^2 dr = 4\pi(2)^2\left(\frac{1}{12}\right) \approx 4.19 \text{ ft}^3$$

97. $P(x, y) = 0.01(-x^2 + 3xy + 160x - 5y^2 + 200y + 2600)$

with $x + y = 280$.

(a) $y = 280 - x$

$$P(x) = 0.01[-x^2 + 3x(280 - x) + 160x - 5(280 - x)^2 + 200(280 - x) + 2600]$$
$$= 0.01(-x^2 + 840x - 3x^2 + 160x - 392{,}000 + 2800x - 5x^2 + 56{,}000 - 200x + 2600)$$

$$P(x) = 0.01(-9x^2 + 3600x - 333{,}400)$$
$$P'(x) = 0.01(-18x + 3600)$$
$$0.01(-18x + 3600) = 0$$
$$-18x = -3600$$
$$x = 200$$

If $x < 200, P'(x) > 0$, and if $x > 200$, $P'(x) < 0$.

Therefore, P is maximum when $x = 200$. If $x = 200$, $y = 80$.

$$P(200, 80) = 0.01[-200^2 + 3(200)(80) + 160(200) - 5(80)^2 + 200(80) + 2600]$$
$$= 0.01(26{,}600) = 266$$

Thus, \$200 spent on fertilizer and \$80 spent on seed will produce a maximum profit of \$266 per acre.

(b) $P(x, y) = 0.01(-x^2 + 3xy + 160x - 5y^2 + 200y + 2600)$

$$P_x(x, y) = 0.01(-2x + 3y + 160)$$
$$P_y(x, y) = 0.01(3x - 10y + 200)$$
$$0.01(-2x + 3y + 160) = 0$$
$$0.01(3x - 10y + 200) = 0$$

These equations simplify to

$$-2x + 3y = -160$$
$$3x - 10y = -200.$$

Solve this system.

$$-6x + 9y = -480$$
$$6x - 20y = -400$$
$$-11y = -880$$
$$y = 80$$

If $y = 80$,

$$3x - 10(80) = -200$$
$$3x = 600$$
$$x = 200.$$

$$P_{xx}(x, y) = 0.01(-2) = -0.02$$
$$P_{yy}(x, y) = 0.01(-10) = -0.1$$
$$P_{xy}(x, y) = 0$$

For (200,80), $D = (-0.02)(-0.1) - 0^2 = 0.002 > 0$, and $P_{xx} < 0$, so there is a relative maximum at (200,80).

$P(200,80) = 266$, as in part (a) Thus, \$200 spent on fertilizer and \$80 spent on seed will produce a maximum profit of \$ 266 per acre.

(c) Maximize $P(x, y) = 0.01(-x^2 + 3xy + 160x - 5y^2 + 200y + 2600)$

subject to $x + y = 280$.

1. $g(x, y) = x + y - 280$

2 $F(x, y, \lambda) = 0.01(-x^2 + 3xy + 160x - 5y^2 + 200y + 2600) - \lambda(x + y - 280)$

3. $F_x = 0.01(-2x + 3y + 160) - \lambda$
$F_y = 0.01(3x - 10y + 200) - \lambda$
$F_\lambda = -(x + y - 280)$

4. $0.01(-2x + 3y + 160) - \lambda = 0$ (1)
$0.01(3x - 10y + 200) - \lambda = 0$ (2)
$x + y - 280 = 0$ (3)

5. Equations (1) and (2) give

$$0.01(-2x + 3y + 160) = 0.01(3x - 10y + 200)$$
$$-2x + 3y + 160 = 3x - 10y + 200$$
$$-5x + 13y = 40.$$

Multiplying equation (3) by 5 gives

$$5x + 5y - 1400 = 0.$$

$$-5x + 13y = 40$$
$$5x + 5y = 1400$$
$$18y = 1440$$
$$y = 80$$

If $y = 80$,

$$5x + 5(80) = 1400$$
$$5x = 1000$$
$$x = 200.$$

Thus, $P(200,80)$ is a maximum. As before,

$$P(200,80) = 266.$$

Thus, \$200 spent on fertilizer and \$80 spent on seed will produce a maximum profit of \$266 per acre.

99. $T(A,W,S) = -18.37 - 0.09A + 0.34W + 0.25S$

(a) $T(65,85,180) = -18.37 - 0.09(65) + 0.34(85) + 0.25(180) = 49.68$

The total body water is 49.68 liters.

(b) $T_A(A,W,S) = -0.09$

The approximate change in total body water if age is increased by 1 yr and mass and height are held constant is -0.09 liter.

$$T_W(A,W,S) = 0.34$$

The approximate change in total body water if mass is increased by 1 kg and height are held constant is 0.34 liter.

$$T_S(A,W,S) = 0.25$$

The approximate change in total body water if height is increased by 1 cm and age and mass are held constant is 0.25 liter.

101. (a) $f(60,1900) \approx 50$

In 1900, 50% of those born 60 years earlier are still alive.

(b) $f(70,2000) \approx 75$

In 2000, 75% of those born 70 years earlier are still alive.

(c) $f_x(60,1900) \approx -1.25$

In 1900, the percent of those born 60 years earlier who are still alive was dropping at a rate of 1.25 percent per additional year of life.

(d) $f_x(70,2000) \approx -2$

In 2000, the percent of those born 70 years earlier who are still alive was dropping at a rate of 2 percent per additional year of life.

103. Let x be the length of each of the square faces of the box and y be the length of the box.

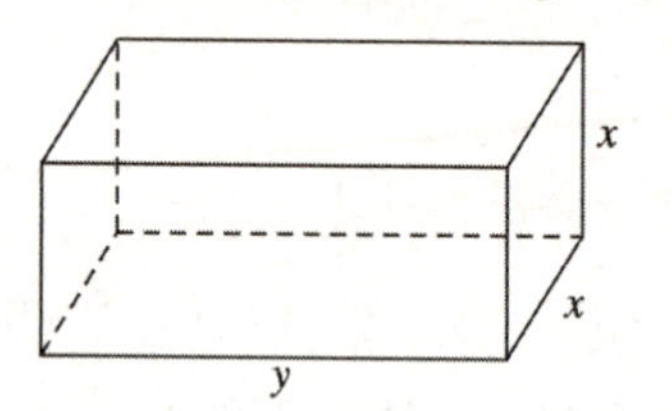

Since the volume must be 125, the constraint is $125 = x^2y$.

$f(x,y) = 2x^2 + 4xy$ is the surface area of the box.

1. $g(x) = x^2y - 125$
2. $F(x,y,\lambda) = 2x^2 + 4xy - \lambda(x^2y - 125)$
3. $F_x(x,y,\lambda) = 4x + 4y - 2xy\lambda$

   $F_y(x,y,\lambda) = 4x - x^2\lambda$

   $F_\lambda(x,y,\lambda) = -(x^2y - 125)$
4. $4x + 4y - 2xy\lambda = 0 \quad (1)$

   $4x - x^2\lambda = 0 \quad (2)$

   $x^2y - 125 = 0 \quad (3)$
5. Factoring equation (2) gives

$$x(4 - x\lambda) = 0$$
$$x = 0 \text{ or } 4 - x\lambda = 0.$$

Since $x = 0$ is not a solution of equation (3), then

$$4 - x\lambda = 0$$
$$\lambda = \frac{4}{x}.$$

Substituting into equation (1) gives

$$4x + 4y - 2xy\left(\frac{4}{x}\right) = 0$$

or

$$4x + 4y - 8y = 0$$
$$x = y.$$

Substituting $x = y$ into equation (3) gives

$$x^2y - 125 = 0$$
$$y^3 = 125$$
$$y = 5.$$

Therefore, $x = y = 5$. The dimensions are 5 inches by 5 inches by 5 inches.

Chapter 18

PROBABILITY AND CALCULUS

18.1 Continuous Probability Models

Your Turn 1

$$f(x) = \frac{2}{x^2} \text{ on } [1, 2]$$

Condition 1 holds because $f(x) \geq 0$ on $[1, 2]$.

Condition 2 holds because
$$\int_1^2 f(x)\,dx = \int_1^2 \frac{2}{x^2}\,dx = -\frac{2}{x}\bigg|_1^2 = -1 - (-2) = 1.$$

Thus $f(x)$ is a probability density function for the interval $[1, 2]$.

$$\begin{aligned} P(3/2 \leq X \leq 2) &= \int_{3/2}^2 \frac{2}{x^2}\,dx \\ &= -\frac{2}{x}\bigg|_{3/2}^2 \\ &= -1 - \left(-\frac{2}{3/2}\right) \\ &= -1 + \frac{4}{3} \\ &= \frac{1}{3} \end{aligned}$$

Your Turn 2

The function $f(x) = kx^3$ will be nonnegative on the interval $[0, 4]$ for any positive k.

$$\begin{aligned} \int_0^4 kx^3\,dx &= \frac{k}{4}x^4\bigg|_0^4 \\ &= \frac{k}{4}4^4 \\ &= 64k \end{aligned}$$

Thus $f(x)$ will be a probability density function on $[0, 4]$ when $64k = 1$, or $k = 1/64$.

Your Turn 3

$$\begin{aligned} P(0 \leq X \leq 1) &= \int_0^1 2xe^{-x^2}\,dx \\ &= \left(-e^{-x^2}\right)\bigg|_0^1 \\ &= -\frac{1}{e} - (-1) \\ &= 1 - \frac{1}{e} \approx 0.6321 \end{aligned}$$

Your Turn 4

The cumulative distribution function from Example 5 is $F(x) = -e^{-x^2} + 1$. The probability that there is a bird's nest within 1 km of the given point is

$$F(1) = -\frac{1}{e} + 1 \approx 0.6321.$$

18.1 Exercises

1. $f(x) = \frac{1}{9}x - \frac{1}{18}$; $[2, 5]$

Show that condition 1 holds.

Since $2 \leq x \leq 5$,

$$\frac{2}{9} \leq \frac{1}{9}x \leq \frac{5}{9}$$
$$\frac{1}{6} \leq \frac{1}{9}x - \frac{1}{18} \leq \frac{1}{2}.$$

Hence, $f(x) \geq 0$ on $[2, 5]$.

Show that condition 2 holds.

$$\begin{aligned} \int_2^5 \left(\frac{1}{9}x - \frac{1}{18}\right)dx &= \frac{1}{9}\int_2^5 \left(x - \frac{1}{2}\right)dx \\ &= \frac{1}{9}\left(\frac{x^2}{2} - \frac{1}{2}x\right)\bigg|_2^5 \\ &= \frac{1}{9}\left(\frac{25}{2} - \frac{5}{2} - \frac{4}{2} + 1\right) \\ &= \frac{1}{9}(8 + 1) \\ &= 1 \end{aligned}$$

Yes, $f(x)$ is a probability density function.

3. $f(x) = \frac{1}{21}x^2$; [1, 4]

Since $x^2 \geq 0$, $f(x) \geq 0$ on [1, 4].

$$\frac{1}{21}\int_1^4 x^2\,dx = \frac{1}{21}\left(\frac{x^3}{3}\right)\Big|_1^4$$
$$= \frac{1}{21}\left(\frac{64}{3} - \frac{1}{3}\right) = 1$$

Yes, $f(x)$ is a probability density function.

5. $f(x) = 4x^3$; [0, 3]

$$4\int_0^3 x^3\,dx = 4\left(\frac{x^4}{4}\right)\Big|_0^3$$
$$= 4\left(\frac{81}{4} - 0\right)$$
$$= 81 \neq 1$$

No, $f(x)$ is not a probability density function.

7. $f(x) = \frac{x^2}{16}$; [−2, 2]

$$\frac{1}{16}\int_{-2}^2 x^2\,dx = \frac{1}{16}\left(\frac{x^3}{3}\right)\Big|_{-2}^2$$
$$= \frac{1}{16}\left(\frac{8}{3} + \frac{8}{3}\right)$$
$$= \frac{1}{3} \neq 1$$

No, $f(x)$ is not a probability density function.

9. $f(x) = \frac{5}{3}x^2 - \frac{5}{90}$; [−1, 1]

Let $x = 0$. Then $f(x) = f(0) = -\frac{5}{90} < 0$.

So $f(x) < 0$ for at least one x-value in [−1, 1].

No, $f(x)$ is not a probability density function.

11. $f(x) = kx^{1/2}$; [1, 4]

$$\int_1^4 kx^{1/2}dx = \frac{2}{3}kx^{3/2}\Big|_1^4$$
$$= \frac{2}{3}k(8 - 1)$$
$$= \frac{14}{3}k$$

If $\frac{14}{3}k = 1$,

$$k = \frac{3}{14}.$$

Notice that $f(x) = \frac{3}{4}x^{1/2} \geq 0$ for all x in [1, 4].

13. $f(x) = kx^2$; [0, 5]

$$\int_0^5 kx^2\,dx = k\frac{x^3}{3}\Big|_0^5$$
$$= k\left(\frac{125}{3} - 0\right)$$
$$= k\left(\frac{125}{3}\right)$$

If $k\left(\frac{124}{3}\right) = 1$,

$$k = \frac{3}{125}.$$

Notice that $f(x) = \frac{3}{125}x^2 \geq 0$ for all x in [0, 5].

15. $f(x) = kx$; [0, 3]

$$\int_0^3 kx\,dx = k\frac{x^2}{2}\Big|_0^3$$
$$= k\left(\frac{9}{2} - 0\right)$$
$$= \frac{9}{2}k$$

If $\frac{9}{2}k = 1$,

$$k = \frac{2}{9}.$$

Notice that $f(x) = \frac{2}{9}x \geq 0$ for all x in [0, 3].

17. $f(x) = kx;\ [1, 5]$

$$\int_1^5 kx\,dx = k\frac{x^2}{2}\Big|_1^5$$

$$= k\left(\frac{25}{2} - \frac{1}{2}\right)$$

$$= 12k$$

If $12k = 1$,

$$k = \frac{1}{12}.$$

Notice that $f(x) = \frac{1}{12}x \geq 0$ for all x in $[1, 5]$.

19. For the probability density function $f(x) = \frac{1}{9}x - \frac{1}{18}$ on $[2, 5]$, the cumulative distribution function is

$$F(x) = \int_a^x f(t)\,dt$$

$$= \int_a^x \left(\frac{1}{9}t - \frac{1}{18}\right) dt$$

$$= \left(\frac{1}{18}t^2 - \frac{1}{18}t\right)\Big|_2^x$$

$$= \frac{1}{18}[(x^2 - x) - (4 - 2)]$$

$$= \frac{1}{18}(x^2 - x - 2),\ 2 \leq x \leq 5.$$

21. For the probability density function $f(x) = \frac{x^2}{21}$ on $[1, 4]$, the cumulative distribution function is

$$F(x) = \int_1^x \frac{t^2}{21}\,dt$$

$$= \frac{t^3}{63}\Big|_1^x$$

$$= \frac{1}{63}(x^3 - 1),\ 1 \leq x \leq 4.$$

23. The value of k was found to be $\frac{3}{14}$. For the probability density function $f(x) = \frac{3}{14}x^{1/2}$ on $[1, 4]$, the cumulative distribution function is

$$F(x) = \int_1^x \frac{3}{14}t^{1/2}\,dt$$

$$= \frac{3}{14} \cdot \frac{2}{3}t^{3/2}\Big|_1^x$$

$$= \frac{1}{7}(x^{3/2} - 1),\ 1 \leq x \leq 4.$$

25. The total area under the graph of a probability density function always equals 1.

29. $f(x) = \frac{1}{2}(1 + x)^{-3/2};\ [0, \infty)$

$$\frac{1}{2}\int_0^\infty (1 + x)^{-3/2}\,dx$$

$$= \lim_{a\to\infty} \frac{1}{2}\int_0^a (1 + x)^{-3/2}\,dx$$

$$= \lim_{a\to\infty} \frac{1}{2}(1 + x)^{-1/2}\left(\frac{-2}{1}\right)\Big|_0^a$$

$$= \lim_{a\to\infty} [-(1 + a)^{-1/2} + 1]$$

$$= \lim_{a\to\infty} \left(\frac{-1}{\sqrt{1 + a}} + 1\right)$$

$$= 0 + 1 = 1$$

Since $x \geq 0$, $f(x) \geq 0$.

$f(x)$ is a probability density function.

(a) $P(0 \leq X \leq 2)$

$$= \frac{1}{2}\int_0^2 (1 + x)^{-3/2}\,dx$$

$$= -(1 + x)^{-1/2}\Big|_0^2$$

$$= -3^{-1/2} + 1$$

$$\approx 0.4226$$

(b) $P(1 \leq X \leq 3)$

$$= \frac{1}{2}\int_1^3 (1 + x)^{-3/2}\,dx$$

$$= -(1 + x)^{-1/2}\Big|_1^3$$

$$= -4^{-1/2} + 2^{-1/2}$$

$$\approx 0.2071$$

(c) $P(X \geq 5)$

$$= \frac{1}{2}\int_5^\infty (1+x)^{-3/2}\,dx$$
$$= \lim_{a\to\infty} \frac{1}{2}\int_5^a (1+x)^{-3/2}\,dx$$
$$= \lim_{a\to\infty} [-(1+x)^{-1/2}]\Big|_5^a$$
$$= \lim_{a\to\infty} [-(1+a)^{-1/2} + 6^{-1/2}]$$
$$= \lim_{a\to\infty}\left(\frac{-1}{\sqrt{1+a}} + 6^{-1/2}\right)$$
$$\approx 0 + 0.4082$$
$$= 0.4082$$

31. $f(x) = \frac{1}{2}e^{-x/2}$; $[0, \infty)$

$$\frac{1}{2}\int_0^\infty e^{-x/2}\,dx$$
$$= \lim_{a\to\infty} \frac{1}{2}\int_0^a e^{-x/2}\,dx$$
$$= \lim_{a\to\infty} \frac{1}{2}\left(\frac{-2}{1}e^{-x/2}\right)\Big|_0^a$$
$$= \lim_{a\to\infty} -e^{-x/2}\Big|_0^a$$
$$= \lim_{a\to\infty}\left(\frac{-1}{e^{a/2}} + 1\right)$$
$$= 0 + 1$$
$$= 1$$

$f(x) > 0$ for all x.

$f(x)$ is a probability density function.

(a) $P(0 \leq X \leq 1) = \frac{1}{2}\int_0^1 e^{-x/2}\,dx$

$$= -e^{-x/2}\Big|_0^1$$
$$= \frac{-1}{e^{x/2}} + 1$$
$$\approx 0.3935$$

(b) $P(1 \leq X \leq 3) = \frac{1}{2}\int_1^3 e^{-x/2}\,dx$

$$= -e^{-x/2}\Big|_1^3$$
$$= \frac{-1}{e^{3/2}} + \frac{1}{e^{1/2}}$$
$$\approx 0.3834$$

(c) $P(X \geq 2) = \frac{1}{2}\int_2^\infty e^{-x/2}\,dx$

$$= \lim_{a\to\infty} \frac{1}{2}\int_2^a e^{-x/2}\,dx$$
$$= \lim_{a\to\infty} (-e^{-x/2})\Big|_2^a$$
$$= \lim_{a\to\infty}\left(\frac{-1}{e^{a/2}} + \frac{1}{e}\right)$$
$$\approx 0.3679$$

33. $f(x) = \begin{cases} \frac{x^3}{12} \text{ if } 0 \leq x \leq 2 \\ \frac{16}{3x^3} \text{ if } x > 2 \end{cases}$

First, note that $f(x) > 0$ for $x > 0$. Next,

$$\int_0^\infty f(x)dx$$
$$= \int_0^2 \frac{x^3}{12}dx + \lim_{a\to\infty}\int_2^a \frac{16}{3x^3}dx$$
$$= \left(\frac{x^4}{48}\right)\Big|_0^2 + \lim_{a\to\infty}\left(-\frac{8}{3x^2}\right)\Big|_2^a$$
$$= \left(\frac{1}{3} - 0\right) + \left[\lim_{a\to\infty}\left(-\frac{8}{3a^2}\right) - \left(-\frac{8}{12}\right)\right]$$
$$= \frac{1}{3} + \frac{2}{3}$$
$$= 1.$$

Therefore, $f(x)$ is a probability density function.

(a) $P(0 \leq X \leq 2) = \int_0^2 f(x)dx$

$$= \left(\frac{x^4}{48}\right)\Big|_0^2$$
$$= \frac{1}{3}$$

(b) $P(X \geq 2) = P(X > 2)$

$$= \int_2^\infty \frac{16}{3x^3}dx$$
$$= \lim_{a\to\infty}\int_2^a \frac{16}{3x^3}dx$$
$$= \lim_{a\to\infty}\left(-\frac{8}{3x^2}\right)\Bigg|_2^a$$
$$= \lim_{a\to\infty}\left(-\frac{8}{3a^2}\right)-\left(-\frac{8}{3\cdot 2^2}\right)$$
$$= 0-\left(-\frac{2}{3}\right)$$
$$= \frac{2}{3}$$

(c) $P(1 \leq X \leq 3)$

$$= \int_1^2 \frac{x^3}{12}dx + \int_2^3 \frac{16}{3x^3}dx$$
$$= \left(\frac{x^4}{48}\right)\Bigg|_1^2 + \left(-\frac{8}{3x^2}\right)\Bigg|_2^3$$
$$= \left(\frac{1}{3}-\frac{1}{48}\right)+\left(-\frac{8}{27}+\frac{2}{3}\right)$$
$$= \frac{295}{432}$$

35. $f(t) = \frac{1}{2}e^{-t/2}$; $[0, \infty)$

(a) $P(0 \leq T \leq 12) = \frac{1}{2}\int_0^{12} e^{-t/2}\,dt$

$$= -e^{-t/2}\Big|_0^{12}$$
$$= \frac{-1}{e^6}+1$$
$$\approx 0.9975$$

(b) $P(12 \leq T \leq 20) = \frac{1}{2}\int_{12}^{20} e^{-t/2}\,dt$

$$= -e^{-t/2}\Big|_{12}^{20}$$
$$= \frac{-1}{e^{10}}+\frac{1}{e^6}$$
$$\approx 0.0024$$

(c) $F(t) = \int_0^t \frac{1}{2}e^{-s/2}ds$

$$= \frac{1}{2}(-2)e^{-s/2}\Big|_0^t$$
$$= -(e^{-t/2}-1)$$
$$= 1-e^{-t/2},\ t \geq 0.$$

(d) $F(6) = 1 - e^{-6/2}$

$$= 1-e^{-3}$$
$$\approx 0.9502$$

The probability is 0.9502.

37. If $f(x)$ is proportional to $(10 + x)^{-2}$, then, for some value of k, $f(x) = k(10 + x)^{-2}$ on [0, 40]. Find k. We know the total probability must equal 1.

$$\int_0^{10} k(10+x)^{-2}\,dx = -k(10+x)^{-1}\Big|_0^{40}$$
$$= -k(50^{-1}-10^{-1})^{-1}$$
$$= -k\left(\frac{1}{50}-\frac{1}{10}\right)$$
$$= \frac{2}{25}x$$

If $\frac{2}{25}k = 1$, then $k = \frac{25}{2}$. Therefore

$$f(x) = \frac{25}{2}(10+x)^{-2}, 0 \leq x \leq 40$$

So the probability distribution function is

$$F(x) = \int_0^x \frac{25}{2}(10+t)^{-2}dt$$
$$= -\frac{25}{2}(10+t)^{-1}\Big|_0^x$$
$$= -\frac{25}{2}[(10+x)^{-1}-10^{-1}]$$
$$= -\frac{25}{2}\left(\frac{1}{10+x}-\frac{1}{10}\right)$$
$$= \frac{25}{2}\left(\frac{1}{10}-\frac{1}{10+x}\right)$$
$$F(6) = \frac{25}{2}\left(\frac{1}{10}-\frac{1}{106}\right)$$
$$\approx 0.47$$

The correct answer choice is **c**.

39. $f(x) = \frac{1}{2\sqrt{x}}$; [1, 4]

(a) $P(3 \le X \le 4) = \int_3^4 \left(\frac{1}{2\sqrt{x}}\right) dx$

$= \frac{1}{2}\int_3^4 x^{-1/2}\,dx$

$= \frac{1}{2}(2)x^{1/2}\Big|_3^4$

$= 2 - 3^{1/2} \approx 0.2679$

(b) $P(1 \le X \le 2) = \int_1^2 \left(\frac{1}{2\sqrt{x}}\right) dx$

$= \frac{1}{2}(2)x^{1/2}\Big|_1^2$

$= 2^{1/2} - 1 = 0.4142$

(c) $P(2 \le X \le 3) = \int_2^3 \left(\frac{1}{2\sqrt{x}}\right) dx$

$= \frac{1}{2}(2)x^{1/2}\Big|_2^3$

$= 3^{1/2} - 2^{1/2} = 0.3178$

41. $f(x) = 1.185 \cdot 10^{-9}x^{4.5222} - 0.049846x$

(a) $P(0 \le X \le 150)$

$= \int_0^{150} 1.185 \cdot 10^{-9}x^{4.5222}e^{-0.049846x}\,dx$

≈ 0.8131

(b) $P(100 \le x \le 200)$

$= \int_{100}^{200} 1.185 \cdot 10^{-9}x^{4.5222}e^{-0.049846x}\,dx$

≈ 0.4901

43. **(a)**

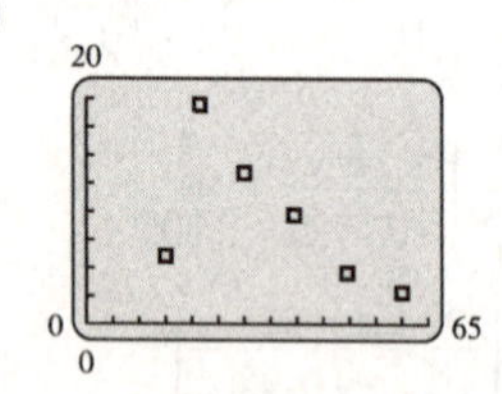

A polynomial function could fit the data.

(b) $N(t) = -0.00007445t^4 + 0.01243t^3 - 0.7419t^2 + 18.18t - 137.5$

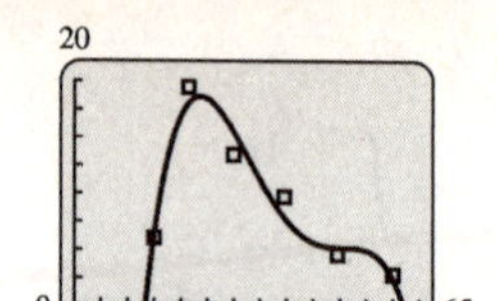

(c) $\int_{13.4}^{62.0} N(t)\,dt \approx 466.26$, as found using a calculator. Thus the density function corresponding to the quartic fit is

$S(t) = \frac{1}{466.26}N(t)$

$= \frac{1}{466.26}(-0.00007445t^4 + 0.01243t^3 - 0.7419t^2 + 18.18t - 137.5).$

(d) Use a calculator to compute the integrals needed in (d).

$P(35 \le \text{age} < 45) = \int_{35}^{45} S(t)\,dt \approx 0.1688;$

actual relative frequency

$= \frac{9.701}{56.065} \approx 0.1730$

$P(18 \le \text{age} < 35) = \int_{18}^{35} S(t)\,dt \approx 0.5896;$

actual relative frequency

$= \frac{19.461 + 13.423}{56.065} \approx 0.5865$

$P(45 \le \text{age}) = \int_{45}^{62} S(t)\,dt \approx 0.1610;$

actual relative frequency

$= \frac{4.582 + 2.849}{56.065} \approx 0.1325$

45. $f(x) = \frac{5.5 - x}{15}$; [0, 5]

(a) $P(3 \le X \le 5) = \int_3^5 \frac{5.5 - x}{15} dx$

$$= \left(\frac{5.5}{15}x - \frac{1}{15} \cdot \frac{x^2}{2}\right)\Big|_3^5$$

$$= \left(\frac{5.5}{15} \cdot 5 - \frac{1}{15} \cdot \frac{5^2}{2}\right) - \left(\frac{5.5}{15} \cdot 3 - \frac{1}{15} \cdot \frac{3^2}{2}\right)$$

$$= 0.2$$

(b) $P(0 \le X \le 2) = \int_0^2 \frac{5.5 - x}{15} dx$

$$= \left(\frac{5.5}{15}x - \frac{1}{15} \cdot \frac{x^2}{2}\right)\Big|_0^2$$

$$= -\left(\frac{5.5}{15} \cdot 2 - \frac{1}{15} \cdot \frac{2^2}{2}\right) - \left(\frac{5.5}{15} \cdot 0 - \frac{1}{15} \cdot \frac{0^2}{2}\right)$$

$$= 0.6$$

(c) $P(1 \le X \le 4) = \int_1^4 \frac{5.5 - x}{15} dx$

$$= \left(\frac{5.5}{15}x - \frac{1}{15} \cdot \frac{x^2}{2}\right)\Big|_1^4$$

$$= \left(\frac{5.5}{15} \cdot 4 - \frac{1}{15} \cdot \frac{4^2}{2}\right) - \left(\frac{5.5}{15} \cdot 1 - \frac{1}{15} \cdot \frac{1^2}{2}\right)$$

$$= 0.6$$

47. $f(t) = \frac{1}{3650.1} e^{-t/3650.1}$

(a) $P(365 < T < 1095)$

$$= \int_{365}^{1095} \frac{1}{3650.1} e^{-t/3650.1}\, dt$$

$$= (-e^{-t/3650.1})\Big|_{365}^{1095}$$

$$= -e^{-1095/3650.1} + e^{-365/3650.1}$$

$$\approx 0.1640$$

(b) $P(T > 7300)$

$$= \int_{7300}^{\infty} \frac{1}{3650.1} e^{-t/3650.1}\, dt$$

$$= \lim_{b\to\infty} \int_{7300}^{b} \frac{1}{3650.1} e^{-t/3650.1}\, dt$$

$$= \lim_{b\to\infty} (-e^{-t/3650.1})\Big|_{7300}^{b}$$

$$= \lim_{b\to\infty} (-e^{-b/3650.1} + e^{-7300/3650.1})$$

$$= 0 + e^{-7300/3650.1} \approx 0.1353$$

49. $f(t) = 0.06049e^{-0.03211t}$; [16, 84]

(a) $P(16 \le T \le 25)$

$$= \int_{16}^{25} f(t)dt$$

$$= \int_{16}^{25} 0.06049e^{-0.03211t}dt$$

$$= \frac{0.06049}{-0.03211}(e^{-0.03211t})\Big|_{16}^{25}$$

$$\approx -1.88384(e^{-0.03211t})\Big|_{16}^{25}$$

$$= -1.88384(e^{-0.03211\cdot 25} - e^{-0.03211\cdot 16})$$

$$\approx 0.2829$$

(b) $P(35 \le T \le 84)$

$$= \int_{35}^{84} 0.06049e^{-0.03211t}dt$$

$$\approx -1.88384(e^{-0.03211t})\Big|_{35}^{84}$$

$$\approx 0.4853$$

(c) $P(21 \le T \le 30)$

$$= \int_{21}^{30} 0.06049e^{-0.03211t}\,dt$$

$$\approx -1.88384(e^{-0.03211t})\Big|_{21}^{30}$$

$$\approx 0.2409$$

(d) $$F(t) = \int_{16}^{t} 0.06049e^{-0.03211s}\,ds$$

$$= 0.06049 \cdot \frac{1}{-0.03211}e^{-0.03211s}\Big|_{16}^{t}$$

$$= -1.8838(e^{-0.03211t} - e^{-0.03211 \cdot 16})$$

$$= 1.8838(0.5982 - e^{-0.03211t}),$$

$$16 \le t \le 84$$

(e) $$F(21) = 1.8838(0.5982 - e^{-0.03211 \cdot 21})$$

$$= 1.8838(0.0887)$$

$$= 0.1671$$

The probability is 0.1671.

51. **(a)**

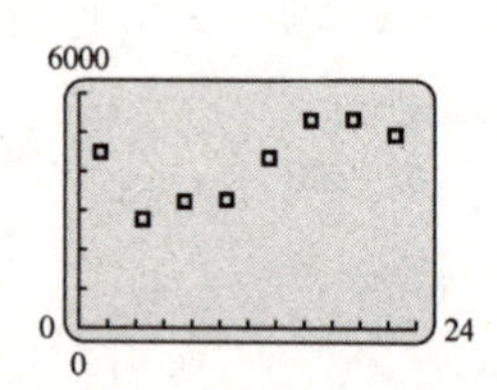

A polynomial function could fit the data.

(b) $$T(t) = -2.564t^3 + 99.11t^2 - 964.6t + 5631$$

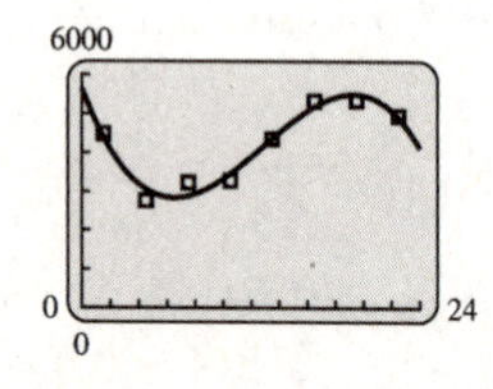

(c) $\int_0^{24} T(t)\,dt \approx 101{,}370$, as found using a calculator. Thus the density function corresponding to the cubic fit is

$$S(t) = \frac{1}{101{,}370}T(t)$$

$$= \frac{1}{101{,}370}(-2.564t^3 + 99.11t^2 - 964.6t + 5631).$$

(d) Use a calculator to compute the integrals needed in (d).

$P(12 \text{ am} \le \text{time} \le 2 \text{ am})$

$$= \int_0^2 S(t)\,dt$$

$$\approx 0.09457$$

$P(4 \text{ pm} \le \text{time} \le 5{:}30 \text{ pm})$

$$= \int_{16}^{17.5} S(t)\,dt$$

$$\approx 0.07732$$

18.2 Expected Value and Variance of Continuous Random Variables

Your Turn 1

Find the expected value and the variance of the random variable X with probability density function $f(x) = \frac{8}{3x^3}$ on [1, 2].

$$\mu = \int_1^2 x f(x)\,dx$$

$$= \int_1^2 x\frac{8}{3x^3}\,dx$$

$$= \int_1^2 \frac{8}{3}x^{-2}\,dx$$

$$= -\frac{8}{3}x^{-1}\Big|_1^2$$

$$= -\frac{8}{6} - \left(-\frac{8}{3}\right)$$

$$= \frac{4}{3}$$

$$\text{Var}(x) = \int_1^2 \left(x - \frac{4}{3}\right)^2 \frac{8}{3x^3}\,dx$$

$$= \int_1^2 \left(x^2 - \frac{8}{3}x + \frac{16}{9}\right)\frac{8}{3x^3}\,dx$$

$$= \int_1^2 \left(\frac{8}{3}x^{-1} - \frac{64}{9}x^{-2} + \frac{128}{27}x^{-3}\right)dx$$

$$= \left(\frac{8}{3}\ln(x) + \frac{64}{9x} - \frac{64}{27x^2}\right)\Big|_1^2$$

$$= \left(\frac{8}{3}\ln 2 + \frac{32}{9} - \frac{16}{27}\right) - \left(\frac{64}{9} - \frac{64}{27}\right)$$

$$= \frac{8}{3}\ln 2 - \frac{16}{9} \approx 0.0706$$

Your Turn 2

Use the alternative formula to find the variance of the random variable X with probability density function $f(x) = \frac{4}{x^5}$ for $x \geq 1$. First find the mean μ.

$$\mu = \int_1^\infty x\left(\frac{4}{x^5}\right)dx$$
$$= \int_1^\infty 4x^{-4}\,dx$$
$$= \lim_{b\to\infty}\int_1^b 4x^{-4}\,dx$$
$$= \lim_{b\to\infty}\left(-\frac{4}{3}x^{-3}\Big|_1^b\right)$$
$$= \lim_{b\to\infty}\left(-\frac{4}{3b^3} - \left(-\frac{4}{3}\right)\right)$$
$$= \frac{4}{3}$$

$$\text{Var}(X) = \int_1^\infty x^2 f(x)\,dx - \mu^2$$
$$= \int_1^\infty x^2\frac{4}{x^5}\,dx - \left(\frac{4}{3}\right)^2$$
$$= 2\int_1^\infty 2x^{-3}\,dx - \frac{16}{9}$$
$$= 2\lim_{b\to\infty}\left(-x^{-2}\Big|_1^b\right) - \frac{16}{9}$$
$$= 2\left(1 - \lim_{b\to\infty} b^{-2}\right) - \frac{16}{9}$$
$$= 2 - \frac{16}{9} = \frac{2}{9}$$

Your Turn 3

Find the median m for the probability density function $f(x) = \frac{4}{x^5}$ for $x \geq 1$.

$$\int_1^m \frac{4}{x^5}\,dx = \frac{1}{2}$$
$$\left(-\frac{1}{x^4}\right)\Bigg|_1^m = \frac{1}{2}$$
$$-\frac{1}{m^4} - (-1) = \frac{1}{2}$$
$$m^4 = 2$$
$$m = \sqrt[4]{2} \approx 1.1892$$

18.2 Exercises

1. $f(x) = \frac{1}{4}$; $[3, 7]$

$$E(X) = \mu = \int_3^7 \frac{1}{4}x\,dx = \frac{1}{4}\left(\frac{x^2}{2}\right)\Bigg|_3^7$$
$$= \frac{49}{8} - \frac{9}{8}$$
$$= 5$$

$$\text{Var}(X) = \int_3^7 (x-5)^2\left(\frac{1}{4}\right)dx$$
$$= \frac{1}{4}\cdot\frac{(x-5)^3}{3}\Bigg|_3^7$$
$$= \frac{8}{12} + \frac{8}{12}$$
$$= \frac{4}{3} \approx 1.33$$
$$\sigma \approx \sqrt{\text{Var}(X)}$$
$$= \sqrt{\frac{4}{3}}$$
$$\approx 1.15$$

3. $f(x) = \frac{x}{8} - \frac{1}{4}$; $[2, 6]$

$$\mu = \int_2^6 x\left(\frac{x}{8} - \frac{1}{4}\right)dx$$
$$= \int_2^6 \left(\frac{x^2}{8} - \frac{x}{4}\right)dx$$
$$= \left(\frac{x^3}{24} - \frac{x^2}{8}\right)\Bigg|_2^6$$
$$= \left(\frac{216}{24} - \frac{36}{8}\right) - \left(\frac{8}{24} - \frac{4}{8}\right)$$
$$= \frac{208}{24} - 4$$
$$= \frac{26}{3} - 4$$
$$= \frac{14}{3} \approx 4.67$$

Use the alternative formula to find

$$\text{Var}(X) = \int_2^6 x^2\left(\frac{x}{8} - \frac{1}{4}\right)dx - \left(\frac{14}{3}\right)^2$$
$$= \int_2^6 \left(\frac{x^3}{8} - \frac{x^2}{4}\right)dx - \frac{196}{9}$$
$$= \left(\frac{x^4}{32} - \frac{x^3}{12}\right)\Bigg|_2^6 - \frac{196}{9}$$
$$= \left(\frac{1296}{32} - \frac{216}{12}\right) - \left(\frac{16}{32} - \frac{8}{12}\right) - \frac{196}{9}$$
$$\approx 0.89.$$
$$\sigma = \sqrt{\text{Var}(X)} \approx \sqrt{0.89} \approx 0.94$$

5. $f(x) = 1 - \frac{1}{\sqrt{x}}$; [1, 4]

$$\mu = \int_1^4 x(1 - x^{-1/2})dx$$
$$= \int_1^4 (x - x^{1/2})dx$$
$$= \left(\frac{x^2}{2} - \frac{2x^{3/2}}{3}\right)\Bigg|_1^4$$
$$= \frac{16}{2} - \frac{16}{3} - \frac{1}{2} + \frac{2}{3}$$
$$= \frac{17}{6} \approx 2.83$$

$$\text{Var}(X) = \int_1^4 x^2(1 - x^{-1/2})dx - \left(\frac{17}{6}\right)^2$$
$$= \int_1^4 (x^2 - x^{3/2})dx - \frac{289}{36}$$
$$= \left(\frac{x^3}{3} - \frac{2x^{5/2}}{5}\right)\Bigg|_1^4 - \frac{289}{36}$$
$$= \frac{64}{3} - \frac{64}{5} - \frac{1}{3} + \frac{2}{5} - \frac{289}{36}$$
$$\approx 0.57$$
$$\sigma \approx \sqrt{\text{Var}(X)} \approx 0.76$$

7. $f(x) = 4x^{-5}$; $[1, \infty)$

$$\mu = \int_1^\infty x(4x^{-5})dx$$
$$= \lim_{a\to\infty} \int_1^a 4x^{-4}\,dx$$
$$= \lim_{a\to\infty} \left(\frac{4x^{-3}}{-3}\right)\Bigg|_1^a$$
$$= \lim_{a\to\infty} \left(\frac{-4}{3a^3} + \frac{4}{3}\right)$$
$$= \frac{4}{3} \approx 1.33$$

$$\text{Var}(X) = \int_1^\infty x^2(4x^{-5})dx - \left(\frac{4}{3}\right)^2$$
$$= \lim_{a\to\infty} \int_1^a 4x^{-3}\,dx - \frac{16}{9}$$
$$= \lim_{a\to\infty} \left(\frac{4x^{-2}}{-2}\right)\Bigg|_1^a - \frac{16}{9}$$
$$= \lim_{a\to\infty} \left(\frac{-2}{a^2} + 2\right) - \frac{16}{9}$$
$$= 2 - \frac{16}{9} = \frac{2}{9} \approx 0.22$$
$$\sigma = \sqrt{\text{Var}(X)} = \sqrt{\frac{2}{9}} \approx 0.47$$

11. $f(x) = \frac{\sqrt{x}}{18}$; [0, 9]

(a) $E(X) = \mu = \int_0^9 \frac{x\sqrt{x}}{18}dx$

$$= \int_0^9 \frac{x^{3/2}}{18}dx$$
$$= \frac{2x^{5/2}}{90}\Bigg|_0^9 = \frac{x^{5/2}}{45}\Bigg|_0^9$$
$$= \frac{243}{45} = \frac{27}{5} = 5.40$$

(b) $\text{Var}(X) = \int_0^9 \frac{x^2\sqrt{x}}{18}\,dx - \left(\frac{27}{5}\right)^2$

$= \int_0^9 \frac{x^{5/2}}{18}\,dx - \left(\frac{27}{5}\right)^2$

$= \left.\frac{x^{7/2}}{63}\right|_0^9 - \left(\frac{27}{5}\right)^2$

$= \frac{2187}{63} - \left(\frac{27}{5}\right)^2 \approx 5.55$

(c) $\sigma = \sqrt{\text{Var}(X)} \approx 2.36$

(d) $P(5.40 < X \le 9)$

$= \int_{5.4}^9 \frac{x^{1/2}}{18}\,dx$

$= \left.\frac{x^{3/2}}{27}\right|_{5.4}^9$

$= \frac{27}{27} - \frac{(5.4)^{1.5}}{27}$

≈ 0.5352

(e) $P(5.40 - 2.36 \le X \le 5.40 + 2.36)$

$= \int_{3.04}^{7.76} \frac{x^{1/2}}{18}\,dx$

$= \left.\frac{x^{3/2}}{27}\right|_{3.04}^{7.76}$

$= \frac{7.76^{3/2}}{27} - \frac{3.04^{3/2}}{27}$

≈ 0.6043

13. $f(x) = \frac{1}{4}x^3;\ [0, 2]$

(a) $E(X) = \mu = \int_0^2 \frac{1}{4}x^4\,dx = \left.\frac{x^5}{20}\right|_0^2$

$= \frac{32}{20} = \frac{8}{5} = 1.6$

(b) $\text{Var}(X) = \int_0^2 \frac{1}{4}x^5 dx - \frac{64}{25}$

$= \left.\frac{x^6}{24}\right|_0^2 - \frac{64}{25}$

$= \frac{8}{3} - \frac{64}{25}$

$= \frac{8}{75} \approx 0.11$

(c) $\sigma = \sqrt{\text{Var}(X)}$

$= \sqrt{\frac{8}{75}}$

≈ 0.3266

≈ 0.33

(d) $P(8/5 < X \le 2) = \int_{8/5}^2 \frac{x^3}{4}\,dx$

$= \left.\frac{x^4}{16}\right|_{8/5}^2$

$= 1 - \frac{256}{625}$

$= \frac{369}{625} \approx 0.5904$

(e) Use a four-place value for the standard deviation.

$P(1.6 - 0.3266 \le X \le 1.6 + 0.3266)$

$= \int_{1.2734}^{1.9266} \frac{x^3}{4}\,dx = \left.\frac{x^4}{16}\right|_{1.2734}^{1.9266}$

$= \frac{1.9266^4}{16} - \frac{0.1.2734^4}{16}$

≈ 0.6967

15. $f(x) = \frac{1}{4};\ [3, 7]$

(a) $m = \text{median}:\ \int_3^m \frac{1}{4}dx = \frac{1}{2}$

$\left.\frac{1}{4}x\right|_3^m = \frac{1}{2}$

$\frac{m}{4} - \frac{3}{4} = \frac{1}{2}$

$m - 3 = 2$

$m = 5$

(b) $E(X) = \mu = 5$ (from Exercise 1)

$P(X = 5) = \int_5^5 \frac{1}{4}dx = 0$

17. $f(x) = \frac{x}{8} - \frac{1}{4}$; [2, 6]

(a) m = median:

$$\int_2^m \left(\frac{x}{8} - \frac{1}{4}\right) dx = \frac{1}{2}$$

$$\left(\frac{x^2}{16} - \frac{x}{4}\right)\Bigg|_2^m = \frac{1}{2}$$

$$\frac{m^2}{16} - \frac{m}{4} - \frac{1}{4} + \frac{1}{2} = \frac{1}{2}$$

$$m^2 - 4m - 4 + 8 = 8$$

$$m^2 - 4m - 4 = 0$$

$$m = \frac{4 \pm \sqrt{16 + 16(1)}}{2}$$

Reject $\frac{4-\sqrt{32}}{2}$ since it is not in [2, 6].

$$m = \frac{4 + \sqrt{32}}{2}$$

$$= 2 + 2\sqrt{2} \approx 4.8284$$

(b) $E(X) = \mu = \frac{14}{3}$ (from Exercise 3)

$$P\left(\tfrac{14}{3} \le X \le 2 + 2\sqrt{2})\right)$$

$$= \int_{14/3}^{2+2\sqrt{2}} \left(\frac{x}{8} - \frac{1}{4}\right) dx$$

$$= \left(\frac{x^2}{16} - \frac{x}{4}\right)\Bigg|_{14/3}^{2+2\sqrt{2}}$$

$$= \frac{\left(2 + 2\sqrt{2}\right)^2}{16} - \frac{2 + 2\sqrt{2}}{4} - \frac{(14/3)^2}{16} + \frac{14/3}{4}$$

$$= \frac{1}{18} \approx 0.0556$$

If you do the integration on a calculator using rounded values for the limits you may get a slightly different answer, such as 0.0553.

19. $f(x) = 4x^{-5}$; $[1, \infty)$

(a) m = median:

$$\int_1^m 4x^{-5}\, dx = \frac{1}{2}$$

$$\frac{4x^{-4}}{-4}\Bigg|_1^m = \frac{1}{2}$$

$$-m^{-4} + 1 = \frac{1}{2}$$

$$1 - \frac{1}{m^4} = \frac{1}{2}$$

$$2m^4 - 2 = m^4$$

$$m^4 = 2$$

$$m = \sqrt[4]{2} \approx 1.189$$

(b) $E(X) = \mu = \frac{4}{3}$ (from Exercise 7)

$$P\left(1.19 \le X \le \tfrac{4}{3}\right) \approx \int_{1.189}^{1.333} 4x^{-5}\, dx$$

$$\approx -x^{-4}\Big|_{1.189}^{1.333}$$

$$\approx -\frac{1}{1.333^4} + \frac{1}{1.189^4}$$

$$\approx 0.1836$$

21. $f(x) = \begin{cases} \frac{x^3}{12} & \text{if } 0 \le x \le 2 \\ \frac{16}{3x^3} & \text{if } x > 2 \end{cases}$

Expected value:

$$E(X) = \mu = \int_0^\infty x\, f(x) dx$$

$$= \int_0^2 x\left(\frac{x^3}{12}\right) dx + \lim_{a\to\infty} \int_2^a x\left(\frac{16}{3x^3}\right) dx$$

$$= \int_0^2 \frac{x^4}{12}\, dx + \lim_{a\to\infty} \int_2^a \frac{16}{3x^2}\, dx$$

$$= \left(\frac{x^5}{60}\right)\Bigg|_0^2 + \lim_{a\to\infty}\left(-\frac{16}{3x}\right)\Bigg|_2^a$$

$$= \left(\frac{8}{15} - 0\right) + \left[\lim_{a\to\infty}\left(-\frac{16}{3a}\right) - \left(-\frac{16}{6}\right)\right]$$

$$= \frac{16}{5}$$

Variance:

$$\text{Var}(X) = \int_0^{\infty} x^2 f(x)dx - \mu^2$$
$$= \int_0^2 x^2\left(\frac{x^3}{12}\right)dx + \int_2^{\infty}\left(\frac{16}{3x^3}\right)dx - \left(\frac{16}{5}\right)^2$$

Examine the second integral.

$$\int_2^{\infty} x^2\left(\frac{16}{3x^3}\right)dx$$
$$= \lim_{a\to\infty}\int_2^a x^2\left(\frac{16}{3x^3}\right)dx$$
$$= \lim_{a\to\infty}\int_2^a \frac{16}{3x}\,dx$$
$$= \lim_{a\to\infty} \frac{16}{3}\ln|a| - \frac{16}{3}\ln|2|$$

Since the limit diverges, neither the variance nor the standard deviation exists.

23. $f(x) = \begin{cases} \frac{|x|}{10} \text{ for } -2 \le x \le 4 \\ 0 \quad \text{otherwise} \end{cases}$

First, note that

$$|x| = \begin{cases} -x \text{ for } -2 \le x \le 0 \\ x \text{ for } 0 \le x \le 4 \end{cases}$$

The expected value is

$$E(X) = \mu = \int_{-2}^{4} x \cdot \frac{|x|}{10}\,dx$$
$$= \int_{-2}^{0} x \cdot \frac{-x}{10}\,dx + \int_0^4 x \cdot \frac{x}{10}\,dx$$
$$= \int_{-2}^{0} -\frac{x^2}{10}\,dx + \int_0^4 \frac{x^2}{10}\,dx$$
$$= -\frac{x^3}{30}\bigg|_{-2}^{0} + \frac{x^3}{30}\bigg|_0^4$$
$$= -\left(0 - \frac{-8}{30}\right) + \left(\frac{64}{30} - 0\right)$$
$$= \frac{56}{30} = \frac{28}{15}$$

The correct answer choice is **d**.

25. $f(t) = \frac{1}{11}\left(1 + \frac{3}{\sqrt{t}}\right); [4, 9]$

(a) From Exercise 6, $\mu = \frac{141}{22}$ yr ≈ 6.409 yr.

(b) $\sigma = 1.447$ yr.

(c) $P\left(T > \frac{141}{22}\right)$

$$= \int_{141/22}^{9} \frac{1}{11}(1 + 3t^{-1/2})\,dt$$
$$= \frac{1}{11}(t + 6t^{1/2})\bigg|_{141/22}^{9}$$
$$= \frac{1}{11}\left[9 + 18 - \frac{141}{22} - 6\left(\frac{141}{22}\right)^{1/2}\right]$$
$$\approx 0.4910$$

27. Using the hint, we have

$$\text{loss not paid} = \begin{cases} x \text{ for } 0.6 < x < 2 \\ 2 \text{ for } x > 2 \end{cases}$$

Therefore, the mean of the manufacturer's annual losses not paid will be

$$\mu = \int_{0.6}^{0} x \cdot f(x)dx + \int_2^{\infty} 2 \cdot f(x)dx$$
$$= \int_{0.6}^{2} x\frac{2.5(0.6)^{2.5}}{x^{3.5}}\,dx$$
$$+ \int_2^{\infty} 2\frac{2.5(0.6)^{2.5}}{x^{3.5}}\,dx$$
$$= 2.5(0.6)^{2.5}\int_{0.6}^{2} \frac{1}{x^{2.5}}\,dx$$
$$+ 5(0.6)^{2.5}\int_2^{\infty} \frac{1}{x^{3.5}}\,dx$$
$$= 2.5(0.6)^{2.5}\left(\frac{1}{-1.5}\right)\frac{1}{x^{1.5}}\bigg|_{0.6}^{2}$$
$$+ 5(0.6)^{2.5}\left(\frac{1}{-2.5}\right)\frac{1}{x^{2.5}}\bigg|_2^{\infty}$$
$$= -\frac{5}{3}(0.6)^{2.5}\left(\frac{1}{2^{1.5}} - \frac{1}{0.6^{1.5}}\right)$$
$$- 2(0.6)^{2.5}\left(0 - \frac{1}{2^{2.5}}\right)$$
$$\approx 0.8357 + 0.0986 \approx 0.93$$

The correct answer choice is **c**.

29. Since the probability density function is proportional to $(1 + x)^{-4}$, we have $f(x) = k(1 + x)^{-4}$, $0 < x < \infty$. To determine k, solve the equation $\int_0^\infty k f(x)dx = 1$.

$$\int_0^\infty k(1 + x)^{-4} dx = 1$$

$$k\left(-\frac{1}{3}\right)(1 + x)^{-3}\Big|_0^\infty = 1$$

$$-\frac{k}{3}(0 - 1) = 1$$

$$\frac{k}{3} = 1$$

$$k = 3$$

Thus, $f(x) = 3(1 + x)^{-4}, 0 < x < \infty$.

The expected monthly claims are

$$\int_0^\infty x \cdot 3(1 + x)^{-4}\, dx = 3\int_0^\infty \frac{x}{(1 + x)^4} dx$$

The antiderivative can be found using the substitution $u = 1 - x$.

$$\int \frac{x}{(1 + x)^4} dx = \int \frac{u - 1}{u^4} du$$

$$= \int \left(\frac{1}{u^3} - \frac{1}{u^4}\right) du$$

$$= -\frac{1}{2u^2} + \frac{1}{3u^3}$$

Resubstitute $u = 1 + x$.

$$3\int_0^\infty \frac{x}{(1 + x)^4} dx$$

$$= 3\left(-\frac{1}{2(1 + x)^2} + \frac{1}{3(1 + x)^3}\right)\Bigg|_0^\infty$$

$$= 3\left[0 - \left(-\frac{1}{2} + \frac{1}{3}\right)\right] = 3\left(\frac{1}{6}\right) = \frac{1}{2}$$

The correct answer choice is **c**.

31. $f(t) = \dfrac{1}{(\ln 20)t}$; $[1, 20]$

(a) $\mu = \displaystyle\int_1^{20} t \cdot \frac{1}{(\ln 20)t} dt$

$$= \int_1^{20} \frac{1}{\ln 20} dt$$

$$= \frac{t}{\ln 20}\Big|_1^{20}$$

$$= \frac{19}{\ln 20} \approx 6.3424 \approx 6.342 \text{ seconds}$$

(b) $\text{Var}(T) = \displaystyle\int_1^{20} t^2 \cdot \frac{1}{(\ln 20)t} dt - \mu^2$

$$= \int_1^{20} \frac{t}{\ln 20} dt - \mu^2$$

$$= \frac{t^2}{2 \ln 20}\Big|_1^{20} - (6.3424)^2$$

$$= \frac{399}{2 \ln 20} - (6.3424)^2$$

$$\approx 26.3687$$

$$\sigma \approx \sqrt{26.3687} \approx 5.1350 \text{ sec}$$

(c) $P(6.3424 - 5.1350 < T < 6.3424 + 5.1350)$

$$= P(1.2074 < T < 11.4774)$$

$$= \int_{1.2074}^{11.4774} \frac{1}{(\ln 20)t} dt$$

$$= \frac{\ln t}{\ln 20}\Big|_{1.2074}^{11.4774}$$

$$= \frac{1}{\ln 20}(\ln 11.4774 - \ln 1.2074)$$

$$\approx 0.7517$$

If you do the integration on a calculator using differently rounded values for the limits you may get a slightly different answer, such as 0.7518.

(d) The median clotting time is the value of m such that $\int_a^m f(t)dt = \frac{1}{2}$.

$$\int_1^m \frac{1}{(\ln 20)t} dt = \frac{1}{2}$$

$$\frac{1}{\ln 20} \ln t\Big|_1^m = \frac{1}{2}$$

$$\frac{1}{\ln 20}(\ln m - 0) = \frac{1}{2}$$

$$\ln m = \frac{\ln 20}{2}$$

$$m = e^{\ln 20/2}$$

$$\approx 4.472$$

33. $f(x) = \dfrac{1}{2\sqrt{x}}$; [1, 4]

(a) $\mu = \displaystyle\int_1^4 x \cdot \frac{1}{2\sqrt{x}}\,dx$

$= \displaystyle\int_1^4 \frac{x^{1/2}}{2}\,dx = \frac{x^{3/2}}{3}\Bigg|_1^4$

$= \dfrac{1}{3}(8 - 1)$

$= \dfrac{7}{3} \approx 2.333$ cm

(b) $\text{Var}(X) = \displaystyle\int_1^4 x^2 \cdot \frac{1}{2\sqrt{x}}\,dx - \left(\frac{7}{3}\right)^2$

$= \displaystyle\int_1^4 \frac{x^{3/2}}{2}\,dx - \frac{49}{9}$

$= \dfrac{x^{5/2}}{5}\Bigg|_1^4 - \dfrac{49}{9}$

$= \dfrac{1}{5}(32 - 1) - \dfrac{49}{9}$

≈ 0.7556

$\sigma = \sqrt{\text{Var}(X)}$

≈ 0.8692 cm

(c) $P(X > 2.33 + 2(0.87))$

$= P(X > 4.07)$

$= 0$

The probability is 0 since two standard deviations falls out of the given interval [1, 4].

(d) The median petal length is the value of m such that $\int_a^m f(x)dx = \frac{1}{2}$.

$\displaystyle\int_1^m \frac{1}{2\sqrt{x}}\,dx = \frac{1}{2}$

$\sqrt{x}\Big|_1^m = \dfrac{1}{2}$

$\sqrt{m} - 1 = \dfrac{1}{2}$

$\sqrt{m} = \dfrac{3}{2}$

$m = \dfrac{9}{4} = 2.25$

The median petal length is 2.25 cm.

35. $f(x) = 1.185 \cdot 10^{-9} x^{4.5222} e^{-0.049846x}$

$E(X) = \displaystyle\int_1^{1000} x\,f(x)dx$

Using the integration function on our calculator.

$E(X) \approx 110.80$

The expected size is about 111.

37. $S(t) = \dfrac{1}{466.26}(-0.00007445t^4 + 0.01243t^3 - 0.7419t^2 + 18.18t - 137.5)$

for t in the interval [13.4, 62.0].

Evaluate the required integrals with a calculator.

$\mu = \displaystyle\int_{13.4}^{62.0} tS(t)\,dt$

≈ 31.75 years

$\text{Var}(X) = \displaystyle\int_{13.4}^{62.0} t^2 S(t)\,dt - (31.75)^2$

≈ 133.44

$\sigma \approx \sqrt{133.44} \approx 11.55$ years

39. $f(x) = \dfrac{5.5 - x}{15}$; [0, 5]

(a) $\mu = \displaystyle\int_0^5 x\left(\frac{5.5 - x}{15}\right)dx$

$= \displaystyle\int_0^5 \left(\frac{5.5}{15}x - \frac{1}{15}x^2\right)dx$

$= \dfrac{5.5}{30}x^2 - \dfrac{1}{45}x^3\Bigg|_0^5$

$= \left(\dfrac{5.5}{30} \cdot 25 - \dfrac{1}{45} \cdot 125\right) - 0$

≈ 1.806

(b) $\text{Var}(X) = \displaystyle\int_0^5 x^2\left(\frac{5.5 - x}{15}\right)dx - \mu^2$

$= \displaystyle\int_0^5 \left(\frac{5.5}{15}x^2 - \frac{1}{15}x^3\right)dx - \mu^2$

$= \left(\dfrac{5.5}{45}x^3 - \dfrac{1}{60}x^4\right)\Bigg|_0^5 - \mu^2$

$= \dfrac{5.5}{45} \cdot 125 - \dfrac{1}{60} \cdot 625 - 0 - \mu^2$

≈ 1.60108

$\sigma = \sqrt{\text{Var}(X)} \approx 1.265$

(c) $P(X \le \mu - \sigma)$

$$= P(X \le 1.806 - 1.265)$$
$$= P(X \le 0.541)$$
$$= \int_0^{0.541} \frac{5.5 - x}{15}\, dx$$
$$= \left(\frac{5.5}{15}x - \frac{1}{30}x^2\right)\Bigg|_0^{0.541}$$
$$= \left(\frac{5.5}{15}(0.541) - \frac{1}{30}(0.541)^2 - 0\right)$$
$$\approx 0.1886$$

41. $f(t) = 0.06049e^{-0.03211t}$; $[16, 84]$

(a) Expected value:

$$E(T) = \mu$$
$$= \int_{16}^{84} t(0.06049e^{-0.03211t})dt$$

Use integration by parts.

$$\int_{16}^{84} t(0.06049e^{-0.03211t})dt$$
$$\approx -1.88384(t + 31.14295)e^{-0.03211t}\Big|_{16}^{84}$$
$$= -1.88384[(84 + 31.14295)e^{-0.03211\cdot 84} - (16 + 31.14295)e^{-0.03211\cdot 16}$$
$$\approx 38.512$$
$$\approx 38.51$$

The expected value is 38.51 years.

(b) $\text{Var}(T)$

$$= \int_{16}^{84} t^2(0.06049e^{-0.03211t})dt - \mu^2$$

Use integration by parts twice.

$$\int_{16}^{84} t^2(0.06049e^{-0.03211t})dt - \mu^2$$
$$\approx -1.88384(t^2 + 62.28589t + 1939.76619)e^{-0.03211t}\Big|_{16}^{84} - 38.512^2$$
$$\approx 308.305$$

$$\sigma = \sqrt{\text{Var}(T)} = \sqrt{308.290} \approx 17.558$$

$s \approx 17.56$ years

$P(16 \le T \le 38.512 - 17.558)$

$$= \int_{16}^{20.954} 0.06049e^{-0.03211t}dt$$
$$= -1.88384e^{-0.03211t}\Big|_{16}^{20.954}$$
$$\approx 0.1657$$

If you do the integration on a calculator using differently rounded values for the limits you may get a slightly different answer, such as 0.1.

(d) To find the median age, find the value of m such that $\int_a^m f(t)dt = \frac{1}{2}$.

$$\int_{16}^{m} 0.06049e^{-0.03211t}dt = \frac{1}{2}$$
$$\frac{0.06049}{-0.03211}e^{-0.03211t}\Big|_{16}^{m} = \frac{1}{2}$$
$$\frac{0.06049}{-0.03211}\left(e^{-0.03211m} - e^{-0.03211(16)}\right) = \frac{1}{2}$$
$$e^{-0.03211m} - e^{-0.51376} = -\frac{0.03211}{2\cdot 0.06049}$$
$$e^{-0.03211m} = e^{-0.51376} - \frac{0.03211}{2\cdot 0.06049}$$
$$-0.03211m = \ln\left(e^{-0.51376} - \frac{0.03211}{2\cdot 0.06049}\right)$$
$$m = \frac{1}{-0.03211}\ln\left(e^{-0.51376} - \frac{0.03211}{2\cdot 0.06049}\right)$$
$$m \approx 34.26$$

The median age is 34.26 years.

43. $S(t) = \dfrac{1}{101{,}370}(-2.564t^3 + 99.11t^2 - 964.6t + 5631)$ for t in $[0, 24]$

$$\mu = \int_0^{24} tS(t)\,dt$$
$$= \frac{1}{101{,}370}\int_0^{24} (-2.564t^4 + 99.11t^3 - 964.6t^2 + 5631t)\,dt$$

Use a calculator to evaluate the integral.

$$\mu \approx 12.964$$

The expected time of day at which a fatal accident will occur is about 1 pm.

18.3 Special Probability Density Functions

Your Turn 1

$42 - 27 = 15$, so the uniform distribution for the maximum daily temperature T is $f(t) = \frac{1}{15}$ for t in $[27, 42]$.

$$\begin{aligned}\mu &= \frac{27+42}{2}\\ &= \frac{69}{2}\\ &= 34.5\end{aligned}$$

The expected maximum daily temperature is 34.5°C.

$$\begin{aligned}\sigma &= \frac{1}{\sqrt{12}}(42-27)\\ &= \frac{15}{\sqrt{12}}\\ &= \frac{5\sqrt{3}}{2}\\ &\approx 4.33\end{aligned}$$

A temperature one standard deviation below the mean is $34.5 - \frac{5\sqrt{3}}{2} \approx 30.16987$. A temperature one standard deviation above the mean is $34.5 + \frac{5\sqrt{3}}{2} \approx 38.83013$.

$$P(\mu - \sigma \le T \le \mu + \sigma) = \int_{30.16987}^{38.83013} \frac{1}{15}dx$$

This integral is 1/15 times the difference of the limits, but this difference is just twice the standard deviation, so the probability is

$$\frac{5\sqrt{3}}{2} \cdot \frac{2}{15} = \frac{\sqrt{3}}{3} \approx 0.5774$$

Your Turn 2

$$f(t) = \frac{1}{25}e^{-t/25} \quad \text{for } t \ge 0$$

(a) The probability that a randomly selected battery has a useful life less than 100 hours is

$$\begin{aligned}P(T \le 100) &= \int_0^{100} \frac{1}{25}e^{-t/25}\,dt\\ &= \frac{1}{25}(-25e^{-t/25})\Big|_0^{100}\\ &= -(e^{-100/25} - e^0)\\ &= 1 - e^{-4}\\ &\approx 0.9817.\end{aligned}$$

(b)

$$\mu = \frac{1}{1/25} = 25$$

$$\sigma = \frac{1}{1/25} = 25$$

(c)

$$\begin{aligned}P(T > 40) &= \int_{40}^{\infty} \frac{1}{25}e^{-t/25}\,dt\\ &= \lim_{b\to\infty} \frac{1}{25}(-25e^{-t/25})\Big|_{40}^{b}\\ &= \lim_{b\to\infty}(-e^{-b/25} + e^{-40/25})\\ &= e^{-8/5}\\ &\approx 0.2019\end{aligned}$$

Your Turn 3

(a) The z-score for an age of 79 is

$$\begin{aligned}z &= \frac{79-\mu}{\sigma}\\ &= \frac{79-75}{16}\\ &= 0.25.\end{aligned}$$

$$\begin{aligned}P(X > 79) &= P(z > 0.25)\\ &= 1 - P(z \le 0.25)\\ &= 1 - 0.5987\\ &= 0.4013\end{aligned}$$

(We find the value 0.5987 in the row for 0.2 and the column for 0.05 in the table giving the area under the normal curve.)

(b) Compute the z-scores for 67 and 83.

$$z = \frac{67-75}{16} = -0.5$$

$$z = \frac{83-75}{16} = 0.5$$

$$\begin{aligned}P(67 \le X \le 83) &= P(-0.5 \le z \le 0.5)\\ &= 0.6915 - 0.3085\\ &= 0.3830\end{aligned}$$

(We find the value 0.6915 in the row for 0.5 and the column for 0.00 in the normal table, and the value 0.3085 in the row for -0.5 and the column for 0.00.)

18.3 Exercises

1. $f(x) = \frac{5}{7}$ for x in $[3, 4.4]$

This is a uniform distribution: $a = 3, b = 4.4$.

(a) $\mu = \frac{1}{2}(4.4 + 3) = \frac{1}{2}(7.4)$
$= 3.7$ cm

(b) $\sigma = \frac{1}{\sqrt{12}}(4.4 - 3)$
$= \frac{1}{\sqrt{12}}(1.4)$
≈ 0.4041 cm

(c) $P(3.7 < X < 3.7 + 0.4041)$
$= P(3.7 < X < 4.1041)$
$= \int_{3.7}^{4.1041} \frac{5}{7}\,dx$
$= \frac{5}{7}x\Big|_{3.7}^{4.1041}$
≈ 0.2886

3. $f(t) = 4e^{-4t}$ for t in $[0, \infty)$

This is an exponential distribution: $a = 4$.

(a) $\mu = \frac{1}{4} = 0.25$ year

(b) $\sigma = \frac{1}{4} = 0.25$ year

(c) $P(0.25 < T < 0.25 + 0.25)$
$= P(0.25 < T < 0.5)$
$= \int_{0.25}^{0.5} 4e^{-4t}\,dt$
$= -e^{-4t}\Big|_{0.25}^{0.5}$
$= -\frac{1}{e^{-2}} + \frac{1}{e^{-1}}$
≈ 0.2325

5. $f(t) = \frac{e^{-t/3}}{3}$ for t in $[0, \infty)$

This is an exponential distribution: $a = \frac{1}{3}$.

(a) $\mu = \frac{1}{\frac{1}{3}} = 3$ days

(b) $\sigma = \frac{1}{\frac{1}{3}} = 3$ days

(c) $P(3 < T < 3 + 3) = P(3 < T < 6)$
$= \int_{3}^{6} \frac{e^{-t/3}}{3}\,dt$
$= e^{-t/3}\Big|_{3}^{6}$
$= -\frac{1}{e^{-2}} + \frac{1}{e^{-1}}$
≈ 0.2325

7. $z = 3.50$

Area to the left of $z = 3.50$ is 0.9998. Given mean $\mu = z - 0$, so area to left of μ is 0.5.

Area between μ and z is

$$0.9998 - 0.5 = 0.4998.$$

Therefore, this area represents 49.98% of total area under normal curve.

9. Between $z = 1.28$ and $z = 2.05$

Area to left of $z = 2.05$ is 0.9798 and area to left of $z = 1.28$ is 0.8997.

$$0.9798 - 0.8997 = 0.0801$$

Percent of total area $= 8.01\%$

11. Since $10\% = 0.10$, the z-score that corresponds to the area of 0.10 to the left of z is -1.28.

13. 18% of the total area to the right of z means $1 - 0.18$ of the total area is to the left of z.

$$1 - 0.18 = 0.82$$

The closest z-score that corresponds to the area of 0.82 is 0.92

19. Let m be the median of the exponential distribution $f(x) = ae^{-ax}$ for $[0, \infty)$.

$$\int_0^m ae^{-ax}dx = 0.5$$
$$-e^{-ax}\Big|_0^m = 0.5$$
$$-e^{-am} + 1 = 0.5$$
$$0.5 = e^{-am}$$
$$-am = \ln 0.5$$
$$m = -\frac{\ln 0.5}{a}$$

or $\quad -am = \ln \frac{1}{2}$

$$-am = -\ln 2$$

$$m = \frac{\ln 2}{a}$$

21. The area that is to the left of x is

$$A = \int_{-\infty}^{x} \frac{1}{\sigma\sqrt{2\pi}} e^{-\frac{(t-\mu)^2}{2\sigma^2}}\, dt.$$

Let $u = \frac{(t-\mu)}{\sigma}$. Then $du = \frac{1}{\sigma}dt$ and $dt = \sigma du$.

If $t = x$,

$$u = \frac{x - \mu}{\sigma} = z.$$

As $t \to -\infty$, $u \to -\infty$.

Therefore,

$$A = \int_{-\infty}^{z} \frac{1}{\sigma\sqrt{2\pi}} e^{(-1/2)u^2}\, \sigma\, du$$

$$= \frac{\sigma}{\sigma}\int_{-\infty}^{z} \frac{1}{\sqrt{2\pi}} e^{-u^2/2} du$$

$$= \int_{-\infty}^{z} \frac{1}{\sqrt{2\pi}} e^{-u^2/2} du.$$

This is the area to the left of z for the standard normal curve.

In Exercise 23, use Simpson's rule with $n = 140$ or the integration feature on a graphing calculator to approximate the integrals. Answers may very slightly from those given here depending on the method that is used.

23. **(a)** $\int_{0}^{35} 0.5e^{-0.5x}\, dx \approx 1.00000$

(b) $\int_{0}^{35} 0.5xe^{-0.5x}\, dx \approx 1.99999$

(c) $\int_{0}^{35} 0.5x^2e^{-0.5x}\, dx = 8.00000$

25. Use Simpson's rule with $n = 40$ and limits of -6 and 6 to approximate the mean and standard deviation of a normal probability distribution.

(a) $\int_{-\infty}^{\infty} \frac{x}{\sqrt{2\pi}} e^{-x^2/2}\, dx$

$$\approx \int_{-6}^{6} \frac{x}{\sqrt{2\pi}} e^{-x^2/2}\, dx$$

$\mu = 0$

(For the integral of an odd function over an interval symmetric to 0, Simpson's rule will give 0; there is no need to do any calculation.)

(b) $\int_{-\infty}^{\infty} \frac{x^2}{\sqrt{2\pi}} e^{-x^2/2}\, dx$

$$\approx \int_{-6}^{6} \frac{x^2}{\sqrt{2\pi}} e^{-x^2/2}\, dx$$

$\sigma \approx 0.9999999224$ (Simpson's rule)

$\sigma \approx 0.9999999251$ (calculator)

27. The probability density function for the uniform distribution is $f(x) = \frac{1}{b-a}$ for x in $[a, b]$.

The cumulative distribution function for f is

$$F(x) = P(X \le x)$$

$$= \int_{a}^{x} f(t)\, dt$$

$$= \int_{a}^{x} \frac{1}{b-a}\, dt$$

$$= \frac{1}{b-a} t \Big|_{a}^{x}$$

$$= \frac{1}{b-a}(x - a)$$

$$= \frac{x-a}{b-a},\ a \le x \le b.$$

29. For a uniform distribution,

$$f(x) = \frac{1}{b-a} \text{ for } [a, b].$$

Thus, we have

$$f(x) = \frac{1}{85 - 10} = \frac{1}{75}$$

for $[10, 85]$.

(a) $\mu = \frac{1}{2}(10 + 85) = \frac{1}{2}(95)$

$= 47.5$ thousands

Therefore, the agent sells \$47,500 in insurance.

(b) $P(50 < X < 85) = \int_{50}^{85} \frac{1}{75} dx$

$$= \frac{x}{75}\Big|_{50}^{85}$$

$$= \frac{85}{75} - \frac{50}{75}$$

$$= \frac{35}{75} = 0.4667$$

31. **(a)** Since we have exponential distribution with $\mu = 4.25$,

$$\mu = \frac{1}{a} = 4.25$$
$$a = 0.235.$$

Therefore, $f(x) = 0.235e^{-0.235x}$ on $[0, \infty)$.

(b) $P(X > 10)$

$$= \int_{10}^{\infty} 0.235e^{-0.235x} dx$$

$$= \lim_{a\to\infty} \int_{10}^{a} 0.235e^{-0.235x} dx$$

$$= \lim_{a\to\infty} (-e^{-0.235x})\Big|_{10}^{a}$$

$$= \lim_{a\to\infty} \left(-\frac{1}{e^{0.235a}} + \frac{1}{e^{2.35}}\right)$$

$$= \frac{1}{e^{2.35}} = 0.0954$$

33. **(a)** $\mu = 2.5, \sigma = 0.2, x = 2.7$

$$z = \frac{2.7 - 2.5}{0.2} = 1$$

Area to the right of $z = 1$ is

$$1 - 0.8413 = 0.1587.$$

Probability $= 0.1587$

(b) Within 1.2 standard deviations of the mean is the area between $z = -1.2$ and $z = 1.2$.

Area to left of $z = 1.2 = 0.8849$

Area to the left of the $z = -1.2 = 0.1151$

$$0.8849 - 0.1151 = 0.7698$$

Probability $= 0.7698$

35. If X has a uniform distribution on $[0, 1000]$, then its density function is $f(x) = \frac{1}{1000}$ for x in $[0, 1000]$. The expected payment with no deductible is

$$E(X) = \int_0^{1000} x \cdot \frac{1}{1000} dx$$

$$= \frac{1}{1000} \cdot \frac{1}{2}x^2\Big|_0^{1000}$$

$$= \frac{1}{2000} \cdot (1000^2 - 0)$$

$$= 500.$$

Now, let the deductible be D. According to the hint,

$$\text{payment} = \begin{cases} 0 & \text{for } x \le D \\ x - D & \text{for } x > D. \end{cases}$$

The expected payment with the deductible is therefore

$$E(X) = \int_0^D 0 \cdot \frac{1}{1000} dx + \int_D^{1000} (x - D) \cdot \frac{1}{1000} dx$$

$$= 0 + \frac{1}{1000} \cdot \left(\frac{1}{2}x^2 - Dx\right)\Big|_D^{1000}$$

$$= \frac{1}{1000} \cdot \left[\left(\frac{1}{2}1000^2 - 1000D\right) - \left(\frac{1}{2}D^2 - D^2\right)\right]$$

$$= 500 - D + \frac{1}{2000}D^2.$$

For this amount to be 25% of the amount with no deductible, we must have

$$500 - D + \frac{1}{2000} \cdot D^2 = 0.25 \cdot 500$$

$$\frac{1}{2000}D^2 - D + 500 = 125$$

$$\frac{1}{2000}D^2 - D + 375 = 0$$

$$D^2 - 2000D + 750{,}000 = 0$$

$$(D - 1500)(D - 500) = 0$$

$$D = 1500 \text{ or } D = 500$$

We reject $D = 1500$ since it is not in $[0, 1000]$.

Therefore, $D = 500$. The correct answer choice is **c**.

37. Let the random variable X be the lifetime of the printer in a years. Then it has exponential distribution $f(x) = ae^{-ax}$ for $x \geq 0$.

If the mean is 2 years, then $\frac{1}{a} = 2$, or $a = \frac{1}{2}$ and the function is $f(x) = \frac{1}{2}e^{-x/2}$ for $x \geq 0$.

We wish to find $P(0 \leq X \leq 1)$ and $P(1 \leq X \leq 2)$.

$$P(0 \leq X \leq 1) = \int_0^1 \frac{1}{2}e^{-x/2}dx$$
$$= -e^{-x/2}\Big|_0^1$$
$$= -e^{-1/2} + 1$$

$$P(1 \leq X \leq 2) = \int_1^2 \frac{1}{2}e^{-x/2}dx$$
$$= -e^{-x/2}\Big|_1^2$$
$$= -e^{-1} + e^{-1/2}$$

If 100 printers are sold, then $(1 - e^{-1/2} + 1)(100)$ will fail in the first year, and $(e^{-1/2} - e^{-1/2})(100)$ will fail in the second year.

The manufacturer pays a full refund on those failing first year and one-half refund on those failing during the second year.

$$\text{Refunds} = (1 - e^{-1/2} + 1)(100)(\$200) + (e^{-1/2} - e^{-1})(100)(\$100) \approx \$10{,}255.90$$

The correct answer choice is **d**.

39. For a uniform distribution,

$$f(x) = \frac{1}{b-a} \text{ for } x \text{ in } [a, b].$$

$$f(x) = \frac{1}{36-20} = \frac{1}{16} \text{ for } x \text{ in } [20, 36]$$

(a) $\mu = \frac{1}{2}(20 + 36) = \frac{1}{2}(56)$
$= 28$ days

(b) $P(30 < X \leq 36)$

$$= \int_{30}^{36} \frac{1}{16}\,dx = \frac{1}{16}x\Big|_{30}^{36}$$
$$= \frac{1}{16}(36 - 30)$$
$$= 0.375$$

41. We have an exponential distribution, with $a = 1$.
$f(t) = e^{-t}, [0, \infty)$

(a) $\mu = \frac{1}{1} = 1$ hr

(b) $P(T < 30 \text{ min})$

$$= \int_0^{0.5} e^{-t}\,dt$$
$$= -e^{-t}\Big|_0^{0.5}$$
$$1 - e^{-0.5} \approx 0.3935$$

43. $f(x) = ae^{-ax}$ for $[0, \infty]$

Since $\mu = 25$ and $\mu = \frac{1}{a}$,

$$a = \frac{1}{25} = 0.04.$$

This, $f(x) = 0.04e^{-0.04x}$.

(a) We must find t such that $P(X \leq t) = 0.90$.

$$\int_0^t 0.04e^{-0.04x}\,dx = 0.90$$
$$-e^{-0.04x}\Big|_0^t = 0.90$$
$$-e^{-0.04t} + 1 = 0.90$$
$$0.10 = -e^{-0.04t}$$
$$-0.04t = \ln 0.10$$
$$t = \frac{\ln 0.10}{-0.04}$$
$$t \approx 57.56$$

The longest time within which the predator will be 90% certain of finding a prey is approximately 58 min.

(b) $P(X \geq 60)$

$$= \int_{60}^{\infty} 0.04e^{-0.04x}\,dx$$
$$= \lim_{b\to\infty} \int_{60}^{b} 0.04e^{-0.04x}\,dx$$
$$= \lim_{b\to\infty} (e^{-0.04x})\Big|_{60}^{b}$$
$$= \lim_{b\to\infty} \left[-e^{-0.04b} + e^{-0.04(60)}\right]$$
$$= 0 + e^{-2.4}$$
$$\approx 0.0907$$

The probability that the predator will have to spend more than one hour looking for a prey is approximately 0.0907.

45. For an exponential distribution, $f(x) = ae^{-ax}$ for x in $[0, \infty)$.

Since $\mu = \frac{1}{a} = 12.3$, $a = \frac{1}{12.3}$.

(a) $$P(X \geq 20) = \int_{20}^{\infty} \frac{1}{12.3} e^{-x/12.3}\, dx$$
$$= \lim_{b\to\infty} \int_{20}^{b} \frac{1}{12.3} e^{-x/12.3}\, dx$$
$$= \lim_{b\to\infty} \left(\frac{1}{12.3} e^{-x/12.3} \Big|_{20}^{b} \right)$$
$$= \lim_{b\to\infty} (-e^{-b/12.3} + e^{-20/12.3})$$
$$= e^{-20/12.3}$$
$$\approx 0.1967$$

(b) $$P(10 \leq X \leq 20) = \int_{10}^{20} \frac{1}{12.3} e^{-x/12.3}\, dx$$
$$= (-e^{-x/12.3}) \Big|_{0}^{20}$$
$$= -e^{-20/12.3} + e^{-10/12.3}$$
$$\approx 0.2468$$

47. We have an exponential distribution, with $a = 0.229$.

So $f(t) = 0.229e^{-0.229t}$, for $[0, \infty)$.

(a) The life expectancy is

$$\mu = \frac{1}{a} = \frac{1}{0.229} \approx 4.37 \text{ millennia.}$$

The standard deviation is

$$\sigma = \frac{1}{a} = \frac{1}{0.229} \approx 4.37 \text{ millennia.}$$

(b) $$P(T \geq 2) = \int_{2}^{\infty} 0.229e^{-0.229t} dt$$
$$= 1 - \int_{0}^{2} 0.229e^{-0.229t} dt$$
$$= 1 + \left(e^{-0.229t} \Big|_{0}^{2} \right)$$
$$= 1 + \left[e^{-0.229(2)} - 1 \right]$$
$$= e^{-0.458} \approx 0.6325$$

49. For an exponential distribution, $f(x) = ae^{-ax}$ for $[0, \infty)$.

Since $\mu = \frac{1}{a} = 8$, $a = \frac{1}{8}$.

(a) $$P(X \geq 10) = \int_{10}^{\infty} \frac{1}{8} e^{-x/8} dx$$
$$= 1 - \int_{0}^{10} \frac{1}{8} e^{-x/8} dx$$
$$= 1 + \left(e^{-x/8} \Big|_{0}^{10} \right)$$
$$= 1 + [e^{-10/8} - 1]$$
$$= e^{-10/8} \approx 0.2865$$

(b) $$P(X < 2) = \int_{0}^{2} \frac{1}{8} e^{-x/8} dx$$
$$= -e^{-x/8} \Big|_{0}^{2}$$
$$= -e^{-2/8} + 1$$
$$\approx 0.2212$$

51. We have an exponential distribution $f(x) = ae^{-ax}$ for $x \geq 0$. Since $a = \frac{1}{90}$, $f(x) = \frac{1}{90} e^{-x/90}$ for $x \geq 0$.

(a) The probability that the time for a goal is no more than 71 minutes is

$$P(0 < X < 71) = \int_{0}^{71} \frac{1}{90} e^{-x/90} dx$$
$$= -e^{-x/90} \Big|_{0}^{71}$$
$$= -e^{-71/90} + 1$$
$$\approx 0.5457.$$

(b) The probability that the time for a goal is 499 minutes or more is

$$P(X \geq 499) = \int_{499}^{\infty} \frac{1}{90} e^{-x/90} dx$$
$$= e^{-x/90} \Big|_{499}^{\infty}$$
$$= 0 + e^{-499/90}$$
$$\approx 0.0039.$$

Chapter 18 Review Exercises

1. True

2. True

3. True

4. False: A density function is always nonnegative.

5. False: If the random variable takes on negative values the expectation may also be negative.

6. True

7. True

8. True

9. False: The normal distribution is symmetrical; the exponential distribution has a long tail to the right.

10. False: The expected value is 0 and the standard deviation is 1.

11. In a probability function, the y-values (or function values) represent probabilities.

13. A probability density function f for $[a, b]$ must satisfy the following two conditions:

(1) $f(x) \geq 0$ for all x in the interval $[a, b]$;

(2) $\int_a^b f(x)dx = 1.$

15. $f(x) = \sqrt{x}$; $[4, 9]$

$$\int_4^9 x^{1/2}dx = \frac{2}{3}x^{3/2}\Big|_4^9$$
$$= \frac{2}{3}(27 - 8)$$
$$= \frac{38}{3} \neq 1$$

$f(x)$ is not a probability density function.

17. $f(x) = 0.7e^{-0.7x}; [0, \infty)$

$$\int_0^\infty 0.7e^{-0.7x}dx = -e^{-0.7x}\Big|_0^\infty$$
$$= \lim_{b\to\infty} (-e^{-0.7b}) + e^0$$
$$= \lim_{b\to\infty}\left(-\frac{1}{e^{0.7b}}\right) + 1$$
$$= 0 + 1 = 1$$

$f(x) \geq 0$ for all x in $[0, \infty)$.

Therefore, $f(x)$ is a probability density function.

19. $f(x) = kx^2$; $[1, 4]$

$$\int_1^4 kx^2dx = \frac{kx^3}{3}\Big|_1^4$$
$$= 21k$$

Since $f(x)$ is a probability density function,

$$21k = 1$$
$$k = \frac{1}{21}.$$

21. $f(x) = \frac{1}{10}$ for $[10, 20]$

(a) $P(10 \leq X \leq 12)$

$$= \int_{10}^{12} \frac{1}{10}dx$$
$$= \frac{x}{10}\Big|_{10}^{12}$$
$$= \frac{1}{5} = 0.2$$

(b) $P\left(\frac{31}{2} \leq X \leq 20\right)$

$$= \int_{31/2}^{20} \frac{1}{10}dx$$
$$= \frac{x}{10}\Big|_{31/2}^{20}$$
$$= 2 - \frac{31}{20}$$
$$= \frac{9}{20} = 0.45$$

(c) $P(10.8 \le X \le 16.2)$

$$= \int_{10.8}^{16.2} \frac{1}{10} dx$$

$$= \frac{x}{10}\Big|_{10.8}^{16.2} = 0.54$$

23. If we consider the probabilities as weights, the expected value or mean of a probability distribution represents the point at which the distribution balances.

25. $f(x) = \frac{2}{9}(x-2)$; $[2, 5]$

(a) $$\mu = \int_2^5 \frac{2x}{9}(x-2)dx$$

$$= \int_2^5 \frac{2}{9}(x^2 - 2x)dx$$

$$= \frac{2}{9}\left(\frac{x^3}{3} - x^2\right)\Bigg|_2^5$$

$$= \frac{2}{9}\left(\frac{125}{3} - 25 - \frac{8}{3} + 4\right) = 4$$

(b) $$\text{Var}(X) = \int_2^5 \frac{2x^2}{9}(x-2)dx - (4)^2$$

$$= \int_2^5 \frac{2}{9}(x^3 - 2x^2)dx - 16$$

$$= \frac{2}{9}\left(\frac{x^4}{4} - \frac{2x^3}{3}\right)\Bigg|_2^5 - 16$$

$$= \frac{2}{9}\left(\frac{625}{4} - \frac{250}{3} - 4 + \frac{16}{3}\right) - 16$$

$$= 0.5$$

(c) $\sigma = \sqrt{0.5} \approx 0.7071$

(d) $$\int_2^m \frac{2}{9}(x-2)dx = \frac{1}{2}$$

$$\frac{1}{9}(m-2)^2\Big|_2^m = \frac{1}{2}$$

$$\frac{1}{9}[(m-2)^2 - 0] = \frac{1}{2}$$

$$m^2 - 4m + 4 = \frac{9}{2}$$

$$m^2 - 4m - \frac{1}{2} = 0$$

$$m = \frac{4 \pm 3\sqrt{2}}{2}$$

$$\approx -0.121, 4.121$$

We reject -0.121 since it is not in $[2, 5]$. So, $m = 4.121$

(e) $$\int_2^x \frac{2}{9}(t-2)dt = \frac{1}{9}(t-2)^2\Big|_2^x$$

$$= \frac{1}{9}[(x-2)^2 - 0]$$

$$= \frac{(x-2)^2}{9}, 2 \le x \le 5$$

27. $f(x) = 5x^{-6}$; $[1, \infty)$

(a) $$\mu = \int_1^\infty x \cdot 5x^{-6}\, dx = \int_1^\infty 5x^{-5}\, dx$$

$$= \lim_{b\to\infty} \int_1^b 5x^{-5}\, dx = \lim_{b\to\infty} \frac{5x^{-4}}{-4}\Bigg|_1^b$$

$$= \lim_{b\to\infty} \frac{5}{4}\left(1 - \frac{1}{b^4}\right) = \frac{5}{4}$$

(b) $$\text{Var}(X) = \int_1^\infty x^2 \cdot 5x^{-6}\, dx - \left(\frac{5}{4}\right)^2$$

$$= \lim_{b\to\infty} \int_1^b 5x^{-4}\, dx - \frac{25}{16}$$

$$= \lim_{b\to\infty} \frac{5x^{-3}}{-3}\Bigg|_1^b - \frac{25}{16}$$

$$= \lim_{b\to\infty} \frac{5}{3}\left(1 - \frac{1}{b^3}\right) - \frac{25}{16}$$

$$= \frac{5}{3} - \frac{25}{16} = \frac{5}{48} \approx 0.1042$$

(c) $\sigma \approx \sqrt{\text{Var}(X)} \approx 0.3227$

(d) $$\int_1^m 5x^{-6}dx = \frac{1}{2}$$

$$-x^{-5}\Big|_1^m = \frac{1}{2}$$

$$-m^{-5} + 1 = \frac{1}{2}$$

$$m^{-5} = \frac{1}{2}$$

$$m^5 = 2$$

$$m = \sqrt[5]{2} \approx 1.149$$

(e) $$\int_1^x 5t^{-6}dt = -t^{-5}\Big|_1^x$$

$$= -x^{-5} + 1$$

$$= 1 - \frac{1}{x^5}, x \ge 1$$

29. $f(x) = 4x - 3x^2$; [0, 1]

(a) $\mu = \int_0^1 x(4x - 3x^2)dx$

$= \int_0^1 (4x^2 - 3x^3)dx$

$= \left(\frac{4x^3}{3} - \frac{3x^4}{4}\right)\Bigg|_0^1$

$= \frac{4}{3} - \frac{3}{4} = \frac{7}{12}$

≈ 0.5833

(b) $\text{Var}(X) = \int_0^1 x^2(4x - 3x^2)dx - \left(\frac{7}{12}\right)^2$

$= \int_0^1 (4x^3 - 3x^4)dx - \left(\frac{7}{12}\right)^2$

$= \left(x^4 - \frac{3x^5}{5}\right)\Bigg|_0^1 - \left(\frac{7}{12}\right)^2$

$= 1 - \frac{3}{5} - \left(\frac{7}{12}\right)^2$

≈ 0.0597

$\sigma \approx \sqrt{\text{Var}(X)}$

≈ 0.2444

(c) $P\left(0 \le X \le \frac{7}{12}\right)$

$= \int_0^{7/12} (4x - 3x^3)dx$

$= (2x^2 - x^3)\Big|_0^{7/12}$

$= 2\left(\frac{7}{12}\right)^2 - \left(\frac{7}{12}\right)^3$

≈ 0.4821

(d) $P(\mu - \sigma \le X \le \mu + \sigma)$

$\approx P(0.339 \le x \le 0.827)$

$= \int_{0.339}^{0.827} (4x - 3x^2)dx$

$= (2x^2 - x^3)\Big|_{0.339}^{0.827}$

$= 2(0.827)^2 - (0.827)^3$

$- 2(0.339)^2 + (0.339)^3$

≈ 0.6114

31. $f(x) = 0.01e^{-0.01x}$ for $[0, \infty)$ is an exponential distribution.

(a) $\mu = \frac{1}{0.01} = 100$

(b) $\sigma = \frac{1}{0.01} = 100$

(c) $P(100 - 100 < X < 100 + 100)$

$= P(0 < X < 200)$

$= \int_0^{200} 0.01e^{-0.01x}\,dx$

$= -e^{-0.01x}\Big|_0^{200}$

$= 1 - e^{-2} \approx 0.8647$

For Exercises 33–40, use the table in the Appendix for the areas under the normal curve.

33. Area to the left of $z = -0.43$ is 0.3336.

Percent of area is 33.36%.

35. Area between $z = -1.17$ and $z = -0.09$ is

$0.4641 - 0.1210 = 0.3431.$

Percent of area is 34.31%.

37. The region up to 1.2 standard deviations below the mean is the region to the left of $z = -1.2$. The area is 0.1151, so the percent of area is 11.51%.

39. 52% of area is to the right implies that 48% is to the left.

$P(z < a) = 0.48$ for $a = -0.05$

Thus, 52% of the area lies to the right of $z = -0.05$.

41. $f(x) = 0.05$ for [10, 30]

(a) This is a uniform distribution.

(b) The domain of f is [10, 30].

The range of f is [0.05].

(c)

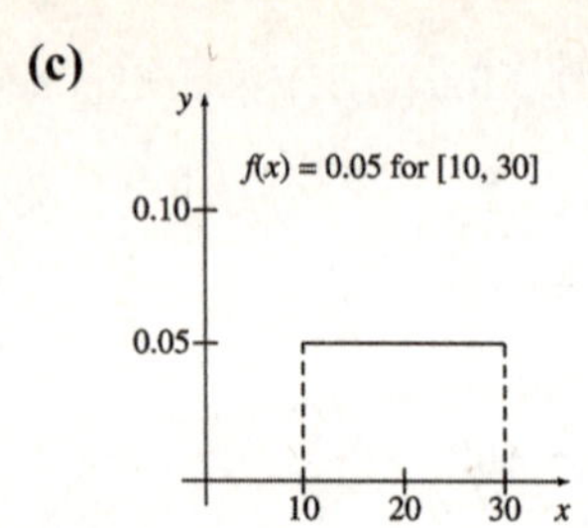

(d) For a uniform distribution,

$$\mu = \frac{1}{2}(b + a) \text{ and}$$

$$\text{Var}(X) = \frac{b^2 - 2ab + a^2}{12}.$$

Thus,

$$\mu = \frac{1}{2}(30 + 10) = \frac{1}{2}(40) = 20$$

$$\text{Var}(X) = \frac{30^2 - 2(10)(30) + 10^2}{12} = \frac{400}{12}.$$

$$\sigma = \sqrt{\frac{400}{12}} \approx 5.77$$

(e) $P(\mu - \sigma \le X \le \mu + \sigma)$

$= P(20 - 5.77 \le X \le 20 + 5.77)$

$= P(14.23 \le X \le 25.77)$

$= \int_{14.23}^{25.77} 0.05\, dx$

$= 0.05x\Big|_{14.23}^{25.77}$

$= 0.05(25.77 - 14.23)$

≈ 0.577

43. $f(x) = \dfrac{e^{-x^2}}{\sqrt{\pi}}$ for $(-\infty, \infty)$

(a) Since the exponent of e in $f(x)$ may be written

$$-x^2 = \frac{-(x-0)^2}{2\left(\frac{1}{\sqrt{2}}\right)^2},$$

and

$$\frac{1}{\sqrt{\pi}} = \frac{1}{\frac{1}{\sqrt{2}}\sqrt{2\pi}},$$

$f(x)$ is a normal distribution with $\mu = 0$ and $\sigma = \frac{1}{\sqrt{2}}$.

(b) The domain of f is $(-\infty, \infty)$.

The range of f is $\left(0, \frac{1}{\sqrt{\pi}}\right]$.

(c)

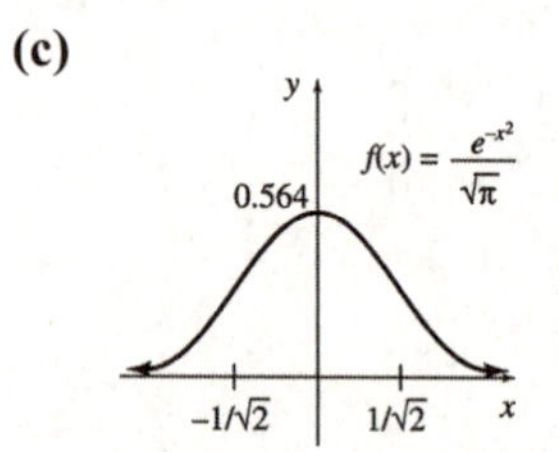

(d) For this normal distribution, $\mu = 0$ and $\sigma = \frac{1}{\sqrt{2}}$.

(e) $P(\mu - \sigma \le X \le \mu + \sigma)$

$= 2P(0 \le X \le \mu + \sigma)$

$= 2P\left(0 \le X \le \frac{1}{\sqrt{2}}\right)$

If $x = \frac{1}{\sqrt{2}}$, $z = \dfrac{\frac{1}{\sqrt{2}} - 0}{\frac{1}{\sqrt{2}}} = 1.00$.

Thus,

$P(\mu - \sigma \le X \le \mu + \sigma)$

$= 2P(0 \le z \le 1.00)$

$= 2(0.3413)$

≈ 0.6826

45. $f(x) = \dfrac{x^{-1/2}e^{-x/2}}{\sqrt{2\pi}}$ for x in $(0, \infty)$

(a) Using integration by parts:

Let $u = e^{-x/2}$ and $dv = x^{-1/2}dx$

Then $du = -\frac{1}{2}e^{-x/2}$ and $v = 2x^{1/2}$.

$$\frac{1}{\sqrt{2\pi}}\int x^{-1/2}e^{-x/2}dx$$

$$= \frac{1}{\sqrt{2\pi}}\left[2x^{1/2}e^{-x/2} - \int 2x^{1/2}\left(-\frac{1}{2}\right)e^{-x/2}dx\right]$$

$$= \frac{1}{\sqrt{2\pi}}\left[2x^{1/2}e^{-x/2} + \int x^{1/2}e^{-x/2}dx\right]$$

Thus,

$P(0 < X \le b)$

$$= \frac{1}{\sqrt{2\pi}}\int_0^b x^{-1/2}e^{-x/2}dx$$

$$= \frac{1}{\sqrt{2\pi}}\left[2x^{1/2}e^{-x/2}\Big|_0^b + \int_0^b x^{1/2}e^{-x/2}dx\right].$$

(b) $P(0 < X \leq 1)$

$$= \frac{1}{\sqrt{2\pi}}\left[2x^{1/2}e^{-x/2}\Big|_0^1 + \int_0^1 x^{1/2}e^{-x/2}dx\right]$$

Notice that

$$2x^{1/2}e^{-x/2}\Big|_0^1 = 2e^{-1/2} - 0 \approx 1.2131.$$

Using Simpson's rule with $n = 12$ to evaluate the improper integral, we have

$$\int_0^1 x^{1/2}e^{-x/2}dx \approx 0.4962$$

Therefore,

$$P(0 \leq X \leq 1) = \frac{1}{\sqrt{2\pi}}\int_0^1 x^{-1/2}e^{-x/2}dx \approx \frac{1}{\sqrt{2\pi}}(1.2131 + 0.4962) \approx 0.6819.$$

(c) $P(0 < X \leq 10)$

$$= \frac{1}{\sqrt{2\pi}}\left[2x^{1/2}e^{-x/2}\Big|_0^{10} + \int_0^{10} x^{1/2}e^{-x/2}dx\right]$$

First,

$$2x^{1/2}e^{-x/2}\Big|_0^{10} = 2\sqrt{10}e^{-5} - 0 \approx 0.0426.$$

Using Simpson's rule with $n = 12$ to evaluate the improper integral, we have

$$\int_0^{10} x^{1/2}e^{-x/2}dx \approx 2.3928$$

Therefore,

$$P(0 \leq X \leq 1) = \frac{1}{\sqrt{2\pi}}\int_0^{10} x^{-1/2}e^{-x/2}dx \approx \frac{1}{\sqrt{2\pi}}(0.0426 + 2.3928) \approx 0.9716.$$

(d) Since $f(x)$ is a probability density function, the limit as b $\to \infty$ should be 1. The previous results do support this conclusion.

47. (a) $f(t) = \frac{5}{112}(1 - t^{-3/2})$; $[1, 25]$

P(No repairs in years 1-3)

= P(First repair needed in years 4-25)

$$= \int_4^{25} \frac{5}{112}(1 - t^{-3/2})\,dt = \frac{5}{112}(t + 2t^{-1/2})\Big|_4^{25} = \frac{5}{112}\left[25 + \frac{2}{5} - 4 - 1\right] = \frac{51}{56} \approx 0.9107$$

(b)
$$\mu = \frac{5}{112}\int_1^{25} t(1 - t^{-3/2})\,dt = \frac{5}{112}\int_1^{25}(t - t^{-1/2})\,dt = \frac{5}{112}\left(\frac{t^2}{2} - 2t^{1/2}\right)\Bigg|_1^{25} = \frac{5}{112}\left(\frac{625}{2} - 10 - \frac{1}{2} + 2\right) = \frac{95}{7} \approx 13.57$$

The expected value for the number of years before the machine requires repairs is 13.57 years.

(c)

$$\text{Var} = \frac{5}{112}\int_1^{25} t^2(1 - t^{-3/2})\,dt - \left(\frac{95}{7}\right)^2$$
$$= \frac{5}{112}\int_1^{25} (t^2 - t^{1/2})dt - \left(\frac{95}{7}\right)^2$$
$$= \frac{5}{112}\left(\frac{t^3}{3} - \frac{2}{3}t^{3/2}\right)\Bigg|_1^{25} - \left(\frac{95}{7}\right)^2$$
$$= \frac{5}{112}\left(\frac{15{,}625}{3} - \frac{250}{3} - \frac{1}{3} + \frac{2}{3}\right) - \frac{9025}{49}$$
$$= \frac{6560}{147}$$
$$\sigma = \sqrt{\frac{6560}{147}}$$
$$\approx 6.68$$

The standard deviation of the number of years before repairs are required is 6.68 years.

49. **(a)** $\mu = 8$

$$\frac{1}{a} = 8$$
$$a = \frac{1}{8}$$
$$f(x) = \frac{1}{8}e^{-x/8} \text{ for } [0, \infty)$$

(b) Expected number $= \mu = 8$

(c) $\sigma = \mu = 8$

(d) $$P(5 \le X \le 10) = \int_5^{10} \frac{1}{8}e^{-x/8}\,dx$$
$$= -e^{-x/8}\Big|_5^{10}$$
$$= -e^{-10/8} + e^{-5/8}$$
$$\approx 0.2488$$

51. Let the random variable X be the number of printers. We have an exponential distribution $f(t) = ae^{-at}$ for t in $[0, \infty)$. Since $\mu = 10, a = \frac{1}{\mu} = 0.1$ so that

$$f(t) = 0.1e^{-0.1t}, t \ge 0.$$

We need to determine the portion of the printers sold in the first year and in the second and third year. In other words, we need to calculate $P(0 \le X \le 1)$ and $P(1 \le X \le 3)$.

$$P(0 \le X \le 1) = \int_0^1 0.1e^{-0.1t}dt$$
$$= -e^{-0.1t}\Big|_0^1$$
$$= -e^{-0.1} + e^0$$
$$= 1 - e^{-0.1}$$
$$P(1 \le X \le 3) = \int_1^3 0.1e^{-0.1t}dt$$
$$= -e^{-0.1t}\Big|_1^3$$
$$= -e^{-0.3} + e^{-0.1}$$
$$= e^{-0.1} - e^{-0.3}$$

Since

$$\text{payment} = \begin{cases} x & \text{for } 0 \le x \le 1 \\ 0.5x & \text{for } 1 \le x \le 3 \\ 0 & \text{for } x > 3 \end{cases}$$

the expected payment will be

$$E(X) = (1 - e^{-0.1})(x) + (e^{-0.1} - e^{-0.3})(0.5x) + 0$$
$$= x - 1.5xe^{-0.1} - 0.5xe^{-0.3}.$$

To determine the level x must be set for this to be 1000, solve

$$E(X) = x - 1.5xe^{-0.1} - 0.5xe^{-0.3} = 1000.$$

Using our calculators, we find $x \approx 5644$. The correct answer choice is **d**.

53. $f(x) = 0.01e^{-0.01x}$ for $[0,\infty)$ is an exponential distribution.

$$P(0 \le X \le 100)$$
$$= \int_0^{100} 0.01e^{-0.01x}\,dx$$
$$= -e^{-0.01x}\Big|_0^{100}$$
$$= 1 - \frac{1}{e}$$
$$\approx 0.6321$$

55. $f(x) = \dfrac{3}{19{,}696}(x^2 + x)$ for x in $[38, 42]$

(a) $$\mu = \frac{3}{19{,}696}\int_{38}^{42} x(x^2 + x)\,dx$$
$$= \frac{3}{19{,}696}\int_{38}^{42} (x^3 + x^2)\,dx$$
$$= \frac{3}{19{,}696}\left(\frac{x^4}{4} + \frac{x^3}{3}\right)\Bigg|_{38}^{42}$$
$$= \frac{3}{19{,}696}\left(\frac{(42)^4}{4} + \frac{(42)^3}{3} - \frac{(38)^4}{4} - \frac{(38)^3}{3}\right)$$
$$\approx 40.07$$

The expected body temperature of the species is 40.07°C.

(b) $P(X \le \mu)$
$$= \frac{3}{19{,}696}\int_{38}^{40.07} (x^2 + x)\,dx$$
$$= \frac{3}{19{,}696}\left(\frac{x^3}{3} + \frac{x^2}{2}\right)\Bigg|_{38}^{40.07}$$
$$= \frac{3}{19{,}696}\left(\frac{(40.07)^3}{3} + \frac{(40.07)^2}{2} - \frac{(38)^3}{3} - \frac{(38)^2}{2}\right)$$
$$\approx 0.4928$$

The probability of a body temperature below the mean is 0.4928.

57. Normal distribution,
$\mu = 2.2$ g, $\sigma = 0.4$ g, X = tension

$P(X < 1.9)$
$$= P\left(\frac{x - 2.2}{0.4} < \frac{1.9 - 2.2}{0.4}\right)$$
$$= P(z < -0.75)$$
$$\approx 0.2266.$$

59. For an exponential distribution,
$f(x) = ae^{-ax}$ for x in $[0, \infty)$.

Since $\mu = \dfrac{1}{a} = 32.5$, $a = \frac{1}{32.5}$.

(a) $$P(X \ge 40) = \int_{40}^{\infty} \frac{1}{32.5}e^{-x/32.5}\,dx$$
$$= \lim_{b\to\infty}\int_{40}^{b} \frac{1}{32.5}e^{-x/32.5}\,dx$$
$$= \lim_{b\to\infty}\left(\frac{1}{32.5}e^{-x/32.5}\Bigg|_{40}^{b}\right)$$
$$= \lim_{b\to\infty}\left(-e^{-b/32.5} + e^{-40/32.5}\right)$$
$$= e^{-40/32.5}$$
$$\approx 0.2921$$

(b) $P(30 \le X < 50)$
$$= \int_{30}^{50} \frac{1}{32.5}e^{-x/32.5}\,dx$$
$$= (-e^{-x/32.5})\Big|_{30}^{50}$$
$$= -e^{-50/32.5} + e^{-30/32.5}$$
$$\approx 0.1826$$

61. $f(t) = \dfrac{1}{3650.1}e^{-t/3650.1}$

This is an exponential distribution with $a = \frac{1}{3650.1}$.

So the expected value is $\mu = \frac{1}{a} = 3650.1$ days.

The standard deviation is $\sigma = \frac{1}{a} = 3650.1$ days.